CONTEMPORARY ISSUES IN SYSTEMS SCIENCE AND ENGINEERING

IEEE Press
445 Hoes Lane
Piscataway, NJ 08854

CONTEMPORARY ISSUES IN SYSTEMS SCIENCE AND ENGINEERING

Edited by

MENGCHU ZHOU
HAN-XIONG LI
MARGOT WEIJNEN

Systems, Man, & Cybernetics Society

IEEE

IEEE PRESS

WILEY

For general information on our other products and services or for technical support, please contact our Customer Care Department within the United States at (800) 762-2974, outside the United States at (317) 572-3993 or fax (317) 572-4002.

Wiley also publishes its books in a variety of electronic formats. Some content that appears in print may not be available in electronic formats. For more information about Wiley products, visit our web site at www.wiley.com.

Library of Congress Cataloging-in-Publication Data:

Contemporary issues in systems science and engineering / [edited by] MengChu Zhou, Han-Xiong Li, Margot Weijnen.
pages cm.
ISBN 978-1-118-27186-5 (hardback)
1. Systems engineering. I. Zhou, MengChu. II. Li, Han-Xiong. III. Weijnen, M. P. C.
TA168.C635 2015
620.001′1—dc23

2014039317

Printed in the United States of America

10 9 8 7 6 5 4 3 2 1

CONTENTS

CONTRIBUTORS

CESARE ALIPPI, Dipartimento di Elettronica e Informazione, Politecnico di Milano, Milano, Italy

CLAUDIA P. ARENAS GUERRERO, Cummins Inc., Columbus, IN, USA

ROBERT ARRABITO, Defence Research and Development of Canada, Toronto, ON, Canada

AMARNATH BANERJEE, Department of Industrial and Systems Engineering, Texas A&M University, College Station, TX, USA

JEAN-PAUL BARTHÈS, UMRCNRS 7253 Heudiasyc, Université de Technologie de Compiègne, Compiègne Cedex, France

ASIMBHATTI, Centre for Intelligent Systems Research, Deakin University, Waurn Ponds, Victoria, Australia

STEPHAN BILLER, General Motors Research & Development Center, Warren, MI, USA

GIACOMO BORACCHI, Dipartimento di Elettronica e Informazione, Politecnico di Milano, Milano, Italy

MICHELE BRISTOW, Conflict Analysis Group, Department of Systems Design Engineering, University of Waterloo, Waterloo, ON, Canada

MARIA PAOLA CABASINO, Department of Electrical and Electronic Engineering, University of Cagliari, Cagliari, Italy

FENG CHU, Laboratoire d'IBISCEA 4526, Université d'Evry Val d'Essonne, EvryCédex, France

PHILIPPE DARONDEAU, Inria Rennes-Bretagne Atlantique, Rennes Cedex, France

GREGORY DITZLER, Department of Electrical and Computer Engineering, Drexel University, Philadelphia, PA, USA

YI DONGYUN, Department of Mathematics and Systems Science, College of Sciences, National University of Defense Technology, Changsha, Hunan, PR China

YUSHUN FAN, Department of Automation, Tsinghua University, Beijing, PR China

LIPING FANG, Conflict Analysis Group, Department of Systems Design Engineering, University of Waterloo, Waterloo, ON, Canada; Department of Mechanical and Industrial Engineering, Ryerson University, Toronto, ON, Canada

ALIREZA FEREIDUNIAN, Smart Grid Lab, Faculty of Electrical Engineering, K. N. Toosi University of Technology, Tehran, Iran

JEFFREY Y. L. FORREST, Institute for Grey System Studies, Nanjing University of Aeronautics and Astronautics, Nanjing, PR China; Department of Mathematics, Slippery Rock University, Slippery Rock, PA, USA

SIMON H. FU, Healthy Aging Research Center, Chang Gung University, Taiwan; IMS Technology and Consultant Inc., Taiwan

ZAIWU GONG, College of Economics and Management, Nanjing University of Information Science and Technology, Nanjing, PR China

NEGAR HASSANPOUR, Systems and Machines Research Lab, Control and Intelligent Processing Center of Excellence, University of Tehran, Tehran, Iran; Computational Intelligence Laboratory, University of Northern British Columbia, Prince George, BC, Canada

ALI G. HESSAMI, Vega Systems, London, UK

KEITH W. HIPEL, Conflict Analysis Group, Department of Systems Design Engineering, University of Waterloo, Waterloo, ON, Canada; Centre for International Governance Innovation, Waterloo, ON, Canada

MING HOU, Defence Research and Development of Canada, Toronto, ON, Canada

FENG HSU, Lockheed Martin Corp., Dallas, TX, USA

YEH-LIANG HSU, Department of Mechanical Engineering, Yuan Ze University, Taiwan; Gerontechnology Research Center, Yuan Ze University, Taiwan

HAN-PANG HUANG, Department of Mechanical Engineering, National Taiwan University, Taiwan

TZU-HAO HUANG, Department of Mechanical Engineering, National Taiwan University, Taiwan

KAO-SHING HWANG, Department of Electrical Engineering, National Sun-Yat-Sen University, Kaohsiung, Taiwan

HAMID JAHANKHANI, GSM London, London, UK

DAWEIJIA, Centre for Intelligent Systems Research, Deakin University, Waurn Ponds, Victoria, Australia

FENG JU, University of Wisconsin, Madison, WI, USA

CHIA-FENG JUANG, Department of Electrical Engineering, National Chung Hsing University, Taichung, Taiwan

GUEYOUNG JUNG, Xerox Research Center, Norwalk, CT, USA

JIUN-YIHKUAN, Department of Mechanical Engineering, National Taiwan University, Taiwan

CHUNG-HSIENKUO, Department of Electrical Engineering, National Taiwan University of Science and Technology, Taiwan

HYUNSOO LEE, School of Industrial Engineering, Kumoh National Institute of Technology, Gumi, Korea

MING-YIH LEE, Healthy Aging Research Center, Chang Gung University, Taiwan; Graduate Institute of Medical Mechatronics, Chang Gung University, Taiwan

WEI-CHEN LEE, Department of Mechanical Engineering, National Taiwan University of Science and Technology, Taiwan

KIN FONG LEI, Healthy Aging Research Center, Chang Gung University, Taiwan; Graduate Institute of Medical Mechatronics, Chang Gung University, Taiwan

HAMID LESANI, Systems and Machines Research Lab, Control and Intelligent Processing Center of Excellence, University of Tehran, Tehran, Iran

JINGSHAN LI, University of Wisconsin, Madison, WI, USA

WEN-YEN LIN, Healthy Aging Research Center, Chang Gung University, Taiwan; Department of Electrical Engineering, Chang Gung University, Taiwan

YI LIN, Department of Mathematics and Systems Science, College of Sciences, National University of Defense Technology, Changsha, Hunan, PR China; School of Business, Slippery Rock University, Slippery Rock, PA, USA

ALAN LIU, Department of Electrical Engineering, National Chung Cheng University, Taichung, Taiwan

SIFENG LIU, Centre for Computational Intelligence, De Montfort University, Leicester, UK

YI-HUNG LIU, Department of Mechanical Engineering, National Taipei University of Technology, Taiwan

JUN-MING LU, Department of Industrial Engineering and Engineering Management, National TsingHua University, Taiwan

JUNZHOU LUO, School of Computer Science and Engineering, Southeast University, Nanjing, PR China

SAÏD MAMMAR, Laboratoired'IBISCEA 4526, Universitéd'Evry Val d'Essonne, EvryCédex, France

SINA SHARIF MANSOURI, Systems and Machines Research Lab, Control and Intelligent Processing Center of Excellence, University of Tehran, Tehran, Iran; Automation & Robotics Lab Dortmund, Dortmund, Germany

THOMAS MILLER, Cognitive Solutions Division, Applied Research Associates Inc., Fairborn, OH, USA

SAEID NAHAVANDI, Centre for Intelligent Systems Research, Deakin University, Waurn Ponds, Victoria, Australia

CHRISTOPHER NEMETH, Cognitive Solutions Division, Applied Research Associates Inc., Fairborn, OH, USA

C. MATTHEW O'CONNOR, Department of the Navy, Washington, DC, USA

MARIA PIAFANTI, Department of Electrical and Information Engineering, Polytechnique of Bari, Bari, Italy

MICHAEL POLIDORO, Basic Commerce and Industries Inc., Dahlgren, VA, USA

ROBI POLIKAR, Electrical and Computer Engineering Department, Rowan University, Glassboro, NJ, USA

CALTON PU, College of Computing, Georgia Institute of Technology, Atlanta, GA, USA

MANUEL ROVERI, Dipartimento di Elettronica e Informazione, Politecnico di Milano, Milano, Italy

CARLA SEATZU, Department of Electrical and Electronic Engineering, University of Cagliari, Cagliari, Italy

MOHAMAD AMIN SHARIFIKOLARIJANI, Systems and Machines Research Lab, Control and Intelligent Processing Center of Excellence, University of Tehran, Tehran, Iran; Bioelectrics Department, Biomedical Engineering Faculty, Amirkabir University of Technology, Tehran, Iran

WEIMING SHEN, National Research Council Canada, Ottawa, ON, Canada

WANN-YUN SHIEH, Healthy Aging Research Center, Chang Gung University, Taiwan; Department of Computer Science and Information Engineering, Chang Gung University, Taiwan

Wei Tan, IBM T. J. Watson Research Center, Yorktown Heights, NY, USA

Ching-Chih Tsai, Department of Electrical Engineering, National Chung Hsing University, Taichung, Taiwan

Wen-Wei Tsai, Healthy Aging Research Center, Chang Gung University, Taiwan; Graduate Institute of Medical Mechatronics, Chang Gung University, Taiwan

Hongguang Wang, Shenyang Institute of Automation, Chinese Academy of Sciences, Shenyang, PR China

Zhikui Wang, HP Labs, Palo Alto, CA, USA

Nai Qi Wu, The Institute of Systems Engineering, Macau University of Technology, Macau, PR China; Department of Industrial Engineering, Guangdong University of Technology, Guangzhou, PR China

Guoxian Xiao, General Motors Research & Development Center, Warren, MI, USA

Jizhong Xiao, The City College, City University of New York, New York, USA; Shenyang Institute of Automation, Chinese Academy of Sciences, Shenyang, PR China

Pengcheng Xiong, College of Computing, Georgia Institute of Technology, Atlanta, GA, USA

Yingjie Yang, Centre for Computational Intelligence, De Montfort University, Leicester, UK; Institute for Grey System Studies, Nanjing University of Aeronautics and Astronautics, Nanjing, PR China

Cang Ye, Department of Systems Engineering, University of Arkansas at Little Rock, Little Rock, AR, USA

Mohammad Ali Zamani, Systems and Machines Research Lab, Control and Intelligent Processing Center of Excellence, University of Tehran, Tehran, Iran; Robotics Lab, Department of Computer Science, Özyeğin University, Istanbul, Turkey

Mengchu Zhou, Department of Electrical and Computer Engineering, New Jersey Institute of Technology, Newark, NJ, USA

Haibin Zhu, Department of Computer Science and Mathematics, Nipissing University, North Bay, ON, Canada

Zhanguo Zhu, College of Economics and Management, Nanjing Agricultural University, Nanjing, PR China

PREFACE

Rapid development and efficient operation of contemporary man-made systems fall into a domain of systems science and engineering. They assume utmost significance in terms of both research and practice worldwide. Man-made systems are highly parallel and distributed, possess uncertainties, and require frequent changes as they evolve in response to varying demands. Mathematically solid systems science and engineering methodologies and system of systems thinking are required to build reliable, sustainable, robust, high quality and high performance man-made systems.

The book is divided into five main parts:

Part I Systems Science and Engineering Methodologies
Part II Learning and Control
Part III Human–Machine System Design
Part IV Cloud and Service-Oriented Computing
Part V Sensing, Networking, and Optimization in Robotics and Manufacturing

Within each part, detailed problems, technical issues, and solution methodologies are presented and illustrated with a series of interesting and significant examples. Individual chapters are written by the leading experts in the field. The book is intended for researchers, graduate students and engineers interested in systems science and engineering. To help readers better focus on their particularly interested topics, we brief each part and chapter as follows.

The first part (five chapters) concerns the fundamentals of system of systems thinking ways and systems engineering approaches. Chapter 1 focuses on an inclusive, generic, and systematic approach to sustainability. It integrates many business and regulated facets of performance, including safety and environmental dimensions.

Sustainability is described as an emergent holistic property of hard, soft, or hybrid systems including large and often complex system of systems.

Chapter 2 reviews the literature on system of systems (SoS) thinking with respect to value systems and ethics, complex adaptive systems, risk, and cooperation. It frames the content in the context of policy development for solving global challenges such as climate change and energy, water, and food crises, which play very significant roles in altering the state of the world as a whole. The global nature of these problems hints at the need for global collective actions. However, how to start the process, monitor the dynamics, and assess the performance of our actions is relatively less obvious. From the holistic perspective of SoS thinking, a unique opportunity exists to tackle global challenges and to design national and global policies and integrate them with local and regional initiatives with deeper insight. A sound policy respects the ethics and values of multiple stakeholders, and harnesses the natural complexities of the interconnected systems to mitigate risks through adaptation, while providing an effective conflict resolution mechanism for resolving unavoidable disputes. With these principles in mind, several policies are discussed in this chapter to exemplify areas of effective policy design and domains for future work.

Chapter 3 is dedicated to systemic yoyos for general systems research. It covers the concept of general systems and its consequent formation of a new dimension for the postal modern science. By investigating the historical successes of calculus and its consequent theories, this chapter identifies the need to examine systems in terms of whole evolutions; and consequently, develops the systemic yoyo model. It discusses the theoretical and empirical justifications of this important model. It then reveals some elementary properties of this model, including four laws on state of motion. It finally introduces some important applications of this model in areas of social science, economics, the foundations of mathematics, and the prediction of extreme weather conditions.

Chapter 4 covers the latest advances in grey system theory research and provides a state-of-the-art overview of this important field. Covering the theoretical foundation, fundamental methods, and main topics in grey system theory, it includes all the elementary concepts: basic principles, grey numbers and their operations, sequence operators and generations of grey sequences, grey incidence analysis, grey cluster evaluations, grey prediction models, grey system prediction, grey models for decision-making, and some of practical applications. At last, the software of grey system modeling is introduced. This chapter will be of interest to those researchers in a wide range of fields including information and systems sciences and management sciences, and to those working in applied areas such as geoscience, engineering, agriculture, medicine, and biosciences.

Chapter 5 focuses on how to build resilience in a practical system. Accounting for individual and team cognitive activities among operating forces can be used to develop requirements and create and evaluate systems that are intended to support decisions. This chapter documents how to incorporate human-centered design into a US Navy's systems acquisition process. It provides a common language for the Chief of Naval Operations and Naval Sea Systems Command staff to use in the acquisition of systems that are intended to support decision-making. Three deliverables make that language evident: a set of requirements for decisions and the displays to support making those decisions that were drawn from extensive interviews and observations; an interactive prototype that demonstrated display organization and operation; a 60-page guidebook that described the approach, for more senior Navy program managers to more junior systems engineers and contractors. The decision support systems that result are expected to enable expeditionary forces to better perform as a complex adaptive system—to improve their resilience.

The second part (five chapters) concerns the learning and control in numerous complex man-made systems.

Chapter 6 is devoted to four significant technical areas: reinforcement learning, evolutionary learning control, intelligent learning control, and learning via case-based reasoning despite many interesting and emerging technologies in the specialized area of intelligent learning in control systems. To illustrate how these methods work well, mobile service robots are used as test platforms. It also illustrates some advances and challenging issues in these technical areas.

Chapter 7 focuses on adaptive classification problem in nonstationary environments. It presents the methods to learn the concept drift and strategies for adapting the classifier to the new working conditions. It introduces the figures of merit to be used to evaluate the classification systems present in the literature. It then describes the methods to deal with supervised information coming from the field during the operational life. It finally gives a number of open problems related to open problems: lack of standard benchmark datasets, lack of real world dataset, and unsupervised or semi-supervised approaches for learning concept drift.

Chapter 8 deals with modeling, analysis, scheduling, and real-time control of one important class of automated systems, that is, cluster tools in wafer fabrication. They are computer-integrated, reconfigurable and widely adopted by semiconductor manufacturers. They can lead to higher yield, shorter cycle time, better utilization of space, and lower fabrication cost. In operating them, one needs to simultaneously arrange the activities of their robots and process modules. Furthermore, there are various constraints and different operational modes, and it is deadlock-prone. Hence, it is challenging to effectively control a cluster tool. This chapter presents a Petri net-based formal technique to model, analyze, schedule, and control cluster tools.

Chapter 9 discusses a Petri net-based approach to the design and dynamic control of large-scale manufacturing process with uncertainties. Representing different types of uncertainties, complicated designs, and interoperable design formats for cooperative modeling can sometimes be challenging while using a Petri net-based approach. This chapter addresses these challenges. Fuzzy colored Petri net with stochastic time delay is proposed to capture different types of uncertainties. This net can incorporate two different types of uncertainties: ambiguity-type uncertainty using fuzzy logic and variation-type uncertainty using stochastic time delay. The chapter then gives a modular-design-based five-stage modeling methodology to represent complicated processes by using a set of parametric Petri net modules to capture and represent commonly occurring complexities in manufacturing processes. It next presents an

Extended Petri Net Markup Language as a new and effective interoperable data format supporting the proposed net model and five-stage modeling methodology. It finally introduces a new and effective learning method via simulation-based optimization for dynamic control of complex manufacturing systems.

Chapter 10 focuses on two important and tightly related problems, that is, identification and synthesis of discrete-event systems. Particular attention is devoted to two main formalisms in this area, that is, finite state automata and Petri nets. This chapter aims to provide a collection of references in this framework and discuss the main research areas where such problems have been investigated. It discusses the most important results in these areas, including the basic ideas related to the theory of regions and the synthesis of labeled Petri nets.

The third part (five chapters) concerns the issues arising from human–machine systems (HMSs) design. Chapter 11 presents a unified agent-based design framework and methodologies to guide operator interface design for complex HMSs, for example, an unmanned aerial vehicle control station. Through the facilitation of operator–agent interaction, an agent-based interface is expected to increase operator situation awareness and reduce workload, thus reducing personnel requirements and enhancing operational performance. This chapter discusses the evolution of interface technologies and describes the concepts and associated conceptual framework of intelligent adaptive interfaces. It reveals the design challenges, alternative solutions, and analytical approaches, along with examples of empirical studies. It introduces the concept of adaptive intelligent agents and an operator–agent interaction model to guide (HMSs) design. It presents hierarchical system architecture to facilitate operator–agent interactions. It illustrates the utility of the proposed framework and design methodologies through an HMSs design example.

Chapter 12 presents a complex adaptive system-of-systems approach to human–automation interaction in smart grids. Human–automation interaction is a typical example of a complex and adaptive phenomenon. The chapter details the core concepts of complexity, complex adaptive systems, system of systems, and complex adaptive system of systems. It gives an introduction to the automation-related notions like the automation itself, human–automation interaction and adaptive autonomy. Subsequently, it introduces the smart grid and its associated human–automation interaction system as complex adaptive systems of systems. Petri nets, as powerful tools for modeling complex adaptive systems, are proposed to manage the mentioned adaptive autonomy in smart grid.

Chapter 13 tackles the challenges in designing effective and usable virtual training systems. It presents user-centered design and evaluation approach and two consecutive case studies with an aim to better understand their efficacy from the end users' perspective. This chapter seeks to answer questions of whether individual users' characteristics, in particular prior experience, lead to varied behavior or affect cognitive learning outcomes, and if so, which of the characteristics is the most influential. It finally discusses the findings from two case studies and the lessons learnt in the development and evaluation of virtual training systems for procedure skills acquisitions.

Chapter 14 focuses on the technologies, systems, and applications of Computer-Supported Collaborative Design (CSCD). Industries and societies require new technologies to address increasingly complex design issues for products, structures, buildings, systems, processes, and services while meeting the high expectation of customers. CSCD emerged in response to this requirement and has been a very active R&D area in the past two decades with dramatic progresses. It is carried out not only among multidisciplinary design teams within the same organization, but also across the boundaries of organizations and time zones, with increased numbers of customers and suppliers involved in the process. This chapter presents an overview of this R&D area, particularly CSCD technologies, systems, and applications. It also discusses CSCD research challenges and opportunities.

Chapter 15 discusses systematically the application of role mechanisms in collaboration or teams. It argues that roles establish trust among team members in collaboration, set up dynamics of a team, facilitate interaction among team members, support adaptation among team members and the team, and enable information sharing among team members. In this chapter, Role-Based Collaboration (RBC) is illustrated with the properties, the process, the principles, and the model called Environments – Classes, Agents, Roles, Groups, and Objects (E-CARGO). A case study is then conducted to explain the application of RBC and the E-CARGO model into analyzing the performance of a team. Conclusions and future challenges are presented to motivate more applications and related investigations.

The fourth part (three chapters) concerns the recently emerging computing technologies, that is, cloud computing and service-oriented computing. Cloud computing is the delivery of computing as a service whereby shared resources, software, and information are provided to computers and other devices as a utility over a network. There are many types of public cloud computing: infrastructure as a service, platform as a service (PaaS), and software as a service. Although cloud computing brings in advantages over traditional computing paradigms, PaaS providers often face a resource sharing dilemma under workload fluctuation where statically configured virtual resources are often either overprovisioned or overloaded. Although much progress has been made in the dynamic resource management using control-based approaches in cloud computing environments, there are two significant research challenges, that is, only single metric-like mean response time and single benchmark application have been considered. Little attention has been paid to the behavior of the applications, including that of the workload generators. Chapter 16 addresses these two issues through experiments on a test bed with multi-tier applications hosted on multiple virtual machines. First, it shows that, when the resource utilization of the applications is maintained at constant levels, the end-to-end performance metrics of the applications that can be perceived by the users can be very different. Second, it shows that, even when the mean response time of the applications is maintained at constant levels, the other end-to-end performance metrics of the application that can be perceived by the users can still be different. Third, it presents insights into different controller performance when the same feedback controller is applied to applications with different workload generators. The results imply that the evaluation of the controller has to be done on multiple applications that can represent different practical scenarios. In reality, the behavior of the applications has to be carefully considered when the controller parameters are configured.

Since the early of this century, service-oriented computing has become the representative direction of distributed computing. The core part of service-oriented computing is web services. When a single web service cannot satisfy the complex functionality required by a user, multiple web services need to be composed together in order to fulfill the request. This trend has triggered a significant number of research efforts on the web services composition both in academia and in industry. As one of the most important problems in service composition, protocol-level mismatch is still a highly complicated problem to solve. The state-of-the-art method to deal with it is to generate adaptors to check deadlock-freeness based on a reachability graph. When this property is violated, the generation process will repeat itself until no deadlock state is found. The main drawback of this method is that it does not take into account the future deadlock state and requires many interactions with a developer. Thus it suffers from low efficiency. Chapter 17 presents a Petri net called composition net (C-net) to model multiple web service interactions. The protocol-level mismatch problem is transformed into the empty siphon problem of a C-net. By this transformation, potential deadlock states are taken into consideration. The optimal solution that involves the fewest interactions with a developer is identified. The proposed method is proved to achieve higher efficiency for resolving protocol-level mismatch issues than traditional ones.

Building workflows using service composition has become an important method to build composite applications and reuse existing resources. Therefore, workflow-based service composition and orchestration is now a hot topic in both academia and industry. Chapter 18 summarizes the research activities in the field of workflow in service-oriented computing. It discusses five major research topics, that is, languages for service orchestration, automatic service composition, mediation-aided service composition, verification of service workflows, and decentralized execution of workflows. It also reveals several topics that are potentially appealing, that is, to combine the power of social network and service computing, more practical and flexible service composition, workflow as a service, and novel applications.

The last part (containing seven chapters) concerns sensing, networking, and optimization issues arising from robotics and manufacturing. Chapter 19 focuses on the design of a rehabilitation robot. To avoid high complexity of the high order dynamic model considering coupling effects and payload variations typically inherited from common robotic systems, a robust dynamic sliding mode control law design based on an independent joint model, tremendously reducing the complexity of the dependable design process, is proposed to control a Multi-DOF rehabilitation robot arm with parameter variations, external disturbances and payload variations. The global asymptotic stability of the controller is guaranteed by the Lyapunov approach, and the input chattering phenomenon is handled by using the augmented system. Through dynamic simulation and practical experiments of a rehabilitation robot, the research results show that the proposed controller is better than the others. A nontarget EMG pattern filtering scheme is used to improve stability of neuroprostheses.

Chapter 20 deals with body sensor networks (BSN). BSN has increasingly become an attractive topic because of the significant demand for health services, as well as related equipment and supplies. It can provide real-time remote monitoring of the health situation of a particular person. A wide range of wearable, miniaturized, and low power consumption sensors has been utilized in the BSN for collecting biological signals. Recently, accelerometer can be fabricated by microelectromechanical system technology that makes sensing of acceleration becomes simple and low cost. Moreover, the sensor itself is miniaturized that can be integrated into various BSN applications. In this chapter, the use of accelerometers for BSN for monitoring the body motions is discussed. Recent advances in the computation of motion identification are described including tilting angle, muscle strength, and gait performance. Based on specific computing algorithms, different interpretations of the signal from accelerometers can be formulated. Moreover, several medical diagnosis assessment and training applications are introduced to demonstrate the capability of an accelerometer-based BSN. Finally, the concept of using biped humanoid robots to develop the BSN simulation system is discussed.

Chapter 21 deals with telepresence robots and their application to medical services and homecare. Over the past few decades, robotics made tremendous progress in saving, protecting, and improving human lives, and is now adopted across a broad range of applications in medicine and homecare services. While technologies are developed at a rapid pace, people are expecting robots to be part of their lives in a more natural way. Among the wide variety of robots, telepresence robots, which allow the user to experience the virtual presence in another place, offer the potential to meet this rising demand better and are thus of great concern. This chapter explores the up-to-date research findings and industry practices in telepresence robots for medical and homecare applications. Moreover, it discusses the key factors contributing to the success of telepresence robots as well to address the future trends and opportunities.

Chapter 22 overviews the research and development of climbing robots. They are special robots with vertical mobility that overcome gravity by using special adhesion methods to perform difficult tasks in dangerous environments or hard-to-access places. This chapter provides a comprehensive review of technological advances in climbing robots since their appearance in the late 1980s. The most difficult challenge for a climbing robot is the efficient adhesion mechanism to ensure that the robot sticks to wall surfaces or vertical structures reliably without sacrificing mobility. This chapter introduces the motivation, history, and applications of climbing robots. It then presents the state-of-the-art adhesion technologies used in climbing robots and summarizes the advantages and limitations of each technology. These adhesion technologies are categorized into six types: magnetic adhesion, vacuum suction techniques, aerodynamic attraction, grasping grippers, bio-mimetic approaches inspired by climbing animals, and compliant electro-adhesion. It next presents some locomotion techniques that enable the climbing robots to operate on vertical surfaces. It finally gives a comparison of successful adhesion technologies used in climbing robots.

Chapter 23 presents a collection of vision data processing methods for small robot autonomy. Two typical robotic perception systems—a LIDAR and Flash LIDAR Camera (FLC)-based system—and the related methods for vision-based navigation are introduced. The former is a complete robot navigation system comprising terrain mapping, terrain traversability analysis, and motion control subsystems for urban navigation. The system registers laser scans captured at different robot positions into a map over time using the robot pose information obtained from a three-axis gyro and wheel odometry. For reliable map building, a filtering method utilizing robot motion constraint and the LIDAR's characteristics is developed for erroneous laser data removal. A motion control subsystem determines the robot motion command and control the robot movement. Experimental results show that the navigation system is able to reliably navigate the robot with smooth motion. In FLC, a camera's visual and range data are used for robot pose estimation and obstacle detection/avoidance. The pose estimation method estimates the robot's ego-motion by co-processing the camera's intensity and range images. The resulting 3D point-pairs are used to compute the robot's attitude and position changes. The method uses feature descriptors to perform feature matching. It works well with large image motion between two frames without the need of spatial correlation search. To enable dual use of the camera's range data for obstacle detection/avoidance, a plane extraction method is proposed. The method encodes the local geometric information and global spatial information of the 3D data into an Enhanced Range Image, which is then clustered into a number of homogeneous groups. The FLC-based system has two distinct advantages. First, due to FLC's consistent depth measurement accuracy, the pose estimation method may achieve a much better accuracy and repeatability than a stereovision-based approach. Second, FLC's range data is complete and therefore can be used for reliable obstacle detection/avoidance. This makes it possible to navigate a mobile robot by using a single perception sensor and thus allow for much higher autonomy in a smaller-sized system.

Chapter 24 overviews the recent advances in automotive battery technologies, their manufacturing processes, and environmental and recycling issues. It presents the main aspects in technology and manufacturing processes for electric vehicle batteries, such as NiMH and Li-ion batteries. It outlines the typical requirements for automobile batteries. It then illustrates the different types of electric vehicles. It next presents the development of vehicle battery technology, focusing on NiMH and Li-ion batteries. Finally, it discusses the challenges faced by battery manufacturers.

The last chapter introduces the recent progress in the area of facility location decision-making. It focuses on capacitated plant location problems with multicommodity flow, a multi-type transshipment point location problem with multicommodity flow, a large-scale new variant of capacitated clustering problems, a location problem with selective matching and vehicles assignment, and competitive facility location and design with reactions of competitors already in the market. It presents some interesting and important ideas for future research.

Acknowledgment: This volume is supported in part by US NSF under Grant No. CMMI-1162482, in part by the PRC Ministry of Science and Technology under Contract No. 2013DFM10100, National Natural Science Foundation of China under Grant No. 61374148, and the Deanship of Scientific Research (DSR), King Abdulaziz University, Jeddah, under Grant No. 23-135-35-HiCi.

PART I

SYSTEMS SCIENCE AND ENGINEERING METHODOLOGIES

CHAPTER 1

A SYSTEMS FRAMEWORK FOR SUSTAINABILITY

ALI G. HESSAMI[1], FENG HSU[2], and HAMID JAHANKHANI[3]
[1]Vega Systems, London, UK
[2]Lockheed Martin Corp., Dallas, TX, USA
[3]GSM London, London, UK

1.1 INTRODUCTION

The ability to maintain a balance in a process or state in a system, whether ecological, technological, or social is currently known as sustainability. This balance is often viewed from multiple perspectives namely, desirability, longevity/survivability, affordability, and social/environmental impact. The UN World Commission on Environment and Development [1] has provided a human focused definition that is now widely accepted as the norm stating that sustainability amounts to "*[meeting] the needs of the present without compromising the ability of future generations to meet their own needs.*" This is a strategic vision and a macro perspective that qualifies the subject matter as a potent candidate for the exploration, development, and progression as a scientific systems framework.

The pervasive significance and prominence of sustainability is typified by the UK Engineering Council in their guidance notes for the entire engineering profession under the Principles for Sustainability [2]. These comprise contributing to building a sustainable society, present and future; applying professional and responsible judgment and taking a leadership role; doing more than just complying with legislation and codes; using resources efficiently and effectively; seeking multiple views to solve sustainability challenges and managing risk to minimize adverse impact to people and environment in all matters related to engineering. Alas, the principles do not elaborate the makeup and constituents of the concept of sustainability itself and mainly amount to a code of behavior and practice for the profession. However, these highlight the emergent sensitivity to the tenets of sustainability in a domain that

Contemporary Issues in Systems Science and Engineering, First Edition.
Edited by MengChu Zhou, Han-Xiong Li and Margot Weijnen.

has traditionally propounded safety, reliability, integrated risk management [3], and social responsibility alone.

The performance of a generic product, process, system, or undertaking can be viewed as a symbiotic balance of many systemic emergent behavioral and structural properties namely:

(1) technical functionality;
(2) cost/schedule, programmatic balance;
(3) environmental impact;
(4) reliability, availability, maintainability;
(5) safety and security;
(6) quality; and
(7) perceived and/or absolute value.

We posit that from a holistic systems perspective [22], all of these are essentially sustainability qualifiers in that compromising any of the above set of requisite performance attributes renders a product, process, system, or undertaking unsustainable.

In this chapter, we approach sustainability as an emergent holistic property of hard, soft, or hybrid systems including large and often complex system of systems. We use the following nomenclature:

Assurance: Increasing confidence and certainty

Gain: Lives saved, improvements made, damages prevented or avoided in the natural habitat, or benefits accrued to a business/society or a combination thereof

Hazard: Object, state or condition which in the absence of adequate detection or containment could lead to an accident

Health: Soundness of body and mind, freedom from illness

Loss: Physical harm to people, detriment to a business/society, or damage/destruction of the natural habitat or a combination thereof

Reward: A forecast for a desirable event or gain. The expected value of a future gain

Risk: A forecast for an accident or loss. The expected value of a future loss

System: A (purposeful) composite of inter-related parts/elements with discernible collective output(s) or emergent property(ies) not manifested by any of the elements

Safety: Freedom of people from (physical) harm

Security: Freedom from vulnerability or loss caused by deliberate and malicious acts

Sustainability: A blend of social, economic, environmental, resources, and technological considerations which render a product, system, or undertaking viable and continually optimal

Systems assurance: The art, science, and technology of ensuring and demonstrating that a system is likely to achieve its objectives without engendering unacceptable levels of loss

Systems safety: The art, science, and technology of ensuring and demonstrating that a system is not likely to lead to unacceptable levels of (physical) harm to people

Systems security: The art, science, and technology of ensuring and demonstrating that a system is not likely to be vulnerable to malicious deliberate acts aimed at engendering unacceptable levels of loss

Vulnerability: Susceptibility to injury, fatality, or loss

Welfare: Wellbeing and quality of life for individuals and the society

1.2 A UNIFIED SYSTEMS SUSTAINABILITY CONCEPT

In the research or debate on many critical issues concerning global climate change, energy and financial system crises in recent years, while there is growing recognition of the importance of system sustainability (SS), there is little or no consensus on the definition of systems sustainability that is widely accepted by the world scientific or policy-making communities. In order to delineate a comprehensive, inclusive yet generically unified concept for system sustainability, we begin by exploring the concept of sustainable systems. It is generally accepted that sustainability applies to integrated systems (or system of systems; SoS) comprising societal and technological infrastructures, humans, and the nature. Beyond the conventional systems concept, a key characteristic of the sustainable systems is that the structures and operation of the human component, in terms of society, economy, governance, law and regulatory apparatus, etc. must be such that these system attributes reinforce or enable the existence, persistence, resilience, and long-term wellbeing of the structures and operation of the natural component.

In the concept of sustainable systems within the realm of Gaia [4], such attributes should also include energy, natural resources and ecosystem dependencies, environmental stability, technology safety and security and dependability, biodiversity, resource cycling, human–nature interactive complexity and biogeochemical cycles, etc. In fact, the term "sustainability" most widely used in recent years, is more of a macro concept. Its strategic perspective usually applies more broadly to entire systems and infrastructures such as the global economy, energy systems, transportation, information, and agriculture systems. However, the true gravity of the term is often loosely defined, and even appears somewhat elusive. Confusion or ambiguity often arises due to many researchers and stakeholders often having their own unique definitions within a wide range of disciplines, in part because the term "sustainability" seems to imply absolute system behaviors, as it is indeed difficult to define or describe a "partially sustainable system" as opposed to a "fully sustainable system," especially given the lack of a precise, clear, and unified definition. So, in an attempt to explore a comprehensive and unified concept for system sustainability which may be referenced or applied to a broad category of systems, based on the principles of the emerging

SoS concept [5], the authors intend to present a sustainability concept by probing into both realms of macro-based (at top layer) and micro-based (at lower layers) lexicons pertaining to sustainability. Relationships of how these concepts converge at different hierarchy of layers (or levels) of clarity are also discussed. In practice, a large and complex system of systems can possess a wide range of sustainability qualities at a macro level. To refer to a system as "more sustainable" is to essentially imply that a system has more resilience to survive or possesses a blend of many other qualities. Meanwhile, in a measurable or metrics-based value judgment system, we could also try to define a sustainable system at a lower level, or in a micro or metrics-based (quantitative) perspective.

Apart from formalizing a comprehensive and inclusive view of the sustainability, we also raise the necessity for evaluation, benchmarking, and aggregating a whole suite of qualities that positively or adversely influence the sustainability of products, processes, systems, and undertakings. However, the treatment of the latter requirement is beyond the scope of the current chapter, and it will be addressed in our future work.

1.3 SUSTAINABILITY ASSURANCE: THE FRAMEWORK

Given the multi-disciplinary and multi-dimensionality of the sustainability as a holistic and macro property, we develop and propose an integrated perspective on the subject through a systems framework. This is intended to clarify the context, components, topology, and the scope of sustainability at any level of perspective from consumable products to a large system of systems such as the ecosystem. The systems framework is developed and represented in weighted factors analysis methodology (WeFA) [6].

1.3.1 Weighted Factors Analysis

The underpinning philosophy for WeFA, the elicitation process, and the representation schema are detailed in the published literature [6, 7–10]. However, a brief account is given here as a quick reference to the methodology in view of the extensive use of WeFA's notation for the elicitation, capture, and encoding of a systems sustainability framework.

WeFA is a group-based knowledge capture, representation, encoding, and evaluation methodology. The expert panel are chosen to represent related but diverse and non-overlapping aspects of the problem being studied. The focal point of a group study in WeFA is an "AIM" (A0) represented graphically by an oval annotated with a brief relevant text. With the active participation of the expert panel, the AIM is debated, formalized, agreed, captured, noted, and decomposed into a number of influencing factors called "GOALs." The GOALs that are deemed to support the attainment of the AIM are classified as Drivers and those opposing the attainment of the AIM are considered Inhibitors. The Driver GOALs are represented by ellipses with bright background linked toward the AIM or other GOALs with green forward arrows implying positive influence. The Inhibitor GOALs are represented by ellipses with a dark background color, linked to the AIM or other GOALs that they influence with red arrows pointing backward (toward themselves) implying negative influence.

All GOALs are annotated with a brief text to indicate their nature and a unique numbering system to simplify referencing. In WeFA, each factor (GOAL) can be in turn decomposed into its localized Drivers and Inhibitors and the process is repeated until the AIM is studied and analyzed at a desirable level of detail. The influences of factors in a WeFA schema, represented by green or red arrows, can be hierarchical as well as lateral. This creates a powerful and inter-related network capable of representing the factors, their influences, dependencies, and relationships with respect to the AIM under scrutiny.

WeFA diagrams are hierarchical and the GOAL numbering system is designed to reflect the hierarchy as well as type, that is, Driver or Inhibitor. The closest layer of GOALs to the AIM of a WeFA diagram is referred to as level 1 and its GOALs are annotated by G followed by a number. The numbering scheme for Driver GOALs is clockwise from 12:00 (top) starting from 1, for example, G1, G2, G3 and anticlockwise for the Inhibitors, for example, G1 and G2, etc. Deeper layers of the hierarchy are annotated as G1.1, G1.1.2, etc. These are referred to as level 2, 3, etc.

The graphical representation of all the factors and their positive and detrimental influences upon an AIM is referred to as a WeFA schema. This form of knowledge representation is principally aimed at ease of comprehension and review due to the graphical representation of the key factors, their influences, and their position within the hierarchy of other factors. This underpins the approach to the development of an objective evaluation and assessment regime for the sustainability assurance framework. In this spirit, WeFA is a knowledge capture, representation, verification, validation, evaluation, and reuse methodology principally suited to strategic complex scenarios such as sustainability.

1.3.2 The Framework

Traditional wisdom about sustainability refers to an amalgam of social, economic, and environmental dimensions in this complex concept [11]. The current bias however is mainly toward ecosystems whilst also assuming a blend of the right ingredients from the social, economic, and environmental dimensions is adequate to give a semblance of sustainability in a given context. By adopting the WeFA methodology, we challenge this unipolar conception of sustainability, thus treating it as a dynamic balance between a host of factors that support and detract from its attainment in a given context. This, we posit, is a more holistic and potent route to understanding [1], representation, evaluation, and benchmarking of this key attribute than the classical hierarchy of summative approaches. To this end, the system sustainability assurance framework, as an extension of the current conventional wisdom is presented in the WeFA schema of Figure 1.1. The framework embodies five key driver and three inhibitor class of goals. This suite of goals in the WeFA methodology essentially constitute the highest level of abstraction for the major influencing factors on the aim of the schema. Furthermore, given the hierarchical and relational nature of the

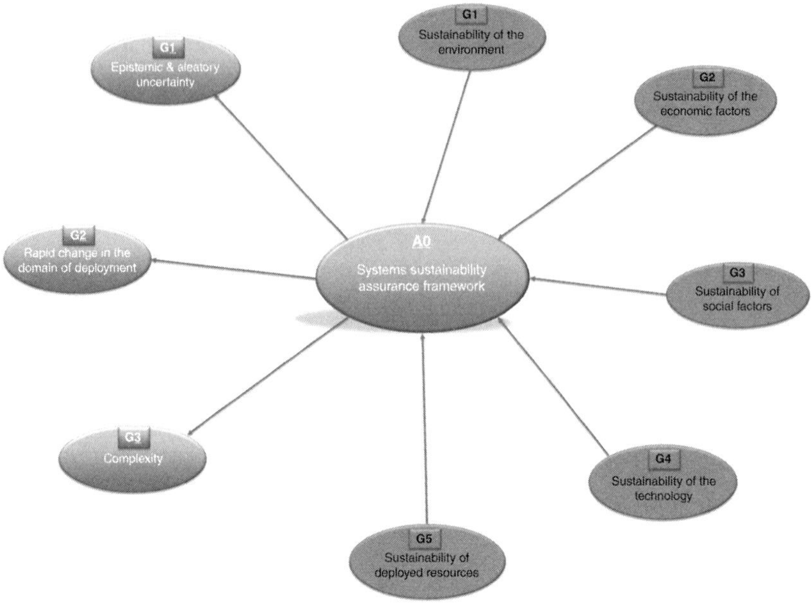

FIGURE 1.1 Systems sustainability framework, level 1 WeFA schema.

WeFA methodology, the so-called Level 1 schema is decomposed into a number of subsequent layers, Level 2 (as shown in Figure 1.2), Level 3 etc. An illustrative level 3 decomposition focuses on the technological facet as shown in Figure 1.3. An outline elaboration of the level 1 schema of Figure 1.1 is given as follows:

G1-Sustainability of the environment. All factors that qualify and underpin sustainability as far as the environmental impact is concerned.

G2-Sustainability of the economic factors. All factors that qualify and underpin sustainability from an economic cost, value, and market response perspective.

G3-Sustainability of the social factors. All factors that qualify and underpin the social acceptance, desirability, and values.

G4-Sustainability of the technology. All factors that influence the long-term viability of the materials and technology employed.

G5-Sustainability of the resource. The factors influencing the availability, desirability, and renewability of the resources employed.

$\underline{G}1$-Uncertainty. The factors influencing the degree of uncertainty in any facet or performance.

$\underline{G}2$-Rapid change in the domain of deployment. The factors underpinning the potential for rapid change internal or external.

$\underline{G}3$-Complexity. The factors indicative of the degree of complexity in time behavior, structure, or composition.

FIGURE 1.2 Systems sustainability framework, level 2 WeFA schema.

Whilst WeFA is a creative and hierarchical knowledge capture, representation, and evaluation methodology, the level 1 schema of Figure 1.1 in principle depicts a taxonomic perspective on sustainability. The suite of dependent and independent influencing factors emerge in level 2 and higher decompositions in a given schema.

Taking the level 3 schema of Figure 1.3, we focus on the technological sustainability (G4) as an illustration of the key concepts of a sustainable system at macro and micro perspectives. This is in turn driven by:

G4.1 Sustainability of the architecture.

G4.2 Sustainability of the structure and build.

The driver goals for G4 are in turn decomposed as an illustration of the underlying concepts.

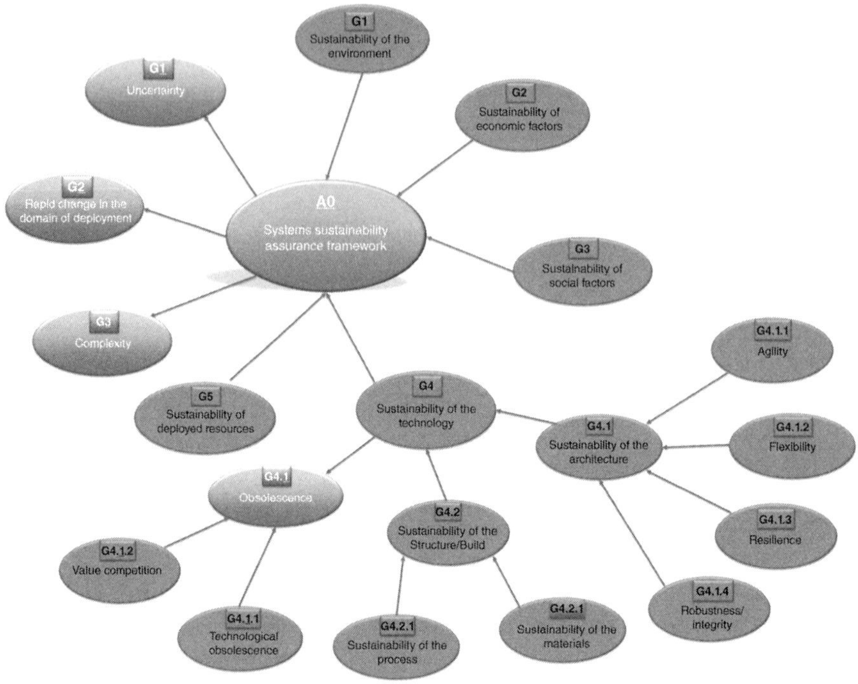

FIGURE 1.3 Systems sustainability framework, technological focus.

1.3.3 The Macro Concept of a Sustainable Architecture (G4.1)

We believe that a system or system of systems is considered architecturally sustainable in a macro perspective, if the following high-level key characteristics are assumed or attained:

(1) Resilience
(2) Robustness
(3) Flexibility
(4) Agility and efficiency

The macro (top-level) concept of a sustainable system can be qualitatively assessed for the sustainability of macro systems, such as the global economic system, environmental system, transportation system, information system, agriculture system, energy and global climate systems. Such a development of the sustainability concept is based on a top-down, goal driven, and attributes-decomposition process [12, 13]. Each of the four major attributes of the sustainable architecture of a system is defined below and also delineated in a graphic form as shown in Figure 1.4:

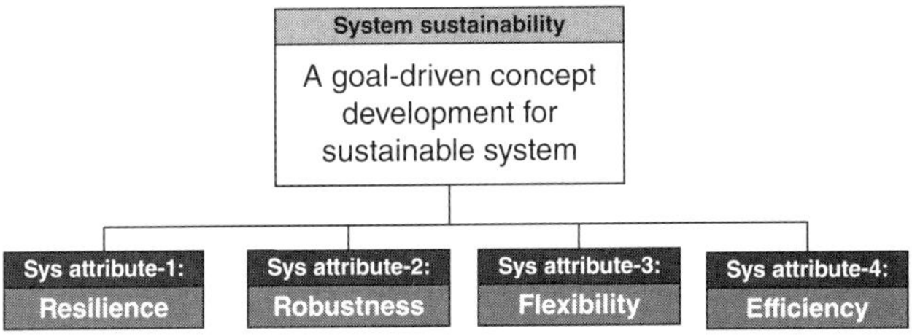

FIGURE 1.4 A macro (top-level) concept delineation of sustainable architectures.

System resilience – system attribute used to imply or indicate the evidence of system operability and survivability within an existing and changing environment or condition.

System robustness – system attribute used to imply or indicate the evidence of system capacity or performance capability given variations in an existing environment.

System flexibility – system attribute used to imply or indicate the evidence of system adaptability, and the ability to reconfigure itself for continued operability under a dynamic or disruptive changing environment (or internal/external conditions).

System agility and efficiency – system attribute used to imply or indicate the evidence of system capability to maximize performance while minimizing waste of resources.

1.3.4 The Micro Concept of a Sustainable System

A system or system of systems is considered sustainable in the micro or low-level sense, if the following low-level attributes or metrics-based characteristics are ensured and assumed:

(1) long-term operability and survivability within an existing and changing environment;
(2) flexibility and capability to adapt to the dynamics of the evolving external–internal conditions;
(3) system structure and interdependencies that support and promote accountability and efficiency;
(4) ability to expand and accept the use of new technologies for increased health and performance;
(5) robustness and capability to handle uncertainty and avoid severe risks of system catastrophe;
(6) system security and safety to resist or tolerate deteriorating or disruptive internal or external threat conditions;

(7) system efficiency and simplicity for satisfying performance expectations while preventing wasteful consumptions of resources; and
(8) long-term system availability or capacity for self-rectification to ensure continued existence and operation for achieving required system performances.

1.3.5 A Top-Down Hierarchy of a Multi-Level Sustainability Concept

As described in the previous two sections, a top-down hierarchical approach is employed to delineate the conceptual framework of a goal/objective-driven and attribute-decomposition process to derive a more comprehensive and unified generic definition of system sustainability. Based on a generic hierarchy (as shown in Figure 1.5), the macro- and micro-levels of system sustainability concepts introduced in this paper is thus logically related to one another.

The term "goal/objective-driven" used here is to explain the thought process of the necessity to define a precise "goal" or purpose and objectives of any sustainable system at the very high level of considerations. Meanwhile, the term "attribute decomposition" is used to describe the top-down process because a low-level sustainability concept (micro concept) is based on the metrics or system performance measures, which are generated from decomposing the high-level system attributes.

For instance, a top-level goal for sustainability of any generic systems could be simply defined as "assurance of system health," or to put it more specifically as "maximizing long-term system operability and successful performance while minimizing resources and system collapse." A top-level goal for sustainability of a globe system for humanity could be defined as "maximize global system survival and long-term security and prosperity." Therefore, under this top goal of achieving the globe system sustainability, a set of key system attributes (secondary or lower-level goals) can be readily developed and defined, and such a macro concept of a global system sustainability is shown in Figure 1.6.

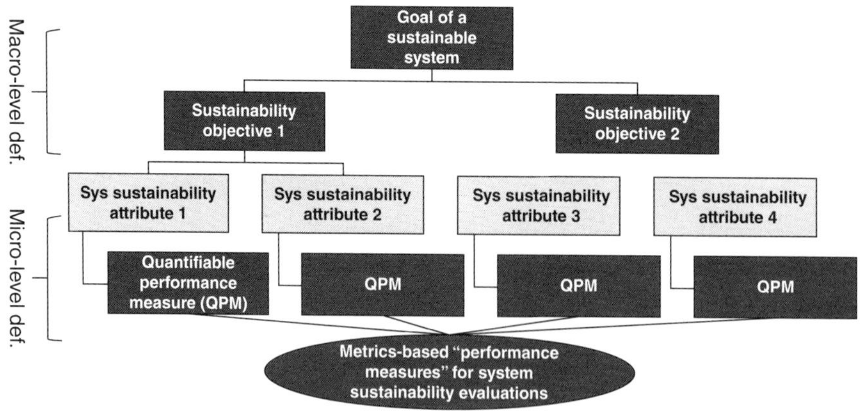

FIGURE 1.5 A goal-driven top-down hierarchy for defining system sustainability.

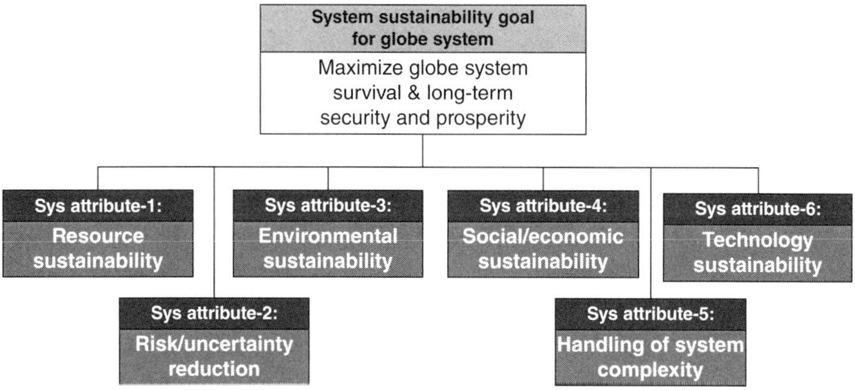

FIGURE 1.6 Illustration of a macro concept for global system sustainability.

In order to advance and properly define the concepts of systems sustainability, we must realize the important nature of sustainability concept with its interdisciplinary and multidisciplinary characteristics. In other words, interdisciplinary research and collaborative efforts with diverse stakeholders need to be carried out to develop and apply life-cycle based system models [14] for the evaluation of system sustainability. Furthermore, a set of complete sustainability metrics or performance measures need to be carefully defined for systems that meet societal needs. The sustainability concept as introduced in this section, can be summarized in the following respective definitions provided for a generic system and for a global system as well:

System sustainability for the generic system – an integrated set of inter-related parts, assemblies, rules, and processes or group of self-contained independent systems working together to deliver functions or services that satisfy a desirable blend of emergent properties pertaining to social, economic, environmental, technological, and resource utilization dimensions.

System sustainability for the global system – an integrated set of industrial, technological, social, and natural processes or systems purposefully working together to deliver products and services that satisfy a desirable blend of emergent properties pertaining to social, economic, environmental, technological, and resource utilization dimensions to ensure the well-being and continued progression of global societies while maintaining the integrity of human and ecosystems over a long-term time horizon.

1.4 TECHNOLOGICAL SUSTAINABILITY CASE STUDY—INFORMATION SYSTEMS SECURITY

We introduce a brief case study in one of the facets of sustainability in the information infrastructure. Even though the case presented does not directly align with the sustainability framework depicted in the schema of Figure 1.1, it highlights an important aspect of sustainability in the cyber-domain context.

Policies and procedures are an important tool in the information security toolbox. The processes upon which sound security engineering management policies, procedures, and practices should be built need to be repeatable, reportable, and auditable (RRA) [15]. A process that is repeatable by two or more people will produce the same desired outcome and thereby provides confidence in the outcome. A process that is reportable ensures that any threat can be identified and assessed before it becomes destructive and allow managers to deal with it. Finally, a process that is auditable allows verification of the process. Processes that meet the criteria expressed as RRA are sustainable and allow cross-skilling and improve the recovery time when information security is threatened.

Organizational design is one of the biggest challenges facing business in the twenty-first century [16]. Even in the information society, most organizational structures were developed at a time when physical assets were managed and processed by physical labor. In today's world many assets are intangible and operate in a mutation of electronic communication networks which has increased complexity and introduced new threats and hence risks to the well-being of nations.

In today's business environment it is difficult to obtain senior management approval for the expenditure of valuable resources to "guarantee" that a potentially disastrous event will not occur that could affect the ultimate survivability of the organization.

Advanced level of network security provides maximum network flexibility as well as an additional layer of protection against unauthorized computer access. Moreover, this advanced security level also makes possible an audit trail of network usage. Another benefit is that a user authorization can be quickly and efficiently rescinded from the network. In general, this advanced security level can help reduce, if not eliminate, the need for costly additional security hardware such as data encryption devices.

Most companies tend to be reactive and respond with quick infrastructure solutions. A strategic approach to computer network security leads to a more efficient plan and a less expensive risk-management strategy. Aligning computer network security to corporate goals provides management with a framework for steering resources, whether it is toward infrastructure, improved controls, training, or insurance, based on a carefully thought-out process that analyses the level of risk the company is willing to absorb.

1.4.1 Network Security as a Business Issue

Many organizations run on information, and a well-planned network circulates this information lifeblood to all parts of an organization as efficiently as possible. Inappropriate network security provisions, however, can reduce network flexibility and still not close the door against unauthorized access and information loss. The ability of a network to blend an advanced level of security with maximum operating flexibility, therefore, must be considered carefully in any network plans.

An effective information security program incorporates a combination of technological and human controls in order to avoid the loss of information, deter accidental or intentional unauthorized activities, prevent unauthorized data access, detect a loss or impending loss, recover after a loss has occurred, and correct system vulnerabilities to prevent the same loss from happening again [17]. Correspondence among businesses, internal or external, is conducted through data transmissions. Data transmissions pass in networks of interconnected portals where parties could get in touch with one another. Networks need to be protected from both outsiders such as hackers and crackers, and insiders such as employees and other individuals with access to the network.

Unprotected information and computer networks can seriously damage a business's future. This happens because of the loss of classified or customer critical information, exposure of trade secrets, unacceptable business interruption, or lawsuits stemming from security breaches. As information and computer network security involves more than technology, companies are now spending more money and man-hours than necessary on cutting-edge technology. Inaccurate analysis of the company's needs can result in greater risk of information loss and higher frequency of security breaches.

Making computer and communication systems more secure is both a technological challenge and a managerial problem. The technology exists to incorporate adequate security safeguards within these systems, but the managerial demand for secure systems is virtually non-existent outside of the defense and financial industries. That so many of our commercial systems provide marginal security at best is a reflection of the lack of managerial awareness and understanding of the need to protect the information stored in, and transmitted between, computers.

Unprotected information and computer networks mean loss of data that are deemed crucial and confidential for the company's own development; loss of confidential third-party data; and business interruption or slowdowns that significantly impact the business as well as other parties.

Senior managers are becoming more and more aware of the need to address security and information technology investments within the context of the corporation's business goals. As Schwartau [18] has observed, security is no longer just about security. This also involves the proper identification of roles. Today, security is about resource and information management and it turns out that good security is a by-product of a well-run organization.

Information systems (IS) executives are most concerned with ensuring that their technology goals are consistent with those of the overall business, believing that an effective organization and usage of the company's data is a critical IS activity.

1.4.2 The Focus of Investment on Network Security

In all organizations the directors are faced with the problem of achieving organizational goals using limited resources available to them. Investing in network security should always seek to move the organization nearer to securing the entire network.

The directors should also keep in mind that investing in network security is an ongoing process as the threats are always being upgraded time and again. This is so due to the fact that the hacking tools are readily available on the market and minimal training is required. Investment decisions are essentially search projects which are worth more than they cost to exploit and which thus create value.

Information security, right now, is a confused and paradoxical business. For example:

(1) You have increased spending significantly, and you are told that this is a good thing, and yet it has had zero effect in mitigating security breaches.

(2) You are constantly warned about "digital Pearl Harbours" and yet the vast majority of incidents you report are relatively small, do not last long and do not cost much.

(3) You are told that aligning security and business strategies is a top priority, and yet those who have fared best in avoiding breaches, downtime, and security-related damages are the least likely to be aligned with the business. But in another sense, you seem to be contributing to the confusion.

(4) Respondents who suffered the most damages from security incidents were two times more likely than the average respondent to plan on decreasing security spending next year.

(5) Those with the most damages were nearly half as likely to list staff training as one of their top three priorities.

(6) A quarter of you neither measured nor reviewed the effectiveness of your information security policies and procedures in the past year.

In short, as much as the emerging information security discipline has grown since its baptism—on September 18, 2001 (one week after the terrorist attacks and the day the Nimda worm hit)—it has not much improved with age.

Network resources allow worldwide access to information, no matter where it resides, or where the users are located. Unfortunately, the physical and logical controls used by an organization to secure its information offer no protection when that information is being electronically transmitted.

As Frankwood [19] observed, a successful financial decision is one that creates value, that is, it is worth more than it costs to implement. The investment decision is perhaps the most important of all decision an organization makes. In any organization there are strategic, technical, and political issues as well as financial aspects to tackle with. It involves more than simply number-crunching or relative financial costs and benefits. Many costs and benefits in network security investment are difficult if not impossible to quantify. Moreover, the numbers are often secondary, what determines whether an investment is accepted or not depends on the strategic direction the organization wants to pursue. Moreover, the approval of investment may hinge on the nature of the decision process within the organization, that is, investment approval in reality is often a question of internal politics. To obtain an approval of investment in network security management, managers must demonstrate that an investment is necessary to maintain security by replacing old or obsolete technology. Many organizations classify investments into various categories in which different amounts of information are required as inputs to the evaluation process. The foregoing discussion highlights the need for a high-level systems framework to address the essential facets of information security. This in essence, can be a facet of the sustainability framework depicted in Figures 1.1.

1.5 CONCLUSIONS

Sustainability is highly desirable yet more complex than is often perceived, hence the need for an integrated systems framework to characterize, verify, validate, and communicate the underpinning factors globally. The secondary gain from a systemic framework is the extension, customization, evaluation, benchmarking, and assessing sustainability properties in products, processes, systems, and undertakings. In the latter context, it is vital that a systems framework for sustainability provides a scalable and adaptive architecture thus rendering it applicable at any level of perspective, from individual components, sub-systems and products to the macro-level consideration of the Gaia. The systems sustainability framework depicted in Figure 1.1 delivers a scalable systemic architecture capable of customization and application at any level of perspective.

However, apart from inclusivity and adaptability, sustainability and its systemic framework should lend themselves to application throughout the life-cycle from concept and feasibility to deployment, maintenance, and disposal. The systems framework depicted in Figures 1.1 and 1.5 is capable of adaptation and transverse application throughout the life of a product, process, system, or undertaking irrespective of the scale and scope of the case.

Given its inclusive and generic nature, the systems sustainability framework developed and proposed here provides a potent candidate for a unification of many emergent properties from safety and quality to security and technical performance. This enables stakeholders, business enterprises, and influential government and private institutions to achieve legal compliance and many other goals alongside other business imperatives.

Whilst the framework provides an image that transcends beyond ecological concerns, it is inevitable that sustainability will gain a more ecological bias should the dire predictions of the climate change begin to surface as real threats to the whole planet. To this end and to ensure risk/reward-based informed decision-making by all stakeholders, an objective set of metrics and benchmarks for sustainability will gain more urgency in an attempt to make this pervasive and increasingly vital facet, an integral aspect of decision-making.

We have developed the case for an inclusive, generic and systematic approach to sustainability which integrates many business and regulated facets of performance namely safety and environmental dimensions. Once verified, validated and adopted across many fields, sustainability may require regulation to ensure that it continues to remain a potent weapon in our quest to preserve the integrity and continued survival of Gaia.

The extension of systems thinking to sustainable systems can span from ethical sustainable decisions in the face of dwindling natural resources [20] to developing sustainable business practices [21]. The scope of the application from soft to hard systems is vast and necessitates adoption of an underlying framework for consistency, transparency, and relative comparability of the potential assessment outcomes. The systemic framework depicted in Figure 1.1 is recommended for such underpinning. As a universal model for systemic sustainability, the framework can be applied to any soft or hard domain through decomposition of the so called level 1 goals in the schema to domain specific factors as we have illustrated in the decompositions depicted in Figures 1.2 and 1.3, respectively.

We should endeavor to progress and prosper whilst assuring the rights of the future generations to enjoy the same privileges. This should be our ultimate goal, beyond short-term gains and losses. Our creativity and ingenuity is the sole weapon in this battle. Systems thinking is the strategic guide and a framework of systemic factors the route-map.

> "Whatever you can do or dream you can, begin it. Boldness has genius, power and magic in it. Begin it now!"
>
> —Johan Wolfgang von Goethe

REFERENCES

[1] United Nations General Assembly. Report of the World Commission on Environment and Development: Our Common Future. Transmitted to the General Assembly as an Annex to Document A/42/427—Development and International Co-operation: Environment, 1987.

[2] Engineering Council UK. Guidance on sustainability for the engineering profession, May 2009. Available at www.engc.org.uk/sustainability, accessed May 2009.

[3] F. Hsu. The triple-triplet conceptual framework for integrated risk management (PSAM-0278). In: Proceedings of the Eighth International Conference on Probabilistic Safety Assessment and Management, New Orleans, LA, May 2006.

[4] J. Lovelock. *Homage to Gaia: The Life of an Independent Scientist*, Oxford University Press, 2001.

[5] S. Popper, S. Bankes, R. Callaway, and D. DeLaurentis. System-of-Systems Symposium: Report on a Summer Conversation, July 21–22, 2004, Potomac Institute for Policy Studies, Arlington, VA, 2004.

[6] A. Hunter and A. G. Hessami. *Formalization of Weighted Factors Analysis, Knowledge-Based Systems*, Elsevier, 2002.

[7] A. G. Hessami. Safety assurance, a systems paradigm. *Hazard Prevention—Journal of System Safety Society*, 35(3), 8–13, 1999.

[8] A. G. Hessami. Risk, a missed opportunity. *Risk and Continuity Journal*, 2: 17–26, 1999.

[9] A. G. Hessami. A systems framework for safety & security, the holistic paradigm. *Systems Engineering—The Journal of the International Council on Systems Engineering*, 7(2): 99–112, 2004.

[10] A. G. Hessami. Risk management a systems paradigm. *Systems Engineering—The Journal of the International Council on Systems Engineering*, 2(3): 156–167, 1999.

[11] Conceptual Framework Working Group of the Millennium Ecosystem Assessment. *Ecosystems and Human Well-being*. London: Island Press, 2003. Chapter 5. ISBN 155634030.

[12] F. Hsu. Managing programmatic risk in strategic decisions—a methodology for programmatic & strategic risk assessment and management. In: Proc. PSAM9, HK, 2008.

[13] F. Hsu. A goal-driven hierarchy process of metrics-based decision assessment for the lunar surface science scenarios, NASA OSEWG Report, 2009.

[14] ISO/IEC 15288, System Life Cycle Processes, October 2002.

[15] P. Maloney. Security zone: opinions and insights from experienced professionals. *Computer Weekly*, December 16, 2008.

[16] J. Holt. A Pragmatic Guide to Business Process Modelling, Second Edition, British Computer Society, 2009, ISBN 1906124124, 9781906124120.

[17] H. Jahankhani, S. Fernando, and M. Z. Nkhoma. Information systems security: cases of network administrator threats. *The International Journal of Information Security and Privacy*, 1(3): 13–25, 2007.

[18] W. Schwartau. Securing the enterprise. Technology alone won't make you safe. Tackle it as a management problem. *Network World*, 42, January 27, 2005.

[19] Frankwood. *Introduction to Accounting*. Prentice Hall, 2000.

[20] P. L. Kunsch, M. Theys, and J. P. Brans. The importance of systems thinking in ethical and sustainable decision-making. *Central European Journal of Operations Research*, 15(3): 253–269, 2007.

[21] H. Bradbury. Sustaining inner and outer worlds: a whole-systems approach to developing sustainable business practices in management. *Journal of Management Education*, 27(2): 172–187, 2003.

[22] A. Waring. *Practical Systems Thinking*. International Thomson Business Press, 1996. ISBN 0-412-71750-6.

CHAPTER 2

SYSTEM OF SYSTEMS THINKING IN POLICY DEVELOPMENT: CHALLENGES AND OPPORTUNITIES

KEITH W. HIPEL[1,2], LIPING FANG[1,3], and MICHELE BRISTOW[1]

[1]Conflict Analysis Group, Department of Systems Design Engineering, University of Waterloo, Waterloo, ON, Canada

[2]Centre for International Governance Innovation, Waterloo, ON, Canada

[3]Department of Mechanical and Industrial Engineering, Ryerson University, Toronto, ON, Canada

2.1 INTRODUCTION

2.1.1 A World in Crisis

The climate crisis [1]. The energy crisis [2]. The water crisis [3]. The food crisis [4]. Despite the fact that the standard of living is rising in many nations of the world, particularly in China and India, the state of the world in the twenty-first century is becoming increasingly vulnerable [5]. High energy and food prices, juxtaposed against a recession [6, 7], are signaling times of great difficulty and distress, where it is not only the poorest of the poor that have been hit hard, but also the urban poor [8, 9]. Although less severely affected, the rich have noticed too. Global warming and ensuing freshwater scarcity still threaten the livelihoods of many. These are matters of international scope. Everyone is affected because everyone is connected through the natural and man-made systems that provide us with the necessities to sustain our living. The connections among systems are complex and these interactions contribute to the emerging crises. As argued by Homer-Dixon [10] and others [5], these complex large-scale system of systems (SoS) problems are highly interconnected and are focusing their highly destructive power on an increasingly vulnerable and fragile world.

Contemporary Issues in Systems Science and Engineering, First Edition.
Edited by MengChu Zhou, Han-Xiong Li and Margot Weijnen.

The climate system, which involves the atmosphere, oceans, land, and ice—an SoS—has served to regulate weather patterns over time and to cultivate the incredible diversity of the natural world that we are only beginning to discover [11]. However, more and more evidence from terrestrial and marine species shows that recent warming is strongly affecting natural biological systems [12]. Shifting habitat ranges, advancing migration dates, altering times of mating, and earlier flowering have been attributed to climate change [12, 13], and although the long-term consequences are still being scientifically and politically debated, the documented observations are real and must be incorporated into our resource management decisions and policy development [14]. Our managed energy, water, and food systems—in and of themselves, systems of systems—are heavily interconnected with the natural environment, as it would simply be physically impossible and uneconomical for them to function without it. Addressing the climate crisis is in effect also considering these other systems of systems, in part.

Factors such as declining oil supply, limited fresh water, and finite arable land are natural constraints to which humans must learn to adapt in managing this energy–water–food supply. However, the bulk of our problems are not imposed by natural systems, but rather by the systems and institutions that have been created by humans—trade and the economy, politics and government, civil law and social norms, and science and technology. Specifically, in the case of the global food system, fundamental factors at the heart of the global food crisis are volatile markets, subsidization, trading cartels, centralized agribusiness, and industrialized agriculture [4]. Paradoxically, it would seem, it is our well-established ways of doing things that are fundamentally the reason why the system is failing. Moreover, considering the food system as a complex adaptive system, the interdependencies that have increased over time among countries now labeled as low income food deficit countries (LIFDC) and other nations have made the system extremely vulnerable to external shocks, such as speculation, increased demand, and decreased supply. The solutions to the global food crisis are not simple. One must study the interactions between and among natural systems and man-made systems, where humans are the fundamental actors who can observe and effect changes (with and without intent) in the dynamical relationships and future outcomes. The climate, food, energy, and water crises are each an SoS problem.

The main barriers that we face, however, which are also the causes of debate over the long-term consequences of global warming, are the unpredictable outcomes, for example, changes in the natural world due to climate change that may have an effect on the quality of human life. Furthermore, there is uncertainty; while it is clear that carbon dioxide concentrations in the atmosphere have increased beyond normal variation, largely due to the burning of fossil fuels, the point of saturation and irreversibility is unknown. In a complex adaptive system, natural variation versus accelerated variation can mean the difference between strengthened resilience and total collapse [10]. Where do we draw the line? The risks due to unpredictability and uncertainty make management very difficult, though all the more necessary. Moreover, in any situation where there are multiple participants, which is an inherent characteristic of systems, there will be multiple conflicting value sets and multiple objectives [15]. Decisions are made largely based on value judgments, which can have a huge impact on whether to escalate a conflict or disarm. The interactions played out through relationships among individuals, groups, governments, non-governmental organizations (NGO), corporations, and the environment can be purely competitive, collaborative, or a mixture of both. With so many stakeholders and interests, coordination of systems is also a major challenge. When following an SoS approach, one desires to mitigate the risks associated with decisions and to resolve conflicts over multiple values and objectives with operational methods.

An SoS approach recognizes that multiple participants and differences in their intrinsic value/survival systems may ignite conflicts, and thereby seeks to operationalize value systems for use in designing policies that are ethically based, such that disputes may be resolved effectively or avoided altogether. Moreover, adaptation in an SoS can serve as a mechanism to reduce the impacts of conflicts and crises, and should therefore be an integral part of policy development. Complex systems modeling is used to understand emergence and to learn how to make use of emergent behavior. In addition to complexity, uncertainty and risk must also be understood in order to provide informative decision-making. In the face of present global crises, what is needed are holistic and operational methods that can integrate multiple stakeholders' values into policies, as well as foster adaptability through policies. SoS thinking seeks to provide key decision makers with tools to model value systems, delineate risk, and coordinate strategies with other decision makers in an attempt to reach win/win resolutions.

Within this chapter, we review the literature on SoS thinking with respect to value systems and ethics, complex adaptive systems, risk, and cooperation and we frame the content in the context of policy development for solving global problems. In addition to climate change, energy, water and food, there are several other fundamental global challenges, such as transnational organized crime, health issues, democratization, and the status of women, that also play very significant roles in altering the state of the world as a whole [5]. The global nature of these problems hints at the need for global collective actions. However, how to start the process, monitor the dynamics, and assess the performance of our actions is relatively less obvious. From the holistic perspective of SoS thinking, a unique opportunity exists to tackle global challenges and to design national and global policies and integrate them with local and regional initiatives with deeper insight. A sound policy respects the ethics and values of multiple stakeholders, and harnesses the natural complexities of the interconnected systems to mitigate risks through adaptation, while providing an effective conflict resolution mechanism for resolving unavoidable disputes. With these principles in mind, several policies are discussed in this chapter to exemplify areas of effective policy design and domains for future work. A briefing of this chapter is also available [16].

2.1.2 System of Systems

A system is a composition of entities, which may be physical, behavioral, or symbolic in nature, that are interrelated, interdependent, and mutually interacting [17–19]. Components connect in a way, whether purposefully or not, to accomplish functions and objectives synergistically, that would not be achievable disjointedly [20]. The point of departure from conventional systems thinking to an SoS perspective is that components in a system are by themselves individual systems that are autonomous, heterogeneous, and complex [21, 22]. Considering that a set of interacting components in a system could be considered a system in its own right is not novel, as Ackoff [23] recognized in stating that "every system can be conceptualized as part of another and larger system." However, Sage and Biemer [22], based on a literature review of SoS concept definitions, point out seven characteristics that differentiate an SoS from a classically engineered system of subsystems. These are: (1) operational independence of the individual systems; (2) managerial independence of the individual systems; (3) geographical distribution; (4) emergent behavior; (5) evolutionary development; (6) self-organization; and (7) adaptation. Alternatively, Boardman and colleagues [24, 25], based on their review of 40 SoS definitions, chose five characteristics: (1) autonomy: encompasses managerial and operational independence; (2) belonging: the right and ability of constituent systems to belong to an SoS; (3) connectivity: the ability to stay connected to other constituent systems; (4) diversity: visible heterogeneity; and (5) emergence: encompasses emergent behavior, evolutionary development, self-organization, and adaptation. Through the advances that have been made in understanding complex adaptive systems in biology, sociology, and computer science [26–29], SoS thinking is simply a natural evolution of systems thinking to incorporate this new knowledge, demonstrating the potentially great scope of systems science across different scales and application areas [30]. For policy development, the purpose of utilizing formalized SoS thinking is to harness the complexity [31, 32] that is inherent in large-scale, multidisciplinary issues and to address the composite SoS problems that arise from interactions among environmental, societal, intelligent, and integrated systems [17, 15].

All systems of systems are inhabited by multiple participants with multiple objectives [15]. For example, water resources management faces many SoS problems that involve energy, services and infrastructure, industrial, and agricultural stakeholders. Furthermore, these systems of systems are intimately reliant on the natural environment, and thus can be adversely affected by natural disasters such as floods and droughts. Note that each individual system is operationally and managerially independent. Moreover, the geographic distribution of the SoS means that numerous people, societies, and nations in our highly interconnected world face a looming water crisis as the world supply of fresh water becomes more and more scarce due to climatic change, pollution, exponential population growth and industrialization of developing countries such as China and India [3, 33–35]. The SoS consists of many different participants or agents who are interacting synergistically for procuring water supply, consumption, and recycling. Some of these participants are public water providers, private water vendors, and consumers ranging from individuals to transnational corporations, who rely upon water treatment technologies, be they natural or human-made. Multiple objectives exist within and among the participants. For policy and decision makers who are tasked with managing water resources, systematic methods that are able to communicate the values and interests of all participants are needed to understand the complex behavior of the overall system [36, 37], and hence, to design a policy that reflects those values and ethically binds all participants to respect it. A deeper discussion into value systems and how they play a central role in addressing SoS problems is furnished in Section 2.2.

All systems of systems, by definition, involve complex adaptive systems [17] that exhibit emergent behavior, evolutionary development, self-organization, and adaptation [22]. Consider the global food system, wherein demonstrations by thousands of Haitians on April 4, 2008, triggered by high food prices, turned into riots [38]. The sharp escalation of food prices world-wide signaled the emergence of a global food security crisis [4, 39, 40]. The causes are proclaimed to be the rising demand of basic staple foods for other purposes such as biofuels and meat production, falling global food stocks, increasing costs of production linked to high oil prices, market speculation, and export bans [4, 41]. Although there is an ongoing debate on what exactly the main factors of the global food crisis are [42, 43]—in itself, a manifestation of the hidden agendas and interests of representative stakeholders—the event demonstrates the nonlinear unpredictability of emergent behavior in the global food system, which has been for thirty years financially distorted by agricultural subsidies in rich developed countries [4]. Moreover, the oligopolistic behavior of agencies in international trade, which is exemplified in the World Wheat Market, creates imperfect competition and enables special interest groups to influence agricultural policy decisions [44]. The vulnerabilities of poor developing countries were not fashioned overnight, but instead evolved over time as their investments moved away from local subsistence agriculture to industries and markets in which they had comparative economic advantage [4]. Accordingly, they became more and more dependent on importing basic staples. Slow processes were in place, unbeknownst to the policy and decision makers, which have caused the global food system to evolve and self-organize into a dysfunctional global agricultural marketplace that exists today. Through the complexity, however, the saving grace is adaptation—a change to an entity in response to other entities and its environment to become fitter for survival. Participants are capable of adapting independently in terms of form (what they are) and function (what they do). Form and function together create a variety of available strategies, which are conditional sequences of actions that participants undertake to achieve their objectives. The adaptation of participants such as farmers, consumers, and transnational food corporations will depend on cues from each other and other stakeholders, as well as from the environment. This can present opportunities as well as barriers. It is important for policy and decision makers to understand the interlocking processes of adaptation in order to guide productive institutional reforms that make use of the complexity. Lessons from complexity research that apply to SoS problems are presented in Section 2.3.

Finally, all systems of systems must manage risks in their uncertain and unpredictable environments. Natural events such as earthquakes and hurricanes occur and are beyond the control of any human-made system. The Asian Tsunami (2004) and Hurricane Katrina (2005) are ranked among the most devastating disasters in the twenty-first century. However, a close look at the case of Hurricane Katrina reveals that the disastrous effects that sunk New Orleans were due to man-made errors in the 100-year flood protection system [45] and faulty engineering solutions that were misguided by conflicting priorities of different stakeholders, such as the US Army

Corps of Engineers, politicians, and other commercial-interest groups in the oil and shipping industry [46].

More frequently, disasters occur and are made worse by societal systems that put people and other systems at greater risk. Thus, in order to fully address vulnerabilities and mitigate risk factors, a more balanced assessment that shifts the dominant attention on technical factors in the natural environmental systems toward greater focus on social, political, and economic factors in societal systems is required [47]. For example, mitigating the risks of natural disasters due to climate change goes far beyond understanding the science behind weather patterns, although that is an important component. It involves understanding the interactions among geophysical, biological, and global human systems, which include food supply, infrastructure, health, and water resources. Human development has significantly reduced the resilience of ecosystems, thereby making ecosystems and their biologically diverse populations more vulnerable to climate change [48]. Moreover, marginalized populations in developing countries, who are caught in social traps such as poverty, are especially vulnerable to the impacts of climate change because of their constrained ability to adapt [49]. The final task of policy and decision makers is to select rules to enforce and strategies to implement policies that are within an acceptable risk tolerance and, in the event of natural hazards or external shocks, are capable of reducing risk of catastrophe and collapse to create opportunities for "catagenesis" through adaptation [10]. As an aid for policy and decision-making in SoS, the quantification of risk with methods that delineate uncertain and unpredictable outcomes is discussed in Section 2.4.

Altogether, SoS thinking is a paradigm shift, succinctly stated by Hipel *et al.* [50], "from a disciplinary to a multidisciplinary outlook, from a mass production to a mass customization focus, from a steady state to a real-time perspective, and from an optimal to an adaptive approach." One should attempt to live with complexity and uncertainty—not to treat global problems in a homogenous and deterministic manner, but rather to continuously search for a state of fairness and balance, which is a dynamic target.

2.2 VALUE SYSTEMS AND ETHICS

While the understanding of systems requires an unbiased, value-neutral science [30], the design of systems requires decisions that are laden with value judgments [51], which if ignored or misunderstood would be an assault on the foundations of systems science, systems engineering and design. Values are ubiquitous in cultures, religions, market economies, institutional laws and policies, and even in science domains. For example, every market transaction presupposes the right to private property and the instituted value of money [52]. The right to private property is believed and prescribed in religious faiths as an undeniable law of God, and is effectively enforced by criminal and civil codes [52]. Without this coercive right and its effective enforcement, there would be no security in the marketplace [52]. Hence, free market exchanges rely on a prescribed value system, which is morally grounded in religious beliefs. In science domains, there are often scientific theories or paradigms that compete with each other to explain natural phenomena, which arise due to the different standards and objectives among the scientific community. When paradigms shift, there are usually significant changes in the criteria determining the legitimacy both of scientific problems and of proposed solutions [53]. However, to be certain that systems science aspires toward neutrality it should be recognized that its application will require deep consideration of the value systems that influence thinking, but also that its application should not be biased toward any singular line of reasoning that stands resolutely on partial truths.

2.2.1 Conflicting Value Systems

Amid the variety of value systems of participants inhabiting an SoS, conflict is an inescapable condition due to the immense diversity. However, important links among their values inextricably connect participants to one another. The environmentalist desires a stable economy and the economist needs a clean environment [11]. To take an extremist stance is to deny the existence of the other and to prescribe the demise of oneself, as each depends on the other. In order to begin understanding global challenges such as climate change, which Wilson [11] notes are too "complicated [and complex] to be solved only by piety and an unyielding clash of good intentions," policy and decision makers need to seek out solutions that put aside moral superiority. Moreover, managed systems of systems must be ethically maintained [15]. Ethics goes beyond ambiguous moral commands that are open to interpretation and are insufficient to protect against the underlying forces that threaten the rights and freedoms of each participant. A universal ethical framework should be able to integrate different values and changing values, reflecting the reality of multiple cultures and beliefs, while maintaining the freedom of each individual. Clearly, current societal systems are not based on a universal ethical framework. Shrybman's [54] commentary on the World Trade Organization (WTO) concludes that "if international investment rules are to foster, rather than undermine, our prospects for achieving environmental goals, they will have to be fundamentally overhauled." Good policies must be firmly based on the values of all stakeholders, not just a privileged minority, which persists in the global market where rights are purposefully given to transnational corporations and no one else [55].

Furthermore, in systems of systems, where direct cause and effect relationships are rarely observable, it is not sufficient to focus on the intentions of participants. It is not intentions that constitute the problem, but rather actions that lead to unintended consequences. Since values drive decision-making, a policymaker must be careful to implement policies that encourage desirable outcomes from decisions and actions that align with participants' values [56], which amidst the complexity is a difficult task. First of all, what would define a desirable outcome? Secondly, how could you ascertain that the desirable outcome was in fact the correct one to encourage? Value assessment and systematic decision-making processes are useful tools to determine clear objectives and criteria when there is no clear solution, which especially abounds in SoS problems. As well, for circumstances in which multiple objectives require some sort of sacrifice (tradeoff), it is all the more important to think deeply about values as a guide to justify the sacrifices that are being asked of each participant. Doing so strengthens the integrity of a policy as in the case of the Montreal protocol on Substances That Deplete the Ozone Layer. Moreover, it aids in uncovering hidden agendas of participants (either obscured by the complexity of the decision or intentionally obscured by a party [57]), which could weaken the enforcement of a policy, as in the case of the Kyoto protocol on reducing greenhouse gases (GHG) in an effort to prevent anthropogenic climate change. Ethical systems and values are made operational through the purposeful analysis of objectives, criteria, and tradeoffs, both qualitatively and quantitatively, to formulate preferences.

2.2.2 Modeling Value Systems

From an alternative-focused perspective, decision-making involves choosing a strategy from a set of identified alternatives [51, 58]. On the other hand, from a value-focused viewpoint, decision-making involves strategy development to meet specified objectives and values [51]. A value system, whether implicitly or explicitly stated, is required in both cases. A value model can be created for any decision-making entity in an SoS, ranging from an individual to a corporation to a computer program to an intelligent robot, entirely designed and hard-wired with value systems by people. Value-focused thinking starts with the development of a value model as opposed to alternative-focused thinking which begins with listing courses of action. In terms of SoS policy development where multiple participants and multiple objectives abound and alternatives have uncertain consequences, a value-focused approach is much more practical, as it can guide the decision-making process along a directed path to creating strategies that align with all participants' value systems. In policy development, a value model may be created for each participant, or an aggregated value model can be constructed to integrate the value systems of all participants. In Keeney's framework of value-focused thinking [51], three key steps in specifying values for a decision situation are laid out as follows:

(1) identifying and structuring objectives;
(2) measuring the achievement of objectives; and
(3) quantifying objectives with a value model.

Objectives are first identified by asking the question: "What would you like to achieve in this situation?" in an open and creative atmosphere, in which the goal is to create a list of objectives that is as complete as possible. For a policymaker who wonders whether the mechanisms in the Kyoto protocol are ineffective [59, 60] or who seeks to create new strategies that would improve international cooperation in reducing GHG emissions, reviewing the value systems of key decision makers and possibly other stakeholders to identify the objectives of a climate change policy is a reasonable first step. Then, after generating a list of objectives, the list is separated into means objectives and fundamental objectives, which is carried out by asking the question for each objective: "Why is this objective important?" A means objective is a way of satisfying other objectives. For example, why is reduction of GHG

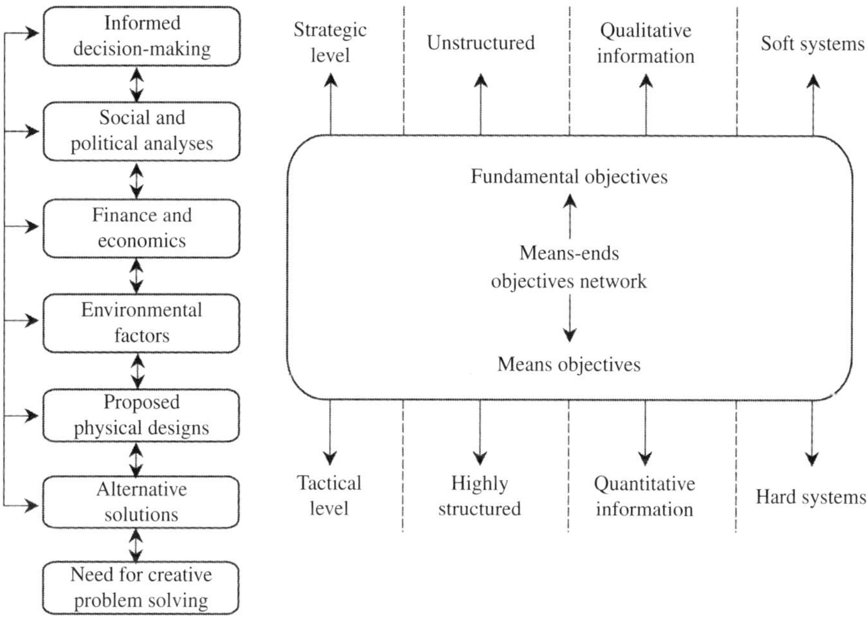

FIGURE 2.1 Fundamental and means objectives in SoS decision-making framework.

emissions important? The response may be to prevent anthropogenic climate change, to which, the question should be asked: why is preventing anthropogenic climate change important? In response, it may be that the reasons are to protect biological diversity, to prevent adverse effects on global water–food supply, and so on. This procedure serves several purposes, which include linking objectives through a means-ends network and uncovering more specific fundamental objectives. A fundamental objective is one whose answer to the "why is this objective important" question is "because it is simply important," and is a stone's throw away from the values of a stakeholder. Means-ends objectives networks and fundamental objectives may differ according to the value systems against which the objectives are considered and they may contain similarities. These observations can provide the policymaker with valuable qualitative insight into potentialities of conflict, as well as opportunities for cooperation.

Fundamental and means objectives can also be draped over the top of an SoS decision-making hierarchical framework [50, 61], where fundamental objectives reside at the strategic level and means-ends objectives chains trickle down to the tactical level, as shown in Figure 2.1. Fundamental objectives are unstructured and associated with strategy, soft systems, and qualitative information. On the other hand, means objectives are highly structured and associated with tactical operations, hard systems, and quantitative information. The methods of measuring the achievement of objectives from the strategic level to the tactical level also range in terms of data and timescale. Strategic level objectives are likely to be longer term and associated with more qualitative data, while tactical level objectives are shorter term and measurable with more quantitative data. Determining an appropriate method of measurement becomes more difficult as one traverses from the tactical level to the strategic level. Attributes or criteria are the yardsticks to determine the degree to which an objective is achieved. Eventually, alternatives are evaluated according to their attributes, which describe the consequences of each alternative.

Keeney [51] groups attributes into three types: natural, constructed, and proxy. Natural attributes are those that fall out from the objective in a fairly direct manner, and are generally straightforward and structured for tactical level objectives. When it is not possible to come up with natural attributes, a constructed attribute which is context specific can be formulated based on a description of impact levels that indicate the degree to which the associated objective is achieved. Moreover, when neither natural nor constructed attributes are identifiable, proxy attributes, which utilize an indirect measure, can be used, which is often the case for fundamental objectives where quantitative data are scarce, and only qualitative information is available.

For all objectives, it is generally desirable for attributes to be measurable, operational, and understandable, which collectively removes ambiguity from the decision context [51], and essentially discloses the value judgments which are used to evaluate alternatives. When objectives are measurable, it becomes possible to utilize mathematical programming or optimization approaches. Wang *et al.* [62] demonstrate how to obtain measurable principles and use them in multi-objective optimization. Measurability places an unbiased scale on the attribute to determine the degree of achieving the objective. Operationality of an attribute is concerned with the details of how measurements are taken, as well as how well the attribute qualifies the consequences of alternatives to specific measurement levels. Finally, understandability means that there should be no ambiguity in describing and interpreting consequences described in terms of attributes. With this in mind, attributes should be value-neutral yardsticks. However, the selection of attributes is based on value judgments. For example, in environmental policies that consider risks to human health, an objective of such policies may be to "minimize the loss of life," for which possible attributes could include an "annual number of fatalities" and "total year of expected life lost" [51]. Both attributes are value-neutral, in that regardless of a decision maker's value system, the measurement of the attribute does not change for different decision makers. Nonetheless, the selection of one attribute over the other requires a value judgment on whether each death is equal or the death of a younger person is more significant than that of an older person. Delineating attributes to their underlying value judgments helps to determine whether the selected attribute is appropriate for the given decision context.

Finally, using the identified objectives and attributes, a value model quantifies the objectives mathematically by assigning either utility (quantitative) [63–65] or conditional ranking (non-quantitative) preference information [66–68]. Put simply, an alternative with a higher utility or higher ranking is a more preferred alternative compared to one with lower utility or lower ranking. A value model describes the preference structure of a decision maker's value system, which is used to determine the relative desirability of consequences, and the overall preferences for alternatives [68]. In general, the foundation for any value model is the set of objectives which are chosen to reflect the decision under consideration [51]. A value model assigns a number to each alternative, where the number corresponds to the measurement of an attribute, which indicates the degree of achieving the corresponding objective [51]. Using preference relations, which may be multiplicative, fuzzy, intuitionistic, linguistic, and possibly incomplete and uncertain, depending on the availability of information, it may be possible to derive a utility function which can be used to calculate the utility of any alternative, or at the least, to rank alternatives. Details on different preference relations are explained well by Xu [69] and Xu and Chen [70]. The general procedure for a policymaker to formulate value models would require one to first determine the value tradeoffs which are encountered in the given set of objectives. Value tradeoffs are necessary when objectives compete for limited resources. Thinking about value tradeoffs requires a judgment on exactly how much a participant is willing to sacrifice in achieving one objective for a specified amount of gain in achieving another objective. Appropriate tradeoffs can then be encapsulated into a value model through the specification of preference relations, for example, with a prioritization algorithm [71]. Preference relations can be used to evaluate alternatives, but more importantly to create alternatives that complement the modeled value systems.

Hence, policy and decision makers can incorporate values into policy development and decision-making in a systematic way. The performance of value assessments should lead to clearer communication of objectives, more transparency of value judgments through the selection of appropriate attributes, and better accounting of value tradeoffs in eliciting utility and preference information. As Hipel and Fang [15] have previously pointed out, it is important to keep in mind that differences in value systems and underlying ethics and beliefs provide the basic fuel for igniting and maintaining conflict. For conflict resolution, it is necessary to determine the value systems of each decision maker, and to subsequently analyze the strategic interactions among the decision makers which result from the differences in values. As in the Graph Model for Conflict Resolution (GMCR) [72], values can be conveniently obtained using an option prioritization approach, which mathematically derives a preference structure based on a set of lexicographic preference statements about options [71]. The derived preference structures of many stakeholders can be utilized in decision support systems (DSS) to model conflicts, which aids the decision maker in making an informed, and thus, better decision [73]. An aggregated preference structure of a group of decision makers can also be generated and integrated into a DSS for use in group decision-making situations. Since a DSS is only as good as the information that is provided to it, it is important to use a systematic method, such as the one presented, to understand and operationalize value systems. However, despite the importance of incorporating all stakeholders' values into policy design, it is a time-consuming process which is often dismissed because of the high immediate costs for obscure long-term payoffs. To be practical, the costs of value assessments need to be reduced and the results have to be relevant in a fast-pace changing environment in which values change and new values emerge. In other words, policies and decisions that need to be made today, should be considered in the context of a complex adaptive system.

2.3 COMPLEX ADAPTIVE SYSTEMS

As we gain more understanding of ourselves and of our world, we become much more powerful and capable of utilizing both the world's natural resources and human resources in SoS architectures to accomplish incredible feats and to overcome insurmountable challenges. Examples include the creation of the Airbus A380, which forged a global network of over a hundred suppliers and partners in more than 20 countries [74]; the institution of the Grameen Bank, which has revolutionized the financial system to include poor people [75]; and the formation of the European Union (EU), which has evolved new forms of political cooperation and economic development and integration [76]. Concurrently, with great power follows great responsibility for the consequences which are brought about by the systemic behavior of technological and societal systems. Consequences can range from observed symptoms such as pollution and climate change, to systemic diseases such as corruption and overexploitation, to diagnosed death—the irreversible collapse of the systems themselves. Thankfully, as in the cases of the A380, the Grameen Bank, and the EU, it is possible to harness the complexity to not only achieve great things, but also to adapt and evolve in such a way that resilience and self-organization are embedded into the system structure and dynamics. In fact, since all systems of systems are complex adaptive systems either in part or whole, dealing with consequences should be simply a matter of turning on the adaptive capabilities of the system. Of course, adaptation itself is not so simple. However, complexity research provides a scientific basis for both designing policies and testing implementations of policies through simulations to identify vulnerabilities, and hence flag opportunities for change.

2.3.1 Emergent Behavior

Complexity research is essentially the study of interactions in a system that give rise to collective behavior, also known as emergent behavior, because the properties of the system are not characteristic or attributable to its constituent parts on their own. It is usually only after the observation of an emerging situation that experts can find an explanation and make further hypotheses based on this new knowledge. This is what happened with chlorofluorocarbons (CFCs) and the ozone system. After determining that the release of CFCs into the atmosphere was the cause of ozone layer depletion, experts have hypothesized that if bromofluorocarbons (BFCs) were used instead of CFCs, it would have likely resulted in the total collapse of the ozone system [1]. Emergence and adaptive learning are central themes of complex systems theory. Emergent behavior is often associated with outcomes that are unpredictable, and thus surprising to observers. In fact, an emergent property is one that cannot be predicted. It is not that it is uncertain, it is that given everything that you know, you would not be able to analytically determine the outcome, either deterministically or stochastically, or both. An example of emergence is the 1997 Asian Financial Markets Crisis, which started when Thailand devalued its currency, triggering a cascade of financial crises in one country after another in East Asia and eventually rippling throughout the global economy. Investors and speculators were stunned when they saw big players like Russia default on its sovereign debt due to the crisis [10]. In the current economic crisis, as of October 2008, financial experts were not sure if Canada was entering a recession, depression, or neither [77, 78].

Another example of emergent behavior is blackouts, which are often surprises, at least for the consumers. In the investigations of power outages that affect many people over a long period of time such as the Northeast Blackout of 2003, it is often found that they are triggered by an event that results in a cascading effect [79], similar to the one observed in the crashing of financial markets. It would be just as ridiculous to blame Thailand for the 1997 financial crisis as it would be to blame a tree that is in contact with a power line for a power outage. Instead, the emergent properties and dynamics of the systems are recognized as products of the interactions between fast- and slow-moving variables and processes, such as, in the case of blackouts: electricity demand, control systems feedback and monitoring, maintenance, and deteriorating infrastructure. As argued by Hipel and Fang [15], the resulting interactions are due to an underlying clash of value systems. Moreover, they are results of decisions made by many individuals, who are governed by value systems that determine human behavior as well as organizational structures and institutional arrangements that mediate the relationship among humans, nature, and technology [80]. Thus, in order to understand emergence so as to avoid catastrophe and to use this property for productive ends, researchers are keen to study the structure and dynamics of complex adaptive systems.

Interactions can be viewed statically in terms of the system structure as well as dynamically with respect to its evolving structure and the flows of influence through the system, be they market crashes or power failures. Networks and graph theory have provided useful tools to model and analyze the structural topology of real-world systems. Current research seeks a dynamical systems viewpoint to analyze the dynamical properties of a system in conjunction with its structure [81]. Several theoretical models such as Erdõs and Rényi's random graphs [82], Watts and Strogatz's concept of small-world networks [83, 84], and Barabási and Albert's derivation of scale-free networks [85] are conceptual abstractions of the complex structures that appear in real-world applications such as social networks, protein structure, air traffic, the Internet, and the World Wide Web. In particular, scale-free network models, due to the ubiquity of this type in real-world systems, are useful in explaining emergent properties such as resilience to breakdowns [86] and attacks [87], and epidemics through the spread of infection [88, 89]. Such models may be useful in gaining more insight into the vulnerabilities of the food, water, and energy supply systems, which must be robust against shortages and failures in the system.

In addition to network models, agent-based modeling and cellular automata are also exhibiting suitability in studying the complexity of systems in which the actions of a large number of participants or "agents" or "cells" are interconnected and patterns of behavior emerge over time and space. Cellular automata and agent-based methods have been proposed to study the complexity of cities, wherein models are used to recreate patterns of land use and urban growth in terms of how individuals behave and the processes they engage in to develop their environment [90]. In an attempt to explain ethnic violence, Lim *et al.* [91] created an agent-based model that simulates a process of global pattern formation of regions differentiated by culture. It is used to predict the occurrence of violence where the boundaries between regions are not sufficiently well defined. The policy implications from modeling these sociological phenomena might be lessons that lead to the creation of strategies that foster desirable emergent properties that counteract the negative ones, or new methods that deal effectively with perpetuating problems [80].

Formulating networks and agent-based and cellular automata models is a relatively straightforward task, but only after the purpose of the analysis and context of the problem are well defined. A model can be constructed fairly quickly based on simple interaction rules and processes. However, modelers and analysts must be cautious of the possibility of equifinality, which is when different processes result in the same outcomes. The real difficulty of dealing with emergence is that even though you may have a model that predicts the situation, it does not mean that it is the right model [90]. Although there is no way to be completely certain about unpredictable behavior, a model is still useful if it fulfills the objectives that it was set out to achieve from the beginning, which under the restrictions of time, may be simply to explore as many new ideas as possible without worrying about the details. Results can lead to more structured and deeper analysis. Perhaps the objective of the analysis is to test an ensemble of policies in multiple scenarios which would require many different model deployments. Simulating multi-agent systems is a way of representing many possible realities based on a single model constructed by the designer [28, 92]. It can be useful in policy analysis in order to understand emergent properties and the consequences of interacting policies. Whatever the case, in order to study the behavior of a complex adaptive SoS or a policy designed to manage it in an integrative and adaptive fashion, one can employ the foregoing and other formal systems engineering tools. In this way, one model is not the end all and be all, but rather hints in the learning process toward a more complete understanding of the evolving organizational structure and dynamical behavior of an SoS. Hence, policy can be thoroughly tested before it is put into place and adaptively improved over time.

2.3.2 Modeling Complex Systems

In modeling complex adaptive systems, the focus is on how local interactions create global structures and patterns. An immense amount of research effort has been and continues to be put into the creation and validation of new computational models. Creating a model is by no means a simple task; however, using a well-defined approach ensures that the constructed model is transparent, clear, and consistent. In an SoS approach, an analyst must first define the problem, and then determine the system perspective of the problem under consideration, which involves identifying the relevant participants, their value systems, and associated processes. An SoS model needs not include everything under the sun, but rather frames the context such that

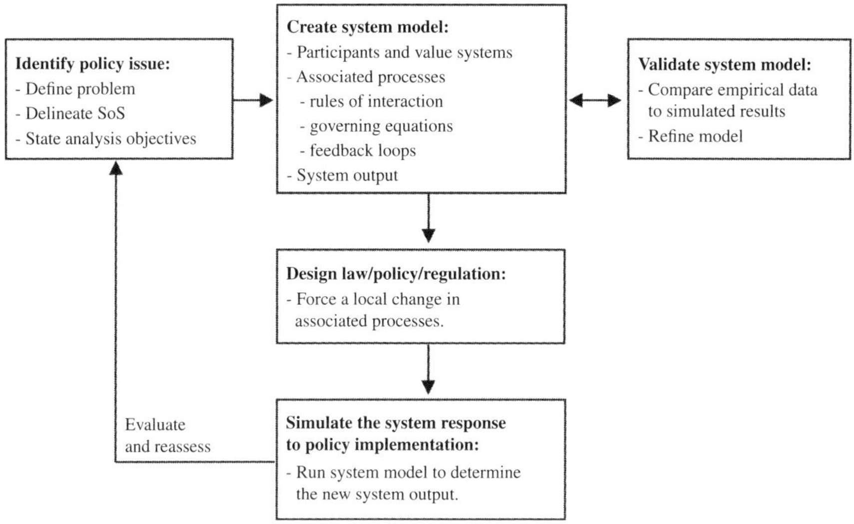

FIGURE 2.2 Main steps in simulating policy design.

the full dimensionality of the problem is brought into the light. Relevant participants may include entities that have influence over the outcomes of the system's behavior as well as those that are affected by the consequences of those outcomes. Participants can be represented abstractly as nodes in a network, cells in a cellular automaton, and decision makers, agents, or actors depending on the terminology of the formalized modeling technique. The relevant processes are the interactions and relationships among participants. Visually, they may be modeled as arcs in a network to expose the communication pathways from one entity to another, which would otherwise be obscured. Processes may be expressed as rules-based strategies for cells and agents to formulate the underlying thinking, or value system, that guides their actions, which have consequences. In particular, the GMCR, which was developed separately from classical game theory and prior to agent-based modeling [66], provides a formalized framework for emulating human behavior in decision-making. Hipel and Fang [15] mention that GMCR solution concepts could be used in agent-based models to dictate agent interactions.

In a computational setting, a complex system can be simulated using very simple rules. The main steps of simulating policies in a complex system are shown in a process diagram in Figure 2.2. The first step is to identify the policy issue with a problem definition, a delineation of the SoS of interest, and a clear statement of what the analysis intends to achieve. Once the correct problem is identified, the next step is to create a system model. The associated processes are formalized with rules of interactions, governing equations, and feedback loops. A modeler needs to specify the system output of interest, which is relevant in characterizing the system behavior or system performance. System output signals are also related to the attributes that are selected to measure the achievement of participants' objectives. The next step is to design a new law, policy, or regulation, which involves the forcing of local changes in the associated processes. A policy, such as a ban on a particular substance or a tax for a certain commodity, would change the strategic interactions among participants. At the local level, the modeler would modify the rules of interactions or the governing equation to reflect the effect of the new policy. The final step is to simulate the system response to the policy implementation, which involves running an analysis or simulation of the model to determine the new system output. As the model output is only as credible as the assumptions that were made, several iterations over a large ensemble of plausible system representations may be necessary in order to observe patterns that are generally true [93]. Hence, the process is iterative to evaluate the simulation results and reassess the original policy issues, objectives, and assumptions, as new information and insights become available.

Understanding complexity means building an intuition of how parts of the system interact and how these interactions affect the overall system behaviors, its responses and outcomes. There are two types of questions based on forward- and reverse-reasoning that our intuition tries to answer. First, given a set of local interaction rules, what is the system's behavior? Second, given a desired system behavior, what are the required local interactions? A computational model should be able to answer the first question. Based on many simulations, correlations between process variables and system outputs can be determined with measurements to assess the validity of a hypothesis. This is the subject of pattern analysis. Furthermore, as shown in Figure 2.2, the model should be validated against empirical data, where available, and appropriately refined when needed. However, policymakers and decision makers are often faced with the second question of how to change the system to attain desired outcomes, which is the essence of policy development. Without an analytical model that can be inverted, building up intuition on this front is very difficult and tends to rely on experience. Where experience is lacking, as occurs when confronted with new problems, this poses a significant handicap for policymakers and decision makers. Also, politicians from a given political party often force their beliefs ("religious value systems") on others without regard to how well the policy will work and meet the objectives of others. In 1997, the Canadian Institute for Environmental Law and Policy (CIELAP) along with other environmental organizations in Ontario presented their concern that the Ontario Government was implementing laws, policies and regulations that undermined its own capacity to deal with environmental problems, which was being sacrificed for short-term economic gain [94]. Hence, it is important to develop a system model such that laws, policies, and regulations are tested to see if they meet certain objectives in advance of implementing them. While a computational model cannot provide a standard solution, it can provide a landscape upon which policymakers can experiment and develop some intuition without bias while avoiding the real impacts of unintended consequences.

Thus, a computational model of a complex system should be flexible and robust, such that changes in the model maintain its applicability and generality. Choosing the appropriate scale for the model is important to observe the dynamical behavior that is of particular interest. Internalizing key evolutionary processes, which are influenced by environmental variables and decisions made by participants, must also be part of the simulation to capture the evolutionary development, self-organization, and adaptation characteristics of complex systems. This means that a participant's values structure must be mutable to allow for values to change and for new values to emerge, which can be captured in complex systems modeling with logic-based programming that is conditional on changing environmental and design parameters. Using a complex systems modeling approach, these capabilities can be embedded into the simulation. However, we must realize that uncertainty is an inescapable fact of reality. Thus, through the model, we must also engage with the world of uncertainty.

2.4 RISK, UNCERTAINTY, AND UNPREDICTABILITY

Understanding risk and managing it are an integral part of SoS thinking and must be undertaken in decision-making and policy development [95]. Failing to assess the risk of a new policy ignores an essential dimension of the problem, which is interconnected with the ethics and complexity of the SoS. Participants' value systems determine what is and what is not an acceptable risk. Also, Haimes [95] stresses the importance of considering multiple objectives in risk management, and hence, uses a multiple criteria decision analysis (MCDA) approach. Moreover, the complexity of an SoS is often intertwined with uncertainty, in that the dynamic behavior of a complex entity is rarely deterministic, and often unpredictable. Fortunately, there exists a plethora of formal mathematical methods that have been developed for risk assessment. Uncertainty can be represented using probability, belief functions, possibility rankings, and plausibility measures [96], as well as fuzzy sets [97], grey sets [98], rough sets [99], information gap theory [100, 101] and other formal methods. Next, we will review the systems-level implications of uncertainty and unpredictability in risk management for policy development.

2.4.1 Risk Management

Leiss [102] identified two facets of handling risk—on the one hand is risk management, and on the other is risk issue management. First, risk management seeks to assess and control the hazards contained in a risk domain and is supported by scientific risk assessments that estimate probable harm to persons and environments [102]. This definition is extended to the control of vulnerability for unpredictable hazards later on. Scientific risk assessments are important for policy development. While they are usually performed at a detailed level in a reductionist manner such that a hypothesized link between a cause and effect is established, they are necessary to understand what risks exist and how they propagate through a system. The current financial crisis exhibits an excellent example of how buyers of "toxic bonds" failed to manage their risk. According to Homer-Dixon, investors did not have any incentive to determine the quality of the original mortgage [103], which was bought and sold so many times over that the original mortgage lenders were no longer truly held accountable to their transactions. The risk associated with these assets were so fragmented and attenuated that what the investors believed was manageable risk was in fact "unbounded uncertainty" [103]. Ultimately, the risk was mismanaged because it was not actually known and there was also no incentive to investigate what the risk entailed. Similar imprudence toward risk economically devastated the Canadian beef industry when bovine spongiform encephalopathy (BSE), or mad cow disease, was discovered in a single cow on an Albertan farm in May 2003 [104]. Hundreds of cattle from BSE-infected countries were being imported into Canada; some were subsequently rendered to produce animal feed, and the estimated probability of at least one infection of BSE occurring in Canada was calculated and judged to be negligible by the Canadian Food Inspection Agency (CFIA) [104]. The mathematical essence of risk, however, is defined as the probability of an occurrence multiplied by its consequences [95]. What the CFIA failed to do was to consider the magnitude of the consequences of a BSE infection, and thus they did not really measure the risk [104]. The incomplete risk assessment was unfortunately catastrophic to the cattle industry. Moreover, the stigma of being the cattle producer to find the next case of BSE created an incentive to cover up any potential discoveries and stay quiet, making risks further unknown. Clearly, these scientific analyses inform high-level decision makers in performing risk management. Furthermore, without a transparent policy environment where stakeholders are encouraged to be vigilant, risks are likely to remain hidden.

Furthermore, risk issue management, which refers to the resolution of risk controversies, requires a system-level perspective to effectively communicate risks to all stakeholders and to design and implement appropriate procedures and policies. At the core of a risk controversy are competing interests of several stakeholders that attempt to define or control the course of social action with respect to perceived hazards [102]. Hatfield and Hipel [105] believe that many cases of controversy around risk-based decisions and assessments can be traced to implicit and undocumented value-based decisions. The various interests of government, industry, and NGOs create public controversy over scientific risk assessments. Each group interprets unbiased data subjectively and if its interests conflict with the information in the data, it may go so far as to create a smoke-screen over it and lobby for a policy that suits its own objectives. Scientific uncertainty can be manipulated to weaken strict regulations or to delay implementation of new policies until a desirable level of certainty about adverse effects is realized. This is the challenge that was faced by public health units with regards to tobacco use; it is the challenge that is faced by governments with regard to climate change [106], and it is the essence of the never-ending saga of the Canadian Environmental Protection Act in controlling toxic substances [102]. What policymakers have to realize and communicate effectively to the public is that uncertainty will always exist.

Effective risk issue management is achieved by confronting the existence and implication of persistent uncertainties in a timely manner, with open dialogue at the interface of science and public policy [102]. The use of the precautionary principle, which states that a lack of full scientific certainty should not be used as a reason for postponing decisions [107], is a method of avoiding unknown consequences and encourages the research community to resolve the uncertainties so that risks are known before they are taken. This method is employed by the European Union in its policies on food safety, which resulted, for example, in banning Canadian and American beef imports due to growth hormones, and implementing strict regulations on genetically modified food [108]. Moreover, the determination of acceptable risk should be divorced from the special interest groups that stand to gain from the risk-taking without bearing the costs of externalities, and should instead be based on the value systems of the public, to whom the costs of the risk-taking are generally transferred. Britain recognized this requirement after its BSE crisis and created an autonomous Food Standards Agency (FSA) in 2000 to represent the public interest [104]. This institutional reorganization relieved food safety monitoring and enforcement tasks from the government department that promotes and supports food commodity producers.

Finally, risk management for SoS also involves being prepared for and resilient to the unpredictable emergent behavior of complex adaptive systems. It means being prepared for a highly unpredictable event, popularly known as the "Black Swan" [109]. This constitutes a slight paradigm shift, in which risk management becomes less about controlling the hazard and more about taking control over the consequences. In this setting, risk is understood as the occurrence of a crisis or disaster, which is expressed in terms of a hazard and one's vulnerability to the effects of exposure to the hazard [47]. Hence, a hazard is comparable to the probability of an event and vulnerability is commensurate to its consequences. Vulnerability is defined as the susceptibility of the system to destabilize itself in the presence of unpredictable disturbances. Conversely, resilience is the ability of the system to absorb random shocks, or rapidly adapt to maintain its basic function and structure [110]. Note that multiple stable states are a characteristic of nonlinear systems and adaptation may cause a jump to a new equilibrium state [10]. Policymakers have to accept that hazards such as floods, droughts, and climate change are essentially unpredictable and unavoidable. Hence, vulnerability, or conversely, resilience of a system is the main variable which policymakers are able to modify in order to manage and mitigate risk. For example, the concept of building resilient and sustainable infrastructure revolves around mitigating risks to unpredictable events, to which large-scale systems within a city are generally predisposed. Critical infrastructure systems, such as water systems, food systems, and energy systems need to be planned and renewed based on SoS risk management principles.

Hashimoto *et al.* [111] developed three risk-related criteria for evaluating the performance of water resource systems which include: (1) reliability: how likely a system is to fail; (2) resiliency: how quickly it recovers from failure; and (3) vulnerability: how severe the consequences of failure may be. A couple decades later, Okada created a conceptual framework for integrated disaster risk management, which is referred to as Vitae System principles, which are: (1) survivability: aim for resilience, defined as surviving shocks with least probable damage or least severe damage; (2) vitality: encourage each individual to be resourceful and to lead a healthy and fulfilling life; and (3) conviviality: emphasize communication and cooperation within and among communities [112, 113]. An important thing to note is that an SoS approach is not only concerned with the physical infrastructure, such as dams, roads, and pipelines, but also with social, financial, and political infrastructure. Munasinghe [114] stresses the importance of social capital in building up resilience to mitigate risk, in comparing responses to Hurricane Katrina and the 2004 Asian Tsunami. Volatile financial markets and unresponsive governments are more responsible for disasters than the environmental systems from which they may originate. Therefore, the goal of risk management for SoS, in addition to hazard control and risk issue communication, is to understand the vulnerabilities of societal–technical systems to unpredictable hazards that propagate through environmental systems, and to build up resilience in both societal and technical capacities.

2.4.2 Modeling Risk and Adaptation Processes

Policymakers should always be asking: "what is the risk?" In order to answer this question, one needs to present the process of an unfolding disaster, which is a dynamic process. This is inherently part of the process of simulating the response of a system to policy implementation, in which risk considerations guide the system designer in assessing the effect of uncertainty and unpredictability of system variables on the system output. Generally known as scenario analysis, the system model is subjected to a variety of model parameter values, initial conditions, and external forces to generate multiple plausible futures. Exploratory modeling [93, 115] surveys multiple hypotheses about the given system by broadening the assumptions of the system model and specifying the uncertainties that are relevant for policy analysis. Patterns of system behavior are explored and displayed graphically and are used to make policy recommendations. The basic steps of risk modeling and management for policy development are as illustrated in Figure 2.3. First, risk and reliability assessments are performed to determine the hazards in the SoS and to search for ways of reducing them. The hazards are delineated with uncertain and unpredictable variables, which are inputs to the SoS model. Then, using the general system model of the SoS

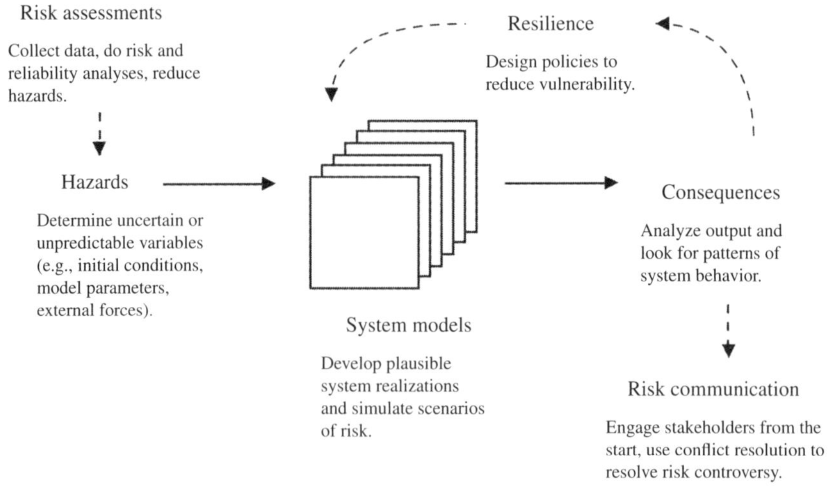

FIGURE 2.3 Risk management in policy development.

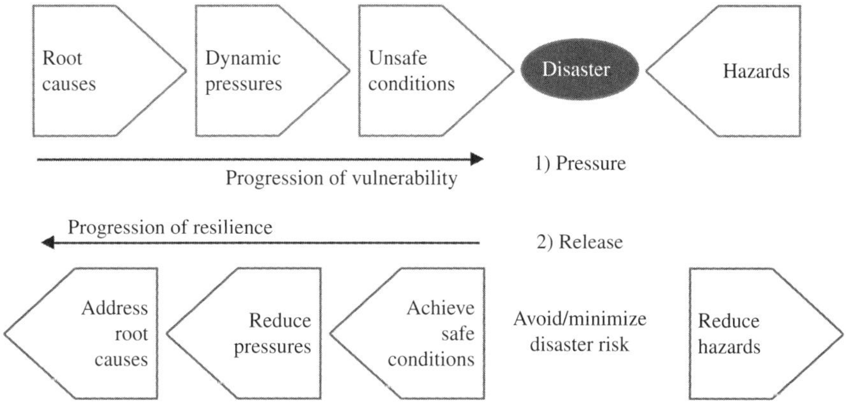

FIGURE 2.4 Pressure and release model (adapted from Wisner *et al.* [47]).

under study, multiple risk scenarios are explored to create a spectrum of specific system realizations. Through the analysis of the system output, consequences are reasoned based on patterns in the system's behavior. Since risk management should be a continuous and adaptive process, a feedback loop is shown to inject resilience into the system as an adaptive and integrative policy design. Here, policy designs are considered adaptive if they are flexible enough to respond effectively to different scenarios. They are considered integrative if they utilize system resources efficiently with cooperation. Risk communication also plays a vital role in risk management to engage stakeholders throughout the process so that uncertain consequences are better understood. Conflict resolution is used to resolve risk controversies.

A key quality that fosters resilience is flexibility, where maintaining flexibility allows for adaptation over time so that better decisions can be made [57, 116] and informed strategic maneuverings can be undertaken [117]. Flexibility is the propensity to change under pressure or stress. Operationally, flexibility can be assessed by determining the ease with which vulnerabilities can be turned around into resilience, with regard to scheduling time and cost [118]. Wisner *et al.*'s pressure and release (PAR) model [47], illustrated in Figure 2.4, can be used to communicate the vulnerabilities that exist in the system and the strategies that may be undertaken to build up resilience. Vulnerability stems from root causes that are amplified by dynamic pressures, which manifest themselves as unsafe conditions. Vulnerability coupled with hazards results in a very likely disaster. An SoS is demonstrably flexible if it can change itself quickly with low cost to achieve safe conditions, self-organize to reduce pressures which are generally self-inflicted, and ultimately address the root causes that are engrained into the core beliefs of its participants. In certain cases, flexibility may be in direct conflict with other desirable system characteristics, such as safety and robustness. Haimes *et al.* [118] identified the need to balance resilience on one hand, and prevention, deterrence and protection on the other. When building up resilience to release the pressures that lead to high-risk scenarios, hardening of critical systems with redundancy and robustness should also be considered. Testing both strategies and combinations of them in high-risk scenarios can be virtually realized through exploratory modeling [93]. Hence, policymakers can develop and realize adaptive capabilities by engaging with the world of uncertainty through exploratory modeling, and by creating a flexible policy environment in which creative problem-solving, or what engineers call design, goes all the way to track down and deal with root causes.

2.5 SYSTEM OF SYSTEMS MODELING AND POLICY DEVELOPMENT

In order to effectively communicate concepts and ideas, SoS thinking requires the use of a common language that transcends domains and disciplines. There are two reasons for this. First, it is because systems science is not bound to a single specialization or group of specialists, but is equally applicable to all fields [119]. Second, an SoS involves more than one field and for multiple specialists to communicate with one another, they need to develop a common understanding using a shared language to transfer their knowledge. Now, more than ever, these dialogues, particularly among scientists, engineers, business leaders, and policymakers, have to take place in order to tackle our global challenges. Modeling is a fundamental skill-set that uses this common language to effectively communicate SoS thinking.

An SoS model's first goal is to help key stakeholders in developing a common understanding of the system, and its second goal is to assist policymakers in developing mechanisms to control, or at least influence, the system in beneficial ways. In a keynote address at the 2008 IEEE International Conference on Systems, Man and Cybernetics, Chuen [120] presented a very simplified SoS model of Singapore and showed how, over a 40-year period of intensive growth and strategic rethinking of Singapore's water system, manufacturing system, and education system, Singapore has become a living example of a human-designed SoS. Looking back on the development of that country's large scale systems, Chuen draws on lessons from the successes and failures of its policy programs, bringing to light the complexities that foster and thwart desired outcomes. The integration of public utilities, drainage, and sewerage, which in most other countries are separately managed and rarely integrated, took 40 years of sustained effort and was a key evolutionary phase in creating the world's first closed water loop system—an incredible feat which has secured Singapore's water supply. On the other hand, in an effort to control the population's growth rate, Singapore's "Two is Enough" policy backfired as the birth rate fell, leading to an overcorrection that could not be reversed. The irreversibility of such a state is why SoS modeling and simulation is so important in designing policies and testing them before they are implemented to avoid unintended consequences. Retrospectively, the model effectively communicates how Singapore functions as a fairly resilient SoS. Using SoS thinking, leaders and system decision makers will continue to lead the evolution and revolution of Singapore's large-scale economic, social, and national security systems.

In this section, an SoS model of the global food system is discussed in the hopes of developing an understanding of the system and to provide insights into which policy mechanisms may work to achieve global food security. It involves the consideration of the value systems of multiple stakeholders, the relationships and interactions among stakeholders and systems, and the risks and uncertainties which threaten to destabilize the system. These components are structured such that the focus is on relations and patterns of large-scale system behavior. This emphasis keeps the analysis manageable. Many assumptions are necessary in order to simplify the real-life system into a model framework. In any type of modeling, researchers should always be forthcoming with the assumptions that are made and both the reasons for and potential consequences of the assumptions should be stated.

2.5.1 Global Food System Model

With the sharp increase in global prices of staple food commodities such as rice, corn, and soybean in April 2008, civil unrest and violence linked to the soaring prices had been occurring around the world [4, 39]. The situation was called a global food crisis. Unlike a natural disaster, experts knew that it could not be turned around quickly due to its complexity [121]. Unfortunately, as the media shifted its attention to the global financial crisis and newly emerging conflicts, the food crisis failed to grasp the attention that it deserved at the year's end. Yet, the United Nations had forecasted that one billion people around the globe would go hungry in 2009—an unprecedented figure [122]. Although that figure decreased to 925 million undernourished in 2010, it is still unacceptably high [123]. It remains as a sustained crisis that should rank high on every country's national security agenda. The former US President George W. Bush hit the point home when he stated: "Can you imagine a country that was unable to grow enough food to feed its people? It would be a nation subject to international pressure. It would be a nation at risk. And so, when we're talking about American agriculture, we're really talking about a national security issue" [124]. In order to secure a nutritious food supply for all nations, fundamental changes to the agribusiness infrastructure, and supporting political and economic systems are required. It is a daunting task and requires the conscious effort of many people over a long period of time. In the same way that Singapore overhauled its education system to fix a 40% failure rate at the primary level of schooling in the 1970s [120], rethinking of the global food system is necessary to fix multiple failings of the food system that manifest themselves in the form of famine and lost livelihoods, environmental degradation and land loss, as well as obesity and human health impacts of an industrially processed food diet and malnutrition [125–128].

In the evolution of natural and man-made systems, the interconnectedness of components, and thus the overall complexity of the system, naturally increases over time. For example, the natural environment and man-made energy and food systems demonstrate very high complexity. Agriculture is essentially the management of a secure food supply. From its simple beginnings of subsistence farming, it has evolved into an SoS that not only exploits environmental systems, but also utilizes large-scale industrial and trade systems. The interactions among these three groups of systems are as illustrated in Figure 2.5. Environmental systems provide resource inputs to our industrial systems, which transform resources into useful products and

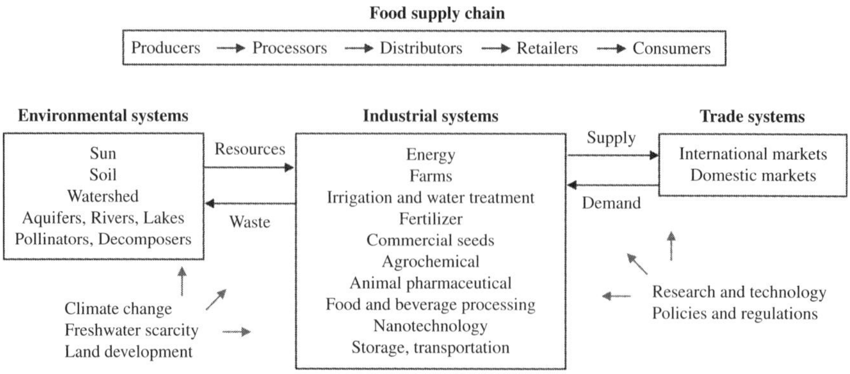

FIGURE 2.5 System of systems model of the global food system.

discharges waste products back into the environment. The industrial systems are kept alive by demand for those useful products, which are supplied to international and domestic markets through sophisticated trade systems. Figure 2.5 also shows the typical food supply chain which has elongated to capitalize markets through value-added processes and centralized distribution networks. The remoteness of the producer from the consumer through the insertion of "middle-men" and industrialized operations has perhaps created more abundance and higher standards of living for some, but it has consequently detached many from the land and according to Shiva, it has stolen sustenance from many [129]. For example, beriberi disease, which was called the national disease of Japan between 1875 and 1885, is directly linked to thiamin deficiency [130]. The process of making white rice significantly decreases the amount of thiamine and other essential vitamins in grain, which means that consumers of white rice need to consume much more compared to consumers of brown rice to obtain the same amount of nutrients [129]. Subsequently, more land is required to grow more rice in order to sufficiently nourish the population.

Agriculture is no longer just about cultivating enough food to live well. Today, it is foremost a business that sells commodities in a market to the highest bidder and seeks to maximize profit rather than equity. Although agriculture also plays a charitable role in food aid because our collective consciousness abhors unfairness and feels some responsibility in feeding the hungry, the global food system is nonetheless dominated by oligopolies consisting of giant agribusiness corporations that control seeds, chemicals, processing, and distribution [2, 127, 131]. Meanwhile, external pressures such as climate change, freshwater scarcity, and land development threaten the stability of the global food system, which if managed adaptively may be mitigated through research and technology and with appropriate policies and regulations.

With this SoS model, the development of the global food crisis can be understood in terms of Holling and Gunderson's [132] model of the adaptive cycle of complex adaptive systems. There are four phases in an adaptive cycle, which are growth or exploitation (r), conservation (K), release (Ω), and reorganization (α). The adaptive

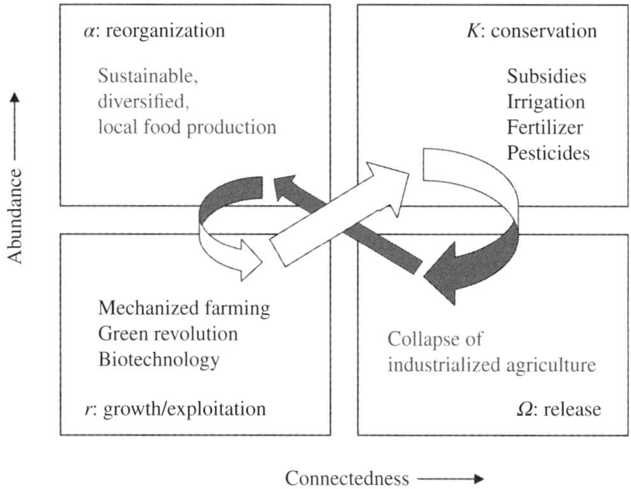

FIGURE 2.6 Current adaptive cycle of the global food system (adapted from Holling and Gunderson [132]).

cycle of the global food system as it exists today is plotted in Figure 2.6. The *X*-axis is connectedness in the system and the *Y*-axis is abundance, which is the inherent potential of the system to feed the world's population. Beginning with the growth phase, it is here that new technologies and processes are introduced with promises of higher yields and better efficiency. The modernization of agriculture through mechanization, refrigeration, the Green Revolution after World War II, and the latest developments in biotechnology, has been the driving force for increased exploitation of the earth's natural resources such as fossil fuels, water, and land, which has resulted in an immense abundance of food. In the next stage of conservation, its idealistic form is when the population stabilizes and food production and distribution is sustained to feed this population. However, the reality is that we are struggling to maintain production levels and increases that were gained through modernization and there is a conceptual fear that food production may not keep up with population growth. To maintain this state of the cycle, countries have created subsidy programs and farmers have become dependent on energy-intensive inputs and operations such as irrigation, chemical fertilizers, and pesticides, just to maintain the same yields that were marketed to them in the growth phase [127]. During this phase, connectedness and abundance are at their highest and it is arguably the phase in which the global food system is currently functioning. The final two phases—release and reorganization—are imminent. Pfeiffer [127] connects the decline of the oil age to the collapse of modern industrial agriculture and points to North Korea as an example of how destructive the collapse will be on society and the irreversibility of its degeneration. It may be argued that the 2008 global food crisis was economically triggered by a whole host of variables from speculation to supply and demand dynamics, but at the core of the global food system, it can be understood that the high connectedness of the industrial and trade systems coupled with the unsustainable resource pathways through the environmental systems resulted in incredible vulnerabilities and sensitivities to economic disturbances, which would not have the same destabilizing effects on a loosely connected or sustainable system. The crisis is but a signal to policymakers that the global food system is both getting ready and is already in the process of transitioning from the conservation phase to the release phase of its adaptive cycle.

Failure of the global food system could very well lead to the total collapse of society. With Jared Diamond's five-point framework of possible contributing factors of collapsed societies [133], which are: (1) environmental damage; (2) climate change; (3) hostile neighbors; (4) friendly trade partners; and (5) society's response, we are given a guide on where policymakers should focus their efforts in order to gracefully handle and respond to the collapse (release) of industrial agriculture. Another important factor, though technically immeasurable, is hope. When faced with a cascading crisis, there is not only the possibility of collapse, but there is also the opportunity to rebound and become even better [2, 10]. As Kingdom recognized [134], crises can open up "windows of opportunity" if they are coupled with available solutions and a favorable political climate in which the time is right for change and the leadership can take action [135]. Friedman [136] also stated: "Only a crisis—actual or perceived—produces real change. When that crisis occurs the actions that are taken depend on the ideas that are lying around. That I believe is our basic function, to develop alternatives to existing policies, to keep them alive and available, until the politically impossible becomes politically inevitable." That the available solutions are the ones that are implemented in a crisis is the reason that they should be tested before they are implemented. Further investigation using SoS thinking in the global food system is done to demonstrate how it may be possible to shape society's response with systematically designed and tested policies. The following sections are illustrative examples to show how SoS thinking may be carried out in practice. It is hoped that this will encourage much more rigorous and in-depth analysis of participants and their value systems, the associated processes, and risks in the global food system, in order to guide policy development and decision-making to achieve global food security.

2.5.1.1 Participants and Their Value Systems On April 29, 2008, at the onset of the global food crisis, Secretary General Ban Ki-Moon established a high-level Task Force on the Global Food Security Crisis. The participants of this task force are heads of the United Nations specialized agencies, funds and programs, Bretton Woods institutions, which are the World Bank and the International Monetary Fund, and relevant parts of the United Nations Secretariat [137, 138]. On a number of occasions, the Secretary General has stressed the need to engage all stakeholders into the decision-making process, including governments of member states, civil society organizations, and the private sector [137]. However, as may be expected, key decisions and policy development are performed by elite members of the society from a high-level perspective with insufficient understanding or regard for the base of society. It is the conventional centralized approach which is efficient in theory, but for on-the-ground implementation, the process is slow and often ineffective. In order to obtain the full SoS perspective and to speed up implementation, elites could fare

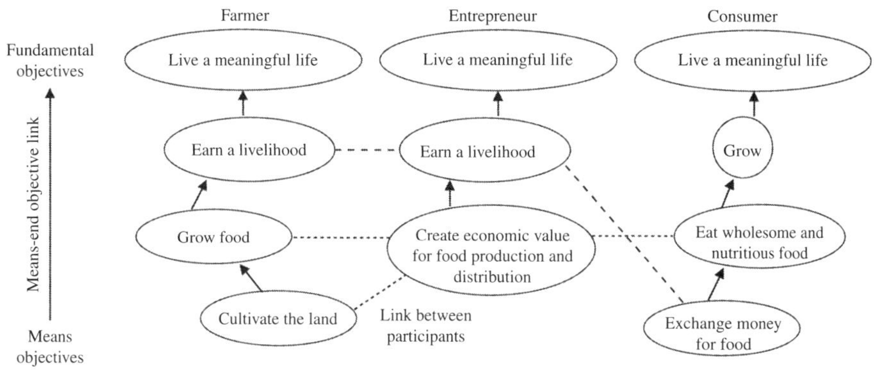

FIGURE 2.7 Identified and structured objectives of participants' value systems.

better by engaging the public, thereby inviting citizen participation. This is certainly not a novel approach and not to be taken lightly. *The Public Participation Handbook* by James L. Creighton [139] provides a comprehensive guide on how to do this with appropriate scope and technique. With this approach, policymakers are able to make more informed decisions and design more effective policies. Participants in the global food system are not only the high-level decision makers who dictate and enforce policies, but rather, the majority are low-level decision makers such as farmers, entrepreneurs, and consumers. They constitute the real value in the food chain from the start to finish. Hence, their value systems need to be understood by policymakers, if policy design is to be legitimized and fruitful.

Let us model the value systems of the archetypal farmer, entrepreneur, and consumer using Keeney's framework of value focused thinking [51]. Note that while a farmer can be considered an entrepreneur and rightly so, for this analysis they are differentiated because of their unique relationship and interaction with key environmental systems. It is assumed that their understanding of the land is greater compared to canners and retailers, which affects the output of the global food system. Also, recall that a model is only a simple representation of the real world, and the following model is intentionally simple to highlight important insights. The values of archetypal farmers, entrepreneurs and consumers are based on the authors' assumptions on what they might likely be. It is recommended that a real policy analysis be based on interviews with real farmers, entrepreneurs, and consumers. Figure 2.7 exhibits the means-ends network of objectives for the three types of participants. As it turns out, the fundamental objectives of all three archetypes are determined to be the following: to live a meaningful life, which is achieved through different means. Farmers cultivate the land to grow food, which they do to earn a livelihood. Entrepreneurs create economic value for food production and distribution [140], which is how they earn their livelihood. Consumers exchange money for food that is hoped to be wholesome and nutritious to eat so that they can grow and survive, which allows them to live.

Farmers, entrepreneurs, and consumers are logically connected and interdependent as the objectives of the others must be fulfilled in order to achieve their own. Farmers and entrepreneurs can only earn a living when consumers have money to buy their products. Without the social and technical innovations of entrepreneurs, farmers and consumers would not have the high levels of safety, diversity, and conveniences that evolved over time and are often taken for granted today. And, of course, without the needs of farmers and consumers, entrepreneurs would have no problems to solve.

From this idealization, what the authors want to emphasize is that the general fundamental objective is the same for all participants and it represents the ethical basis upon which the global food system stands. It mirrors the vitality principle of a vitae system, which is to encourage each individual to be resourceful and to lead a healthy and fulfilling life [112, 113]. As much as the elites recognize this and have stated their intention of creating such an environment for smallholder farmers, they have in general failed to follow through in a complete manner. The Uruguay Round and current Doha Round of WTO negotiations exemplify how difficult and destructive policies can be when world leaders lose sight of the ethical dimension of their decisions. The Uruguay Round paved the way for unintended consequences of a North–South divide and the rise in profile of multinational enterprises [141]. Entrepreneurs, who are not by nature malevolent, have unfittingly attained too much power compared to their capacity of responsibility. The loosely regulated control of agricultural inputs and processed food outputs by private transnational corporations undermines the means by which farmers and consumers can meet their own objectives. The Doha Round of negotiations collapsed, specifically due to sharp divisions between the United States, India, and China about access to agricultural markets in the developing world [142]. As a matter of national security, whoever controls world agricultural markets has basically attained ultimate influence in all matters of security. World domination appears to be the fundamental objective of the world's elites. On the other hand, this is generally not on the agenda of a typical Canadian wheat farmer. For the global food system to function and for policies to achieve global food security, high-level decision makers are encouraged to step out of their sphere of power and examine the value systems of the people whom they are seeking to help. A translation of a teaching of Lao Tzu, the father of Taoism, encompasses exactly how to do this in practice: "Go to the People; Live among them; Love them; Learn from them; Start from where they are; Work with them; Build on what they have" [143]. Here, SoS thinking has been used to operationalize this process. Understanding what motivates people, based on their values, may lead to insights on the appropriate incentives (carrots) and regulations (sticks) to put in place.

2.5.1.2 Associated Processes and Risks The process of agriculture is not inherently complicated, nor is it easy work. Over the past 11,000 years, human culture and agriculture has evolved simultaneously, to the point that local cuisine is an inseparable part of one's culture. Given the historical length of agricultural practice, it has for the most part been a sustainable development. Save for a few exceptions throughout history, agriculture along with other key components of development is a dismal mess. Take Haiti, for example, which Jared Diamond discusses in his book, *Collapse* [133]. It is grossly over populated, suffers from irreversible environmental damage, and had responded violently to the food crisis [133]. The systemic breakdown

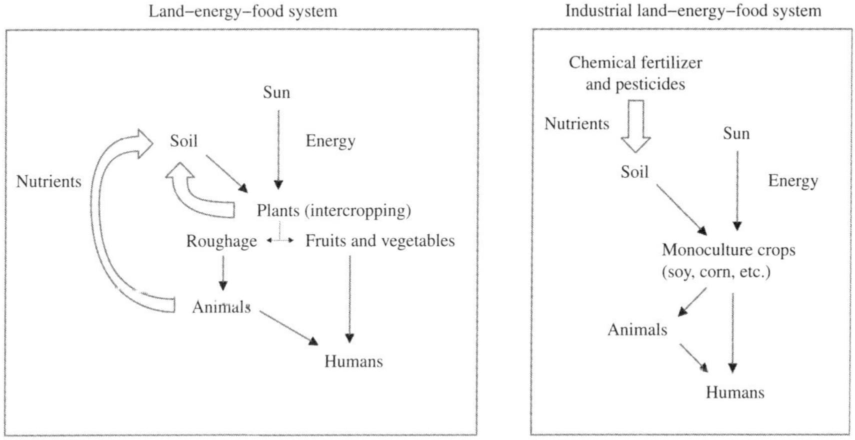

FIGURE 2.8 Comparison of the energy-nutrient circulation in the traditional and industrial land–energy–food systems.

of all parts of Haitian society demonstrates how interrelated our natural and man-made systems are. To understand the complexity of an SoS and how it evolves, the underlying processes and related risks are studied in the context of dynamic interactions. For this part of the analysis, we focus on the interactions between humans and environmental systems. The reason is because this side of the agriculture debate is often pushed aside by special interest groups. Also, the exploitation of environmental resources without due diligence creates high risks. Societal–technical interactions through trade systems, which are equally important, are saved for another discussion.

Based on the books *Stolen Harvest* by Vandana Shiva [129] and *Eating Fossil Fuels* by Dale A. Pfeiffer [127], the processes of energy-nutrient circulation through traditional and industrial land–energy–food systems are as illustrated in Figure 2.8. The traditional process is exemplified through common agricultural practices in India and those which mainly existed elsewhere prior to the Green Revolution. Organic and biodynamic agricultural practices attempt to emulate traditional agriculture. The key interactions that foster sustainable relationships are the return of nutrients to the soil through animal manure and plant decomposition, and the dissociated sources of animal feed and human food. On the other hand, industrial agriculture is the process that was adopted and developed since the Green Revolution after World War II. Figure 2.8 shows a crude representation, but is honest in its message. The stark differences from the traditional process are the dependence on fossil-fuel-based inputs, and competition between animals and humans for food. Actual agricultural practices throughout the world likely cast a spectrum between these two models, but due to the intensive output of industrialized systems, they tend to dominate the global food system.

Clapp [4, 144, 145] argues that these large-scale industrialized agricultural systems are not performing well in light of the food crisis, and she brings attention to solutions, which include more food aid money, investment in agriculture in poor countries, and ultimately, a major overhaul of global agriculture to small-scale sustainable farming and subsidy reduction in rich countries. Global leaders, such as those in the United Nations, are keen to catalyze change, and have outlined a Comprehensive Framework for Action [146], which is a combination of immediate actions to meet urgent needs and long-term strategies to build up resilience against future shocks. However, the gap which exists between theory and application, in addition to the mismatch between policies and values, is a misunderstanding of the dynamics of complex systems at relevant spatial and temporal scales. That is, we are struggling to harness the complexity of the global food system to our advantage. Complexity is manifested through the different relationships which people around the world had developed for thousands of years, with the land that they cultivated and the food that they ate. Whether farmers produce food to feed a family, a community, a country, or the world, it begins with dynamic interactions between them and the environment. The effectiveness of distributing food aid, giving smallholder farmers patented seeds, fertilizer and pesticides, and encouraging market-based economic development tactics in solving the global food security crisis should be judged based on how well they utilize and promote advantageous complexity in the land–energy–food system, wherein sustained relationships between humans and the environment is a vital component. Testing specific policies and activities in a virtual simulation of the complex system dynamics, or in a small-scale low-risk experiment, will facilitate better decision-making before policies are implemented by governments.

Secretary General Ban Ki-Moon stressed that the Comprehensive Framework for Action is more than just a document—it is a process [137]. It encompasses a deliberate approach that is needed to reconfigure the global food system into a system that is sustainable and capable of feeding the growing world's population in a dignified manner. Such a process must be continuously mindful of risks, while progressing steadfastly toward a visible goal. The agriculture sector of an economy faces an immense amount of uncertainty and unpredictability. Agricultural subsidies were originally paid to farmers in the industrialized world to ensure that their nation's food security was not compromised by low crop yields, or low prices in the marketplace which would cause farmers' output to contract. However, the perverse effects of agricultural subsidies on international markets resulted in poor developing countries increasing their vulnerability to food insecurity. Developing countries that could not afford to compete were encouraged to switch their food crops to other commodity crops in which they had a comparative advantage [4]. Over the course of 30 years, investments in agriculture diminished and these countries came to rely on food imports [4]. The history of agricultural subsidies can help us to appreciate that the threat of food insecurity always exists, a fact which industrialized countries were well aware. However, the progression to an increasingly vulnerable state is a dynamic process, which policymakers have the ability to influence, and industrialized countries took advantage of this. In order to build up resilience in the global food system, risks that contribute to food insecurity need to be identified, and specific policies must be judged based on their ability to mitigate the risk or adapt the system to avoid risk altogether.

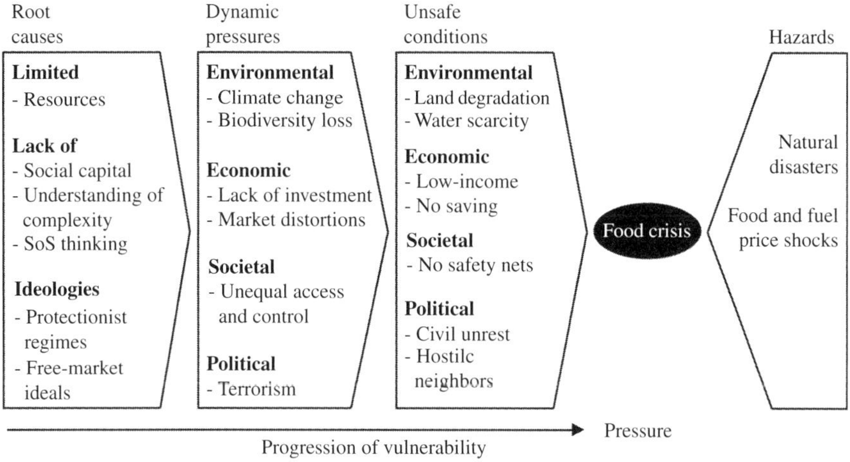

FIGURE 2.9 Pressure model of vulnerabilities that contributed to the global food crisis.

For example, an illustrative set of unsafe conditions, dynamic pressures, and root causes of the global food crisis are identified in Figure 2.9. Here, vulnerability is defined as participants' inability to absorb price shocks. Notwithstanding that food and fuel prices are a result of human systems and actions, price shocks are considered here as hazards rather than contributors to participants' vulnerability. Unsafe conditions include factors such as land degradation, low-income households, a lack of social safety nets, and civil unrest. The dynamic pressures that sustain the unsafe conditions are, for example, loss of biodiversity, and unequal access to and control of agricultural inputs and food products. The ultimate root causes are traced back to natural resource constraints and engrained beliefs such as free-market ideals that are in conflict with protectionist regimes. Out of this diagram, a policymaker obtains a basis upon which to perform a systematic risk assessment of different policies. It also encourages policymakers to look for vulnerabilities, which a policy may create or worsen unintentionally.

Now, much more in-depth analysis using computational models and experimental simulations is called for to fully understand the intricate behavior of the global food SoS. The paper and pen models of participants' value systems, human–environment interactions, and vulnerabilities that led to the food crisis presented in this section provide a starting point on how to use SoS thinking in policy development. To carry them forward, a policy development framework is proposed in the following section.

2.5.2 Policy Implications

The current food system is a product of globalization, which has generated great advancements and discontent [147]. The globalized economy has technically enabled countries to import food from around the world, thereby empowering people to enjoy a diverse diet and higher standard of living. However, free-trade agreements

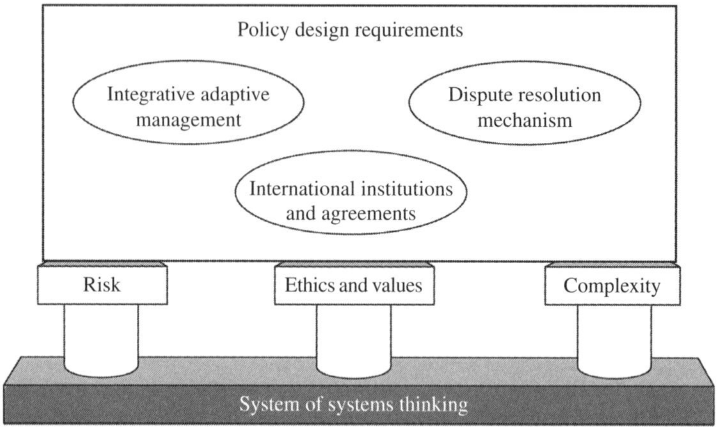

FIGURE 2.10 Policy development framework supported by SoS thinking.

and international institutions have placed a large amount of power in the hands of transnational corporations, and, consequently, have weakened the abilities of nation states to govern and regulate matters of national security, such as financial or food security. In general, as Adam Harmes explains in *The Return of the State*, corporations operate at the global level, while governments are constrained to the national level and cannot regulate international activities to ensure accountability to the public good [148]. What is needed, in order to restore the balance of powers and effectively address our global challenges, are national and international policies that are systematically conceived to harness complexity rather than eliminate it, to mitigate and avoid risks instead of increasing them, and to respect the rights and freedoms of every human being, which are constantly being violated throughout the world today.

Current policy development is generally lacking in an approach that is cooperative, integrative, adaptive, firm, and compassionate. The proposed policy development framework, illustrated in Figure 2.10, seeks to fill this gap. In this framework, policy development is supported by the three pillars of SoS thinking: ethics and values, complexity, and risk. This approach to policy development will require the modification or creation of international institutions and agreements that reflect good governance and democratic principles to enforce regulations that protect the public good. Effective policies should create and sustain an environment for integrative and adaptive management, so that participants in the system can meet their objectives and manage their risks according to their own value systems. Lastly, since disputes or differences of opinion can arise even when you have good policies, a dispute resolution mechanism, designed to lead conflict in a positive direction, is also a design requirement of all policies.

In the following sections, it is argued that SoS analysis methods are required in policy development. The negative consequences of not using SoS thinking are proliferating as emerging crises create panic and perverse phenomena, such as "disaster capitalism," which, as Klein [149, 150] argues, can sustain a market for terrorism and, as pointed out by Saul [151], undermines world efforts to establish good governance, fairness, and peace. In order to understand and manage our global challenges, both technical and social ingenuity gaps need to be filled [152]. The way that we do this will play a large role in determining society's responses to problems in an evolving and self-organizing SoS environment.

2.5.2.1 Institutions and Agreements One of the policy recommendations of the Canadian Food Security Policy Group is to ensure that any changes to multilateral food/agriculture institutions reflect a well-designed institutionally integrated response to food insecurity [153]. The policy brief mentions the relevant institutions as being the Food and Agriculture Organization (FAO) of the United Nations, the United Nations World Food Programme (WFP), the International Fund for Agricultural Development (IFAD), and the Consultative Group on International Agricultural Research (CGIAR). Each of these agencies serves a role of paramount importance in the distribution of food aid and the advancement of agriculture through investment and innovation. However, the SoS model of the global food system recognizes that trade systems play a direct role in the functioning of the global food system. This indicates that changes must also be extended to multi-lateral trade institutions such as the WTO and the International Monetary Fund (IMF), which are criticized for being antiquated [154], if a systems approach is to be realized. These international bodies have significantly more clout and influence in resolving the persistent root causes of the global food crisis, the urgency of which are often played down. While the policy brief recognizes that a supportive and non-threatening international trade environment is crucial for sustainable local food protection, a serious concerted effort should be demanded from food and agriculture and international trade institutions. Integrative institutional reform of these organizations is surely needed.

A similar integrative approach is needed to solve other global challenges such as climate change, water resources management, and energy security, with international institutions and agreements [155]. In the arena of global environmental problems, the indisputable link between the environment and the economy necessitates international collective action across conventionally defined discipline boundaries [49, 156]. The success of the 1987 Montreal protocol on Substances that Deplete the Ozone Layer is credited, among several factors, to the use of trade restrictions on non-compliant parties [157]. For certain countries, such as China and India, with large production capacities of CFCs and strong domestic markets for CFC-based products, trade restrictions had too small a bite to convince those countries to join the agreement. Instead the creation of the Multilateral Fund to assist developing countries in technology transfer and provision of alternative substances was a key incentive to facilitate widespread participation [157]. This combination of "carrot and stick" incentives—trade restrictions and provisions of adequate financial resources and technology—likely worked because it reflected the value systems of the different stakeholders, and participants were able to negotiate terms that supported their own objectives or interests. Furthermore, the flexibility of the trade restrictions, which could be relaxed for countries that were compliant but not part of the agreement, or conversely, were imposed on former parties that became non-compliant, was essential in maintaining the integrity of the protocol. Risk considerations of non-compliance, such as illegal trade and parties that are unable to meet their commitments, must be closely monitored and managed by the institutions of the agreement. In order to reduce these risks, institutions still need to facilitate increased cooperation, which Axelrod [158] sees as an evolutionary process, specifically the emergence of norms to solve dilemmas of collective action. The result is a highly integrated system in which the complexities of cooperation emerge from an ethical framework to avoid the consequences of a depleted ozone layer that is beyond an irreversible threshold.

The experience of the Montreal protocol is a lesson for the negotiation of other international environmental agreements, such as the United Nations Framework Convention on Climate Change (UNFCCC) and the Kyoto protocol [159]. Unfortunately, the lessons are not being applied well. While mitigation of and adaptation to climate change due to the impact of human activities is clearly on the international agenda, the Kyoto protocol, as of January 1, 2015, still awaits ratification by the United States [159]. Moreover, few parties seem to be on track in fulfilling their voluntary commitments. Failing to obtain the participation of one of the largest GHG producers—past and present—makes it difficult to have any kind of binding effect, which, as Stiglitz [160] points out, is the key weakness of the Kyoto protocol. But more fundamentally, it misses the mark on setting targets for national emission reductions. The objective of reducing GHG emissions some percentage points below 1990 levels, without due consideration to a country's values system, causes a misalignment between what the country commits to and what the country actually does. Canada, which committed to a 6% reduction below 1990 levels, was up 24% in 2006 [161]. Discordant policies and values, coupled with no enforcement mechanism, are likely contributing factors for why the Kyoto protocol is not meeting its intended objectives. A policy regime that is based upon uniform treatment of the parties is unlikely to receive much support [156]. The Montreal protocol was able to avoid this through flexible mechanisms and differential treatment for industrialized and developing countries. The Kyoto protocol has yet to achieve a regime that suits every country's individual development goals. Essentially, cap and trade systems will not work until the conflict between values systems and the Kyoto protocol is resolved.

Moreover, to seriously curb global warming which is happening much faster than expected by climate scientists [162], carbon would have to be exorbitantly priced which would result in both economic and political backlashes [163]. While Shellenberger *et al.* [163] advocate for massive public investments into energy research to cut the proverbial "Gordian knot" of global warming, Engel and Orbach perceive a much more profound solution through the development of state and local initiatives [164]. They looked at regional, state, and local initiatives to analyze the seemingly irrational behavior of citizens accepting the high costs of climate change mitigation and the little benefit gained if others do nothing. They show that once the motivations of voters and local decision makers are understood, their actions are in fact in line with maximum utility theory. They also speculate that local actions could lead to changes in global patterns, a behavioral trait of a complex adaptive system. Engel and Orbach's observations bring to light the power of local buy-in, local engagement, and local decision-making in solving global challenges, which international institutions and agreements need to foster. Similarly, a global food system that promotes small-scale sustainable agricultural systems will require regional and international institutions and agreements that are amenable to local conditions.

2.5.2.2 Integrative Adaptive Management The connectedness of environmental, industrial, and trade systems in the global food system with the variability of weather, crop yields, pests, and human behavior requires a complementary management framework to efficiently and effectively produce and distribute food according to the needs of diverse people. Such a framework is called integrative adaptive management, which Tien identifies as synonymous to real-time customized management (RTCM) [165]. In essence, integration and adaptation are required to manage variable supply and variable demand interrelationships within an SoS at a sufficiently fast response rate [165]. In our policy development framework, a policy is designed to foster an environment conducive to integrative and adaptive management. For example, integration and adaptation in the global food system may be encouraged with the following policy goals, which are based on the same three criteria of Hashimoto *et al.* [111]: (1) Reliability: to ensure that people can produce, distribute and procure food according to their and others' needs in an ecologically, financially, and politically stable climate; (2) Resiliency: to enable systems to respond sufficiently quickly to ecological, financial, and political disturbances such that people maintain access to the resources that they need in order to adapt; and (3) Vulnerability: to minimize the consequences of ecological, financial, and political crises such that people see the situation as an opportunity to change, rather than to remain imprisoned in an irreversible state of degradation. In order to achieve these goals, we refer to the three pillars of the policy development framework which are ethics and values, risk, and complexity, portrayed in Figure 2.10.

Essentially, integrative adaptive management that benefits the whole SoS is not possible in a corrupt environment. Policies are first and foremost put into place so that ethical behavior may trump unethical conduct that tears away at the fabric of society. In other words, policies should allow all people to lead healthy and fulfilling lives according to their individual and communal value systems. Jacobs deliberated on the idea that guardians of society such as policymakers, judges, and enforcement officers, enable commerce [56]. If industry, business and markets are left to their own devices, or in other words, are deregulated completely or unscrupulously, then, as demonstrated by the global financial crisis, they can no longer manage themselves. This reflects the classic case of the fox guarding the chicken coop—ultimately, the fox devours all the chickens. Hence, policy design always needs to be evaluated against ethical and value systems of all participants in order to assess its ability to enhance integrative adaptive management.

The increasing complexity of problems due to highly integrated systems and processes is why integrative management is needed to efficiently solve emerging problems. For integrative management to be possible, policy needs to open up communication pathways among stakeholders to enhance group negotiation and decision-making. Holling *et al.* [80] argue that for deeper understanding, an integrative framework and theories that combine insights and people from diverse disciplines leads to more robust and sustainable decision-making. Naturally, large scale systems of systems span jurisdictional levels and areas. For example, effective and efficient infrastructure management needs to involve all levels of government to bring consistency and uniformity to infrastructure renewal and rehabilitation [166]. In Canada, environmental protection is both the jurisdiction of the provincial and the federal governments. On transboundary waters, the United States and Canada autonomously manage their own developments, yet both are collectively responsible for monitoring its management and resolving disputes through the International Joint Commission (IJC) according to the 1909 Boundary Waters Treaty [167]. Effective responses to global warming and climate change require international cooperation and action. The integrative component of integrative adaptive management relies on policy mechanisms that are able to bring stakeholders to the table and enhance communication and cooperation among them.

The nonlinear and unpredictable behavior of complex systems and processes create many unknown risks, which is why adaptive management is necessary to respond quickly to failures and to pre-emptively reduce expected damages in cases of failure. The framework of adaptive management, which was initially developed by Holling [168] and several other colleagues (e.g., Walters [169]) from studies on the resilience of ecological systems [170], has been applied to real-world systems problems, such as water resources management [37] and environmental management [171]. Olsson *et al.* [135] extend the framework to adaptive governance of socio-ecological systems, using rules of thumb to build leadership and social networks that facilitate successful transformations of the system. Among the rules of thumb is the recommendation to "try to facilitate adaptive governance by allowing just enough flexibility in institutions and politics" [135]. Hence, adaptation is hindered by rigidly structured institutions and agreements. Flexibility in institutions and agreements means that policies have to be constructed with due consideration to unknown risks that may arise and need to be tested thoroughly before their implementation to ensure that the policies maintain their integrity under adaptive pressures. A good policy should not have to change to changing circumstances, but should facilitate innovation in and evolution of management practices to handle risk efficiently and effectively.

2.5.2.3 Dispute Resolution Invariably, conflicting value systems, different attitudes toward risks, and complex social and political interactions among decision makers warrant the requirement of a dispute resolution mechanism in all treaties. Debates, such as those on genetically engineered seeds and patenting hybrid seeds, industrialized agriculture and its environmental impacts, and transnational food corporations and their influence on health and nutrition, demonstrate the ubiquitous presence of conflict. Evidence-based, formal policy appraisals are needed to resolve conflict in policies. As well, a clearly defined and transparent process on how to resolve future disputes should be designed into a policy and its effectiveness should be evaluated. Ma *et al.* [167] show how to perform a formalized appraisal of several transboundary water treaties with respect to their enforcement capabilities, treaty implementations, and dispute settlement mechanisms, using a flexible MCDA technique known as the elimination method [172]. (For descriptions of MCDA approaches, see, for instance, publications by Hipel [61], Saaty [173], Roy [174], Hobbs and Meier [175], Belton and Stewart [176], Chen *et al.* [177], Chankong and Haimes [178], and Hipel *et al.* [179]). The policy analysis leads to important insights into the improvement of the current treaties.

Because of the universal nature of conflict, valuable contributions to the development of a rich range of negotiation and conflict resolution methodologies have been made by researchers working in many different fields such as systems engineering, sociology, operational research, psychology, political science, law and business [180, 181]. The GMCR [66, 182] is a useful tool for tracing the evolution of conflicts from the status quo state to a final equilibrium [183], as well as for testing out the outcomes of new policies and their stability [184], based on modeled values systems, taking into account preference uncertainty [185], strength of preferences [186], and the influences of attitudes [187] and emotions [188]. Moreover, in addition to investigating the strategic consequences of pure competition or non-cooperative behavior under conflict, GMCR can ascertain if cooperation via coalition formation can lead to more desirable win–win resolutions [189–191]. Zeng *et al.* [192] utilize graph model concepts of equilibrium and generalized metarationalities to perform policy analysis for multiple participant conflicts. Fang *et al.* [71, 193] outline a formalized process for developing a DSS [194] from model formulation to analysis. In addition, special algorithms are being devised to improve the computational efficiency of the analytical calculations [195]. Developing an advanced DSS for everyday policy making is becoming a more and more pressing requirement as the complexity of policy problems increases [196] and governments are expected to pre-empt where possible and react quickly to detrimental crises. Moreover, the importance of formalized methods for designing, policies and having them tested by independent and disinterested experts cannot be understated [196]. The power imbalances among stakeholders in an SoS, as in the WTO, makes it very difficult to settle conflicts with win–win resolutions. Hence, a policy should be designed to separate the two sides of the decision making and conflict-resolution process, the scientific process from the political process, while at the same time ensuring that both sides complement each other, as yin and yang.

For an SoS to thrive, competition and conflict are natural and required to motivate continuous innovation and evolution. At the same time, cooperation as a response to conflict can in some cases create a more preferred win/win resolution, which in itself represents a type of innovative and evolutionary development [158]. Hence, policies need to strike a balance between competition and cooperation; however, when disputes arise, they should facilitate the scientific and political processes of decision-making to inspire an all-win resolution. This requires serious consideration of values and tradeoffs, calculated risk-taking and insurances, and concerted coordinated effort on the part of all stakeholders. As an example, consider the relationship between modern industrialized agribusiness and local small-scale/organic farmers. Disagreements have arisen on dumping and agricultural subsidies in the WTO, the ownership of seeds, and use of genetic engineering. The clash between small-scale and large-scale agriculture is pictured as a classic David versus Goliath battle, which was exemplified in the case of Monsanto versus Schmeiser [129]. The Monsanto

Corporation has a history of controversies (see Reference 105 for a discussion on the controversy surrounding Alachlor-based herbicides). International and national policies are needed to creatively rein in big business, and at the same time, bring small-scale operations, which generally have very few regulatory standards, under an integrated and flexible SoS policy umbrella to enhance healthy competition and mutually beneficial cooperation among different scale farmers. In this way, competitors are on a more level playing field, and dispute resolution is more likely going to result in win–win outcomes. Ensuring that dispute settlement mechanisms guide stakeholders into a positive direction is a non-trivial, complex task. For how to do this, policymakers are again encouraged to practice the three pillars of policy development depicted in Figure 2.10, modeling values and ethics, risks, and complexity, to understand the participants, their associated processes and the relevant interactions among participants, processes, and systems. Hence, new policies can be explored from an SoS perspective.

2.6 CONCLUSIONS

Current and future generations are faced with complex global challenges in their journey toward achieving general well-being and security for all. The interconnectedness of natural, societal and technological systems means that decision makers and policymakers must understand the SoS which they influence, as a whole, in order to achieve their desired outcomes and avoid unintended consequences. The aims of a holistic understanding are respect for different value systems within a universal ethical framework, acceptance of complexity and utilization of complexity toward productive ends, and engagement with uncertainty and unpredictability to plan and be prepared for shocks in order to respond gracefully to crises. Regardless of one's role in an SoS, every decision maker, from the lowest to the highest level, should be cognizant of the system perspective in order to fully understand one's role and how to best act in that role. That is a challenge for all, especially for decision makers whose decisions in policy development invariably trickle down to affect the everyday lives of many others. This responsibility should be undertaken with an approach that is cooperative, integrative, adaptive, firm, and compassionate.

In this chapter, SoS thinking and methods are put into perspective for policymakers who face increasingly complex policy challenges in an SoS environment. Formalized procedures for modeling values systems, complex adaptive systems, and uncertain and unpredictable risks are presented. Several practical uses of these tools are presented to deepen our understanding of the global food system and the global food crisis. The strong links among environmental, industrial, and trade systems provide an argument for acknowledging the interrelationships among trade and agricultural policies and their effects on the environment and people's livelihoods. The analysis also traces the causes of the global food crisis to various factors, from low-income conditions to the free-market ideology, which increases people's vulnerability to food price shocks. In order to address complex global challenges, such as the global food and financial crises, a policy development framework based on values and ethics, risks, and complexity is introduced. The framework is a guide for policymakers in reforming institutions and agreements, fostering an environment for integrative and adaptive management, and designing effective dispute resolution mechanisms.

As we tackle the world's problems, there is a tremendous opportunity to generate solutions, including national and international policies and institutions, which are formally tested and evaluated before they are implemented. They are then made available, so that when a window of opportunity presents itself, we will be ready.

REFERENCES

[1] T. Flannery. *The Weather Makers*. Toronto: Harper Collins, 2005.

[2] R. Heinberg. *The Party's Over: Oil, War and the Fate of Industrial Societies*, 2nd edn. Gabriola Island, British Columbia: New Society Publishers, 2005.

[3] M. Barlow. *Blue Covenant: The Global Water Crisis and the Coming Battle for the Right to Water*. Toronto: McClelland & Stewart, 2007.

[4] J. Clapp. The global food crisis: causes, consequences, solutions? Food for Thought Lecture, CIGI, May 13, 2008.

[5] J. C. Glenn, T. J. Gordon, and E. Florescu (Editors). *2008 State of the Future*. Washington, DC: Millenium Project, World Federation of United Nations Associations, 2008.

[6] BBC. Timeline: global credit crunch: a quick guide to the origins of the global financial crisis. *BBC News*. Available at http://news.bbc.co.uk/2/hi/business/7521250.stm (accessed December 31, 2008).

[7] S. C. Johnson, US recession deepens, countries boost spending. New York: Reuters. Available at http://www.reuters.com/article/ousiv/idUSTRE4B70ME20081224 (accessed December 24, 2008).

[8] IRIN. Somalia: no longer able to live on 100 dollars a month. Available at http://www.irinnews.org/Report.aspx?ReportId=79301 (accessed July 17, 2008).

[9] G. York. Food crisis squeezing Vietnamese poor. *Globe and Mail*. Available at http://www.theglobeandmail.com/servlet/story/RTGAM.20080429.wvietnamfood-0429/BNStory/sp (accessed April 30, 2008).

[10] T. F. Homer-Dixon. *The Upside of Down: Catastrophe, Creativity and the Renewal of Civilization*. Toronto: Random House of Canada, 2006.

[11] E. O. Wilson. *The Future of Life*. New York: Alfred A. Knopf, 2002.

[12] C. Rosenzweig, G. Casassa, D. J. Karoly, A. Imeson, C. Liu, A. Menzel, S. Rawlins, T.L. Root, B. Seguin, and P. Tryjanowski. 2007: Assessment of observed changes and responses in natural and managed systems. In: *Climate Change 2007: Impacts, Adaptation and Vulnerability. Contribution of Working Group II to the Fourth Assessment Report of the Intergovernmental Panel on Climate Change*, edited by M. L. Parry, O. F. Canziani, J. P. Palutikof, P. J. van der Linden, and C. E. Hanson. Cambridge, UK: Cambridge University Press, 2007, pp. 79–131.

[13] F. Montaigne. No room to run. *National Geographic*, 206(3): 34–57, 2004.

[14] D. Glick. The big thaw. *National Geographic*, 206(3): 12–33, 2004.

[15] K. W. Hipel and L. Fang. Multiple participant decision making in societal and technological systems. In: *Systems and Human Science—For Safety, Security, and Dependability*, edited by T. Arai, S. Yamamoto, and K. Makino. Amsterdam, The Netherlands: Elsevier, 2005, Chapter 1, pp. 3–31. Selected Papers of the 1st International Symposium, SSR2003, Osaka, Japan.

[16] K. W. Hipel, L. Fang, and M. Heng. System of systems approach to policy development for global food security. *Journal of Systems Science and Systems Engineering*, special issue on Strategic Decision Making for Global Security from a Systems Engineering Perspective in the Post-911 Environment, 19(1): 1–21, 2010.

[17] K. W. Hipel, A. Obeidi, L. Fang, and D. M. Kilgour. Sustainable environmental management from a system of systems engineering perspective. In: *System of Systems – Innovations for the 21st Century*, edited by M. M. Jamshidi. New York: John Wiley & Sons, 2008, Chapter 18, pp. 443–481.

[18] W. B. Rouse. Engineering complex systems: implications for research in systems engineering. *IEEE Transactions on Systems, Man, and Cybernetics–Part C: Applications and Reviews*, 33, 154–156, 2003.

[19] A. P. Sage. *Systems Engineering*. Hoboken, New Jersey: John Wiley & Sons, 1992.

[20] R. L. Ackoff. Science in the systems age: beyond IE, OR and MS. *Operations Research*, 21, 661–671, 1973.

[21] M. M. Jamshidi. *System of Systems Engineering—Innovations for the 21st Century*. New York: John Wiley & Sons, 2009.

[22] A. P. Sage and S. M. Biemer. Processes for system family architecting, design, and integration. *IEEE Systems Journal*, 1(1): 5–16, 2007.

[23] R. L. Ackoff. Towards a system of systems concepts. *Management Science*, 17, 661–671, 1973.

[24] J. Boardman and B. Sauser. System of systems—the meaning of OF. In: Proceedings of the 2006 IEEE/SMC International Conference on System of Systems Engineering, Los Angeles, CA, April 24–26, 2006, pp. 118–123.

[25] A. Gorod, B. Sauser, and J. Boardman. System-of-systems engineering management: a review of modern history and a path forward. *IEEE Systems Journal*, 2(4): 484–499, 2008.

[26] Y. Bar-Yam. *Making Things Work: Solving Complex Problems in a Complex World*. Cambridge, MA: Springer, 2005.

[27] D. Braha, A. A. Minai, and Y. Bar-Yam (Editors). *Complex Engineered Systems: Science Meets Technology*. Cambridge, MA: Springer, 2006.

[28] J. L. Casti. *Would-be Worlds: How Simulation Is Changing the Frontiers of Science*. New York: John Wiley & Sons, 1997.

[29] J. H. Miller and S. E. Page (Editors). *Complex Adaptive Systems: An Introduction to Computational Models of Social Life*. Princeton, NJ: Princeton University Press, 2007.

[30] J. N. Warfield. *An Introduction to Systems Science*. Hackensack, NJ: World Scientific, 2006.

[31] R. Axelrod and M. D. Cohen. *Harnessing Complexity: Organizational Implications of a Scientific Frontier*. New York: Basic Books, 2000.

[32] J. Gharajedaghi. *Systems Thinking: Managing Chaos and Complexity: A Platform for Designing Business Architecture*, 2nd edn. Burlington, MA: Elsevier, 2006.

[33] K. W. Hipel, L. Wang, and L. Fang. Systems thinking in fair water resources allocation. In: *Proceedings of the International Conference on Water, Environment, Energy and Society (WEES-2009), Volume II: Statistical and Systems Analysis Techniques*, edited by S. K. Jain, V. P. Singh, V. Kumar, R. Kumar, R. D. Singh, and K. D. Sharma, conference held in New Delhi, India, January 12–16, 2009. New Delhi, India: Allied Publishers Pvt. Ltd., 2009, pp. 937–952.

[34] P. Rogers, Facing the freshwater crisis. *Scientific American*, 299(2): 46–53, 2008.

[35] M. W. Rosegrant, X. Cai, and S. A. Cline. *World Water and Food to 2025: Dealing with Scarcity*. Washington, DC: International Food Policy Research Institute, 2002.

[36] L. Hermans. *Actor Analysis for Water Resources Management: Putting the Promise into Practice*. Delft, The Netherlands: Eburon, 2005.

[37] K. W. Hipel, A. Obeidi, L. Fang, and D. M. Kilgour. Adaptive systems thinking in integrated water resources management with insight into conflict over water exports. *Information Systems and Operational Research*, 46(1): 51–69, 2008.

[38] BBC. Food riots turn deadly in Haiti. *BBC News*, April 5, 2008. Available at http://news.bbc.co.uk/2/hi/americas/7331921.stm (accessed January 18, 2015).

[39] Food and Agriculture Organization of the United Nations. Soaring food prices: facts, perspectives and actions required. In: HLC/08/INF/1, High Level Conference on World Food Security: The Challenges of Climate Change and Bioenergy, Rome, June 3—5, 2008. Available at http://www.fao.org/foodclimate/conference/doclist/en/ (accessed January 18, 2015).

[40] R. Sheppard. Rice riots and empty silos: is the world running out of food? *CBC News*, 30 April 30, 2008. Available at http://www.cbc.ca/news/background/food/prices.html (accessed February 20, 2009).

[41] A. Evans. Rising food prices: drivers and implications for development. Chatham House Briefing Paper, April 2008. Available at http://www.chathamhouse.org.uk/publications/papers/view/ /id/612/ (accessed February 20, 2009).

[42] M. Grant. Is food commodity securitization worsening the world's food problem? The Conference Board of Canada Briefing, July 2008. Available at http://www.conferenceboard.ca/documents.asp?rnext=2662 (accessed January 18, 2015).

[43] M. Liebreich. Biofuels not main player in the food price heist. May 2008. Available at http://www.ren21.net/forum/forum.asp?id=11 (accessed February 20, 2009).

[44] P. L. Paarlberg and P. C. Abbott. Oligopolistic behaviour by public agencies in international trade: the world wheat market. *American Journal of Agricultural Economics*, 68(3): 528–543, 1986.

[45] I. Van Heerden and M. Bryan. *The Storm: What Went Wrong and Why During Hurricane Katrina—The Inside Story from One New Orleans Scientist*. New York: Penguin, 2006.

[46] M. Grunwald. The threatening storm—Hurricane Katrina: two years later. *Time Magazine, Special Report*, 17(7): 18–29, 2007.

[47] B. Wisner, P. Blaikie, T. Cannon, and I. David. *At Risk*, 2nd edn. London: Routledge, 2004.

[48] S. H. Schneider, S. Semenov, A. Patwardhan, I. Burton, C. H. D Magadza, M. Oppenheimer, A. B. Pittock, A. Rahman, J. B. Smith, A. Suarez, and F. Yamin. 2007: Assessing key vulnerabilities and the risk from climate change. In: *Climate Change 2007: Impacts, Adaptation and Vulnerability. Contribution of Working Group II to the*

Fourth Assessment Report of the Intergovernmental Panel on Climate Change, edited by M. L. Parry, O. F. Canziani, J. P. Palutikof, P. J. van der Linden, and C. E. Hanson. Cambridge, UK: Cambridge University Press, 2007, pp. 779–810.

[49] N. Stern (Editor). *The Economics of Climate Change: The Stern Review*. Cambridge, UK: Cambridge University Press, 2007.

[50] K. W. Hipel, M. M. Jamshidi, J. M. Tien, and C. C. White. The future of systems, man, and cybernetics: application domains and research methods. *IEEE Transactions on Systems, Man, and Cybernetics, Part C: Applications and Reviews*, 37(5): 726–743, 2007.

[51] R. L. Keeney. *Value-Focused Thinking: A Path to Creative Decisionmaking*. Cambridge, MA: Harvard University Press, 1992.

[52] J. McMurtry. *Unequal Freedoms: The Global Market as an Ethical System*. Toronto: Garamond Press, 1998.

[53] T. Kuhn. *The Structure of Scientific Revolutions*. Chicago, IL: University of Chicago Press, 1962.

[54] S. Shrybman. *The World Trade Organization: A Citizen's Guide*, 2nd edn. Toronto: Canadian Centre for Policy Alternatives and James Lorimer and Company Ltd., 2001.

[55] J. McMurtry. *Value Wars: The Global Market versus the Life Economy*. London: Pluto Press, 2002.

[56] J. Jacobs. *Systems of Survival: A Dialogue on the Moral Foundations of Commerce and Politics*. New York: Vintage Books, 1992.

[57] J. S. Hammond, R. L. Keeney, and H. Raiffa. *Smart Choices: A Practical Guide to Making Better Decisions*. Cambridge, MA: Harvard Business School Press, 1999.

[58] B. Malakooti and Y. Q. Zhou. Feedforward artificial neural networks for solving discrete multiple criteria decision making problems. *Management Science*, 40, 1542–1561, 1994.

[59] S. Barrett. *Environment and Statecraft: The Strategy of Environmental Treaty-Making*. Oxford: Oxford University Press, 2003.

[60] D. G. Victor. *The Collapse of the Kyoto Protocol and the Struggle to Slow Global Warming*. Princeton, NJ: Princeton University Press, 2001.

[61] K. W. Hipel (Editor). Multiple objective decision making in water resources. In: *Monograph Series No. 18*, Vol. 28. Middleburg, VA: American Water Resources Association, 1992.

[62] L. Wang, L. Fang, and K. W. Hipel. On achieving fairness in the allocation of scarce resources: measurable principles and multiple objective optimization approaches. *IEEE Systems Journal*, 1(1): 17–28, 2007

[63] R. L. Keeney and H. Raiffa. *Decisions with Multiple Objectives: Preferences and Value Tradeoffs*. Cambridge, MA: Cambridge University Press, 1993.

[64] J. Von Neumann and O. Morgenstern. *Theory of Games and Economic Behaviour*. 1st edn. Princeton, NJ: Princeton University Press, 1944.

[65] J. Von Neumann and O. Morgenstern. *Theory of Games and Economic Behaviour*. 3rd edn. Princeton, NJ: Princeton University Press, 1953.

[66] L. Fang, K. W. Hipel, and D. M. Kilgour. *Interactive Decision Making: The Graph Model for Conflict Resolution*. New York: John Wiley & Sons, 1993.

[67] M. Fraser and K. W. Hipel. *Conflict Analysis: Models and Resolutions*. New York: North Holland, 1984.

[68] N. Howard. *Paradoxes of Rationality: Theory of Metagames and Political Behaviour*. Cambridge, MA: MIT Press, 1971.

[69] Z. Xu. A survey of preference relations. *International Journal of General Systems*, 36(2) 179–203, 2007.

[70] Z. Xu and J. Chen. Some models for deriving the priority weights from interval fuzzy preference relations. *European Journal of Operational Research*, 184, 266–280, 2008.

[71] L. Fang, K. W. Hipel, D. M. Kilgour, and X. Peng. A decision support system for interactive decision making—Part I: model formulation. *IEEE Transactions on Systems, Man, and Cybernetics, Part C: Applications and Reviews*, 33(1): 42–55, 2003.

[72] K. W. Hipel, D. M. Kilgour, and L. Fang. The graph model for conflict resolution. In: *Wiley Encyclopedia of Operations Research and Management Science*, edited by J. J. Cochran (Editor-in-Chief) with L. A. Cox, P. Keskinocak, J. P. Kharoufeh, and J. C. Smith (Area Editors). New York: John Wiley & Sons, Vol. 3 of 8, 2011, pp. 2099–2111.

[73] K. W. Hipel, L. Fang, and D. M. Kilgour. Decision support systems in water resources and environmental management. *ASCE Journal of Hydrologic Engineering*, 13(9): 761–770, 2008.

[74] G. Norris and M. Wagner. *Airbus A380: Superjumbo of the 21st Century*. Wisconsin: Zenith Press, 2005.

[75] M. Yunus. *Creating a World Without Poverty: Social Business and the Future of Capitalism*. New York: PublicAffairs, 2008.

[76] M. Telò. *European Union and New Regionalism: Regional Actors and Global Governance in a Post hegemonic Era*. Surrey, UK: Ashgate Publishing, 2007.

[77] E. Beauchesne. Close, but no recession for Canada, forecasters agree. *National Post*, Oct 8, 2008. Available at http://www.financialpost.com/story.html?id=868581 (accessed February 20, 2009).

[78] CBC. Canada heading for a recession, say economists. *CBC News*, Oct 6, 2008.

[79] G. Andersson, P. Donalek, R. Farmer, N. Hatziargyriou, I. Kamwa, P. Kundur, N. Martins, J. Paserba, P. Pourbeik, J. Sanchez-Gasca, R. Schulz, A. Stankovic, C. Taylor, and V. Vittal. Causes of the 2003 major grid blackouts in North America and Europe, and recommended means to improve system dynamic performance. *IEEE Transactions on Power Systems*, 20(4): 1922–1928, 2005.

[80] C. S. Holling, L. Gunderson, and D. Ludwig. In quest of a theory of adaptive change. In: *Panarchy: Understanding Transformations in Human and Natural Systems*, edited by L. H. Gunderson and C. S. Holling. Washington, DC: Island Press, 2002, pp. 3–24.

[81] M. Newman, A.-L. Barabási, and D. J. Watts (Editors). *The Structure and Dynamics of Networks*. Princeton, NJ: Princeton University Press, 2006.

[82] P. Erdõs and A. Rényi. On random graphs. I. *Publicationes Mathematicae*, 6: 290–297, 1959.

[83] L. A. N. Amaral, A. Scala, M. Barthélémy, and H. E. Stanley. Classes of small-world networks. *Proceedings of the National Academy of Sciences*, 97(21): 11149–11152, 2000.

[84] D. J. Watts and S. H. Strogatz. Collective dynamics of "small—world" networks. *Nature*, 393(6684): 440–442, 1998.

[85] A.-L. Barabási and R. Albert. Emergence of scaling in random networks. *Science*, 286(5349): 509–512, 1999.

[86] R. Cohen, K. Erez, D. Ben Avraham, and S. Havlin. Resilience of the internet to random breakdowns. *Physical Review Letters*, 85(21): 4626–4628, 2000.

[87] R. Albert, H. Jeong, and A.-L. Barabási. Error and attack tolerance of complex networks. *Nature*, 406(6794): 378–382, 2000.

[88] R. M. May and A. L. Lloyd. Infection dynamics on scale-free networks. *Physical Review E*, 64(6): 066112, 2001.

[89] R. Pastor-Saltorras and A. Vespignani. Epidemic spreading in scale-free networks. *Physical Review Letters*, 86(14): 3200–3203, 2001.

[90] M. Batty. *Cities and Complexity*. Cambridge, MA: MIT Press, 2005.

[91] M. Lim, R. Metzler, and Y. Bar-Yam. Global pattern formation and ethnic/cultural violence. *Science*, 317, 1540–1544, 2007.

[92] J. Ferber. *Multi-agent Systems: An Introduction to Distributed Artificial Intelligence*. Harlow, UK: Addison-Wesley, 1999.

[93] D. B. Agusdinata. *Exploratory Modeling and Analysis: A Promising Method to Deal with Deep Uncertainty*. Delft, The Netherlands: Next Generation Infrastructures Foundation, 2008.

[94] D. Macdonald, R. Nadarajah, and M. Winfield. Our future, our health: a statement of concern by environmental organizations in Ontario. Centre for International Environemtnal Law and Policy (CIELAP), March 1997. Available at http://www.cielap.org/pdf/ourfutureourhealth.pdf (accessed January 18, 2015).

[95] Y. Y. Haimes. *Risk Modelling, Assessment, and Management*, 3rd edn. Hoboken, NJ: John Wiley & Sons, 2009.

[96] J. Y. Halpern. *Reasoning about Uncertainty*. Cambridge, MA: MIT Press, 2005.

[97] L. A. Zadeh. Outline of a new approach to the analysis of complex systems and decision processes. *IEEE Transactions on Systems, Man and Cybernetics*, 3(1): 28–44, 1973.

[98] J. L. Deng. Introduction to grey system theory. *The Journal of Grey System*, 1(1): 1–24, 1989.

[99] S. Greco, B. Matarazzo, and R. Slowinski. Rough sets theory for multicriteria decision analysis. *European Journal of Operational Research*, 129: 1–47, 2001.

[100] Y. Ben-Haim. *Info-gap Decision Theory: Decisions under Severe Uncertainty*. 2nd edn. London: Elsevier, 2006.

[101] K. W. Hipel and Y. Ben-Haim. Decision making in an uncertain world: information gap modelling in water resources management. *IEEE Transactions on Systems, Man and Cybernetics, Part C: Applications and Reviews*, 29(4): 506–517, 1999.

[102] W. Leiss. *In the Chamber of Risks: Understanding Risk Controversies*. Montreal: McGill-Queen's University Press, 2001.

[103] T. F. Homer-Dixon. Unbounded uncertainty. *Globe and Mail,* Sept 26, 2008.

[104] I. MacLachlan. Betting the farm: food safety and the beef commodity chain. In: *Feeding the Future: From Fat to Famine: How to Solve the World's Food Crises*, edited by A. Heintzman and E. Solomon. Toronto: House of Anansi Press, 2004, pp. 37–69.

[105] A. J. Hatfield and K. W. Hipel. Risk and systems theory. *Risk Analysis*, 22(6): 1043–1057, 2002.

[106] D. A. Stainforth, M. R. Allen, E. R. Tredger, and L. A. Smith. Confidence, uncertainty and decision-support relevance in climate predictions. *Philosophical Transactions of the Royal Society London, Series A (Mathematical, Physical and Engineering Sciences)*, 365(1857): 2145–2161, 2007.

[107] UN Conference on Environment and Development. Rio Declaration on environment and development. Principle 15, 1992. Available at http://www.un.org/documents/ga/conf151/aconf15126-1annex1.htm (accessed January 18, 2015).

[108] T. Van de Haegen. EU view of precautionary principle in food safety. European Union in the US, October 23, 2003. Available at http://www.eurunion.org/news/speeches/2003/031023tvdh.htm (accessed February 20, 2009).

[109] N. N. Taleb. *The Black Swan: The Impact of the Highly Improbable*. Toronto: Random House of Canada, 2007.

[110] B. H. Walker, D. A. Salt, and W. V. Reid. *Resilience Thinking: Sustaining Ecosystems and People in a Changing World*. Washington, DC: Island Press, 2006.

[111] T. Hashimoto, J. R. Stedinger, and D. P. Loucks. Reliability, resiliency, and vulnerability criteria for water resource system performance evaluation. *Water Resources Research*, 18(1): 14–20, 1982.

[112] K. W. Hipel, D. M. Kilgour, and L. Fang. Systems methodologies in vitae systems of systems. *Journal of Natural Disaster Science*, 33: 63–77, 2011.

[113] N. Okada. A methodology for disaster risk management and its relationships to economic analysis. In: *Economic Analysis for Disaster Prevention: Risk Management Policies and their Evaluation*, edited by H. Tatano and A. Takagi. Japan: Keiso Shobo Publishing Corp., 2005 (in Japanese).

[114] M. Munasinghe. The importance of social capital: comparing the impacts of the 2004 Asian Tsunami on Sri Lanka, and Hurricane Katrina 2005 on New Orleans. *Ecological Economics*, 64(1): 9–11, 2007.

[115] S. Bankes. Exploratory modelling for policy analysis. *Operations Research*, 41(3): 435–449, 1993.

[116] M. W. Merkhofer. The value of information given decision flexibility. *Management Science*, 23: 716–727, 1977.

[117] J. S Evans. Strategic flexibility for high technology manoeuvres: a conceptual framework. *Journal of Management Studies*, 28: 69–89, 1991.

[118] Y. Y Haimes, K. Crowther, and B. M. Horowitz. Homeland security preparedness: balancing protection with resilience in emergent systems. *Systems Engineering*, 11(4): 287–308, 2008.

[119] R. L. Flood and E. R. Carson. *Dealing with Complexity: An Introduction to the Theory and Applications of Systems Science*. 2nd edn. New York: Plenum Press, 1993.

[120] L. P. Chuen. Singapore—an example of a system of systems. Keynote address presented at the IEEE International Conference on Systems, Man, and Cybernetics, Singapore, October 12–15, 2008.

[121] B. Ki-moon. Remarks to general assembly meeting on the global food and energy crisis, New York, July 18, 2008. http://www.un.org/apps/news/infocus/sgspeeches/search_full.asp?statID=296 (accessed February 20, 2009).

[122] G. Lean. Year of the hungry: 1,000,000,000 affected. *The Independent*, Dec 28, 2008. http://www.independent.co.uk/news/world/politics/year-of-the-hungry-1000000000-afflicted-1213843.html (accessed January 18, 2015).

[123] *The State of Food Insecurity in the World*. Economic and Social Development Department, Food and Agriculture Organization of the United Nations, Rome, October 2010.

[124] G. W. Bush. Remarks to the future farmers of America. July 27, 2001, Washington, DC.

[125] M. Nestle. *Food Politics: How the Food Industry Influences Nutrition and Health*. Berkley, CA: University of California Press, 2007.

[126] T. F. Pawlick. *The End of Food: How the Food Industry Is Destroying Our Food Supply—And What You Can Do About It*. Vancouver: Greystone Books, 2006.

[127] D. A. Pfeiffer. *Eating Fossil Fuels: Oil, Food and the Coming Crisis in Agriculture*. Gabriola Island, British Columbia: New Society Publishers, 2006.

[128] M. Pollan. *In Defense of Food: An Eater's Manifesto*. New York: Penguin Press, 2008.

[129] V. Shiva. *Stolen Harvest: The Hijacking of the Global Food Supply*. Cambridge, MA: South End Press, 2000.

[130] K. J. Carpenter. *Beriberi, White Rice, and Vitamin B*. Berkley, CA: University of California Press, 2000.

[131] ETC Group. Oligopoly, Inc. 2005: concentration in corporate power. *Communiqué*, Issue 91, 2005. Available at http://www.etcgroup.org/upload/publication/pdf_file/44 (accessed February 20, 2009).

[132] C. S. Holling and L. Gunderson. Resilience and adaptive cycles. In: *Panarchy: Understanding Transformations in Human and Natural Systems*, edited by L. H. Gunderson and C. S. Holling. Washington, DC: Island Press, 2002, pp. 25–62.

[133] J. Diamond. *Collapse: How Societies Choose to Fail or Succeed*. New York: Penguin, 2005.

[134] J. W. Kingdom. *Agendas, Alternatives, and Public Policies*. New York: Harper Collins, 1995.

[135] P. Olsson, L. H. Gunderson, S. R. Carpenter, P. Ryan, L. Lebel, C. Folke, and C. Holling. Shooting the rapids: navigating transitions to adaptive governance of social-ecological systems. *Ecology and Society*, 11(1): article 18, 2006. Available at http://www.ecologyandsociety.org/vol11/iss1/art18/ (accessed January 18, 2015).

[136] M. Friedman. *Capitalism and Freedom*. 2nd edn. Chicago, IL: University of Chicago Press, 1982.

[137] UN High-Level Task Force on the Global Food Security Crisis Website, Available at http://www.un.org/issues/food/taskforce/ (accessed February 20, 2009).

[138] O. Ward. UN task force to take aim at food crisis facing the world's poor. *Toronto Star*, April 30, 2008. Available at http://www.thestar.com/News/World/article/419817 (accessed January 18, 2015).

[139] J. L. Creighton. *The Public Participation Handbook: Making Better Decisions Through Citizen Involvement*. San Francisco, CA: John Wiley & Sons, 2005.

[140] D. Wheeler and J. Thomson. Brand barons and the business of food. In: *Feeding the Future: From Fat to Famine: How to Solve the World's Food Crises*, edited by A. Heintzman and E. Solomon. Toronto: House of Anansi Press, 2004, pp. 191–235.

[141] S. Ostry. Between feast and famine: fixing global trade. In: *Feeding the Future: From Fat to Famine: How to Solve the World's Food Crises*, edited by A. Heintzman, E. Solomon. Toronto: House of Anansi Press, 2004, pp. 237–257.

[142] A. Beattie and F. Williams. Doha trade talks collapse. *Financial Times*, July 29, 2008. Available at http://www.ft.com/cms/s/0/0638a320-5d8a-11dd-8129-000077b07658.html?nclick_check=1 (accessed January 18, 2015).

[143] Lao Tzu, *Tao Te Ching* (translation: *The Book of the Way*), 600 BC–531 BC. Available at http://cec.vcn.bc.ca/cmp/modules/emp-go.htm (accessed January 19, 2015).

[144] J. Clapp. Stepping up to the plate: rich nations must hand over more cash for food aid and reduce subsidies to their farmers. *Globe and Mail*, April 30, 2008. Available at http://www.theglobeandmail.com/servlet/story/RTGAM.20080430.wcofood30/BNStory/National (accessed February 20, 2009).

[145] C. Walker. Jennifer Clapp of CIGI talks about the current global food crisis. *Cambridge Now!*, May 2008. Available at http://www.cambridgenow.ca/npps/story.cfm?id=1114 (accessed February 20, 2009).

[146] UN High-Level Task Force on the Global Food Security Crisis. Comprehensive framework for action. July 2008. Available at http://www.un.org/issues/food/taskforce/Documentation/FINAL008.pdf (accessed February 20, 2009).

[147] J. E. Stiglitz. *Globalization and Its Discontents*. New York: W.W. Norton & Company, 2002.

[148] A. Harmes. *The Return of the State: Protestors, Power-brokers, and the New Global Compromise*. Vancouver: Douglas & McIntyre, 2004.

[149] N. Klein. Disaster capitalism: the new economy of catastrophe. *Harper's Magazine*, 315(1889): 47–58, 2007.

[150] N. Klein. *The Shock Doctrine: The Rise of Disaster Capitalism*. Toronto: Random House of Canada, 2007.

[151] J. R. Saul. *A Fair Country: Telling Truths about Canada*. Toronto: Penguin, 2008.

[152] T. F. Homer-Dixon. *The Ingenuity Gap: Facing the Economic, Environmental, and other Challenges of an Increasingly Complex and Unpredictable World*. New York: Alfred A. Knopf, 2000.

[153] Canadian Council for International Co-operation. Canada and the global food crisis. Policy Brief from the Canadian Food Security Policy Group, May 2008. Available at www.ccic.ca/e/docs/003_food_2008-05_global_food_crisis.pdf (accessed February 20, 2009).

[154] B. Carin, J. Sewell, W. Wilson, C. Abbott, and L. Innis. Breaking global deadlocks: the future of international governance architecture (meeting report #3). Centre for Global Studies and CIGI, Mexico City, March 10–11, 2008.

[155] E. T. Loehman and D. M. Kilgour (Editors). *Designing Institutions for Environmental and Resource Management*. Cheltenham, UK: Edward Elgar Publishing Ltd., 1998.

[156] T. Swanson and S. Johnston. *Global Environmental Problems and International Environmental Agreements: The Economics of International Institution Building*. Northampton, MA: Edward Elgar, 1999.

[157] J. Krueger. Trade restrictions for the global environment: the case of the Montreal Protocol. In: *The Environment and International Trade Negotiations: Developing*

Country Stakes, edited by D. Tussie. Ottawa: International Development Research Centre, 2000.

[158] R. Axelrod. *The Complexity of Cooperation*. Princeton, NJ: Princeton University Press, 1997.

[159] UN Framework Convention on Climate Change (UNFCCC), Available at www.unfccc.int (accessed September 1, 2011).

[160] J. E. Stiglitz. A new agenda for global warming. *The Economists' Voice*, 3(7), Article 3, July, 2006.

[161] CBC. Canada–Kyoto timeline. *CBC News*, Last updated: February 14, 2007. Available at http://www.cbc.ca/news/background/kyoto/timeline.html (accessed February 20, 2009).

[162] C. Campbell. If every home in the US put in one compact fluorescent light bulb...what it will really take to stop global warming. *Maclean's*, March 26, 2008. Available at http://www.macleans.ca/science/environment/article.jsp?content=20080326_994_994 (accessed February 20, 2009).

[163] M. Shellenberger, T. Nordhaus, J. Navin, T. Norris, and A. Van Noppen. Fast, clean, & cheap: cutting global warming's Gordian knot. *Harvard Law and Policy Review*, 2: 93–118, 2008. Available at http://www.hlpronline.com/Shellenberger_HLPR.pdf (accessed January 18, 2015).

[164] K. H. Engel and B. Y. Orbach. Micro-motives and state and local climate change initiatives. *Harvard Law and Policy Review*, 2: 119–137, 2008. Available at http://www.hlpronline.com/Engel_Orbach_HLPR.pdf (accessed January 18, 2015).

[165] J. M. Tien. On integration and adaptation in complex service systems. *Journal of Systems Science and Systems Engineering*, 17(4): 385–415, 2008.

[166] N. Axworthy. An innovative approach to sustainable infrastructure. *Engineering Dimensions*, 28(4): 26–29, 2008.

[167] J. Ma, K. W. Hipel, M. De, and J. Cai. Transboundary water policies: assessment, comparison and enhancement. *Water Resources Management*, 22: 1069–1087, 2008.

[168] C. S. Holling (Editor). *Adaptive Environmental Assessment and Management*. New York: John Wiley & Sons, 1978.

[169] C. J. Walters. *Adaptive Management of Renewable Resources*. New York: MacMillan, 1986.

[170] C. S. Holling. Resilience and stability of ecological systems. *Annual Review Ecology and Systematics*, 4: 1–23, 1973.

[171] B. F. Nobel. Applying adaptive environmental management. In: *Resource and Environmental Management in Canada*, 3rd edn, edited by B. Mitchell. Toronto: Oxford University Press, 2004, pp. 442–446.

[172] K. R. MacCrimmon. An overview of multiple objective decision making. In: *Multiple Criteria Decision Making*, edited by J. L. Cochrance, M. Zeleny. Columbia: University of South Carolina Press, 1973, pp. 18–44.

[173] T. L. Saaty. *Decision Making for Leaders: The Analytical Hierarchy Process for Decisions in a Complex World*. Belmont, CA: Lifetime Learning Publications, 1982.

[174] B. Roy. *Multicriteria Methodology for Decision Aiding*. Dordrecht: Kluwer, 1996.

[175] B. R. Hobbs and P. Meier. *Energy Decisions and the Environment: A Guide to the Use of Multicriteria Methods*. Kluwer, 2000.

[176] V. Belton and T. J. Stewart. *Multicriteria Decision Analysis: An Integrated Approach*. Dordrecht: Kluwer, 2002.

[177] Y. Chen, K. W. Hipel, and D. M. Kilgour. Multiple criteria sorting using case-based distance models with an application in water resources management. *IEEE Transactions on Systems, Man, and Cybernetics, Part A: Systems and Humans*, 37(5): 680–691, 2007.

[178] V. Chankong and Y. Y. Haimes. *Multiobjective Decision Making: Theory and Methodology*. New York: North Holland, 1983.

[179] K. W. Hipel, D. M. Kilgour, S. Rajabi, and Y. Chen. Chapter 27—Operations research and refinement of courses of action. In: *Handbook of Systems Engineering and Management*, 2nd edn, edited by A. P. Sage and W. B. Rouse. New York: John Wiley & Sons, 2009, pp. 1171–1222.

[180] K. W. Hipel (Editor). *Conflict Resolution*. Oxford, UK: EOLSS Press, 2009, Vol. 1.

[181] K. W. Hipel (Editor). *Conflict Resolution*. Oxford, UK: EOLSS Press, 2009, Vol. 2.

[182] K. W. Hipel. Competition and cooperation in societal and technological systems of systems. Keynote address presented at the IEEE International Conference on Systems, Man, and Cybernetics, Montreal, October 7–10, 2007.

[183] K. W. Li, D. M. Kilgour, and K. W. Hipel. Status quo analysis in the graph model for conflict resolution. *Journal of the Operations Research Society*, 56: 699–707, 2005.

[184] A. Obeidi, D. M. Kilgour, and K. W. Hipel. Perceptual stability analysis of a graph model system. *IEEE Transactions on Systems, Man, and Cybernetics, Part A: Systems and Humans*, 39(5): 993–1006, 2009.

[185] K. W. Li, K. W. Hipel, D. M. Kilgour, and L. Fang. Preference uncertainty in the graph model for conflict resolution. *IEEE Transactions on Systems, Man, and Cybernetics, Part A: Systems and Humans*, 34(4): 507–520, 2004.

[186] L. Hamouda, D. M. Kilgour, and K. W. Hipel. Strength of preference in the graph model for conflict resolution. *Group Decision and Negotiation*, 13: 449–462, 2004.

[187] T. Inohara, K. W. Hipel, and S. Walker. Conflict analysis approaches for investigating attitudes and misperceptions in the War of 1812. *Journal of Systems Science and Systems Engineering*, 16(2): 181–201, 2007.

[188] A. Obeidi, K. W. Hipel, and D. M. Kilgour. The role of emotions in envisioning outcomes in conflict analysis. *Group Decision and Negotiation*, 14: 481–500, 2005.

[189] T. Inohara and K. W. Hipel. Coalition analysis in the graph model for conflict resolution. *Systems Engineering*, 11(4): 343–359, 2008.

[190] T. Inohara and K. W. Hipel. Interrelationships among noncooperative and coalition stability concepts. *Journal of Systems Science and Systems Engineering*, 17(1): 1–29, 2008.

[191] D. M. Kilgour, K. W. Hipel, L. Fang, and X. Peng. Coalition analysis in group decision support. *Group Decision and Negotiation*, 10(2): 159–175, 2001.

[192] D.-Z. Zeng, L. Fang, K. W. Hipel, and D. M. Kilgour. Policy equilibirum and generalized metarationalities for multiple decision-maker conflicts. *IEEE Transactions on Systems, Man, and Cybernetics, Part A: Systems and Humans*, 37(4): 456–463, 2007.

[193] L. Fang, K. W. Hipel, D. M. Kilgour, and X. Peng. A decision support system for interactive decision making—Part II: analysis and output interpretation. *IEEE Transactions*

on Systems, Man, and Cybernetics, Part C: Applications and Reviews, 33(1): 56–66, 2003.

[194] A. P. Sage. *Decision Support Systems Engineering*. Hoboken, NJ: John Wiley & Sons, 1991.

[195] H. Xu, K. W. Hipel, and D. M. Kilgour. Matrix representation of solution concepts in multiple decision maker graph models. *IEEE Transactions on Systems, Man, and Cybernetics, Part A: Systems and Humans*, 39(1): 96–108, 2009.

[196] M. Nilsson, A. Jordan, J. Turnpenny, J. Hertin, B. Nykvist, and D. Russel. The use and non-use of policy appraisal tools in public policy making: an analysis of three European countries and the European Union. *Policy Sciences*, 41(4): 335–355, 2008.

CHAPTER 3

SYSTEMIC YOYOS: AN INTUITION AND PLAYGROUND FOR GENERAL SYSTEMS RESEARCH

YI LIN[1,2], YI DONGYUN[1], and ZAIWU GONG[3]

[1]Department of Mathematics and Systems Science, College of Sciences, National University of Defense Technology, Changsha, Hunan, PR China
[2]School of Business, Slippery Rock University, Slippery Rock, PA, USA
[3]College of Economics and Management, Nanjing University of Information Science and Technology, Nanjing, PR China

3.1 INTRODUCTION

Von Bertalanffy [1] challenged the research of biology by stating provokingly that because the fundamental character of living things is its organization, the customary investigation of individual parts and processes cannot provide a complete explanation of the phenomenon of life. This investigation gives us no information about the coordination of parts and processes. Thus the chief task of biology must be to discover the laws of biological systems (at all levels of organization). We believe that the attempts to find a foundation at this theoretical level point at fundamental changes in the world picture. This view, considered as a method of investigation, we call "organismic biology" and, as an attempt at an explanation, "the system theory of the organism."

By carefully pondering over this statement, where the concept of systems is formally introduced, it becomes quite clear that such a challenge indeed exists in every corner of the spectrum of modern science as soon as one replaces "living things" by "physical matters" or "social events," where each physical matter or social event does have a beginning, followed by a period of relatively stable existence, and ended in a moment of disappearance so that the phenomenon of life can be loosely observed in many areas of scientific research and human endeavor. As a matter of

Contemporary Issues in Systems Science and Engineering, First Edition.
Edited by MengChu Zhou, Han-Xiong Li and Margot Weijnen.

fact, the existence of such a challenge facing the entire modern science is evidenced by our inabilities in making reliable predictions about the future, no matter whether what we are about to foretell is regarding the nature or regarding the next stages of development of social events. The difficulty to foretell the forthcoming changes in nature indicates that modern science still cannot deal with transitional changes beyond the theories established on the concept of continuity. Our inability to foresee the next stages of development of social events implicitly indicates that when many forces interact nonlinearly with each other, the current linear science simply loses its validity and strength. To this end, see Reference 2 for detailed discussions.

From this realization, in the past 80 plus years, studies in systems science and systems thinking have brought forward brand new understandings and discoveries to some of the major unsettled problems in conventional science. For more details, see References 3, 4 and the references therein. Due to these studies of wholes, parts, and their relationships, a forest of interdisciplinary explorations has appeared, revealing the overall trend of development in modern science and technology of synthesizing all areas of knowledge into a few major blocks, and the boundaries of conventional disciplines have become blurred [5]. Underlying these explorations, we can see the united effort of studying similar problems in different scientific fields on the basis of wholeness and parts, and of understanding the world in which we live by employing the point of view of interconnectedness. As having been tested in the past 80 plus years, the concept of systems has been widely accepted by the entire spectrum of science and technology [6, 7].

This chapter consists of eight sections. In the rest of this first section, we will look at the concept of general systems and its consequent formation of a new dimension for the postal modern science. By investigating the historical successes of calculus and its consequent theories, we will see the need to look at systems in terms of whole evolutions; and consequently, the systemic yoyo model is naturally established. In Section 3.2, we focus on the theoretical and empirical justifications of this important model. In Section 3.3, some elementary properties of this model are developed, including four laws on state of motion. In Sections 3.4–3.7, we will introduce the reader to some of the wonderful applications of this model in areas of social science, economics, the foundations of mathematics, and the prediction of extreme weather conditions. This chapter is concluded with some open questions that are crucial for the further development of this systemic model.

3.1.1 The Concept of General Systems

Similar to how numbers are introduced the concept of systems can be abstractly proposed out of any object, event, and process. For instance, behind a collection of objects, say, apples, there is a set of abstract numbers, such as 1, 2, 3, ... Behind each organization or structure there is an abstract system within which the relevant whole, component parts, and the related interconnectedness are emphasized. In other words, when the internal structure is ignored, numbers are useful; otherwise the world will be dominated by systems. That is, the abstract concept lying beneath each organization or structure stands for a system, which brings together all the components into a meaningful relationship which acts as a whole. Now, a natural question is how to describe the specific system of interest. In practice, this is done according to what the researcher wants to see as a system [3]. That is, a system is anything one sees as an entity with internal structure.

Now, let us look at what systems methodology entails. To this end, different scholars in the area of systems science have different takes. Even so, the fundamental points underlying these different understandings are roughly the same. For instance, Quastler [8] wrote that

> Generally speaking, systems methodology is essentially the establishment of a structural foundation for four kinds of theories of organization: cybernetics, game theory, decision theory, and information theory. It employs the concepts of a black box and a white box (a white box is a known object, which is formed in certain way, reflecting the efficiency of the system's given input and output), showing that research problems, appearing in the afore-mentioned theories on organizations, can be represented as white boxes, and their environments as black boxes. The objects of systems are classified into several categories: motivators, objects needed by the system to produce, sensors, and effectors. Sensors are the elements of the system that receive information, and effectors are the elements of the system that produce real reactions. Through a set of rules, policies, and regulations, sensors and effectors do what they are supposed to do.

By using these objects, Quastler proved the following laws:

(1) Interactions are between systems and between systems and their environments.
(2) A system's internal movements and the reception of information about its environment stimulate its efficiency.

Zadeh [9] listed the following problems as important for systems science: systems characteristics, systems classifications, systems identification, signal representation, signal classification, systems analysis, systems synthesis, systems control and programming, systems optimization, learning and adaptation, systems liability, stability, and controllability. To Zadeh, the main task of systems science is the study of general properties of systems without considering their physical specifics; and systems methodology is that systems science is an independent scientific endeavor whose abstract foundation contains concepts and frameworks that are useful in the study of various behaviors of different kinds of systems.

Although the concept of systems was first introduced formally in the second decade of the twentieth century in biology [1] and has been a fashionable topic of discussion in modern science and technology, like other concepts of science, its idea and thinking logic can be traced as far back in time as the recorded history goes. For example, Chinese traditional medicine, treating each human body as a whole, can be dated to the time of Yellow Emperor about 4800 years ago; Aristotle's statement that "the whole is greater than the sum of its parts" represents a fundamental problem in systems science. That is, throughout the history, the mankind has been studying and exploring nature by using the thinking logic of systems with various time-specific terms. Only in modern times have some new contents been added to the ancient systems thinking. In other words, the concept of systems we are studying today is not simply a product of yesterday. Instead, it is a reappearance of some ancient thought and a modern form of an ancient quest. This quest has been recognized in the human struggle for survival with nature, and has been studied at various points in time by using the languages of different historical moments.

At the same time when modern science and technology continuously employ Descartes' second principle and Galileo's method, with which wholes and processes are divided into small parts, systems methodology is established to deal with problems of order, structure, and organization. It is because scholars have gradually realized that the world is not a pile of infinitely many isolated objects, where not every problem or phenomenon can be simply described by causal relationships. The fundamental characteristics of the physical world are its organizational structure and connections of interior and exterior relations of different matters. The study of either isolated parts or separate causalities of problems can hardly explain completely or relatively globally the surrounding world [2].

Technologically, systems research has appeared to meet the challenges imposed by modern conveniences and demands, such as energies produced by various devices, self-controlled or self-directed devices, information highways, and new construction materials. That is, modern technological advances have been forcing the mankind to consider not only a single machine or matter or phenomenon, but also systems of machines, systems of matter and phenomena, and systems of men, matter, and machines. In particular, science and technology have been so maturely developed that each rational combination of different pieces of the available knowledge could result in unexpected consumable products.

As a matter of fact, all areas of human activity have faced with complexity, totality, and systemality. This tendency betokens a sharp change in scientific thinking. By comparing Descartes' second principle and Galileo's method with systems methodology and thinking logic, it is not hard to see that because of the introduction of systemic concepts, another new scientific and technological revolution is in the making. For example, each application of systems thinking points out the fact that the relevant classical theory needs to be modified somehow, see [10]. As always the case in scientific history, not all scientific workers have the same optimistic outlook [11, 12]. Some scholars believed that this phenomenon is an omen that systems research itself is facing a crisis [13]. Due to the lack of an adequate language, intuition, and playground, most of the conclusions and predictions derived earlier in systems research were ordinary language based [11] so that to the scientific minds, their validities are questionable. Contrary to this situation, some recent works have shown some very significant progress in this front [2, 7, 14].

In 2001, Klir [7] looked at the controversial phenomenon from a different angle. Because systems thinking focuses on those properties of systems and associated problems that emanate from the general notion of systemhood, while the divisions of the classical science have been done largely on properties of thinghood, systems science and research would naturally transcend all the disciplines of the classical science and becomes a force making the existing disciplinary boundaries totally irrelevant and superficial. The cross-disciplinary feature of the systems science and research implies that

(1) Researches, done in systems science, can be applied to virtually all disciplines of the classical science.
(2) Issues involving systemhood, studied in individual specialization of the classical science, can be studied comprehensively and thoroughly.
(3) A unifying influence on the classical science, where a growing number of narrow disciplines appear, is created.

Therefore, the classical and systems sciences can be viewed as a genuine two-dimensional (2D) science, where with the advantage of the newly discovered second dimension, we should be able to resolve some of the unsettled problems that have been difficult for modern science to conquer.

In the rest of this chapter, we will show how this second dimension—the systems science—can be successfully employed to address some of the age-old problems in diverse areas of knowledge. That is, we will show how systems concepts and methodology, in particular, the systemic yoyo model, can be equally applied to natural and social sciences.

3.1.2 A Look at the Success of Calculus-Based Theories

From the primeval to modern civilizations, the mankind has gone through a history of over several millions of years. However, a relatively well-recorded history goes back only as far as about 3000 years. During this time frame, the development of science mirrors that of human civilizations. And, each progress of pursuing after the ultimate truth is a process of getting rid of the stale and taking in the fresh and of making new discoveries and new creativities. Each time when the authority was repudiated, science was reborn again. Each time when the "ultimate truth" was questioned, a scientific prosperity appeared. That is, each time when people praise authorities, they are in fact praising ignorance.

When the mankind enters the twenty-first century, it faces new challenges along with age-old unsettled problems. Some of these challenges impose an urgent need to expand the boundary of science in order to address additional problems, while the age-old problems require the man to scrutinize the available science more closely. That is, along with history turning to another new page, it is once again a golden opportunity for us to think about the limitations of and problems existing within science, which we inherit from the generations before us. It is found [15–17] that in terms of the formalism, nonlinear evolution models are the singularity problems of mathematical blown-ups of uneven formal evolutions; and in terms of physical objectivity, nonlinear evolution models describe mutual reactions of uneven structures of materials, which is no longer a problem of formal quantities. Since uneven structures stand for eddy sources, leading to eddy motions instead of waves, the mystery of nonlinearity, which has been bothering the mankind for over 300 years, is resolved at once both physically and mathematically. What is shown is that the essence of nonlinear evolutions is the destruction of the initial-value automorphic structures and appearance of discontinuity. It provides a tool of theoretical analysis for studying objective transitional and reversal changes of materials and events.

In the early stages of the Western civilization, there existed two opposite schools on the structure of materials [18, 19]. One school believed that materials were made up of uncountable and invisible particles, named "atoms"; the other school that all materials were continuous. A representative of the former school is the ancient Greek philosopher Democritus (about 460–370 BC) and the later the ancient philosopher Aristotle (384–322 BC). Since abundant existence of solids made it easy for people to accept the Aristotelian continuity, the theory of "atoms" was not treated with any validity until the early part of the nineteenth century when J. Dalton (1766–1844) established relevant evidence. In principle, Leibniz and Newton's calculus was originated in the Aristotelian thoughts. Along with calculus, Newton constructed his laws of motion on the computational basis of calculus and accomplished the first successes in applications in celestial movements under unequal quantitative effects. With over 200 years of development, the classical mechanics has gradually evolved into such a set of analysis methods based on continuity that even nearly a century after quantum mechanics and relativity theory were established, the thinking logic and methods developed on continuity are still in use today [20, 21].

Note: Because this chapter will touch on social sciences, there is one natural concern at this junction regarding these two schools on the structure of materials because societies and individuals are not conventionally seen as material things. The key and unconventional concept we will establish here is that all matters and materials, either living or nonliving, either biological or physical, in the universe are systems, each of which can be legitimately investigated as a rotational pool of fluids. More specifically, the physical world is indeed an ocean of constantly changing flows of spinning pools of fluids, which will be discussed in Section 3.2. So, neither of the ancient schools on the structure of materials when seen as systems and in turn spinning pools of fluids has provided a comprehensive portrait for the structure of materials, be they physical objects or human organizations. It is because in the universe, as an abstract ocean of flowing spinning pools of fluids, both continuity and atomic particles are regional, local phenomena, where when isolated particles flow in a uniform direction, the sense of continuity appears if the flow is seen from a distance. And what are most common in the ocean are the interactions of rotational pools of fluids, and these interactions should be more righteously represented by chaotic discontinuities. For further deliberation, please refer to the discussion about Figure 3.1 below.

Due to differences in the environmental conditions and living circumstances, where the West was originated from castle-like environments and the East from big river cultures with agriculture and water conservation as the foundation of their national prosperities, the ancient Eastern civilizations were different from those of the West. So, naturally, Chinese people have been more observing about reversal and transitional changes of weathers and rivers. Since fluid motions are irregular and difficult to compute exactly, that was why *Book of Changes* (Yi Ching, in Chinese

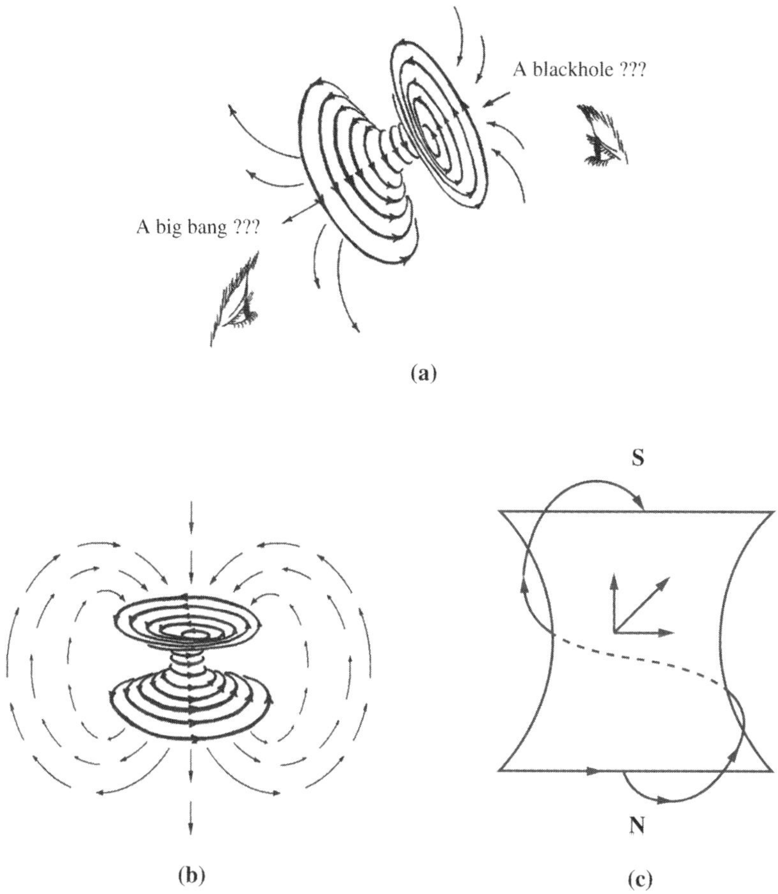

FIGURE 3.1 The eddy motion model of the general system.

name) and *Lao Tzu* (Tao De Ching, in Chinese name) appeared in China [22–25]. The most important characteristic of the first book is its way of knowing the world through materials' images and analyzing materials' changes through figurative structures with an emphasis placed on materials' irregularities, discontinuities, transitional and reversal changes.

(As a side note for those readers who are not familiar with these two Chinese classics. *Book of Changes* can be traced back to more than 3000 years ago when it was used to foretell the future. And throughout the history of the Far East, this book has been employed in many different settings, such as battle field strategies, coding method for sending secret messages, and political coups. On the other hand, Lao Tzu has been the book by a human that has been translated most among all books ever written, only second to the Holy Bible, which is supposed to be written by God. Different from the first book, this small but rich-in-content book has been read by various generations as a philosophy, a military strategy book, a book on how to rule, etc. Some of the mysteries surrounding these books include: There is no record about who wrote the first book. Why could the world be seen through the binary Ying and Yang system? What was the purpose of Lao Tzu? Who was supposed to be the audience?)

As what is pointed out in the blown-up theory [15, 17], in terms of mathematical symbolism, due to escapes in uneven forms from continuity, the evolution of any nonlinear evolution model is no longer a problem of simply expanding the given initial values. What is significant here is that through nonlinear evolutions, the concept of blown-ups can represent Lao Tzu's teaching that "all things are impregnated by two alternating tendencies, the tendency toward completion and the tendency toward initiation, which, acting together, complete each other" [22–24], and agrees with noninitial-value automorphic evolutions as what the *Book of Changes* describes: "At extremes, changes are inevitable. As soon as a change occurs, things will stabilize and stability implies long lasting" [25].

Since nonlinearity describes eddy motions, there must be different eddy vectorities and consequent irregularities. That is, the phenomenon of orderlessness, as observed in the study of chaos theory [26], is inevitable. When looking at fluid motions from the angle of eddies, one can see that the corresponding quantitative irregularities, orderlessness, multiplicities, complexities, etc., are all about the multiplicity of rotating materials. Therefore, there exist underlying reasons for the appearance of quantitative irregularities, multiplicities, and complexities. Those underlying reasons are the unevenness of time and space of the evolutionary materials' structures. For a detailed discussion on the concept of time, please refer Reference 14.

3.1.3 Whole Evolution and Yoyo Fields

With the conclusions of blown-up theory, one can see that all eddy motions, as described with nonlinearities, are irregular. And, regularized mathematical methods become powerless in front of the challenge of solving discontinuously quantified deterministic problems of nonlinear evolution models. In addition, one important concept studied in the blown-up theory is that of equal quantitative effects, which describes the conclusion of quantitative analysis under quasi-equal acting forces. Although this concept is initially introduced in the study of fluid motions, it represents the fundamental and universal characteristic of materials' movements. What is important about this concept is that it reveals the fact that nonlinearities originate from the structures of materials instead of nonstructural quantities.

On the basis of the blown-up theory and discussion on whether or not the world can be seen from the point of view of systems [27, 28], the concepts of black holes, big bangs, and converging and diverging eddy motions are coined together in the model shown in Figure 3.1. This model was established in Reference 17 for each object and every system imaginable. In particular, each system or object considered in a study is a multidimensional entity that spins about its invisible axis. If we fathom such a spinning entity in our three-dimensional (3D) space, we will have a structure as shown in Figure 3.1a. The side of black hole sucks in all things, such as materials, information, and energy. After funneling through the short narrow neck, all things are spit out in the form of a big bang. Some of the materials, spit out from the end of big bang, never return to the other side and some will (Figure 3.1b). For the sake of convenience of communication, such a structure, as shown in Figure 3.1a, is called a (Chinese) yoyo due to its general shape. More specifically, what this model says is that each physical entity in the universe, be it a tangible or intangible object, a living being, an organization, a culture, a civilization, etc., can all be seen as a kind of realization of a certain multidimensional spinning yoyo with an invisible spin field around it. It stays in a constant spinning motion as depicted in Figure 3.1a. If it does stop its spinning, it will no longer exist as an identifiable system. What Figure 3.1c shows is that due to the interactions between the eddy field, which spins perpendicularly to the axis of spin, of the model, and the meridian field, which rotates parallel to the axis of spin, all the materials returning to the black-hole side travel along a spiral trajectory.

To satisfy as many readers as possible, let us address the following natural questions: How and why a social system may have a structure similar to the yoyo model? Why each social entity would be spinning about an invisible axis? And what might be a black hole in a given social system?

First, each social entity is an objectively existing system that is made up of objects, such as people and other physical elements, and some specific relations between the objects, where it is these relations that make the objects emerge as an organic whole, called a social system. For example, each university consists of people, buildings, equipment, etc. However, it will not be a university (system) without the specific setup of organizational whole (relationships). Now, what the yoyo model says is that each imaginable system, which is defined as the totality of some objects and some relationships between the objects [4], possesses the yoyo structure so that each chosen social system, as a specific system involving people, has its own specific multidimensional yoyo structure with a rotational field.

Second, there are many different ways to see why each social entity spins about an invisible axis. In particular, let us imagine a business entity. As is well known in management science, each firm has its own particular organizational culture. Differences in organizational cultures lead to varied levels of productivity. Now, the basic components of an organizational culture change over time. These changes constitute the evolution of the firm and are caused by inventing and importing ideas from other organizations and consequently modifying or eliminating some of the existing ones. The concept of spin beneath the systemic yoyo structure of the firm comes from what ideas to invent, which external ideas to import, and which existing ones to eliminate. If idea A will likely make the firm more prosperous with higher level of productivity, while idea B will likely make the firm stay as it has been, then these ideas will form a spin in the organizational culture. Specifically, some members of the firm might like additional productivity so that their personal goals can be materialized in the process of creating the extra productivity, while some other members might like to keep things as they have been so that what they have occupied, such as income, prestige, and social status, will not be adversely affected. These two groups will fight against each other to push for their agendas so that theoretically, ideas A and B actually "spin" around each other. For one moment, A is ahead; for the next moment B is leading. And at yet another moment no side is ahead when the power struggle might very well return to the initial state of the affair. In this particular incidence, the abstract axis of spin is invisible, because no one is willing to openly admit his underlying purpose for pushing for a specific idea (either A or B or other ones).

Thirdly, the black-hole and big-bang structure in a social organization can be seen relatively clearly, because each social organization is an input–output system, no matter whether the organization is seen materially, holistically, or spiritually. The input mechanism is naturally the "black hole," while outputs the "big bang." Again, when the organization is seen from different angles, the meanings of "black hole" and "big bang" are different. But, together these different "black holes" and "big bangs" make the organization alive. Without the totality of "black holes" and that of "big bangs," no organization can be physically standing. Other than intuition, to this end, the existing literature on civilizations, business entities, and individual humans readily testify.

At this junction, it should be clearly pointed out that each system, be it physical or biological, stands for a whole that is made up of the physical body, its internal structure, and its interactions with the environment. This whole, according to the systemic yoyo model, is a high dimensional spin field. Considering the fact that the body is the only carrier of all other (such as cultural, philosophical, spiritual, and psychological) aspects of the structure, in theory each body of human organization is a pool of fluid realized through human sensing organs in the 3D space. The word "fluid" here is an abstract term totalling the flows of energy, information, materials, etc., circulating within the inside of, going into, and giving off from the body. And in all published references we have searched these flows are studied widely in natural and social science using continuous functions, which in physics and mathematics mean flows of fluids and are widely known as flow functions. On the other hand, as will be shown in this chapter the universe is a huge ocean of eddies, which changes and evolves constantly. That is, the totality of the physically existing world can be legitimately studied as fluids.

Notes: (1) Historically, although the name for this model was officially introduced in Reference 29, the particular yoyo structure as given in Figure 3.1a appeared on the front cover of Reference 17. The spirit of the model is spin, which has been well documented in Reference 15, where the focus of attention was the discussion of Lorenz's chaos and the predictability of disastrous weather conditions. As detailed in Reference 15, the realization of the inherent constant spin of general systems was initially discovered by Shoucheng OuYang in the 1960s when he was forced to look at how to predict major convective weather conditions. As for how OuYang successfully fulfilled his 1960s' task, the interested reader is encouraged to refer References 15 and 17 and the works listed there by OuYang.

(2) The concept of systems: The yoyo model presented hereafter is entirely based on a comprehensive understanding of the concept of (general) systems. As

shown earlier, this concept has been investigated in different ways using various languages throughout the recorded history. However, like the concept of Tao [22], whenever the concept of systems is defined in a specific fashion, what is described is not exactly the concept. This fact means that each specific technical definition of systems only captures a particular aspect of the desired concept. So, to achieve the desired level of understanding for our purpose here, one needs to combine several available versions of the concept in order to achieve the required mastery. To this end, the listed versions of definition of systems in this chapter will be adequate.

(3) After the fact that all systems, be they physical objects or organizations of biological beings, can be investigated as rotational pools of fluids is established, the rest of this presentation becomes quite intuitive and is developed on analytic analysis, laboratory experiments, and formal logic reasoning. To this end, scholars from social sciences might prefer to lay the theory on a more thorough analysis of human phenomena themselves. However, the truth is that if reliable thorough analysis of human phenomena already existed, we would not have much trouble explaining many organizational behaviors and events, such as the financial crisis of the late 2000s. In other words, there is no reliable thorough analysis of human phenomena especially those developed for large scale ones. Because of this reason, this work could not be seen as secondary as developed based on the existing scientific literature.

3.2 THEORETICAL AND EMPIRICAL JUSTIFICATIONS

In this section, we will look at justifications for the validity of the afore-described yoyo structure underneath each and every system under consideration.

3.2.1 Transitional Changes in Whole Evolutions

When we treat everything we see as a system [7], many systems in nature seem to evolve in concert. When one thing changes, many other seemingly unrelated matters alter their states of existence accordingly. If we attempt to make predictions regarding an evolving event, such as the price of a stock, which is dependent on the behaviors of the market participants, or the weather system, which is not affected by how man thinks, we may find ourselves somewhere in between having not enough information or too much information. It has been shown [26, 30, 31] that no matter which situation we are in, too much information or too little or anywhere in between, we face uncertainties. That is why the evolution of the system that we need to predict about its future needs to be seen as a whole [32]. In whole evolutions, other than continuities, as well studied in modern mathematics and science, what is more important and common is discontinuity with which transitional changes (or blown-ups) occur. These blown-ups reflect not only the singular transitional characteristics of the whole evolutions of nonlinear equations, but also the changes of old structures being replaced by new structures.

Thousands of case studies have shown [14] that transitional changes are the central issue and extremely difficult open problem of prediction science, since the method of linearity, which extends the past rise-and-fall into the future, does not have the ability to predict forthcoming transitional changes and what will occur after the changes. In terms of nonlinear evolution models, blown-ups reflect destructions of old structures and establishments of new ones. By borrowing the form of calculus, we can write the concept of blown-ups as follows: For a given (mathematical or symbolic) model, that truthfully describes the physical situation of concern, if its solution $u = u\left(t; t_0, u_0\right)$, where t stands for time and u_0 the initial state of the system, satisfies

$$\lim_{t \to t_0} |u| = +\infty, \tag{3.1}$$

and at the same time when $t \to t_0$, the underlying physical system goes through a transitional change, then the solution $u = u(t; t_0, u_0)$ is called a blown-up solution and the relevant physical movement expresses a blown-up. Analysis of thousands of real-life cases of various evolutionary systems [14] indicates that disastrous events appear at the moments of blown-ups in the whole evolutions of systems. For nonlinear models in independent variables of time (t) and space (x, x and y, or x, y, and z), the concept of blown-ups are defined similarly, where blow-ups in the model and the underlying physical system can appear in time or in space or in both.

Based on the evolutionary behaviors of the system before and after the transition, we can classify blown-ups into two categories: transitional and nontransitional. A blown-up is transitional, if the development of the physical system after the special blow-up moment in time or space is completely opposite to that before the blow-up. For example, if the evolution of the system grows drastically before the blow-up and the development of the system after the blow-up starts from nearly ground "zero," then such blown-up is transitional (Figure 3.2). Otherwise, it is called nontransitional (Figure 3.3).

The mathematical characteristics of blown-ups of the following constant-coefficient differential equation,

$$\begin{aligned} \dot{u} = F &= a_0 + a_1 u + \ldots + a_{n-1} u^{n-1} + u^n \\ &= (u - u_1)^{p_1} \ldots (u - u_r)^{p_r} (u^2 + b_1 u + c_1)^{q_1} \ldots (u^2 + b_m u + c_m)^{q_m} \end{aligned} \tag{3.2}$$

where u is the state variable, $\dot{u}$ derivative of u with respect to time t, as how Newton used to write, a_0, a_1, ..., and a_{n-1} are constants, p_i and q_j, $i = 1, 2, \ldots, r$ and $j = 1, 2, \ldots, m$, are positive whole numbers satisfying $n = \sum_{i=1}^{r} p_i + 2\sum_{j=1}^{m} q_j$, $\Delta = b_j^2 - 4c_j < 0, j = 1, 2, \ldots, m$, (without loss of generality, assume that $u_1 \geq u_2 \geq \ldots \geq u_r$), are described qualitatively in the following result.

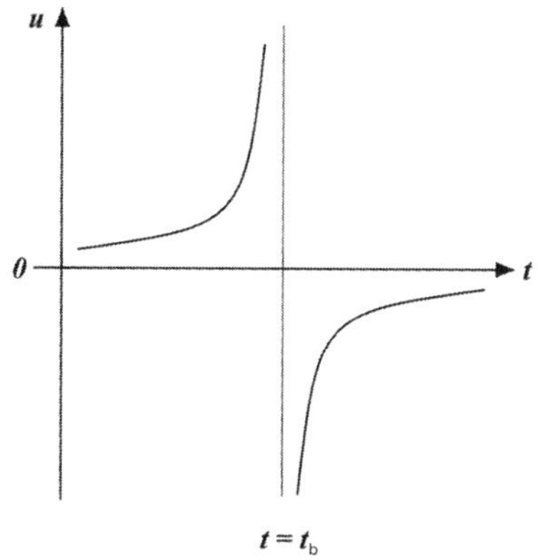

FIGURE 3.2 A transitional blown-up.

Theorem 1 For each initial-value problem of Equation (3.1), the following hold:

(1) When u_i, $i = 1, 2, \ldots, r$, does not exist, that is, $F = 0$ does not have any real solution, the solution u of the initial problem contains blown-up(s).
(2) If $F = 0$ does have real solutions u_i, $i = 1, 2, \ldots, r$, satisfying $u_1 \geq u_2 \geq \ldots \geq u_r$,
   (a) when n is an even number, if $u > u_1$, then the solution u of the initial problem contains blown-up(s); and
   (b) when n is an odd number, no matter whether $u > u_1$ or $u < u_r$, there always exist blown-ups in the solution u of the initial problem.

A detailed proof of this theorem can be found in Reference 17, pp. 65–66. □

3.2.2 Quantitative Infinity and Equal Quantitative Effects

One feature of blown-ups is the quantitative ∞, a mathematical indeterminacy, which in applications causes instabilities and calculation spills that have stopped every

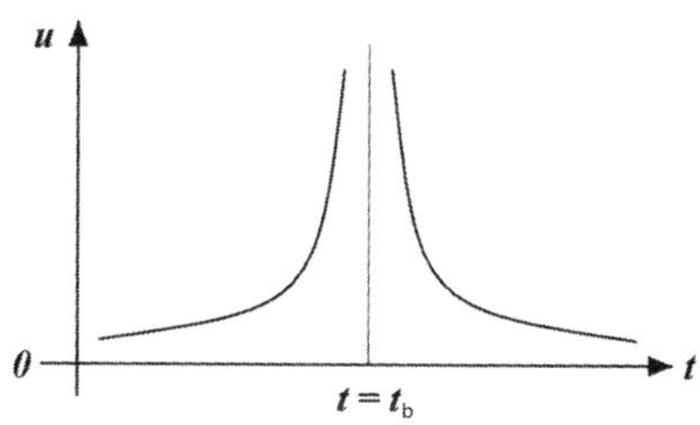

FIGURE 3.3 A nontransitional blown-up.

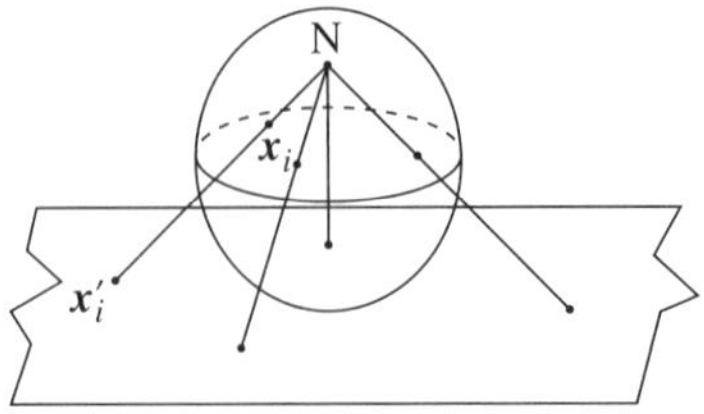

FIGURE 3.4 The Riemann ball—relationship between planar infinity and three-dimensional North Pole.

working computer. To comprehend ∞, let us look at the mapping relation of the Riemann ball, which is well studied in complex functions (Figure 3.4). This ball, a curvature space, illustrates the relationship between the infinity ∞ on the plane and the North Pole N of the ball. Such a mapping relation connects $-\infty$ and $+\infty$ through a blown-up. In other words, when a dynamic point x_i travels through the North Pole N on the sphere, the corresponding image x_i' on the plane of the point x_i shows up as a reversal change from $-\infty$ to $+\infty$ through a blown-up. So, treating the planar points $\pm\infty$ as indeterminacy can only be a product of the thinking logic of a low dimensional observation and control, because, speaking generally, these points stand implicitly for direction changes of one dynamic point on the sphere at the polar point N. Or speaking differently, the phenomenon of directionless, as shown by blown-ups of a lower dimensional space, represents exactly a direction change of movement in a higher dimensional curvature space.

From this illustration, one can see that the concept of blown-ups can represent implicit transformations of spatial dynamics. Through blown-ups, indeterminacy of a distorted low dimensional space (in this case, the Euclidean space) is transformed into determinant situations of a curvature space.

Summarizing what has been discussed above, we can see that nonlinearity mathematically stands (mostly) for singularities in Euclidean spaces; and in terms of physics, it represents eddy motions, the movements on the curvature space. Such motions represent a problem of structural evolutions, which are a natural consequence of uneven evolutions of materials. So, nonlinearity accidentally describes discontinuous singular evolutionary characteristics of eddy motions (in curvature spaces) from the angle of a special, narrow observe–control system, the Euclidean spaces. To support this end, please go to Bjerknes' Circulation Theorem discussed later for details.

As for how general dynamic system is related to eddy motions, let us look at Newton's second law of motion:

$$m\frac{d\vec{v}}{dt} = \vec{F}. \tag{3.3}$$

Based on Einstein's concept [33] of uneven time and space of materials' evolutions, we can assume

$$\vec{F} = -\nabla S\left(t, x, y, z\right), \tag{3.4}$$

where $S = S(t, x, y, z)$ stands for the time–space distribution of the external acting object, and the symbol ∇ stands for the gradient of the multivariable function. Let $\rho = \rho(t, x, y, z)$ be the density of the object being acted upon. Then, the kinematic Equation (3.3) for a unit mass of the object being acted upon can be written as

$$\frac{d\vec{u}}{dt} = -\frac{1}{\rho(t, x, y, z)} \nabla S(t, x, y, z), \tag{3.5}$$

where $\vec{u}$ is used to replace the original $\vec{v}$ to represent the meaning that each movement of some materials is a consequence of mutual reactions of materials' structures. Evidently, if ρ is not a constant, then Equation (3.5) becomes

$$\frac{d\left(\nabla x \times \vec{u}\right)}{dt} = -\nabla x \times \left[\frac{1}{\rho} \nabla S\right] \neq 0, \tag{3.6}$$

which stands for an eddy motion because of the cross-products involved. In other words, a nonlinear mutual reaction between materials' uneven structures and the unevenness of the external forcing object will definitely produce eddy motions.

Another important concept studied in the blown-up theory is that of equal quantitative effects, which stands for the eddy effects with nonuniform vortical vectorities existing naturally in systems of equal quantitative movements due to the unevenness of materials. Here, equal quantitative movements are mainly characterized by the kind of measurement uncertainty that when I observe an object, the object is constrained by me; when an object is observed by another object, the two objects cannot really be separated apart. More specifically, by equal quantitative movements, it is meant to be the movements with quasi-equal acting and reacting objects or under two or more quasi-equal mutual constraints. For example, the relative movements of two or more planets of approximately equal masses are considered equal quantitative movements. In the microcosmic world, an often seen equal quantitative movement is the mutual interference between the particles to be measured and the equipment used to make the measurement. Many phenomena in daily lives can also be considered equal quantitative effects, including such events as wars, politics, economies, chess games, races, and plays.

The concept of equal quantitative effects not only possesses a wide range of applications, but also represents an important omission of modern science of the past 300 plus years. To intuitively see why equal quantitative effects are so difficult for modern science to handle, let us first look at why all materials in the universe are in rotational movements. According to Einstein's uneven space and time, we can assume that all materials have uneven structures. Out of these uneven structures, there naturally exist gradients. With gradients, there will appear forces. Combined with uneven arms of forces, the carrying materials will have to rotate in the form of moments of forces.

Now, we can imagine that the natural world is entirely composed of eddy currents, where eddies exist in different sizes and scales and interact with each other. That is, the universe is a huge ocean of eddies, which change and evolve constantly. One of

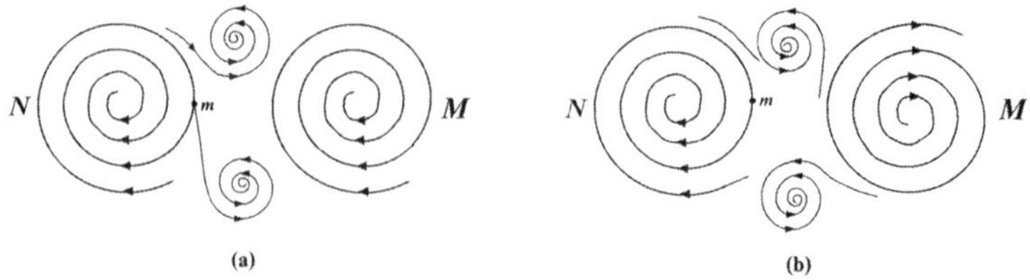

FIGURE 3.5 Appearance of sub-eddies.

the most important characteristics of spinning fluids, including spinning solids, is the difference between the structural properties of inwardly and outwardly spinning pools and the discontinuity between these pools. Due to the stirs in the form of moments of forces, in the discontinuous zones, there exist sub-eddies and sub-sub-eddies (Figure 3.5, where sub-eddies are created naturally by the large eddies *M* and *N*). Their twist-ups (the sub-eddies) contain highly condensed amounts of materials and energies. Since evolution is about changes in materials' structures, it is mainly about transformations of rotations in the form of moments of forces ignited by irregularities. The enclosed areas in Figure 3.6 stand for the potential places for equal quantitative effects to appear, where the combined pushing or pulling is small in absolute terms. However, it is generally difficult to predict what will come out of the power struggles. In general, what comes out of the power struggle tends to be drastic and unpredictable by using the theories and methodologies of modern science.

3.2.3 Fluid Circulation, Informational Infrastructure, and Human Communications

Continuing on what was done previously, we will in this subsection study several empirical evidences and observations that underline the existence of the yoyo structure behind each and every system, which either tangibly exists or is intellectually imaginable.

3.2.3.1 Bjerknes' Circulation Theorem At the end of the nineteenth century, V. Bjerknes (1898) discovered the eddy effects due to changes in the density of the

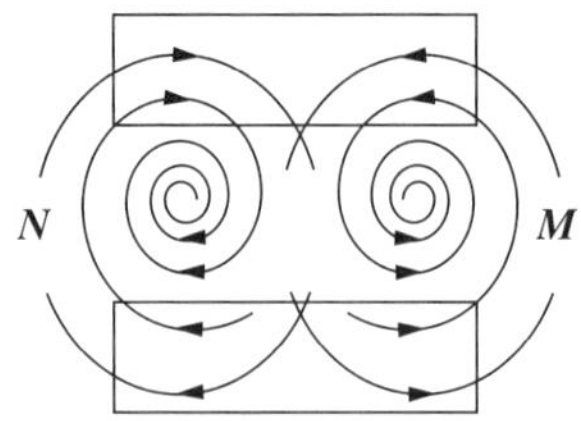

FIGURE 3.6 Structural representation of equal quantitative effects.

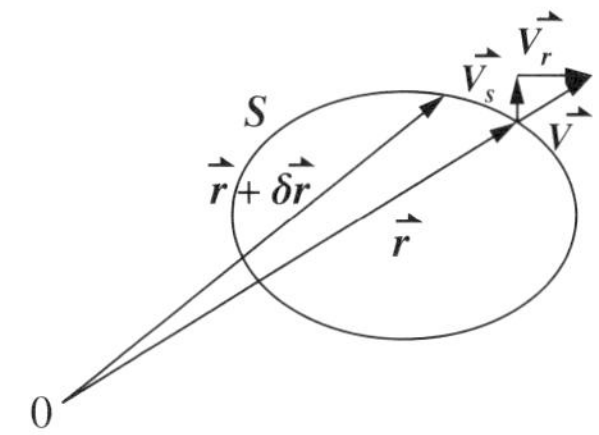

FIGURE 3.7 The definition of a closed circulation.

media in the movements of the atmosphere and ocean. He consequently established the well-known circulation theorem, which was later named after him.

By a circulation, it means a closed contour in a fluid. Mathematically, each circulation Γ is defined as the line integral about the contour of the component of the velocity vector locally tangent to the contour. In symbols, if $\vec{V}$ stands for the speed of a moving fluid, S an arbitrary closed curve, $\delta\vec{r}$ the vector difference of two neighboring points of the curve S (Figure 3.7), then a circulation Γ is defined as follows:

$$\Gamma = \oint_S \vec{V}\delta\vec{r}. \tag{3.7}$$

By applying a sequence of ingenious operations, Bjerkes obtained the following circulation theorem:

$$\frac{d\vec{V}}{dt} = \times(-\nabla p) \cdot \delta\sigma - 2\Omega\frac{d\sigma}{dt}, \tag{3.8}$$

where t stands for time, σ the projection area on the equator plane of the area enclosed by the closed curve S, p the atmospheric pressure, ρ the density of the atmosphere, and Ω the earth's rotational angular speed.

The left-hand side of Equation (3.8) represents the acceleration of the moving fluid, which according to Newton's second law of motion is equivalent to the force acting on the fluid. On the right-hand side, the first term is called a solenoid term in meteorology. It is originated from the interaction of the p- and ρ-planes due to uneven density ρ so that a twisting force is created. Consequently, materials' movements must be rotations with the rotating direction determined by the equal p- and ρ-plane distributions (Figure 3.8). The second term in Equation (3.8) comes from the rotation of the earth.

To see the importance of this result, we need to realize that the common form of materials' movements is eddies, such as the eddy motions of the solar system, galaxies, and nebula, in the cosmic level; polar eddies, cyclones, and anticyclones, on earth at the mesoscale level; at the microscopic level, various rotational movements are found in atomic structures. That is why Kuchemann [34] once claimed that "the tendon of moving fluids is eddies." What OuYang said is [17] that the field of eddies is both a place with high concentration of fluids' kinetic energies and a place with

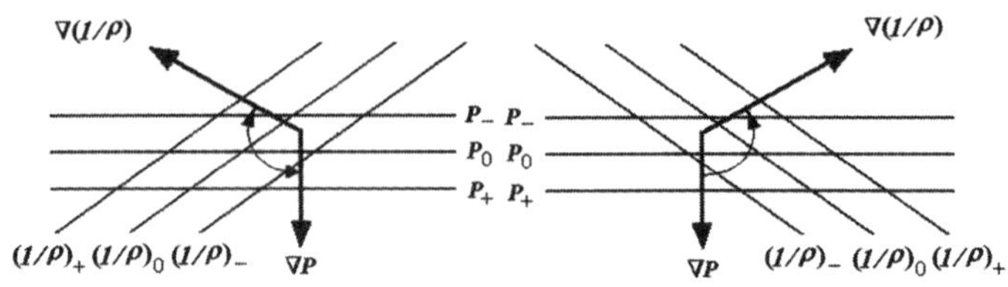

FIGURE 3.8 A diagram for solenoid circulations.

efficient transformations of kinetic energies into heat energies. In this sense, people are reminded that without eddy motions, there would not be any transformation of kinetic energies [14]. Through concentration of kinetic energies and through transformations of these energies into heat energies, eddies actually consume kinetic energies. Such a rise-and-fall of energies determines the equilibrium of the heat–kinetic forces, internal to the eddies, and reflects the close relationship between the quantities of eddies and transformations of energies.

Because uneven densities create twisting forces, fields of spinning currents are naturally created. Such fields do not have uniformity in terms of types of currents. Clockwise and counterclockwise eddies always co-exist, leading to destructions of the initial smooth, if any, fields of currents. When such evolutions are reflected in the continuous analytical methodology of any calculus-based narrow observe–control system, discontinuous singularities in solutions will naturally appear. Conversely, discontinuities of nonlinear differential equations offer us an analytic method for predicting discontinuous transitional changes of materials and for predicting the appearance of fields of eddy currents.

3.2.3.2 *Informational Infrastructure* A historical fact is that because of the various laws of conservation, some branches of modern science have become "exact." By thinking alone the same line, based on an intuitive understanding of the concept of general systems, Reference 35 established a law of conservation emphasizing on the uniformity of "spatial structures" of various "systems."

In particular, from a practical point of view, a system is what is distinguished as a system [3]. From a mathematical point of view, a (general) system S is defined [36] as an ordered pair (M, R) of sets, where M is the set of objects of the system S and R a set of some relations on the set M. The sets M and R are called the object set and the relation set of the system S, respectively. For those readers who are sophisticated enough in mathematics, for each relation r in R, it implies that there exists an ordinal number $n = n(r)$, a function of r, such that r is a subset of the Cartesian product of n copies of the set M. The idea of using an ordered pair of sets to define the general system is to create the convenience of comparing systems. In particular, when two systems S_1 and S_2 are given, by writing each of these systems as an ordered pair (M_i, R_i), $i = 1, 2$, we can make use of the known facts of mathematics to show that $S_1 = S_2$, if, and only if $M_1 = M_2$ and $R_1 = R_2$. When two systems S_1 and S_2 are not equal, then with their ordered pair structures, we can readily investigate their comparisons, such as how and when they are similar, congruent, or one is structurally less than the other, and other properties between systems. For more details about this end, please refer Reference 4.

By combining these two understandings of general systems, we can intuitively see the following. Each thing that can be imagined in human minds is a system according to Klir's definition so that this thing would look the same as that of an ordered pair (M, R) according to Lin's definition. Furthermore, relations in R can be about some information of the system, its spatial structure, its theoretical structure, etc. That is, there should exist a law of conservation that reflects the uniformity of all tangible and imaginable things with respect to

(1) the content of information;
(2) spatial structures; and
(3) various forms of movements.

With this intuition in place, Reference 35 searched several areas of scientific research, covering particle physics, Dirac's Large Number Hypothesis, electromagnetic interactions, angular momentum of the solar system, celestial mechanics, and movements of the earth's atmosphere, and proposed the following law of conservation of informational infrastructure in terms of structures:

$$AT \times BS \times CM \times DE = a \tag{3.9}$$

or more generally,

$$AT^{\alpha} \times BS^{\beta} \times CM^{\gamma} \times DE^{\varepsilon} = a, \tag{3.10}$$

where α, β, γ, ε, and a are constants, T, S, M, E and A, B, C, D are respectively time, space, mass, energy, and their coefficients. These two formulas hold true for various conservative systems at the cosmic, macroscopic, and microscopic levels. The constants α, β, γ, ε, and a are determined by the initial state and properties of the natural system of interest.

This law carries significant theoretical and practical consequences. For instance, in the widest domain of human knowledge of our modern time, this law reveals the structural unification of different layers of the universe so that it might provide a clue for the unification of the four basic forces in physics. This law can be a new evidence for the big-bang theory. It supports the model for the infinite universe with border and the oscillation model for the evolution of the universe, where the universe evolves as follows:

$$\cdots \to \text{explosion} \to \text{shrinking} \to \text{explosion} \to \text{shrinking} \to \cdots$$

This law shows that there indeed exists some kind of uniformity in terms of time, space, mass, and energy among different natural systems of various scales under either macroscopic or microscopic conditions or relativity conditions.

Philosophically, this law indicates that same as mass, energy is also a characteristic of physical entities. Time and space can be seen as the forms of existence of physical entities with motion as their basic characteristics. This law connects time, space, mass, and motion closely into an inseparable whole so that time, space, mass, and energy can be seen as attributes of physical entities. With this understanding, the concept of mass is generalized and the wholeness of the world is further proved and the thoughts of never diminishing mass and never diminishing universes are evidenced.

3.2.3.3 Human Communications In this subsection, we look at how the systemic yoyo model is manifested in different areas of life. Based on the systemic yoyo model, each human being is a 3D realization of a spinning yoyo structure of a certain dimension higher than three. To illustrate this end, let us consider two simple and easy-to-repeat experiments.

Experiment 1: Imagine we go to a swim meet. The area of competition contains a pool of the Olympic size and about 200 seats along one long side of the pool available for spectators. It is enclosed with a roof and walls all around the space.

As soon as we enter the pool area, we immediately fall into a boiling pot of screaming and jumping spectators, cheering for their favorite swimmers competing in the pool. After we settle down in our seat a distance away from the pool deck anywhere in the seating area, let us purposelessly pick a voluntary helper on the pool deck for whatever reason, either for her beauty or for his strange body posture, and stare at him intensively. Here is what will happen next: Magically enough, before long, our stare will be felt by the person from quite a good distance; she/he will turn around and locate us in no time out of the reasonably sized and boiling audience.

In this silent communication, each side, either the person being stared at or us, is a high dimensional spinning yoyo. Although we are separated by space and by informational noise, the stare of one party on the other has directed that party's spin field of the yoyo structure into the spin field of the yoyo structure of the other party. Even though the latter party initially did not know the forthcoming stare, when her/his spin field is interrupted by the sudden intrusion of another unexpected spin field, the person surely senses the exact direction and location where the intrusion is from. That is the underlying mechanism for the silent communication to be established.

Experiment 2: Let us look at human relationship. When an individual A has a good impression about another individual B, magically, individual B also has a similar and almost identical impression about A. When A does not like B and describes B as a dishonest person with various undesirable traits, it has been clinically proven in psychology that what A describes about B is exactly who A is himself [37].

Once again, the underlying mechanism for such a quiet, unspoken evaluation of each other is based on the fact that each human being stands for a high dimensional spinning yoyo and its rotational field. Our feelings toward each other are formed through the interactions of our invisible yoyo structures and their spin fields. So, when person A feels good about another person B, it generally means that their underlying yoyo structures possess the same or very similar attributes, such as spinning in the same direction, both being either divergent or convergent at the same time, both having similar intensities of spin, etc. When person A does not like B and lists many undesirable traits B possesses, it fundamentally stands for the situation that the underlying yoyo fields of A and B are fighting against each other in terms of either opposite spinning directions, or different degrees of convergence, or in terms of other attributes.

3.3 ELEMENTARY PROPERTIES OF YOYO FIELDS

To prepare the theoretical foundation for the rest of this chapter, in this section, based on the basic attributes of the yoyo model, we introduce the structure of meridian fields that helps to hold the dynamic spin field of the yoyo model together. With the spin field of a systemic yoyo structure fully established, we study various properties of general systems, which will soon be fully employed in the following sections. This section is based on Reference 38.

3.3.1 Eddy and Meridian Fields

Because each yoyo spins as in Figure 3.1a, other than the eddy field, which is perpendicular to the axis of rotation of the yoyo structure, there is also a meridian field accompanying each yoyo (see Figure 3.1b). The invisible meridians go into the center of the black hole, through the narrow neck, and then out the big bang. They travel through the space and return to the center of the black hole. Somehow we can imagine that these meridians help to hold different layers of the spin field of the yoyo structure together.

Here, the word “spin” is used to capture the concept of angular momentum or the presence of angular momentum intrinsic to a body as opposed to orbital angular momentum of angular momentum that is the movement of the object about an external point. For example, the spin of the earth stands for the earth’s daily rotation about its polar axis. The orbital angular momentum of the earth is about the earth’s annual movement around the sun. In general, a 2D object spins around a center (or a point), while a 3D object rotates around a line called an axis. Here, the center and the axis must be within the body of the object.

Mathematically, the spins of rigid bodies have been understood quite well. If a spin of a rigid body around a point or axis is followed by a second spin around the same point (respectively axis), a third spin results. The inverse of a spin is also a spin. Thus, all possible spins around a point (respectively axis) form a group of mathematics. However, a spin around a point or axis and a spin around a different point (respectively axis) may result in something other than a rotation, such as a translation. Spins around the x-, y-, and z-axes in the 3D Euclidean space are called principal spins. Spin around any axis can be performed by taking a spin around the x-axis, followed by a spin around the y-axis, and then followed by a spin around the z-axis. That is, any 3D spin can be decomposed into a combination of principal spins.

In astronomy, spin (or rotation) is a commonly observed phenomenon. Stars, planets, galaxies all spin around on their axes. The planetary speeds of spin in the solar system were first measured by tracking visible features. This spin induces a centrifugal acceleration, which slightly counteracts the effect of gravity and the phenomenon of precession, a slight "wobble" in the movement of the axis of a planet.

In social science, spin appears in the study of many topics. That is why we have the old saying: Things always "go around and come around." As an example, the theory and practice of public relations heavily involve the concept of spin, where a person, such as a politician, or an organization, such as a publicly traded company, signifies his often biased favor of an event or situation. While traditional public relations may rely on creative presentation of the underlying facts, "spin" tends to imply disingenuous, deceptive and/or highly manipulative tactics used to influence public attitudes and opinions [39, 40].

In quantum mechanics, spin is particularly important for systems at atomic length scales, such as individual atoms, protons, or electrons. Such particles and the spin of quantum mechanical systems possess unusual or nonclassical features. For such systems, spin angular momentum cannot be associated with the concept of rotation precisely, but instead, refers to the presence of angular momentum [41].

So, each yoyo structure contains spins of materials in two perpendicular directions. The spin field perpendicular to the axis of rotation of the entire structure is referred to as the eddy field and the other field, which is quasi-parallel to the axis, is named the meridian field. For the convenience of our communication, for any given yoyo structure, the black-hole side will be referred to as the south pole of the structure and the big-bang side the north pole.

Also, in theory, we can think of the totality of all materials that can be physical, tangible, intangible, or epistemological and that are contained in a systemic yoyo, if this yoyo is situated in isolation from other yoyo structures. So, the concept of mass for a systemic yoyo can be defined as in the conventional physics. Since systems are of various kinds and scales, the universe can be seen as an ocean of eddy pools of different sizes, where each pool spins about its center or axis. At this junction, one good example in our 3D physical space is the spinning field of air in a tornado. In the solenoidal structure, at the same time when the air within the tornado spins about the eye in the center, the systemic yoyo structure continuously sucks in and spits out air. In the spinning solenoidal field, the tornado takes in air and other materials, such as water or water vapor at the bottom, lifts up everything it took in into the sky, and then it continuously spays out the air and water from the top of the spinning field. At the same time, the tornado also breathes in and out with air in all horizontal directions and elevations. If the amounts of air and water taken in by the tornado are greater than those given out, then the tornado will grow larger with increasing effect on everything along its path. That is the initial stage of formation of the tornado. If the opposite holds true, then the tornado is in its process of dying out. If the amounts of air and water that are taken in and given out reach an equilibrium, then the tornado can last for at least a while. In general, each tornado (or a systemic yoyo) experiences a period of stable existence after its initial formation and before its disappearance.

Similarly, for the general systemic yoyo model, it also constantly takes in and spits out materials. For the convenience of our discussion, we assume that the spinning of the yoyo structures follows the Left-Hand Rule 1 below:

> **Left-Hand Rule 1**: When holding our left hand, the four fingers represents the spinning direction of the eddy plane and the thumb points to the north pole direction along which the yoyo structure sucks in and spits out materials at its center (the narrow neck). (*Note:* It should be clear that in the physical world, systemic yoyos do not have to comply with this left-hand rule.)

As influenced by the eddy spin, the meridian directional movement of materials in the yoyo structure is actually slanted instead of being perfectly vertical. In Figure 3.1c, the horizontal vector stands for the direction of spin on the yoyo surface toward the reader and the vertical vector the direction of the meridian field, which is opposite of that in which the yoyo structure sucks in and spits out materials. Other than breathing in and out materials from the black-hole (the south pole) and big-bang (the north pole) sides, the yoyo structure also takes in and gives out materials in all horizontal directions and elevations, just as in the case of tornadoes discussed earlier.

In the process of taking in and giving out materials, an outside surface of materials is formed that is mostly imaginary of our human mind; this surface holds most of the contents of the spinning yoyo. The density of materials of this surface decreases as one moves away from the yoyo structure. The maximum density is reached at the center of the eddy field. As the spin field, which is the combination of the eddy and meridian fields, constantly takes in and gives out materials, there does not exist any clear boundary between the yoyo structure and its environment, which is analogous to the circumstance of a tornado that does not have a clear-cut separation between the tornado and its surroundings.

Next, let us look at the quark structure [42] of yoyo structures. Using the so-called three-jet event from particle physics [43] as our supporting evidence, all microscopic systemic yoyos can be classified on the basis of laboratory experiments into classes using the number of quarks involved [42, p. 41]. In particular, each spinning yoyo, as shown in Figure 3.1a, is seen as a 2-quark structure, where we imagine the yoyo is cut through its waist horizontally in the middle, then the top half is defined as an absorbing quark and the bottom half a spurting quark. Now, let us study 3-quark yoyos by looking at a proton P and a neutron N. At this junction, the three jet events are employed as the evidence for the structure of 3-quark yoyos, where there are two absorbing and one spurting quarks in the eddy field. The proton P has two absorbing u-quarks and one spurting d-quark (Figure 3.9), while the neutron N has two spurting d-quarks and one absorbing u-quark (Figure 3.10). Figure 3.10b are the simplified flow charts with the line segments indicating the imaginary axes of rotation of each local spinning column. Here, in Figure 3.9, the absorbing u-quarks stand for local spinning pools while together they also travel along in the larger eddy field in which they are part of. Similarly in Figure 3.10, the spurting d-quarks are regional spinning pools. At the same time when they spin individually, they also travel along in the large yoyo structure of the neutron N. In all these cases, the spinning directions of these

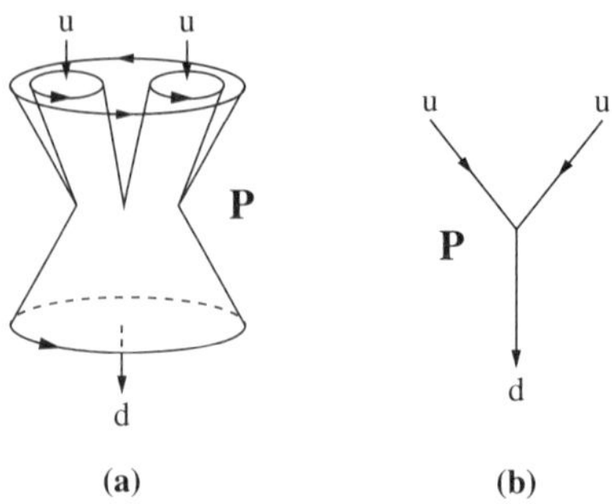

FIGURE 3.9 The quark structure of a proton P.

u- and d-quarks are the same except that each u-quark spins convergently (inwardly) and each d-quark divergently (outwardly).

Different yoyo structures have different numbers of absorbing u-quarks and d-quarks. And, the u-quarks and d-quarks in different yoyos are different due to variations in their mass, size, spinning speed and direction, and the speed of absorbing and spurting materials. This end is well supported by the discovery of quarks of various flavors, two spin states (up and down), positive and negative charges, and colors. That is, the existence of a great variety of quarks has been firmly established.

Now, if we fit Fultz's dishpan experiment [44] to the discussion above by imagining both the top and the bottom of each yoyo as a spinning dish of fluids, then the patterns as observed in the dishpan experiment suggest that in theory, there could exist such a yoyo structure that it has n u-quarks and m d-quarks, where $n \geq 1$ and $m \geq 1$ are arbitrary natural numbers, and each of these quarks spins individually and along with each other in the overall spinning pool of the yoyo structure.

3.3.2 Interactions Between Systemic Yoyos

When more than two yoyo structures are concerned with, the interactions between these entities, a multibody problem, become impossible to describe analytically, This

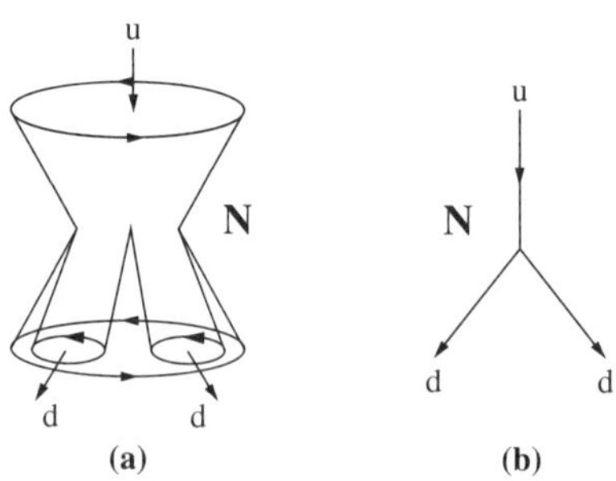

FIGURE 3.10 The quark structure of a neutron N.

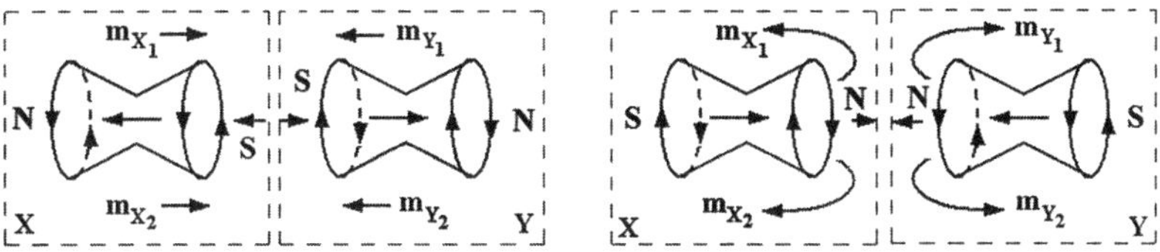

(a) (b)

FIGURE 3.11 Repulsion of like yoyo fields.

end is witnessed by the difficulties scholars have faced in the investigation of the three-body problem since about 300 years ago when Newton introduced his laws of motion regarding two-body systems. So, to achieve anything meaningful in terms of interactions of systemic yoyos, let us now focus on figurative expressions.

Just as electric/magnetic fields, for yoyo fields, the like-kind ends (the north poles or south poles) repel and the opposite attract, Figures 3.11 and 3.12.

The yoyo's (vertical) absorbing/spurting and (all directional) taking in/giving out along the meridian field direction create an active field that can be visibly seen as a field with source (or active field). In the physical world, gravitational fields, magnetic fields, electric fields, and nuclear fields are some of the examples of both active fields and eddy fields. In particular, in a given yoyo, if its eddy field is identified as its magnetic field, then the active (the meridian) field will correspond to the electric field. In this case, the so-called N and S poles of the magnetic field correspond respectively to the diverging and the converging sides of the yoyo structure, while the positive and negative electric fields to the converging and diverging eddy fields. Similarly, the eddy field can be identified as an electric field and so correspondingly the meridian field as the magnetic field. Now, if the eddy field is seen as a gravitational field, then the meridian field will be the corresponding magnetic field. In this case, we can see that the gravitational field can be both attraction and repulsion that exist side by side, the former corresponds to the converging side of the eddy field, and the latter the diverging side. From this discussion, it can be seen that all forms of fields must have two opposite effects, such as the N and S poles of a magnetic field, where the opposite effects have to coexist at all times and no one side effect can exist without the other opposite effect, even though the opposite effects do not have to be visible at the same time or both of them do not have to be visible at all.

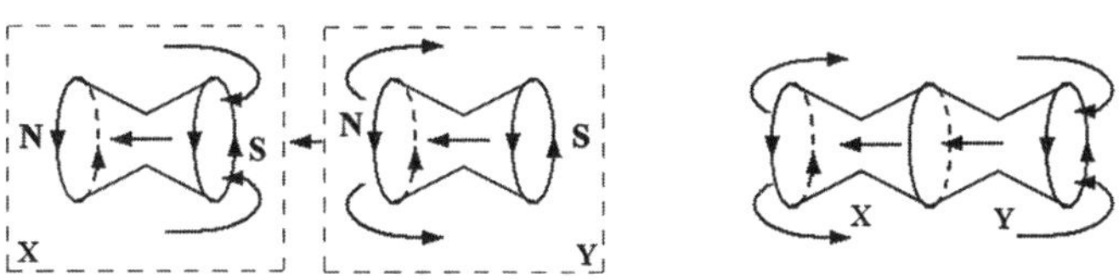

(a) (b)

FIGURE 3.12 How two yoyos can potentially become one yoyo.

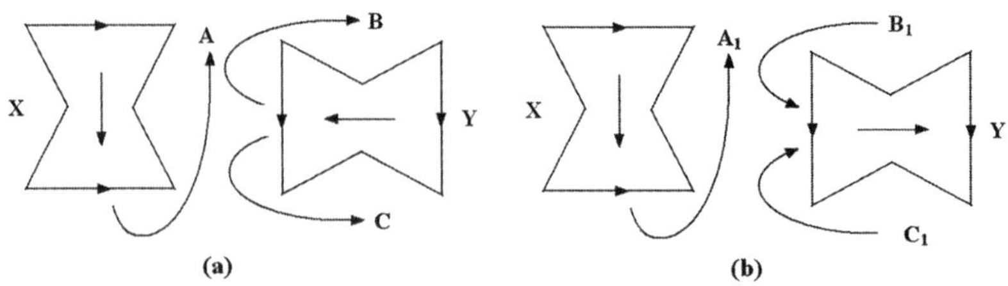

FIGURE 3.13 The tendency for yoyos to line up.

Now, let us look at the relative positioning of two yoyo structures. Assume that two spinning yoyos X and Y are given as shown in Figure 3.13. Then the meridian field A of X fights against C of Y so that both X and Y have the tendency to realign themselves in order to reduce the conflicts along the meridian directions. Similarly in Figure 3.13b, the meridian field A_1 of yoyo X fights against B_1 of Y. So, the yoyos X and Y also have the tendency to realign themselves as in the previous case.

Of the two yoyos X and Y above, let us assume that one of them is mighty and huge and the other so small that it can be seen as a tiny particle. Then, the tiny particle yoyo m will be forced to line up with the much mightier and larger spinning yoyo M in such a way that the axis of spin of the tiny yoyo m is parallel to that of M and that the polarities of m and M face the same direction. For example, Figure 3.14 shows how the particle yoyo m has to rotate and reposition itself under the powerful influence of the meridian field of the much mightier and larger yoyo structure M.

Now, let us imagine a tiny particle yoyo m, say, an electron in the conductive circuit, or a unit in the supply chain of a business activity, where either the circuit or

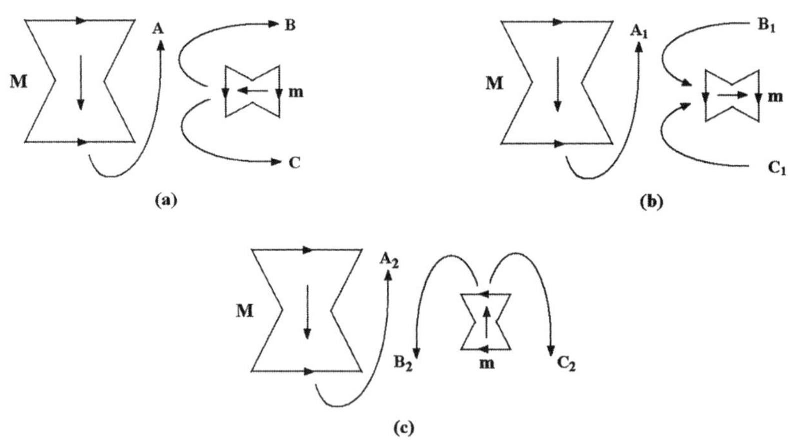

FIGURE 3.14 How mighty spinning yoyo M bullies particle yoyo m.

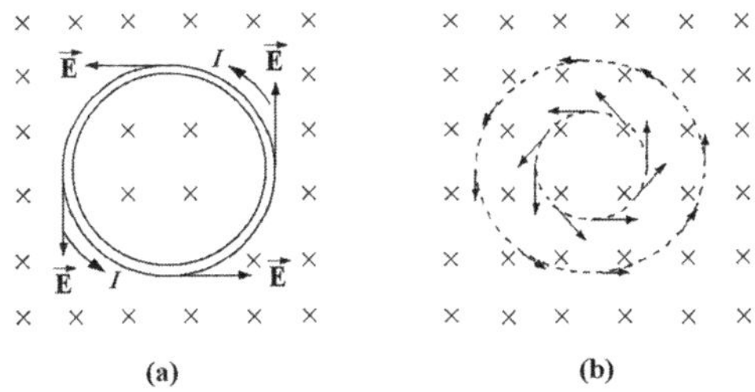

FIGURE 3.15 The manifestation of an induced yoyo field.

the supply chain is theoretically seen as the circular loop in Figure 3.15a, where the Xs stand for an overriding yoyo field of the environment going into the page. Then, the axis of spin and its polarities of the tiny yoyo m have to live up with those of an abstract mighty spinning yoyo, part of whose eddy field is symbolized by the Xs in Figure 3.15a. Based on the Left-Hand Rule 1 and the north direction of the yoyo field outside the circuit or supply chain, (the reason why we use the direction outside the circuit instead of the inside is because in the eddy field of the mighty spinning yoyo, the circuit would be only the size of a dot), we can see how the particle yoyos, such as electrons or different units in the supply chain, line up inside the circuit as shown in Figure 3.16. This line up of particle yoyos inside the circuit also determines the north direction of the superpositioned magnetic eddy field along the circuit, which fights against the increase in the intensity of the original magnetic yoyo field. This end in fact provides a systems theory support for the validity of the following generalized Lenz's Law, where Lenz's Law is well known in the theory of electromagnetism.

Generalized Lenz's Law: There is an induced yoyo current in any closed, conducting loop if and only if the yoyo flux, that is, a yoyo field through the loop is changing. The direction of the induced current is such that its induced yoyo field opposes the change in the yoyo flux.

To make sense out of the generalized Lenz's Law beyond the traditional understanding in the field of electromagnetism, let us imagine a person, who lives in an

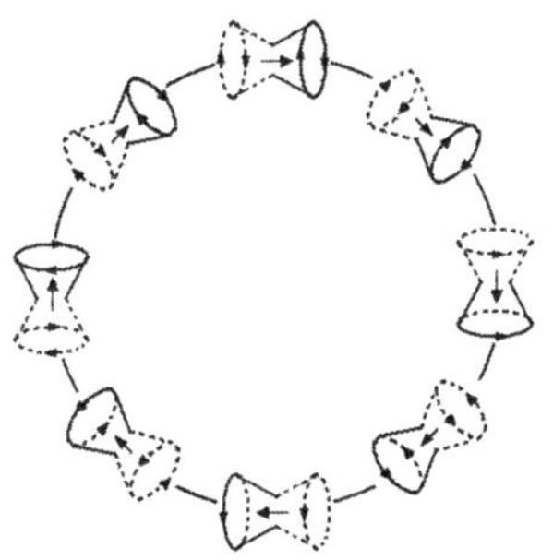

FIGURE 3.16 How particle yoyos line up inside the circuit.

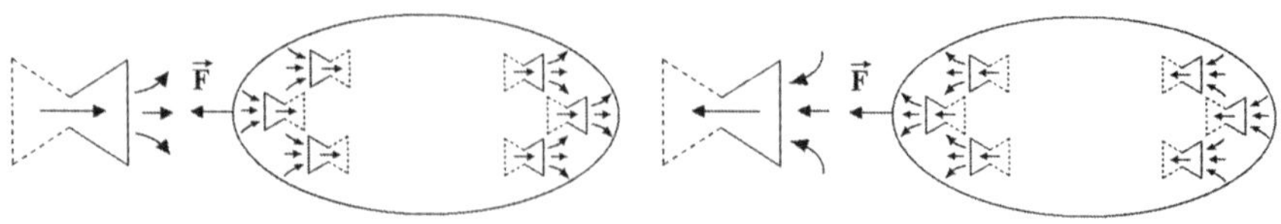

FIGURE 3.17 Formation of yoyo dipoles.

established environment. If one day, he suddenly plans to change his own behavior completely, what will happen to him is that he will find that everyone else in the environment instantly become united in the fight against his planned changes. The more he is determined to carry out his planned changes, the more resistance he will experience from the united environmental front. This example is one case of application of the generalized Lenz's Law in social sciences.

Similar to electric charge, if an object, when seen as a spinning yoyo, possesses an overall attraction or repulsion, we say that the object has a yoyo charge. Now, it is well known that electrically charged plastic or glass rods can attract certain nonmetal insulators, such as scraps of paper. Does this phenomenon indicate certain properties of yoyo charges? As a matter of fact, when a yoyo charge neutral object is affected by an external yoyo charge, the object can experience the phenomenon of polarization, just as the scraps of paper being affected by an electric charge. In particular, when a positive yoyo charge is placed near the object, the negative component yoyos inside the object have the potential to move to the side closer to the positive charge, while positive component yoyos move to the side that is furthest away from the external charge. Similarly, if the external charge is negative, the charged component yoyos inside the object will move in directions opposite to those described above. That is, under the influence of an external powerful yoyo charge, a yoyo dipole can be formed, as shown in Figure 3.17. Because opposite yoyo poles attract, the external yoyo charge attracts the created yoyo dipole.

3.3.3 Laws on State of Motion

From Reference 17, it follows that spins are the fundamental evolutionary characteristic of materials. In this subsection, we study the figurative analysis method of the systemic yoyo model and how to apply it to establish laws on state of motion by generalizing Newton's laws of motion. Corresponding to what is known that Newton's laws of motion are one of the reasons why physics is an "exact" science, in the rest of this chapter, we will show that these generalized laws will be equally applicable to social sciences and humanity areas as their classic forms in natural science. For all the omitted details of this subsection, please refer Reference 29.

Newton's first law says: An object will continue in its state of motion unless compelled to change by a force impressed upon it. This property of objects, their natural resistance to changes in their state of motion, is called inertia. By using the theory of blown-ups, how yoyo fields are created, and by attempting to address such

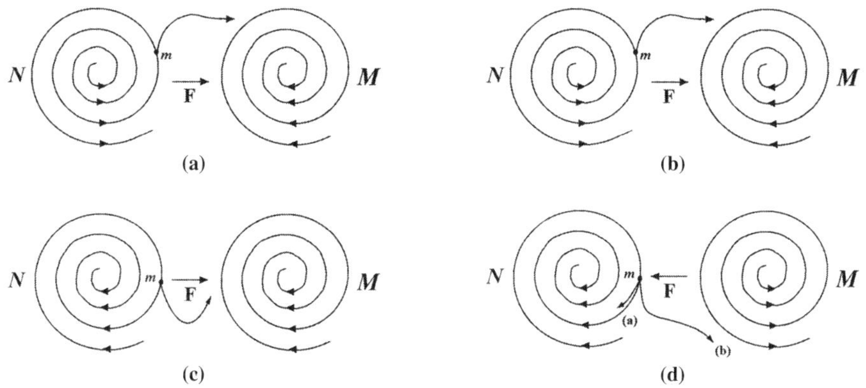

FIGURE 3.18 Acting and reacting models with yoyo structures of harmonic spinning patterns. (a) Object *m* is located in a diverging eddy and pulled by a converging eddy *M*. (b) Object *m* is located in a diverging eddy and pulled or pushed by a diverging eddy *M*. (c) Object *m* is located in a converging eddy and pulled by a converging eddy *M*. (d) Object *m* is located in a converging eddy and pulled or pushed by a diverging eddy *M*.

a question as *how can such resistance to change be considered natural*, the following is established [29]:

> *The First Law on State of Motion:* Each imaginable and existing entity in the universe is a spinning yoyo of a certain dimension. Located on the outskirt of the yoyo is a spin field. Without being affected by another yoyo structure, each particle in the said entity's yoyo structure continues its movement in its orbital state of motion.

Because for Newton's first law to hold true, one needs an external force, when people asked Newton where such an initial force could be from, he answered (jokingly?): "It was from God. He waved a bat and provided the very first blow to all things he created [19]." If such an initial blow is called the *first push*, then the yoyo model in Figure 3.1a and the stirring forces naturally existing in each "yoyo" created by uneven densities of materials' structures will be called the second stir.

Newton's second law of motion says that when a force does act on an object, the object's velocity will change and the object will accelerate. In symbols, the second law is written in Equation (3.3). Now, if we consider how an object *m* is acted upon by another object *M*, Figures 3.18 and 3.19, we have the following result:

> *The Second Law on State of Motion:* When a constantly spinning yoyo structure M does affect an object m, which is located in the spin field of another object N, the velocity of the object m will change and the object will accelerate. More

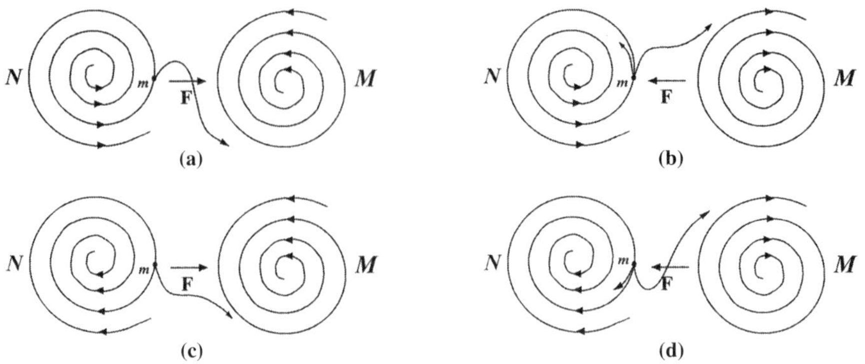

FIGURE 3.19 Acting and reacting models with yoyo structures of inharmonic spinning patterns. (a) Object *m* is located in a diverging eddy and pulled by a converging eddy. (b) Object *m* is located in a diverging eddy *M* and pushed or pulled by a diverging eddy *M*. (c) Object *m* is located in a converging eddy and pulled by a converging eddy *M*. (d) Object *m* is located in a converging eddy and pushed or pulled by a diverging eddy *M*.

specifically, the object m experiences an acceleration $\vec{a}$ toward the center of M such that the magnitude of $\vec{a}$ is given by

$$a = \frac{v^2}{r}, \tag{3.11}$$

where r is the distance between the object m and the center of M and v the speed of any object in the spin field of M about distance r away from the center of M. And, the magnitude of the net pulling force $\overrightarrow{F_{net}}$ that M exerts on m is given by

$$F_{net} = ma = m\frac{v^2}{r}. \tag{3.12}$$

Newton's third law is commonly known as that to every action, there is an equal, but opposite, reaction. By analyzing the situation in two different angles: two eddy motions act and react to each other's spin field, as shown in Figures 3.20 and 3.21, and one spinning yoyo is acted upon by an eddy flow of a higher level and scale (Figure 3.22), the following laws are established [29]:

The Third Law on State of Motion: When the spin fields of two yoyo structures N and M act and react on each other, their interaction falls in one of the six scenarios as shown in Figure 3.20a–3.20c and Figure 3.21a–3.20c. And, the following are true:

(1) For the cases in (a) of Figures 3.20 and 3.21, if both *N* and *M* are relatively stable temporarily, then their action and reaction are roughly equal but in opposite directions during the temporary stability. In terms of the whole

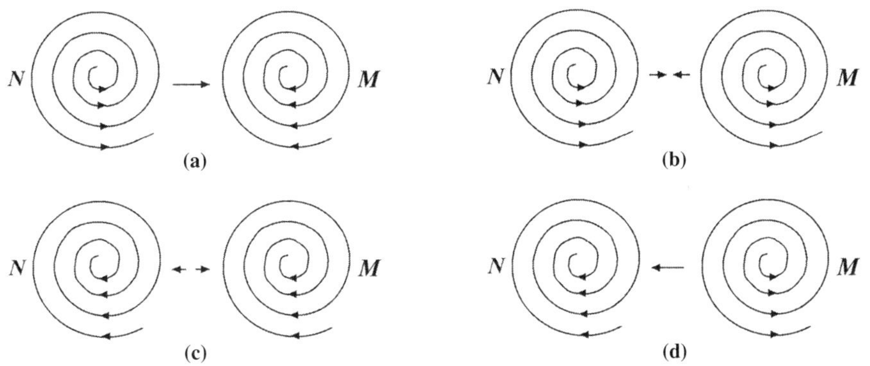

FIGURE 3.20 Same scale acting and reacting spinning yoyos of the harmonic pattern. (a) *N* diverges and *M* converges. (b) Both *N* and *M* diverges. (c) Both *N* and *M* converge. (d) *N* converges and *M* diverges.

evolution involved, the divergent spin field (*N*) exerts more action on the convergent field (*M*) than *M*'s reaction peacefully in the case of Figure 3.20a and violently in the case of Figure 3.21a.

(2) For the cases (b) in Figures 3.20 and 3.21, there are permanent equal, but opposite, actions and reactions with the interaction more violent in the case of Figure 3.20b than in the case of Figure 3.21b.

(3) For the cases in (c) of Figures 3.20 and 3.21, there is a permanent mutual attraction. However, for the former case, the violent attraction may pull the two spin fields together and have the tendency to become one spin field. For the latter case, the peaceful attraction is balanced off by their opposite spinning directions. And, the spin fields will coexist permanently.

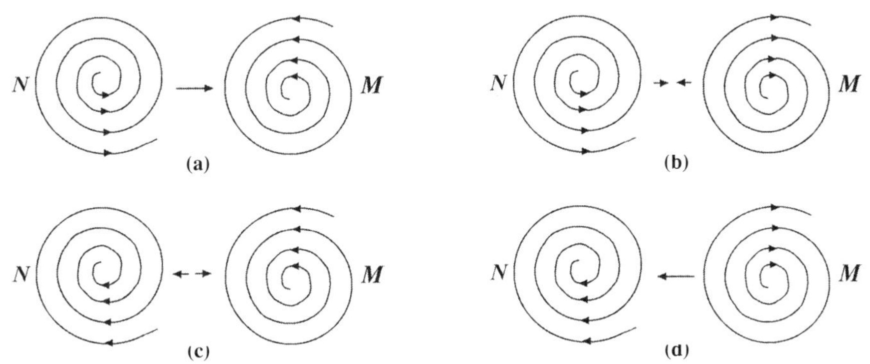

FIGURE 3.21 Same scale acting and reacting spinning yoyos of inharmonic patterns. (a) *N* diverges and *M* converges. (b) Both *N* and *M* diverge. (c) Both *N* and *M* converge. (d) *N* converges and *M* diverges.

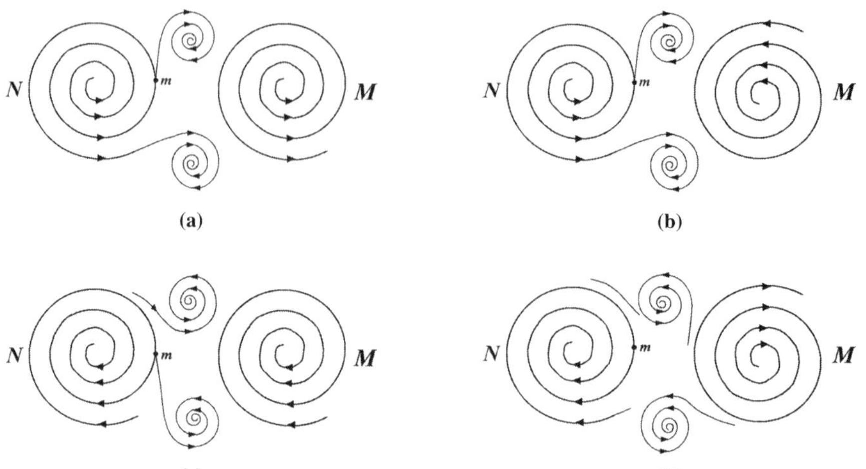

FIGURE 3.22 Object *m* might be thrown into a sub-eddy created by the spin fields of *N* and *M* jointly. (a) Object *m* is situated as in Figure 18b. (b) Object *m* is situated as in Figure 19a. (c) Object *m* is situated as in Figure 18c. (d) Object *m* is situated as in Figure 19d.

That is to say, Newton's third law holds true temporarily for cases (a), permanently for cases (b) and partially for cases (c) in Figures 3.20 and 3.21.

The Fourth Law on State of Motion: When the spin field M acts on an object m, rotating in the spin field N (3.22), the object m experiences equal, but opposite, action and reaction, if it is either thrown out of the spin field N and not accepted by that of M (Figures 3.18a and 3.18d, Figures 3.19b and 3.19c) or trapped in a sub-eddy motion created jointly by the spin fields of N and M (Figures 3.18b and 3.18c, Figures 3.19a and 3.19d). In all other possibilities, the object m does not experience equal and opposite action and reaction from N and M.

3.4 APPLICATIONS IN SOCIAL SCIENCES

As the title of this chapter suggests, we will in this chapter look at how the yoyo model has been applied to address some of the most important problems in social sciences.

3.4.1 Systemic Structures of Civilizations

In this subsection, we will see how this systemic model and relevant properties [29] can be readily employed in the study of civilizations by providing a series of novel explanations for some of the basic observational facts that have been employed as granted in sociology [45]. For more details, please refer Reference 46. Due to the sound conclusions derived on solid scientific foundations, it is expected that this work produces such results that can be truly useful for policy makers at national and international levels.

History is mainly composed of stories of civilizations from ancient Sumerian and Egyptian to Classical and Mesoamerican to Christian and Islamic civilizations and through successive manifestations of Sinic and Hindu civilizations. Many distinguished historians, sociologists, and anthropologists have well studied the causes, emergence, rise, interactions, achievements, decline, and fall of civilizations [47].

The concept of civilization was initially developed by eighteenth century French thinkers. German thinkers of the nineteenth century distinguished civilizations by mechanics, technology, and material factors, the cultures that involved values and ideals, and the higher intellectual, artistic, moral qualities of a society. Some anthropologists conceived cultures as characteristic of primitive unchanging nonurban societies and complex developed urban and dynamic societies as civilizations. In general, civilization and culture both refer to the overall way of life of people with a civilization being larger than a culture. They both involve the values, norms, institutions, and modes of thinking to which successive generations in a given society have attached primary importance [48]. By a civilization, it is meant to be a space, a cultural area, a collection of cultural characteristics, and phenomena [49, pp. 177, 202], a particular concatenation of worldview, customs, structures, and culture (both material culture and high culture) which forms some kind of historical whole and which coexists with other varieties of this phenomenon [50, p. 215], and a kind of moral milieu encompassing a certain number of nations, each national culture being only a particular form of the whole [51, p. 811). For example, blood, language, religion, and way of life were what the ancient Greeks had in common and what distinguished them from the Persians and other non-Greeks. When people are divided by cultural characteristics, they are grouped into civilizations, which are comprehensive, just as the concept of general systems [4], none of their constituent units (or elements) can be fully understood in isolation without reference to the encompassing whole (civilization). A civilization is a maximal whole or totality in terms of its fundamental values and philosophical assumptions. Civilizations comprehend without being comprehended by others [52, p. 455]. Civilizations do not seem to have clear-cut boundaries, precise endings or beginnings. Throughout history people have redefined their identities and the composition and shape of civilizations have gone through changes over time. Civilizations interact and overlap and are meaningful entities with ambiguous but real dividing line between them. They tend to be long lived by appropriately evolving and adapting to the environment, and represent some of the most enduring forms of human associations and possess long historical continuity, representing the longest story of all [49, pp. 209–210].

Now, let us employ the yoyo model to address such questions as: How do civilizations form? What factors influence the determination of the ambiguous but real dividing line between civilizations? Why do different civilizations have different underlying assumptions and values of philosophy?

With the yoyo model in place, we can answer these and other related questions as follows. At the start of time when still living in the primitive conditions, due to the existing natural conditions and available resources within the environment, people formed their elementary beliefs, basic values, and fundamental philosophical assumptions. Since the population density was low, tools available for production, conquering surroundings, so on, were extremely limited and inefficient, minor obstacles of nature in today's standard easily divided people on the same land into small tribes. Since these individual and separated tribes in reality lived in the same natural environment, they naturally held an identical value system and an identical set of philosophical assumptions, on which they reasoned in order to explain whatever inexplicable, developed approaches to overcome hardships, and established methods to administrate members in their individual tribes. As time went on, better tools for production and transportation and better practices of administration were designed and employed in various individual tribes. The natural desire for better living conditions paved the way for the inventions of new tools, discovery of new methods of reasoning, and the introduction of more efficient ways of administration to pass around the land through word of mouth. So, a circulation of people with special skills started to form. Along with these talented people, information, knowledge, and commercial goods were also parts of the circulation. As a circulation started to appear, Bjerknes' Circulation Theorem guarantees the appearance of abstract eddy motions over the land consisting of migration of people, spread of knowledge and information, and transportation of goods. That explains how civilizations are initially formed and why different civilizations have different underlying assumptions and values of philosophy.

Our yoyo model analysis and discussion lead naturally to explanations on how civilizations are separated from each other and how civilizations remain and survive political, social, economic, even ideological upheavals, although empires rise and fall, and governments come and go. In particular, a good indicator for telling different civilizations apart from each other is the natural environment and geographic conditions in which people live. For example, let us think of some islanders who are cut off from other varieties of environmental conditions by large bodies of water. If throughout history their closest neighboring lands are always occupied by well-formulated civilizations (vigorously spinning yoyos), then the islanders (constituting a small and barely spinning yoyo) have no alternative other than forming their own civilization while periodically absorbing useful and beneficial elements from the neighboring civilizations. And when the neighboring civilizations (as seen as spinning fluids in the dishpan experiment) experience internal chaos (that is when the spinning fluids contain traveling eddy leaves), the small civilizational yoyo of the islanders might have the opportunity to expand temporarily onto the land occupied by the organizationally chaotic people.

As for political, social, economic, even ideological upheavals, they only appear on the surface of civilizations without fundamentally altering the underlying systems of values and philosophical assumptions. They represent some of the eddy leaves periodically appearing in the "whirlpools" of civilizations. So, they are natural phases of the evolution of civilizations. As for the rise and fall of empires, it is like the situation of an ocean of spinning yoyos, the appearance of which is similar to what is shown on the charts of airstreams of the earth as seen from above the North Pole, where one spinning pool, helped by some invisible joint forces from other whirlpools, attempts to conquer many other pools of various sizes by forcing each of them to "spin" in exactly the same way as itself. When the forces that combine internal and external strengths become exhausted, the most likely consequence is that the conquering whirlpool structure is greatly weakened, while the other pools either recover back to their original spinning motion or spins more energetically than before, because their basic sets of values and philosophical assumptions have been just renewed and/or expanded by adopting some of the useful elements from the failed conquering whirlpool. Similar to the situation of political, social, economic, and ideological upheavals, governments are just temporary centers of cultural yoyo structures. As cultures evolve, their centers also move from one "location" to another, similar to the case of the magnetic poles of the earthly yoyo structure.

Civilizations move through seven stages: mixture, gestation, expansion, age of conflict, universal empire, decay, and invasion [53, 146ff.]; Each civilization arises as a response to challenges and then goes through a period of growth involving increasing control over its environment produced by a creative minority, followed by a time of troubles, the rise of a universal state, and then disintegration [52, 569ff.]. The political composition of civilizations varies from one civilization to another and varies over time within a civilization. A civilization may contain a few or many political units. These units may be city states, empires, federations, confederations, nation states, multinational states, all of which may have varying forms of government. As a civilization evolves over time, changes normally occur in the number and nature of its constituent political units [45, p. 44].

To investigate why some civilizations contain few political units and other many units, let us place civilizations in the historical flow of time and see each civilization as a spinning yoyo coexisting with many other civilizational yoyos. By doing so, the earlier systemic yoyo analysis indicates that as long as a civilization has uneven internal organization, there must be uneven moments of forces. Combining with the naturally existing uneven gradient forces produced by the uneven internal organization, the abstract but realistic yoyo structure of the civilization will have to spin. That is, the uneven internal organization of an entity produces eddy motion for the entity. And, the more uneven internal organization the entity possesses, the greater the naturally existing uneven gradient forces will be, and the faster the entity will spin. In a fast spinning civilization, where the high speed of rotation represents a high degree of unity of the underlying civilization, the number of political units changes over time just as what is shown in the dishpan experiment. This analysis also indicates that it is also possible for a civilization to have many political units as long as the overall internal organization of the civilization is relatively even. This end combined with what is obtained earlier leads to the following conclusions.

(1) Civilizations that are situated in such lands that have severely, unevenly distributed natural resources tend to have higher degrees of unity than those that

occupy lands with relatively even distribution of necessities for the existence of ordinary lives.

(2) Civilizations existing on rich lands with relatively even distributions of natural resources tend to have multiple centers (known as core states).

In addition, the phenomenon of political units in a civilization can be modeled by using the concept of centralized systems. In particular, Hall and Fagen [54] introduce the concept of centralized systems, where a system is centralized if one object or a subsystem of the system plays a dominant role in the system's operation. The leading part can be thought of as the center of the system, since a small change in it would affect the entire system, causing considerable changes across the entire spectrum. Now, let a system be an ordered pair [36] of sets, $S = (M, R)$, such that M is the set of all objects and R a set of some relations defined on M. The sets M and R are called the object and relation set of S, respectively. Here, each $r \in R$ is defined as follows: There is an ordinal number $n = n(r)$, a function of r, called the length of the relation r, such that $r \subseteq M^n$, where

$$M^n = \underbrace{M \times M \times \ldots \times M}_{n \text{ times}} = \{f : n \to M \text{ is a mapping}\} \tag{3.13}$$

is the Cartesian product of n copies of M. A system $S = (M, R)$ is trivial, if $M = R = \emptyset$. Given two systems $S_i = (M_i, R_i)$, $i = 1, 2$, S_1 is a partial system of S_2 if either (1) $M_1 = M_2$ and $R_1 \subseteq R_2$ or (2) $M_1 \subset M_2$ and there exists a subset $R' \subseteq R_2$ such that $R_1 = R'|M_1 = \{f: f$ is a relation on M_1 and there is $g \in R'$ such that f is the restriction of g on $M_1\}$. In symbols,

$$R'|M_1 = \{f : g \in R' \ (f = g|M_1)\}, \tag{3.14}$$

where $g|M_1 \equiv g \cap M_1^{n(g)}$. A system $S = (M, R)$ is called a centralized system, if each object in S is a system and there is a nontrivial system $C = (M_C, R_C)$ such that for any distinct elements x and $y \in M$, say, $x = (M_x, R_x)$ and $y = (M_y, R_y)$, then $M_C = M_x \cap M_y$ and $R_C \subseteq R_x|M_C \cap R_y|M_C$. The system C is called a center of S. The following has been shown.

Theorem 2 (Lin [55]). Assume that we are in the ZFC axiomatic set theory. Suppose that S = (M, R) is a system such that |M|, the cardinality of M $\geq$ c, where c is the cardinality of the set of all real numbers, and that each object in S is a system with a finite object set. If there exists such an element that belongs to at least c objects in M, there then exists a partial system B of S with an object set of cardinality $\geq$c and B forms a centralized system. □

This result indicates that as long as a civilization is seen as a system, where different parts of the civilization are closely connected by various relationships, and some special elements exist such that each of these elements transcends through a great number of the parts of the civilization, then in this civilization, at least one center (or political unit) will appear. And, when dictated by nature, such as severely uneven distribution of resources, various existing centers (political units) will naturally fight for control of greater power and influence. In the process of power struggle, this theorem indicates that many political units along with their spin fields could be eliminated without damaging the underlying structure of the civilization. This theoretical result might be the explanation for the phenomenon of Sinic single center of political power and the historical fact that after each unification of China, all oppositions were mercilessly eliminated, which, according to our yoyo model analysis, created a uniform motion in the civilizational spin field.

To conclude this section, let us apply the following result of microeconomics to see why each moment of transition or blown-up that it bridges a transition between organizational expansion and contraction represents a weakest link in the evolution of a social entity.

Theorem 3 (The Theorem of Never-Perfect Value Systems [56]) In a family of at least two members, one member h is the head who is benevolent and altruistic toward all other members. The head establishes a value system for all members of the family to follow so that in his eyes, every selfish member will be better off both for now and for the future. If a selfish member k measures up well to the value system, he will then be positively rewarded by the head h. Unfortunately, the more effort member k puts in to measure up to the value system, the more he will be punished by the reward system. □

As depicted in Figure 3.23 and as in practical situations, what is described in this theorem is that the benevolent head transfers his resources to other family members in a periodic fashion over time without an end in sight, where although the word "benevolent" is used, based on the proof of this theorem it only means that he is a person responsible to distribute wealth around, also the phrase of "his resources" does not necessarily mean that he is the person who has made all the wealth. The idea of value systems implies that the head tells other members at some time moment along the time line that starting at a certain pay period, each member's behavior will affect the amount he will receive from the head at the end of that period. And, the head will design his response to each member's behavior in such a way as to maximize his own utility. That is, Theorem 3 explains a sequential behavioral change and reflection over time between the head and his family members. In Figure 3.23, the scale marks on the time line represent the moments of individual asset distributions and are given for reference purposes without much practical implications. Mark 0 can be located anywhere on the line as the beginning of our discussion or focus. Negative scale

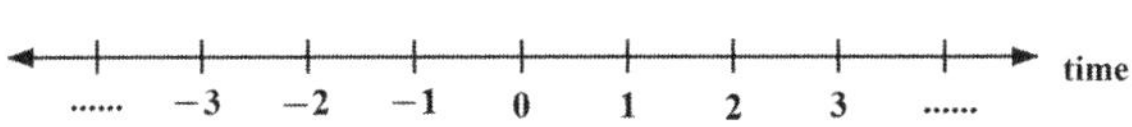

FIGURE 3.23 Time periods of the head's benevolent giving to selfish members.

marks represent the moments of distribution of the past and the positive marks the future moments of distribution.

What Theorem 3 implies is that in each civilization, which functions and evolves on a fundamental set of values and philosophical assumption, has to go through moral decline over time, because the reward system of the culture in the long run punishes those who actually try to measure up to the requirements and to promote the civilizational values. This fact very well illustrates why each of the transitional changes or blown-ups between the civilizational expansions and contractions represents a weakest link in the whole evolution of the civilization, since the overall moral of the society is at its low during these time periods.

3.4.2 Systemic Structures Beneath Business Organizations

In this section, using the yoyo model, Figure 3.1, let us look at business organizations, where human desires always alter the evolution of the outcome, leading to unpredictable, chaotic consequences [57]. For more details about this subsection, please refer Reference 14.

For the interacting economic forces of demand and supply, let us consider a market controlled completely by free competitions, where the price P of a consumer good is closely related to and determined by the demand D and the supply S. Assume that the changes in P is directly proportional to the difference of the demand and supply:

$$\frac{dP}{dt} = k(D - S),\ k > 0, \tag{3.15}$$

where t stands for time and k a constant. With all other variables fixed, assume that both the demand D and the supply S are functions of the price P such that $D = D(P)$ and $S = S(P)$. The relationship between D and P is generally linear, which we can write

$$D(P) = -\lambda P + \beta,\ \lambda, \beta > 0, \tag{3.16}$$

where λ is the rate of change of the demand with respect to the price P and β the saturation constant of the demand.

As for the relationship between S and P, it is generally nonlinear, because when the price of a consumer good lowers, the number of buyers will increase and the demand consequently increases. The increased demand stimulates the production of the good so that the supply is increased. If the price gradually increases, the demand will accordingly decrease so that the supply will consequently be lowered. Since the demand and supply are not correlated directly, when the price reaches certain height, even though the demand continues to drop, the supply might be increased because the increased price can stimulate the production (Figure 3.24). Therefore, the relation

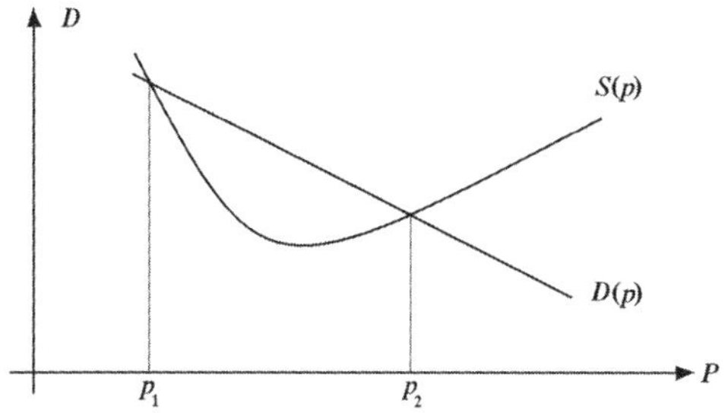

FIGURE 3.24 The functional relationships of the demand and supply on the price.

of the supply with respect to the price is generally nonlinear. This relationship can be written symbolically as (at least by means of local approximations)

$$S(P) = \delta + \alpha P + \gamma P^2, \tag{3.17}$$

where $\delta > 0$ is a constant, α and γ are respectively the linear and nonlinear intensities of the supply, satisfying $\alpha > 0$ and $\gamma < 0$.

Substituting Equations (3.16) and (3.17) into Equation (3.15) produces

$$\frac{dP}{dt} = AP^2 + BP + C, \tag{3.18}$$

where $A = -k\gamma > 0$, $B = -k(\lambda + \alpha) < 0$ and $C = k(\beta - \delta) < 0$. Since the price evolution model (3.18) is quadratic, the price in general changes discontinuously and has the characteristic of singular reversal transitions or blown-ups. See Theorem 1 for more details. That is, the concept of demand and supply is in fact about mutual restrictions or mutual reactions of different forces under equal quantitative effects.

Because various economic forces, including, but not limited to those of demand and supply, are underneath each business organization, it can be reasonably expected that the yoyo model will find its wide-ranging applications in the investigations of business organizations.

3.4.3 Systemic Structure in Human Mind

The totality of each human being is four dimensional: body, mind, heart, and spirit. It is physically made up of flesh, bone, blood, hair, and brain cells, and systemically of self-awareness, imagination, conscience, and free will [58, p. 70]. By using self-awareness, he is able to examine his own thoughts and has the freedom to choose his response to whatever he comes across or whatever is imposed on him. With imagination, he is able to create a (fantasy) world in his mind beyond the present reality. With conscience, he is deeply aware of what is right and wrong, of the principles that govern his behavior, and a sense of the degree to which his thoughts and actions are in harmony with the principles. And he has to act based on his self-awareness, free of all other influences. In this section, we use the yoyo model to provide new insights as for what the human endowments—self-awareness, imagination, conscience, and free will—are and how they function. For more details and related discussions, please refer References 59, 60.

By self-awareness it stands for a person's self sense that he exists as an individual separated from other people and objects and that he possesses his own private thoughts and individual rights [61, p. 106]. It also includes the understanding that other people are similarly self-aware. Self-awareness is a self-conscious state in which attention focuses on oneself. It makes people more sensitive to their own attitudes and dispositions [62, p. 41]. With the utilization of self-awareness, people examine their own thoughts and have the power to choose their responses to whatever situations they are faced with. It includes such elements as self-discipline, carefulness, thoroughness, organization, deliberation (the tendency to think carefully before acting), and need for achievement [63].

Our discussions above indicate that a person's self-awareness is really a natural consequence of his underlying multidimensional systemic yoyo structure. In particular, because each system breaths in and out materials throughout each part of its surface area, between different systems, there is a constant battle for (1) pushing against each other when materials that are emitted outward are thrown against each other. This is how each person feels that he is separate from other people and objects, and (2) attracting toward each other when different systems try to absorb a piece of material of common interest. This is how the feeling of individual rights and entitlements is created.

These constant battles between systems collectively make them aware of their own existence, the existence of others, and their private thoughts, be they human beings or objects. The reason why each person focuses on himself, is sensitive to his own attitudes and dispositions is because his very own viability is at stake, where his viability is determined by how strong his yoyo field spins and how much and how effectively the field can absorb materials from the environment.

By imagination, it means the faculty of the human mind to imagine, to form mental images or concepts of what is not actually present to the biological senses, and the action or process of forming such images or concepts. With this faculty, one derives meaning to experience and understanding to knowledge. It is a fundamental human endowment through which people make sense of the world [64, pp. 1–2] and plays a key role in the learning process [65, p. 50].

Going along with the systemic yoyo model of self-awareness, imagination is also a natural consequence of the human yoyo structure. Because human sense organs are simply some 3D realizations of the underlying multidimensional spinning fields, there are many field structures of the surrounding world these sense organs cannot really pick up. From the fact that each human yoyo constantly interacts with other spin fields in the forms of pushing against each other or grabbing over materials of common interests, it follows that the experience and knowledge gained from the human field interactions are much richer than what is known based on the information collected by the sense organs. This end provides a systemic model for human imagination. That is, the so-called imagination is in fact the collection of all the conscious (meaning through the sense organs) and unconscious (meaning not through any of the sense organs) records of the interactions between the underlying yoyo fields. When one needs to establish an abstract image or concept, he relies on his various trainings to tap into this reservoir of experience and knowledge collected both consciously and unconsciously from his field interactions with others. Here, the level of training on his self-awareness determines how deeply he can reach into this reservoir. Consequently, it determines how thought provoking his established concepts and images will be. When imagination is called for action, it simply matches what is given, be it a difficult problem, a challenging situation, or a difficult task, with some of the relevant knowledge or experience from the reservoir. If the match is nearly perfect, the action will be considered successful. If the match is not good or no match is found, then new experience or knowledge will be added into the reservoir. This reservoir consists of two parts: the collection of all learned knowledge and experience through the sense organs, and the other the collection of all knowledge and information collected not through the sense organs. With adequate training on self-consciousness, one can "magically" apply a great deal of materials out of the reservoir of his imagination.

By conscience, it means the ability with which one distinguishes whether his actions are right or wrong, and is deeply aware of the principles that govern his behavior and a sense of the degree to which his thoughts and actions are in harmony with the principles. When one does things not in agreement with his moral values, his conscience will make him feel remorse; when his actions confirm to his moral values, his conscience brings him feelings of rectitude or integrity. Conscience also represents the attitude which informs a person's moral judgment before performing any act. Many authors from different angles, including religious views, secular views, and philosophical views, have studied the concept of conscience and its role in ethical decision-making [66–68]

Along with the systemic models of self-awareness and imagination, it can be seen that the so-called conscience is simply a partial function with two output values such that the partial function is defined on some of the spin patterns and interactions of these patterns stored in the reservoir of imagination. Here, certain kinds of field flows and interactions of the flows are assigned the value of +, known as being right or moral, and some other kinds of flows and interactions assigned the value of −, known as being wrong or immoral. The reason why this function is only partial is because other than the flow patterns and their interactions that are assigned either a + value or a value −, there are still plentiful of other flow patterns and interactions in the reservoir of imagination that are not assigned with any + or − value. By the domain of this ± function, we mean the totality of all the flow patterns and the flow interactions each of which has been assigned either a + or − value. The assignments of + and/or − values of certain flow patterns and interactions of flows start when the person is still an infant and continue throughout the entire life of the person. Initially, the domain of the ± function is the empty set. As the person grows older, the domain starts to expand. And, if some dramatic event happens during his lifetime, the assignments of the + and/or − values may be altered.

As suggested by the formation of the sub-eddies in Figure 3.22, some of the very first assignments of + values should be given to the flows heading in the same direction; and − values to currents flowing in opposite directions. It is because same directional flows feed into the existence of the sub-eddies, while opposite directional currents slow down or even attempt to stop the rotations of the sub-eddies. In daily language, the initial + values are given to behaviors that help strength the well being of all children involved, and the initial − values to the behaviors that damage or destroy the well being. For instance, when two children fight over a toy, most likely the parents would demand the children to share, reinforcing a + value on making everybody happy, even though the parents know very well which child actually owns the toy. When two little children are hitting each other, the parents would simply stop the fight forcefully, indicating the assignment of a − value to violence. After a person reaches a certain level of maturity, his ± function will be quite well defined deeply in his head. So, whatever action he takes or thought he thinks of, he would unconsciously compare it to the elements in the domain of his ± function. If the action or thought has a + value, he senses the feeling of rectitude as the consequence of being constantly appraised by adults for such occasions; if the action or thought has a − value, he feels remorse, since similar situations have been cursed regularly by grownups and he has let his care takers down one more time. If the action or the thought does not have a well assigned + or − value, the person will feel afraid of the potentially uncertain reactions from others. Afterward, he receives either a + or − value, or he will start his journey to explore the potential value for his specific action or thought by pushing this action further or thinking of the thought deeper until he reaches a certain outcome or he is attracted to some other more interesting, or urgent, or important topic before reaching any meaningful + or − value.

So, our discussion indirectly says that one's capacity for conscience is genetically determined, because it is completely established on imagination, which in turn is dependent on self-awareness, while self-awareness is an innate ability. On the other hand, the subject matter of conscience, the specific content of the domain of the ± function and the assignments of the + or − values, is learned.

By free will, it means [69–72] the human ability to keep the promises one make to himself and others. It is the human ability to make decisions and choices and to act in accordance with those choices and decisions. The extent to which personal free will is developed is tested in day-to-day lives in the form of personal integrity. It stands for one's ability to give meaning to his words and walk the walk. It is an integral part of how much value is placed on oneself.

Based on what has been discussed, the so-called free will is the human ability to predict at least for the short term what one can or cannot accomplish and what choices are better or the best for the situation involved. In particular, with self-awareness, one forms his reservoir of imagination and his ± function of moral values with its constantly expanding domain. What matters here is where his self, as identified by his self-awareness, is located, inside the domain of the ± function or outside the domain. If his identified self is inside the domain of his ± function with a specific value assigned to it, then the assignment of a + value to his self will force him to make as accurate prediction as possible on what he can or cannot accomplish and what choices are better or the best for the situation involved. If he were uncertain with respect to a situation, his promise either to himself or others would be that he would try his best without any guarantee for success. On the other hand, if a − value is assigned to his self, he will still force himself to make as honest predictions as possible on what he can or cannot accomplish and what choices are better or the best for the situation involved. However, in this case, most likely he will make promises opposite to what he foresees based on his imagination and his ± function. Now, when the third scenario occurs, where his identified self is not inside the domain of his ± function, then whatever appears, he would make his random promise, because no matter whether or not he keeps his promise, the outcome would not bear any conscientious consequence to him.

3.5 APPLICATIONS IN ECONOMICS

Based on what has been discussed in the previous sections, it is expected that the yoyo model can be widely employed in the analysis of economic systems.

3.5.1 Becker's Rotten Kid Theorem

People interact with each other for different purposes. Due to differences in personalities and preferences, there exist conflicting interests. Then how can one keep all these people on the same track, working for a common purpose? To address this problem, Becker [73], in *A Theory of Social Interactions*, published his Rotten Kid Theorem. This theorem suggests that family members, even those who are selfish, will contribute to efforts to raise family income, if there is a benevolent head who voluntarily bequeaths gifts to other members periodically. The importance of this result has been well described by Bergstrom in Reference 74:

> If it is generally correct, the Rotten Kid theorem must be one of the most remarkable and important results in the theory of incentives. For it tells us that a sufficiently benevolent household head would *automatically* internalize all the external effects that family members have on each other. Benevolent parents of intelligent, though selfish, children can breathe more easily. In the family there will be no free riders or principal-agent problems. Elaborate incentive schemes and detection devices are unnecessary. All that is needed is to explain the Rotten Kid theorem to each family member and they will all (except possibly for a few irrational lapses) behave in the common interest. Not only would this be remarkable good news for parents, it would suggest a promising way of avoiding the incentive problems that bedevil firms and other social organizations. Shouldn't it be possible to find group incentive structures similar to those of families with benevolent heads?

Due to the theorem's wide-ranging importance, Bergstrom [74] notes that it is worthwhile to find the limits of generality for theorems with strong and interesting conclusions, the Rotten Kid Theorem included. Subsequently, he constructed three examples, each showing where this theorem could go wrong. Along the same idea, after developing a two-time-period model for the study of child labor, which encompasses the human capital dimension, Baland and Robinson [75] show that the Rotten Kid Theorem does not hold true when parental savings are at a corner. In addition, Bommier and Dubois [76] use a similar model to reveal that a child's disutility, a dimension overlooked by previous researchers, considerably affects the validity of this important theorem. Even though 30 years have elapsed since the Rotten Kid Theorem was initially published in 1974, no one has formalized a condition under which the theorem holds true in general until 2008 [77], on which this presentation is based.

In this subsection, we see how to apply the yoyo model [29] to establish a sufficient and necessary condition under which Becker's Rotten Kid Theorem (the 1974 version) holds true in general. As a matter of fact, when the yoyo model is applied to the study of the family [78], it is obvious that Becker's Rotten Kid Theorem can only hold true conditionally. By specifying several relevant yoyos and how they interact, we can intuitively see the needed condition under which Becker's theorem holds true in general.

Theorem 4 (Becker's Rotten Kid Theorem [73]) If a family has a head who cares about all other members so much that he transfers his resources to them automatically, then any redistribution of the head's income among members of the household would not affect the consumption of any member, as long as the head continues to contribute to all. Additionally, other members are also motivated to maximize the family income and consumption, even if their welfare depends on their own consumption alone. □

This version of the Rotten Kid Theorem is original cited from Reference 73. In [78], a slightly different version is given. In 1989, Bergstrom established his version of the Rotten Kid Theorem. And, Dijkstra [79] generalizes Bergstrom's results to a higher level. Not only so, he also establishes a condition under which in the Samaritan's dilemma, the altruist reaches her first best when she moves after the parasite. The reason why we base our work on Becker's original result published in 1974 is because this version of the Rotten Kid Theorem provides a sense of evolution of the head's benevolence and the selfish kids' unselfish behaviors and tolerance toward potential redistributions of the head's resources. In particular, if we assume, as in practical situations, that the head transfers his resources to other family members periodically at fixed time intervals, we can depict the implied evolution in Figure 3.25. The scale marks on the time line represent the moments of individual asset distributions and are given for reference purposes without much practical implications. Mark 0 can be located anywhere on the line as the beginning of our discussion. Negative scale marks represent the moments of distribution of the past and the positive marks the future moments of distribution.

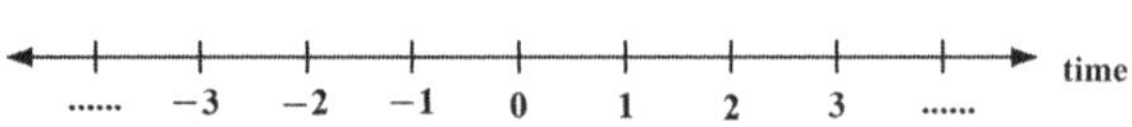

FIGURE 3.25 Time periods of the head's benevolent giving to selfish members.

As indicated by Reference 73, let us assume that at each past moment of distribution, the head transferred $\$R_j$ to selfish member j. If for a reason such as that member k suffers from a temporary financial setback in time period $(-1, 0)$ and the head has to (because of his consideration of others) bail out k by providing k extra monetary support at $t = 0$ and maybe a few following pay periods, then Theorem 4 states that the head can adjust his amount of distribution to member j and others in order to finance his increased support for k. And this adjustment in his transfers will not affect the consumption of any member, *as long as* the head continues to contribute to all. Intuitively, because of such potential adjustments in transfers, all selfish members feel the sense and existence of the family and are willing to help maximize the family income and consumption, since each of them might one day need such an increased financial support.

On the other hand, Bergstrom's 1989 version [74] of the Rotten Kid Theorem focuses on one time period as $(k, k + 1)$, and Dijkstra's work [79] looks at such time periods as $(k, k + 1)$ and $[k, k + 1]$, respectively and individually. Specifically, Bergstrom's result focuses on the case that the head's transfer is made after whatever the selfish kids do beforehand, and Dijkstra establishes the conditions under which the head reaches his first best when he makes transfers, respectively, either before or after however the selfish kids behave. Note that it is Hirshleifer [80] who first points out the importance of the order of moves. Indeed, both Bergstrom's and Dijkstra's results can be employed to successfully analyze practical situations. On the other hand, if we look at the payment cycles of a work environment, where employees get paid periodically at fixed time intervals as depicted in Figure 3.25, then Becker's original version of the Rotten Kid Theorem is more appropriate. Here, because of potential future transfers, both the benevolent head and selfish kids will have to adjust their expectations and behaviors from one time period to the next in order to reach their individual Pareto optimums over time. That is, there are fundamental differences between one-shot deals and repeated businesses. Countless examples have been seen in the business world where people behave in one way for one-shot deals and in completely different manners in potentially repeating business opportunities. Because of this reason, we see the need to look at Becker's 1974 result and see under what general conditions, this result holds true.

Now, if we look at Becker's Rotten Kid Theorem in the light of the yoyo model, we can easily see that the correctness of this result must be conditional, not as simply stated as in the theorem. In particular, the family can be seen in two different fashions:

(1) The entire family is a spinning yoyo, where the family consumption is seen as the materials spitting out of the big-bang side and the family income as the materials sucked in from the black-hole side.

(2) The family is a collection of individually spinning yoyos that interact with each other, where each individual's consumption is seen as the materials spit out of his yoyo's big-bang side and his income as the materials sucked in from his yoyo's black-hole side.

For case (1), if we stand above the yoyo and look at the big-bang side, similar to the situation where scientists look at the earth from above its north pole, then the flows of the materials spit out of the big-bang side can be simulated by the fluid patterns in the dishpan experiment. In particular, Dave Fultz [44] and his colleagues of University of Chicago partially fill a cylindrical vessel with water, place it on a rotating turntable, and subject to heating near the periphery and cooling near the center. In this experiment design, the bottom of the container simulates one hemisphere of the earth's surface, the water the air above this hemisphere, the rotation of the turntable the earth's rotation, the heating and cooling the excess external heating of the atmosphere in low latitudes and the excess cooling in high latitudes.

In order to observe the pattern of flows at the upper surface of the water, which is intended to simulate atmospheric motion at high elevations, Fultz sprinkles some aluminum powder. A special camera that effectively rotates with the turntable took time exposures so that a moving aluminum particle would appear as a streak and sometimes each exposure ends with a flash, which can add an arrowhead to the forwarded end of each streak. The turntable generally rotates counterclockwise, as does the earth when viewed from the North Pole. Even though everything in the experiment is arranged with perfect symmetry about the axis of rotation, such as no impurities are added in the water, the bottom of the container is flat, Fultz and his colleagues observe more than they bargain for. First, both expected flow patterns as shown in Figure 3.26 to appear, and the choice depends on the speed of the turntable's rotation and the intensity of the heating. Briefly, with fixed heating, a transition from circular symmetry (Figure 3.26a) takes place as the rotation increases past a critical value. With the sufficiently rapid, but fixed rate of rotation, a similar transition occurs when the heating reaches a critical strength, while another transition back to the symmetry occurs when the heating reaches a still higher critical strength.

With this experiment in place, it is safe for us to conclude that for case (1) where the entire family is seen as a spinning yoyo, sooner or later the spin field of the family yoyo will have local eddies, which in some locations may not necessarily spin in the same direction of the overall family yoyo. That is, case (1) turns into the situation described in case (2).

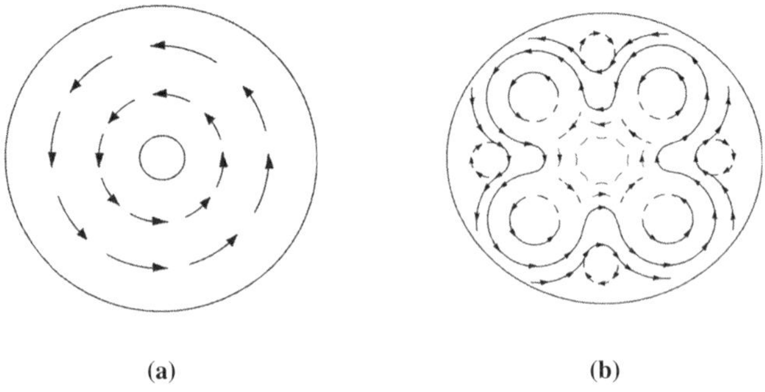

(a) (b)

FIGURE 3.26 Patterns observed in Fultz's dishpan experiment. (a) Symmetric flow at the upper surface. (b) Asymmetric flow at the upper surface.

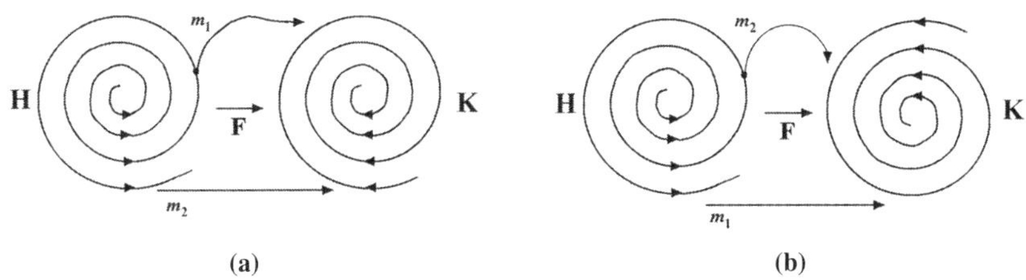

FIGURE 3.27 Interactions between benevolent head H and a selfish kid K.

For case (2), to make the situation easier to analyze, let us assume that there are only two members, the benevolent head H and a selfish kid K, in the family. In this case, the interactions between the yoyos can be depicted in Figure 3.27, where the benevolence of the head H is represented as a divergent whirlpool and the selfishness of the kid K as a convergent whirlpool. If both m_1 and m_2 stand for the voluntary money transfers from the head, the spin field of K in Figure 3.27 will accept the gift m_1 happily and the transfer m_2 unwillingly (Figure 3.27b) or even reject such a monetary gift (Figure 3.27a). Here, m_1 is given to the selfish kid K without violating his own preference of consumption, while m_2 is forced on K against his wills or personal preferences. That is, we have seen the following result.

Theorem 5 Becker's Rotten Kid Theorem holds true, if and only if the distribution of the benevolent head's resources is not in conflict with the consumption preferences of any selfish member. □

For the detailed proof of this result, please refer Reference 77 or 14.

3.5.2 Interindustry Wage Differentials

Over the years, numerous scholars have tried to provide various reasons for the existence of the interindustry wage pattern. It has been found that high wages can be explained in part by

(1) unpleasant and unsafe working environment;
(2) the purpose of hiring better workers for their both measurable and immeasurable labor quality;
(3) compensation differences naturally existing between industries;
(4) each employee's effort $e(w)$ being an increasing function of his wage rate w;
(5) firms' engagement in monitoring their workers' performance so that those who are caught shirking will be fired;
(6) the purpose of reducing the rate of employees' quits, since hiring and training workers can be expensive [81];
(7) making employees feel that they are paid fairly;
(8) the firms' ability to pay;
(9) labor union density, so on.

The uniformity of wage differentials across occupations works against both explanations (1) and (2) [82]. While explanation (3) is undoubtedly true to a degree [83], it cannot explain yet why such compensation differences exist from one industry to another. Murphy and Topal [84] support explanation (2), believing that the unexplained variance is due to employees' unobserved abilities. By identifying "the unobserved abilities" as intelligence, in particular, as IQ test scores, Blackburn and Neumark [85] find that there is a negative relationship between an industry's wage and the average IQ scores of its employees.

Explanations (4)–(7) are the fundamental premises of the so-called efficiency wage models using the mathematical rigor [86]. These models have attracted a great deal of attention and tend to prove that higher than competitive wages can be profitable. By specifying the positive effort–wage relationship, as described in (4) above, four main classes of efficiency wage models have been established. More specifically, shirking models [87] are developed to describe how hard employees work at their jobs where piece rates are impractical due to either the difficulty of counting "pieces," or high costs of monitoring, or both. By paying above market wages, firms engage in monitoring their workers and fire those who are caught shirking. As expected, these models indicate that high wage industries are those with high monitoring costs or those bearing high costs of employee shirking. About shirking models, Thaler [82] casts the following natural questions: Do employees work harder when they think they are in danger of losing a high paying job? Do employees work enough harder to justify the higher wages? Are the firms that pay high wages those who would gain the most from an increase in workers' effort? The analysis in this chapter will address these important questions from a new angle completely and definitely.

The so-called turnover models are established on the assumption that explanation (6) holds true [88, 89]. Consequently, these models predict that high wage industries are those with the highest costs associated with quits. Since data on quit rates are published, it has been confirmed that paying high wages does decrease quit rates. Adverse selection models [90, 91] are established on the explanations (2) and (4) by assuming that the average quality of the applicant pool increases with the wage rate. These models imply that industries that are sensitive to labor quality or spend a lot on measuring quality offer high wages. Evidently, these models do not provide a convincing explanation for the interindustry wage patterns. On explanation (7), fair-wage models are created [92, 93] by assuming that employees will exert more effort if they think they are being paid fairly. Under this premise, firms pay wages above competitive levels whenever their workers' perceived fair wage exceeds the competitive wage. As a result, these models lead to such conclusions as that industries with high profits and/or where team work and employee cooperation are particularly important will pay high wages [94]. Even though both common sense and social psychological research on "equity theory" suggest that workers are more productive if the morale is high, the following practical problem remains open: Is the true efficiency wage which sets the marginal gains from increased morale equal to marginal costs [82]? To this end, Raff and Summers [95] evaluated the Ford's decision in 1913 to double wages.

Evidently, explanation (8) does not make sense, because each dollar wage increase means a dollar dividend less for the stockholders. As for explanation (9), our analysis below will show that labor unions are attracted to high wage industries instead of the other way around, even though studies find that industry wage rates are corrected to union density (the percentage of workers in an industry who belong to a union) and that the unionization rate increases wages for both union and nonunion members in an industry.

Now, let us assume that a company produces and sells a line of special product. Then, the company has two sources of income: (1) producing and selling the product and (2) hiring each worker, because the expense of hiring a worker is expected to be lower than the income the worker produces for the company. In particular, from each unit of the product produced and sold, the company makes as much profit as $p_s^p - p_p^p$, where p_s^p is the selling price per unit and p_p^p the unit cost. And for each worker it hires, the company generates as much profit as $p_s^W - p_p^W$, where p_s^W is the average revenue the worker is expected to make and p_p^W the average cost associated with the worker.

To help us uncover the underlying mechanism of the interindustry wage differentials, we will analyze the company in two situations.

(A) The company has strong financial backings.

(B) The company's financial network and resources are limited.

For (A), for the company's product, the market demand is given by

$$n_p\left(p_s^p\right) = n_p\left(p_{s0}^p\right)\frac{p_{s0}^p - p_p^p}{p_s^p - p_p^p}\ \text{(units)} \tag{3.19}$$

and the total profit from selling its product, is

$$P^p = \text{profit of its product} = n_p\left(p_{s0}^p\right)\left(p_{s0}^p - p_p^p\right), \tag{3.20}$$

where $n_p\left(p_{s0}^p\right)$ is the initial market demand for the product selling at $\$p_{s0}^p$ each unit. Equation (3.19) implies that when $p_s^p \to (p_p^p)^+$, the demand will approach infinity. Equation (3.20) indicates that if the unit profit $(p_{s0}^p - p_p^p)$ stays constant, the lower the p_{s0}^p-value the greater the demand $n_p\left(p_{s0}^p\right)$ and the greater the total profit P^p.

Similarly, the company's staffing need is given by

$$n_W\left(p_s^W\right) = n_W\left(p_{s0}^W\right)\frac{p_{s0}^W - p_p^W}{p_s^W - p_p^W}\ \text{(persons)} \tag{3.21}$$

and the total profit from hiring $n_W(p_s^W)$ employees is given by

$$P^W = \text{profit of personnel} = n_W\left(p_{s0}^W\right)\left(p_{s0}^W - p_p^W\right), \qquad (3.22)$$

where $n_W\left(p_{s0}^W\right)$ stands for the company's initial need for additional personnel, hired at the initial expected revenue $\$p_{s0}^W$ per worker. Then, we have the following:

Proposition 1 If a business venture is well funded, and all other aspects of the operation stay the same, then

(1) The market demand for the product increases as the unit selling price drops close to the unit cost basis, while the total profit increases drastically;
(2) The company taking on the venture will hire additional employees at higher than competitive wage rates with the total profit soaring. □

Proposition 2 Assume that the company has all necessary funds for its operation. If at the expected revenue level $p_s^P = p_{s0}^P$ a new hire can create for the company and at a total cost $p_p^W = p_{s0}^W$, the total profit $P^W(p_s^W, p_p^W)$ all employees are expected to generate satisfies

(i) $\left.\frac{\partial P^W}{\partial p_s^W}\right|_{p_{s0}^W} < -n_W\left(p_{s0}^W\right)$, then the company can reduce its expected per-employee revenue p_{s0}^P to increase its total profit; or

(ii) $\left.\frac{\partial P^W}{\partial p_p^W}\right|_{p_{p0}^W} > n_W\left(p_{p0}^W\right)$, then the company can raise its expected per-employee cost basis p_{p0}^P to reap in additional profit; or

(iii) $\left.\frac{\partial P^W}{\partial p_s^W}\right|_{p_{s0}^W} < -n_W\left(p_{s0}^W\right)$ and $\left.\frac{\partial P^W}{\partial p_p^W}\right|_{p_{p0}^W} > n_W\left(p_{p0}^W\right)$, then the company can both reduce its expected per-employee revenue p_{s0}^P and raise its per-employee cost basis p_{p0}^P to maximize its overall profit. □

For situation (B) where the company's resources are limited, the total profit is given by

$$P_{\text{total}} = \mathrm{P}^{\mathrm{P}} + P^W = n_p\left(p_s^P\right)\left(p_s^P - p_p^P\right) + n_W\left(p_s^W\right)\left(p_s^W - p_p^W\right), \qquad (3.23)$$

subject to the budget constraint:

$$n_p\left(p_s^P\right)p_p^P + n_W\left(p_s^W\right)p_p^W = I, \qquad (3.24)$$

where I stands for the total funds available to the company. By solving the maximization problem of Equation (3.23) subject to Equation (3.24), we establish the following results:

$$n_{\mathrm{p}}\left(p_{\mathrm{s}}^{\mathrm{p}}\right) = \frac{n_{\mathrm{p}}\left(p_{\mathrm{s}0}^{\mathrm{p}}\right) p_{\mathrm{s}0}^{\mathrm{p}}}{p_{\mathrm{s}}^{\mathrm{p}}}, \tag{3.25}$$

$$n_{\mathrm{W}}\left(p_{\mathrm{s}}^{\mathrm{W}}\right) = \frac{n_{\mathrm{W}}\left(p_{\mathrm{s}0}^{\mathrm{W}}\right) p_{\mathrm{s}0}^{\mathrm{W}}}{p_{\mathrm{s}}^{\mathrm{W}}}, \tag{3.26}$$

and

$$P_{\mathrm{total}} = \mathrm{P}^{\mathrm{p}} + P^{\mathrm{W}} = \frac{n_{\mathrm{p}}\left(p_{\mathrm{s}0}^{\mathrm{p}}\right) p_{\mathrm{s}0}^{\mathrm{p}}}{p_{\mathrm{s}}^{\mathrm{p}}}\left(p_{\mathrm{s}}^{\mathrm{p}} - p_{\mathrm{p}}^{\mathrm{p}}\right) + \frac{n_{\mathrm{W}}\left(p_{\mathrm{s}0}^{\mathrm{W}}\right) p_{\mathrm{s}0}^{\mathrm{W}}}{p_{\mathrm{s}}^{\mathrm{W}}}\left(p_{\mathrm{s}}^{\mathrm{W}} - p_{\mathrm{p}}^{\mathrm{W}}\right) \tag{3.27}$$

$$= n_{p}\left(p_{s0}^{p}\right) p_{s0}^{p}\left(1 - \frac{p_{p}^{p}}{p_{s}^{p}}\right) + n_{W}\left(p_{s0}^{W}\right) p_{s0}^{W}\left(1 - \frac{p_{p}^{W}}{p_{s}^{W}}\right), \tag{3.28}$$

where $p_{\mathrm{s}0}^{\mathrm{p}}\, n_{\mathrm{p}}\left(p_{\mathrm{s}0}^{\mathrm{p}}\right)$ and $p_{\mathrm{s}0}^{\mathrm{W}}\, n_{\mathrm{W}}\left(p_{\mathrm{s}0}^{\mathrm{W}}\right)$ are defined the same as in the analysis of situation (A).

Equation (3.26) indicates that to hire more employees the company has to lower the average expected per-worker revenue. Since the capital markets are not perfect to the company, this result implies that the company has to limit how many workers it could afford to hire. To maximize its profit in the dimension of human resources, Equation (3.28) implies that $p_{\varepsilon}^{\mathrm{W}} \gg p_{\mathrm{p}}^{\mathrm{W}}$. So, in such a company with limited resources, it can either hire a relatively large number of employees at low wage rates or hire a relatively small number of high quality workers at a relatively higher wage rates. For the latter case to occur, the workers' productivity must be very high, which in general means that the company needs to invest a great deal in technology. And, this end might not be possible. Together with the company's weak financial standing, high quality worker option may never be practically possible for the company to take.

When Equation (3.28) is compared to Equation (3.22), it can be seen that

(1) Financially resourceful companies can spend extra money on workers' retraining programs to lower the average per-worker cost basis $p_{\mathrm{p}}^{\mathrm{W}}$, while firms with limited resources cannot. It is because in the later case, extra spending on workers' training programs increases both $p_{\mathrm{s}}^{\mathrm{W}}$- and $p_{\mathrm{p}}^{\mathrm{W}}$-values. So, the ratio $p_{\mathrm{p}}^{\mathrm{W}}/p_{\mathrm{s}}^{\mathrm{W}}$ may not change in the favorable direction to the firms.

(2) Similar reasoning indicates that financially resourceful companies can afford to invest in programs to make their workers feel good and to raise their morale, while firms with limited resources just cannot afford such luxuries.

Consequently, workers in financially strong firms produce more and are less likely to quit when compared to those hired by financially strained companies.

(3) While financially resourceful companies hire large numbers of workers so that these companies can easily reduce their per-employee benefit costs, companies with few workers have to pay the inflated market prices for the same benefit packages. That is, volume savings are not available to firms with limited resources.

This analysis explains why in a firm, if an occupation is high paid, so do all other occupations. In particular, if one occupation is paid at a level above the competitive wage rates, then all other occupations within the firm must play supporting roles. The increased productivity from the central occupation will be more than enough to finance the supporting occupations so that their respective wages are higher than those occupations' competitive market rates. For more in-depth discussion along this line, please refer Reference 96.

3.5.3 Price Behaviors of Projects

In terms of corporate governance, corporations need capitals to fund their ventures or provide the founders with cash-out opportunities, and suppliers of finance want to be assured that they get a return on their investment. In this subsection, we will look at the interactions between a firm's board of directors, which is dominated by long-term, large shareholders, and the CEO who is not a long-term, large shareholder.

Let us imagine each corporation and each supplier of finance as an economic yoyo, where the black-hole side sucks in funds from whatever available sources, while the big-bang side spends the money in various ways. More specifically, if a yoyo models a corporation, then its black-hole side absorbs funds, while part of the materials spit out of the big-bang side contains such a portion of the corporation's profits that it is returned to the investors as capital gains or dividends. If a yoyo represents an investor, then the black-hole side stands for ways and amounts he could get his return on his investments, and the big-bang side contains opportunities of future finance provided to each particular corporation.

According to Alchian [97] and Stigler [98], we have the following evolutionary theory: No one should worry about corporate governance reform, since in the long run market competition would force firms to minimize costs. And as part of this cost minimization, firms would be able to raise external capital at the lowest cost. If we imagine that all the yoyos, representing firms and investors, coexist side-by-side, interacting with each other, then it is ready to see that due to differences in spinning directions and angles, some yoyos will be destroyed and some will combine into greater (yoyo) structures with enhanced ability to absorb more investors' funds. To prevent abusive competitions and maximize the cash flows (so that the governments could potentially collect the most taxes), regulatory entities must be involved in order to help those tiny (yoyos) investors from being crushed so that when necessary, firms can still attract capital for their venture needs from all potential sources, including these investors. That is, even though market competition is a powerful force toward economic efficiency, it alone is not sufficient to solve the problem of corporate governance. That is, the afore-mentioned evolutionary theory holds true conditionally in such a sense that the field of competition must be level for all competitors (yoyos) to participate (spin). To this end, countries like the United States, Britain, Germany, and Japan, have well-introduced sophisticated laws to protect investors.

For a publicly traded firm, since financers generally are either too small or not qualified or not informed enough to make detailed operational decisions, they rely on the manager to run the firm. Consequently, the manager ends up with substantial control rights and discretion to allocate funds as he chooses. He can expropriate the funds [99], or spend the funds on his private benefits of control, such as building an empire for himself [100], spending on consumption perquisites [101], expanding the firm irrationally, and pursuing pet projects that do not benefit the investors [102]. And, worst of all, managers can expropriate investors by staying on the job even if they are no longer competent enough to run the firm [103]. To prevent any of these possibilities of how the manager could misuse the investors' investments—an agency problem, the systemic yoyo model suggests that one possible solution would be make sure to place the manager's personal yoyo in the same spin field of the yoyos of each investors—earn a return for his finance and/or effort in the firm. That is, tie the financial interests of the manager with those of the financers [104]. One method of tying the manager's interest with those of the investors is to employ incentive contracts, which have been in use since at least the time of Reference 105. However, evidence [106] exists, showing that incentive contracts create enormous opportunities for self-dealing for the managers. As suggested by the enclosed areas in Figure 3.28, no solution would work to solve this agency problem completely, as long as the manager is not the same as the numeraire investor, an abstract investor representing the common characteristics and desires of all financers together, there will be conflicts. At the same time, the dishpan experiment shows that as long as the manager is not the same as the only investor, the agency problem will appear in one form or another. In particular, individual investors have a single goal—earn a return on their investment. However, the manager in general has more goals to achieve and more benefits, such as any of the afore-mentioned private benefits of control, to enjoy beyond simply making an earning or return on his labor effort and talent for being the manager. Our analysis implies that in terms of corporate governance, as long as there exist investors who are not the manager, there will be an unsolvable agency

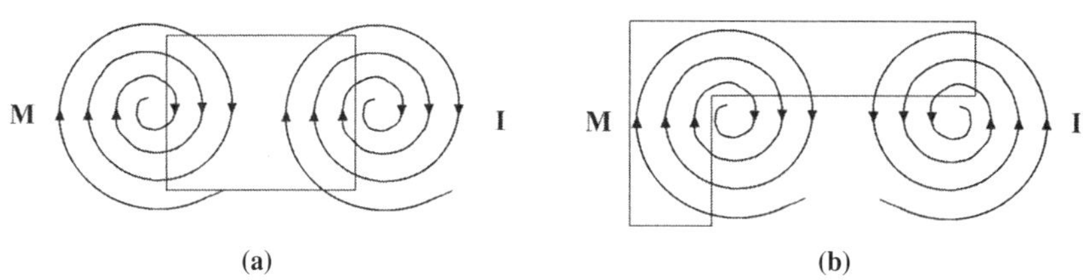

FIGURE 3.28 Conflicts of interests between the manager M and the numeraire investor I.

problem. The best one can do is to reduce its severity. For more in-depth analysis of this model, please refer Reference 14.

Our conclusion above points directly to the following question: Why do investors leave their money to managers who have enormous discretion on how to expropriate much of it for their personal gains [107]. To this end, the systemic yoyo model provides the answer. That is, based on the fact that eddy motions exist in pairs, one can see that when there is an opportunity, there will be risk takers. Among the risk takers, some are prudent, others are not and are gullible and get taken. The reason why in developed market economies, enormous outside finance exists is because, it also explains why these economies are more developed than others, the societal structures (yoyos), such as the local and national governments, underlying the economic yoyos of the developed economies, spin in the same directions and angles as the yoyos of these economies. In particular, the fiscal welfare of the governments is closely tied to the cash flows in the economies through taxation. So, to make all the yoyos of large scales, such as different levels of the governments and firms, do well financially, the governments introduced laws to protect investors. Because of these laws, many small investors, as the tiny particles evolving about in the spin fields of the underlying yoyos, become gullible and get taken either sometimes or most of the times. On the other hand, the laws established are not simply for protecting investors whose financial gains in general do not contribute much to the fiscal welfare of the governments, but also play the magic role of monitoring the behaviors of firms. Every time when investors get taken, some noise will be made and heard, which very likely brings additional monetary awards to the governments through the use of the legal system.

The systemic yoyo model suggests that the board of a firm exists as the concentrated area of interests of all involved parties, including the management of the firm, investors and creditors, specifically those large shareholders and significant creditors. The board serves as the narrow neck of the economic yoyo of the firm. It connects the activities of the firm with all the parties whose interests are tied to the success of the firm. Without the neck (or the board), the firm would not exist both financially and physically. The existence of a functioning board induces investors to trust the manager with their money.

For large shareholders and investors, to protect their investment, guarantee their return, and prevent the managers from expropriating funds and spending on their private benefits of control, they would monitor the managers' moves and decision-makings closely. To show how the board of a publicly traded company and the CEO interact, let us look at such a problem as how the CEO chooses between long-term and short-term projects (assets). In this subsection, projects and assets are interchangeably used, because for investors, they mainly care about when mispricings of their holdings disappear so that they can close their positions with profits. By short-term projects, we mean such projects that they cannot stay mispriced for long. Otherwise, these projects would be called long term.

Assume that the board is effectively controlled by large, long-term shareholders. The controlling interest wants long-term growth, even though it could mean that they sometimes have to suffer from some occasional temporary losses. And, assume that the CEO is not a long-term, large shareholder of the company. That is, there is a conflict of interest here between the board and the CEO. That is, the board desires the CEO to devote all his talent and labor effort to work on the existing long-term projects and initiate new ones so that the market value of the company would grow steadily over time. On the other hand, the CEO needs short-term equity performance, because poor equity performance raises the likelihood of replacement by the board or through a hostile takeover. In either case, the CEO looses his job and all the privilege of control [108]. In the following analysis, we assume that there is no discount of the future over time.

To look at the price behavior of different investment projects, assume that a short-term asset is traded at $p_s < v_s$ a share in time period 1, where v_s is the fundamental share value of the asset. In this period, an investor buys $n(p_s)$ shares at the market price p_s a share with a total cost of $I_s = n(p_s)p_s$. Since this asset is a short-term investment, the mispricing disappears in period 2 with the trading price the same as the fundamental value v_s. So, the profit from this short-term asset is

$$n\left(p_s\right) v_s - I_s R = \frac{I_s v_s}{p_s} - I_s R = I_s \left(\frac{v_s}{p_s} - R\right), \tag{3.29}$$

where $R > 1$ is the gross interest spent on the total investment amount I_s.

Assume that for a long-term asset, the same investor buys in period 1 $n(p_L)$ shares at p_L a share satisfying $p_L < v_L$, where v_L is the fundamental share value of the asset. So, the total cost to the investor is $I_L = n(p_L)p_L$. For the convenience of recognizing the subscripts and distinguishing 1 and L, we choose to use lower case s and upper case L. Since the asset is a long-term investment, its mispricing does not disappear until period 3 when the total profit is

$$n(p_L)v_L - I_L R^2.$$

Discounting this amount to period 2 provides

$$\frac{1}{R}\left[n\left(p_L\right) v_L - I_L R^2\right] = \frac{1}{R}\left[\frac{I_L v_L}{p_L} - I_L R^2\right] = I_L\left[\frac{v_L}{p_L R} - R\right]. \tag{3.30}$$

If the investor wants to produce the same return on his investments from the short-term and the long-term assets, we have from Equations (3.29) and (3.30) that

$$I_s\left[\frac{v_s}{p_s} - R\right] = I_L\left[\frac{v_L}{p_L R} - R\right].$$

From the assumption that $I_s = I_L$, meaning that the investor allocated the same amount of fund to each opportunity, we have

$$\frac{v_s}{p_s} = \frac{v_L}{p_L} \cdot \frac{1}{R}, \quad R > 1. \tag{3.31}$$

That is, for the investor to put his money with the long-term asset in period 1, this asset must be more mispriced than the short-term asset. This result is the same as that obtained by Shleifer and Vishny [107], where they establish a different model. Our results in the following paragraphs are opposite of those predicted by these authors.

Similar to the analysis above, we can compare the mispricings between short-term and long-term assets in three additional different cases.

(1) Both the long-term and short-term assets are traded in period 1 at a per-share price $p_i > v_i$, $i \in \{s, L\}$;

(2) The investor is risk averse and the long-term asset is more risky than the short-term investment, and the trading prices $p_i < v_i$, $i \in \{s, L\}$; and

(3) The same situation as in (2) above holds true except that the trading prices $p_i > v_i$, $i \in \{s, L\}$.

For case (1) the same equation as (3.31) holds. However, this equation means something completely opposite of that of Equation (3.31). In particular, for this current situation, Equation (3.31) implies that for the investor to put his money in the short-term asset, the asset must be more mispriced than the long-term asset. This end confirms the practical experience that going short can make quicker and handsome profit in relatively short period of time than going long [57]. For case (2), we obtain

$$\frac{v_{\mathrm{s}}}{p_{\mathrm{s}}} < \frac{v_{\mathrm{L}}}{p_{\mathrm{L}}} \cdot \frac{1}{R_{\mathrm{L}}}, \ R_{\mathrm{L}} > 1, \tag{3.32}$$

where R_{L} is the long-term gross interest rate. For case (3), we have

$$\frac{v_{\mathrm{L}}}{p_{\mathrm{L}}} \cdot \frac{1}{R_{\mathrm{L}}} = \frac{v_{\mathrm{s}}}{p_{\mathrm{s}}} + \left(R_{\mathrm{L}} - R_{\mathrm{s}}\right) > \frac{v_{\mathrm{s}}}{p_{\mathrm{s}}}, \ R_{\mathrm{s}} > 1, \tag{3.33}$$

where $R_{\mathrm{s}} < R_{\mathrm{L}}$ stands for the short-term gross interest rate.

Summarizing what is obtained, it can be seen that when assets are undervalued, the mispricing of the long-term asset in equilibrium is worse than that of the short-term asset. When the assets are overvalued, the mispricing of the short-term asset in equilibrium is worse than that of the long-term asset. And, when risky long-term asset is involved, mispricings become even worse for the long-term asset if both assets are undervalued, and for the safer short-term asset if both assets are overvalued, than when risk-free assets are considered.

Since it takes longer time for fundamental uncertainties to resolve, CEOs are typically averse to severely under pricing their long-term equity. Considering their job security, they also try to avoid overvalued short-term projects and all long-term projects, even though some of these projects could be detrimentally important to the long-term health and growth of their companies. Combining this reasoning with the fact that CEOs' compensation depends typically in part on short-term equity performance, CEOs should prefer undervalued short-term projects, if the CEOs are not large, long-term shareholders [14]. To satisfy the curiosity of the reader, all main results of Reference 107 are also derived using our model here.

3.6 APPLICATIONS IN THE FOUNDATIONS OF MATHEMATICS

Now, let us model the history of mathematics mentally as an abstract systemic yoyo, where the black-hole side sucks in empirical data and puzzles man collects and experiences in daily lives. Through human biological capability, man processes the data with rational thinking (the narrow neck of the yoyo), then man produces (out of the big-bang side) all kinds of concepts, theories, and products that man can physically use in daily lives. Now, some concepts, theories, and products become obsolete and disregarded over time, while the rest are recycled back into the black-hole side. These recycled concepts, theories and products, combined with new observations and newly collected empirical data, through the tunnel of rational thinking again, lead to new or renewed concepts, theories, and products, which help to bring the quality of life to another level.

Now, if we stand at a distance away from this yoyo structure and look at the spinning field from either the black-hole side or the big-bang side, what we see is like a huge whirlpool of data, observations, puzzles, concepts, theories, products, etc. What the dishpan experiment suggests is that this huge whirlpool goes through pattern changes periodically, alternating between uniform and symmetric flows to chaotic currents with local eddies formed. If the chaotic currents with local eddies is identified with one of the crises in the history of mathematics, and the uniform and symmetric flow pattern with one of the quiet periods in the history between two consecutive crises, then we can expect that after the relatively quiet twentieth century, one should be able to see another major crisis of mathematics in the making. As the previous crises, this forthcoming event will also be as devastating, as frustrating as any of the known crises. Its impact will be felt throughout most major corners of the modern mathematics. Here, we notice that some of the major concepts and puzzles, such as those of infinitesimals, limits, and summations of numbers, played a role in each of the past crises. So, they can be seen as some of those knowledge, concepts, theories, puzzles, etc., which are sucked into the black hole of the yoyo structure of the modern mathematics. After being processed by human thinking and reasoning, they are spit out of the big-bang side in their renewed forms. The following subsections indicate that the impact of the hidden contradictions existing in modern mathematics and the return of the Berkeley paradox should be seen and felt in all major corners of modern mathematics in the years and decades to come. Speaking differently, the following discussions imply that instead of the ultimate truth, mathematics will forever become an area of knowledge like physics where the truthfulness of each statement will only be conditional to the given ranges.

To be clear, what we try to say is that as predicted by the yoyo model, established mentally for the structure of modern mathematics, the fourth crisis in the foundations of mathematics has appeared. For more details about this end, please refer References 109–114. And, the presentation of this section is based on Reference 115.

3.6.1 Historical Crises in the Foundations of Mathematics

With an invitation from Ronald Mickens in 1987, Yi Lin had the honor to join some of the very well-known scholars from around the globe, such as Wendell Holladay (Vanderbilt University, USA), Saunders Mac Lane (University of Chicago, USA), John Polkinghorne (Cambridge, UK), Robert Rosen (Dalhousie University, Canada), and others, to express his opinion from the angle of systems research on Nobel Laureate Eugene P. Wigner's assertion about "the unreasonable effectiveness of mathematics." His paper, entitled "A few systems-colored views of the World," was eventually published in 1990 in Reference 116. This work led him to ponder over the following question: Does the human way of thinking have the same structure as that of the material world? To this end, the difficulty is: what is meant by "structure", and how can this structure be uncovered? After many years of struggle both theoretically and empirically, finally in 2002, Lin and colleagues discovered the structure named the yoyo model that underlies all physical and imaginary systems [29, 17]. In this subsection, we will use the history of mathematics to uncover the systemic yoyo structure in human thoughts by assuming that mathematics is an organic whole of thoughts created on the bases of experience, intuition, and logic thinking. In terms of references, the history of mathematics presented herein without mentioning the source is from either Reference 117 or 19. And the presentation in this entire section is mainly based on Reference 118.

In the fifth century B.C., unexpected discovery of irrational numbers gives rise to the first crisis in the foundations of mathematics. It is one of the greatest achievements of the Pythagoreans. This finding implies that there are points on the number line that are not corresponding to any rational number. The rigorous proof for the existence of irrational numbers was surprising and disturbing to the Pythagoreans, since it was a mortal blow to the Pythagorean philosophy that all in the world depend on the whole numbers. The final resolution of this crisis in the foundations of mathematics was achieved in about 370 B.C. by Eudoxus. His treatment of incommensurables coincides essentially with the modern exposition of irrational numbers that was first given by Richard Dedekind in 1872. This crisis in the foundations of mathematics seems to be largely responsible for the subsequent formulation and adoption of the axiomatic method in mathematics.

Toward the end of the seventeenth century, after many other mathematicians had prepared the way, the creation of calculus was made by Isaac Newton and Gottfried Wilhelm Leibniz. This theory proved to be remarkably powerful and capable of attacking problems quite unassailable in earlier days. It was the wide range and astonishing applicability of the theory that attracted the bulk of the mathematics researchers of the day with the result that papers were published in great profusion with little concern for the unsatisfactory foundations of the theory. The processes employed were justified largely on the ground that they worked. It was not until the nineteenth century was almost elapsed, after a number of absurdities and contradictions had crept into mathematics, that mathematicians felt the need to examine and to establish rigorously the basis of their work. The effort to place calculus on a logically sound foundation proved to be a difficult task. Various ramifications occupied the better part of the next 100 years. A result of this careful work in the foundations of calculus was that it led to equally careful work in foundations of all branches of mathematics and to the refinement of many important concepts. Chronically, calculus was created in the latter half of the seventeenth century. The eighteenth century was largely spent in exploiting the new and powerful methods of calculus. And, the nineteenth century was largely devoted to the effort of establishing calculus on a firm, logical foundation.

Since the early calculus was established on the ambiguous and vague concept of infinitesimals without a solid foundation, many operations and applications of the theory were reproached and attacked from various angles. Among all the reproaches, the most central and the ablest about the faulty foundation of the early calculus was brought forward by Bishop George Berkeley (1685–1753). In 1734, he published a book, entitled *The Analyst: A Discourse Addressed to an Infidel Mathematician*. This book was a direct attack on the foundation and principles of calculus, specifically on Newton's notion of fluxion and on Leibniz's notion of infinitesimal change. The following passage has been most frequently quoted from *The Analyst*:

> And what are these fluxions? The velocities of evanescent increments? And what are these same evanescent increments? They are neither finite quantities nor quantities infinitely small, nor yet nothing. May we not call them the Ghosts of departed Quantities?

To understand what Berkeley was talking about, let us look at an example from a paper by Newton on how to compute the area a region with a curved boundary. In his work, Newton claimed [19] that he avoided using infinitesimals by going through the following steps: Give an increment to x, expand $(x + 0)^n$, subtract x^n, divide by 0, compute the ratio of the increment in x^n over that in x. Then by throwing away the 0 term, he obtained the fluxion of x^n. Berkeley said that the variable was first given an increment, then let the increment to be 0. This involved the fallacy of a shift in the hypothesis. As for the derivative seen as the ratio of the disappeared increments in y and x, that is, the ratio of dy and dx, Berkeley called these disappeared increments "neither finite quantities nor quantities infinitely small, not yet nothing. May we not call them the Ghosts of departed quantities?"

With the passage of time, a serious crisis in the foundations of mathematics became evident. It was Joseph Louis Lagrange (1736–1813) who was the earliest to attempt a rigorization of calculus. A great stride was made in 1821 when Augustin-Louis Cauchy (1789–1857) successfully developed an acceptable theory of limits and then defining continuity, differentiability, and definite integral in terms of the concept of limits.

Specifically, on the basis of Cauchy's theory of limits, Dedekind proved the fundamental theorems in the theory of limits using the rigorized real number theory. With the combined efforts of many mathematicians, the methods of $\varepsilon - N$ and $\varepsilon - \delta$ became widely accepted so that infinitesimals and infinities could be successfully avoided. Therefore, it has been a common belief that due to the satisfactory establishment of the theory of limits, calculus has ever since been constructed on a solid, rigorous theoretical foundation. So, the second crisis in the foundations of mathematics is considered resolved successfully.

In 1874, Georg Cantor (1845–1918) commenced his revolutionary work on set theory and the theory of the infinite, where he developed a theory of transfinite numbers based on the actual infinite. Since so much of mathematics is permeated with set concepts, the superstructure of mathematics can actually rest upon set theory as its foundation. However, in 1897, Burali-Forti brought to light the first publicized paradox of set theory. Cantor, 2 years later, found a nontechnical description of a very similar paradox. In his theory of set, Cantor proved that for any given set X, there is always another set Y such that the cardinality of X is less than that of Y. What is shown is similar to the situation that there is no greatest natural number. There also is no greatest transfinite number. Now, consider the set that contains all sets as its elements. Surely no set can have more elements than this set of all sets. But, if this is the case, how can there be another set whose cardinality is greater than the cardinality of this set of all sets? In 1902, Bertrand Russell discovered another paradox, which involves nothing more than just the concept of sets itself.

Now a simple analysis of the Russell paradox suggests that it directly touches on the entire foundation of mathematics so that the common belief that mathematics and logic are two most rigorous and exact scientific disciplines is severely challenged. The chaotic situation, caused by the appearance of these set-theoretic paradoxes, is adequately called the third crisis in mathematics.

To resolve the third crisis, mathematicians focused on modifying the naïve set theory into a plausible theory. As of the present day, there have been two plans to remold the set theory. The first is to employ the theory of types, developed by Whitehead and Russell. The second is to continue Zermelo's axiomatic set theory. Walking along the idea of Russell's "theory of extensionality," Zermelo in 1908 established his system of axioms for his new set theory. After several rounds of modifications, Fraenkel and Skolem (1912–1923) provided a rigorous interpretation and formed the present-day ZF system. Since the ZF system accepts the axiom of choice, it is often written as ZFC system. The specific route of thinking of the ZFC system is the following: By introducing the axioms of the empty set and of infinity, the legality of the set of all natural numbers is warranted. The legality of the set of all real numbers is derived by using the axiom of power sets. Then, the legality of each subset of those elements, satisfying a given property P, of the real numbers is based on the axiom of subsets. Therefore, as long as the ZFC system is consistent, the theory of limits, as a rigorous theoretical foundation of calculus, can be satisfactorily constructed on the ZFC axiomatic system. However, the remaining problem is that even though various paradoxes of the two-value logic will not reappear in the ZFC system, as of this writing, we still cannot show in theory that no paradox of whatever kind will ever be constructed in the ZFC system. That is, the consistency of the ZFC system or other presently available axiomatic set-theoretic systems still cannot be shown. To this end, Poincare commented about a century ago:

> We set up a fence to protect our flock of sheep from potential attacks of wolves. However, at the time when we installed the fence, there might have already been a wolf in sheep's clothing being enclosed inside the flock of our sheep. So, how can we guarantee that in the future there will not be any problems?

After several paradoxes were publicized in the naïve set theory, Hausdorff [119] felt deeply graceful and reminded the community of mathematicians with the following comment:

> These paradoxes made people feel unease. It is not because of the appearance of the paradoxes. Instead, it is because we did not expect these contradictions would ever exist. The set of all cardinalities seems so empirically indubitable just as the set of all natural numbers is that naturally acceptable. So, the following uncertainty is created. That is, is it possible that other infinite sets, all of them, are this kind contradictory, specious non-sets?

However, what attracted the most attention and what was most surprising are the following comments [120]:

> In terms of the foundations of mathematics, my position is based on the following two main principles: (1) No matter which semantics is applied, infinite sets do not exist (both in practice and in theory). More precisely, any description about infinite sets is simply meaningless. (2) However, we still need to conduct mathematical research as we have used to. That is, in our work, we should still treat infinite sets as if they realistically exist.

3.6.2 Actual and Potential Infinities

By sorting through the ideas studied in the history of mathematics, it is discovered that some concepts, such as those of infinitesimals, infinities and limits, underlie each crisis in the foundations of mathematics. To this end, in this subsection we will study the concepts of actual and potential infinities. Then, in the following subsection, we will uncover some hidden contradictions in the system of modern mathematics and see the return of the Berkeley paradox.

By actual infinity, it is meant to be a nonfinite process that definitely reaches the very end. For example, when a variable x approaches the endpoint b from within the interval $[a, b]$, x actually reaches its limit b. In this case, this limit process reflects the spirit of an actual infinity. In history, Plato was the first scholar to clearly recognize the existence of actual infinities [121]. By potential infinity, it is meant to be a nonfinite, nonterminating process that will never reach the very end. For example, consider a variable x defined on an open interval (a, b) that approaches the endpoint b from within this open interval. Then, the process for x to get indefinitely close to its limit b is a potential infinity, since even though x can get as close to b as any pre-determined accuracy, the variable x will never reach its limit b. In history, Aristotle was the first scholar to acknowledge the concept of potential infinities. And, he never accepted the existence of actual infinities. Similarly, Plato did not believe in the existence of potential infinities [121].

Although these two kinds of infinites are different, where actual infinity stands for a perfect tense and potential infinity a present progressive tense, in modern mathematics system, they have been treated as the same. For example, in the theory of limits, in order to avoid the Berkeley paradox, the definitions of limits, either infinite or finite, using $\varepsilon - N$ and $\varepsilon - \delta$ methods, are completely based on the thinking logic of potential infinities. In the naïve set theory and the modern (ZFC) axiomatic set theories, from Cantor to Zermelo, the existence and construction of infinite sets have been established on the concept of actual infinities. In many areas of the mathematics system, the existence of specific subsequences of a given sequence is guaranteed on the basis of mathematical induction and the recognition that actual infinities are the same as potential infinities. For example, it is well known in mathematical analysis that any bounded sequence $\{a_i\}_{i=1}^{\infty}$ of real numbers has a convergent subsequence $\{a_{k_n}\}_{n=1}^{\infty}$. A proof of this fact is based on inductively constructing a convergent subsequence $\{a_{k_j}\}_{j=1}^{\infty}$ term by term; then applying the assumption that actual infinity = potential infinity, the mathematical induction implies that the convergent subsequence is obtained. However, based on the descriptive definitions of actual and potential infinities, as detailed in Reference 122, all proven above is that for any natural number j, a term a_{k_j} of $\{a_i\}_{i=1}^{\infty}$ can be picked to satisfy a set of desirable conditions. So, the nonterminating process of getting one more term a_{k_j} out of $\{a_i\}_{i=1}^{\infty}$ can be carried out indefinitely. That is, the existence of a potential infinity is shown. After that what is claimed is that the actual infinity, the subsequence $\{a_{k_j}\}_{j=1}^{\infty}$, has been obtained. In other words, in modern mathematics actual infinities are seen as the same as potential infinities.

At this junction, a natural question is: Other than some scholars in the history, such as Aristotle and Plato, who did not recognize both kinds of infinities at the same time, is there any evidence that these two kinds of infinities can in fact lead to different consequences? The answer to this question is YES. To this end, let us consider the following vase puzzle, which was first published in Reference 4.

3.6.3 Vase Puzzle and the Fourth Crisis

Here is the vase puzzle: Suppose that a vase and infinitely many pieces of paper are available. The pieces of paper are labeled with natural numbers 1, 2, 3, … , so that each piece has only one label. Now, let us perform the following recursive procedure:

Step 1: Put the pieces of paper, labeled from 1 to 10, into the vase; then remove the piece labeled 1.

Step 2: Put the pieces of paper, labeled from $10n - 9$ through $10n$ into the vase; then remove the piece labeled n, where n is an arbitrary natural number 1, 2, 3, …

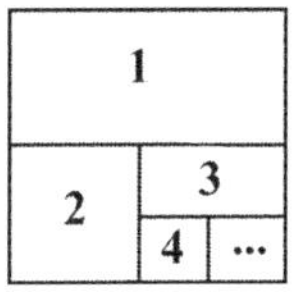

FIGURE 3.29 Obtain as many pieces of paper as needed out of a chosen area.

Question: After the recursive procedure is finished, how many pieces of paper are left in the vase?

First of all, some comments are needed here to make the situation practically doable. (1) The vase need not be infinitely large; actually any size will be fine. (2) The total area of the infinite number of pieces of paper can also be any chosen size. For example, Figure 3.29 shows how an infinite number of pieces of paper can be obtained. (3) The number labeling can be done according to the steps in the puzzle. (4) The recursive procedure can be finished within any chosen period of time. The details here are similar to that of any chosen size for the total area of the pieces of paper above.

An Elementary Modeling: To answer the question, define the following using mathematical induction:

$$f(n) = 9n,$$

which tells how many pieces of paper are left in the vase right after step n, where $n = 1, 2, \ldots$ Therefore, if the recursive procedure can be finished, the number of pieces of paper left in the vase should be equal to the limit of $f(n)$ as n approaches ∞. So, the answer is infinitely many pieces of paper are left in the vase.

What we should note here is that this entire modeling process is based on the concept of potential infinities. The recursive procedure cannot really be finished The step of taking limit only represents the progressive growth in the number of pieces of paper in the vase and the (imaginary) limit state.

A Set-Theoretic Modeling: Based on this modeling, the answer to the question of the vase puzzle is "no piece of paper is left in the vase." This, of course, contradicts the conclusion derived in the elementary modeling above.

In particular, for each natural number n, define the set M_n of the pieces of paper left in the vase right after the nth step of the recursive procedure:

$$M_n = \{x | x \text{ has a label between } n \text{ and } 10n + 1 \text{ exclusively}\}.$$

Then mathematical induction implies that an actual infinity, in particular, the sequence $\{M_n\}_{n=1}^{\infty}$ exists. And as soon as this actual infinity is assumed to exist, one realizes that no piece of paper could have been left in the vase, because otherwise he could pick out a piece of paper from the vase. By the convention, this piece has a natural number label k, which is impossible, because k has been drawn out at step k, a contradiction. So, there is no piece of paper is left in the vase, an answer completely opposite of that of the elementary modeling.

This vase puzzle vividly shows the difference between actual and potential infinities. Of course, this example is theoretical in nature. One might very well claim that in applications of mathematics, he will never run into such a problem. To this end, we should not be so sure, because in the study of quality control of the products out of an automated assembly line, for example, the concept of infinities is involved. In particular, when studying quality, we draw a random sample of the products and use the sample statistics to make inferences on the continually expanding population, which is theoretically the collection of all the products that have been and will be produced from the assembly line. To make the inferences more reliable, we often treat the ever-expanding population as an infinite population, either actually infinite or potentially infinite. For more impacts of the vase puzzle, please refer Reference 4, where a discussion about the connections between the vase puzzle with methodology, epistemology, and philosophy of science are given. And some comments on methodological indication of the fundamental structure of general systems, mathematical induction, and the knowability of the physical world can be found in Reference 123.

What is shown by the vase puzzle is that in general actual infinities are different from potential infinities. However, in modern mathematics, they have been assumed to be the same. For instance, when using mathematical induction, the inductive step says that as long as a property P holds true for natural number n, one can show in theory that the property P holds true for $n + 1$. That stands for a present progressive tense—a potential infinity. However, when mathematical induction is used in existence proofs, the positive conclusion is always drawn as that the property P holds true for all natural numbers n—an actual infinity, which is a perfect tense. Since such applications of mathematical induction appear all over the entire system of mathematics, we conclude that in modern mathematics, the convention that potential and actual infinities are the same has been implicitly assumed.

On the other hand, the convention that actual infinities are not the same as potential infinities is also implicitly assumed in modern mathematics. In particular, in formal logic, the concept of contrary opposites and contradictory opposites are distinguished since the time of Aristotle. Assume that P is a predicate. The contrary opposite of P is written ⌉P and the contradictory opposite ¬P. Let us now limit our attention to modern mathematics, where the two-value logical calculus is the tool of deduction. The law of proof by contradiction can be expressed as: $\Gamma, \neg A \vdash B, \neg B \Rightarrow \Gamma \vdash A$. This law is the logical foundation of the often-used method of deduction, called proof by contradiction. Within the system of two-value logical calculus, from the law of proof by contradiction, one can prove the principle of excluded middle: $\vdash A \vee \neg A$, and the principle of no contradiction: $\vdash \neg(A \wedge \neg A)$. Now, in the two-value logical system, both contrary opposites (P, ⌉P) and contradictory opposites (P, ¬P) coincide. So, ⌉P = ¬P. Therefore, potential infinities (*poi*) and actual infinities (*aci*), as the opposites in a pair of contrary opposites (P, ⌉P), must now be a pair of contradictory opposites (P, ¬P). That is, the *poi* and *aci* must respectively represent the affirmative and the negative aspect of a concept and have to satisfy the principles of the excluded middle and no contradiction:

$$\vdash poi \vee aci \text{ and} \vdash \neg\, (\mathrm{poi} \wedge \mathrm{aci}).$$

That is, we can show $poi \neq aci$. For a more detailed discussion along this line, please refer Reference 124.

What is shown is that an inherent part of modern mathematics allows both of the contradictory conventions: actual infinities ≠ potential infinities; and actual infinities = potential infinities. However, the previous discussion has clearly shown that these two kinds of infinities are two different concepts. They might lead to completely opposite outcomes. So, in the rest of this discussion, let us treat these infinities as different concepts.

If we have $a\uparrow b \wedge a \top b$, where ↑ means "approaches" and ⊤ "reaches", then we say that the variable a approaches the limit b in the fashion of actual infinities. If we have $a\uparrow b \wedge a \not\top b$, then we say that the variable a approaches the limit b in the fashion of potential infinities. It can be seen that for any variable a, it approaches its limit b, if it exists, in the fashion of either actual or potential infinities, but not both. For the conventional expressions of limits,

$$\lim_{x\to x_0} f(x) = A, \quad \lim_{x\to 0} f(x) = A, \quad \text{or} \quad \lim_{x\to\infty} f(x) = A,$$

there are two sets of variables in each of these expressions: $x \to x_0$ (respectively, $x \to 0$ or ∞), and $f(x) \to$ A. If the variable x approaching its limit x_0 (respectively, $x \to 0$ or ∞), in the fashion of actual infinities causes trouble or leads to contradictions, that is, $x\uparrow x_0 \wedge x \top x_0 \vdash \mathrm{B}, \neg\mathrm{B}$, then the variable x has to approach its limit x_0 in the fashion of potential infinities. That is, it must become that $x\uparrow x_0 \wedge x \not\top x_0$. With these preparations, the Berkeley paradox of the past actually returns.

In particular, the well-known distance formula for a free falling object is given by

$$S = \tfrac{1}{2}gt^2,$$

where g is the gravitational constant. When $t = t_0$, the distance the object has fallen through is $S_0 = \frac{1}{2}gt_0^2$. When $t = t_0 + h$, the distance of falling is $S_0 + L = \frac{1}{2}g(t_0 + h)^2$. This end implies that during the h seconds, the object has fallen through the distance L given by

$$L = \tfrac{1}{2}g(t_0 + h)^2 - S_0 = \tfrac{1}{2}g(t_0 + h)^2 - \tfrac{1}{2}gt_0^2 = \tfrac{1}{2}g(2t_0 + h)h.$$

So, within the h seconds, the average speed of the falling object is

$$\mathrm{V} = \frac{L}{h} = gt_0 + \tfrac{1}{2}gh.$$

Evidently, the smaller the time interval h is, the closer the average speed is to the instantaneous speed at $t = t_0$. However, no matter how small h is, as long as $h \neq 0$, the average speed is not the same as the instantaneous speed at $t = t_0$. When $h = 0$, there is no change in the falling distance. So, $V = \frac{0}{0}$ becomes meaningless. Hence, it is impossible to compute the instantaneous speed of the free falling object at $t = t_0$.

To this end, both Newton and Leibniz provided several explanations in order to get rid of this difficulty. One of the explanations goes as follows: Assume that h is an infinitesimal. So, $h \neq 0$ and the ratio

$$\frac{L}{h} = \frac{\frac{1}{2}g\left(2t_0 + h\right)h}{h}$$

is meaningful. And, this ratio can be simplified to $\frac{L}{h} = gt_0 + \frac{1}{2}gh$. Since the product of the infinitesimal h and a positive, bounded value $\frac{1}{2}g$ can be ignored, the term $\frac{1}{2}gh$ can be erased, producing gt_0. That is, the instantaneous speed of the falling object at $t = t_0$ is $V|_{t=t_0} = gt_0$.

As we can see, this explanation, like all others provided by Newton and Leibniz, cannot really resolve the following contradictions:

$$\begin{aligned}&\text{(A)}\quad \text{To make } \tfrac{L}{h} \text{ meaningful, one must have } h \neq 0;\\ &\text{(B)}\quad \text{To obtain } gt_0 \text{ as the outcome, one must assume } h = 0.\end{aligned} \tag{3.34}$$

This is the so-called Berkeley paradox in the history of mathematics. After the theory of limits is established, this problem of computing the instantaneous speed of the free falling object is solved as follows:

$$\begin{aligned} V|_{t=t_0} &= \lim_{\Delta t \to 0} \frac{\Delta S}{\Delta t} = \lim_{\Delta t \to 0}\left(gt_0 + \tfrac{1}{2}g\Delta t\right) \\ &= gt_0 + \tfrac{1}{2}g \lim_{\Delta t \to 0} \Delta t = gt_0 + \tfrac{1}{2}g \cdot 0 = gt_0 + 0 = gt_0. \end{aligned} \tag{3.35}$$

Within the current theory of limits, the Berkeley paradox is successfully resolved. However, for now, we have shown the need to separate the concept of a variable approaching its limit in the fashion of potential infinities from that in the fashion of actual infinities, we can expect a return of the Berkeley paradox to the theory of limits. To this end, the limit expression in Equation (3.35) consists of two important limits:

(1) $\lim_{\Delta t \to 0} \frac{\Delta S}{\Delta t} = gt_0$; and
(2) $\lim_{\Delta t \to 0} \Delta t = 0$.

The first variable $\Delta t \to 0$ is the same in both expressions. However, the limit expression in (1) is not realizable, meaning that $\Delta t \uparrow 0 \wedge \Delta t \,\overline{\uparrow}\, 0$. That is, the variable

Δt must approach its limit 0 in the fashion of potential infinities. Now, the limit expression in (2) is realizable, meaning that $\Delta t \uparrow 0 \wedge \Delta t \top 0$. That is, the variable Δt must approach its limit 0 in the fashion of actual infinities. So, a natural question is: For the same process of solving the same problem and for the same variable $\Delta t \to 0$, how can both $\Delta t \top 0$ and $\Delta t \overline{\top} 0$ be allowed at the same time?

What is suggested in Equation (3.35) is that to make $\lim_{\Delta t \to 0} \frac{\Delta S}{\Delta t} = gt_0$ meaningful, we took the assumption $\Delta t \uparrow 0 \wedge \Delta t \overline{\top} 0$. On the other hand, to obtain $\frac{1}{2}g \cdot \lim_{\Delta t \to 0} \Delta t = \frac{1}{2}g \cdot 0$, we changed our mind and allowed $\Delta t \uparrow 0 \wedge \Delta t \top 0$. This kind of deduction is difficult for people to feel comfortable. To resolve this dissatisfaction, can we suggest permitting Δt to approach its limit 0 in the fashion of actual infinities? It is impossible, since it would make $\lim_{\Delta t \to 0} \frac{\Delta S}{\Delta t} = \frac{0}{0}$ meaningless. Since it is so, can we unify the situation by allowing $\Delta t \to 0$ in the fashion of potential infinities? This end is also impossible, since under the assumption that $\Delta t \uparrow 0 \wedge \Delta t \overline{\top} 0$, it must cause the limit expression $\lim_{\Delta t \to 0} \Delta t = 0$ to be not realizable (the expression $\lim_{x \to x_0} f(x) =$ A is said to be realizable, if (i) $x \uparrow x_0 \wedge x \top x_0$, $f(x) \uparrow \mathrm{A} \wedge f(x) \top \mathrm{A}$, and (ii) $f(x_0) =$ A. Otherwise, the limit expression $\lim_{x \to x_0} f(x) = \mathrm{A}$ is not realizable). But, it can be checked that the limit $\lim_{\Delta t \to 0} \Delta t = 0$ is realizable. From the definition for a limit expression to be realizable, it follows that each limit expression has to be either realizable or not realizable, but not both. So, the expression $\lim_{\Delta t \to 0} \Delta t = 0$ should not be an exception.

What is discussed implies that as soon as one introduces and distinguishes the concepts of approaching limits in the fashion of either actual or potential infinities, the Berkeley paradox comes back. For more detailed discussion along this line, please refer Reference 125.

As we have seen earlier, the ultimate goal of reconstructing the naïve set theory is to lay the foundation of the modern mathematics on it so that the new theory of sets has to be rich enough and powerful enough to handle infinities. However, that is exactly where new contradictions and paradoxes are found and where the Berkeley paradox made its way back.

3.7 APPLICATIONS IN EXTREME WEATHER FORECAST

After having seen how the yoyo model and its methodology can be beautifully applied to study natural and social sciences, economics, and mathematics, in this section, we will see how the yoyo model has been successfully employed in the prediction of disastrous weather conditions. The forecasting of these nearly zero-probability disastrous events have been world-class hard problems in modern science.

3.7.1 V-3θ Graphs: A Structural Prediction Method

The set of V-3θ graphs, a structural prediction method, was established [17, 126] to forecast disastrous weather conditions. This method has been shown to be effective in terms of the prediction of disastrous conditions like torrential rains, strong convections, windstorms and sandstorms, heavy fogs with low visibility, and abnormally high temperatures. Currently, many meteorological stations across China employ this method for their practical forecasts of weathers. Since it has helped to greatly improve the forecasting accuracies, when compared to the conventional methods, definite social and economic benefits have been materialized. In particular, the difference between this and the conventional methods contains the following:

(1) The concept of predicting disastrous weathers. This new method emphasizes on confrontational movements instead of the first push system initially established by Isaac Newton. In other words, investigations on evolution are about how moving materials "fight" against each other instead of being stepped over passively by a super being.
(2) Converting quantitative forces back to the original uneven structures of materials with a focus on materials' irregular or peculiar structures. It is because: (a) well-established computational schemes in the Newtonian system cannot handle irregularities; (b) materials' structures are prior to the existence of quantitative forces; so by using structures of materials or events, one can gain valuable lead time of forecasts.
(3) To represent confrontational movements, one has to find out how they move, which is exactly what was missed from Newton's laws of motion. On materials' uneven structurality, confrontational movements have to produce rotations with differences in rotational directions. So, discontinuities in movements have to appear, making the concept of continuity in Newton's system not appropriate for the analysis of evolutions and prediction of disasters.
(4) The order structure. In the Newtonian system, forces have been seen as the reason for materials to move. However, when dealing with practical problems, it involves how to recognize forces and how to apply them. For example, even if the atmospheric movement is treated as an acting force, one still has to deal with the rotational force of the earth and the stirring effect of the uneven atmospheric structure, where there exist at least two kinds of air streams, one cold and the other warm. Because whatever moves on Earth changes its direction, directions have to appear prior to quantities. In addition, rotations, either inwardly or outwardly, produce different effects such that directional structures appear. Besides, quantities cannot deal with irregular structures, while realistic rotational structures are exactly irregular. In particular, when two materials rotate in the same direction, the quantities representing their combined forces are relatively large as shown by the enclosed area in Figure 3.30a. When the materials rotate in different directions, the quantities representing their combined forces will be relatively small as shown in Figure 3.30b. Traditionally, these small quantities might very well be ignored. However, in terms of structures and confrontations, it is the latter case with small quantities that the stirring energy is greater due to the acting moments of forces so that evolution is caused.

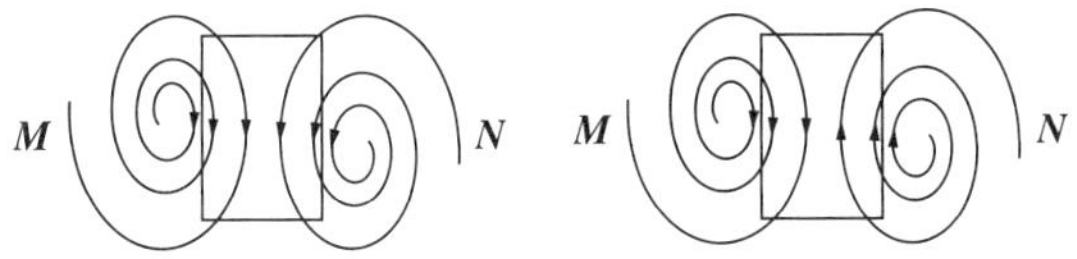

FIGURE 3.30 Interactions of spinning yoyos.

Presently, the widely applied method of weather forecast is developed on the basis of the pushes of barometric pressures, and utilizes the information of pressure, temperature, humidity, and wind, with the barometric pressure system as the core. However, in practical applications, the observations of pressures are statically modified, while no information on atmospheric densities are available, making the pressure information post-event that lags behind what is forthcoming. That explains why actual disastrous weathers often occur before the observed pressure systems. Therefore, in the structural prediction method, the order of information becomes: wind (its direction and speed), humidity, temperature, and pressure, with emphasis placed on wind directions. That is where the yoyo model was initially germinated. In other words, this new method does not employ the regularized quantitative computational schemes of Newton's first push. Instead, it uses an analysis method of irregular structures of rotations.

In the designs of the traditional weather forecasting methods, the impacts of eddy motions are already known. However, due to historical reasons, only horizontal vortices are considered for practical purposes without clearly understanding the transformational roles of vortices in evolutions. In particular, vertical vortices (or called horizontal vorticities) have not even been seriously looked at both theoretically and practically. Although in the study of dynamics, vorticity equations have involved horizontal vorticities, the functions of these vorticities and how to practically apply them are not clearly pointed out. In addition, the weather maps used in weather forecasting, consist of only four maps for the altitudes of 500 hPa, 700 hPa, 850 hPa, and the ground surface, produced after smoothing treatments with small-probability (or called irregular) information greatly damaged. That is, in the current weather forecasting, not only are vertical vortices not practically used, but also did not even get noticed, wasting all the available irregular information beneath 500 hPa and all available meteorological data above 500 hPa.

What is most important is that at the upper layer 300 $\sim$ 100 hPa of the troposphere, about 9–12 km above the sea level, there might exist an ultra low temperature that is closely related to disastrous weathers. This phenomenon is an important weather factor ignored and missed out by the traditional meteorological science. It involves a drop in temperature of the troposphere and touches on the problem of under what physical and photochemical mechanisms the temperature slides suddenly. Evidently, only with a temperature drop at the upper layer of troposphere, the rising water vapors will condensate. And, a severe drop in the temperature naturally leads to increased condensation of water vapors, causing windstorm and/or torrential rain weathers. Hence, for such weathers conditions as strong convections and torrential rains, the existence of such a severe ultra low temperature becomes a necessary requirement. That is why in employing the $\mathbf{V}$-3θ graphs, we make use of irregular information collected from the ground level to 100 hPa.

Historically, the discoveries of ultra low temperatures and functions of rolling currents were the initial motivations of recognizing that mutual reactions are not a problem of the first push, that forces are originated from materials' structures, that forces under Newton's quantitative manipulations cannot lead to the actual forces existing in uneven structures of materials, and the eventual establishment of the yoyo model [17]. Also, the quantitative forces, studied in the Newtonian system of the first push, come into being after the existence of materials' structures. What can be known are only the forces of the present moment without involving any knowledge about potential changes in the forces and their future. That is, these quantitative forces do not have the leadability of time, which is badly needed in prediction practices. In particular, mutual reactions or confrontations operate in the form of stirring motion of the structures of moments of forces. So, these operations are no longer the simple pushes of particles (without any size and volume) of the inertial system.

Since realistic movements of materials are almost entirely rotational under confrontational stirrings, the corresponding disasters only represent the most intensive and hottest spots of confrontation. So, to comprehend disastrous changes, one must investigate confrontational movements, where evolutions and rolling currents need to be studied.

Practical experiments show that atmospheric rolling currents play extremely important roles in transitional changes in weather evolutions. It can even be said that by knowing the existing rolling currents, one can basically foretell transitional changes in weathers. What is practically significant is that changes in the direction of rolling currents occur ahead of weather changes. When combined with the distribution and transportation of water vapors, one can basically foretell torrential rains, severe convections, and other disastrous weathers, and transitional changes in weather systems.

For more detailed illustrations on how the $\mathbf{V}$-3θ graphs are designed and practically employed in weather forecasts, along with case studies, please refer Reference 14.

3.7.2 Digitization of Irregular Information

Facing the countless failures of predicting the unknown future, it is realized [15] that materials or events cannot really be studied by replacing them with numbers. Since quantities cannot substitute for events, a natural question is: Can quantities be employed to substitute for information? For information, the before-and-after order of time can be clearly seen. However, without change in materials, the concept of time would become meaningless. It has been an indisputable fact that modern science has not resolved the problem of what time is. At the very basic level, modern science has avoided the fundamental problem of change in materials and events. Because change is the fundamental characteristic of the material world, one naturally sees the need for studying changes in materials and relevant processes. To this end, the yoyo model indicates naturally that materials exist first before any post-event quantity so that from quantities one will not be able to locate the causes of evolutions. This subsection is based on Reference 2.

Because weather stands for a problem of constant change, in order to improve the accuracy of predicting disastrous weather conditions, one has to question what causes modern science's inability to deal with changes. What is discovered is that quantitative analysis cannot handle irregular information. And, because quantities cannot deal with irregular information, such information cannot be simply ignored as what has been done throughout the entire spectrum of modern science. This end leads to the problem of how to comprehend irregular information and the realization that irregular information is in fact about changes. Because quantitative analysis cannot deal with information of change, an innovative method has to be introduced to deal with information of change in order to meaningfully produce forecasts of the future.

Newton's second law of motion stands for a system of quantitative average numbers. Even if we treat averages as the generality, they still cannot substitute for the specifics of events. Theories of statistics are also being limited to stable series so that peculiarities are eliminated. So, from the logical reasoning, it follows that if the generality studied by using average numbers represents invariance, then the peculiarities have to represent variability. Correspondingly, irregular information is that of change. That is why in the investigation of problems of evolution, there should not be any randomness, and events and information cannot be processed or modified in the fashion of numbers. From this understanding, we introduced the method of digitization of irregular information.

It should be noted that digitization is different from quantification, although events can be denoted using the symbols of quantities. However, these quantitative symbols of events are only employed to distinguish the events from each other; they are not the same as the underlying quantities. What is important is that even when events are labeled with quantitative symbols, these quantities do not in general comply with the quantitative calculus of the formal logic. That is the essential difference between digitization and quantification. Also, quantities can be employed to compute nonexistent matters more accurately than those that actually exist. This fact reveals the nonrealisticity characteristic of quantities and explains why by using quantities, one loses the specifics of events. Without specifics there would not be any practically meaningful accuracy.

On the other hand, digitization is about a refinement of events with regard to their differences. The significance of digitization is about walking out of the realm of quantification so that digital forecast is not the same as numerical forecast. Digits do not comply with the rules of operations of quantities. Similar to the symbols used in the "Book of Changes," digits cannot be directly added, subtracted, multiplied, or divided from each other. They are merely labels used to truthfully pass on the information of concern. Digital information is a further development from symbolic information elevated to refined information. The latter is not the same as the earlier stage symbolization of information. Although the current digitization is mainly about visual, figurative structuralization, it did not walk out of the category of visible figures of the Euclidean spaces and can only be seen as at the elementary stage of development. Based on this discussion, the method of digital (or digitized) forecasts is exactly about how to predict digitized events. For the future of digitization, it should also include the structuralization in non-Euclidean spaces of the dynamic senses of seeing, hearing, smelling, and touching. It should reflect the structural characteristics of recognizing information from multiple angles so that it can delicately describe the specifics and change of events without pursuing after the formal unit in the name of quantification. In particular, information of change is about the digitization of processes. Its applications and future development will alter the system of modern science, leading to resolutions of many currently unsolvable problems.

In terms of what time is, the core to the question is at the evolutionality of materials and about changes in events. Modern science avoids changes in materials so that time becomes a parameter that has nothing to do with change. The yoyo model indicates that time is given an expression in materials and in the changes of events; time is neither a static parameter nor a static physical quantity. It is embedded in the changes of events. No matter whether the underlying materials change or not, such concepts as force, mass, and other physical quantities exist on top of the materials, while "time" only exists along with the changes in the underlying events. Without change, "time" will not exist. So, "time" exists only in the differences between the before and after of events. That is why "time" cannot be treated as a physical quantity or parametric dimension as it has been in modern science. Correspondingly, timely differences of events or materials contain the irregularities of information of change. That is why the method of digitization can be employed to reveal the changeability of materials and events, providing a brand new way of analyzing digitized irregular information.

The digitization of time series data is designed to uncover the significance and effects of the specifics of events. If errors are controlled within the allowed range, the observational data will represent a realization of change in the underlying event, while smoothing or filtering the data only eliminates the essential irregularities of the data. As long as the distribution of changes in events is not even, distributional gradient has to exist so that one can use the azimuth of the gradient as a variable and employ the mathematical tool of reciprocating functions to represent the changes in the information of events without cooking the information of change using the quantitative calculus of the formal logic. By doing so, one transforms a time series data into digitized figures, reflecting the changes in the underlying events, while the realisticity of the information is not injured and the changeability of events is revealed. In other words, the purpose of this method is to make use of irregular information as much as possible so that the underlying changeability can be realistically reflected without damaging the invariance of static events.

To confirm the practical validity of this method of digitization, sets of data from many disciplines have been tested. What is found is that as long as the observational information is realistic, the outputs will reveal the general laws of change. For details about how this method has been successfully employed in forecasting disastrous weather conditions, earthquakes, etc., please refer Reference 2.

3.8 CONCLUSIONS

From the presentation in this chapter, it can be seen that the yoyo model has been successfully employed in the study of a wide range of different scientific disciplines and provides the desperately needed intuition and playground for systems research. From the perspective of evolution the importance of this model can be more adequately understood. Historically, the current state of systems science is in a crucial moment of development. In particular, since the time when the concept of general systems was initially hinted by von Bertalanffy in the 1920s, the systems movement has experienced over 80 years of ups and downs. After the highs of the movement in the 1960s and 1970s, the heat wave started to cool gradually in the past 20, 30 years. And in the past few years, we witnessed fast disappearance of major systems science programs from around the United States. By critically studying the recent and drastic cooling of the systems movement in the United States, we can see the following problems facing systems research:

(1) With over 80 years of development systems research has mainly stayed at the abstract level of philosophy without developing its own clearly visible and tangible methodology and consequent convincingly successful applications in the traditional disciplines. This end is extremely important, because systems scientists are humans too, they also need a means to make a living. And, only when some of the unsolvable problems in the traditional science, together with the capability of solving new problems that arise along with the appearance of systems science, can be resolved using systems methods, systems research will be embraced by the world of learning as a legitimate branch of knowledge.

(2) Those specific theories, claimed to be parts of systems science by systems enthusiasts, can be and have been righteously seen as marbles developed within their individual conventional areas of knowledge without any need to mention systems science. This fact partially explains why programs in systems science from around the United States are disappearing in recent years.

(3) Systems engineering considers various practical systems that do not really have much in common even at the level of abstract thinking. This end constitutes a real challenge to the systems movement. Specifically, when a systems project or idea needs public support, such as locating reputable reviewers for a research grant application, the organizer in general has a hard time to locate the base of supporters other than a few possible personal contacts. Due to this reason, it generally takes a long time, if it even happens, for a new rising star in systems research to be recognized. Without a steady supply of new blood, the effort of systems research will surely stay on the sideline and as secondary to the traditional sciences.

Comparing this state of systems science to that of, for instance, calculus, one can see the clear contrast. The former does not have a tightly developed system of theory, which new comers can feel firstly excited about and consequently strongly identify themselves with and scientific practitioners can simply follow procedures to produce their needed results, while the latter gives one the feeling of holistic body of thoughts where each concept is developed on the previous ones in a well-accepted playground, the Cartesian coordinate system. Beyond this, calculus possesses a high level of theoretical beauty and contains a large reservoir of procedures scientific practitioners can follow to obtain their desired consequences. In other words, by working on further developing calculus and related theories or using these theories, thousands of people from around the world in the generations before us and after us in the foreseeable future have made and will continue to make their satisfactory living.

It is on the basis of this understanding of the history of the systems movement that the yoyo model is initially introduced and established to play several crucial roles as expected:

(1) It will be the intuition and playground for all systemic thinking in a similar fashion as that of Cartesian coordinate systems in modern science. To this end, of course each of the characteristics of the model, such as the black hole, big bang, narrow neck, and spin, needs to be specified in each individual scenario of study. Hopefully, the presentation of this chapter has clearly shown how this end could be done.

(2) It will provide a means and ground of logic thinking for one to see how to establish models in the traditional field, the first dimension of science, in an unconventional sense in order to resolve the problem at hand.

(3) It will help to produce understandings of nature from an angle that is not achievable from the traditional science alone. Comparing to the items (1) and (2) above, it is this line that will provide the needed strength of life for systems science to survive the current slowdown of development and to blossom again to truly become an established second dimension of science, as claimed by George Klir.

Based on what is said above, let us now outline briefly what has been achieved with the yoyo model and its methodology as of this writing. For the relevant references, please refer Reference 115. In particular, since the idea of the yoyo model was initially felt in the 1960s,

(i) The fundamental thinking of the yoyo model was used to establish two methods for predicting (nearly) zero-probability disastrous weather conditions, which represent a world level difficult problem in modern science, see Section 3.7 and the references listed there. So, by using the yoyo model, one can make a living and produce economic benefits to the society in at least the area of weather forecasting.

(ii) The basis of the yoyo model led to the initial discovery of how human sense organs are limited in terms of understanding the physical world. This discovery implies that current methods of science must have their inherent limitations and need to be expanded in maybe the dimension of systems science.

(iii) More than a decade after the previous discovery was published, the fourth crisis in the foundations of mathematics was discovered once again on the basis of the yoyo model, see Section 3.6. This discovery is extremely important for the development of systems science, because when systems research is done on the basis of the traditional mathematics, such as the case so far in history as described in previous problem (ii) facing systems science, the conclusions of the research will be righteously and conventionally seen as part of the traditional science instead of systems science. To break out this vicious cycle, systems science needs its own theory of mathematics that complements the traditional mathematics. To this end, the newly discovered fourth crisis of mathematics provides the motivation for both systems theorists and mathematicians alike to develop mathematical theories and procedures useful for solving problems involving systems. At the same time, this discovery shows the world of learning that systemic thinking can indeed lead to unconventional consequences.

(iv) Newton's laws of motion have been generalized so that they can be employed to the study of human organizational behaviors, see Sections 3.4 and 3.5 and relevant chapters in Reference 14. This fact shows that in the coming years systemic thinking can potentially permeate the research of social sciences, while making these sciences more "exact." If this becomes a reality, it will be another magnificent achievement of systems science.

(v) Based on the study of the yoyo model, the conventional concept of energy was enriched by the establishment of stirring energy. This new concept explains why some of the natural phenomena could not be explained successfully by using the conservation of the conventional energies.

(vi) A civil engineering design was developed for the purpose of reducing the damaging consequences of natural disasters, while economic savings could be duly achieved.

As expected, along the lines of theoretical and applied successes of the yoyo methodology, there are some very important aspects of the model that need to be addressed.

First of all, to make the model more practically applicable so that scientific practitioners can employ this model in their real-life related works, relevant mathematical theories are needed for n-dimensional rotations, for any natural number n. To accomplish this, one needs to generalize the concept of cross-product of vectors to the general n-dimensional space (so far this is a bottleneck for the mathematicians). Then, important theorems of calculus that deal with rotations will need to be generalized accordingly.

Second, different yoyo fields need to be classified into equivalent classes so that those that belong to the same class would share similar properties.

Thirdly, to better study different natural systems using this model, there is a need to identify the underlying yoyo structures of various systems so that from the differences in the yoyo field properties, one can tell roughly what system he is investigating.

As a matter of fact, the yoyo model has currently been well used as a roadmap for applying systemic thinking and reasoning when dealing with specific problems.

However, to materialize the expected goal of establishing this model as the intuition and playground in different systemic investigations as the role Euclidean spaces have played in modern science [14], there is still a long way to go. That is, the introduction of this model has opened up a wide and fertile field for brand new scientific investigation that will truly bring forward theoretically significant and practically useful theories and methods in the years and decades to come.

REFERENCES

[1] L. von Bertalanffy. Einfuhrung in Spengler's werk. Literaturblatt Kolnische Zeitung, May, 1924.

[2] Y. Lin and S. C. OuYang. *Irregularities and Prediction of Major Disasters*. New York: Auerbach Publications, an imprint of Taylor and Francis, 2010.

[3] G. Klir. *Architecture of Systems Problem Solving*. New York: Plenum Press, 1985.

[4] Y. Lin. *General Systems Theory: A Mathematical Approach*. New York: Kluwer Academic and Plenum Publishers, 1999.

[5] Mathematical sciences: a unifying and dynamic resource. *Notices of the American Mathematical Society*, 33: 463–479, 1985.

[6] I. V. Blauberg, V. N. Sadovsky, and E. G. Yudin. *Systems Theory, Philosophy and Methodological Problems*. Moscow: Progress Publishers, 1977.

[7] G. Klir. *Facets of Systems Science*. New York: Springer, 2001.

[8] H. Quastler. General principles of systems analysis. In: *Theoretical and Mathematical Biology*, edited by T. H. Waterman and H. J. Morowitz. New York: Blaisdell Publishing, 1965.

[9] L. Zadeh. From circuit theory to systems theory. *Proceedings of the IEEE*, 50: 856–865, 1962.

[10] G. Klir. *An Approach to General Systems Theory*. Princeton, NJ: Van Nostrand, 1970.

[11] D. Berlinski. *On Systems Analysis*. Cambridge, MA: MIT Press, 1976.

[12] D. Lilienfeld. *The Rise of Systems Theory*. New York: John Wiley & Sons, 1978.

[13] A. T. Wood-Harper and G. Fitzgerald. A taxonomy of current approaches of systems analysis. *The Computer Journal*, 25: 12–16, 1982.

[14] Y. Lin. *Systemic Yoyos: Some Impacts of the Second Dimension*. New York: Auerbach Publications, an imprint of Taylor and Francis, 2008.

[15] Y. Lin (guest editor). Mystery of nonlinearity and Lorenz's chaos [A special double issue]. *Kybernetes: The International Journal of Cybernetics, Systems and Management Science*, 27(6/7): 605–854, 1998.

[16] S. C. OuYang. *Break-Offs of Moving Fluids and Several Problems of Weather Forecasting*. Chengdu: Press of Chengdu University of Science and technology, 1994.

[17] W. Yong and Y. Lin. *Beyond Nonstructural Quantitative Analysis: Blown-Ups, Spinning Currents and the Modern Science*. River Edge, NJ: World Scientific, 2002.

[18] G. Jarmov. *Development History of Physics* [Chinese translation]. Beijing: Commercial Press, 1981.

[19] M. Kline. *Mathematical Thought from Ancient to Modern Times*. Oxford: Oxford University Press, 1972.

[20] I. Prigogine. Structure, dissipation and life. In: First International Conference of the Theoretical Physics and Biology, Versailles, June 26–30, 1967.

[21] R. Thom. *Structural Stability and Morpho-Genesis*. Reading, MA: Benjamin, 1975.

[22] J. English and Feng G. F. *Tao De Ching*. New York: Vintage Books, 1972.

[23] T. Lao. *Tao Te Ching: A New Translation by Gia-fu Feng and Jane English*. New York: Vintage Books, 1972.

[24] H. M. Liang. *Lao Tzu* [in Chinese]. Liaoning: Liaoning People's Press, 1996.

[25] R. Wilhalm and C. Baynes. *The I Ching or Book of Changes*, 3rd edn. Princeton, NJ: Princeton University Press, 1967.

[26] E. N. Lorenz. *The Essence of Chaos*. Seattle, WA: Washington University Press, 1993.

[27] Y. Lin. Can the world be studied in the viewpoint of systems? *Mathematical and Computer Modelling*, 11: 738–742, 1988.

[28] Y. Lin, Y. Ma, and R. Port. Several epistemological problems related to the concept of systems. *Mathematical and Computer Modelling*, 14: 52–57, 1990.

[29] Y. Lin. Systemic yoyo model and applications in Newton's, Kepler's laws, etc. *Kybernetes: The International Journal of Cybernetics, Systems and Management Science*, 36(3–4): 484–516, 2007.

[30] S. Liu and Y. Lin. *Grey Information: Theory and Practical Applications*. London: Springer, 2006.

[31] G. Soros. *The Alchemy of Finance*. New York: John Wiley & Sons, 2003.

[32] S. C. OuYang, Y. G. Chen, and Y. Lin. *Digitization of Information and Prediction*. Beijing: China Meteorological Press, 2009.

[33] A. Einstein. *Collected Papers of Albert Einstein*. Princeton: Princeton University Press, 1997.

[34] D. F. Kuchemann. Report on the I.U.T.A.M. symposium on concentrated vortex motions in fluids. *Journal of Fluid Mechanics*, 21: 1–20, 1965.

[35] Z. Q. Ren, Y. Lin, and S. C. OuYang. Conjecture on law of conservation of informational infrastructures. *Kybernetes: The International Journal of Systems and Cybernetics* 27: 543–552, 1998.

[36] Y. Lin. A model of general systems. *Mathematical Modeling: An International Journal*, 9: 95–104, 1987.

[37] H. Hendrix. *Getting the Love You Want: A Guide for Couples*. New York: Owl Books, 2001.

[38] Y. Lin. Research studies: systemic yoyos and their applications. *Kybernetes: The International Journal of Systems, Cybernetics, and Management Science*, 39(2): 174–378, 2010.

[39] E. Bernays. *Public Relations*. Boston, MA: Bellman Publishing Co., 1945.

[40] L. Stoykov and V. Pacheva. *Public Relations and Business Communications*. Sofia: Ot Igla Do Konetz, 2005.

[41] D. J. Griffiths. *Introduction to Quantum Mechanics*, 2nd edn. New Jersey: Prentice Hall, 2004.

[42] G. R. Chen. *The Original State of the World: The Theory of Ether Whirltrons*. Hong Kong: Tianma Books Limited, 2007.

[43] R. Brandelik, W. Braunschweig, K. Gather, V. Kadansky, K. Lübelsmeyer, P. Mättig, H.-U. Martyn, G. Peise, J. Rimkus, H. G. Sander, *et al.* Evidence for planar events in e^+e^- annihilation at high energies. *Physics Letters B*, 86: 243–249, 1979.

[44] D. Fultz, R. R. Long, G. V. Owens, W. Bohan, R. Kaylor, and J. Weil. Studies of thermal convection in a rotating cylinder with some implications for large-scale atmospheric motion. *Meteorological Monographs* (American Meteorological Society), 21(4): 1959.

[45] S. P. Huntington. *The Clash of Civilizations and the Remaking of World Order*. New York: Simon & Schuster, 1996.

[46] Y. Lin and B. Forrest. The state of a civilization. *Kybernetes: The International Journal of Cybernetics, Systems and Management Science*, 39(2): 343–356, 2010.

[47] H. A. Kissinger. *Diplomacy*. New York: Simon & Schuster, 1994.

[48] A. B. Bozeman. Civilizations under stress. *Virginia Quarterly Review*, 51(winter): 1–18, 1975.

[49] F. Braudel. *On History*. Chicago, IL: University of Chicago Press, 1980.

[50] I. M. Wallerstein. *Geopolitics and Geoculture: Essays on the Changing World-System*. Cambridge: Cambridge University Press, 1991.

[51] E. Durkheim and M. Mauss. Note on the notion of civilization. *Social Research*, 38: 808–813, 1971.

[52] A. Toynbee. *A Study of History*, 12 volumes. Oxford: Oxford University Press, 1934–1961.

[53] C. Quigley. *Evolution of Civilizations: An Introduction to Historical Analysis*. New York: Macmillan, 1961.

[54] A. D. Hall and R. E. Fagen. Definitions of systems. *General Systems*, 1: 18–28, 1956.

[55] Y. Lin. An application of systems analysis in sociology. *Cybernetics and Systems: An International Journal*, 19: 267–278, 1988.

[56] Y. Lin and D. Forrest. Economic yoyos and never-perfect value system. *Kybernetes: The International Journal of Cybernetics and Systems*, 37(1): 149–165, 2008.

[57] G. Soros. *The Crisis of Global Capitalism: Open Society Endangered*. New York: Public Affairs, 1998.

[58] S. R. Covey. *The 7 Habits of Highly Effective People: Powerful Lessons in Personal Change*. New York: Free Press, 1989.

[59] Y. Lin and B. Forrest. Nature, human, and the phenomenon of self-awareness. *Kybernetes: The International Journal of Cybernetics, Systems and Management Science*, in press.

[60] Y. Lin and B. Forrest. The mechanism behind imagination, conscience, & free will. *Kybernetes: The International Journal of Cybernetics, Systems and Management Science*, in press.

[61] E. F. Cooke. *A Detailed Analysis of the Constitution*. Littlefield Adams & Co, 1974.

[62] N. Branden. *The Psychology of Self-esteem*. Nash Publishing Corp, 1969, p. 41.

[63] J. F. Salgado. The five factor model of personality and job performance in the European community. *Journal of Applied Psychology*, 82(1): 30–43, 1997.

[64] R. Norman. Cultivating imagination in adult education. In: Proceedings of the 41st Annual Adult Education Research, Vancouver, Canada, June 2–4, 2000.

[65] K. Egan. *Imagination in Teaching and Learning*. Chicago, IL: University of Chicago Press, 1992.

[66] D. Buss. *Evolutionary Psychology: The New Science of the Mind*. Boston, MA: Pearson Education, Inc., 2004.

[67] D. W. Pfaff. *The Neuroscience of Fair Play: Why We (Usually) Follow the Golden Rule*. New York: Dana Press, The Dana Foundation, 2007.

[68] N. Tinbergen. *The Study of Instinct*. New York: Oxford University Press, 1951.

[69] T. Benditt. *Philosophy Then and Now*, edited by Arnold and Graham. Oxford: Blackwell Publishing, 1998.

[70] D. Dennett. *Elbow Room: The Varieties of Free Will Worth Wanting*. Bradford Books, 1984.

[71] J. Greene and J. Cohen. For the law, neuroscience changes nothing and everything. *Philosophical Transactions of the Royal Society of London B*, 359, 1775–1785, 2004.

[72] D. Hume. *A Treatise of Human Nature*, 1967 edn. Oxford: Oxford University Press, 1740.

[73] G. Becker. A theory of social interactions. *Journal of Political Economy*, 82(6): 1063–1093, 1974.

[74] T. Bergstrom. A fresh look at the rotten kid theorem and other household mysteries. *Journal of Political Economy*, 97(5): 1138–1159, 1989.

[75] J. M. Baland and J. Robinson. Is child labor inefficient? *Journal of Political Economy*, 108: 663–679, 2000.

[76] A. Bommier and P. Dubois. Rotten parents and child labor. *Journal of Political Economy*, 112(1): 240–248, 2004.

[77] Y. Lin and D. Forrest. Economic yoyos and Becker's rotten kid theorem. *Kybernetes: The International Journal of Cybernetics and Systems*, 37(2): 297–314, 2008.

[78] G. Becker. *A Treatise of the Family*. Cambridge, MA: Harvard University Press, 1991.

[79] B. R. Dijkstra. Samaritan vs. rotten kid: another look. *Journal of Economic Behavior and Organization*. To appear, 64(1): 91–110, 2007.

[80] J. Hirshleifer. Shakespeare vs. Becker on altruism: the importance of having the last word. *Journal of Economic Literature*, 15: 500–502, 1977.

[81] D. Hamermesh. *Labor Demand*. Cambridge, MA: MIT Press, 1993.

[82] R. Thaler. Interindustry wage differentials. *Journal of Economic Perspectives*, 3: 181–193, 1989.

[83] S. Rosen. The theory of equalizing differences. In: *Handbook of Labor Economics*, Vol. 1, edited by O. Ashefelter and R. Layard. New York: Elsevier Science Publishers BV, 1986.

[84] K. Murphy and R. Topal . Unemployment, risk, and earnings: testing for equalizing wage differences in labor market. In: *Unemployment & The Structure of Labor Markets*, edited by L. Lang and J. S. Leonard. Oxford: Basil Blackwell, 1987.

[85] M. Blackburn and D. Neumark. Efficiency wages, inter-industry wage differentials, and the returns to ability. Finance and Economics Discussion Series, Federal Reserve Board, 1988.

[86] J. Yellen. Efficiency wage models of unemployment. *American Economic Review*, 74: 200–205, 1984.

[87] C. Shapiro and J. Stiglitz. Equilibrium unemployment as a worker discipline device. *American Economic Review*, 27: 433–444, 1984.

[88] S. Salop. A model of the natural rate of unemployment. *American Economic Review*, 69: 117–125, 1979.

[89] J. E. Stiglitz. Alternate theories of wages determination and unemployment in L.C.D.'s: the labor turnover model. *Quarterly Journal of Economics*, 88: 194–227, 1974.

[90] J. E. Stiglitz. Prices and queues as screening devices in competitive markets. IMSSS Technical Report No. 212, Stanford University, August, 1976.

[91] A. Weiss. Job queues and layoffs in labor markets with flexible wages. *Journal of Political Economy*, 88: 526–538, 1980.

[92] G. Akerlof and J. Yellen. The fair wage-effort hypothesis and unemployment. *Quarterly Journal of Economics*, 105: 255–284, 1990.

[93] R. M. Solow. Another possible source of wage stickiness. *Journal of Macroeconomics*, 1: 79–82, 1979.

[94] D. Kahneman, J. Knetsh, and R. Thaler. Fairness as a constraint on profit seeking: entitlements in the market. *American Economic Review*, 76: 728–741, 1986.

[95] M. Raff and L. Summers. Did Henry Ford pay efficiency wages? *Journal of Labor Economics*, 5: S57–S86, 1987.

[96] (Yi Lin, guest editor) Research Studies: Systemic Yoyos and Their Applications, *Kybernetes: The International Journal of Systems, Cybernetics, and Management Science*, 39(2): 174–378.

[97] A. Alchian. Uncertainty, evolution, and economic theory. *Journal of Political Economy*, 58: 211–221, 1950.

[98] G. Stigler. The economics of scale. *Journal of Law and Economics*, 1: 54–71, 1958.

[99] L. Zingales. The value of the voting right: a study of the Milan stock exchange experience. *The Review of Financial Studies*, 7: 125–148, 1994.

[100] T. Owen. *The Corporation under Russian Law, 1800–1917: A Study in Tsarist Economic Policy*. New York: Cambridge University Press, 1991.

[101] B. Burrough and J. Helyar. *Barbarians at the Gate: The Fall of RJR Nabisco*. New York: Harper & Row, 1990.

[102] M. Jensen. Agency costs of free cash flow, corporate finance, and takeovers. *American Economic Review*, 76: 323–329, 1986.

[103] A. Shleifer and R. Vishny. Management entrenchment: the case of manager-specific investments. *Journal of Financial Economics*, 25: 123–140, 1989.

[104] M. Jensen and W. Meckling. Theory of the firm: managerial behavior, agency costs, and ownership structure. *Journal of Financial Economics*, 3: 305–360, 1976.

[105] A. Berle and G. Means. *The Modern Corporation and Private Property*. New York: Macmillan, 1932.

[106] D. Yermack. Good timing: CEO stock option awards and company news announcements. *Journal of Finance*, 52: 449–476, 1997.

[107] A. Shleifer and R. Vishny. Equilibrium short horizons of investors and firms. *American Economic Review Papers and Proceedings*, 80: 148–153, 1990.

[108] S. Kaplan and B. Minton. Appointments of outsiders to Japanese boards: departments and implications for managers. *Journal of Financial Economics*, 36: 225–257, 1994.

[109] W. J. Zhu, Y. Lin, G. P. Du, and N. S. Gong. Mathematical system of potential infinities (I)—preparation. *Kybernetes: The International Journal of Systems and Cybernetics*, 37(3/4): 489–493, 2008.

[110] W. J. Zhu, Y. Lin, G. P. Du, and N. S. Gong. Mathematical system of potential infinities (II)—formal system of logical basis. *Kybernetes: The International Journal of Systems and Cybernetics*, 37(3/4): 494–504, 2008.

[111] W. J. Zhu, Y. Lin, G. P. Du, and N. S. Gong. Mathematical system of potential infinities (III)—metatheory of logical basis. *Kybernetes: The International Journal of Systems and Cybernetics*, 37(3/4): 505–515, 2008.

[112] W. J. Zhu, Y. Lin, G. P. Du, and N. S. Gong. Mathematical system of potential infinities (IV)—set theoretic foundation. *Kybernetes: The International Journal of Systems and Cybernetics*, 37(3/4): 516–525, 2008.

[113] W. J. Zhu, Y. Lin, N. S. Gong, and G. P. Du. Problem of infinity between predicates and infinite sets. *Kybernetes: The International Journal of Systems and Cybernetics*, 37(3/4): 526–533, 2008.

[114] W. J. Zhu, *et al.* Intension and structure of actually infinite, rigid sets. *Kybernetes: The International Journal of Systems and Cybernetics*, 37(3/4): 534–542, 2008.

[115] Y. Lin (guest editor). Systematic studies: the infinity problem in modern mathematics. *Kybernetes: The International Journal of Cybernetics, Systems and Management Sciences*, 37(3/4): 385–542, 2008.

[116] R. E. Mickens. *Mathematics and Science*. Singapore: World Scientific, 1990.

[117] J. Eves. *An Introduction to the History of Mathematics with Cultural Connections*, 6th edn. Fort Worth, TX: Saunders College Publishing, 1992.

[118] Y. Lin, W. J. Zhu, N. S. Gong, and G. P. Du. Systemic yoyo structure in human thoughts and the fourth crisis in mathematics. *Kybernetes: The International Journal of Systems*, Cybernetics, and Management, in press, 2008.

[119] F. Hausdorff. *Mengenlehve*. Watter ac. Hruyler, 1935.

[120] A. Robinson. Formalism 64, logic, methodology, and philosophy of science. In: *Proceedings of the 1964 International Congress*, edited by Y. Bar-Hillel. North-Holland and Pubco., 1964, pp. 228–246.

[121] IFHP (Institute of Foreign History of Philosophy). *The Philosophy of Ancient Greeks [in Chinese]*. Beijing: Commercial Press, 1962.

[122] W. J. Zhu, Y. Lin, N. S. Gong, and G. P. Du. Descriptive definitions of potential and actual infinities. *Kybernetes: The International Journal of Systems and Cybernetics*, 37(3/4): 424–432, 2008.

[123] Y. Lin and FanT. H. The fundamental structure of general systems and its relation to knowability of the physical world. *Kybernetes: The International Journal of Systems and Cybernetics*, 26: 275–285, 1997.

[124] W. J. Zhu, Y. Lin, N. S. Gong, and G. P. Du. Modern system of mathematics and a pair of hidden contradictions in its foundation. *Kybernetes: The International Journal of Systems and Cybernetics*, 37(3/4): 438–445, 2008.

[125] W. J. Zhu, Y. Lin, N. S. Gong, and G. P. Du. New Berkeley paradox in the theory of limits. *Kybernetes: The International Journal of Systems and Cybernetics*, 37($^3/_4$): 474–481, 2008.

[126] S. C. OuYang. *Weather Evolutions and Structural Forecasting*. Beijing: Meteorological Press, 1998.

CHAPTER 4

GREY SYSTEM: THINKING, METHODS, AND MODELS WITH APPLICATIONS

SIFENG LIU[1], JEFFREY Y.L. FORREST[2,3], and YINGJIE YANG[1,2]
[1]Centre for Computational Intelligence, De Montfort University, Leicester, UK
[2]Institute for Grey System Studies, Nanjing University of Aeronautics and Astronautics, Nanjing, PR China
[3]Department of Mathematics, Slippery Rock University, Slippery Rock, PA, USA

4.1 INTRODUCTION

4.1.1 Inception and Growth of Grey System Theory

At the same time when scientific and technological endeavors are compartmented into finer sections, the development of science has shown the tendency of synthesis at a higher level. At this higher level have appeared various studies of systems science, which reveals intrinsic connections of objects and events and has greatly enriched the overall progress of science and technology. Many of the historically difficult scientific problems have been resolved successfully along with the inception of specific branches of systems science [1–5]. Because of the emergence of various new areas in systems science, human understanding of nature and the laws that govern objective evolutions has been deepened. At the end of the 1940s, there appeared general systems theory [6], information theory [7], cybernetics [8], and operations research. During the 1950s and the 1960s, systems dynamics and the theory of dissipative structures [9] have been put forward. During the 1970s, there appeared one by one such new transfield and interfield theories of systems science as the synergetics [10], catastrophe theory [11], hypercycle theory [12], genetic algorithms [13], chaos theory [14], and fractal theory [15].

Contemporary Issues in Systems Science and Engineering, First Edition.
Edited by MengChu Zhou, Han-Xiong Li and Margot Weijnen.

Due to various reasons, such as the limitation of our sensing organs, the available information tends to contain uncertainty. In the recent history of science, understandings of uncertain systems have been gradually deepened and reached a new height. The second half of the twentieth century has witnessed the emergence of various theories of uncertain systems. For instance, L. A. Zadeh established fuzzy set theory in the 1960s [16], Julong Deng developed grey system theory [17–20], and Z. Pawlak proposed rough set theory in the 1980s [21]. These works represent some of the most important efforts in the research of uncertain systems of this period and provide the foundation for dealing with uncertainties.

In terms of grey system theory, it is a new methodology that focuses on the problems involving small samples and poor information. It deals with uncertain systems with partially known information through generating, excavating, and extracting useful information from what is available. Because in the natural world, uncertain systems with small samples and poor information exist commonly, it implies the wide range of applicability of grey system theory [22]. In particular, the first paper in grey system theory was written by Julong Deng and appeared in 1982 in *Systems & Control Letters*, a journal published by North-Holland. As soon as this work was published, it caught the attention of many scholars and scientific practitioners from across the world. Numerous well-known scientists strongly supported the validity and livelihood of such research. Many young scholars actively participated in the investigation of grey system theory. Currently, a great number of scholars from China, United States, England, Germany, Japan, Australia, Canada, Austria, Russia, Turkey, The Netherlands, Iran, and others have been involved in the research and application of grey system theory. In 1989, the British journal, *The Journal of Grey System*, was launched. In 1997, a Chinese publication, named *Journal of Grey System*, was launched in Taiwan. It was later in 2004 that this publication becomes all English. In 2011, Emerald launched a new journal, entitled *Grey system: Theory and application*. As of this writing, over 1000 different professional journals in the world have accepted and published papers in grey system theory. Some of these journals have respectively published special issues on grey system theory.

Many universities around the world offer courses in grey system theory. For instance, Nanjing University of Aeronautics and Astronautics (NUAA) not only offers such courses to PhD level and master level students, but also provides a service course on grey system to all undergraduate students in different majors.

Huazhong University of Science and Technology, NUAA, Wuhan University of Technology, Fuzhou University, De Montfort University, Bogazici University, University of Cape Town, Bucharest University of Economics, Kanagawa University, and several universities in Taiwan recruit and produce PhD students focusing on the research in grey systems. It is estimated that thousands upon thousands of graduate students from around the world employ the thinking logic and methodology of grey systems in their research and the writing of their dissertations.

Many publishers from around the world, such as Science Press, Press of National Defense Industry, Literature Press of Science and Technology, Taiwan Guaoli Books Limited, Science and Engineering Press of Japan, Taylor and Francis Group (USA), and Springer-Verlag (Germany), have published over 100 different kinds of monographs in grey system.

A whole array of brand new hybrid branches of study, such as grey hydrology, grey geology, grey theory and methods of breeding, grey medicine, and grey system analysis of regional economics, have appeared along with the opportunity presented by grey system theory. Agencies at national, provincial, and local governments and various foundations actively sponsored research works in grey system. It is estimated that throughout China more than 200 research outcomes of grey system were recognized officially by national, provincial, or ministerial agencies.

Based on our incomplete statistics, it was found that such internationally recognized indexing sources as SCI, EI, ISTP, SA, MR, and MA, have collected research works in grey system theory published by Chinese scholars alone more than 10,000 times. According to Chinese Science Citation Data base (CSCD), Science Time of China, November 26, 1997, the works by Julong Deng of Huazhong University of Science and Technology were cited the most times consecutively for many years among all publications by Chinese scholars. In 1993, the Blue Book of Chinese Science and Technology (no. 8), edited and published by Chinese Ministry of Science and Technology, recognized grey system theory as a new methodology of soft science established by Chinese scholars. In 2008, the Reports of Scientific Developments and Researches (2007–2008), edited and published by Chinese Association of Science and Technology, used a good amount of space to introduce grey system theory as one of the major innovative achievements in management science and engineering disciplines.

In 2006, the conference on grey system theory and applications was held successfully in Beijing with financial sponsorship provided by Chinese Center for Advanced Science and Technology, headed by T. D. Lee (a Nobel Laureate), Guangzao Zhou, and Yongxian Lu (academicians of Chinese Academy of Sciences). In addition, grey system theory has caught the attention of many important international conferences and become a center of discussion at many international events, which no doubt will play an important and active role for the world of systems researchers to get better acquainted with grey system theory. Since November 2007, the biennial IEEE International Conference on Grey System and Intelligent Services has been successfully held in Nanjing, China, attracting scholars from different parts of the world.

For a detailed comparison between grey system theory and other theories of uncertain systems, such as probability and statistics, rough set theory, and fuzzy mathematics, please see Reference 23. Also, since the late 1950s, there has appeared a study named interval analysis in the attempt of producing accurate numerical computations. Although the interval analysis is developed on the same premises as those of grey system theory and is more than 20 years older than grey system theory, the former primarily focuses on producing accurate numerical results while the latter on how to deal with uncertainty. Grey system theory has grown into a much more fertile field of theoretical and applied research than the former. For details on this comparison, please see Reference 24.

4.1.2 Basics of Grey System

Many systems under study are named after their features of the research objects, while grey systems are labeled using the color of the systems of concern. For instance,

Ashby refers to objects with unknown internal information as black boxes. Another example, as a society moves toward democracy, the citizens want to have an increased degree of transparency. We use "black" to indicate unknown information, "white" the completely known information, and "grey" the partially known and partially unknown information. Accordingly, systems with completely known information will be regarded as white, those with completely unknown information as black, and those with partially known and partially unknown information as grey. Here, the term "system" is employed to indicate the study of the structure and functions of the object of concern through analyzing the existing organic connections between the object, relevant factors, and its environment and the related laws of change [18, 25, 26].

The research of grey system theory consists of uncertain systems that are only partially known with small samples and poor information. The theory focuses on the generation and excavation of the partially known information to materialize the accurate description and understanding of the material world. Generally, incomplete system information can exist in one of the following four cases, where incompleteness in information is the fundamental meaning of being "grey" [18, 27, 28]:

(i) the information about the elements (parameters) is incomplete;
(ii) the information about the structure of the system is incomplete;
(iii) the information about the boundary of the system is incomplete; and
(iv) the information on the system's behaviors is incomplete.

Through nearly 30 years of history, grey system theory has been developed with its very own theoretical structure consisting of systems analysis, evaluation, modeling, prediction, decision-making, control, and techniques of optimization. Its main contents contain [27, 29]

(a) the theory established on the basis of grey algebraic system, grey equations, grey matrices, etc.;
(b) the methodology developed on sequence operators and generations of grey sequences;
(c) the analysis system constructed on grey incidence spaces and grey cluster evaluations;
(d) the prediction method centered around GM(1, 1);
(e) the decision-making theory represented by multi-attribute intelligent grey target decision models;
(f) combined grey models developed for producing new and practically useful results [30]; and
(g) the optimization method consisting of grey programming, grey input–output analysis, grey game theory, and grey control.

Grey algebraic system, grey matrices, grey equations, etc., constitute the foundation of grey system theory. The so-called interval analysis fits into this foundation.

4.2 SEQUENCE OPERATORS

A number, instead of its range, whose exact value is unknown, is referred to as a *grey number*, denoted using the symbol "$\otimes$". If a grey number $\otimes$ has only a lower bound $\underline{a}$, it is written $\otimes \in [\underline{a}, \infty]$ or $\otimes(\underline{a})$, where $[\underline{a}, \infty]$ is referred to as the field of $\otimes$. If a grey number $\otimes$ has only an upper bound $\bar{a}$, it is written $\otimes \in (-\infty, \bar{a}]$ or $\otimes(\bar{a})$. If a grey number $\otimes$ has both a lower $\underline{a}$ and an upper bound $\bar{a}$, it is known as an interval grey number, written $\otimes \in [\underline{a}, \bar{a}]$. If a grey number takes only a finite number or a countable number of potential values, then it is known as discrete. If the grey number can potentially take any value within an interval, then it is known as continuous. When $\otimes \in (-\infty, +\infty)$ has neither an upper nor a lower bound, then $\otimes$ is a black number. When $\otimes \in [\underline{a}, \bar{a}]$ and $\underline{a} = \bar{a}$, $\otimes$ is a white number. An essential grey number is one that temporarily cannot be represented by a white number, while a nonessential grey number can be represented by a white number obtained either through experience or certain methods. The definite white number is referred to as the whitenization (value) of the grey number, denoted $\tilde{\otimes}$, and $\otimes(a)$ is used to represent the grey number with a as its whitenization.

Given grey numbers $\otimes_1 \in [a, b]$, $a < b$, and $\otimes_2 \in [c, d]$, $c < d$, the corresponding operations of addition, subtraction, multiplication, reciprocal of $\otimes_1$, division of $\otimes_1$ by $\otimes_2$, scalar multiplication by a positive real number k, and exponent are as follows [18, 31]:

$$\otimes_1 + \otimes_2 \in [a + c, b + d], \ \otimes_1 - \otimes_2 = \otimes_1 + (-\otimes_2) \in [a - d, b - c],$$

$$\otimes_1 \cdot \otimes_2 \in [\min\{ac, ad, bc, bd\}, \max\{ac, ad, bc, bd\}],$$

$$\otimes_1^{-1} \in \left[\frac{1}{b}, \frac{1}{a}\right], \text{ if } a < b, \ a \neq 0, b \neq 0, \ ab > 0,$$

$$\otimes_1/\otimes_2 = \otimes_1 \times \otimes_2^{-1} \in \left[\min\left\{\frac{a}{c}, \frac{a}{d}, \frac{b}{c}, \frac{b}{d}\right\}, \max\left\{\frac{a}{c}, \frac{a}{d}, \frac{b}{c}, \frac{b}{d}\right\}\right], \quad \text{if } c \neq 0, d \neq 0, cd > 0,$$

$$k \cdot \otimes_1 \in [ka, kb], \text{ and } \otimes_1^k \in [a^k, b^k], \text{if } k \text{ is a positive real number.}$$

Any algebraic equation, differential equation, or matrix that contains grey parameter(s) (or called grey element(s)) is called a *grey algebraic equation*, *grey differential equation*, or *grey matrix*, respectively.

A grey number could possibly vibrate around a certain fixed number. The *whitenization* of such a grey number is relatively easy. In particular, one can simply use value a as its whitenization. This grey number can be written as $\otimes(a) = a + \delta_a$ or $\otimes(a) \in (-, a, +)$, where δ_a stands for the vibration. For an interval grey number $\otimes \in [a, b]$, its whitenization value $\tilde{\otimes}$ can be taken as $\tilde{\otimes} = \alpha a + (1 - \alpha)b, \alpha \in [0, 1]$, which is known as the equal-weight whitenization. If the weight $\alpha = 1/2$, the resultant whitenization is known as *equal-weight mean whitenization* when the distribution of an interval grey number is unknown.

For two given interval grey numbers $\otimes_1 \in [a, b]$, $\otimes_2 \in [a, b]$, such that $\tilde{\otimes}_1 = \alpha a + (1 - \alpha)b$, $\alpha \in [0, 1]$, and $\tilde{\otimes}_2 = \beta a + (1 - \beta)b$, $\beta \in [0, 1]$, if $\alpha = \beta$, both $\otimes_1$ and $\otimes_2$ are said to take the same value. If $\alpha \neq \beta$, $\otimes_1$ and $\otimes_2$ are said to take different

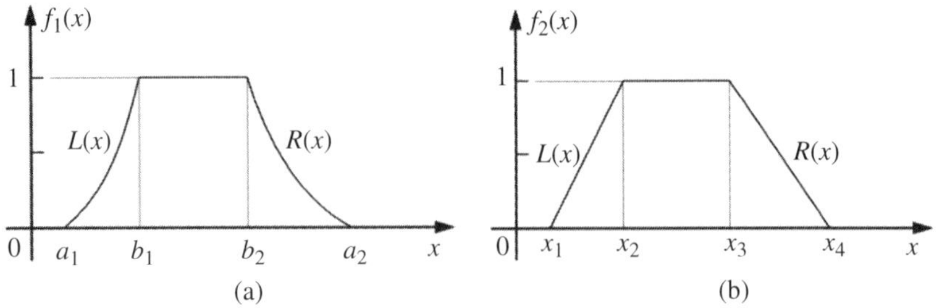

FIGURE 4.1 Typical whitenization weight functions.

values. If $\otimes_1$ and $\otimes_2$ take their values in the same interval $[a, b]$, then only when they take the same value, one can possibly have $\otimes_1 = \otimes_2$.

When the distribution of an interval grey number is known, it is natural to take such a whitenization that is generally not equal weight. For instance, a certain person's age is possibly between 30 and 45 years. Thus, $\otimes \in [30, 45]$ is a grey number. What is also known is that the person finished his 12 years of pre-college education and entered college in the 1980s. Hence, the chance for the person to be 38 years old or in the interval from 36 to 40 years in age in 2005 is quite good. For this grey number, it will not be reasonable for us to employ equal-weight whitenization. To describe what is known, we use a *whitenization weight function* to describe the preference the grey number has over the potential values it might take [33, 34, 35].

Generally speaking, the whitenization weight function of a grey number is designed based on what is known to the researcher. So, it does not have a fixed form. The start and end of the curve should have its significance. For instance, in a trade negotiation, there is a process of changing from a grey state to a white state. The eventual agreed upon deal will be somewhere between ask and bid. Hence, the relevant whitenization weight function should start at the level of ask (or bid) and end at the level of bid (or ask). The typical whitenization weight function is a continuous function with fixed starting and ending points such that the left-hand side increases and right-hand side decreases as shown in Figure 4.1(a). In practice, for simplicity, the curves are changed to strait lines as shown in Figure 4.1(b).

4.2.1 Buffer Operators

One task of grey system theory is to uncover mathematical relationships between different system variables and the laws of change of certain variables. The theory looks at each stochastic variable as a grey quantity that varies within a fixed region and within a certain timeframe, and each stochastic process as a grey process.

In a study, what is available is often a sequence of several white numbers. Although no substantial difference exists between whether the sequence is treated as a trajectory or actualization of a stochastic process or whitenized values of a grey process, different methods have been developed using different thinking logics to uncover the laws of systems evolution. For instance, the stochastics theory investigates statistical laws on the basis of probabilities established on prior knowledge. This method generally requires a large sample and can be difficult to apply to deal with nontypical distribution processes. In comparison, a number of grey system theories uncover laws of change by excavating and organizing the available raw data, representing an approach to finding data out of data, which is referred to as grey sequence generation. It is assumed that although any precise expression of the system might be complicated, and its data is chaotic, the system possesses its overall functions and properties. Therefore, there must be laws that govern the existence and operation of the system. The key is how investigators choose their appropriate method to reveal the laws. For any given grey sequence, its implicit pattern can always be revealed through weakening the explicit randomness.

Let $X^{(0)} = (x^{(0)}(1), x^{(0)}(2), \ldots, x^{(0)}(n))$ be a sequence of a system's true behavior. The observed system behavior $X = (x(1), x(2), \ldots, x(n)) = (x^{(0)}(1) + \varepsilon_1, x^{(0)}(2) + \varepsilon_2, \ldots, x^{(0)}(n) + \varepsilon_n) = X^{(0)} + \varepsilon$, where $\varepsilon = (\varepsilon_1, \varepsilon_2, \ldots, \varepsilon_n)$ represents a shocking disturbance, is a disturbed sequence. To correctly recognize the true behavior sequence $X^{(0)}$ from the disturbed sequence X, one first has to go over the hurdle ε. If we directly establish a model and make the predictions using the severely affected data X without first cleaning up the disturbance, then our predictions would most likely fail, because what the model describes is not the true state $X^{(0)}$ of change of the underlying system [36].

If the sequence $X = (x(1), x(2), \ldots, x(n))$ satisfies that $\forall\, k = 2, 3, \ldots, n$, $x(k) - x(k-1) > 0$, then X is referred to as a *monotonic increasing sequence*; if the previous inequality is replaced by $x(k) - x(k-1) < 0$, then X is referred to as a *monotonic decreasing sequence*; otherwise, X is referred to as a random vibrating or fluctuating sequence. In this case, $M - m = \max\{x(k) | k \in N_n^+\} - \min\{x(k) | k \in N_n^+\}$ is referred to as the *amplitude* of the sequence X.

Let D be a sequence operator to work on X such that after applying D, the sequence X becomes $XD = (x(1)d, x(2)d, \ldots, x(n)d)$, where XD is referred to as the *first-order sequence* (of the operator D). If D_1, D_2, and D_3 are all sequence operators, then D_1D_2 is referred to as a second-order sequence operator; its application on X produces $XD_1D_2 = (x(1)d_1d_2, x(2)d_1d_2, \ldots, x(n)d_1d_2)$, a second-order sequence of D_1D_2. Similarly, $D_1D_2D_3$ is referred to as a third-order sequence operator and $XD_1D_2D_3 = (x(1)d_1d_2d_3, x(2)d_1d_2d_3, \ldots, x(n)d_1d_2d_3)$ a third-order sequence of $D_1D_2D_3$, and so forth.

Any sequence operator D satisfying the following conditions is referred to as a *buffer operator*: (1) $x(n)d = x(n)$; (2) each entry of X is sufficiently applied in D; and (3) each entry $x(k)d$ of XD is expressed by a uniform, elementary analytic representation of $x(1), x(2), \ldots, x(n)$. If the buffered sequence XD increases, decreases, or fluctuates slower or with smaller amplitude, respectively, than the original sequence X, then D is referred to as a *weakening operator*. On the other hand, if XD increases, decreases, or fluctuates faster or with larger amplitude, respectively, than the original sequence X, then D is referred to as a *strengthening operator* [36].

Theorem 1 Assume that X is a monotonic increasing (respectively, decreasing) sequence. If D is a weakening operator, then $x(k)d \geq x(k)$ (respectively, $x(k)d \leq x(k)$); if D is a strengthening operator, then $x(k)d \leq x(k)$ (respectively, $x(k)d \geq x(k)$) [36].

Theorem 2 Assume that X is a fluctuating sequence and XD a buffered sequence. Then, if D is a weakening operator, then $\max_{1\le k\le n}\{x(k)\} \ge \max_{1\le k\le n}\{x(k)d\}$ and $\min_{1\le k\le n}\{x(k)\} \le \min_{1\le k\le n}\{x(k)d\}$; if D is a strengthening operator, then $\max_{1\le k\le n}\{x(k)\} \le \max_{1\le k\le n}\{x(k)d\}$ and $\min_{1\le k\le n}\{x(k)\} \ge \min_{1\le k\le n}\{x(k)d\}$.

For the proofs of these theorems, see Reference 23. As for the construction of practically useful buffer operators, see Reference 24.

4.2.2 Generation of Grey Sequences

When in the available data sequences there are entries that are either missing or out of ordinary, a need to fill the missing entries or to correct the abnormal entries arises when one processes the data. Generation by average is one of the often employed methods to create new data, fill a vacant entry, and construct new sequences. In particular, let $X = (x(1), x(2), \ldots, x(k), x(k+1), \ldots, x(n))$ be a sequence of data. If X has a blank entry at location k, denoted $\o(k)$, then the filling $x^*(k) = \alpha x(k) + (1 - \alpha)x(k-1)$, for $\alpha \in [0, 1]$, of this blank entry is referred to as generated by the new $x(k)$ and old information $x(k-1)$ under the generation coefficient (weight) α. If the blank entry $\o(k)$ is filled by using $x^*(k) = 0.5x(k-1) + 0.5x(k+1)$, then the resultant sequence is referred to as generated by using the nonadjacent neighbor mean. If X has no blank entry, then $x^*(k) = 0.5x(k) + 0.5x(k-1)$ is referred to as generated by adjacent neighbor mean. The sequence $Z = (z(2), z(3), \ldots, z(n))$ consisting of adjacent neighbor means is referred to as generated by adjacent neighbor means.

By borrowing the idea of functions, the smoothness of a discrete sequence is defined as follows: For a given $X = (x(1), x(2), \ldots, x(k), x(k+1), \ldots, x(n), x(n+1))$, let $Z = (z(1), z(2), \ldots, z(n))$ be the sequence of adjacent neighbor means. Assume that X^* is a representative differentiable function and d a distance function of $\boldsymbol{R}^n$. Let the same symbol X represent the sequence X with the entry $x(n+1)$ deleted. If X satisfies: (1) when k is sufficiently large, $x(k) < \sum_{i=1}^{k-1} x(i)$; and (2) $\max_{1\le k\le n} |x^*(k) - x(k)| \ge \max_{1\le k\le n} |x^*(k) - z(k)|$, then the sequence X is referred to as *a smooth sequence*. That is, the greater a pattern appears in X, the closer X is to a smooth continuous function and the greater its degree of smoothness. The ratio

$$\rho(k) = \frac{x(k)}{\sum_{i=1}^{k-1} x(i)}; k \in N_n^+ - \{1\} \tag{4.1}$$

is referred to as the *smoothness ratio* of the sequence X [18]. This concept reflects the smoothness of sequences from a different angle. It uses the ratio $\rho(k)$ of the kth data value $x(k)$ over the sum $\sum_{i=1}^{k-1} x(i)$ of the previous values to check whether or not the changes in the data points of X are stable.

If a sequence X satisfies

(1) $\frac{\rho(k+1)}{\rho(k)} < 1, k \in N_{n-1}^+ - \{1\}$ and

(2) $\rho(k) \in [0, \varepsilon], k \in N_n^+ - \{1, 2\}$, where $\varepsilon < 0.5$,

then X is referred to as a *quasi-smooth sequence*. If X has missing entries and a new sequence Y generated based on X is quasi-smooth, then Y is referred to as quasi-smooth generated.

If the first entry $x(1) = \o(1)$ and last entry $x(n)\ \o(n)$ of a sequence X are blank, the method of *stepwise ratio generation* is often employed. Based on the *stepwise ratios* of X,

$$\sigma(k) = \frac{x(k)}{x(k-1)}; k \in N_n^+ - \{1\} \tag{4.2}$$

the missing entries $x(1) = \o(1)$ and $x(n) = \o(n)$ can be generated respectively by using the stepwise ratio of its right-hand side and left-hand side neighbors. The sequence obtained by filling all its missing entries using stepwise ratios is referred to as stepwise ratio generated [18].

For an original sequence $X^{(0)} = (x^{(0)}(1), x^{(0)}(2), \ldots, x^{(0)}(n))$, the *accumulating (generation) operator D* [18, 27], when applied to $X^{(0)}$ to obtain $X^{(0)}D = (x^{(0)}(1)d, x^{(0)}(2)d, \ldots, x^{(0)}(n)d)$, is defined as follows:

$$x^{(0)}(k)d = \sum_{i=1}^{k} x^{(0)}(i); k \in N_n^+. \tag{4.3}$$

Correspondingly, the *inverse accumulating operator D*, when applied to $X^{(0)}$ to obtain $X^{(0)}D = (x^{(0)}(1)d, x^{(0)}(2)d, \ldots, x^{(0)}(n)d)$, is defined as follows:

$$x^{(0)}(k)d = x^{(0)}(k) - x^{(0)}(k-1); \quad k \in N_n^+ - \{1\}. \tag{4.4}$$

Proposition 1 Assume that $X^{(0)} = (x^{(0)}(1), x^{(0)}(2), \ldots, x^{(0)}(n))$ is a nonnegative sequence satisfying $x^{(0)}(k) \in [a, b]; k \in N_n$. If $X^{(r)} = (x^{(r)}(1), x^{(r)}(2), \ldots, x^{(r)}(n))o = X^{(0)}D^{(r)}$ is the rth accumulated sequence of $X^{(0)}$, then when r is sufficiently large, for any $\varepsilon > 0$, there is natural number N such that for any k, $N < k \leq n$,

$$\frac{x^{(r)}(k)}{\sum_{i=1}^{k-1} x^{(r)}(i)} < \varepsilon.$$

Proposition 2 Assume that $X^{(0)}$ is the same as in Proposition 1. Let $X^{(0)}D = X^{(1)} = (x^{(1)}(1), x^{(1)}(2), \ldots, x^{(1)}(n))$ and $Z^{(1)} = (z^{(1)}(2), \ldots, z^{(1)}(n))$ the sequence generated by adjacent neighbor means of $X^{(1)}$. Then, for any $\varepsilon > 0$ there is a positive natural number $N = N(\varepsilon)$ such that for any k, $N < k \leq n$,

$$\frac{x^{(0)}(k)}{z^{(1)}(k)} < \rho(k) = \frac{x^{(0)}(k)}{\sum_{i=1}^{k-1} x^{(0)}(i)} < \varepsilon.$$

For the proofs of these propositions, see References 23 and 24.

4.2.3 Exponentiality of Accumulating Generations

Let $X^{(0)} = (x^{(0)}(1), x^{(0)}(2), \ldots, x^{(0)}(n))$ be a sequence of data and its inverse accumulation-generated sequence $\alpha^{(1)}X^{(0)} = (\alpha^{(1)}x^{(0)}(1), \alpha^{(1)}\ x^{(0)}(2), \ldots, \alpha^{(1)}x^{(0)}(n))$. If there is k such that $\alpha^{(1)}x^{(0)}(k) = x^{(0)}(k) - x^{(0)}(k-1) > 0$, then $X^{(0)}$ increases at step k. If $\alpha^{(1)}x^{(0)}(k) < 0$, $X^{(0)}$ decreases at step k. If there are k_1 and k_2 such that $\alpha^{(1)}x^{(0)}(k_1) > 0$ and $\alpha^{(1)}x^{(0)}(k_2) < 0$, then $X^{(0)}$ is said to be a *random sequence*. If $X^{(0)}$ is either monotonic increasing or decreasing, while $\alpha^{(1)}X^{(0)}$ is random, then $X^{(0)}$ is referred to as a *first-order* weak random sequence. If for each $i \in N_{r-1}$, $\alpha^{(i)}X^{(0)}$ is monotonic (either increasing or decreasing), while $\alpha^{(r)}X^{(0)}$ is random, then $X^{(0)}$ is referred to as an rth-*order* weak random sequence. If for any $r \in N$, $\alpha^{(r)}X^{(0)}$ is monotonic, then $X^{(0)}$ is referred to as a nonrandom sequence.

Theorem 3 Assume that $X^{(0)}$ is a positive sequence and $X^{(r)}$ the rth accumulation-generated sequence of $X^{(0)}$, then $X^{(r)}$ must be an rth-order weak random sequence.

The exponential function $X(t) = ce^{at} + b, c, a \neq 0$, is *homogeneous*, if $b = 0$, nonhomogeneous, if $b \neq 0$. If a sequence $X = (x(1), x(2), \ldots, x(n))$ satisfies $x(k) = ce^{ak}$, c, $a \neq 0$, $\forall k \in N_n^+$, then X is referred to as a *homogeneous exponential sequence*. If $x(k) = ce^{ak} + b$, c, a, $b \neq 0$, $\forall k \in N_n^+$, then X is referred to as a *nonhomogeneous sequence*.

Theorem 4 A sequence X is homogeneous exponential, if and only if $\sigma(k)\ \forall k \in N_n^+$ is a constant.

For the proof of this result, see Reference 23.

For the given sequence $X = (x(1), x(2), \ldots, x(n))$, if $\forall k$, $\sigma(k) \in (0, 1)$, then X is said to satisfy the *law of negative grey exponent*; if $\forall k$, $\sigma(k) \in (1, b)$, for some $b >$ 1, then X is said to satisfy the *law of positive grey exponent*; if $\forall k$, $\sigma(k) \in [a, b]$, $b - a = \delta$, then X is said to satisfy the *law of grey exponent* with the absolute degree of greyness δ; and if $\delta < 0.5$, then X is said to satisfy the *law of quasi-exponent* [37, 38].

Theorem 5 If $X^{(0)}$ is a nonnegative quasi-smooth sequence, the sequence $X^{(1)}$, generated by applying accumulating generation once on $X^{(0)}$, satisfies the law of quasi-exponent.

For the proof, see Reference 23.

Theorem 6 Assume that $X^{(0)}$ is a nonnegative sequence. If $X^{(r)}$ satisfies a law of exponent and the stepwise ratio of $X^{(r)}$ is given by $\sigma^{(r)}(k) = \sigma$, then

(1) $\sigma^{(r+1)}(k) = \frac{1-\sigma^k}{1-\sigma^{k-1}}$;
(2) When $\sigma \in (0, 1)$, $\lim_{k\to\infty} \sigma^{(r+1)}(k) = 1$; and for each k, $\sigma^{(r+1)}(k) \in (1, 1+\sigma]$;
(3) When $\sigma > 1$, $\lim_{k\to\infty} \sigma^{(r+1)}(k) = \sigma$; and for each k, $\sigma^{(r+1)}(k) \in (\sigma, 1+\sigma]$.

This theorem says that if the rth accumulating generation of $X^{(0)}$ satisfies an obvious law of exponent, additional application of the accumulating generator will destroy the pattern. In practical applications, if the rth accumulating generation of $X^{(0)}$ satisfies the law of quasi-exponent, one generally stops applying the accumulating generator any further. To this end, Theorem 5 implies that only one application of the accumulating generator is needed for nonnegative quasi-smooth sequences before establishing an exponential model. For the proof of Theorem 6, see Reference 23.

4.3 GREY INCIDENCE ANALYSIS

In the study of a complex system, many factors are involved; and it is the mutual interactions of these factors that determine its behavior. It is generally interesting to investigate which factors are primary with the dominant effect, while others exert less influence, on the development of the system. To this end, regression analysis, difference equations, main component analysis, etc., are commonly employed methods. However, these methods suffer from the following weaknesses: (1) Large samples are needed to produce reliable conclusions; (2) The available data need to satisfy some typical type of probability distribution; linear relationships between factors and system behaviors are often assumed, while between the factors it is required to have no interactions (these requirements in generally are difficult to satisfy); (3) The amount of computation is large and generally done by using computers; and (4) It is possible that quantitative conclusions do not agree with qualitative analysis outcomes such that the laws governing the system of concern are distorted or misunderstood. What is especially important is that when the available data contain a relatively large amount of greyness and experience large rises and major falls without following any conventional probability distribution, it will be extremely difficult to apply the traditional methods of statistics to analyze these data [18].

Grey incidence analysis provides a new method to analyze systems when conventional methods do not seem appropriate. This method can be relied upon no matter what the sample size is and no matter whether the data satisfy any assumed distribution or not. What is more advantageous is that the amount of computation involved is small and can be carried out conveniently without the problem of disagreement between quantitative and qualitative conclusions [39, 40, 41].

4.3.1 Grey Incidence Factors and Set of Grey Incidence Operators

For a system factor X_i, let $x_i(k)$ be its observational value at the ordinal position $k \in N_n^+$, then $X_i = (x_i(1), x_i(2), \ldots, x_i(n))$ is seen as the *behavioral sequence* of X_i. For example, if X_i represents an economic factor, k time, and $x_i(k)$ the observed value of X_i at the time k, then $X_i = (x_i(1), x_i(2), \ldots, x_i(n))$ is a time series of economic behavior. If k is the ordinal number of an index, then $X_i = (x_i(1), x_i(2), \ldots, x_i(n))$ is the index sequence of economic behavior. If k represents the ordinal number of different economic regions or departments, then $X_i = (x_i(1), x_i(2), \ldots, x_i(n))$ is a horizontal sequence of economic behavior.

Let X_i be the behavioral sequence of X_i as described above and D_j a sequence operator such that $X_iD_j = (x_i(1)d_j, x_i(2)d_j, \ldots, x_i(n)d_j), j \in N_5^+$. If $x_i(k)d_1 = x_i(k)/x_i(1)$, $x_i(1) \neq 0$, $\forall k \in N_n^+$, then D_1 is known as an *initialing operator* with X_iD_1 called *initial image* of X_i. If $x_i(k)d_2 = \frac{x_i(k)}{\overline{X_i}}$, $\overline{X_i} = \frac{1}{n}\sum_{k=1}^{n} x_i(k)$, $\forall k \in N_n^+$, then D_2 is known as an *averaging operator* with X_iD_2 called the *average image* of X_i. If $x_i(k)d_3 = \frac{x_i(k)-\min_k x_i(k)}{\max_k x_i(k)-\min_k x_i(k)}$, $k = 1, 2, \ldots, n$, then D_3 is known as an *interval operator* with X_iD_3 called the *interval image* of X_i. If $x_i(k) \in [0, 1]$, and $x_i(k)d_4 = 1 - x_i(k)$, $\forall k \in N_n^+$, then D_4 is known as a *reversing operator* with X_iD_4 called the *reverse image* of X_i. If $x_i(k)d_5 = 1/x_i(k)$, $x_i(k) \neq 0$, $\forall k \in N_n^+$, then D_5 is known as a *reciprocating operator* with X_iD_5 called the *reciprocal image* of X_i.

The set $D = \{D_j | j \in N_5^+\}$ is known as the *set of grey incidence operators*. If X stands for the set of all system factors and D the set of grey incidence operators, then (X, D) is referred to as the *space of grey incidence factors* of the system.

4.3.2 Degrees of Grey Incidences

Given a sequence $X = (x(1), x(2), \ldots, x(n))$, without causing confusion the same symbol X is also used to represent the corresponding *zigzagged line* of the plane

$$X = \{x(k) + (t - k)(x(k + 1) - x(k)) | k \in N_{n-1}^+; t \in [k, k + 1]\}.$$

Let X_0 be the sequence of a system's characteristic behavior, which is increasing, and X_i the behavioral sequence of a relevant factor. If X_i is also an increasing sequence, then we say that both X_i and X_0 have a *positive* or *direct incidence relationship*. If X_i is a decreasing sequence, then we say that both X_i and X_0 have a *negative* or *inverse incidence relationship*. Without loss of generality, let us focus on the study of positive incidence relationships.

Let $X_0 = (x_0(1), x_0(2), \ldots, x_0(n))$ be a data sequence of a system's characteristic and $X_i = (x_i(1), x_i(2), \ldots, x_i(n)) \ldots$, $i \in N_m^+$, relevant factor sequences. For given real numbers $\gamma(x_0(k), x_i(k))$, $i \in N_m^+$, $k \in N_n^+$, if $\gamma(X_0, X_i) = \frac{1}{n}\sum_{k=1}^{n} \gamma(x_0(k), x_i(k))$ satisfies

(1) *Normality*: $0 < \gamma(X_0, X_i) \leq 1$, $\gamma(X_0, X_i) = 1 \Longleftrightarrow X_0 = X_i$
(2) *Wholeness*: $\forall X_i, X_j \in X = \{X_s | s \in N_m; m \geq 2\}$, $\gamma(X_i, X_j) \neq \gamma(X_j, X_i)$, $i \neq j$
(3) *Pair Symmetry*: $\forall X_i, X_j \in X$, $\gamma(X_i, X_j) = \gamma(X_j, X_i) \Longleftrightarrow X = \{X_i, X_j\}$
(4) *Nearness*: the smaller the $|x_0(k) - x_i(k)|$ is, the greater the $\gamma(x_0(k), x_i(k))$,

then $\gamma(X_0, X_i)$ is referred to as a *degree of grey incidence* between X_i and X_0, $\gamma(x_0(k), x_i(k))$ the *incidence coefficient* of X_i and X_0 at point k [18].

Theorem 7 Given a system's behavioral sequences $X_0 = (x_0(1), x_0(2), \ldots, x_0(n))$ and $X_i = (x_i(1), x_i(2), \ldots, x_i(n))$, $i \in N_m^+$, for $\xi \in (0, 1)$, define

$$\gamma(x_0(k), x_i(k)) = \frac{\min_i \min_k |x_0(k) - x_i(k)| + \xi \max_i \max_k |x_0(k) - x_i(k)|}{|x_0(k) - x_i(k)| + \xi \max_i \max_k |x_0(k) - x_i(k)|} \tag{4.5}$$

and

$$\gamma(X_0, X_i) = \frac{1}{n} \sum_{k=1}^{n} \gamma(x_0(k), x_i(k)), \tag{4.6}$$

then $\gamma(X_0, X_i)$ is a *degree of grey incidence* between X_0 and X_i, where ξ is known as the *distinguishing coefficient* [18].

For the proof, see Reference 23.

4.3.3 General Grey Incidence Models

4.3.3.1 Absolute Degree of Grey Incidence Let X_i, $i \in N_2^+$, be two sequences of the same length that is defined as the sum of the distances between two consecutive time moments, and $s_i = \int_1^n (X_i - x_i(1))dt$, $i \in N_2^+$. Then,

$$\varepsilon_{12} = \frac{1 + |s_1| + |s_2|}{1 + |s_1| + |s_2| + |s_1 - s_2|} \tag{4.7}$$

satisfies the properties of normality, pair symmetry, and nearness, and is referred to as the *absolute degree of* (grey) incidence between X_1 and X_2.

As for sequences of different lengths, the concept of absolute degree of incidence can be defined by either shortening the longer sequence or prolonging the shorter sequence with appropriate methods to make the sequences have the same length. However, by doing so, the ultimate value of the absolute degree of incidence will be affected.

Theorem 8 The absolute degree of incidence ε_{12} satisfies the following properties.

(1) $0 < \varepsilon_{12} \leq 1$.
(2) ε_{12} is only related to the geometric shapes of X_1 and X_2, and has nothing to do with the spatial positions of these sequences.
(3) The more X_1 and X_2 are geometrically similar, the greater the ε_{12}.
(4) If X_1 and X_2 are parallel or X_1 fluctuates around X_2 with the area of the parts of X_1 located above X_2 equal to that of the parts with X_1 located underneath X_2, then $\varepsilon_{12} = 1$.
(5) When one of the observed values of X_1 and X_2 changes, so does ε_{12}.
(6) When the lengths of X_1 and X_2 changes, so does ε_{12}.
(7) $\varepsilon_{11} = \varepsilon_{22} = 1$.
(8) $\varepsilon_{12} = \varepsilon_{21}$ [29].

4.3.3.2 Relative Degree of Grey Incidence Let X_i and X_j be two sequences of the same length with nonzero initial values and X_i' and X_j' respectively the initial images of X_i and X_j. The absolute degree of grey incidence of X_i' and X_j' is referred to as the *relative degree* of (grey) incidence of X_i and X_j, denoted r_{ij}.

Theorem 9 The relative degree r_{ij} of incidence satisfies the following properties:

(1) $0 < r_{ij} \leq 1$.
(2) The value of r_{ij} has something to do with only the rates of change of the sequences X_i and X_j with respect to their individual initial values and has nothing to do with the magnitudes of other entries. In other words, scalar multiplication does not change the relative degree of incidence.
(3) The closer the individual rates of change of X_i and X_j with respect to their initial values, the greater r_{ij} is.
(4) If $X_j = aX_i$, or when the images of zero initial points of the initial images of X_i and X_j satisfy that $X_i'^0$ fluctuates around $X_j'^0$ and that the area of the parts where $X_i'^0$ is located above $X_j'^0$ equals that of the parts where $X_i'^0$ is located underneath $X_j'^0$, $r_{ij} = 1$.
(5) When an entry in X_i or X_j changes, so does r_{ij}.
(6) When the length of X_i or X_j changes, so does r_{ij}.
(7) $r_{jj} = r_{ii} = 1$; and
(8) $r_{ij} = r_{ji}$ [29].

4.3.3.3 Synthetic Degree of Grey Incidence Let X_i and X_j be two sequences of the same length with nonzero initial entries, ε_{ij} and r_{ij} respectively the absolute and relative degrees of incidence between X_i and X_j, and $\theta \in [0, 1]$. Then

$$\rho_{ij} = \theta\varepsilon_{ij} + (1 - \theta)r_{ij} \tag{4.8}$$

is known as the *synthetic degree of* (grey) incidence between X_i and X_j. The concept of synthetic degree of incidence reflects the degree of similarity between the zigzagged lines of X_i and X_j as well as the nearness between the rates of changes of X_i and X_j with respect to their individual initial values.

Theorem 10 The synthetic degree ρ_{ij} of incidence satisfies the following properties:

(1) $0 < \rho_{ij} \leq 1$;
(2) The value of ρ_{ij} is related to the individual observed values of the sequences X_i and X_j as well as the rates of changes of these values with respect to their initial values;
(3) When the lengths of X_i and X_j change, so does ρ_{ij};

(4) When $\theta = 1$ or 0, $\rho_{ij} = r_{ij}$;
(5) $\rho_{jj} = \rho_{ii} = 1$; and
(6) $\rho_{ij} = \rho_{ji}$ [29].

4.3.4 Grey Incidence Models Based on Similarity and Nearness

Let X_i and X_j be two sequences of the same length and $X_i^0 = (x_i^0(1), x_i^0(2), \ldots x_i^0(n))$ and $X_j^0 = (x_j^0(1), x_j^0(2), \ldots, x_j^0(n))$ their respective images of the zero starting point. Define $s_i - s_j = \int_1^n (X_i^0 - X_j^0)dt$, $S_i - S_j = \int_1^n (X_i - X_j)dt$. Then

$$\varepsilon_{ij} = \frac{1}{1 + |s_i - s_j|} \tag{4.9}$$

and

$$\rho_{ij} = \frac{1}{1 + |S_i - S_j|} \tag{4.10}$$

are known as the *similitude degree* and nearness *degree* of (grey) incidence between X_i and X_j, respectively. It can be seen that the more similar the geometric shapes of X_i and X_j, the greater the value ε_{ij} takes, and vice versa; and that the closer X_i and X_j are, the greater the value ρ_{ij} takes, and vice versa [27] .

Theorem 11 The similitude degree ε_{ij} of incidence satisfies the following properties:

(1) $0 < \varepsilon_{ij} \leq 1$.
(2) The value of ε_{ij} is determined only by the geometric shape of the sequences X_i and X_j without anything to do with their relative spatial positions. In other words, horizontal translations of X_i and X_j will not change the value of ε_{ij}.
(3) The more geometrically similar the sequences X_i and X_j, the greater the value of ε_{ij}; and vice versa.
(4) When X_i and X_j are parallel or when X_i^0 fluctuates around X_j^0 and the area of the parts where X_i^0 is located above X_j^0 equals that of the parts where X_i^0 is located beneath X_j^0, $\varepsilon_{ij} = 1$.
(5) $\varepsilon_{ii} = 1$.
(6) $\varepsilon_{ij} = \varepsilon_{ji}$ [27].

Theorem 12 The nearness degree ρ_{ij} of incidence satisfies the following properties:

(1) $0 < \rho_{ij} \leq 1$.

(2) The value of ρ_{ij} is determined not only by the geometric shape of the sequences X_i and X_j but also by their relative spatial positions. In other words, horizontal translations of X_i and X_j will change the value of ρ_{ij}.

(3) The closer X_i and X_j, the greater the value of ρ_{ij}; and vice versa.

(4) If X_i and X_j coincide or X_i fluctuates around X_j with the area of the parts where X_i is located above X_j equals that of the parts where X_i is located beneath X_j, $\rho_{ij} = 1$.

(5) $\rho_{ii} = 1$.

(6) $\rho_{ij} = \rho_{ji}$ [29].

It is through degrees of incidence that grey incidence analysis investigates the mutual influences and connections between sequences. Its focus is about the relevant order relations instead of the specific magnitudes of the values of the degrees of incidence. For instance, when computing the similitude degrees or nearness degrees of incidence, when the absolute values of the sequence data are relatively large, the values of both $|s_i - s_j|$ and $|S_i - S_j|$ might be large too, creating the situation that the resultant similitude and nearness degrees of incidence are relatively small. This scenario does not have any substantial impact on the analysis of order relationships. If the situations to be dealt with demand relatively large numerical magnitudes in the degrees of incidence, one can consider either replacing the number 1 appearing in the numerators or denominators in (4.9) and (4.10) by a relevant constant, or using the grey absolute degree of incidence model or other appropriate models [42].

4.4 GREY CLUSTER EVALUATION MODELS

Grey clustering is developed for classifying observation indices or objects into definable classes using grey incidence matrices or grey whitenization weight functions. Each cluster can be seen as a set consisting of all the observational objects of a same kind. In practice, it is often the situation that each observational object possesses quite a few characteristic indices, which are difficult to classify accurately. Based on the objects to be clustered, grey clustering consists of two methods: clustering using grey incidence, and clustering using grey whitenization weight functions. The first method is mainly applied to group the same kinds of factors into their individual categories such that a complicated system can be simplified. By using the clustering method of grey incidence, one can examine whether or not some of the factors under consideration really belong to the same kind such that a synthetic index of these factors or one of these factors can be used to represent all these factors without losing any part of the available information carried by them. This problem is about the selection of variables to be used in the study of the system of concern. Before conducting a large-scale survey, which generally costs a lot of money and man power, by using the clustering method of grey incidence on a typical sample data, one can reduce the amount of data collection to a minimal level by eliminating the unnecessary variables such that tangible savings can be materialized. On the other hand, the clustering method of grey whitenization weight functions is mainly applicable for checking whether or not the observational objects belong to predetermined classes such that they can be treated differently [43, 44].

4.4.1 Grey Incidence Clustering

Assume that m characteristic data points $X_i = (x_i(1), x_i(2), \ldots, x_i(n))$, $i \in N_m^+$, are collected for each of the n observational objects. From these data, construct the following *incidence matrix* of the characteristic variables: $A = [\varepsilon_{ij}]_{m\times m}$, where ε_{ij} is the absolute degree of incidence between X_i and X_j, and $\varepsilon_{ii} = 1$, for all $i < j$, and $\varepsilon_{ij} = 0$, when $i > j$, $i, j \in N_m^+$. For a chosen threshold value $r \in [0, 1]$, which in general satisfies $r > 0.5$, if $\varepsilon_{ij} \geq r$, $i \neq j$, the variables X_j and X_i are seen as having the same charactcristics, resulting in a so-called *clustering of r grey incidence*. When studying a specific problem, the particular r value is determined based on the circumstances involved. The closer r is to 1, the finer the clustering is with relatively fewer variables in each cluster. Conversely, the smaller the r is, the coarser the clustering becomes with relatively more variables in each cluster. For examples on how to practically employ this clustering method, please refer see Reference 23.

4.4.2 Grey Variable Weight Clustering

Consider clustering n objects according to m criteria into s different grey classes. Classifying the ith object into the kth grey class according to the observed value x_{ij} of the ith object judged against the jth criterion, $i \in N_n^+$, $j \in N_m^+$, is known as *grey clustering*. Classifying n objects into s grey classes using the jth criterion is referred to as the jth criterion subclassification with the whitenization weight function of the jth criterion kth subclass denoted as $f_j^k(\bullet)$ [29]. If the whitenization weight function $f_j^k(\bullet)$ takes the typical form as shown in Figure 4.2, then $x_j^k(1)$, $x_j^k(2)$, $x_j^k(3)$, and $x_j^k(4)$ are referred to as turning points of $f_j^k(\bullet)$. Such a typical whitenization function is written as $f_j^k[x_j^k(1), x_j^k(2), x_j^k(3), x_j^k(4)]$. If the whitenization weight function $f_j^k(\bullet)$ does not have the first and second turning points $x_j^k(1)$ and $x_j^k(2)$, as shown in Figure 4.3, then

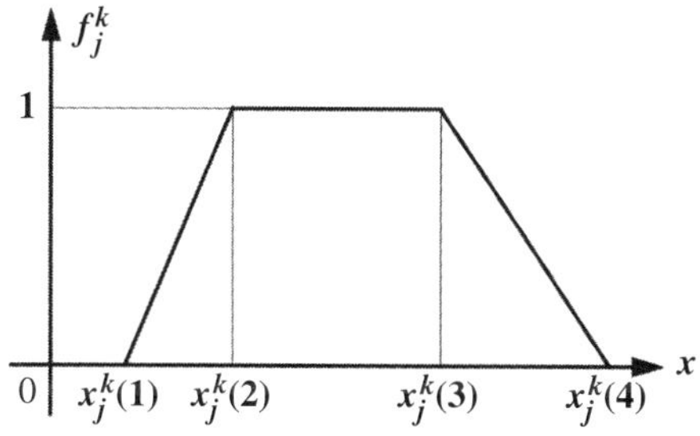

FIGURE 4.2 The typical jth criterion kth subclass whitenization weight function.

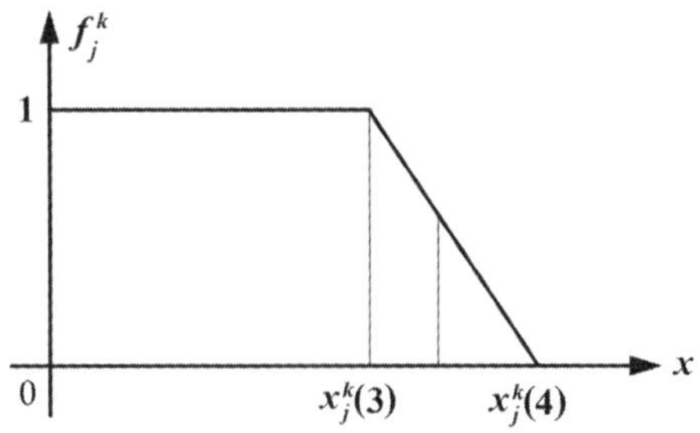

FIGURE 4.3 A whitenization weight function of lower measure.

$f_j^k(\bullet)$ is referred to as a *whitenization weight function of lower measure*, and denoted $f_j^k[-, -, x_j^k(3), x_j^k(4)]$. If the second and third turning points $x_j^k(2)$ and $x_j^k(3)$ of $f_j^k(\bullet)$ coincide, as shown in Figure 4.4, then $f_j^k(\bullet)$ is referred to as a *whitenization weight function of moderate* (or middle) *measure,* denoted $f_j^k[x_j^k(1), x_j^k(2), -, x_j^k(4)]$. If $f_j^k(\bullet)$ does not have the turning points $x_j^k(3)$ and $x_j^k(4)$, as shown in Figure 4.5, then this function is referred to as a *whitenization weight function of upper measure*, denoted $f_j^k[x_j^k(1), x_j^k(2), -, -]$ [18].

For the kth subclass of the jth criterion, if its whitenization weight function is given as in Figures 4.2, 4.3, 4.4 or 4.5, respectively, let $\lambda_j^k = \frac{1}{2}(x_j^k(2) + x_j^k(3))$, $\lambda_j^k = x_j^k(3)$, or $\lambda_j^k = x_j^k(2)$, accordingly. This λ_j^k value is referred to as the *critical value* and $\eta_j^k = \lambda_j^k / \sum_{j=1}^m \lambda_j^k$ the weight of the kth subclass of the jth criterion [18].

Assume that x_{ij} is the observed value of object i with regard to the jth criterion, $f_j^k(\bullet)$ the whitenization weight function, and η_j^k the weight of the kth subclass of the jth criterion. Then $\sigma_i^k = \sum_{j=1}^m f_j^k(x_{ij}) \cdot \eta_j^k$ is referred to as the (grey) cluster coefficient of variable weight for object i to belong to the kth grey class, $\sigma_i = (\sigma_i^1, \sigma_i^2, \ldots, \sigma_i^s)$ the cluster coefficient vector of object i, and $\sum = (\sigma_i^k)_{n\times s}$ the cluster coefficient matrix. If $\max_{1\le k\le s}\{\sigma_i^k\} = \sigma_i^{k^*}$, one then clusters object i into grey class k^*.

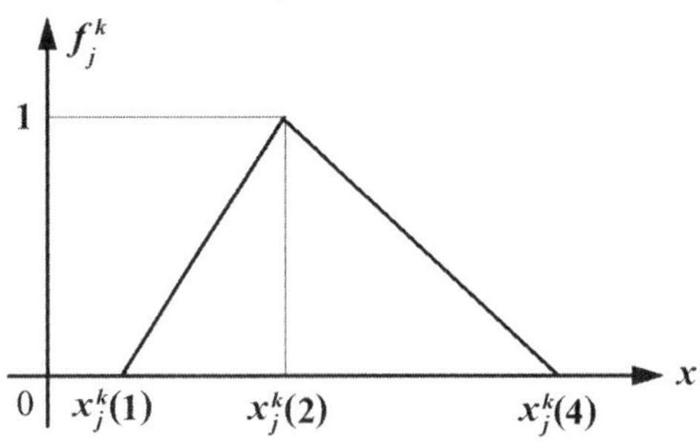

FIGURE 4.4 A whitenization weight function of moderate (or middle) measure.

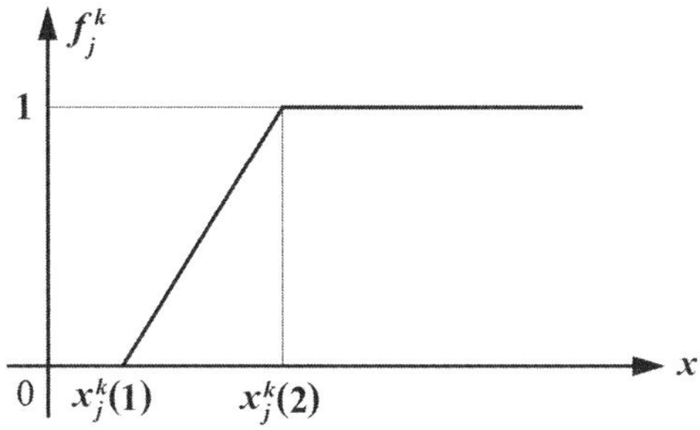

FIGURE 4.5 A whitenization weight function of upper measure.

Variable weight clustering method is useful to study problems with such criteria that have the same meanings and units. Otherwise, it is not appropriate to employ this method. Also, if the numbers of observed values of individual criteria are greatly different from each other, this clustering method should not be applied. In terms of how to determine the whitenization weight function for the kth subclass of the jth criterion, can it generally be done by using the background information of the problem of concern. When resolving practical problems, one can determine the whitenization weight functions from either the angle of the objects that are to be clustered or by looking at all the same type objects in the whole system, not just the ones involved in the clustering. For more details, see References 18 and 23.

4.4.3 Grey Fixed Weight Clustering

When the criteria for clustering have different meanings, dimensions (units), and drastically different numbers of observed data points, applying variable weight clustering method can lead to the problem that the effects of some of the criteria are very weak. There are two ways to solve this problem:

(1) First transform the sample of data values of all the criteria into nondimensional values by applying either the initiating operator or averaging operator; then cluster the transformed data. When employing this method, all the criteria are treated equally so that no difference played by the criteria in the process of clustering is reflected.
(2) Assign each clustering criterion a weight ahead of the clustering process. In this subsection, we will mainly study this second method.

Assume that x_{ij} is the observed value of object i with regard to criterion j, $i \in N_n^+, j \in N_m^+$, and $f_j^k(\bullet)$ the whitenization weight function of the kth subclass of the jth criterion, $k \in N_s^+$, $j \in N_m^+$. If the weight η_j^k of the kth subclass of the jth criterion is not a function of k, $k \in N_s^+, j \in N_m^+$, that is, for any $k_1, k_2 \in N_s^+$, we always have $\eta_j^{k_1} = \eta_j^{k_2}$. Then the symbol η_j^k can be written as η_j with the superscript k removed, $j \in N_m^+$. In this case, $\sigma_i^k = \sum_{j=1}^m f_j^k(x_{ij})\eta_j$ is referred to as the fixed weight clustering coefficient for object i to belong to the kth grey class. Furthermore, $\eta_j = \frac{1}{m}$, $\forall j \in N_m^+$, then $\sigma_i^k = \sum_{j=1}^m f_j^k(x_{ij}) \cdot \eta_j = \frac{1}{m}\sum_{j=1}^m f_j^k(x_{ij})$ is referred to as the *equal-weight clustering coefficient* for object i to belong to the kth grey class. The method of clustering objects by using grey fixed weight clustering coefficients is known as *grey fixed weight clustering*. The method of using grey equal-weight clustering coefficients is known as *grey equal-weight clustering*. Grey fixed weight clustering can be carried out according to the following steps [29, 45]

Step 1: Determine the whitenization weight function $f_j^k(\bullet)$ for the kth subclass of the jth criterion, $k \in N_s^+, j \in N_m^+$.

Step 2: Determine a clustering weight η_j for each criterion $j = 1, 2, \dots, m$.

Step 3: Based on the whitenization weight functions $f_j^k(\bullet)$ obtained in step 1, the clustering weights η_j obtained in step 2, and the observed data value x_{ij} of object i with respect to criterion j, calculate the fixed weight clustering coefficients $\sigma_i^k = \sum_{j=1}^m f_j^k(x_{ij})$, $i \in N_n^+, j \in N_m^+, k \in N_s^+$.

Step 4: If $\max_{1\le k\le s}\{\sigma_i^k\} = \sigma_i^{k^*}$, then object i belongs to grey class k^*.

4.4.4 Grey Evaluation Using Triangular Whitenization Functions

4.4.4.1 Evaluation Model Using Endpoint Triangular Whitenization Functions Assume that n objects are to be clustered into s different grey classes according to m criteria. The observed value of object i in terms of criterion j is x_{ij}, $i \in N_n^+, j \in N_m^+$. One needs to evaluate and diagnose object i based on x_{ij}. The particular computational steps of the *grey evaluation model* based on endpoint triangular whitenization functions are as follows [29, 46]:

Step 1: Based on the predetermined number s of grey classes for the planned evaluation, divide the individual ranges of the criteria into s grey classes. For example, let $[a_1, a_{s+1}]$ be the range of the values of criterion j. Now, divide $[a_1, a_{s+1}]$ into s grey classes as follows, $[a_1, a_2], \dots, [a_{k-1}, a_k], \dots, [a_{s-1}, a_s], [a_s, a_{s+1}]$.

Where $a_k, k \in N_{s+1}^+$, in general, can be determined based on specific requirements of a situation or relevant qualitative analysis.

Step 2: Calculate the geometric midpoint between the various cells, $\lambda_k = (a_k + a_{k+1})/2, k \in N_s^+$.

Step 3: Let the whitenization weight function value for λ_k to belong to the kth grey class be 1.When $(\lambda_k, 1)$ is connected to the starting point λ_{k-1} of the $(k-1)$th class grey and the ending point λ_{k+1} of the $(k+1)$th grey class, one obtains a triangular whitenization weight function $f_j^k(\cdot)$ in terms of criterion j about the kth grey class, $j \in N_m^+, k \in N_s^+$. For $f_j^1(\cdot)$ and $f_j^s(\cdot)$, the range of criterion j can be extended to the left and the right to a_0 and a_{s+2}, respectively (as shown in Figure 4.6).

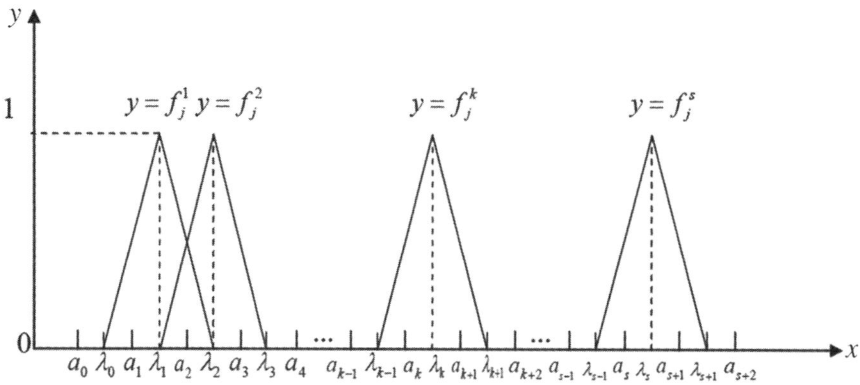

FIGURE 4.6 The general endpoint triangular whitenization weight function.

For an observed value x of criterion j, its degree $f_j^k(x)$ of membership in the kth grey class, $j \in N_m^+, k \in N_s^+$, can be computed as follows:

$$f_j^k(x) = \begin{cases} 0, & x \notin [a_{k-1}, a_{k+2}] \\ \dfrac{x - a_{k-1}}{\lambda_k - a_{k-1}}, & x \in [a_{k-1}, \lambda_k] \\ \dfrac{a_{k+2} - x}{a_{k+2} - \lambda_k}, & x \in [\lambda_k, a_{k+2}] \end{cases} . \tag{4.11}$$

Step 4: Compute the comprehensive clustering coefficient for object i with respect to grey class k:

$$\sigma_i^k = \sum_{j=1}^{m} f_j^k(x_{ij}) \cdot \eta_j, \ i \in N_n^+, k \in N_s^+, \tag{4.12}$$

where $f_j^k(x_{ij})$ is the whitenization weight function of the kth subclass of the jth criterion, and η_j the weight of criterion j in the comprehensive clustering.

Step 5: From $\max_{1 \le k \le s}\{\sigma_i^k\} = \sigma_i^{k^*}$, it follows that object i belongs to grey class k^*. When several objects belong to the same k^* grey class, one can further determine the order of preference of these objects in grey class k^* by using the magnitudes of their clustering coefficients.

4.4.4.2 Evaluation Model Using Center-Point Triangular Whitenization Functions When deciding on the grey classes, the point within a grey class with the maximum degree of greyness is referred to as the center of the class. The specific steps of employing the grey evaluation model based on center-point triangular whitenization weight functions are given below [27, 29]:

Step 1: Based on the number s of grey classes required by the evaluation task, respectively determine the centers $\lambda_1, \lambda_2, \ldots, \lambda_{s-1}$, and λ_s of grey class 1, 2, ..., $s-1$, and s. The centers stand for the particular points for the observed values to belong to particular grey classes; they could be either their individual interval's centers or not). Also, correspondingly partition the field of each criterion into s grey classes, which are represented respectively by using their centers $\lambda_k, k \in N_s^+$.

Step 2: Expand the grey classes in two different directions by adding a 0 and $(s+1)$ grey classes with their centers λ_0 and λ_{s+1} determined. Thus, we have a new sequence of centers: $\lambda_0, \lambda_1, \lambda_2, \ldots, \lambda_s$ and λ_{s+1}. By respectively connecting the point $(\lambda_k, 1)$ with the center $(\lambda_{k-1}, 0)$ and $(\lambda_{k+1}, 0)$ of the $(k-1)$th and the $(k+1)$th small grey classes, we obtain the triangular whitenization weight function $f_j^k(\cdot)$ for the kth grey class of the jth criterion, $j \in N_m^+$, $k \in N_s^+$, as shown in Figure 4.7.

For an observed value x of criterion j, one can employ the formula

$$f_j^k(x) = \begin{cases} 0, & x \notin [\lambda_{k-1}, \lambda_{k+1}] \\ \dfrac{x - \lambda_{k-1}}{\lambda_k - \lambda_{k-1}}, & x \in (\lambda_{k-1}, \lambda_k] \\ \dfrac{\lambda_{k+1} - x}{\lambda_{k+1} - \lambda_k}, & x \in (\lambda_k, \lambda_{k+1}) \end{cases} \tag{4.13}$$

to compute its degree of membership $f_j^k(x)$ in grey class k, $k \in N_s^+$.

Step 3: Compute the comprehensive clustering coefficient σ_i^k for object i, $i \in N_n^+$, with respect to grey class k, $k \in N_s^+$, $\sigma_i^k = \sum_{j=1}^m f_j^k(x_{ij}) \cdot \eta_j$, where $f_j^k(x_{ij})$ is the whitenization weight function of the kth subclass of the jth criterion, and η_j the weight of criterion j in the comprehensive clustering.

Step 4: From $\max_{1\le k\le s}\{\sigma_i^k\} = \sigma_i^{k^*}$, it is decided that object i belongs to grey class k^*. When there are several objects in grey class k^*, these objects can be ordered according to the magnitudes of their comprehensive clustering coefficients.

The gray cluster evaluation model based on end-point triangular whitenization weight functions is applicable to the case that the boundary of each grey class is clear but the most possible point of each grey class is unknown; the gray cluster evaluation based on center-point triangular whitenization weight functions is applicable to the case that the most possible point of each grey class is known but the boundary of each grey class is unclear.

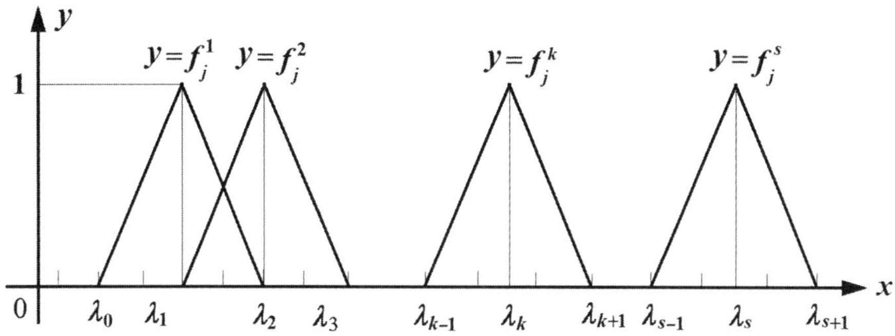

FIGURE 4.7 Center-point whitenization weight functions.

4.4.5 Practical Applications

All the degrees of grey incidence presented earlier are about how closely related two sequences are. For any chosen grey incidence operator, the values of the absolute and relative degrees of grey incidence are uniquely determined. When the grey incidence operator and the θ-value are fixed, the synthetic degree of grey incidence is also unique. These kinds of conditional uniqueness do not affect how investigators analyze their problems. As a matter of fact, when analyzing a system, they investigate the relationship between the system's characteristic of interest and relevant factors. His main focus is on ordering the degrees of incidence between the characteristic sequence and the individual factor sequences without much attention given to the specific magnitudes of the degrees of incidence.

Let X_0 be a system's characteristic sequence, $X = \{X_1, X_2, \ldots, X_m\}$ a set of relevant factor sequences, and γ the degree of grey incidence. If $\gamma_{0i} \geq \gamma_{0j}$, factor X_i is said to be superior to factor X_j, written $X_i \succ X_j$. The order relation "$\succ$" is known as the *grey incidence order* derived from degrees of grey incidence. That is, the general grey incidence order contains such special cases as *grey absolute incidence order*, *grey relative incidence order*, and *grey synthetic incidence order*. It can be shown [23] that grey incidence order, absolute incidence order, relative incidence order, and synthetic incidence order are all partial ordering on the set X. Furthermore, if the factor sequences in X all have the same length as X_0, then both grey incidence order and grey absolute incidence order are linear orderings on X. If the initial values of all the sequences are nonzero, then both grey relative incidence order and grey synthetic incidence order are also linear orders on X.

Assume that $Y_1, Y_2, \ldots$, and Y_s are a system's characteristic behavioral sequences and $X_1, X_2, \ldots$, and X_m relevant factor behavioral sequences with the same length. Let γ_{ij} be the degree of grey incidence between Y_i and X_j, $i \in N_s^+$, $j \in N_m^+$. Then $\Gamma = (\gamma_{ij})_{s\times m}$ is known as the *grey incidence matrix* of the system, where the ith row is made up of the degrees of grey incidence between the characteristic sequence $Y_i (i \in N_s^+)$ and each of the factor sequences $X_1, X_2, \ldots$, and X_m; and the jth column the degrees of grey incidences between each of the characteristic sequences $Y_1, Y_2, \dots,$ and Y_s and $X_j (j \in N_m^+)$.

Similarly, general grey incidence matrices, such as the grey *absolute incidence matrix* $A = (\varepsilon_{ij})_{s\times m}$, the grey *relative incidence matrix* $B = (r_{ij})_{s\times m}$, and the grey *synthetic incidence matrix* $C = (\rho_{ij})_{s\times m}$ are defined.

Now, for the grey incidence matrix $\Gamma = (\gamma_{ij})_{s\times m}$, if there are $k, i \in N_s^+$ such that $\gamma_{kj} \geq \gamma_{ij}$, $\forall j \in N_m^+$, then the system's characteristic Y_k is said to be more favorable than the system's characteristic Y_i, written $Y_k \succ Y_i$. If $\forall i \in N_s^+, i \neq k$, $Y_k \succ Y_i$ always holds true, then Y_k is said to be the *most favorable characteristic*. In addition, if there are $l, j \in N_m^+$ such that $\gamma_{il} \geq \gamma_{ij}$, $\forall i \in N_s^+$, then we say that the system's factor X_l is more favorable than the factor X_j, written $X_l \succ X_j$. If $\forall j \in N_m^+, j \neq l$, $X_l \succ X_j$ always holds true, then is said to be the *most favorable factor*.

If there are $k, i \in N_s^+$ satisfying $\sum_{j=1}^{m} \gamma_{kj} \geq \sum_{j=1}^{m} \gamma_{ij}$, then the system's characteristic Y_k is said to be more quasi-favorable than Y_i, denoted $Y_k \underline{\succ} Y_i$. If there are $l, j \in N_m^+$ satisfying $\sum_{i=1}^{m} \gamma_{il} \geq \sum_{i=1}^{m} \gamma_{ij}$, then the system's factor X_l is said to be more *quasi-favorable* than X_j, denoted $X_l \underline{\succ} X_j$.

If there is $i \in N_s^+$ such that $\forall i \in N_s^+, i \neq k$, $Y_k \underline{\succ} Y_i$, then the system's characteristic is said to be quasi-preferred. If there is $l \in N_m^+$ such that $\forall j \in N_m^+, j \neq l$, $X_l \underline{\succ} X_j$, then the system's factor is said to be a *quasi-preferred.*

Proposition 3 In a system of s characteristics and m relevant factors, there may not be any most favorable characteristic and most favorable factor. However, there must be quasi-preferred characteristic and factor.

When practical problems are investigated, the analyses of the three incidence orders may not provide cohesive conclusions. It is because the absolute incidence order looks at the relationship between absolute quantities, the relative incidence order focuses on the rates of change with respect to the initial values of the observed sequences, while the synthetic incidence order combines both the relationships between absolute quantities and between the rates of change. Considering the background of the problem of concern, one can choose one of the incidence orders. For the purpose of simplicity, after a particular grey incidence operator is applied to the system's characteristic behavioral sequences and relevant factor sequences, one only needs to employ the absolute incidence order on the processed data. Please see Reference 23 for detailed case studies.

4.5 GREY PREDICTION MODELS

4.5.1 GM(1,1) Model

Let $X^{(0)} = (x^{(0)}(1), x^{(0)}(2), \dots, x^{(0)}(n))$ and $X^{(1)} = (x^{(1)}(1), x^{(1)}(2), \dots, x^{(1)}(n))$ be respectively a sequence of raw data and its accumulation-generated sequence. Then $x^{(0)}(k) + ax^{(1)}(k) = b$ is referred to as the *original form* of the GM(1,1) model, where the symbol GM(1,1) stands for "first-order grey model in one variable."

Let $Z^{(1)} = (z^{(1)}(2), z^{(1)}(3), \ldots, z^{(1)}(n))$ be the sequence generated from $X^{(1)}$ by adjacent neighbor means, where $z^{(1)}(k) = \frac{1}{2}(x^{(1)}(k) + x^{(1)}(k-1)), k = 2, 3, \ldots, n$. Then, $x^{(0)}(k) + az^{(1)}(k) = b$ is referred to as the *basic form* of the GM(1,1) model [18].

Theorem 13 Assume that $X^{(0)}$ is nonnegative. If $\hat{a} = (a, b)^{\mathrm{T}}$ is a sequence of parameters, and

$$Y = \begin{bmatrix} x^{(0)}(2) \\ x^{(0)}(3) \\ \vdots \\ x^{(0)}(n) \end{bmatrix}, \quad B = \begin{bmatrix} -z^{(1)}(2) & 1 \\ -z^{(1)}(3) & 1 \\ \vdots & \vdots \\ -z^{(1)}(n) & 1 \end{bmatrix}, \tag{4.14}$$

then the least square estimate sequence of the basic form GM(1,1) model satisfies $\hat{a} = (B^T B)^{-1} B^T Y$.

For the proof, see References 18 and 23.

Continuing all the notations from Theorem 13, if $[a, b]^{\mathrm{T}} = (B^{\mathrm{T}} B)^{-1} B^{\mathrm{T}} Y$, then $\frac{dx^{(1)}}{dt} + ax^{(1)} = b$ is known as a *whitenization* (or *image*) *equation* of the GM(1,1) model of the basic form.

Theorem 14 Let B, Y, and $\hat{a}$ be the same as those in Theorem 13. If $\hat{a} = [a, b]^T = (B^T B)^{-1} B^T Y$, then

(1) The solution, also known as a time response function, of the whitenization equation $\frac{dx^{(1)}}{dt} + ax^{(1)} = b$ is given by $x^{(1)}(t) = (x^{(1)}(1) - \frac{b}{a}) e^{-at} + \frac{b}{a}$.
(2) The time response sequence of the GM(1,1) model of the basic form is given below:

$$\hat{x}^{(1)}(k+1) = \left(x^{(0)}(1) - \frac{b}{a}\right) e^{-ak} + \frac{b}{a}, \quad k \in N_n^+.$$

(3) The restored values of $x^{(0)}(k)$'s are given as follows:

$$\begin{aligned} \hat{x}^{(0)}(k+1) &= \alpha^{(1)} \hat{x}^{(1)}(k+1) = \hat{x}^{(1)}(k+1) - \hat{x}^{(1)}(k) \\ &= (1 - e^{a}) \left(x^{(0)}(1) - \frac{b}{a}\right) e^{-ak}, \quad k \in N_n^+. \end{aligned}$$

4.5.2 Improvements on GM(1,1) Models

When a GM(1,1) model does not meet the predetermined accuracy requirement, one can establish another GM(1,1) model using the error sequence to remedy the original model to improve the accuracy. In particular, let $X^{(0)}$ be a sequence of raw data,

$X^{(1)}$ be the accumulation-generated sequence based on $X^{(0)}$, and the time response formula of the GM(1,1) model be

$$\hat{x}^{(1)}(k+1) = \left(x^{(0)}(1) - \frac{b}{a}\right) e^{-ak} + \frac{b}{a}.$$

Then

$$d\hat{x}^{(1)}(k+1) = (-a)\left(x^{(0)}(1) - \frac{b}{a}\right) e^{-ak} \tag{4.15}$$

is known as the *restored value* through derivatives. Because in general $d\hat{x}^{(1)}(k+1) \neq \hat{x}^{(0)}(k+1)$, $\hat{x}^{(0)}(k+1) = \hat{x}^{(1)}(k+1) - \hat{x}^{(1)}(k)$ stands for the restored value through inverse accumulation. It implies that the GM(1,1) is neither a differential equation nor a difference equation. And, when $|a|$ is sufficiently small, the fact that $1 - e^{a} \approx -a$ implies that $d\hat{x}^{(1)}(k+1) \approx \hat{x}^{(0)}(k+1)$, meaning that the results of differentiation and difference are quite close. Therefore, in this case the GM(1,1) model can be seen as both a differential equation and a difference equation.

To reduce the possible errors caused by reciprocating operators, the errors of $X^{(1)}$ are often used to improve the simulated values $\hat{x}^{(1)}(k+1)$ of $X^{(1)}$. To this end, assume that $\varepsilon^{(0)} = (\varepsilon^{(0)}(1), \varepsilon^{(0)}(2), \ldots, \varepsilon^{(0)}(n))$ is the error sequence of $X^{(1)}$, where $\varepsilon^{(0)}(k) = x^{(1)}(k) - \hat{x}^{(1)}(k)$. If there is k_0 satisfying that $n - k_0 \geq 4$ and $\forall k \geq k_0$, the signs of $\varepsilon^{(0)}(k)$ stay the same, then $(|\varepsilon^{(0)}(k_0)|, |\varepsilon^{(0)}(k_0+1)|, \ldots, |\varepsilon^{(0)}(n)|)$ is known as the *error sequence of modelability*, still denoted as $\varepsilon^{(0)} = (\varepsilon^{(0)}(k_0), \varepsilon^{(0)}(k_0+1), \ldots, \varepsilon^{(0)}(n))$. In this case, let $\varepsilon^{(1)} = (\varepsilon^{(1)}(k_0), \varepsilon^{(1)}(k_0+1), \ldots, \varepsilon^{(1)}(n))$ be accumulation generated on $\varepsilon^{(0)}$ with the following GM(1,1) time response formula

$$\hat{\varepsilon}^{(1)}(k+1) = \left(\varepsilon^{(0)}(k_0) - \frac{b_\varepsilon}{a_\varepsilon}\right) \exp\left[-a_\varepsilon(k-k_0)\right] + \frac{b_\varepsilon}{a_\varepsilon}, \ k \geq k_0.$$

Then the simulation sequence of $\varepsilon^{(0)}$ is given by $\hat{\varepsilon}^{(0)} = (\hat{\varepsilon}^{(0)}(k_0), \hat{\varepsilon}^{(0)}(k_0+1), \ldots, \hat{\varepsilon}^{(0)}(n))$, where

$$\hat{\varepsilon}^{(0)}(k+1) = (-a_\varepsilon)\left(\varepsilon^{(0)}(k_0) - \frac{b_\varepsilon}{a_\varepsilon}\right) \exp\left[-a_\varepsilon(k-k_0)\right], \ k \geq k_0.$$

If $\hat{\varepsilon}^{(0)}$ is used to improve $\hat{X}^{(1)}$, the modified time response formula

$$\hat{x}^{(1)}(k+1) = \begin{cases} \left(x^{(0)}(1) - \frac{b}{a}\right) e^{-ak} + \frac{b}{a}, & k < k_0 \\ \left(x^{(0)}(1) - \frac{b}{a}\right) e^{-ak} + \frac{b}{a} \pm a_\varepsilon \left(\varepsilon^{(0)}(k_0) - \frac{b_\varepsilon}{a_\varepsilon}\right) e^{-a_s(k-k_0)}, & k \geq k_0 \end{cases} \tag{4.16}$$

is known as the GM(1,1) *model with error modification* or *remnant* GM(1,1) for short, where the sign of the error modification value

$$\hat{\varepsilon}^{(0)}(k+1) = a_\varepsilon \times \left(\hat{\varepsilon}^{(0)}(k_0) - \frac{b_\varepsilon}{a_\varepsilon}\right) \exp\left[-a_\varepsilon(k-k_0)\right]$$

needs to stay the same as those in $\varepsilon^{(0)}$.

If a model of the error sequence $\varepsilon^{(0)} = (\varepsilon^{(0)}(k_0), \varepsilon^{(0)}(k_0+1), \ldots, \varepsilon^{(0)}(n))$ of $X^{(0)}$ and $\hat{X}^{(0)}$ is used to modify the simulation value $\hat{X}^{(0)}$, then different methods of restoration from $\hat{X}^{(1)}$ to $\hat{X}^{(0)}$ can produce different time response sequences of error modification. In particular, if

$$\hat{x}^{(0)}(k) = \hat{x}^{(1)}(k) - \hat{x}^{(1)}(k-1) = \left(1-e^{a}\right)\left(x^{(0)}(1) - \frac{b}{a}\right)e^{-a(k-1)},$$

then the corresponding time response sequence of error modification

$$\hat{x}^{(0)}(k+1) = \begin{cases} (1-e^{a})\left(x^{(0)}(1) - \dfrac{b}{a}\right)e^{-ak}, & k < k_0 \\ (1-e^{a})\left(x^{(0)}(1) - \dfrac{b}{a}\right)e^{-ak} \pm a_\varepsilon\left(\varepsilon^{(0)}(k_0) - \dfrac{b_\varepsilon}{a_\varepsilon}\right)e^{-a_\varepsilon(k-k_0)}, & k \geq k_0 \end{cases}$$

is known as the *error modification model* of inverse accumulation restoration.If $\hat{x}^{(0)}(k+1) = (-a)(x^{(0)}(1) - \frac{b}{a})e^{-ak}$, the corresponding time response sequence of error modification

$$\hat{x}^{(0)}(k+1) = \begin{cases} (-a)\left(x^{(0)}(1) - \dfrac{b}{a}\right)e^{-ak}, & k < k_0 \\ (-a)\left(x^{(0)}(1) - \dfrac{b}{a}\right)e^{-ak} \pm a_\varepsilon\left(\varepsilon^{(0)}(k_0) - \dfrac{b_\varepsilon}{a_\varepsilon}\right)e^{-a_\varepsilon(k-k_0)}, & k > k_0 \end{cases}$$

is known as the *error modification model* of derivative restoration.

In the previous discussion, all the error simulation terms in remnant GM(1,1) have been taken as the derivative restoration. Of course, they can be taken as inverse accumulation restoration. As long as $|a_\varepsilon|$ is sufficiently small, the effects of different error restoration methods on the modified $\hat{x}^{(0)}(k+1)$ are almost the same.

In practical applications, one does not have to use all the available data in the modeling. Each subsequence of the original data can be employed to establish a model. Generally, different subsequences lead to different models [47, 48]. Even when a same kind GM(1,1) is applied, different subsequences lead to different a and b values. These changes reflect the fact that varied circumstances and conditions have different effect on the system of our concern [49, 50]. For a given sequence $X^{(0)} = (x^{(0)}(1), x^{(0)}(2), \ldots, x^{(0)}(n))$, if $x^{(0)}(n)$ is the origin of the time axis, then $t < n$ is the past, $t = n$ the present, and $t > n$ the future. If $\hat{x}^{(0)}(k+1) = (1-e^{a})(x^{(0)}(1) - \frac{b}{a})e^{-ak}$ is the restored values of inverse accumulation of the GM(1,1) time responses of $X^{(0)}$, then for $t \leq n$, $\hat{x}^{(0)}(t)$ is referred to as the *simulated value*; $\hat{x}^{(0)}(t)$ the *prediction* one when $t > n$. The main purpose of modeling is to make predictions. To improve the prediction accuracy, one might want to have sufficiently high accuracy in simulation, especially for the moment $t = n$. In general, the data, including $x^{(0)}(n)$, used for modeling should be an *equal time interval sequence*.

For the given sequence $X^{(0)}$, the GM(1,1) model established using the entire sequence $X^{(0)}$ is known as *all-data* GM(1,1). For $\forall k_0 > 1$, the GM(1,1) model established on the tail sequence $X^{(0)} = (x^{(0)}(k_0), x^{(0)}(k_0+1), \ldots, x^{(0)}(n))$ is known as *partial-data* GM(1,1). If $x^{(0)}(n+1)$ stands for a piece of new information, then the GM(1,1) model established on the prolonged sequence $X^{(0)} = (x^{(0)}(1), x^{(0)}(2), \ldots, x^{(0)}(n), x^{(0)}(n+1))$ is known as *new information* GM(1,1); the GM(1,1) model established on $X^{(0)} = ((x^{(0)}(2), \ldots, x^{(0)}(n), x^{(0)}(n+1))$ with the new information added and the oldest piece $x^{(0)}(1)$ of information removed is known as *metabolic* GM(1,1). For detailed empirical analysis of these GM(1,1) models, please see References 18 and 23.

4.5.3 Applicable Ranges of GM(1,1) Models

Studies have revealed the fact that GM(1,1) only holds true conditionally. In particular, it is shown [24, 51] that

(1) when $(n-1)\sum_{k=2}^{n}[z^{(1)}(k)]^2 \to [\sum_{k=2}^{n} z^{(1)}(k)]^2$, the GM(1,1) becomes invalid; and

(2) when the development coefficient a of the GM(1,1) model satisfies $|a| \geq 2$, the GM(1,1) model becomes invalid.

4.5.4 Discrete Grey Models

Any equation of the form $x^{(1)}(k+1) = \beta_1 x^{(1)}(k) + \beta_2$ is known as a *discrete grey model* or a *discretization of the GM(1,1) model*.

Theorem 15 Let $X^{(0)} = \{x^{(0)}(1), x^{(0)}(2), \ldots, x^{(0)}(n)\}$ and $X^{(1)} = \{x^{(1)}(1), x^{(1)}(2), \ldots, x^{(1)}(n)\}$ be respectively a nonnegative sequence and its accumulation generation. If $\hat{\beta} = (\beta_1, \beta_2)^{\mathrm{T}}$ is the parametric sequence and

$$Y = \begin{bmatrix} x^{(1)}(2) \\ x^{(1)}(3) \\ \vdots \\ x^{(1)}(n) \end{bmatrix}, \quad B = \begin{bmatrix} x^{(1)}(1) & 1 \\ x^{(1)}(2) & 1 \\ \vdots & \vdots \\ x^{(1)}(n-1) & 1 \end{bmatrix},$$

then the least square estimates of the parameters of the discrete model $x^{(1)}(k+1) = \beta_1 x^{(1)}(k) + \beta_2$ satisfy $\hat{\beta} = (B^{\mathrm{T}}B)^{-1}B^{\mathrm{T}}Y$.

Theorem 16 Let B, Y, and $\hat{\beta}$ be the same as those defined in Theorem 15, and $\hat{\beta} = [\beta_1, \beta_2] = (B^T B)^{-1} B^T Y$.

(1) If $x^{(1)}(1) = x^{(0)}(1)$, the recurrence relation is

$$\hat{x}^{(1)}(k+1) = \beta_1^k x^{(0)}(1) + \frac{1-\beta_1^k}{1-\beta_1} * \beta_2; k \in N_{n-1}^+;$$

or

$$\hat{x}^{(1)}(k+1) = \beta_1^k \left(x^{(0)}(1) - \frac{\beta_2}{1-\beta_1} \right) + \frac{\beta_2}{1-\beta_1}; k \in N_{n-1}^+.$$

(2) The restored values are

$$\hat{x}^{(0)}(k+1) = \alpha^{(1)} \hat{x}^{(1)}(k+1) = \hat{x}^{(1)}(k+1) - \hat{x}^{(1)}(k); k \in N_{n-1}^+.$$

For the proofs of these results, see References 24 and 52.

The discrete grey model, as studied in the previous section, uses $x^{(0)}(1)$ as the basis of iteration. However, in practical applications, this basis of iteration can be adjusted according to the specific circumstances involved in order to improve the degree of accuracy. As a matter of fact, any term in the given sequence can be employed as the basis of iteration. According to different choices of the initial value of iteration, there are three different kinds of discrete grey models. In particular, assume that the observed values of a system's behavioral characteristic are given in a sequence $X^{(0)} = \{x^{(0)}(1), x^{(0)}(2), \dots, x^{(0)}(n)\}$, its accumulation generation is $X^{(1)} = \{x^{(1)}(1), x^{(1)}(2), \dots, x^{(1)}(n)\}$. Based on different choices of the iteration basis, that is the sequence data value that is assumed at the time of modeling to be the same as the simulated value, the consequent discrete grey models can take one of the following three forms.

(1) $\begin{cases} \hat{x}^{(1)}(k+1) = \beta_1 \hat{x}^{(1)}(k) + \beta_2 \\ \hat{x}^{(1)}(1) = x^{(1)}(1) = x^{(0)}(1) \end{cases}$,

where $\hat{x}^{(1)}(k)$ stands for a simulated value of the original sequence value with β_1 and β_2 to be determined and $\hat{x}^{(1)}(1)$ the basis of iteration. This form of the discrete grey model is referred to that *with fixed starting points*, written in short as *starting points discrete grey model* (SPDGM).

(2) $\begin{cases} \hat{x}^{(1)}(k+1) = \beta_1 \hat{x}^{(1)}(k) + \beta_2 \\ \hat{x}^{(1)}(m) = x^{(1)}(m) = \sum_{i=1}^{m} x^{(0)}(i), \quad 1 < \mathrm{m} < n \end{cases}$,

where $\hat{x}^{(1)}(k)$, $k \in N_n^+$, stands for a simulated value of the original sequence with β_1 and β_2 to be determined and $\hat{x}^{(1)}(m)$ the basis of iteration. This kind of the discrete grey model is referred to as that *with a middle point fixed*, denoted in short as *middle point discrete grey model* (MPDGM).

(3) $\begin{cases} \hat{x}^{(1)}(k+1) = \beta_1 \hat{x}^{(1)}(k) + \beta_2 \\ \hat{x}^{(1)}(n) = x^{(1)}(n) = \sum_{i=1}^{n} x^{(0)}(i) \end{cases}$,

where $\hat{x}^{(1)}(k)$, $k = 1, 2, \ldots, n$, stand for simulated values of the original sequence with β_1 and β_2 to be determined and $\hat{x}^{(1)}(n)$ the basis of iteration. This kind of the discrete grey model is referred to as that *with the end point fixed*, denoted in short as *end point discrete grey model* (EPDGM).

4.5.5 GM(r,h) Models

Let $X_i^{(0)} = (x_i^{(0)}(1), x_i^{(0)}(2), \ldots, x_i^{(0)}(n))$, $i \in N_N^+$, be given, where $X_1^{(0)}$ stands for a sequence of a system's characteristic data, $X_i^{(0)}$ data sequences of relevant factors, $X_i^{(1)}$ the accumulated generation of $X_i^{(0)}$, $i \in N_N^+$, and $Z_1^{(1)}$ the adjacent neighbor mean generated sequence of $X_1^{(1)}$. Then,

$$x_1^{(0)}(k) + az_1^{(1)}(k) = \sum_{i=2}^{N} b_i x_i^{(1)}(k) \tag{4.17}$$

is known as the GM(1,*N*) *model.* The constant $(-a)$ is known as the system's development coefficient, $b_i x_i^{(1)}(k)$ the *driving term*, b_i the *driving coefficient*, and $\hat{a} = [a, b_1, b_2, \ldots, b_N]^{\mathrm{T}}$ the sequence of parameters.

Theorem 17 For the previously defined terms $X_i^{(0)}$, $X_i^{(1)}$, and $Z_1^{(1)}$, $i \in N_N^+$, let

$$B = \begin{bmatrix} -z_1^{(1)}(2) & x_2^{(1)}(2) & \cdots & x_N^{(1)}(2) \\ -z_1^{(1)}(3) & x_2^{(1)}(3) & \cdots & x_N^{(1)}(3) \\ \cdots & \cdots & \ddots & \cdots \\ -z_1^{(1)}(n) & x_2^{(1)}(n) & \cdots & x_N^{(1)}(n) \end{bmatrix}, \quad Y = \begin{bmatrix} x_1^{(0)}(2) \\ x_1^{(0)}(3) \\ \vdots \\ x_1^{(0)}(n) \end{bmatrix},$$

then the least square estimate of the sequence $\hat{a} = [a, b_1, b_2, \ldots, b_N]^{\mathrm{T}}$ of parameters satisfies

$$\hat{a} = (B^{\mathrm{T}}B)^{-1}B^{\mathrm{T}}Y.$$

For $\hat{a} = [a, b_1, b_2, \ldots, b_N]^{\mathrm{T}}$,

$$\frac{dx^{(1)}}{dt} + ax_1^{(1)} = b_2 x_2^{(1)} + b_3 x_3^{(1)} + \cdots + b_N x_N^{(1)} \tag{4.18}$$

is known as the *whitenization* (or *image*) *equation* of the GM(1,*N*) model in (4.17).

Theorem 18 Let $X_i^{(0)}$, $X_i^{(1)}$, $Z_1^{(1)}$, B, and Y, $i \in N_N^+$, be defined as in Theorem 17, and $\hat{a} = [a, b_1, b_2, \ldots, b_N]^{\mathrm{T}} = (B^{\mathrm{T}}B)^{-1}B^{\mathrm{T}}Y$. Then,

(1) The solution of the whitenization equation $\frac{dx_1^{(1)}}{dt} + ax_1^{(1)} = \sum_{i=2}^{N} b_i x_i^{(1)}$ is

$$x^{(1)}(t) = e^{-at}\left[\sum_{i=2}^{N}\int b_i x_i^{(1)}(t)e^{at}dt + x^{(1)}(0) - \sum_{i=2}^{N}\int b_i x_i^{(1)}(0)dt\right]$$
$$= e^{-at}\left[x_1^{(1)}(0) - t\sum_{i=2}^{N} b_i x_i^{(1)}(0) + \sum_{i=2}^{N}\int b_i x_i^{(1)}(t)e^{at}dt\right]. \quad (4.19)$$

(2) When $X_i^{(1)}$, $i \in N_N^+$, vibrate in small magnitudes, $\sum_{i=2}^{N} b_i x_i^{(1)}(k)$ can be seen as a grey constant. In this case, an approximate time response sequence of the GM(1,N) model in (4.17) is given by

$$\hat{x}_1^{(1)}(k+1) = (x_1^{(1)}(0) - \frac{1}{a}\sum_{i=2}^{N} b_i x_i^{(1)}(k+1))e^{-ak} + \frac{1}{a}\sum_{i=2}^{N} b_i x_i^{(1)}(k+1), \quad (4.20)$$

where $x_1^{(1)}(0)$ is taken as $x_1^{(0)}(1)$.

(3) The restored sequence of inverse accumulation is

$$\hat{x}_1^{(0)}(k+1) = \alpha^{(1)}\hat{x}_1^{(1)}(k+1) = \hat{x}_1^{(1)}(k+1) - \hat{x}_1^{(1)}(k).$$

(4) The GM(1,N) difference model is $\hat{x}_1^{(0)}(k) = -az_1^{(1)}(k) + \sum_{i=2}^{N} b_i \hat{x}_i^{(1)}(k)$.

Let $X_1^{(0)}$ be a data sequence of a system's characteristic, $X_i^{(0)}$ the data sequences of relevant factors, and $X_i^{(1)}$ the accumulation-generated sequence of $X_i^{(0)}$, $i \in N_N^+$-{1}, then

$$x_1^{(1)}(k) = a + b_2 x_2^{(1)}(k) + b_3 x_3^{(1)}(k) + \cdots + b_N x_N^{(1)}(k) \quad (4.21)$$

is known as the GM(0,N) *model*. Although its form looks like a multivariate linear regression model, it is essentially different from any of the statistical models. In particular, the general multivariate linear regression model is established on the basis of the original data sequences, while the GM(0,N) model is constructed on the accumulation generation of the original data.

Theorem 19 With the same sequences $X_i^{(0)}$ and $X_i^{(1)}$ as given above, let

$$B = \begin{bmatrix} 1 & x_2^{(1)}(2) & x_3^{(1)}(2) & \cdots & x_N^{(1)}(2) \\ 1 & x_2^{(1)}(3) & x_3^{(1)}(3) & \cdots & x_N^{(1)}(3) \\ \cdots & \cdots & \cdots & \cdots & \cdots \\ 1 & x_2^{(1)}(n) & x_3^{(1)}(n) & \cdots & x_N^{(1)}(n) \end{bmatrix}, \quad Y = \begin{bmatrix} x_1^{(1)}(2) \\ x_1^{(1)}(3) \\ \vdots \\ x_1^{(1)}(n) \end{bmatrix},$$

then the least square estimate of the parametric sequence $\hat{a} = [a, b_1, b_2, \ldots, b_N]^{\mathrm{T}}$ is given by $\hat{a} = (B^{\mathrm{T}}B)^{-1}B^{\mathrm{T}}Y$.

For a given sequence of raw data $X^{(0)} = (x^{(0)}(1), x^{(0)}(2), \ldots, x^{(0)}(n))$, let its sequences of accumulation generation and inverse accumulation generation, respectively, be $X^{(1)} = (x^{(1)}(1), x^{(1)}(2), \ldots, x^{(1)}(n))$ and $\alpha^{(1)}X^{(0)} = (\alpha^{(1)}x^{(0)}(2), \ldots, \alpha^{(1)}x^{(0)}(n))$, where $\alpha^{(1)}x^{(0)}(k) = x^{(0)}(k) - x^{(0)}(k-1)$, $i \in N_n^+ - \{1\}$, and the sequence of adjacent neighbor mean generation of $X^{(1)}$ be $Z^{(1)} = (z^{(1)}(2), z^{(1)}(3), \ldots, z^{(1)}(n))$. Then,

$$\alpha^{(1)}x^{(0)}(k) + a_1x^{(0)}(k) + a_2z^{(1)}(k) = b \tag{4.22}$$

is known as the GM(2,1) *model*; and

$$\frac{d^2x^{(1)}}{dt^2} + \alpha_1\frac{dx^{(1)}}{dt} + \alpha_2x^{(1)} = b \tag{4.23}$$

the *whitenization equation* of the GM(2,1) model.

Theorem 20 For the sequences $X^{(0)}$, $X^{(1)}$, $Z^{(1)}$, and $\alpha^{(1)}X^{(0)}$, as defined above, let

$$B = \begin{bmatrix} -x^{(0)}(2) & -z^{(1)}(2) & 1 \\ -x^{(0)}(3) & -z^{(1)}(3) & 1 \\ \cdots & \cdots & \cdots \\ -x^{(0)}(n) & -z^{(1)}(n) & 1 \end{bmatrix}, \quad Y = \begin{bmatrix} \alpha^{(1)}x^{(0)}(2) \\ \alpha^{(1)}x^{(0)}(3) \\ \cdots \\ \alpha^{(1)}x^{(0)}(n) \end{bmatrix} = \begin{bmatrix} x^{(0)}(2) - x^{(0)}(1) \\ x^{(0)}(3) - x^{(0)}(2) \\ \cdots \\ x^{(0)}(n) - x^{(0)}(n-1) \end{bmatrix}$$

then the least square estimate of the parametric sequence $\hat{a} = [a_1, a_2, b]^{\mathrm{T}}$ of the GM(2,1) is given as $\hat{a} = (B^{\mathrm{T}}B)^{-1}B^{\mathrm{T}}Y$.

Theorem 21 For the solution of the GM(2,1) whitenization equation, the following hold true:

(1) If $X^{(1)*}$ is a special solution of $\frac{d^2x^{(1)}}{dt^2} + \alpha_1\frac{dx^{(1)}}{dt} + \alpha_2x^{(1)} = b$ and $\overline{X}^{(1)}$ the general solution of the corresponding homogeneous equation $\frac{d^2x^{(1)}}{dt^2} + \alpha_1\frac{dx^{(1)}}{dt} +$

$\alpha_2 x^{(1)} = 0$, then $X^{(1)^*} + \overline{X}^{(1)}$ is the general solution of the GM(2,1) whitenization equation.

There are the following three cases for the general solution of the homogeneous equation above:

(i) when the characteristic equation $r^2 + \alpha_1 r + \alpha_2 = 0$ has two distinct real roots r_1, r_2,

$$\overline{X}^{(1)} = c_1 e^{r_1 t} + c_2 e^{r_2 t}; \tag{4.24}$$

(ii) when the characteristic equation has a repeated root r,

$$\overline{X}^{(1)} = e^{rt}(c_1 + c_2 t); \tag{4.25}$$

(iii) when the characteristic equation has two complex conjugate roots $r_1 = \alpha + i\beta$ and $r_2 = \alpha - i\beta$,

$$\overline{X}^{(1)} = e^{\alpha t}(c_1 \cos \beta t + c_2 \sin \beta t). \tag{4.26}$$

(2) A special solution of the whitenization equation may take one of the three possibilities: (i) when 0 is not a root of the characteristic equation, $X^{(1)^*} = C$; (ii) when 0 is one of the two distinct roots of the characteristic equation, $X^{(1)^*} = Cx$; and (iii) when 0 is the only root of the characteristic equation, $X^{(1)^*} = Cx^2$.

Assume that $X^{(0)}$ is a sequence of raw data, $X^{(1)}$ the sequence of accumulation generation of $X^{(0)}$, and $Z^{(1)}$ the adjacent neighbor mean generation of $X^{(1)}$. Then,

$$x^{(0)}(k) + az^{(1)}(k) = b(z^{(1)}(k))^{\alpha} \tag{4.27}$$

is known as the GM(1,1) *power model*. And,

$$\frac{dx^{(1)}}{dt} + ax^{(1)} = b(x^{(1)})^{\alpha} \tag{4.28}$$

is known as the *whitenization equation* of the GM(1,1) power model.

Theorem 22 The solution of the whitenization equation of the GM(1,1) power model is

$$x^{(1)}(t) = \left\{ e^{-(1-a)at} \left[(1-a) \int be^{(1-a)at} dt + c \right] \right\}^{\frac{1}{1-a}}. \tag{4.29}$$

Theorem 23 Let $X^{(0)}$, $X^{(1)}$, and $Z^{(1)}$ be defined above. Let

$$B = \begin{bmatrix} -z^{(1)}(2) & (z^{(1)}(2))^{\alpha} \\ -z^{(1)}(3) & (z^{(1)}(3))^{\alpha} \\ \vdots & \vdots \\ -z^{(1)}(n) & (z^{(1)}(n))^{\alpha} \end{bmatrix}, \quad Y = \begin{bmatrix} x^{(0)}(2) \\ x^{(0)}(3) \\ \vdots \\ x^{(0)}(n) \end{bmatrix}$$

then the least square estimate of the parametric sequence $\hat{a} = [a, b]^{\mathrm{T}}$ of the GM(1,1) power model is $\hat{a} = (B^{\mathrm{T}}B)^{-1}B^{\mathrm{T}}Y$.

When the power in the GM(1,1) power model $\alpha = 2$, the resultant model $x^{(0)}(k) + az^{(1)}(k) = b(z^{(1)}(k))^2$ is known as the *grey Verhulst model*; and $\frac{dx^{(1)}}{dt} + ax^{(1)} = b(x^{(1)})^2$ is known as the *whitenization equation* of the grey Verhulst model.

Theorem 24

(1) The solution of the Verhulst whitenization equation is

$$\begin{aligned} x^{(1)}(t) &= \frac{1}{e^{at}\left[\frac{1}{x^{(1)}(0)} - \frac{b}{a}(1 - e^{-at})\right]} = \frac{ax^{(1)}(0)}{e^{at}\left[a - bx^{(1)}(0)(1 - e^{-at})\right]} \\ &= \frac{ax^{(1)}(0)}{bx^{(1)}(0) + (a - bx^{(1)}(0))e^{at}}. \end{aligned} \tag{4.30}$$

(2) The time response sequence of the grey Verhulst model is

$$\hat{x}^{(1)}(k+1) = \frac{ax^{(1)}(0)}{bx^{(1)}(0) + (a - bx^{(1)}(0))e^{ak}}. \tag{4.31}$$

Assume that $X_i^{(0)} = (x_i^{(0)}(1), x_i^{(0)}(2), \ldots, x_i^{(0)}(n)), i \in N_h^+$, is given, where $X_1^{(0)}$ stands for a data sequence of a system's characteristic, and $X_i^{(0)}$, $i \in N_h^+$, data sequences of relevant factors. Let

$$\alpha^{(1)}\hat{x}_1^{(1)}(k) = \hat{x}_1^{(1)}(k) - \hat{x}_1^{(1)}(k-1) = \hat{x}_1^{(0)}(k)$$

$$\alpha^{(2)}\hat{x}_1^{(1)}(k) = \alpha^{(1)}\hat{x}_1^{(1)}(k) - \alpha^{(1)}\hat{x}_1^{(1)}(k-1) = \hat{x}_1^{(0)}(k) - \hat{x}_1^{(0)}(k-1)$$

$$\cdots \ \cdots \ \cdots \ \cdots \ \cdots \ \cdots$$

$$\alpha^{(r)}\hat{x}_1^{(1)}(k) = \alpha^{(r-1)}\hat{x}_1^{(1)}(k) - \alpha^{(r-1)}\hat{x}_1^{(1)}(k-1) = \alpha^{(r-2)}\hat{x}_1^{(0)}(k) - \alpha^{(r-2)}\hat{x}_1^{(0)}(k-1)$$

and $z^{(1)}(k) = \frac{1}{2}(x^{(1)}(k) + x^{(1)}(k-1))$, then

$$\alpha^{(r)}x_1^{(1)}(k) + \sum_{i=1}^{r-1} a_i\alpha^{(r-i)}x_1^{(1)}(k) + a_r z_1^{(1)}(k) = \sum_{j=1}^{h-1} b_j x_{j+1}^{(1)}(k) + b_h \tag{4.32}$$

is known as the GM(r,h) *model*, an rth-order grey model in h variables. In such a model, $-\hat{a} = [-a_1, -a_2, \ldots, -a_h]^{\mathrm{T}}$ is known as the *development coefficient vector*, $\sum_{j=1}^{h-1} b_j x_{j+1}^{(1)}(k)$ the *driving term*, and $\hat{b} = [b_1, b_2, \ldots, b_h]^{\mathrm{T}}$ the *vector of driving coefficients*.

Theorem 25 Let $X_1^{(0)}$ be a data sequence of a system's characteristic, $X_i^{(0)}$, $i = 2, 3, \ldots, N$, the data sequences of relevant factors, $X_i^{(1)}$ the accumulation-generated sequence of $X_i^{(0)}$, $Z_1^{(1)}$ the adjacent neighbor mean generated sequence from $X_1^{(1)}$, and $\alpha^{(r-i)}X_1^{(1)}$ the $(r-i)$th-order inverse accumulation sequence of $X_1^{(1)}$. Define

$$B = \begin{bmatrix} -\alpha^{(r-1)}x_1^{(1)}(2) & -\alpha^{(r-2)}x_1^{(1)}(2) & \cdots & -\alpha^{(1)}x_1^{(1)}(2) & -z_1^{(1)}(2) & x_2^{(1)}(2) & \cdots & x_h^{(1)}(2) & 1 \\ -\alpha^{(r-1)}x_1^{(1)}(3) & -\alpha^{(r-2)}x_1^{(1)}(3) & \cdots & -\alpha^{(1)}x_1^{(1)}(3) & -z_1^{(1)}(3) & x_2^{(1)}(3) & \cdots & x_h^{(1)}(3) & 1 \\ \cdots & \cdots & \cdots & \cdots & \cdots & \cdots & \cdots & \cdots & \cdots \\ -\alpha^{(r-1)}x_1^{(1)}(n) & -\alpha^{(r-2)}x_1^{(1)}(n) & \cdots & -\alpha^{(1)}x_1^{(1)}(n) & -z_1^{(1)}(n) & x_2^{(1)}(n) & \cdots & x_h^{(1)}(n) & 1 \end{bmatrix},$$

$$Y = \begin{bmatrix} \alpha^{(r)}x_1^{(1)}(2) \\ \alpha^{(r)}x_1^{(1)}(3) \\ \vdots \\ \alpha^{(r)}x_1^{(1)}(n) \end{bmatrix}.$$

Then the parametric sequence $\hat{a} = [a, b_1, b_2, \ldots, b_N]^{\mathrm{T}}$ of the least square estimate satisfies $\hat{a} = (B^{\mathrm{T}}B)^{-1}B^{\mathrm{T}}Y$.

The GM(r,h) model is the general form of grey system models. In particular, (1) when $r = 1$ and $h = 1$, the previous becomes

$$\frac{dx_1^{(1)}}{dt} + a_1 x_1^{(1)} = b_1 \text{ and } \alpha^{(1)}x_1^{(1)}(k) + a_1 z_1^{(1)}(k) = b_1$$

which is the GM(1,1) model. (2) When $r = 1$ and $h = N$, the previous model takes the form of

$$\frac{dx_1^{(1)}}{dt} + a_1 x_1^{(1)} = b_1 x_2^{(1)} + b_2 x_3^{(1)} + \cdots + b_{N-1}x_N^{(1)} + b_N$$
$$\alpha^{(1)}x_1^{(1)}(k) + a_1 z_1^{(1)}(k) = b_1 x_2^{(1)}(k) + b_2 x_3^{(1)}(k) + \cdots + b_{N-1}x_N^{(1)}(k) + b_N$$

which is the GM(1,*N*) model. (3) When $r = 0$ and $h = N$, the previous model is

$$x_1^{(1)}(k) = b_1 x_2^{(1)}(k) + b_2 x_3^{(1)}(k) + \cdots + b_{N-1} x_N^{(1)}(k) + b_N$$

which is the GM(0,*N*) model. (4) When $r = 2$ and $h = 1$, the previous model reduces to

$$\frac{d^2 x_1^{(1)}}{dt^2} + a_1 \frac{dx_1^{(1)}}{dt} + a_2 x_1^{(1)} = b_1$$
$$\alpha^{(2)} x_1^{(1)}(k) + a_1 \alpha^{(1)} x_1^{(1)}(k) + a_2 z_1^{(1)}(k) = b_1$$

which is the GM(2,1) model. That is, such models as GM(1,1), GM(1,*N*), GM(0,*N*), and GM(2,1) are all special cases of the GM(*r*,*h*) model.

For all the relevant proofs of the results presented in this subsection, see References 23 and 24.

4.5.6 Grey System Predictions

Prediction is foretelling the possible future development of societal events, political matters, economic ups and downs, so on, using scientific methods and techniques based on attainable historical and present information such that appropriate actions can be planned and carried out. In short, prediction is about making scientific inferences about the evolutions of materials and events ahead of time. Prediction includes not only *static inference* about the unknown matters based on what is known within the same time frame, but also *dynamic inference* about the future based on the history and the present state of affairs of a certain matter. A particular prediction is only about a dynamic forecast within which a scientific inference about the future evolution of a certain event is given.

Grey prediction is about making scientific, quantitative forecasts about the future states of systems based on unascertained characteristics of the systems by making use of sequence operators on the original data sequences in order to generate, treat, and excavate the hidden laws of system evolution so that grey system models can be established and the future predicted. In particular, let $X^{(0)} = (x^{(0)}(1), x^{(0)}(2), \ldots, x^{(0)}(n))$ be some raw data, $\hat{X}^{(0)} = (\hat{x}^{(0)}(1), \hat{x}^{(0)}(2), \ldots, \hat{x}^{(0)}(n))$ the simulated data out of a chosen prediction model, $\varepsilon^{(0)} = (\varepsilon(1), \varepsilon(2), \ldots, \varepsilon(n)) = (x^{(0)}(1) - \hat{x}^{(0)}(1), x^{(0)}(2) - \hat{x}^{(0)}(2), \ldots, x^{(0)}(n) - \hat{x}^{(0)}(n))$ the error sequence, and

$$\Delta = \left(\left| \frac{\varepsilon(1)}{x^{(0)}(1)} \right|, \left| \frac{\varepsilon(2)}{x^{(0)}(2)} \right|, \ldots, \left| \frac{\varepsilon(n)}{x^{(0)}(n)} \right| \right) = \{\Delta_k\}_1^n$$

the relative error sequence. For $k \leq n$, $\Delta_k = \left|\frac{\varepsilon(k)}{x^{(0)}(k)}\right|$ is known as *relative error* of the simulation at point k, and $\bar{\Delta} = \frac{1}{n}\sum_{k=1}^{n} \Delta_k$ the *mean relative error*; $1 - \bar{\Delta}$ is known as the *mean relative accuracy*, and $1 - \Delta_k$ the *simulation accuracy* at point k, $k \in N_n^+$; for a given α, when $\bar{\Delta} < \alpha$ and $\Delta_n < \alpha$ hold true, the prediction model is said to be *error satisfactory*. Let ε stand for the absolute degree of incidence between the raw data $X^{(0)}$ and the simulated values $\hat{X}^{(0)}$. If for a given $\varepsilon_0 > 0$ the absolute degree ε of incidence satisfies $\varepsilon > \varepsilon_0$, then the simulation model is said to be *incidence satisfactory*.

Based on the sequences $X^{(0)}$, $\hat{X}^{(0)}$, and $\varepsilon^{(0)}$, consider the relevant means and variances

$$\bar{x} = \frac{1}{n}\sum_{k=1}^{n} x^{(0)}(k),\ S_1^2 = \frac{1}{n}\sum_{k=1}^{n}(x^{(0)}(k) - \bar{x})^2$$

and

$$\bar{\varepsilon} = \frac{1}{n}\sum_{k=1}^{n} \varepsilon(k),\ S_2^2 = \frac{1}{n}\sum_{k=1}^{n}(\varepsilon(k) - \bar{\varepsilon})^2.$$

If for a given $C_0 > 0$, the ratio of variances $C = \frac{S_2}{S_1} < C_0$, then the model is said to be *variance ratio satisfactory*. If $p = P(|\varepsilon(k) - \bar{\varepsilon})| < 0.6745S_1)$ is seen as a small error probability and for a given $p_0 > 0$, when $p > p_0$, then the model is said to be *small error probability satisfactory*.

What discussed above are three different ways to test a chosen model. Each of them is based on observations of the error to determine the accuracy of the model. For both the mean relative error $\bar{\Delta}$ and the simulation error, the smaller, the better. For the degree of incidence ε, the greater, the better. For the variance ratio C, the smaller the value is the better. It is because a small C indicates that S_2 is relatively small while S_1 is relatively large, meaning that the error variance is small while the variance of the original data is relatively large. Therefore, for better simulation results, the smaller S_2 is, when compared to S_1, the better. For the case of small error probability p, as soon as a set of α, ε_0, C_0, and p_0 values are chosen, a scale of accuracy for testing models is determined. The most commonly used scales of accuracy for testing models are listed in Table 4.1.

A *sequence prediction* is an attempt at foretelling the future behaviors of a system's variables. To this end, the GM(1,1) model is commonly employed. After making sure that the model passes the accuracy test, it is applied to make predictions. For related practical applications, see Reference 23.

4.5.6.1 *Interval Predictions*

When dealing with chaotic data sequences, it is generally difficult for any chosen model to pass the accuracy test, leading to the

TABLE 4.1 Commonly Used Scales of Accuracy for Model Testing

Threshold / Accuracy scale	Relative error α	Degree of incidence ε_0	Variance ratio C_0	Small error probability p_0
1st level	0.01	0.90	0.35	0.95
2nd level	0.05	0.80	0.50	0.80
3th level	0.10	0.70	0.65	0.70
4th level	0.20	0.60	0.80	0.60

trouble of making questionable quantitative predictions. In this case, one can consider providing a range for the future values to fall within. In particular, let $X(t)$ be a zigzagged line. If there are smooth and continuous curves $f_u(t)$ and $f_s(t)$, satisfying that for any t, $f_u(t) < X(t) < f_s(t)$, then $f_u(t)$ is known as the *lower bound function* of $X(t)$ and $f_s(t)$ the *upper bound function*, and $S = \{(t, X(t)) | X(t) \in [f_u(t), f_s(t)]\}$ the *value band* of $X(t)$. If the upper and lower bounds of $X(t)$ are the same kind of functions, then S is known as a uniform band. When S is a uniform band with exponential functions as its upper and lower bounds $f_u(t)$ and $f_s(t)$, then S is known as a *uniform exponential band*. If a uniform band S has linear upper and lower bound functions $f_u(t)$ and $f_s(t)$, then S is known as a *uniform linear band* or a *straight band* for short. If for $t_1 < t_2$, $f_s(t_1) - f_u(t_1) < f_s(t_2) - f_u(t_2)$ always holds true, then S is known as a *trumpet-like band*.

Let $X^{(0)} = (x^{(0)}(1), x^{(0)}(2), \ldots, x^{(0)}(n))$ be a sequence of raw data, and its accumulation generation be $X^{(1)} = (x^{(1)}(1), x^{(1)}(2), \ldots, x^{(1)}(n))$. Define $\sigma_{\max} = \max_{1 \le k \le n}\{x^{(0)}(k)\}$ and $\sigma_{\min} = \min_{1 \le k \le n}\{x^{(0)}(k)\}$, respectively, the upper and lower bound functions of $X^{(1)}$ be taken as

$$f_u(n + t) = x^{(1)}(n) + t\sigma_{\min}$$

and

$$f_s(n + t) = x^{(1)}(n) + t\sigma_{\max}.$$

Then $S = \{(t, X(t)) | t > n, X(t) \in [f_u(t), f_s(t)]\}$ is known as the *proportional band*.

For a sequence $X^{(0)}$ of raw data, let $X_u^{(0)}$ be the sequence corresponding to the curve that connects all the low points of $X^{(0)}$ and $X_s^{(0)}$ the sequence corresponding to the curve of all the upper points of $X^{(0)}$. Assume that

$$\hat{x}_u^{(1)}(k + 1) = \left(x_u^{(0)}(1) - \frac{b_u}{a_u}\right) \exp(-a_u k) + \frac{b_u}{a_u}$$

and

$$\hat{x}_s^{(1)}(k + 1) = \left(x_s^{(0)}(1) - \frac{b_s}{a_s}\right) \exp(-a_s k) + \frac{b_s}{a_s}$$

are respectively the GM(1,1) time response sequences of $X_u^{(0)}$ and $X_s^{(0)}$. Then $S = \{(t, X(t)) \, | X(t) \in [\hat{X}_u^{(1)}(t), \hat{X}_s^{(1)}(t)]\}$ is known as a *wrapping band (see Figure 4.8)*.

For a given sequence $X^{(0)}$ of raw data, let us take m different subsequences to establish m GM(1,1) models with the corresponding parameters $\hat{a}_i = [a_i, b_i]^T; i \in N_m^+$. Let $-a_{\max} = \max_{1 \le i \le m}\{-a_i\}$, $-a_{\min} = \min_{1 \le i \le m}\{-a_i\}$,

$$\hat{x}_u^{(1)}(k + 1) = \left(x_u^{(0)}(1) - \frac{b_{\min}}{a_{\min}}\right) \exp(-a_{\min} k) + \frac{b_{\min}}{a_{\min}}$$

$$\hat{x}_s^{(1)}(k + 1) = \left(x_s^{(0)}(1) - \frac{b_{\max}}{a_{\max}}\right) \exp(-a_{\max} k) + \frac{b_{\max}}{a_{\max}}$$

then $S = \{(t, X(t)) | X(t) \in [\hat{X}_u^{(1)}(t), \hat{X}_s^{(1)}(t)]\}$ is known as a *development band*.

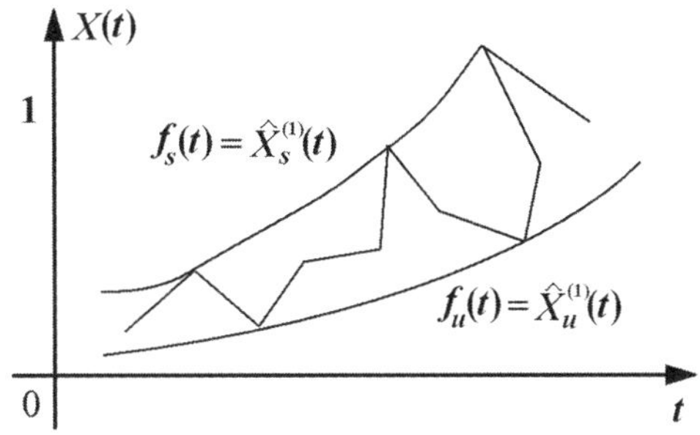

FIGURE 4.8 A wrapping band.

For a sequence $X^{(0)} = (x^{(0)}(1), x^{(0)}(2), \ldots, x^{(0)}(n))$ of raw data, let $f_u(t)$ and $f_s(t)$ be upper and lower bound functions of the accumulation generation $X^{(1)}$ of $X^{(0)}$. For any $k > 0$, $\hat{x}^{(0)}(n+k) = \frac{1}{2}[f_u(n+k) + f_s(n+k)]$ is known as *basic prediction value*, and $\hat{x}_u^{(0)}(n+k) = f_u(n+k)$ and $\hat{x}_s^{(0)}(n+k) = f_s(n+k)$, respectively, the *lowest and highest predicted values*.

4.5.6.2 Disaster Predictions The idea of *grey disaster predictions* is the foretelling of abnormal values. As for what kinds of values are considered abnormal, it is commonly determined based on individuals' experiences. The task of grey disaster predictions is to provide the time moments of the forthcoming abnormal values such that relevant parties can prepare for the worst ahead of time. In particular, let $X = (x(1), x(2), \ldots, x(n))$ be a sequence of raw data. For a given upper abnormal (or catastrophe) value ξ, the subsequence $X_\xi = (x[q(1)], x[q(2)], \ldots, x[q(m)]) = \{x[q(i)] | x[q(i)] \geq \xi; i \in N_m^+\}$ of X is known as the *upper catastrophe sequence*. For a given lower abnormal (or catastrophe) value ζ, the subsequence $X_\zeta = (x[q(1)], x[q(2)], \ldots, x[q(l)]) = \{x[q(i)] | x[q(i)] \leq \zeta; i \in N_l^+\}$ is known as the *lower catastrophe sequence*. These upper and lower catastrophe sequences are referred to as *catastrophe sequences*. Because the idea behind the discussion of catastrophe sequences is the same, in the following presentation, we will not distinguish upper and lower catastrophe sequences.

Assume that $X_\xi = (x[q(1)], x[q(2)], \ldots, x[q(m)]) \subset X$ is a catastrophe sequence. Then, $Q^{(0)} = (q(1), q(2), \ldots, q(m))$ is known as the *catastrophe date sequence*. In grey system theory, disaster prediction is about finding patterns, if any, through the study of catastrophe date sequences to predict future dates of occurrences of catastrophes by establishing GM(1,1) models for relevant catastrophe date sequences. In particular, if the accumulation generation $Q^{(1)} = (q(1)^{(1)}, q(2)^{(1)}, \ldots, q(m)^{(1)})$ of the catastrophe date sequence $Q^{(0)}$ is employed to establish the GM(1,1) model $q(k) + az^{(1)}(k) = b$, where $Z^{(1)}$ is the adjacent neighbor mean generated sequence of $Q^{(1)}$, then this model is known as a catastrophe GM(1,1) model. For a data sequence $X = (x(1), x(2), \ldots, x(n))$, if n stands for the present and the last entry $q(m)(\leq n)$ in the corresponding catastrophe date sequence $Q^{(0)}$ represents when the last catastrophe occurred, then the predicted value $\hat{q}(m+1)$ means when the next catastrophe will happen and for any $k > 0$, $\hat{q}(m + k)$ stands for the predicted date for the kth catastrophe to occur in the future.

4.5.6.3 Stock-Market-Like Predictions

When the available data sequence vibrates widely with large magnitudes, it is often difficult to find an appropriate simulation model. In this case, one can consider making use of the pattern of fluctuation of the data to predict the future development of the wavy movement. This kind of prediction is known as a *stock-market-like prediction*.

In particular, let $X = (x(1), x(2), \ldots, x(n))$ be a data sequence. Then $x_k = x(k) + (t - k)[x(k + 1) - x(k)]$ is known as a *k-piece zigzagged line* of the sequence X, and $\{x_k = x(k) + (t - k)[x(k + 1) - x(k)] \,|k = 1, 2, \ldots, n - 1\}$ the zigzagged line, still denoted by X. Let $\sigma_{\max} = \max_{1\le k\le n}\{x(k)\}$ and $\sigma_{\min} = \min_{1\le k\le n}\{x(k)\}$. For any $\forall\xi \in [\sigma_{\min}, \sigma_{\max}]$, $X = \xi$ is known as the *ξ-contour* (line); and the solutions $(t_i, x(t_i))(i = 1, 2, \ldots)$ of system of equations

$$\begin{cases} X = \{x(k) + (t - k)[x(k + 1) - x(k)] \,|k = 1, 2, \ldots, n - 1\} \\ X = \xi \end{cases}$$

the *ξ-contour points*. These points are the intersections of the zigzagged line X and the ξ-contour line.

Proposition 4 If there is a ξ-contour point on the ith segment of X, then the coordinates of this point are given by $(i + \frac{\xi - x(i)}{x(i+1) - x(i)}, \xi)$.

Let $X_\xi = (P_1, P_2, \ldots, P_m)$ be the sequence of ξ-contour points of X such that point P_i is located on the ith segment. Let $q(i) = t_i + \frac{\xi - x(t_i)}{x(t_i+1) - x(t_i)}$, $i \in N_m^+$. Then $Q^{(0)} = (q(1), q(2), \ldots, q(m))$ is known as the *ξ-contour time moment sequence*. By establishing a GM(1,1) model using this ξ-contour moment sequence, one produces the predicted values for future ξ-contour time moments: $\hat{q}(m + 1), \hat{q}(m + 2), \ldots,$ and $\hat{q}(m + k)$. The lines $X = \xi_i(i \in N_s)$, where $\xi_0 = \sigma_{\min}$, $\xi_1 = \frac{1}{s}(\sigma_{\max} - \sigma_{\min}) + \sigma_{\min}, \ldots$ $\xi_i = \frac{i}{s}(\sigma_{\max} - \sigma_{\min}) + \sigma_{\min}, \ldots,$ $\xi_{s-1} = \frac{s-1}{s}(\sigma_{\max} - \sigma_{\min}) + \sigma_{\min}$, $\xi_s = \sigma_{\max}$, are known as *equal time distanced contours*. When taking contour lines, one needs to make sure that the corresponding contour moments satisfy the conditions for establishing valid GM(1,1) models.

Let $X = \xi_i(i \in N_s^+)$ be s different contours, $Q_i^{(0)} = (q_i(1), q_i(2), \ldots, q_i(m_1), i \in N_s^+$, the sequence of ξ_i-contour time moments, and $\hat{q}_i(m_i + 1), \hat{q}_i(m_i + 2), \ldots, \hat{q}_i(m_i + k_i)$, $i \in N_s^+$, the GM(1,1) predicted ξ_i-contour time moments. If there are $i \ne j$ such that $\hat{q}_i(m_i + l_i) = \hat{q}_j(m_j + l_j)$, then these values are known as a pair of *invalid moments*.

Proposition 5 Let $\hat{q}_i(m_i + j), j \in N_{k_i}^+$, $i \in N_s^+$, be the GM(1,1) predicted ξ_i-contour time moments. After deleting all invalid predictions, order the rest in terms of their magnitudes as follows: $\hat{q}(1) < \hat{q}(2) < \ldots < \hat{q}(n_s)$, where $n_s \leq k_1 + k_2 + \cdots + k_s$. If $X = \xi_{\hat{q}(k)}$ is the contour line corresponding to $\hat{q}(k)$, then the predicted wavy curve of $X^{(0)}$ is given by $X = \hat{X}^{(0)} = \{\xi_{\hat{q}(k)} + [t - \hat{q}(k)][\xi_{\hat{q}(k+1)} - \xi_{\hat{q}(k)}] | k = 1, 2, \ldots, n_s\}$.

4.6 GREY MODELS FOR DECISION-MAKING

Determining what to do based on the given circumstances to actualize pre-determined goals is known as *decision-making*. It not only plays an important part of various kinds of management activities, but also appears throughout every person's daily life. The concept of decision-making can be divided into two categories: general and specific. For the general case, each decision-making stands for the entire process of activities, including posting questions, collecting the data, establishing the goal, making, analyzing, and evaluating the plan of action, implementing the plan, looking back, and modifying the plan. For the specific case, decision-making only represents the step of choosing a specific plan of action out of the entire decision-making process. In addition, some scholars understand decision-making as choosing and picking a plan of action under uncertain conditions. For this case, the choice can be most likely influenced by the decision maker's prior experience, attitude, and willingness to take a certain amount of risk. Grey decision-making is about making decision using such decision models that involve grey elements or combine the general decision model and grey system models. Its focus of study is on the problem of choosing a specific plan. In the rest of this section, the problem waiting to be resolved, the event needing to be handled, a current state of a system's behavior, etc., is known as an *event*, on which we begin our presentation.

The concepts of events, *countermeasures*, *objectives*, and *effects* are the four key elements of decision-making. The set of all events within the range of a study is denoted $A = \{a_1, a_2, \ldots, a_n\}$ with a_i being the ith event, $i \in N_n^+$ The set of all possible countermeasures is denoted $B = \{b_1, b_2, \ldots, b_m\}$ with b_j, $j \in N_m^+$, be the jth countermeasure. The Cartesian product $S = A \times B = \{(a_i, b_j) | a_i \in A, b_j \in B\}$ is known as the *situation set*, where each ordered pair $s_{ij} = (a_i, b_j)$, for any $a_i \in A$ and $b_j \in B$, is known as a *situation*. For a given situation $s_{ij} \in S$, evaluating the effects under a set of pre-determined objectives and deciding on the basis of the evaluation on what to take and what to let go is the decision-making we will look at in this section.

4.6.1 Grey Target Decisions

Let $S = \{s_{ij} = (a_i, b_j) | a_i \in A, b_j \in B\}$ be the situation set, $u_{ij}^{(k)}$ the *effect value of situation* s_{ij} with respect to objective k, R the set of all real numbers. Then $u_{ij}^{(k)}$: $S \mapsto R$, defined by $s_{ij} \mapsto u_{ij}^{(k)}$, is known as the *effect mapping* of S with respect to object k. If $u_{ij}^{(k)} = u_{ih}^{(k)}$, then countermeasures b_j and b_h of event a_i are said to be equivalent with respect to objective k, written $b_j \cong b_h$; and $B_i^{(k)} = \{b|b \in B, b \cong b_h\}$ is known as the effect *equivalence class* of countermeasure b_h of event a_i with respect to objective k.

If k is such an objective that the greater the effect value the better, and $u_{ij}^{(k)} > u_{ih}^{(k)}$, then the countermeasure b_j is considered *superior* to b_h in terms of event a_i with respect to objective k, written $b_j \succ b_h$. $B_{ih}^{(k)} = \{b|b \in B, b \succ b_h\}$ is known as the superior set of countermeasure b_h of event a_i with respect to objective k. Similarly, the concepts of superior classes of countermeasures can be defined for the situations that the closer to a fixed moderate value the effect value is the better, and that the smaller the effect value is the better.

If $u_{ij}^{(k)} = u_{ih}^{(k)}$, then events a_i and a_j are considered *equivalent* in terms of countermeasure b_h with respect to objective k, written $a_i \cong a_j$. The set $A_{jh}^{(k)} = \{a|a \in Aa \cong a_i\}$ is known as the *effect equivalence class* of events of countermeasure b_h with respect to objective k. If k is such an objective that the greater the effect value, the better, and $u_{ih}^{(k)} > u_{jh}^{(k)}$, then event a_i is *superior* to event a_j in terms of countermeasure b_h with respect to objective k, denoted $a_i \succ a_j$. The set $A_{jh}^{(k)} = \{a|a \in A, a \succ a_j\}$ is known as the superior class of event a_j in terms of countermeasure b_h with respect to objective k. Similarly, the concepts of superior classes can be defined for the situations that the closer to a fixed moderate value the effect value is, the better and that the smaller the effect value, the better.

If $u_{ij}^{(k)} = u_{hl}^{(k)}$, then situation s_{ij} is *equivalent* to situation s_{hl} under objective k, denoted $s_{ij} \cong s_{hl}$. $S^{(k)} = \{s|s \in S, s \cong s_{hl}\}$ is known as the *effect equivalence class* of situation s_{hl} under objective k. If k is such an objective that the greater the effect value, the better, and $u_{ij}^{(k)} > u_{hl}^{(k)}$, then situation s_{ij} is said to be *superior* to situation s_{hl} under objective k, denoted $s_{ij} \succ s_{hl}$. $S_{hl}^{(k)} = \{s|s \in S, s \succ s_{hl}\}$ is known as the *effect superior class* of situation s_{hl} under objective k. Similarly, the concepts of superior classes for a situation effects can be defined for the scenarios that the closer to a fixed moderate value the effect value of a situation is, the better and that the smaller the effect value of situation, the better.

Proposition 6 Assume that $S = \{s_{ij} = (a_i, b_j)|a_i \in A, b_j \in B\} \neq \emptyset$ and $U^{(k)} = \{u_{ij}^{(k)}|a_i \in A, b_j \in B\}$ is the set of effects under objective k, and $\{S^{(k)}\}$ the set of effect equivalence classes of situations under objective k. Then mapping $u^{(k)} : \{S^{(k)}\} \to U^{(k)}$, defined by $S^{(k)} \mapsto u_{ij}^{(k)}$, is bijective.

Let $d_1^{(k)}$ and $d_2^{(k)}$ be the upper and lower threshold values of the situational effects under objective k. Then $S^1 = \{r|d_1^{(k)} \le r \le d_2^{(k)}\}$ is known as the one-dimensional *grey target* of objective k, $u_{ij}^{(k)} \in [d_1^{(k)}, d_2^{(k)}]$ a satisfactory effect under objective k, the corresponding s_{ij} a desirable situation with respect to objective k, and b_j a desirable countermeasure of event a_i with respect to objective k [18, 29].

Proposition 7 Assume that $u_{ij}^{(k)}$ stands for the effect value of situation s_{ij} with respect to objective k. If $u_{ij}^{(k)} \in S^1$, that is, s_{ij} is a desirable situation with respect to objective k. Then for any $s \in S_{ij}^{(k)}$, s is also a desirable situation. That is, when s_{ij} is desirable, all situations in its effect superior class are desirable.

Assume that $d_1^{(1)}$ and $d_2^{(1)}$ are the threshold values of situational effects of objective 1, $d_1^{(2)}$ and $d_2^{(2)}$ the threshold values of situational effects of objective 2. Then $S^2 = \{(r^{(1)}, r^{(2)}) | d_1^{(1)} \le r^{(1)} \le d_2^{(1)}, d_1^{(2)} \le r^{(2)} \le d_2^{(2)}\}$ is known as a grey target of two-dimensional decision-making. If the effect vector of situation s_{ij} satisfies $u_{ij} = \{u_{ij}^{(1)}, u_{ij}^{(2)}\} \in S^2$, then s_{ij} is seen as a *desirable situation* with respect to objectives 1 and 2, and b_j a *desirable countermeasure* for event a_i with respect to objectives 1 and 2.

Generally, assume that $d_1^{(1)}, d_2^{(1)}$; $d_1^{(2)}, d_2^{(2)}$; $\cdots$; and $d_1^{(s)}, d_2^{(s)}$ are respectively the threshold values of situational effects under objectives 1, 2, … , and s. Then the following region of the s-dimensional Euclidean space

$$S^s = \left\{ (r^{(1)}, r^{(2)}, \ldots, r^{(s)}) \left| d_1^{(1)} \le r^{(1)} \le d_2^{(1)}, d_1^{(2)} \le r^{(2)} \le d_2^{(2)}, \ldots, d_1^{(s)} \le r^{(s)} \le d_2^{(s)} \right. \right\}$$

is known as a *grey target of an s-dimensional decision-making*. If the effect vector of situation s_{ij} satisfies $u_{ij} = (u_{ij}^{(1)}, u_{ij}^{(2)}, \ldots, u_{ij}^{(s)}) \in S^s$, where $u_{ij}^{(k)}$ stands for the effect value of situation s_{ij} with respect to objective k, $k \in N_s^+$, then s_{ij} is known as a *desirable situation* with respect to objectives 1, 2, … , and s, and b_j a *desirable countermeasure* of event a_i with respect to objectives 1, 2, … , and s.

$$R^s = \left\{ (r^{(1)}, r^{(2)}, \ldots, r^{(s)}) \left| (r^{(1)} - r_0^{(1)})^2 + (r^{(2)} - r_0^{(2)})^2 + \cdots + (r^{(s)} - r_0^{(s)})^2 \le R^2 \right. \right\}$$

is known as an s-dimensional spherical grey target centered at $r_0 = (r_0^{(1)}, r_0^{(2)}, \ldots, r_0^{(s)})$ with radius R. The vector $r_0 = (r_0^{(1)}, r_0^{(2)}, \ldots, r_0^{(s)})$ is seen as the *optimum effect vector*. For $r_1 = (r_1^{(1)}, r_1^{(2)}, \ldots, r_1^{(s)}) \in R$,

$$|r_1 - r_0| = \left[(r_1^{(1)} - r_0^{(1)})^2 + (r_1^{(2)} - r_0^{(2)})^2 + \cdots + (r_1^{(s)} - r_0^{(s)})^2 \right]^{1/2}$$

is known as the *bull's-eye distance* of vector r_1. The values of this distance reflect the superiority of the corresponding situational effect vectors.

Let s_{ij} and s_{hl} be two different situations, and $u_{ij} = (u_{ij}^{(1)}, u_{ij}^{(2)}, \ldots, u_{ij}^{(s)})$ and $u_{hl} = (u_{hl}^{(1)}, u_{hl}^{(2)}, \ldots, u_{hl}^{(s)})$, respectively, their effect vectors. If $|u_{ij} - r_0| \ge |u_{hl} - r_0|$, then situation s_{hl} is said to be *superior* to s_{ij}, denoted $s_{hl} \succ s_{ij}$. When the equal sign holds true, they are said to be *equivalent*, written $s_{hl} \cong s_{ij}$. If for $i \in N_n^+$ and $j \in N_m^+$, $u_{ij} \ne r_0$ always holds true, then the optimum situation does not exist, and the event does not have any optimum countermeasure. If the optimum situation does not exist, however, there are such h and l that for any $i \in N_n^+$ and $j \in N_m^+$, $|u_{hl} - r_0| \le |u_{ij} - r_0|$ holds true, that is, for any $s_{ij} \in S$, $s_{hl} \succ s_{ij}$ holds, then s_{hl} is known as a *quasi-optimum situation*, a_h a *quasi-optimum event*, and a *quasi-optimum countermeasure*.

Theorem 26 Let $S = \{s_{ij} = (a_i, b_j) | a_i \in A, b_j \in B\}$ be the set of situations,

$$R^s = \left\{ (r^{(1)}, r^{(2)}, \ldots, r^{(s)}) \left| (r^{(1)} - r_0^{(1)})^2 + (r^{(2)} - r_0^{(2)})^2 + \cdots + (r^{(s)} - r_0^{(s)})^2 \le R^2 \right. \right\}$$

an s-dimensional spherical grey target. Then S becomes an ordered set with "superiority" as its order relation $\prec$.

Theorem 27 There must be quasi-optimum situations in the situation set $(S, \succ)$.

The bull's-eye distance between a situational effect vector and the center of the target measures the superiority of the situation in comparison with other situations. At the same time, the degree of incidence between the effect vector of a situation and the optimum effect vector can be seen as another way to evaluate the superiority of the situation. In particular, let $S = \{s_{ij} = (a_i, b_j) | a_i \in A, b_j \in B\}$ be the set of situations, and $u_{i_0j_0} = \{u_{i_0j_0}^{(1)}, u_{i_0j_0}^{(2)}, \ldots, u_{i_0j_0}(s)\}$ the optimum effect vector. If the situation corresponding to $u_{i_0j_0}$ satisfies $u_{i_0j_0} \notin S$, then $u_{i_0j_0}$ is known as an *imagined optimum effect vector*, and $s_{i_0j_0}$ the *imagined optimum situation*.

Proposition 8 Let S be the same as above and the effect vector of situation s_{ij} is $u_{ij} = \{u_{ij}^{(1)}, u_{ij}^{(2)}, \ldots, u_{ij}^{(s)}\}$, for $i \in N_n^+, j \in N_m^+$.

(1) When k is an objective such that the greater its effect value, the better, let $u_{i_0j_0}^{(k)} = \max_{1 \le i \le n, 1 \le j \le m} \{u_{ij}^{(k)}\}$.
(2) When k is an objective such that the closer to a fixed moderate value u_0 its effect value is, the better, let $u_{i_0j_0}^{(k)} = u_0$.
(3) When k is an objective such that the smaller its effect value, the better, let $u_{i_0j_0}^{(k)} = \min_{1 \le i \le n, 1 \le j \le m} \{u_{ij}^{(k)}\}$.

Then $u_{i_0j_0} = \{u_{i_0j_0}^{(1)}, u_{i_0j_0}^{(2)}, \ldots, u_{i_0j_0}^{(s)}\}$ is the imagined optimum effect vector.

Proposition 9 Assume the same as in Proposition 8 and let $u_{i_0j_0} = \{u_{i_0j_0}^{(1)}, u_{i_0j_0}^{(2)}, \ldots, u_{i_0j_0}^{(s)}\}$ be the imagined optimum effect vector, ε_{ij} the absolute degree of grey incidence between u_{ij} and $u_{i_0j_0}$, for $i \in N_n^+, j \in N_m^+$. If $\forall i \in N_n^+, j \in N_m^+$ satisfying $i \ne i_1$ and $j \ne j_1$, $\varepsilon_{i_1j_1} \ge \varepsilon_{ij}$ always holds true, then $u_{i_1j_1}$ is a quasi-optimum effect vector and $s_{i_1j_1}$ a quasi-optimum situation.

Grey development decision-making is done based on the development tendency or the future behavior of the situation of concern. It does not necessarily place specific emphasis on the current effect of the situation. Instead it focuses more on the change of the situational effect over time. This method of decision-making can be and has been employed for long-term planning and the decision-makings of large-scale engineering projects and urban plannings. It looks at problems from the angle of development while attempting to make feasible arrangements and avoid repetitious constructions such that great savings of capital and manpower can be materialized.

Assume that $A = \{a_1, a_2, \ldots, a_n\}$ and $B = \{b_1, b_2, \ldots, b_m\}$ are respectively a set of events and a set of countermeasures, and $S = \{s_{ij} = (a_i, b_j)|a_i \in A, b_j \in B\}$ the set of situations. Then, $u_{ij}^{(k)} = (u_{ij}^{(k)}(1), u_{ij}^{(k)}(2), \ldots, u_{ij}(k)(h))$ is known as the *situational effect time series* of situation s_{ij} with respect to objective k.

Proposition 10 Let the situational effect time series of situation s_{ij} with respect to objective k be $u_{ij}^{(k)} = (u_{ij}^{(k)}(1), u_{ij}^{(k)}(2), \ldots, u_{ij}^{(k)}(h))$, and $\hat{a}_{ij}^{(k)} = [a_{ij}^{(k)}, b_{ij}^{(k)}]^{\mathrm{T}}$ the least square estimate of the parameters of the GM(1,1) model of $u_{ij}^{(k)}$. Then the inverse accumulation restoration of the GM(1,1) time response of $u_{ij}^{(k)}$ is given by

$$\hat{u}_{ij}^{(k)}(l+1) = \left[1 - \exp\left(a_{ij}^{(k)}\right)\right] \cdot \left[u_{ij}^{(k)}(1) - \frac{b_{ij}^{(k)}}{a_{ij}^{(k)}}\right] \exp\left(-a_{ij}^{(k)} \cdot l\right).$$

Assume that the restored sequence through inverse accumulation of the GM(1,1) time response of the situational effect time series of the situation s_{ij} with respect to objective k is

$$\hat{u}_{ij}^{(k)}(l+1) = \left[1 - \exp\left(a_{ij}^{(k)}\right)\right] \cdot \left[u_{ij}^{(k)}(1) - \frac{b_{ij}^{(k)}}{a_{ij}^{(k)}}\right] \exp\left(-a_{ij}^{(k)} \cdot l\right).$$

When objective k satisfies that the greater the effect value, the better, if

(1) $\max_{1\le i\le n, 1\le j\le m}\{-a_{ij}^{(k)}\} = -a_{i_0j_0}^{(k)}$, then $s_{i_0j_0}$ is known as the *optimum situation of development coefficients* with respect to objective k.

(2) $\max_{1\le i\le n, 1\le j\le m}\{\hat{u}_{ij}^{(k)}(h+l)\} = \hat{u}_{i_0j_0}^{(k)}(h+l)$, then $s_{i_0j_0}$ is known as the *optimum situation of predictions* with respect to objective k.

Similarly, the concepts of optimum situations of development coefficients and predictions can be defined for the cases of objectives satisfying that the smaller the effect value, the better, and that the closer to a moderate value the effect value is, the better, respectively.

In practical applications, one may face the scenarios of either that both the optimum situations of development coefficients and predictions are the same or that they are different. Even so, the following theorem tells that eventually these optimum situations would converge to one.

Theorem 28 Assume that k is such an objective that the greater its effect value, the better, $s_{i_0j_0}$ the optimum situation of development coefficients, that is, $-a_{i_0j_0}^{(k)} = \max_{1\leq i\leq n, 1\leq j\leq m}\{-a_{ij}^{(k)}\}$, and $\hat{u}_{i_0j_0}^{(k)}(h+l+1)$ the predicted value for the situation effect of $s_{i_0j_0}$. Then there must be $l_0 > 0$ such that

$$\hat{u}_{i_0j_0}^{(k)}(h+l_0+1) = \max_{1\leq i\leq n, 1\leq j\leq m}\left\{\hat{u}_{ij}^{(k)}(h+l_0+1)\right\}.$$

That is, in a sufficiently distant future, $s_{i_0j_0}$ will also be the optimum situation of predictions.

Similar results hold true for those objectives satisfying either that the smaller the effect value, the better or that the closer to a fixed moderate value the effect value is, the better.

Grey cluster decision is useful for synthetic evaluations about some objects with respect to several different criteria such that decisions can be made about whether or not an object meets the given standards for inclusion or exclusion. This method has been often employed for the classification decision-making regarding objects or people. Assume that there are n objects to make decisions on, m criteria, s different grey classes, the quantified evaluation value of object i with respect to criterion j is x_{ij}, $f_j^k(*)$ the whitenization weight function of the kth grey class with respect to the jth criterion, w_j the synthetic decision-making weight of criterion j such that $\sum_{j=1}^m w_j = 1$, $i \in N_n^+$, $j \in N_m^+$, $k \in N_s^+$, then $\sigma_i^k = \sum_{j=1}^m f_j^k(x_{ij})w_j$ is known as the *decision coefficient* for the object i to belong to grey class k; $\sigma_i = (\sigma_i^1, \sigma_i^2, \ldots, \sigma_i^s)$ the *decision coefficient vector* of object i, $i \in N_n^+$; and $\sum = (\sigma_i^k)_{n\times s}$ the decision coefficient matrix. If $\max_{1\leq k\leq s}\{\sigma_i^k\} = \sigma_i^{k^*}$, then the decision is that object i belongs to grey class k^*.

When one needs to order objects to decide which objects to take and which to delete from his consideration, because decision coefficient vectors are generally not constructed against a same background or dimension, it is often difficult to actually compare these vectors. To this end, first normalize all the decision coefficient vectors by letting $\delta_i^k = \sigma_i^k / \sum_{k=1}^s \sigma_i^k$, which is known as the *normalized decision coefficient* for object i to belong to grey class k. Then, construct for object i its *normalized decision coefficient vector* $\delta_i = (\delta_i^1, \delta_i^2, \ldots, \delta_i^s)$, $i \in N_n^+$, and its *normalized decision coefficient matrix* $\prod = (\delta_i^k)_{n\times s}$.

Proposition 11 When decision object i is respectively clustered using its decision coefficient σ_i^k and its normalized decision coefficient δ_i^k, the outcome is the same.

Within a grey class, after all decision coefficients are normalized, the magnitudes of their coordinates may well determine which vector should be ordered first. For instance, given $\delta_1 = (0.4, 0.35, 0.25)$ and $\delta_2 = (0.41, 0.2, 0.39)$, the corresponding objects belong to the first grey class. If one compares directly their comprehensive decision coefficients, since $0.41 > 0.4$, object 2 will be ordered ahead of object 1. However, if we closely compare in coordinates, we might well be convinced that object 1 should be ordered ahead of object 2. This difference in opinion is caused by the fact that when we compare the normalized decision coefficient vectors, we do not look at the vectors holistically. To resolve this problem, let us look at the concept of synthetic measures.

Assume that there are s different grey classes. Then

$$\begin{aligned}
&\eta_1 = (s, s-1, s-2, \ldots, 1)\\
&\eta_2 = (s-1, s, s-1, s-2, \ldots, 2)\\
&\eta_3 = (s-2, s-1, s, s-1, \ldots, 3)\\
&\cdots,\\
&\eta_k = (s-k+1, s-k+2, \ldots, s-1, s, s-1, \ldots, k)\\
&\cdots,\\
&\eta_{s-1} = (2, 3, \ldots s-1, s, s-1)\\
&\text{and } \eta_s = (1, 2, 3, \ldots s-1, s)
\end{aligned}$$

are known as the *adjustment coefficients* of the first, second, ... , sth grey class, respectively.

If there are n decision objects, s different grey classes, and object i belongs to grey class k, then $\omega_i = \eta_k \cdot \delta_i^{\mathrm{T}}$ is known as the *synthetic decision measure* of the object i. If $\max_{1\le k\le s}\{\delta_{i_1}^k\} = \delta_{i_1}^{k^*}$ and $\max_{1\le k\le s}\{\delta_{i_2}^k\} = \delta_{i_2}^{k^*}$ hold with $\omega_{i_1} > \omega_{i_2}$, then in grey class k^*, decision object i_1 is seen as *superior* to decision object i_2. If $\max_{1\le k\le s}\{\delta_{i_1}^k\} = \delta_{i_1}^{k^*}$, $\max_{1\le k\le s}\{\delta_{i_2}^k\} = \delta_{i_2}^{k^*}$, ... , $\max_{1\le k\le s}\{\delta_{i_l}^k\} = \delta_{i_l}^{k^*}$ hold, that is, decision objects $i_1, i_2, \ldots,$ and i_l all belong to grey class k^*, and if $\omega_{i_1} > \omega_{i_2} > \cdots > \omega_{i_l}$ and the decision grey class k^* is only allowed to have l_1 objects, then the objects $i_1, i_2, \ldots,$ and i_{l_1} are admitted into class k^* with the remaining ordered objects being the candidates on the waiting list of grey class k^*. In general, a grey cluster decision can be made by going through the following steps [18, 29]:

Step 1: Decide on s grey classes based on the requirements of the comprehensive evaluation; accordingly, also divide the individual fields of the criteria into s grey classes, and define the whitenization weight function $f_j^k(*)$ for the kth subclass of the jth criterion, $j \in N_m^+$, $k \in N_s^+$.

Step 2: Determine the clustering weight w_j of the criteria, $j \in N_m^+$;

Step 3: Compute the decision coefficient $\sigma_i^k = \sum_{j=1}^{m} f_j^k(x_{ij})w_j$ for object i to belong to grey class k.

Step 4: Calculate the normalized decision coefficient δ_i^k for object i to belong to grey class k.

Step 5: From $\max_{1\le k\le s}\{\delta_i^k\} = \delta_i^{k^*}$, it is determined that object i belongs to grey class k^*.

Step 6: Compute the synthetic decision measure $\omega_i = \eta_k \cdot \delta_i^{\mathrm{T}}$ of object i with respect to grey class k.

Step 7: According to the synthetic decision measures, order all the objects that belong to the same grey class k.

As for the problem of how to make decisions for multiple-target situations to meet a synthesized criterion, assume that $A = \{a_1, a_2, \ldots, a_n\}$ is a set of events, $B = \{b_1, b_2, \ldots, b_m\}$ a set of countermeasures, $S = \{s_{ij} = (a_i, b_j)|a_i \in A, b_j \in B\}$ the relevant set of situations, and $u_{ij}^{(k)}$ the observed effect value of situation $s_{ij} \in S$ with respect to objective k, $i \in N_n^+$, $j \in N_m^+$. Then, $r_{ij}^{(k)} = u_{ij}^{(k)}/\max_{i,j}\{u_{ij}^{(k)}\}$ is known as an *upper effect measure*; $r_{ij}^{(k)} = \min_{i,j}\{u_{ij}^{(k)}\}/u_{ij}^{(k)}$ a *lower effect measure*; and $r_{ij}^{(k)} = u_{i_0j_0}^{(k)}/u_{i_0j_0}^{(k)} + |u_{ij}^{(k)} - u_{i_0j_0}^{(k)}|$ a *moderate effect measure*, where $u_{i_0j_0}^{(k)}$ is a chosen moderate effect value with respect to objective k.

When making situational decisions with a synthesized target, for the kinds of objectives where the greater or the more the effect sample values, the better, one can make use of the concept of upper effect measure. For the kinds of objectives where the smaller or fewer the effect sample values, the better, one can utilize the concept of lower effect measure. As for the kinds of objectives that require "neither too large nor too small" and/or "neither too many nor too few," one can apply the concept of moderate effect measure.

Proposition 12 The three effect measures $r_{ij}^{(k)}(i \in N_n^+, j \in N_m^+)$ as defined above satisfy: (1) $r_{ij}^{(k)}$ is nondimensional; (2) $r_{ij}^{(k)} \in [0, 1]$; and (3) the more ideal the effect, the greater the $r_{ij}^{(k)}$.

For the given set S of situations, $R^{(k)} = (r_{ij}^{(k)})_{n\times m}$ is known as the *matrix of uniform effect measures* of S with respect to objective k. For $s_{ij} \in S$, $r_{ij} = (r_{ij}^{(1)}, r_{ij}^{(2)}, r_{ij}^{(s)})$ is known as the *vector of uniform effect measures* of the situation s_{ij}. If η_k stands for the decision weight of objective k, $k \in N_s^+$, satisfying $\sum_{k=1}^{s} \eta_k = 1$, then $\sum_{k=1}^{s} \eta_k \cdot r_{ij}^{(k)}$ is known as the *synthetic effect measure of the situation* s_{ij}, which is still denoted as $r_{ij} = \sum_{k=1}^{s} \eta_k \cdot r_{ij}^{(k)}$; and $R = (r_{ij})_{n\times m}$ is known as the *matrix of synthetic effect measures*. If $\max_{1\le j\le m}\{r_{ij}\} = r_{ij_0}$, then b_{j_0} is known as the *optimum countermeasure* of event a_i. If $\max_{1\le i\le n}\{r_{ij}\} = r_{i_0j}$, then a_{i_0} is known as the *optimum event* corresponding to countermeasure b_j. If $\max_{1\le i\le n, 1\le j\le m}\{r_{ij}\} = r_{i_0j_0}$, then $s_{i_0j_0}$ is known as the *optimum situation*.

Each synthesized situation decision can be made by following the steps below:

Step 1: Based on set $A = \{a_1, a_2, \ldots, a_n\}$ of events and set $B = \{b_1, b_2, \ldots, b_m\}$ of countermeasures, construct the set of situations $S = \{s_{ij} = (a_i, b_j) | a_i \in A, b_j \in B\}$.

Step 2: Determine decision objectives $k \in N_s^+$.

Step 3: For each objective $k \in N_s^+$, compute the corresponding observed effect matrix $U^{(k)} = (u_{ij}^{(k)})_{n\times m}$.

Step 4: Calculate matrix $R^{(k)} = (r_{ij}^{(k)})_{n\times m}$ of uniform effect measures of objective $k \in N_s^+$.

Step 5: Determine the decision weights $\eta_1, \eta_2, \ldots$, and η_s of the objectives.

Step 6: From $r_{ij} = \sum_{k=1}^{s} \eta_k \cdot r_{ij}^{(k)}$, compute the matrix of synthetic effect measures $R = (r_{ij})_{n\times m}$.

Step 7: Determine the optimum situation $s_{i_0 j_0}$.

4.6.2 Multi-Attribute Intelligent Grey Target Decision Models

For a benefit-type objective k, that is, for k the greater the effect sample value, the better, and the decision grey target of objective k is $u_{ij}^{(k)} \in [u_{i_0 j_0}^{(k)}, \max_{i,j}\{u_{ij}^{(k)}\}]$, that is, $u_{i_0 j_0}^{(k)}$ stands for the threshold effect value of objective k. Then $r_{ij}^{(k)} = (u_{ij}^{(k)} - u_{i_0 j_0}^{(k)}) \div (\max_{i,j}\{u_{ij}^{(k)}\} - u_{i_0 j_0}^{(k)})$ is known as the *effect measure of a benefit-type objective*. If k is a cost-type objective, that is, for k, the smaller the effect value, the better, and the decision grey target of objective k is $u_{ij}^{(k)} \in [\min_{i,j}\{u_{ij}^{(k)}\}, u_{i_0 j_0}^{(k)}]$, that is, $u_{i_0 j_0}^{(k)}$ stands for the threshold effect value of objective k. Then $r_{ij}^{(k)} = (u_{i_0 j_0}^{(k)} - u_{ij}^{(k)}) \div (u_{i_0 j_0}^{(k)} - \min_{i,j}\{u_{ij}^{(k)}\})$ is known as the *effect measure of cost-type objective*. If k is a moderate-value type objective, that is, for k, the closer to a moderate value A the effect value is, the better, and the decision grey target of objective k is $u_{ij}^{(k)} \in [A - u_{i_0 j_0}^{(k)}, A + u_{i_0 j_0}^{(k)}]$, that is, both $A - u_{i_0 j_0}^{(k)}$ and $A + u_{i_0 j_0}^{(k)}$ are respectively the lower and upper threshold effect values of objective k. Then, (i) when $u_{ij}^{(k)} \in [A - u_{i_0 j_0}^{(k)}, A]$, $r_{ij}^{(k)} = (u_{ij}^{(k)} - A + u_{i_0 j_0}^{(k)}) \div u_{i_0 j_0}^{(k)}$ is known as the *lower effect measure of moderate-value type objective*; (ii) when $u_{ij}^{(k)} \in [A, A + u_{i_0 j_0}^{(k)}]$, $r_{ij}^{(k)} = (A + u_{i_0 j_0}^{(k)} - u_{ij}^{(k)}) \div u_{i_0 j_0}^{(k)}$ is referred to as the *upper effect measure of moderate-value type objective*.

The effect measures of benefit-type objectives reflect the degrees of both how close the effect sample values are to the maximum sample values and how far away they are from the threshold effect values of the objectives. Similarly, the effect measures of cost-type objectives represent the degrees of how close the effect sample values are to the minimum effect sample values and how far away the effect sample values are from the threshold effect values of the objectives; and the lower effect measures of moderate-value type objectives indicate how far away the effect sample values that are smaller than the moderate value A are from the lower threshold effect value, and the upper effect measures how far away the effect sample values that are greater than the moderate value A are from the upper threshold effect values of the objectives [29, 53].

For the situation of missing the target, it can be considered in the following four different possibilities:

(1) The effect value of a benefit-type objective is smaller than the threshold value $u_{i_0 j_0}^{(k)}$, that is, $u_{ij}^{(k)} < u_{i_0 j_0}^{(k)}$.
(2) The effect value of a cost-type objective is greater than the threshold value $u_{i_0 j_0}^{(k)}$, that is, $u_{ij}^{(k)} > u_{i_0 j_0}^{(k)}$.
(3) The effect value of a moderate-value type objective is smaller than the lower threshold effect value $A - u_{i_0 j_0}^{(k)}$, that is, $u_{ij}^{(k)} < A - u_{i_0 j_0}^{(k)}$.
(4) The effect value of a moderate-value type objective is greater than the upper threshold effect value $A + u_{i_0 j_0}^{(k)}$, that is, $u_{ij}^{(k)} > A + u_{i_0 j_0}^{(k)}$.

In order for the effect measures of each type objective to satisfy the condition of normality, that is, $r_{ij}^{(k)} \in [-1, 1]$, without loss of generality, it can be assumed that for a benefit-type objective, $u_{ij}^{(k)} \geq -\max_{i,j}\{u_{ij}^{(k)}\} + 2u_{i_0 j_0}^{(k)}$; for a benefit-type objective, $u_{ij}^{(k)} \leq -\min_{i,j}\{u_{ij}^{(k)}\} + 2u_{i_0 j_0}^{(k)}$; for the situation where the effect value of a moderate-value type objective is smaller than the lower threshold effect value $A - u_{i_0 j_0}^{(k)}$, $u_{ij}^{(k)} \geq A - 2u_{i_0 j_0}^{(k)}$; and for the situation where the effect value of a moderate-value type objective is greater than the upper threshold effect value $A + u_{i_0 j_0}^{(k)}$, $u_{ij}^{(k)} \leq A + 2u_{i_0 j_0}^{(k)}$. With these assumptions, the following holds true.

Proposition 13 The effect measures $r_{ij}^{(k)} (i \in N_n^+, j \in N_m^+, k \in N_s^+)$, as defined above, satisfy the following properties: (1) $r_{ij}^{(k)}$ is nondimensional; (2) the more satisfied the effect, the greater the $r_{ij}^{(k)}$ is; and (3) $r_{ij}^{(k)} \in [-1, 1]$.

4.7 PRACTICAL APPLICATIONS

4.7.1 To Analyze the Time Difference of Economic Indices

In order to effectively monitor the performance of macro-economic systems and provide timely warning, there is a need to investigate the time relationship of various economic indices with respect to the economic cycles in terms of the peaks and valleys by addressing such questions as: Which indices can provide warning ahead of time? Which ones would be synchronic with the evolution of the economic systems? And which ones tend to be lag behind the economic development? In other words, we need to divide economic indices into three classes: leading indicators, synchronic indices, and stagnant representations. To this end, the method of grey incidence analysis provides an effective tool for classifying economic indices.

Through careful research and analysis, we selected the following 8 major classes and 17 criteria as indices for economic performance:

(1) the class of energy and raw materials: the total production of energy;
(2) the class of investment: the total investment in real estate;
(3) the class of production: increase in industry output, increase in light industry output, increase in heavy industry output;
(4) the class of revenue: national income, national expenditure;
(5) the class of currency and credit: currency in circulation, savings at various financial institutions, amount of loans issued by financial institutions, cash payout in the form of salary and wages, net amount of currency in circulation;
(6) the class of consumption: the gross retail amount of the society;
(7) the class of foreign trade: gross amount of imports, gross amount of exports, direct investments by foreign entities; and
(8) the class of commodity prices: the consumer price index.

By applying the following standards, we classify the previous criteria into three classes: leading indicators, synchronic indices, and stagnant representations.

The standards for determining leading indicators:

(1) The indicated appearance of economic cyclic peaks needs to be at least 3 months ahead of the actual occurrence. Such leading relationship needs to be relatively stable with few exceptions.
(2) The indicated cycles and the historical cycles are nearly one-to-one corresponded to each other. And, for the most recent three economic cycles, at least two times the indicated cycles are ahead of the actual occurrences with at least 3 months of lead time.
(3) The economic characteristics of the indices provide relatively definite and clear leading relationships with respect to the background economic cycles.

The standards for determining both synchronic indices and stagnant representations are similar to what are outlined above with the exception that for synchronic indices, the time differences between the indicated appearances and the actual occurrences of economic cycles need to be within plus or minus 3 months, while for stagnant representations, the indicated appearances of economic cycles are behind the actual occurrences by at least 3 months.

As expected, in practical selections it is almost impossible to actually find an index that meets all the stated standards. So, what can be done is that based on the recorded reference cycles, we look for the statistical indices that meet the previously stated standards as closely as possible. Even so, the actual situation is that an identified leading indicator can also sometimes lag behind the actual economic development, while an identified stagnant representation can also provide nice lead time in its forecast of a specific economic evolution. Similar scenarios also occur with regard to synchronic indices. However, theoretically, if among the one-to-one correspondences between an index and the actually recorded cycles over two-thirds of times the index is leading the actual occurrences, then we treat such an index as leading. Similar treatments are applied to synchronic indices and stagnant representations.

Because for Chinese economy the increase in industry output has played a very important role, as a synchronic index it has a quite high quality. So, it can be employed as the basic index in our grey incidence analysis. We will compute not only the absolute degree of grey incidence between each criterion and the increase in industry output, but also the absolute degrees of grey incidence of other 16 criteria with their data translated 1–12 months along the time axis either left or right. When the data are translated to the left, the months will take negative values, when translated to the right, the months will take positive values. The amount of horizontal translation is denoted by L. That is, we compute the absolute degrees of grey incidence between all the 16 individual criteria, not including that of increase in industry output, and that of the increase in industry output for $L = -12, \dots, 12$. For each L-value, we order the obtained absolute degrees of grey incidence from the smallest to the largest with the criterion listed in the front chosen as candidate criterion for that specific L-value. For instance, when $L = 0$, the absolute degrees of grey incidence of the criteria are listed in Table 4.2.

Synchronic indices should be selected from those with larger absolute degrees of grey incidence, because the large degrees of incidence indicate that these criteria have greater similarities in comparison with that of increase in industry output, which we employ as the basic standard of the Chinese economic cycles. However, we still do not have theoretical evidence to support that an index with large absolute degree of

TABLE 4.2 The Absolute Degrees of Grey Incidence of the Criteria When $L = 0$

Index	Absolute degree of grey incidence	Index	Absolute degree of grey incidence
Increase in heavy industry output	0.979810	National income	0.559540
Increase in light industry output	0.972655	Gross amount of exports	0.544870
Gross retail amount of the society	0.862105	Total production of energy	0.541044
Cash payout as salaries	0.789278	Net amount of currency in circulation	0.525936
Currency in circulation	0.753681	Loans issued by financial institutions	0.507958
Total investment in real estate	0.726366	Savings at financial institutions	0.505226
Gross amount of imports	0.598248	Consumer price index	0.500173
National expenditure	0.566914	Direct investments by foreign entities	0.500002

TABLE 4.3 The Absolute Degrees of "Cash Payout as Salaries" When $L \neq 0$

L	Absolute degree of grey incidence	L	Absolute degree of grey incidence
−12	0.664615	1	0.877090
−11	0.705983	2	0.867859
−10	0.733564	3	0.857366
−9	0.752740	4	0.832260
−8	0.753598	5	0.825027
−7	0.732221	6	0.806787
−6	0.723942	7	0.806782
−5	0.731232	8	0.820384
−4	0.742249	9	0.803771
−3	0.752628	10	0.806649
−2	0.770216	11	0.805679
−1	0.800838	12	0.836308
0	0.789278		

grey incidence must be synchronic. To this end, we also need to consider whether or not the related absolute degrees of grey incidence will be even greater when $L \neq 0$. If when $L = 0$ the value of the absolute degree of grey incidence of a certain index is ranked quite in the front, and when $L = -4$, its value is even greater, it means that after this index is translated to 4 months earlier, it is more similar to the pattern of the increase in industry output. So, in this case, this specific index can be seen as one leading the economic cycle by as much as about 4 months. By using these two standards, we can not only classify indices as synchronic, leading, or stagnant, but also the specific amount of leading or staggering amount of time.

When $L = 0$, the index of "cash payout as salaries" is ranked relatively in the front. So, it is a natural candidate for being a synchronic indicator. When L-value changes, the relevant changes in its absolute degrees of grey incidence are given in Table 4.3.

From Table 4.3, it follows that when $L = 1$, the absolute degree of grey incidence reaches its maximum. So, speaking strictly, this specific index should be seen as one that is lagging the economic cycle by as much as 1 month. According to the custom, leading or lagging no more than 2 months is seen as synchronic, while more than this range of time will it be treated as either a leading or staggering index.

As a second example, for the index of "gross retail amount of the society," the computational results are provided in Table 4.4.

From Table 4.4, it can be seen that when $L = -6$ the absolute degree of grey incidence of the particular index reaches its maximum. Therefore, it can be seen as a leading indicator.

By using this method, we can compute the L-values corresponding to the maximum absolute degrees of grey incidence of each of the indices of our interest. The results are listed in Table 4.5.

Table 4.5 indicates that we can classify the 16 indices of the eight major classes into three classes as leading, synchronic, and stagnant indices as shown in Table 4.6.

TABLE 4.4 The Absolute Degrees of "Gross Retail Amount of the Society"

L	Absolute degree of grey incidence	L	Absolute degree of grey incidence
−12	0.914466	1	0.856944
−11	0.915117	2	0.866789
−10	0.918527	3	0.876758
−9	0.887243	4	0.882430
−8	0.888258	5	0.889590
−7	0.928151	6	0.895899
−6	0.948684	7	0.900899
−5	0.939351	8	0.900130
−4	0.923900	9	0.895977
−3	0.909621	10	0.894374
−2	0.884610	11	0.892662
−1	0.846814	12	0.889532
0	0.862105		

4.7.2 The Evaluation of Science and Technology Park

The indexes for the evaluation of the science and technology park development in Table 4.7 are explained as follows:).

X_1—the proportion of R&D funds in gross income, measures in percentage

X_2—the products with self-determination knowledge property right or patent right, it can be derived as "number of products/100"

X_3—the proportion of R&D personnel in the whole employees, measured by %

TABLE 4.5 *L*-Values Corresponding to Maximum Absolute Degrees of Incidence of the Indices

Index	L	Absolute degree	Index	L	Absolute degree
Currency in circulation	−6	0.983452	National income	+12	0.718998
Increase in heavy industry output	0	0.979810	Gross amount of imports	−9	0.606556
Increase in light industry output	0	0.972655	Gross amount of exports	+10	0.560054
Gross retail amount of the society	−6	0.948684	Total production of energy	−6	0.555035
Cash payout as salaries	+1	0.877090	Direct investments by foreign entities	−11	0.510016
National expenditure	+12	0.800533	Loans issued by financial institutions	−5	0.508375
Net amount of currency in circulation	+8	0.796688	Savings at financial institutions	−6	0.505588
Total investment in real estate	−11	0.769778	Consumer price index	+11	0.503235

TABLE 4.6 Classifications of Leading, Synchronic, and Stagnant Indices

	Leading index	Synchronic index	Stagnant index
Energy and raw materials	Total production of energy (−6)[a]		
Investment	Total investment in real estate (−11)		
Production		Increase in light industry output (0) Increase in heavy industry output (0)	
Finance			National income (+12) National expenditure (+12)
Currency and credit	Currency in circulation (−6) Savings at financial institutions (−6) Loans issued by financial institutions (−5)	Cash payout as salaries (+1)	Net amount of currency in circulation (+8)
Consumption	Gross retail amount of the society (−6)		
Foreign trade	Gross amount of imports (−9) Direct investments by foreign entities (−11)		Gross amount of exports (+10)
Commodity price			Consumer price index (+11)

[a]Numbers in parentheses stand for the time difference between indicated cycles and reference cycles.

X_4—the proportion of the employees with a Master's or Doctor's degree in the whole employees, measures in percentage

X_5—the organizations engage in R&D, measured by number

X_6—the number of whole employees, measured by 1000 persons

X_7—the policy for science and technology development, marked by experts

X_8—management and impelling mechanism, marked by experts

X_9—venturous capital or seed funds, measured by 100 million yuan

X_{10}—basic installations, marked by experts

X_{11}—information network, marked by experts

X_{12}—environment, marked by experts

X_{13}—gross income, measured by 10 billion yuan

X_{14}—the average income of each employee, measured by 10,000 yuan

X_{15}—the average income of each ha. occupied, measured by 10 million yuan

X_{16}—the total tax and fee turned over to the government, measured by 100 million yuan

X_{17}—international projects, measured by number

TABLE 4.7 The Index System to Appraising Science and Technology Park Development

Indexes		Weight	Weaker class	General class	Stronger class
Capability for innovation 31	X_1	8	$1 \le X_1^1 < 4$	$4 \le X_1^2 < 8$	$8 \le X_1^3 < 12$
	X_2	6	$2 \le X_2^1 < 5$	$5 \le X_2^2 < 10$	$10 \le X_2^3 < 20$
	X_3	4	$2 \le X_3^1 < 5$	$5 \le X_3^2 < 10$	$10 \le X_3^3 < 20$
	X_4	4	$1 \le X_4^1 < 3$	$3 \le X_4^2 < 6$	$6 \le X_4^3 < 10$
	X_5	4	$1 \le X_5^1 < 3$	$3 \le X_5^2 < 6$	$6 \le X_5^3 < 10$
	X_6	5	$2 \le X_6^1 < 5$	$5 \le X_6^2 < 10$	$10 \le X_6^3 < 30$
Environment for innovation 28	X_7	6	$40 \le X_7^1 < 60$	$60 \le X_7^2 < 75$	$75 \le X_7^3 < 90$
	X_8	5	$40 \le X_8^1 < 60$	$60 \le X_8^2 < 75$	$75 \le X_8^3 < 90$
	X_9	5	$5 \le X_9^1 < 10$	$10 \le X_9^2 < 20$	$20 \le X_9^3 < 40$
	X_{10}	4	$40 \le X_{10}^1 < 60$	$60 \le X_{10}^2 < 75$	$75 \le X_{10}^3 < 90$
	X_{11}	3	$40 \le X_{11}^1 < 60$	$60 \le X_{11}^2 < 75$	$75 \le X_{11}^3 < 90$
	X_{12}	5	$40 \le X_{12}^1 < 60$	$60 \le X_{12}^2 < 75$	$75 \le X_{12}^3 < 90$
Economic benefit 25	X_{13}	8	$3 \le X_{13}^1 < 5$	$5 \le X_{13}^2 < 8$	$8 \le X_{13}^3 < 15$
	X_{14}	6	$10 \le X_{14}^1 < 30$	$30 \le X_{14}^2 < 60$	$60 \le X_{14}^3 < 100$
	X_{15}	5	$1 \le X_{15}^1 < 3$	$3 \le X_{15}^2 < 6$	$6 \le X_{15}^3 < 10$
	X_{16}	6	$10 \le X_{16}^1 < 20$	$20 \le X_{16}^2 < 50$	$50 \le X_{16}^3 < 100$
Internationalization 16	X_{17}	4	$30 \le X_{17}^1 < 50$	$50 \le X_{17}^2 < 100$	$100 \le X_{17}^3 < 200$
	X_{18}	3	$3 \le X_{18}^1 < 5$	$5 \le X_{18}^2 < 8$	$8 \le X_{18}^3 < 15$
	X_{19}	4	$10 \le X_{19}^1 < 30$	$30 \le X_{19}^2 < 60$	$60 \le X_{19}^3 < 100$
	X_{20}	5	$10 \le X_{20}^1 < 30$	$30 \le X_{20}^2 < 60$	$60 \le X_{20}^3 < 100$

X_{18}—real receipted Foreign Direct Investment (FDI), measured by 10 million USD

X_{19}—the number of returned students and foreign experts to be engaged, measured by person

X_{20}—export income, measured by 10 million USD

Twenty indexes in Table 4.1 are wholly positive alteration indexes, that is to say: the bigger the index's figure is, the stronger the science and technology park will be.

Four synthetic indexes' weight of the capability for innovation, environment for innovation, economic benefit, and internationalization are 31, 28, 25, and 16, respectively. Each synthetic index's weight can distribute to each relevant branch index again. People can calculate the synthetic capability index from each branch index directly, and may not have to calculate four synthetic indexes separately [54].

The results in Table 4.2 are suited to evaluate science and technology park at the province (city) level, but cannot be directly applied to the county or country level. At county or country level, the index, requirements, and stresses will be changed according to the alternation of regional level. And the index system, weight of the subindex and its field must be clarified again.

TABLE 4.8 The Data of the Science and Technology Park Indexes in Jiangsu Province in 2006

Index	X_1	X_2	X_3	X_4	X_5	X_6	X_7	X_8	X_9	X_{10}
values	4.6	3.1	5.4	3.3	3.52	16.7	81	72	7.6	86
Index	X_{11}	X_{12}	X_{13}	X_{14}	X_{15}	X_{16}	X_{17}	X_{18}	X_{19}	X_{20}
values	90	85	12.5	73.1	7.29	90.6	122	9.4	26	68.5

On the basis of the data of the science and technology park indexes in Jiangsu province in 2006 (see Table 4.8), the synthetic clustering coefficients σ_i^k of the science and technology park belonged to class $k(k = 1, 2, 3)$ can be computed by the evaluation model using endpoint triangular whitenization functions which was introduced in Section 4.4.4.

$$\sigma_{\text{STP}}^1 = \sum_{j=1}^{20} f_j^1\left(x_j\right)\eta_j = 20.85$$

$$\sigma_{\text{STP}}^2 = \sum_{j=1}^{20} f_j^2\left(x_j\right)\eta_j = 25.97$$

$$\sigma_{\text{STP}}^3 = \sum_{j=1}^{20} f_j^3\left(x_j\right)\eta_j = 47.79$$

As $\max_{1<k<3}\{\sigma_{\text{STP}}^k\} = 47.79 = \sigma_{\text{STP}}^3$, the science and technology park in Jiangsu province can be sorted into " stronger" class.

From Table 4.3, we can get the conclusion that the sticking points that interfere with the science and technology park are X_2, X_3, X_9, X_{19} which have something to do with the products with self-determination knowledge property right or patent right, the proportion of R&D personnel and venturous capital or seed funds.

The whitenization weight functions are suited to evaluate provincial science and technology park in China. If used for evaluation of other levels of region or other country, it must be clarified accordingly.

4.7.3 To Select the Supplier of Key Components of Large Commercial Aircrafts

In China, the production of large commercial aircrafts is managed using the model of main manufacturers–suppliers, where a great amount of key components comes from international suppliers. So, the scientificality of the decision-making on the selection of relevant suppliers is a key link that directly determines the success or failure of the operation. As a typical decision-making problem involved in the production process of sophisticated products, the selection of suppliers is generally accomplished through the form of public bidding. What has been done is that the main manufacturer first lists his clear demands, and then each potential supplier puts together his proposal that meets the needs of the manufacturer. After collecting all the proposals, the manufacturer selects the optimum proposal and signs the purchase agreement by comprehensively evaluating all the submitted proposals of the potential suppliers.

As for what factors actually affect the manufacturer's decision, it is an extremely complicated matter. In order to arrive at educated and scientifically sound decisions, there is a need to analyze all involved factors closely and holistically.

During the selection of international suppliers for a specific key component of the production of large commercial aircrafts, there were there suppliers accepted into the second round of competition. To decide on the eventual decision, let us go through the following steps.

Step 1: Establish the sets of events, countermeasures, and situations. Let us define event a_1 to be the selection of a supplier for the said component for the production of large commercial aircrafts. So, the set of events is $A = \{a_1\}$. Define selecting supplier 1, supplier 2, or supplier 3 to be our countermeasures b_1, b_2, and b_3 so that the set of countermeasures is $B = \{b_1, b_2, b_3\}$. Therefore, the set of situations of our concern is $S = \{s_{ij} = (a_i, b_j) | a_i \in A, b_j \in B, i = 1; j = 1, 2, 3\} = \{s_{11}, s_{12}, s_{13}\}$.

Step 2: Determine the decision objectives. Through three rounds of surveys of relevant experts, the following five objectives are considered: quality, price, time of delivery, design proposal, and competitiveness.

Among these objectives, competitiveness, quality, and design proposal are qualitative. They are scored by relevant experts' evaluations satisfying that the higher the evaluation scores are the better. That is, they are benefit-type objectives. Let us take the threshold value $u_{i_0j_0}^{(k)} = 9, k = 1, 4, 5$. For the objective of cost, the lower the cost, the better. So, it is a cost-type objective. Let us take the threshold value $u_{i_0j_0}^{(2)} = 15$. The objective of "time of delivery" is one of moderate-value type. The main manufacturer desires the delivery at the end of the 16th month with 2 months deviation allowed. That is, $u_{i_0j_0}^{(3)} = 2$, the lower threshold effect value is $16 - 2 = 14$, and the upper threshold effect value is $16 + 2 = 18$.

Step 3: Determine the decision weights of the objectives. To this end, we apply the AHP method to determine the weights, see Table 4.9 for details.

Step 4: Determine the effect sample vectors of each of the objectives:

$$U^{(1)} = (9.5, 9.4, 9),\ U^{(2)} = (14.2, 15.1, 13.9),\ U^{(3)} = (15.5, 17.5, 19),$$
$$U^{(4)} = (9.6, 9.3, 9.4),\ U^{(5)} = (9.5, 9.7, 9.2).$$

Step 5: Assign the threshold effect values for the objectives. Because competitiveness, quality, and design proposal are all benefit-type objectives, let us take the threshold values $u_{i_0j_0}^{(k)} = 9, k = 1, 4, 5$. Because price is a cost-type objective, let us take the threshold value $u_{i_0j_0}^{(2)} = 15$. Because time of delivery is a moderate-value type objective and the main manufacturer desires the delivery at the end of the 16th month with a tolerance of +/− 2 months, we set $u_{i_0j_0}^{(3)} = 2$, the lower threshold effect value $16 - 2 = 14$, and the upper threshold effect value $16 + 2 = 18$.

TABLE 4.9 The Objectives' Evaluation System

Objective	Quality	Price	Delivery	Design	Competitiveness
Unit	Qualitative	Million US$	Month	Qualitative	Qualitative
Order number	1	2	3	4	5
Weight	0.25	0.22	0.18	0.18	0.17

Step 6: Calculate the vectors of uniform effect measures. For the three qualitative objectives, competitiveness, quality, and design proposal, we employ the effect measures of benefit type. For the objective of price, we utilize the effect measures of cost type. For the objective of time of delivery, we apply the lower and upper effect measures. So, we obtain the following vectors of uniform effect measures:

$$R^{(1)} = [1, 0.8, 0], \; R^{(2)} = [0.73, -0.09, 1], \; R^{(3)} = [0.75, 0.25, -0.5],$$
$$R^{(4)} = [1, 0.5, 0.67], \text{ and } R^{(5)} = [0.71, 1, 0.29].$$

Step 7: From $r_{ij} = \sum_{k=1}^{5} \eta_k \cdot r_{ij}^{(k)}$, we compute the following vector of synthetic effect measures:

$$R = [r_{11}, r_{12}, r_{13}] = [0.8463, 0.4852, 0.2999].$$

Step 8: Make the final decision. Because $r_{11} > 0, r_{12} > 0, r_{13} > 0$, it means that all these three suppliers have hit the target. This result implies that it is reasonable for these suppliers to have entered the second round of competition. From $\max_{1 \le j \le 3}\{r_{1j}\} = r_{11} = 0.08463$, it follows that the main manufacturer should sign agreement with supplier 1.

4.8 INTRODUCTION TO THE SOFTWARE OF GREY SYSTEM MODELING

Currently, grey system theory has been widely applied in different scientific areas, while providing solutions to a large number of practical problems and challenges met in daily lives and day-to-day works. To this end, various versions of grey system modeling software have played a very important role.

A computer software on most of grey systems modeling can be downloaded from the web site of Institute for Grey System studies of NUAA (http://ieeegsss.nuaa.edu.cn/institute) for free.

4.8.1 Features and Functions

The software system [24, 29, 55] combines the capabilities of the C/S (client/server) and B/S (browser/server) modules, where the C/S part completes computational functions, while the B/S part handles the relevant operations that serve users and their communication with the server. This package has the following characteristics: (1) Data entry is convenient and fast. For data sequences of the same kind, the package provides a rectangular window, into which users can simply copy

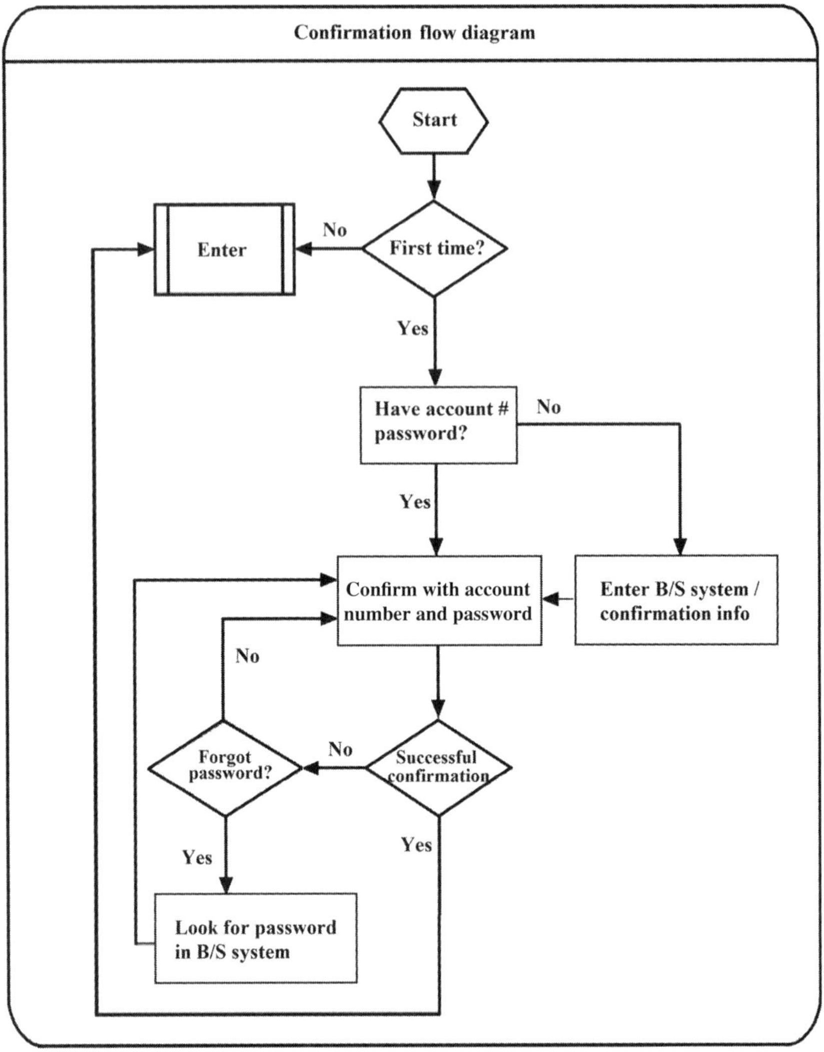

FIGURE 4.9 The confirmation flow chart.

the sequences with one operation. For large sets of data, the user can enter data in an EXCEL document and then open the data file into this package system. (2) Modules are designed according to functionalities. The package contains a group of independent procedures, each of which has its own external characteristics, such as its own name, label, and interfaces. (3) This system provides operational details as well as periodic results such that the user can monitor changes of the data in each computational step such that he can further understand how the model operates. Also, to remind the user of the relevant formulas employed in the model, the software interface provides the relevant information.

4.8.2 Operation Guide

4.8.2.1 The Confirmation System To guarantee the legal ownership of the user and maintain the simplicity of operation of the system, the software applies the XML-based client programming technique. When the user attempts to run the program the first time, the system will prompt him to provide the needed account number and password. The provided data will then be delivered through the Internet to the data base located at the server to verify their legality. When the user attempts to use the program at different, subsequent occasions, he can directly enter the main interface window to avoid the annoyance of having to enter account number and password every time. Figure 4.9 is the flow chart of confirmation.

4.8.2.2 Using the Software Package After users successfully confirm their legal ownership, they will enter the system's main interface window. Various modules (and their submodules) of grey system theory are mainly administrated through menus. Figure 4.10 provides the flow chart of various submodules of the system.

Before running the program, one needs to first enter data into the software package and specify the system parameters. There are two ways to input data. One can directly enter data into the provided text box or import data from an external EXCEL document. For those modules that require large amounts of data, the only way provided for entering data is through importing from EXCEL documents.

Let us look at several examples of applications.

Example 1 Sequence operators

Click on "Sequence generation." From the pull-down menu that appear, select the module according to the practical modeling need. The corresponding detailed modeling interface appears. In particular, what is shown in Figure 4.11 is the interface of the mean weakening buffer operator.

This interface window contains three main areas: the first shows the original data sequence, which is the area for data entry or importing data; the second area the "order and outcome precision," in which one adjusts the order of the operator being applied and the corresponding precision of the computational outputs based on users' need of modeling; and the third the area in which the computational results are

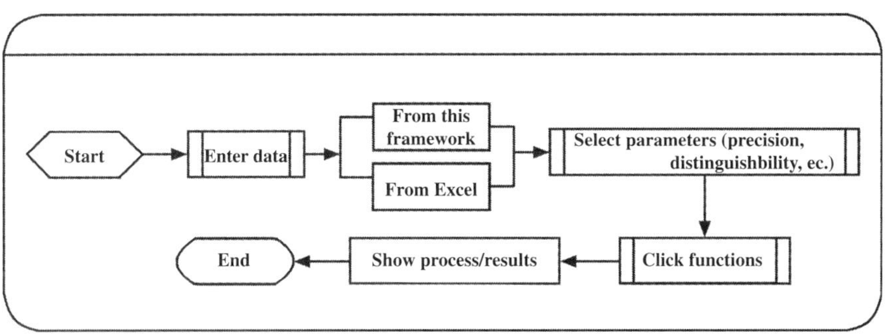

FIGURE 4.10 Submodules.

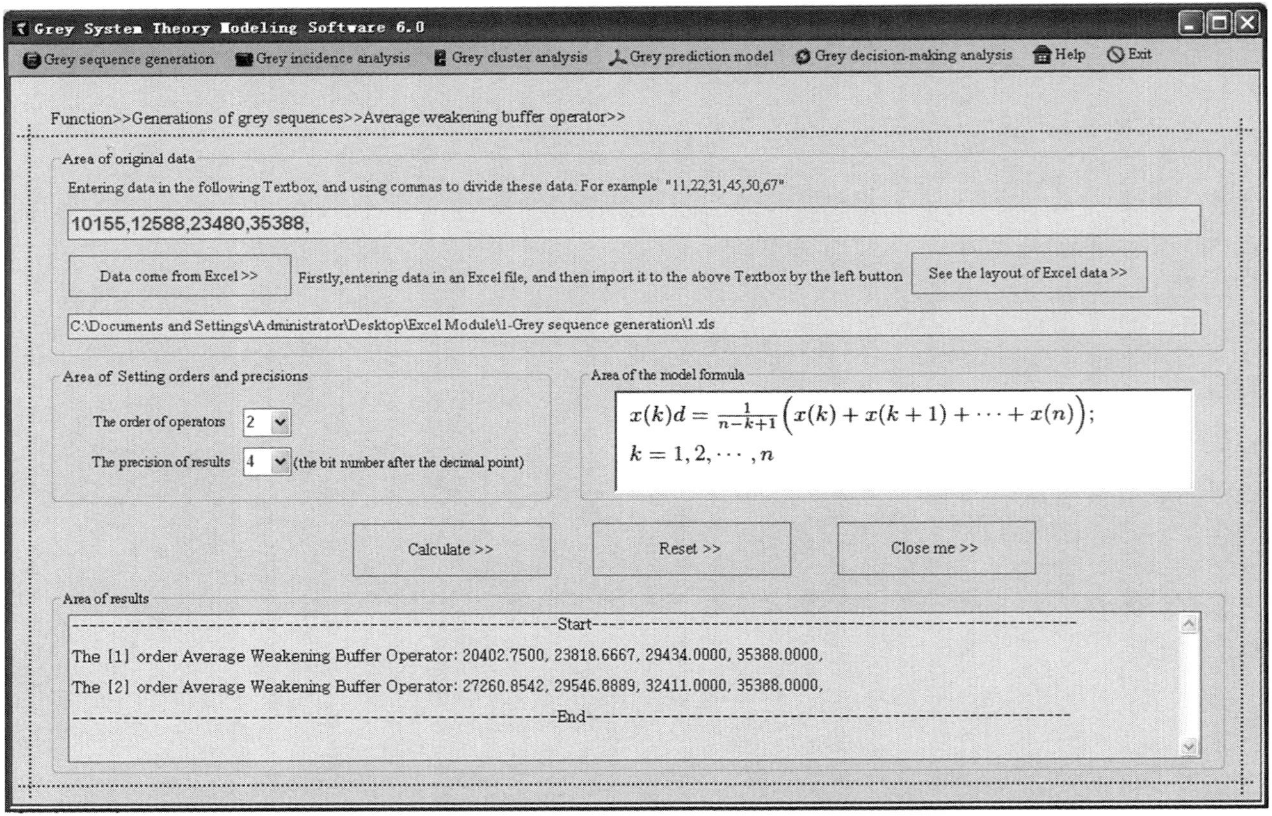

FIGURE 4.11 The interface of the mean weakening buffer operator.

Grey_sequence_generation

	A	B	C	D	E
1	the first data	the second data	the third data	the fourth data	...
2	10155	12588	23480	35388	...
3					

Sheet1 / Sheet2 / Sheet3

FIGURE 4.12 The required EXCEL file format.

shown. After the data entry is completed, click on "mean weakening buffer operator (AWBO)" button. Immediately, the generated sequence appears in the generated sequence window. Figure 4.12 shows a work sheet of an EXCEL document. When applying this capability of EXCEL and deciding on importing data from EXCEL, the user has to follow this shown format exactly.

Example 2 Grey incidence analysis.

Similar to what is shown in Example 1, all parts of incidence analysis, except that of Deng's degree of grey incidence, provide two ways to input data. For Deng's degree of grey incidence, due to the need of large amount of data, this software system provides only the method of data entry through EXCEL documents. Figure 4.13 shows the editing format of an EXCEL document, while Figure 4.14 shows the complete work interface.

Example 3 Grey clustering analysis.

Grey clustering analysis generally requires a large amount of original data with the exception that in grey clustering analysis, the data may have several different types, including objective criterion data, whitenization weight functions, and criterion weights. Therefore, for grey clustering analysis, the software only provides one way to enter data by importing from EXCEL files. The key of using this part of the functions is to correctly edit the different types of data in the EXCEL documents. Sheet1 contains the objective.criteria data (Figure 4.15), Sheet2 the corresponding whitenization weight functions (Figure 4.16), and Sheet3 the weights of the criteria (Figure 4.17).

Figure 4.18 shows the operating interface of grey clustering analysis. As for how to apply grey variable weight clustering and that based on center-point triangular whitenization weight functions, the operational details are similar and omitted.

Deng's incidence degree

	A	B	C	D	E	F
1	Sequences\Data	the first data	the second data	the third data	the fourth data	...
2	the first sequence	45.8	43.4	42.3	41.9	...
3	the second sequence	39.1	41.6	43.9	44.9	...
4	the third sequence	3.4	3.3	3.5	3.5	...
5	the fourth sequence	6.7	6.8	5.4	4.7	...
6	...	...	...	...	...	...
7						

Sheet1 / Sheet2 / Sheet3

FIGURE 4.13 The required EXCEL file format.

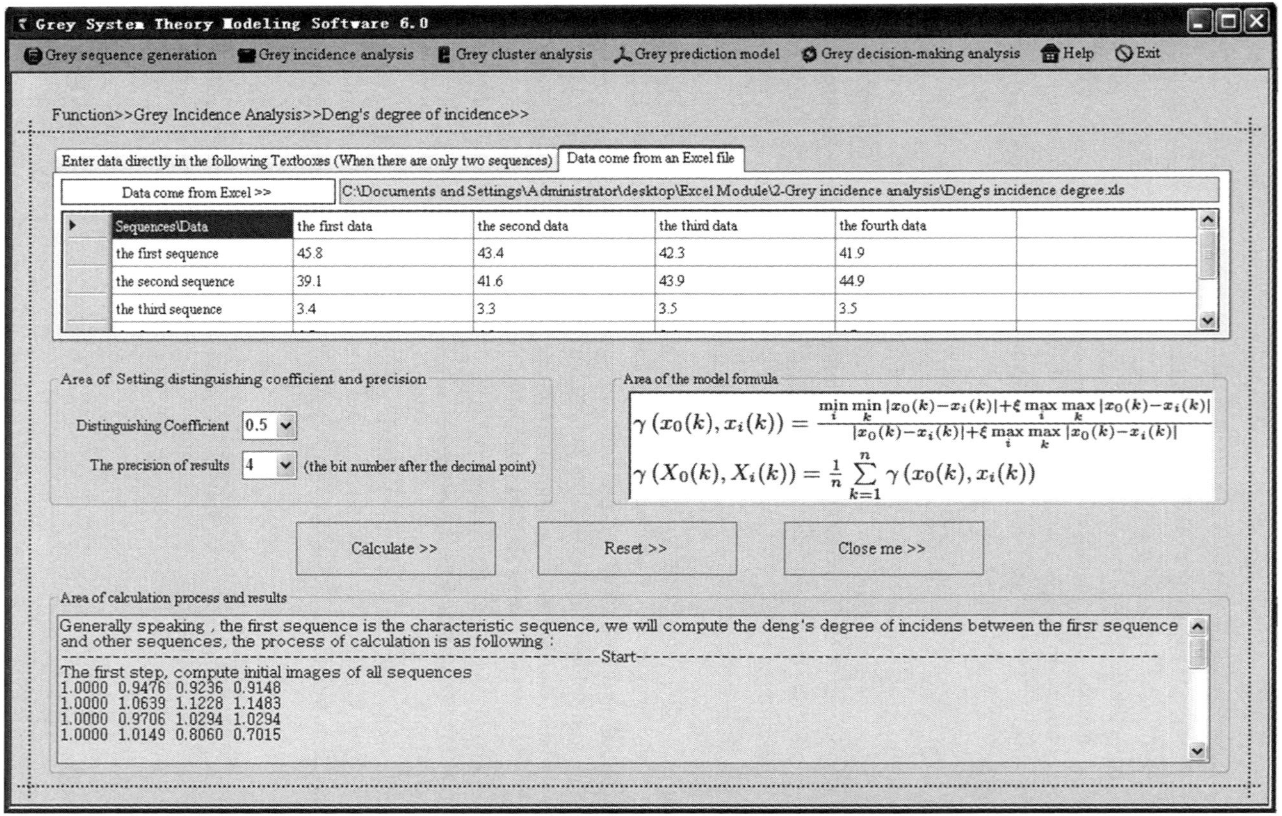

FIGURE 4.14 The complete work interface of Deng’s degree of grey incidence.

Grey fixed clustering [兼容模式]

	A	B	C	D	E	F
1		Parameter values	Parameter values	Parameter values	Parameter values	...
2	Object	22.5	4	0	0	...
3	Object	79.37	6	600	0.75	...
4	Object	144	7	300	0.75	...
5	Object	300	6.1	189	12	...
6	Object	456	12	250	12	...
7	Object	189	8	700	1.5	...
8	Object	369	8	1300	2.25	...
9	Object	1127.11	16.2	550	3	...
10	Object	260	11	600	1	...
11	Object	200	8	600	1.25	...
12	Object	475	10	1000	0.75	...
13	Object	314.1	8	900	0.75	...
14	Object	282.8	7.4	1300	0.5	...
15	Object	240	8	1200	0.5	...
16	Object	160	5	1000	0.25	...
17	Object	270	8	1200	0.25	...
18	Object	9	1	200	0	...
19	...	...	...	...	...	...
20						

Sheet1 Sheet2 Sheet3

FIGURE 4.15 The objective criteria data.

Grey fixed clustering [兼容模式]

	A	B	C	D	E	F
1		the first parameter	the second parameter	the third parameter	the forth parameter	...
2	Whitenization weight function of the first grey class	100,300,-,-	3,10,-,-	200,1000,-,-	0.25,1.25,-,-	...
3	Whitenization weight function of the second grey class	50,150,-,250	2,6,-,10	100,600,-,1100	0,0.5,-,1	...
4	Whitenization weight function of the third grey class	-,-,50,100	-,-,4,8	-,-,300,600	-,-,0.25,0.5	...
5	...	...	...	...	...	...

Sheet1 Sheet2 Sheet3

FIGURE 4.16 The corresponding whitenization weight functions.

Grey fixed clustering [兼容模式]

	A	B	C	D	E	F	G	H
1		the first parameter	the second parameter	the third parameter	the forth parameter	...		
2	Weight	0.3	0.25	0.25	0.2	...		

Sheet1 Sheet2 Sheet3

FIGURE 4.17 The weights of the criteria.

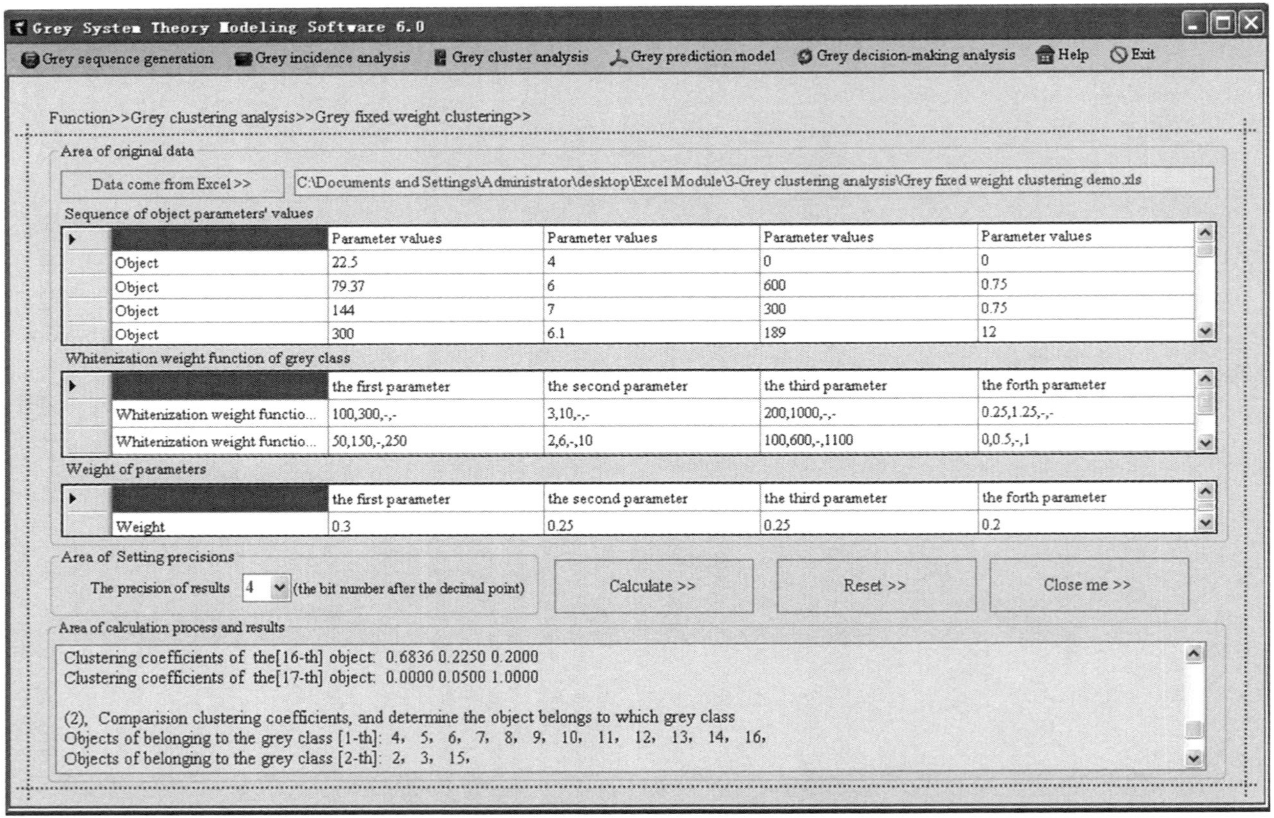

FIGURE 4.18 The operating interface of a grey clustering analysis.

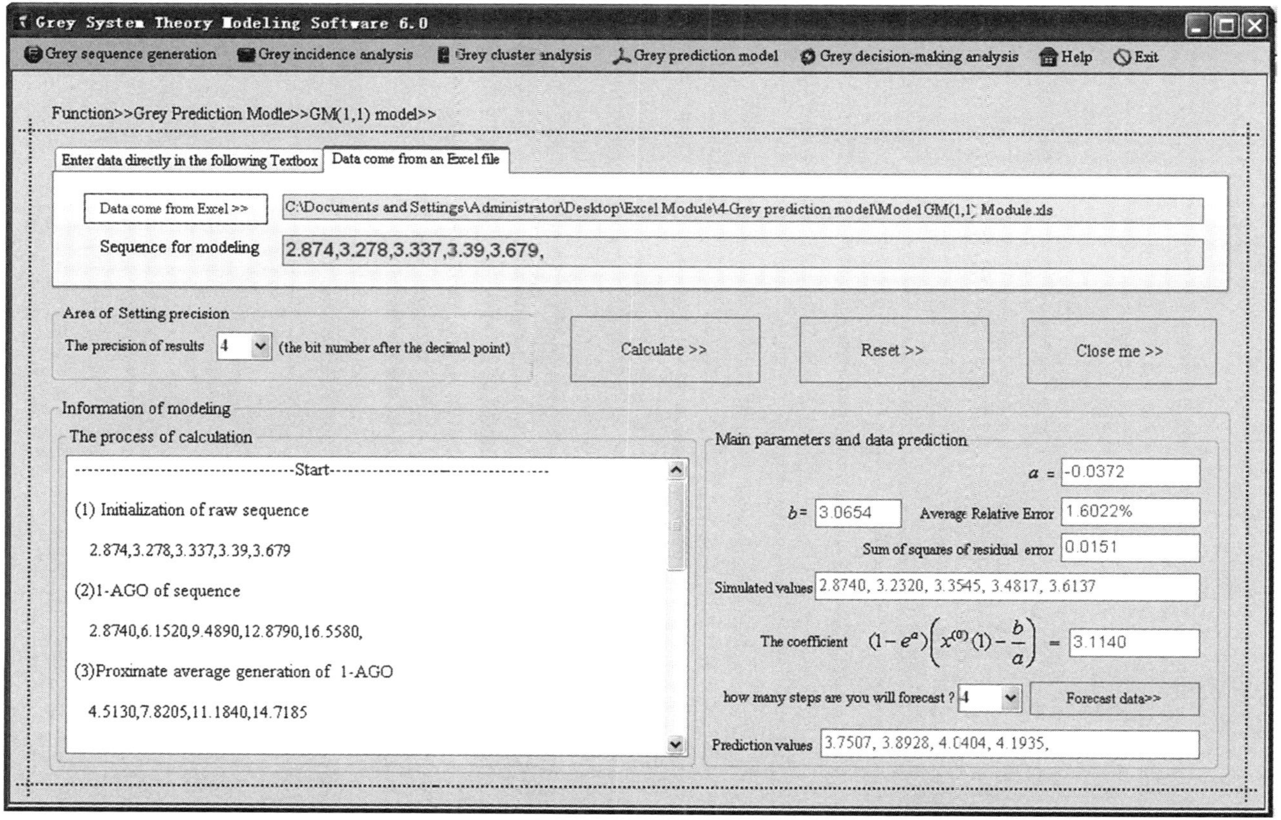

FIGURE 4.19 The operational interface of the GM(1,1) model.

Decision-making of intelligence grey target demo [兼容模式]

	A	B	C	D	E	F	G	H
1		Situation-11	Situation-12	Situation-13	...	Critical Value	Weight of parameter	Type of measure
2	Target1	9.5	9.4	9	...	9	0.25	max
3	Target2	14.2	15.1	13.9	...	15	0.22	min
4	Target3	15.5	17.5	19	...	14,18	0.18	moderate
5	Target4	9.6	9.3	9.4	...	9	0.18	max
6	Target5	9.5	9.7	9.2	...	9	0.17	max
7	...	...	...	...	...	...	...	...
8								

Sheet1 Sheet2 Sheet3

FIGURE 4.20 The exact layout of data page for intelligent grey target decision-making.

Example 4 Grey prediction models

The operation of each individual grey prediction model is roughly the same. So, let us use the GM(1,1) model to illustrate how to use the software system. The main steps include: enter or import data; click the computation, simulation, prediction button to compute the model parameters and the simulated values, and select the simulation accuracy; by entering the desired number of predicted values, then click "prediction results." Figure 4.19 shows the operational interface of the GM(1,1) model.

Example 5 Grey target decision-making.

This part of the software contains two modules, one for synthesized objective decision-making, and the other intelligent grey target decision-making. Similar to grey clustering analysis, one needs a large amount of data, including the comprehensive evaluation matrices of situations, measurement types of criteria and relevant weights, etc. So, for synthesized objective decision-making, there is also only one way to enter data through importing from EXCEL sources. The key of applying this part of the functions is to correctly edit the data files in EXCEL.

The data layout of an intelligent grey target decision-making is the same as that of any synthesized objective decision-making except that there is an additional column of threshold values, as shown in Figure 4.20, where interval [56, 57] means that the lower effect threshold value is 14 and the upper effect value is 18. Figure 4.21 shows the entire operating page of an intelligent grey target decision-making.

ACKNOWLEDGMENTS

This work was supported by a Marie Curie International Incoming Fellowship within the 7th European Community Framework Programme (Grant No. FP7-PIIF-GA-2013-629051), a project of the Leverhulme Trust International Network entitled "Grey Systems and Its Applications" (IN-2014-020), the National Natural Science Foundation of China (91324003), the joint research project of both the NSFC (71111130211) and the RS of UK, the major project and key project of Social Science Foundation of the China (10zd&014, 12AZD102), At the same time, the authors would like to acknowledge the partial support of the Fundamental Research Funds for the Central

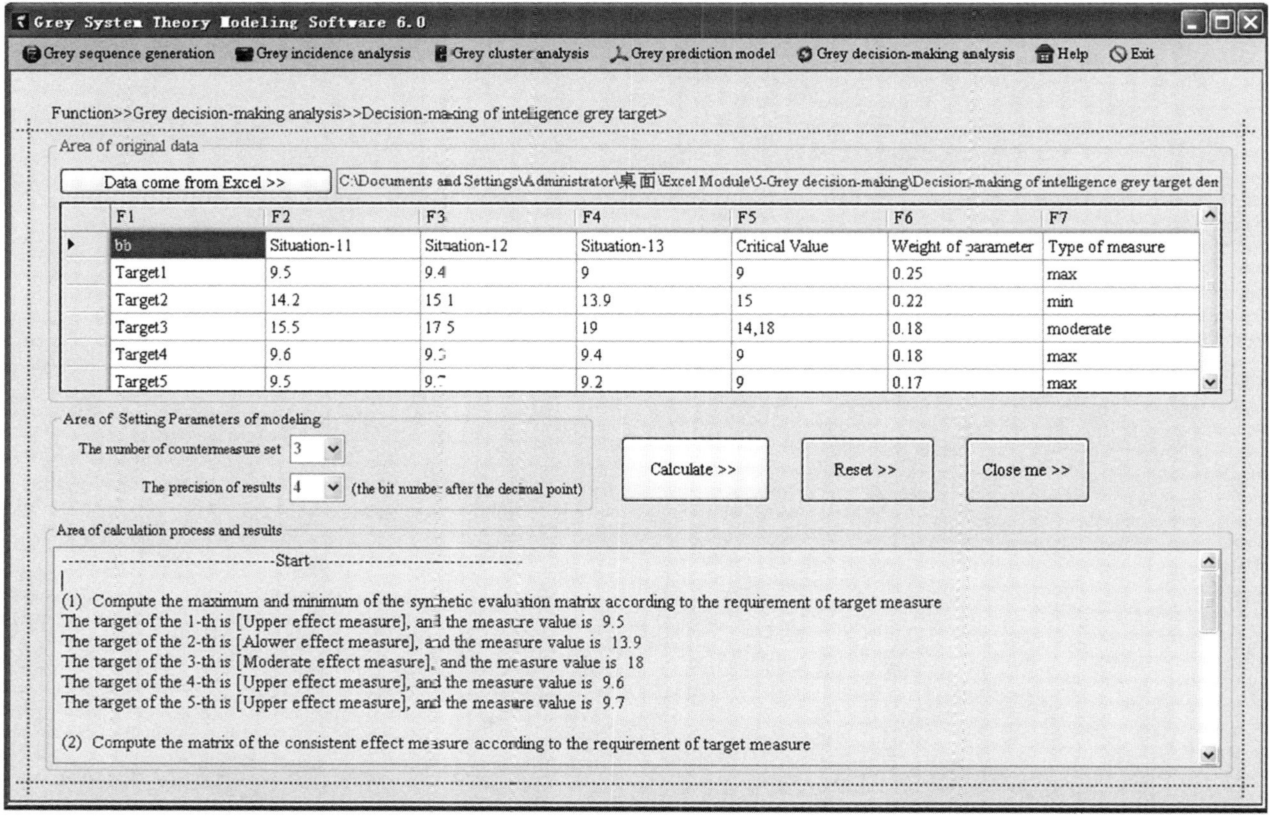

FIGURE 4.21 The entire operating page of grey target decision-making.

Universities (NP2015208), the Research Funds for the Think Tanks for Science and Technology Development (BR2014100), the Foundation for Key Research Base of Philosophy and Social Science in Universities of Jiangsu Province (2014JDXM015), and the Foundation for national outstanding teaching group of China (No. 10td128).

REFERENCES

[1] A. M. Andrew. Why the world is grey. *Grey Systems: Theory and Application*, 1(2): 112–116, 2011.

[2] J. F. Gu and G. Z. Xu. *Preface for Grey Systems Theory and Its Application* [in Chinese], 2nd edn. Beijing: Science Press, 1999.

[3] K. W. Hipel. Book reviews: Grey systems: theory and applications. *Grey Systems: Theory and Application*, 1(3): 274–275, 2011.

[4] R. Vallee. Cybernetics and systems, from past to future. In: Proceedings of the Joint Conference of WOSC and IIGSS, Pittsburgh, Pennsylvania, PA, 2002, pp. 1–15.

[5] R. Vallee. Book reviews: grey information: theory and practical applications [J]. *Kybernetes*, 37(1): 89, 2008.

[6] L. V. Bertranffy. General Systems Theory: Foundation, Development, Applications, New York: George Braziller, 1968.

[7] C. E. Shannon. *A mathematical theory of communication, Bell System Technical Journal*, 27: 379–423 & 623–656, 1948.

[8] N. Wiener. Cybernetics: Or Control and Communication in the Animal and the Machine. New York: John Wiley, 1948.

[9] I. Prigogine and G. Nicolis. *Self-Organization in Non-Equilibrium Systems: From Dissipative Structures to Order Through Fluctuations.* J. Wiley & Sons, New York, 1977.

[10] H. Haken. *Synegetics*, 2nd edn. Berlin: Springer, 1978.

[11] R. Thom. Structural Stability and Morphogenesis. Benjamin-Addison-Wesley, 1975.

[12] M. Eigen and P. Schuster. “Part A: Emergence of the Hypercycle.” *Naturwissenschaften*, 65: 7–41, 1978.

[13] J. H. Holland. *Adaptation in Natural and Artificial Systems.* Oxford, England: U Michigan Press, 1975.

[14] T. K. Alligood, T. D. Sauer, and J. A. Yorke. Chaos: An Introduction to Dynamical Systems, Springer, 2000.

[15] B. Mandelbrot. The Fractal Geometry of Nature. W. H. Freeman, 1983.

[16] L. A. Zadeh. Fuzzy sets. *Information and Control*, 8: 338–353, 1965.

[17] J. L. Deng. Control problems of grey systems. *Systems & Control Letters*, 1(5): 288–294, 1982.

[18] J. L. Deng. *Grey Control Systems* [in Chinese]. Wuhan: Press of Huazhong University of Science and Technology, 1985.

[19] J. L. Deng. *A Course in Grey Systems Theory* [in Chinese]. Wuhan: Press of Huazhong University of Science and Technology, 1990.

[20] J. L. Deng. *Several Problems on the Research and Development of Grey Systems, New Developments in Grey Systems Research* [in Chinese], edited by S. F. Liu and Z. X. Xu. Wuhan: Press of Huangzhong University of Science and Technology, 1996, pp. 1–12.

[21] Z. Pawlak. Rough sets. *International Journal of Computer & Information Sciences*, 11(5): 341–356, 1982.

[22] S. F. Liu and Y. Lin. *An Introduction to Grey Systems Theory*. IIGSS Academic Publisher, 1998.

[23] S. F. Liu and Y. Lin. *Grey Information: Theory and Practical Applications [M]*. Springer-Verlag London Ltd, 2006, pp. 64–67, 82, 83, 97, 119–120, 139–153, 199–202, 224–228, 277–281.

[24] S. F. Liu and Y. Lin. *Grey Systems: Theory and Applications [M]*. Springer, 2011, pp. 119–122 & 149–152.

[25] Y. Lin and S. F. Liu. A systemic analysis with data (I). *International Journal of General Systems*, 29(6): 989–999, 2000.

[26] Y. Lin and S. F. Liu. A systemic analysis with data (II). *International Journal of General Systems*, 29(6): 1001–1013, 2000.

[27] S. F. Liu, N. M. Xie, and J. Forrest. Novel models of grey relational analysis based on visual angle of similarity and nearness. *Grey Systems: Theory and Application*, 1(1): 8–18, 2011.

[28] Y. Zheng. On the optimization concept of grey systems. *Applied Mathematical Modelling*, 17(7): 388–392, 1993.

[29] S. F. Liu, Y. J. Yang, and L. F. Wu. *Grey System Theory and Application* [in Chinese], 7th edn. Science Press, 2014.

[30] S. F. Liu and Y. D. Zhu. Grey-econometrics combined model. *The Journal of Grey System*, 8(1): 103–110, 1996.

[31] Y. J. Yang and S. F. Liu. Reliability of operations of grey numbers using kernels. *Grey Systems: Theory and Application*, 1(1): 57–71, 2011.

[32] S. F. Liu and N. Xie. A new grey evaluation method based on reformative triangular whitenization weight function. *Journal of Systems Engineering*, 26(2): 244–250, 2011.

[33] S. F. Liu. On measures of information content of grey numbers. *Kybernetes: The International Journal of Systems & Cybernetics*, 35(6): 899–904, 2006.

[34] S. F. Liu. Axiom on grey degree. *The Journal of Grey System*, 8(4): 397–400, 1996.

[35] K. L. Wen, T. C. Chang, and M. L. You. Grey entropy and its application in weighting analysis. In: Proceedings of the IEEE International Conference on Systems, Man and Cybernetics, San Diego, California, USA, 1998, pp. 1842–1844.

[36] S. F. Liu. The three axioms of buffer operator and their application. *The Journal of Grey System*, 3(1): 39–48, 1991.

[37] Y. Lin and S. F. Liu. Law of exponentiality and exponential curve fitting. *Systems Analysis Modelling Simulation (SAMS)*, 38(1): 621–636, 2000.

[38] Y. Wei. Two effective methods of solving approximately the parameter of GM(1,1). *Mathematics in Practice and Theory*, 37(7): 108–112, 2007.

[39] S. F. Liu, Z. G. Fang, and Y. Lin. Study on a new definition of degree of grey incidence. *Journal of Grey System*, 9(2): 115–122, 2006.

[40] C. Sun, J. Li, and H. Zheng. A new method of faulty insulation diagnosis in power transformer based on degree of area incidence analysis [J]. *Power System Technology*, 26(7): 24–29, 2002.

[41] Q. S. Zhang. Genetic algorithm-based optimization of the distinguishing coefficient in relational analysis. *Chinese Journal of Management Science*, 11(5): 76–79, 2003.

[42] E. Scarlat and C. Delcea. Complete analysis of bankruptcy syndrome using grey systems theory. *Grey Systems: Theory and Application*, 1(1): 19–32, 2011.

[43] J. Li, C. Sun, and W. Cheng. Study on fault diagnosis of insulation of oil-immersed transformer based on grey cluster theory [J]. *Transactions of China Electrotechnical Society*, 17(4): 80–83, 2002.

[44] G. P. Wu and Z. X. Xu. Grey ART dynamic clustering evaluation for oil and gas information in oxido-reducing potential [in Chinese]. *Geophyscial Prospecting for Petroleum*, 40(3): 107–111, 2001.

[45] S. F. Liu. Fixed weight grey cluster analysis and evaluation. In: New Methods of Grey Systems, Beijing: Agriculture Press, pp. 178–184, 1993.

[46] S. F. Liu and Y. D. Zhu. Study on Triangular Model and Indexes in Synthetic Evaluation of Regional Economy. *Transactions of the Chinese Society of Agricultural Engineering*, 9(2): 8–131, 1993.

[47] Z. L. Wang and S. F. Liu. Trend testing of grey dynamic models. In: 2004 IEEE International Conference on Systems, Man and Cybernetics, The Hague, Netherlands, 2004, pp. 2438–2442.

[48] X. P. Xiao and K. K. Peng. Research on generalized non-equidistance GM(1,1) model based on matrix analysis. *Grey Systems: Theory and Application*, 1(1): 87–96, 2011.

[49] H. Morita, T. Kase, Y. Tamuray, and S. Iwamoto. Interval prediction of annual maximum demand using grey dynamic model. *International Journal of Electrical Power and Energy System*, 18(7): 409–413, 1996.

[50] Z. M. Song and S. H. Zhang. Separated modeling method in grey system. *Systems Engineering—Theory & Practice*, 22(5): 103–107, 2002.

[51] S. F. Liu and J. L. Deng. The applicability range of GM(1,1) model. *Systems Engineering Theory and Practice*, 20(5): 121–124, 2000.

[52] N. M. Xie and S. F. Liu. Discrete GM (1,1) and the modeling mechanism of grey prediction model. *Systems Engineering Theory and Practice*, 25(1): 93–99, 2005.

[53] S. F. Liu, B. Xu, J. Forrest et al. On Uniform Effect Measure Functions and a Weighted Multi-attribute Grey Target Decision Model. *The Journal of Grey System*, 25(1): 1–11, 2013.

[54] S. F. Liu. On index system and mathematical model for evaluation of scientific and technical strength. *Kybernetes: The International Journal of Systems & Cybernetics*, 35(7–8): 1256–1264, 2006.

[55] B. Zeng, S. F. Liu, and W. Meng. Development and Application of MSGT6.0 (Modeling System of Grey Theory 6.0) Based on Visual C# and XML. *The Journal of Grey System*, 23(2): 145–154, 2011.

[56] C. R. Lin and J. L. Deng. *On Grey Prediction of Gas Pool, New Developments in Grey Systems Research* [in Chinese], edited by S. F. Liu and Z. X. Xu. Wuhan: HUST Press, 1996, pp. 25–31.

[57] Y. Lin and S. F. Liu. Several programming models with unascertained parameters and their application. *Journal of Multi-Criteria Decision Analysis*, 8(1): 206–220, 1999.

CHAPTER 5

BUILDING RESILIENCE: NAVAL EXPEDITIONARY COMMAND AND CONTROL

CHRISTOPHER NEMETH[1], THOMAS MILLER[1], MICHAEL POLIDORO[2], and C. MATTHEW O'CONNOR[3]
[1]Cognitive Solutions Division, Applied Research Associates Inc., Fairborn, OH, USA
[2]Basic Commerce and Industries Inc., Dahlgren, VA, USA
[3]Department of the Navy, Washington, DC, USA

5.1 INTRODUCTION

This chapter describes an 18-month project to demonstrate how to incorporate human-centered design [1] into the Navy's systems acquisition process. Our research team used cognitive systems engineering (CSE) methods [2] including cognitive task analysis (CTA) [3] to understand the cognitive work and information needs up and down the C2 (command and control) hierarchy of two Navy Expeditionary Combat Command (NECC) forces. Maritime Expeditionary Security Force (MESF) provides port and harbor defense, protects US high value assets, and performs humanitarian missions. The Riverine Force ensures the security of inland waterways and performs peacekeeping missions. Both forces have the highest operations tempo and most complex tactical coordination requirements of any in NECC.*

The project provides a common language for the Chief of Naval Operations and the Naval Sea Systems Command staff to use in the acquisition of systems that are intended to support decision-making. Three deliverables our team produced made that language evident.

*In 2012, The U.S. Navy reorganized the Maritime Expeditionary Security Force and the Riverine Force into the Coastal Riverine Force.

Contemporary Issues in Systems Science and Engineering, First Edition.
Edited by MengChu Zhou, Han-Xiong Li and Margot Weijnen.

- A set of requirements for decisions, and the displays to support making those decisions, were drawn from extensive interviews and observations.
- An interactive prototype demonstrated display organization and operation.
- A 60-page guidebook that described the approach, from more senior Navy program managers to more junior systems engineers and contractors.

Project results make it possible to account for individual and team cognitive activities among operating forces, and to use those requirements to create and evaluate systems that are intended to support decisions. We anticipate that this will produce decision support systems that enable expeditionary forces to better perform as a complex adaptive system—to improve their resilience [4].

5.2 EXPEDITIONARY OPERATIONS COMMAND AND CONTROL

The 2000 attack on USS Cole in Aden, Yemen demonstrated the need for Navy force protection. In 2006, the Navy responded by establishing the NECC to manage, train, and equip forces in order to conduct maritime security and joint contingency operations in novel, evolving missions. These missions include coastal warfare and patrol, riverine warfare, maritime port security and naval construction, expeditionary diving and combat salvage, and explosive ordnance disposal. Depending on the mission, portions of 11 NECC commands are assembled and deployed in what are termed "bolt-on" forces to support other US Navy forces, joint forces, and coalition partners.

NECC requirements for C2 are complex for a number of reasons. NECC forces are routinely configured differently to serve a range of different and novel missions across multiple theatres of operation. The force's concepts of operations, organization, and procedures have continued to evolve since its inception. The force continues to define its role in the larger Navy organization, its capabilities, and how to best support naval expeditionary missions around the world. NECC command roles differ widely, from MESF's coastal patrol and harbor defense mission to the Naval Mobile Construction Battalion's mission.

When NECC forces are deployed, a three-tier command structure handles expeditionary mission tactics and planning. As Figure 5.1 shows, MESF and Riverine Force patrol leaders, watch supervisors, and senior commanders collaborate with allied and coalition forces, often at a high operations tempo.

At the lowest level of the command structure, crews aboard boats conduct operations along coasts and harbors (MESF) as well as rivers (Riverine Force). A watch staff at the next senior level monitors the seaward and landward tactical situation in an assigned area of responsibility and manages boat operations. If terrain effects on radio signal impede radio communications between the tactical operations center and Riverine boats, a mobile armored vehicle may be assigned to relay communications between them. The senior staff oversees the tactical watch and handles planning,

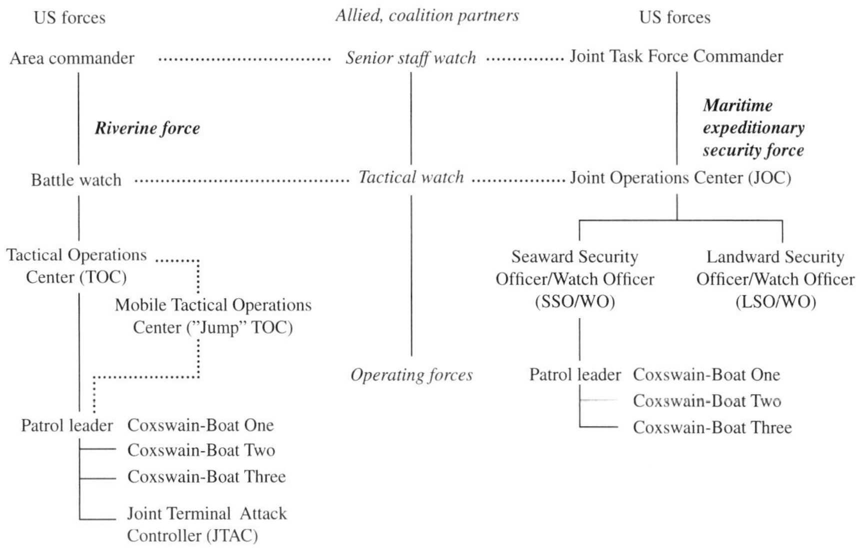

FIGURE 5.1 Deployed expeditionary command hierarchy [5]. Copyright © 2012 Applied Research Associated, Inc. Used by permission.

communications support, logistics, and coordination with the next senior command echelon. Each level also coordinates operations with coalition and host country forces.

5.2.1 Systems Acquisition

The US Department of Defense follows a standard process to acquire systems for military forces to use in the performance of missions that vary widely in nature, scale, and location. The acquisition process follows a prescribed series of steps according to Defense Architecture Framework [6] that govern enterprise architecture and systems architecture. The enterprise architecture is a coherent set of principles, methods, and models. These are used to design and realize the organizational structure, business processes, information system, and infrastructure of an enterprise (collection of organizations with common goals/bottom line) [7]. The systems architecture is the organization of a system that is embodied in its components, their relationships to each other and the environment, and the principles that govern its design and evolution [8]. The Defense Architecture Framework is a reference model that the Department of Defense has developed in order to organize the enterprise architecture and systems architecture to be complementary and consistent. In the Navy, the Chief of Naval Operations (OPNAV) staff and Naval Sea Systems Command (NAVSEA) staff work together to acquire new surface systems using the Defense Architecture Framework model. While the OPNAV staff accounts for operator requirements, the NAVSEA staff develops solutions to meet those requirements.

Acquisition projects range in size from the largest (ACAT I) to the smallest (ACAT IV). A further smaller scale program, the Abbreviated Acquisition Process (AAP), is an *ad hoc* approach that has been developed to quickly field systems by implementing only necessary ACAT IV steps. NECC C2 systems are typically ACAT IV and AAP projects. In these projects the Navy purchases commercial off-the-shelf (COTS) technology and modifies it to fit the military need and specifications.

The Navy has traditionally used the Joint Capabilities Integration Development System to acquire the smaller scale systems that the NECC forces use. The process accounts for simpler tasks that operators perform such as sending a radio message. However, US Navy officers and enlisted personnel routinely perform more complex activities such as decision-making, planning, replanning, sensemaking, and risk management during operational missions. This is cognitive work that a C2 system must support [9].

5.3 HUMAN-CENTERED SYSTEM DEVELOPMENT

The function of C2 relies on individual and team decision-making that have traditionally been studied at the individual, group, and organizational levels. This is human behavior that typically results in a choice among options, as an individual or group considers alternatives, estimates outcomes, elects a course of action, and assesses the results. In complex environments such as those in which NECC forces operate, decision makers continually engage in a series of behaviors that are designed to assess the current situation, reduce uncertainty to the extent possible, anticipate choice consequences, execute courses of action, and then assess the effects of those actions to make further decision–action–assessment sequences possible. Previous research has identified additional macrocognitive behaviors [3] in addition to these cognitive behaviors. Macrocognitive behaviors are thought of as a collection of cognitive behaviors that characterize how people think in real-world settings, including:

- *Planning*—Change in actions, in order to change a current state into a desired state.
- *Replanning/adaptation*—Change to, or replacement of, a plan that is already in progress.
- *Problem detection*—Recognition of anomalies in time to avoid or deflect consequences.
- *Coordination*—The manner in which team members sequence actions to perform a task.
- *Naturalistic decision-making*—Reliance on experience to identify plausible courses of action that can be evaluated through mental simulation.
- *Sensemaking and situation assessment*—Diagnosis of how things came to be the way they are and anticipation of how it will develop in the future.

These processes are typically related to each other although they are typically studied separately. In real-world settings such as expeditionary missions we studied, situations can require operators to engage in any and all of these processes; sometimes in succession, other times in combination [9].

5.3.1 Envisioned World Problem

Envisioned world [10] studies explore how people will operate in a world that has yet to be created, and how to support the way the future world is expected to work. Design for an envisioned world is the design of technology to fit the cognitive demands in an anticipated work setting. The need to foresee a future world presents a number of challenges for a project such as this. Our approach had to anticipate what would occur in a work context that the introduction of new technology would substantially change. In such a project, the research team uses analyses of operators and the cognitive work that they perform in the present to develop hypotheses about ways to improve future performance [11]. This makes it possible to manage the difference between current and expected circumstances. The hypotheses are embodied in design prototypes and used to discover additional support requirements.

5.3.2 Cognitive Systems Engineering

CSE is a research approach that is used to understand work domains, operator goals and work, obstacles that operators confront, as well as to develop support that will enable them to perform more reliably, efficiently, and safely. It includes Cognitive Task Analysis (CTA), which is a structured interview method that is used to elicit knowledge from subject-matter experts (SMEs). The accumulation of experience over time produces expertise: the ability to see what others who are less experienced do not [12]. The expert decision maker is able to detect what is typical, or able to notice when things are not going as they do normally, because he or she has seen the circumstances many times before. CSE research methods such as CTA can be used to elicit expertise, then to develop information systems that support decision-making. As Figure 5.2 illustrates, CSE follows an iterative five-phase process to design a C2 system that is based on decision-making.

- *Preparation*—Develop an understanding of the work domain, tasks, and users in order to identify cognitively complex tasks.
- *Knowledge elicitation*—Understand critical decisions and identify key decisions through the use of CTA methods.
- *Analysis and representation*—Decompose data into discrete elements to identify decision requirements and central themes and issues for which the solution will make a difference.
- *Application design*—Use decision requirements to drive design requirements. Creation of rough and refined prototypes gives the solution a physical form.
- *Evaluation*—Use criteria developed during data analysis to determine how and how well the solution improves user performance.

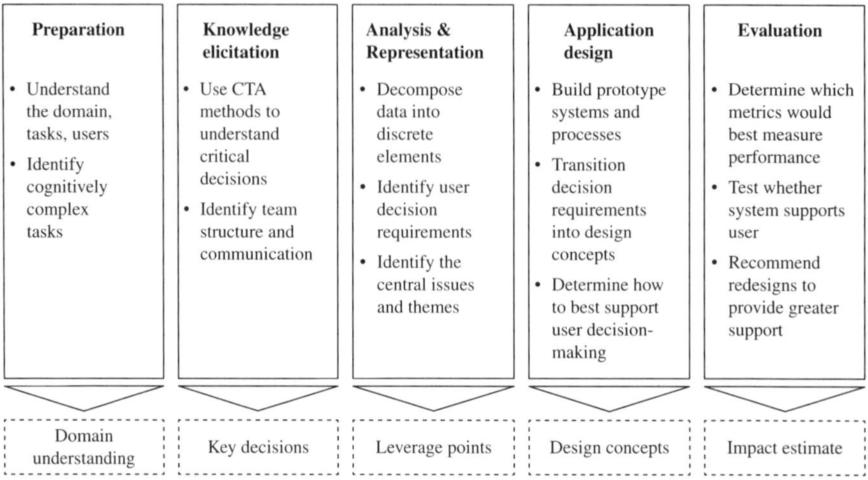

FIGURE 5.2 Human-centered design five-phase research process. (With permission from Reference 3; Copyright © 2006 MIT Press).

Other researchers have applied CSE methods to Navy applications. For example, Bisantz *et al.* [13] reported the use of CSE methods to support large-scale system design of a new combatant surface ship. They used cognitive work analysis modeling and CTA interview methods to develop an abstraction hierarchy and decision ladder models. Hutton *et al.* [14] incorporated CTA to develop design concepts that would support decisions made during damage control activities aboard surface ships. Dominguez *et al.* [15], as well as Moundalexis *et al.* [16], describe the use of CTA to develop concepts that address critical cognitive issues that submarine commanding officers and command teams face. Despite these and other reported uses of CSE methods, both Potter *et al.* [17] and Roth *et al.* [18] noted that transitions from research data into concepts for systems to support cognition need to be stronger. Hoffman and Deal [10] proposed steps to address those issues from both the CTA practitioner and the systems engineering point of view.

5.3.3 Application: Navy Expeditionary Combat Command

To prepare for our project, the research team reviewed relevant materials such as manuals and instructions, and conducted phone interviews with five different SMEs. The phase oriented the project team to NECC operations, requirements, and planning. To elicit expertise of operators who would use the C2 system, team members made four field visits to MESF and Riverine Force commands. The research team conducted over 43 structured CTA interviews, and observed work settings, equipment, and an operational exercise. CTA interviews included background information on experience, descriptions of operations and issues, and finished with Critical Decision Method [19]. For analysis and representation, the team identified common and

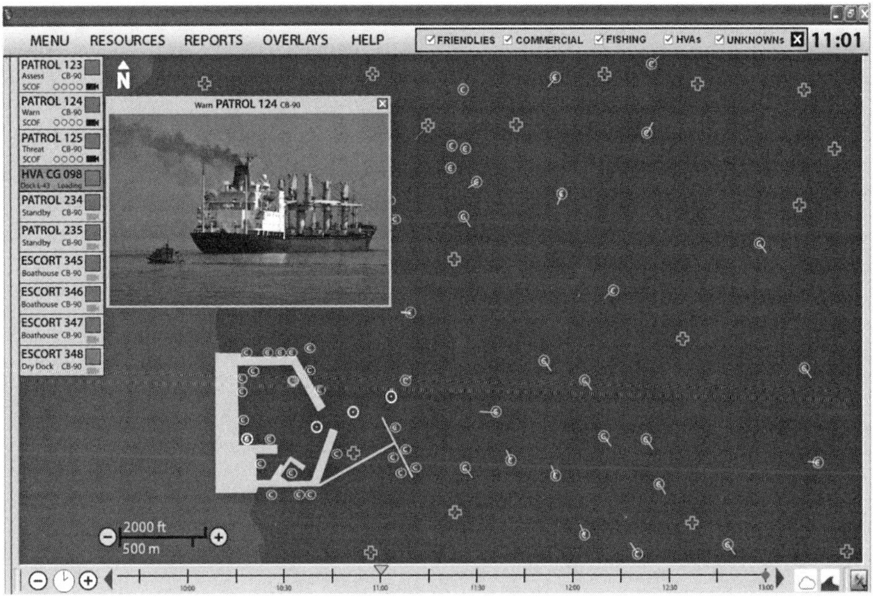

FIGURE 5.3 Sample screen from NECC C2 prototype. Copyright © 2012. Applied Research Associates, Inc. Used by permission.

shared operator requirements for information between MES and Riverine forces, as well as other needs that included training and information system architecture. The team developed initial display prototypes based on requirements, then evaluated them by conducting a "cognitive wall walk." This was an in-person session with SMEs from both forces to verify display prototype features, using scenarios to determine how well display design concepts fit operator experience. The team used results of the research and the wall walk review to program an interactive information display (Figure 5.3), using a previously validated JEC3 software development platform. Final expert review sessions with SMEs determined that the display features provided what the operators needed while performing their mission on deployments.

5.3.4 Reasonable Scientific Criteria

Both more senior as well as more junior staff will use this human-centered design process. Pentagon-level Navy program managers, resource officers, and the materiel Integrated Product Team will use it to manage acquisition. Staff members at Naval Surface Warfare Centers and contractors will use it to develop user operational evaluation systems. Few are likely to be familiar with how to conduct research into cognitive performance. Providing a simple description would not be sufficient to ensure that the process produces a quality result. Some means would be needed to set a standard for how well to perform each step in human-centered design. In order to provide that standard, our research team developed what we termed "reasonable scientific criteria." The criteria can guide new practitioners through CSE in a manner that is both scientifically rigorous, and link design recommendations directly to operator needs. Using steps in the CSE process shown in Figure 5.2, the team considered three questions.

- What reliability/validity criteria are important and reasonable to apply to CTA data?
- What are our standards of practice, and what do we need to do to meet those standards?
- How do we establish and adhere to a rigorous process while also remaining open to the discovery of unexpected processes and outcomes?

Answers to these questions identified a set of quality standards for each stage of the CSE process (Table 5.1).

5.4 DISCUSSION

5.4.1 Resilience Engineering

Our research in this project found various aspects of the NECC command structure and work setting include circumstances that might be considered sources of brittleness [21]. These are vulnerabilities that are the result of an inability to adapt to demand that a common C2 operating picture can benefit. For example, barriers that stand in the way of performing necessary cognitive work include:

- communications that can be impeded by terrain, making coordination erratic;
- disruptions to information sharing due to verbal relays of technical information;
- suboptimal cognitive workload that can result in operators missing information that may be relevant;
- tools and equipment that are a poor match for mission performance;
- organizational issues such as lack of standardization; and
- training issues, such as the need to cross-train individuals in more than one job.

Even the systems acquisition process that was mentioned early in the chapter can itself be considered to be a source of brittleness. The lack of human-centered design up to this point has created systems that do not support the cognitive work of its operators, which limits their ability to adapt to mission demands.

The complexity of activities that expeditionary forces perform requires cognitive aiding to support individual and team performance. Aiding is needed to effectively assign and manage resources to perform the missions we have described. Officers and enlisted personnel at each level of the command hierarchy need an accurate, real-time operational picture. That picture needs to integrate relevant information from the perspectives of patrol boats and patrol leaders and the staff watches in the

TABLE 5.1 Reasonable Scientific Criteria for CSE [20]

CSE step	Standards
Preparation and framing	Clear statements of: • Issue or problem • Framing activities outcome • Method, settings, project participant selection rationale
Knowledge elicitation	Use of multiple knowledge elicitation methods Use of interview and observation guides Purposeful sampling of participants and settings Qualified, prepared data collectors Quality control protocols (specified format to document data) Manage the dual requirements for rigor and flexibility
Analysis and representation	Systematic, purposeful, and documented analysis process Audit trail to connect data elements to findings to design elements Multiple analysis processes and multiple passes through the data Qualified analysis team members Validity checks on findings Goal-driven selection of qualitative versus quantitative analysis Use of reliability indices
Application: design	Iterative design–build–evaluate process SMEs for credibility checks Audit trail to connect data elements, to findings, to design
Evaluation	Clear assessment criteria Review evaluation results systematically and purposefully Evaluation methods reflect key cognitive components, behaviors Outcomes reflect cognitive and behavioral issues critical for cognitive work Verify whether the design/changes improve performance

maritime and tactical operations centers. Without a shared operating picture, patrols can miss information that is crucial to understanding their tactical situation. Critical information can be omitted or distorted through multiple relays that are required to transmit it by voice from node to node. Radar or visual contacts such as surface vessels that may be potential threats may go undetected, or be detected too late. Insufficient time to evaluate or react can compromise the force's ability to engage a threat.

Our research revealed a number of requirements for information that is needed to provide an accurate, integrated, real-time contact picture. These include the status and capability of assets such as the status of US boats, potential threats, and schedules and movement of other crafts. These elements can be developed into display prototypes and evaluated using criteria that we developed during the data collection and analysis phases. Results of display evaluation indicate how well the sources of brittleness have been addressed.

5.4.2 The Data Hub

During the analysis phase, team members developed an internal work-in-process tool in order to keep track of the link from collected data, through data analyses, to requirements for decisions, requirements for displays, and finally display features. They termed this the "data hub" and used two tables to organize the information. One accounts for barriers to NECC mission performance such as unreliable communications, inadequate tools and equipment, boredom and cognitive underload, organizational issues, gaps in training, and information flow up and down the command hierarchy. The other accounts for cognitive demands the operators must meet in order to perform their mission. These demands include understanding the operational picture, developing/repairing/maintaining common ground, information sharing, collaboration and coordination, detection and assessment of potential threats, management of the communication process and equipment, anticipation and planning, problem detection and solving, and the maintenance of vigilance and attention. Table 5.2 is part of the cognitive demand data hub that accounts for information that operators need to understand the operational picture. The table organizes the data starting with the operational need (e.g., accurately interpret what the patrol leader reports about the status of boats), the cost in case the need is not met (e.g., the watch officer cannot provide guidance/direction to the patrol leaders regarding mission accomplishment), barriers that get in the way of mission performance (e.g., delayed, distorted information from boats), and C2 display requirements (e.g., need to understand all patrol craft capabilities and status with respect to current environmental conditions such as sea state). The continuity that this structure provides ensures that the C2 display is valid: based soundly on research team direct observations and operator needs that interviews revealed.

5.4.3 Implementation Challenges

The human-centered design process in this project faces a number of challenges in order to make the transition to actual use. Human-centered research is new to technical staff members who acquire systems for NECC. The variety of their professional abilities and experience may produce a variety of interpretations on how to perform it. A guidebook by itself may not be enough to enable them to successfully perform this process. In addition, the potential exists for technical staff to follow the process, yet overlook the "reasonable scientific criteria" quality standard. For example, it is possible to perform interviews and related activity, yet miss insights that are essential to understand operator cognitive work. That is the reason why the team made quality standards explicit.

5.4.4 Limitations

Our research team was obliged to respect NECC operational requirements. This made it necessary to adjust schedules, and limited our access to some of the operators who were either deployed or had recently returned from deployment. As a demonstration

TABLE 5.2 Portion of the NECC C2 Data Hub

CGD TABLE A: Understanding the Operational Picture
This individual-level activity involves **building and maintaining an understanding of the operational situation, including status and capability of assets; potential threats; and schedules and movement of other craft.** It is a sensemaking activity that involves tracking, filtering, and integrating information from multiple sources, and constantly updating one's understanding.

A-1 Cognitive Demands Inventory

Need to be able to *Understand operational picture* in order to . . .	If cannot do this then . . . (costs)	Things that get in the way . . . (glitches/challenges)	Requirements—in order to do this they need . . .
(from MOC perspective) • Provide information relay between the JOC and the water • Understand implications of information they are receiving • Accurately interpret what the PL reports about the status of boats • Know what action to take to ensure mission continues • Provide "oversight"—grasp implications of specific events for the overall mission picture • Step in and drive the situation if/when the PL actions are inconsistent with what the MOC watch officer thinks needs to	• Watch officer cannot make timely, effective, appropriate decisions • WO cannot provide guidance/direction to the PLs re: accomplishing the mission • Watch officer will not know implications of patrol craft status for security plan and ongoing mission • WO are "behind the curve"—response time is shorter; lose ability to anticipate or react • JOC Watch officer not able to provide CO/XO	• Unreliable radio communications • No sensors • Delayed/distorted information from boats • Vessels not submitting their movements • Delayed (or lack of) communication from host nation Port Authority about schedule and vessels coming in/out of port • Difficulty finding the person who has the information you need	• Need reliable radio communications for information sharing and updating • Need accurate, timely info from other links in the C2 chain • Need information that is real time (not time delayed), or capability that allows operators to check recency of information update • Need integration of both perspectives—boat/water perspective and bigger picture (MOC) • Need understanding of capabilities and status of all patrol crafts, with respect to current environmental conditions (e.g., sea state) • Need to know what are the available human resources

project, we developed display concepts to the stage of a functional prototype but not a completed product. Because of this, evaluations we performed were exploratory and assessment usability tests, not validation tests [1]. If the C2 system had been taken to the stage of a completed product, validation tests would have made it possible to collect data on individual and team performance.

5.5 FUTURE WORK

Three areas provide an opportunity for future research related to what we have discussed in this chapter: the inclusion of human performance research in system development, the transition from qualitative research to design, and resilience engineering.

5.5.1 Human Performance Research

Qualitative methods such as the interviews and observations that are described in this chapter are powerful means to study how individuals and teams perform cognitive work. Use of these methods from sociology and cognitive psychology rests on the three features of Harold Garfinkel's theoretical framework of the 1960s: indexicality (activities and events are unique), reflexivity (actions are inseparable from their immediate context), and sequentiality (actions and events occur in relation to each other) [22]. Successful design depends on understanding users' ordinary work practices, tasks, and situational requirements [23]. As a result, system development needs to continue to build on the methods that are used to understand and incorporate human performance.

5.5.2 Transition from Qualitative Research to Design

The relationship between human performance studies and the design and development of new product concepts needs to be more direct. Sociologists and psychologists who study human behavior capture the richness of human experience and their accounts can often produce extensive and detailed ("thick") descriptions. By contrast, engineers decompose a problem into manageable elements, work on them, and reassemble them into a finished product. This difference makes it difficult to judge "what contributions the reports and findings make to the improvement of the artifact" [24]. Further study of how to best develop requirements based on operator needs can make the relationship between researcher and engineer stronger and more efficient.

5.5.3 Resilience Engineering

High hazard sectors such as the military must be resilient—able to mount a robust response to unforeseen, unpredicted, and unexpected demands, and to resume normal operations or develop new ways to achieve operational objectives [25]. Resilience engineering (RE) provides tools to launch safety improvements by assessing changes in the adaptive capacity of an organization as it confronts disruptions, change, and pressures. RE can enable an organization to cope with and recover from unexpected developments. RE accounts for the manner in which people at all levels of an organization can try to anticipate paths that may lead to failure, to create and sustain strategies that are resistant to failure, and adjust tasks and activities to maintain margins in the face of pressure to do more and to do it faster. Understanding resilience makes the difference between organizations that unintentionally create complexity and miss signals that risks are increasing, and those that can manage high hazard processes well.

A resilient organization maintains the ability to adapt when demands go beyond an organization's customary operating boundary. This can enable it to cope with and recover from unexpected developments. Operators are the most adaptable element of a system, and are routinely assigned tasks that are beyond the adaptive ability of hardware or software. For that reason, understanding human performance as this chapter described is essential to the development of systems that have resilient characteristics.

5.6 CONCLUSIONS

Results of this research are expected to contribute to the Navy's resilience in two ways. The first, and most evident, way is to enable each level in the command hierarchy of a deployed expeditionary force to share a common view of operations through the C2 display. Through a shared awareness of and ability to reliably communicate about the capabilities of one's own force and threats, the display will enable expeditionary forces to adjust and continue operations in the face of unexpected challenges. The second (and less evident benefit) will be to integrate human-centered design into Navy systems acquisition. Use of the human-centered approach as part of the systems acquisition process will account for the human performance of actual work. As a result, future systems that are acquired are more likely to reflect and support cognitive and macrocognitive functions and thereby improve expeditionary force resilience.

ACKNOWLEDGMENTS

Beth Crandall, Cynthia Dominguez, Anna Grome, Robert Strouse, Michelle Ter-Maath, and Sterling Wiggins comprised the project team led by one of the co-authors (CN) who performed the research this chapter describes. The authors are grateful to the men and women of Navy Expeditionary Combat Command for their interest and support. We also thank the Department of the Navy (OPNAV N-857) for guidance and funding, and to the NECC Futures Working Group for support and guidance.

REFERENCES

[1] C. Nemeth. *Human Factors Methods for Design*. Boca Raton, FL: Taylor and Francis/CRC Press, 2004.

[2] D. Woods and E. Roth. Cognitive systems engineering. In: *Handbook of Human–Computer Interaction*, edited by M. Helander. Amsterdam: North-Holland, 1988, pp. 3–43.

[3] B. Crandall, G. Klein, and R. R. Hoffman. *Working Minds: A Practitioners Guide to Cognitive Task Analysis*. Cambridge, MA: MIT Press, 2006.

[4] E. Hollnagel, D. Woods, and N. Leveson. *Resilience Engineering: Concepts and Precepts*. Aldershot, UK: Ashgate Publishing, 2006.

[5] C. Nemeth, R. Strouse, M. TerMaath, and M. O'Connor. Acquiring cognition: adding human-centered design to Navy system development. In: Proceedings of the IEEE Systems, Man and Cybernetics Society 2011 International Symposium, Anchorage, AL. 2011, October.

[6] DoD. Architecture Framework, Version 1.5. Washington, DC: U.S. Department of Defense, 2007.

[7] M. Lankhorst. *Enterprise Architecture at Work: Modeling, Communication and Analysis*. 2nd edn. New York: Springer, 2009, p. 3.

[8] IEEE. 1471-2000—IEEE recommended practice for architectural description for software-intensive systems. Accessed on 20 September, 2011. Available at IEEE web site http://standards.ieee.org/findstds/standard/1471-2000.html. Institute of Electrical and Electronic Engineers, 2011.

[9] C. Nemeth and G. Klein. The naturalistic decision making perspective. In: *Wiley Encyclopedia of Operations Research and Management Science*, edited by J. J. Cochran. New York: John Wiley & Sons, 2011.

[10] R. Hoffman and S. Deal. Informing vs influencing design, part 1: A gap analysis. In: *IEEE Intelligent Systems*, September/October, 2008, pp. 72–75.

[11] D. D. Woods and S. W. A. Dekker. Anticipating the effects of technology change: a new era of dynamics for human factors. *Theoretical Issues in Ergonomics Science*, 1(3): 272–282, 2001.

[12] G. A. Klein. *Sources of Power*. Cambridge, MA: The MIT Press, 1998.

[13] A. Bisantz, E. Roth, B. Brickman, L. L. Gosbee, L. Hettinger, and J. McKinney. Integrating cognitive analyses in a large-scale system design process. *International Journal of Human-Computer Studies*, 58: 177–206, 2003.

[14] R. T. B. Hutton, T. E. Miller, and M. L. Thordsen. Decision centered design: leveraging CTA in design. In: *Handbook of Cognitive Task Design*, edited by E. Hollnagel. Mahwah, NJ: Lawrence Erlbaum & Associates, Inc., pp. 383–416, 2003.

[15] C. O. Dominguez, W. G. Long, T. E. Miller, and S. L. Wiggins. Design directions for support of submarine commanding officer decision making, Undersea HSI Symposium. Mystic, CT, 2006.

[16] M. Moundalexis and J. A. McKneely. Exceptional Expertise For Submarine Command Team Decision Making (E2SCDM) Project. In: 2008 Undersea Human System Integration Symposium, Bangor, WA, July 30, 2008.

[17] S. Potter, E. Roth, D. Woods, and W. Elm. Bootstrapping multiple converging cognitive task analysis techniques for system design. In: *Cognitive Task Analysis*, edited by J. M. C. Schraagen, S. F. Chipman, and V. L. Shalin. Mahwah, NJ: Lawrence Erlbaum Associates, Inc., pp. 317–340, 2000.

[18] E. Roth, J. Gualtieri, J. Easter, S. Potter, and W. Elm. Bridging the gap between cognitive analysis and cognitive engineering. *Gateway*, Volume XII: Number 1. Human Systems Information Analysis Center, 2001.

[19] G. A. Klein, R. Calderwood, and D. MacGregor. Critical decision method for eliciting knowledge. *IEEE Transactions on Systems, Man and Cybernetics—Part A*, 19(3): 462–472, 1989.

[20] C. Nemeth, C. Dominguez, A. Grome, B. Crandall, S. Wiggins, and M. O'Connor. Setting the bar: performance standards in naturalistic decision making research. In: Proceedings of the 10th International Conference on Naturalistic Decision Making (NDM2011). Orlando, FL, 2011.

[21] D. Woods, N. Sarter, and C. Billings. Automation surprises. In: *Handbook of Human Factors and Ergonomics*, edited by G. Salvendy. New York: John Wiley & Sons, 1997, pp. 1926–1943.

[22] M. Bonzoni and C. Heath. *The Interactional Organisation of Workplace Activities*, 1997.

[23] C. Heath and P. Luff. Work, interaction and technology: empirical studies of social ergonomics. (Technical Report EPC-1991-108). In: *Proceedings of the 11th International Conference on Ergonomics, Paris, FR.* London: Taylor and Francis, 1991.

[24] R. Anderson. Where the rubber hits the road: notes the deployment problem in workplace studies. In: *Workplace Studies: Recovering Work Practice and Informing System Design*, edited by P. Luff, J. Hindmarsh, and C. Heath. New York: Cambridge University Press, 2000, pp. 215–229.

[25] C. Nemeth, R. Wears, D. D. Woods, E. Hollnagel, and R. I. Cook. Minding the gaps: creating resilience in healthcare. In: *Advances in Patient Safety: New Directions and Alternative Approaches*, Vol. 3, edited by K. Henricksen, J. B. Battles, M. A. Keyes, and M. L. Grady. Performance and Tools, AHRQ Publication No. 08-0034-3. Rockville, MD: Agency for Healthcare Research and Quality, 2008.

PART II

LEARNING AND CONTROL

CHAPTER 6

ADVANCES AND CHALLENGES ON INTELLIGENT LEARNING IN CONTROL SYSTEMS

CHING-CHIH TSAI[1], KAO-SHING HWANG[2], ALAN LIU[3], and CHIA-FENG JUANG[1]

[1]Department of Electrical Engineering, National Chung Hsing University, Taichung, Taiwan

[2]Department of Electrical Engineering, National Sun-Yat-Sen University, Kaohsiung, Taiwan

[3]Department of Electrical Engineering, National Chung Cheng University, Taichung, Taiwan

6.1 INTRODUCTION

In recent years, a trend has emerged in which techniques of machine learning, evolutionary computations, and intelligent control have been integrated into autonomous systems on a variety of scales to meet the needs of implementation at the angle of products. Many learning methods, such as expert systems, reinforcement learning, mimetic algorithms, evolutionary computations, and support vector machines and so on, have been investigated by a finite number of researchers. On the other hand, several intelligent control approaches, including fuzzy control, neural networks, real-coded genetic algorithm, ant colony optimization (ACO), artificial immune networks, and swarm intelligence, have gained successful applications in many engineering fields. In light of this emerging trend, the technical committee on intelligent learning in control systems has been timely established to promote the advanced theory, practice, and interdisciplinary aspects of integration of intelligent learning and control in control systems.

Contemporary Issues in Systems Science and Engineering, First Edition.
Edited by MengChu Zhou, Han-Xiong Li and Margot Weijnen.

Although there are many interesting and emerging technologies in the specialized area of intelligent learning in control systems, this chapter is devoted to four significant technical areas: reinforcement learning, evolutionary learning control (ELC), intelligent learning control, and learning via case-based reasoning (CBR). To illustrate how the main methods in the four areas work well, mobile service robots are used throughout the chapter as test platforms. With these interesting platforms, some advances and challenging issues in the four technical areas are respectively reviewed and discussed.

Programming mobile robots can be a sophisticated and time-consuming process. It often takes much computation power to refine the low level mapping from sensors to actuators. Besides, the process always needs affine domain knowledge about how to complete a task into terms that are useful for the robot. Therefore, the idea of providing some high level specification of the task and using machine learning techniques to bridge the gap is an appealing one. It seems like it would take less time to write this high level instructions and the learned action policy should yield less tendency to programmer bias than handwritten programming does, if it uses empirical reception from the working field.

However, writing high level instructions for robotic control is still a profound task, because of complexity and diversity between the system and working environment. Recently there is much interest in using machine learning or rule-based techniques to help deal with this complexity. Reinforcement learning is one of the prevailing resolution mechanisms proposed to overcome the difficulty [1]. Eventually, two strategies are always applied to reinforcement learning problems, either searching in the space of behaviors to find one that performs the best in the environment, or using statistical techniques and dynamic programming methods to estimate the utility of taking action in states of the world. The former approach has been taken by genetic algorithms and genetic programming as well as some more novel search technologies [2]. This chapter will concentrate on the latter because they take advantage of the special structure of a reinforcement learning problem that is not available in optimization problems [3].

The advent of bio-inspired evolutionary computation techniques has inspired new methods for solving control problems. Evolutionary algorithms (EAs) and swarm intelligence (SI) are two well-known categories of bio-inspired evolutionary optimization algorithms. These optimization algorithms are heuristic and stochastic. They are less likely to get stuck at a local minimum than gradient-based learning algorithms do, and are derived from populations of individuals with specified behavior, similar to biological phenomena. Because the design of control systems can be regarded as an optimization problem, researchers have proposed controller design using various bio-inspired evolutionary optimization algorithms. The ELC approaches generally show the advantages of avoiding the calculation of complicated mathematical equations of controlled plants and good control performance.

Recently, much attention has been directed toward intelligent control methods by combining fuzzy logic control and neural networks, thus forming many kinds of fuzzy-neural-network (FNN) control structures. With the fused features of the fuzzy logics and neural networks, FNN control has been developed to cope with the increasing demands for controlling highly complex, uncertain, and nonlinear systems in many applications. FNN-based control approaches have been regarded as enabling technologies for many control systems that allow online adaptation to uncertainties. The online adaptation laws for stable FNN-based controllers are usually derived using Lyapunov stability theory, thereby achieving regulation and tracking control. In this chapter, some interesting and useful FNNs are briefly introduced, some intelligent learning control approaches based on these FNNs are described, and some challenging issues are also mentioned.

CBR is a decision-making and learning method widely used in artificial intelligence. The purpose of using CBR is to provide a way to accumulate past experiences for future use, and CBR is sometimes considered as using analogy to solve problems. A problem along with its corresponding solution is stored in a case, and such cases are accumulated in a case base as more problems are solved. As a learning method for a service robot, a case may contain the user request as a problem and the applicable service as the solution. Moreover, the preference of a user can be stored in a case to provide a preferred service for the user. For providing user-friendly commands to be used in human–robot interaction, a case may contain a high level command as a problem and the corresponding sequence of robot operations as a solution.

The rest of the chapter is constructed as follows. Section 6.2 surveys some contemporary issues on the application of reinforcement learning to mobile robots. In Section 6.3, some new concepts and techniques of bio inspired ELC are presented. Section 6.4 discusses some stable intelligent learning approaches using Lyapunov stability theory for various FNN control systems. The concept of learning through CBR is introduced in Section 6.5. Section 6.6 states some challenges for future work.

6.2 REINFORCEMENT LEARNING

The resolving framework about reinforcement learning is well recognized as a machine learning paradigm especially well-suited to learning action policies for mobile robots. In what follows, some pitfalls and an introduction to a framework for efficiency using reinforcement learning on mobile robots are brought into discussion. In addition, discussions will go on with some experimental results from applying this framework to a mobile robot control task: behavior learning of soccer robots.

6.2.1 Reinforcement Learning

Reinforcement learning (RL) is a machine learning paradigm that is particularly well-suited for use in mobile robots. Basically, most general reinforcement learning includes five parts [4]: a decision-making agent, decisions (actions) made by the agent, environment interacting with the agent, rewards reflecting the decisions made by the agent, and situations (state) at which the agent is being. The relationship among these five parts is shown in Figure 6.1. The agent and the environment interact at discrete time steps. The agent observes the state at step $t : s_t \in S$, produces the action at step t, $a_t \in A(s_t)$, and receives the reward: $r_{t+1} \in R$, while its action results

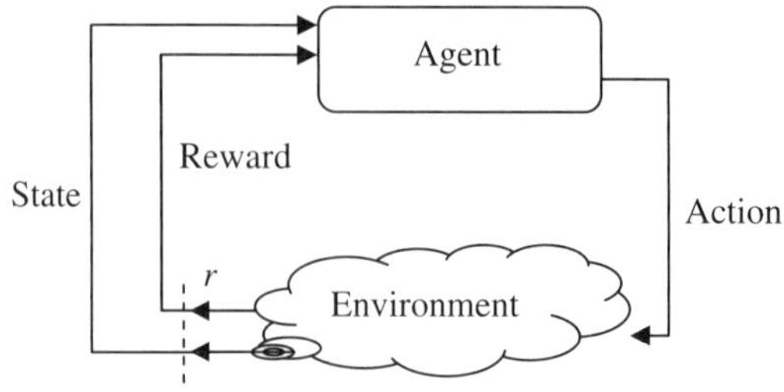

FIGURE 6.1 The architecture of reinforcement learning.

in the next state: s_{t+1}. The agent makes a decision to take an action depending on the state and the experience; this action brings the environment into a new state, makes estimate according to the consequence made by the action, and sends the result back to the agent at the same time. The agent accumulates experience from the feedback of returned rewards for the next decision. Briefly speaking, the basic reinforcement learning model consists of a set of environment states $s_{(t)}$, a set of actions $a_{(t)}$, and a set of scalar "rewards" or a scalar reward function in R. At each time t, the agent perceives its state $s_i \in S$ and the set of possible actions $A(s_t)$. It chooses an action $a \in A(s_i)$ and receives from the environment the new state s_{t+1} and a reward r_{t+1}. Based on these interactions, the reinforcement learning agent must develop or search a policy $\pi : S \to A$ which maximizes the quantity $R = r_0 + r_1 + \cdots + r_\pi$ for the Markov decision processes (MDPs) that have a terminal state, or the quantity $R = \Sigma \gamma^i r_i$ for MDPs without terminal states (where $0 \leq \gamma \leq 1$ is some "future reward" discounting factor). Thus, reinforcement learning is particularly well-suited to problems that include a long-term versus short-term reward trade-off.

After repeating this cycle continuously, the agent can make most appropriate and consistent decision for long-term benefits. In other words, reinforcement learning algorithms attempt to find a policy that maps states of the world to the actions the agent ought to take in those states. Reinforcement learning differs from supervised learning in that neither correct input/output pairs are presented, nor suboptimal actions explicitly corrected. Further, there is a focus on online performance, which involves finding a balance between exploration (of uncharted territory) and exploitation (of current knowledge). The exploration versus exploitation trade-off in reinforcement learning has been mostly studied through the multi-armed bandit problem.

The agent predicts expected reward in state s before or after taking action a by an evaluative function. Two kinds of evaluative functions are defined here: state-value function V^π and action-value function Q^π, where π denotes the policy applied at present. The state-value function for policy π is given by

$$V^\pi(s) = E_\pi\{R_t | s_t = s\} = E_\pi\left\{\sum_{k=0}^{\infty} \gamma^k r_{t+k+1} | s_t = s\right\}, \tag{6.1}$$

where $E_\pi\{\}$ denotes the expected reward based on policy π, and t represents a time step. The value of taking an action in a state under policy π is the expected return starting from that state with that action and the action-value function for policy π:

$$Q^\pi(s,a) = E_\pi\{R_t | s_t = s, a_t = a\} = E_\pi\left\{\sum_{k=0}^{\infty} \gamma^k r_{t+k+1} | s_t = s, a_t = a\right\}. \quad (6.2)$$

The purpose of reinforcement learning is to learn a set of optimal policy π^*. The optimal policy shares the same optimal state-value function

$$V*(s) = \max_\pi V^\pi(s) \quad \forall s \in S. \quad (6.3)$$

The optimal policy also shares the same optimal action-value function

$$Q*(s,a) = \max_\pi Q^\pi(s) \quad \forall s \in S, a \in A(s). \quad (6.4)$$

According to the optimal policy $\pi*$, the expected reward under state s when the agent takes action a is

$$Q*(s,a) = E\{r_{t+1} + \gamma V*(s_{t+1}) | s_t = s, a_t = a\}. \quad (6.5)$$

6.2.2 Q-Learning Algorithm

Q-learning is a form of model-free reinforcement learning method based on stochastic dynamic programming and a set of Q-values associated with the current state and the action to be taken. It accumulates the experience from each action the agent has taken and provides the agent with the capability of learning to act optimally in a Markovian environment. Each Q-value approaches an optimal $Q*$, but needs long-term learning to approach it. According to the Q-learning algorithm, with the increase in the number of learning times, $\max_{a'} Q(s', a')$ is closer and closer to $Q*$, and the Q-value takes $\max_{a'} Q(s', a')$ as a goal to update the value. As a result, the Q-value tends to be more reliable in each state. Based on the above argument, the action it takes based on the Q-value is more appropriate as well.

As illustrated by the architecture of Q-learning shown in Figure 6.2, the functions are separated into four parts: memory, evaluation function, update function, and reward function. The "memory" function can be used to save the Q-value under every state and the corresponding action in a form called Q-table. The "reward function" in the environment returns a scalar value as an indicator used for strengthening or weakening the agent's decision-making capability. The "update function" is used for updating the Q-value, based on the reinforcement signal and current state. An "evaluation function" is used to evaluate the agent's current state, to select an action to achieve the optimal Q-value as output, according to the Q-value recorded in the memory.

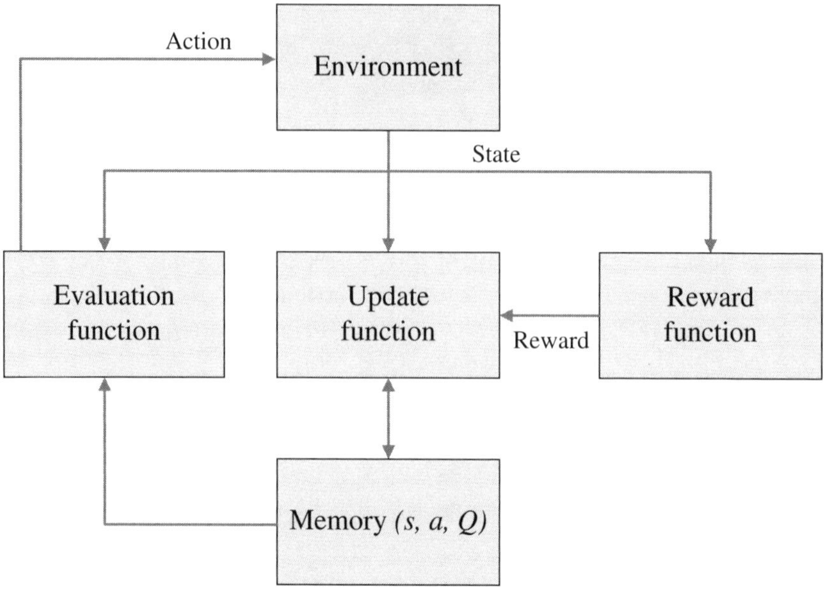

FIGURE 6.2 The architecture of Q-learning.

The agent is assumed to be able to discriminate the set S of distinct world states, and can take the set A of actions on the world. The following is a procedure of the one-step Q-learning algorithm.

Initialize $Q(s, a)$ to be 0 for all states s and actions a.

Perceives a current state s.

Choose an action a according to the action-value function.

Carry out an action a in the environment. Let s' be a next state and r be an immediate reward.

Update the action-value function in terms of state s and action as a pair.

Update the action-value

$$Q_{t+1}(s, a) \leftarrow (1 - \alpha) Q_t(s, a) + \left[r + \gamma V\left(s'\right)\right], \tag{6.6}$$

where α is a learning rate parameter and γ is a fixed discount factor between 0 and 1.

Obtain the optimal Q

$$Q^* = \max_{a'} Q(s', a'). \tag{6.7}$$

Under certain conditions this is guaranteed to converge to the optimal Q-function, $Q^*(s, a)$. One of the most important features of Q-learning is the so-called off-policy algorithm. This means that the distribution from which the training samples are drawn has no effect on the policy learned.

6.2.3 Reinforcement Learning in Robots

Reinforcement learning, in particular, seems to be a natural choice for learning action policies in mobile robots. Instead of designing an action policy to drive robots, we can design a much higher level task description in the form of the reward function, $R(s, a)$. Designing a sparse reward function is generally easier than designing the low level mapping from observations to actions. Often, for robot tasks, rewards correspond to the physical events in the world. This makes it easy to come up with simple reward functions for many tasks. For example, for an obstacle avoidance task, the robot might get a reward of 1 for reaching the goal, and -1 for hitting an obstacle. In theory, this is all that is necessary for the robot to learn the optimal policy. However, there are a number of problems with simply using standard Q-learning techniques. This type of reward functions is called sparse reward functions, which hold zero everywhere, except for a few places (the obstacles and goal in the our example). Its contrast is the dense reward functions that give nonzero rewards most of the time. A dense reward function for obstacle avoidance might be, for example, the sum of distances to the obstacles divided by the distance to the goal state. It gives more information after each action, but is much more difficult to construct than sparse reward functions do. In this work, we will mostly be interested in the sparse ones, since they are generally much simpler to design.

One problem with RL in mobile robots is that the state-space description of mobile robots is often best expressed in terms of vectors of real values, while Q-learning requires discrete states and actions. One approach to overcoming this problem, known as value-function approximation, replaces the tabular representation of $Q(s, a)$ with a general-purpose function approximator. However, it has been shown that this will not work for general function approximators, even in seemingly benign situations. In the previous work [5], an algorithm is presented to address the problems associated with value-function approximation. The algorithm is based on an aggregating method where the cerebellar model articulation controller (CMAC) method is used as a function approximator [6].

The other main problem that should be tackled is how to transfer prior knowledge into the learning system. The only way that Q-learning can extract information about its environment is to take actions and observe their effects. In the early stages of learning, the system knows nothing about the environment, or how to act in it. It is, therefore, forced to choose more-or-less arbitrary actions. As more information, in the form of rewards arrives, Q-learning can iteratively improve its approximation of the value function. However, if no rewards are observed, the value-function approximation will never change. One of the major hurdles to implementing RL systems in real robots is the inclusion of prior knowledge in the learning system. Using a sparse reward function only, and without some prior knowledge of the environment, the learning system seems doomed to fail.

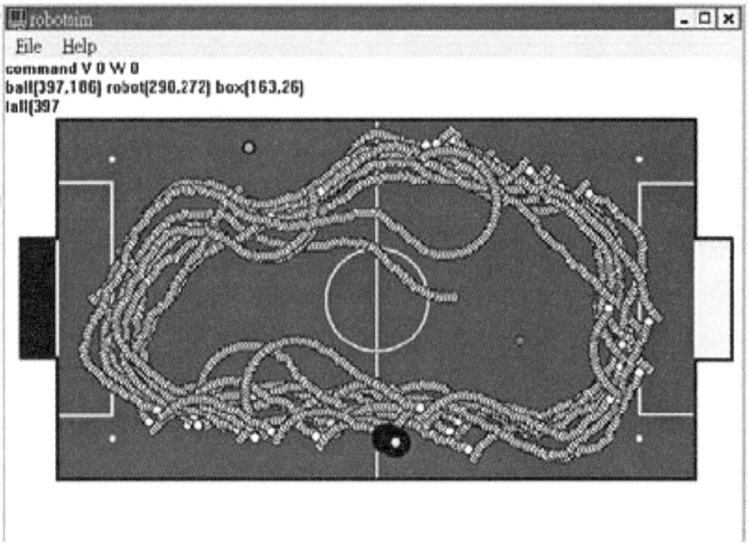

FIGURE 6.3 Obstacle avoidance behavior.

6.2.4 Soccer Robot Behaviors

A soccer robot needs to learn three basic types of behaviors, that is, ball shooting, obstacle avoidance, and target reaching. These primitive behaviors can be combined as more complex behavior, such as defending and attacking. Ball shooting is a simple but essential action. Through this action, before a robot comes to the ball position, it follows an imaginary shooting line, and switches the kicking motor on. The imaginary shooting line is a line passing through the ball and the target door. Obstacle avoidance is necessary because during the game, a robot is not allowed to collide with other robots. The temporal information about other robots' positions and moving directions is provided by a vision system. The robot moves in a direction to avoid collision according to the distance between the other robots and the current position by means of a reinforcement learning mechanism. The outputs of all available actions of the learning mechanism are regarded as driving reference signals to motorize wheel. The target reaching action drives a robot to the desired position and orientation.

The simulation environment to emulate a soccer field is defined by the Federation of International Robot–Soccer Association (FIRA) [7]. A robot is initially placed at the center but slightly to the right. The robot motion track is shown by a series of white dots. Figure 6.3 shows the behavior of a robot when performing obstacle avoidance. The robot walks randomly on the field without passing the boundary, which is regarded as an obstacle. Figure 6.4 shows a robot's capability of moving to a target. In this figure, the robot moves as closely as possible to the assignment posture (position with a certain orientation) with a slight tolerance represented by a hidden region centered by the assigned position. Figure 6.5 shows the learnt ball

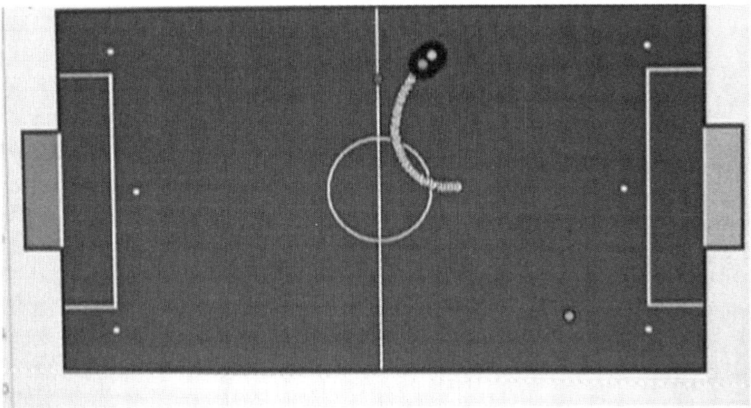

FIGURE 6.4 Target moving behavior.

shooting behavior. The robot moves to an aiming position where the robot aligns the line between the center of the opponent's gate and ball, but with a certain distance from the ball in preparation for kicking.

6.2.5 Concluding Remarks

Programming mobile robots in hard coding can be a tedious and time-consuming process. Specifying the low level structure along the data flow from sensors to actuators is prone to programmer misconceptions, and constructing such a mapping can be profound. Therefore, the idea of having a robot learns how to accomplish a task with certain self-awareness capability, rather than being told explicitly, has attracted many research efforts. It seems easier and much more intuitive for filed

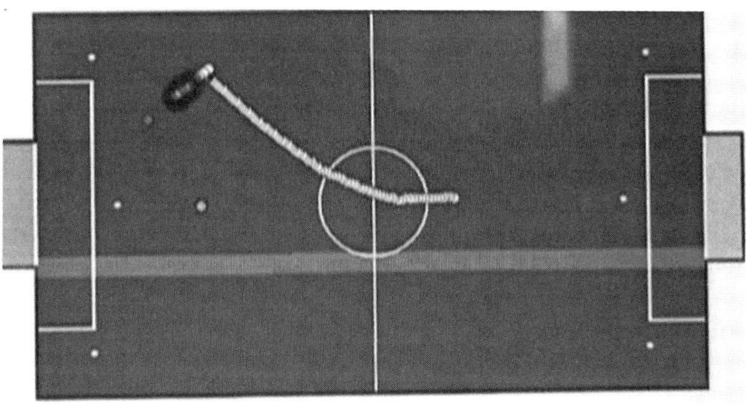

FIGURE 6.5 Ball shooting behavior.

operators or users to specify what the robot should be doing, instead of detailing how the robot should do the job. In this section, a framework for reinforcement learning on mobile robots has been introduced and the experimental outcomes by applying it to learning some simple behaviors have been shown as well. The results actually show the feasibility of a leaning algorithm in realizing autonomy and intelligence of mobile robots.

6.3 BIO-INSPIRED EVOLUTIONARY LEARNING CONTROL

This section introduces ELC using evolutionary algorithms (EAs) [8] and swarm intelligence (SI) [9]. EAs involve populations of individuals with a specific behavior similar to a biological phenomenon, like crossover or mutation. The most well-known EA is genetic algorithms (GAs). GAs are stochastic search procedures based on the mechanics of natural selection, genetics, and evolution. They simultaneously evaluate many points in the search space and are more likely to find the global solution of a given problem. The SI technique is a relatively new optimization algorithm compared to EAs. The SI technique studies collective behavior in decentralized systems. Its development is based on mimicking the social behavior of animals or insects in an effort to find the optima in the problem space. Two well-known SI models are particle swarm optimization (PSO) and ACO [10]. The development of PSO is based on observations of the social behavior of animals such as bird flocking and fish schooling. The technique of ACO is inspired by real ant colony observations. EAs and SI use only a simple scalar performance measure that does not require or use derivative information, and thus they are suitable for solving controller optimization problems.

6.3.1 Bio-Inspired Evolutionary Learning Control

ELC has drawn much attention in recent years. Its studies can be divided into two categories. The first one applies bio-inspired optimization algorithms to the design of traditional controllers. Its typical representative is optimization of PID controllers using GAs [11] or PSO [12]. The idea of designing sliding mode controllers or H_2/H_∞ controllers using GAs has also been proposed [13, 14].

The second one applies bio-inspired optimization algorithms to the design of intelligent controllers, such as fuzzy or neural controllers. One main advantage of fuzzy controllers in contrast to the latter is transparency in how they comprise fuzzy if–then rules. Manual derivation of fuzzy control rules is often difficult and time consuming, and requires expert knowledge. To ease the design task and improve control performance, GA- or PSO-based fuzzy control has been extensively studied. Figure 6.6 shows the general learning configuration of this ELC approach. According to the IEEE Explorer database, more than 2000 papers in this topic have been published after the year 2000. ACO-based fuzzy control is a relatively new ELC approach compared to the GA- or PSO-based fuzzy control approaches, and previous studies have demonstrated its potential in solving controller optimization problems [15–17]. In

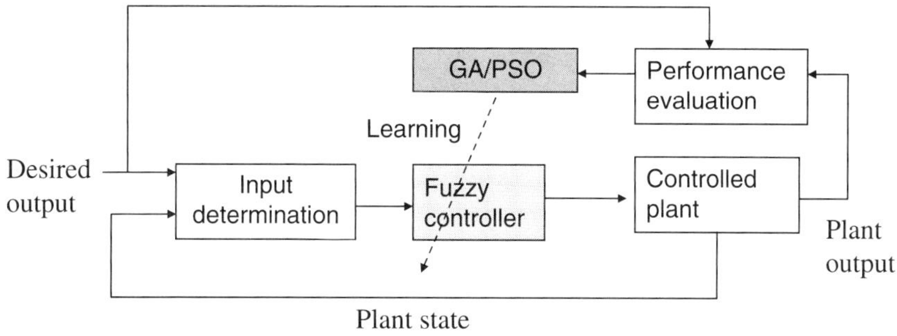

FIGURE 6.6 GA- or PSO-based fuzzy control configuration.

the ACO technique, artificial ant colonies cooperate to find good solutions to difficult discrete or continuous optimization problems. Discrete ACO is applied to reinforcement fuzzy control problems in Reference 16, where the consequent parameters in each fuzzy rule are optimized in a discrete space and the antecedent parameters are manually assigned in advance. The Q-value in reinforcement learning is incorporated with ACO in order to improve learning efficiency. For problems where high control accuracy is a major concern, such as tracking control, continuous ACO that finds solutions in a continuous space is a better choice than discrete ACO. Continuous ACO is applied to optimize all of the free parameters in a fuzzy controller in Reference 10, and the control performance is shown to be better than various advanced GAs and PSO do.

6.3.2 Bio-Inspired Evolutionary Robots

One promising application of ELC is evolutionary robots. The technique of evolutionary robotics has been proposed for the automatic creation of autonomous robots. Evolutionary fuzzy control of a wheeled robot for wall-following or obstacle avoidance is proposed in References 15,18, and 19. In Reference 18, a robot automatically evolves with wall-following behavior using a Q-value aided ant optimized fuzzy controller under reinforcement learning environments. Figure 6.7 shows the learning configuration. It is unnecessary to collect precise input–output data in advance for robot training. All fuzzy control rules are generated and optimized during the online learning process. Several techniques for the ELC of a legged robot have also been proposed. Evolution of a neural controller through EA or PSO for the gait control of quadruped and hexapod robots is proposed in References 20–22. In Reference 22, a symbiotic species-based PSO is applied to train a neural controller for the gait generation of a hexapod robot. Evolution of the gait sequence is based on the total walking distance of each neural controller (represented as a particle) in a certain time period. Simulation and experimental results of the hexapod gait control show the success of this ELC approach.

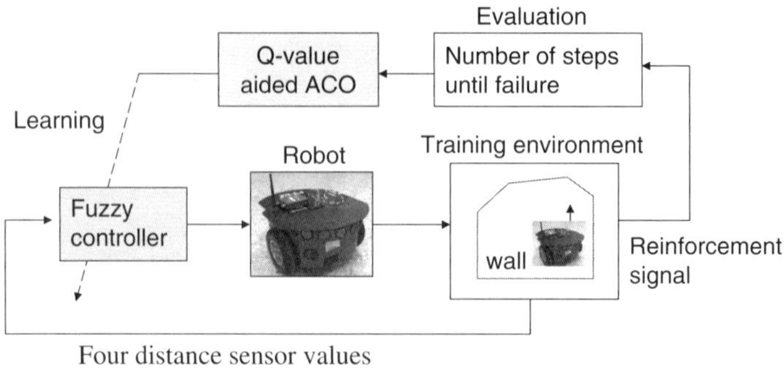

FIGURE 6.7 Evolutionary robot wall-following control using reinforcement ant optimized fuzzy control.

6.4 INTELLIGENT LEARNING CONTROL USING FUZZY NEURAL NETWORKS

6.4.1 Introduction

During the past two decades, the nonlinear control of uncertain dynamical systems using neural networks (NNs) attracted tremendous interest in the intelligent control community. An artificial neural network comprises many interconnected nonlinear processing elements called neurons to map input–output relations. The general description of a neuron function is that its weighted inputs are summed to yield an output through a nonlinear activation function if the weighted sum exceeds an internal threshold. An NN function demonstrates a universal approximation capability and many outstanding features such as learning, fault toleration, parallelism, and generalization, thus it is usually implemented to compensate for the inversion error or the unknown dynamics [23, 24]. When an exact dynamic model is difficult to obtain due to its uncertainty and complexity, the conventional NN-based control techniques are less ineffective to meet the requirement of system performance. Because of this reason, much attention has recently been directed toward intelligent control methods by combining fuzzy logic control and neural networks to form many kinds of FNN control structures. The fusion of both advantages from fuzzy logics, neural networks and nonlinear control theory provides FNN-based adaptive controllers with the ability to adapt to the environment uncertainties. FNN control has been developed to cope with the increasing demands for controlling highly complex, uncertain, and nonlinear systems in industrial applications [25–31]. Several researchers have presented several FNN-based controllers that have emerged as enabling technologies for many control systems that allow online adaptation to uncertainties. The online adaptation laws for stable FNN-based controllers are usually derived using Lyapunov stability theory, thereby achieving regulation and tracking control.

Two important issues must be solved while an FNN-based controller is used. First, the network structure needs to be determined to permit a minimum processing time in real-time control applications, that is, the number of nodes in each level and the connection. Second, online learning techniques for the FNN parameters and connecting weights should be developed with fast convergence of approximation errors. Usually, the stability of the FNN-based control system mainly depends on the selection of the learning rate and the initial values of the weights. The overall stability of the FNN-based control systems has been investigated by Lyapunov stability theory.

6.4.2 Intelligent Learning Control Using FNNs

Researchers have proposed several FNN-based controllers that can be classified into two types: feedforward and recurrent (or feedback). Fuzzy basis function neural networks (FBFN), wavelet fuzzy neural networks (WFNN), fuzzy cerebellar model articulation controller (FCMAC), and interval type 2 fuzzy neural networks (IT2FNN) represent four useful categories of the feedforward FNNs. They have been increasingly receiving attention in solving complex and practical problems. They have also been adopted widely for nonlinear system modeling and control because they possess a simple structure, good local approximating performance, particular resolvability, and function equivalence to a class of nonlinear functions. Along with the multiplayer perceptions, FBFN [30,32] and FWNN [31] hold much interest in the current FNN literature due to its universal approximation property and straightforward computation using linearly weighted combination of single hidden layer neurons. For example, the architecture of a three-layered FBFN [30, 32] usually has an input layer, a membership layer with a flexible number of neurons, and an output layer with specified number of neurons. The architecture of a three-layered FWNN [31] has also three layers, including an input layer, a membership layer with a flexible number of wavelet functions. Their performance depends on centers and widths of the basis and membership functions, shapes, and Lyapunov methods used for learning the input–output mapping. Many different methods have been proposed to determine and train their centers and widths, roughly classified as supervised and unsupervised. The most common method of supervised learning is gradient descent or search methods. However, supervised learning methods may have slow convergence or tracking due to nonlinearity and nonstationarity of the physical signals. In addition, the gradient-based searching may be trapped at the local minima of the cost function surface because the global minimum point may not be found if the input has a large bounded disturbance. On the other hand, an unsupervised learning approach does not search for the global minimum point along the cost function surface in the parameter space, but aims at constructing and shaping an energy surface with a single global minimum point in the time domain through the adjustment of the weight parameters in the Lyapunov sense. The intelligent learning control using feedforward FNNs have been well applied to motion control of several uncertain wheeled mobile platforms, including electric unicycle platforms, self-balancing two-wheeled mobile platforms [32], nonholonomic mobile platforms, three-wheeled and four-wheeled omnidirectional mobile platforms, robotic manipulators [33], and voice coil motors [34]. Figure 6.8 shows an interesting example of using FBFN to achieve self-balancing and motion control of an electric unicycle.

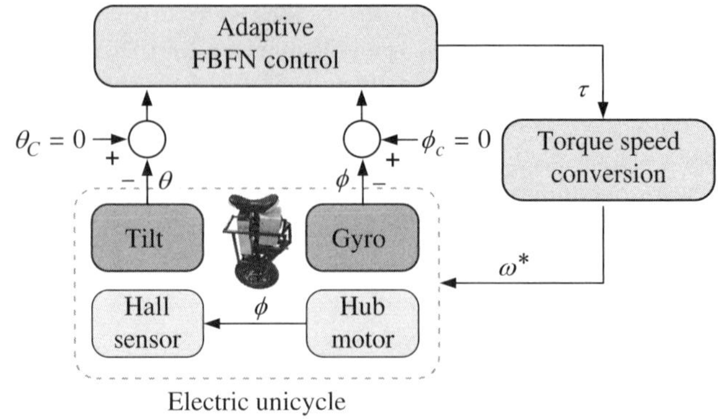

FIGURE 6.8 Self-balancing and motion control of an electric unicycle using FBFN.

In addition to these feedforward FNN-based controllers, there has been considerable interest in exploring the applications of recurrent fuzzy neural networks (RFNN) to deal with nonlinearities and uncertainties of the control systems. RFNN [25–29], recurrent wavelet fuzzy neural network (RWFNN) [35], recurrent fuzzy cerebellar model articulation controller (RFCMAC) [36], and recurrent IT2FNN (RIT2FNN) [37] can be four useful and important types of FNNs with feedback. The hidden layer neurons in these FNN structures sum their weighted inputs and yield an output through a nonlinear activation function. Dynamic RFNN, RWFNN, RFCMAC, and RIT2FNN have been developed to extend the static neurons through adding state feedback and provide more computational advantages than a backpropagation neural network does. These RFNNs are composed of both feedback and feedforward connections between layers and neurons forming complicated dynamics with which they demonstrate good control performance in the presence of uncertainties. Their structures are often composed of three or five layers in which the hidden layer neurons have a dynamic self-recurrent connection relationship using a nonlinear activation function. The recurrent neurons with internal feedback loops serve to capture the dynamic response of a system without external feedback through delays. Thus RFNN, RWFNN, RFCMAC, and RIT2FNN have superior capabilities to other static mapping FNNs, such as storing the past information for later use and approximating dynamic characteristics. Of particular interest are their abilities to deal with time-varying input or output through their own natural temporal operation. These dynamic RFNNs may be treated as subsystems of the adaptive system when they are used to approximate and control unknown nonlinear systems through an online learning process. The weights of the networks need to be updated using a dynamic learning algorithm during the control process. All the online learning laws can be constructed using Lyapunov stability theory. It is worthwhile to mention that the above-mentioned RFNN, RWFNN, RFCMAC, and RIT2FNN can be applied to achieve servo control for induction motors [28], identification and predictive control for some industrial plants [35], tracking control [36], and dynamic system processing [37].

6.5 CASE-BASED REASONING AND LEARNING

The goal of CBR is to provide a way to build up a knowledge base through experiences, and the experiences will be used in solving similar problems in the future. It is also considered as using analogy to solve problems. CBR is used in wide varieties of applications in which non-algorithmic problems are solved [38]. The basic concept comes from our daily lives, such as a doctor diagnosing a patient based on experiences, a mechanic fixing cars with experiences, and a sales person approaching clients according to experiences. These experiences may be recorded and categorized as notes or cases. In CBR, each individual problem is a case that the expert needs to solve, and that particular case may be used in the future if an identical or a similar problem occurs.

6.5.1 Case-Based Reasoning Process

The cases stored in the case base may include either successful or unsuccessful experiences from the past. The successful cases can be applied to the new problems for providing solutions while the failure cases can be the warnings to avoid applying seemingly useful solutions. A case consists of a part called problem description which distinguishes a case from another. The problem description contains a set of features and their values. When a system is to process a problem, the problem is defined with those features, and the values of those futures are matched with the features of the cases in the case base to find the most similar cases. When such similar cases are identified, they are further evaluated accordingly and a single case is proposed as the solution to the problem.

If the problem and the proposed case have an exact match in features and values, then the solution is applied directly. Otherwise, adaptation is needed before the solution can be applied to the problem. There are many adaptation strategies and heuristics proposed. The adaptation procedures are usually domain dependent. After case adaptation, the concept of learning is applied by considering the case to be stored into the case base. This general procedure of a CBR is shown in Figure 6.9, where four Rs are used in the process, namely Retrieve, Reuse, Revise, and Retain [39].

6.5.2 Case Design and Reuse

Many kinds of case representation methods like the structural case representation have been proposed [40]. As the number of cases grows, the major challenge of a case base is the inefficiency in search time. Research has been done on the organizational case base such as the dynamic memory model and the category and exemplar model, which use the hierarchical structure and the network structure framework [41]. The use of indexing reduces the cost of searching a case base. For search and storage efficiency, a case can be represented with several levels of abstraction.

The CBR systems rely on the similarity assessment to retrieve the most similar cases or the most adaptable cases [42]. Surface similarity can be defined as the distance between two cases with weights. Structural similarity uses the representation

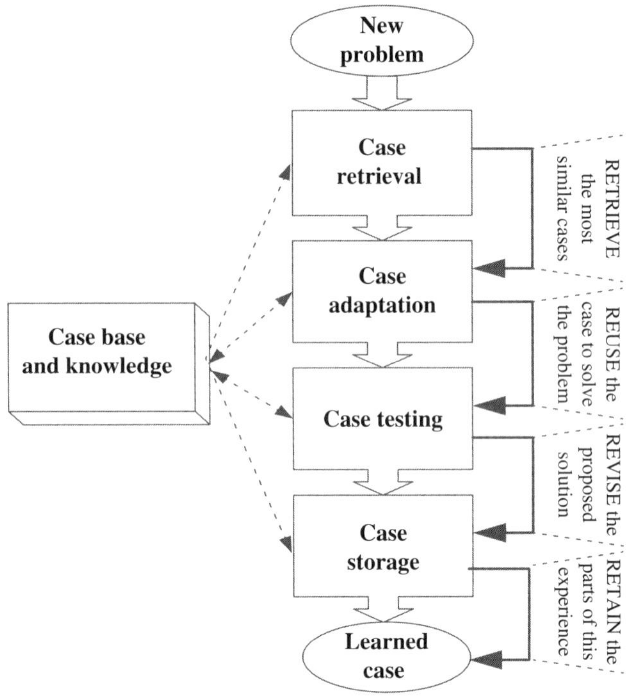

FIGURE 6.9 Process of case-based reasoning.

of the structural case and organizes the conceptual graphs before the case retrieval. Explanation-based similarity uses the explanation case to estimate the similarity in case retrieval and uses domain knowledge to support an explanation of a case.

Among the phases in the CBR process, how to develop adaptation knowledge is the most difficult task, because the expert knowledge in the problem domain is needed. For this reason, machine learning is used as a supportive role in CBR systems [43]. Researchers focus on using neural networks to assist the case retrieval or case adaptation and integrate rule induction and CBR to improve case matching.

6.5.3 Hybrid Learning Method Architectures in CBR

Combining different learning methods is considered promising [44, 45]. One example of using multiple cases is robot soccer to present a teamwork strategy [46]. Using a CBR method in solving the multirobot soccer cooperation problem, the soccer robot is first trained by using a suitable machine learning algorithm to develop the adaptation knowledge base. It is important for a CBR system to be adaptable during dynamic situation changes through receiving online feedback to achieve online learning.

Figure 6.10 shows a possibility of combining offline and online learning, in which the offline learning method is placed there to increase the accuracy of the solution

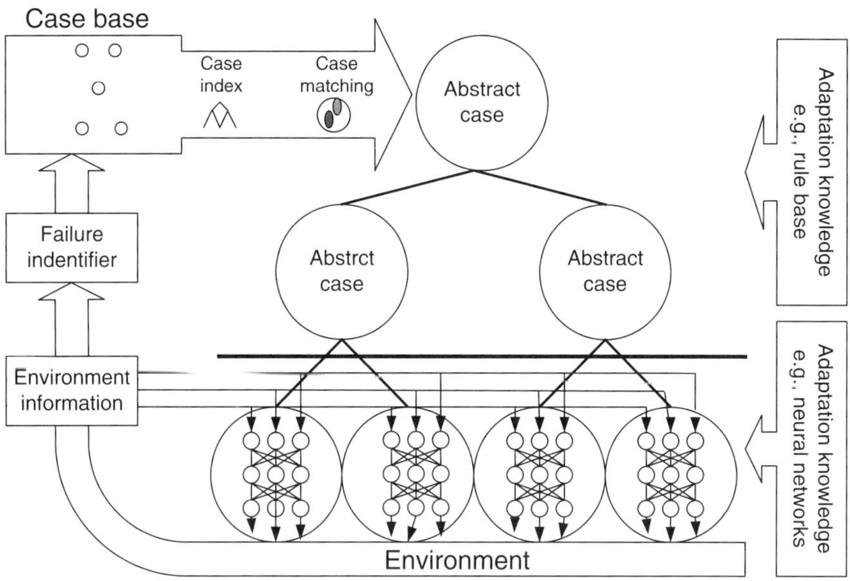

FIGURE 6.10 Framework of combining offline learning and online learning into CBR.

provided by cases while employing online learning method for solving a problem in a dynamic situation. Hierarchical case representation with two steps of case adaptation is used in this framework. In the high level of abstraction, the case adaptation belongs to quality adaptation, and in the low level, the case adaptation belongs to quantity adaptation. This framework uses a rule-based method for the high level adaptation knowledge and neural network systems for the low level adaptation knowledge.

6.5.4 Applications in Human–Robot Interaction

The main goal of a service robot is to provide services to a user. In order to deliver a correct service, the robot needs to know what the user wants. Thus, interaction between the user and the robot becomes a critical issue. This also contributes to the acceptance of robots in a household setting. When considering that each user may have different ways of interacting with the robot and there are different ways of interpreting user requests, the concept of personalization needs to be studied. For such special needs in interaction and interpreting someone's requests, human beings learn from experiences and use analogy. Thus, CBR is considered as a suitable method for meeting such demand.

Human–robot interaction has been an important research topic and especially the interaction between the users and social robots has been one of the core research topics in service robots [47]. As for using CBR in human–robot interaction, there is a project team working with Emergent Behavior Theory to organize abstract behaviors which consist of simpler and reflective behaviors [48]. The lower layer in the behavior definition relies on the relationship among different sensors and dynamic behaviors. Thus, the reflective behaviors and visional information are combined as a case and stored in a case base. The relationship between the commands and sensor data is also represented as a case, and this relationship is then combined with the reflective behaviors to help learning. Another approach, S-learning, uses a case base to record action sequences to produce a sequence library. By using reinforcement learning, no model needs to be constructed [49]. Its advantage is the possibility of having reflective response. However, training still takes a long time to complete.

The STARhome project uses the concept, called case-driven ambient intelligence (C-AmI), to use the past experiences for solving context, prediction, reasoning, and executing tasks, in which the daily activities of a user are stored as cases [50]. Similarly, the SHARE-it project integrates agent technologies for assisting the elderly or handicapped [51]. The technologies include safety and assistance in regular daily lives. The goal is to rapidly integrate different technologies with intelligent data acquisition and interpretation.

Research in smart home settings requires the interaction between the user and the robots. The robots can be passive and wait for the user's commands, but they can also actively initiate interaction. The user intention is studied in order to let the robot understand more abstract user commands [52]. At the same time, the robots are not only required to communicate with human users, but also with the devices in the execution environment. A service-oriented architecture integrates the robots with human users and also devices in the execution environment. With this environment for achieving assisted living or care, CBR can be used for allowing the users to have different preferences and ways of communication with the robot and the environment. Personalization and personal attention are also other research topics that can be benefited from learning. By knowing the user intention and user preferences, one can predict what the user wants currently and also in the near future if the user's habits and expectation are known.

6.6 CONCLUSIONS

This chapter has presented some contemporary issues in the technical field of intelligent learning in control systems. Particular attention has been paid to the four important, ever-prevalent and growing technical areas: reinforcement learning, ELC, intelligent learning control, and learning via CBR. Below are the statements of some interesting but challenging topics for future research.

How Q-learning approaches can be implemented in a successful way has been shown to deal with the problems of mobile robots in behavior learning. In practice, it always encounters two major problems due to an excessively large state space. First, learning Q-values in a tabular form may be infeasible because of the excessive amount of memory needed to store the table. Second, rewards in the state space may be so sparse that the policies can be discovered extremely slowly with some exploration schemes. In this aspect, the examples for mobile robots address the issue of individual behavior design for a soccer robot using a simple Q-learning method, where a robot learns how to accomplish a task with certain self-awareness capability, rather than being told explicitly. However, there are some major limitations of current reinforcement learning methods, such as their reliance on human expertise to design a representation for the agent solutions, planning for a task-specific model, and so forth. Nevertheless, along with the examples in this chapter showing the possibility of how an existing behavior can be imitated by a learning method, some areas of intriguing future research including multiagent learning, hierarchical action presentation, aggregation of receptive fields on state spaces, and so on, can eventually inspire the dream of intelligent robots with fully autonomous learning capability.

ELC approaches help optimize parameters in traditional or intelligent controllers via learning. They ease design effort and improve control performance. However, EC and SI algorithms usually take more than thousands of control evaluations to find a good parameter set for a controller. Such long learning period may be inefficient when learning is applied to a real system instead of a simulated one. In addition, the evolution of a controller may be too slow to handle changes in a time-varying plant. Future research directions on ELC may focus on addressing these challenges.

Several types of forward and recurrent FNNs have been introduced as a powerful means to achieve intelligent learning control for a class of nonlinear systems in which motion control of mobile robots are used for illustration. The well-known Lyapunov stability theory can be used to find not only membership functions and weights of the used FNNs, but also control laws, in order to ensure the overall system stability of the controlled systems or robots. In this research direction, some challenging issues for future research would be to explore how the recurrent FNNs can be used to synthesize effective controllers for complicated nonlinear plants and mobile robots, in order to obtain excellent control performance in the presence of unknown system models, parameter variations, and unpredictable external disturbances.

Learning by experience seems natural for human beings, but it may not be easily implemented with algorithms. Thus, CBR is a promising method for retaining the knowledge from experience. Such knowledge is especially important in providing services by robots and also interacting with different users. In order to take the human–robot interaction to the next level, understanding user intention and satisfying user preferences will be challenging research topics. CBR would be beneficial as a learning method for meeting such requirements.

REFERENCES

[1] Y. Takahashi, K. Noma, and M. Asada. Efficient behavior learning based on state value estimation of self and others. *Advanced Robotics*, 22(12): 1379–1395, 2008.

[2] J. Schmidhuber. A general method for multi-agent learning and incremental self-improvement in unrestricted environment. In: *Evolution Computation: Theory and Application*. Singapore: Scientific Publishing, 1996.

[3] L. Kaelbling, M. Littman, and A. Moore. Reinforcement learning: a survey. *Journal of Artificial Intelligent Research*, 4, 237–285, 1996.

[4] A. G. Barto, R. S. Sutton, and C. W. Anderson. Neuron-like adaptive elements that can solve difficult learning control problems. *IEEE Transactions on System, Man, and Cybernetics*, 13(5): 834–846, 1983.

[5] K. S. Hwang and J. S. Lin. Smoothing trajectory tracking of three-link robot: a self-organizing CMAC approach. *IEEE Transactions on System, Man, and Cybernetics*, 28(5): 680–692, 1998.

[6] K. S. Hwang, C. Y. Lo, and W. L. Liu. Modular agent architecture for an autonomous robot. *IEEE Transactions on Instrumentation & Measurement*, 58(8): 2797–2806, 2009.

[7] http://www.fira.com (accessed 7th January, 2015)

[8] X. Yao (Ed.), *Evolutionary Computation—Theory and Applications*. World Scientific, 1999.

[9] J. Kennedy, R. Eberhart and Y. Shi. *Swarm Intelligence*. Morgan Kaufmann, 2001.

[10] M. Dorigo and T. Stutzle. *Ant Colony Optimization*, MIT, 2004.

[11] R. A. Krohling and J. P. Rey. Design of optimal disturbance rejection PID controllers using genetic algorithms. *IEEE Trans. Evolutionary Computation*, 5(1): 78–82, 2001.

[12] Z. L. Gaing. A particle swarm optimization approach for optimum design of PID controller in AVR system. *IEEE Transactions on Energy Conversion*, 19(2): 384–391, 2004.

[13] X. Yu and O. Kaynak. Sliding-mode control with soft computing: a survey. *IEEE Transactions on Industrial Electronics*, 56(9): 3275–3285, 2009.

[14] S. J. Ho, S. Y. Ho, M. H. Hung, L. S. Shu, and H. L. Huang. Designing structure-specified mixed H_2/H_∞ optimal controllers using an intelligent genetic algorithm IGA. *IEEE Transactions on Control Systems Technology*, 13(6): 1119–1124, 2005.

[15] M. Mucientes and J. Casillas. Quick design of fuzzy controllers with good interpretability in mobile robotics. *IEEE Transactions on Fuzzy Systems*, 15(4): 636–651, 2007.

[16] C. F. Juang and C. M. Lu. Ant colony optimization incorporated with fuzzy Q-learning for reinforcement fuzzy control. *IEEE Transactions on System, Man, and Cybernetics, Part A: Systems and Humans*, 39(3): 597–608, 2009.

[17] C. F. Juang and P. H. Chang. Designing fuzzy rule-based systems using continuous ant colony optimization. *IEEE Transactions on Fuzzy Systems*, 18(1): 138–149, February, 2010.

[18] C. F. Juang and C. H. Hsu. Reinforcement ant optimized fuzzy controller for mobile-robot wall-following control. *IEEE Transactions on Industrial Electronics*, 56(10): 3931–3940, October, 2009.

[19] C. F. Juang and Y. C. Chang. Evolutionary group-based particle swarm-optimized fuzzy controller with application to mobile robot navigation in unknown environments. *IEEE Transactions on Fuzzy Systems*, 19(2): 379–392, April, 2011.

[20] D. P. Krasny and D. E. Orin. Generating high-speed dynamic running gaits in a quadruped robot using an evolutionary search. *IEEE Transactions on System, Man, and Cybernetics B: Cybernetics*, 34(4): 1685–1696, 2004.

[21] D. K. Pratihar, K. Deb, and A. Ghosh. Optimal path and gait generations simultaneously of a six-legged robot using GA-fuzzy approach. *Robotics and Autonomous Systems*, 41: 1–20, 2002.

[22] C. F. Juang, Y. C. Chang, and C. M. Hsiao. Evolving gaits of a hexapod robot by recurrent neural networks with symbiotic species-based particle swarm optimization. *IEEE Transactions on Industrial Electronics*, 58(7): 3110–3119, 2011.

[23] N. Hovakimyan, F. Nardi, A. Calise, and N. Kim. Adaptive output feedback control of uncertain nonlinear systems using single-hidden-layer neural networks. *IEEE Transactions on Neural Networks*, 13(6): 1420–1431, 2002.

[24] K. K. Tan, S. N. Huang, and T. H. Lee. Adaptive backstepping control for a class of nonlinear systems using neural network approximations. *International Journal of Robust and Nonlinear Control*, 14, 643–664, 2004.

[25] J. Zhang and A. J. Morris. Recurrent neuro-fuzzy networks for nonlinear process modeling. *IEEE Transactions on Neural Networks*, 10(2): 313–326, 1999.

[26] Z. Huaguang, L. Cai, and Z. Bien. A fuzzy basis function vector-based multivariable adaptive controller for nonlinear systems. *IEEE Transactions on Systems, Man, and Cybernetics, Part B*, 30(1): 210–217, 2000.

[27] C. H. Lee and C. C. Teng. Identification and control of dynamic systems using recurrent fuzzy neural networks. *IEEE Transactions on Fuzzy Systems*, 8(4): 349–366, 2000.

[28] F. J. Lin and R. J. Wai. Hybrid control using recurrent fuzzy neural network for linear induction motor servo drive. *IEEE Transactions on Fuzzy Systems*, 9(1): 102–115, 2001.

[29] C. F. Juang. A TSK-type recurrent fuzzy network for dynamic systems processing by neural network and genetic algorithms. *IEEE Transactions on Fuzzy Systems*, 10(2): 155–170, 2002.

[30] C. K. Lin. Adaptive critic autopilot design of bank-to-turn missiles using fuzzy basis function networks. *IEEE Transactions on Systems, Man, and Cybernetics, Part B*, 35(2): 197–207, 2005.

[31] C. K. Lin. Nonsingular terminal sliding mode control of robot manipulators using fuzzy wavelet networks. *IEEE Transactions on Fuzzy Systems*, 14(6): 849–859, 2006.

[32] C. C. Tsai, H. C. Huang, and S. C. Lin. Adaptive neural network control of a self-balancing two-wheeled scooter. *IEEE Transactions on Industrial Electronics*, 57(4): 2010.

[33] C. H. Lu. Wavelet fuzzy neural networks for identification and predictive control of dynamic systems. *IEEE Transactions on Industrial Electronics*, 58(7): 3046–3058, 2011.

[34] C. M. Lin, and Hsin-Yi Li. A novel adaptive wavelet fuzzy cerebellar model articulation control system design for voice coil motors. *IEEE Transactions on Industrial Electronics*. 59, 2021–2033, 2011.

[35] C. H. Lu. Design and application of stable predictive controller using recurrent wavelet neural networks. *IEEE Transactions on Industrial Electronics*, 56(9): 3733–3742, 2009.

[36] J. Peng, Y. Wang, H. Zhang. Recurrent fuzzy cerebellar model articulation controller and its application on robotic tracking control. In: *Proceedings of 2009, WRI World Congress on Intelligent Systems*, Xiamen, China, , May 19–21, 2009, vol. 2, pp. 293–297.

[37] C.F. Juang, R. B. Huang, and Y. Y. Lin. A recurrent self-evolving interval type-2 fuzzy neural network for dynamic system processing. *IEEE Transactions on Fuzzy Systems*, 17(5): 1092–1105, 2009.

[38] J. Kolodner. *Case-Based Reasoning*. Morgan Kaufmann, 1993.

[39] A. Aamodt and E. Plaza. Case-based reasoning: foundational issues, methodological variations, and system approaches. *AI Communication*, 7(1): 39–59, 1994.

[40] R. Bergmann, J. Kolodner, and E. Plaza. Representation in case-based reasoning. *Knowledge Engineering Review*, 20(3): 209–213, 2005.

[41] B. Smyth, M. T. Keane, and P. Cunningham. Hierarchical case-based reasoning integrating case-based and decompositional problem-solving techniques for plant-control

software design. *IEEE Transactions on Knowledge and Data Engineering*, 13(5): 793–812, 2001.

[42] P. Cunningham. A taxonomy of similarity mechanisms for case-based reasoning. *IEEE Transactions on Knowledge and Data Engineering*, 21(11): 1532–1543, 2009.

[43] F. Fernandez-Riverola, F. Diaz, and J. M. Corchado. Reducing the memory size of a fuzzy case-based reasoning system applying rough set techniques. *IEEE Transactions on Systems, Man, and Cybernetics, Part C: Applications and Reviews*, 37(1): 138–146, 2007.

[44] R. S. Michalski, I. Bratko, and M. Kubat. *Machine Learning and Data Mining*. John Wiley & Sons, 1998.

[45] N. Cassimatis, P. Bignoli, M. Bugajska, S. Dugas, U. Kurup, A. Murugesan, and P. Bello. An architecture for adaptive algorithmic hybrids. *IEEE Transactions on Systems, Man, and Cybernetics, Part B: Cybernetics*, 40(3): 903–914, 2010.

[46] R. Ros, J. L. Arcos, R. L. de Mantaras, and M. Veloso. A case-based approach for coordinated action selection in robot soccer. *Artificial Intelligence*, 173(9–10): 1014–1039, 2009.

[47] F. Harashima and S. Suzuki. State-of-the-art intelligent mechatronics in human–machine interaction [guest introduction]. *IEEE Industrial Electronics Magazine*, 4(2): 9–13, 2010.

[48] J. M. Peula, C. Urdiales, I. Herrero, I. Sanchez-Tato, and F. Sandoval. Pure reactive behavior learning using case based reasoning for a vision based 4-legged robot. *Robotics and Autonomous Systems*, 57(6): 688–699, 2009.

[49] B. Rohrer. S-learning: a model-free, case-based algorithm for robot learning and control. In: *Case-Based Reasoning Research and Development*, Vol. 5650, Berlin/Heidelberg: Springer, 2009, pp. 285–297.

[50] F. Zhou, J. Jiao, S. Chen, and D. Zhang. A case-driven ambient intelligence system for elderly in-home assistance applications. *IEEE Transactions on Systems, Man, and Cybernetics, Part C: Applications and Reviews*, 41(2): 179–189, 2011.

[51] U. Cortes, C. Barrue, A. B. Martinez, C. Urdiales, F. Campana, R. Annicchiarico, C. Caltagirone. Assistive technologies for the new generation of senior citizens. the SHARE-it approach. *International Journal of Computers in Healthcare*, 1(1): 35–65, 2010.

[52] C. H. L. Lee and A. Liu. A goal-driven approach for service request modeling. *International Journal of Intelligent Systems*, 25(8): 733–756, 2010.

CHAPTER 7

ADAPTIVE CLASSIFIERS FOR NONSTATIONARY ENVIRONMENTS

CESARE ALIPPI[1], GIACOMO BORACCHI[1], MANUEL ROVERI[1], GREGORY DITZLER[2], and ROBI POLIKAR[3]

[1]Dipartimento di Elettronica e Informazione, Politecnico di Milano, Milano, Italy
[2]Department of Electrical and Computer Engineering, Drexel University, Philadelphia, PA, USA
[3]Electrical and Computer Engineering Department, Rowan University, Glassboro, NJ, USA

7.1 INTRODUCTION

Current deployments of sensor/actuator networks designed to address industrial and environmental applications make possible the online acquisition of a large set of data from a monitored environment. On such data a designer either builds an application or assesses its performance. In most cases the designer implicitly assumes that a process generating the data is stationary, that is, the probabilistic description of the process does not change over time. However, real systems naturally evolve in time due to ageing effects, soft or hard faults affecting the composing parts; at the same time the phenomenon might evolve itself. The consequence is that the stationary hypothesis is no longer valid [1] and the effectiveness of the application generally degrades with the intensity of the change.

Predictive models developed to work in nonstationary environments must then be able to detect changes, characterize and isolate faults and, if requested, adapt itself to track the change to maximize performance.

The same comments hold in applications endowed with a classification system where the stationarity hypothesis is implicitly assumed. Violation of the stationarity assumption can be induced by perturbations affecting the knowledge base of the classifier (e.g., the weights of neural networks) which, if not suitably designed to address such changes, rapidly impairs its performance abilities [2].

Contemporary Issues in Systems Science and Engineering, First Edition.
Edited by MengChu Zhou, Han-Xiong Li and Margot Weijnen.

Classification systems designed to work in nonstationary environments must therefore implement methods to handle the occurrence of unpredictable changes, or *concept drifts*, in a data generating process [3]. In particular, by relying on supervised information coming from the field during the operational life, they must adapt themselves to accommodate the occurrence of both gradual and abrupt variations. In addition, these classifiers should be able to deal with recurrent concepts [4], distinguish between true changes and noise [3] and maintain or improve the classification accuracy whenever stationary conditions are detected [1].

Algorithms designed to work in nonstationary environments can be characterized in several ways: online versus batch algorithms depending on the number of instances used to update the algorithm; a single classifier versus ensemble algorithms depending on the number of envisaged classifiers; active versus passive approaches, depending on whether the change detection is explicitly sought by the algorithm or not [4].

Against this background, the structure of this chapter is as follows. Section 7.2 defines the classification problem in nonstationary environments. Section 7.3 expands the adaptive classifier study by presenting methods to learn the concept drift, leaving to Section 7.4 the presentation of strategies for adapting the classifier to the new working conditions. Section 7.5 introduces the figures of merit to be used to evaluate the classification systems present in the literature, and Section 7.6 describes the approaches to deal with supervised information coming from the field during the operational life.

7.2 DEFINITION OF THE PROBLEM

Let X be the data generating process providing, at time t, an independent and identically distributed random variable $X(t) \in \mathbb{R}^d$ classified as $y(t) \in \{c_1, \ldots, c_c\}$. Without loss of generality we consider a two-class problem with labels $\{c_1, c_2\}$ Due to perturbation $\delta(t)$ affecting the system at unknown time τ, the probability density function (pdf) $p(X(t))$ of $\boldsymbol{X}$ changes to $p(X(t), \delta(t))$.

Perturbations affecting the system can be grouped into two main families according to the behavior of $p(X(t), \delta(t))$:

- *Time-independent variations*: The process moves from a stationary state into a new stationary state. At time instant τ the pdf changes from p(X(t)) to $p'(X(t), \delta)$ with δ being a time independent function (i.e., $\delta(t) = \delta, t > \tau$). This kind of variation is known as *abrupt concept drift* (also sudden or instantaneous), and can be mostly associated with hard faults affecting the system.
- *Time-dependent variations:* The pdf after the change is time dependent, that is, $p'(X(t), \delta(t))$. Examples of such variations are *gradual concept drifts*, where $p'(X(t), \delta(t))$ slowly changes over time so as to represent aging effects, *slow change-of-state concept drifts,* where a variation slowly drives the data-generating process from a stationary condition to a different one, and *cyclical*

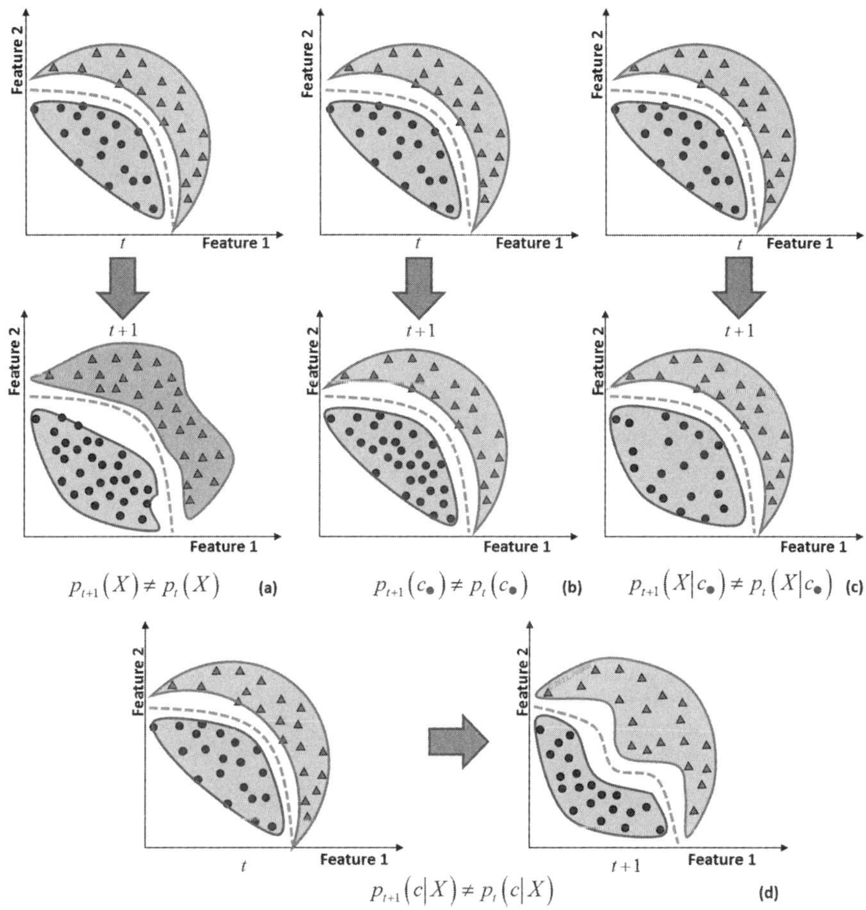

FIGURE 7.1 Change in (a) data distribution; (b) class priors; (c) likelihoods; (d) posteriors.

concept drifts, where a periodical variation affects the probability density function of the data generating process (e.g., the perturbation is related to seasonal phenomena such as the day–night cycle).

If we look at the concept drift problem from a more pragmatic classification perspective; however, we see that concept drift can affect any of the three terms of the posterior probability: the probability of the data distribution itself, the prior probability of individual classes, and/or the likelihood of class conditional probabilities. For simplicity, we refer to these cases as $p_{t+1}(X) \neq p_t(X)$, $p_{t+1}(c) \neq p_t(c)$; and $p_{t+1}(X|c) \neq p_t(X|c)$, respectively, where $\boldsymbol{X}$ represents the data, and c represents a particular class label. These three cases are illustrated in Figure 7.1a–7.1c, where—and when relevant—the circle indicates that change in probabilities are observed in that class, and the dashed line indicates the true (but unknown) decision boundary. We also note that, from a classification perspective, we only have a real concept drift if any of these terms actually causes the posterior probability to change, that is, if $p_{t+1}(c|X) \neq p_t(c|X)$. This is illustrated in Figure 7.1d.

Adaptive classification systems designed to work in nonstationary environments have to deal with a generic perturbation, and the scenario that causes the nonstationarity is typically unknown. Of course, exploiting a priori information allows us to use a more appropriate mechanism hence improving accuracy both in detection and classification.

7.3 LEARNING CONCEPT DRIFTS

Recently, learning concept drift (CD) has been receiving an increasing attention, in part due to many practical applications, such as spam/fraud detection, or longitudinal medical studies, where data distributions inherently change over time. However, much of early work in learning concept drifts (NSL) have primarily been on the definition of the problem, identifying the types of nonstationary environments [5–7] and the conditions under which they can be learned [8]. Characterizing such an environment is not trivial: as mentioned previously, the change can be abrupt or gradual, slow or fast, rare, random or systematic, cyclical or otherwise.

A variety of approaches have recently been proposed which can be categorized into the taxonomy mentioned in the previous section, that is, active versus passive approaches, online versus batch approaches, one single classifier versus ensemble-based approaches.

Regardless of whether there is an explicit mechanism to monitor detecting changes, CD algorithms typically employ a procedure to adjust the learner's parameters for incorporating the change in the environment, and/or forget what is no longer relevant. Earliest examples of CD algorithms include STAGGER [5] and FLORA [7] which use a sliding window to choose a block of (new) instances to train a new classifier. Such approaches are also called *instance selection* approaches (Figure 7.2). The window size is modified via "window adjustment heuristic," based on how fast the environment is changing.

FLORA has a built-in forgetting mechanism with the implicit assumption that those instances that fall outside of the window are no longer relevant, and the information carried by them can be forgotten. More recently, there have been several additions to this window-based approach, each introducing its own heuristics on drift detection

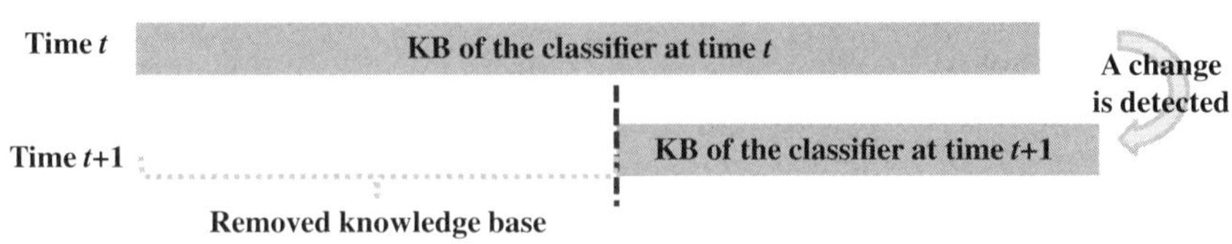

FIGURE 7.2 Instance selection approach.

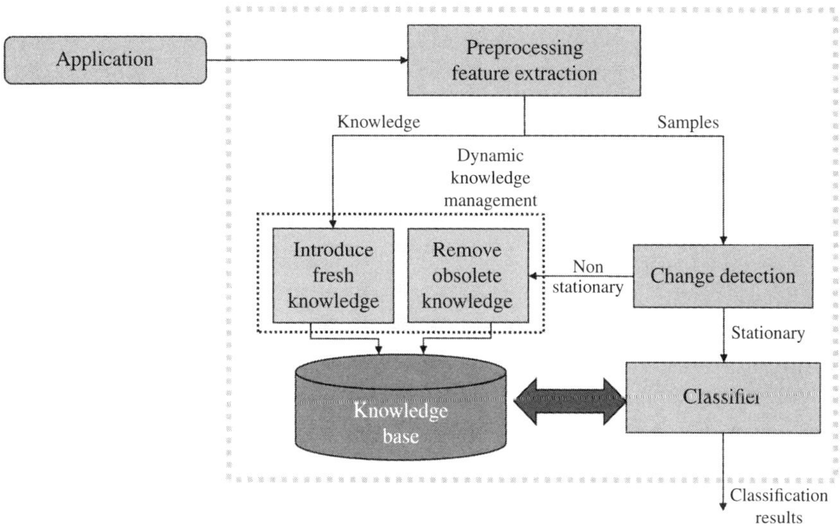

FIGURE 7.3 Just-in-time approach.

[9, 10], choice of classifier (such as decision trees, fuzzy rules, and kNN) [11–14], or establishing error thresholds. The primary shortcoming of these approaches is that they are usually not incremental (i.e., they need access to old data).

Other active approaches include change (anomaly) detection mechanisms to determine when changes occur [14–18]. More recent examples of these active approaches include those based on control charts, such as cumulative sum (CUSUM). Alippi and Roveri's *just in time* (JIT) classifiers (Figure 7.3) [19–21], and their more recent solution exploiting a change detection test (CDT) based on the *intersection of confidence intervals* (ICI) rule [22] are examples of such approaches. Most drift detection approaches are generally quite successful in detecting abrupt changes, but may struggle with gradual drift.

As an online active approach, the Early Drift Detection Method (EDDM) is specifically designed for gradual drift [23], which monitors the distance between the errors of a classifier and compares the mean of the errors to a threshold. EDDM has the ability to flag not only drift, but also issue a warning if necessary.

Other approaches for learning concept drift, or more generally, learning in nonstationary environments, include treating CD as a prediction problem used with a classification (e.g., PNN) [24], or tracking algorithm (such as Kalman filters) [25–27], or even deriving online update rules based on the minimization of a penalty function for the perceptron [28]. Another group of approaches use information theoretic measures, for example, entropy, mutual information, or Hoeffding's bounds of individual features for detecting drift and updating a decision tree [29–32]. These approaches also include a FLORA-like windowing mechanism, so do Hulten *et al.*'s concept adapting very fast decision tree (CVFDT) [11] or Cohen *et al.*'s incremental online-information network (IOLIN) [18, 33] algorithms.

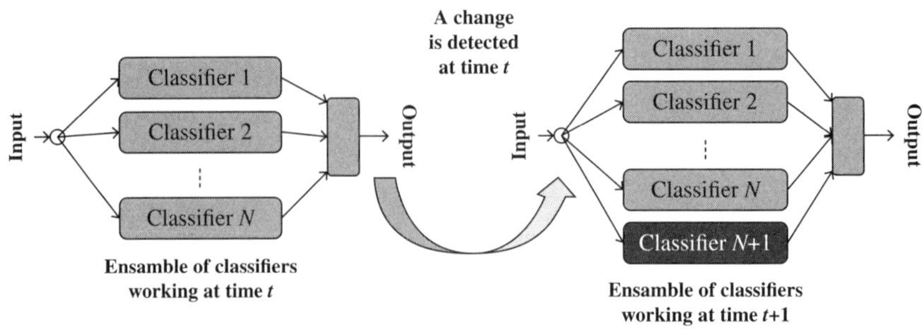

FIGURE 7.4 Ensemble of classifiers approach.

Bayesian approaches for online update of model parameters have also been proposed, for example, for regression problems [34, 35], or linearly separable rules [28].

The ensemble-based approaches that combine multiple classifiers constitute a new breed of CD algorithms. Kuncheva puts ensemble-based approaches into one of three general categories [36]: given new data, those that (i) update the combination rules or voting weights of a fixed ensemble, such as [37–39]; the origins of which can be traced to Littlestone's winnow [40], weighted majority algorithm [41], and Freund and Schapire's hedge (a precursor of AdaBoost) [42]; (ii) update the parameters of the (existing) members of the ensemble, using an online learner [14, 43]; and/or (iii) add new members to grow an ensemble with each incoming dataset as shown in Figure 7.4. The latter category approaches typically use a fixed number of classifiers in the ensemble, where the oldest (as in Street's Streaming Ensemble Algorithm (SEA) [44]) or the least contributing ensemble members are replaced with a new one (as in Tsymbal's Dynamic Integration [45], Kolter and Maloof's Dynamic Weighted Majority (DWM) [46]). Other efforts that follow a similar approach include [47–50]. Alternatively, Scholz and Klinkenberg's use a boosting-based ensemble creation approach which maintains two ensembles—one trained on the current data, and one trained on a cache of previously seen data, and chooses the better of the two ensembles in each time step. Classifier weights are based on the LIFT of each classifier, which measures the correlation between the classifier's decision and the true class based on conditional probabilities. Changing values of LIFT for each classifier across time indicate the existence of drift [51]. The way in which the LIFT values are computed, however, restricts the algorithm to binary classification problems only.

Some of the abovementioned methods can also fall into multiple categories; for example, DWM updates all of the (online) classifiers every time there is new data while adding new ones [46], and so does the approach in Reference 47. Many still use some form of a window either to limit the number of instances or to pick old instances still thought to be relevant (and hence these are not incremental learning algorithms) [11, 49]. Other efforts that follow similar ensemble approaches include using random forests with entropy [52] and Bifet's Hoeffding tree with Kalman-filter-based active change detection using adaptive sliding window (ADWIN) [53, 54]. ADWIN is also available within the WEKA-like software suite, Massive Online Analysis (MOA) in Reference 55.

More recently, Elwell and Polikar introduced Learn^{++}.NSE (for incremental learning in nonstationary environments), a new member of the Learn^{++} family of algorithms [56, 57]. Like all members of the Learn^{++} family of algorithms, Learn^{++}.NSE is a truly incremental algorithm that only uses currently available data. It is an ensemble-based batch learning algorithm with a passive drift detection mechanism: it generates a new classifier with each batch of data that becomes available, and combines the classifiers using weighted majority voting. Its novelty is with its mechanism for determining the voting weights, which are dynamically updated with respect to the classifiers' time-adjusted errors on current and past environments. A particular advantage of Learn^{++}.NSE is its ability to handle a variety of nonstationary environments, including sudden concept change, or concept drift that is slow or fast, gradual or abrupt, cyclical, or even variable rate drift. It is also one of the few algorithms that can handle concept addition (new class) or deletion of an existing class [4, 58].

While concept drift itself is a difficult problem on its own, it becomes a particularly challenging problem when encountered along with imbalanced data. A dataset is said to be unbalanced if the classification categories are not (approximately) equally represented. Real-world datasets are often predominantly composed of *normal* examples (also called *negative* or *majority* class) with only a small percentage of *abnormal* or *interesting* examples (also called *positive* or *minority* class).

A comprehensive literature review of learning from imbalanced data can be found in Reference 59. The most obvious solution is, of course, under sampling the majority class data and/or oversampling the minority class data. However, each has potentially serious drawbacks: under-sampling throws away data from the majority class, whether they are useful or not. Over-sampling, on the other hand, creates exact replicates of the minority instances, which may cause the classifier to overfit the minority class instances. A more novel sampling method, however, is followed by SMOTE [60], which enriches the minority class space by creating strategically placed synthetic examples located on the line segment connecting two minority instances. SMOTE has been shown to improve the classification accuracy on the minority class over other standard approaches.

Approaches that attempt to address the joint problem of learning concept drift from imbalanced data are far and few in between, perhaps due to difficulty of the problem. One notable algorithm capable of learning concept drift from imbalanced data has been proposed by Gao *et al.* [61, 62]. Their algorithm employs uncorrelated bagging to train classifiers on a subset of the majority class instances and the union of all minority class instances seen thus far (current + previous positive examples). With each iteration providing new data, the minority class instances are saved and accumulated to be used in training the classifiers at the future iterations. However, this approach implicitly assumes that the minority class data are stationary, potentially (and quite likely) an incorrect assumption. Chen and He's Selectively Recursive Approach (SERA) algorithm uses a similarity measure to select previous minority examples that are most similar to those in the most recent dataset [63]. This algorithm is less prone to issues of the minority class drifting compared to uncorrelated bagging presented in Reference 61. SERA can be implemented in two different ways: one method simply generates a single classifier on each dataset while the second method implements a form of biased bagging known as *BBagging*, which manually increases the sampling weights of the minority class instances. SERA effectively discards what it considers as irrelevant accumulated instances from the current training set by employing a Mahalanobis distance-based metric. SERA is modified to employ an ensemble-based approach in References 64 and 65. More recently, Ditzler and Polikar proposed two additional members to the Learn^{++} family of algorithms for learning imbalanced data in nonstationary environments. The first is Learn^{++}.CDS (*C*oncept *D*rift with *S*mote) which combines the previously mentioned Learn^{++}.NSE and SMOTE algorithms: SMOTE is used first to rebalance the data distribution, followed by Learn^{++}.NSE used to learn the concept drift. The second algorithm is Learn^{++}.NIE (*N*onstationary and *I*mbalanced *E*nvironment), which uses an ensemble of ensembles approach by generating a subensemble for each dataset that becomes available. The subensemble uses the classifiers that are trained with all minority class examples of the current batch (hence avoiding the problem with accumulating minority class instances) along with subsampled examples of the majority class. Learn^{++}.NIE also recognizes that the algorithms that address class imbalance cannot be evaluated on raw classification accuracy (RCA), and attempts to minimize an objective function that includes a weighted average of recall and precision [66, 67]. The alternate figures of merit for assessing algorithm performance are described in Section 7.5.

7.4 CHANGE DETECTION

In active adaptive classifiers following the detect-and-react philosophy the change detection mechanism is the key issue aiming at detecting the instant of time τ associated with the change and is carried out by designing suitable CDTs.

A CDT inspects the pdf $p(X(t))$, builds an initial hypothesis and checks over time whether the hypothesis is still valid or must be rejected. In doing this, the CDT has to minimize the detection delay time yet maximizing the detection accuracy, that is, reducing false positives and negatives.

Traditional CDTs detect a change by inspecting the measurements produced by X as shown in Figure 7.5a. Differently, detect-and-react classifiers either check the occurrence of changes in $\boldsymbol{X}$ (similar to the traditional change detection approach) and then react as shown in Figure 7.5b or monitor the output of the classifiers as shown in Figure 7.5c. These two approaches, which will be detailed in Sections 7.4.1 and 7.4.2, are characterized by different working conditions.

In particular, adaptive classifiers following the first approach do not require supervised samples to monitor the process and are able to detect only variations affecting the pdf of the process generating the data by inspecting the pdf of $\boldsymbol{X}$ (a swap of the pdfs of the two classes cannot be detected). Differently, adaptive classifiers following the latter approach are able to deal with variations affecting the classifier (e.g., the

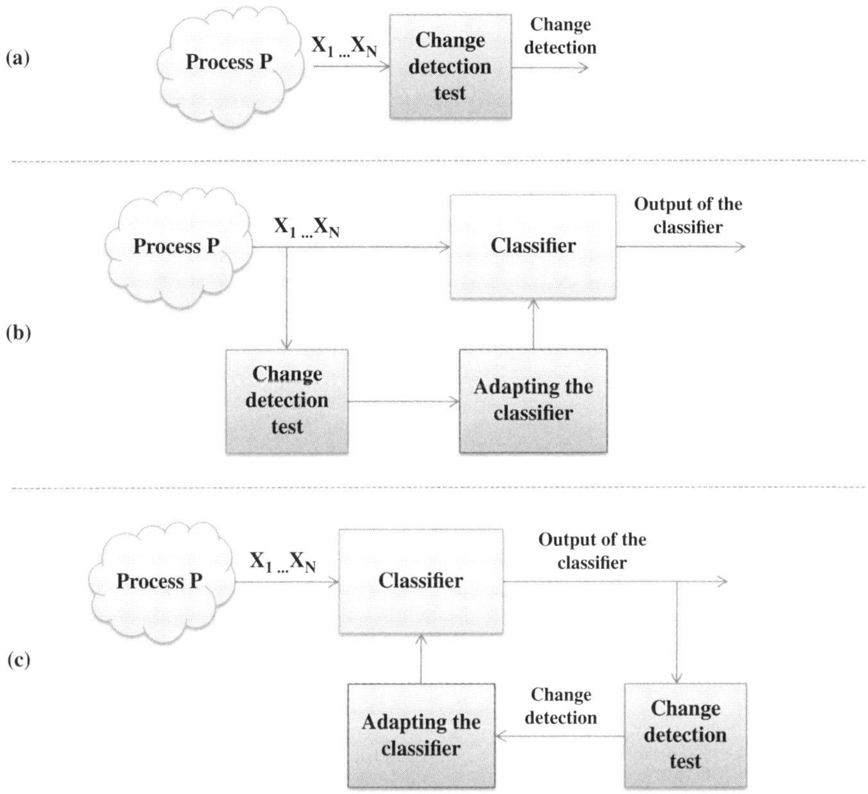

FIGURE 7.5 Change-detection tests in adaptive classifiers. (a) Traditional change detection activity. (b) Change detection enabling update of the classifier. (c) Change acting by inspecting the output of the classifier.

swap of the classes would be detected) but require the availability of supervised samples to accurately train the classifier and estimate its accuracy over time.

Section 7.4.1 summarizes the most common CDTs present in the literature to address variation in the process X (detailed reviews can be found in References 68 and 69), while Section 7.4.2 describes the use of these tests to inspect the output of adaptive classifiers.

7.4.1 Change-Detection Tests: A Review

CDTs can be grouped into three main families: statistical hypothesis tests, sequential hypothesis tests, and change-point detection tests.

Statistical hypothesis tests verify a generated hypothesis by inspecting a finite set of measurements. In particular, they either partition the available samples into two disjoint subsets to evaluate statistical discrepancies of features estimated from them (e.g., the differences in the mean values) or exploit all samples to verify the validity of an assumption on the pdf (e.g., the mean is equal to a predefined value). Z-test and t-test [70] are two parametric tests aimed at verifying the mean of a set of samples equal to a constant value (or comparing the means of two sets of samples) extracted from a normal distribution of known and unknown variance, respectively.

The Mann–Whitney U-test [71] (or Wilcoxon test) is a nonparametric test evaluating whether two sets of data have been generated from the same distribution. The test operates by ordering the available samples to verify if they are randomly ordered (i.e., the samples in the two sets have been generated by the same pdf) or not. Similarly, the Kolmogorov–Smirnov test [72] is a nonparametric test able to detect changes in the shape of two cumulative distributions by evaluating the maximum difference between the empirical and the true cumulative distribution (or between two empirical cumulative distributions). This test is largely used to assess a given hypothesis regarding a distribution, for example, to test if a distribution is normal. The Kruskal–Wallis test [73] extends the Mann–Whitney U-test, which is limited to the comparison of two sets, to three or more sets. As stated in Reference 74, the partitioning of the available samples affects not only the change detection accuracy, but also becomes a critical issue for an online classification system. In fact, the need to continuously detect changes makes it infeasible to exhaustively evaluate all possible partitioning and requires, even by considering a sliding window, a priori knowledge about the changing rate to suitably define the window size.

Differently from statistical hypothesis tests where the number of samples taken into account for the change detection is fixed, in *sequential hypothesis tests* the total number of samples depends on available data and is thus a random variable [75]. These tests, which have been developed during World War II to provide a more efficient testing of anti-aircraft gunnery, sequentially analyze incoming samples (one at the time) until a decision to accept or refuse the “no-change” hypothesis can be taken; unnecessary samples are not considered after the decision is taken. Reference 76 presents a sequential probability ratio test (SPRT) that reformulates the Neyman–Pearson lemma [77] for sequential analysis. In particular, similar to statistical hypotheses tests, SPRT tests the “no-change” hypothesis (P_1) versus the “change” hypothesis (P_2) by assessing the value of a likelihood-ratio-based figure of merit over time. When this value exceeds the threshold related to the “no-change” hypothesis, P_2 is rejected. On the contrary, when the figure of merit drops below the threshold related to the “change” hypothesis, P_2 is accepted. Detailed analysis of these two thresholds (and their relationship with the desired level of false positives and negatives) is presented in Reference 76.

The stopping time (i.e., when the test has enough significant information to be able to make a decision) is defined as the time instant in which the figure of merit overcomes one of the two thresholds. The extension of the SPRT test to composite hypotheses can be found in References 78 and 80. In the field of signal detection, a different approach is suggested in References 81 and 82, where the analysis of the sign is considered: the main advantages of this sign-based approach are the simplicity and their nonparametric property. An alternative to the SPRT test is the Repeated

Significance Test proposed in References 83 and 84. Initially developed for medical statistics, its main feature is the need to repeat statistical tests on larger and larger datasets until the statistical process is characterized with enough confidence (and the null hypothesis can be accepted or rejected). As stated in Reference 80, the repetition of tests on accumulating data may inflate the overall statistical significance level; the issue can be addressed by considering smaller significance levels for individual tests to guarantee that the overall statistical significance of the test is lower than a prescribed value.

Sequential hypothesis tests are theoretically well founded but suffer intrinsically from the need to make a decision about the null hypothesis (i.e., after the stopping time the null hypothesis is accepted or rejected). This is not a drawback in medical statistic applications, where a decision about the hypothesis is required (i.e., a new drug is/is not effective), but is a strong limitation in statistical quality control where the aim of the CDT is to identify when the process under monitoring departs from a predefined working state.

This aspect is tackled instead by *sequential change-point detection tests* that focus on the online monitoring and evaluation of the statistical behavior of a data generation process [68]. These tests keep on monitoring the process until a change is detected (i.e., the null hypothesis is rejected) but, differently from sequential hypothesis tests, do not interrupt the monitoring to assess the null hypothesis.

The concept of "state of statistical control" was introduced by Shewhart [85] in 1931 to model the statistical behavior of some chosen quality characteristic at a specific instant of time. Any change from this state, which results in a variation in the pdf of the characteristic under monitoring, is detected when the current sample exceeds a predefined threshold. To increase the robustness (i.e., reduce false positives) of the Shewhart chart, several moving average mechanisms have been proposed [86]. In particular, the exponentially weighted moving average (EWMA) [86] aims at emphasizing the weight of present observations with respect to past ones, while finite moving average (FMA) relies on the sample mean of a finite running window. A change is detected when the moving average exceeds a predefined threshold. A different approach is suggested by the CUSUM [69, 87] and the Shiryaev–Roberts [88, 89] tests, which have been developed by modifying the sequential hypothesis theory (i.e., the SPRT test) proposed by Wald. The CUSUM tests, which is based on the log-likelihood ratio between two pdfs (i.e., the null and the alternative pdf), is one of the most well-known CDTs and, in case of i.i.d. samples, it minimizes the worst mean detection delay among the tests providing the same false-alarm mean time [90].

By relying on a Bayesian approach, Shiryaev [88] presents a CDT that assumes the geometric prior distribution on the change time and addresses the change detection problem as an optimal stopping problem. Authors of [89] suggest an extension of the Shiryaev rule (i.e., the Shiryaev–Roberts test) that does not assume a specific apriori distribution about the change time. Both the non-Bayesian approach of the CUSUM test and the Bayesian approach of the Shiryaev–Roberts test provide excellent detection performance but require the knowledge of the pdf before and after the change (which is impractical in many real-world applications).

Several nonparametric extensions of the CUSUM and the Shiryaev–Roberts test have been presented in the literature (e.g., [91–94]). In particular, Reference 91 suggests a distribution-free CUSUM test aiming at detecting abrupt changes in the median of a distribution, Reference 92 presents a multi-dimensional nonparametric CUSUM-based test for the detection of changes in the mean value, and Reference 93 proposes a self-configuring and nonparametric CUSUM test following a computational-intelligent approach. Reference 94 proposes a nonparametric Shiryaev–Roberts test that relies on the likelihood ratios of signs and ranks of absolute values. Differently, Reference 95 presents a novel nonparametric sequential change-point detection test whose uniqueness is the use of the ICIs rule to monitor the evolution of the data generating process. These CDTs are summarized in Table 7.1.

7.4.2 Change-Detection Tests in Adaptive Classifiers

As presented in Figure 7.5c, detecting changes in the process by analyzing the output of the classifier requires monitoring the accuracy of the classifier itself. This is a key point since, differently from CDTs that work directly on the measurements coming from the process, assessing the accuracy of the classifier requires to operate on supervised information. Besides the need to have enough supervised samples to adequately train the model (to avoid bias in the trained model which could lead to false positive detections), the approach has two main effects. First, the amount of supervised samples coming from the field during the operational life must be adequate to allow the online estimation of the classification accuracy. In other words, classification systems following this approach generally rely on the availability of a batch of supervised samples to effectively estimate the accuracy (e.g., [4, 9, 64]). Other approaches (e.g., in References 7, 14, and 96] require one supervised sample at each time instant, but, to estimate accuracy, a weighted average of samples acquired at different time instants is considered. Second, the number of supervised samples coming from the field in a working condition is generally limited. This has a fundamental effect when comparing performances (i.e., the accuracies computed on subsequent batches of supervised samples) since the confidence of the accuracy estimate depends on the amount of samples on which the accuracy is computed [97]. Thus, if the number of available supervised samples in each batch is very small, the confidence on the respective estimates is small and the comparison among the accuracies might be very inaccurate.

Adaptive classifiers following the detect-and-react approach can be grouped into two main families according to the techniques used to assess the accuracy over time.

Adaptive classifiers belonging to the first approach [7, 14, 96] rely on a CDT (as the one presented in the previous section) to monitor accuracy during the operational life of the classifier. For example, Reference 7 exploits a Shewhart-based approach to monitor accuracy. In particular, when the accuracy of the classifier estimated on recent supervised samples decreases below a predefined threshold, a change is detected and subsequent adaptive procedures are then performed. A similar

TABLE 7.1 Change-Detection Tests

	Test family	Type (P,NP)	Change (Ab/Dr)	Entity under test	1D/ND	Online/offline	Training set/a priori information	Notes
Z-test	Statistical hypothesis testing	Parametric	Abrupt	Mean	1D	Offline	Parameters	Assume normality and known variance [70]
t-test	Statistical hypothesis testing	Parametric	Abrupt	Mean	1D	Offline	None	Assume normality [70]
Mann–Whitney U test	Statistical hypothesis testing	Nonparametric	Abrupt	Median	1D	Offline	None	Rank test [71]
Kolmogorov–Smirnov test	Statistical hypothesis testing	Nonparametric	Abrupt	Pdf	1D	Offline	None	Also goodness of fit test [72]
Kruskal–Wallis test	Statistical hypothesis testing	Nonparametric	Abrupt	Median	1D	Offline	None	Mann–Whitney-based, multiple subsets [73]
SPRT	Sequential hypothesis testing	Parametric	Abrupt	Pdf	1D	Online	Parameters	Minimize the stopping time nonparametric extensions [76]
CUSUM	Sequential change-point detection	Parametric	Abrupt	Pdf	1D	Online	Pdf and parameters	Minimize the worst detection delay [69]
CUSUM extension	Sequential change-point detection	Nonparametric	Abrupt	Median	1D	Online	None	Nonparametric extension of the CUSUM test [91]
CUSUM extension	Sequential change-point detection	Nonparametric	Abrupt	Median	1D/ND	Online	Parameters or training sequence	Detection of intrusion system [92]
CUSUM extension	Sequential change-point detection	Nonparametric	Abrupt/Drift	PDF/sample moments	1D/ND	Online	Training sequence	Computational intelligence approach [93]
ICI change-detection test	Sequential change-point detection	Nonparametric	Abrupt/Drift	Mean and variance	1D	Online	Training sequence	Exploits the intersection of confidence interval (ICI) rule [95]
Shiryaev–Roberts extension	Sequential change-point detection	Nonparametric	Abrupt	Median	1D	Online	None	Nonparametric extension of the Shiryaev–Roberts test [94]

Shewhart-based approach is presented in Reference 14, where accuracy is compared with thresholds that, differently from Reference 7, are not fixed a priori but adapted to the process by means of confidence intervals on a reference minimum accuracy. To overcome limitations of the Shewhart test, the approach presented in Reference 96 relies on an EWMA chart on the error observations. As stated by the authors, a Bernoulli CUSUM chart could also be considered but the EWMA chart on error observations natively provides, together with the change detection ability, the current probability of error of the classifier (which is then used in a weighting mechanism).

Approaches belonging to the first family are committed to monitor the output of the classifier by means of CDTs. Differently, adaptive classifiers belonging to the latter one [4, 10, 64, 67] more generally speculate about the classification accuracy to decide how and when to modify the classifier. For example, one of the approaches discussed in Reference 4 suggests ranking the classification accuracies of an ensemble of classifiers and pruning the classifier with the lowest accuracy. The works in References 64, and 67 perceive the occurrence of changes in the data generating process as well as in weights used to aggregate the final decision of an ensemble of classifiers. Finally, the approach presented in Reference 10 ranks the classification accuracies of classifiers configured on different training sets to identify which training set provides the highest accuracy (and hence using it to carry out the classification phase).

7.5 ASSESSING THE PERFORMANCE: FIGURES OF MERIT

There are several figures of merit commonly used in assessing classification algorithms. Each figure of merit provides a different perspective in evaluating the classifier performance, and in general, multiple figures of merit should ideally be used—and compared—in assessing the performance of a given algorithm. These metrics need to be used with care, particularly in concept drift and perhaps even more in imbalanced data scenarios, as blind use of any given metric can lead to incorrect interpretations.

In this section, we provide an overview of these approaches, while pointing to potential pitfalls in their usage. We use the following setting and terminology in this section: at the current evaluation time t, the classification algorithm returns a hypothesis $H^{(t)}$ as its classification rule, which is then provided with an evaluation dataset that consists of N instances $\mathbf{x}_i = 1, 2, \dots, N$ with their corresponding true labels $y \in \{c_1, \dots, c_c\}$. For two class problems, we define one of the classes (the more interesting one, or the minority one) as the positive class, whereas the other one as the negative class. In such cases, we can further define *true positive* (*TP*) and *true negative* (*TN*) as correct classifications of positive and negative classes, respectively; and *false positive* (*FP*) and *false negative* (*FN*) as incorrect classifications of positive class as negative, and negative class as positive class, respectively.

7.5.1 Raw Classification Accuracy

Raw classification accuracy (RCA), the ratio of the number of correct classifications to the total number of instances evaluated, is by far the most commonly used—and perhaps rather most misused and abused—evaluation metric:

$$\mathrm{RCA}^{(t)} = \frac{1}{N}\sum_{i=1}^{N}\left[\left[H^{(t)}(\mathbf{x}_i) = y_i\right]\right],$$

where $[[\cdot]]$ indicates the predicate that evaluates to 1 when its argument is true, and to 0, otherwise.

For two class problems, RCA can also be written as

$$\mathrm{RCA}^{(t)} = \frac{\mathrm{TP}^{(t)} + \mathrm{TN}^{(t)}}{\mathrm{TP}^{(t)} + \mathrm{TN}^{(t)} + \mathrm{FP}^{(t)} + \mathrm{FN}^{(t)}}.$$

Note that the above performance metrics are only evaluated at the current time t. For a global view of the algorithm's performance, the figure of merit can (whether it is RCA, or any of the following ones) be averaged over time, giving

$$\mathrm{RCA} = \frac{1}{T}\sum_{t=1}^{T}\mathrm{RCA}^{(t)}$$

However, with such averaging, the temporal characteristics of the algorithm performance is lost. In a nonstationary environment, such temporal characteristics can be very important. Therefore, a better approach is to plot the figure of merit against time.

7.5.2 Confusion Matrix

The primary problem with RCA is that it provides no information on which classes, if any, is particularly difficult to classify. When an algorithm correctly classifies certain classes, but consistently misclassifies others, such information can be extremely useful in troubleshooting and designing a better classifier. One of the easiest ways to assess the classifier's performance on each class is to observe its confusion matrix. The ith row and jth column element of this matrix denotes the number of class c_i instances classified as class c_j. Therefore, the trace of this matrix provides the total number of correct classifications, whereas the sum of off-diagonal elements is the total number of misclassifications.

While the confusion matrix provides all the information necessary to evaluate a classifier, it is a relatively inadequate tool for evaluating concept drift algorithms, since a new matrix needs to be generated for each time stamp. While these matrices can be averaged across time stamps, such an approach, as mentioned above, leads to the loss of temporal information. On the other hand, the confusion matrix does not lend itself naturally to plotting against time.

7.5.3 Geometric Mean

Geometric mean provides an alternative figure of merit that uses class-specific performances. It is defined as follows:

$$\mathrm{GM}^{(t)} = \prod_{j=1}^{C} \left(\frac{1}{N_j} \sum_{\mathbf{x}_i \in X_j} \left[H^{(t)}(\mathbf{x}_i) = y_i \right] \right)^{\frac{1}{C}},$$

where X_j is the set of instances from class c_j and N_j is the number of instances from class c_j. Note that the expression inside the product term is simply the class-specific RCA of the algorithm on class c_j instances. Also note that these performances are raised to the power of $1/C$ (where C is the total number of classes), and then multiplied. Therefore, an algorithm must perform well on *all* classes in order to obtain a good geometric-mean-based performance. $\mathrm{GM}^{(t)}$ can then be plotted against t to observe the temporal characteristics of the performance.

7.5.4 Precision and Recall

Precision and recall are important figures of merit particularly for imbalanced datasets. Precision (also known as *positive predictive value* in medical context) is the ratio of correctly identified positive instances to all instances identified as positive by the classifier. Precision can be interpreted as the probability of an instance identified as positive by the classifier to be truly a positive instance (or, in the medical context, the probability of a person identified as diseased by a test to be truly diseased), and it is defined as

$$\mathrm{Precision}^{(t)} = \frac{\mathrm{TP}^{(t)}}{\mathrm{TP}^{(t)} + \mathrm{FP}^{(t)}}.$$

Recall (also known as the *sensitivity*) is the ratio of correctly identified positive instances to all positive instances in the dataset. Recall can be interpreted as the probability of the classifier to truly identify a positive class instance, and it is defined as

$$\mathrm{Recall}^{(t)} = \frac{\mathrm{TP}^{(t)}}{\mathrm{TP}^{(t)} + \mathrm{FN}^{(t)}}.$$

As mentioned above, precision and recall are important for imbalanced datasets, where RCA can be particularly misleading. Consider an imbalanced dataset generated from the application of financial transaction fraud detection, where only 0.1% of all transactions are fraudulent, and the remaining are legitimate. In such applications, we are more interested in the classifier's ability to detect the minority class (here, the fraudulent transactions). A classifier that blindly classifies all instances as legitimate will achieve an *RCA* of 99.9%, while misclassifying all fraudulent cases. On the other hand, this classifier would have a zero recall and zero precision accuracy (as the number of true positives would be zero), a better reflection of the true performance of the classifier for this application.

Once computed for all time stamps, $t = 1, 2, \ldots, T$, both of these measures can either be averaged across all times, or more appropriately, plotted against t.

7.5.5 *F*-measure

The F-measure, or more accurately the F_1 score, is an attractive figure of merit because it is the harmonic mean of the precision and recall, and hence incorporates both these metrics in one measure. The F_1 score is the special case of the F_β measure, for the specific value of $\beta = 1$:

$$F_\beta^t = \frac{(1+\beta^2)\,Precision^{(t)}\,Recall^{(t)}}{\beta^2\,Precision^{(t)} + Recall^{(t)}}$$

$$F_1^t = \frac{2\,Precision^{(t)}\,Recall^{(t)}}{Precision^{(t)} + Recall^{(t)}}.$$

We can also show that the F_1 score is related to the geometric mean:

$$F_1^{(t)} = \frac{2(\mathrm{GM}^{(t)})^2}{Precision^{(t)} + Recall^{(t)}}.$$

As with previous measures, once computed for all time stamps, $t = 1, 2, \ldots, T$, F-measure can either be averaged across all times, or more appropriately, plotted against t.

7.5.6 Receiver Operator Characteristic Curve and Area Under the Curve

Originally developed for signal detection applications, receiver operator characteristics (ROC) curves provide a trade-off between the true positive rate (TPR, which is equivalent to recall) and false positive rate (FPR), defined as $\mathrm{FPR} = \mathrm{FP}/(\mathrm{FP}+\mathrm{TN})$. ROC curves are generally generated for binary classification problems, and require a classifier with a probabilistic output. A threshold θ is chosen to determine the label output of the classifier. If the probabilistic output, interpreted as the posterior probability of the positive class, is greater than θ, the positive label is assigned. Otherwise, the negative label is assigned. The ROC curve is generated by varying this threshold. For one extreme value of the threshold all positive instances are correctly classified, whereas all negative instances are correctly classified for the other extreme. A classifier with a random output displays an ROC that is diagonal, with an area (under this "curve") equal to 0.5. The larger the area under the ROC curve (and closer to 1), the better the classifier.

While the ROC curve itself is not that useful for concept drift applications, the area under this curve (AUC) is. It can be measured for each time stamp and plotted against time, providing a very informative evaluation metric for the concept drift tracking classifier.

7.6 CONCLUSIONS

A significant portion of recent and current research on machine learning and computational intelligence have been on developing new classification algorithms that make a fundamental assumption on the nature of data: the data come from a fixed but unknown distribution, an assumption that is clearly violated in many applications of current practical interest. These applications include tracking user interests for placing appropriate advertisements on web pages; spam filtering, financial data analysis, long-term epidemiological studies, energy demand and pricing, among many others. In such cases, the algorithms developed with the aforementioned assumption fail after a period of time when the decision boundaries generated based on previous data no longer describe the current decision boundaries. A new framework is required to develop algorithms that can track changing environments and take appropriate corrective actions.

In this chapter, we present an overview of a new breed of algorithms, commonly referred to as *concept drift* algorithms, which are designed to work in such nonstationary environments.

While now there is a growing body of work currently available, there are many open research problems. Some of these open problems can be listed as follows:

- *Lack of standard benchmark datasets*: The UCI machine learning datasets, a popular repository for testing and benchmarking new machine learning algorithms, does not include many (if any) datasets that include known drift. Therefore, different researchers have developed synthetic datasets to evaluate their algorithms. These include the STAGGER dataset [5], the SEA dataset [44], rotating checkerboard and travelling Gaussians datasets [58], rotating hyperplane dataset [61], among others. However, there are no standard benchmark datasets with controlled drift that can be used to compare algorithms.
- *Lack of real world dataset*: While the number of real-world applications that generate nonstationary applications are many, there are relatively few datasets that are available.
- *Change detection*: There are now several change detection mechanisms proposed in the literature; however, there is no benchmark either, or even a standard figure of merit for evaluating these algorithms.
- *Unsupervised or semi-supervised approaches for learning concept drift*: Often times, obtaining labeled data is costly, difficult or even impossible. Such cases call for semi-supervised or unsupervised approaches, and to the best of our knowledge such approaches for learning in nonstationary environments is severely underexplored.

REFERENCES

[1] C. Alippi and M. Roveri. Just-in-time adaptive classifiers–part II: designing the classifier. *IEEE Transactions on Neural Networks*, 19(11): 2053–2064, 2008.

[2] C. Alippi. Selecting accurate, robust and minimal feedforward neural networks. *IEEE-Transactions on Circuits and Systems: Part I, Fundamental Theory and Applications*, 49(12): 1799–1810, 2002.

[3] A. Tsymbal. Technical Report: The Problem of Concept Drift: Definitions and Related Work. Trinity College, Dublin, Ireland, TCD-CS-2004–15, 2004.

[4] R. Elwell and R. Polikar. Incremental learning in nonstationary environments with controlled forgetting. In: IJCNN 2009. International Joint Conference on Neural Networks, 2009, Atlanta, June 14–19, 2009, pp. 771–778.

[5] J. C. Schlimmer and R. H. Granger. Incremental learning from noisy data. *Machine Learning*, 1(3): 317–354, 1986.

[6] D. P. Helmbold and P. M. Long. Tracking drifting concepts by minimizing disagreements. *Machine Learning*, 14(1): 27–45, 1994.

[7] G. Widmer and M. Kubat. Learning in the presence of concept drift and hidden contexts. *Machine Learning*, 23(1): 69–101, 1996.

[8] J. Case, S. Jain, S. Kaufmann, A. Sharma, and F. Stephan. Predictive learning models for concept drift. *Theoretical Computer Science*, 268(2): 323–349, 2001.

[9] R. Klinkenberg and T. Joachims. Detecting concept drift with support vector machines. In: 17th International Conference on Machine Learning, California, June 29—July 2, 2000, pp. 487–494.

[10] R. Klinkenberg. Learning drifting concepts: example selection vs. example weighting. *Intelligent Data Analysis, Special Issue on Incremental Learning Systems Capable of Dealing with Concept Drift*, 8(3): 281–300, 2004.

[11] G. Hulten, L. Spencer, and P. Domingos. Mining time-changing data streams. In: Conference on Knowledge Discovery in Data, 2001, pp. 97–106.

[12] M. Nunez, R. Fidalgo, and R. Morales. Learning in environments with unknown dynamics: towards more robust concept learners. *Journal of Machine Learning Research*, 8: 2595–2628, 2007.

[13] P. Wang, H. Wang, X. Wu, W. Wang, and B. Shi. A low-granularity classifier for data streams with concept drifts and biased class distribution. *IEEE Transactions on Knowledge and Data Engineering*, 19(9): 1202–1213, 2007.

[14] J. Gama, P. Medas, G. Castillo, and P. Rodrigues. Learning with drift detection. *Advances in Artificial Intelligence—SBIA 2004 in Lecture Notes in Computer Science*, 3171: 286–295, 2004.

[15] M. Markou and S. Singh. Novelty detection: a review–part 1: statistical approaches. *Signal Processing*, 83(12): 2481–2497, 2003.

[16] M. Markou and S. Singh. Novelty detection: a review–part 2: neural network based approaches. *Signal Processing*, 83(12): 2499–2521, 2003.

[17] Y. Liao, V. R. Vemuri, and A. Pasos. Adaptive anomaly detection with evolving connectionist systems. *Journal of Network and Computer Applications*, 30(1): 60–80, 2007.

[18] L. Cohen, G. Avrahami-Bakish, M. Last, A. Kandel, and O. Kipersztok. Real-time data mining of non-stationary data streams from sensor networks. *Information Fusion*, 9(3): 344–353, 2008.

[19] C. Alippi and M. Roveri. Just-in-time adaptive classifiers; part I: detecting nonstationary changes. *IEEE Transactions on Neural Networks*, 19(7): 1145–1153, 2008.

[20] C. Alippi and M. Roveri. Just-in-time adaptive classifiers; part II: Designing the classifier. *IEEE Transactions on Neural Networks*, 19(12): 2053–2064, 2008.

[21] C. Alippi, G. Boracchi, and M. Roveri. Just in time classifiers: managing the slow drift case. In: International Joint Conference on Neural Networks, Atlanta, 2009, pp. 114–120.

[22] C. Alippi, G. Boracchi, and M. Roveri. Adaptive classifiers with ICI-based adaptive knowledge base management. In: International Conference on Artificial Neural Network (ICANN 2010), September 15–18, 2010, Thessaloniki, Greece, 2010, pp. 458–467.

[23] M. Baena-Garcia, J. del Campo-vila, R. Fidalgo, and A. Bifet. Early drift detection method. In: ECML PKDD International Workshop on Knowledge Discovery from Data Streams, Berlin (Germany), 2006, pp. 77–86.

[24] L. Rutkowski. Adaptive probabilistic neural networks for pattern classification in time-varying environment. *IEEE Transactions on Neural Networks*, 15(4): 811–827, 2004.

[25] G. Cheng-Kui, W. Zheng-Ou, and S. Ya-Ming. Encoding a priori information in neural networks to improve its modeling performance under non-stationary environment. In: International Conference on Machine Learning and Cybernetics (ICMLC 2004), Shanghai, China, 2004, vol. 5, pp. 3068–3072.

[26] A. Bifet and R. Gavalda. Kalman filters and adaptive windows for learning in data streams. In: Proceedings of 9th International Conference on Discovery Science (DS 2006), Barcelona, Spain, 2006, vol. 4265, pp. 29–40.

[27] S. Bai, S. Yi-Dong, and X. Wei. Modeling concept drift from the perspective of classifiers. In: 2008 IEEE Conference on Cybernetics and Intelligent Systems, Chengdu, China, 2008, pp. 1055–1060.

[28] E. A. de Oliveira. The Rosenblatt Bayesian algorithm learning in a nonstationary environment. *IEEE Transactions on Neural Networks*, 18(2): 584–588, 2007.

[29] Z. Yan, M. S. Mulekar, and P. Nerellapalli. Adaptive spam filtering using dynamic feature space. In: 17th IEEE International Conference on Tools with Artificial Intelligence (ICTAI 05), Hong Kong, 2005, p. 8.

[30] P. Vorburger and A. Bernstein. Entropy-based concept shift detection. In: Sixth International Conference on Data Mining (ICDM '06), Hong Kong, 2006, pp. 1113–1118.

[31] S. Hoeglinger and R. Pears. Use of Hoeffding trees in concept based data stream mining. In: International Conference on Information and Automation for Sustainability, ICIAFS 2007, Melbourne, Australia, 2007, pp. 57–62.

[32] C. J. Tsai, C. I. Lee, and W. P. Yang. Mining decision rules on data streams in the presence of concept drifts. *Expert Systems with Applications*, Part 1, 36(2): 1164–1178, 2009.

[33] L. Cohen, G. Avrahami, M. Last, and A. Kandel. Info-fuzzy algorithms for mining dynamic data streams. *Applied Soft Computing*, 8(4): 1283–1294, 2008.

[34] C. Jen-Tzung and C. Jung-Chun. Recursive Bayesian regression modeling and learning. In: IEEE International Conference on Acoustics, Speech and Signal Processing (ICASSP 2007), Honolulu, 2007, vol. 2, pp. 554–557.

[35] S. Srinivasan, J. Samuelsson, and W. B. Kleijn. Codebook-based Bayesian speech enhancement for nonstationary environments. *IEEE Transactions on Audio, Speech and Language Processing*, s15(2): 441–452, 2007.

[36] L. I. Kuncheva. Classifier ensembles for changing environments. In: *Multiple Classifier Systems (MCS 2004), Lecture Notes in Computer Science*, 2004, Vol. 3077, pp. 1–15.

[37] A. Blum. Empirical support for winnow and weighted-majority algorithms: results on a calendar scheduling domain. *Machine Learning*, 26(1): 5–23, 1997.

[38] C. Fang, W. Yizhou, and C. Zaniolo. An adaptive learning approach for noisy data streams. In: Fourth IEEE International Conference on Data Mining (ICDM 2004), Brighton (UK), 2004, pp. 351–354.

[39] Z. Xingquan, W. Xindong, and Y. Ying. Dynamic classifier selection for effective mining from noisy data streams. In: Fourth IEEE International Conference on Data Mining (ICDM '04), Brighton, UK, 2004, pp. 305–312.

[40] N. Littlestone. Learning quickly when irrelevant attributes abound: a new linear-threshold algorithm. *Machine Learning*, 2(4): 285–318, 1988.

[41] N. Littlestone and M. Warmuth. Weighted majority algorithm. *Information and Computation*, 108, 212–261, 1994.

[42] Y. Freund and R. E. Schapire. Decision-theoretic generalization of on-line learning and an application to boosting. *Journal of Computer and System Sciences*, 55(1): 119–139, 1997.

[43] N. Oza. Online ensemble learning. PhD Dissertation, University of California, Berkeley, 2001.

[44] W. N. Street and Y. Kim. A streaming ensemble algorithm (SEA) for large-scale classification. In: Seventh ACM SIGKDD International Conference on Knowledge Discovery & Data Mining (KDD-01), San Francisco, 2001, pp. 377–382.

[45] A. Tsymbal, M. Pechenizkiy, P. Cunningham, and S. Puuronen. Dynamic integration of classifiers for handling concept drift. *Information Fusion*, 9(1): 56–68, 2008.

[46] J. Z. Kolter and M. A. Maloof. Dynamic weighted majority: an ensemble method for drifting concepts. *Journal of Machine Learning Research*, 8: 2755–2790, 2007.

[47] H. Wang, W. Fan, P. Yu, and J. Han. Mining concept-drifting data streams using ensemble classifiers. In: Ninth ACM SIGKDD International Conference on Knowledge Discovery and Data Mining, Washington, DC, 2003, pp. 226–235.

[48] J. Rushing, S. Graves, E. Criswell, and A. Lin. A coverage based ensemble algorithm (CBEA) for streaming data. In: 16th IEEE International Conference on Tools with Artificial Intelligence (ICTAI 2004), Boca Raton, 2004, pp. 106–112.

[49] K. Nishida, K. Nishida, and K. Yamauchi. Adaptive classifiers-ensemble system for tracking concept drift. In: *2007 International Conference on Machine Learning and Cybernetics, Hong Kong*, edited by K. Yamauchi, 2007, vol. 6, pp. 3607–3612.

[50] H. He and S. Chen. IMORL: incremental multiple-object recognition and localization. *IEEE Transactions on Neural Networks*, 19(10): 1727–1738, 2008.

[51] M. Scholz and R. Klinkenberg. Boosting classifiers for drifting concepts. *Intelligent Data Analysis, Special Issue on Knowledge Discovery from Data Streams*, 11(1): 3–28, 2007.

[52] H. Abdulsalam, D. Skillicorn, and P. Martin. Classification using streaming random forests. *IEEE Transactions on Knowledge and Data Engineering*, vol. (in press) 2011.

[53] A. Bifet. Adaptive learning and mining for data streams and frequent patterns. PhD Dissertation, Universitat Politècnica de Catalunya, 2009.

[54] A. Bifet, E. Frank, G. Holmes, and B. Pfahringer. Accurate ensembles for data streams: combining restricted Hoeffding trees using stacking. In: 2nd Asian Conference on Machine Learning in Journal of Machine Learning Research, 2010, vol. 13.

[55] A. Bifet. MOA: Massive Online Analysis. 2010.

[56] R. Polikar, L. Upda, S. S. Upda, and V. Honavar. Learn++: an incremental learning algorithm for supervised neural networks. *IEEE Transactions on Systems, Man and Cybernetics, Part C*, 31(4): 497–508, 2001.

[57] M. D. Muhlbaier, A. Topalis, and R. Polikar. Learn++.NC: combining ensemble of classifiers with dynamically weighted consult-and-vote for efficient incremental learning of new classes. *IEEE Transactions on Neural Networks*, 20(1): 152–168, 2009.

[58] R. Elwell and R. Polikar. Incremental learning of variable rate concept drift. In: International Workshop on Multiple Classifier Systems, Reykjavik, Iceland, 2009, vol. 5519, pp. 142–151.

[59] H. He and E. A. Garcia. Learning from imbalanced data. *IEEE Transactions on Knowledge and Data Engineering*, 21(9): 1263–1284, 2009.

[60] N. V. Chawla, K. W. Bowyer, L. O. Hall, and W. P. Kegelmeyer. SMOTE: synthetic minority oversampling technique. *Journal of Artificial Intelligence*, 16: 321–357, 2002.

[61] J. Gao, W. Fan, J. Han, and P. S. Yu. A general framework for mining concept-drifting data streams with skewed distributions. In: SDM07 Vienna, Austria, 2007.

[62] J. Gao, B. Ding, W. Fan, J. Han, and P. S. Yu. Classifying data streams with skewed class distributions and concept drifts. *IEEE Internet Computing*, 12(6): 37–49, 2008.

[63] S. Chen and H. He. SERA: Selectively recursive approach towards nonstationary imbalanced stream data mining. In: IJCNN09, Atlanta, 2009, pp. 552–529.

[64] S. Chen, H. He, K. Li, and S. Desai. MuSERA: multiple selectively recursive approach towards imbalanced stream data mining. In: World Congress on Computational Intelligence, Barcelona, Spain, 2010, pp. 2857–2864.

[65] S. Chen and H. He. Towards incremental learning of nonstationary imbalanced data stream: a multiple selectively recursive approach. *Evolving Systems*, 2(1): 35–50, 2011.

[66] G. Ditzler, R. Polikar, and N. Chawla. An incremental learning algorithm for nonstationary environments and class imbalance. In: 20th International Conference on Pattern Recognition (ICPR), Istanbul, Turkey, 2010, pp. 2997–3000.

[67] G. Ditzler and R. Polikar. An ensemble based incremental learning framework for concept drift and class imbalance. In: The 2010 International Joint Conference on Neural Networks (IJCNN), Barcelona, Spain, 2010, pp. 1–8.

[68] T. L. Lai. Sequential analysis: some classical problems and new challenges. *Statistica Sinica*, 11(2): 303–350, 2001.

[69] Michèle Basseville and Igor V. Nikiforov. *Detection of Abrupt Changes: Theory and Application*. Englewood Cliffs, NJ: Prentice-Hall, 1993.

[70] A. M. Mood, F. A. Graybill, and D. C. Boes. *Introduction to the Theory of Statistics*, 3rd edn. New York: McGraw-Hill, 1974.

[71] H. B. Mann and D. R. Whitney. On a test of whether one of two random variables is stochastically larger than the other. *The Annals of Mathematical Statistics*, 18(1): 50–60, 1947, JSTOR.

[72] F. J. Massey Jr. The Kolmogorov-Smirnov test for goodness of fit. *Journal of the American Statistical Association*, 46(253): 68–78, 1951, JSTOR.

[73] W. H. Kruskal and W. A. Wallis. Use of ranks in one-criterion variance analysis. *Journal of the American Statistical Association*, 47(260): 583–621, 1952, JSTOR.

[74] S. Muthukrishnan, E. van denBerg, and W. Yihua. Sequential change detection on data streams. In: Seventh IEEE International Conference on Data Mining, Omaha, 2007, pp. 551–550

[75] Y. Mei. Asymptotically optimal methods for sequential change-point detection. PhD Dissertation, Calif. Inst. Technol., Pasadena, CA, 2003.

[76] A. Wald. Sequential tests of statistical hypotheses. *Annals of Mathematical Statistics*, 16(2): 117–186, 1945.

[77] N. Jerzy and P. Egon. On the problem of the most efficient tests of statistical hypotheses. *Philosophical Transactions of the Royal Society of London. Series A, Containing Papers of a Mathematical or Physical Character*, 231: 289–337, 1933.

[78] Z. Govindarajulu. *Sequential Statistical Procedures*. Academic Press, 1975

[79] J. Kiefer and J. Sacks. Asymptotically optimum sequential inference and design. *The Annals of Mathematical Statistics*, 34: 705–750, 1963.

[80] D. Siegmund. *Sequential Analysis: Tests and Confidence Intervals*. New York: Springer, 1985.

[81] R. Chandramouli and N. Ranganathan. A generalized sequential sign detector for binary hypothesis testing. *Signal Processing Letters, IEEE*, 5(11): 295–297, 1998.

[82] S. Tantaratana and J. Thomas. On sequential sign detection of a constant signal. *IEEE Transactions on Information Theory*, 23(3): 304–315, 1977.

[83] P. Armitage. *Sequential Medical Trials*, 2nd edn. Oxford, UK: Blackwell, 1975.

[84] H. R. Lerche. An optimal property of the repeated significance test. *Proceedings of the National Academy of Sciences of the United States of America*, 83: 1546–1548, 1986

[85] Walter A. Shewhart. *Economic Control of Quality of Manufactured Product*. New York: D. Van Nostrand Company, 1931, pp. 501, LCCN 31–32090.

[86] S.W. Roberts. Control chart tests based on geometric moving averages. *Technometrics*, 42(1): 97–101, 1959.

[87] E. S. Page. Continuous inspection schemes. *Biometrika*, 41: 100–114, 1954.

[88] A. N. Shiryaev. On optimum methods in quickest detection problems. *Theory of Probability and Its Applications*, 8: 22–46, 1963.

[89] S. W. Roberts. A comparison of some control chart procedures. *Technometrics*, 8(3): 411–430, 1966, JSTOR

[90] G. Lorden. Procedures for reacting to a change in distribution. *The Annals of Mathematical Statistics*, 42(6): 1897–1908, 1971, JSTOR

[91] C. A. McGilchrist and K. D. Woodyer. Note on a distribution-free CUSUM technique. *Technometrics*, 17(3): 321–325, 1975.

[92] A. G. Tartakovsky, B. L. Rozovskii, R. B. Blazek, and H. Kim. Detection of intrusions in information systems by sequential change-point methods. *Statistical Methodology*, 3(3): 252–293, 2006.

[93] C. Alippi and M. Roveri. Just-in-time adaptive classifiers. Part I. detecting non-stationary changes. *IEEE Transactions on Neural Networks*, 19(7): 1145–1153, 2008.

[94] L. Gordon and M. Pollak. An efficient sequential nonparametric scheme for detecting a change of distribution. *The Annals of Statistics*, 22(2): 763–804, 1994, JSTOR

[95] C. Alippi, G. Boracchi, and M. Roveri. Change detection tests using the ICI rule. In: Proceedings of IEEE IJCNN 2010—International Joint Conference on Neural Networks, Barcelona, Spain, July 18—23, 2010.

[96] K. Nishida and K. Yamauchi. Learning, detecting, understanding, and predicting concept changes. In: IJCNN 2009—International Joint Conference on Neural Networks, June 14–19, 2009, pp. 2280–2287.

[97] C. Alippi and P. Braione. Classification methods and inductive learning rules: what we may learn from theory. *IEEE Transactions on Systems, Man, Cybernetics: Part C*, 36(5): 649–655, 2006.

CHAPTER 8

MODELING, ANALYSIS, SCHEDULING, AND CONTROL OF CLUSTER TOOLS IN SEMICONDUCTOR FABRICATION

NAI QI WU[1,2], MENGCHU ZHOU[1,3], FENG CHU[4], and SAÏD MAMMAR[4]

[1]The Institute of Systems Engineering, Macau University of Technology, Macau, PR China

[2]Department of Industrial Engineering, Guangdong University of Technology, Guangzhou, PR China

[3]Department of Electrical and Computer Engineering, New Jersey Institute of Technology, Newark, NJ, USA

[4]Laboratoire d'IBISC EA 4526, Université d'Evry Val d'Essonne, Evry Cédex, France

8.1 INTRODUCTION

Most wafer fabrication processes are chemical processes, and traditionally wafers are processed in batches for high throughput. However, for batch processing it is difficult to assure the quality of wafers. One of the important innovations in semiconductor fabrication is the adoption of single-wafer processing by using cluster tools. With single-wafer processing technology, cluster tools provide a flexible, reconfigurable, and efficient environment for semiconductor manufacturing [1, 2] resulting in higher yield [3], shorter cycle time [3–5], better utilization of space [2, 5], and lower capital cost [5]. Thus, cluster tools are increasingly adopted by the semiconductor fabrication industry and, consequently, their effective operation is receiving more and more attention.

A cluster tool integrates several processing modules (PMs) within a closed environment such that the wafers are processed one by one in a cassette without going out of the tool. Cluster tools are a kind of robotic cells. Extensive studies have been done to schedule robotic cells [6–17]. These studies mainly focus on robotic flow-shop with a single PM for each processing step and a completed part can stay in

Contemporary Issues in Systems Science and Engineering, First Edition.
Edited by MengChu Zhou, Han-Xiong Li and Margot Weijnen.

the machine for unlimited time. Cluster tools in semiconductor fabrication can be differentiated from the robotic flow-shop by: (1) there may be time constraints on the wafer processing; (2) multiple PMs can be configured for a step; (3) the robot moving times from one PM to another are much shorter than the wafer processing times [18]; and (4) there are wafer revisiting for some wafer fabrication processes. The hoist system is somehow similar to the flow-shop robot cells except that there is a part of processing window constraint [19–21]. However, a hoist system is different from a dual-arm cluster tool in the following ways: (1) a hoist can hold one part at a time, whereas the robot in a dual-arm cluster tool can hold two parts at a time; (2) in hoist system, there is only one PM for each processing step, whereas in a cluster tool more than one PMs can be configured for a step; and (3) because of the radial layout, the movement of the robot in a cluster tool is much efficient than a hoist. Hence, different techniques for modeling, analysis, scheduling, and control of cluster tools are necessary.

Extensive studies on the modeling, performance evaluation, control, and scheduling of cluster tools have been conducted. This chapter presents the advancements made in this field, analyzes the open problems, and points out the future research directions.

8.2 CLUSTER TOOLS AND THEIR OPERATIONS

8.2.1 Architecture of Cluster Tools

A cluster tool consists of a number of PMs, an aligner module, a wafer-handling robot, and one or two loadlocks for wafer cassette loading/unloading as shown in Figure 8.1. There are no more than six PMs in it and these PMs are arranged in a

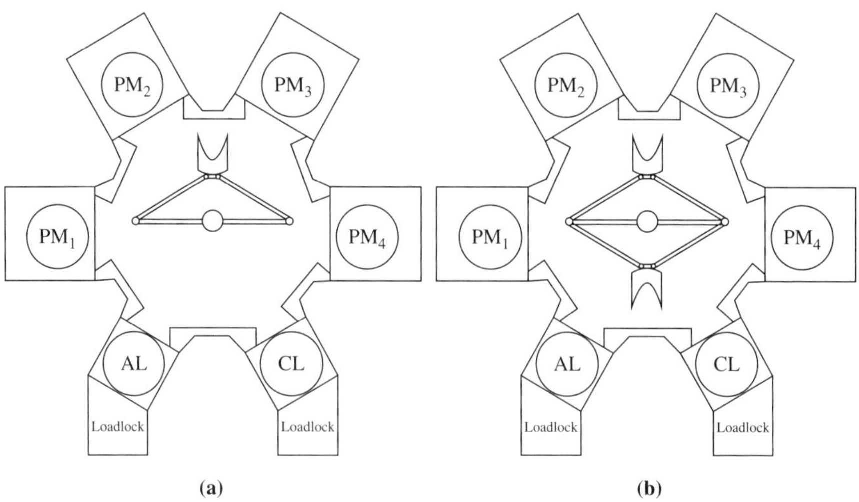

FIGURE 8.1 Cluster tools: (a) single-arm robot; (b) dual-arm robot.

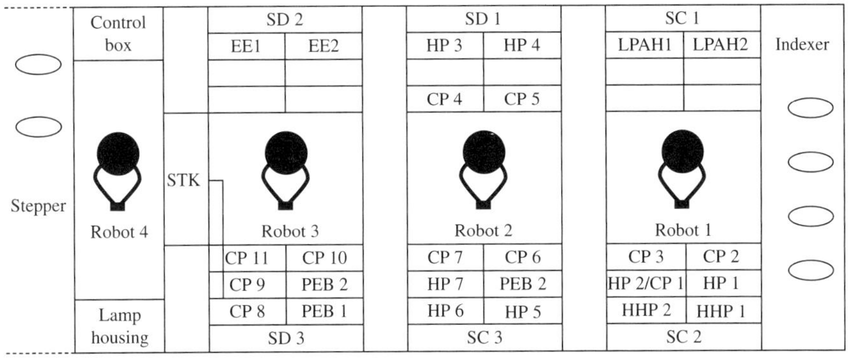

FIGURE 8.2 Layout of a typical semiconductor track system. CP, cool plate; HP or HHP, hot plate; PEB, post exposure bake hot plate; EE, edge exposure; SC, spin coater; SD, spin developer; LPAH, low-pressure adhesion.

radial way such that the robot can efficiently move among them. Raw wafers enter the system through a loadlock and visit one or more PMs in a specific order and return to the loadlock. The robot unloads a wafer from a loadlock, loads it into the PMs, unloads it after processing, moves it between the PMs, and finally returns the completed wafer into the loadlock. The robot can be a single-arm (see Figure 8.1a) or dual-arm one (Figure 8.1b). As shown in Figure 8.1b, the two arms of the dual-arm robot point to opposite directions and at any time only one of them can pick up or drop down a wafer.

A system that integrates a number of PMs and multiple wafer handling robots is called a track system. For example, a track system is used for photolithography processes. There are 5–10 parallel PMs for a step [22]. A typical track system is shown in Figure 8.2.

In a radial cluster tool, the robot makes only radial movements and the time taken from one PM to another PM is small and hence is regarded identical. However, in track systems, a robot should make both radial and vertical movements with different moving time, which has an effect on the operation.

Several cluster tools can be connected by intermediate buffers to form a multi-cluster tool [23–28]. A multi-cluster tool can be configured as a linear or tree-like topology, and a linear topology multi-cluster tool is shown in Figure 8.3. In a track system, a robot and the PMs served by the robot can be seen as cluster tool. Thus, a track system is in fact a multi-cluster tool.

8.2.2 Wafer Flow Patterns

Wafer flow patterns take an important role in operating cluster tools. Because multi-cluster tools are composed of a number of single-cluster tools, we just present the wafer flow patterns for a single-cluster tool. Raw wafers enter into a cluster tool through the loadlocks cassette by cassette. In general, a cassette has 25 wafers with

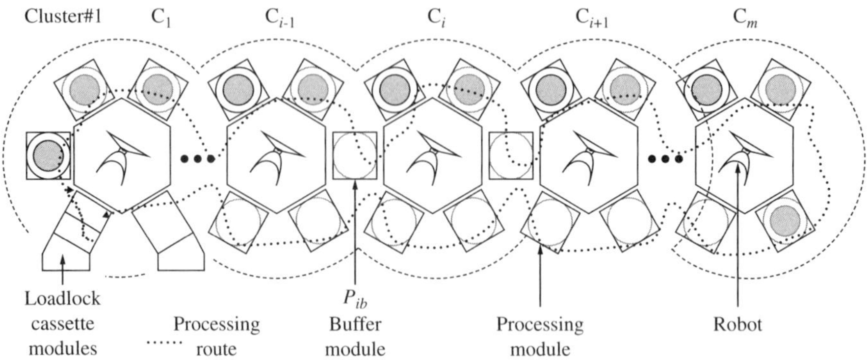

FIGURE 8.3 A linear multi-cluster tool.

an identical recipe [18, 29]. When a cassette is loaded into a loadlock, it is pumped into vacuum and then the wafers in the cassette are ready for processing. The robot unloads a wafer from the loadlock and moves it to the aligner. A vision system at the aligner checks the wafer alignment and adjusts the wafer holding position on the robot arm. After an unloaded wafer is aligned, it is moved and loaded into a PM. When a PM completes the processing, the chuck holding the wafer comes down to the slot valve position. As soon as a robot arm is available, the slot valve opens and the just processed wafer is unloaded onto the robot arm through the slot valve, then the wafer is moved to the next step for processing. A wafer visits a number of PMs in the order according to a prespecified recipe. After a wafer is completed, it is cooled down in the cooling module. Then, it comes back to the same loadlock from where it came from.

Different types of wafers have different processing routes that determine the wafer flow patterns. A cluster tool is reconfigurable because a PM in a tool can be configured to complete a required fabrication process. A number of PMs can be configured for the same fabrication process and these PMs are called parallel PMs for the process. To describe the wafer flow pattern, we define a fabrication process as a fabrication step. By this definition, different fabrication steps have different fabrication processes. The fabrication processes for a wafer type are composed of a number of operations. However, different operations may require the same step for processing. In this way, a general wafer flow pattern without wafer revisiting is denoted by $(m_1, m_2, \ldots, m_n)$ [18], where n is the number of operations. Because there is no wafer revisiting, each operation should be processed by a different step. Thus, it needs n steps to process the wafers and m_i is the number of parallel PMs at process Step i. When $m_1 = m_2 = \ldots = m_n = 1$ holds, it is called a serial wafer flow pattern. Such a wafer pattern is shown in Figure 8.4a. However, if any $m_i > 1$, $1 \leq i \leq n$, it is called a parallel wafer flow pattern. An example is shown in Figure 8.4b, where the first step has two PMs, that is, $m_1 = 2$.

Some wafer fabrication processes are repeated processes. For such processes, the wafers need to visit some fabrication steps again. For example, in atomic layer deposition (ALD) processes, the deposition thickness is controlled by repeating extremely

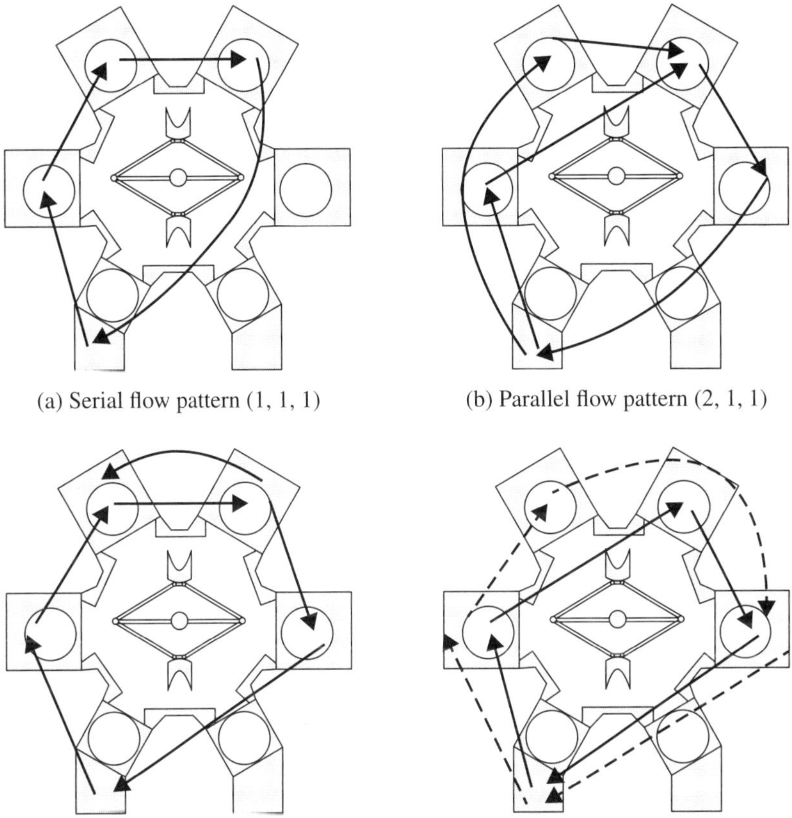

(a) Serial flow pattern (1, 1, 1)

(b) Parallel flow pattern (2, 1, 1)

(c) Flow pattern with revisiting $(1, (1, 1)^k, 1)$

(d) Two-product flow pattern with dotted one (1, 1, 1) corresponding to (PM_1, PM_2, PM_4) and solid one (1, 1, 1) corresponding to (PM_1, PM_3, PM_4)

FIGURE 8.4 Typical wafer flow patterns.

thin deposition multiple times. The general wafer flow pattern with wafer revisiting is denoted by $(m_1, \ldots, (m_i)_f^h, \ldots, (m_g, \ldots, m_j)^k, \ldots, m_n)$, where $n > j > g$. Notation $(m_i)_f^h$ denotes that after executing Step f, Step i is repeated once, and then the wafer is delivered to Step h for the next processing. Note that if it requires repeating a single step, it repeats only once. When consecutive fabrication steps need to be repeated, there is no confusion. Hence, $(m_g, \ldots, m_j)^k$ means that after executing Step $g - 1$, it needs to repeat k times the processing of steps g to j, and then the next step is $j + 1$. For example, the revisiting wafer flow pattern $(1, (1, 1)^{\boldsymbol{k}}, 1)$ is shown in Figure 8.4c.

A cluster tool can be used to process multiple wafer types simultaneously as shown in Figure 8.4d. The wafer flow pattern for such a process can be denoted by the wafer flow pattern of the individual wafer type.

8.2.3 Operation Requirements

With two loadlocks, when the wafers of a cassette in a loadlock are being processed, the completed wafers in the other loadlock can be unloaded and a new cassette can be loaded into this loadlock. Thus, a cluster tool can continue its operation without interruption. Often, the wafers in a cassette have the same recipe such that a cluster tool mainly repeats the operation most of the time, in other words, a cluster is in its steady state most of the time. A transient state appears only when the system starts from the idle state or when it switches from processing one type of wafers to another type. Therefore, to effectively operate a cluster tool is to schedule it such that the productivity is maximized, the automation operation can be continued, and the processing quality is assured.

In a cluster tool, the space is very limited and there is no intermediate buffer. Thus, some wafer flow may block the flow of others, leading to deadlocks. When deadlock occurs, the automation operation would be stopped. It is crucial to avoid deadlock and conflict in operating a cluster tool.

In a cluster tool, there is a robot for wafer delivery only. When a wafer is processed in a PM, it remains in the PM until the robot arrives and unloads it. Thus, there is a time duration for which the processed wafer waits in the PM. This wafer waiting time is called wafer delay or wafer residency time. Due to the residual gases and heat in a PM, a wafer waiting in it after being completed results in surface quality problems. Most of wafer fabrication processes require that after being completed a wafer should be unloaded from the PM within a given time range, which is called wafer residency time constraints [30]. For some processes, there are very strict wafer residency time constraints. For instance, in low pressure chemical vapor deposition (LPCVD), when the wafer residency time exceeds an upper limit, the wafer is scrapped [18]. In a wet station, it requires that a wafer cleaned in a chemical bath should be immediately unloaded and rinsed in a water bath [31]. Coating and developing processes in a photolithography process also have strict wafer residency time constraints [32]. Thus, it is very important to make the wafer residency time constraints satisfied in operating a cluster tool.

In cluster tool operations, wafer processing times and robot task times are rather deterministic and can be regarded as constant in the normal condition. However, some abnormal events can occur, leading to the variation of activity times. In particular, unpredictable variation arises from wafer alignment failure and retrial, PM processing delay, communication delay, and computer processing delay. Hence, the process times and robot task times are subject to certain variation and sometimes abrupt random disturbances [18, 33]. They result in the wafer residency time fluctuation and make an otherwise feasible and optimal schedule obtained under a deterministic activity time assumption infeasible. Thus, activity time variation should be considered in operating such a cluster tool.

One more issue in operating a cluster is that processing tools in semiconductor manufacturing are prone to failures and the downtimes are quite lengthy, ranging in the order of several hours [34, 35]. Thus, techniques are necessary to deal with the tool failures.

8.3 MODELING AND PERFORMANCE EVALUATION

For the effective operation of cluster tools, important efforts are made for modeling and performance analysis. Cluster tools are reconfigurable and their different configurations lead to different performance results, as analyzed in References 26, 27, and 36. When the configuration is given, two types of properties are concerned: quantitative and qualitative properties. The most concerned qualitative property is deadlock. The most concerned quantitative property is throughput or cycle time performance. There are mainly two types of models: timing diagram and Petri net.

8.3.1 Analysis Based on Timing Diagram Model

At the early stage of studies on cluster tools, the main concern focused on the throughput. It is assumed that the activity times are deterministic and there is no wafer residency time constraint. As discussed above, most of the time a cluster tool is in its steady state and operates in a periodic way. Thus, in the steady state the throughput is characterized by its cycle time. The analysis done in References 37 and 38 is the representative work. It uses so-called timing diagram to model the behavior of a cluster tool. A timing diagram model for a single-arm cluster tool with four steps is shown in Figure 8.5, where each step has only one PM. In the model, it defines two types of activities: robot moving and wafer processing. A robot moving activity denotes the robot moving from a PM to another including picking a wafer up from a PM or dropping a wafer down into a PM. Pictorially, a robot moving activity is represented by an arrow in the model and the time taken by it is a multiple of T. The wafer processing is represented by a shaded rectangle whose length indicates the time taken.

With the timing diagram model and backward scheduling strategy, the throughput of a single-arm cluster tool without wafer revisiting is analyzed in References 37 and 38. For a four-step process, the backward scheduling strategy operates as follows: the robot moves to Step 4 → unloads a completed wafer there → moves to the loadlocks → loads the completed wafer into a loadlock → moves to Step 3 → unloads a processed wafer there → moves to Step 4 → loads the wafer into a PM there → moves to Step 2 → unloads a processed wafer there → moves to Step 3 → loads the wafer into a PM there → moves to Step 1 → unloads a processed wafer there → moves to Step 2 → loads the wafer into a PM there → moves to the loadlocks → unloads a raw wafer there → moves to Step 1 → loads the wafer into a PM there →

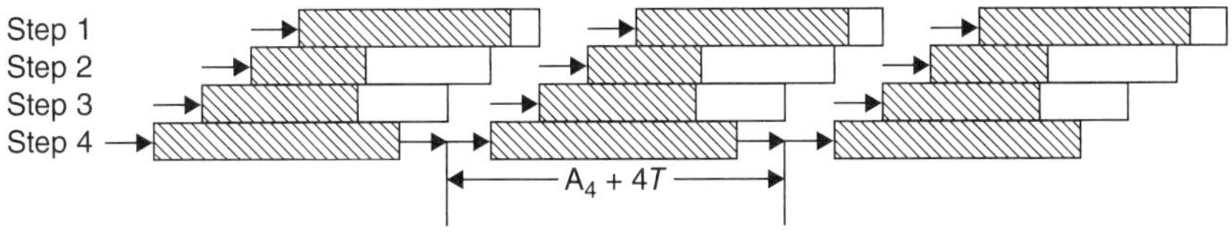

FIGURE 8.5 Timing diagram for a single-arm cluster tool with four steps.

moves to Step 4 again. In this way, a period is completed. The cycle time is defined as the fundamental period, as denoted by Π in this chapter. Let n denote the number of steps, A_i be the wafer processing time at Step i, Π_i the fundamental period for Step i, and ψ the time taken by the robot in a cycle. Then it is found that

$$\Pi_i = \frac{A_i + 4T}{m_i} \tag{8.1}$$

and

$$\psi = 2T(n+1). \tag{8.2}$$

Then, $\Pi = \max[\psi, \Pi_1, \Pi_2, \ldots, \Pi_N]$. If $\Pi = \psi$, it is said that the tool operates in the transport-bound region. If $\Pi > \psi$, it is said that the tool operates in the process-bound region. In the former, the robot is always busy and the system cycle time is determined by the time for robot tasks, whereas in the latter, the robot has idle time and the processing times in PMs determine the cycle time. With the fact that for cluster tools in semiconductor fabrication the robot moving time is much less than the wafer processing time, backward scheduling is optimal for single-arm cluster tools.

Based on the model and the results obtained in References 37 and 38, the throughput of dual-arm cluster tools is analyzed by Venkatesh *et al.* [39]. It defines the so-called swap strategy. The swap strategy operates as follows: the robot with both arms empty moves to loadlocks → unloads a raw wafer there by one arm → the idle arm moves to Step 1 and unloads a processed wafer there → the other arm moves to Step 1 and unloads the wafer held by the arm into a PM there → the idle arm moves to Step 2. This process is repeated until a completed wafer at the last step is unloaded and returned to the loadlocks. It is shown that, by using a swap strategy, while ψ is the same as that of single-arm cluster tools, Π_i can be calculated by

$$\Pi_i = \frac{A_i + 2T}{m_i}. \tag{8.3}$$

Thus, it shows that the throughput can be increased by using dual-arm cluster tools. Also, it indicates that swap is an efficient scheduling strategy for dual-arm cluster tools.

Perkinson *et al.* [37] also analyzed single-arm cluster tools with wafer revisiting. It shows that the wafer processing time needed by the revisiting step has great affect on the throughput and different schedules may yield different results. However, they provided no scheduling method.

8.3.2 Analysis Based on Marked Graph

Although the timing diagram model can be used to analyze the throughput of cluster tools, it cannot describe the dynamic behavior of the tools. Especially, it cannot be

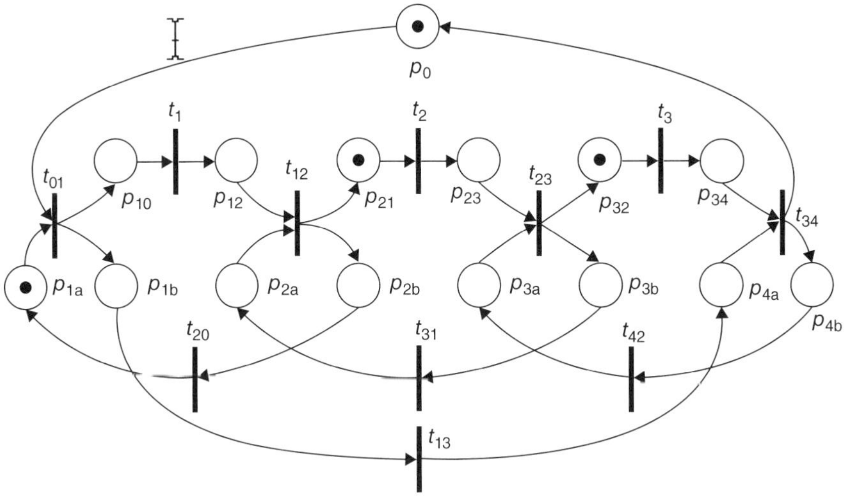

FIGURE 8.6 A marked graph for single-arm cluster tool with three steps.

used to analyze the qualitative properties. Petri nets are powerful in modeling and performance evaluation of manufacturing systems. Thus, Petri nets are used to model cluster tools for performance evaluation [40–43].

Notice that by backward scheduling for single-arm cluster tools or swap strategy for dual-arm cluster tools, the activity sequence is deterministic. A marked graph is a good choice for modeling such a system. The marked graph to model the steady state of a single-arm cluster tool with three steps is developed in Reference 42, as shown in Figure 8.6. It is a timed Petri net and the time is associated with its transitions. The interpretation of the transitions is given in Table 8.1.

According to Petri net theory, the cycle time of the system can be calculated as follows: (1) find all the circuits in the net; (2) calculate the time needed for each circuit by adding the times taken by the transitions on the circuit; and (3) the largest one among these circuits is the cycle time of the system. In Figure 8.6, four circuits are identified:

$$C_1 = \{t_{01}, p_{10}, t_1, p_{12}, t_{12}, p_{2b}, t_{20}, p_{1a}, t_{01}\},$$
$$C_2 = \{t_{12}, p_{21}, t_2, p_{23}, t_{23}, p_{3b}, t_{31}, p_{2a}, t_{12}\},$$
$$C_3 = \{t_{23}, p_{32}, t_3, p_{34}, t_{34}, p_{4b}, t_{42}, p_{3a}, t_{23}\},$$

and $C_4 = \{t_{01}, p_{1b}, t_{13}, p_{4a}, t_{34}, p_{4b}, t_{42}, p_{3a}, t_{23}, p_{3b}, t_{31}, p_{2a}, t_{12}, p_{2b}, t_{20}, p_{1a}, t_{01}\}$.

Circuits $C_{1\text{–}3}$ correspond to the wafer processing processes of $PM_{1\text{–}3}$, whereas C_4 corresponds to the robot tasks in a cycle. Let $f(t)$ denote the time associated with transition t, it can be obtained that $\Pi_1 = f(t_1) + f(t_{01}) + f(t_{12}) + f(t_{20})$, $\Pi_2 = f(t_2) + f(t_{12}) + f(t_{23}) + f(t_{31})$, $\Pi_3 = f(t_3) + f(t_{23}) + f(t_{34}) + f(t_{42})$, and $\psi = f(t_{01}) + f(t_{12}) +$

TABLE 8.1 Interpretation of Transitions in Figure 8.6

Transition	Operation	Time taken
t_1	PM_1 processes a wafer	A_1
t_2	PM_2 processes a wafer	A_2
t_3	PM_3 processes a wafer	A_3
t_{01}	Robot picks up a wafer from the LL, moves to PM_1, and loads the wafer into PM_1	$2\alpha + \mu$
t_{13}	Robot moves to PM_3	μ
t_{34}	Robot picks up a wafer from PM_3, moves to the LL, and drops it into the LL	$2\alpha + \mu$
t_{42}	Robot moves to PM_2	μ
t_{23}	Robot picks up a wafer from PM_2, moves to PM_3, and loads it into PM_3	$2\alpha + \mu$
t_{31}	Robot moves to PM_1	μ
t_{12}	Robot picks up a wafer from PM_1, moves to PM_2, and loads it into PM_2	$2\alpha + \mu$
t_{20}	Robot moves to the LL	μ

$f(t_{23}) + f(t_{34}) + f(t_{13}) + f(t_{20}) + f(t_{31}) + f(t_{42})$. Then the cycle time for the system is $\Pi = \max[\Pi_1, \Pi_2, \Pi_3, \psi]$.

Let μ denote the time needed for the robot to move from a PM to another and α the time for the robot to load a wafer into a PM/loadlock or unload a wafer from a PM/loadlock. Then, for this case, we have

$$\begin{aligned}\Pi_1 &= A_1 + 3\mu + 4\alpha,\\ \Pi_2 &= A_2 + 3\mu + 4\alpha,\\ \Pi_3 &= A_3 + 3\mu + 4\alpha,\\ \psi &= 8\,(\mu + \alpha)\,.\end{aligned}$$

By this method, for the general wafer pattern $(m_1, m_2, \ldots, m_n)$ without wafer revisiting, the throughput of a single-arm cluster tool with n steps is calculated as

$$\Pi_i = \frac{A_i + 3\mu + 4\alpha}{m_i} \tag{8.4}$$

and

$$\psi = 2\,(n + 1)\,(\mu + \alpha)\,. \tag{8.5}$$

Obviously, the throughput obtained by Equations (8.4) and (8.5) is more accurate than that obtained by Equations (8.1) and (8.2). In Reference 42, a marked graph model is also developed for dual-arm cluster tools with swap strategy. It is believed that the time taken by a swap is not a simple summation of the times taken by the three activities. Let λ denote the time for a swap. By using the marked graph model, we obtain the throughput for dual-arm cluster tools as follows:

$$\Pi_i = \frac{A_i + \lambda}{m_i} \tag{8.6}$$

and

$$\psi = (n+1)\,\mu + 2\alpha + n \times \lambda. \tag{8.7}$$

It is seen that if a schedule is decided, the robot task sequence is known. Then, a marked graph can be developed. The system and the throughput of the system can be obtained by finding all the circuits in the net. Following the idea, Zuberek [43] analyzes the throughput of cluster tools with wafer revisiting. For single-arm cluster tools, it presents some deadlock-free schedules and develops Petri nets for them such that throughput analysis can be done. He also presents such a model to describe dual-arm cluster tools with revisiting. However, with wafer revisiting there are lots of schedules and the throughput analysis is done based on given schedules in Reference 43. Their optimality in terms of throughput is not discussed.

All the aforementioned analyses belong to quantitative analysis, but there is no qualitative analysis. In fact, the timing diagram model cannot describe the dynamic behavior of the system. However, the marked graph models developed in References 42 and 43 are choice-free. Thus, such models cannot be used for qualitative analysis.

8.3.3 Analysis Based on Resource-Oriented Petri Nets

To analyze both the qualitative and quantitative properties of cluster tools, based on the colored resource-oriented Petri net (CROPN) modeling methods proposed in References 44 and 45, CROPN models are developed to describe the dynamic behavior of cluster tools in References 41, 46, and 47. A CROPN is a kind of finite-capacity Petri net. The robot in a cluster tool is modeled by two places r_1 and r_2 with their capacity being one as shown in Figure 8.7a. Initially, if there is a token in only one of the two places, it means that it is a single-arm robot. If both places have a token, it is a dual-arm robot. The delivery of a wafer from a PM/loadlock to another PM/loadlock is modeled by a macro-transition t_{ij} as shown in Figure 8.7a. Here a_i represents a wafer unloading from Step i, a_j a wafer loading into Step j, and q_{ij} the robot moving from Step i to Step j. Then, a cluster tool is modeled as follows. Step i is modeled by place p_i with capacity being m_i and p_0 is for the loadlocks with infinite capacity, indicating that there are always wafers for processing. A macro-transition t_{ij} is used to connect p_i and p_j if there is wafer flow from p_i to p_j. The obtained model is shown in Figure 8.7b for a four-PM cluster tool. Table 8.2 shows the interpretation of all places and transitions. In the model, it uses dashed lines to represent transition t_{32} and its relative arcs, indicating that they are necessary when there is a wafer revisiting. Based on the net structure shown in Figure 8.7b, it defines the transition enabling and firing rules in Reference 41.

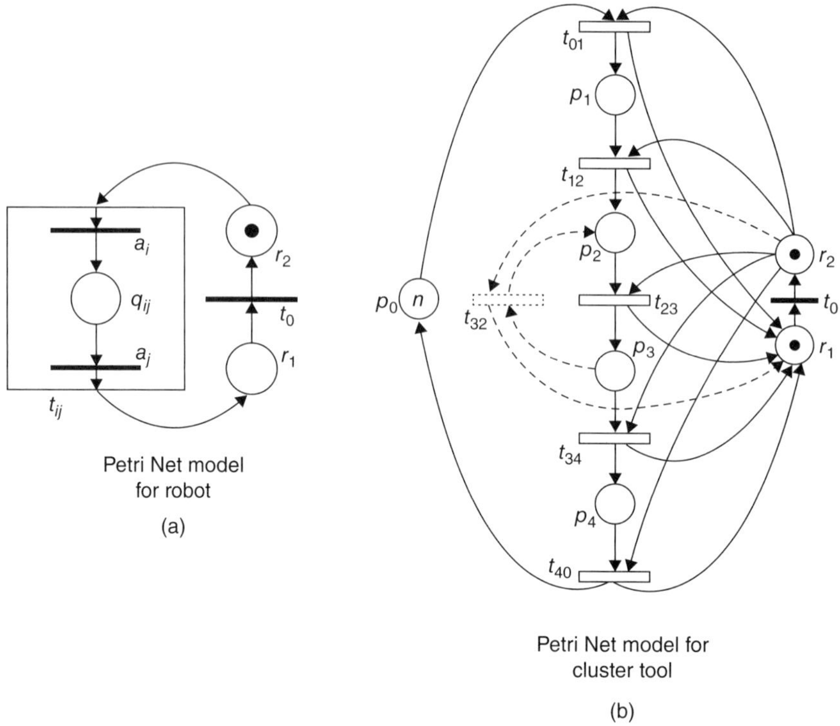

FIGURE 8.7 ROPN model for cluster tools.

This model describes all possible activities such that it can implement any applicable schedule by controlling the transition firing order. Also, it can be used to describe both single-arm and dual-arm ones with or without wafer revisiting.

With the model shown in Figure 8.7, it is shown in Reference 41 that a single-arm cluster tool governed by this model is deadlock-free if there is no wafer revisiting,

TABLE 8.2 Interpretation of Places and Transitions in Figure 8.7

Place or transition		Activity or state
Macro-transition t_{ij}	a_i	Robot unloads a wafer from Step i or a loadlock
	q_{ij}	Robot moves from Steps i to j
	a_j	Robot loads a wafer to Step j or a loadlock
r_1, r_2		A robot arm available
t_0		A transition that connects r_1 and r_2
p_0		Raw wafers in the loadlocks
p_1		Wafers that are being processed at Step 1
p_2		Wafers that are being processed at Step 2
p_3		Wafers that are being processed at Step 3
p_4		Wafers that are being processed at Step 4

whereas it is deadlock-prone if there is wafer revisiting. However, a control policy is developed to make the system deadlock-free and it is a maximally permissive policy. It is also shown that the system for dual-arm cluster tools is deadlock-free, no matter whether there is wafer revisiting or not.

Based on the model, throughput analysis is also conducted in Reference 41 and the same results as that obtained based on the timing diagram and marked graph models are obtained. In fact, all the aforementioned throughput analyses are done to find the bottleneck of the fabrication processes for a given schedule and to calculate the workload for the bottleneck. Then, the cycle time is obtained based on the workload for the bottleneck. In fact, the time taken by the bottleneck is the lower bound of cycle time for the system. The question is that whether such a lower bound can be reached by a schedule.

In References 47 and 48, for dual-arm cluster tools, the atomic layer deposition (ALD) process that is a typical wafer revisiting process is studied. For this process, the wafer flow pattern is $(1, (1, 1)^2)$. It is found that under a swap strategy this process is composed of three local cycles and three global cycles. Under some conditions, this process never reaches its steady state. Let ψ_1 and ψ_2 denote the robot cycle time for the global and local cycles, respectively. Then, the cycle time for the process is presented in Reference 47 by the following properties:

Property 1 For a dual-arm cluster tool with wafer revisiting under a swap strategy,

(a) if $\Pi_1 \leq 3\Pi_{\text{local}} + \psi_1$, $\Pi_3 > \Pi_2$, $\psi_1 > \Pi_1$, and $\Pi_3 < \psi_1$, the cycle time is

$$\Pi = \begin{cases} \Pi_3 + \psi_1, \text{ if } \Pi_3 \geq \psi_2 \\ \psi_2 + \psi_1, \text{ if } \Pi_3 < \psi_2 \end{cases}; \tag{8.8}$$

(b) if $\Pi_1 \leq 3\Pi_{\text{local}} + \psi_1$, $\Pi_3 > \Pi_2$, $\psi_1 < \Pi_1$, and $\Pi_3 < \psi_1$,

$$\Pi = \begin{cases} \Pi_3 + \Pi_1, \text{ if } \Pi_3 \geq \psi_2 \\ \psi_2 + \Pi_1, \text{ if } \Pi_3 < \psi_2 \end{cases}; \tag{8.9}$$

(c) if $\Pi_1 \leq 3\Pi_{\text{local}} + \psi_1$, $\Pi_3 > \Pi_2$, $\Pi_1 - \Pi_3 = \chi$, and $\Pi_3 > \psi_1$,

$$\Pi = \begin{cases} 2\Pi_3, \text{ if } \Pi_3 - \psi_1 \geq 2\chi \\ 2\Pi_3 + \left(2\chi + \psi_1 - \Pi_3\right)/3, \text{ if } \Pi_3 - \psi_1 < 2\chi \end{cases}. \tag{8.10}$$

For all the above cases, the lower bound of the cycle time is $2\Pi_3$. The lower bound cannot be reached by a swap strategy in general. However, if the methods presented in References 41 and 43 are used, it would take $2\Pi_3$ as the cycle time, while in fact it is not. The problem is that, by the methods presented in References 41 and 43, it is taken for granted that the system operates in its steady state. However, in fact, it does not operate in the steady state as shown in studies in References 47 and 48. This gives rise to a question whether there exists a schedule such that the lower bound can be reached, and if so how such a schedule can be obtained.

8.3.4 Discussion

It follows from the above discussion that extensive studies have been made on the modeling and performance analysis of cluster tools. Although these studies are done on single-cluster tools, the performance evaluation of multi-cluster tools can be done by extending the existing results. Thus, we can say that the future research direction will be mainly on the operation of cluster tools.

8.4 SINGLE-CLUSTER TOOL SCHEDULING

It follows from the above discussion that the performance of a cluster is dependent on its scheduling. Thus, to effectively operate a cluster tool, it is necessary to search for effective scheduling approaches. With two loadlocks, a cluster tool can be operated in an uninterrupted way for a batch of identical wafers. Thus, the key is to obtain a cyclic schedule to maximize the throughput. A cyclic schedule is called a k-unit schedule if exactly k parts are released into the system and at the same time k parts are completed [16]. When $k = 1$, it is a one-unit cyclic schedule. Because it is easy to understand, implement, and control one-unit cyclic schedule, it is attractive in practice and is widely used in industry [11, 49]. For cluster tools, one-wafer cyclic schedule is desired.

For a single-cluster tool, if there is no wafer residency time constraint and no wafer revisiting, the backward scheduling is optimal for single-arm cluster tools [27, 50] and swap strategy is optimal for dual-arm cluster tools [39, 51]. Thus, in these cases, the scheduling technique is well developed and the earliest starting strategy is applied [31, 52, 53]. However, for practical applications, wafer residency constraint, wafer revisiting, and activity time variation should be considered. Next, we review the advancements made by considering these requirements.

8.4.1 Scheduling with Wafer Residency Time Constraints

The first study on scheduling cluster tools with wafer residency time constraints was conducted by Rostami *et al.* [54]. Their cluster tool was a dual-arm cluster tool and the goal was to find a cyclic schedule. In scheduling it, they proposed to use one of the two arms as temporary buffer. An iterative algorithm is proposed to make the constraints satisfied. It first finds the cycle time Π without considering the wafer residency time constraints. Then, it searches if there exists a feasible solution such that the cycle time is Π. If so, a feasible solution is found. If not, it increases Π by one time unit and repeats the search. To reduce the search space, three heuristics are used. This method is further extended to dual-arm cluster tool scheduling with both wafer and robot residency time constraints [55] and dual-arm cluster tool scheduling with intermediate buffer [56]. Although it can find the optimal solution, it is not a systematic method and it may not be computationally efficient.

A systematic scheduling method for dual-arm cluster tools with wafer residency time constraints is presented in Reference 18. Whether the wafer residency time

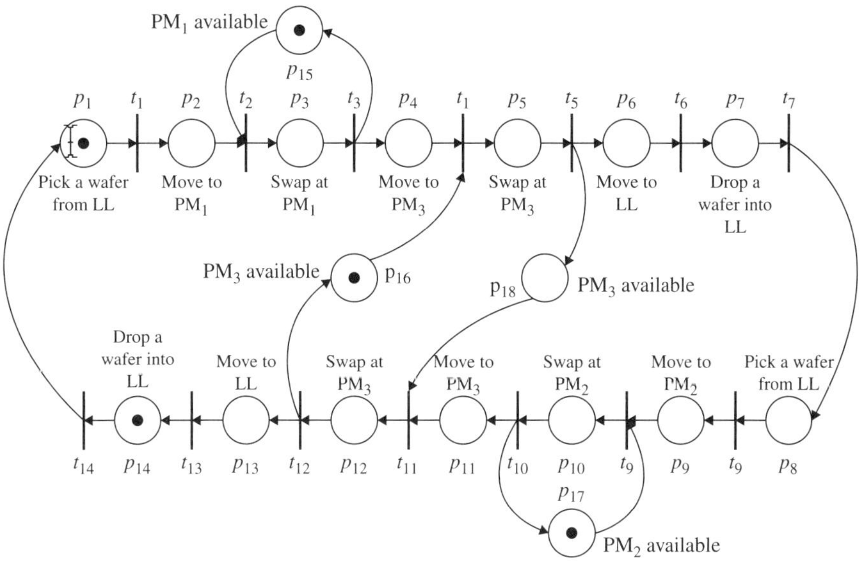

FIGURE 8.8 Timed event graph (TEG) for wafer flow pattern (2, 1).

constraints can be satisfied greatly depends on the extent to which workloads balance among the steps. To ensure a certain level of balance, a concept called all-wait-swap is presented. In all-wait-swap strategy, during the swap operation, after the completed wafer is unloaded from a PM, the wafer to be loaded into the PM can wait on the robot arm for some time before it is loaded into the PM. The system is modeled by a kind of Petri net called timed event graph (TEG) or timed marked graph. The difference between the TEG and the models in References 42 and 43 is that TEG associates the time with places but not transitions. A TEG for dual-arm cluster tools with wafer flow pattern (2, 1) is shown in Figure 8.8, where PM_1 and PM_2 are configured for Step 1 and PM_3 for Step 2.

Based on the TEG, the scheduling problem is formulated as a linear programming problem. It includes the schedulability condition as its constraints. This work studies, for the first time, the schedulability problem in scheduling cluster tools with wafer residency time constraints. If schedulable, an optimal schedule is found by solving the linear programming model. However, the schedulability condition obtained is just a sufficient but not necessary one.

The TEG model is extended by using negative places and tokens in Reference 29. With this model, the scheduling problem is formulated as a mixed-integer programming model. By using the mixed-integer programming model, necessary and sufficient schedulability condition is obtained. If schedulable, an optimal schedule can be found. The problem with this approach is that to formulate the mixed-integer programming model, it needs to find all the circuits in the net. Theoretically, the number of circuits in such a net is exponential to the size of net. Also, it is difficult to

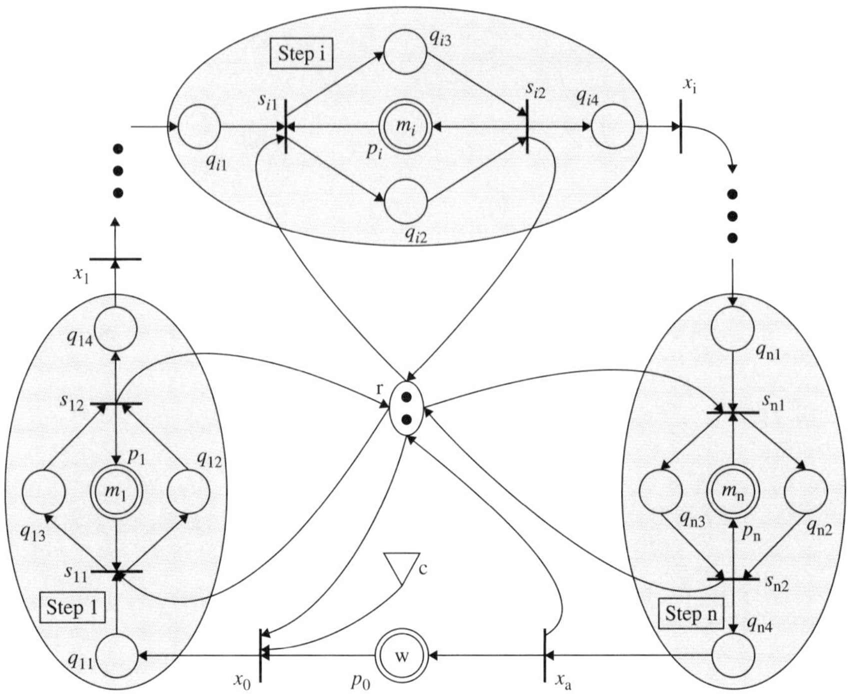

FIGURE 8.9 A generic Petri net model for dual-arm cluster tools.

solve a mixed-integer programming model. Thus, the approach faces a computational complexity problem.

For both models in References 18 and 29, they are not generic, but case-dependent. It implies that, for a different wafer flow pattern, a new model should be developed which limits the application of the proposed methods. For dual-arm cluster tools, all-wait-swap can be used to offset the workload unbalance to some extent, but for single-arm cluster tools, all-wait-swap is not applicable. Hence, their method is not applicable to single-arm cluster tools.

To schedule dual-arm cluster tools with wafer residency time constraints, Wu and Zhou [57] proposed a generic Petri net model, which is shown in Figure 8.9. In this model, a step is modeled by a subnet as follows. It uses place p_i for the PMs at Step i with capacity m_i indicating that there are m_i PMs. Transitions s_{i1} and s_{i2} model unloading a wafer from a PM and loading a wafer into a PM during the swap operation, respectively. Places q_{i2} and q_{i3} model the robot waiting with both arms holding a wafer during the swap operation, and q_{i1} models the robot waiting before swapping. The loadlocks are just treated as a step. Thus, to model the wafer fabrication process by a dual-arm cluster tool is to connect these steps in a serial way, no matter what the wafer flow pattern is. In this model, times are associated with both transitions and places.

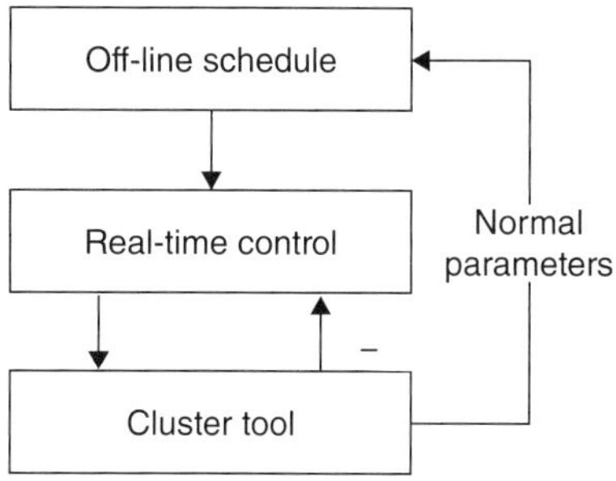

FIGURE 8.10 Architecture for cluster tool scheduling with activity time variation.

With this model, to find a schedule with wafer residency time constraints is simply to determine the token sojourn times (robot waiting times) at places q_{i1}, q_{i2}, and q_{i3}. Because the robot waiting time at q_{i2} and q_{i3} must be the same, it needs to determine one of them. Based on the model, by analyzing the temporal properties, analytical schedulability conditions are presented. If schedulable, the optimal schedule is found analytically. Hence, it is computationally efficient.

The results obtained in Reference 57 are extended to schedule single-arm cluster tools with wafer residency time constraints [58].

The above studies for scheduling cluster tools with wafer residency time constraints are based on the assumption that there is only one type of wafer to be processed. In practice, it needs to process multiple types simultaneously. This problem is studied in Reference 59, and an online scheduling algorithm with polynomial complexity is presented. However, the schedule obtained is not a cyclic schedule.

8.4.2 Scheduling with Both Wafer Residency Constraints and Activity Time Variation

With activity time variation, the wafer sojourn time in a PM fluctuates such that the wafer residency time constraints may be violated when a feasible schedule obtained by the methods discussed above is applied. Thus, for practical application the activity time variation should be taken into account in scheduling cluster tools.

Kim and Lee [33] pioneered in studying the schedulability problem when bounded activity time variations are considered. Under the earliest starting scheduling strategy, for dual-arm cluster tools, they identify so-called *always schedulable* and *never schedulable* cases by using Petri nets and a branching technique. If it is always schedulable, a schedule can be found such that the residency time constraints can be satisfied. If not, it is an open problem whether a feasible schedule can be found or not.

Motivated by the work in Reference 33, for dual-arm cluster tools, this problem is studied in References 58 and 60–64. The activity time variation is treated as a disturbance to the normal operation and a two-level real-time scheduling architecture is proposed as shown in Figure 8.10. At the upper level, it assumes that the system operates normally and an offline cyclic schedule is found such that the wafer residency time constraints are satisfied. At the lower level, a real-time control policy is applied to offset the activity time disturbance based on real-time observation. A Petri net model is developed to find the offline schedule and real-time control policy.

Based on the architecture and the Petri net model, a real-time control policy is presented in References 62 and 64. It is shown that, by using the control policy, some systems identified as never schedulable in Reference 33 are in fact schedulable. Then, Wu and Zhou [60] analyzed the effect of activity variation on the wafer sojourn time by analytically presenting the upper bound of wafer sojourn time fluctuation. With the wafer sojourn time fluctuation analysis, the schedulability and scheduling methods are studied in References 61, 63, and 64. These references present the schedulability conditions and develop the scheduling algorithms. The schedulability conditions and scheduling algorithms are presented in analytical expressions. If schedulable, an optimal schedule is found. This is a significant advancement in this field.

8.4.3 Scheduling with Wafer Revisiting

When there is wafer revisiting, analysis is made in References 41, 43, and 46–48. However, the scheduling problem is not discussed. Perkinson *et al.* [37] just discuss the effect of different schedules on the throughput by using an example and present no scheduling method.

The scheduling problem of single-arm cluster tools with wafer revisiting is studied and a systematic method is proposed in Reference 65. The Petri net model developed in Reference 18 is extended to describe the system. Then, after analyzing the deadlock-free condition, the problem is formulated as a mixed-integer programming model. Thus, an optimal solution can be found by solving the mixed-integer programming model. Nevertheless, it may involve prohibitive computation.

For the ALD, a typical wafer revisiting process, the scheduling problem is studied in Reference 46. The wafer flow patterns addressed include $(1, (1, 1)^k)$, $(1, (1, 2)^k)$, and $(1, (2, 1)^k)$. To obtain an efficient scheduling method, a multi-level architecture is proposed, as shown in Figure 8.11. The Petri net model developed in Reference 41 is extended to describe the system. With this model, a large number of infeasible states are avoided such that the search space is greatly reduced. Furthermore, based on the model, a necessary and sufficient deadlock-control policy is obtained. Thus, one needs to solve the scheduling problem in the feasible space.

After analyzing the properties of the system, Wu *et al.* [46] show that for wafer flow pattern $(1, (1, 1)^k)$, the optimal solution is unique and it can be obtained by an analytical expression. For wafer flow patterns $(1, (1, 2)^k)$ and $(1, (2, 1)^k)$, the optimal solution is one of the k schedules. Thus, it needs to compare the k schedules only. In this way, the scheduling problem for such a process is efficiently solved.

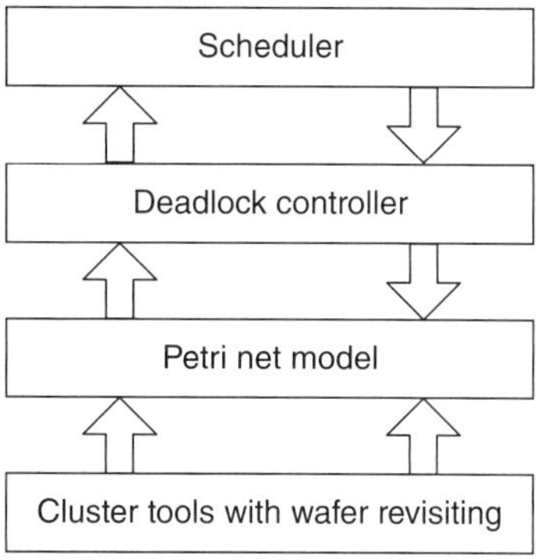

FIGURE 8.11 Architecture for single arm cluster tool scheduling with wafer revisiting.

8.4.4 Schedule Implementation

When the earliest starting scheduling strategy is applied, a schedule is easy to implement. However, as pointed out in Reference 18, an optimal schedule for cluster tools with wafer residency time constraints, the state at which the system enters its steady state, should be compatible with the schedule. This is not an easy task.

In Reference 57, a simple but effective method is presented for the scheduling implementation. If initially m_i virtual tokens called W_0 are put into place p_i in the Petri net model, this Petri net model can describe the initial transient process. Thus, with the virtual tokens in the p_i's, initially the system can be treated just as in the steady state and the system operates as in the steady state according to the schedule obtained. When all the virtual tokens are out of the system, it enters the steady state. At this time, the state is compatible with the desired optimal schedule.

8.4.5 Discussion

It follows from the above discussion that extensive studies have been conducted for single-cluster tool scheduling and great advancement has been made. However, there are still some topics to be explored.

As the market demands are changing, wafer fabrication batch is continuously decreased. Thus, it requires processing multiple products simultaneously in a single cluster tool. Up to now, the scheduling problem for this case is studied only in Reference 59, where a noncyclic schedule is found. A cyclic schedule is more preferable as it is easy to implement.

Revisiting is a property in wafer fabrication. Up to now, research on scheduling such processes has been done for single-arm cluster tools only. As discussed in Section 8.3, it is unknown if there exists a schedule such that the lower bound cycle time can be reached. Also, there are various wafer revisiting modes and only a few of them have been explored. Although the method presented in Reference 65 is a systematic one and can be applied for different revisiting modes, it involves prohibitive computation. It may be useful to explore other efficient methods for different revisiting modes that are in practical use.

One of the challenges in scheduling cluster tools comes from the wafer residency time constraints. In the existing studies, such constraints are considered only for non-revisiting processes. Notice that to make the wafer residency time constraints satisfied, the workload imbalance among the steps should not be too large. As pointed out in Reference 37, with wafer revisiting, the imbalance may become large. Although the methods presented in Reference 65 may be able to solve this problem, they may not able to find a feasible solution without significant computational resources. Thus, further research is necessary.

Activity time variation has an important impact on both the system performance and schedule feasibility. For the applicability of methods obtained, this issue should not be ignored. It follows from the above discussion that there are many open problems to deal with.

When a PM fails, response should be made to maximally reduce the loss in both throughput and quality. Up to now, there is little work on this topic and research attention should be paid on it.

8.5 SCHEDULING OF MULTI-CLUSTER TOOLS

With multiple cluster tools, the system is much more complex than a single-cluster tool in both structure and the fabrication processes. Thus, it is more difficult to schedule a multi-cluster tool than a single-cluster tool. Nevertheless, there is research advancement in this field.

8.5.1 Deadlock Control and Scheduling of Track Systems

A track system, such as photolithography equipment, often needs to process multiple products simultaneously. The recipes of these products may be different, or they have different processing routes. Without enough intermediate buffers, the system is deadlock-prone. As an automated system, to make the automation possible it is necessary to resolve deadlock.

A deadlock-free scheduling method is presented by Yoon and Lee [66, 67]. The method is based on the architecture shown in Figure 8.12. With the product routes known, it divides the PMs into groups: potential deadlock and deadlock-safe PMs. Then, a deadlock avoidance policy is developed. Consider that PMs may fail and the deadlock avoidance policy may not avoid all deadlocks, such as the second-level deadlock, so a deadlock detection-and-recovery method is included. Hence, a dispatching rule is presented such that an online deadlock-free schedule can be obtained. Essentially, it finds a feasible schedule but pays no attention on the throughput.

This problem is later studied by Wu and Zhou [68, 69]. It takes the advantage of the routing flexibility of the products, and an effective deadlock avoidance policy is

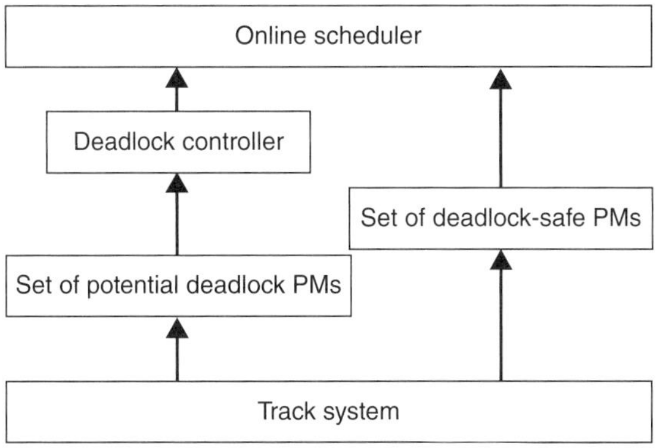

FIGURE 8.12 Architecture for track system scheduling.

developed. Based on the control policy, then a dispatching rule is presented to improve the throughput. Thus, while freedom from deadlock is guaranteed, the performance is improved.

8.5.2 Schedule Optimization

When a system is deadlock-prone, such as a track system, the key problem is deadlock resolution. However, the situation changes when a multi-cluster tool copes with only one type of wafer and, most of the time operates in the steady state. In such situation, more attention is paid to the system performance.

In Reference 70, methods are presented to optimize the wafer lot sequencing for a multi-cluster tool. However, little detail is given about the activity scheduling of the tool itself. Heuristics are presented for multi-cluster tool scheduling in References 71 and 72. However, the performance is difficult to evaluate. Significant advancement in scheduling multi-cluster tools is made in References 23–25, 28, and 73.

In multi-cluster tools, one or more processing PMs are used as a buffer to link two single-cluster tools. Wafers should flow between these two cluster tools through the buffer. The order in which the wafers take the buffer affects the scheduling performance. Thus, there are multiple choices in scheduling the system, which complicates the scheduling problem.

In Reference 25, an event graph model is used to describe the dynamic behavior of the system. With this model, feasible action sequences are found and an undirected network is formed by using these sequences. The extended critical path in the network corresponds to the fundamental period of the process, or the cycle time. Hence, a schedule is found. To reduce the search space, a method is proposed to avoid searching the redundant sequences.

To reduce the computational complexity, a decomposition method is proposed in References 28 and 73. It decomposes the M-cluster tool into M single-cluster tools as follows. A buffer that connects cluster tools C_i and C_{i+1} is treated as a PM for C_i, but as a loadlock for C_{i+1} as shown in Figure 8.13. With the decomposition, the fundamental period for each single-cluster tool can be obtained as done for scheduling single-cluster tools. Then, by analyzing time delays needed for loading and unloading the connecting buffers, the global fundamental period, or the cycle time, for the system is determined. In this way, a schedule is found.

In References 28 and 73, it is assumed that the robot moving from one PM to another takes no time. This assumption is relaxed in References 23 and 24. To obtain the optimal schedule for a multi-cluster tool, the key is to determine the number of wafers, n_i, for each single-cluster tool C_i that are being concurrently processed. By defining the concept of basic activity set, a polynomial algorithm is presented to find n_i. Then, the optimal schedule can be obtained by a polynomial algorithm [23, 24]. This stands for a significant advancement in this field.

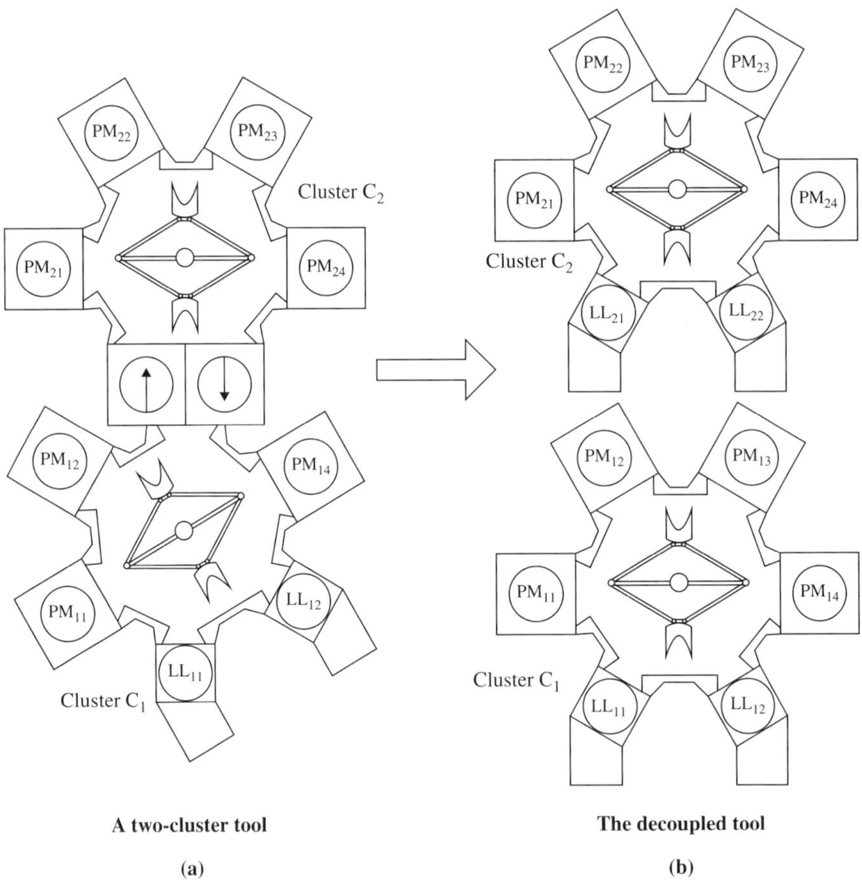

FIGURE 8.13 Decomposition of a two-cluster tool.

8.5.3 Discussion

Scheduling a multi-cluster tool is much more complex than scheduling a single-cluster tool. Although advancement is made in this field, it is still in its infant stage. There is no research reported on scheduling a multi-cluster tool such that wafer residency time constraints are satisfied. It is not known that under what conditions a feasible schedule exists if such constraints are considered. It is more difficult to deal with activity time variation, because a real-time scheduling technique is necessary. Thus, there is much work to be done.

8.6 CONCLUSIONS

Cluster tools in semiconductor fabrication adopt the single-wafer processing technique and represent one of the important innovations in the semiconductor industry. Their use leads to high productivity and quality, and cost reduction. However, as a kind of reconfigurable and automated manufacturing system, it is challenging to effectively operate them. Thus, research is necessary to find methodologies for their modeling, analysis, simulation, scheduling, and control. During the past two decades, extensive studies were performed on modeling, performance evaluation, control, and scheduling of cluster tools and significant advancements are made. This chapter has reviewed the existing work and pointed out the future research directions. It is hoped that it will motivate more researchers and practitioners to work on various open problems in this important area.

REFERENCES

[1] M. Bader, R. Hall, and G. Strasser. Integrated processing equipment. *Solid State Technology*, 33: 149–154, 1990.

[2] P. Burggraaf. Coping with the high cost of wafer fabs. *Semiconductor International*, 38: 45–50, 1995.

[3] B. Newboe. Cluster tools: a process solution. *Semiconductor International*, 13(8): 82–88, 1990.

[4] T. K. McNab. Cluster tools, part 1: emerging processes. *Semiconductor International*, 13(9): 58–63, 1990.

[5] P. Singer. The driving forces in cluster tool development. *Semiconductor International*, 18(8): 113–118, 1995.

[6] N. Brauner. Identical part production in cyclic robotic cells: concepts, overview and open questions. *Discrete Applied Mathematics*, 156: 2480–2492, 2008.

[7] Y. Crama, V. Kats, J. van de Klundert, and E. Levner. Cyclic scheduling in robotic flowshops. *Annals of Operations Research: Mathematics of Industrial Systems*, 96: 97–124, 2000.

[8] Y. Crama and J. van de Klundert. Cyclic scheduling of identical parts in a robotic cell. *Operations Research*, 45: 952–965, 1997.

[9] M. Dawande, H. N. Geismar, M. Pinedo, and C. Sriskandarajah. Throughput optimization in dual-gripper interval robotic cells. *IIE Transactions*, 42(1): 1–15, 2009.

[10] M. Dawande, H. N. Geismar, S. P. Sethi, and C. Sriskandarajah. *Throughput Optimization in Robotic Cells*. New York: Springer, 2007.

[11] M. Dawande, C. Sriskandarajah, and S. Sethi. On throughput maximization in constant travel-time robotic cells. *Manufacturing and Service Operations Management*, 4(4): 296–312, 2002.

[12] H. N. Geismar, L. M. A. Chan, M. Dawande, and C. Sriskandarajah. Approximations to optimal *k*-unit cycles for single- and dual-gripper robotic cells. *Production and Operations Management*, 17(5): 551–563, 2008.

[13] H. N. Geismar, M. Dawande, and C. Sriskandarajah. Throughput optimization in constant travel-time dual-gripper robotic cells with parallel machines. *Production Operations Management*, 15: 311–328, 2006.

[14] L. Lei and T. J. Wang. Determining optimal cyclic hoist schedules in a single-hoist electroplating line. *IIE Transactions*, 26(2): 25–33, 1994.

[15] S. Sethi, J. Sidney, and C. Sriskandarajah. Scheduling in dual-gripper robotic cells for productivity gains. *IEEE Transactions on Robotics and Automation*, 17: 324–341, 2001.

[16] S. Sethi, C. Sriskandarajah, G. Sorger, J. Blazewicz, and W. Kubiak. Sequencing of parts and robot moves in a robotic cell. *International Journal of Flexible Manufacturing Systems*, 4: 331–358, 1992.

[17] Q. Su and F. Chen. Optimal sequencing of double-gripper gantry robot moves in tightly-coupled serial production systems. *IEEE Transaction on Robotics and Automation*, 12: 22–30, 1996.

[18] J.-H. Kim, T.-E. Lee, H.-Y. Lee, and D.-B. Park. Scheduling analysis of timed-constrained dual-armed cluster tools. *IEEE Transactions on Semiconductor Manufacturing*, 16(3): 521–534, 2003.

[19] A. D. Che and C. B. Chu. Multicyclic hoist scheduling with constant processing times. *IEEE Transactions on Robotics and Automation*, 18(1): 69–80, 2002.

[20] A. Che and C. Chu. Cyclic hoist scheduling in large real-life electroplating lines. *OR Spectrum*, 29(3): 445–470, 2007.

[21] V. Kats, E. Levner, and L. Mayzin. Multiple-part cyclic hoist scheduling using a sieve method. *IEEE Transactions on Robotics and Automation*, 15: 704–713, 1999.

[22] H. J. Yoon and D. Y. Lee. Real-time scheduling of wafer fabrication with multiple product types. In: Proceedings of IEEE International Conference on Systems, Man, and Cybernetics, 1999, pp. 835–840.

[23] W. K. V. Chan, J. Yi, and S. Ding. Optimal scheduling of multicluster tools with constant robot moving times, part I: two-cluster analysis. *IEEE Transactions on Automation Science and Engineering*, 8(1): 5–16, 2011.

[24] W. K. V. Chan, S. Ding, J. Yi, and D. Z. Song. Optimal scheduling of multicluster tools with constant robot moving times, part II: tree-like topology configuration. *IEEE Transactions on Automation Science and Engineering*, 8(1): 17–28, 2011.

[25] S. Ding, J. Yi, and M. T. Zhang. Multicluster tools scheduling: an integrated event graph and network model approach. *IEEE Transactions on Semiconductor Manufacturing*, 19(3): 339–351, 2006.

[26] M. J. Lopez and S. C. Wood. Systems of multiple cluster tools: configuration and performance under perfect reliability. *IEEE Transactions on Semiconductor Manufacturing*, 11(3): 465–474, 1998.

[27] M. J. Lopez and S. C. Wood. Systems of multiple cluster tools: configuration, reliability, and performance. *IEEE Transactions on Semiconductor Manufacturing*, 16(2): 170–178, 2003.

[28] J. Yi, S. Ding, and D. Song. Steady-state throughput and scheduling analysis of multi-cluster tools for semiconductor manufacturing: a decomposition approach. In: Proceedings of 2005 IEEE International Conference on Robotics and Automation, 2005, pp. 292–298.

[29] T.-E. Lee and S.-H. Park. An extended event graph with negative places and tokens for timed window constraints. *IEEE Transactions on Automation Science and Engineering*, 2(4): 319–332, 2005.

[30] T.-E. Lee. A review of scheduling theory and methods for semiconductor manufacturing cluster tools. In: *Proceedings of 2008 Winter Simulation Conference*, 2008, pp. 2127–2135.

[31] T.-E. Lee, H.-Y. Lee, and S.-J. Lee. Scheduling a wet station for wafer cleaning with multiple job flows and multiple wafer-handling robots. *International Journal of Production Research*, 45(3): 487–507, 2007.

[32] H. L. Oh. Conflict resolving algorithm to improve productivity in single-wafer processing. In: Proceedings of the International Conference on Modeling and Analysis of Semiconductor Manufacturing, 2000, pp. 55–60.

[33] J.-H. Kim and T.-E. Lee. Schedulability analysis of timed-constrained cluster tools with bounded time variation by an extended Petri net. *IEEE Transactions on Automation Science and Engineering*, 5(3): 490–503, 2008.

[34] P. K. Johri. Practical issues in scheduling and dispatching in semiconductor wafer fabrication. *Journal of Manufacturing Systems*, 12: 474–485, 1993.

[35] H. Toba. Segment-based approach for real-time reactive rescheduling for automatic manufacturing control. *IEEE Transactions on Semiconductor Manufacturing*, 13(3): 264–272, 2000.

[36] E. J. Koehler and M. S. Seppanen. Evaluation of cluster tool throughput for thin film head production. In: Proceedings of 1999 Winter Simulation Conference, 1999, pp. 714–719.

[37] T. L. Perkinson, R. S. Gyurcsik, and P. K. MacLarty. Single-wafer cluster tool performance: an analysis of the effects of redundant chambers and revisitations sequences on throughput. *IEEE Transactions on Semiconductor Manufacturing*, 9: 384–400, 1996.

[38] T. L. Perkinson, P. K. MacLarty, R. S. Gyurcsik, and R. K. Cavin, III. Single-wafer cluster tool performance: an analysis of throughput. *IEEE Transactions on Semiconductor Manufacturing*, 7: 369–373, 1994.

[39] S. Venkatesh, R. Davenport, P. Foxhoven, and J. Nulman. A steady state throughput analysis of cluster tools: dual-blade versus single-blade robots. *IEEE Transactions on Semiconductor Manufacturing*, 10: 418–424, 1997.

[40] R. S. Srinivasn. Modeling and performance analysis of cluster tools using Petri nets. *IEEE Transactions on Semiconductor Manufacturing*, 11(3): 394–403, 1998.

[41] N. Q. Wu and M. C. Zhou. Colored timed Petri nets for modeling and analysis of cluster tools. *Asian Journal of Control*, 12(3): 253–266, 2010.

[42] W. M. Zuberek. Timed Petri nets in modeling and analysis of cluster tools. *IEEE Transactions on Robotics and Automation*, 17(5): 562–575, 2001.

[43] W. M. Zuberek. Cluster tools with chamber revisiting—modeling and analysis using timed Petri nets. *IEEE Transactions on Semiconductor Manufacturing*, 17(3): 333–344, 2004.

[44] N. Q. Wu. Necessary and sufficient conditions for deadlock-free operation in flexible manufacturing systems using a colored Petri net model. *IEEE Transactions on Systems, Man, and Cybernetics, Part C*, 29(2): 192–204, 1999.

[45] N. Q. Wu and M. C. Zhou. Avoiding deadlock and reducing starvation and blocking in automated manufacturing systems based on a Petri net model. *IEEE Transactions on Robotics and Automation*, 17(5): 658–669, 2001.

[46] N. Q. Wu, F. Chu, C. B. Chu, and M. C. Zhou. Petri net-based scheduling of single-arm cluster tools with reentrant atomic layer deposition processes. *IEEE Transactions on Automation Science and Engineering*, 8(1): 42–55, 2011.

[47] N. Q. Wu, F. Chu, C. B. Chu, and M. C. Zhou. Petri net-based cycle time analysis of dual-arm cluster tools with wafer revisiting and swapping strategy. In: Proceedings of 2011 IEEE International Conference on Robotics and Automation, Shanghai, China, May 9–13, 2011, pp. 5499–5504.

[48] N. Q. Wu, C. B. Chu, F. Chu, and M. C. Zhou. Petri net modeling and cycle time analysis of dual-arm cluster tools with wafer revisiting. *IEEE Transactions on Systems, Man, and Cybernetics: Systems*, 43(1): 196–207, 2013.

[49] I. Drobouchevitch, S. P. Sethi, and C. Sriskandarajah. Scheduling dual-gripper robotic cells: 1-unit cycles. *European Journal of Operational Research*, 171: 598–631, 2006.

[50] T.-E. Lee, H.-Y. Lee, and Y.-H. Shin. Workload balancing and scheduling of a single-armed cluster tool. In: Proceedings of the 5th APIEMS Conference, Gold Coast, Australia, 2004, pp. 1–15.

[51] J.-H. Paek and T.-E. Lee. Optimal scheduling of dual-armed cluster tools without swap restriction. In: Proceedings of 2008 IEEE International Conference on Automation Science and Engineering, 2008, pp. 1–6.

[52] T.-E. Lee. Stable earliest starting schedules for periodic job shops: a linear system approach. *International Journal of Flexible Manufacturing Systems*, 12(1): 59–80, 2000.

[53] Y.-H. Shin, T.-E. Lee, J.-H. Kim, and H.-Y. Lee. Modeling and implementing a real-time scheduler for dual-armed cluster tools. *Computers in Industry*, 45(1): 13–27, 2001.

[54] S. Rostami, B. Hamidzadeh, and D. Camporese. An optimal periodic scheduler for dual-arm robots in cluster tools with residency constraints. *IEEE Transactions on Robotics and Automation*, 17: 609–618, 2001.

[55] S. Rostami and B. Hamidzadeh. Optimal scheduling techniques for cluster tools with process-module and transport-module residency constraints. *IEEE Transactions on Semiconductor Manufacturing*, 15(3): 341–349, 2002.

[56] S. Rostami and B. Hamidzadeh. An optimal residency-aware scheduling technique for cluster tools with buffer module. *IEEE Transactions on Semiconductor Manufacturing*, 17(1): 68–73, 2004.

[57] N. Q. Wu and M. C. Zhou. A closed-form solution for schedulability and optimal scheduling of dual-arm cluster tools with wafer residency time constraint based on steady schedule analysis. *IEEE Transactions on Automation Science and Engineering*, 7(2): 303–315, 2010.

[58] N. Q. Wu, C. B. Chu, F. Chu, and M. C. Zhou. A Petri net method for schedulability and scheduling problems in single-arm cluster tools with wafer residency time constraints. *IEEE Transactions on Semiconductor Manufacturing*, 21(2): 224–237, 2008.

[59] H. J. Yoon and D. Y. Lee. On-line scheduling of integrated single-wafer processing tools with temporal constraints. *IEEE Transactions on Semiconductor Manufacturing*, 18(3): 390–398, 2005.

[60] N. Q. Wu and M. C. Zhou. Analysis of wafer sojourn time in dual-arm cluster tools with residency time constraint and activity time variation. *IEEE Transactions on Semiconductor Manufacturing*, 23(1): 53–64, 2010.

[61] N. Q. Wu and M. C. Zhou. Petri net-based scheduling of time-constrained dual-arm cluster tools with bounded activity time variation. In: Proceedings of 6th IEEE Conference on Automation Science and Engineering, Toronto, Canada, August 21–24, 2010, pp. 465–470.

[62] N. Q. Wu, M. C. Zhou, S. S. Peng, F. Chu, and C. B. Chu. Petri net modeling and real-time control of dual-arm cluster tools with residency time constraint and activity time variations. In: Proceedings of the 4th IEEE Conference on Automation Science and Engineering, Washington, DC, August 23–26, 2008, pp. 109–114.

[63] N. Q. Wu and M. C. Zhou. Schedulability analysis and optimal scheduling of dual-arm cluster tools with residency time constraint and activity time variation. *IEEE Transactions on Automation Science and Engineering*, 9(1): 203–209, 2012.

[64] N. Q. Wu and M. C. Zhou. Modeling, analysis and control of dual-arm cluster tools with residency time constraint and activity time variation based on Petri nets. *IEEE Transactions on Automation Science and Engineering*, 9(2): 446–454, 2012.

[65] H.-Y. Lee and T.-E. Lee. Scheduling single-armed cluster tools with reentrant wafer flows. *IEEE Transactions on Semiconductor Manufacturing*, 19(2): 224–240, 2006.

[66] Yoon, H. J. and D. Y. Lee. Identification of potential deadlock set in semiconductor track systems. In: Proceedings of 2001 IEEE International Conference on Robotics and Automation, Seoul, Korea, 2001, pp. 1820–1825.

[67] Yoon, H. J. and D. Y. Lee. Deadlock-free scheduling of photolithography equipment in semiconductor fabrication. *IEEE Transactions on Semiconductor Manufacturing*, 17(1): 42–54, 2004.

[68] N. Q. Wu and M. C. Zhou. Deadlock modeling and control of semiconductor track systems using resource-oriented Petri nets. *International Journal of Production Research*, 45(15): 3439–3456, 2007.

[69] N. Q. Wu and M. C. Zhou. Real-time deadlock-free scheduling for semiconductor track systems based on colored timed Petri nets. *OR Spectrum*, 29(3): 421–443, 2007.

[70] M. A. Dummler. Using simulation and genetic algorithms to improve cluster tool performance. In: Proceedings of the 1999 Winter Simulation Conference, 1999, pp. 875–879.

[71] D. Jevtic. Method and apparatus for managing scheduling a multiple cluster tool. European Patent 1,132,792 (A2), December, 2001.

[72] D. Jevtic and S. Venkatesh. Method and aparatus for scheduling wafer processing within a multiple chamber semiconductor wafer processing tool having a multiple blade robot. U.S. Patent 6,224,638, May, 2001.

[73] J. Yi, S. Ding, D. Song, and M. T. Zhang. Steady-state throughput and scheduling analysis of multi-cluster tools: an decomposition approach. *IEEE Transactions on Automation Science and Engineering*, 5(2): 321–336, 2008.

CHAPTER 9

DESIGN, SIMULATION, AND DYNAMIC CONTROL OF LARGE-SCALE MANUFACTURING PROCESS WITH DIFFERENT FORMS OF UNCERTAINTIES

HYUNSOO LEE[1] and AMARNATH BANERJEE[2]
[1]School of Industrial Engineering, Kumoh National Institute of Technology, Gumi, Korea
[2]Department of Industrial and Systems Engineering, Texas A&M University, College Station, TX, USA

9.1 INTRODUCTION

Modeling and simulation of processes and systems have been popular approaches to estimate or verify the performance of processes and systems using metrics such as cycle time, throughput, or work in progress (WIP). In addition, these approaches have been used to identify inefficiencies and potential problems in processes or systems, such as bottlenecks. During the analyzing process of existing systems, they are often modeled using one of several available modeling methodologies. Simulation software tools are used to simulate current operating conditions in the modeled system to identify limitations in the existing systems and possibly isolate the causes and locations of the limitations. Modeling and simulation are often used to estimate the effectiveness of a new system during its planning phase. The versatility of modeling and simulation tools and the statistical analysis approach receive considerable emphasis from designers and engineers when the need for modeling and analyzing a system arises. In the past few decades, there was a big increase in the number of software-based modeling and simulation tools with newer versions capable of performing complex modeling, simulation, and analysis. Even

Contemporary Issues in Systems Science and Engineering, First Edition.
Edited by MengChu Zhou, Han-Xiong Li and Margot Weijnen.

with the advent of newer modeling and simulation tools, there still exist several limitations, particularly, in the design and analysis of large-scale manufacturing systems.

The characteristics and challenges of contemporary manufacturing systems and processes can be categorized in terms of size, complexity, uncertainty, control, and interoperability/integration contexts. Table 9.1 summarizes some of these characteristics and challenges in the problems that have been studied in these contexts and some of the observed trends in literature, and the high-level activity that they fall under (modeling, simulation, and control).

The characteristics of manufacturing environments are shown in Table 9.1. They can be described using several adjectives: large, complicated, uncertain, dynamic, cooperative, and decentralized. This implies that it is quite difficult to model a large-scale manufacturing process. In addition, even though a process model is modeled successfully, the simulation and control of the model might be another issue. Due to these reasons, current control strategies are dependent on decentralized control techniques and merge-and-conquer strategies without fully considering global control and optimization. Sections 9.1.1 and 9.1.2 discuss issues in modeling, simulation, and control for contemporary manufacturing environments. This is followed by a description of a new and effective modeling framework in Sections 9.3, 9.4, and 9.5 with an example of performing dynamic control.

9.1.1 Issues in Design of Large-Scale Manufacturing Processes

The increase in size and degree of complexity of manufacturing processes has resulted in a number of modeling problems. The primary objective of a process model is not only to identify the elements in processes and their static interactions but also to use itself meaningfully in subsequent activities: simulation and control. The elements in a process model include inputs, outputs, control strategies and algorithms, mechanism, device, and human resource with connected relationship between preceding and subsequent processes. As the scale of manufacturing environments grows, there is manifold increase in the number of elements to capture the complex interactions. In addition, one of the observed trends in manufacturing system modeling and analysis is the requirement to increasingly model other processes associated with the system. It implies that a process model may cover other areas such as R&D and logistics/supply chain management (SCM), in addition to production and manufacturing areas.

It is frequently observed that there is some level of trade-off that exists between modeling scale and complexity in existing modeling techniques. Such trade-offs have a significant effect on the ability to develop in-depth models. For example, there are commonly used modeling tools such as UML diagrams or IDEF 0/3 that can capture a system at a high level, but fail to adequately capture processes with complicated structures and special conditions. Furthermore, one of the other limitations is that such models are difficult to use in subsequent activities (e.g., monitoring, simulation, and control) automatically. In order to overcome such limitations, it is observed that the commercial software products have tried to closely integrate the modeling

TABLE 9.1 Characteristics of Contemporary Manufacturing Processes

Activity	Issues	Characteristics and related studies
Modeling	Large scale of processes	• Increasing size of facilities and number of resources from mass production [1] • Integration among processes covering all product life cycle [2] (e.g., marketing, R&D, manufacturing, logistics, disposal)
	High degree of complexities	• Increase of regenerative processes [3] (e.g., recycling processes due to quality problems) • Volatile manufacturing environments [4] (e.g., change of manufacturing processes from fluctuating demands and planning)
	Existence of uncertainties	• Variance type uncertainties [5] (e.g., variance in processing time and arrival intervals) • Ambiguity type uncertainties [6] (e.g., Unfixed operation rules)
	Integration issues	• Decrease of "islands of automation" from highly advanced automation technologies [7] • Integration between human and facilities [8] (e.g., human-oriented factory modeling)
Simulation and control	Dynamic environment	• Occurrence of unexpected events [9] (e.g., breakdown or problems with resource operation) • Changes in operation rules for increasing efficiency [10]
	Real-time control	• Quick recovery from facilities' problems [11] • Development of early warning detection techniques and real-time controllers [12]
	Interoperability issues	• Necessities of seamless manufacturing operations [13] • Increase of cooperative manufacturing environments [14]

tools with simulation functionalities using customized code, functionalities, and objects that are often difficult, if not impossible, to transfer information from and to other products. These products exist in niche application areas where they are able to handle domain-specific special conditions, providing the ability to develop detailed models under special environments and conditions. However, most of these products encounter difficulties covering all product lifecycle models in large-scale manufacturing environments.

The above problem is directly related to interoperability and cooperative modeling issues. As the modeling scale grows, process designers tend to participate in cooperative modeling to develop the desired overall model. In this case, the requirements and modeling techniques may be considerably different. As a final output, one unified manufacturing process model is expected from the integration of sub-process models possibly developed in different formats by different designers. The task of generating a unified process model is sometimes time consuming and difficult requiring in-depth technical knowledge of the modeled systems. This is one of the reasons that make a strong case for an interoperable format for modeling in a large-scale modeling framework.

Capturing different types of uncertainties prevailing in a process or system is another challenge for most modeling tools. Most of the commonly used modeling tools (e.g., IDEF tools, UML diagrams, and business process execution language (BPEL)) capture and describe inputs/controls/outputs/mechanism (ICOM) and linking information only. There are a few advanced modeling tools that support the capture of process uncertainties mainly in the form of variances. The modeling tools that are classified as "advanced" refer to the family of tools that are able to be used in the subsequent simulation activities. The ability to capture process uncertainties is a critical activity to seamlessly simulate process operations during simulation and control. Most of the advanced process design tools capture only variance type uncertainties, mainly in the form of data that are used to identify the underlying probability density functions (PDF) and the related parameters using statistical data–fitting tools. These distributions and parameters are used to generate random variates in the simulation model to simulate the random behaviors.

Although variance type uncertainties are most frequently observed and modeled, there is another form of uncertainty that often exist in systems and processes—ambiguity type uncertainty. For example, the unfixed operation rules, ambiguous criteria or definitions in manufacturing environments can be considered as ambiguity type uncertainties. Lee and Banerjee have classified and modeled several ambiguity type uncertainties in manufacturing environments [6] and logistics [15]. The requirement of capturing ambiguity type uncertainties increases with the development of intelligent control techniques.

9.1.2 Simulation Model for Dynamic Control

Following the creation of a process model of a manufacturing system, the next logical activity is to check for feasibilities and inconsistencies of the designed model and an analysis of system performance using several metrics. For example, a number of different process models can be designed for a target process. In this situation, simulation can be used as a tool to determine the best process model representation from the available model choices. This is one of the reasons that designers and analysts consider modeling and simulation as inseparable activities. It is seen that most simulation models simulate the dynamics based on existing routes that are part of the process model. The entity flow inside a simulation model is driven by the constraints of the manufacturing system environment. A simulation model allows for selection of alternative paths in a model, mostly based on using probabilistic data or condition-based rules. The alternative paths are already part of the existing process model. This sometimes poses a challenge particularly in the event of a breakdown in a process. Most simulation models use concepts such as mean time to failure (MTTF), mean time to recover (MTTR), or mean time between failure (MTBF) to model failure and recovery without emphasizing the creation of dynamic routes for entities to flow as it might occur in a real-world situation. This makes it difficult sometimes to use a simulation model as a dynamic control model as it may not be a true replica of the real-world environment that is being modeled and analyzed. Several research studies have been proposed to overcome this limitation using intelligent holonic modules [16], agent-based concepts, and evolution techniques.

Another challenge that is frequently encountered is the way simulation models can be used for controlling processes. Most often, it is observed that the outputs of a simulation are point and interval estimates of parameters such as cycle time, work in process (WIP), throughput, waiting time, and resource utilization. The inputs that drive most of the simulation models are arrival rates of entities into the system, service time estimates, resource availability, and failure rate models. Equation (9.1) attempts to show a conceptual formula of a simulation model:

$$Y(\text{CT}, \text{WIP}, \text{TH}, \ldots) = \otimes f(\lambda_1, \lambda_2, \ldots, \mu_1, \sigma_1, \ldots, \alpha_1, \beta_1, \ldots), \tag{9.1}$$

where $Y(\cdot)$ is simulation output function; CT, WIP, and TH are the elements of simulation output (e.g., cycle time, WIP, or throughput); $\otimes f(\cdot)$ the simulation mechanism; $\lambda_1, \lambda_2, \mu_1, \sigma_1, \alpha_1, \beta_1$ the elements of simulation input (e.g., inter-arrival times or processing mean/standard deviation).

In most simulation models, the input distributions and parameters are known a priori or estimated based on existing data making up the elements that drive the model. The output data generated from the simulation models are used to determine $Y(\cdot)$ using $\otimes f(\cdot)$. However, a control mechanism can be considered from an opposite perspective. From a control standpoint, the known values are not input elements such as $\lambda_1, \lambda_2, \mu_1, \sigma_1, \alpha_1$, and β_1 but target output metrics such as CT, WIP, and TH. For example, the control target may be set to a desired time for overall cycle time. When the target cycle time is set, the objective of control is to determine the appropriate combination of input parameters such as $\lambda_1, \lambda_2, \mu_1, \sigma_1, \alpha_1$, and β_1 to meet the output goals using $\otimes f(\cdot)^{-1}$. From a control model perspective, the main problem of a simulation model is that it may be very difficult to generate or estimate the inverse function $\otimes f(\cdot)^{-1}$. In most situations, the parameter settings for obtaining the target output is performed using a trial-and-error-based parameter setting approach or design of experiment (DOE) technique. However, these approaches are insufficient for controlling processes. To overcome this limitation, some control logic or optimization formulations have been added to simulation models. These additions play a role in guiding $\otimes f(\cdot)$ to meet the control targets. This chapter introduces a new and efficient modeling framework for designing large-scale manufacturing processes to address some of the issues discussed in this section. It also shows how the modeling and simulation framework can be used to effectively perform dynamic process control.

9.2 BACKGROUND AND LITERATURE REVIEW

As discussed in Section 9.1, there are a number of challenges in the process of modeling large-scale manufacturing processes and their simulation covering dynamic control. Even though several good modeling frameworks and tools exist, there are only few frameworks that couple both modeling and simulation to address some of the challenges. Table 9.2 shows characteristics and limitations of several existing modeling frameworks/tools.

As the need for modeling and simulation of large-scale manufacturing processes grows, several criteria are required for a good modeling and simulation framework or tool. These criteria are categorized and shown in Figure 9.1.

As shown in Figure 9.1, the radar chart with five scales has eight criteria. Among them (1) generalization of modeling and (2) easy representation factors are considered as basic criteria which are needed for a modeling framework. Then, (3) interoperability and (4) expansion/flexibility criteria are required for modeling a large-scale manufacturing process. The use of interoperable formats and support of modular concepts for expansion are useful for designing large-scale manufacturing processes.

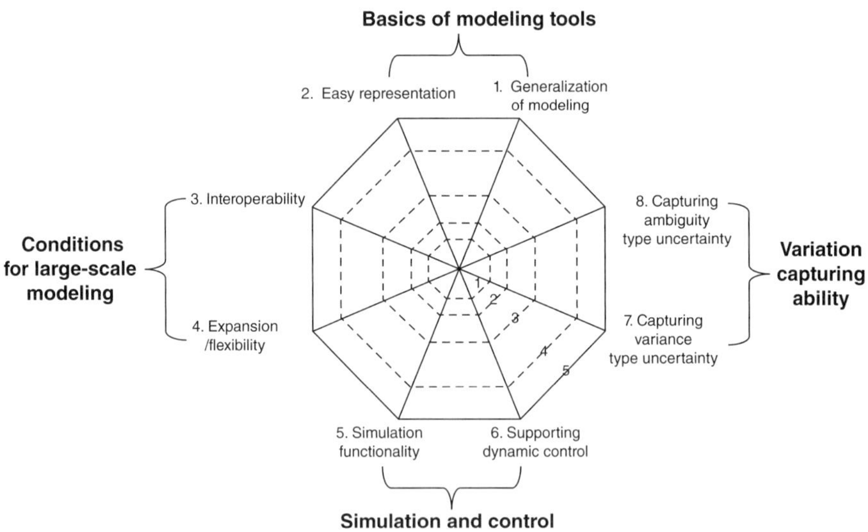

FIGURE 9.1 Criteria to judge modeling and simulation tools for large-scale manufacturing processes.

TABLE 9.2 Characteristics and Limitations of Several Existing Modeling Tools

Modeling frameworks/tools		Characteristics	Limitations
General-purpose framework	UML [17][a]	• Supports several diagrams (class diagram, use case diagram, sequence diagram, and so on) • Effective modeling using software and programming codes	• Non-executable modeling tool • Absence of mechanism to capture uncertainties • Unsuitable for large-scale process modeling
	IDEF [18]	• Supports various modeling mechanisms –IDEF 0 (function modeling) and IDEF 3 (process modeling) • Effective in designing algorithms and modeling processes	• Non-executable modeling tool • Absence of mechanism to capture uncertainties
	ALPS/Process plan graph [19]	• Variation of IDEF 3 –Advanced version of IDEF 3 • Representation of alternative production routings using "AND" node and "OR" node	• Non-executable modeling tool • Absence of mechanism to capture uncertainties
	IEC 61499 [16, 20, 21]	• Modeling using function block (FB) • Executable modeling framework • Supports real-time control interface	• Absence of mechanism to capture uncertainties
	Petri Nets [22]	• Executable modeling tools: several evolved types (colored Petri Net, stochastic Petri Net, Fuzzy Petri Net, and so on) • Supports mechanism to capture uncertainties	• High complexity/unsuitable for large-scale process modeling

(continued)

TABLE 9.2 ***(continued)***

Modeling frameworks/tools		Characteristics	Limitations
	Process specification language (PSL) [23]	• Supports interoperability –XML-formatted process description tool • Supported by the international standards committee	• Non-executable modeling tool
Modeling tools for specific/custom purpose	Production and quality activity model (PQAM) [24]	• Manufacturing process diagnosis-oriented modeling framework • Sharing of IDEF0 and SIMAN	• Very specialized modeling framework • Does not have an interoperable format
	Chemical-mechanical polishing modeling tool (CMP) [25]	• Mathematical and chemistry-based modeling tools • Custom modeling tool for a specific process (polishing process)	• Very specialized modeling framework • Does not have an interoperable format
Commercial simulation tools	ARENA, Witness, ExtendSim, etc.	• General simulation software package • Support modeling tools which can be used in simulation directly	• Unsuitable for large-scale process modeling • Cannot capture ambiguity type uncertainties

[a]Related research studies.

Criteria (5) and (6) provide the mechanism to transition a framework from modeling to dynamic simulation and control. The other criteria (7) and (8) are related to capturing degrees of uncertainties embedded in a process model. Among modeling frameworks shown in Table 9.2, the general frameworks tend to have weaknesses in criteria (5), (6), (7), and (8). For modeling tools meant for custom purposes, they typically have a low score in criteria (1), (3), (4), and (8). Even popular commercial simulation tools have limitations in criteria (3), (4), and (8). Even though there are limitations for a modeling framework to meet all requirements, several modeling frameworks have attempted to meet these conditions for designing large-scale manufacturing systems.

Among them, Lee and Banerjee have proposed a Petri Net-based modeling and simulation framework—*Fuzzy colored Petri Net with stochastic time delay*

(FCPN-std) [26]. Petri Net is a type of bipartite graph with static content (model) of a system captured in place and transition nodes, and dynamic content of the system captured through the movement of tokens through the nodes. Petri Net supports dynamic system analysis such as reachability, boundedness, liveness, and reversibility [27–31]. Since it was proposed by Carl A. Petri in 1964, many research scholars and engineers have used and developed Petri Nets for modeling, simulation, and control of different types of system. FCPN-std is an inherited model from colored Petri Nets (CPNs) [32], stochastic Petri Nets (SPNs) [33], and Fuzzy Petri Nets (FPNs) [34]. The radar chart shown in Figure 9.2a summarizes the characteristics of FCPN-std, and Figure 9.2b shows a relationship among other types of Petri Nets. This chapter uses FCPN-std as a modeling and simulation framework for designing large-scale manufacturing processes. Section 9.3 provides a detailed definition, advantages, and characteristics of FCPN-std.

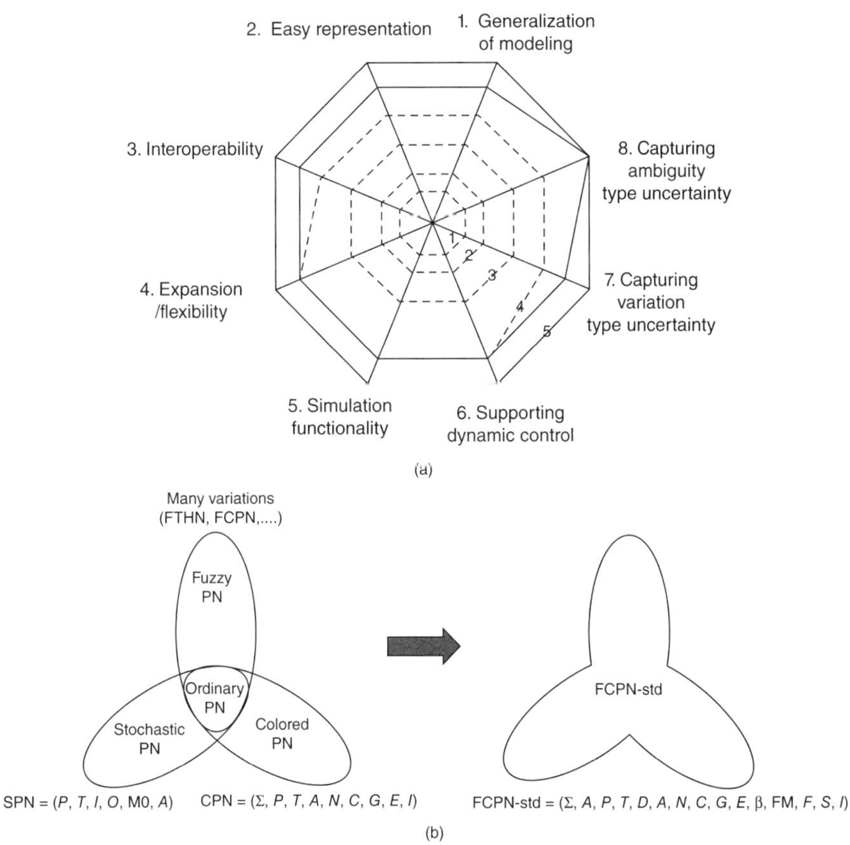

FIGURE 9.2 Characteristics of FCPN-std and relationship among other Petri Nets. (a) Characteristics of FCPN-std; (b) relationship between other types of Petri Nets and FCPN-std. (Reproduced from Lee and Banerjee [26]. © 2009 IEEE.)

As discussed in Section 9.1, interoperability is one of the important factors for modeling large-scale manufacturing processes. One of the ways interoperability can be achieved is by using a common model authoring format. As the FCPN-std framework is considered as a modeling and simulation framework, there are two representing common formats to capture the system information—PNML [35] and xPNML [13]. Petri Net Markup Language (PNML) is the first standard format for representing Petri Net models. However, in spite of their expansion ability and convenience, their meta-model and schema cover mainly the modeling part. With the integration of simulation and analysis modules within Petri Nets, information (e.g., warm-up period, number of replications, simulation runs, stochastic distributions, and parameters) supporting simulation and analysis becomes an important factor to successfully perform the simulation and analysis. The absence or limited availability of such information may generate different results and interpretations among designers and process modelers. The eXtensible Petri Net Markup Language (xPNML) overcomes these limitations with addition of these parameters/attributes in its meta-model and schema.

Figure 9.3 shows the meta-model of xPNML. It has several parameters for simulation setting and analysis. Stochastic distribution and its parameters are described as attributes in the *label* node. Section 9.4 discusses how an FCPN-std model can be represented using the xPNML format.

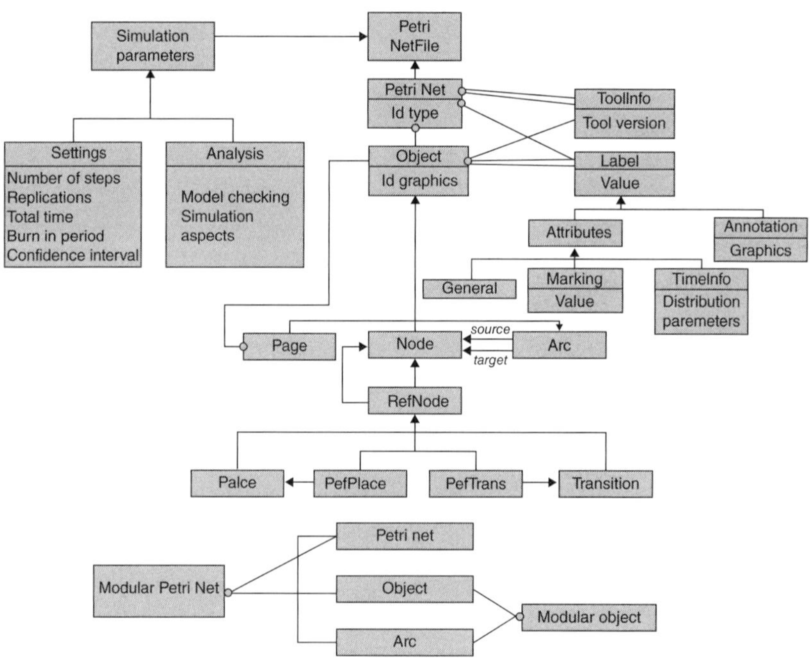

FIGURE 9.3 Meta-model of xPNML. (Reproduced from Lee *et al.* [13]. © 2007 IEEE.)

9.3 DIFFERENT TYPES OF UNCERTAINTIES AND FCPN-STD

This section discusses the two main types of uncertainties that can be effectively captured in the FCPN-std framework in a modular fashion. The section starts with the definition of FCPN-std followed by a description and example of modeling and simulation using FCPN-std.

9.3.1 Definition of FCPN-std

This section provides the definition of FCPN-std framework and its modular concept supporting large-scale processes modeling. An FCPN-std is a type of Petri Net model which can incorporate different types of uncertainties in its model. It means that FCPN-std can capture two main types of uncertainties—variation type uncertainty and ambiguity type uncertainty. Moreover, this framework shares the characteristics and advantages of timed Petri Nets (TPNs), SPNs, and FPNs.

FCPN-std is defined as a 15-tuple FCPN = (Σ, Λ, P, T, D, A, N, C, G, E, β, FM, F, S, I). Table 9.3 defines each element in the FCPN definition.

In the FCPN-std framework, variation type uncertainty can be captured using probability distribution functions (Λ) and mapped to each transition through stochastic mapping function (S). The ambiguity type uncertainty is incorporated using fuzzy membership functions (FM). A set of fuzzy membership functions (FM) consists of fuzzy descriptions (D). These fuzzy descriptions are connected with each place/transition using bijective mapping functions (β). The mapping function (F) connects these descriptions (D) to fuzzy membership functions (FM). The detailed definition and descriptions are provided in References 16, 26.

Equation (9.2) gives the condition that an FCPN-std model is converted to a general stochastic Petri Net (GSPN):

$$\{F(t) \rightarrow \phi\} \cap \{\beta(p|t) \rightarrow \phi\} \cap \{|\Sigma| = 1\}, \tag{9.2}$$

where p means places t means transition.

Similarly, the following equations show the conditions when an FCPN-std can be converted to an ordinary Petri Net, CPN, and FPN, respectively.

$$\{F(t) \rightarrow \phi\} \cap \{\beta(p|t) \rightarrow \phi\} \cap \{|\Sigma| = 1\} \cap \{S(t) \rightarrow \phi\}, \tag{9.3}$$

$$\{F(t) \rightarrow \phi\} \cap \{\beta(p|t) \rightarrow \phi\} \cap \{S(t) \rightarrow \phi\}, \tag{9.4}$$

$$\{|\Sigma| = 1\} \cap \{S(t) \rightarrow \phi\}. \tag{9.5}$$

This extension ability of FCPN-std is useful for modeling a large-scale manufacturing process with uncertainties. Furthermore, the use of mapping functions (β, F, S) and color tokens (Σ) overcomes some of the main complexities of a Petri Net model

TABLE 9.3 Definition of FCPN-std [26]

Notation	Meaning	Relations	Example
$\sum$	Set of color tokens	$\sum = \{\sigma_1, \sigma_2, \ldots\}$, where, σ_i = ith colored token	$\sum$ = {"sedan," "truck," …}
Λ	Set of distribution	$\Lambda = \{\text{Dist}_1, \text{Dist}_2, \ldots\}$	Λ = {4, expo(5.5), unif(2,3)}
P	Set of places	$P = P_g \cup P_f$, where P_g is the set of normal places and P_f is the set of fuzzy places	$P_g = \{P_1, P_4, P_5, P_6\}$ $P_f = \{P_2, P_3\}$
T	Set of transitions	$T = T_g \cup T_f$, where T_g is the set of normal transitions and T_f is the set of fuzzy transitions; normal transition can have stochastic time delay and fuzzy transition can have fuzzy membership values	$T_g = \{T_1, T_4, T_7\}$ $T_f = \{T_2, T_3, T_5, T_6\}$
D	Set of fuzzy description/propositions	$D = \{\text{Des}_1, \text{Des}_2, \ldots\}$	Des_1 = "if token type is σ_1 and process in P_5 is busy, then 2~3 resources are needed"
A	Set of arcs	$P \cap T = P \cap A = A \cap T = \emptyset$	$A = \{a_1, a_2, \ldots\}$
N	Node function	$A \rightarrow P \times T \cup T \times P$	$N(a_1) : a_1 \rightarrow P_1 \times T_1$
C	Color function	$P \cup T \rightarrow \sum$	$C(P_1) = \{\sigma_1, \sigma_2\}$, $C(T_7) = \{\sigma_2\}$
G	Guard function	$\forall t \in T$: [Type($G(t)$) = Boolean $\wedge$ Type(Var($G(t)$)) $\subseteq \sum$]	T_4 is fired if σ_2 goes to T_4
E	Arc expression function	$\forall a \in A$: [Type($E(a)$) = $(C(p),R)$ $\wedge$ Type(Var($E(a)$)) $\subseteq \sum$], where R is real number to represent a capacity of arc	$C(a_3) = \{(\sigma_1,2), (\sigma_2,1))$
β	Bijective function	$P_f \cup T_f \rightarrow D$	If place is P_3, follow Des_1 If transition is T_2, follow Des_4
FM	Set of Fuzzy membership functions	$\text{FM} = \{m_1, m_2, \ldots\}\}$	m_2 is a trapezoid membership function with parameters (s_1,s_2,s_3,s_4) [36]
F	Fuzzy mapping function	$D \rightarrow \text{FM}$	$F(\text{Des}_1) = \{m_1, m_2\}$
S	Stochastic mapping function	$\forall t \in T_g$: [Type($S(t)$) = Λ $\wedge$ Type(Var($S(t)$)) $\subseteq \sum$]	If σ_2 goes to T_2, T_2's firing time follows expo(4)
I	Initial marking	$\forall p \in P$: [Type($\delta(p)$) = $C(p)$] where δ: $P \rightarrow \sum$	$\{(\sigma_1, \sigma_2),(\sigma_2), \ldots\}$

encountered during the design of a large-scale manufacturing process. The following section describes how to design an FCPN-std model with a modular concept and five-stage modeling methodology.

9.3.2 Modular Design and Five-Stage Modeling Methodology

Despite many advantages of different Petri Net frameworks, the disproportionate growth in the number of places and transitions with increasing complexity of a system often leads to hesitation in using Petri Nets as a modeling and simulation framework. In addition, another issue with Petri Nets is the possibility of creating different Petri Net models to represent the same process, which can often lead to inconsistent modeling. For example, one designer can model a sub-activity as a place. On the other hand, another designer can design the same activity as a transition. This problem is a result of the degree of design freedom which a Petri Net framework offers as a general bipartite graph.

Both issues—disproportional growth in number of places and transitions, and multiple ways of expressing a model—can be overcome with the use of a modular concept. Initially, modular concept in Petri Nets was introduced as a mechanism to represent several structures (e.g., sequence, concurrency, parallel choice, circuit/self-loop, choice-synchronization, and mutual exclusion) described in References 28 and 29. Lee and Banerjee have developed this concept as a standard Petri Net modular concept [16, 26]. In their research, a Petri Net module (PNM) is defined as place, transition, and arc along with their parameters. The characteristics of several commonly occurring manufacturing environments are captured and represented as PNMs. Table 9.4 shows several manufacturing PNMs.

Using these PNMs can prevent redundancies in cooperative modeling with a number of designers and play a role in standardized design while modeling large-scale manufacturing systems. In addition, PNMs can be used as reconfigurable modules for designing reconfigurable manufacturing processes (RMP) and systems (RMS). The combination of PNMs improves the design process. A recent article [16] has shown the effectiveness of FCPN-std as a tool for modeling RMP/RMS with a fuzzy Q-learning algorithm.

Another strong point of FCPN-std is the ability to incorporate different types of uncertainties. The degree of complexities and uncertainties is classified into five stages/degrees. Table 9.5 shows the different types of Petri Nets with their corresponding degree of uncertainties.

The uncertainty level of a manufacturing process is reviewed in advance prior to starting a modeling task. This provides the basis for selection of the appropriate Petri Net framework. When a manufacturing process has the two types of uncertainties—ambiguity type uncertainty and variation type uncertainty—the FCPN-std framework is selected and used as a modeling and simulation framework with PNMs. This methodology is called the five-stage modeling methodology. The use of the five-stage modeling methodology helps decrease the complexity of an FCPN-std model. Figure 9.4 shows a simple FCPN-std model which has both types of uncertainties.

TABLE 9.4 Several Manufacturing PNMs [16]

No	Module	Description	Parameters
1	General process module	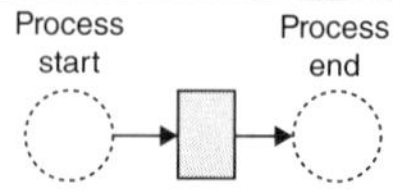	1. Processing time
2	Batch process	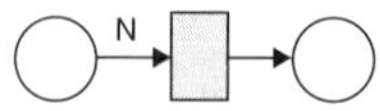	1. Batch size
3	Process module with resource	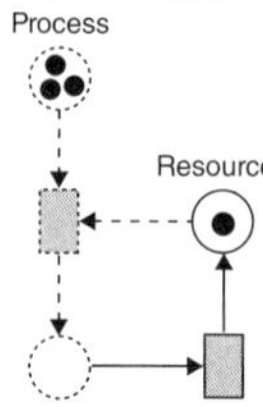	1. Processing time 2. Number of resources
4	Resource sharing module	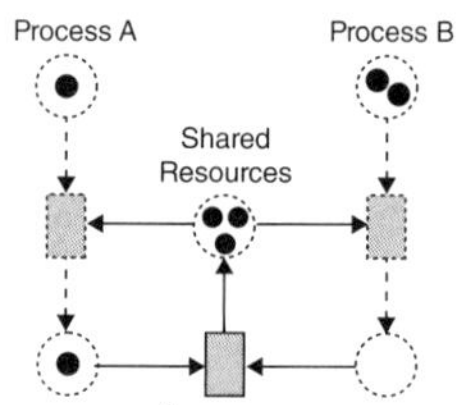	1. Processing time 2. Number of processes 3. Number of shared resources
5	Switching module	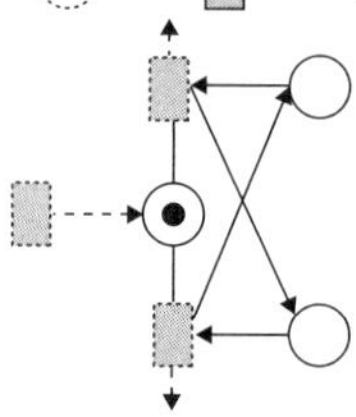	1. Shared Resource : 0, 1
6	Break/recovery module	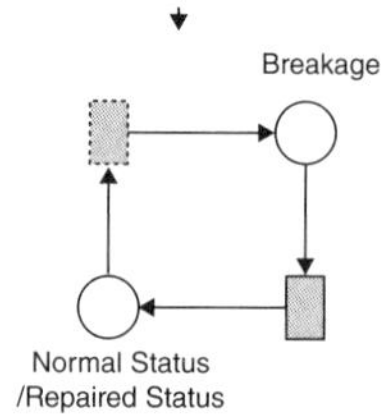	1. Break frequency/distribution 2. Recovery parameters/ distribution
7	Finite buffer Module	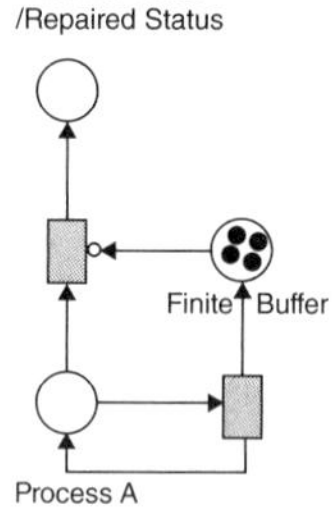	1. Size of finite buffer
8	Counter module	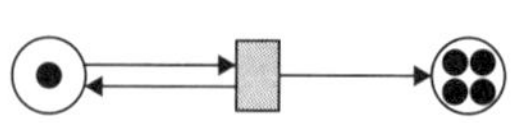	1. Number of tokens per each counting
9	User-defined module	–	1. User-defined parameters

TABLE 9.5 Types of Petri Nets with Corresponding Degree of Uncertainties [26]

Stage	Degree of complexities/ uncertainties	Modeling activities and characteristics	Related types of PN
1	Low (no uncertainties)	Describe a sequence of processes	Ordinary PN/timed PN
2	Low (no uncertainties)	Stage 1 + Design using parametric PN module	Ordinary PN/timed PN
3	Medium (no uncertainties)	Stage 2 + Describe multiple product and related operation rules	CPNs
4	High (variation)	Stage 3 + Describe variation using stochastic information	SPNs
5	Very high (variation and ambiguity)	Stage 4 + Describe ambiguity using fuzzy membership function	FCPN-std

This model describes a G/G/1 case with two types of products and one PNM—a process module with one resource ($p2 \rightarrow t2 \rightarrow p3 \rightarrow t3 \rightarrow p4 \rightarrow t4 \rightarrow p2$) as shown in Table 9.4. The two products are represented using a "*circle*"-shaped token and "*plus*"-shaped token. A "*rectangle*"-shaped token represents a resource. In this example, the average inter-arrival rate of "*circle*"-shaped product is 10 time units and the other product's average inter-arrival rate is 100 unit times. The resource (*machine #1*)'s processing time follows *Gaussian* ($150,10^2$).

When these conditions are embedded in the process model, it can be seen that the PNM belongs to stage 4 degree of complexity and can be represented using SPNs. However, in this example there is an ambiguous operation rule associated

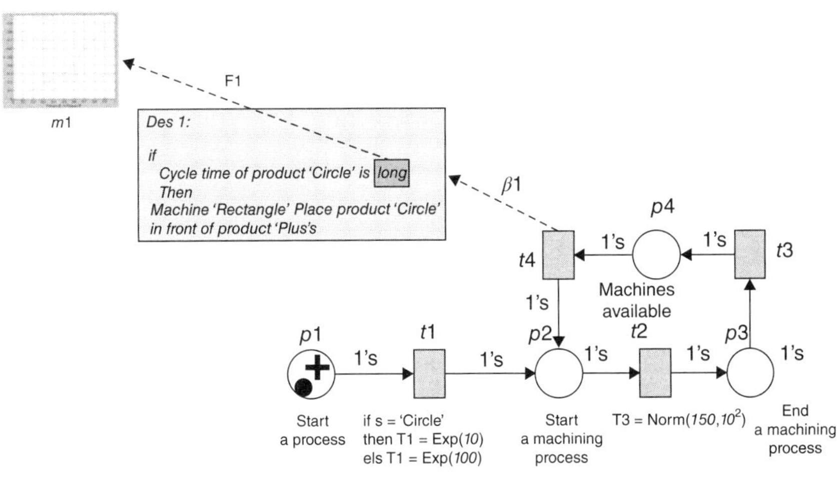

FIGURE 9.4 Example of an FCPN-std model.

with *machine #1*—"*if cycle time of product 'circle' is* long, *machine #1 changes the working order of waiting product at p2.*" The ambiguity is a result of the definition or interpretation of "*long*" in a product's cycle time. In a general case, this type of situation may occur from unfixed operation rules using experts' knowledge and know-how. As it is difficult for these types of knowledge to be captured in a model, the modeling of this process requires a stage 5 Petri Net framework: FCPN-std.

As shown in Figure 9.4, the FCPN-std model uses fuzzy logic [36] to capture the ambiguity type uncertainty— "*long*." The experts' experience or feelings for a qualitative value can be represented using a fuzzy membership function ($m1$). Then, it is mapped to a fuzzy description (Des 1) and is connected to $t4$ using a bijective mapping ($\beta 1$). The use of FCPN-std framework guarantees the incorporation of the two types of uncertainties. The following section describes how an FCPN-std model allows modeling the dynamic component of a system model in the form of simulation and analysis.

9.3.3 Simulation Using FCPN-std

The incorporation of fuzzy descriptions (D in Table 9.3) can play a role as control rules in an FCPN-std model. It implies that the handling of fuzzy control rules have an influence on the outcome of a simulation of an FCPN-std model. As it shares characteristics of SPNs, the general simulation of FCPN-std follows the simulation of GSPN [33]. In addition, the execution of fuzzy control is added to the simulation module to enable the execution of both the variance and ambiguity types of uncertainties. A transition t with fuzzy descriptions can be fired using four steps: (1) fuzzy matching, (2) inference, (3) combination, and (4) defuzzification step, as illustrated in Figure 9.5.

When a dependent variable's (parameter in fuzzy descriptions) value is detected in step 1, the mapping fuzzy utility value μ_a is obtained using a predefined fuzzy membership function. Then, the related fuzzy utility value μ^* $\left(|\mu^*| = \mu_\alpha\right)$ is found in the fuzzy membership function describing a transition's firing action. This process is called an inference step. If a fuzzy description consists of several fuzzy conditions, the minimum fuzzy utility value among fuzzy utility values is selected as a common descriptor shown as follows:

$$\mu^* = \mathrm{Min}(\mu_1, \mu_2, \ldots, \mu_n), \tag{9.6}$$

where n is the number of fuzzy conditions (fuzzy membership function).

This step is called a combination step. After finding μ^* through combination and inference steps, the final decision c is determined through a defuzzification process as follows:

$$c = \frac{\int\limits_x x \cdot \tilde{f}_{\mathrm{o}}\,(\mathrm{x})\,\mathrm{d}x}{\int\limits_x \tilde{f}_{\mathrm{o}}\,(x)\,\mathrm{d}x}, \tag{9.7}$$

where $\tilde{f}_{\mathrm{o}}\,(\cdot)$ is the output fuzzy membership function, and $x = \{x \in \tilde{f}_{\mathrm{o}}^{-1} | \tilde{f}_{\mathrm{o}}(x) \leq \mu^*\}$.

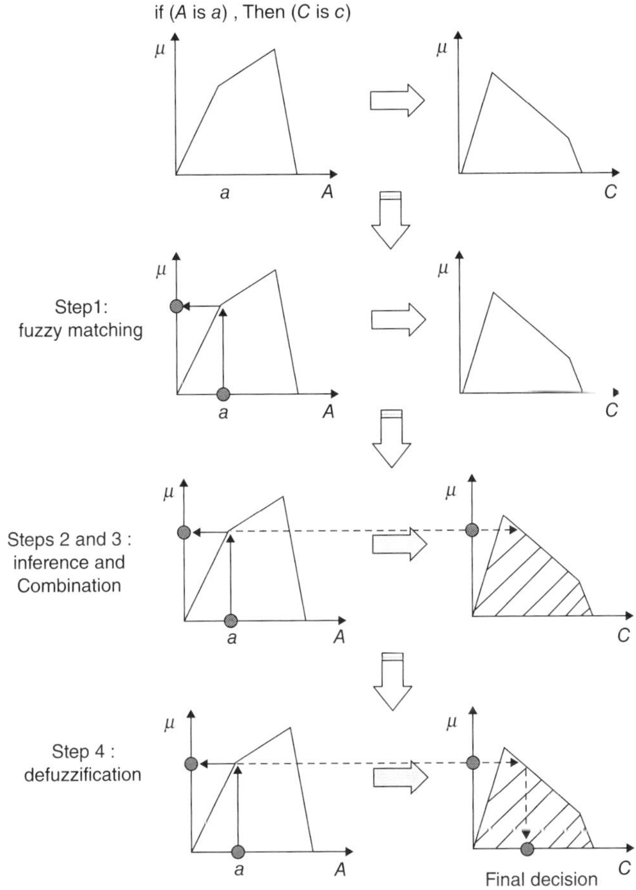

FIGURE 9.5 Determination of a final decision of fuzzy control in an FCPN-std framework.

The use of Equation (9.7) guarantees to decide the most common action among all alternatives, which not only coincides with the same fuzzy utility value μ^*("=" relationship) but also can satisfy fuzzy number μ^* ("<" relationship). Figure 9.5 shows the determination process of a final action using fuzzy logic. The simulation of an FCPN-std model is followed using the GSPN simulation method and this execution process.

9.4 DESIGN OF LARGE-SCALE MANUFACTURING PROCESSES

The Petri Net modules shown in Table 9.4 and the use of the five-stage modeling methodologies can contribute to the decrease in the degree of complexity along with an increase of design consistency at the same time.

However, there is one remaining issue—interoperability. As discussed in Section 9.2, the design of large-scale manufacturing processes can be achieved by a number of designers using different modeling tools. Each designer may be responsible for the design of their assigned subprocesses. On the other hand, several designers can be involved in the design of the same process with different views. The demand for cooperative process modeling keeps increasing in such conditions. In this situation, the use of an interoperable format is an important criterion for integrating different subprocess models. This section discusses the underlying format of the FCPN-std framework that addresses the interoperability issue.

The interoperability of process models can be satisfied using an interoperable format. As the FCPN-std model is a type of Petri Nets, one of the representative interoperable formats is PNML. However, as discussed in Section 9.2, the PNML format offers limited support of parameters and attributes for simulation and analysis of a process model. xPNML [13] was developed as an alternative solution to overcome these limitations. The meta-model of xPNML can be extended using the integration of xPNML *common type definition* (CTD) and *user-defined module definition* (UDMD). CTD has a role of data type definition (DTD) in an XML document. UDMD is similar to plug-ins reflecting designers' modeling objectives and characteristics of a modeling framework. Lee and Banerjee proposed the meta-model for FCPN-std framework [6]. Figure 9.6 shows the meta-model of xPNML supporting the FCPN-std framework.

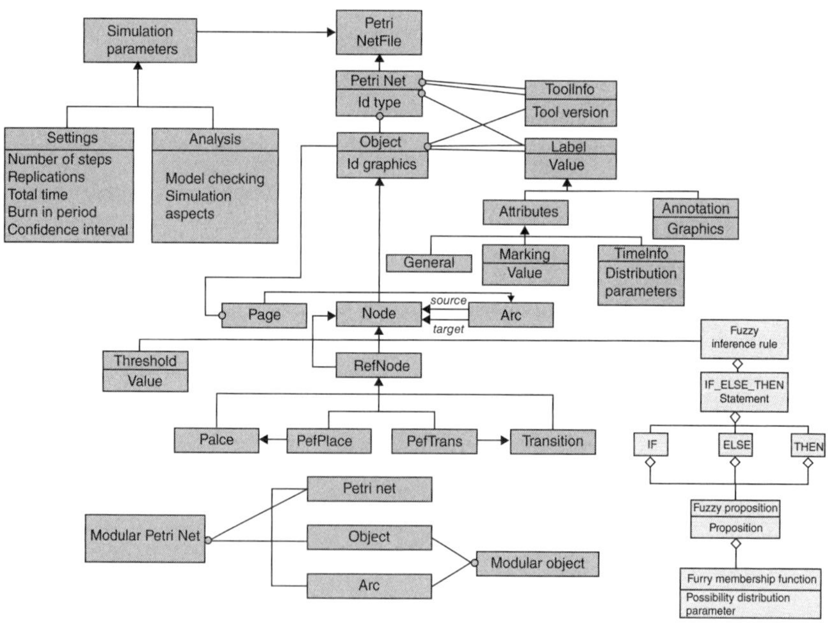

FIGURE 9.6 Meta-model of xPNML supporting FCPN-std framework. (Reproduced from Lee and Banerjee [6]. © 2009 IEEE.)

```
- <xPNML>
 - <place id="P15">
  - <Fuzzy_Inference_Rule>
   - <IF_ELSE_THEN_Statement id="2">
    - <IF>
       resource availability in lens holder mold design is
     - <Fuzzy_proposition>
        high
      - <fuzzy_membership_function>
          <possibility_distribution id="trapezoid" />
          <parameter id="param 1">7</parameter>
          <parameter id="param 2">8</parameter>
          <parameter id="param 3">10</parameter>
          <parameter id="param 4">10</parameter>
        </fuzzy_membership_function>
      </Fuzzy_proposition>
     </IF>
    - <IF>
       resource availability in lens mold design is
     - <Fuzzy_proposition>
        low
      - <fuzzy_membership_function>
          <possibility_distribution id="trapezoid" />
          <parameter id="param 1">0</parameter>
          <parameter id="param 2">0</parameter>
          <parameter id="param 3">0</parameter>
          <parameter id="param 4">2</parameter>
        </fuzzy_membership_function>
      </Fuzzy_proposition>
     </IF>
    - <IF>
       resource availability in cover/body mold design is
     - <Fuzzy_proposition>
        low
      - <fuzzy_membership_function>
          <possibility_distribution id="trapezoid" />
          <parameter id="param 1">0</parameter>
          <parameter id="param 2">0</parameter>
          <parameter id="param 3">0</parameter>
          <parameter id="param 4">2</parameter>
        </fuzzy_membership_function>
      </Fuzzy_proposition>
```


FIGURE 9.7 Example of an FCPN-std model using xPNML format. (Reproduced from Lee and Banerjee [6]. © 2009 IEEE.)

This meta-model is drawn using Unified Modeling Language (UML) 2.0 class diagram. Compared with the general meta-model of xPNML (Figure 9.3), it has a *Fuzzy_Inference_Rule* class and a *Threshold* class. *Fuzzy_Inference_Rule* class consists of several *IF_THEN_ELSE_Statement* classes. Each class can have several threshold values representing fuzzy utility values and other parameters for representing control logic. Finally, *Fuzzy_Inference_Rule* class and *Threshold* class can be related to the general node, *Node* class. *Node* class consists of *Place* and *Transition* nodes. It implies that a place or transition can have several fuzzy inference rules with fuzzy utility values. Figure 9.7 shows an example of an xPNML-formatted FCPN-std model.

The use of xPNML with the FCPN-std framework makes it possible to successfully execute cooperative process modeling. The following section describes an example which shows the effectiveness of the FCPN-std framework in a dynamic manufacturing environment.

9.5 DYNAMIC CONTROL OF MANUFACTURING PROCESSES

This section shows the development of an FCPN-std model of a small manufacturing system and the use of FCPN-std to simulate the dynamic content of the system with the goal of performing some control functions. An ingot purification process, which

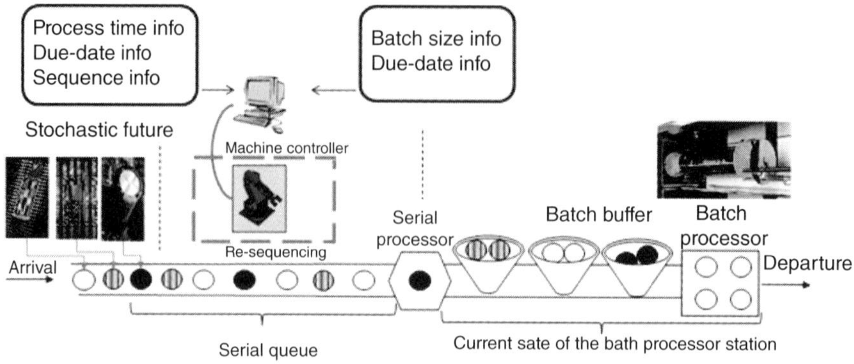

FIGURE 9.8 Serial batch process with a re-sequencing function.

is part of semiconductor or photovoltaic cell manufacturing process, is considered as an example here. Broadly, semiconductor or photovoltaic cell manufacturing processes consist of four subprocesses: (1) poly-silicon processes (purification processes), (2) ingot/wafer processes, (3) cell manufacturing processes, and (4) module manufacturing processes. Among these processes, the second process (Ingot and wafer processes) requires additional purification using laser and furnace devices. The components arrive to this process individually, but they are processed in batches. Therefore, they are referred to as a serial batch process.

Figure 9.8 shows a typical serial batch process. The scheduling of serial batch processes has been studied by many researchers due to the dynamic characteristics of processes. The decision at every epoch is to decide whether to start a partial batch or wait for a future arrival. There are many strategies that have been studied to improve the system performance. For example, the re-sequencing of jobs can potentially decrease the delivery delay of a product thus improving on the tardiness metric. Many research studies including Abdullah and Banerjee [37] suggest several mathematical programming or heuristics approaches to solve this problem. Even though their approaches improve the system performance, one of their main limitations is that the suggested mathematical model is set up only under a fixed situation. When the routing number of products or re-sequencing criteria are changed, the mathematical model have to be reconstructed and solved.

One alternative solution to overcome this limitation is the use of a simulation model to analyze the process dynamics. However, it is often difficult to capture accurate control logic and insert them into the simulation model. The FCPN-std framework can be a good simulator for this situation. The ambiguous or unfixed control rule can be captured using fuzzy logic and incorporated in the FCPN-std. A numerical example is used below to show the use of FCPN-std.

A serial batch processor handles three types of products: $\{\sigma_1, \sigma_2, \sigma_3\}$. The product inter-arrival time follows an exponential distribution with parameters $\lambda_1 = 5$, $\lambda_2 = 7$, and an uniform distribution $U = [3, 5]$, respectively. Each product's batch size is 3, 2, and 3, respectively. The serial processor that feeds the components to the batch

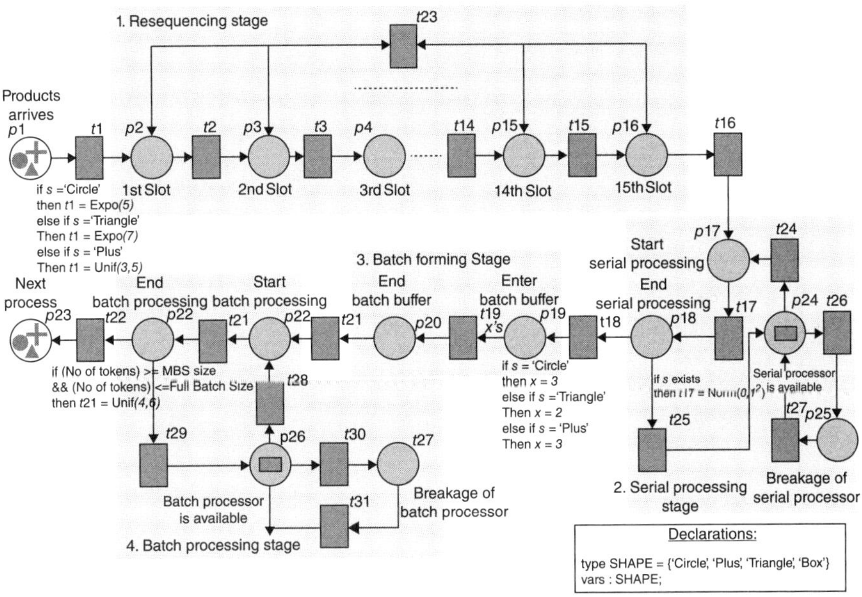

FIGURE 9.9 An SPN model for a serial batch process (the degree of complexity = 4).

processor has a processing time that follows a Gaussian distribution with $\mu = 6$ and $\sigma = 1$. The batch processor's processing time follows a uniform distribution [4, 6]. The batch processor rule is a minimum batch size (MBS) rule. Figure 9.9 shows an SPN model reflecting this manufacturing scenario.

With the simulation parameters (*warm-up period = 1440 time units; run time = 28,800 time units, and number of replications = 30*), the cycle time of the three products is estimated as 44.88, 46.02, and 37.46 time units, respectively.

Then, a series of fuzzy rules is introduced as re-sequencing control logic (Table 9.6) to determine if they can improve this process. The re-sequencing order is determined by the priority of products in terms of cycle time (e.g., second product σ_2 has a higher priority than σ_1 and σ_3). For example, the seventh fuzzy rule signifies that if cycle time status of σ_1, σ_2, and σ_3 are "Good," "Bad," and "Medium," respectively, re-sequencing is achieved with the input order of $(\sigma_2, \sigma_1, \sigma_3)$.

The determination of fuzzy status is calculated using the initial simulation result. When the histograms of the three products' cycle time are measured (Figure 9.10a), the degree of status ("Good"/"Medium"/"Bad") is determined and captured using a fuzzy triangle-shaped membership function:

$$\tilde{S}_{ij}(a_{ij}, b_{ij}, c_{ij}), \tag{9.8}$$

where $\tilde{S}_{ij}(\cdot)$ is the ith product's jth fuzzy utility value (if $j = 1/2/3$, the status of cycle time is good/medium/bad) and a_{ij}, b_{ij}, and c_{ij} are corner points of a fuzzy triangle function.

TABLE 9.6 Re-sequencing Strategy

	Fuzzy status of cycle time				
Rule	Product no. 1 (σ_1)	Product no. 2 (σ_2)	Product no. 3 (σ_3)	Action	Sequencing order
1	Good	Good	Good	No re-sequencing	–
2	Good	Good	Medium	No re-sequencing	–
3	Good	Good	Bad	No re-sequencing	–
4	Good	Medium	Good G	Sequencing	(σ_2,σ_1,σ_3)
5	Good	Medium	Medium	Sequencing	(σ_2,σ_1,σ_3)
6	Good	Medium	Bad	Sequencing	(σ_2,σ_3,σ_1)
7	Good	Bad	Good	Sequencing	(σ_2,σ_1,σ_3)
8	Good	Bad	Medium	Sequencing	(σ_2,σ_1,σ_3)
9	Good	Bad	Bad	Sequencing	(σ_2,σ_3,σ_2)
10	Medium	Good	Good	No re-sequencing	–
...	...	...	...	...	...

For example, $\tilde{S}_{12}(35, 45, 55)$ means that the fuzzy membership function of the first product's "Medium" status is

$$\tilde{S}_{12}(35, 45, 55) = \begin{cases} 0 & \mathrm{CT}_1 \leq 35 \\ 1/10^* \ \mathrm{CT}_1 - 3.5, & 35 < \mathrm{CT} \leq 45 \\ -1/10^* \ \mathrm{CT}_1 - 5.5, & 45 < \mathrm{CT} \leq 55 \\ 0 & \mathrm{CT}_1 > 35 \end{cases},$$

where CT_1 is cycle time.

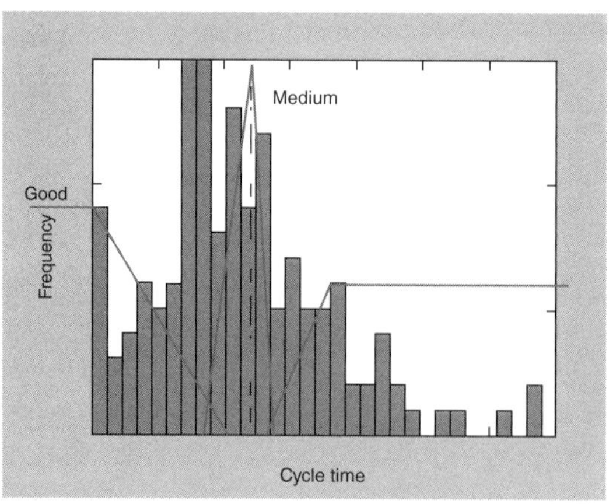

(a)

$$\tilde{s}_{11} = \begin{cases} 1 & ,ct_1 = 0 \\ -1/40^* \ ct_1 + 1 & ,0 < ct_1 \leq 40 \\ 0 & ,ct_1 > 40 \end{cases}$$

$$\tilde{s}_{12} = \begin{cases} 0 & ,ct_1 \leq 35 \\ 1/10^* \ ct_1 - 3.5 & ,35 < ct_1 \leq 45 \\ -1/10^* \ ct_1 + 5.5 & ,45 > ct_1 \leq 55 \\ 0 & ,ct_1 > 55 \end{cases}$$

$$\tilde{s}_{12} = \begin{cases} 0 & ,ct_1 \leq 50 \\ 1/15^* \ ct_1 - 10/3 & ,50 < ct_1 \leq 65 \\ 1 & ,ct_1 > 65 \end{cases}$$

(b)

FIGURE 9.10 A fuzzy membership function describing fuzzy utilities in terms of product #1 cycle time. (a) Histogram of first product's cycle time; (b) fuzzy membership function.

TABLE 9.7 Fuzzy Membership Functions for Products No. 1, No. 2, and No. 3

i	j	a_{ij}	b_{ij}	c_{ij}
1	1	0	0	40
	2	35	45	55
	3	50	65	65
2	1	0	0	40
	2	35	45	60
	3	50	80	80
3	1	0	0	30
	2	20	35	50
	3	45	50	50

TABLE 9.8 Simulation Result of the FCPN-std Model

			Cycle time	
Product	Batch size	Throughput	Case with "no fuzzy rules"	Case with "fuzzy rules"
Product no. 1	3	166	44.88	45.01
Product no. 2	2	161	46.02	41.51
Product no. 3	3	166	37.46	41.52

Warm-up period—1440 time units; simulation run length—28,800 time units; number of replications—30.

Table 9.7 shows the extracted fuzzy membership function parameters.

Then, these fuzzy rules are connected to the transition node "$t23$" in Figure 9.9 and the FCPN-std model is designed. Finally, the simulation of this FCPN-std model is achieved using the execution rule described in Section 9.3.3. Table 9.8 shows the simulation result of this FCPN-std model.

As shown in Table 9.8, the cycle time of product no. 2 decreases using the specified fuzzy rule. This illustrates the power of FCPN-std framework where the ambiguous rules can potentially improve the dynamic performance of systems. Thus, an FCPN-std model with the use of xPNML format can address not only design issues but also simulation and control issues for a large-scale manufacturing process.

9.6 CONCLUSIONS

This chapter discusses how a large-scale manufacturing process can be designed including different forms of uncertainties. The trend in current manufacturing environment can be summarized with the terms: "large scale," "general uncertainty," "cooperation/integration," and "dynamic control." The use of the proposed FCPN-std framework and xPNML format can make the design of large-scale manufacturing processes easier while maintaining design consistencies.

The FCPN-std framework is a general modeling framework incorporating two types of uncertainties: variation type uncertainty and ambiguity type uncertainty. Variation type uncertainty is captured using stochastic distribution or data, and ambiguity type uncertainty can be embedded using fuzzy logic. The incorporation of both types of uncertainties can reduce the number of modeling assumptions that are made. Moreover, the FCPN-std framework adheres to the "modular Petri Net" concept and "five-stage modeling methodology." Considering the degree of uncertainty, the use of Petri Net modules streamlines the design of large-scale manufacturing processes.

The xPNML is an advanced version of PNML and provides an interoperable format for Petri Net models. The embedded simulation parameters and attributes assist in the sharing of the consistent design viewpoints during simulation and analysis. Finally, the xPNML-formatted FCPN-std model can be used as a general framework for modeling large-scale manufacturing processes.

The theoretical studies of the suggested framework are currently being done and will be reported in the future. The incorporation of ambiguity type uncertainty makes it difficult to analyze the characteristics of Petri Nets such as boundedness, liveness, and others. As illustrated in Section 9.5, an FCPN-std model can play a role of a dynamic controller through its simulation component. This can play an important role in improving the performance of large-scale manufacturing systems without having to solve the underlying mathematical programming problem repeatedly for every situation.

REFERENCES

[1] R. V. Mohan, K. K. Tamma, D. R. Shires, and A. Mark. Advanced manufacturing of large-scale composite structures: process modeling, manufacturing simulations and massively parallel computing platforms. *Advances in Engineering Software*, 29: 249–263, 1998.

[2] C. C. Bozarth, D. P. Warsing, B. B. Flynn, and E. J. Flynn. The impact of supply chain complexity on manufacturing plant performance. *Journal of Operations Management*, 27: 78–93, 2009.

[3] D. Tranfield and S. Smith. The strategic regeneration of manufacturing by changing routines. *International Journal of Operations & Production Management*, 18: 114–129, 1998.

[4] B. Tan and S.B. Gershwin. Production and subcontracting strategies for manufacturers with limited capacity and volatile demand. *Business and Economics*, 125: 205–232, 2004.

[5] J. Mula, R. Poler, J. P. Garcia-Sabater, and F. C. Lario. Models for production planning under uncertainty: a review. *International Journal of Production Economics*, 103: 271–285, 2006.

[6] H. Lee and A. Banerjee. Modeling and representation of manufacturing process with uncertainties using parametric modular design and xPNML. In: The 2009 International Symposium on Collaborative Technologies and Systems, Baltimore, MD, 2009, pp. 512–519.

[7] M. Hardwick. On STEP-NC and the complexities of product data integration. *Journal of Computing and Information Science in Engineering*, 4: 60–68, 2004.

[8] X. Yang, E. Deines, C. Lauer, and J. C. Aurich. A human-centered virtual factory. In: 2011 International Conference on Management Science and Industrial Engineering, 2011, pp. 1138–1142.

[9] H. Lee, B. Sharda, and A. Banerjee. A closed-loop control architecture for CAM accounting for shop floor uncertainties. *Journal of Manufacturing Systems*, 27: 166–175, 2008.

[10] G. E. Vieira, J. W. Herrmann, and E. Lin. Rescheduling manufacturing systems: a framework of strategies, policies, and methods. *Journal of Simulation*, 6: 39–62, 2003.

[11] M. Schiebe and S. Pferrer. *Real-Time Systems: Engineering and Applications*. Norwell, MA: Kluwer Academic Publishers, 1992.

[12] H. Meyer, F. Fuchs, and K. Thiel. *Manufacturing Execution Systems: Optimal Design, Planning, and Deployment*. New York: McGraw-Hill, 2009.

[13] H. Lee, B. Sharda, and A. Banerjee. Modeling and representation of stochastic models using xPNML. In: Proceedings of the 2007 Winter Simulation Conference, Washington, DC, 2007, pp. 1063–1071.

[14] J. Kim, H. Lee, and A. Banerjee. Cooperative modeling and design history tracking using design tracking matrix. In: 2009 Industrial Engineering Research Conference (IERC), Miami, FL, 2009, pp. 1–8.

[15] H. Lee and A. Banerjee. Design, simulation and analysis of logistics with uncertainties. In: International Conference on Value Chain Sustainability, Louisville, KY, 2009, pp. 1–6.

[16] H. Lee and A. Banerjee. Executable design and control framework using reconfigurable manufacturing holons and its evolution algorithm. *International Journal of Production Research*, 49: 1405–1423, 2011.

[17] X. Gao and Z. Li. Business process modeling and analysis using UML and polychromatic sets. *Production Planning and Control*, 17: 780–791, 2006.

[18] J. A. B. Montevechi, F. Leal, A. F. d. Pinho, R. F. d. S. Costa, M. L. M. d. Oliveira, and A. L. F. d. Silva. Conceptual modeling in simulation projects by mean adapted IDEF: an application in a Brazilian tech company. In: The Proceedings of 2010 Winter Simulation Conference, Baltimore, MD, 2010, pp. 1624–1635.

[19] B. A. Catron and S. R. Ray. ALPS–a language for process specification. *International Journal of Computer Integrated Manufacturing*, 4: 105–113, 1991.

[20] G. Black and V. Vyatkin. On practical implementation of holonic control principles in baggage handling system using IEC 61499. Lecture Notes in Artificial Intelligence, 4659: 314–325, 2007.

[21] R. W. Brennan and J. Wang. A simulation-based approach to developing real-time distributed control systems. In: Proceedings of Summer Computer Simulation Conference, Montreal, Canada, 2003, pp. 24–29.

[22] W. Tan, Y. Fan, and M. Zhou. A Petri Net-based method for compatibility analysis and composition of web services in business process execution language. *IEEE Transactions on Automation Science and Engineering*, 6: 94–106, 2009.

[23] L. Qiao, S. Kao, and Y. Zhang. Manufacturing process modelling using process specification language. *International Journal of Advanced Manufacturing Technology*, 55: 549–563, 2011.

[24] J. C. Hernandez-Matias, A. Vizan, J. Perez-Garcia, and J. Rios. An integrated modelling framework to support manufacturing system diagnosis for continuous improvement. *Robotics and Computer-Integrated Manufacturing*, 24: 187–199, 2008.

[25] S. R. Runnels, I. Kim, J. Schleuter, C. Karlsrud, and M. Desai. A modeling tool for chemical-mechanical polishing design and evaluation. *IEEE Transactions on Semiconductor Manufacturing*, 11: 501–510, 1998.

[26] H. Lee and A. Banerjee. A modular Petri Net based architecture to model manufacturing systems exhibiting resource and timing uncertainties. In: 5th Annual IEEE Conference on Automation Science and Engineering, Bangalore, India, 2009, pp. 525–530.

[27] B. Hruz and M. C. Zhou. *Modeling and Control of Discrete Event Dynamic Systems*. London: Springer, 2007.

[28] S. B. Joshi and J. S. Smith. *Computer Control of Flexible Manufacturing Systems: Research and Development*. London: Chapman & Hall, 1994.

[29] M. C. Zhou, and F. DiCesare. *Petri Net Synthesis for Discrete Event Control of Manufacturing Systems*. London: Kluwer Academic Publishers, 1993.

[30] M. C. Zhou and M. D. Jeng. Modeling, analysis, simulation, scheduling and control of semiconductor manufacturing systems: a Petri Net approach. *IEEE Transactions on Semiconductor Manufacturing*, 11: 333–357, 1998.

[31] M. C. Zhou and K. Venkatesh. *Modeling, Simulation and Control of Flexible Manufacturing Systems: A Petri Net Approach*. Singapore: World Scientific, 1998.

[32] K. Jensen and L. M. Kristensen. *Coloured Petri Nets*. Heidelberg, Germany: Springer, 2009.

[33] P. J. Haas and P. J. Haas. *Stochastic Petri Nets: Modelling, Stability, Simulation*. New York: Springer, 2002.

[34] C. G. Looney. Fuzzy Petri nets and applications. *Fuzzy Reasoning in Information, Decision and Control Systems*, 11: 511–527, 1994.

[35] J. Billington, S. Christensen, K. v. Hee, E. Kindler, O. Kummer, L. Petrucci, R. Post, C. Stehno, and M. Weber. The Petri net markup language: concepts, technology, and tools. Lecture Notes in Computer Science, 2679: 484–505, 2003.

[36] J. Yen and R. Langari. *Fuzzy Logic: Intelligence, Control, and Information*. New Delhi, India: Prentice Hall, 1999.

[37] C. Abdullah and A. Banerjee. Dynamic control of the batch processor in a serial-batch processor system with mean tardiness performance. *International Journal of Production Research*, 48: 1339–1359, 2010.

CHAPTER 10

MODEL IDENTIFICATION AND SYNTHESIS OF DISCRETE-EVENT SYSTEMS*

MARIA PAOLA CABASINO[1], PHILIPPE DARONDEAU[2], MARIA PIA FANTI[3], and CARLA SEATZU[1]

[1]Department of Electrical and Electronic Engineering, University of Cagliari, Cagliari, Italy

[2]Inria Rennes-Bretagne Atlantique, Rennes Cedex, France

[3]Department of Electrical and Information Engineering, Polytechnique of Bari, Bari, Italy

10.1 INTRODUCTION

System identification is a classical problem in system theory: it deals with choosing mathematical models from a known model set to characterize the input–output behavior of an unknown system from finite data [1].

In the discrete-event system (DES) framework, input data are usually given in terms of behavioral descriptions (e.g., transition system and language) and the set of behavioral sequences may be fixed or may be increased in the course of the identification process by performing new experiments. The *identification* problem aims to address two main issues. First, it decides whether for the given behavioral specification there exists a DES (e.g., a Petri net of a given class) that generates the specified behavior. Secondly, it provides a constructive procedure to determine such a DES (usually an automaton or a Petri net).

A problem strictly related to identification is *synthesis*. While only a partial description of the system is assumed in identification, the synthesis problem starts from a complete description of the system. This does not imply that bisimulation is necessarily a goal to achieve, but a requirement of exactness or approximation should

*This work has been partially supported by the European Community's Seventh Framework Programme under project DISC (Grant Agreement no. INFSO-ICT-224498).

Contemporary Issues in Systems Science and Engineering, First Edition.
Edited by MengChu Zhou, Han-Xiong Li and Margot Weijnen.

be specified as an input parameter of the synthesis problem. In simple words, the essential differences between synthesis and identification can be summarized in the following two items.

- While in many synthesis approaches there are both examples and counterexamples, that is, respectively sequences of events that have to be generated by the system and strings that have to be avoided by the system, in identification methods counterexamples are not considered.
- In identification only a fraction of all possible behaviors has been observed, that is, processes allow for much more traces than the ones that are observed.

It is mandatory to highlight that the above distinction between synthesis and identification is not universally recognized in the literature. Thus it often occurs that the two terms are used equivalently.

A rich literature related to identification and synthesis of DES exists. The first contributions appeared in the 1960s and have been formulated in terms of language identification from sets of accepted words, whose solutions are commonly denoted as *learning techniques* [2]. A series of other contributions appeared later in the context of finite-state automata (FSA) and Moore or Mealy machines. They are surveyed in Section 10.3.

In the Petri net (PN) framework, different problem formulations have been proposed since the early 1990s, and several solutions have been given in quite different settings. Besides, it has also occurred that the same problem formulation has been studied by people from two different areas, namely computer science and automatic control. The use of different terminologies and notations, as well as the separate reference journals and conferences for these two areas, lead to the fact that very similar solutions to the same problem—either identification or synthesis—have been developed independently by different research groups. In Section 10.4 we first recall the basic problem formulations, and the main solutions. Then we discuss a series of tightly related problems, mentioning the main contributions appeared in the literature.

Finally, in Section 10.5 we discuss some approaches that basically solve identification problems but are mainly devoted to workflow management processes. Such approaches are usually referred to in the literature as *process mining* or *workflow mining*.

10.2 BACKGROUND ON FINITE STATE AUTOMATA AND PETRI NETS

In this section we briefly recall some basic definitions and notations used in the sequel.

10.2.1 Finite State Automata

FSA are one of the most important formalisms to model DES.

Formally, a (deterministic) FSA can be defined as follows.

Definition 1 An FSA is a 5-tuple $G = (X, E, \delta, x_0, X_m)$ where

- X is a finite set of states;
- E is a finite set of events, called the alphabet of symbols;
- $\delta : X \times E \to X$ is a partial transition function; $\delta(x, e)$ denotes the state reached by the automaton when the event e occurs in state x;
- $x_0 \in X$ is the initial state; and
- $X_m \subseteq X$ is the set of final states (or marked states). □

An FSA is usually described by an oriented graph whose nodes are the states, where the initial state is denoted by an arrow, and where the final states are indicated by a double circle. If $\bar{x} = \delta(x, e)$, then there is an arc from node x to node $\bar{x}$ labeled with e.

An FSA may be seen as a *generator*, that is, as a device that *spontaneously* generates events starting from the initial state. Alternatively, it may be seen as an *acceptor*, i.e., as a device that receives a sequence of events from the environment and distinguishes accepted sequences (those leading to a final state) from rejected sequences (those leading to a nonfinal state). By definition, the languages (i.e., sets of sequences) accepted by FSA are the *regular* languages, i.e., a language is regular if it is accepted by some FSA.

There exist two important classes of DES that generalize FSA by taking a sequence of input events and producing a sequence of output events, namely the *Moore machines* and the *Mealy machines*.

Definition 2 A Moore machine is a 6-tuple $G_{mo} = (X, E, \Theta, \delta, \lambda, x_0)$ where:

- X, E, δ, and x_0 are defined as in the case of FSA (now E is called input alphabet),
- Θ is the output alphabet; and
- $\lambda : X \to \Theta$ is the output function, that is, $\lambda(x)$ denotes the output event produced when the machine is in the state x. □

Note that in a Moore machine, no set of final states is given. In FSA, the states are classified as marked and unmarked. In a Moore machine the output function allows to classify the states in a much more general way, using as many classes as the number of output symbols. Thus an FSA can be seen as a particular Moore machine with $\Theta = \{m, \bar{m}\}$ where $\lambda(x) = m$ if x is final, while $\lambda(x) = \bar{m}$ if x is not final.

Definition 3 A Mealy machine is a 6-tuple $G_{me} = (X, E, \Theta, \delta, \lambda, x_0)$, where:

- X, E, δ, and x_0 are defined as in the case of FSA (now E is called input alphabet);
- Θ is the output alphabet;

- $\lambda : X \times E \to \Theta$ is the output function, that is, $\lambda(x, e)$ denotes the output event produced when the machine executes from x the transition to $\delta(x, e)$ under the effect of the input event e. □

Therefore, both Moore and Mealy machines allow us to give a description of a DES in terms of input–state–output. However a Mealy machine is a more general model being the output function λ not only dependent on the current state x but on the produced event e as well.

10.2.2 Petri Nets

A *Place/Transition net* (*P/T* net) is a structure $N = (P, T, \text{Pre}, \text{Post})$. P is a set of m places. T is a set of n transitions. $\text{Pre} : P \times T \to \mathbb{N}$ and $\text{Post} : T \times P \to \mathbb{N}$ are the pre- and post-incidence matrices that specify flow arcs with integer weights from places to transitions and from transitions to places, respectively. Matrix $C = \text{Post} - \text{Pre}$ is the incidence matrix.

A *marking* is a vector $M : P \to \mathbb{N}$ that assigns to each place a nonnegative integer number of tokens, represented by black dots. We denote $M(p)$ the marking of place p. A P/T *system* or *net system* $\langle N, M_0 \rangle$ is a net N with an initial marking M_0.

A transition t is *enabled* at M if $M \geq \text{Pre}(\cdot, \text{t})$ and may *fire* yielding the marking $M' = M + C(\cdot, t)$. We write $M\ [\sigma\rangle$ to denote that the sequence of transitions σ is enabled at M, and we write $M\ [\sigma\rangle\ M'$ to denote that the firing of σ yields M'. In the following we let $\vec{\sigma} \in \mathbb{N}^{|T|}$ denote the firing vector associated with sequence σ, where the entry of $\vec{\sigma}$ relative to transition t is equal to the number of occurrences of t in σ, and $|T|$ denotes the cardinality of the set T, that is, the number of transitions.

A marking M is *reachable* in $\langle N, M_0 \rangle$ if there exists a firing sequence σ such that $M_0\ [\sigma\rangle\ M$. The set of all markings reachable from M_0 defines the *reachability set* of $\langle N, M_0 \rangle$ and is denoted $R(N, M_0)$.

Given a PN system $\langle N, M_0 \rangle$, its *free language* is the set of its firing sequences

$$L(N, M_0) = \{\sigma \in T^* \mid M_0[\sigma\rangle\}.$$

We also define the set of firing sequences of length less than or equal to $k \in \mathbb{N}$:

$$L_k(N, M_0) = \{\sigma \in L(N, M_0) \mid |\sigma| \leq k\}.$$

When observing the evolution of a net, it is common to assume that each transition t is assigned a label $\varphi(t)$ and the occurrence of t generates an observable output $\varphi(t)$. If $\varphi(t) = \varepsilon$, that is, if the transition is labeled with the empty string, its firing cannot be observed. This leads to the definition of labeled nets.

Definition 4 Given a PN N with set of transitions T, a *labeling function* $\varphi : T \to E \cup \{\varepsilon\}$ assigns to each transition $t \in T$ a symbol from a given alphabet E or the empty string ε. A *labeled* PN system is a 3-tuple $G = \langle N, M_0, \varphi \rangle$ where $N = (P, T, \text{Pre}, \text{Post})$, M_0 is the initial marking, and $\varphi : T \to E \cup \{\varepsilon\}$ is the labeling function. □

Given a labeled PN system $G = \langle N, M_0, \varphi \rangle$ we define its *λ-free labeled language* as the set of words in E^* generated from the initial marking M_0, namely,

$$L^E(N, M_0) = \{w \in E^* \mid \exists \sigma \in T^* : M_0[\sigma\rangle, \varphi(\sigma) = w\}.$$

We also denote $L_k^E(N, M_0)$ the set of words in $L^E(N, M_0)$ of length less than or equal to $k \in \mathbb{N}$, that is,

$$L_k^E(N, M_0) = \{w \in L^E(N, M_0) \mid |w| \leq k\}.$$

10.3 IDENTIFICATION AND SYNTHESIS OF LANGUAGES AND FINITE STATE AUTOMATA

The problem of identifying a DES has been first formulated in terms of language identification from sets of accepted words. The proposed solutions are commonly called *learning techniques* [3].

One of the first contributions in this framework dates back to 1967 and is due to Gold [4]. The study in [4] derives its motivation from artificial intelligence. In particular, the following situation is considered: a class of possible languages is specified, together with a method of presenting information to the learner about an unknown language, which is to be chosen from the class. The problem is to establish if the information is sufficient to determine exactly which of the possible languages is the unknown language. The class of languages is called *learnable* with respect to the given method of presenting information if learners can use for making their guesses some algorithm with the following property: for any language in the class, there is a finite amount of information, counted in terms of input strings, after which the guesses made by the algorithm will be the same forever and will be correct. If that is the case, the language is said to be *learnable in the limit*.

Many contributions based on the idea of *learning in the limit* have been proposed since then. A good survey has been done by Jain and Kinber in [5]. Let us mention the works of Case and Lynes [6], and Osherson and Weinstein [7] who introduce a more powerful type of learning for languages, where learners behave correctly if they almost always output correct (but not necessarily the same) grammar for the target language. In both cases, the authors also considered a much stronger (and less realistic) model of learning languages in the presence of full positive (examples) and negative data (counterexamples). In [8] Baliga *et al.* consider an intermediate model, where a learner gets full positive data and a finite number of negative data. However, negative data are preselected, which dramatically affects learning capabilities.

Another important contribution in the framework of learning has been made by Angluin [9]. The author considers inductive inference of formal languages, as defined by Gold [4], in the case of positive data, that is, when the examples are successive elements in an arbitrary enumeration of the unknown language. Moreover, in [4] Gold proves that the question of whether there is an automaton with the minimum number of states which agrees with a finite set of positive data is NP-complete. However, in [10] Gold proposes an automaton identification rule that allows the identification-in-the-limit in polynomial time from positive data. Successively, in [9] Angluin characterizes when an indexed family of nonempty recursive formal languages is inferable from positive data. From this theorem other useful conditions for inference from positive data can be obtained. Later, in [11] Angluin introduces another important learning paradigm, that is, learning from queries to a teacher (oracle), that has been extensively investigated by many other authors as discussed in detail in [5].

In the context of FSA, the contributions to the identification problem are based on *Moore machines* and *Mealy machines*. A very good survey on the developed techniques has been presented by Estrada-Vargas *et al.* in [2]. The end of the section draws much inspiration from that survey (italicized paragraphs are almost excerpts from [2]).

One of the first contributions in this framework has been given by Booth in [12] who proposed a method to identify a Moore or Mealy machine. The system under investigation is placed within a test bed and connected to a so-called experimenter, which generates the input signals and records the output signals of the system. The identification can be started considering a very few number of states. If, at some point of the experiment, it is impossible to find a correct machine with the assumed number of states, then the identification is started again considering a machine with one more state.

The method proposed by Kella in [13] obtains models representing Mealy machines. The presented method does not require any a priori knowledge of the system, and only a single observed sequence is available. The algorithm lists all reduced machines which may produce the given input-output sequence. The construction principle is the merging of equivalent states.

In [14] Biermann and Feldman modify the Nerode realization technique for synthesizing finite-state machines from their associated right-invariant equivalence relations [15] and give a method for identifying nondeterministic Moore machines from finite subsets of their input–output behavior. The synthesis procedure includes a parameter that may be adjusted to obtain machines that represent the given input–output behavior with varying degrees of accuracy (and consequently with varying complexities).

In [16] Veelenturf presented an algorithm for finding a minimal, though not a smallest, Mealy machine with a behavior comprising a given finite sample of the input-output behavior of some unknown sequential machine. The automaton is built step by step. At each step, the already available machine is examined and completed by adding transitions and possibly new states.

Later, in [17] Veelenturf proposed an algorithm to identify a Moore machine generating from the initial state a behavior observed during a certain number of steps. The learning procedure operates in three phases: induction, contradiction, and discrimination. A state can never be deleted, and only transitions between states can be modified. This method has been improved by Richetin et al. in [18] *where two algorithms are proposed for identifying multiple automata as well as automata that may not be initialized.*

A series of interesting and efficient identification algorithms based on neural networks, and in particular on recurrent neural networks (RNNs) [19] have also been presented. Among the contributions in [20, 21, 22] that only deal with short input and output sequences because the training algorithm fails to effectively deal with long sequences. Recently, online identification with a single pair of input and output sequences has been proposed in [23]; the online identification procedure, nonetheless, requires a pair of overly long input and output sequences to train the RNN in the online fashion. Finally in [24], Won *et al.* propose a class of RNNs and show that the class possesses the capability of identifying any discrete time dynamical system. Moreover, since the states of the proposed network represent those of the system being modeled, the encoding of an FSA in the proposed network becomes quite simple.

While the main goal of identification is to construct an automaton based on the results of experimenting with a "black box," that is, an implementation, synthesis consists of constructing an automaton from given specifications of the needed, possible, and forbidden behavior [25]. Significant contributions to both problems have been given in the framework of the theory of experiments with automata. An experiment with an automaton is generally considered as the process of supplying input sequences for the automaton, observing its behavior in response to these inputs, and reaching some conclusions on its functioning and properties, by exploiting both the observations done and the a priori information available about the automaton. The main objective of the considered theory is to design effective experiments that would allow one to obtain partial knowledge of the structure of the automaton and the properties of information-transformation processes it realizes [25]. An excellent survey is given in the recent paper by Kudryavtsev *et al.* that appeared in 2010 [25].

We conclude this section by mentioning a problem strictly related to identification, known as *testing*: in a testing problem we have a machine about which we lack some information, for example, the initial state; we would like to deduce this information by providing a sequence of inputs to the machine and observing the outputs produced. An excellent survey on this problem, even if not very recent, has been written by Lee and Yannakakis in [26]. For the sake of brevity, we do not address this problem here.

10.4 IDENTIFICATION AND SYNTHESIS OF PETRI NETS

The first solutions to the synthesis problem of PNs date back to the early 1990s and are based on the *theory of regions*. Most contributions in this line have been proposed by people from the computer science area. Among these we mention the works by Ehrenfeucht and Rozenberg [27], Badouel *et al.* [28], Badouel and Darondeau [29], and Cortadella *et al.* [30]. A series of more recent contributions based on the theory of regions have been brought by Bergenthum *et al.* [31], Carmona *et al.* [32], Lorenz *et al.* [33], and Lorenz and Juhás [34].

Independently, a series of other contributions have been proposed by people from the automatic control community. Among these we mention the works by Meda-Campaña and López-Mellado [35, 36], Cabasino *et al.* [37], and Dotoli *et al.* [38].

Three main formulations of the synthesis/identification problem have been considered.

- *Synthesis from graphs:* given a set of transitions T and a finite directed graph with arcs labeled by transitions in T, decide whether there exists a PN whose reachability graph is isomorphic to the given graph, and if such PN exists, then construct it [28, 29, 37, 39].
- *Identification and synthesis from finite or regular languages over T*: given a prefix-closed language $\mathcal{L} \subset T^*$, where T is a set of transitions, synthesize a PN such that $L(N, M_0) = \mathcal{L}$, or $L(N, M_0) \supseteq \mathcal{L}$, or $L_k(N, M_0) \supseteq \mathcal{L}$ for some finite k [28, 33, 37, 40–42].
- *Identification and synthesis from finite languages over E*: given a prefix-closed language $\mathcal{L} \subset E^*$, where E is an alphabet and $\varphi : T \rightarrow E$ is a labeling function that associates a symbol in E with each transition, synthesize a PN labeled by φ, such that $L^E(N, M_0) \supseteq \mathcal{L}$ (or $L^E_k(N, M_0) \supseteq \mathcal{L}$ for some finite k) [37].

Basically, we talk about synthesis when the relationships in the two items above are equalities, and otherwise we talk about identification. Obviously, the complexity of the problem depends highly on whether the goal is $L(N, M_0) = \mathcal{L}$ (synthesis) or $L(N, M_0) \supseteq \mathcal{L}$ (identification). Indeed, the goal $L(N, M_0) = \mathcal{L}$ imposes additional constraints for preventing the net from producing forbidden words, so-called counterexamples [33, 37]. Analogously, for the third problem, the complexity highly depends on whether the goal is $L^E(N, M_0) = \mathcal{L}$ or $L^E(N, M_0) \supseteq \mathcal{L}$.

To the best of our knowledge, the third problem has only been investigated in [37, 38], and only the work in [37] handles the case where the goal is $L^E(N, M_0) = \mathcal{L}$, for which no solution based on the theory of regions has been proposed yet.

10.4.1 Synthesis from Graphs

One of the major contributions to the synthesis problem from graphs is given by Badouel *et al.* in [28] and by Badouel and Darondeau in [29]. A simplified interpretation of the theory of regions using the basic notions of PNs is presented later by Ghaffari *et al.* in [39]. Although these authors use this new approach to design a control policy, we report here their interpretation because it explains in a very easy way how the theory of regions synthesizes a PN with a set of transitions T, starting from a graph whose arcs are labeled by transitions in T.

Problem 1 Assume that given a set of transitions $T = \{t_1, \ldots, t_n\}$ and a finite directed graph $\mathcal{G}$ whose arcs are labeled by transitions in T. Assume that in the graph $\mathcal{G}$ all nodes may be reached by directed paths from a fixed node S_0. We want to identify a pure PN system $\langle N, M_0\rangle$ whose set of transitions coincides with T and whose reachability graph is isomorphic to $\mathcal{G}$, such that M_0 is the image of S_0 under this isomorphism.

The unknowns we want to determine are a set of places $P = \{p_1, \ldots, p_m\}$, the elements of the incidence matrix

$$C = \{c_{ij}\} \in \mathbb{N}^{m\times n}$$

and the elements of the vector

$$M_0 = \left[m_{0,1}\ m_{0,2} \cdots m_{0,m}\right]^{\mathrm{T}} \in \mathbb{N}^m.$$

Note that in the above formulation the incidence matrix C completely characterizes the structure of the net N since by assumption, this net is pure (i.e., loop free: $\mathrm{Pre}(p,t) \neq 0 \Rightarrow \mathrm{Post}(p,t) = 0$ and $\mathrm{Post}(p,t) \neq 0 \Rightarrow \mathrm{Pre}(p,t) = 0$).

The following theorem states necessary and sufficient conditions for the existence of a solution to the above identification problem. □

Theorem 1 [39] Problem 1 has some PN system solution $\langle N, M_0\rangle$ iff there exists a set of places P such that

(1) each place $p \in P$ satisfies the *cycle equation*

$$\sum_{t\in T} C(p,t)\cdot \vec{\gamma}(t) = 0, \quad \forall \gamma \in U, \tag{10.1}$$

where U is the set of nondirected elementary cycles of $\mathcal{G}$, and for any $\gamma \in U$, $\vec{\gamma}$ is the firing count vector of γ, that is, if $\gamma = e_1 \ldots e_n$ is a sequence of edges e_i connecting corresponding nodes s_i and s_{i+1} (or s_1 if $i = n$) by a direct transition from s_i to s_{i+1} labeled t_i (let $\varphi(e_i) = t_i$ in this case) or by a reverse transition from s_{i+1} to s_i labeled t_i (let $\varphi(e_i) = t_i$ in that case), then $\vec{\gamma}(t) = \sum\{\varphi(e_i) \mid 1 \leq i \leq n \wedge t_i = t\}$ for every transition $t \in T$;

(2) each place $p \in P$ satisfies the *reachability conditions*

$$M_0(p) + C(p,\cdot)\cdot \vec{\Gamma} \geq 0, \quad \forall\, \Gamma \in V, \tag{10.2}$$

where V is a fixed set of nondirected initial paths in $\mathcal{G}$ containing exactly one path from s_0 to s for every node s, and $\vec{\Gamma}$ is the firing count vector of Γ;

(3) For any two distinct nodes s and s' of $\mathcal{G}$ reached by paths $\Gamma, \Gamma' \in V$, or equivalently, for any two distinct paths $\Gamma, \Gamma' \in V$, there exists a place p that satisfies the *state separation condition* of (s, s'), that is,

$$M_0(p) + C(p,\cdot)\cdot \vec{\Gamma} \neq M_0(p) + C(p,\cdot)\cdot \vec{\Gamma}'; \tag{10.3}$$

(4) for each node s of $\mathcal{G}$ and for each transition t not firable from s in $\mathcal{G}$, there exists a place p that satisfies the *event separation condition* of (s, t), that is,

$$M_0(p) + C(p, \cdot) \cdot \vec{\Gamma} + C(p, t) \leq -1, \tag{10.4}$$

where $\Gamma \in V$ is the path from s_0 to s in $\mathcal{G}$. □

The validity of the above theorem, that is formally proved in [39], is quite intuitive given the structure of the constraints. One remark needs to be done concerning constraint (2). For a given node s, there may certainly exist several paths Γ from s_0 to s in $\mathcal{G}$. However, under the cycle equations (10.1), the product $C(p, \cdot) \cdot \vec{\Gamma}$ is the same for all these paths. As a result the path Γ may be chosen arbitrarily.

We conclude this subsection mentioning another approach presented by Cabasino *et al.* [37] and not based on the theory of regions, that has been quite recently proposed to solve the problem of synthesizing a bounded PN system whose reachability graph is isomorphic to a given graph $\mathcal{G}$. The authors derive a similar algebraic characterization of the set of feasible constraints. However, the main feature of their approach is in the selection of one among the set of possible solutions. In particular, they define a performance index and compute one feasible PN system that minimizes this index via the solution of a linear integer programming problem.

10.4.2 Identification and Synthesis from Finite Languages Over *T*

Most of the contributions in this framework are region-based approaches. The basic idea behind region-based approaches can be summarized in simple words as follows [33].

Given a prefix-closed language $\mathcal{L}$ over some alphabet T, one wants to determine a finite PN system $\langle N, M_0 \rangle$ with set of transitions in one-to-one correspondence with the symbols in the alphabet, such that all words in the language are generated as firing sequences. If one considers the net with the set of transitions T and with the empty set of places, then clearly this net generates as firing sequences all words in $\mathcal{L}$ and many more since it generates fully T^*. The idea is then to add *feasible* places, namely places whose incorporation into the net restricts the behavior although still guaranteeing that $\mathcal{L} \subseteq L(N, M_0)$, where $\langle N, M_0 \rangle$ is the augmented net system. Adding *all* feasible places in one shot amounts to *minimize* the language $L(N, M_0)$ while guaranteeing that $\mathcal{L} \subseteq L(N, M_0)$, but more progressive techniques may also be considered. The theory of regions provides a method to calculate the feasible places, based on the following definition of *regions of a language*.

Definition 5 Given a finite prefix-closed language $\mathcal{L}$ over an alphabet T, define two matrices A and $A' \in \mathbb{N}^{|\mathcal{L}| \times |T|}$ as follows.

The transitions in T index the columns of A and A'. The words $w \in \mathcal{L}$ index the rows of A. The entry $A(w, t)$ of A is the number of occurrences of t in word w. Thus, each row of A is the firing vector of some sequence $w \in \mathcal{L}$. Matrix A' is defined similarly, except that its rows are indexed by the words in the language $\mathcal{L}'$ obtained from $\mathcal{L}$ by removing the last event symbol from each sequence. A region defining an associated place p is a triple $(M_0(p), \text{Post}(p,\cdot), \text{Pre}(p,\cdot))$ with $\text{Post}(p,\cdot), \text{Pre}(p,\cdot) \in \mathbb{N}^{|T|}$ and $M_0(p) \in \mathbb{N}$, such that

$$M_0(p) \cdot \vec{1} + A' \cdot \text{Post}(p,\cdot) - A \cdot \text{Pre}(p,\cdot) \geq \vec{0}. \tag{10.5}$$

The set of all regions of a language is denoted $\mathcal{R}(\mathcal{L})$ and the region $(0, \vec{0}, \vec{0})$ is called the *trivial* region. □

In simple words the above equation (10.5) guarantees that place p does not disable any sequence in $\mathcal{L}$. Note that each region defines the initial marking of a feasible place and its set of input and output arcs to or from transitions in T. Such regions have been used by Badouel *et al.* [28], Bergenthum *et al.* [40], Darondeau [42], and Lorenz *et al.* [33] among others.

Let us observe that, since the set of feasible places is in general *infinite* [33], a finite subset should be appropriately selected. Two main solutions have been presented in this respect, denoted as *basis representation* and *separating representation* [33, 40].

In the basis representation, the set of places is chosen such that it is a basis for the nonnegative integer solution space of the linear inequality system (10.5). The main limitation of this approach is that the basis representation is worst-case exponential in the number of sequences in $\mathcal{L}$ thus leading to a PN with an enormous number of places. In [40] Bergenthum *et al.* also provide some heuristics on how to remove redundant places from the basis. However this still requires the basis to be fully constructed.

The separating representation aims to reproduce exactly the language $\mathcal{L}$ [33, 40], or more precisely to impose that $L_k(N, M_0) = \mathcal{L}$ where $\langle N, M_0 \rangle$ is the computed net system, i.e., it provides a solution to what is commonly known as a synthesis problem. The basic principle is to add only those places needed to prevent the firing of the last transition t in some sequence $wt \notin \mathcal{L}$ such that $w \in \mathcal{L}$ and $t \in T$. This leads to a PN whose number of places is no longer exponential, but polynomial in the size of the language. This solution, as well as all other solutions based on the theory of regions, leads to a linear algebraic characterization of the net structure and initial marking, in which the unknowns are integer numbers. However, the constant terms of the equations or inequalities found in such algebraic characterizations are always equal to 0, which means that solutions can nevertheless be computed in polynomial time.

A similar remark is at the base of the approach recently proposed by Cabasino *et al.* in [41] for solving the synthesis problem $L_k(N, M_0) = \mathcal{L}$, where k is a fixed integer. It is also observed there that the set of constraints characterizing the solutions of the identification problem is ideal and rational [41]. Hence, the integrality constraints on the unknowns are removed. In the considered work, a different place is constructed for each sequence $w \in \mathcal{L}$ and for each transition symbol $t \in T$ such that $wt \notin \mathcal{L}$, hence the number of places of the net is exactly equal to the number of disabling constraints. In particular, there may exist redundant places. To overcome this issue, in [41] the authors provide a criterion to check if the computed solution has a minimal number of places, and, if that is not the case, they propose two methods to reduce this number, called *prereduction* and *postreduction*. In the first approach, the set of constraints that characterize feasible solutions is written in a modified form, using a reduced number of places. In the second approach, one first computes a net with as many places as disabling constraints, and then identifies the redundant places that can be removed without affecting the correctness of the result.

Another approach that aims at solving the synthesis problem, or equivalently at computing a separating representation, has been proposed by Cabasino *et al.* in [37]. In that work, the net structure and the initial marking are computed by solving an integer linear programming problem. The main feature of the approach is that the authors associate a performance index, and among all feasible solutions, they select one with minimal index. Minimizing a performance index cannot be done without integer linear programming, but the incurred algorithmic complexity is compensated by the expected quality (conciseness for instance) of the synthesized net. The approach has been extended in several ways in [37], by adding different information about the model (structural constraints, conservative components, and stationary sequences) or about its initial marking. Finally, as discussed in detail in Section 9, Cabasino *et al.* in [37] also show that the approach can be generalized to the case of labeled PNs, where two or more transitions may share the same label.

We conclude this section by mentioning a series of other approaches focused on the synthesis of restricted classes of PNs proposed in the 1980s and 1990s.

Murata presents in [43] a method for synthesizing or augmenting live and safe marked graph models of decision-free concurrent computations. The approach is modular in the sense that subsystems, that is, marked subgraphs, are added one by one without redesigning the entire system. The following properties or parameters can be prescribed for the synthesized marked graphs: liveness (absence of deadlocks), safeness (absence of overflows), number of reachability classes, maximum resource (temporary storage) requirement, computation rate (performance), as well as the numbers of arcs and states.

Czaja [44] proposes a method for synthesizing nets by using a concurrency operator, Esparza *et al.* [45] give synthesis rules for live and bounded free choice nets, and Graubmann [46] proposes a method for synthesizing nets from given trace behavior.

In [47] Hiraishi presents an interactive method for constructing a special class of safe marked PNs from the knowledge of a finite set of their firing sequences. The proposed algorithm consists of two phases. In the first phase, a language is identified from the given firing sequences in the form of a finite-state automaton. In a second phase, the dependency relation is extracted from the regular language obtained, and the structure of a PN is guessed. Provided that the language is generated by the special class of safe nets under consideration, the algorithm identifies uniquely the original net if a sufficiently large set of firing sequences is given as input to the algorithm.

The problem of synthesizing ordinary interpreted PNs has been more recently investigated by Meda-Campaña and López-Mellado in [35]. In this work the marking of a subset of places is assumed to be observable, and the part of the net related to unobservable places is computed from cyclic output data streams; this allows the online operation of the identification procedure that infers a succession of PN models, which acquire progressively more specific structure; thus at each time, the currently computed PN model represents abstractly the behavior of the system observed so far.

Later, in [36], Meda-Campaña and López-Mellado investigate a problem tightly related to the previous one. They provide an approach to determine the set of transition sequences which are really needed to identify an interpreted PN modeling the complete behavior of a DES. These representative transition sequences play a crucial role because, if one can compute them from the observed output signals of the system, then it is guaranteed that the complete behavior of the system is captured in the synthesized PN model, even though some other transition sequences of the system have not been directly revealed by the output signals.

10.4.3 Identification and Synthesis from Finite Languages Over *E*

The problem of synthesizing a labeled PN system from finite languages over an alphabet E has also been investigated in the literature. To the best of our knowledge no solution has been proposed based on the theory of regions.

In the sequel we first discuss in detail an approach based on linear integer programming that provides a separating representation of the solution to the synthesis problem in its most general formulation, that is, without any simplifying assumption on the structure and labeling function of the synthesized net [37]. In the second part of the section, we survey several other similar approaches.

10.4.3.1 A Solution to the Synthesis Problem Based on Linear Integer Programming

Problem 2 [37] Assume that a set of places $P = \{p_1, \dots, p_m\}$ and a set of transitions $T = \{t_1, \dots, t_n\}$ are given. Let

$$T = \bigcup_{e \in E} T_e$$

and $\varphi : T \to E$ be a labeling function over E, such that $T_e = \varphi^{-1}(e)$ for all $e \in E$ (i.e., $\varphi(t) = e$ iff $t \in T_e$). Let $\mathcal{L} \subset E^*$ be a given finite prefix-closed language over E^*, and

$$k = \max_{w \in \mathcal{L}} |w|$$

be the length of the longest word in $\mathcal{L}$.

We want to synthesize the structure of a *deterministic* net[1] $N = (P, T, \text{Pre}, \text{Post})$ labeled by φ and an initial marking M_0 such that

$$L_k^E(N, M_0) = \mathcal{L}.$$

The unknowns we want to determine are the elements of the two matrices

$$\text{Pre} = \{e_{i,j}\} \in \mathbb{N}^{m \times n} \quad \text{and} \quad \text{Post} = \{o_{i,j}\} \in \mathbb{N}^{m \times n}$$

[1] A net labeled over E is deterministic if, for each $e \in E$ and for each reachable marking M, at most one transition t with $\varphi(t) = e$ can be fired in M.

and the elements of the vector

$$M_0 = \left[m_{0,1}\ m_{0,2}\ \cdots\ m_{0,m}\right]^T \in \mathbb{N}^m. \quad \square$$

Note that, as discussed in [37], this problem may be solved without assuming that the labeled net to be synthesized is deterministic.

The following theorem provides a linear algebraic characterization of the deterministic labeled PN systems with m places, n transition,s and labeling function φ such that $L_k^E(N, M_0) = \mathcal{L}$.

Theorem 2 [37] A solution to the synthesis Problem 2 satisfies the following set of linear algebraic constraints

$\mathcal{G}(m, T, \mathcal{L}, \varphi) \triangleq$

$$\begin{cases}
\left\{\begin{array}{l}
M_w - \mathrm{Pre}(\cdot, \mathrm{t}_1^{\mathrm{e}}) \geq -\mathrm{z}_1^{\mathrm{e,w}} \cdot \mathbf{K} \\
\quad\vdots \\
M_w - \mathrm{Pre}(\cdot, \mathrm{t}_{\mathrm{n_e}}^{\mathrm{e}}) \geq -\mathrm{z}_{\mathrm{n_e}}^{\mathrm{e,w}} \cdot \mathbf{K} \\
M_{we} - M_w - \mathrm{Post}(\cdot, \mathrm{t}_1^{\mathrm{e}}) + \mathrm{Pre}(\cdot, \mathrm{t}_1^{\mathrm{e}}) \leq \mathrm{z}_1^{\mathrm{e,w}} \cdot \mathbf{K} \\
M_{we} - M_w - \mathrm{Post}(\cdot, \mathrm{t}_1^{\mathrm{e}}) + \mathrm{Pre}(\cdot, \mathrm{t}_1^{\mathrm{e}}) \geq -\mathrm{z}_1^{\mathrm{e,w}} \cdot \mathbf{K} \\
\quad\vdots \\
M_{we} - M_w - \mathrm{Post}(\cdot, \mathrm{t}_{\mathrm{n_e}}^{\mathrm{e}}) + \mathrm{Pre}(\cdot, \mathrm{t}_{\mathrm{n_e}}^{\mathrm{e}}) \leq \mathrm{z}_{\mathrm{n_e}}^{\mathrm{e,w}} \cdot \mathbf{K} \\
M_{we} - M_w - \mathrm{Post}(\cdot, \mathrm{t}_{\mathrm{n_e}}^{\mathrm{e}}) + \mathrm{Pre}(\cdot, \mathrm{t}_{\mathrm{n_e}}^{\mathrm{e}}) \geq -\mathrm{z}_{\mathrm{n_e}}^{\mathrm{e,w}} \cdot \mathbf{K} \\
z_1^{e,w} + \ldots + z_{n_e}^{e,w} = n_e - 1 \\
z_1^{e,w}, \ldots, z_{n_e}^{e,w} \in \{0, 1\}
\end{array}\right\} \quad \forall (w,e) \in \mathcal{E} & (a) \\
-K\bar{S}(w, t_j^e) + M_w - Pre(\cdot, t_j^e) \leq \mathbf{-1} \qquad \forall (w,e) \in \mathcal{E} \,:\, |T_e| > 1,\ \forall t_j^e \in T_e & (b) \\
\vec{1}^T \bar{S}(w, t_j^e) \leq m - z_j^{e,w} \qquad \forall (w,e) \in \mathcal{E} \,:\, |T_e| > 1,\ \forall t_j^e \in T_e & (c) \\
-KS(w, t_j^e) + M_w - \mathrm{Pre}(\cdot, \mathrm{t}_j^{\mathrm{e}}) \leq \mathbf{-1} \qquad \forall (\mathrm{w,e}) \in \mathcal{D},\quad \forall \mathrm{t}_j^{\mathrm{e}} \in \mathrm{T_e} & (\mathrm{d}) \\
\vec{1}^{\,T} S(w, t_j^e) \leq m - 1 \qquad \forall (w,e) \in \mathcal{D},\quad \forall t_j^e \in T_e & (e) \\
M_w \in \mathbb{N}^m, \quad \forall w \in \mathcal{L} & (f) \\
\mathrm{Pre, Post} \in \mathbb{N}^{\mathrm{m \times n}} & (\mathrm{g}) \\
S(w, t_j^e) \in \{0, 1\}^m & (h) \\
\bar{S}(w, t_j^e) \in \{0, 1\}^m & (i)
\end{cases} \tag{10.6}$$

where

$$T_e = \{t_1^e, \ldots, t_{n_e}^e\}$$

$$\mathcal{E} = \{(w, e) \mid w \in \mathcal{L}, |w| < k, we \in \mathcal{L}\},$$

$$\mathcal{D} = \{(w, e) \mid w \in \mathcal{L}, |w| < k, we \notin \mathcal{L}\},$$

and $M_\varepsilon = M_0$. □

Assume that K is an arbitrarily large integer and $\vec{K}$ is a vector with all entries equal to K. Also assume that for all $w \in \mathcal{L}$, M_w denotes the marking such that $M_0[w\rangle M_w$ in the synthesized net. Then the above set of constraints can be explained as follows.

- Let $we \in \mathcal{L}$ with $e \in E$. Then *at least* one transition $t_j^e \in T_e$ should be enabled at M_w, or equivalently, for *at least* one $t_j^e \in T_e$ it should hold:

$$M_w \geq \mathrm{Pre}(\cdot, \mathrm{t}_\mathrm{j}^\mathrm{e}).$$

Thus, following a quite standard approach to convert the logical *or* operator in terms of linear constraints [48], we can write

$$\begin{cases} M_w - \mathrm{Pre}(\cdot, \mathrm{t}_1^\mathrm{e}) \geq -\mathrm{z}_{1,\mathrm{w}}^\mathrm{e} \cdot \mathbf{K} \\ \vdots \\ M_w - \mathrm{Pre}(\cdot, \mathrm{t}_{\mathrm{n_e}}^\mathrm{e}) \geq -\mathrm{z}_{\mathrm{n_e},\mathrm{w}}^\mathrm{e} \cdot \mathbf{K} \\ z_{1,w}^e + \ldots + z_{n_e,w}^e = n_e - 1 \\ z_{1,w}^e, \ldots, z_{n_e,w}^e \in \{0, 1\} \end{cases}$$

If $z_{j,w}^e = 0$ it means that $t_j^e \in T_e$ may fire at M_w, and the marking M_{we} reached after this firing is

$$M_{we} = M_w + \mathrm{Post}(\cdot, \mathrm{t}_\mathrm{j}^\mathrm{e}) - \mathrm{Pre}(\cdot, \mathrm{t}_\mathrm{j}^\mathrm{e})$$

and the following set of linear inequalities are therefore satisfied:

$$\begin{cases} M_{we} - M_w - \mathrm{Post}(\cdot, \mathrm{t}_1^\mathrm{e}) + \mathrm{Pre}(\cdot, \mathrm{t}_1^\mathrm{e}) \leq \mathrm{z}_{1,\mathrm{w}}^\mathrm{e} \cdot \mathbf{K} \\ M_{we} - M_w - \mathrm{Post}(\cdot, \mathrm{t}_1^\mathrm{e}) + \mathrm{Pre}(\cdot, \mathrm{t}_1^\mathrm{e}) \geq -\mathrm{z}_{1,\mathrm{w}}^\mathrm{e} \cdot \mathbf{K} \\ \vdots \\ M_{we} - M_w - \mathrm{Post}(\cdot, \mathrm{t}_{\mathrm{n_e}}^\mathrm{e}) + \mathrm{Pre}(\cdot, \mathrm{t}_{\mathrm{n_e}}^\mathrm{e}) \leq \mathrm{z}_{\mathrm{n_e},\mathrm{w}}^\mathrm{e} \cdot \mathbf{K} \\ M_{we} - M_w - \mathrm{Post}(\cdot, \mathrm{t}_{\mathrm{n_e}}^\mathrm{e}) + \mathrm{Pre}(\cdot, \mathrm{t}_{\mathrm{n_e}}^\mathrm{e}) \geq -\mathrm{z}_{\mathrm{n_e},\mathrm{w}}^\mathrm{e} \cdot \mathbf{K} \end{cases}$$

Let $z_k^{e,w} = 0$ in the above constraints, entailing that $t_k^e \in T_e$ is enabled at M_w. If we want the net to be deterministic, in case when $|T_e| > 1$, we must impose that no other transition $t_j^e \in T_e$ with $j \neq k$ is enabled at M_w. As $z_k^{e,w} = 0$ entails $z_j^{e,w} = 1$ for $j \neq k$ (because $z_{1,w}^e + \cdots + z_{n_e,w}^e = 1$), this amounts to requiring that, for every transition $t_j^e \in T_e$ with $z_j^{e,w} = 1$, it holds that

$$M_w - \mathrm{Pre}(\cdot, t_j^e) \not\geq \vec{0}.$$

In order to impose such constraints, for all $t_j^e \in T_e$ we introduce a vector of binary variables $\bar{S}(w, t_j^e)$ that should satisfy the following set of linear inequalities:

$$\begin{cases} -K\bar{S}(w, t_j^e) + M_w - \text{Pre}(\cdot, t_j^e) \leq -\vec{1} \\ \vec{1}^T \bar{S}(w, t_j^e) \leq m - z_j^{e,w} \end{cases}$$

If $z_j^{e,w} = 0$ (i.e., $j = k$ where t_k^e is the e-labeled transition enabled at M_w), then all entries of $\bar{S}(w, t_j^e)$ may be unitary, thus adding no additional constraint (the corresponding inequality is trivially verified). On the contrary, if $z_j^{e,w} = 1$, then at least one entry of $\bar{S}(w, t_j^e)$ is null, thus making t_j^e disabled at M_w (by the first inequality). Thus, only one transition labeled e is enabled at M_w.

- Let $w \in \mathcal{L}$ and $we \notin \mathcal{L}$ with $e \in E$. Then for all $t_j^e \in T_e$ the following set of linear constraints should be satisfied:

$$\begin{cases} -K \cdot S(w, t_j^e) + M_w - \text{Pre}(\cdot, \mathrm{t_j^e}) \leq -\mathbf{1}_\mathrm{m} \\ \vec{1} \cdot S(w, t_j^e) \leq m - 1 \\ S(w, t_j^e) \in \{0, 1\}^m. \end{cases}$$

Indeed, if these constraints are satisfied, then the vector of binary variables $S(w, t_j^e)$ has at least one null entry, hence t_j^e is disabled at M_w (by the first inequality). Note that, for the above characterization to be valid, it not necessary that the net be K-bounded. In fact, one can take any value

$$\begin{aligned} K &\geq \max_i M_0(p_i) + k \cdot \max_{i,j} \text{Post(i, j)} \\ &\geq \max_i M(p_i) + k \cdot \max_{i,j} \text{Post(i, j)} \quad , \\ &\geq \max_i M_w(p_i) \end{aligned}$$

where k is the maximum of $|w|$ for $w \in \mathcal{L}$.

The above linear algebraic characterization enables us to solve by integer linear programming net synthesis problems including moreover a performance index to be minimized over the set of feasible solutions, thus obtaining, for example, a solution with a minimal number of tokens in the initial marking and/or minimal arcs with nonunit weight.

Note that in [37] the authors also relax the assumption that the number of places is fixed in advance.

The main limitation of the approach is an exponential complexity with respect to k. Nevertheless, this high complexity originates from the fact that we want to reconstruct exactly the language $\mathcal{L}$ (constraints (d) and (e)). The complexity becomes polynomial in k if this requirement is weakened to $\mathcal{L} \subseteq L_K^E(N, M_0)$, that is, if we solve a simple identification problem instead of a synthesis problem.

10.4.3.2 Other Important Contributions Other important contributions in the framework of labeled net identification and synthesis are due to Ru and Hadjicostis who present in [49] an approach for the state estimation of DES modeled by labeled PNs. More specifically, given an unknown (nondeterministic) labeled PN with a known set of places and a known initial marking, they show that the number of markings possibly reachable from this initial marking is at most polynomial in the length of the observation sequence, even though the set of possible firing sequences can be exponential in the length of the observation sequence.

In [38] Dotoli *et al.* present a procedure to identify a labeled PN system modeling the DES from the real time observation of its dynamical evolution. Assuming that the DES can be detected and distinguished by the identifier and can be modeled by a labeled PN, the proposed approach is based on the construction of a set of integer linear constraints and of a metric as an indicator of the size of the PN. Moreover, an ILP problem is defined in order to minimize the selected metric and to obtain an optimal PN modeling the observed DES behavior. The set of places, of transitions and the labeling function are unknown and only an upper bound of the cardinality of the place set is given. To identify the PN system modeling the DES, the authors propose an identification algorithm that waits until a new event occurs and stores the corresponding output symbols. At each observation the algorithm updates the labeling function, the transition and the place sets, redefines and solves the ILP problem. This approach does not require the complete knowledge of the system evolution in the form of the reachability graph or of the positive examples and counterexamples of the DES language. On the contrary, the proposed method identifies the PN modeling the DES behavior on the basis of the observation of a sequence of events and of the corresponding output symbols. Moreover, only limited knowledge of the system structure is necessary since the sets of transitions and of places are obtained by the proposed identification algorithm on the basis of a limited online DES observation. In addition, the identification algorithm is general and does not impose any restrictive hypotheses on the succession of the observed events as in the incremental synthesis method proposed by Meda-Campaña and E. López-Mellado [35]. Finally, from the computational complexity point of view, the authors point out that the number of the unknowns of the ILP problem is linear in the number of places and transitions and in the length of the observed firing sequence.

A problem related to the PN identification is studied in [50] by Sreenivas. Given a measure of size of a labeled PN, the author questions the existence of a procedure that takes as input a description of an arbitrary, labeled PN, and returns a description of a language equivalent but possibly different with the smallest possible size. In [50], the author investigates the existence of such minimization procedures for a variety of measures but concludes that these procedures cannot exist for the considered measures. However, for families of PN languages in which controllability is decidable, there can exist minimization procedures for a restricted class of measures. Unfortunately, such procedures are intractable. It is shown in [50] that this is already the case for the family of bounded PN languages, that coincides with the regular languages (for which there exist minimal automata), and a fortiori for any family of PN languages including the regular languages.

Other issues related to PN identification have been recently studied by Ru and Hadjicostis in [49] and by Li *et al.* in [51]. In particular, Li *et al.* [51] assume that each transition in the given net has a nonnegative cost which could represent its likelihood (e.g., in terms of the amount of workload or power required to execute the transition). A recursive algorithm is proposed that, given the structure of a labeled PN and the observation of a sequence of labels, finds the transition firing sequence(s) that has (have) the least total cost and is (are) consistent with both the observed label sequence and the PN. The complexity of the procedure is polynomial in the length of the observed label sequence and is thus amenable to online event estimation and monitoring.

10.4.4 Related Problems in the PN Framework

A series of problems related to the identification and synthesis of PNs and based on the theory of regions, have appeared in the literature and deserve to be mentioned.

In [30] Cortadella *et al.* present a method which, given a finite state model, namely a deterministic transition system, synthesizes a safe, place-irredundant labeled PN with a reachability graph bisimilar to the original transition system.

In [32] Carmona *et al.* provide an efficient synthesis approach for concurrent systems. An algorithm for the synthesis of bounded PNs from deterministic transition systems, based on the theory of general regions, is presented. A bounded PN is always provided in case when the given transition system is isomorphic to the reachability graph of some bounded PN with injectively labeled transitions. Otherwise, the algorithm always succeeds to synthesize a labeled PN with bisimilar behavior, where each event may be split to several transitions. Starting from the theory and algorithms for synthesizing safe PNs in [30], the authors extend them to k-bounded PNs by generalizing the notion of excitation closure from sets of states to multisets of states. The extension covers the case of k-bounded PNs with weighted arcs, as well as the heuristics needed to guide event splitting in case when the synthesis of labeled PNs is actually needed.

In [34] given a set of firing sequences, Lorenz and Juhás answer the question whether this set equals the set of all executions of a PN. They propose a definition of regions for a partial language. Moreover, given a partial language of firing sequences, they prove a necessary and sufficient condition (based on regions) for the partial language to be the partial language of executions of a PN.

In [40] Bergenthum *et al.* present an algorithm to synthesize a finite PN from a finite set of labeled partial orders (a finite partial language). This PN has minimal nonsequential behavior including the specified partial language. Consequently, either this net has exactly the nonsequential behavior specified by the partial language, or there is no such PN. They finally develop an algorithm to test whether the synthesized net has exactly the nonsequential behavior specified by the partial language.

In [31] Bergenthum *et al.* present an algorithm to synthesize a finite unlabeled PN from a possibly infinite partial language, which is given by a term over a finite set of labeled partial orders using operators for union, iteration, parallel composition and sequential composition. The synthesis algorithm is based on the theory of regions for partial languages presented in [34] and produces a PN having minimal net behavior including the given partial language. The algorithm uses linear programming techniques that were already successfully applied in [40].

10.5 PROCESS MINING AND WORKFLOW PROBLEMS

The term *process mining* refers to approaches that capture the behavior of a structured process from a set of real executions, called *logs*. Actually the problem statement is very similar to, or even undistinguishable from, the one described in the identification approaches. However the application field is different since process mining is mainly devoted to workflow management processes, namely processes that are case driven. This is the reason why the more specific term of *workflow mining* is often used. The aim of workflow mining is to derive "good" workflow models using as little information as possible.

The workflow log on which the process mining is based can provide either only minimal information on the events traced in the log, namely the case identifier and the task identifier, or provide additional detailed information such as the time stamp for each event, the type of events, or the identity of the user generating the event (if any).

In [52–54] a process mining approach dealing with a special class of PNs specifically tailored toward workflow processes, called workflow nets (WF-nets), has been presented. Under the assumptions of no noise and "sufficient" information in the workflow log, the authors investigate whether it is possible to rediscover the workflow process, namely for which class of workflow models it is possible to accurately construct the model by just looking at the logs. It is shown that it is impossible to rediscover the class of all WF-nets, however an algorithm, called the α algorithm, is proposed that can successfully rediscover a large class of practically relevant WF-nets. The α algorithm is based on four ordering relations which can be derived from the log and constructs the "simplest" WF-net that generates the desired behavior.

The assumptions made in [54] are quite restrictive. In fact, in real cases the workflow model is very large and the workflow log can never contain all possible combinations. Moreover, the detection of some sequences of events with low probability is problematic. Finally, the workflow logs typically contain noise, thus the data they provide can be incorrect or incomplete, or can refer to exceptions. To overcome these problems, some heuristic mining techniques, less sensitive to noise and to the possible incompleteness of logs, have been presented. In [55–57] the heuristic approaches consist of three mining steps: (i) starting from the workflow log, one builds a dependency/frequency table (D/F-table) that contains the frequency and the degree of dependency of a task with respect to the other tasks; (ii) using simple heuristics, a basic relations table (R-table) is extracted from the D/F-table; and (iii) starting from the R-table, a WF-net is built using the α-algorithm introduced in [52] and [53]. The testing of these heuristic approaches has given good results in the presence of four different types of noise generating the following situations: (1) delete the head of an event sequence, (2) delete the tail of an event sequence, (3) delete a part of the body, and (4) interchange two randomly chosen events.

In [58] van der Werf *et al.* consider the problem of process discovery. Given a finite set of execution traces of an information system, that is, a log, they construct a model (in their case a PN) of the processes within this information system. They restrict their attention to the control flow, that is, they focus on the ordering of activities executed, rather than on the data recorded. The approach consists of applying results obtained from the theory of regions. However, in the framework of process discovery, the authors want a PN whose behavior does not correspond exactly to the given log, but may be larger since the log does not represent all possible behaviors. The authors first show that a straightforward application of the theory of regions would lead to a PN with a set of places as large as the log. Then they discuss how this method can be merged with generally accepted concepts from the field of process discovery in order to generate smaller PNs. Finally they present an implementation of their approach in ProM (Process Mining framework) [59], a completely plugable environment for process discovery and related topics.

In [60] Carmona *et al.* present a new method for the synthesis of PNs from event logs in the area of process mining. The method derives a bounded PN that overapproximates the behavior of the event log. The algorithm for the synthesis of bounded PNs is based on the theory of regions. The most important property is that it produces (mines) a net with the smallest possible behavior including the event log, that is, all traces in the event log are feasible in the PN (but the PN may accept extra traces not in the log). In addition, a minimality property is demonstrated on the mined PN: no other net exists that both covers the log and accepts less traces than the mined PN. The methods presented can mine a particular k-bounded PN, for a given bound k. Finally, the theory has been incorporated in a synthesis tool, used to test the approach.

A paper that shares common goals with [60] has been written by Bergenthum *et al.* [40]. In this work, they present two process mining strategies based on regions of languages. The strategy differs from the one adopted in [60]: integer linear systems are solved first in order to precompute a set of special places, called feasible places, namely all places that do not forbid the firing of any trace included in the event log. Then places are picked from this bag and progressively incorporated to the synthesized net. The more places are added, the more traces are forbidden in the net. If all feasible places are incorporated to the net, then the minimality property is obtained (the net has the smallest possible behavior including the event log). However, the set of feasible places might be infinite. On the contrary, in [60], given a maximal bound k for the mining of a k-bounded PN, minimal regions of the transition system are enough to demonstrate the minimality property with respect to this bound.

We finally mention the very recent contribution by Esparza *et al.* [3] where the authors present a learning approach that relaxes the assumption that the event log is complete. In particular, their learning algorithm automatically derives queries about the executability of some event sequences. If a teacher answers these queries, the algorithm always terminates with a correct model. The authors also provide matching upper and lower bounds on the number of queries required by the algorithm, and report on the application of an implementation to some examples.

We conclude this section by mentioning a good survey on process mining approaches that can be found in [61].

10.6 CONCLUSIONS

This chapter surveys the most significant contributions on identification and synthesis of DES in the framework of automata and PNs. Particular attention is devoted to the approaches based on the theory of regions and to a more powerful approach proposed by some of the authors to deal effectively with the most general formulation of the identification or synthesis problem. A series of other surveys and related problems are also presented.

In ongoing research on model identification and synthesis of DES, a major trend is to develop computationally efficient techniques that can be applied to highly complex systems, for example, multifunction copy machines and medical diagnostic equipments, or to very large systems, for example, for constructing the workflow of an hospital. Experiments in such fields are reported in a recent book [62].

REFERENCES

[1] O. Nelles. *Nonlinear System Identification*, Berlin Heidelberg, Germany: Springer, 2001.

[2] A. P. Estrada-Vargas, E. Lòpez-Mellado, and J.-J. Lesage. A comparative analysis of recent identification approaches for discrete-event systems. *Mathematical Problems in Engineering*, vol. 2010, Article ID 453254, 21 pages, 2010. doi:10.1155/2010/453254

[3] J. Esparza, M. Leucker, and M. Schlund. Learning workflow Petri nets. *Applications and Theory of Petri nets, Lecture Notes in Computer Science*, 6128: 206–225, 2010. doi:10.1016/j.conengprac.2010.12.010

[4] M. Gold. Language identification in the limit. *Information and Control*, 10: 447–474, 1967.

[5] S. Jein and E. Kinber. Learning languages from positive data and a finite number of queries. *Information and Control*, 204: 123–175, 2006.

[6] J. Case and C. Lynes. Machine inductive inference and language identification. *Lecture Notes in Computer Science*, 140: 107–115, 1982.

[7] D. Osherson and S. Weinstein. Criteria of language learning. information and control. *Information and Control*, 52: 123–138, 1982.

[8] G. Baliga, J. Case, and S. Jain. Language learning with some negative information. *Journal of Computer and System Sciences*, 51(5): 273–285, 1995.

[9] D. Angulin. Inductive inference of formal languages from positive data. *Information and Control*, 45: 117–135, 1980.

[10] E. M. Gold. Complexity of automaton identification from given data. *Information and Control*, 37(3): 302–320, 1978.

[11] D. Angluin. Queries and concept learning. *Machine Learning*, 2: 319–342, 1988.

[12] T. L. Booth. *Sequential Machines and Automata Theory*, New York: John Wiley & Sons, 1967.

[13] J. Kella. Sequential machine identification. *IEEE Transactions on Computers*, 20(3): 332–338, 1971.

[14] A. W. Biermann and J. A. Feldman. On the synthesis of finite-state machines from samples of their behavior. *IEEE Transactions on Computers*, 21(6): 592–597, 1972.

[15] A. Nerode. Linear automaton transformations. *Proceeding of Americal Mathematical Society*, 9: 541–544, 1958.

[16] L. P. J. Veelenturf. Inference of sequential machines from sample computations. *IEEE Transactions on Computers*, 27(2): 167–170, 1978.

[17] L. P. J. Veelenturf. An automata-theoretical approach to developing learning neural networks. *Cybernetics and Systems*, 12(1–2): 179–202, 1981.

[18] M. Richetin, M. Naranjo, and P. Luneau. Identification of automata by sequential learning. *Pattern Recognition Letters*, 2(6): 379–385, 1984.

[19] C.-C. Ku and K. Y. Lee. Diagonal recurrent neural networks for dynamic systems control. *IEEE Transactions on Neural Networks*, 6(1): 144–156, 1995.

[20] S. Das and R. Das. Induction of discrete state-machine by stabilizing a continuous recurrent network using clustering. *Computer Science and Informatics*, 21(2): 35–40, 1991.

[21] S. Das and M. Mozer. Dynamic on-line clustering and state extraction: an approach to symbolic learning. *Neural Networks*, 11(1): 53–64, 1998.

[22] Z. Zeng, R. M. Goodman, and P. Smythe. Learning finite state machines with self-clustering recurrent neural networks. *Neural Computation*, 5(6): 976–990, 1993.

[23] I. Gabrijel and A. Dobnikar. On-line identification and reconstruction of finite automata with generalized recurrent neural networks. *Neural Networks*, 16(1): 101–120, 2003.

[24] S. H. Won, I. Song, S. Y. Lee, and C. H. Park. Identification of finite state automata with a class of recurrent neural networks. *IEEE Transactions on Neural Networks*, 21(9): 1408–1421, 2010.

[25] V. B. Kudryavtsev, I. S. Grunskii, and V. A. Kozlovskii. Analysis and synthesis of abstract automata. *Journal of Mathematical Sciences*, 169(4): 481–532, 2010.

[26] D. Lee and M. Yannakakis. Principles and methods of testing finite state machines—a survey. *Proceedings of IEEE*, 48(8): 1090–1123, 1996.

[27] A. Ehrenfeucht and G. Rozenberg. Partial (Set) 2-Structures—Part 1 and Part 2. *Acta Informatica*, 27(4): 315–368, 1989.

[28] E. Badouel, L. Bernardinello, and P. Darondeau. Polynomial algorithms for the synthesis of bounded nets. *Lecture Notes in Computer Science*, 915: 647–679, 1995.

[29] E. Badouel and P. Darondeau. Theory of regions. *Lecture Notes in Computer Science*, 1491: 529–586, 1998.

[30] J. Cortadella, M. Kishinevsky, L. Lavagno, and A. Yakovlev. Deriving Petri nets from finite transition systems. *IEEE Transactions on Computers*, 47(8): 859–882, 1998.

[31] R. Bergenthum, J. Desel, R. Lorenz, and S. Mauser. Synthesis of Petri nets from infinite partial languages. In: Proceedings of the 8th International Conference on Application of Concurrency to System Design, Xian, China, 2008, pp. 170–179.

[32] J. Carmona, J. Cortadella, A. Kishinevsky, L. Lavagno, A. Kondratyev, and A. Yakovlev. A symbolic algorithm for the synthesis of bounded Petri nets. In: Proceedings of the

International Conference on Application and Theory of Petri Nets and Other Models of Concurrency, Xi'an, China, 2008, pp. 92–111.

[33] R. Lorenz, G. Juhás, and S. Mauser. How to synthesize nets from languages – a survey. In: Proceedings of 2007 Winter Simulation Conference, Washington DC, 2007, pp. 637–647.

[34] R. Lorenz and G. Juhas. Towards synthesis of Petri nets from scenarios. *Lecture Notes in Computer Science*, 4024: 302–321, 2006.

[35] M. E. Meda-Campaña and E. López-Mellado. Incremental synthesis of Petri net models for identification of discrete event systems. In: Proceedings of the 41th IEEE Conference on Decision and Control, Las Vegas, NV, 2002, pp. 805–810.

[36] M. E. Meda-Campaña and E. López-Mellado. Required event sequences for identification of discrete event systems. In: Proceedings of the 42th IEEE Conference on Decision and Control, Maui, HI, 2003, pp. 3778–3783.

[37] M. P. Cabasino, A. Giua, and C. Seatzu. Identification of Petri nets from samples of their languages. *Discrete Events Dynamical Systems*, 17(4): 447–474, 2007.

[38] M. Dotoli, M. P. Fanti, and A. M. Mangini. Real time identification of discrete event systems using Petri nets. *Automatica*, 44(5): 1209–1219, 2008.

[39] A. Ghaffari, N. Rezg, and X. Xie. Design of a live and maximally permissive Petri net controller using the theory of regions. *IEEE Transactions on Robotics and Automation*, 19(1): 137–142, 2003.

[40] R. Bergenthum, J. Desel, R. Lorenz, and S. Mauser. Process mining based on regions of languages. *Lecture Notes in Computer Science*, 4714: 375–383, 2007.

[41] M. P. Cabasino, A. Giua, and C. Seatzu. Linear programming techniques for the identification of Place/Transition nets. In: Proceedings of the 47th IEEE Conference on Decision and Control, Cancun, Mexico, 2008, pp. 514–520.

[42] P. Darondeau. Deriving unbounded Petri nets from formal languages. In Proceedings of the 9th International Conference on Concurrency Theory, London, UK, 2008, pp. 533–548.

[43] T. Murata. Synthesis of decision-free concurrent systems for prescribed resources and performance. *IEEE Transactions on Software Engineering*, SE-6(6): 525–530, 1980.

[44] L. Czaja. Marking nets abstract and structured. *Lecture Notes in Computer Science*, 222: 181–202, 1986.

[45] J. Esparza and M. Silva. Top-down synthesis of live and bounded free choice nets. *Lecture Notes in Computer Science*, 524: 118–139, 1991.

[46] P. Graubmann. The construction of EN systems from a given trace behaviour. *Lecture Notes in Computer Science*, 340: 133–153, 1989.

[47] K. Hiraishi. Construction of a class of safe Petri nets by presenting firing sequences. *Lecture Notes in Computer Science*, 616: 244–262, 1992.

[48] A. Bemporad and M. Morari. Control of systems integrating logic, dynamics and constraints. *Automatica*, 35(3): 407–429, 1999.

[49] Y. Ru and C. N. Hadjicostis. State estimation in discrete event systems modeled by labeled Petri nets. In: Proceedings of the 45th IEEE Conference on Decision and Control, San Diego, CA, 2006, pp. 6022–6027.

[50] R. S. Sreenivas. On minimal representations of Petri net languages. *IEEE Transactions on Automatic Control*, 51(5): 799–804, 2006.

[51] L. Li, Y. Ru, and C. N. Hadjicostis. Least-cost firing sequence estimation in labeled Petri nets. In: Proceedings of the 45th IEEE Conference on Decision and Control, San Diego, CA, 2006, pp. 416–421.

[52] W. M. P. van der Aalst and B. F. van Dongen. Discovering workflow performance models from timed logs. *Lecture Notes in Computer Science*, 2480: 45–63, 2002.

[53] W. M. P. van der Aalst, A. J. M. M. Weijters, and L. Maruster. Workflow mining: which processes can be rediscovered? BETA Working Paper Series, WP 74, Eindhoven University of Technology, Eindhoven, 2002.

[54] W. M. P. van der Aalst, A. J. M. M. Weijters, and L. Maruster. Workflow mining: discovering process models from event logs. *IEEE Transactions on Data & Knowledge Engineering*, 16(9): 1128–1142, 2004.

[55] L. Maruster, A. J. M. M. Weijters, W. M. P. van der Aalst, and A. van den Bosch. Process mining: discovering direct successors in process logs. *Lecture Notes in Artificial Intelligence*, 2534: 364–373, 2002.

[56] A. J. M. M. Weijters and W. M. P. van der Aalst. Process mining: discovering workflow models from event-based data. In: Proceedings of the 13th Belgium-Netherlands Conference on Artificial Intelligence, Amsterdam, The Netherlands, 2001, pp. 283–290.

[57] A. J. M. M. Weijters and W. M. P. van der Aalst. Rediscovering workflow models from event-based data. In: Proceedings of the 11th Dutch–Belgian Conference on Machine Learning, Amsterdam, The Netherlands, 2001, pp. 93–100.

[58] J. M. E. M. van der Werf, B. F. van Dongen, B. F. Hurkens, and C. A.Serebrenik. Process discovery using integer linear programming. In: Proceedings of the 29th International Conference on Applications and Theory of Petri Nets, Xi'an, China, 2008, pp. 368–387.

[59] W. M. P. van der Aalst, B. F. van Dongen, C. W. Gunther, R. S. Mans, A. K. Alves de Medeiros, A. Rozinat, V. Rubin, M. Song, H. M. W. Verbeek, and A. J. M. M. Weijters. ProM 4.0: Comprehensive support for real process analysis. *Application and Theory of Petri Nets and Other Models of Concurrency*, 4546: 484–494, 2007.

[60] J. Carmona, J. Cortadella, and A. Kishinevsky. A region-based algorithm for discovering Petri nets from event logs. In: Proceedings of the 6th International Conference on Business Process Management, Milan, Italy, 2008, pp. 358–373.

[61] W. M. P. van der Aalst, B. F. van Dongen, J. Herbst, L. Maruster, G. Schimm, and A. J. M. M. Weijters. Workflow mining: a survey of issues and approaches. *IEEE Transactions on Data & Knowledge Engineering*, 47(2): 237–267, 2003.

[62] W. M. P. van der Aalst. *Process Mining: Discovery, Conformance and Enhancement of Business Processes*, Springer, 2011.

PART III

HUMAN–MACHINE SYSTEMS DESIGN

CHAPTER 11

ADVANCES AND CHALLENGES IN INTELLIGENT ADAPTIVE INTERFACE DESIGN

MING HOU[1], HAIBIN ZHU[2], MENGCHU ZHOU[3], and ROBERT ARRABITO[1]

[1]Defence Research and Development Canada, Toronto, ON, Canada
[2]Department of Computer Science and Mathematics, Nipissing University, North Bay, ON, Canada
[3]Department of Electrical and Computer Engineering, New Jersey Institute of Technology, Newark, NJ, USA

11.1 INTRODUCTION

With advances in robotics, computer science, and artificial intelligence (AI), more and more automated and complex systems have been adopted. One recent technology introduced and explored actively for military and commercial applications is the unmanned aerial vehicle (UAV). Due to the UAV's rapid deployability, increased communication and joint operations capability, and revolutionary potential in battlefields and environmental surveillance (in the air, sea, and on the ground), UAV systems are effective tools for command, control, communications, computers, intelligence, surveillance, and reconnaissance [1, 2]. With the greater intelligence and autonomy of unmanned vehicles used in battlefields, these systems can supplement traditional manned forces as force multipliers. The United States Congress has mandated that up to one-third of future military systems should be unmanned by 2015 [3]. A key benefit of UAVs is their ability to carry out missions without casualties that would otherwise endanger human life (e.g., suppression of enemy air defense or reconnaissance in contaminated areas) [4].

Although current UAV systems have the characteristics of autonomous systems, they are in fact remotely controlled by human operators through various control

Contemporary Issues in Systems Science and Engineering, First Edition.
Edited by MengChu Zhou, Han-Xiong Li and Margot Weijnen.

stations [1–7]. The functions of control stations include: (1) receiving wireless telemetry data from the UAV; (2) processing incoming data; (3) displaying the UAV's status; and (4) supervising navigation at waypoints. A control station can be put on the ground or on another mobile carrier (e.g., ground vehicle, airborne platform, or vessel). A control system has a variety of features and settings and therefore requires an operator to be aware of all relevant data and be able to exercise control at all times [8, 9]. Clearly, a knowledge-intensive UAV control system involves extremely high workloads for operators. This situation can be exacerbated by system time delays and reduced operator situation awareness due to extensive information processing demands. For example, the RQ-1 Predator UAV mishap on April 18, 1999 partially resulted from losing situation awareness. In that mishap, the Predator experienced aircraft icing, leading to a loss of engine power. Although the UAV pilots performed recovery procedures, they became too focused on rarely encountered severe weather conditions to be able to recover, resulting in the loss of the UAV [5, 10].

Incidents such as the RQ-1 Predator UAV mishap indicate that current UAV control stations cannot provide adequate information about system states and operator needs due to the complex and dynamic nature of information management for UAV control. This situation presents the challenge of how to truly integrate knowledge of human capabilities and limitations into human–machine systems (HMSs) design, and affects systems engineering and concept development processes. Since operator interfaces play a critical role in the management of enormous amounts of data and information in complex and dynamic environments, a natural alternative solution to the challenge is to develop innovative and advanced interface technologies and associated design techniques and methods. A properly designed interface can increase operator situation awareness [5, 11] and thus provide the right information to the right person in the right format at the right time. One way to increase situation awareness is to employ software agents to take over some of the operator's tasks. These agents can act as automation assistants and acquire knowledge of mission goals, operator goals and states, and environmental states. With regard to agent autonomy, an agent may perform a function automatically, ask permission to act, provide needed information and suggestions, ask for clarification, or relinquish control to operators. Through the interface, operators can delegate responsibilities to these automation agents and focus exclusively on the crucial tasks that need human supervision. In a more general context, an agent can be compared to a representative authorized to act on one's behalf (e.g., a travel agent will book flight tickets and hotels for a customer). It is argued that an agent-based architecture increases operator situation awareness and facilitates the processes of knowledge acquisition, attention, reasoning, and decision-making through adaptation, allocation, and automation delegation [12, 13]. Agent systems have been widely used in the software engineering and AI fields, and an agent-based approach is well suited to assist operators using complex and dynamic HMSs.

How operators interact with various levels of automation through an interface to enhance overall HMS performance is domain dependent [14, 15]. Sheridan and Verplank [16] proposed a human–automation interaction taxonomy from a human factors (HF) perspective that categorizes interaction into 10 levels from no automation assistance to partial assistance or full automation. Sheridan and Verplank's taxonomy does not address who monitors the interaction. On the other hand, Endsley and Kaber's [17] taxonomy discusses not only who monitors the interaction, but also who initiates it (i.e., agent or human). Their taxonomy considers interaction in four stages of information processing: acquisition, analysis, decision-making, and action. However, from a human–computer interaction (HCI) perspective, the process of decision-making stretches as a continuum between adaptive systems and adaptable systems. According to the spectrum of adaptivity [18], adaptive systems consist of agent/machine-initiated adaptivity and adaptable systems consist of human-initiated adaptivity. The spectrum has five levels of human involvement and authority. Interaction levels are based on the amount of system status information given to the user and how much control the user and agent have over the initiation of interaction. It is obvious that there are overlapping areas in human–automation interaction among these taxonomies, demonstrating a lack of research integration between the HF and HCI communities. Therefore, a pressing need exists to develop a coherent body of knowledge that integrates these conceptual and design approaches using consistent and unambiguous terminologies. Advances in one domain can benefit the other by working together to address the lack of established design guidance for advanced operator interfaces.

As illustrated in Figure 11.1, a typical HMSs development procedure involves a four step system development processes: conception, analysis, design, and implementation. First, theoretical frameworks need to be reviewed and modified to meet the needs of a project plan. Specific design concepts can then be developed to capture the scope. Second, analytical approaches need to be selected in order to capture the interface display, communication and control requirements, and the task and functional decomposition of the domain conceived for the system. This step also involves identification of means to capture more detailed knowledge from

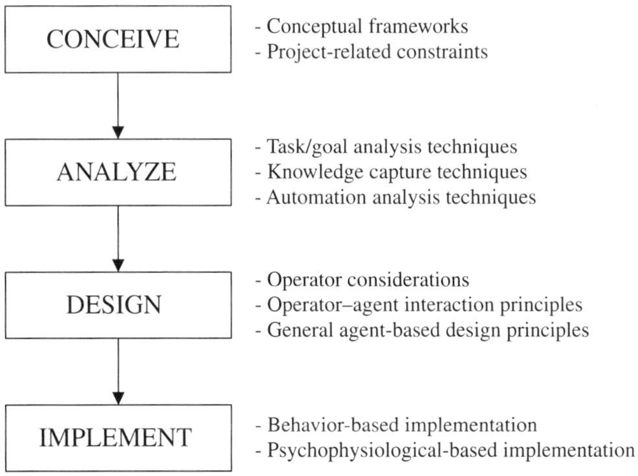

FIGURE 11.1 Overview of a human–machine systems design procedure.

subject matter experts for function allocation between automation and users. Function allocation assigns the performance of each function to the element (e.g., humans, hardware or software) that is best suited to perform the function [19]. Third, when considering the improvement of human–automation interaction quality, the system design should follow general design principles from a multiagent perspective, and consider how to facilitate operator–agent interaction and thus maximize overall HMS performance. Fourth, in order to provide information about the objective and subjective states of the user within a mission or task context, behavior-based and/or psychophysiological-based implementation techniques need to be identified for the analysis of the psychological, physiological, and behavioral states of the user. The goal of the system is to adapt to changes in mission status, system status, environmental conditions, and user states. As with knowledge of the external context, information about the internal context (i.e., the user) provides the basis for an interface to support the user and achieve system goals.

This chapter presents a unified agent-based design framework and methodologies to guide operator interface design for complex HMSs (e.g., a UAV control station). Through the facilitation of operator–agent interaction, an agent-based interface is expected to increase operator situation awareness and reduce workload, thus reducing personnel requirements and enhancing operational performance.

The remainder of this chapter is arranged as follows: Section 11.2 discusses the evolution of interface technologies and describes the concepts and associated conceptual framework of intelligent adaptive interfaces (IAIs), while defining relevant nomenclature to minimize cross-discipline term confusion. Section 11.3 discusses IAI design challenges, alternative solutions, and analytical approaches, along with examples of empirical studies. Section 11.4 introduces the concept of adaptive intelligent agents (AIAs) and an operator–agent interaction model to guide IAI system design by understanding the differences among human–human, human–machine, and operator–agent interactions. Section 11.5 describes a hierarchical IAI system architecture with different levels of AIAs to facilitate operator–agent interactions. Section 11.6 illustrates the utility of the IAI framework and associated design methodologies through an IAI design example. Finally, Section 11.7 concludes the chapter and discusses future work.

11.2 EVOLUTION OF INTERFACE TECHNOLOGIES AND IAI CONCEPT

The maturation of computers, communication, AI, and other advanced technologies has led to increased automation in nearly all aspects of society. For instance, through the use of AI technology, software systems are gaining the ability to reason and make decisions on their own. This trend has triggered a shift in the human operator's role from largely perceptual/motor tasks (e.g., controlling a vehicle by directly manipulating a stick or wheel) to tasks that require more cognitive processing (e.g., monitoring, reasoning, judging, decision-making, and managing automated or semi-automated system operations). Consequently, the design of effective human–machine interfaces becomes ever more critical to overall system performance if both operators and the system have to adapt to shifting roles. However, conventional interfaces do not meet the interaction requirements imposed by complex HMSs and often fail to properly reflect users' tasks, goals, plans, and problem domains [20–22]. In addition, there lacks clarity about what an interface requires for aiding decision-making and facilitating interaction with users. Also, there is very little guidance on how to evaluate the effectiveness of interface design. This is especially true for complex HMSs [23] such as UAV control stations, where information flow is dynamic, operators have excessive workload, and situation awareness is crucial to the accomplishment of the mission [24]. This section discusses the evolution of interface technologies, defines the IAI concept in the human–machine interaction domain, and identifies the essential components of a generic IAI system.

11.2.1 Evolution of Interface Technologies

Before we discuss interface technologies, it is necessary to examine the overlaps existing between interface terminologies. Human–computer interface, also known as user interface, human–machine interface (HMI), also known as man–machine interface (MMI) and user interface, and operator–machine interface (OMI), also known as operator interface, all have different formal definitions, but are used loosely to describe the same interface concepts. The following are the formal definitions of "interface" that can be applied to IAI design:

- A human–computer interface can be described as the point where the human interacts with the computer. It is also known as a "user interface". Human–computer interface design seeks to discover the most efficient way to design understandable electronic messages [25, 26]. A user interface provides an input and output. An input allows the user to enter information into the system, whereas an output indicates the effects to the user as a result of user input. For example, a keyboard is an input device, and a Windows™ computer application that displays typed text is an output device.
- An HMI allows interaction between the user and the machine. According to Yu *et al.* [27], the terms HMI and MMI are typically used in industrial and process control applications, whereas "user interface" is the common term in business information systems.
- OMI is the aggregate of means by which a human operator interacts with the machine. The OMI provides the means of input (i.e., allowing the operator to manipulate the machine) and output (i.e., allowing the machine to produce the effects of the operator's manipulation) [28].
- Despite these different definitions of "interface," the term has a singular meaning (i.e., the means by which humans can interact with technology/tool). Because each definition arises from a different domain, definitions are derived from different contexts. Within an HMS, the OMI is one component of the overall system. An OMI is a means that allows the human to interact with the system.

For example, a computer mouse, keyboard, display, and operating system are individual components of a computer interface. As a whole, it is also known as an interface. Interface technologies exist in a number of forms, such as static interfaces, adaptive/intelligent interfaces, and IAIs. Now let us describe the evolution of interface technologies.

11.2.1.1 Conventional Operator–Machine Interfaces In their most general form, conventional OMIs are mediums that support operator interaction with a machine, device, computer program, or other complex tools. The OMI facilitates interaction between the operator and the machine. For example, the Windows operating system is a computer interface that allows one to manipulate the system and perform tasks. Windows takes a user's inputs and produces outputs to respond to the inputs. To do so, Windows needs to have knowledge relating to users' tasks (i.e., task model). Tasks are composed of activities that are made up of operations. A *task model* is the system's knowledge of likely operations an operator performs. As shown in Figure 11.2, task models are technology centered, and focus on the machine rather than the human. A technology-centered OMI design approach emphasizes how technology advances can help allocate more tasks to automation. The purpose is to reduce workload, increase task efficiency and accuracy, and bring in more convenience (e.g., ATM and mouse). In a task model, automation functions are pre-planned and pre-designed. Automation capability derives from two principles: the *left-over* principle and the *compensatory* principle. The former states that operators are left with functions that have not been automated, or could not be automated. The latter states that function allocation is based on the strengths and weaknesses of the human and machine [29]. With the advance of computerized aiding technologies, more and more tasks are being allocated to automation. However, an operator may have difficulty maintaining situation awareness due to the static nature of conventional automation and OMI [30, 31]. For example, when driving with cruise control for long distances, a driver's mind may wander. The driver may become relaxed and less sensitive to his/her surroundings.

To maintain and increase situation awareness, an HMS needs to know how a human operator is doing. Interface and automation technologies need to be flexible to adapt to changing operator states. The need to assist the operator in a flexible fashion has fostered intelligent interface and adaptive interface technologies. To assist the operator, the system must trigger automated aid based on one, or the combination of three conditions: (1) evaluate a task condition, (2) access operator stress levels, and (3) access operator work levels. These three conditions depend on a *user model*. A user model is the system's knowledge of the capabilities, limitations, needs, and demands of the human. Thus, interface technology evolves into intelligent interface and/or adaptive interface technology by combining the task and user models. The design of such technologies is based on user-centered design (UCD) principles. Designers must have a deep and detailed knowledge of their potential users. UCD ensures that HMSs do work and that people can use them. Good UCD can lead to a safer HMS that will mitigate potential risks.

11.2.1.2 Intelligent Interfaces and Adaptive Interfaces With embedded task and user models, an interface can be personalized to enhance user interaction with a system by making the system more efficient, effective, and easy to use [32, 33]. An intelligent interface can match its content to changing task-related circumstances. For example, depending on the type of a menu, each menu a user selects in a word processor will have different contents. The system controls whether an adaptation occurs, and how it occurs, along with the amount of adaptation that occurs. In addition, the user has control over how the system adaptation is initially configured. Keeble and Macredie [34] define an intelligent interface as "an interface where the appearance, function or content of the interface can be changed by the interface itself in response to the user's interaction."

However, an intelligent interface does not always make decisions based on external contexts. For example, a Global Positioning Satellite (GPS) navigation device for vehicles takes drivers through the shortest possible route to reach their destinations. Without knowing changes in external events and without built-in intelligence, many GPSs cannot process external environmental factors such as traffic jams and constructions to automatically prescribe a new route for a driver. The driver has to manually select an alternate route through the GPS interface. In other words, an intelligent interface does not have a model for external environments, nor the adaptation intelligence to deduct logical actions based on an external context.

Originally, intelligent interfaces were known as adaptive user interfaces [35–39]. Within the context of human–machine interaction, Powers *et al.* [40] and Sternberg [41] defined intelligence as the ability to perceive, understand, reason, and infer relevant changes in environments, and adapt accordingly. Tomlinson *et al.* [42] defined adaptive interfaces as interfaces that seek to predict what features are desirable and customizable by users. In other words, an adaptive interface is a flexible tool that intends to customize systems for individual users [43]. Customization enhances usability between human operators and systems by matching contents to changing circumstances, which results in increased efficiency and effectiveness. For example, speech recognition tools convert spoken words into computer-readable language. They are commonly used in telephony services as a way for customers to interact and receive help from an organization. By uttering commands such as "account balance" or "billing inquiries" to a machine through a telephone, it is possible to receive appropriate information pertaining to the subject at hand. The system is able to recognize and adapt to the user's spoken input in order to respond back with relevant information. However, there are many times when speech recognition fails due to noisy environments or speakers' accents. Without knowledge of the noisy external environment and the ability to discriminate between language and noise, the speech recognition system cannot recognize speech input in a noisy environment. Thus, filtering out external noise is necessary for the interface to provide seamless instructions.

Despite the differing definitions of an intelligent interface and an adaptive interface, their shared goal is to satisfy users' needs by making the interface more effective, efficient, and easy to use. In other words, they use a user-centered approach

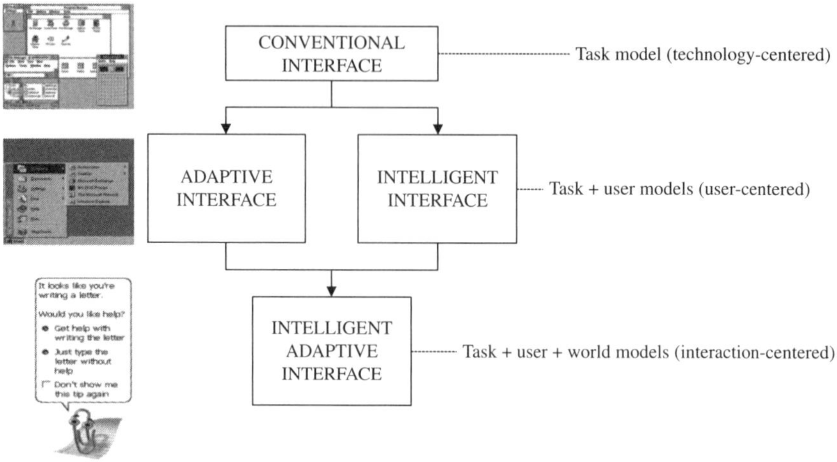

FIGURE 11.2 Evolution of interface technologies.

to interface design, as illustrated in Figure 11.2. Where one adjusts the content output based on changing situations of the task, user, and environment (i.e., adaptive), the other decides when, what, how, and the amount of adaptation that is needed (i.e., intelligent). To do so, more models (e.g., world models) beyond task and user models are needed to reflect the entire picture of the relations among the task, user, system, and environment. With the knowledge of the task, user, system, and environment and built-in adaptation mechanisms, an HMS can intelligently adapt to the changes in all four of these domains.

11.2.1.3 Intelligent Adaptive Interface (IAI) Obviously, the environment is another critical element of a HMS. As such, the dynamic nature of complex environments makes deciding the correct type of interface a challenge. The right type of interface must adapt to rapidly changing tasks (e.g., UAV route replanning), operator cognitive states (e.g., information overload), and working environments (e.g., weather changes). For example, a jetfighter radar system must communicate with satellites, ground control, the Internet, and all personnel monitoring the radar. Communication capabilities must interlink the pilot, co-pilot, and airbase like a spider's web. When technology becomes intelligent, it is able to behave dynamically, and make decisions based on its environment. As seen in Figure 11.2, the progression from conventional interface, to adaptive (i.e., intelligent) interface, to IAI is possible through combining task, user, and world models. An HMS needs more knowledge than that of task and user models in order to intelligently and dynamically allocate tasks between automation and human operators. The world model encompasses the system's knowledge of the HMS working environment. In other words, it contains knowledge about the external world, such as the activities and objects that exist in the world, their properties, and the rules that govern them [44]. For example, a world model allows the HMS to monitor mission situations and environmental changes, calculate real-time decisions, and then relay the external status to the human operator.

Interface technologies with both intelligent and adaptive capability needs evolve into unified IAIs. As indicated in Figure 11.2, IAIs seek to enhance HMS performance by using technologies that intelligently adapt to interface and task automation. IAIs also support the operator within external and internal contexts. A system monitoring the task environment is an example of an external context. A system monitoring system status and the operator's mental state is an example of an internal context.

Thus, an IAI can be defined as an interface that dynamically changes its control, and displays characteristics that react in real time to task, user, system, and environmental states [45]. IAI design takes an interaction-centered approach, with the goal being to facilitate users' interaction with an HMS and improve overall system performance and efficiency. The Microsoft Office Assistant epitomizes IAIs. This feature was included in Microsoft Office 97 and subsequent versions until Office 2007. The most well-known office assistant was named "Clippy" after its animated paperclip representation. Clippy presented various help search functions and offered advice based on mathematical algorithms (i.e., intelligence). It would be turned on when the program thought the user required assistance. Also, Clippy used adaptation to modify the formatting of document and menu content accordingly. For example, typing an address followed by "Dear" would prompt Clippy to open and state, "It looks like you're writing a letter. Would you like help?" The embedded algorithms used a combination of task models (e.g., how a letter is formatted), user models (e.g., how many mistakes the operator makes when writing a letter), and world models (e.g., what the context of the task and what the user is trying to do) to modify the interface to match the user's needs and requirements. Clippy can be deactivated by the user. Otherwise, the system dynamically controls the amount of adaptation that occurs.

Understanding the user's capabilities and limitations within assigned tasks and their context is important for IAI design. A typical IAI is driven by software agents (i.e., automation) that intelligently aid decision-making and operator actions under different levels of workload and task complexity. IAIs should present the right information, present action sequence proposals, or perform actions in the correct format at the right time. Clearly, the emphasis in this definition is on intelligent adaptation.

11.2.2 A Conceptual Framework of IAI Systems

When starting an IAI design, a generic framework is needed to conceptualize high level system requirements. A framework provides a comprehensive and efficient means for IAI designs, and shows major components and their relationships. As illustrated in Figure 11.3, there are four components that make up the IAI system framework: (1) Situation Assessment, (2) Operator State Assessment, (3) Adaptation Engine, and (4) Operator–Machine Interface. Collectively, these components form the backbone of a UAV's IAI system.

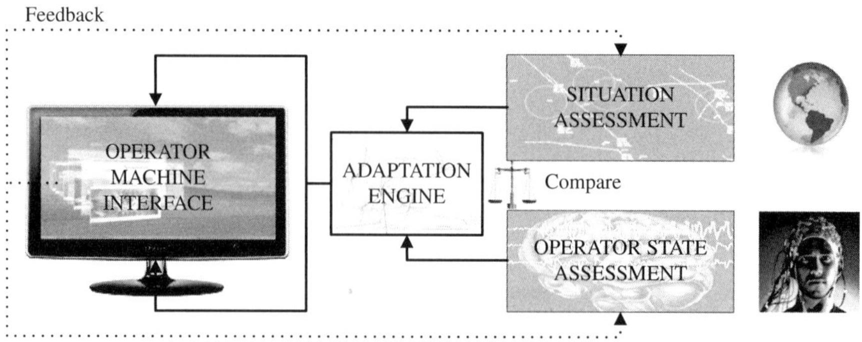

FIGURE 11.3 A conceptual framework of an intelligent adaptive interface system.

11.2.2.1 Situation Assessment The Situation Assessment component conducts real-time analysis on mission goals, system status, and the external environment. It monitors and tracks the state of the mission, whether the mission has been accomplished or changed, and changes in system status and the operational environment. For example, it will know that an aircraft has completed the mission and needs to refuel. It calculates the closest and safest landing strip and reports the results to the pilots. Other factors that the Situation Assessment component can monitor include heading, altitude, and potential threats to the aircraft. This component is also capable of analyzing tactical and operational factors. Lastly, it provides information about the state of the system within the context of a specific mission. It uses a knowledge-based system to automate tasks and support the user, which includes information about the user, the external environment, and the goals of the mission.

11.2.2.2 Operator State Assessment Perhaps one of the most critical functions in an IAI system is the ability to monitor operator state and adjust system settings accordingly. As operators become stressed, the system has to make adjustments to better accommodate them. An Operator State Assessment component monitors and analyzes the psychological, physiological, and behavioral state of the operator. Its primary function is the continuous monitoring of workload, attention and focus, visual and verbal processing loads, and intentions. From these data, it determines the level of stress and mental load placed on the operator. The component also monitors extreme high or low levels of arousal. Overall, it provides information about the internal and external states of the operator within a specific mission context. For example, if a pilot were to fly through a dangerous zone, it would anticipate increased heart rate, stress, and so forth. From this anticipation of increased heart rate and stress, it will accommodate the operator by making adjustments to the system. Possible adjustments could include pre-planning an escape route or emergency maneuvers to evade an enemy attack.

11.2.2.3 Adaptation Engine The Adaptation Engine uses higher order outputs from the Operator State and Situation Assessment components. It maximizes the match between the system states, tactical assessment, and the operator states provided by the two other components with built-in models of the external world, task/mission goals, and operator states. It compares them with the new inputs from the Operator State and System Assessment components, and then provides adaptation solutions for the system that react to internal and external changes. It also presents information and updates to the operator through an OMI.

11.2.2.4 Operator–Machine Interface An OMI is the gateway that the operator uses to interact with the system in order to satisfy mission goals and tasks. An interface is the bridge between the operator and the system, as discussed in Section 11.2.1. An OMI maintains communication between the operator and the system, and continually updates the status of the operator, system, and environment. Based on the information provided by the OMI, the operator maintains situation awareness and takes actions accordingly. An OMI is the only physical component of a UAV system that enables an operator to interact with the platform by remote control.

All the four components operate within a closed-loop IAI system that resamples operator state and situation assessment after the adaptation of the OMI and automation. The concept of a closed-loop system originates from perceptual control theory (PCT). In PCT, the goal is to adjust the level of adaptation so that optimal operator states (e.g., performance and workload) are maintained [8, 46]. The reality of intelligent interface technology is "indirect management" of information rather than "direct manipulation" [47]. Since an IAI is multifaceted in purpose and nature, it should include capabilities for multimedia input analysis (e.g., mouse, keyboard, spoken language, eye and head tracking, and three-dimensional gesture), and multimedia presentation generalization (e.g., windows, menus, dialogue boxes for typed or spoken language, graphics, and gesture), and it should then apply user, system, and task models to personalize and enhance interactions. To recognize indirect information management and improve the efficiency, effectiveness, and intuitiveness of human–machine interaction, it is necessary to represent, reason, and act on models of the organization, knowledge (e.g., the user, domain, task, system, and world), agent, communication (e.g., graphic, natural language, and gesture), and design. Therefore, essential models must be specified and constructed to support the development of the four components. These models are widely used in knowledge-based system design techniques and the intelligent interface technology community [44, 48–50]. They are as follows.

11.2.2.5 Domain Model A domain model provides abstract representations of domain knowledge relevant to the purpose of the mission/task. It represents the complexity, criticality, uncertainty, and environmental constraints that are particularly relevant to the Operator State Assessment component. It can also describe various entities, their attributes, their roles and relationships, and the constraints that govern the integrity of the model elements comprising that problem domain. A domain model can be thought of as a conceptual model of a domain of interest.

11.2.2.6 Organizational Model An organizational model incorporates knowledge relating to the organizational context in which the knowledge-based system is intended to operate. For example, an organizational model implemented into a system will determine the proper knowledge model to incorporate.

11.2.2.7 Knowledge Model Knowledge models are abstract representations of knowledge in contexts such as domain, task, user, inference, and goals. A knowledge model incorporates a detailed record of the knowledge required to perform a task. For example, it may include information on operator behavior, knowledge, preferences, and needs. It also enables the system to provide advice to the operator, automate tasks, and adapt to the OMI [44].

11.2.2.8 Dialogue/Communication Model A dialogue/communication model incorporates knowledge of how the operator communicates with the system. This allows the system to understand the commands issued by the human operator. For example, if a system had voice recognition implemented, the model would be responsible for recognizing human voice commands, and turning human voice commands into actions.

11.2.2.9 Design Model A design model determines the hardware and software requirements for the creation of an IAI. It also specifies how the operator state is monitored. Its aim is to develop a detailed design for constructing IAI software using specifications from the knowledge and dialogue/communication models.

11.2.2.10 Task Model A task model incorporates knowledge relating to the tasks and functions by all agents, including the operator. It contains knowledge about the tasks being pursued, and organizes the tasks in a hierarchy of high level tasks to subtasks [44]. For example, in an aircraft cockpit, the navigation system may have a task model built into the system. The navigation system's task model can include mission and pilot parameters, the duration of the mission, the altitude the aircraft is flying at, and so on. Task models are vital to complex systems because they outline the parameters of the operator and the mission.

11.2.2.11 Agent Model An agent model identifies task participants, allowing participant responsibilities to be incorporated into the knowledge system. It incorporates knowledge relating to system participants (e.g., human and computer) and their respective roles and responsibilities. For example, it will know the specifications of the system and the limitations of the operator. From that information, it can determine the limits of the human and the system.

11.2.2.12 User Model A user model incorporates the system's knowledge about the user's behavior. It also monitors the user's knowledge, abilities, needs, and preferences. An effective user model allows the system to have a thorough understanding of the user's needs and demands, and develop a user profile. A user model also enables physiological monitoring of the user [44]. Physiological monitoring may be required to develop an effective IAI.

11.2.2.13 System Model A system model consists of the system's knowledge about itself and its ability to assist users. For example, it will determine how the system can help a user through means such as offering advice, automation, or interface adaptation.

11.2.2.14 World Model A world model contains the system's knowledge about the external world. This includes the objects that exist in the world, their properties, and the rules that govern them [44]. For example, it will specify the principles of flight control and human behavioral principles under stress. By monitoring external factors through the world model, the system can make suggestions to the operator depending on external factors. Note that the "world" model is actually a model about the external environment of the system and operators.

Although the above models consider most aspects of IAI design and implementation requirements, maximizing overall HMS performance through the interactions of an operator and agents has not been actively explored.

11.3 CHALLENGES OF IAI DESIGN, ALTERNATIVE SOLUTIONS, AND EMPIRICAL INVESTIGATIONS

The area of IAIs covers a variety of topics concerned with applying AI and knowledge-based techniques to issues of human–machine interaction. The main issues addressed by IAI research are: (a) making interactions more clear and efficient; (b) supporting users' tasks, goals, and plans; (c) presenting information effectively; and (d) designing and implementing interfaces effectively. To address these issues, this section discusses IAI design challenges, two related interface design approaches, analytical methodologies, and ends with a summary of three empirical studies.

11.3.1 Challenges of IAI Design

IAI technology is an effective means to overcome the problems that conventional interfaces cannot handle (i.e., information overflow and real-time cognitive overload). In the past decade, a number of theories were developed and tested (see Sections 11.2.1.2 and 11.2.1.3). However, there are still a number of problems that prevent this technology from being broadly applied. The challenges for IAI designs include the following [43]:

- What are the most reliable and cost-efficient IAI development methods?
- What are the universal usability principles that do not lead users' expectations astray?
- How and when can intelligence substantially improve interaction?
- How do we evaluate whether the system supports users' real tasks?
- How do we design authoring tools to enable easy development and maintenance of the intelligent parts of the system (i.e., scalability)?

Within the context of multiple-UAV control in a dynamic and complex environment, an IAI design demands better usability principles, easier ways to interact with the interface, and better tools to support the full life cycle of an HMS. In addition, the problem becomes more complicated in a networked environment during joint operations where interoperability is important and many human operators use the same information from a single UAV. A systematic approach is needed to address these issues and guide IAI design toward effective interaction between the operator and the system.

Furthermore, research on agent technologies has increased in prominence in the systems design field, which includes the use of agents to display system data and status via facial displays, multimodal communication between animated agents, and standards and open architectures for building agent-based multimodal interfaces. However, the key questions are: what can and should agents help with, and how, when and why should they interact with users? In other words, what intelligence for adaptation resides in an HMS? To answer these questions, user requirements need to be assessed, in order to understand what tasks can and cannot be allocated to agents. Before looking into the suitable tasks for agents, a UCD approach should be considered.

11.3.2 User-Centered Design Approach

UCD was initiated by Norman and Draper [51–53]. It is based on the active involvement of users to improve the understanding of user and task requirements, and the iteration of design and evaluation [54].

A typical representation of the user-centered philosophy is the intelligent interface architecture proposed by Rouse *et al.* [55]. First, the operator state is central to the functioning of interface components. The relevant elements include: activities, awareness, intentions, resources, and performance. The other component is the interface manager, similar to executive assistants who zealously guard their superior's time and resources. The user-centered principle is used as a key component in the IAI framework for guiding interface design. However, the concept of using intelligent agents only as a back-up is not applicable in a dynamic, complex, and interactively networked system such as the control of multiple UAVs. When dealing with a large amount of information in a complex mission environment, delegating automation tasks is essential for robust operator control of multiple UAVs. Otherwise, overloading of the operator occurs that will likely lead to undesirable outcomes. In this example, instead of acting as a mere back-up, automation agents should be more of a partner that proactively prevents anything from going wrong. Thus, to maximize overall system performance the operator and the intelligent agents should work as a team and share responsibilities. In other words, operator–agent interaction should resemble the behavior of human–human interaction. In this way, the IAI does not act like a passive "listener" but an active partner. The operator will not be a passive observer but an active controller. With specific task division done beforehand, mission-critical tasks will be neither missed nor duplicated.

11.3.3 Agent-Based Interface Design Approaches

For very complex systems, interface development requires high skill levels and massive time commitments [55, 56]. Thus, interface development techniques were originally conceived as software development environments [57]. These environments include software components such as windows, menus, and dialogue boxes developed using toolkits (e.g., Visual Basic and OpenGL). Unfortunately, using these toolkits often results in the mixing of interface and application code. To resolve this issue, model-based systems were introduced to divide applications into multiple layers (i.e., application actions, dialogue control, specifications of presentation and behavior, and primitive toolkit objects) to support a more declarative development process. Model-based systems (e.g., MAYA) draw upon automated input analysis and output generation techniques, and promise automated design critique, refinement, and implementation. However, these design environments neither emphasize human–computer interaction for the final interface, nor provide enough means to facilitate interaction between operators and interfaces.

On the other hand, an agent-based interface design approach emphasizes the use of AI techniques to create software that performs information filtering and other autonomous tasks for users. An agent is an entity capable of autonomous and goal-oriented behavior [58]. In other words, an agent can act as an assistant to the operator. In an HMS, agents are software applications that can take action on behalf of an operator [44]. Agents can improve situation awareness by reducing the mental load on operators. Also, they can acquire knowledge about mission goals, operator goals, operator states, and environmental states. Rogers [10] described a UAV mishap where pilots crashed a UAV due to lack of situation awareness regarding their fuel, and thus loss of engine power. It was discovered that the pilots became too focused on flying the Predator (UAV) in weather conditions they had rarely encountered. Incorporating agents into the UAV control system may have reduced operator workload and prevented the aircraft from crashing.

The essential function of embedded agents is to act as a bridge between user goals and computer capabilities. This function can be supported by making the interface more intuitive and by encouraging types of interactions that may be difficult to evoke when compared to an interface developed in a model-based environment. There are many agent-based techniques to help with interface design in the systems engineering and AI fields. Among them, the common knowledge acquisition and design structuring (CommonKADS) and explicit models design (EMD) [48] approaches have the potential to facilitate operator–agent interaction [59].

CommonKADS is a knowledge management and engineering approach for developing knowledge-based systems. It provides the guidelines to analyze and design a knowledge management system using a formalized representation of knowledge and an associated inference mechanism with the capabilities of human activities such as diagnosis, planning, and design [60]. A multiagent system extension of the CommonKADS methodology has been proposed by Iglesias *et al.* [61]. The extension adds specific agent-related constructs to the system, including those relating to: (a) interagent communication; (b) the division of tasks among individual agents; and (c) the implementation of multiagent systems.

Although these methodologies offer a framework for approaching the design and implementation of an agent-based system (e.g., an IAI), they do not provide a means of identifying and subdividing the knowledge required by IAI systems. Integrated Computer-Aided Manufacturing (ICAM) Definition (IDEF) was developed by the US Air Force [62, 63] to look at methods for analyzing and improving manufacturing operations. IDEF language allows symbolic representations of complex constraints that cannot be depicted using schematic language (e.g., unified modeling language (UML)) alone. Thus, IDEF permits flexible modeling of temporal concepts and formal logical representations of process constraints, and allows for precise specification of event timing and duration. Consequently, integrating the above approaches and combining the strengths of the individual components will provide an opportunity to form a comprehensive approach for IAI system designs and solve the above mentioned problem of CommonKADS.

As another design approach, EMD promises to offer a means of subdividing the knowledge content of CommonKADS into five distinct and interacting components: task, user, system, dialogue, and world [64]. By making explicit the knowledge required by IAI systems, EMD has the potential to determine the goals an operator is trying to achieve, the plans for achieving those goals, and how the system can most effectively assist the operator. Its challenges lie in how to decompose knowledge into the various models and how to coordinate knowledge among the models to build effective supporting systems. To tackle these challenges, a technique called hierarchical goal analysis (HGA) provides the mechanisms for decomposing user and system goals and recognizing user plans based on observed actions [65]. Thus, these techniques have been integrated in EMD for programmatic implementation of the method.

Another human–computer interface design technique, ecological interface design (EID), is well suited to problems of controlling and monitoring complex systems such as the control stations of UAVs. EID was developed at the Risø National Laboratory in Denmark by Vicente and Rasmussen [66] as a framework for problem domain analysis and the design of human–machine interfaces in complex work environments. The name reflects the incorporation of elements from ecological psychology, particularly emphasizing the importance of human interaction with their environment. While most traditional interface design approaches confine their attention to human characteristics, EID also examines how humans interact with their surroundings; taking into account both physical and cognitive factors in the context of the complex system under control. To achieve this, EID offers concrete guidelines on interface design, with the aim of producing optimal usability and safety. EID is a set of techniques for constructing safe and reliable user interfaces based on levels of cognitive control. It improves fault diagnosis speed and accuracy and creates more flexible and robust control strategies [67].

EID has been applied in a variety of areas, including nuclear power plants, industrial process control, and in the medical and air domains [68–73]. It recognizes that system reliability depends on not only the engineering of that system, but also how humans interact with it. For that reason, the content and structure of the interface must be designed to facilitate the operator's understanding of the system being controlled. Traditional ergonomic design techniques govern how controls are situated, clearly displayed, and made intuitive to use; an ergonomic approach can reduce the likelihood of execution errors by the operator, such as manipulating one control when the intention was to adjust a different one. However, without considering the cognitive processes of the operator, ergonomic techniques fall short at preventing errors of intention, such as when a user deliberately adjusts a control for the wrong reason. EID aims to minimize errors of intention by incorporating cognitive factors into the interface that take into account how people make decisions and analyze problems. Thus, EID can support collaboration with multiagent systems and contribute to an IAI design that improves situation awareness and HMS effectiveness in complex situations.

Since the details of these techniques are beyond the scope of this chapter, they will not be discussed here. However, the methodologies for identifying and subdividing the required knowledge for IAI system design provide a foundation for an IAI framework that focuses on operator interaction with the machine and the working environment. The integration of the above techniques and methods into a comprehensive, cross-disciplinary interface design approach should serve the goals of reducing operator workload and manning requirements while generating a robust, maintainable, and reliable system.

11.3.4 Analytical Methodologies

To build a complex system like an IAI, high level design principles must be combined with thorough analytical methodologies. These methodologies are used to analyze and build various models (as discussed in Section 11.2.2). The models represent system components or functions at different operational levels, and should enable system functions to:

- Modify OMIs to handle the interaction and dialogue between an operator and system components or software agents.
- Track the operator's goals/plans/intents (and progress toward them).
- Monitor the operator state.
- Monitor the world state.
- Provide knowledge of the effects of system advice, automation, and/or adaptation on the operator and world states (i.e., closed-loop feedback).

In addition, the relationships among various methodologies and models (Figure 11.4) can be mapped back onto the four components and 10 models of a generic IAI system (Section 11.2.2). These models are necessary to effectively facilitate operator–machine interaction, and provide the following benefits: (1) The user model enables the monitoring of the operator's behavior, physiological, and psychological states. (2) The task, system, organization, domain, and world models enable the

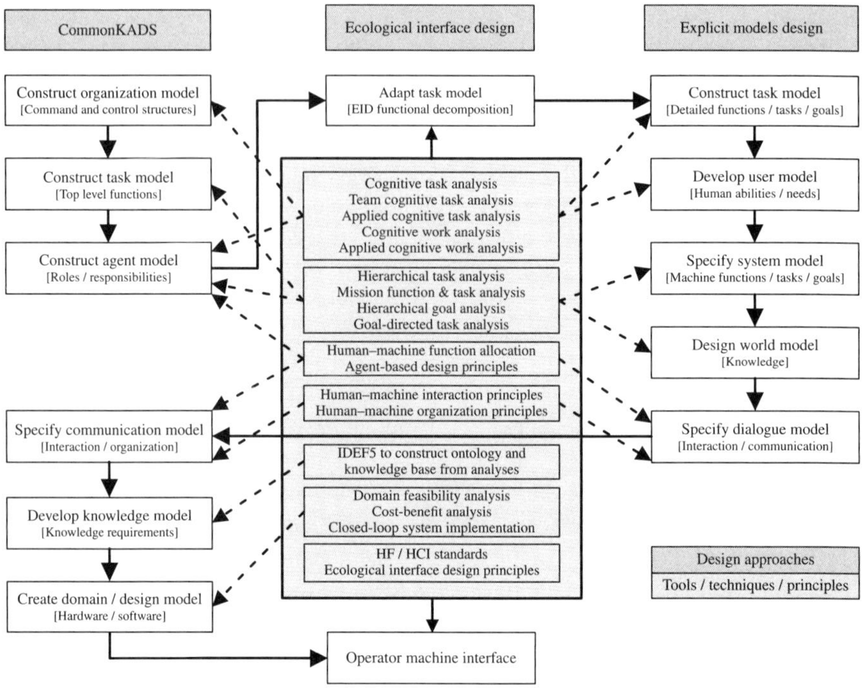

FIGURE 11.4 The analytical process for the development of intelligent adaptive interfaces.

monitoring of mission plan/goal completion, tasks/activities, and entities/objects in the external environment. (3) The design and knowledge models enable the machine to provide advice to the operator, automate tasks, or adapt to the OMI. (4) The dialogue/communication model enables effective interaction between the system and the operator. The generation of these models is one of the implementation steps of the generic conceptual framework. Figure 11.4 also illustrates various analytical approaches that can contribute to the creation of different models, where solid arrows mean process flows and dashed arrows mean tasks for the specified process. The various analytical approaches and their contributions are as follows:

- *Cognitive Analysis Methodologies* contribute to the construction of organization, task, agent, and user models.
- *Task Analysis Methodologies* contribute to the construction of task, agent, system, and world models.
- *Human–Machine Function Allocation and Agent-Based Design Principles* contribute to the construction of agent and dialogue/communication models.
- *Human–Machine Interaction and Organization Principles* contribute to the construction of dialogue/communication models.

- *IDEF5 Guidelines* contribute to the construction of the ontology and knowledge base. This is then used to enumerate the knowledge captured by the analysis process.
- *Domain Feasibility, Cost–Benefit Analysis, and Principles for Closed-Loop Implementation* contribute to the construction of the domain and design models, including the means by which operator state is monitored.
- *Human Factors and Human–Computer Interaction Principles* contribute to the construction of the OMI and related systems. The design process might also include principles from EID.

Most analytical approaches are generic (i.e., context independent) and scalable. Thus, selection of these analytical tools is less critical as they can be (and sometimes must be) modified to suit the domain. Furthermore, these analytical approaches can be combined to present their strengths and mitigate their weaknesses. Several criteria can be used to determine which analytical and design tools, techniques, and methodologies should be used to design a specific IAI.

- *Project constraints*: schedule and budget.
- *Domain*: complexity, criticality, uncertainty, and environmental constraints (particularly relevant to the choice of operator state monitoring systems).
- *Operator*: consequences of error and overload, what kind and quantity of support is needed, and who needs to be in control (particularly relevant in combat domains).
- *Tasks*: suitability for adaptation, assistance, or automation.

11.3.5 Empirical Investigations

Although the concept of IAI has been around since the early 1990s [74], validating the advantages of IAI over other interfaces is still needed because there have been few empirical investigations of IAI technologies [75]. Here, three relevant studies are considered as investigations of IAI technologies and are summarized below.

Bennett and Hettinger [76] conducted a preliminary study of a dynamically adaptive interface (DAI) in the domain of aviation (i.e., precision, low level navigation). According to the study, a DAI should change display and/or control characteristics of a system in real time, anticipate the information needs of users, and provide that information without requiring explicit user control input. Standard (i.e., conventional controls and displays), candidate (i.e., alternative controls and displays), and adaptive (i.e., dynamic between the standard and candidate displays) interfaces were evaluated. Results indicated that significant performance advantages in the quality of route navigation were obtained with the candidate and adaptive interfaces compared to the standard one. However, there were no significant differences between the former two. The implication of this study is that adaptive interfaces have the potential to improve overall HMS performance if they are properly designed. But if designed improperly, adaptive interfaces have the potential to degrade system performance by:

(a) preventing the development of automatic processes in the operator; (b) presenting irrelevant information; or in the worst-case scenario; and (c) eliminating information that is currently needed. The implications of these results raised the issue of using IAI technologies for automation and adaptation.

To address the above issue, the key research questions that need to be answered before any interface is designed are: What can and should an agent do? Who (i.e., which agent) should interact with users? When, where, why, and how should an agent do it? These questions were called the W5 question set by Duric *et al.* [77] and W5+ by Maat and Pantic [78]. The research discussed IAIs by integrating perceptual and cognitive models through both theoretical analysis and empirical investigations. Although these authors [77, 78] concentrated on the effectiveness of tools for the interpretation of perceptual processes, including lower arm movements, facial data processing, eye–gazing tracking, and mouse gestures, other IAI technologies (i.e., behavioral and cognitive tools) were also advocated in interface design for IAIs where human cognitive, perceptual, motor, and affective factors can be modeled and used in the interface to adapt to the changes of users. The essence of the research was to monitor affective or emotional behavior, or nonverbal information, to answer the W5 and change display characteristics according to user behavior. This method for interface design is more human like, as the interface or the embodied agent/automation is regarded as a human assistant that monitors perceptual and cognitive states and understands users as partners or teammates, thus facilitating more effective collaboration. The idea is to emphasize team collaboration, which is true in human–human interaction as shown in Table 11.2 (see Section 11.4.4). In human–human interaction, since certain people may not work well with others, individuals will have to adapt to each other in order to achieve better overall collaboration performance.

Another set of experiments investigated how different design choices and interactions affect the success of adaptive graphical user interfaces (GUI) [79]. Three adaptive user interfaces were implemented: split, moving, and visual pop out. A split interface copies important functions onto a toolbar in a spatially stable manner. Users can choose either to continue using the original (i.e., unmodified) interface or the adaptive toolbar. A moving interface *moves* promoted functionality from inside popup panes onto the main toolbar, causing the remaining elements in the popup pane and the existing buttons on the toolbar to shift to make space for the promoted button. A visual pop out interface highlights promoted buttons in magenta. If a promoted button resides inside a popup menu, both the button invoking the popup menu and the menu item are highlighted.

Gajos *et al.* [79] computed an estimate of participants' perceived benefits and costs associated with adaptive interfaces. They used the average of responses to the efficiency and performance questions as a measure of benefit. The mental demand, physical demand, frustration, and confusion questions were used to compute the cost. The split interface was found to be the most beneficial and least costly despite having lower theoretical benefits than the moving interface. The visual pop out interface was found to confer little benefit, as expected, but participants found it very distracting and assigned it a higher than expected cost. Participants were also observed to perceive adaptations as incurring a cost. Thus, this study shows that purely mechanical properties of an adaptive interface are not a good predictor of a user's performance or satisfaction. A balance between perceived cost and benefit would be a better predictor of user acceptance.

The most important implication from these studies is not only that computer technology is needed to make novel interfaces a reality, but also that users adapt to the interface presented by the computer. At the end, both the operator and the computer must understand each other's intentions and/or motivations, provide feedback to each other as necessary, and eventually adapt to each other. Although the above empirical studies have shown the advantages of IAI over other conventional interfaces, there is still a lack of design guidance for IAIs, especially from the perspective of operator interaction with intelligent agents.

11.4 MULTIAGENT-BASED DESIGN AND OPERATOR–AGENT INTERACTION

As indicated in Section 11.3.3, little research has focused on operator–agent interaction because agent-based technology has not yet been adopted in many complex and dynamic system designs. Particularly, from a HF perspective, lack of theoretical development and empirical studies makes many designs costly and ineffective. Many existing frameworks focus on only individual models (e.g., user, task, and domain) rather than operator–interface interaction. To address this issue for maximizing overall HMS performance, this section introduces the concept of an adaptive intelligent agent (AIA), a model of operator–agent interaction, the use of AIAs, and the model to facilitate operator–agent interactions.

11.4.1 AIA Concept

Improvements in AI techniques and the growing need for better interface metaphors have led to two converging yet often overlapping areas of research on agents [80]. Generally, an agent refers to a representative authorized to act for another. For example, a travel agent makes an individual's vacation plans, and a stockbroker selects an opportune time to purchase stock for clients. In software technology, the use of the term "agent" denotes that they can take action on behalf of a user. Research on software agents has been active since the early 1990s, when the technology showed promise as an effective computing paradigm for implementing users' delegations and making the interface an active partner to the users [81, 82].

The term "agent" has been used differently in different research communities, but definitions of agents in the user interface domain (i.e., "interface agents") are most relevant to this study [83–85]. Although there are still different definitions for this term [86], there are special types of agents that share several common characteristics. According to Finin *et al.* [87] and Wooldridge [88, 89], their most

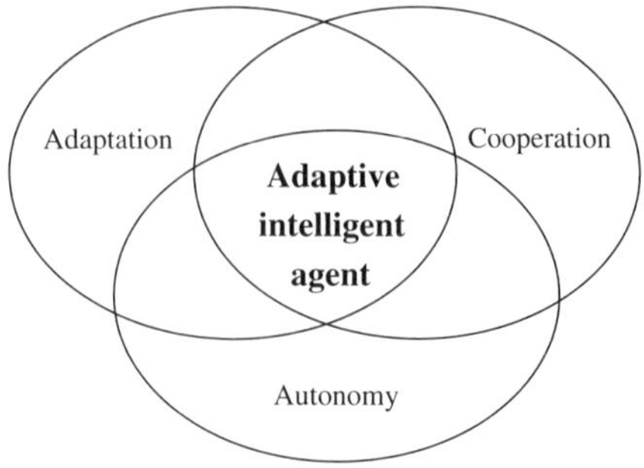

FIGURE 11.5 Programs that exhibit characteristics of an adaptive intelligent agent (AIA).

prevalent characteristics are adaptation (i.e., reactivity to the changes in users and environments), autonomy (i.e., taking certain tasks from users automatically), and cooperation (i.e., social ability among agents themselves), as illustrated in Figure 11.5. Thus, these three characteristics become the defining attributes of an AIA. Hence, an AIA can be defined as

> *A personification of computer code and algorithms so as to mimic human behavior, perception, and cognition to some extent, which can cooperate with other agents, automatically take actions on the user's behalf, and adapt to the changes in the user, system, and environment.*

Therefore, an AIA exhibits adaptation, autonomy, and cooperation in accomplishing tasks for the user. Also, an AIA should be able to adapt autonomously to facilitate collaboration between operators and the interface. It can be regarded as a partner to a human operator that autonomously takes over certain tasks and adapts itself to changes in the operator, system, and environment.

Within the context of UAV control, interface embedded AIAs are well suited to take over select tasks (e.g., monitoring fuel state or altitude level) delegated by the operator. This type of task automation serves to optimize operator–agent interaction, which in turn increases overall HMS performance in complex and dynamic environments. Thus, AIAs are capable of planning or reacting to achieve goals, archiving and retrieving information, sensing and acting, modeling the environment, coordinating with other agents, and resolving conflicts.

Note that using this definition permits a broad range of applications to be classified as AIAs, but excludes the "normal" range of computer programs. For example, a standard macro (i.e., a small program embedded within a larger application) would not be classified as an AIA. Macros automate tasks for the user, but they are usually input and environment dependent. Any deviation from the initial inputs or changes to the environment can cause the macro to fail.

11.4.2 Operator–Agent Interaction Model

Ten models have been discussed in Section 11.2 that explicitly represent the domains of applications, task constraints, and the flexibility inherent in human interaction with a complex IAI system. However, they are not sufficient to describe the interaction between an operator and AIAs to reflect the work environment and its dynamic nature. As illustrated in Figure 11.2, IAI technologies are interaction centered. Only after Benyon and Murray [47] defined the interaction model was the importance of interaction between operator and agents emphasized. A theory and associated strategies are needed to facilitate operator–agent interactions in a particular IAI system in which AIAs are embedded.

According to Beaudouin-Lafon [90], "An interaction model is defined as a set of principles, rules, and properties that guide the design of an interface. It describes how to combine interaction techniques in a meaningful and consistent way and defines the 'look and feel' of the interaction from a user's perspective. Properties of the interaction model can be used to evaluate specific interaction designs." Thus, an interaction model should represent at least three properties of the control and controlled systems as well as the operator: (1) what changes to the system the operator wants to make; (2) why the changes should be made with respect to system goals and the current state; and (3) how the needed changes to the system can be made (i.e., the operator activities undertaken to achieve the desired state). In addition, if there are concurrent activities, the model should present their nature, and provide choices to the operator given the current system state. To be useful to the design, an effective interaction model must be both descriptive and prescriptive. It should aid decision-making by describing what an operator actually does and specify what the operator should do next.

Since an IAI contains many AIAs to understand operators, act for operators, and explain and specify adaptations to operators, operator–interface interaction is essentially operator–agent interaction. With assistance from various automation agents, an operator's tasks can be diverted from direct information manipulation to indirect information management. An operator's focus will be on the supervision of overall operation and performance of a system, instead of the detailed functions of the system that should be delegated to various embedded automation agents. These agents should inherit all the characteristics of an AIA as described in Figure 11.5 (i.e. adaptation, autonomy, and cooperation). AIAs should be able to tell operators what is happening (i.e., be descriptive) and what should or will be done next (i.e., be prescriptive). AIAs should be able to learn operators' intentions, monitor their cognitive workload and performance, and guard their resources and time (i.e., be intelligent). They should also be able to learn from past experiences and change how they behave in any given situation (i.e., be adaptive). In addition, these AIAs should enable themselves to communicate and cooperate with each other and act according to the results of communications (i.e., be cooperative). Therefore, the interface is an entity that gathers all the AIAs that are not only descriptive/prescriptive, but also adaptive and cooperative, as illustrated in Figure 11.6.

When comparing the way operators interact with interface agents in an HMS with the way a management team runs a corporation, it is apparent that they have similar

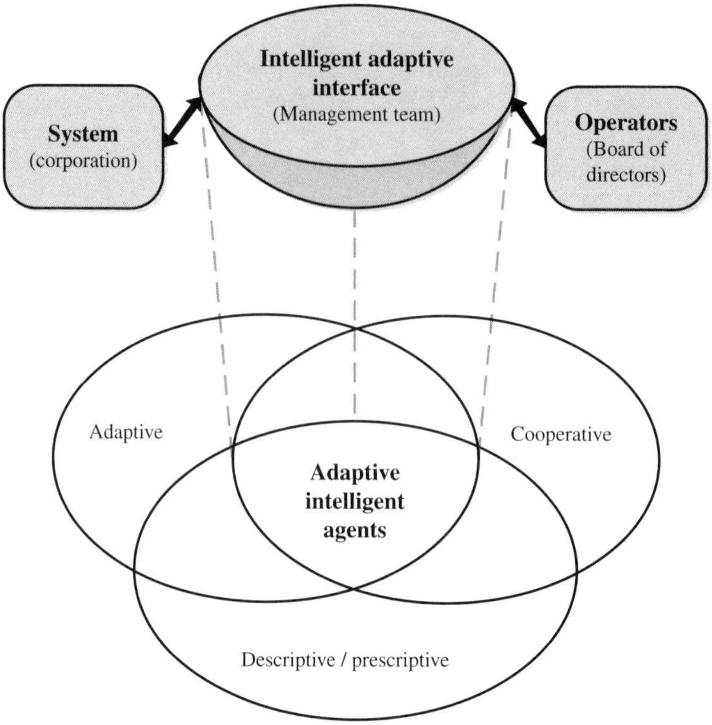

FIGURE 11.6 Operator–agent interaction model.

fashions of interaction. As shown in Figure 11.6, operators can be regarded as a board of directors of a corporation with the purpose of making profits (i.e., maximizing overall HMS performance). The board must trust its management team (i.e., components of an interface) and delegate them to run the business. Through the management team (i.e., a group of agents), the interface will then act as a bridge between the board of directors (i.e., the operators) and the corporation (i.e., the system) to run the business (i.e., execute the mission) and represent the interest of the operators. Thus, an IAI can be regarded as a management team or communication channel that consists of a group of assistants (i.e., AIAs), including a chief executive officer (CEO) and his/her assistants. Through the communication channel (i.e., IAI), the management team (i.e., AIAs) automatically conducts business and keeps the board and shareholders (i.e., operators) updated on what is happening and what the corporation (i.e., the system) should or will do next. The management team will handle any issues between itself and the system, including correcting any mistakes, without any interference from the board, except for emergencies to which they must draw the board's attention. Whenever that happens, the board must understand what the team is doing through interactive communications, as well as why something happened, how to be directly involved in the business (i.e., take over system control), and/or how to instruct the management team to handle situations. In other words, the board or shareholders should only play a supervisory role. To do so, the management team should have the ability to adapt to not only changes of business environment (i.e., system/environment changes), but also changes in the board's directions/intentions. At the same time, the management team should be able to communicate effectively and work collaboratively to make profits for the corporation (i.e., maximize overall HMS performance). Therefore, the IAI consists of various AIAs that are descriptive/prescriptive, adaptive, and cooperative.

To facilitate operator–agent interaction, the IAI should also include a combination of algorithmic and symbolic models for assessing and predicting an operator's activities, awareness, intention, resources, and performance. The IAI should support more rapid task completion by operators with less work (i.e., efficiency), provide the right information at the right time, tailor the content and form of the interaction to the context of the operator, tasks, and dialogue (i.e., effectiveness), and facilitate spoken, written, and gestural interactions, ideally as if interacting with a human interlocutor (i.e., naturalness).

11.4.3 Difference Between Human–Human Interaction, Human–Machine Interaction, and Operator–Agent Interaction

IAI technology focuses on the interaction between operators and agents and maximizing overall HMS performance. To analyze operator–agent interaction, understanding human–machine and human–human interactions serves as a starting point because many HF issues in these areas have been extensively studied for decades and should be effectively applied in an IAI design process.

First, the differences in their definitions should be reviewed as follows:

- *Human–Machine Interaction*—interplay between a human and any device under defined/known environments [91].
- *Human–Computer Interaction*—interaction between humans and computers that is usually in a sedentary setting.
- *Human–Human Interaction*—communication between humans where beliefs and trust are of significant importance as the context changes.
- *Operator–Agent Interaction*—exchanges between humans and an agent system in a dynamic environment.

Both human–machine interaction and human–computer interaction typically have static or well-defined contexts based on assumptions, rules, and procedures. The difference between them is subtle, except that the former may involve more direct manipulations than the latter. They are both different from human–human interaction and operator–agent interaction as the latter are dynamic and their contexts are sometimes unknown a priori. Although there are some overlaps between human–machine interaction and operator–agent interaction, there are still some major differences between them as highlighted in Table 11.1. Because of the differences, existing

TABLE 11.1 Differences Between Human–Machine and Operator–Agent Interactions

Interaction aspects	Human–machine interaction	Operator–agent interaction
Predictability	Plans, actions, and system states are known within limits	Plans and actions are not known a priori, and may produce unexpected system states
Procedures	It is specific, systematic, and often associated with standard operating procedures	It is fuzzy and there may be many means to achieve the same end
Belief	The human has beliefs (assumptions) about the machine and task. The machine design takes into consideration certain assumptions about the human and task	The operator has beliefs about the agent and task. The agent may have equivalent beliefs about the operator and task
Trust	Typically, trust is binary—the machine works or does not	Trust must be built over time since there are few definite ways to judge whether the agent is performing well
Levels of autonomy	Typically, there are two levels of autonomy—completely manual or completely automatic	There are multiple levels of autonomy
Awareness	The human knows how the system will process/display information	The operator does not necessarily know how the system will process/display information
Context	The context is typically static or well defined	The context is dynamic and typically unknown a priori

human–machine interface design standards are probably still relevant but not sufficient for an agent-based IAI system design. If designers are using current standards to design intelligent agent systems, users may find these systems frustrating. A fresh look at operator–agent interaction may capture new principles or guidelines that do not appear in current standards. Thus, there is a strong need to develop them to address IAI design challenges.

Second, since AIAs aim to help users by automatically taking over some tasks and adapting themselves to changes in users and the IAI, they should be more human like and resemble human behavior for effective and natural interactions. Thus, operator–agent interaction should adopt some human behaviors that include physiological attributes (e.g., eyes and other body parts), intellectual characteristics (e.g., capacity, recognition, learning, and decision), knowledge basis (e.g., knowing

TABLE 11.2 Behavior of Human–Machine and Operator–Agent Interactions

Behavior	Component
Observation/perception	Aural, taste, vision, touch, olfactory
Communication	Mouth, eyes, body, scent
Cognitive process	Learning, understanding, reasoning, reference, trust
Adaptation	Belief influence, behavior change
Collaboration	Reduce workload and improve situation awareness and operational effectiveness

the environment, system, task, and user) and psychological states (e.g., concentration, vigilance, fatigue, and patience), as illustrated in Table 11.2.

An example of an agent paradigm resembling the cognitive and adaptation processes of human behavior is the belief–desire–intention (BDI) approach [92–95]. An agent's beliefs refer to the knowledge the agent has about the world, which may be incomplete or incorrect. An agent's desires correspond to the tasks allocated to it. In other words, goals in the system are required to be logically consistent. However, even if they are consistent, an agent will not be able to achieve all the system goals. An agent must thus fixate on select subsets of available desires and commit resources to achieving them. These chosen desires are then defined as intentions. An agent will typically continue to achieve an intention until the intention is believed to be either satisfied or no longer available [96]. The key data structure in a BDI model is a plan library, which is a set of plans for an agent. This set of plans (i.e., recipes) specifies courses of actions that an agent may take to achieve its intentions. An agent's plan library represents its procedural knowledge about producing system states.

The BDI agent model only resembles cognitive and adaptation processes of human behavior. There should be more human-like agents to imitate other human behavior as listed in Table 11.2. Only when a number of these agents are effectively implemented and deployed in IAIs can the operator–agent interaction promise better human–machine interaction with more benefits, including:

- More effective interaction—doing the right thing at the right time, and tailoring the content and form of the interaction to the context of users, tasks, system, and communications.
- More efficient interaction—enabling more rapid task completion with less work.
- More natural interaction—supporting spoken, written, and gestural interaction, ideally as if interacting with a human interlocutor.

Given the difference between human–machine and operator–agent interactions, the most effective way to design an IAI is to incorporate many proactive and personalized AIAs. AIAs can act as partners and take responsibilities delegated from operators to optimize operator–agent interactions in an intuitive fashion, thus maximizing overall HMS performance.

11.4.4 Optimization of Operator–Agent Interaction

As pointed out by Lajos [97], more complex adaptation schemes for an intelligent interface should require and/or allow:

- multiple levels of interaction;
- multiple dialogues and modes (content and form) of interaction;
- involvement of many personal shaping factors into adaptation;
- wide choice of skill, rule, and knowledge-based interaction elements; and
- multiple modularity/hierarchical levels in the interface structure and functions.

Besides visual inputs, other communication channels between operators and interfaces should also be considered, such as auditory (e.g., verbal and aural) inputs. Multimodal inputs are effective for supporting human–computer interaction [78]. Furthermore, to help reduce operator workload and accompanying stress due to an overloaded visual output channel, multimodal outputs (e.g., auditory, tactile/haptic) should be considered. Relative to unimodal cue presentation, multimodal cues can facilitate effective presentation of information in interfaces that require complementary, redundancy, substitution, and suitability modes of sensory cues [98]. For example, redundancy gains, such as faster response times to an incident, can be observed when the same information is mapped to multiple modalities [99]. Also, there is a synergistic benefit when the related information is mapped to different modalities [98]. The appropriate level and channel of input/output and interaction should depend on the task itself and the current state of interaction. The reliability and accuracy of operator/user models and algorithms are also critical for system design and interaction optimization. In addition, the optimization should also consider trust and transparency issues. With trust, an AIA can accomplish delegated tasks without operator intervention. A dynamic adaptive interface [100] that has cognitive inference aiding systems should automatically provide information without requiring control input from the operator. With transparency, some agents (e.g., some dialogue windows) effectively disappear when necessary, thus enabling the operator to interact directly with objects of interest in the domain and to achieve effective interaction with minimal cognitive effort.

An IAI is tightly coupled with an HMS. The task distribution and collaboration between humans and system agents are essential to IAI design. Human–automation teaming is a promising approach to enhancing operator–agent interaction. Cuevas *et al.* [101] presented a theoretical framework that outlined how information exchange and updating between humans and automation technologies affect the cognitive processes of both individual humans and human–agent teams. They proposed that non-human team members should aid their human counterparts (similar to human–human interaction as discussed in Section 11.4.3 and Table 11.2) by

- maintaining awareness of member actions and status;
- updating team members regarding important changes;

- monitoring intra-team and inter-team communications;
- tracking task execution and completion and progress toward the achievement of the team's goals; and
- providing timely feedback on performance and guidance on correcting team errors.

11.5 A GENERIC IAI SYSTEM ARCHITECTURE AND AIA COMPONENTS

This section proposes a generic IAI architecture of an AIA-based system for optimizing operator–agent interaction that considers the aspects discussed above. This is followed by a description of AIA structure and adaptation processes from an interface design perspective. Figure 11.7 gives a more detailed illustration of the IAI interface structure shown in Figure 11.6 (i.e., an IAI is composed of various AIAs). Figure 11.6 provides an external view of an operator–agent interaction model, whereas Figure 11.7 illustrates the internal IAI system architecture and depicts inter-relations among AIAs.

11.5.1 Generic IAI System Architecture

The hierarchical IAI system architecture consists of many AIAs working at different levels. The IAI architecture is a cooperative multiagent structure that can keep track of all the resources in the system and coordinate agent activities. As such, agent technology plays a crucial role in enabling operator–agent interaction.

The foundation of this architecture is to classify AIAs into four different function groups and manage them in a hierarchical way. Typically, four function groups of an IAI system represent a hierarchy of various AIAs: executive, senior, working, and junior. Except for the executive AIA, each AIA group has a number of agents communicating with each other and working collaboratively. The Managing Agent is the only executive agent in the first group. This agent is responsible to the operator with whom it directly communicates. It acts like a CEO of an organization as a part of a Management Team (Figure 11.6). It receives instructions from the operator and advice from four other senior AIAs to adapt to the changes of the operator, system, and environment. It is a decision-making AIA who makes changes to the interface through its own working and junior level AIAs.

The second group has four senior AIAs to manage higher level tasks for achieving higher level goals of the system (i.e., as part of the management team for running a corporation—IAIs in Figure 11.6). It includes an Interaction Agent, Modeling Agent, Tasking Agent, and Sensing Agent. These senior AIAs play the role of managing and coordinating tasks, information flow, communications, and feedback among the operator, agents, tasks, system, and environment. They communicate with each other and with their own working AIAs. They update the Managing Agent about changes in the operator states and the status of the system and environment. They give advice to

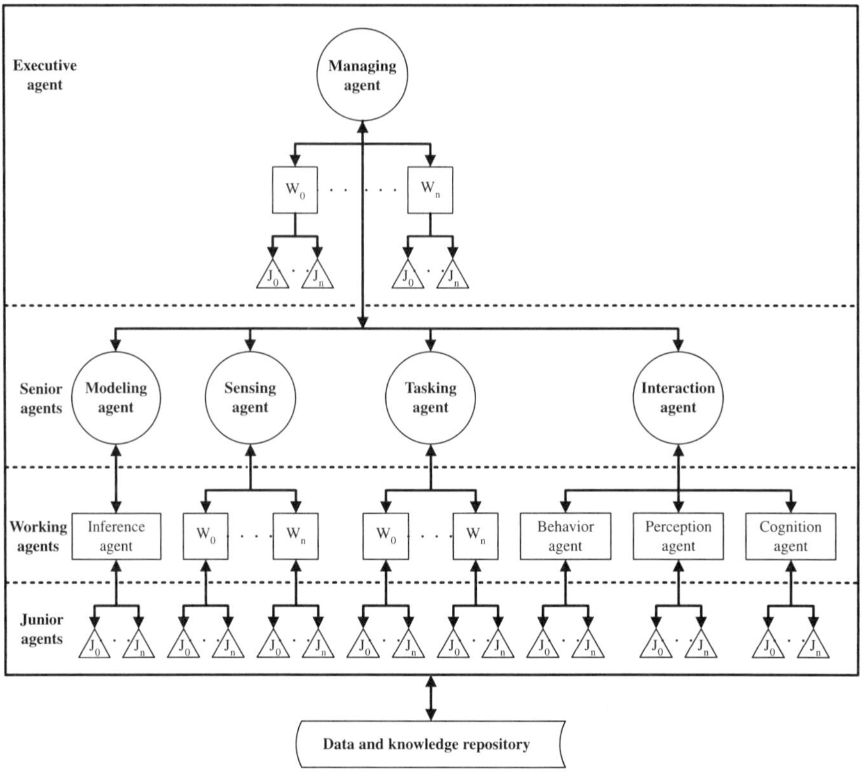

FIGURE 11.7 A generic IAI system architecture (the bidirectional arrows represent communications and feedback).

the latter about the adaptation strategies. They also communicate with other working and junior AIAs through their superior AIAs.

The third function group consists of many working AIAs (i.e., working level employees of an organization—AIAs in Figure 11.6) that work on individual aspects of major tasks so a senior AIA can achieve individual goals. They communicate within the group and report to their superior senior AIAs, and are depicted as W_0 to W_n in Figure 11.7. They also communicate with other working or junior AIAs in other groups through their superior AIAs. For the purpose of optimizing operator–agent interaction in this discussion, four working AIAs are needed and outlined in Figure 11.7: Inference Agent (which works for the Modeling Agent), Behavior Agent, Perception Agent, and Cognition Agent (the three of which work under the Interaction Agent). The Inference Agent plays a reasoning role in order to interpret information gathered from other AIAs. The Behavior Agent expresses an operator's manual control of input devices; the Perception Agent covers interaction aspects such as visual, verbal, aural, tactile, and force; and the Cognition Agent covers the operator's workload, situation awareness, complacency, skill degradation, fatigue, and frustration.

Junior AIAs (i.e., assistants to the working level employees in an organization—AIAs in Figure 11.6) are the fourth function group of the IAI hierarchy that works on the details of individual tasks for the working level AIAs. For example, an eye-gaze tracking agent is a junior AIA that monitors an operator's eye movement and delivers an operator's eye-gaze tracking information to its working level AIA (i.e., the Perception Agent). Junior AIAs communicate with each other in the same group and report to their working AIAs. Junior AIAs are labeled as J_0 to J_n in Figure 11.7. They also communicate with other AIAs through their superior working and senior AIAs.

Due to the conceptual nature of this research, the discussions here focus on the AIAs contributing to operator–agent interactions at senior and working levels. The functions and communications between the executive agent and the other four senior AIAs in Figure 11.7 are

- A *Managing Agent* that communicates and coordinates with the other four senior AIAs to collect information about the states (i.e., knowledge) of the operator, tasks, external environment, and the system itself, including error monitoring and emergency control. Through its own working level AIAs, it manages information flow and data output, shares knowledge with other four senior AIAs, controls the display characteristics, prioritizes system reactions to any emergencies, decides adaptation strategies to the changes, assigns tasks, and decides automation levels, etc. In other words, it communicates with the operator directly at the top level and manages all AIAs in the IAI hierarchy.
- A *Modeling Agent* that gathers information from Sensing, Tasking, and Interaction senior AIAs about current states of the operator, tasks, system, and environment. It compares the newly collected data with embedded models of the operator, tasks, system, and environment. It then provides updated models to senior and other agents and the database (i.e., Data and Knowledge Repository in Figure 11.7). This entity has a working agent (i.e., Inference Agent) that collects information from Sensing, Tasking, and Interaction AIAs through the Modeling Agent for the processes of comparisons and inferences.
- A *Sensing Agent* that gathers information from sensors and data link to keep the Managing Agent and Modeling Agent updated on current states of internal and external sensors, assets, environments, etc. It also provides feedback on the effect of system adaptation elicited in response to system changes. This entity has a number of working level AIAs.
- A *Tasking Agent* that gathers information from internal and external tasks to keep the Managing Agent and Modeling Agent updated on the current states of internal and external tasks. It also provides feedback on the effect of system adaptation on internal and external tasks. This entity has a number of working level AIAs.
- An *Interaction Agent* that collects and integrates information from its three working level AIAs (i.e., Behavior Agent, Perception Agent, and Cognition Agent) about an operator's state when interacting with the system. It provides

feedback on an operator's physiological attributes, intellectual characteristics, and psychological states. It coordinates with the Modeling Agent to process information and communicates with the Managing Agent. It also makes recommendations to the Managing Agent to manage interface output. These recommendations facilitate operator–agent interaction by deciding optimal modality, controlling the display characteristics, assigning tasks and automation levels to different agents, prioritizing the emergency system, so on.

The functions and communications among the four AIAs in the working level agent group (Figure 11.7) are

- A Behavior Agent that collects an operator's input data from a keyboard, mouse, joystick, or any other input device. The data are then fed into the Interaction Agent for integration with perception and cognition data. The Interaction Agent coordinates with the Modeling Agent to process the integrated data set and inform the Managing Agent about the current behavioral state of the operator.
- A Perception Agent that collects data of an operator's low arm movements, facial expression, eye-gaze characters, and gestural changes. The data are then fed into the Interaction Agent for integration with behavior and cognition data. The Interaction Agent coordinates with the Modeling Agent to process the integrated data set and inform the Managing Agent about the operator's attention, fatigue, frustration, and even fear or excitement.
- A Cognition Agent that collects an operator's physiological data through biofeedback devices such as electroencephalography, electrocardiography, and functional near-infrared imaging. The data are then fed into the Interaction Agent for integration with behavior and perception data. The Interaction Agent coordinates with the Modeling Agent to process the integrated data set and informs the Managing Agent of the operator's workload, situation awareness, complacency, and skill (i.e., performance) degradation.
- An Inference Agent that collects information from the Sensing Agent, Tasking Agent, and Interaction Agents through the Modeling Agent. After capturing information about the states of the operator, tasks, system, and environment, it systematically organizes and categorizes the information. It then processes the information and turns it into knowledge through reasoning. The knowledge is disseminated to agents relevant for decision-making through the Modeling and Managing agents.

It is noted in Figure 11.7 that the executive agent and each senior AIA has its own working level agents, and each working level agent has its own junior agents. All agents constantly update their own status and send the collected information to their superiors and the Data and Knowledge Repository. The bidirectional arrows in Figure 11.7 represent communication and feedback between agents, the interface, and the operator, and interface and external systems.

In the IAI hierarchy, working agents take major responsibilities for detailed work, and they not only work independently, but also cooperate with other AIAs under the supervision of a senior AIA. For instance, the Behavior, Perception, and Cognition Agents work independently to collect data about an operator's behavioral, perceptive, and cognitive states, and send the data to the Interaction Agent. They also communicate with each other through the Interaction Agent for coordination purposes to reflect current states of the operator (i.e., description). Other descriptive information comes from the Sensing Agent and the Tasking Agent about current system states, tasks, and goals. Based on the information described by these three senior AIAs, the Modeling Agent communicates with its working agent (i.e., Inference Agent) to reason and model the operator, tasks, system, and environment. The Inference Agent classifies, compares, and evaluates all the data sets in the system database. The result is then sent back to the Modeling Agent to inform it of the possible state changes of the operator, tasks, system, and environment. The Modeling Agent updates and informs the other four senior agents about the states and possible changes. Then, according to the algorithms and criteria built into the Managing Agent and Interaction Agent, they decide how to make necessary adjustments (i.e., adaptation) when an operator's goals, the tasks, the system status, and the environmental situation (i.e., prescription) change. These changes include decisions on adaptation methods, combination of modality, automation levels, and actions to be taken.

The following example illustrates the communication and collaboration of various AIAs in the framework. If an input is captured by an eye-tracking junior AIA, the data (and other possible inputs from speech recognition, arm movement, facial expression, and gesture change AIAs) will be sent to the Perception Agent for processing. The output of the Perception Agent will be the knowledge of where an operator is looking at in the interface at the moment. This output will be then fed into its senior Interaction Agent. The Interaction Agent combines this output with other outputs from the Behavior Agent and Cognition Agent. It then processes these data to understand the operator's behavior, situation awareness, workload, so on. In this case, the operator is found not looking at the area in the interface she/he is supposed to focus on. Based on the assessment, the Interaction Agent will describe the situation and make recommendations (e.g., send a reminder message to the operator) to the senior Modeling Agent and Managing Agent. The Modeling agent will pack the data with other possible data from other AIAs through the Sensing Agent and Tasking Agent to its working level Inference Agent. The Inference Agent will compare all these inputs with the current state records of the operator, task, system, and environment stored in the Data and Knowledge Repository. After the data are processed, it passes its outputs to the Interaction Agent and the Managing Agent about the operator's current attention and other related information (e.g., losing a data link to an asset on the scene) through its Modeling Agent. According to its built-in protocol, the Modeling Agent provides a new and enhanced action plan (e.g., augment the visual modality with auditory modality because an asset might have been lost and the operator's attention is needed immediately). The Managing Agent will then assess action recommendations provided by the Interaction Agent and the Modeling Agent and make a decision. The Managing Agent may ask its own working and junior AIAs to change the output of the interface.

This proposed agent-based IAI hierarchical architecture focuses on interaction between the human operator and various AIAs. It optimizes operator–agent interaction by perceiving, reasoning, interpreting, and predicting the current and future states of the operator, tasks, system, and environment. The prescriptive aiding system also assists in decision-making on task allocation and automation level assignment as needed. This paradigm provides a foundation for AIAs to automatically take certain tasks from an operator.

11.5.2 AIA Structure

An AIA can be regarded as a standard computer program with arguments and return values. Figure 11.8 illustrates an example of a working AIA (i.e., the Perception Agent) structure. It has six arguments (i.e., inputs) that are the return values (i.e., outputs) of six junior AIAs. The latter are responsible for collecting facial data, states of haptics, arm movement data, speech recognition data, eye-gaze tracking data, and gesture changes, respectively. Each junior AIA has its own inputs from its own sensors. Based on embedded algorithms and models, it compares new data with its previous states after decoding. The return values are delivered to the Perception Agent after encoding. The AIA gathers all outputs (i.e., knowledge) of these junior AIAs and compares them with the results from its own embodied models. Its outputs

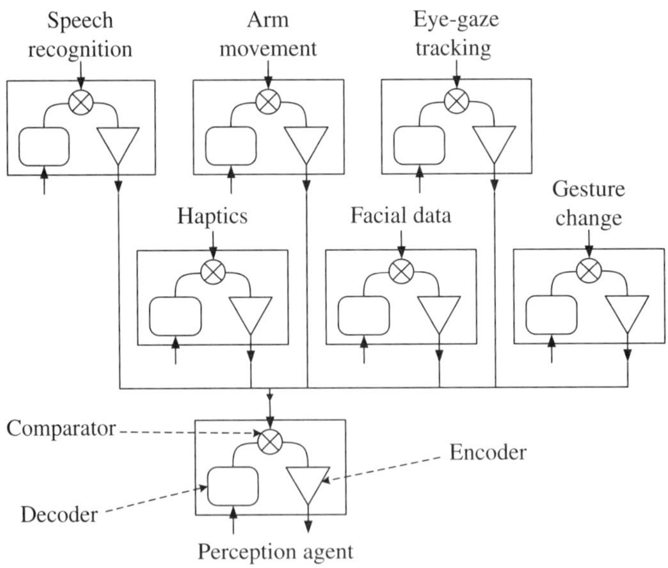

FIGURE 11.8 Computer program as a standard AIA ("⊗" means a comparator, "▢" means a decoder, and "∇" means an encoder).

consist of information about levels of workload, situation awareness, frustration, fear, excitement, etc. The outputs become inputs and are fed into the Interaction Agent for the interpretation of the operator's states. The Interaction agent passes the information to the Inference Agent through the Modeling Agent. The three AIAs work together, comparing the current states with previously stored data. They then infer the future status of workload, situation awareness, complacency, and skill degradation according to cognitive models. Through the Managing Agent, the system and the operator are informed of why and what events are happening and what events need to be prioritized according to pre-defined rules.

Similarly, the Managing Agent can also be regarded as a main function of a computer program. It has arguments (i.e., inputs) from four other senior agents: Sensing, Tasking, Modeling, and Interaction Agents. The four AIAs have knowledge of the system (i.e., interface) itself, task processes, task environments, an operator's physical, emotional, and cognitive states, current automation levels of all agents, emergency situations, so on. The returns (i.e., outputs) of the Managing Agent are adjusted to other actions accordingly. The returns include: (1) the most effective modality for system–operator interaction; and (2) the new models chosen by the Modeling Agent to adapt to the operator, tasks, system, and interaction (refer back to the example in the above section). For the Sensing Agent, the inputs include all the data from internal and external sensors and data links. The outputs are the current and next states of the system and task environments. For the Interaction Agent, the inputs are the states of an operator's behavior, perception, and cognition. The outputs are the possible next stages of the process and recommended actions that the operator and the interface will take. For the Modeling Agent, the inputs are the information from the Sensing, Tasking, and Interaction senior AIAs and its own working Inference Agent. The outputs are adaptation strategies and new models of the operator, tasks, system, and environment for the other four senior AIAs.

11.5.3 Adaptation Processes

To optimize operator–agent interaction and thus maximize HMS performance, AIAs enable the IAI to be aware of activities, resources, and the operator's intention, plan, and performance [102]. Junior AIAs gather basic information of the operator, system, environment, and processes of various tasks. Working AIAs interpret and process the information into the knowledge of the system parts. Then, senior AIAs build adaptation strategies into the interface and react to the changes. As a result, AIAs interact with different aspects of an operator's behavior (refer to Table 11.2) including physiological attributes (e.g., eye, mouth, and body), intellectual characteristics (e.g., capacity, recognition, learning, reasoning, decision, and trust), knowledge basis (the environment, system, operator, tasks, etc.), and psychological states (concentration, situation awareness, vigilance, fatigue, patience, etc.).

From a management perspective, the interface AIA (e.g., the Managing Agent) has the control to allocate appropriate tasks to the operator and other AIAs. The function allocation follows certain philosophies of adaptation in terms of answering the W5+ (i.e., what, who, where, why, when, and how) questions that are addressed

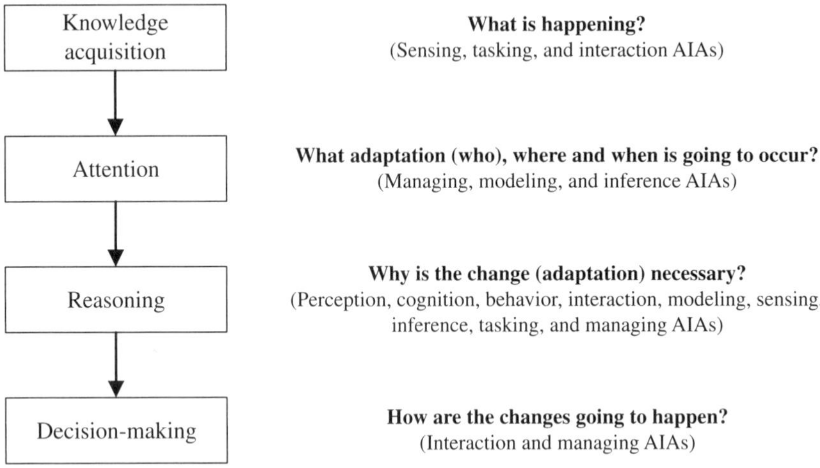

FIGURE 11.9 AIA adaptation process.

in Section 11.3.5. As illustrated in Figure 11.9, a process for the function allocation and adaptation can then be proposed. The AIA adaptation process includes four steps:

- *Knowledge Acquisition* gathers information through various junior and working AIAs through the Sensing, Tasking, and Interaction Agents about the states of an operator, tasks, system, and environment. The purpose is to inform the system and the operator of what is currently happening.
- *Attention* involves various junior and working AIAs through two senior AIAs (e.g., the Managing and Modeling Agents) and one working AIA (e.g., the Inference Agent). The two senior AIAs enable operators to be aware of what should or will be happening next. They indicate to the operator what changes (i.e., adaptations) are needed, and where and when the changes will occur. The two senior AIAs also provide an operator with feedback and alerts.
- *Reasoning* involves various junior and working AIAs through all five senior AIAs (i.e., the Managing, Sensing, Tasking, Interaction, and Modeling Agents) and four working AIAs (i.e., the Behavior, Perception, Cognition, and Inference Agents). Junior AIAs collect information on the states of an operator, tasks, system, and environment. The information is provided to the working AIAs to classify, interpret, reason, and compare. The Inference Agent compares the current states of these variables with the existing models in the database. Then, it judges whether changes are necessary and suggests what models should be used for adaptation through the Modeling Agent. The generated knowledge is then disseminated to other senior AIAs with the understanding of why all changes are necessary.

- *Decision-Making* involves the Interaction, Tasking, and Managing Agents that evaluate proposed methods of changes from the reasoning process by applying two criteria. The primary criterion looks at human performance consequences (e.g., mental workload, situation awareness, complacency, and skill degradation). The secondary criterion includes automation reliability and costs of consequences [103]. Once confirmed with the criteria, this process chooses a method of adaptation and identifies automation levels for various agents. In the meantime, proper feedback or alerts to an operator are presented.

The primary goal of facilitating these processes is to optimize operator–agent interaction through adaptation by providing the states of an operator, tasks, system, and environment, including activities, awareness, intentions, resources, performance, and timeline.

11.6 AN IAI AND AIA DESIGN: CASE STUDY

To optimize operator–agent interaction in an IAI-based system, the above theoretical issues need to be addressed in empirical investigations and thus provide solid guidelines for IAI design. This section illustrates an example of agent-based interface design using the IAI conceptual framework, analytical methodologies, AIA concept, operator–agent interaction model, and IAI hierarchical system architecture. The example provided is a case study for an IAI design that will control multiple UAVs from an airborne platform for a counter-terrorist mission. IAIs are applied here to reduce operator workload and improve decision-making effectiveness in the employment and operation of UAVs. In the context of UAV control, an IAI is driven by AIAs that support the decision-making and action requirements of operators under different levels of workload and task complexity. The IAI manifests itself by presenting the right information and action sequence proposals, or by performing actions at the right time. In addition to reducing workload for humans involved in UAV missions, an IAI also provides an opportunity to reduce manning requirements (e.g., moving from ratio of 10 operators controlling 1 UAV to 1 operator controlling 10 UAVs).

As illustrated in Figure 11.10, a mixed-factor experimental design is applied to evaluate the role of AIAs facilitating operator–agent interaction in an IAI-based system. It includes three domain variables of human–machine interfaces: level of task complexity, level of operator temporal workload, and level of interface intelligence/automation. Within the context of UAV control, task complexity is determined by the cognitive nature of the task a UAV operator has at a specific position (e.g., UAV Pilot (UP), UAV Payload Operator (UO), and UAV Tactical Navigator (TN)) with low, medium, and high levels. Operator temporal workload is determined by the extent of operator time pressure and/or the amount of information processing at low, medium, and high levels. Interface intelligence is determined by the agent technology, and should cover all aspects of human perception, behavior, and cognition. Intelligence is also directly related to fully manual, semiautomatic,

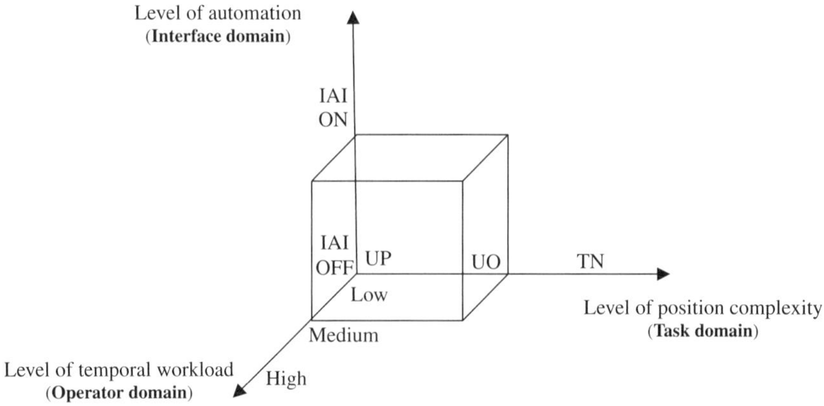

FIGURE 11.10 Depiction of the mixed-factor experimental design.

and fully autonomous automation levels. Since automation changes the level of an operator's involvement, how will AIA assistance impact an operator's perception, cognition, behavior, and goals when undertaking tasks at different levels of automation and task complexity? The question of how task complexity, operator workload, or interface intelligence affects overall HMS performance is the driving force that will optimize operator–agent interaction and thus IAI design guidelines. The following case study illustrates how these methodologies have been applied for IAI system design.

11.6.1 Interface Design Requirements for the Control of Multiple UAVs

UAV control is operator intensive and can involve high levels of workload. As the quantity and variety of data collected increases, the workload for UAV operators also increases significantly. Moreover, the allocated data must be integrated and/or converted into information and then disseminated to those operators who make decisions. Recently, data collection, data fusion, information management and distribution, intelligence collecting, and data-related decision-making threaten to become a bottleneck. This situation has been made even more complex by increasing joint operations, and rapid and flexible warfare. Feedback from UAV operators indicates that improvements in the operator interface aspect of these emerging systems will likely reap significant gains in system performance and effectiveness. This applies to effective control of UAVs, management of data, and efficient dissemination of the associated information. The level of automation to be applied to decision-making processes is a key concern for both tactical commanders and UAV system managers. As a result, supporting technologies such as IAI systems that combine operators with automation to satisfy mission requirements need to be investigated.

11.6.2 Issues

The control of multiple UAVs is a complex cognitive task with high workload. Applying the above mentioned IAI system design methodologies into the design of these advanced HMSs will likely provide real potential for improvement in the effectiveness of UAV control. To investigate the efficacy of the IAI concept, IAI design principles were applied into the operator interface design for the control of multiple UAVs from an airborne platform. The issues to be addressed in this case study include the components of the IAI system that are required to analyze scenarios, missions, and goals, and whether the design methodologies are effective in conducting the development of the UAV control system.

11.6.3 How the IAI Design Method Was Used

As a standard systems engineering approach, the IAI framework was used from the starting point (i.e., analysis) of the operator interface design. For systems where human functions are predominantly cognitive, the method of analysis should capture this essential human activity. The IAI framework was used in the following design processes:

Operational mission scenario generation. A fictitious counter-terrorism scenario was developed for a maritime patrol task along the east coast of Canada. The scenario involved a UAV crew controlling multiple UAVs from airborne platform CP140. The UAV crew consists of UP, UO, and TN operators. The UP was responsible for the safe and appropriate conduct of all UAVs under the crew's control. The UO was responsible for managing the information being returned by the sensors onboard the UAVs, and for relating findings based on that sensor data back to the rest of the crew as appropriate. The TN was ultimately responsible for overall operations. According to the nature of the tasks each crew member performs, three levels of task complexity were assigned: low (UP), medium (UO), and high (TN). Depending on the number of UAVs to control, workload was also assigned at three levels: low (one UAV), medium (two UAVs), and high (five UAVs).

Operational sequence analysis. To understand the logical interconnection of operators' tasks and the flow of information throughout the system during the conduct of a mission scenario, operational sequence diagrams (OSDs) were used to show the flow of information and operator functions through the system in relation to the mission timeline. The visual representation of OSDs indicated actions, inspections, data manipulation (i.e., transmission, reception, and storage), time delays, and decisions of the mission scenario. Thus, OSDs are particularly useful for the analysis of highly complex systems that require many time-critical information–decision–action functions by multiple users, which was the case for this study. With the scenario written, a series of OSDs were prepared to facilitate a function and goal/task analysis of

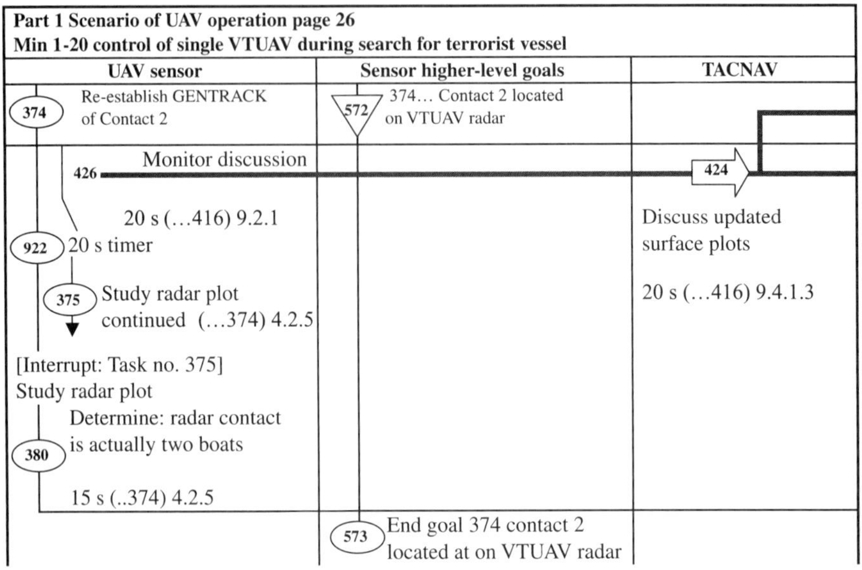

FIGURE 11.11 A sample of operational sequence diagrams.

the envisaged system and the development of a model from which performance predictions could be made and the potential for IAI automation agents could be identified. These OSDs also created an inventory of all bottom-up lowest level tasks in a temporal sequence. Figure 11.11 illustrates a small portion of an OSD as an example of the operational network among three UAV operators, where VTUAV is a naval UAV and TACNAV is a navigation system.

Hierarchical goal analysis (*HGA*). Research on IAIs in the military research and development community was limited when this work was conducted. As a result, the work began with a HGA within which standard mission, operation, and goal analysis procedures were followed to gain a more detailed understanding of implementation issues and opportunities for automated IAI tasks. A particular interest in this work was identifying goals that were candidates for IAI automation agents. Using the AIA principles and IAI framework, the agent, user, system, and world models were created. Upon the completion of the scenario and OSDs, a hierarchical decomposition of the envisaged system goals was conducted. The goal decomposition for all three UAV operators took place according to a means–end hierarchy, and analysis needs were typically satisfied at the fourth or fifth level of this exercise. The goal decomposition was first performed in a top-down fashion from the highest level (e.g., GOAL= counter-terrorist mission is completed) down to lower levels (e.g., GOAL= vertical take-off UAV sector search is planned). Then, a more stringent bottom-up approach was completed by studying the detailed mission activities in the OSDs. As a result, more goals were added to the top-down analysis list to complete the generation of the HGA inventory. An example of a small portion of the HGA results is illustrated in Table 11.3.

AIA identification. To reduce operator workload, goals and associated tasks identified by HGA that can be implemented automatically were selected as appropriate AIA candidates. For example, the system could track keystrokes and cursor movement and would be able to deduce that an operator is attempting to complete multiple concurrent tasks, such as UAV flight control and surface plot manipulation. In this case, the interface could adapt to the situation and provide partial or complete assistance to the busy operator. For instance, if an operator is unable to assign all airborne UAVs a task due to high workload demands, the interface could automatically evoke potential UAV search or loiter patterns. Thus, this type of AIA was chosen as the UAV route planning agent. The feasibility of implementing selected AIAs has to be confirmed by military subject matter experts following pre-defined operational rules. Similarly, many other AIAs were chosen and designed to work together at senior, working, and junior levels (as shown in Figure 11.7) and implemented in the operator interface.

11.6.4 Task Network Modeling and Simulation

After the OSD generation and HGA, the tasks related to a system that can be automatically operated (with automation intelligence or expert knowledge) were chosen to be appropriate AIAs. Goals for systems that have no information on automated operations could not be identified as AIA tasks. With the identification of these AIAs, a task network model was developed in an integrated performance modeling environment (IPME). IPME is a discrete event-simulation framework used to model and assess human and system performance [104]. It comprises an environment model, a crew model, a task network model, a performance-shaping model, and optional external models. Combined with IPME's scheduling algorithm, these models can help an analyst predict workload and operator performance. An IPME database was also generated while producing the OSD and HGA tasks. In the database, each task was allocated to an operator or a system with a descriptive label, and an operational network model was developed. Although the network model was a UAV operator model, external events (other aircrew activities, UAV activities, and other unit activities) had to be established to allow the network to function as a closed-loop feedback system. These were prepared and included in the network and were used to define task behavior, operator assignment, and interactions between tasks and operators. By linking together various networks or tasks, the model attempted to replicate the behavior of a real system. The task network model was run 10 times in IPME to simulate the mission. The collected simulation data were used to analyze operator behavior and determine the effect of IAI agents on operator performance. The IAI models provided not only an effective means to assess the merits of incorporating new automation technologies, but also a clear indication of the most fruitful areas for further research and development.

TABLE 11.3 An Example of HGA Results

Number	Level / Goal/objective and subgoals/subobjectives: 1	2	3	4	5	AIA candidate	Influenced variable	Assignment	Completion time (s)
Top	I want to perceive the (. . .) conduct of the terrorist patrol mission								
9	. . . *communications are conducted and maintained*								
9.1		. . . directions (instructions) are received							
9.1.1			. . . directions are received from other crew members			Yes	Directions	UAV pilot	4–25
9.1.2			. . . directions are received from other units			Yes	Directions	UAV pilot	5–14
9.2		. . . information is received							
9.2.1			. . . information is received from other crew members						
9.2.1.1				. . . VTUAV refueling location			Location	Tactical navigator	5
9.2.1.2				. . . VTUAV calculated time on task		Yes	Time	Tactical navigator	5
9.2.1.3				. . . the flight crew's message that contact is identified			Message	Sensor operator	5

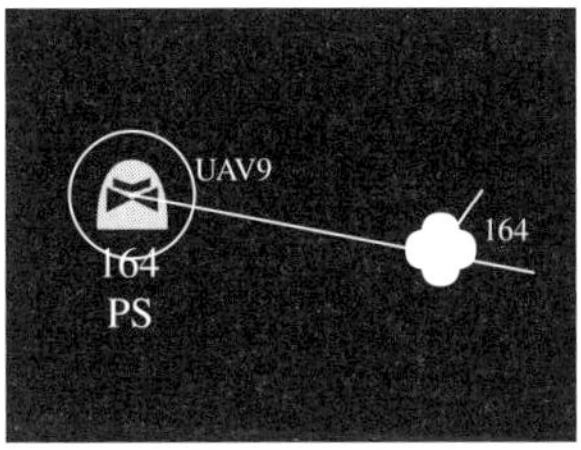

FIGURE 11.12 IAI readouts in TACPLOT.

11.6.5 AIA Implementation

For illustration purposes, only six working level AIAs are listed below to demonstrate the utility of an operator–agent interaction model and IAI framework.

The *Screen Manager* is a working AIA that serves the senior Managing Agent. According to pre-defined rules, this AIA manages a shared tactical plot among three operators whenever new high priority events occur. This includes automatically panning the tactical plot to locations of interest and zooming in or out. On the tactical plot, text was added to the UAV icon as illustrated in Figure 11.12, where TACPLOT is a tactical plot manager. The text helps identify the contact designed by a junior AIA and indicates when this AIA is exercising UAV flight control (P) and/or sensor geo-tracking (S).

The *Interaction Communicator* is a working AIA subservient to the senior Interaction Agent. To optimize operator–agent interaction, this AIA relays all information and knowledge generated by the Sensing, Tasking, Modeling, and Managing Agents. It communicates events and actions and provides feedback to operators through a message window. The window is designed to allow all messages to be displayed in the primary display, thus attaining the crew's attention faster and more reliably. Figure 11.13a illustrates the Interaction Communicator message window that is located at the bottom left corner of the primary displays for all three operators. As depicted in Figure 11.13b, the message window continuously shows all active UAVs, allocated tracks, and information about how the AIAs support the UAVs.

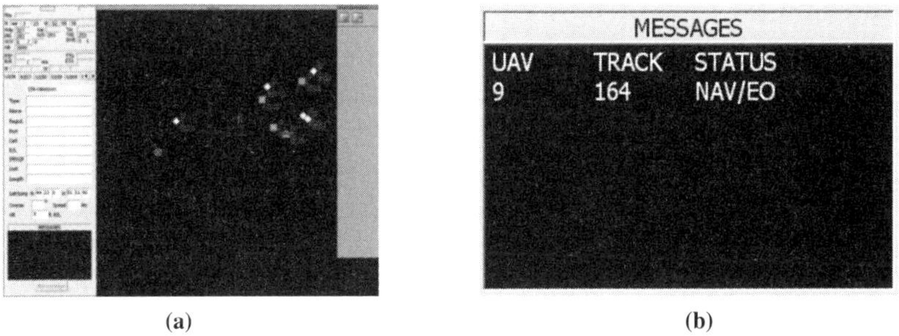

FIGURE 11.13 Interaction communicator message windows: (a) an operator's interface and (b) IAI message window.

The *Route Planner* is a working AIA subservient to the senior Modeling Agent. If a UAV is used to investigate an unknown or hostile contact, this AIA will work with the Inference Agent to compute the most direct route, and activate that route for the UAV. The allocation of tracks to UAVs is based on a search for the closest unknown or hostile contact. Additional logic ensures that no more than one UAV can be engaged on a single unknown contact. However, more than one UAV can be engaged on hostile contacts. Once a route is planned, the route will be passed onto the Tasking Agent through the Modeling Agent.

The *Route Follower* is a working AIA subservient to the senior Tasking Agent. This AIA will pilot a UAV on the active route that is provided by the Route Planner. This includes flight altitude, speed management, and self-preservation in close proximity to the track. The AIA enters an orbital flight pattern around the track once the UAV reaches the sensor identification range. The AIA will update the IAI about the UAV status and its tracks in the message window through the Interaction Communicator.

The *Sensor Manager* is a working AIA subservient to the senior Sensing Agent. Once a UAV is close enough to a track to engage an electronic optic sensor, the AIA will take over sensor management. This includes pointing the sensor and establishing a stable lock on the moving target once the track is within the visual range. The AIA will pass the information to the Screen Manager to manage the shared tactical plot. The information will also be passed on to the Route Planner through the Modeling Agent for route planning, and to the Interaction Communicator through the Interaction Agent for display in the message window.

The *Data-Link Monitor* is a working AIA subservient to the senior Sensing Agent. The AIA monitors the flight pattern (and other status of the UAVs) to determine whether the data links are working. If not, the AIA will immediately inform operators. It works with the Inference Agent through the Modeling Agent to update the system about current communications with all assets in the theatre. The AIA also keeps communication with the Interaction Communicator through the Interaction Agent and displays red emergency text in the message window if anything goes wrong. It will also turn on an alarm to draw the immediate attention of operators. This is a good example of operator–agent interaction optimization using multiple modalities.

All AIAs developed for the UAV control stations follow this AIA adaptation sequence:

- Knowledge Acquisition—gathering status information about all active UAVs, their tracks, the states of operators, and the current display configuration.
- Attention—analyzing information with respect to pre-defined operational rules (i.e., models) and determining which events have occurred.
- Reasoning—explaining to the system and the operator why the events are happening and prioritizing the events according to pre-defined rules (i.e., models).
- Decision-Making and Action—executing pre-defined tasks for each event following the prioritization order.

11.6.6 Human-in-the-Loop Experimentation

The IAI framework and related AIA principles were used to analyze mission scenarios, identify IAI tasks, and generate task network models. This case study was undertaken because simulation results suggested that operator–machine interaction would augment decision-making. However, further analysis was required to identify the most beneficial IAI tasks to modify. In addition, considerable effort was still needed to apply the IAI framework and related AIA principles to the design of prototype systems. Strong empirical evidence was also required to substantiate this effort.

Thus, an experimental synthetic environment (SE) was developed to conduct a human-in-the-loop experiment and validate the task network modeling method used in the simulation. Consistent with the UAV crew positions used in the performance modeling phase, the SE had three control consoles replicating CP140 tactical compartment multifunction workstations. The workstations were designed to communicate with virtual UAVs through fully functional, real-world software interfaces. Each of them had a set of appropriate displays and controls for the UP, UO, and TN as illustrated in Figure 11.14. The experimental environment also had the ability to collect and integrate audio and video data, thus enabling empirical assessments of IAI concepts developed in the first phase.

11.6.7 AIA Evaluation

The aim was to use the IAI hierarchy with various AIAs to support the decision-making and action requirements of operators under three different levels of workload and task complexity. The AIAs were functional components of the UAV control SE developed for this research. They supported the participants in accomplishing the assigned mission tasks of the experiment by providing decision support to the crew and by taking over certain high workload crew tasks. A $3 \times 3 \times 2$ (i.e., Operator Workload: one UAV vs. two UAVs vs. three UAVs; Operator Position Complexity: UP vs. UO vs. TN; IAI Interface Condition: ON vs. OFF) mixed-factor design was used in the experiment, as illustrated in Figure 11.10. The IAI for each crewmember was tailored to suit his/her individual needs. Both Operator Workload and IAI Interface Condition were within-subjects factors, meaning that participants were tested under all levels of those factors, whereas Position Complexity is a between-subjects factor (i.e., crewmembers remain at one position throughout the experiment).

11.6.8 Discussions and Implications

Although the IAI implemented as a prototype system in this case study was only a small subset of a future, more extensive, and fully optimized agent system, the experimental findings indicated that a dynamic and complex system such as multiple disparate UAV control from an airborne platform can be improved through the use of a multiagent IAI suite. More importantly, through the discussions and observations made during the conduct of the study, experience and knowledge were gained

FIGURE 11.14 Experimental environment showing three UAV control stations for UAV pilot, sensor operator, and tactical navigator.

regarding the use of the IAI framework and associated AIA principles for the design of IAI agents, the implementation of synthetic IAI prototype environments, and the conduct of the experiments. Many thoughts are also given to the details of IAI design issues as discussed in Hou *et al.* [45].

This case study on IAI design for the control of multiple UAVs demonstrates that the proposed IAI framework and associated AIA principles provide useful guidance for IAI development. The framework is *complete*: the principles, the models and the analytical methodologies cover all the tasks related to the entire design and evaluation cycle of an HMS; *consistent*: the principles, the models and the analytical methodologies keep a consistent track from abstract guidelines to concrete models and methods; *effective*: the framework speeds up the design and execution of the experiment in both time and personal management because of the clear division between the principles and models; and *user friendly*: the analytical methodologies in Figure 11.2 provide clear processes for the system design.

In summary, the generic IAI framework and system architecture were conceived at the beginning of the design process. The operational scenario analysis and HGA helped identify the types and number of different AIAs to be designed and implemented in the system. Finally, the selected AIAs optimized operator–agent interaction by providing assistance to operators with reduced workload and improving operational effectiveness [44, 45].

11.7 CONCLUSIONS

Interface technologies have evolved from traditional task-centered OMIs for supporting operator interaction with tools, to more user-centered adaptive or intelligent interfaces for facilitating operator interaction with HMSs, to interaction-centered IAIs for optimizing operator–agent interactions. Combined with AI, automation technologies, and AIAs, IAI systems are effective operator decision-making aids. IAIs represent a class of technology that is sufficiently mature to increase human information processing capabilities in all of the most critical applications. This is especially true for UAV control, which involves a variety of workload-intensive activities that potentially impose severe constraints and excessive workload on operators. In the UAV context, IAIs help pilots, sensor operators, and those involved in integrated intelligence, surveillance, and reconnaissance activities. However, designing IAIs is extremely challenging because there are no established design guidelines for this type of advanced interface.

This chapter has introduced currently available theoretical approaches and the requirements for designing an IAI, a generic IAI conceptual framework, and associated methodologies that guide interface design and design verification.

First, different definitions of interfaces from different domains were reviewed. The facilitation of operator interaction with technologies is a common goal of interfaces. OMIs are an aggregate of means that allow the machine to produce the effects of an operator's inputs. In other words, an OMI is a system of an HMS. Interface technologies have evolved from conventional to adaptive or intelligent interfaces to IAIs as a result of a shift in design focus, from a task-centered approach to a task- and user-centered approach, to an interaction-centered approach. With the recognition that an IAI is a closed-feedback loop system, a generic conceptual IAI framework was proposed that consists of four essential components of an IAI system and 10 associated construction models.

Second, the challenges of IAI design were recognized from different aspects: usability principles, development methods, design practice, and scalability. To address these challenges, UCD and agent-based design approaches were introduced as an integral part of the interaction-centered IAI design approach. CommonKADS, EID, and EMD were recommended as baseline interface design methods along with IDEF and HGA for analyzing the models of organization, system, domain, tasks, goals, and various structures of agents for the purpose of implementing an IAI. Through the review of two empirical investigations on IAI technologies, the lack of guidance to optimize operator–agent interaction was revealed as one of the issues for IAI system design.

To facilitate operator–agent interaction, software agent technology and the definition of agent were examined. The concept of AIA was then defined within the context of human–machine interaction, in order to develop a coherent body of interface design knowledge and avoid confusion. An operator–agent interaction model was proposed that may optimize operator–agent interaction based on the philosophy of the proactive use of AIAs and UCD principles.

The primary purpose of AIAs is to take over certain responsibilities from users (who are still in control) and maintain situation awareness. Thus, both users and IAIs can work actively and collaboratively as partners to maximize overall performance. The objective of designing such a system is to provide operators with a descriptive and prescriptive decision-making aid that provides insight into current actions and what actions should be executed next. The similarities and differences among human–human, human–machine, and operator–agent interactions were also reviewed. This helps one understand the benefits of operator–agent interaction optimization as a means to maximize HMS performance.

To achieve the goal of helping operators increase situation awareness and making correct decisions in a timely manner, a generic IAI system hierarchy was proposed. This architecture has executive, senior, working, and junior levels of AIAs that monitor operator needs, task progress, and system status. The executive AIA communicates with the operator by providing decision aids in the right format at the right time. It intelligently changes interface display and control characteristics based on information and knowledge supplied by the senior agents. The four senior agents oversee the internal and external system and environment states, manage the information flow and operation routines, and communicate with internal and external sensors, the operator, and other embedded AIAs. The working agents gather information from both senior and junior level agents, then reason and model the states of the operator, tasks, system, and environment, and then provide feedback and decision-making aids to the senior level agents when necessary. The junior level agents act on direct inputs from the operator, sensors, system, and environment to provide information to the working AIAs for further analysis and decision-making.

To optimize operator–agent interaction, the communication between an operator and AIAs has to be as natural as possible (i.e., resembling human–human interaction). To keep an effective and efficient partnership when AIAs actively take some tasks, an operator should not be a passive observer but an active controller involved in the adaptation processes of knowledge acquisition, attention, reasoning, and decision-making. The partnership was illustrated by an IAI design example for the control of multiple UAVs. Reduced workload and improved situation awareness were empirical evidence that six working level AIAs can facilitate the optimization of operator–agent interaction. Thus, the IAI framework and associated methodologies are useful for guiding IAI design.

The IAI design case study was a small subset of a more extensive system of fully optimized UAV AIAs. As the level of complexity increases with the system, the degree to which AIAs outperform humans may increase. More designs and empirical research need to be conducted to enhance the usefulness of IAI design methodologies and strengthen the robustness of the design models. To do so, the operator–agent interaction model and IAI framework are being used to design operator interfaces on handheld devices for the control of a micro-UAV [44]. The feasibility of implementing the framework will also be carried out with the context of multimodal interface design for the control of UAVs and UAV swarming. The framework will be further developed with research progress in the area of augmented cognition, where an operator's workload and stress can be measured by physiological states [105]. Investigating the connections between workload and physiological states will provide more empirical evidence to optimize operator–agent interaction and improve performance.

ACKNOWLEDGMENTS

The authors thank Paul Hillier and Chunhui Liu for their editing work.

REFERENCES

[1] A. E. Gil, K. M. Passino, and S. Ganapathy. Cooperative task scheduling for networked uninhabited air vehicles. *IEEE Transactions on Aerospace and Electronic Systems*, 44(2): 561–581, 2008.

[2] H. A. Ruff, S. Narayana, and M. H. Draper. Human interaction with levels of automation and decision-aid fidelity in the supervisory control of multiple simulated unmanned air vehicles. *Presence*, 11(4): 335–351, 2002.

[3] National Research Council. Building unmanned group vehicles requires more funding, greater focus. *News from the National Academies*. Available at http://www8.nationalacademies.org/onpinews/newsitem.aspx?RecordID=10592 (accessed January 9, 2003).

[4] NATO Air Force Armaments Group Air Group 7 on Unmanned Aerial Vehicles. Review of UAV Procurement Activities in NATO. September, 2003.

[5] S. D. Manning, C. E. Rash, P. A. LeDuc, R. K. Noback, and J. McKeon. *The Roles of Human Causal Factors in U.S. Army Unmanned Aerial Vehicle Accidents (2004–11)*. Fort Detrick, MD: U.S. Army Aviation Branch Safety Office, U.S. Army Aeromedical Research Laboratory, and U.S. Army Safety Center, 2004.

[6] A. Schulte, C. Meitinger, and R. Onken. Human factors in the guidance of uninhabited vehicles: oxymoron or tautology?. *Cognition, Technology and Work*, 11(1): 71–86. 2009.

[7] G. F. Wilson and C. A. Russell. Performance enhancement in an uninhabited air vehicle task using psychophysiologically determined adaptive aiding. *Human Factors*, 49(6): 1005–1018, 2007.

[8] W. T. Powers. *Behavior: The Control of Perception*. New York: Aldine De Gruyter, 1973.

[9] W. B. Rouse. Adaptive aiding for human/computer control. *Human Factors*, 30(4): 431–443, 1988.

[10] K. Rogers. Crash result of pilot losing track of another jet. *Las Vegas Review-Journal*, 3B, 1999.

[11] P. Guyot and S. Honiden. Agent-based participatory situations: merging multi-agent systems and role-playing games. *Journal of Artificial Societies and Social Simulation*, 9(4). Retrieved July 29, 2012 from this website. Available at http://jasss.soc.surrey.ac.uk/9/4/8.html.

[12] C. A. Miller. Delegation architectures: playbooks and policy for keeping operators in charge. In: *Proceedings of the Workshop on Mixed-Initiative Planning and Scheduling*. Monterey, CA: ICAPS, 2005, pp. 28–35.

[13] C. A. Miller and R. Parasuraman. Designing for flexible interaction between humans and automation: delegation interfaces for supervisory control. *Human Factors*, 49(1): 57–75, 2007.

[14] T. Inagaki. Design of human–machine interactions in light of domain-dependence of human-centered automation. *Cognition, Technology and Work*, 8(3): 161–167, 2006.

[15] T. B. Sheridan and R. Parasuraman. Human-automation interaction. *Reviews of Human Factors and Ergonomics*, 1(41): 89–129, 2005.

[16] T. B. Sheridan and W. L. Verplank. *Human and Computer Control of Undersea Teleoperators*. Cambridge, MA: MIT Man-Machine Systems Laboratory, 1978.

[17] M. R. Endsley and D. B. Kaber. Level of automation effects on performance, situation awareness and workload in a dynamic control task. *Ergonomics*, 42(3): 462–492, 1999.

[18] R. Oppermann and H. Simm. *Adaptive User Support: Ergonomic Design of Manually and Automatically Adaptable Software*. Hillside, NJ: Lawrence Erlbaum, 1994, pp. 14–64.

[19] C. P. Nemeth. *Human Factors Methods for Design: Making Systems Human-Centered*. New York: CRC Press, 2004.

[20] D. A. Bowan and L. F. Hodges. Formalizing the design, evaluation, and application of interaction techniques for immersive virtual environments. *Journal of Visual Languages and Computing*, 10(1): 37–53, 1999.

[21] P. Sukaviriya. From user interface design to the support of intelligent and adaptive interfaces: an overhaul of user interface software infrastructure. *Knowledge-Based Systems*, 6(4): 220–229, 1993.

[22] A. Sutcliffe, S. Kurniawan, and J. E. Shin. A method and advisor tool for multimedia user interface design. *International Journal of Human-Computer Studies*, 64(4): 375–392, 2006.

[23] P. M. Jones, R. W. Chu, and C. M. Mitchell. A methodology for human-machine systems research: knowledge engineering, modelling and simulation. *IEEE Transactions on Systems, Man, and Cybernetics*, 25(7): 1025–1038, 1995.

[24] K. Button. The mac attack: multi-aircraft control of UAVs by a single pilot won't come easily, experts warn. *C4ISR Journal*, 8(9): 30–32, 2009.

[25] D. A. Norman. *The Design of Everyday Things*. New York: Basic Books, 1988.

[26] B. Shneiderman. *Designing the User Interface: Strategies for Effective Human-Computer Interaction*, 3rd edn. Reading, MA: Addison Wesley Longman, 1998.

[27] Z. Yu, Y. Wei, B. Ramkumar, and Y. Guan. An efficient scheme for securing XOR network coding against pollution attacks. In: *Proceedings of INFOCOM 2009*. Rio de Janeiro, Brazil: IEEE, 2009, pp. 406–414.

[28] S. Banbury, M. Gauthier, A. Scipione, and M. Hou. *Literature Research of Design Guidance for Intelligent Adaptive Automation and Interfaces* (CR 2007-075). Toronto, Canada: Defence Research and Development Canada, 2007.

[29] P. M. Fitts. *Human Engineering for an Effective Air Navigation and Traffic Control System*. Washington, DC: National Research Council, Committee on Aviation Psychology, 1951.

[30] C. E. Billings. *Human-Centered Aircraft Automation: A Concept and Guidelines* (NASA Technical Memorandum, 103885). Moffet Field, CA: NASA Ames Research Center, 1991.

[31] M. R. Endsley and E. O. Kiris. The out-of-the-loop performance problem and the level of control in automation. *Human Factors*, 37(2): 3181–3194, 1995.

[32] P. A. Hancock and M. H. Chignell. Mental workload dynamics in adaptive interface design. *IEEE Transactions on Systems, Man, and Cybernetics*, 18(4): 647–658, 1988.

[33] M. Maybury. Intelligent user interfaces: an introduction. In: *Proceedings of the 1999 International Conference on Intelligent User Interfaces*. Redondo Beach, Los Angeles, CA: ACM, 1993, pp. 3–4.

[34] R. J. Keeble and R. D. Macredie. Assistant agents for the worldwide web: intelligent interface design challenges. *Interacting With Computers*, 12(4): 357–381, 2000.

[35] J. Abascal, I. F. D. Castro, A. Lafuene, and J. M. Cia. Adaptive interfaces for supportive ambient intelligence environments. In: *ICCHP '08 Proceedings of the 11th International Conference on Computers Helping People with Special Needs*, edited by K. Miesenberger, J. Klaus, W. Zagler, and A. Karshmer. Heidelberg, Germany: Springer, 2008, pp. 30–37.

[36] D. R. Benyon and D. M. Murray. Adaptive systems: from intelligent tutoring to autonomous agents. *Knowledge-Based Systems*, 6(4): 197–219, 1993.

[37] D. Browne, P. Totterdell, and M. Norman. *Adaptive User Interfaces*. London, UK: Academic Press, 1990.

[38] P. R. Innocent. Towards self-adaptive interface systems. *International Journal of Man-Machine Studies*, 16(3): 287–299, 1982.

[39] N. Mitrovic, J. A. Royo, and E. Mena. Performance analysis of an adaptive user interface system based on mobile agents. In: *Engineering Interactive Systems*, edited by

J. Gulliksen, M. Harning, P. Palanque, G. Van Der Veer, and J. Wesson. Berlin-Heidelberg, Germany: Springer, 2007, pp. 1–18.

[40] W. T. Powers, R. K. Clark, and R. L. McFarland. A general feedback theory of human behavior: part 1. *Perceptual and Motor Skills*, 11: 71–88, 1960.

[41] R. J. Sternberg. *Cognitive Psychology*. Orlando, FL: Harcourt Brace College Publishers, 1996.

[42] B. Tomlinson, E. Baumer, M. L. Yau, P. MacAlpine, L. Canales, A. Correa, B. Hornick, and A. Sharma. Dreaming of adaptive interface agents. In: The ACM Conference on Human Factors in Computing Systems, San Jose, CA, 2007.

[43] K. Hook. Steps to take before intelligent user interfaces become real. *Interacting with Computers*, 12(4): 409–426, 2000.

[44] M. Hou and H. Zhu. Intelligent adaptive interface: a design tool for enhancing human-machine system performance. In: NATO Research and Technology Organisation Human Factors and Medicine Panel Symposium, Sofia, Bulgaria (RTO-HFM-181), 2009.

[45] M. Hou, R. D. Kobierski, and M. Brown. Intelligent adaptive interfaces for the control of multiple UAVs. *Journal of Cognitive Engineering and Decision Making*, 1(3): 327–362, 2007.

[46] K. C. Hendy, D. Beevis, F. Lichacz, and J. L. Edwards. Analyzing the cognitive system from a perceptual control theory point of view. In: *Cognitive Systems Engineering in Military Aviation Environments: Avoiding Cogminutia Fragmentosa* (A report produced under the auspices of The Technical Cooperation Programme Technical Panel HUM TP-7 Human Factors in Aircraft Environments), edited by M. D. McNeese and M.A. Vidulich. Wright-Patterson AFB, OH: Human Systems Information Analysis Center, 2002, pp. 201–252.

[47] D. R. Benyon and D. M. Murray. Special issue on intelligent interface technology: editor's introduction. *Interacting with Computers*, 12(4): 315–322, 2000.

[48] J. L. Edwards. *A Generic Agent-Based Framework for the Design and Development of UAV/UCAV Control Systems* (CR 2004-062). Toronto, ON, Canada: Defence Research and Development Canada, 2004

[49] H. Zhu and M. Zhou. Role-based collaboration and its kernel mechanisms. *IEEE Transactions on Systems, Man, and Cybernetics*, 36(4): 578–589, 2006.

[50] H. Zhu and M. Zhou. Roles in information systems: a survey. *IEEE Transactions on Systems, Man, and Cybernetics*, 38(3): 377–396, 2008.

[51] D. A. Norman and S. W. Draper. *User-Centered System Design*. Hillside, NJ: Lawrence Erlbaum, 1986.

[52] J. Karat. Evolving the scope of user-centered design. *Communications of the ACM*, 40(7): 33–38, 1997.

[53] T. Keinonen. User-centered design and fundamental need. In: *NordiCHI '08 Proceedings of the 5th Nordic Conference on Human-computer Interaction: Building Bridges*, edited by K. Tollmar and B. Jönsson. New York: ACM, 2008, pp. 211–220.

[54] J. Y. Mao, K. Vredenburg, P. W. Smith, and T. Carey. The state of user-centered design practice. *Communications of the ACM—The Disappearing Computer*, 48(3): 105–109, 2005.

[55] W. B. Rouse, N. D. Geddes, and R. E. Curry. An architecture for intelligent interfaces: outline of an approach to supporting operators of complex systems. *Human-Computer Interaction*, 3(2): 87–122, 1987.

[56] Y. Xiao, P. Milgram, and D. J. Doyle. Planning behavior and its functional role in interactions with complex systems. *IEEE Transactions on Systems, Man, and Cybernetics*, 37(6): 313–324, 1997.

[57] M. Maybury. Intelligent user interfaces: an introduction. In: *Proceedings of the 4th International Conference on Intelligent User Interfaces*, edited by M. Maybury, P. Szekely, and C. G. Thomas. New York: ACM, 1999, pp. 3–4.

[58] O. Etzioni and D. S. Weld. Intelligent agents on the Internet: fact, fiction, and forecast. *IEEE Expert*, 10(4): 44–49, 1995.

[59] A. T. Schreiber, J. Akkermans, A. Anjewierden, R. De Hoog, N. Shadbolt, W. Van De Velde, and B. Wielinga. *Knowledge Engineering and Management: The Commonkads Methodology*. Cambridge, MA: The MIT Press, 2000.

[60] W. C. Yoon and J. M. Hammer. Deep-reasoning fault diagnosis: an aid and a model. *IEEE Transactions on Systems, Man, and Cybernetics*, 18(4): 659–676, 1988.

[61] C. A. Iglesias, M. Garijo, J. C. Gonzalez, and J. R. Velasco. Analysis and design of multi-agent systems using MAS-commonkads. In: *ATAL '97 Proceedings of the 4th International Workshop on Intelligent Agents IV, Agent Theories, Architectures, and Languages*, edited by M. P. Singh, A. Rao, and M. J. Wooldridge. London, UK: Springer, 1998, pp. 313–328.

[62] T. J. Lydiard. Using idef3 to capture the air campaign planning process. Artificial Intelligence Applications Institute, University of Edinburgh, UK. Retrieved July 29, 2012 from this URL. Available at http://www.aiai.ed.ac.uk/~arpi/ACP-MODELS/ACP-OVERVIEW/96-MAR/HTML-DOC/idef3.html, 1995.

[63] R. J. Mayer, C. P. Menzel, M. K. Painter, P. S. DeWitte, T. Blinn, and B. Perakath. *Information Integration for Concurrent Engineering (IICE) idef3 Process Description Capture Method Report* (AL-TR-1995-XXXX). College Station, TX: Knowledge Based Systems, Inc., 1995.

[64] J. L. Edwards and K. C. Hendy. A testbed for intelligent aiding in adaptive interfaces. In: The Spring Symposium on Adaptive User Interfaces, Stanford University, March 20–22, 2000. Retrieved July 29, 2012 from this URL. Available at https://www.aaai.org/Papers/Symposia/Spring/2000/SS-00-01/SS00-01-008.pdf, 2000.

[65] K. C. Hendy, D. Beevis, F. Lichasz, and J. Edwards. Analyzing the cognitive system from a perceptual control theory point of view. In: *Cognitive Systems Engineering in Military Aviation Environments: Avoiding Cogminutia Fragmentosa!*, edited by M. D. McNeese and M. A. Vidulich. Dayton, OH: WPAFB, 2002.

[66] K. J. Vicente and J. Rasmussen. Ecological interface design: theoretical foundations. *IEEE Transactions on Systems, Man, and Cybernetics*, 22(4): 589–606, 1992.

[67] K. Vicent. Ecological interface design: progress and challenges. *Human Factors*, 44(1): 62–78, 2002.

[68] C. M. Burns and J. R. Hajdukiewicz. *Ecological Interface Design*. Boca Raton, FL: CRC Press, 2004.

[69] C. M. Burns, G. Skraaning, N. Lau, G. Jamieson, J. Kwok, R. Welch, and G. Andresen. Evaluation of ecological interface design for nuclear process control: situation awareness effects. *Human Factors*, 50(4): 663–679, 2008.

[70] T. C. Davies and C. M. Burns. Advances in cognitive work analysis and the design of ecological visual and auditory displays. *Cognitive Technology*, 13(2): 17–23, 2008.

[71] H. Furukawa and R. Parasuraman. Supporting system-centered view of operators through ecological interface design: two experiments on human-centered automation. *Proceedings of the 47th Annual Meeting of the Human Factors and Ergonomics Society,* 47(3): 567–571, 2003.

[72] N. Lau, G. Skraaning, G. Jamieson, and C. M. Burns. Ecological interface design in the nuclear domain: an empirical evaluation of ecological displays for the secondary subsystems of a boiling water reactor plant simulator. *IEEE Transactions on Nuclear Science*, 55(6): 3597–3610, 2008.

[73] P. P. Morita, C. M. Burns, and S. J. Calil. The influence of strong recommendations, good incident reports and a monitoring system over an incident investigation system for healthcare facilities. In: Proceedings of the 53rd Annual Meeting of the Human Factors and Ergonomics Society, 2009, pp. 1679–1683.

[74] D. M. Lamberti and W. A. Wallace. Intelligent interface design: an empirical assessment of knowledge presentation in expert systems. *MIS Quarterly*, 14(3): 279–311, 1990.

[75] J. Eklund and K. Sinclair. An empirical appraisal of the effectiveness of adaptive interfaces for instructional systems. *Educational Technology and Society*, 3(4): 165–177, 2000.

[76] K. B. Bennett and S. Hettinger. A theoretical analysis and preliminary investigation of dynamically adaptive interfaces. *The International Journal of Aviation Psychology*, 11(2): 169–195, 2001.

[77] Z. Duric, W. D. Gray, R. Heishman, F. Li, A. Rosenfeld, M. J. Scholles, C. Schunn, and H. Wechsler. Integrating perceptual and cognitive modeling for adaptive and intelligent human-computer interaction. *Proceedings of IEEE*, 90(7): 1272–1289, 2002.

[78] L. Maat and M. Pantic. Gaze-x: adaptive, affective, multimodal interface for single-user office scenarios. In: *Proceedings of the ICMI 2006 and IJCAI 2007 International Conference on Artificial Intelligence for Human Computing*, edited by T. S. Huang, A. Nijholt, M. Pantic, and A. Pentland. Heidelberg, Germany: Springer, 2007, pp. 251–271.

[79] K. Z. Gajos, M. Czerwinski, D. S. Tan, and D. S. Weld. Exploring the design space for adaptive graphical user interfaces. In: *Proceedings of the Working Conference on Advanced Visual Interfaces*. New York: ACM, 2006, pp. 201–208.

[80] H. Lieberman. Interfaces that give and take advice. In: *Human-Computer Interaction in the New Millennium*, edited by J. Carroll. New York: ACM, 2002, pp. 475–485.

[81] B. Shneiderman and P. Maes. Direct manipulations vs. interface agents. *Interactions*, 4(6): 42–61, 1997.

[82] J. Grossklags and C. Schmidt. Software agents and market (in) efficiency: a human trader experiment. *IEEE Transactions on Systems, Man, and Cybernetics*, 36(1): 56–67, 2006.

[83] A. L. Baylor, R. B. Rosenberg-Kima, and E. A. Plant. Interface agents as social models: the impact of appearance on females' attitude toward engineering. In: *CHI '06 Extended Abstracts on Human Factors in Computing*. New York: ACM, 2006, pp. 526–531.

[84] J. J. Gómez-Sanz, J. Pavón, and J. P. F. Garijo. Intelligent interface agents behaviour modelling. In: *Lecture Notes in Computer Science—Mexican International Conference on Artificial Intelligence*, edited by O. Cairó, L. Sucar, and F. Cantu. Heidelberg, Germany: Springer, 2000, pp. 598–609.

[85] B. Laurel. Interface agents. In: *The Art of Human-Computer Interface Design*, edited by B. Laurel. UK: Addison-Wesley, 1990, pp. 355–365.

[86] J. Bradshaw. *Software Agents*. Cambridge, MA: The MIT Press, 1997.

[87] T. Finin, C. Nicholas, and J. Mayfield. Tutorial: agent-based information retrieval. In: *Proceedings of the Second ACM International Conference on Digital Libraries*. New York: ACM, 1997.

[88] M. Wooldridge. *Intelligent Agents*. Cambridge, MA: The MIT Press, 1999.

[89] M. Wooldridge. *An Introduction to Multiagent Systems*, 1st edn. Chichester, West Sussex, UK: John Wiley & Sons, 2002.

[90] M. Beaudouin-Lafon. Instrumental interaction: an interaction model for designing post-wimp user interfaces. In: *CHI '00 Proceedings of the SIGCHI Conference on Human factors in Computing Systems*. New York: ACM, 2000, pp. 446–453.

[91] J. Rasmussen. *Information Processing and Human-Machine Interaction: An Approach to Cognitive Engineering*. North-Holland Series in System Science and Engineering Edition, Vol. 12. New York: Elsevier Science, Inc., 1986.

[92] M. E. Bratman, D. J. Israel, and M. E. Pollack. Plans and resource-bounded practical reasoning. *Computational Intelligence*, 4(3): 349–355, 1988.

[93] M. Georgeff, B. Pell, M. Pollack, M. Tambe, and M. Wooldridge. The belief-desire-intention model of agency. In:*Proceedings of the 5th International Workshop on Intelligent Agents V: Agent Theories, Architectures, and Languages (ATAL-98)*, edited by J. Müller, M. P. Singh, and A. S. Rao. Lecture Notes in Artificial Intelligence, 1555. Berlin, Germany: Springer, 1999, pp. 1–10.

[94] B. Jarvis, D. Corbett, and L. C. Jain. Beyond trust: a belief-desire-intention model of confidence in an agent's intentions. In: *Proceedings of the 10th International Conference on Knowledge-Based Intelligent Information and Engineering Systems*, edited by R. Khosla, R. J. Howlett, and L. C. Jain. 3681. Berlin, Germany: Springer, 2005, pp. 844–850.

[95] J. Sudeikat, L. Braubach, A. Pokahr, and W. Renz. On the validation of belief–desire–intention agents. In: *ProMAS'06 Proceedings of the 4th International Conference on Programming multi-Agent Systems*, edited by R. H. Bordini, M. Dastani, J. Dix, and A. E. F. Seghrouchni. Heidelberg, Germany: Springer, 2007, pp. 185–200.

[96] P. R. Cohen and H. J. Levesque. Intention is choice with commitment. *Artificial Intelligence*, 42(2–3): 213–261, 1990.

[97] B. Lajos. Adaptive human-computer interfaces for man-machine interaction in computer-integrated systems. *Computer Integrated Manufacturing Systems*, 8(2): 133–142, 1995.

[98] N. B. Sarter. Multimodal information presentation: design guidance and research challenges. *Journal of Industrial Ergonomics*, 36(5): 439–445, 2006.

[99] V. Santangelo, C. Ho, and C. Spence. Capturing spatial attention with multisensory cues. *Psychonomic Bulletin and Review*, 15(2): 398–403, 2008.

[100] S. F. Scallen, P. A. Hancock, and J. A. Duley. Pilot performance and preference for short cycles of automation in adaptive function and allocation. *Applied Ergonomics*, 26(6): 397–403, 1995.

[101] H. M. Cuevas, S. M. Fiore, B. S. Caldwell, and L. Strater. Augmenting team cognition in human-automation teams performing in complex operational environments. *Aviation, Space, and Environmental Medicine*, 78(5, Sec. II): B63–B70, 2007.

[102] B. Tomlinson, E. Baumer, M. L. Yau, P. M. Alpine, L. Canales, A. Correa, B. Hornick, and A. Sharma. Dreaming of adaptive interface agents. In: *Proceedings of the Computer/Human Interaction 2007 Extended Abstracts on Human Factors in Computing Systems*. San Jose, CA: CHI, 2007, pp. 2007–2012.

[103] R. Parasuraman, T. B. Sheridan, and C. D. Wickens. A model for types and levels of human interaction with automation. *IEEE Transactions on Systems, Man, and Cybernetics*, 30(3): 286–297, 2000.

[104] D. Dahn and K. R. Laughery. The integrated performance modeling environment – simulating human-system performance. In: *Proceedings of the 1997 Winter Simulation Conference*, edited by S. Andradottir, K. J. Healy, D. H. Withers, and B. L. Nelson. Atlanta, GA: ACM 1997, pp. 1141–1145.

[105] D. Schmorrow, D. Nicholson, J. Drexler, and L. Reeves. *Foundations of Augmented Cognition*, 4th edn. Arlington, VA: Strategic Analysis, Inc., 2007.

CHAPTER 12

A COMPLEX ADAPTIVE SYSTEM OF SYSTEMS APPROACH TO HUMAN–AUTOMATION INTERACTION IN SMART GRID

ALIREZA FEREIDUNIAN[1], HAMID LESANI[2], MOHAMMAD ALI ZAMANI[2,3], MOHAMAD AMIN SHARIFI KOLARIJANI[2,4], NEGAR HASSANPOUR[2,5], and SINA SHARIF MANSOURI[2,6]

[1]Smart Grid Lab, Faculty of Electrical Engineering, K. N. Toosi University of Technology, Tehran, Iran
[2]Systems and Machines Research Lab, Control and Intelligent Processing Center of Excellence, University of Tehran, Tehran, Iran
[3]Robotics Lab, Department of Computer Science, Özyeğin University, Istanbul, Turkey
[4]Bioelectrics Department, Biomedical Engineering Faculty, Amirkabir University of Technology, Tehran, Iran
[5]Computational Intelligence Laboratory, University of Northern British Columbia, Prince George, BC, Canada
[6]Automation & Robotics Lab Dortmund, Dortmund, Germany

12.1 INTRODUCTION

Human–automation interaction (HAI) is a typical example of a complex and adaptive phenomenon in two senses: firstly, the collaboratively performed tasks of humans and automation systems are complex and should be adapted to the changing environment; and secondly, the human–automation collaboration itself is a complex phenomenon and should be adapted to the changing environment. Thus, an HAI system can be regarded as a complex adaptive system (CAS).

An HAI scheme primarily consists of the following systems: a core task process (depending on the application field that is electric Smart Grid, in this chapter), a

Contemporary Issues in Systems Science and Engineering, First Edition.
Edited by MengChu Zhou, Han-Xiong Li and Margot Weijnen.

human system (as both user and operator), an automation (and control) system, an information technology (IT) system, an HAI sphere, an HAI regulator, and a surrounding environment.

A conceptual outline of this chapter is depicted in Figure 12.1, expressing three cognitive genres representing this chapter as: descriptive genre, normative/prescriptive genre, and know-how genre. The descriptive genre (the second row of the table in Figure 12.1) introduces and discusses on the following disciplines: the core concepts of complexity, CASs and system of systems (SoS) (Sections 12.2–12.5), the application domains of HAI and adaptive autonomy (Section 12.6), especially in Smart Grid (Section 12.7), and two implementation ideas of expert systems (Section 12.9) and Petri nets (Section 12.8). Going upward, the third row in table of Figure 12.1 is related to the normative/prescriptive genre of this chapter, prescribing the HAI in Smart Grids as being a CAS of systems (Section 12.7). The upmost row in the table of Figure 12.1 is relevant to the know-how genre of this chapter, presenting a realization methodology (know how) for adaptive autonomy using hierarchical Petri nets (Sections 12.9 and 12.10).

The rest of this chapter is organized as follows: Sections 12.2–12.5 discuss on the core concepts of complexity, CASs, SoS and complex adaptive system of systems (CASoS). Section 12.6 describes the automation related notions, that is, automation and HAI, followed by an investigation of the HAI models' evolution from both perspectives of dimensions and dynamism. The idea of adaptive autonomy (AA) is then introduced as a dynamic HAI scheme, followed by a classification of AA implementation methods. Section 12.7 is dedicated to a discussion on HAI in Smart Grid as a CAS and as an SoS. Petri nets are introduced as powerful tools for modeling complex systems (CxS) in Section 12.8. Section 12.9 introduces a model-based implementation methodology for adaptive autonomy concept. Finally, a Petri net realization of the adaptive autonomy expert system (AAES) is presented, followed by a performance evaluation study in Section 12.10. Section 12.11 sums up the chapter and enumerates a couple of future prospective and open questions regarding to the subject of the chapter for future research.

12.2 COMPLEXITY IN SYSTEMS SCIENCE AND ENGINEERING

12.2.1 The Nature of Complexity

Human brain is associated with the Cartesian[1] approach of decomposing problems to simpler and easier subproblems [1]; therefore, a *complex problem* is a problem which cannot easily be decomposed (reduced) to simpler subproblems.

While being indecomposable to simpler subproblems, complexity can be attributed to two different sources:

(1) The complex entity (innate complexity).
(2) The observer (cognitive complexity).

[1]Cartesian: pertaining to René Descartes (1596–1650), to whom the reductionism approach is attributed.

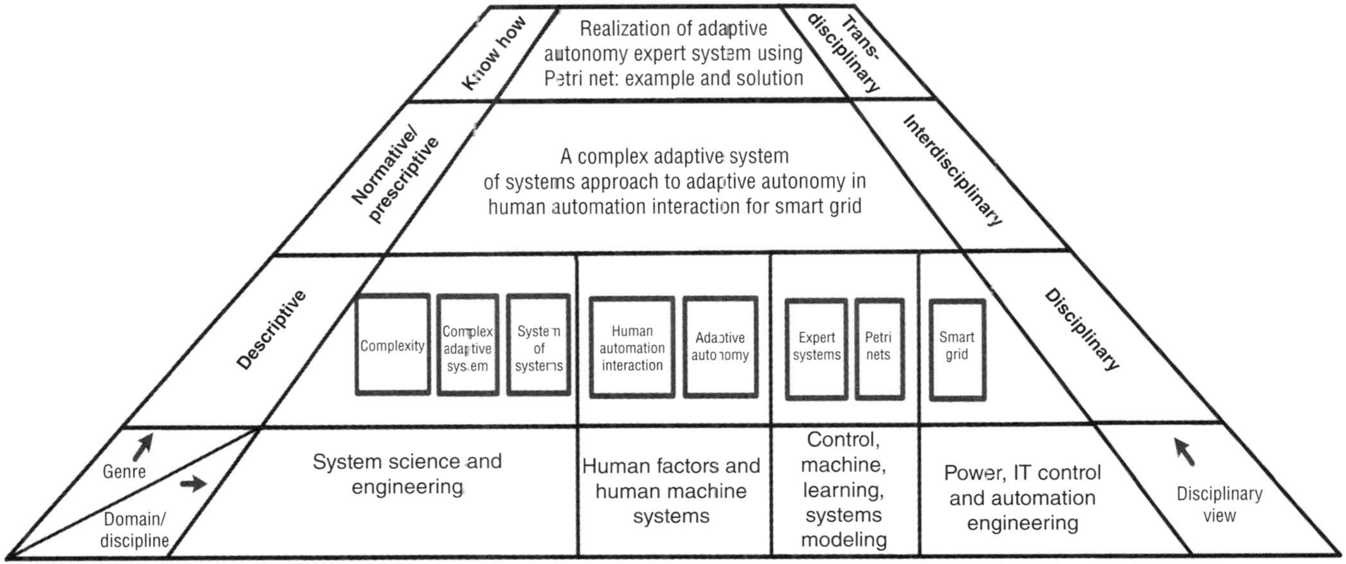

FIGURE 12.1 Conceptual outline of this chapter: genres versus disciplines/domains.

The innate complexity is the one that is inherently sourced by the complex entity itself; while the cognitive complexity is the one sourced by the observer's cognitive limits during observing an entity that is not necessarily inherently complex. The former notion is considered regardless of an observer's cognitive limits; while the latter depends on an observer's level of expertise and intelligence; thus, it does not represent a genuine expression of complexity. The former can be simply called complexity and described as a state of the world; whereas the cognitive complexity can be called as complicatedness (i.e., making confused) and described as a state of mind when responding to complexity[2] [2]. In fact, many real-world problems are complex, nonlinear and unpredictable; thus, science (and engineering) is expected to devise methods to explore and tame the complexity.

In line with efforts to explore the notion of complexity and complex problems, Warren Weaver categorized the scientific problems into three classes [3]:

(1) *Problems of simplicity*: concerning two (say few, like: three, four, or five) variables, approached by Newtonian physics and calculus mathematics, belonging to the nineteenth century.
(2) *Problems of disorganized complexity*: concerning astronomical number of variables, approached by statistical or probabilistic mathematics and quantum mechanics, belonging to the twentieth century.
(3) *Problems of organized complexity*: concerning an intermediate number of variables, expectedly approached by complexity management methods, belonging to the twenty-first century.

Apparently, few-variable problems in physical sciences can be tackled by simplicity methods. They can be decomposed by a Cartesian approach to easier subproblems. However, the multivariable complex problems could not be tackled by the same simplicity methods as few variables. Weaver asserts that this weakness of simplicity methods has led the complex problems (especially in life sciences) to "*become highly quantitative or analytical in character.*" [3]

On problems of complexity, Weaver regarded the problems of complexity variables as two different scales: astronomical number of variables (problems of disorganized complexity) and large yet sizeable number of problems (problems of organized complexity).

Problems of disorganized complexity are problems in which the variables are too many and each of these many variables "*has a behavior which is individually erratic, or perhaps totally unknown.*" Weaver continues then: "... *in spite of unknown behavior of all the individual variables, the system as a whole possesses certain orderly and analyzable average properties.*" Due to the very high number of variables in problems of disorganized complexity, statistical and probabilistic methods are utilized to predict the average behavior of the whole problem. Application examples of

[2]This approach to complexity sources (innate or cognitive) is a matter of controversy; however, more investigation on this issue might be furthered in future research.

disorganized complexity can be found in insurance, telephone, and quantum physics [3].

The other class of complex problems is the class of organized complexity, which deals with a sizable number of variables that are interrelated into an organic whole [3]. Problems of organized complexity lay on a spectrum, somewhere between few-variable methods of simplicity and astronomical number of variables methods of disorganized complexity. Application examples of organized complexity include: "*what is the description of aging in biochemical terms? Do complex proteins 'know how' to reduplicate their pattern? On what does the price of wheat depend? How can currency be wisely and effectively stabilized?*" [3]. Reference 4 associates the organized complexity with a *significant* emergent behavior in the complex problem elements.

12.2.2 Complex Systems

Systems—like problems—can be complex, and CxSs are systems whose study leads to complex problems. A system can be complex from one point of view, while not complex from another point of view; as Bouwmans *et al.* states "*systems that have the potential to be 'complex' (by any formal definition we would adopt), do not necessarily show complex behaviour [behavior] under all conditions.*" [5] Similar to a complex problem, a *complex system* is a system that cannot easily be decomposed to simpler subsystems [1].

CxSs, as most definitions of CxS suggest, are unpredictable in their behavior. However, the complex system components' behavior might be either known (predictable or deterministic) or unknown (unpredictable or stochastic). This unknown behavior of system components might be caused by either unpredictable or stochastic nature of a finite number of individual components or by astronomical number of deterministically natured components. Thus, we can argue that the unpredictable system components only can be associated with disorganized complexity, while predictable system components can be associated with organized complexity. Table 12.1 represents a source–problem description of CxS classification in which a system may belong to one (or two) of the four classes of A–D.

As in Table 12.1, the complexity classes can be described as follows:

- Complexity Class A: systems with innate source of complexity and organized complexity problem.

TABLE 12.1 Source–Problem Representation of Complex Systems Classes

	Source of complexity →		
↓Relevant problem of complexity	↓System components	Innate	Cognitive
Organized	Predictable	Class A	Class B
Disorganized	Unpredictable	Class C	Class D

- Complexity Class B: systems with cognitive source of complexity and organized complexity problem.
- Complexity Class C: systems with innate source of complexity and disorganized complexity problem.
- Complexity Class D: systems with cognitive source of complexity and disorganized complexity problem.

We defined a *complex system* as *a system which cannot easily be decomposed (reduced) to simpler subsystems*. Nevertheless, many different definitions of complexity have been presented, based on different perceived sources and types of complexity that depend on the discipline and application field. A couple of useful and concise definitions of complexity are presented here, followed by a discussion on its view on source–problem of complexity.

One of the most concise and most well-known definitions of complexity[3] is: "*The complexity of a system is the degree of difficulty in predicting the properties of the system, given the properties of the system's component.*" This definition of complexity is credited to Weaver [1, 6], although Weaver's paper itself does not directly (or even indirectly) point to the given definition. Even we could assert that Weaver's description for disorganized complexity is opposite to that given in this definition, where he states: "*one in which each of the many variables has a behavior which is individually erratic or perhaps totally unknown*" [3]. Therefore, we conclude that the above mentioned definition of complexity points to the organized complexity problems, disregarding the complexity source (i.e., Classes A and B, in Table 12.1). In fact, according to major writings in complexity, the organized complexity is the main type of complexity in the twenty-first century.

In Reference 7, Edmonds reviewed several definitions of complexity and finally proposed a concise definition as "*that property of a language expression which makes it difficult to formulate its overall behaviour [behavior] even when given almost complete information about its atomic components and their inter-relations.*" This definition also points to the organized complexity problem, disregarding the complexity source (i.e., Classes A and B, in Table 12.1).

Price gives an interesting definition of complexity: "*[life is] … a property of improbable complexity possessed by an entity that works to keep itself out of equilibrium with its environment.*" [8] (Quoted from [9]). Based on Dawkins's notion, Price states that a measure of system complexity is the degree in which it defies the thermodynamics equilibrium. For instance, if a dead bird is thrown into the air, it obeys the laws of mechanics, going up and falling down in a parabola; while a live thrown bird runs away, disregarding laws of mechanics. Price's definition points to the innate source of complexity and the organized complexity problem ("*improbable complexity*"), pointing to Class A of Table 12.1.

[3]When we define complexity, the definition of complex system is in hand: Complexity is "the quality or condition of being complex" (Oxford English Dictionary). A complex system is a system which embodies the attribute of complexity.

Kinsner presents another view on definition of complexity: "*a large number of interacting elements with many degrees of freedom whose individual behaviour [behavior] could not be traced back or predicted*" [10]. Kinsner's definition regards unpredictable behavior of individual components (disorganized complexity). However, he has not mentioned that "*could not be traced back or predicted*" is sourced to the observer (cognitive complexity) or to the complex entity itself (innate complexity). Thus, Kinsner's definition points to Classes C and D of Table 12.1.

From another point of view, J. N. Warfield defines complexity as a state of mind: "*that sensation experienced in the human mind when, in observing or considering a system, frustration arises from lack of comprehension of what is being explored*" [11, 12]. Warfield's definition of complexity as a state of mind applies to the cognitive source of complexity, yet disregarding the problem of complexity (organized or disorganized), i.e., mentioning Classes B and D in Table 12.1.

To sum up, D. K. Hitchins has commented as: "*Most definitions have something to say about an inability to predict the behaviour [behavior] of the whole system, knowing the behaviours [behaviors] of the interacting subsystems*" [13]. Further, on the characteristics of a complex system, he has added: "*We are talking about non-linear interactions, emergence, open systems (which adapt as they exchange energy, information and material with their environment). We also appear to be suggesting degrees of self similarity, with subsystems in systems and sub-subsystems within subsystems*" [13]. Hitchin's description on CxS belongs to Class A of Table 12.1.

We will discuss on characteristics of CxS more in the next subsections.

12.2.3 Complexity Measures

In science and engineering, we study classifications and taxonomies in order to attribute certain relevant characteristics to certain classes. Of course, almost all of those taxonomies and classifications are not crisp classifications; on the contrary, they are mostly fuzzy ones. Consider the class of tall people: why we might be interested in classifying people into tall and short sets? Because we can attribute some features to those classes of tall and short people: tall people need bigger spaces in cars, higher chairs or tables in offices, etc. This classification obviously helps us in facilitating peoples' lives; however, the tall people class is rather fuzzy and we cannot strictly say who is tall. But should we stop thinking about facilitating tall people, by an excuse of the tall people class being fuzzy? We would say no. Complexity is similar in the way that it is fuzzy too; we can talk about degrees of complexity. Some systems are strongly complex; whereas, some systems are weakly complex. So we accept that the complex system boundary is fuzzy, and acknowledge the sophistication of the definition of complexity and its characteristics.

In fact, cognition and management of systems are considered as main gains of the complexity theory. Therefore, complexity measures could lead to understanding the emergence phenomena in physical, biological, and societal systems and clarifying the level of randomness and correlation. Moreover, it can contribute to enhancing

TABLE 12.2 Measuring Complexity Type Classification by Reference 10

Measuring Complexity Classification	Description (difficulty of describing . . .)
Structural	m-dimensional object
Dynamical	The patterns of systems trajectory
Functional	The functionality of the system
Synergetic	The level of societal development
Design	Difficulty of designing, embodiment, verification, testing, and maintaining

the perceptual and cognitive processes and it could also determine self-organization [10].

In Reference 10, Kinsner introduced five types of complexity: structural, dynamic, functional, organizational, and design complexity which are summarized in Table 12.2. He also introduced different ways to express complexity which include: (i) local or global, (ii) single scale or multiscale, (iii) algorithmic or probabilistic, (iv) absolute, differential, or relative, (v) static or dynamic, (vi) average or asymptotic, and (vii) arithmetic or logical. Therefore, according to Reference 10, complexity can be measured in five types and expressed in seven ways. This two-dimensional (2D) classification of complexity types and ways is represented in Figure 12.2 in which, the seven ways of expressing complexity are shown in the horizontal plane, while

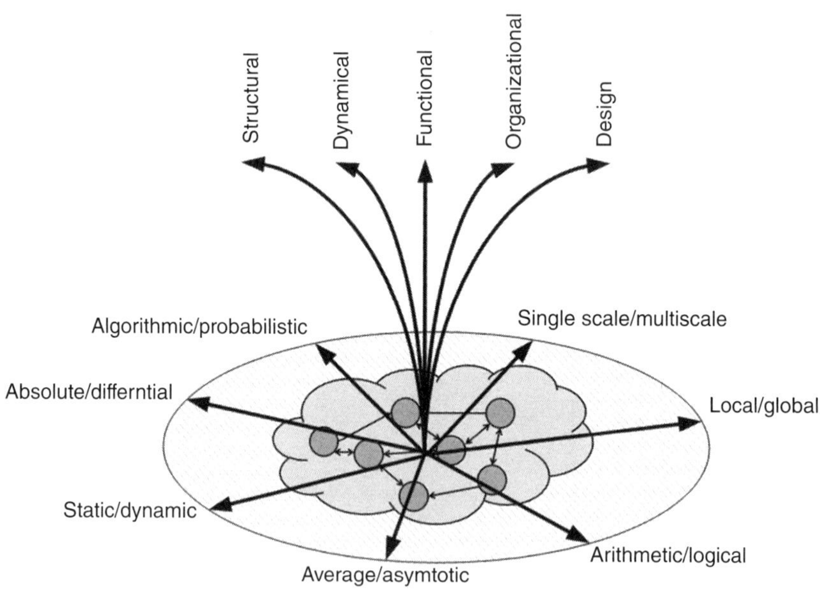

FIGURE 12.2 Two-dimensional classification of complexity types and ways according to Reference 10.

the five types of complexity are shown in vertical axes. A comprehensive survey on complexity measurement is presented in References 10, 14–17.

Alongside with definitions provided for complexity in relevant subsections (12.2.1 and 12.3.2), it is worthy to mention some forms of complexity categorized by Reference 18 "*as:*

(a) *Lack of knowledge in characterizing the behavior of process (Unit Behavioral Complexity).*
(b) *Complexity of computational engine associated with a subprocess (Computational Complexity).*
(c) *Difficulties in characterizing the interconnection topology (Interconnection Topology Complexity).*
(d) *Organizational alternatives for decision making (Organizational Complexity).*
(e) *Variability, uncertainty and multi-level couplings in the system's organization in describing the overall system organization (System of Systems Complexity).*
(f) *Large scale dimensionality impacting on methodologies (Large Scale Complexity)*
(g) *Heterogeneous nature of sub-processes, resulting in behavior (Hybrid Behavioral Complexity).*
(h) *Variability and/or uncertainty on the system's environment during the lifecycle requiring flexibility in organization and operability (Lifecycle Complexity)*".

12.2.4 Complexity-Related Terms in Literature

Figures 12.3 and 12.4 show the number of articles including *complexity* and *complex system* in their titles, by decade since 1950 from the two sources of *ISI Web of*

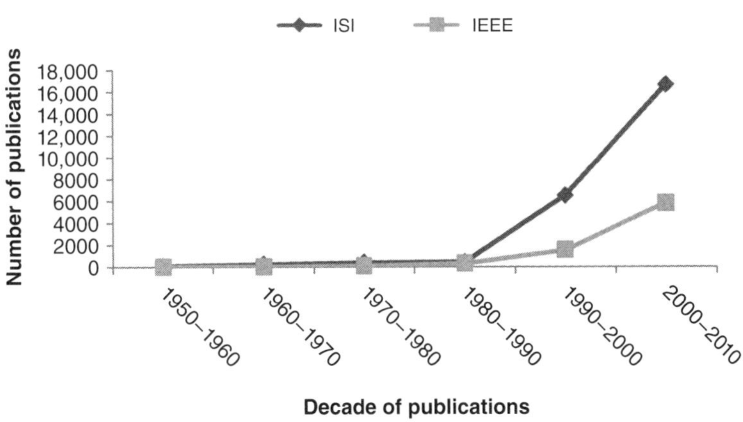

FIGURE 12.3 Number of titles containing *Complexity* by decade in ISI Web of Knowledge and IEEEXplore.

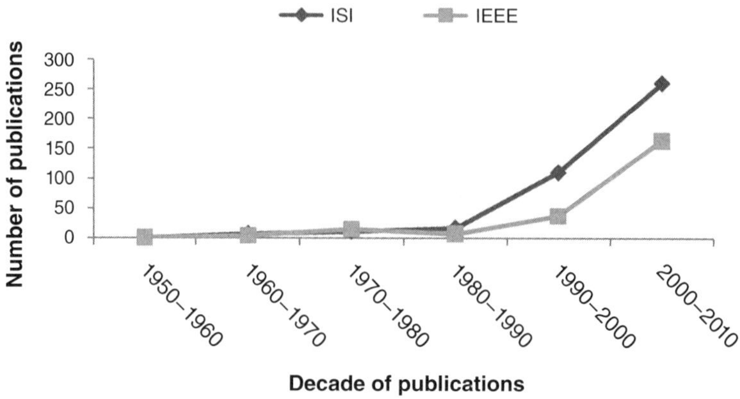

FIGURE 12.4 Number of titles containing *Complex System* by decade in ISI Web of Knowledge and IEEEXplore.

Knowledge and *IEEEXplore*. As shown, these concepts were introduced before 1950s; however, it was during 1980s that researchers showed considerable interest in these areas. Since then, the number of publications on *complexity* and *complex system* has been exponentially increasing due to growing application of these concepts in a variety of disciplines.

The industrial need, as well as research on complexity in systems science and engineering, has led to the introduction of several complexity-related terms: CASs and SoS, to be discussed in the next sections. Figures 12.5 and 12.6 show the number of articles including *CAS* and *SoS* titles respectively by decade since 1950 in *ISI Web*

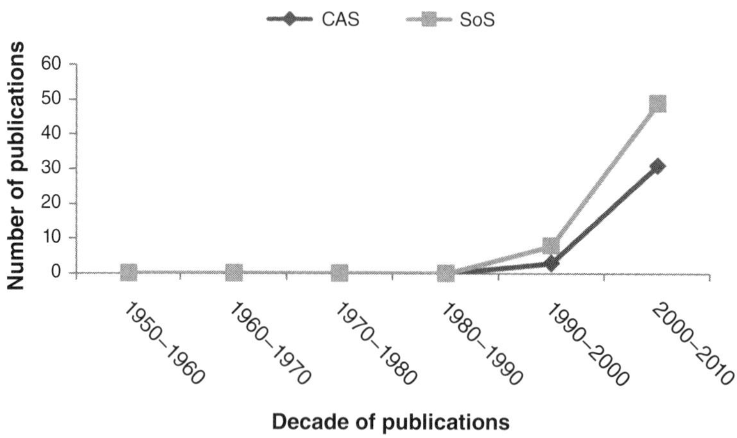

FIGURE 12.5 Number of titles containing *Complex Adaptive System* and *System of Systems* respectively by decade in ISI Web of Knowledge.

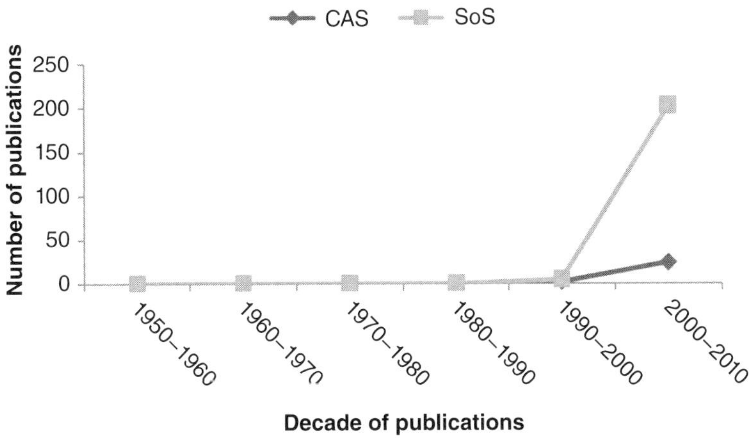

FIGURE 12.6 Number of titles containing *Complex Adaptive System* and *System of Systems* respectively by decade in IEEEXplore.

of Knowledge and *IEEEXplore*. As seen, both CAS and SoS concepts are rather new and it was after 1990 that research interests in these concepts increased drastically, specially on SoS.

Figures 12.7 and 12.8 show the number of articles including *CAS* and *SoS* titles respectively by year during last two decades in *ISI Web of Knowledge* and *IEEEXplore*. As seen, both CAS and SoS concepts are rather new and it was after 1990 that research interests in these concepts increased drastically, specially on SoS.

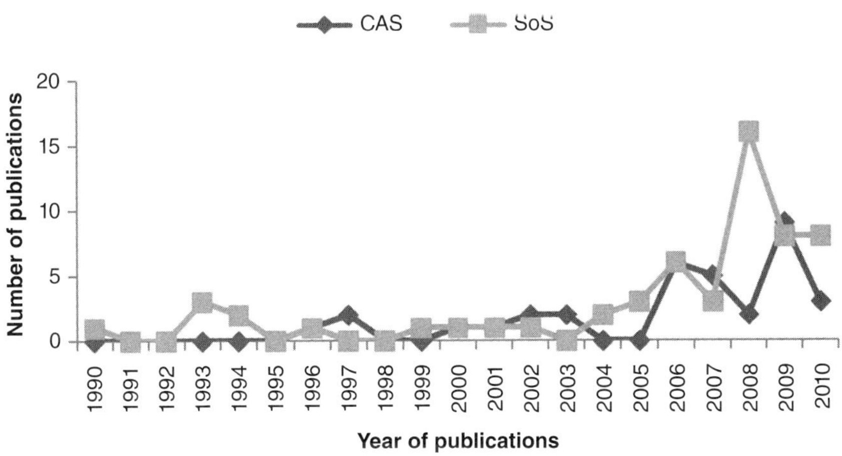

FIGURE 12.7 Number of titles containing *Complex Adaptive System* and *System of Systems* respectively by year in ISI Web of Knowledge.

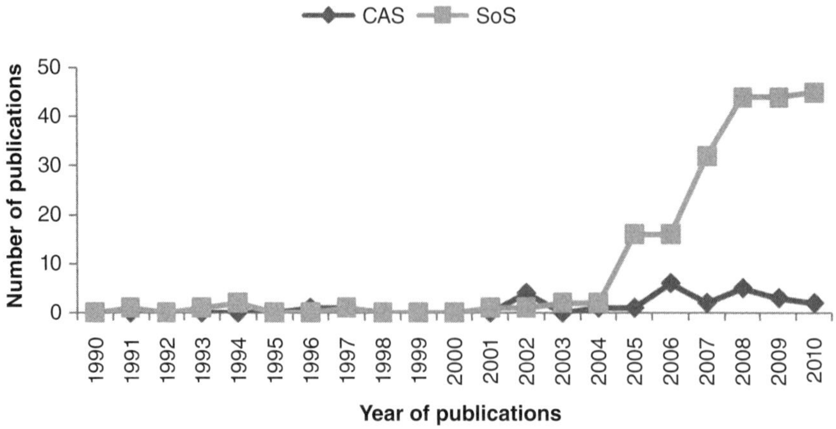

FIGURE 12.8 Number of titles containing *Complex Adaptive System* and *System of Systems* respectively by year in IEEEXplore.

12.3 COMPLEX ADAPTIVE SYSTEMS

12.3.1 What are Complex Adaptive Systems?

The term *Complex Adaptive Systems* (CASs) was coined at the *Santa Fe Institute* by J. H. Holland (father of Genetic Algorithms), M. Gell-Mann (Physics Nobel laureate), S. Kauffman (Biologist) and K. Arrow (Economics Nobel laureate) [19–24]. As a result of efforts at the Santa Fe Institute, a "*common theoretical framework for complexity*" was developed. This framework was built based on the previous works in many different disciplines such as biology, artificial intelligence, cybernetics, neural networks, ecology, economics, and chaos theory [20]. The basic idea of CAS is believed to be introduced in 1960s by the Belgian Nobel laureate I. Prigogine who was studying self-organizing structures in the nature [20].

Apparently, complex system analysis techniques differ from those of the conventional (classical) techniques as compared in Table 12.3 [25, 26]. The left column in Table 12.3 is associated with the problems of simplicity, while the right column is associated with the problems of complexity as described in Section 12.2.1 [3].

A CAS is defined as a dynamic network of heterogeneous agents that adapt or learn as they interact [20, 27]. These agents (representing cells, species, individuals, firms, nations) act in parallel, constantly exchange information and influence each other and the environment in a conditional manner—using IF/THEN structures—based on signals they receive. They also have the ability to adapt their behavior as a result of their experience [27, 28]. The overall behavior of the system is the result of the cooperation of agents. Since the relationships between causes and effects are nonlinear and actions of some parts affect actions of other parts, the system as a whole shows emergent properties which cannot be understood referring to the individual behaviors of the agents [20, 29, 30].

TABLE 12.3 Classical Versus Complex System Analysis Techniques [26]

Classical problems	Complex problems
Mechanistic, linear, separate parts, events, moments	Holistic, nonlinear, integrated
Whole is defined as sum of parts	Whole is greater than sum of parts
Reality is predictable; laws determine the outcome	Reality is full of possibility; nothing is predetermined
Work with building blocks; those in control dictate what is done.	Work with networks; system is emergent and self-referencing.
Chaos is suppressed; structures are taken apart to examine and control	Natural order emerges from chaos; self-organization
Science is objective; what is not observed does not exist	No objective reality; our observation evolves—we cannot avoid having an impact

The entire system never reaches equilibrium, due to continuous interaction of the agents [29]. Yoffee also mentions that "*There is no optimum state of the system performance and the system can always surprise, as when a small initial perturbation can result in a large outcome*" [29]. Gell-Mann notes that a CAS acts in the real world based on a set of regularities drawn from the information that system acquires about environment and its own interactions [22]. Dooley mentions that these schemata go through a rule discovery process in which high level schemata evolve from smaller and more basic schemata [31]. Furthermore, Dodder and Dare define CAS as a "*network of many elements gathering information, learning and acting in parallel in an environment produced by the interactions of these agents*" [32]. They also argue about the co-evolution of these agents with their environment, as well as the fact that states of agents lie between orders and anarchy at the edge of chaos.

12.3.2 Characteristics of Complex Adaptive Systems

Different authors have counted various numbers of (4–13) characteristics for CASs. For instance, Holland enumerates four characteristics of nonlinearity, aggregation, flows, and diversity for CASs [21], in 1995. In 2006, he himself mentions parallelism, conditional action, modularity, adaptation, and evolution as main characteristics for CASs. Wildberger also names the following four characteristics for CASs: emergence, strategic learning and adaptation, nonlinearity (and a potential for chaotic behavior), and feedback [33]. On the other hand, Grus *et al.* listed 13 characteristics for CASs as: openness, components, nonlinearity, feedback, emergence, adaptability, multi-understanding, self-organization, dynamism, unpredictability, sensitivity to initial condition, and fractal building [25].

Some of the mentioned characteristics are almost identical, for example, emergence and aggregate behavior sound to indicate the same behavior of the CASs [33]. Schema and internal model also seem to be conceptually identical [22].

In the following part, some of the most cited characteristics of a CAS are briefly introduced.

12.3.2.1 Emergence/Aggregate Behavior *Emergence* and *aggregate* behavior are interchangeably used to point out the same concept [33]. Merriam Webster Dictionary defines aggregation as "*formed by the collection of units or particles into a body*" [19]. In a CAS, the systematic behavior of the whole system emerges from the interdependent activities of the agents, that is, it is not simply the sum of the behaviors of its agents [34]. This property results in unexpected patterns in the whole system which cannot be produced by the components individually [35, 36]. The emergent behavior in CAS is reverted to the interconnections between the components rather than being originated from the inherent characteristics of the components [37]. Rotmans recalls this characteristic as *spontaneous* growth of patterns from inside the system [38].

This aggregate behavior can hardly be predicted by thoroughly knowing each component. *Organized complexity* is the name Weaver gave to a system with a *significant* emergent behavior (Section 12.2.1) [3]. If the emergent behavior is not *significant*, then the system does not exhibit organized complexity [3, 4].

Emergence, however, like any other feature can have both good and bad effects. Despotou *et al.* state that because of its dynamic reconfiguration character, the CAS can lead to beneficial emergence, increasing the robustness of the system [4]. Decentralized control and collaboration, on the other hand, harden system behavior prediction, since the number of internal interactions in the system has been increased.

Weijnen *et al.* exemplify electricity and IT infrastructures as CASs which were not firstly designed to form integrated systems, though they have emerged over time to become so [39]. Emergent behavior of these systems is due to the unpredictable consequences of system operators' actions and disturbances. For instance, cascading blackouts are a result of emergence occurred due to the operator's inability in directing the flows over the network. As another instance of emergent behavior, Epstein notes the neural system of human memory: "*people can have happy memories of childhood while, presumably, individual neurons cannot*" [40].

Emergence can be viewed from two philosophical aspects: Epistemological view, concerning the original nature of the subject–the study of knowledge—and Ontological view, discussing on reality or existence of emergence and its qualities [41]. O'Connor and Wong define the epistemological emergence as "*systemic features of complex systems which could not be predicted (practically speaking; or for any finite knower; or for even an ideal knower) from the standpoint of a pre-emergent stage, despite a thorough knowledge of the features of, and laws governing, their parts*" [42]. On the other hand, they introduce the ontological emergence as: "*see the physical world as entirely constituted by physical structures, simple or composite. But composites are not (always) mere aggregates of the simples. There are layered strata, or levels, of objects, based on increasing complexity*" [42].

12.3.2.2 Complexity The concept of complexity is discussed in Section 12.2.1. Here, we present some definitions of complexity regarding CAS. Waldrop believes that the complexity of CAS arises from the simultaneous interactions of many simple agents within the system [20]. From the *control* point of view, the complexity of CAS leads to a distributed control system in which there is no single governing rule that controls the whole system and each of the interacting parts is governed by its own rules. These rules may influence the outcome of the system or the actions of other parts [21, 33, 43].

An immediate consequence of complexity is dynamism. The only definite trait of the CASs is change: equilibrium and stasis are equivalent to death for CAS [8, 9, 44–46]. This is because of the number of agents, their interdependence, and their openness to external influence, that is, the system learns and explores its environment to create new structures and new patterns of the relationships [20, 47].

12.3.2.3 Adaptability The future behavior of a CAS is based on its past and current interactions with the environment, that is, it adjusts itself to deal with the changes in the environment. As an instance, the language structure adapts itself during its emergence of interrelated patterns of experience, social interaction, and cognitive mechanism [48]. Besides, changes in the environment bring the need for adaptation inside the CAS, that is, the agents' behavior changes during the adaptation process which leads to changes in the overall behavior of the system [49, 50].

Holland mentions that for a system to adapt, it should be able to change its rules which bring up two computational procedures: credit assignment and rule discovery. The former procedure includes *rewarding* those parts of the system which lead to a better situation. In a rule-based system, rules with good contributions to the system's aggregate performance are assigned credit, meaning that these rules will have more influence in future. That is "*rules with good outcome in some situations in past are more likely to be used in similar situations in future*" [46]. A rule discovery procedure is for dealing with new situations in which the system needs new rules where the credit assignment procedure cannot work properly. For the newly discovered rules to be plausible, the rules are thought to be made up of building blocks and strong rules identified via the credit assignment procedure [46].

Wildberger mentions two types of adaptation for CAS: passive and active [33]. In a passive adaptation the CAS only responds to the environmental changes; while, in an active adaptation the CAS influences the environment, trying to improve its adaptation power by modifying the environment.

Moreover, it is worth knowing that adaptability is about being adapted to the environment; while self-organization (another characteristic of the CASs) is about adaptation without being influenced by the environment (internal adaptation).

12.3.2.4 Nonlinearity The CASs show nonlinear dynamism due to the nonlinear interactions within the system components. The nonlinear behavior of the CASs makes the future of the system unpredictable. Cilliers mentions that although the interactions between the components are known (even well defined), their strengths change over time [37].

12.3.2.5 Unpredictability As stated in Section 12.2.1, the behavior of a complex system is unpredictable; while, this unpredictability can either root back to the nonlinearity of the interconnections between the system components (organized complexity), or to the unsizeable number of components (disorganized complexity) [3].

In order to predict the output of a CAS, we must know its mechanism and component interactions, while the system changes constantly and these changes do not follow constant patterns, thus it may bring surprising outcomes at any point in time [20]. Since interactions and operations in a CAS are neither linear nor fixed, there is no agreed upon pattern that governs them [32]. The unpredictability of CASs does not imply the randomness of its output; for instance, the weather pattern, as a CAS, is very difficult to predict in detail; however, the weather does not change randomly [21, 25, 29, 37, 44, 51].

12.3.2.6 *Sensitivity to Initial Conditions* Sensitivity to initial condition, also known as *butterfly effect*, was discovered by Lorenz in 1961 [52, 53]. It states that an infinitesimal change in an initial action may cause large unpredictable consequences in the future. Weather changes can explain this concept: "*slight changes in wind velocity or minor difference in temperature could produce sunshine one day or rain the next*" [52]. As seen in Figure 12.9, the same system with two slightly different initial conditions lead to a highly different final position.

12.3.2.7 *Openness* The openness property of CASs has two aspects: first, a CAS is open to external influences in a way that CAS and its environment constantly exchange influence; second, the boundaries between CAS and environment are hard to determine [20, 21, 25, 30, 37, 38, 51].

It should be noted that openness is different from adaptability. Openness of a CAS states that the CAS and its environment exchange influence; while the property of adaptability provides the CAS with the ability to adapt itself to changes in the environment. In other words, for the system to be adaptable, it should be open [21, 25, 37, 38, 43, 46].

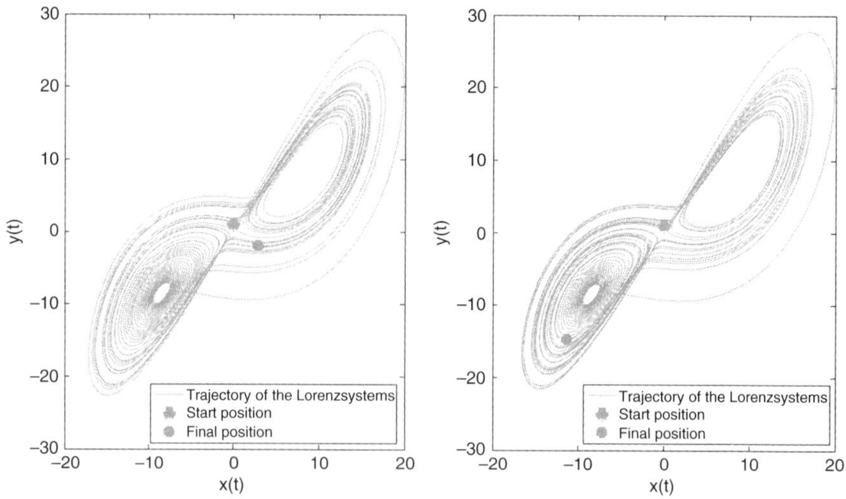

FIGURE 12.9 Sensitivity to initial condition: the initial conditions are very similar, while the future behavior of the system differs drastically.

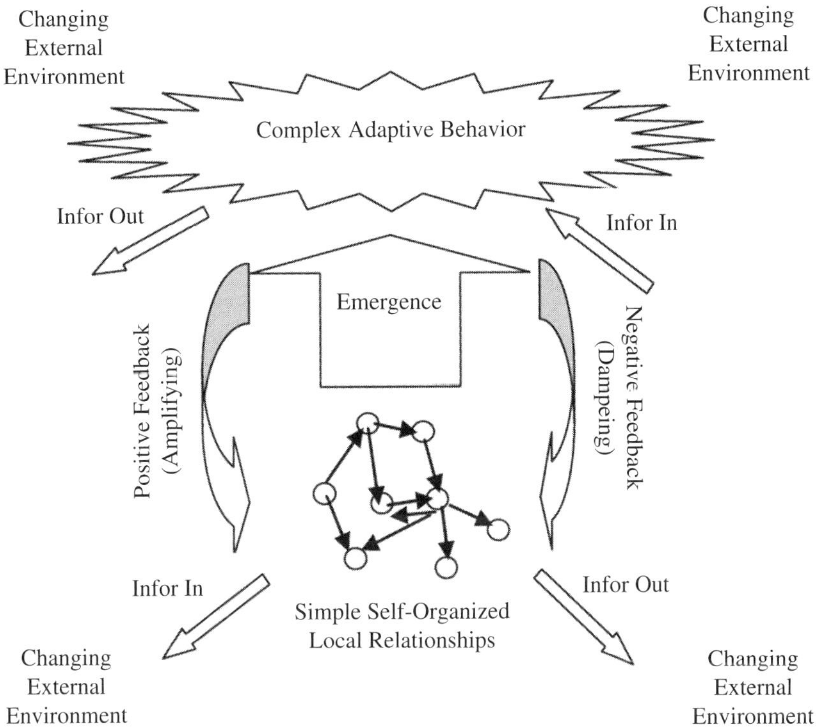

FIGURE 12.10 An illustration of the feedback mechanisms in a CAS [54].

12.3.2.8 Feedback Loops Feedback means feeding the output to input; thus, feedback loops give the CAS the ability of using the output of the previous process as the input of the next. From one point of view, the feedback loops are categorized into internal and external. Internal feedback loops connect agents within the CAS and exchange resources (materials, information, and energy) between them. These transforming feedback loops provide CAS with stability and changeability [20]. The external feedback loops help system to adapt; while the CAS and environment influence on each other.

From another point of view, feedbacks in a CAS are positive or negative. A positive (amplifying) feedback means that the CAS is learning, while a negative (damping) feedback means that the CAS is discouraging the process in order to regulate or damp the output, as shown in Figure 12.10. [25, 32–34, 51, 54].

FIGURE 12.11 Self-organization of leaf structure as a CAS.

12.3.2.9 Self-Organization The self-organization is the ability of a CAS to develop new structures, as shown in Figure 12.11 as leaf structure development. [38, 55–58] or as Levin mentions "*system assumes shape through a process of self-organization*" [55]. For instance, in human systems, spontaneous group activity, dissenting factions or clique are examples of self-organization [25]. Note that the difference between adaptability and self-organization is that self-organization is the development of new structures in the internal architectures of the system; while adaptability is the response to the changes in the environment.

Note that three characteristics of adaptability, feedback loops, and self-organization provide the system with the ability of evolution.

12.3.2.10 Scale Independence According to Merriam Webster Dictionary, scale independence or fractal structure or self-similarity means: "*any of various extremely irregular curves or shapes for which any suitably chosen part is similar in shape to a given larger or smaller part when magnified or reduced to the same size*" [19]. A well-known example of scale independence can be seen in plants: the same structure of angles exists between veins in leaves, twigs, branches, and roots [51].

12.3.2.11 Flows All CASs have nodes connected to each other by connectors. The resources within a CAS move from one node to another through connector (actions between agents involving the exchange of information and resources); this phenomenon is known as flow. For instance, in a financial system banks are nodes, electronic transfers are connectors, and money is the resource; hence, the movement of money by electronic transfers between banks is flow [21, 31, 52]. Simon uses ecosystems as an example of CAS to mention that flows provide the interconnections between agents and make it possible for the system to evolve from random collection of species into an integral whole [55].

CAS characteristics extracted out of different texts and the authors mentioning them are listed in Table 12.4.

12.4 SYSTEM OF SYSTEMS

12.4.1 Necessity and Definition

An SoS, as the name implies, is a system whose components are systems as well. Apparently, an SoS is developed only if the required application is not performable by a single system alone [41, 71, 72].

The notion of SoS has recently found its way through many application fields[4] (Figures 12.5 and 12.6); thus, converging to a unique definition for an SoS, or

[4] As Lane and Valerdi state: "In the business domain, an SoS is the enterprise-wide integration and sharing of core business information across functional and geographical areas . . . In the military domain, an SoS is a dynamic communications infrastructure to support operations in a constantly changing, sometimes adversarial environment . . . For some, an SoS may be a multi-system architecture that is planned up-front by a Lead System Integrator (LSI) . . . For others, an SoS is an architecture that evolves over time, often driven by organization needs, new technologies appearing on the horizon, and available budget and schedule . . . " [83].

TABLE 12.4 CAS Characteristics and Authors Who Have Mentioned Them

Characteristic	Explanation	References
Emergence/aggregate behavior	Behavior of the whole system emerges from the interdependent activities of the agents, i.e., it is not simply the summation of the behavior of the agents	21, 22, 24, 25, 29, 33–38, 40, 44, 46, 52, 59–63
Complexity	Simultaneous interaction of many simple agents within the system results in complexity which is the reason why the overall behavior of system differs from sum of its parts	21, 22, 25, 28, 33, 51, 64–67
Adaptability	Changes in the environment brings the need for adaptation for CAS, i.e., the behavior of agents changes during the process of adaptation to the environmental changes leading to changes in the overall behavior of the system	25, 27, 33, 37, 38, 43, 46, 48–50
Nonlinearity	The CASs show nonlinear dynamic systems due to the nonlinear interactions within the system. This nonlinear behavior of CAS makes the future of the system unpredictable	20, 21, 25, 33, 34, 37, 44, 51
Unpredictability	The system changes constantly and these changes does not follow constant pattern and may bring surprising outcomes at every point in time	20, 21, 25, 29, 33, 37, 44, 51
Sensitivity to initial conditions	An infinitesimal change in an initial action may cause large unpredictable consequences in future	25, 29, 34, 51, 52, 68
Openness	CAS and its environment constantly exchange influence in a way that the bounding between CAS and the environment is hard for determining	20–22, 25, 30, 37, 38, 51
Feedback loops	Feedback loops give CAS the ability to use the previous process's output as the input of the next	20, 25, 29, 32–34, 51
Self-organization	The self-organization is the ability of CAS to develop new structures	21, 24, 25, 29, 38, 44, 46, 55, 69, 70
Scale independence	Repeated irregular shapes in the system in different levels	25, 51
Flows	Movement of resources within the CAS from one node to another through connector actions between agents involving the exchange of information and resources	21, 31, 35, 52, 55

according to Reference 73 a "*multiple integrated complex system*" seems scarcely achievable [74–77].

An SoS is defined as a large-scale system composed of heterogeneous, independent, and self-organizing systems, each providing useful services in its own right [78, 79]. These component systems, together as a whole (SoS), are managed for a common goal and show various characteristics of emergence, complexity, evolutionary development, and synergy [78–80].

SoS itself is subjected to controversies regarding its distinction against *systems*. D. K. Hitchins, as it appears from his writings, may not accord with the current usage of the term SoS [81]. Describing SoS as "*an open set of complementary, interacting systems with properties, capabilities and behaviours [behaviors] of the whole SoS emerging both from the systems and from their interaction*", and system as "*an open set of complementary, interacting parts with properties, capabilities and behaviours [behaviors] of the whole set emerging both from the parts and from their interactions*", he deduces that these two terms, SoS and System, sound to be identical with a "*simple hierarchy shift*" [82].

Another important issue is that whether all SoSs are human made or natural—or specifically speaking, biological. Bar-Yam regards a living organism as an SoS for the following reasons: firstly, each of its constituent cells consists of a reproducing system in order to pass along their DNA information to survive, which is due to their operational and managerial independence. Secondly, as a result of environmental needs, higher effectiveness of survival is achieved for this living organism by evolutionary development, emergent behavior of the whole system, and a location distribution of the cells along the body. This representation of living organisms satisfies the expected characteristics of an SoS according to Bar-Yam's own definition of SoS [80].

Sheard and Mostashari, on the other hand, assert that a biological system is complex but cannot be considered as an SoS: "*Complex systems that consist of a large number of elementary particles or are biological systems not related to engineering would not be considered systems-of-systems*" [43].

12.4.2 Characteristics of System of Systems

As it appears from the definitions in section 12.4.1, most experts in the field define SoS in terms of its characteristics or applications. Here, a survey is performed on all of the founded definitions and some of the more cited characteristics are extracted out of them. Some of these characteristics seem to be almost identical or subset of one another.

12.4.2.1 Geographical Distribution Constituent systems of an SoS are often located widely dispersed. Therefore, one of the most challenging characteristics of SoS is the communication capability in order to secure the collaboration toward their common goal. Eisner, Maier, Kotov, Clare, Sage and Cuppan, DeLaurentis and Callaway, DeLaurentis, Purdue, and Jolly and Muirhead have enumerated geographical distribution as an SoS characteristic, and Shenhar has mentioned this characteristic as being largely widespread [71, 74, 84–91]. Samad and Parisini have exemplified Smart Grid as a geographically distributed system [92].

12.4.2.2 Complexity To define complexity, some tend to emphasize on the behavior of the system while others define it via the system components' intricate interconnections [93, 94]. Complexity in SoS, therefore, has two sides: It is either based on each constituent system's inherent behavior, or the number of entangled interconnections and information contained in those interconnections among constituent systems. As Bouwmans *et al.* assert, increasing interconnectedness between infrastructures can lead to "*new vulnerabilities, as changes or failures in one infrastructure may affect other infrastructures as well*" [5].

Despotou *et al.* [4], on the other hand, elaborate that complexity is actually a combination of three properties of autonomy, decentralized control, and collaboration among the constituent systems.

Gell-Mann, Kotov, Clare, Anderson *et al.*, Parks *et al.*, Bar-Yam, and Stevens have pointed to complexity as a major characteristic of SoS [21, 80, 86, 87, 95–97].

12.4.2.3 Emergence As stated in Section 12.3.2.1, emergence can be viewed from two philosophical aspects: epistemological view, concerning the original nature of the subject and ontological view, discussing on reality or existence of emergence and its qualities [41].

In the SoS field, emergence is regarded mostly from an ontological point of view: This behavior is an overall result of interactions among system components and cannot necessarily be predicted by just knowing each component thoroughly. *Organized complexity* is the name Weaver gives to a system with *significant* emergent behavior [3]. If the emergent behavior is not *significant*, then the system does not exhibit organized complexity [4].

Emergence seems to be the most important characteristic of an SoS, since Stoudt claims the name system-of-systems was coined "*to describe the emergent behavior of new mega-systems created by the tight integration of previously distinct and independent systems*" [98]. He also mentions that emergence cannot be seen when the constituent systems are separate [98].

Maier, Sage and Cuppan, Periorellis and Dobson, Stoudt, Despotou *et al.*, Bar-Yam, Purdue, and Sauser and Boardman have counted emergent behavior as one of SoS's characteristics [4, 72, 74, 80, 89, 98–100].

12.4.2.4 Heterogeneity/Diversity SoS is considered to have the ability to work with heterogeneous constituent systems; that is, the constituent systems may be designed and used in different contexts and be made from distinct elements and qualities. Since heterogeneity compels the SoS to have standardized protocols and interfaces for communication among its constituent systems, the immediate consequence of heterogeneity is for constituent systems to be open. That is, the constituent systems of SoS, as Azani states, must be able to exchange *energy, material, and information with outside world* and with each other [101]. As a consequence, reinforcing a new system into SoS would not be a problem [4]. This is important because is the only way for an SoS to "*achieve higher purpose(s) by leveraging the diversity of its constituent systems*" [72].

Keating *et al.*, Despotou *et al.*, DeLaurentis and Callaway, DeLaurentis, Purdue, Sauser and Boardman, and Jamshidi have listed heterogeneity/diversity as one of SoS's characteristics [4, 71, 72, 78, 88, 89, 102].

12.4.2.5 Connectivity Connectivity is a key concept in SoS, since all of the constituent systems require to exchange information or even substantial qualities of mass or energy [85]. Sauser and Boardman have defined this characteristic as the ability of a system to link with other systems, which is regarded as an important issue due to the fact that constituent systems may be highly heterogeneous and diverse [72]. Connectivity takes place via networking.

Shenhar, Purdue, DeLaurentis, Sauser and Boardman, and Jamshidi have mentioned connectivity as a characteristic of SoS [72, 78, 88, 89, 91].

12.4.2.6 Synergy Synergism, according to Reference 101, is the collaborative interaction among constituent parts of a system, while their combined effect is greater than the sum of their individual effects. In the SoS approaches, synergy between independent constituent systems is desirable in order to achieve the desired overall performance [103].

Boardman and Sauser, Bar-Yam, and Saunders *et al.* have stated synergy as SoS characteristics [79, 80, 104].

12.4.2.7 Large Scale McGraw-Hill Encyclopedia of Science and Technology presents three commonly accepted definitions of a large-scale system based on concepts of decomposition, complexity, and centrality [105]. It notes that a system is large scale if it is (i) decomposable into small-scale subsystems, (ii) complex, or (iii) geographically distributed. It also mentions that geographical distribution may be due to "*a lack of either centralized computing capability or a centralized information structure*", so the conventional control systems including "*components and information grouped in one geographical location or center*" are not applicable.

Jamshidi, Kotov, and Stevens have stated that an SoS must be large scale while all the three notions by which large scale is defined have already been regarded as SoS characteristics independently [78, 86, 97]. Thus, the *large-scale* feature of SoS can be viewed as a dependent characteristic.

12.4.2.8 Operational Independence Maier was the first to notice independency in SoS [99]. As he states, each constituent system of a SoS has its own purpose in its own right and is capable of operating independently to fulfill that purpose if separated from other constituent systems of the SoS [85].

Sage and Cuppan, Crossley, and DeLaurentis have also stated operational independence as one of SoS characteristics [74, 88, 106].

12.4.2.9 Managerial Independence Constituent systems of an SoS operate independently and "*are separately acquired and integrated but maintain a continuing* *operational existence independent of the SoS, i.e., they are managed partly for their own purposes*" [85].

Sage and Cuppan and DeLaurentis have also stated managerial independence as one of SoS characteristics [74, 88].

12.4.2.10 Autonomy Autonomy means "*the ability to act and make decisions without being controlled by anyone else*" [107]. Samad and Parisini state that the word *independence* in operational and managerial independence implies inherent autonomy in SoS [92]. Thus, operational and managerial independence are autonomy feature's subsets and may not be regarded as distinct characteristics of SoS. However, the autonomous system not only needs to have its independent operation, but also should take proper reactions against external stimuli and should make sure to accomplish the SoS purpose [73]. This characteristic leads to semi-intelligent actions of SoS.

Clough declares that the concept of autonomy resides between automation and intelligence. He illustrates that an automatic system will exactly follow a program, while an intelligent system is capable of discovering knowledge; thus, an autonomous system having "*the free will to make its own choices*" resides somewhere between two edges of automation and intelligence [108].

Despotou *et al.* (quoted from Reference 109) identify "*ten levels of autonomy, so called autonomous control levels (ACL): (1) Remotely Controlled Systems (conventional 'dull' systems), (2) Real time health diagnosis (self-awareness), (3) Adaptation to failures/weather (data loss tolerance), (4) Execution replanning (e.g. route for UAVs, intelligence), (5) Group coordination (emergent behavior), (6) Group tactical replanning (shared awareness state), (7) Group tactical goals, (8) Distributed control, (9) Group strategic goals, (10) Fully autonomous systems*" [4].

It is good to note that the above mentioned *ten levels of autonomy* differ from Sheridan's *ten levels of automation* (LOA) introduced in References 110–112 which will be described in Section 12.6.4. Periorellis and Dobson, Keating *et al.*, and Despotou *et al.* have stated autonomy as an SoS characteristic [4, 100, 102].

12.4.2.11 Self-Organization Self-organization is the process by which the system finds its way through planning without being imposed by outside stimuli or inside central authority. Although Bjelkemyr *et al.* claim that self-organization can be decomposed into operational and managerial independence—which by definition seems to be true [113]. Bar-Yam has counted self-organization as one of SoS characteristics [80].

12.4.2.12 Adaptability Adaptability pertains to the ability of a system to make changes in itself to deal with changes in its environment. It involves environmental change recognition, realization of the proper modifications inside the system toward the environmental change, and the ability to make the decided modification happen [114].

Despotou *et al.*, Holland, Bar-Yam, and Carney *et al.* state adaptability as a characteristic of SoS [4, 21, 80, 115].

12.4.2.13 Dynamic Reconfiguration In order to improve the reliability of the SoS, it should have the ability to reconfigure itself if any of its constituent systems or elements become faulty. SoS must find new resources to compensate for the loss. This process is called graceful degradation [4]. Adaptability should not be confused with dynamic reconfiguration. Adaptability is the changes in the system in response to the environment; while, dynamic reconfiguration is a reflection to a loss or failure inside the system [114].

12.4.2.14 Evolutionary Development SoS is not fully designed and formed at the beginning. It runs an evolutionary development via adding, removing and modifying its functions and purposes [85]. Electric power systems can be regarded as an example of evolutionary development. Power systems deliver the electric power to the end consumers, using interconnections which must evolve "*over time to meet the needs of an ever growing demand for electricity.*" [116]

Maier, Sage and Cuppan, Bar-Yam, Purdue, Stoudt, and Carney *et al.* have enumerated evolutionary development as a characteristic of SoS [74, 80, 85, 89, 98, 99, 115].

Characteristics extracted out of the SoS definitions and the authors mentioning them are listed in Table 12.5.

12.4.3 System of Systems Types

Maier assumes that systems of systems with similar complexity and extent should not be regarded as equivalent; and based on managerial control, he suggests three types of SoS: Directed, Collaborative, and Virtual [99]. Moreover, from system architecture point of view, Chen and Clothier classify SoS into two types: Dedicated and Virtual [118]. Recently, Dahmann and Baldwin have introduced a new type of SoS, Acknowledged SoS, which is mostly growing in military context [119]. Different types of SoS and their initiatives' definition are described in Table 12.6.

Note that due to the *specific* purposes which *directed* SoSs are designed to serve—perhaps for a long term—and their centrally managed control system, these SoSs may not be distinguished from a system per se [120].

12.4.4 A Taxonomy of Systems Family

12.4.4.1 SoS versus Complex Systems Determining the type of a system is essential in system analysis and design. Nevertheless, Sage and Cuppan have asserted that "*What distinguishes a system of systems from other systems does not, at this point, have a definitive answer*" [74]. They continued with an example: most of the systems today are made up of systems but can we really call them all systems of systems? A PC for example is composed of several systems but monolithic in its purpose; Internet, on the other hand, serves several purposes and makes wider communications possible. Converging to a clear distinction between systems requires first bringing a definition of each system and then comparing them with SoS characteristics, context, and applications.

TABLE 12.5 SoS Characteristics and Authors Who Have Mentioned Them

SoS Characteristics	Explanation	References
Geographical distribution	Wide dispersion of component systems regarding geographical locations	71, 74, 84, 86–90, 99
Complexity	Combination of autonomy (inherent behavior of component systems), decentralized control, and collaboration of systems (intricate interconnections)	4, 21, 80, 86, 87, 95–97
Emergence	The overall behavior of the system cannot be predicted by knowing each component thoroughly due to complicated and nonlinear interactions among components	4, 72, 74, 80, 89, 99, 100
Heterogeneity/diversity	Constituent systems may be designed and used in different contexts, and be made from distinct elements and qualities	4, 71, 78, 88, 89
Connectivity	The ability of system to link with other systems	72, 78, 88, 89, 91
Synergy	The combined effect of component systems is greater than the sum of their individual effects	79, 80, 104, 117
Large scale	A system which is decomposable into small-scale subsystems, complex, or geographically distributed	78, 86, 97
Operational independence	Each constituent system of an SoS has its own purpose in its own right and is capable of operating independently	74, 85, 88, 99, 106
Managerial independence	Each constituent system of an SoS is managed to fulfill its own purpose	74, 85, 88, 99
Autonomy	*"The ability to act and make decisions without being controlled by anyone else"* [107]	4, 72, 100, 102
Self-organization	The process in which the system finds its way through planning, without being imposed by outside stimuli or inside central authority	80
Adaptability	The changes system make in itself to deal with changes in the environment	4, 21, 80, 115
Dynamic reconfiguration	The ability to reconfigure if any of constituent systems or elements becomes faulty	4
Evolutionary development	SoS runs an evolutionary development via adding, removing, and modifying its functions and purposes	74, 80, 89, 98, 99, 115

TABLE 12.6 Different Types of SoS and Their Examples

	Definition—exact quotation
Directed	"*Directed systems are those in which the integrated system-of-systems is built and managed to fulfill specific purposes. It is centrally managed during long term operation to continue to fulfill those purposes*" [99] "*If the system is developed through formal organizations to fulfill a common purpose, it is a directed SoS*" [121]
Collaborative	"*Collaborative systems are distinct from directed systems in that they do not have coercive power to run the system. The component systems must, more or less, voluntarily collaborate to fulfill the agreed upon central purposes*" [99] "*If the system is developed through the collaboration of its participants, it is a collaborative SoS*" [121]
Dedicated	"*These SoS are consciously engineered and operated to fulfill an evolving need, term them dedicated SoS. Examples of such dedicated SoS are air traffic control systems and the Internet*" [122] "*If the component systems are architected so that they can be integrated to work together to fulfill a goal, it is a dedicated SoS*" [121].
Virtual	"*Virtual systems lack a central management authority. Indeed, they lack a centrally agreed upon purpose for the system-of-systems. Large scale behavior emerges, and may be desirable, but the supersystem must rely upon relatively invisible mechanisms to maintain it*" [99] "*Virtual SoSs take forms that are rarely envisaged at design time and that they frequently comprise elements that were never designed to be integrated*" [122] (quoted from References 123 and 124) "*If subsystems are previously existing architectures that are integrated to meet an immediate mission requirement, it is a virtual SoS*" [121]
Acknowledged	"*Acknowledged SoS, like directed ones, has recognized objectives, a designated manager, and resources for the SoS; however, the constituent systems retain their independent ownership, objectives, funding, and development and sustainment approaches. And like collaborative SoS, changes in the systems are based on collaboration between the SoS and the system*" [119] "*In acknowledged SoS an organization is responsible for the SoS and supporting SoS systems engineering while independent organizations and SE teams are responsible for the constituent systems that support the SoS capability objectives*" [125]

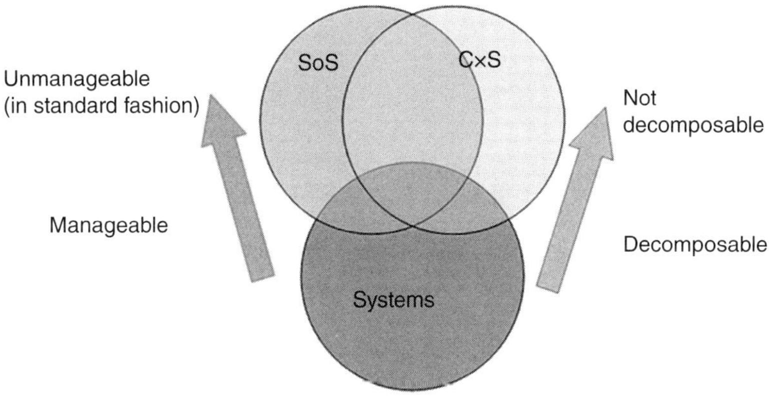

FIGURE 12.12 An illustration of complex systems and systems of systems in terms of decomposability and manageability [43].

Not all systems of systems are considered as CxS. Sheard and Mostashari compare these two as shown in Figure 12.12. They state that systems of systems are related to "*program acquisition context*" and "*unmanageable using standard top-town systems engineering*"; while, CxS are related to "*analytical or scientific context*" and "*described as being not decomposable*" [43].

12.4.4.2 Monolithic Systems Mostafavi *et al.* have recognized differences between a monolithic system and an SoS. These differences are mentioned in Table 12.7 [126].

12.4.4.3 Families of Systems We suggest the Von diagram of Figure 12.13, to represent the interrelation of families of systems, followed by a short explanation of each member of the families of systems, including: family of systems (FoS), SoS, federation of systems, and coalition of systems.

TABLE 12.7 Comparing Monolithic Systems and System of Systems [126]

	Monolithic system analysis	System of systems analysis
Focus	Single system	Integrated systems
Boundaries	Static	Dynamic
Problem	Defined	Emergent
Structure	Hierarchical	Network
Goals	Unitary	Pluralistic
Approach	Process	Methodology
Timeframe	System lifecycle	Continuous
Centricity	Platform	Network

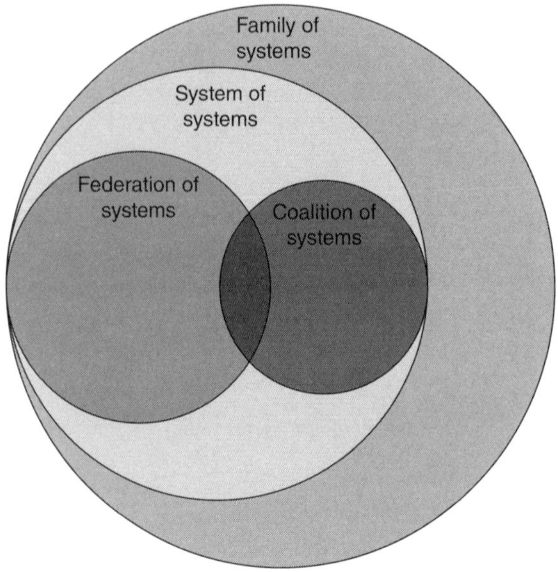

FIGURE 12.13 Von diagram rendering interrelation of taxonomy of systems family.

12.4.4.4 Family of Systems An FoS is said to be a set of systems and "*not considered to be a system per se*" [120]. To distinguish it from SoS, Clark has mentioned some differences: While SoS's constituent systems have been integrated and it is their interconnections that create a capability beyond the sum of each individual's capabilities, FoS's constituent systems are not integrated and therefore, there is no synergism among them [127]. The same deduction could be made about emergence behavior of SoS causing new and unpredictable properties that FoS cannot achieve. Furthermore, one of the main characteristics of SoS is heterogeneity, but considering the mentioned definition, FoS's constituent systems should possess some common characteristics, such as being in the same domain or product line. In other word, "*the member systems may not be connected into a whole*" [120].

12.4.4.5 Federations of Systems To distinguish federation of systems from systems, Krygiel introduces three aspects to be investigated: autonomy, heterogeneity, and dispersion [128]. Federations of systems, as Wells and Sage assert, are more diverse (this diversity especially addresses to the *transcultural and transnational sociopolitical* aspects) and therefore, managed autonomously in a way that each component system fulfills its own objectives, and are geographically more distributed [129].

12.4.4.6 Coalition of Systems Coalitions of systems are "*a class of system similar to systems-of-systems but they differ in that they interact to further overlapping self-interests rather than an overarching mission*" [130]. Shared interests and continuing operation dependence may be coalition of systems' Achilles heel, since there always is the possibility that "*coalition partners renege on their responsibilities to provision parts of a service or application*" [130]. Consequently, a proactive risk management is essential.

12.4.4.7 Other Types of Systems' Relations

Composite systems: Karcanias and Hessami recognize SoS "*as an evolution of the standard notion in engineering of Composite Systems (CoS)*" [131]. However, they have mentioned that SoS and CoS are different regarding independence and autonomy of their constitute systems; while subsystems of SoS are autonomous and satisfy their own goals, in a CoS, subsystems are subjected to "*the rules of the interconnection topology*" and do not have independent goals [131].

Cyber physical systems (CPS): CPS are composed of systems with tightly combined and conjoined computational and physical elements. Their "*ability to interact with and expand the capabilities of the physical world through computation, communication, and control is a key enabler for future technology developments*" [132].

These systems, according to Reference 133, are developed to achieve "*systems that:*

(a) *respond more quickly (e.g., autonomous collision avoidance),*
(b) *are more precise (e.g., robotic surgery and nano-tolerance manufacturing),*
(c) *work in dangerous or inaccessible environments (e.g., autonomous systems for search and rescue, firefighting, and exploration),*
(d) *provide large-scale, distributed coordination (e.g., automated traffic control),*
(e) *are highly efficient (e.g., zero-net energy buildings),*
(f) *augment human capabilities, and enhance societal wellbeing (e.g., assistive technologies and ubiquitous healthcare monitoring and delivery).*"

Samad and Parisini have mentioned two differences between CPS and SoS. Firstly, considering that CPS totally interacts with physical world while SoS does not "*necessarily require closing the loop in the real world*"; for instance, applications that are purely in the information space are not necessarily outside SoS realm and secondly, a CPS does not necessarily require distributed and hierarchal systems; even a Single Input–Single Output Proportional–Integral–Derivative Controller (SISO PID) could be regarded as CPS [92]. Security of CPSs is of a great concern nowadays, due to recent cyber-attacks and cyber-intrusion trials [134, 135].

12.5 COMPLEX ADAPTIVE SYSTEM OF SYSTEMS

Holland claims that "*it is feasible to understand any System of Systems as an artificial complex adaptive system. It is manufactured to achieve a predefined mission and will involve a large number of interacting entities with persistent movement and reconfiguration, changing based on changes in context, ordered through self-organization, with local governing rules for entities and increasing complexity as those rules become more sophisticated"* [21].

CASoS—firstly defined in the Sandia National Laboratories—are systems which exhibit four qualities: (1) be a system; a system consists of some interacting components placed in an environment which is its context of use; (2) be an SoS; since the overall operation is not achievable by a single system; (3) be complex; both inherent complexity in each constituent system and interconnections between them. They are large and irreducible, so "*interpretation, modification, and quantifying the impacts of modification are difficult*," and (4) be adaptive; behavior of component systems change during connection to the environment [136, 137].

Sandia researchers have stated some examples of CASoS including tropical rain forests, agro-ecosystems, cities and megacities (and their network on the planet), interdependent infrastructures (local to regional to national to global), government and political systems, educational systems, health care systems, financial systems, economic systems and their supply networks (local to regional to national to global), the global energy system, and global climate [138].

12.6 HUMAN–AUTOMATION INTERACTION

12.6.1 Automation

The term *automation* was first used in the meaning as is now accepted in industry by Ford Motor Company's VP, *D. S. Harder* around 1946 [139]; while, in academia, E. Nagel used it for the first time in a *Scientic American* article [140].

Automation is defined in Britannica Encyclopedia as: "*a wide variety of systems in which there is a significant substitution of mechanical, electrical, or computerized action for human effort and intelligence*" [139]. In a narrower sense of industrial context, automation can be described as: "*the application of sensors, control systems, and information technologies to reduce the need for human work in the production and delivery of goods and service*" [141].

Historically, mechanical machines were the first automation tools that substituted the human labor; whereas, computers and IT (information technology) systems were widely used then to assist humans in information and decision tasks. Therefore, in this chapter, the terms *automation*, *computer*, *machine* and *IT*, are interchangeably used as a general instance for the automation system or agent "*who executes the function, task or job previously performed or conceivably could be accomplished by a human"* [142]. The latter is our adopted definition of automation. Machines, especially computers, are now capable of accomplishing many functions that at one time could only be performed by humans. Machine execution of such functions (automation) has also been extended to functions that humans do not wish to perform, or cannot perform as accurately or reliably as machines do [110].[5]

[5]Due to the centrality of Reference 110 in our approach to human–automation interaction, we write this reference in its complete form, to stress on the role of all of its authors.

Etymologically, the term automation roots back to Greek word *automatos* [143]; thus, full automation (i.e., complete substitution of humans by machines) might be taken for granted for the term automation. Nof, for instance, argues: "*automation, in general, implies operating or acting or self-regulating, independently, without human intervention*" [143]. Whereas, automation, in a wider sense, can be regarded within a spectrum of no automation (manual) to full automatic (automate), since practically many tasks are performed in a collaboration of humans and automation systems, that is, partial automation or semi-automation [111, 143].

The main reasons to apply automation systems in industry are higher efficiency, avoiding human from hard/hazardous situations, assisting human as a(n) (intelligence/analysis/decision/action) supporting agent and succeeding in critical tasks [110, 144–147].

12.6.2 HAI: Where Humans Interact with Automation

Humans might passively utilize the automation (or its products); however, this can hardly be regarded as an instance for human interaction with automation. Instead, Sheridan and Parasuraman confine the HAI concept to the situations in which humans "*(a) specify to the automation the task goals and constraints (do X but avoid doing Y) and trade-offs between the goals and constraints; (b) control the automation to start or stop or modify the automatic task execution; and (c) receive from the automation information, energy, physical objects, or substances*" [111].

The *automate-as-possible* philosophy, which is the traditional approach to automation systems design, mostly rely on Paul M. Fitts' list [148], in which, a human operator is ironically expected to be responsible for the weaknesses of automation, which itself has been developed for covering human operators' pitfalls: "*functions better performed by automation are automated and the operator remains responsible for the rest, and for compensating for the limits of the automation*" [149].

On the contrary, human-centered automation (HCA) approach promises the notion of more humane automation systems [150, 151], although there might be a controversy on different interpretations of its meaning [152]. HCA seeks the optimum function allocation between the humans and the computers instead of the older attitude of substituting human by computer, and consequently, leaving the un-automatable jobs for humans [153]. The idea of "*eliminating the human to eliminate the human errors*" now appears obsolete amongst most of the automation and computer engineers, and of course, within the human factors engineers as well [111]. Research shows that the moderate combination of job sharing between the humans and the computers might provide better performance and situation awareness (SA) than that of extremist *automate-as-possible* philosophy [154–156].

Bainbridge articulated the ironies of automation to stress on the lack of attention to the role of human in automation [153], and Sheridan ruminated on automation, linking the human decision-making to his well-known taxonomy of automation levels [112, 157]. Moreover, Billings popularized the phrase *human-centered automation (HCA)*, highlighting the importance of human–computer collaboration in the automation system design [111, 150–152, 157]. The term *human-centered computing (HCC)* is also used to demonstrate the significant contribution of human factors to the computer system development [158].

12.6.3 HAI and Function Allocation

A well-crafted automation solution is expected to consider a *manifesto* in which the job sharing between the humans and the machines is completely clarified. This job sharing could aim to optimize the total system performance, mitigate risks, minimize costs, maximize return, and eliminate the operational errors.

The human–machine function allocation has been considered as a classic problem since the early days of very basic automation. The Fitts' list of MABA-MABA, that is, "*men are better at - machines are better at*" is amongst the most well-known solution methodologies [110, 111]. Fitts' MABA-MABA slogan is furthered by T.J. Watson, as IBM Pollyanna principle by "*machines should work- people should think*". Table 12.8 shows an exemplary list of MABA-MABA.

Fitts' list supports automation system designers with a basic idea for static allocation of functions to humans and machines; however, some weaknesses are attributed to that objective. One major drawback of the list, as Dekker and Woods declare, is the false idea of fixed strengths and weaknesses of humans and machines [159]. Whereas, Fitts writes: "*the performance capacity of the human motor system plus its associated visual and proprioceptive feedback mechanisms, when measured in information units, is relatively constant over a considerable range of task conditions*" [148, 160]. In fact, the performance of both humans and machines varies in different situations which implies that the function allocation can scarcely be static, as the MABA-MABA list changes over time. Sheridan also enumerated seven problems of Fitts' approach to function allocation; specifically, he argues that increased autonomy causes function allocation (FA) to be much different from anything Fitts could have imagined [161].

Being a classic problem, function allocation is still a matter of controversy among the scientists and engineers; even on the definitions, concepts and its practical solvability as an either scientific or artistic design problem [152, 161–166]. Moreover, the introduction of intelligent computerized machines and software agents has raised the need for more sophisticated HAI models, capable of providing human–automation

TABLE 12.8 An Exemplary List of MABA-MABA [148]

Men are better at (MABA)	Machines are better at (MABA)
Perceiving patterns	Responding quickly to control tasks
Improvising and using flexible procedures	Repetitive and routine tasks
Recalling relevant facts at the appropriate time	Handling simultaneous complex tasks
Reasoning inductively	Reasoning deductively
Exercising judgment	Fast and accurate computation

collaborative environments rather than simple models that perform a straightforward job allocation. As Proud *et al.* declared: *"The question, 'How autonomous should the system be?' is of primary importance to the designers of the next generation human spaceflight vehicle"* [167]. Endsley and Kaber wrote *"very little experimental work has been conducted to examine the benefits of applying the intermediate LOA (levels of automation) in complex tasks"* [154].

The HCI and HAI methods are developed either by introducing various models and approaches [110–112, 154, 168–175], or by performing practical experiments on implementation of the pre-introduced HAI models [155, 156, 167, 176–188]. Moreover, ergonomists, cognitive engineers, and applied psychologists presented valuable works on the human side of the human–computer systems [142, 185, 187, 189–191].

12.6.4 Evolution of HAI Models: Dimensions

Figure 12.14 reveals the chronological evolution of the HAI models, indicating four breakthroughs in 1951, 1978, 1999/2000, and 2006. The first milestone belongs to Fitts' manual or automate model. P. M. Fitts' list of MABA-MABA can be regarded as one of the first models of HAI [148].

As for the second milestone, the concept of 10 autonomy levels for human–computer (or human–automation) systems is initially proposed by Sheridan and Verplank for teleoperation applications, which Sheridan himself modestly states "*was taken more seriously than were expected*" [112, 157]. However, the considerable

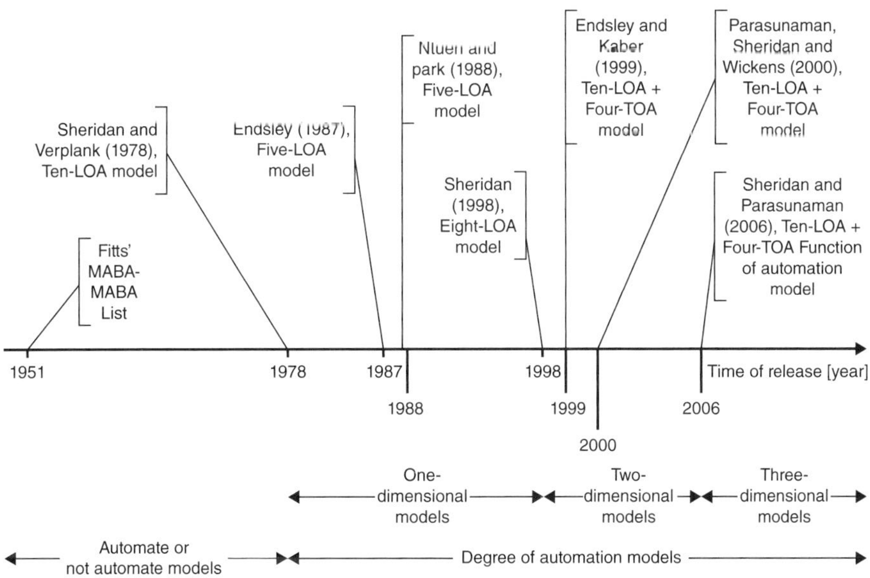

FIGURE 12.14 The chronological evolution of human–computer interaction models.

number of citations to their presented taxonomy shows that it deserves to be taken seriously [155–167, 174, 178, 182, 188]. (LOAs represent degrees or grades of automation for an automatable task, ranging from the lowest (manual) to the highest (full automatic) LOA. We refer to this 10-level model as one-dimensional (1D) model of LOAs in this chapter which is shown as the vertical axis of Table 12.9.

Albeit Endsley introduced an LOA taxonomy for decision support applications; and, Ntuen and Park presented another hierarchy of LOAs for teleoperation applications as well [154, 192]. Although many researchers develop HAI (HCI) models after Sheridan, the 10-level taxonomy was named after Sheridan, as Sheridan's LOA or Sheridan's model of HAI (HCI), since he has initially introduced the idea of degrees of automation in Reference 112, instead of Fitts' automate or non-automate notion.

Subsequently, Endsley and Kaber and Parasuraman, Sheridan and Wickens expanded the 1D LOA model to a 2D model by offering another dimension as type (or stage) of automation (TOA) [110, 154]. TOAs are presented as four types (or stages) for performing a single task, as in horizontal axis of Table 12.9. We refer to this model as 2D model of HAI in this chapter. Table 12.9 compares the 2D HAI models, based on the contributions of Endsley and Kaber and Parasuraman, Sheridan and Wickens [110, 154]. The other dimension of this model (TOA) roots back to the idea that "*proper function allocation differs by process stage*" [161]. It is not clearly stated in their publications which group of authors firstly introduced the second dimension (TOA); however, with respect to both groups of authors, we call the four-stage classification (the second dimension) as "*type of automation (TOA)*" in this chapter. Fereidunian *et al.* discussed on TOAs, questioning it as it is actually *stages of automation* or rather *types of automation* [193, 194].

Another milestone in HAI model development is the introduction of a three-dimensional model by Sheridan and Parasuraman [111]. They added another dimension to the former 2D model as *functions of automation* to make a more comprehensive model.

12.6.5 Evolution of HAI Models: Dynamism

Figure 12.15 shows another aspect of HAI model evolution, in terms of function allocation dynamism, in a chronological order from up to down. As mentioned in the previous section, the first stage in HAI modeling was Fitts' list of MABA-MABA in which a fixed model is considered for manual or automatic function allocation between humans and automation systems as shown in the upper part picture in Figure 12.15 [148].

A step ahead, Sheridan and Verplank furthered the Fitts' two-level—either manual or automatic—model to a 10-degree LOA taxonomy, asserting that an automating job should be considered as incremental degrees of automation, shown in the middle picture in Figure 12.15 [112]. The LOA of HAI model is still a static one, that is, the functions (tasks) are once divided between human and automation in more or less 10 LOAs. This approach was called *static automation* by Parasuraman *et al.* [195].

The performance of human–automation systems is affected by environmental conditions; therefore, the fixed determination of LOA fails to maintain full advantages

TABLE 12.9 The TOA–LOA Model of HAI

LOA	Description [110]↓		Name of the LOA given by Endsley and Kaber [154]↓		
10	The computer decides everything, acts autonomously, ignoring the human		Full automation (FA)		
9	Informs the human only if it, the computer, decides to		Supervisory control (SC)		
8	Informs the human only if asked, or		Automated decision-making (ADM)		
7	Executes automatically, then necessarily informs the human, and		Rigid system (RS)		
6	Allows the human a restricted time to veto before automatic execution, or		Blended decision-making (BDM)		
5	Executes that suggestion if the human approves, or		Decision support (DS)		
4	Suggests one alternative		Shared control (SC)		
3	Narrows the selection down to a few		Batch processing (BP)		
2	The computer offers a complete set of decision/action alternatives, or		Action support (AS)		
1	The computer offers no assistance: human must take all decisions and actions		Manual control (MC)		
	Stages (or types) of automation (TOA) (Parasuraman, Sheridan and Wickens [110]**):** →	I: Information acquisition	II: Information analysis	III: Decision selection	IV: Decision Action
	Stages (or types) of automation (TOA) (Endsley and Kaber [154] **)**→	Monitoring	Generating	Selecting	Implementing

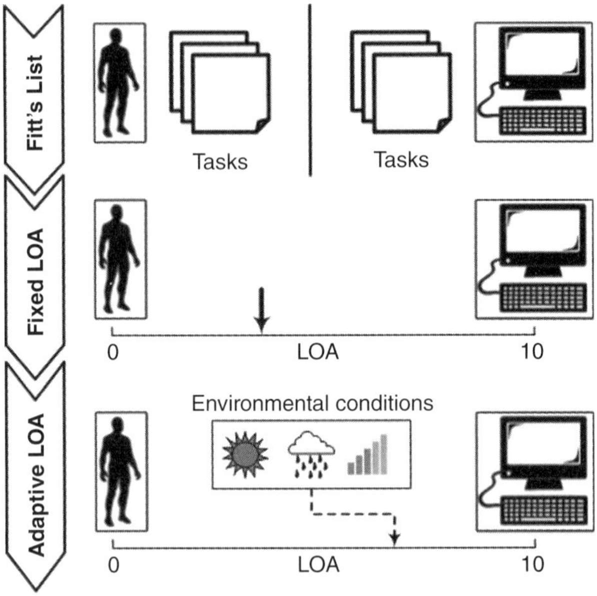

FIGURE 12.15 Development of the HAI in the literature from up to down: Fitts' list (fixed manual or automatic function allocation), fixed LOA (static LOA or fixed 10-level LAO) and adaptive LOA (AA or dynamic LOA).

of the basic idea of automation degrees. As a result, LOAs should be adapted to the environmental conditions. This adaptation necessity paves the way toward further development of the HAI, which according to Sheridan and Parasuraman in Reference 111, is a concept rooting back to Rouse [197]; however, its practical model has been introduced as *adaptive automation* (AA) in References 110 and 155, *adjustable automation* in Reference 169, *dynamic automation* in Reference 111, or *adaptive autonomy* in References 194, 196, and 198, as shown in the lower part picture of Figure 12.15. Improvement of HAI system performance has been reported, as a result of implementing adaptive autonomy, comparing to that of static LOAs [155, 192, 199].

12.6.6 Adaptive Autonomy Implementation

Although the notion of adaptive autonomy is simple, plausible, understandable, and even old (197), it can scarcely be implemented straightforwardly [110, 111, 196, 166]. Parasuraman *et al.* categorized the adaptive autonomy implementing techniques as: "*critical events, operator performance measurement, operator physiological assessment, modeling, and hybrid methods combining one or more of these techniques*" [195] (quoted in Reference 111). These five classes of adaptive autonomy

TABLE 12.10 Summary of Comparisons of Adaptive Autonomy Implementation Methods

Class of adaptive autonomy implementation method	Advantages	Disadvantages
Environment monitoring	• Responsive to the unpredicted changes in environmental conditions	• Ignoring the unpredicted changes in humans' performance and physiological conditions • Sensitivity to the environmental monitoring system
Human monitoring	• Responsive to the unpredicted changes in humans' condition	• Ignoring the unpredicted changes in environmental conditions • Sensitivity to the human performance or physiological monitoring system
Human–automation interaction modeling adaptive autonomy (model based)	• Implementable as an offline adaptive autonomy expert system (AAES) • Less dependence on online monitoring systems	• Dependent on correctness, preciseness and complexity of the models • Delayed response to the changes in environmental or human conditions, due to lack of online monitoring

Source: Taken and Summarized with adaptation from References 111 and 195

implementation methods can be summarized as the three classes of environment monitoring methods, human monitoring methods and HAI modeling (model-based) methods, as shown and compared in Table 12.10.

In the class of environment monitoring adaptive autonomy methods, the external environment (including the under-control plant and other influential factors) is continuously monitored to identify a certain event, like a fault or an abnormal condition in the operation. The automation system intervenes in a human task if those specific events occur, according to some predetermined signs. Parasuraman *et al.* explain the critical event technique (applicable to its general form: environment monitoring) as "*automation is invoked if certain external events occur, but not otherwise*" [195]. Automatic safety management and emergency shutdown (ESD) systems are examples of this class of methods: the safety management system intervenes if certain events occur. For example, consider a safety management system in a robotic manufacturing line; if an operator's hand or head would be recognized entering the operation area of the material handling or welding robots by light guards or other presence sensors, the safety management system will shutdown the manufacturing process, followed by the safety restoration process in an automatic way. Electrical protection and relaying systems and automatic autopilot mechanisms are also good examples for the environmental monitoring adaptive autonomy methods. As Sheridan and Parasuraman argue, the possible insensitivity to actual systems and human operator performance is a disadvantage of the environmental monitoring methods [111]. For instance, in the environment monitoring AA methods, the emergency shutdown system will shutdown the whole process in any prospective recognition of emergency signatures which can be false or not as critical as making the system shutdown.

Class of human monitoring adaptive autonomy methods, on the contrary, continuously monitors the human operator or supervisor of the system to identify any changes in mental workload, fatigue or even improper intention [111]. This can be achieved by assessing operators' performance or measuring operators' physiological condition. If the human performance decrease exceeds a certain limit, more tasks shall be assigned to the automation system, that is, migrating to a higher LOA. For example, Kaber and Riley utilize a measurement technique to determine operator workload in a complex control system. Sheridan and Parasuraman suggest that measuring physiological signals, such as EEG, could potentially prevent operators' extreme fatigue or workload to provide computer aiding to considerably alleviate the potential danger [111]. EEG signals, event-related potentials (ERPs) and eye scanning could be mentioned as examples of human monitoring in the class of AA methods reported in the literature [111, 189, 195, 200–202].

Class of HAI modeling adaptive autonomy (model-based) methods consider both human and automation sides of the interaction. The privileges of model-based methods include its potential for offline implementation, as it could easily be modeled by expert systems. On the other hand, necessity of validation, consideration of all aspect of human operator performance, and divergence of different model results are accepted to be the limitations of this class of methods [111]. Consequently, hybrid methods are introduced to overcome the limitations [111].

12.7 HAI IN SMART GRID AS A CASOS

12.7.1 Smart Grid

Smart Grid, as the future vision of electric power systems, is expected to be intelligent, reliable, optimized, self-healing, and adaptive. This vision leads to significant extension prior to the traditional power system's features. Decentralization of generation, high penetration of renewable energies resources, high contribution of information technologies, power flows from all conceivable places, possibility of anyone to control connections, and adaptive behavior of the Smart Grid feature main differences from the traditional grid [203, 204].

Electric power delivery is a complex phenomenon, due to technical difficulties of storing electric energy. In fact, power distribution system is a collective system, rather than being a delivery system: collecting the power demands from the customers, and distributing the requested energy. This scheme can be considered as a signaling system. Thus, the information availability and quality is one of the most important factors in success of an electric utility company. Electric utilities are committed to perform their delivery job successfully, in terms of legal, technical, and ethical devotions. Any unbalance condition between demand and delivery may cause severe technical issues, which in turn causes economic hazards.

Electric power systems have been subject to many changes in the last two decades. They have emerged from a regulated governmentally owned infrastructure, to a privatized market in capitalized economies. Power systems have migrated from making large power generation plants to encouraging the private sector to contribute in power generation as distributed generations.

Information technology (IT), on the other hand, has become one of the most considerable features of the twenty-first century. All industries devote an effort to adapt themselves to this rapid growing field, to gain more and more from its capabilities. Information technology not only provides an infrastructure to help the industries to overcome their shortages and solve their problems in a more efficient way, but also enables them to do businesses in innovative ways. These industries will be called IT enabled. In power systems, information technology is extensively used to support supervisory control and data acquisition (SCADA), automated operations and *maneuvers* within the network, asset management/facility management (AM/FM), automated meter readings (AMRs) and billings, customer relation management (CRM), and customer information systems (CIS). This extensive use of IT can shift the whole paradigm in electric power utilities, on both technical and managerial sides.

The Smart Grid notion is the fruit of such a paradigm shift: gaining from IT leverages for managing the grid, as well as using the market mechanism for a better demand response, mixed up with the most recent innovations in distributed energy resources (DERs), especially the renewable ones.

Figure 12.16 shows the major subsystems of a Smart Grid scheme. As shown in Figure 12.16, it consists of the following subsystems: power system, automation, and control system, IT infrastructure, operational human (operator, supervisor, controller, or manager), consumer human, HAI system, adaptive AAES, and the surrounding environment.

As the core technology of the Smart Grid, the power system is responsible for electric energy processing and its delivery to the end consumer. The automation and control system manages the operation of the power system. The IT infrastructure provides the data processing and communication between the different nodes and hubs of the system, that is, the interconnections among automation and control, power system, and human [203, 205–212].

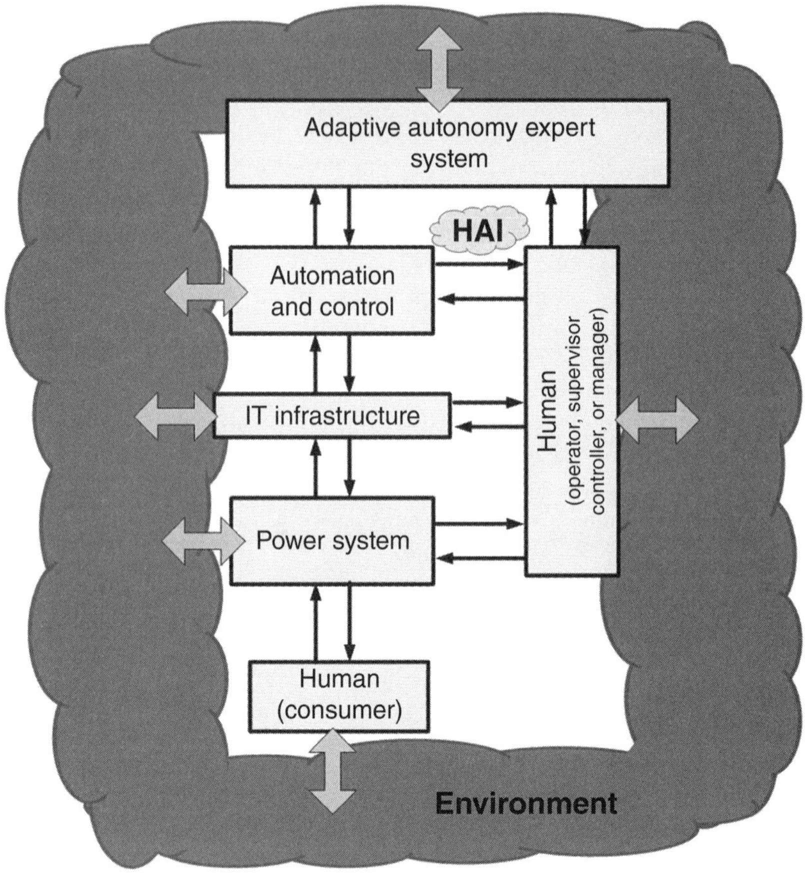

FIGURE 12.16 Smart Grid and its constituent subsystems.

As in Figure 12.16, human participation in Smart Grid is considered as playing two roles: consumption and operation. The consumer human deals with the Smart Grid by consuming the electric energy. Whereas, the operational human is involved in operation, supervision, control or management of the power system, by interacting with the power system, IT infrastructure, and automation and control system with the purpose of design, decision-making, operation, and maintenance.

The HAI system is the sphere in which the operational humans and the automation system collaborate in operation and management of the power system. Finally, the AAES (the up-most block in Figure 12.16) regulates the collaboration of the operational human and the automation system, that is, it adapts the autonomy level (LOA) of the humans and the automation system to the environmental changes. The surrounding environment (the surrounding cloud in Figure 12.16) is the external and internal environment that encompasses the whole Smart Grid system.

In the succeeding subsections the concepts and definitions from previous sections—CAS and SoS—are utilized to investigate the HAI as a CASoS.

12.7.2 HAI in Smart Grid as a CAS

Systems are natural or artificial (human made[6] or engineered) [64]. An artificial system is usually designed and engineered to achieve some specific objectives and to behave according to a predetermined scheme. A conventional control system, a computer program, and a building are examples of artificial systems. Artificial systems are expected to express deterministic behavior, thus, some do not normally regard them asCxS. Unpredictable behaviors may occur in artificial systems—like instability in the control system, divergence in the computer program, or dynamic instability in the building—however, these behaviors are exceptions to regular norms.

The power system, automation and control, and IT infrastructure subsystems of the Smart Grid are artificial systems that are engineered to serve specific goals. Some researchers regard these systems as Information/Decision/Action systems (IDA systems). IDA systems are abstractions of systems that all involve human decision-making, like: individuals and groups of humans, piloted vehicles, companies, governments, air traffic control, finance and banking, management information, command and control information [13, 213]. All of these systems include sensing, communication, assessment of the sensed information (Information), humanistic decision-making (Decision), and consequent controlled action (Action) [13]. IDA systems are claimed to not necessarily be categorized as CASs, since "*a feature missing from some of these IDA systems, for instance, is a degree of regulation and control which tries to prevent them from self-organizing*" [13].

Natural systems, on the other hand, are not engineered by humans. The natural systems evolve during the time from different systems, as they may evolve to other different systems. A lake may evolve to a desert and a tree may evolve to a garbage wood. Science might predict futures for the natural systems based on the empirical experiences from the similar systems. Nevertheless, humans are not personally aware of the goal and target of the natural systems, thus the natural systems are unpredictable in their behavior.

However, between these two extremes, there are systems that are artificial, yet they are larger and older than to be engineered by a team of humans. Infrastructures are one of the most crystal instances of such systems. Weijnen *et al.* exemplify electricity and IT infrastructures as CASs which were not firstly designed to form integrated systems, though they have emerged over time to become so [39, 214]. Emergent behavior of these systems is due to the unpredictable consequences of system operators' actions and disturbances. For instance, cascading blackouts are a result of emergence occurred due to the operator's inability in directing the flows over the network.

[6]In accordance with gender neutrality, here we call the man-made systems as human-made systems.

Now let us apply the above discussion to our Smart Grid case:

> As a holistic view, in a Smart Grid system, although the behavior of each component, for instance a bus-bar, a data link, or a measuring device within different sub-systems is known, the overall behavior of the system is unpredictable. That is, the summation of behaviors of the agents does not simply lead us to the aggregated behavior of the system's components. In other words, the overall behavior of the system is the result of cooperation of its agents. Furthermore, since the relationships between causes and effects are nonlinear, and the action of some parts always affects the action of other parts; the system as a whole shows emergent properties. This nonlinear quality exists both within different subsystems (power system and human operators/consumers) and in the interconnections between them. Moreover, one of the most important qualities of Smart Grid is its ability to adapt to the environmental changes. Furthermore, the subsystems within the Smart Grid system adapt themselves to the changes in the context provided by other subsystems. This internal adaptation is called self-organization (See Section 12.3).

Due to the above discussion, and according to the definitions and characteristics of CASs given in Section 12.3, humans—both operational humans and consumer humans—are obviously CASs, as natural systems. The same applies to the surrounding environment, as it is unpredictable and adaptive.

Power systems are artificial systems that are initially engineered by humans to behave in a deterministic manner; however, they express complex, emergent, adaptive, nonlinear, and even chaotic behavior during their lifecycle. For instance, as a clear example of emergence behavior, a single machine system (i.e., a power generator connected to the bus-bar) is designed and installed to produce electricity with a high level of static, dynamic, and transient stability. However, interconnection of a multimachine system is shown to be prone to dynamic instabilities, causing oscillations in frequency and power, leading to potential cascading trips and even blackouts [215]. As another example for adaptive behavior, consider a situation in which a load varies its consumption. In such a case, the other nodes of the electric network change their voltages and currents to meet the power flow equations [216], thus behaving adaptively.

Similar to the power system, the IT infrastructure and the automation and control subsystems of the Smart Grid are artificial, and are initially engineered by humans to behave deterministically. As stated, these IT infrastructures and automation and control systems are categorized in the class of IDA systems [213]. However, this is provided to that they are conventional ones, that is, they do not include intelligent agents. Otherwise, if the IT infrastructure and the automation and control system include artificial intelligence, the intelligent agents adapt themselves to the changes and learn from their experiences during operation. Thus, apparently, they are complex and adaptive. An example for the actions of automation and control in power systems is restoration by reconfiguration, in response to a fault in the power delivery system by the automation and control system and the employment of the proper tool by IT infrastructures [217–219]. Restoration by reconfiguration is performed while dealing with changes in environment, which can be considered as an adaptive behavior. Nevertheless, even the conventional automation and control systems (i.e., non-intelligent) they also express complex and adaptive behaviors, as discussed earlier quoting Reference 39.

The HAI system (as well as its governing AAES system) is the sphere in which the humans and the automation systems collaborate. The human side is obviously a CAS, and the automation side also behaves as a CAS, as discussed earlier. The HAI is a CAS phenomenon in two senses: firstly, the collaboratively performed tasks of humans and automation systems are complex and should be adapted to the changing environment; and secondly, the human–automation collaboration itself is a complex phenomenon and should be adapted to the changing environment. Thus, an HAI system can be regarded as a CAS.

12.7.3 HAI in Smart Grid as an SoS

HAI can also be regarded as an SoS. In addition to the above mentioned characteristics for a CAS, it is needed to be some complementary characteristics to form an SoS. These characteristics must depict the dispersion aspect of constituent systems: geographical distribution, heterogeneity, and connectivity.

Power Systems—and consequently Smart Grids—are geographically dispersed due to the geographically dispersion of their consumers of electrical power. North American power system is considered to be largest ever implemented human-made system [33].

Smart Grid consists of heterogeneous subsystems that are diverse in purpose, technologies, and context. For instance, the IT infrastructure is responsible for running the required communications among different subsystems. The power system, on the other hand, is responsible for generation, transmission, and distribution of the electrical power.

Connectivity is an inseparable characteristic of SoS, since constituent systems require to communicate with each other to gain the ability of synchronization and pursue their common goal if necessary. Connectivity in Smart Grid is implemented via the IT infrastructure subsystem.

To summarize, a Smart Grid contains a set of geographically dispersed and heterogeneous systems including a large number of agents with nonlinear interconnections—complexity—plus the ability of adaptation to changes in the context produced by other agents or the environment—adaptability. These systems are collaborating together to accomplish a common goal and therefore, Smart Grid can be regarded as a CASoS.

12.8 PETRI NETS FOR COMPLEX SYSTEMS MODELING

Petri nets, introduced in 1962 by C. A. Petri, are basically known as powerful tools for modeling and analysis of systems with concurrent, distributed, nondeterministic, and/or asynchronous behavior. Some examples of application areas of Petri nets are manufacturing systems [220–225], robot planning [226–231], financial systems

[232–234], computational biology [235–240], transportation systems [241–245], and work flow analysis [246–248].

This section is intended to briefly introduce the main concepts of Petri nets. We use notations, definitions, and properties as given by Wu and Zhou in Reference 241.

12.8.1 Definition

A Petri net is a particular kind of bipartite directed graph together with an initial marking depicting the initial state of net. The net includes *places*, *transitions*, and *directed arcs*. Directed arcs connect places to transitions or transitions to places. A *marking* is an assignment of *tokens* (nonnegative integers) to the places of the net. The dynamic behavior of a Petri net is shown by flow of tokens from some places to others by firing transitions resulting in a different marking (state).

A Petri net is formally defined as a 5-tuple $PN = (P, T, I, O, M_0)$, where

(1) $P = \{p_1, p_2, \ldots, p_m\}$ is a finite set of places;
(2) $T = \{t_1, t_2, \ldots, t_n\}$ is a finite set of transitions, $P \cup T \neq \emptyset$, $P \cap T = \emptyset$;
(3) $I : P \times T \rightarrow N$ is an *input function* that defines directed arcs from places to transitions where N is a set of nonnegative integers;
(4) $O : T \times P \rightarrow N$ is an *output function* that defines directed arcs from transitions to places; and
(5) $M_0 : P \rightarrow N$ is the initial marking.

12.8.2 Graph Representation of Petri Nets

A Petri net graph has two types of nodes, *circles* and *bars (boxes)* representing places and transitions, respectively. *Directed arcs (arrows)*, labeled with their multiplicity (weight), connect places and transitions. *Dots* resided in the circles represent tokens in places (as shown in Figure 12.17).

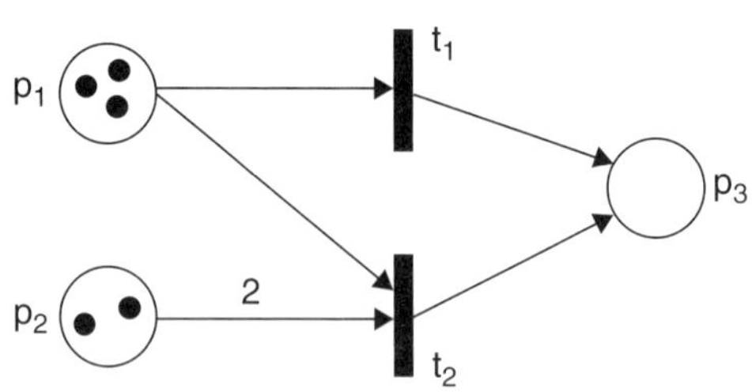

FIGURE 12.17 A simple Petri net graph.

For Petri net of Figure 12.17:

$$\begin{aligned}
&P = \{p_1, p_2, p_3\};\\
&T = \{t_1, t_2\};\\
&I(p_1, t_1) = 1, I(p_2, t_1) = 0, I(p_1, t_2) = 1, I(p_2, t_2) = 2, I(p_3, t_i) = 0;\\
&O(t_1, p_3) = 1, O(t_2, p_3) = 1, O(t_i, p_j) = 0;\\
&M_0 = (3, 2, 0)^T;\\
&(i, j = 1, 2).
\end{aligned} \tag{12.1}$$

12.8.3 Transition Firing

The execution of a Petri net is controlled by the number and distribution of tokens in places. *Enabling rule* and *firing rule* of a transition which control the flow of tokens in places are as follows (see Figure 12.18):

(1) Enabling rule: A transition t is enabled if $\forall p \in P : M(p) \geq I(p, t)$.
(2) Firing rule: The firing of an enabled transition t removes from each input place p the number of tokens equal to the weight of arc (WOA) connecting p to t; and deposits in each output place the same number of tokens equal to the WOA connecting t to p.

Mathematically, firing t at M yields a new marking M' determined:

$$\forall p \in P : M'(p) = M(p) - I(p, t) + O(t, p) \tag{12.2}$$

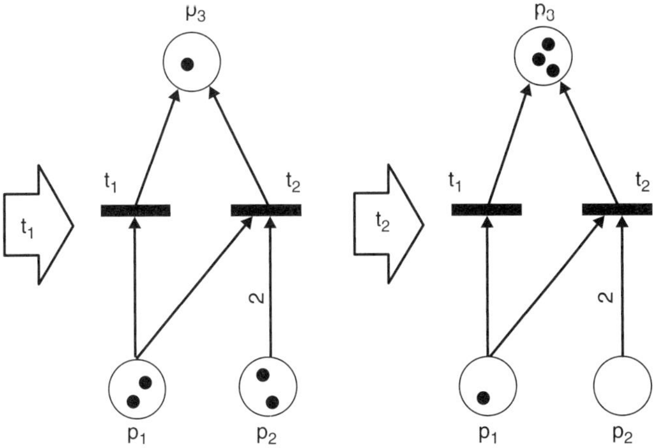

FIGURE 12.18 Transition firing and new markings.

12.8.4 Reachability

A marking M_1 is said to be *immediately reachable* from M_0 if firing an enabled transition in M_0 results in M_1. Reachability is generalized in the way that a marking M_2 is said to be reachable from M_0 if firing a sequence of transitions in T, starting from M_0, results in M_2. The set of all reachable markings of a graph Z from initial marking M_0 is denoted by $R\left(Z, M_0\right)$ [249].

12.8.5 Incidence Matrix and State Equation

The incidence matrix of a Petri net with m places and n transitions is $A = [a_{ij}]_{n\times m}$ with typical entry $a_{ij} = a_{ij}^{+} - a_{ij}^{-}$ where $a_{ij}^{+} = O(t_i, p_j)$and $a_{ij}^{-} = I(p_j, t_i)$. According to firing rule, a_{ij} represents change in the markings in place p_j when transition t_i fires once.

Suppose M_k as an $m \times 1$ column vector whose jth entry denotes the marking in place p_j immediately after the kth firing in some firing sequence, and x_k as the kth firing vector with only one nonzero entry, a 1 in the ith position for the ith transition to be fired at the kth firing. The state equation for a Petri net is as follows (250):

$$M_k = M_{k-1} + A^T.x_k; k = 1, 2, \ldots. \tag{12.3}$$

Now, suppose that destination marking M_d is reachable from M_0 through a firing sequence $\{x_1, x_2, \ldots, x_d\}$. The state equation can be generalized as follows [250]:

$$M_d = M_0 + A^T. \sum_{k=1}^{d} x_k. \tag{12.4}$$

12.8.6 Inhibitor Arc

An *inhibitor arc* connects an input place to a transition and changes the transition enabling condition in a way that there should be no tokens in each input place connected to the transition by the inhibitor arc (see Figure 12.19). Inhibitor arcs are used to model priority in the system.

12.8.7 IF–THEN Rules by Petri Net

An IF–THEN rule can be modeled as a transition whose input places and output places represent antecedent portion and consequence portion of the rule respectively, in a way that each proposition in the antecedent portion is modeled as an input place and each proposition in the consequence portion is modeled as an output place [251]. For instance, the following IF–THEN rule can be modeled as shown in Figure 12.20:

$$R_1 : if\ ((A\ or\ B)\ and\ C)\ then\ ((D\ or\ E)\ and\ F). \tag{12.5}$$

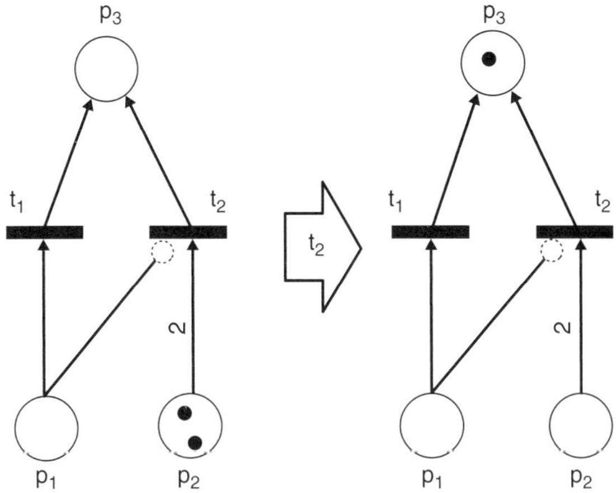

FIGURE 12.19 Inhibitor arc in Petri net graph.

12.9 MODEL-BASED IMPLEMENTATION OF ADAPTIVE AUTONOMY

12.9.1 The Implementation Framework

An implementation framework for adaptive autonomy is proposed by Fereidunian *et al.,* which belongs to the model-based class of adaptive autonomy implementation methods (see Table 12.10 for classification), as shown in Figure 12.21 [193, 194, 198]. The upper loop in Figure 12.21 sequentially checks for the changes in performance shaping factors (PSF). PSFs are used to introduce the environmental conditions to the adaptive autonomy implementation process. PSFs represent the most influential factors that shape the performance of humans and the automation system (intelligent electronic devices (IEDs)). PSFs are used in this framework to tackle the issue of quantitative representation of the SCADA system environmental conditions [193, 198].

Each time the environmental conditions change—according to the monitored field dynamic data—PSFs are updated. Afterward, the new autonomy level is determined

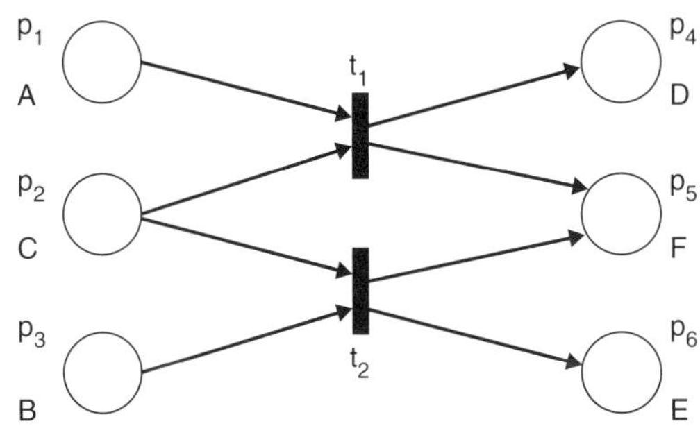

FIGURE 12.20 Petri net modeling of IF–THEN rule.

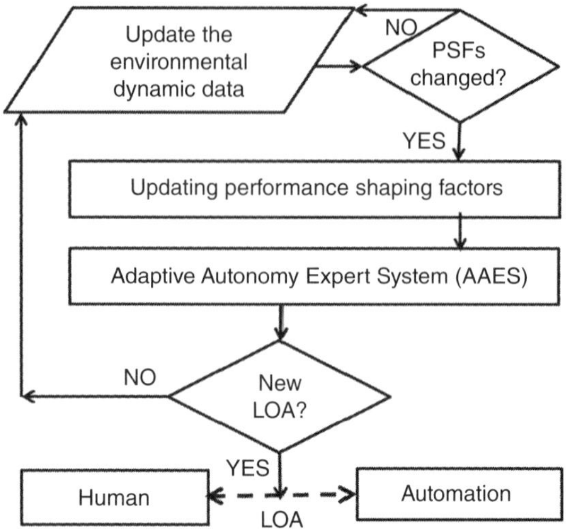

FIGURE 12.21 The proposed framework for implementation of adaptive autonomy.

by a trade-off analysis between the humans and the automation in light of the subjective knowledge of Experts' Judgment. Since the aim of the human performance evaluation is to compare it to those of automation, this research does not seek to focus on direct methods of human mental models, that is, the objective methods; instead, we referred to the subjective knowledge of the filed experts. The Experts' Judgment is gathered through interviews with the GTEDC's (Greater Tehran Electric Distribution Company) SCADA and dispatching experts, based on a standardized questionnaire, where each interview took at least 1 hour time [193, 194, 198].

An extended version of the LOA–TOA model of Parasuraman, Sheridan, and Wickens is used in our proposed method: level 1 of the original Sheridan's taxonomy is shifted down to form a new level 0*; and a new level of 1* is introduced as a new LOA [110]. Hence, our adaptive autonomy framework deals with 11 LOAs. The justification for the necessity of the new level is given in [193], and the definitions of the LOAs, TOAs, and HAI model can also be found in References 110, 193, and 194.

Up until now, this implementation framework has been realized using the following eight technologies: AAES using weighted summation fusion [196], AAFES using fuzzy systems [252], AAGLMES using generalized linear models [208], AALRES using logistic regression [253], AAHES using hybrid neural network [254], AAPNES using Petri nets [255, 256], AAHPNES using hierarchical Petri nets [257], and AAFGES using fuzzy gradient descent [258].

12.9.2 Case Study: Adaptive Autonomy in Smart Grid

Utility management automation (UMA), as a subsystem of Smart Grid, acts as a SCADA system for the electric utility in which human operators and automation systems work collaboratively. In this section, an expert system (referred to as

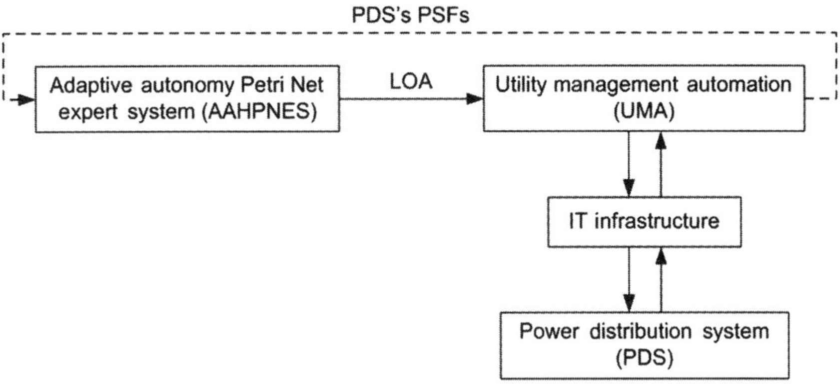

FIGURE 12.22 Position of Petri net adaptive autonomy expert system in power distribution system.

AAHPNES) is employed to adapt the autonomy level (LOA) of the UMA system to the changes in the PSFs. In other words, the AAHPNES controls the LOA of the UMA system.

AAHPNES is implemented to one of the power distribution automation functions, referred to as feeder reconfiguration function of utility management automation (UMA-FRF). The UMA-FRF system—which has been introduced in References 207, 219, 259–261—automatically restores the electric energy for the affected customers (electric power delivery load points) by reconfiguring the distribution network topology after a failure in the distribution network [217, 218, 262, 263]. Figure 12.22 shows the proposed expert system role in relation with the other subsystems of the UMA. The dashed arrow from the UMA conveys the PSFs to the AAHPNES where the other solid line arrows command the LOA recommended by AAHPNES to the UMA.

12.10 ADAPTIVE AUTONOMY REALIZATION USING PETRI NETS

12.10.1 Implementation Methodology

Here the implementation method of the AAHPNES is presented.

12.10.1.1 IF–THEN Rules of AAHPNES IF–THEN rules and their representation in Petri net are the primary concerns in realization of the AAHPNES. In this section, the general form of the extracted rules from the experts' judgment is presented.

There are two kinds of rules that are employed in the AAHPNES: *Comparative Rules* and *Combinational Rules*. The comparative rules suggest an LOA when only *one* PSF changes from its normal condition, while combinational rules suggest an LOA when *multiple* PSFs change from their normal conditions. In other

TABLE 12.11 PSFs' Values and Their Corresponding WoAs

PSF	PSF's value (a_i)	WoA (b_i)
Time	Day*	1
	Night	2
Service area	Uncrowded urban*	1
	Crowded urban	2
	Rural	3
Customer type	Residential*	1
	Commercial/industrial	2
	VIP	3
Number of faults per 2 hours	Few*	1
	More	2
	Much more	3
Network age	New*	1
	Middle aged	2
	Old	3
Load	Low*	1
	High	2

a * denotes a basic PSF state.

words, combinational rules describe the effect of change in one PSF's value on LOA at a time, while combinational rules describe effects of changes in two or three PSFs' values on LOA together at the same time. The list of practical PSFs and their values is shown in Table 12.11. For example, according to experts' judgment for an *old* network (PSF_5 = Old), while other PSFs are normal, the LOA is 3—comparative rule; and for a *highly* loaded network with *much more* (10) faults per 2 hours ($PSF_4 = Muchmore$and$PSF_6 = High$), independent from other PSFs, the LOA is 7—combinational rule. To put it in other words, the first rule demonstrates that an *old* network, alone and in comparison with normal condition, decrease two units in LOA; while, the second rule implies that the effects of *much more* faults per 2 hours combined with high load increase two units in LOA.

12.10.1.2 *Petri Net Representation of Rules* In this part, rules are customized to be applied to the AAHPNES. According to GTEDC's experts, LOA depends on six main PSFs which are shown in Table 12.11 [252]. Using these PSFs, rules are represented in the following general form:

$$R_1 : if((PSF_1 \ is \ a_1) \ and \ldots and \ (PSF_6 \ is \ a_6)) \ then \ (LOA \ is \ c). \qquad (12.6)$$

In order to model this rule by Petri net, antecedent and consequent portions of the rule have to be transformed to Weight of Arcs (WoAs) to the corresponding transition and from it, respectively (see Figure 12.23). To do so, the proposition "$PSF_i isa_i$" is transformed to a WoA (b_i), connecting the input place corresponding to PSF_i to transition corresponding to this rule. Table 12.11 shows the corresponding WOAs

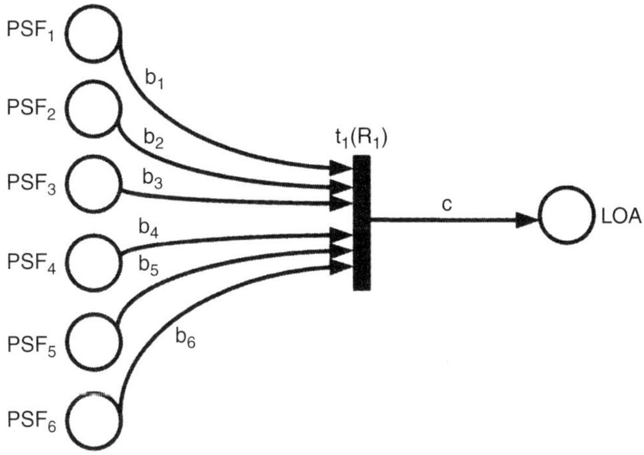

FIGURE 12.23 Petri net modeling of rule of Equation (12.6).

(b_i) for different PSFs' values (a_i) of each PSF. As shown in Table 12.11, the concept in determining WOAs is that for each PSF, the normal PSF's value (marked by * in Table 12.11) gets $WOA = 1$ and for other PSFs' value, WOA increases as the abnormality of PSF's value, relative to the normal PSF's value increases. Note that all PSFs' values are linguistic values: for the first three PSFs (Time, Service area, and Customer type) these values describe different types of the corresponding PSF considered in our modeling and for the last three PSFs (Number of faults per 2 hour, Network age, and Load) these values describe different levels of numerical values of the corresponding PSF, that is, the practical range of numerical values of each PSF is quantized into two or three levels and each quantization level, labeled with a specific linguistic value, is transformed into a nonnegative integer (WoA).

For consequent portion of the rule, "*LOAisc*," *b* is suggested as WOA between the corresponding transition and output place.

12.10.2 Realization of AAHPNES

AAHPNES recommends the proper LOA in the presence of different PSFs. The proposed Petri net expert system has two layers, *Comparison layer* and *Combination layer*, corresponding to two sorts of rules.

12.10.2.1 *Comparison Layer* This layer of the Petri net model is constructed using *11 comparative rules*. The antecedent portion of these rules is in the form of$PSF = [1, \ldots, i, \ldots, 1], i \neq 1$, that is, all PSFs are normal except one of them (Table 12.11). To model these rules by a Petri net, the six PSFs are divided in three groups based on their effect on LOA [260]. These groups and their effects are:

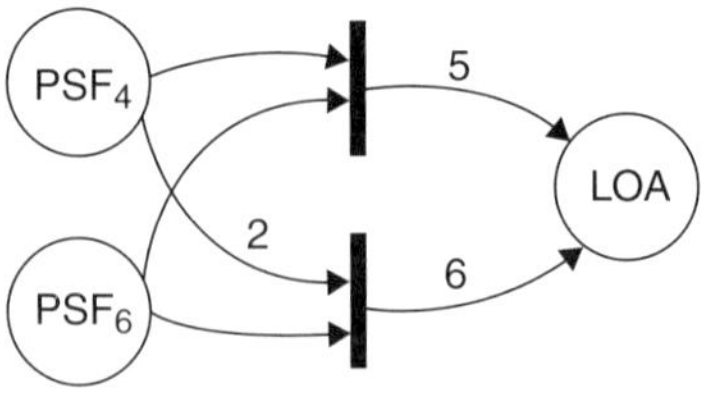

FIGURE 12.24 Simple Petri net modeling of rules of Equation (12.7), each rule is modeled by a transition.

(a) PSF_1 and PSF_2, tend to increase the LOA when their value deviates from normal;

(b) PSF_3 and PSF_5, tend to highly decrease the LOA when their value deviates from normal; and,

(c) PSF_4 and PSF_6, tend to highly increase the LOA when their value deviates from normal.

For each group, using the corresponding rules, a hierarchical Petri net model is derived. To illustrate the hierarchical method of modeling, suppose the following rules:

$$R_1 : PSF = [1, 1, 1, 1, 1, 1] \rightarrow LOA = 5$$
$$R_2 : PSF = [1, 1, 1, 2, 1, 1] \rightarrow LOA = 6. \tag{12.7}$$

These two rules can simply be modeled as shown in Figure 12.24 in which each rule is modeled by a transition using WOAs from Table 12.11; however, this simple model, since there is no priority between transitions, cannot guarantee the correct output for some combinations of inputs. Therefore, it needs a complicated analysis to determine the final LOA; for instance, suppose that PSF_4has two tokens, it can also fire the upper transition without firing the lower transition. This problem was resolved in AAPNES of Zamani *et al.* in References 255 and 256 by introducing priority to the model using deterministic timed Petri nets (DTPN). Here in AAHPNES of Fereidunian *et al.* in Reference 257, these rules are modeled in a hierarchal organized Petri net, as shown in Figure 12.25. This hierarchical modeling technique is based on experts' judgment procedure while determining the proper LOA for real input PSFs. Moreover, in the final modeling, instead of the LOA itself, its deviation from the normal condition ($LOA = 5$) is considered. Figure 12.26 shows this concept for the model of Figure 12.25. The concept of determining deviation from normal LOA as the output of the net instead of the real LOA is employed to make the judgment procedure simpler for experts.

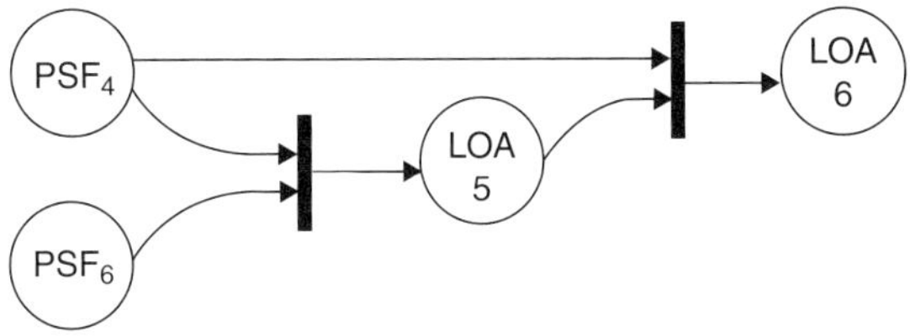

FIGURE 12.25 Hierarchical Petri net modeling of rules of Equation (12.7).

Now that the hierarchical modeling method is explained, we explain the complete procedure of deriving a Petri net model for the third PSFs group (PSF_4 and PSF_6). This model is derived from four comparative rules listed below:

$$\begin{aligned} &R_1 : PSF = [1,1,1,1,1,1] \rightarrow LOA = 5 \\ &R_2 : PSF = [1,1,1,2,1,1] \rightarrow LOA = 6 \\ &R_3 : PSF = [1,1,1,3,1,1] \rightarrow LOA = 7. \\ &R_4 : PSF = [1,1,1,1,1,2] \rightarrow LOA = 6 \end{aligned} \tag{12.8}$$

As stated above, the first rule indicates the normal condition, R_2 and R_3, are the comparative rules regarding $\text{PSF}_4$and $\text{R}_4$is the comparative rule regarding PSF_6. Nonetheless, to be able to derive a model, there is a lack of information about two other feasible conditions: $PSF = [1,1,1,2,1,2]$ and $PSF = [1,1,1,3,1,2]$. For the later one, using R_3 and the fact that the highest practical level of LOA for this application is 7, it is concluded that

$$R_5 : PSF = [1,1,1,3,1,2] \rightarrow LOA = 7. \tag{12.9}$$

For the former one, we have to use a combinational rule—which is actually derived from the nature of the system:

$$R_6 : PSF = [1,1,1,2,1,2] \rightarrow LOA = 6. \tag{12.10}$$

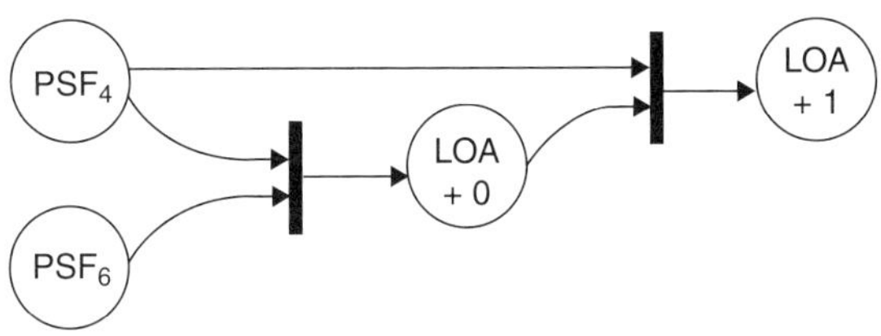

FIGURE 12.26 Hierarchical Petri net modeling of rules of Equation (12.7), output places show the change in LOA from normal condition.

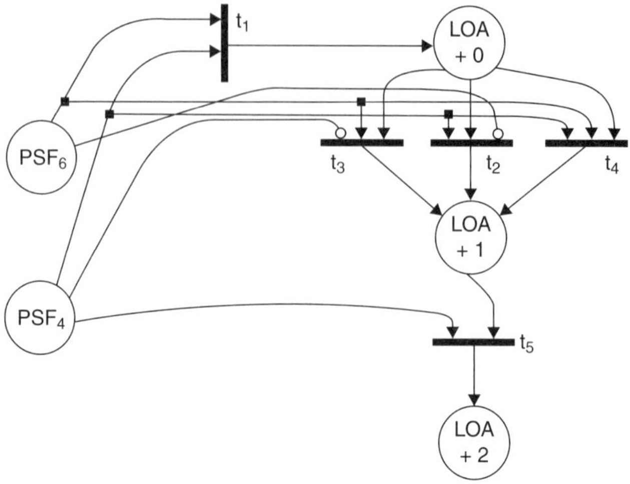

FIGURE 12.27 Hierarchical Petri net modeling of group c includes PSF_4 and PSF_6.

Note that in all of the rules above, the other PSFs (PSF_1, PSF_2, PSF_3, and PSF_5) have their normal values. Now that all needed rules are available, the hierarchical Petri net model can be derived as shown in Figure12.27. In this model t_1, t_2, t_3, and t_4 correspond to R_1, R_2, R_4, and R_6respectively; andt_5 corresponds to R_3 and R_5.

Using the same procedure, the Petri net models of the other groups of PSFs are derived as shown in Figure 12.28. Note that the notation "$\frac{1}{2}$+" in Figure 12.28a indicates that the corresponding input PSFs cannot change the LOA individually and their effect is determined in combination with other input PSFs (this task is performed by the combination layer of modeling).

12.10.2.2 Combination Layer The combination layer of the Petri net model is constructed using *11 combinational rules*. In order to apply these rules, the antecedent portions of them are translated to make them applicable to the outputs of the comparison layer. For instance, the rule

$$R_{12} :, \mathrm{PSF}_4 = 3, \mathrm{PSF}_6 = 2 \rightarrow LOA = 7 \tag{12.11}$$

is translated to

$$R_{12} : (\mathrm{LOA2+}) \,\&\, (LOA2+) \rightarrow LOA = 7. \tag{12.12}$$

The modeling method for these rules is transforming each rule to a transition; like the method applied in Figure 12.24 for modeling the rules of Equation (12.7). These rules determine the final LOA (Figure 12.30).

In addition, during the execution of the overall Petri net, in order to give priority to the comparison layer over the combination layer, a transition is added as

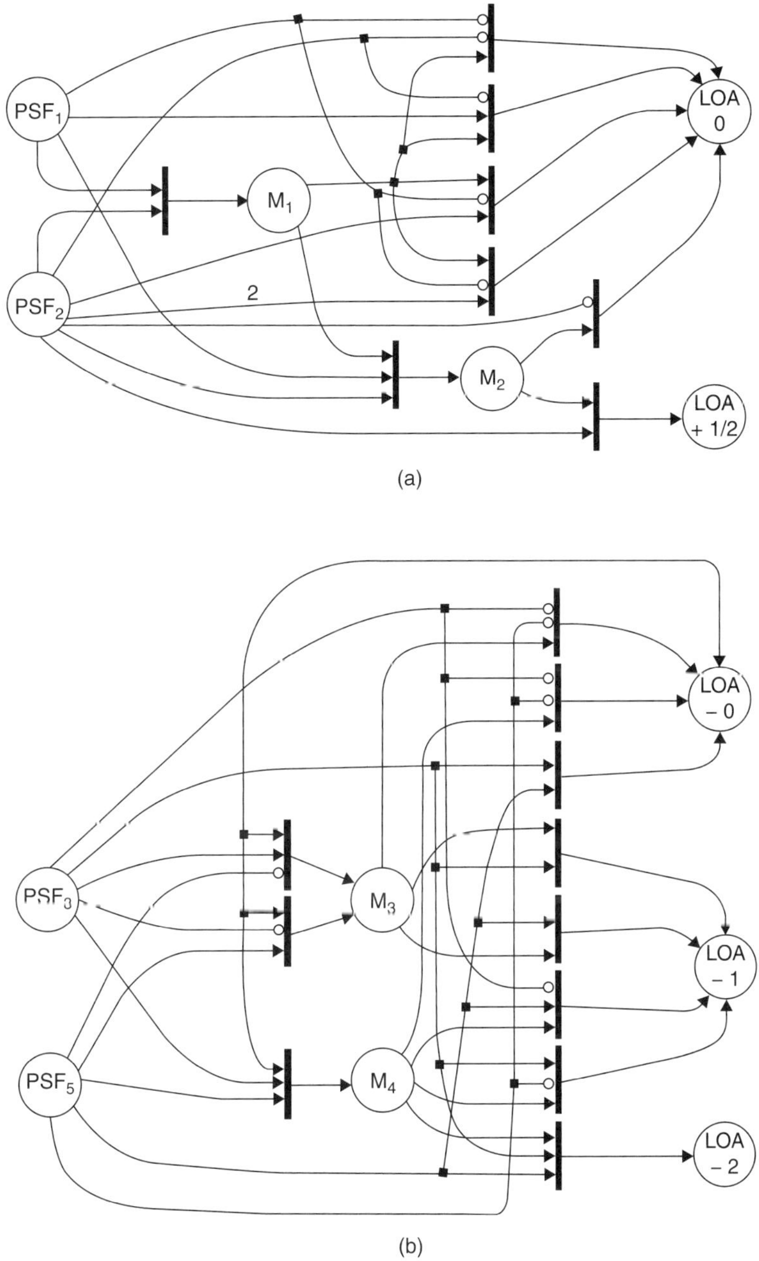

FIGURE 12.28 Hierarchical Petri net modeling of: (a) group a includes PSF_1 and PSF_2, (b) group b includes PSF_3 and PSF_5.

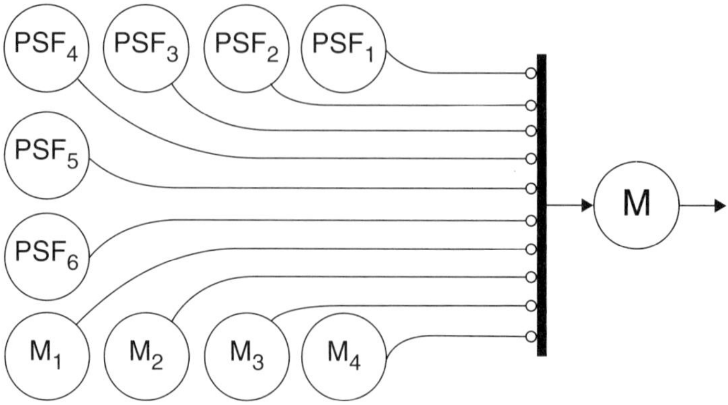

FIGURE 12.29 Trigger for giving propagation priority to comparison layer over combination layer.

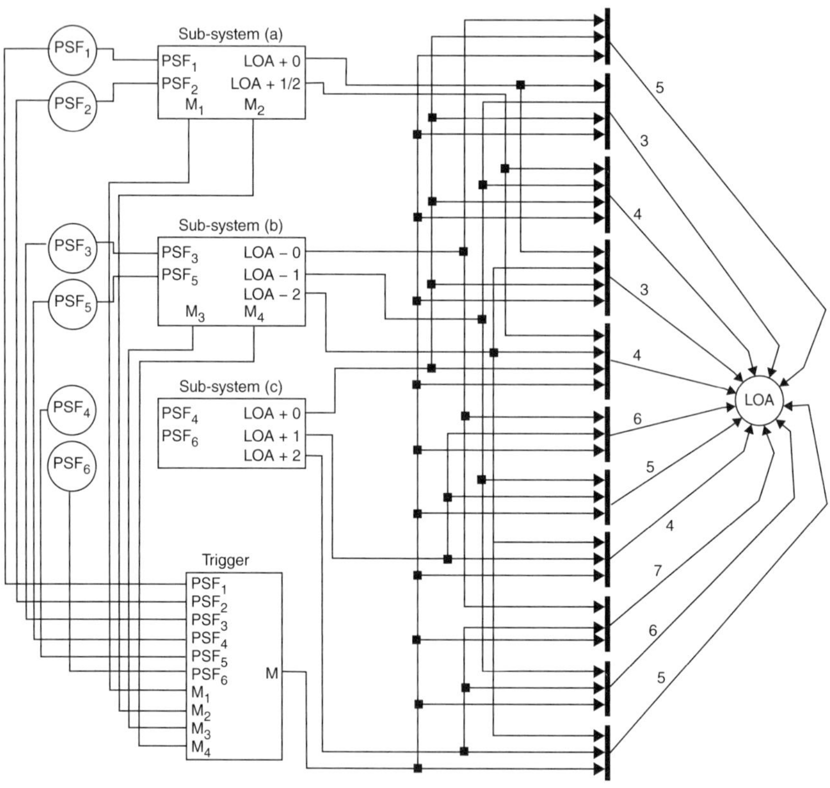

FIGURE 12.30 The overall Petri net expert system.

a trigger. Figure 12.29 shows the structure of the trigger. As shown, the transition will be enabled when all the input and medium places of the comparison layer are empty, that is, the execution of the comparison layer is finished. The output place of this transition (M) is added as an input place to all transitions of the combination layer; thus, the combination layer's transitions would not be enabled, unless the process of the comparison layer is finished.

Figure 12.30 shows the overall Petri net expert system. As can be seen, the hierarchical Petri net modeling of grouped PSFs in elementary layer and the trigger are shown as blocks (subsystems); while, the secondary layer of the modeling is shown in detail.

To apply the Petri net model, the input condition (PSF vector) is transformed to an initial marking according to Table 12.11, that is, the same mechanism for determining WOAs from rules holds for determining the initial marking from the input PSF vector. Afterward, using reachability analysis, the dead-end marking from a firing sequence, in which there is no repeated transition, is calculated.

Note that all of the reachable markings from any initial marking (in which only input places,PSF_{1-6}, include a nonzero number of tokens) lead to a dead-end marking. To understand this, note that in the overall Petri net of Figure 12.30, tokens only propagate from the left-hand side of the net to the right-hand side. Based on the modeling method, there is only one such marking and only this marking can be used for determining LOA, because other dead-end markings, even with a nonzero number of tokens in the place "LOA," are not reasonable due to firing some transition(s) more than once. In this dead-end marking, number of tokens is the place "LOA" demonstrates the final LOA. To illustrate, consider the subsystem (a) of Figure 12.30, which is shown in detail in Figure 12.27. As shown in Figure 12.27, for any initial marking (number of tokens in places PSF_4 and PSF_6), only transition t_1, is enabled at first, since other places has no token. Thus, t_1 is the first transition to be fired. Subsequently, since t_1 cannot be fired anymore and regarding the number of tokens remained in $PSF_4$and PSF_6, only one of the transitions t_2, t_3, or t_4 is enabled and it is the next entity of the firing sequence. Finally, if still a token remains in PSF_4, since none of the transitions t_1, t_2, t_3, and t_4 can be fired, even if they are enabled, t_5 will be enabled and fired. Figure 12.31 shows all of the feasible initial markings and corresponding firing sequences. As you can see, they all end in a marking with only one place with nonzero (one) number of tokens—this place is one of the three output places of "LOA + 0," "LOA + 1," and "LOA + 2."

Using the same procedure, we can also show that for the Petri nets in Figure 12.28 (subsystems b and c of Figure12.30), any feasible initial marking leads to a dead-end marking in which there is only one place with nonzero (one) number of tokens and that place is one of their output places. Now that one of the output places of each subsystem is set (has only one token) and all other places within the subsystems are empty, the combination layer is triggered. Based on the distribution of tokens in output places of subsystems, only one transition among 14 transitions of combination layer will be enabled. Firing this transition deposits in place "LOA" a number of tokens equal to the final LOA—note that each distribution of tokens in output places of subsystems of comparison layer corresponds to only one of the

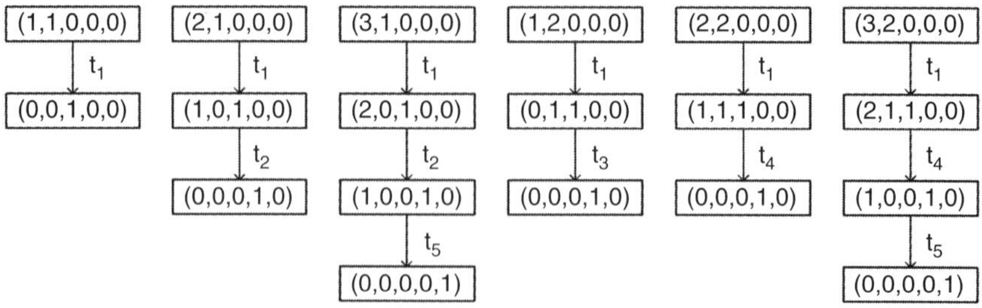

FIGURE 12.31 Firing sequences and corresponding markings for all feasible initial markings for Petri net of Figure 12.27 (the order of places in given markings is PSF_4, PSF_6, LOA + 0, LOA + 1, and LOA + 2).

transitions in the combination layer and only one transition can be fired since output places of subsystems of comparison layer include only one token.

To summarize, AAHPNES paves the steps shown in Figure 12.32, to determine LOA for a HAI system.

12.10.3 Results and Discussions

The intelligence of an expert system intensively relies on including appropriate rules. Furthermore, to evaluate the performance of the expert system, it is needed to verify whether the system can simulate an expert opinion or not. Therefore, both rule set and test set are asked from a superior expert in various PSFs combinations. The superior experts are experts whose superiority (in higher and more reliable expertise) has been verified according to consistency for their expert judgments [252]. All feasible conditions include 324 states which are used to determine correct classification rate (CCR) of the system.

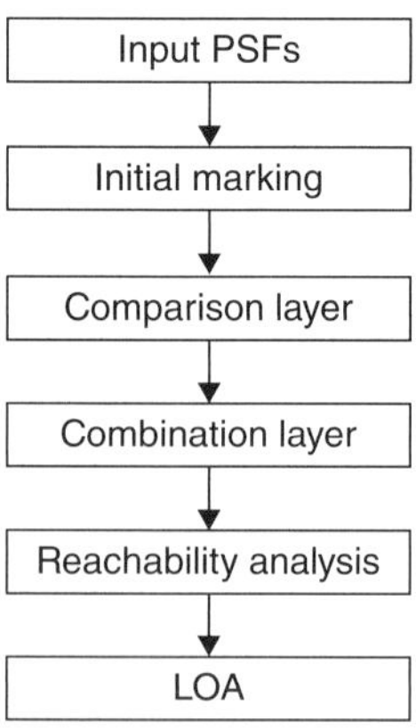

FIGURE 12.32 Steps of expert system in determining hybrid LOA from input PSFs.

The CCR of the proposed AAHPNES is 78% which indicates the high overall performance of the expert system. Moreover, the CCR of the proposed expert system for $LOA = 3 or 7$ is 94% which demonstrates that the system almost exactly simulates an expert opinion in critical situations. This illustrates higher performance of proposed AAHPNES, both generally and in critical situations, in comparison with its ancestor, AAPNES, introduced in References 255 and 256.

In comparison with the other model-driven systems of References 196 and 252, AAHPNES not only shows a higher CCR—especially in complicated situations, but also employs a systematic method that is more like a humanistic decision-making in terms of hierarchical reasoning. To illustrate, while expert systems of References 196 and 252 apply all extracted rules from the experts' judgment at once, the presented AAHPNES first determines the effect of PSFs separately, then combines them to determine to proper LOA. This method is based on experts' attitude during interviews. Moreover, the introduced AAHPNES, like its ancestor AAPNES of References 255 and 256 has the unique quality of employing the priority introduced in experts' judgments in modeling. This priority is used in the second level of modeling in the process of combining outputs of the first level. Although the expert systems of References 208, 253, and 261 may show higher performance regarding CCR, they are all data driven and require a large amount of data to be able to determine a proper LOA, while the proposed AAPNES is based on 22 general rules.

12.11 CONCLUSIONS

After introducing the systems-related concepts of complexity, CASs and SoS, different views regarding these concepts were compared. Especially, the characteristics of the CASs and SoSs were enumerated and explained. Subsequently, the automation-related notions of automation and HAIs were described, and then, evolution of HAI models from both viewpoints of dimensions and dynamism were investigated. Furthermore, the idea of adaptive autonomy (AA) was introduced as a dynamic HAI scheme, followed by a classification of AA implementation methods. Afterward, the HAI system was studied as a CAS and an SoS. Petri nets were introduced as powerful tools for modeling CxS. Finally, the Petri net realization of AAES was expressed and its performance was evaluated.

Besides introducing and explaining the topic itself, a couple of open questions were elaborated in this chapter. One of the most important questions is the source of complexity: either the complex entity itself or the observer (Section 12.2.1). The complexity that is sourced back to the complex entity is referred to as innate complexity, and described as a state of world. Whereas, the complexity that is sourced by the observer is referred to as cognitive complexity or complicatedness (i.e., making confused), and described as a state of mind when responding to complexity [2]. This is a controversial issue, as some argue that if there was no human (as observer), we could not regard any entity as complex. Others, on the other hand, believe that complex entities are inherently complex, regardless of the human observer. An interesting advocate for the innate complexity is that the human mind is a complex phenomenon itself. Thus when talking about cognitive complexity, one should remember that the observer's mind evolve, as a complex entity. This controversy needs more discussions in a proper situation and more investigation on this issue might be furthered in future research.

The 2D classification of Table 12.1 (Section 12.2.2), or as we called it source–problem representation of CxS, has been introduced in this chapter for the first time. Thus it needs more investigation, and perhaps more development.

Definitions of complexity are qualitative ones, while, we can talk about degrees of complexity: two systems might be complex; nevertheless, one of the two could be much more complex than the other one. Many complexity measures have been introduced for different types of complexity (in Section 12.2.3); however, a unified measure of complexity is needed. Moreover, more quantitative work on complexity needs to be performed. Although fuzziness of the complexity concept is acknowledged, little research has been conducted on this, to the best of our knowledge.

We enumerated 11 characteristics for CASs, out of the almost 20 found in the literature (Section 12.3.2). This can be reduced to fewer characteristics by extracting the most principal characteristics. If it is not theoretically possible, a practical list of minimum requirement for CAS' characteristics can be agreed upon. For example, it can be said that a system is a CAS, if and only if it exposes the four principal characters of emergent behavior, complexity, adaptability, and nonlinearity. It is quite plausible that the other characteristics can be deduced from those principal characteristics. Similarly, 14 characteristics were counted for SoSs, out of many (in Section 12.4.2). This can also be reduced to fewer by extracting the most principal ones or by agreeing upon a criteria list for SoS characteristics. The latter two studies—that is, criteria of systems for being CAS and SoS—are more expected from the professional organizations and standardization bodies like IEEE.

A preliminary taxonomy of systems family was developed in this chapter (Section 12.4.4.3) and depicted in Figure 12.13. Some other members of the systems family should be added to it and perhaps some editions required, which can be done in future works.

In this chapter, the qualitative HAI model of Parasuraman, Sheridan, and Wickens in Reference 110 (see Section 12.6.4) was engineered according to the model-based adaptive autonomy framework of Fereidunian *et al.,* in Reference 198 (see Sections 12.6.6, 12.9.1 and 12.9.2), and then realized using a hierarchical Petri net expert system (Sections 12.10.1 and 12.10.2), referred to as AAHPNES [257]. The HAI model of Reference 110 may be implemented by different methods and the adaptive autonomy framework of Figure 12.21 can be improved. Both models can also be implemented to different automation application domains, to compare the context-specific results with our results in Smart Grid.

Furthermore, according to Fereidunian *et al.,* in Reference 194, implementation of the HAI model faces to the following challenges: quantification of a qualitative model, type or stage, the lost LOA, and independency or interdependency of the TOA and LOA. The latter challenge is still an important open question and is a niche in the future research.

Moreover, the performance of the proposed AAHPNES system can be improved by applying additional rules; however, the issue is that a large number of rules (transitions) are needed to provide noticeable improvement. In other words, there is a trade-off between system performance and number of rules. A colored Petri net might show better performance compared to that of the hierarchical Petri net presented in this chapter.

ACKNOWLEDGMENTS

The authors would like to thank Dr. Bashir Mahdi Ebrahimi of University of Tehran. Furthermore, collaboration of Mr. Babak Eskandarnezhad and Mr. Arash Mehrjou is acknowledged.

Contributors of INCOSE Discussion Forum are thankfully acknowledged for their insightful and informative discussions and comments.

REFERENCES

[1] A. W. Wymore. *Model-Based Systems Engineering: An Introduction to the Mathematical Theory of Discrete Systems and to the Tricotyledon Theory of System Design*. Boca Raton, FL: CRC Press, 1993.

[2] D. A. Norman. *Living with Complexity*. MIT Press, 2011.

[3] W. Weaver. Science and complexity. *American Scientist*, 36: 536–544, 1948.

[4] G. Despotou, R. Alexander, and M. Hall-May. Key concepts and characteristics of Systems of Systems, (SoS). Defense and Aerospace Research Partnership, (DARP-HIRTS) Public Document, 2003.

[5] I. Bouwmans, M. P. C. Weijnen, and A. Gheorghe. Chapter 3: infrastructures at risk. In: *Critical Infrastructures at Risk* edited by A. V. Gheorghe, M. Masera, M. Weijnen. L. J. De Vries, Springer, 2006.

[6] WikiComplexity. Complexity. Available at http://en.wikipedia.org/wiki/Complexity (accessed July 29, 2011).

[7] B. Edmonds. What is complexity?—The philosophy of complexity per se with application to some examples in evolution. In: *The Evolution of Complexity*, edited by F. Heylighen and D. Aerts. Dordrecht: Kluwer, 1999.

[8] I. Price. Complexity, complicatedness and complexity: a new science behind organizational intervention?. *E:CO*, 6(1–2): 40–48, 2004.

[9] R. Dawkins. *The Blind Watchmaker: Why the Evidence of Evolution Reveals a Universe Without Design*. London: Longmans, 1988.

[10] W. Kinsner. Complexity and its measures in cognitive and other complex systems. In: 7th IEEE International Conference on Cognitive Informatics, 2008. ICCI 2008, August 14–16, 2008, pp. 13–29.

[11] J. N. Warfield. *Understanding Complexity: Thought and Behavior*. Palm Harbor, FL: Ajar Publishing Company, 2002.

[12] J. N. Warfield. Available at www.jnwarfield.com/ (accessed July 29, 2011).

[13] D. K. Hitchins. Characteristics of CASs. INCOSE Discussion Forum, 2011.

[14] C. Mario. Complexity and chaos: state of-the-art formulations and measures of complexity. Technical Note TN 2006-451, Defence R&D Canada-Valcartier, ON, 62 pp. Available at http://pubs.drdc.gc.ca/PDFS/unc65/p528160.pdf (accessed August 10, 2011).

[15] E. Bruce What is complexity? The philosophy of complexity *per se* with applications to some examples in evolution. In: *The Evolution of Complexity*, Kindle Edition, edited by H. Francis, B. Johan, and R. Alexander. New York: Springer, 1999, pp. 296.

[16] E. Bruce. Syntactic measures of complexities, PhD Thesis. Manchester, UK: University of Manchester, 254 pp. Available at http://bruce.edmonds.name/thesis/ (accessed August 2011, 1999).

[17] L. Seth. Ultimate physical limit to computation. *Nature*, 406(6799): 1047–1054, 2000. Available at http://arXiv.org/abs/quant-ph/9908043 (accessed August 10, 2011).

[18] N. Karcanias and A. G. Hessami. Complexity and the notion of system of systems: part (I): general systems and complexity. In: World Automation Congress (WAC), 2010, 2010, pp. 1–7.

[19] Merriam Webster's Online Dictionary, 2011.

[20] M. Waldrop. Complexity: *The Emerging Science at the Edge of Order and Chaos, Simon & Schuster*. New York, 1992.

[21] J. H. Holland. *Hidden Order: How Adaptation Builds Complexity*. New York: Addison-Wesley, 1995.

[22] M. Gell-Mann. *The Quark and the Jaguar: Adventures in the Simple and the Complex*. London: ABACUS, 1994.

[23] S. A. Kauffman. *The Origins of Order: Self-organization and Selection in Evolution*, New York/Oxford: Oxford University Press, 1993.

[24] S. A. Kauffman. *At Home in the Universe: The Search for the Laws of Self-organization and Complexity*. Oxford University Press, 1995.

[25] L. Grus, A. Bregt, and J. Crompvoets. Defining national spatial data infrastructures as complex adaptive systems. In: Proceedings of GSDI-9 Conference, November 6–10, 2006, Santiago, Chile, 2006.

[26] O. P. Rud. Book highlight—adaptability: A key to business intelligence success. *Global Business and Organizational Excellence*, 29: 76–86.

[27] J. H. Holland. Studying complex adaptive systems. *Journal of Systems Science and Complexity*, 19(1): 1–8, 2006.

[28] R. Chiva-Gomez. Repercussions of complex adaptive systems on product design management. *Technovation*, 24: 707–711, 2004.

[29] N. Yoffee. Understanding complexity in the prehistoric southwest. In: *SFI Studies in the Sciences of Complexity*, edited by G. Gumerman and M. Gell-Mann. Proceedings vol. XVI, Addison-Wesley, 1993.

[30] M. Barnes, E. Matka, and H. Sullivan. Evidence, understanding and complexity: evaluation in non-linear systems. *Evaluation*, 9(3): 265–284, 2003.

[31] K. Dooley. A nominal definition of complex adaptive systems. *The Chaos Network*, 8(1): 2–3, 1996.

[32] R. Dodder and R. Dare. Complex adaptive systems and complexity theory: interelated knowledge domains, In: ESD.83 Research Seminar in Engineering Systems, Massachusetts Institute of Technology, 2000. Available at

http://web.mit.edu/esd.83/www/notebook/ComplexityKD.PDF (accessed July 29, 2011).

[33] A.M. Wildberger. Complex adaptive systems: concepts and power industry applications. In: *Automation, Control and Complexity, An Integrated Approach*, edited by T. Samad and J. Weyrauch. West Sussex, England: John Wiley & Sons, Ltd, 2000.

[34] G. H. Eoyang. Complexity and Dynamics of Organizational Change. In *The Sage Handbook of Complexity and Management*, edited by P. Allen, S. Maguire, and B. McKeley. SAGE Publications Ltd: CA, 2012.

[35] K. Dooley, and S. Corman. Agent-based genetic and emergent computational models of complex systems. In: *UNESCO-Encyclopedia of Life Support Systems (EOLSS)*, Available at http://www.eolss.net/sample-chapters/c15/E1-29-02-00.pdf (accessed August 10, 2011).

[36] F. Nilsson, and V. Darley. On complex adaptive systems and agent-based modeling for improving decision-making in manufacturing and logistics settings: experiences from a packaging company. *International Journal of Operations & Production Management*, 26(12): 1351–1373, 2006.

[37] P. Cilliers. Complexity, deconstruction and relativism. *Theory, Culture and Society*, 22:255–267, 2005.

[38] J. Rotmans. *Societal Innovation: Between Dream and Reality Lies Complexity*. Erasmus Research Institute of Management, Erasmus University Rotterdam, 2005.

[39] M. P. C. Weijnen, Z. Lukszo, and G. Deconinck. Chapter 1: Introduction. In: *Securing Electricity Supply in the Cyber Age:* edited by Z. Lukszo, G. Deconinck, and M. P. C. Weijnen. Springer, 2010.

[40] J. M. Epstein. Agent-based computational models and generative social science. *Complexity*, 4(5): 41–60, 1999.

[41] B. G. McCarter and B. E. White. Emergence of SoS, sociocognitive aspects. In: *System of Systems Engineering—Principles and Applications*, edited by M. Jamshidi. Boca Raton, FL: Taylor Francis CRC Publishers, 2008.

[42] T. O'Connor and H. Y. Wong. The Metaphysics of Emergence. *Noûs*, 39(4): 658–678, 2005.

[43] S.A. Sheard and A. Mostashari. Principles of complex systems for systems engineering. *Systems Engineering*, 12(4): 295–311, 2009.

[44] T. Y. Choi, K. J. Dooley, and M. Rungtusanatham. Supply networks and complex adaptive systems: control versus emergence. *Journal of Operations Management*, 19(3): 351–366, 2001.

[45] K. Dooley. A complex adaptive systems model of organizational change. *Nonlinear Dynamics, Psychology, and Life Science*, 1(1): 69–97, 1997.

[46] J. H. Holland. Complex adaptive systems. *Daedalus*, 121(1): 17–30, 1992.

[47] G. Nicolis and I. Prigogine. *Exploring Complexity: An Introduction*. New York: W. H. Freeman, 1989.

[48] C. Beckner, R. Blythe, J. Bybee, M. H. Christiansen, W. Croft, N.C. Ellis, J. Holland, J. Ke, D. Larsen-Freeman, and T. Schoenemann. Language is a complex adaptive system: position paper. *Language Learning*, 59: 1–26, 2009.

[49] R. Lewin and B. Regine. The core of adaptive organizations. In: *Complex Systems and Evolutionary Perspectives on Organisations: The Application of Complexity*

Theory to Organizations, edited by E. Mitleton-Kelly. Amsterdam: Pergamon, Elsevier, pp. 167–184, 2003.

[50] C. W. Furneaux. Engineering asset procurement: a complex adaptive system perspective. In: "*Contemporary Issues in Public Management,*" *The Twelfth Annual Conference of the International Research Society for Public Management, (IRSPM XII), March 26–28*, edited by K. A. Brown, M. Mandell, C. W. Furneaux, and S. Beach. Quorum Books, Brisbane, Australia, 2008.

[51] G. H. Eoyang and T. Berkas. Evaluation in a complex adaptive system. In: *Managing Complexity in Organizations*: A View in Many Directions, edited by M. Lissack and H. Gunz. Quorum Books, Westport, USA, 1998.

[52] M. S. G. Nitzschke. Vietnam: a complex adaptive perspective. Master's Thesis, Command and Staff College, Marine Corps University, 1997.

[53] WikiButterfly. Butterfly effect. Available at http://en.wikipedia.org/wiki/Butterfly_effect (accessed August 10, 2011).

[54] R. Chuan-Jun, H. Hong-bing, and J. Shi-yao. Specification of Agent in Complex Adaptive System. In: IEEE International Symposium on Computer Science and Computational Technology (ISCSCT '08), pp. 210–216, 20–22 December, 2008.

[55] S. A. Levin. Ecosystems and the Biosphere as Complex Adaptive Systems. *Ecosystems*, 1(5): 431–436, 1998.

[56] S. A. Levin. Self-organization and the Emergence of Complexity in Ecological Systems. *BioScience*, 55(12): 1075–1079, 2005.

[57] E. F. Keller. Ecosystems, organisms, and machines. *BioScience*, 55: 1069–1074, 2005.

[58] J. T. Bonner. First Signals: The Evolution of Multicellular Development. Princeton University Press, Princeton (NJ), 2000.

[59] Y. Bar-Yam. *Dynamics of Complex Systems*, 1st edn. Reading, MA: Perseus Books, 1997.

[60] B. Goodwin. Out of control into participation. *Emergence*, 2(4): 40–49, 2000.

[61] M. R. Lissack. Complexity: the science, its vocabulary, and its relation to organizations. *Emergence*, 1(1): 110–126, 1999.

[62] R. D. Stacey, D. Griffin, and P. Shaw. *Complexity and Management: Fad or Radical Challenge to Systems Thinking*. London: Routledge, 2000.

[63] E. Bonabeau. Predicting the unpredictable. *Harvard Business Review*, 80(3): 109–116, 2002.

[64] H. A. Simon. *The Sciences of the Artificial*. MIT Press, 1996.

[65] G. H. Eoyang. *Complex Adaptive Systems*, The Kellogg Foundation, 2004.

[66] H. Sherman and R. Schultz. *Open Boundaries*. New York: Perseus Books, 1998.

[67] K. Kontopoulos. *The Logics of Social Structure*. New York: Cambridge University Press, 1993.

[68] E. N. Lorenz. Deterministic nonperiodic flow. *Journal of the Atmospheric Sciences*, 20(2): 130–141, 1963.

[69] P. Anderson. Complexity theory and organization science. *Organization Science*, 10: 216–232, 1999.

[70] R. Axelrod and M. D. Cohen. *Harnessing Complexity: Organizational Implications of a Scientific Frontier*. New York: Free Press, 1999.

[71] D. DeLaurentis and R. K. Callaway. A system-of-systems perspective for public policy decisions. *Review of Policy Research*, 21(6): 829, 2004.

[72] B. Sauser and J. Boardman. From prescience to emergence: taking hold of system of systems management. In: American Society for Engineering Management National Conference, Huntsville, AL, 2006.

[73] A. Gorod, B. Sauser, and J. Boardman. Paradox: holarchical view of system of systems engineering management. In: IEEE International Conference on System of Systems Engineering, 2008.SoSE '08. pp. 1–6, 2–4, 2008.

[74] A. P. Sage and C. D. Cuppan. On the systems engineering and management of system of systems and federations of systems. *Information, Knowledge, Systems Management*, 2(4): 325–345, 2001.

[75] A. Meilich. System of systems (SoS) engineering & architecture challenges in a net centric environment. In: 2006 IEEE/SMC International Conference on System of Systems Engineering, April 24–26, 2006, 5 pp.

[76] R. Abbott. Open at the top; open at the bottom; and continually, (but slowly) evolving. In: 2006 IEEE/SMC International Conference on System of Systems Engineering, 2006, pp. 6.

[77] M. Jamshidi. Introduction to system of systems. In: *System of Systems Engineering—Principles and Applications*, edited by M. Jamshidi. Boca Raton, FL: Taylor Francis CRC Publishers, 2008.

[78] M. Jamshidi. Introduction to system of systems. In: *System of Systems Engineering—Innovations for the 21st Century*, edited by M. Jamshidi. Hoboken, NJ: John Wiley & Sons, Inc., 2009.

[79] T. Saunders, C. Croom, W. Austin, J. Brock, N. Crawford, M. Endsley, E. Glasgow, D. Hastings, A. Levis, R. Murray, D. Rhodes, M. Sambur, H. Shyu, P. Soucy, and D. Whelan. System-of-systems engineering for air force capability development: executive summary and annotated brief. SAB-TR-05–04, United States Air Force Scientific Advisory Board, 2005.

[80] Y. Bar-Yam. The characteristics and emerging behaviors of system of systems. NECSI: Complex Physical, Biological and Social Systems Project, 2004

[81] D. K. Hitchins. Family, system, federation, coalition of systems. INCOSE Discussion Forum, 2011.

[82] D. K. Hitchins. What's in a system-of-systems? Available at http://www.hitchins.net/SystemofSystems.pdf (accessed August 3, 2011).

[83] J. A. Lane and R. Valerdi. Synthesizing SoS concepts for use in cost estimation. In: IEEE 2005 Conference on Systems, Man, and Cybernetics, 2005.

[84] H. Eisner. RCASSE: Rapid computer-aided system of systems engineering. In: Proceedings of the 3rd International Symposium of the National Council of Systems Engineering, 1993, Vol. 1, pp. 267–273.

[85] M. Maier. Architecting principles for system-of-systems. *Systems Engineering*, 1(4): 267–284, 1998.

[86] V. Kotov. *System of Systems as Communicating Structures*. Hewlett-Packard Company, pp. 1–14, 1997.

[87] T. Clare. The engineering and acquisition of systems of systems. US Department of Defense, Research Development and Acquisition Office, 2000.

[88] D. DeLaurentis. Understanding transportation as a system of systems design problem. In: The 43rd AIAA Aerospace Sciences Meeting, Reno, NV, January 10–13, 2005.

[89] Purdue. *System of Systems Research Report*. West Lafayette, IN: Purdue University, 2005.

[90] S. D. Jolly and B. K. Muirhead. System of systems engineering in space exploration. In: *System of Systems Engineering—Innovations for the 21st Century*, edited by M. Jamshidi. Hoboken, NJ: John Wiley & Sons, Inc., 2009.

[91] A. J. Shenhar. One size does not fit all projects: exploring classical contingency domains. *Management Science*, 47(3): 394–414, 2001.

[92] T. Samad and T. Parisini. Systems of systems. In: *IEEE Control Systems Society Report: The Impact of Control Technology: Overview, Success Stories, and Research Challenges*, edited by T. Samad and A. Annaswamy. Available at http://ieeecss.org/general/impact-control-technology (accessed August 11, 2011).

[93] J. Moses. Flexibility and its relation to complexity and architecture. In: *Proceedings of the First International Conference on Complex Systems Design and Management*, edited by M. Aiguier, F. Bretaudeau, and D. Krob. Springer, 2010, pp. 197–206.

[94] J. M. Sussman. Ideas on complexity in systems: twenty views. Available at web.mit.edu/esd.83/www/notebook/20ViewsComplexity.PDF (accessed August 4, 2011).

[95] D. J. Anderson, J. E. Campbell, and L. D. Chapman. Evaluating a complex system of systems using state modeling and simulation. In: National Defense Industrial Association Systems Engineering Conference, San Diego, CA, 2003.

[96] R. C. Parks, R. A. Jung, and K. O. Ramotowski. Attacking agent-based systems. In: The 2004 IEEE 1st Symposium on Multi-agent Security and Survivability, 2004.

[97] R. Stevens. Systems engineering in the information age: the challenge of mega-systems. In: Presentation to the Naval Post Graduate School Systems Engineering Management Course, 2005.

[98] C. A. Stoudt. A systems perspective on current ATC trends. *IEEE Aerospace and Electronic Systems Magazine*, 17(9): 28, 2002.

[99] M. Maier. Architecting principles of systems-of-systems. In: The Sixth Annual International Symposium of the International Council on Systems Engineering, Boston, MA, 1996.

[100] P. Periorellis and J. Dobson. Organizational failures in dependable collaborative enterprise systems. *Journal of Object Technology*, 1(3): 107–117, 2002, Special Issue: TOOLS USA 2002 Proceedings.

[101] C. Azani. An open systems approach to system of systems engineering. In: *System of Systems Engineering—Innovations for the 21st Century*, edited by M. Jamshidi. Hoboken, NJ: John Wiley & Sons, Inc., 2009.

[102] C. Keating, R. Rogers, R. Unal, D. Dryer, A. Sousa-Poza, R. Safford, W. Peterson, and G. Rabadi. System of systems engineering. *Engineering Management Journal*, 15(3): 36, 2003.

[103] H. Azarnoosh, B. Horan, P. Sridhar, A. M. Madni, and M. Jamshidi. Towards optimization of a real-world robotic-sensor system of systems. In: Proceedings of World Automation Congress, (WAC) 2006, Budapest, Hungary, 2006.

[104] J. Boardman and B. Sauser. System of Systems – the meaning of of. In: IEEE/SMC International Conference on System of Systems Engineering: 24–26, 2006.

[105] *McGraw-Hill Encyclopedia of Science and Technology*, 5th edn. USA: McGraw-Hill, 2005.

[106] W. A. Crossley. System of systems: an introduction of Purdue University schools of engineering's signature area. In: The Engineering Systems Symposium, Cambridge, MA, March 31, 2004.

[107] Oxford Dictionary, 2004.

[108] B. T. Clough. Metrics, schmetrics! How the heck do you determine a UAV's autonomy anyway? In: Proceedings of the Performance Metrics for Intelligent Systems Workshop, Gaithersburg, MD, 2002.

[109] DoD. Unmanned aerial vehicles (UAV) Roadmap 2000–2025. 2001.

[110] R. Parasuraman, T. B. Sheridan, and C. D. Wickens. A model for types and levels of human interaction with automation. *IEEE Transactions on Systems, Man, and Cybernetics—Part A*, 30(3): 286–297, 2000.

[111] T. B. Sheridan and R. Parasuraman. Human–automation interaction. In: *Review of Human Factors and Ergonomics*, Vol. 1, edited by R. S. Nickerson. HFES/SAGE Publications, 2006, pp. 89–129.

[112] T. B. Sheridan and W. L. Verplank. *Human and Computer Control of Undersea Teleoperators*. Cambridge, MA: MIT Man-Machine Systems Lab., 1978.

[113] M. Bjelkemyr, D. T. Semere, and B. Lindberg. Definition, classification, and methodological issues of system of systems. In: *System of Systems Engineering—Principles and Applications*, edited by M. Jamshidi. Boca Raton, FL: Taylor Francis CRC Publishers, 2008.

[114] N. Subramanian and L. Chung. Software architecture adaptability: an NFR approach. In: International Workshop on Principles of Software Evolution, September 2001.

[115] D. Carney, D. Fisher, and P. Place. *Topics in Interoperability: System-of-Systems Evolution*. Pittsburgh, PA: Software Engineering Institute, Carnegie Mellon University, 2005, pp. 1–14.

[116] M. D. Ilić. The future of electricity systems: general trends, developments. In: *Securing Electricity Supply in the Cyber Age:* edited by Z. Lukszo, G. Deconinck, and M. P. C. Weijnen. Springer, 2010.

[117] W. H. J. ManthorpeJr. The emerging joint system of systems: a systems engineering challenge and opportunity for APL. *Johns Hopkins APL Technical Digest (Applied Physics Laboratory)*, 17(3): 305, 1996.

[118] P. Chen and J. Clothier. Advancing systems engineering for systems-of-systems challenges. *Systems Engineering*, 6(3): 170–183, 2003.

[119] J. S. Dahmann and K. J. Baldwin. Understanding the current state of US defense systems of systems and the implications for systems engineering. In: 2nd Annual IEEE Systems Conference, 2008, pp. 1–10.

[120] D. Swinney. Family, system, federation, coalition of systems, INCOSE Discussion Forum, 2011.

[121] C. H. Dagli and N. Kilicay-Ergin. System of systems architecting. In: *System of Systems Engineering—Innovations for the 21st Century*, edited by M. Jamshidi. Hoboken, NJ: John Wiley & Sons, Inc., 2009.

[122] S. C. Cook. On the acquisition of system of systems. In: The 11th Annual Symposium of the INCOSE, Melbourne, Australia, 2001.

[123] W. A. Owens. The emerging US systems of systems. In: *Dominant Battlespace Awareness*, edited by S. Johnson and M. Libicki. Institute for National Strategic Studies, National Defense University, ISSN 1071-7552, 1997.

[124] W. A. Owens. The emerging U.S. systems-of systems. Institute for National Strategic Studies, Strategic Forum, Number 63, February 1996.

[125] J. Dahmann, G. Rebovich, R. Lowry, J. Lane, K. Baldwin. An implementers' view of systems engineering for systems of systems. In: The 2011 IEEE International Systems Conference, Montreal, Canada, April 2011, pp. 212–217.

[126] A. Mostafavi, D. M. Abraham, D. DeLaurentis, and J. Sinfield. Exploring the dimensions of systems of innovation analysis: a system of systems framework. *IEEE Systems Journal,* 5(2): 256–265, 2011.

[127] J. O. Clark. System of systems engineering and family of systems engineering from a standards, V-Model, and Dual-V model perspective. In: 3rd Annual IEEE Systems Conference, 2009, pp. 381–387.

[128] A. J. Krygiel. *Behind the Wizard's Curtain: An Integration Environment for a System of Systems*. Vienna, VA: CCRP Publication Series, 1999.

[129] G. D. Wells and A. P. Sage. Engineering of a system of systems. In: *System of Systems Engineering—Innovations for the 21st Century*, edited by M. Jamshidi. Hoboken, NJ: John Wiley & Sons, Inc., 2009.

[130] D. Greenwood and I. Sommerville. Responsibility modeling for identifying sociotechnical threats to the dependability of coalitions of systems. In: 6th International Conference on System of Systems Engineering, (SoSE), 2011, pp. 173–178.

[131] N. Karcanias and A. G. Hessami. Complexity and the notion of system of systems: part, (II): general defining the notion of system of systems. In: World Automation Congress (WAC), 2010, pp. 1–7.

[132] R. Bahet and H. Gill. Cyber-physical systems. In: *IEEE Control Systems Society Report: The Impact of Control Technology: Overview, Success Stories, and Research Challenges*, edited by T. Samad and A. Annaswamy. Available at http://ieeecss.org/general/impact-control-technology (accessed August 11, 2011).

[133] US National Science Foundation. Available at http://www.nsf.gov/pubs/2011/nsf11516/nsf11516.pdf (accessed August 14, 2011).

[134] D. Wei, Y. Lu, M. Jafari, P. M. Skare, and K. Rohde. Protecting Smart Grid automation systems against cyberattacks. *IEEE Transactions on Smart Grid,* online first at IEEE *Xplore*, 2011.

[135] F. Boroomand, A. Fereidunian, M. A. Zamani, M. Amozegar, H. R. Jamalabadi, H. Nasrollahi, M. Moghimi, H. Lesani, and C. Lucas. Cyber security for smart grid: a human-automation interaction framework. In: Proceedings of the IEEE PES ISGT, Gothenburg, Sweden, October 11–13, 2010.

[136] R. J. Glass, A. L. Ames, W. A. Stubblefield, S. H. Conrad, S. L. Maffitt, L. A. Malczynski, D. G. Wilson, J. J. Carlson, G. A. Backus, M. A. Ehlen, K. B. Vanderveen, and D. Engi. A roadmap for the complex adaptive systems of systems (CASoS) engineering initiative. Sandia National Laboratories, SAND 2008–4651, September, 2008.

[137] R. J. Glass, A. L. Ames, T. J. Brown, S. L. Maffitt, W. E. Beyeler, P. D. Finley, T. W. Moore, J. M. Linebarger, N. S. Brodsky, S. J. Verzi, A. V. Outkin, and A. A. Zagonel. Complex adaptive systems of systems (CASoS) engineering: mapping aspirations to problem solutions. In: 8th International Conference on Complex Systems, 2011.

[138] Sandia. CASoS engineering website at Sandia National Laboratories. Available at http://www.sandia.gov/CasosEngineering/ (accessed August 3, 2011).

[139] Britannica Encyclopedia Online. Available at www.britannica.com (accessed July 22, 2011).

[140] E. Nagel. Automatic control. *Scientific American*, 187(3): 44–47, 1952.

[141] L. Floyd. Opportunities in automation and control. *Editorial to IEEE Industry Applications Magazine*, 17(2): 2, 2011.

[142] R. Parasuraman and V. Riley. Humans and automation: use, misuse, disuse, abuse. *Human Factors*, 39(2): 230–253, 1997.

[143] S. Y. Nof. Automation: what it means to us around the world. In: *Handbook of Automation*, edited by S. Y. Nof. Berlin: Springer, 2009, pp. 13–52.

[144] J. Frohm. Levels of automation in production systems. PhD Thesis, Chalmers University of Technology, Göteborg, Sweden, 2008.

[145] P. M. Satchell. *Innovation and Automation*. Ashgate Publishing Ltd., 1998.

[146] R. Parasuraman and M. Mouloua. *Automation and Human Performance: Theory and Applications*. Lawrence Erlbaum Associates, Inc., 1996.

[147] S. A. Reveliotis. Accommodating FMS operational contingencies through routing flexibility. *IEEE Transactions on Robotics and Automation*, 15(1): 3–19, 1999.

[148] M. L. Cummings and K. M. Thornburg. Paying attention to the man behind the curtain. *IEEE Pervasive Computing*, 10(1): 58–62, 2011.

[149] J. D. Lee and B. D. Seppelt. Human factors in automation design. In: *Handbook of Automation*, edited by S. Y. Nof. Berlin: Springer, 2009, pp. 417–436.

[150] C. E. Billings. Human-centered aircraft automation: a concept and guidelines. NASA Technical Memorandum 103885, Moffett Field, CA: NASA—Ames Research Center, 1991.

[151] C. E. Billings. *Aviation Automation: The Search for a Human-Centered Approach*. Mahwah, NJ: Lawrence Erlbaum Associates, 1997.

[152] T. B. Sheridan. Human-centered automation: oxymoron or common sense? In: Proceedings of IEEE International Conference on SMC95, 1995, vol. 1, pp. 823–828

[153] L. Bainbridge. Ironies of automation. *Automatica*, 19(6): 775–779, 1983.

[154] M. R. Endsley and D. B. Kaber. Level of automation effects on performance, situation awareness and workload in a dynamic control task. *Ergonomics*, 42(3): 462–492, 1999.

[155] D. B. Kaber, M. C. Wright, L. J. PrinzelIII, and M. P. Clamann. Adaptive automation of human-machine systems information processing functions. *Human Factors*, 47(4): 730–741, 2005.

[156] D. B. Kaber, M. C. Wright, and M. A. Sheik-Nainar. Investigation of multi-modal interface features for adaptive automation of a human–robot system. *International Journal of Human-Computer Studies*, 64(6): 527–540, 2006.

[157] T. B. Sheridan. Rumination on automation. 1998, *Annual Reviews in Control*, vol. 25, 2001, Elsevier Science Ltd, Pergamon. First presented at the IFAC-MMS Conference, Kyoto, Japan, 1998.

[158] J. Zhang, V. L. Patel, A. J. Kathy, J. W. Smith, and J. Malin. HCC at NASA: designing the human-centered distributed information systems. *IEEE Intelligent Systems*, 17(5): 42–47, 2002.

[159] S. W. A. Dekker and D. D. Woods. MABA-MABA or abracadabra? Progress on human automation coordination. *Cognition, Technology, and Work*, 4(4): 240–244, 2002.

[160] Y. Guiard and M. Beaudouin-Lafon. Fitts' law 50 years later: applications and contributions from human–computer interaction. *International Journal of Human-Computer Studies*, 61: 747–750, 2004.

[161] T. B. Sheridan. Function allocation: algorithm, alchemy or apostasy? *International Journal of Human-Computer Studies*, 52(2): 203–216, 2000.

[162] T. B. Sheridan. *Humans and Automation: System Design and Research Issues*. Santa Monica, CA: HFES, Wiley Inter-science, 2002.

[163] R. B. Fuld. The fiction of function allocation revisited, *International Journal of Human-Computer Studies*, 52(2): 217–233, 2000.

[164] J. C. McCarthy, E. Fallon, and L. Bannon. Dialogues on function allocation. *International Journal of Human-Computer Studies*, 52(2): 191–201, 2000.

[165] P. Chen and W. Wu. Toward a human nature based automation system with dynamic function allocation model supported for pervasive systems. In: International Symposium on Parallel Architectures, Algorithms, and Networks, 10th International Symposium on Pervasive Systems, Algorithms, and Networks, 2009, pp. 129–134.

[166] R. Parasuraman and C. D. Wickens. Humans: still vital after all these years of automation. *Human Factors*, 50(3): 511–520, 2008.

[167] R. W. Proud, J. J. Hart, and R. B. Mrozinski. Methods for determining the level of autonomy to design into a human spaceflight vehicle: a function specific approach. NASA Johnson Space Center Report, NASA Road, Houston, TX, 2003.

[168] A. M. Bisantz, E. M. Roth, B. Brickman, L. Gosbee, L. Hettinger, and J. McKinney. Integrating cognitive analyses in a large-scale system design process. *International Journal of Human Computer Studies*, 58: 177–206, 2003.

[169] J. M. Bradshaw, M. Sierhuis, A. Acquisti, P. Feltovich, R. Hoffman, R. Jeffers, D. Prescott, N. Suri, A. Uszok, and R. van Hoof. Adjustable autonomy and human-agent teamwork in practice: an interim report on space applications. In: Agent Autonomy, edited by H. Hexmoor, C. Castelfranchi, and R. Falcone, Chapter 11: 243–280, Springer, 2003.

[170] A. Bye, E. Hollnagel, and T. S. Brendeford. Human-machine function allocation: a functional modeling approach. *Reliability Engineering and System Safety*, 64: 291–300, 1999.

[171] A. Dearden, M. Harrison, and P. Wright. Allocation of function: scenarios, context and the economics of effort. *International Journal of Human-Computer Studies*, 52(52): 289–318, 2000.

[172] E. Hollnagel and A. Bye. Principles for modeling function allocation. *International Journal of Human-Computer Studies*, 52(2): 253–265, 2000.

[173] A. G. Sutcliffe and A. Gregoriades. Automating scenario analysis of human and system reliability. *IEEE Transactions on Systems, Man, and Cybernetics—Part A*, 37(2): 249–261, 2007.

[174] H. Wandke. Assistance in human-machine interaction: a conceptual framework and a proposal for taxonomy. *Theoretical Issues in Ergonomics Science*, 6(2): 129–155, 2005.

[175] Z. Wei, A. Macwan, and P. Wierenga. A quantitative measure for degree of automation, and its relation to system performance. *Human Factors*, 40: 277–295, 1998.

[176] A. R. Crocker. Operational considerations in the development of autonomy for human spaceflight. In: 2005 AIAA Proceedings of 1st Space Exploration Conference, Orlando, FL, 2005.

[177] M. L. Cummings. Designing decision support systems for revolutionary command and control domains. Ph.D. dissertation, University of Virginia, 2004.

[178] F. Di Nocera, B. Lorenz, and R. Parasuraman. Consequences of shifting from one level of automation to another: main effects and their stability. In: *Human Factors in Design, Safety, and Management*, edited by D. de Waard, K. A. Brookhuis, R. van Egmond, and T. H. Boersema. Maastricht, The Netherlands: Shaker Publishing, 2005, pp. 363–376.

[179] S. Guerlaine, G. A. Jamieson, P. Bullemer, and R. Blair. The MPC Elucidator: a case study in the design for human-automation interaction. *IEEE Transactions on System, Man, and Cybernetics—Part A*, 32(1): 25–40, 2002.

[180] H. C. Hu, R. W. Proud, and J. J. Hart. Spacecraft mission assessment and replanning tool, (SMART). In: 2004 42nd AIAA Aerospace Science Meeting and Exhibit, Reno, NV, 2004.

[181] M. Itoh and T. Inagaki. A microworld approach to identifying issues of human-automation systems design for supporting operator's situation awareness. *International Journal of Human-Computer Interaction*, 17(1): 3–24, 2004.

[182] D. B. Kaber and M. R. Endsley. The effect of level of automation and adaptive automation on human performance, situation awareness and workload in a dynamic control task. *Theoretical Issues in Ergonomics Science*, 5(2): 113–153, 2004.

[183] R. S. McCann and L. Spirkovska. Human factors of integrated systems health management on next generation spacecraft. In: Proceedings of ISHEM 2005, NASA Ames Research Center, 2005.

[184] U. Metzger and R. Parasuraman. Automation in future air traffic management: effects of decision aid reliability on controller performance and mental workload. *Human Factors*, 47(1): 35–50, 2005.

[185] R. Parasuraman. Designing automation for human use: empirical studies and quantitative models. *Ergonomics*, 43(7): 931–951, 2000.

[186] M. Sierhuis, J. M. Bradshaw, A. Acquisti, P. Feltovich, R. Hoffman, R. Jeffers, D. Prescott, N. Suri, A. Uszok, and R. van Hoof. Human-agent teamwork and adjustable autonomy in practice. In: Proceedings of the 7th International i-SAIRAS 2003, NARA, Japan, 2003.

[187] C. D. Wickens and S. R. Dixon. Is there a magic number 7? University of Illinois, Technical Report, AHFD-05-01/MAAD-05-1, 2005.

[188] M. Wilson and M. Neal. Diminishing returns of engineering effort in Telerobotic systems. *IEEE Transactions on System, Man, and Cybernetics—Part A*, 31(5): 459–465, 2001.

[189] E. A. Byrne and R. Parasuraman. Psychology and adaptive automation. *Biological Psychology*, 42: 249–268, 1996.

[190] R. Parasuraman. Neuroergonomics: research and practice. *Theoretic Issues in Ergonomic Science*, 4(1–2): 1–5, 2003.

[191] J. V. Nickerson and R. R. Reilly. A model for investigating the effects of machine autonomy on human behavior. In: Proceedings of the 37th Hawaii International Conference on System Sciences, 2004.

[192] C. A. Ntuen and E. H. Park. Human factors issues in teleoperated systems. In: *Ergonomics of Hybrid Automated Systems I*, edited by W. Karwowski, H.R. Parsaei, and M.R. Wilhelm. Amsterdam: Elsevier, 1988, pp. 210–230.

[193] A. Fereidunian, M. Lehtonen, H. Lesani, C. Lucas, and M. Nordman. Adaptive autonomy: smart cooperative cybernetic systems for more humane automation solutions. In: Proceedings of the IEEE International Conference of SMC07, Montreal, QC, Canada, 2007, pp. 202–207.

[194] A. Fereidunian, C. Lucas, H. Lesani, M. Lehtonen, and M. Nordman. Challenges in implementation of the human-automation interaction models. In: Proceeding of 15th IEEE-MED'07 Conference, Athens, Greece, 2007, pp. 1–6.

[195] R. Parasuraman, T. Bahri, J. E. Deaton, J. G. Morrison, and M. Barnes. *Theory and design of adaptive automation in aviation systems*. Technical Report, Code 6021. Naval Air Development Center, Warminster, PA, 1992.

[196] A. Fereidunian, M. A. Zamani, H. Lesani, C. Lucas, and M. Lehtonen. An expert system realization of adaptive autonomy in electric utility management automation. *Journal of Applied Sciences*, 8: 1524–1530, 2009.

[197] W. B. Rouse. Monitoring behavior and supervisory control. In: *Adaptive Allocation of Decision Making Responsibility Between Supervisor and Computer*, edited by T. B. Sheridan and G. Johannsen. New York: Plenum, 1976, pp. 295–306.

[198] A. Fereidunian, H. Lesani, C. Lucas, and M. Lehtonen. A framework for implementation of adaptive autonomy for intelligent electronic devices. *Journal of Applied Sciences*, 8: 3721–3726, 2008.

[199] B. Hilburn, R. Molloy, D. Wong, and R. Parasuraman. Operator versus computer control of adaptive automation. In: *Proceedings of the 7th International Symposium on Aviation Psychology*, edited by R. S. Jensen and D. Neumeister. NASA Technical Reports Server (NTRS): 161–166, 1993.

[200] M. W. Scerbo, F. G. Freeman, P. J. Mikulka, R. Parasuraman, F. Di Nocero, and L. J. Prinzel. *The Efficacy of Psychophysiological Measures for Implementing Adaptive Technology* (NASA TP-2001-211018). Hampton, VA: NASA Langley Research Center, 2001.

[201] L. J. Prinzel, F. G. Freeman, M. W. Scerbo, P. J. Mikulka, and A. T. Pope. A closed-loop system for examining psychophysiological measures for adaptive task allocation. *The International Journal of Aviation Psychology*, 10(4): 393–410, 2000.

[202] G. Wilson and C. A. Russell. Operator functional state classification using multiple psychophysiological features in an air traffic control task. *Hum Factors*, 45(3): 381–389, 2003.

[203] W. Wang, Y. Xu, and M. Khanna. A survey on the communication architectures in smart grid. *Computer Networks*, 55: 3604–3629, 2011.

[204] M. Amin. Scanning the technology: energy infrastructure defense system. *Proceedings of the IEEE*, 93(5): 861–875, 2005.

[205] V. C. Gungor and F. C. Lambert. A survey on communication networks for electric system automation. *Computer Networks*, 50: 877–897, 2006.

[206] D. J. Marihart. Communications technology guidelines for EMS/SCADA systems. *IEEE Transactions on Power Delivery*, 16(2): 181–188, 2001.

[207] A. Fereidunian, M. Nordman, H. Lesani, C. Lucas, and M. Lehtonen. A systems approach to information technology, (IT) infrastructure design for utility management automation systems. *Iranian Journal of Electrical and Electronic Engineering*, 2(3&4): 91–105, 2006.

[208] A. Fereidunian, F. Boroomand, M. A. Zamani, H. Jamalabadi, H. Lesani, C. Lucas, and S. Afkhami. AAGLMES: an intelligent expert system realization of adaptive autonomy using generalized linear models. In: Proceedings of 6th Annual IEEE Conference on Automation Science and Engineering, (CASE 2010), Toronto, ON, Canada, August 21–24, 2010.

[209] M. A. Zamani, A. Fereidunian, H. R. Jamalabadi, F. Boroomand, P. Sepehri, H. Lesani, and C. Lucas. Smart Grid IT infrastructure selection: a T3SD fuzzy DEA approach. In: Proceedings of the IEEE PES ISGT, Gothenburg, Sweden, October 11–13, 2010.

[210] M. A. Zamani, A. Fereidunian, S. Sharif Mansouri, H. Lesani, and C. Lucas. A fuzzy-T3SD policy driven method for IT infrastructure selection in smart grid. In: Proceedings of the 5th IEEE International Symposium on Telecommunications, (IST), Kish-Tehran, Iran, December 4–6, 2010.

[211] M. A. Zamani, A. Fereidunian, S. Sharif Mansouri, M. A. Sharifi K., F. Boroomand, and H. Lesani. A Petri net-T3SD policy driven method for IT infrastructure selection in smart grid. In: Proceedings of IEEE Systems Conference, (SysCon), Montreal, QC, Canada, April 4–7, 2011.

[212] M. Rezagholizadeh, A. Fereidunian, B. Dehghan, B. Moshiri, and H. Lesani. Multi criteria decision making, (MCDM): a modified partial order theory, (POT) approach. In: Proceedings of the 3rd IEEE International Conference on Advanced Computer Control, (ICACC 2011), Harbin, China, January 18–20, 2011.

[213] D. K. Hitchins. Managing system creation. *Physical Science, Measurement and Instrumentation, Management and Education—Reviews, IEE Proceedings A*, 133(6): 343–354, 1986.

[214] M. Masera, Y. Wijnia, L. De Vries, C. Kuenzi, M. Sajeva, and M. Weijnen. Chapter 5: governing risks in the European critical electricity infrastructure. In: *Critical Infrastructures at Risk: Topics in Safety, Risk, Reliability and Quality*, Vol. 9, A. V. Gheorghe, M. Masera, M. Weijnen, and L. J. De Vries. 2006.

[215] J. Machowski, J. Bialek, and J. Bumby. *Power System Dynamics: Stability and Control.* Wiley-IEEE Press, 2008.

[216] A. Bergen and V. Vital. *Power Systems Analysis*, Prentice-Hall, 2000.

[217] R. E. Brown and A. P. Hanson. Impact of two-stage service restoration on distribution reliability. *IEEE Transactions on Power Systems*, 16(4): 624–629, 2001.

[218] S. Curcic, C. S. Ozveren, L. Crowe, and P. K. L. Lo. Electric power distribution network restoration: a survey of papers and a review of the restoration problem. *Electric Power Systems Research*, 35: 73–86, 1996.

[219] A. Fereidunian, H. Lesani, and C. Lucas. Distribution system reconfiguration using pattern recognizer neural networks. *International Journal of Engineering-B*, 15(2): 135–144, 2002.

[220] H. Hu, M. Zhou, and Z. Li. Supervisor optimization for deadlock resolution in automated manufacturing systems with Petri nets. *IEEE Transactions on Automation Science and Engineering*, 8(4): 794–804, 2011.

[221] H. Hu, M. C. Zhou, and Z. Li. Low-cost and high-performance supervision in ratio-enforced automated manufacturing systems using timed Petri nets. *IEEE Transactions on Automation Science and Engineering*, 7(4): 933–944, 2010.

[222] S. Peng and M. Zhou. Production cycle-time analysis based on sensor-based stage Petri nets for automated manufacturing systems. In: IEEE International Conference on

Robotics and Automation, 2003. Proceedings of ICRA '03, 2003, vol. 3, pp. 4300–4305.

[223] Z. W. Li and M. C. Zhou. Elementary siphons of Petri nets and their application to deadlock prevention in flexible manufacturing systems. *IEEE Transactions on Systems, Man and Cybernetics, Part A: Systems and Humans,* 34(1): 38–51, 2004.

[224] Z.W. Li and M.C. Zhou. Two-stage method for synthesizing liveness-enforcing supervisors for flexible manufacturing systems using Petri nets. *IEEE Transactions on Industrial Informatics,* 2(4): 313–325, 2006.

[225] M. C. Zhou and K. Venkatesh. *Modeling, Simulation and Control of Flexible Manufacturing Systems: A Petri Net Approach*. Singapore: World Scientific, 1998.

[226] T. Murata. Some recent applications of high-level Petri nets. *IEEE International Symposium on Circuits and Systems*, 2: 818–821, 1991.

[227] W. Zhang. Automatic robot planning by marked Petri net. In: Proceedings of the International Workshop on Artificial Intelligence for Industrial Applications, 1988. IEEE AI '88, 1988, pp. 358–361.

[228] W. Zhang. Representation of assembly and automatic robot planning by Petri net. *IEEE Transactions on Systems, Man and Cybernetics,* 19(2): 418–422, 1989.

[229] J. Rosell, N. Munoz, and A. Gambin. Robot tasks sequence planning using Petri nets. In: Proceedings of the IEEE International Symposium on Assembly and Task Planning, 2003, pp. 24–29.

[230] K. Kobayashi, A. Nakatani, H. Takahashi, and T. Ushio. Motion planning for humanoid robots using timed Petri net and modular state net. In: 2002 IEEE International Conference on Systems, Man and Cybernetics, 2002, vol. 6, p. 6.

[231] H. Costelha and P. Lima. Modelling, analysis and execution of robotic tasks using Petri nets. In: IEEE/RSJ International Conference on Intelligent Robots and Systems, 2007, IROS 2007, 2007, pp. 1449–1454.

[232] B. Mikolajczak and C. A. Sefranek. Integrating object oriented design with concurrency using Petri nets—a case study of a banking system and the Syroco-Macao environment. In: IEEE International Conference on Systems, Man, and Cybernetics, 2001.

[233] J. Lin, H. Pu, and L. Cui. Petri Net based business rule modeling of intelligent taxation system in pervasive environment. In: 2010 5th International Conference on Pervasive Computing and Applications (ICPCA), 2010, pp. 398–401.

[234] T. S. Staines. Using a timed Petri net (TPN) to model a bank ATM. In: 13th Annual IEEE International Symposium and Workshop on Engineering of Computer Based Systems, 2006. ECBS 2006, 2006, pp. 8, 158.

[235] I. Koach, W. Reising, and F. Schreiber (Eds). *Modeling in Systems Biology: The Petri Net Approach*. London: Springer, Computational Biology Series, 2011.

[236] G. Chalkidis, M. Nagasaki, and S. Miyano. High performance hybrid functional Petri net simulations of biological pathway models on CUDA. *IEEE/ACM Transactions on Computational Biology and Bioinformatics,* 8(6): 1545–1556, 2010.

[237] S. Troncale, J. P. Comet, and G. Bernot. Validation of biological models with temporal logic and timed hybrid Petri nets. In: Engineering in Medicine and Biology Society, 2007. EMBS 2007. 29th Annual International Conference of the IEEE, 2007, pp. 4603–4608.

[238] C. Chaouiya. Qualitative modeling of biological regulatory networks combining a logical multi-valued formalism and Petri nets. In: 9th International Workshop on Discrete Event Systems, 2008. WODES 2008, 2008, pp. 263–268.

[239] R. I. Hamed, S. I. Ahson, and R. Parveen. From fuzzy logic theory to fuzzy Petri nets predicting changes in gene expression level. In: Proceeding of International Conference on Methods and Models in Computer Science, 2009. ICM2CS 2009, 2009, pp. 1–7.

[240] Y. Yang and H. Lin. Reachability analysis based model validation in systems biology. In: 2010 IEEE Conference on Cybernetics and Intelligent Systems (CIS), 2010, pp. 14–19.

[241] N. Q. Wu and M. C. Zhou. *System Modeling and Control with Resource-Oriented Petri Nets*. New York: CRC Press, 2010.

[242] N. Wu, F. Chu, C. Chu, and M. Zhou. Hybrid Petri net modeling and schedulability analysis of high fusion point oil transportation under tank grouping strategy for crude oil operations in refinery. *IEEE Transactions on Systems, Man, and Cybernetics, Part C: Applications and Reviews*, 40(2): 159–175, 2010.

[243] N. Wu, F. Chu, S. Mammar, and M. Zhou. Petri net modeling of the cooperation behavior of a driver and a copilot in an advanced driving assistance system. *IEEE Transactions on Intelligent Transportation Systems*, 12(4): 977–989, 2011.

[244] H. Abouaissa, J. C. Nicolas, A. Benasser, and E. Czesnalowicz. Formal specification of multi-agent systems: approach based on meta-models and high-level Petri nets - case study of a transportation system. In: IEEE International Conference on Man and Cybernetics, 2002, vol. 5, p. 6.

[245] A. Nait Sidi Moh, M. A. Manier, and A. El Moudni. A control policy for a public transport network modelled by Petri nets and max-plus algebra. In: Automation Congress, 2002 Proceedings of the 5th Biannual World, 2002, vol. 14, pp. 59–64.

[246] B. Annappa, P. Jiju, K. Chandrasekaran, and K. C. Shet. Petri net based verification of a cooperative work flow model. In: First International Conference on Networked Digital Technologies, 2009. NDT '09, 2009, pp. 82–87.

[247] Y. Song and G. Zeng. Formalized description of workflow exception handlings based on automation. In: 2011 Third Pacific-Asia Conference on Circuits, Communications and System (PACCS), 2011, pp. 1–4.

[248] L. Li, X. Yue, B. Yan, and K. Zhou. Workflow model based on colored and timed Petri nets. In: 2011 International Conference on Consumer Electronics, Communications and Networks (CECNet), 2011, pp. 3795–3798.

[249] C. T. Leondes (Ed.): Computer aided design, engineering, and manufacturing: systems techniques and applications. In: *Optimization Methods for Manufacturing*, New York: CRC Press, 2001, vol. IV,.

[250] B. Hruz and M. C. Zhou. *Modeling and Control of Discrete Event Dynamic Systems*. London, UK: Springer, 2007.

[251] A. Mehrez, M. Muzumdar, W. Acar, and G. Weinroth. A Petri net model view of decision making: an operational management analysis. *Omega*, 23(1): 63–78, 1995.

[252] A. Fereidunian, M. A. Zamani, H. Lesani, C. Lucas, and M. Lehtonen. AAFES: An intelligent fuzzy expert system for realization of adaptive autonomy concept in utility management automation. In: Proceedings of IEEE-ISDA'09, Pisa, Italy, 30 November–2 December, 2009.

[253] A. Fereidunian, M. A. Zamani, F. Boroomand, H. R. Jamalabadi, H. Lesani, and C. Lucas. AALRES: expert system realization of adaptive autonomy, using logistic regression, in electric utility management automation. In: Proceedings of the 15th IEEE Mediterranean Electrotechnical Conference, Valletta, Malta, 2010.

[254] A. Fereidunian, M. A. Zamani, F. Boroomand, H. R. Jamalabadi, M., Lesani, H. Lucas, C. S. Shariat-Torbaghan, M. Meydani. AAHES: a hybrid expert system realization of adaptive autonomy for Smart Grid. In: Proceedings of the IEEE PES ISGT, Gothenburg, Sweden, October 11–13, 2010.

[255] M. A. Zamani, A. Fereidunian, M. A. Sharifi K, and H. Lesani. AAPNES: a Petri net expert system realization of adaptive autonomy in smart grid. In: Proceedings of Fifth International Symposium on Telecommunication, (IST), Tehran, Iran, December 4–6, 2010.

[256] M. A. Zamani, M. A. Sharifi K, A. Fereidunian, H. Lesani, and C. Lucas. Deterministic timed Petri nets as adaptive autonomy expert system in smart grid realization of adaptive autonomy expert system in smart grid. *International Journal of Information & Communication Technology (IJICT)*, 2(3): 49–59, 2010.

[257] A. Fereidunian, M. A. Zamani, M. A. Sharifi K, and H. Lesani. AAHPNES: a hierarchical Petri net expert system realization of adaptive autonomy in Smart Grid. In: *Proceedings of IEEE PowerTech Conference*, Trondheim, Norway, July 19–23, 2011.

[258] M. A. Zamani, A. Fereidunian, M. A. A. Akhoundi, C. Lucas, and H. Lesani. An intelligent fuzzy expert system realization of adaptive autonomy using gradient descent algorithm. In: Proceedings of IEEE PowerTech Conference, Trondheim, Norway, June 19–23, 2011.

[259] A. Fereidunian, H. R. Zare, H. Lesani, and D. Moazami. Graph theory approach to distribution systems management by reconfiguration. *Journal of Faculty of Engineering, (JFE) of University of Tehran*, 38(1): 9–20, 2004.

[260] A. Fereidunian, H. Lesani, C. Lucas, and T. C. Green. A systems approach to design of a utility management automation system. In: Proceedings of the IRCE2005 Conference, Leeds, UK, 2005.

[261] A. Fereidunian, C. Lucas, H. Lesani, R. Rahmani, and A. W. Wymore. A policy-driven method for IT infrastructure selection in power distribution automation system. *International Review of Electrical Engineering, (IREE), Part B*, 5(2): 671–682, 2010.

[262] C. M. Huang. Multiobjective service restoration of distribution systems using fuzzy cause-effect networks. *IEEE Transactions on Power Systems*, 18(2): 867–874, 2003.

[263] M. Y. Huang, C. S. Chen, and C. H. Lin. Innovative service restoration of distribution systems by considering short-term load forecasting of service zones. *Electrical Power and Energy Systems*, 27: 417–427, 2005.

CHAPTER 13

VIRTUAL TRAINING FOR PROCEDURAL SKILLS DEVELOPMENT: CASE STUDIES AND LESSONS LEARNT

DAWEI JIA, ASIM BHATTI, and SAEID NAHAVANDI
Centre for Intelligent Systems Research, Deakin University, Waurn Ponds, Victoria, Australia

13.1 INTRODUCTION

Virtual environments (VEs) represent a computer-generated three-dimensional (3D) environment and are perceived to facilitate human perception and the comprehension of complex concepts and/or enhance the human ability to perform complex psychomotor tasks [1]. Enhanced perception is achieved by using displays that provide the rich visual, auditory, and haptic sensory information that allows users to easily immerse themselves in and engage and interact with learning tasks [2]. A significant amount of recent research on VEs has been dedicated to the simulation of spatial manipulation tasks in the context of mechanical design, assembly planning, and assembly evaluation. For instance, in manufacturing settings, VEs have been used for object assembly operations, and the aerospace industry embrace VEs for prefight training activities. Reaching far beyond entertainment, VEs offer a route to training [3] in a manner that is intuitive, interactive, engaging, and fun. Virtual training (VT) systems utilizing virtual reality technologies are distinguished from VE where the former represents advanced computer-assisted training systems capable to provide haptic, audio, and visual feedbacks simultaneously, whereas the latter is considered as a subset or component of the VT system. Significant adoption of VT systems has been reported in domains including medicine, biotechnology, manufacturing, automotive, and aerospace. Despite the current advancements in VT systems, the

Contemporary Issues in Systems Science and Engineering, First Edition.
Edited by MengChu Zhou, Han-Xiong Li and Margot Weijnen.

difficulties of replicating real user–environment interaction, accurate simulation of multisensory feedback, and effective design of useful and usable VT systems have yet to be overcome. Moreover, regardless of the increasing demand to use VT systems, a lack of understanding of human–computer interaction in VE [4, 5], especially human–VE interaction (HVEI), has not helped designers to ameliorate the situation of how to better and effectively design such a class of applications. This has led to further problems as to what actually contributes to learning in a technology-mediated learning context, and how to better design a VE-mediated learning environment to facilitate and enhance learning. Consequently, questions of how virtual training and learning could be made more effective and efficient are needed to be addressed. Evaluation of the efficacy of VT systems may be a solution, but it is a difficult task. In particular, due to the lack of empirically derived validations of interaction and learning in VT systems, training effectiveness is hardly ever systematically tested [5, 6].

13.2 RELATED WORK

13.2.1 Background

There are many VE applications available and each may possess unique design features, for instance, the virtual surgery training that provides the basis for retraining patients with neurological deficits resulting from multiple sclerosis, spinal cord injury, and stroke [7]. Other VEs can be of substantial benefit in compensating for sensory deficits, particularly of vision and hearing. Regardless of their application domains, they provoke distinct interaction styles, metaphors, and/or different purposes. A generic immersive VE possesses the features of 3D immersion, multisensory cues, and frame of reference (FOR). Specifically, *3D immersion* enables users to develop a subjective impression of "being in" a computer-generated synthetic environment that is comprehensive and realistic enough to induce the willing suspension of disbelief. *Multisensory cues* refer to high-end user interfaces or displays of visual, auditory, and haptic cues to facilitate users in gathering information to navigate and control objects in the synthetic environment. FOR or perspectives are spatial metaphors which can enhance the meaningfulness of data and provide qualitative insights. For instance, VEs can enable users to experience learning tasks directly through the involvement of "whole being"—brain, body, and the surrounding environment [8]. Additionally, they enable users to interact with spatial representations from various FORs that may improve performance and deepen learning through different and complementary insights [9]. These distinctive features of VEs have led to an increasing demand to use VEs to facilitate the familiarization of inaccessible environments, facilitate task mastery through practice of dangerous or expensive tasks, improve transfer by situating learning in a realistic context, and improve motivation [10].

Designing a human–computer interface that is intuitive, interactive, engaging, and fun is a challenging task and requires designs with a strong sense of presence and immersion as well as system comfort [11]. In particular, the nature of human perception requires VEs to have a rich interaction that utilizes various input modalities, as humans are capable of making use of a combination of sensory stimuli and prefer parallel information input [12]. In the context of training, the design of effective VEs does not merely require a human–computer interface that is intuitive, engaging, and supports strong sense of presence and immersion, but they also need to be effective if users are to achieve their intended learning outcomes. By "effective" we mean that a VE should provide an enjoyable user experience, support quality interaction (high level of interactivity), and facilitate learning. In other words, the design must facilitate users to be proficient at new skills and/or tasks. At the same time, the VE should be easy to learn and use (needing little learning curve), with little or no cognitive overload and should result in minimum side-effects. Therefore, user experiences and learning outcomes are key factors in the study of the effectiveness of VE.

Additionally, making the interface intuitive of little or no learning requires careful attention to the interaction between human activities and the VE features designed to support these activities. Previous research has shown that the more intuitive and interactive the interface is, the less a learning curve a user needs to master the learning tasks mediated by computing technology [3]. Furthermore, the more obtrusive the interface is, the more easily it supports the information flow and HVEI, and thus the better the learning outcomes it facilitates a user to achieve.

13.2.2 Human Side of VT System Efficacy—Issues and Concerns

Previous research implies that user characteristics, task characteristics, and environment characteristics or VE design features are the critical determinants of the effectiveness of VE training systems and that various factors relating to VE effectiveness fall into these categories [3, 9, 13, 14]. User characteristics including age, gender, spatial ability, perceptual style, concentration, and experience, all serve to mediate the relationship between VE design features and learning and they may influence the learning and interaction experience [9]. VE design features or environmental characteristics, which include 3D immersion, FORs, multisensory cues, field-of-view (FOV), fidelity, and resolution, have proven to affect the quality of HVEI, user experience, and learning. Task characteristics include degree of control, level of difficulty (LOD)/complicity, and interactivity requirements, for example, head movement and posture.

It has been established that VE features are highly influential on the level of immersion and presence, and this can lead to different levels of user performance. For instance, system features such as visual complexity are found to have a direct impact on user experience, performance, and memory [15]. Several psychological variables, such as concentration, previous experience with a VE, previous experience with required tasks, expectations regarding the mediated experience (that relate to individual users), all play an active part in this regard [13]. Furthermore, each of these variables can have an influence on the extent to which a user becomes involved in the tasks required. A recent study on a meta-analysis of the training effectiveness of

VE surgical simulators [16] confirmed that there is a clear difference between experienced trainees and less experienced ones on task performance. The experiences are distinguished on the bases of domain knowledge and skills (rather than on experience with VEs). In this sense, regardless of users' prior experience with the VE training simulator, their performance is influenced by their familiarity or domain knowledge of the procedural tasks.

Likewise, research in educational psychology suggests that an individual's prior knowledge is the most important single factor influencing learning [17]. Other characteristics such as age, gender, spatial ability, and experience (domain knowledge and skills) are also found to influence task performance in 2D computer systems [14]. For example, a research study [18] in information visualization shows that individual characteristics such as spatial ability and associative memory have a variety of relationships with user behavior and performance in a spatial-semantic VE. It also suggests that factors such as prior experience with computers and the knowledge of the task domain are more dominant predictors of user performance in the VE than the effect of spatial ability. Human–computer interaction research also supports the claim that individual differences and system characteristics have significant effects on user perceptions, such as the perceived ease of use and the perceived usefulness of computer systems based on empirical evidence [19, 20]. Thus, it makes intuitive sense to explore the effects of individual characteristics and VE system characteristics on user interaction, perception, and learning in these more advanced and complex user interfaces.

In the context of user-centered design for training systems, individual differences between users can be considered to be different in the resources that users bring with them to training tasks [21]. These differences are likely to involve age, gender, prior experience with VEs, domain knowledge (task related), cognitive abilities, learning styles, etc. The effects of individual difference on task performance can vary with the extent to which these resources are required during the performance process. Research has shown that individual difference in user characteristics can be related to a variety of motor-skill learning tactics [22], and the tendency to use certain VE system features [1, 13]. For example, past research found that users with prior experience using force-reflective haptic interfaces exhibited higher performance levels in a task of sphere size identification. Such a task requires users to operate a PHANToM (SensAble Technologies Inc., Cambridge, MA) to explore an accurate and efficient way of estimating human performance [23, 24]. Additionally, research on information visualization has also found a variety of relationships between individual characteristics and information search behaviors [21] and suggests that it is important to match users with system configurations that will optimize their performance. Similarly, Conati and Maclaren [25], exploring whether a user's individual differences can be used as predictors to select the most effective visualization for different users, have found that individual differences do affect visualization effectiveness. In the same vein, research has found women to be less good at spatial reasoning than men in a VE, and that their spatial reasoning is found to be positively correlated with performance [26]. On the other hand, Waller [27] have found no gender effect or differences in training effectiveness in a VE.

The research [18] has been able to isolate the role of individual differences, that is, spatial ability and associative memory, in searching through a spatial-semantic VE. Although no significant effect is found for spatial ability on performance, it suggests that users who are more experienced with the task environment will achieve a higher recall score. A complex interaction is also found between individuals' cognitive abilities, domain knowledge, and direct manipulation skills in the VE [4]. However, the knowledge gathered through any of the research on individual differences on user performance and memory is limited. In particular, there is a lack of research associated with such information as user affect. Therefore, we hope to further the understanding of designing haptic–audio–visual interfaces to best suit individual users from a diverse background of experiences and expertise, in particular to examine the effects of prior experience in manipulating 3D object (LOE3D) on user performance, perceptions, and memory.

13.3 PRESENT STUDY

13.3.1 Motivation and Aims

Since the establishment of Embodied and Embedded Cognition (ECC) [8] and Cognitive Affective Learning with Media (CATLM) [28], technologies such as haptic–audio–visual environments have been designed, developed, and used with respect to individual differences and VT system affordance for training and education. Accompanied by the trend of increasingly adopting and developing such technology into training, various methods for assessing the efficacy of these VT systems are increasing dramatically. Nevertheless, limited research has been done on how the human side of VT system efficacy is expressed by these technologies and the impact of this on the user is evident in terms of performance, cognition, and memory. In addition, users vary in terms of physiological and psychological reactions to VTs due to individual characteristics (e.g., age, gender) and prior experience (e.g., domain knowledge, VT exposure) [1, 13]. Although the above discussion indicates the clear impact of individual characteristics and prior experience on performance, just how they influence user perception and the ability of recognition and memory recall are not reported. Further, the extent to which these individual characteristics and prior experiences influence a user's perception, performance, and memory is unclear (as can be seen from Salzman's statement below). To effectively design VTs for different user groups, they need to be better understood.

> "Individual learner characteristics may serve to moderate the relationship between VR's features and learning and they may influence the learning and interaction experiences."
>
> Salzman *et al.* [1]

Therefore, the main research questions of the Case study 1 are

Can an individual user's characteristics, in particular prior experience, lead to varied performance, perception, and memory in VTs? Which of these characteristics shows the most influence in varying a user's levels of performance?

In general terms, it is expected that task performance, user perception, and memory are significantly affected by individual's prior experience. In addition, it is understandable that an effective system design that does lead to higher performance, perception, and memory would be associated with a higher level of prior experience, for example, 3D object manipulation. The broad aim of the second study is to explore further whether an individual user's characteristics, specifically prior experience with VTs, lead to differences in performance, perception, and memory; and which of the characteristics is the most influential in this regard. Moreover, it is of interest to see whether a higher quality of user experience and interaction can lead to better learning outcomes and vice versa. It is these questions that prompt us to investigate further. A widely recognized problem in the field of VT implies that only some aspects of user experiences have been explored, and few VTs have been tested in formal user studies [29–32]. In fact, most VTs have never gone through a thorough testing with real or representative user samples [6, 33]. Consequently, they run the risk of ineffective information delivery and ineffective training outcomes.

13.3.2 System Architecture and Human–Machine Interface

The system architecture and screenshots of object assembly training are shown in Figure 13.1. The overall VT system architecture uses a modular approach where different software modules process information independently. This modular approach makes the system highly scalable as new modules can be added into the system or discarded at any time with minor changes in the central processing module.

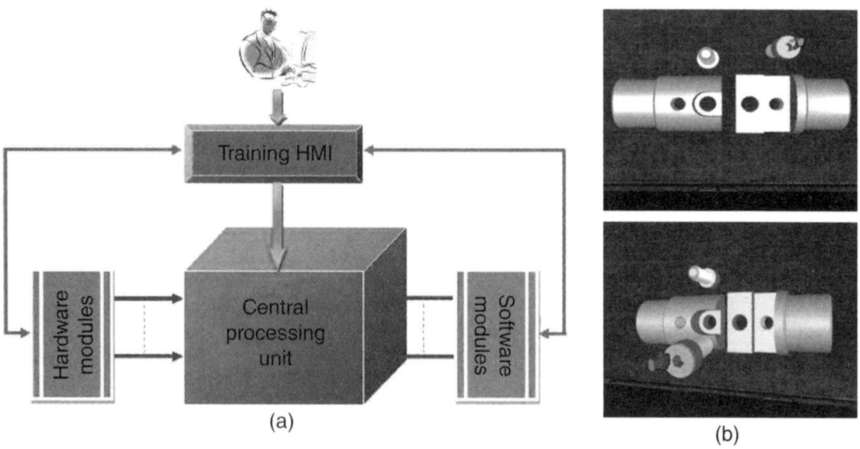

FIGURE 13.1 System architecture (a) and screenshots of object assembly training (b).

Furthermore, independent processing modules take advantage of the current multi-core architectures of the computer processors by running operations in parallel if processes are completely independent. Moreover, the functional aspects of the VT system are event-driven where communications between system modules are encapsulated as events that are propagated to the appropriate destinations. This event-driven approach provides a framework of assessment and evaluation of the user performance. It also portrays an outlook similar to computer games, keeping the user motivated to keep progressing throughout the simulation. This event-driven system design considers the repository, object interaction, and user interface aspects of the system. The repository is needed to provide for the storage and retrieval of geometric models representing virtual worlds and assembly parts as well as the information models encapsulating relevant assembly sequences.

The overall system can be divided into two broad classes: software modules and hardware equipment. Hardware includes input and output (I/O) devices such as a Phantom® haptic device, 5DT® data glove, Flock of Birds, and visualization equipment such as Emagin Z800 Head-mounted display (HMD) or Stereo projectors. Software is responsible for providing interactive functionality to the user. The hardware modules used to provide a complete immersive and interactive training environment can be divided into two broad categories, namely the devices to provide immersion and the devices responsible for interaction. For display purposes, two different stereoscopic modes are provided, namely the stereo projection system mode and the HMD mode. The display of the GUI of the VE system can be selected from any of the aforementioned display modes. Both of the modes provide depth perception to the user.

For the developed setup of the VT, a NEC® stereo projection system and Emagin Z800 HMD are used. A software training environment consists of different information processing modules separated on the basis of information availability to the user and the interaction required from the user. The software modules developed within the functionality of a VT training system are the Central Information Processing Module, Registration Module, Physics Engine, Data Acquisition Module, Collision Detection engine, and Evaluation Module. The hardware components of the VE in this study included an intergraph workstation that runs a SensAble PHANToM haptic device, an HMD, and a 3D mouse. These hardware components are used to provide users with force feedbacks, 3D object perception, and 3D environment manipulation. Software components include a user interface that consists of a series of user menus and 3D visual models of assembly objects.

Moreover, HMDs are used for immersive visualization equipped with 6DOF trackers to keep the virtual view synchronized with the human vision; and PHANToM devices are used to impose physical movement constraints. In addition, 5DT data gloves are used to provide human hand representation within the virtual world. The user is also provided with audio-visual feedback [34] to keep him/her engaged with the training progress, as well as keeping the user informed about different events that occur during the training operation.

The events that trigger audio-visual cues include task completion, requested assistance, violation of recommended practices, wrong operational sequences, etc. Visual feedback involves a change of color of the objects on assembly task completion, graphical animations to assist the user to accomplish the task at hand, and visual cues to highlight the right assembly sequence, assembly path, and correct procedure. Similarly, audio feedback involves three distinct audio cues linked to right, wrong, and forbidden actions emulating the positive and negative reinforcement learning mechanisms used in computer games. Audio cues linked to correct actions and successful completion of tasks create an enhanced feeling of engagement and motivation, whereas audio cues linked to wrong or forbidden actions help to promote best operational practice.

Procedural tasks, such as object assembly, require users to comprehend an assembly sequence presented in the VT, recognize correct objects for specific task procedures, and utilize various VT I/O devices to achieve learning. The VT system can support the learning process of general assembly operators as well as provide an intuitive training platform to enable assembly operators to perform their learning practices repeatedly, until they are proficient with their assembly tasks and sequences. It aims to provide an interactive training platform where users can explore their targeted assembly sequences through experiential learning in 3D virtual space. Users are able to interact with virtual objects directly and experience the effects of their interactions. Through direct manipulation, implicit and explicit learning modes can be induced [35]. Consequently, user motivation is increased and concepts become more readily internalized, as the result of such direct manipulation of virtual objects [36].

Procedural tasks training utilizes all aspects of the unique features a VT possesses, that is, 3D interaction via advanced I/O devices (that many educational VTs lack). In this sense, a 3D interaction is a unique learning experience that is not restricted to the visual aspect but also includes the audio and haptic aspect of interaction. Specifically, the experimental tasks used in this research require object assembly for a car cockpit. Each participant is required to perform seven object assembly tasks in the VE via four main task sequences. As the performance results are strongly dependent on the tasks to be performed, in the task design, both static objects, that is, those physically constrained to move within the prescribed limits of the VT, and dynamic objects, that is, those without any constraints being placed upon their special behavior [37], were included. For example, the car cockpit is a static object that users cannot move around whereas other objects, such as the radio box, screw driver, stereo, and power connector are dynamic objects that users could manipulate and maneuver through the VT.

13.3.3 Measures

A multidimensional user-centered systematic training evaluation (MUSTe) method [38] is adopted to assess cognitive, affective, and skill-based learning in the VT system. This method represents an effort from us to establish a way to quantify human side of VT efficacy. Two self-report user perception measures, the SE scale and the PVE scale, an integral part of the MUSTe method, are utilized. These scales allow us to gather users' subjective perceptions of the VT system. For example, on a 10-point semantic differential rating scale (from 0 to 100 with 0 being the lowest rating), participants rate their capability to perform a training test with similar types of tasks in terms of accuracy, efficiency, and effectiveness (SE). PVE is used to measure the individual's beliefs in the effectiveness of the VE to assist them in learning the object assembly tasks. A seven-point Likert scale is used to gather participants' rating for each item, ranging from 1 (strongly disagree) to 7 (strongly agree). Sample statements include "I was able to focus my attention on learning assembly procedures rather than the input control tools (e.g., haptic device)"; "the input control tools (e.g., haptic device, data glove, and 3D mouse) were comfortable to operate together in unison"; and "I have a strong sense of 'being there' (sufficiently immersed) in the VT environment." Higher ratings are considered to indicate a higher perception of the VT efficacy.

Objective measures of task performance and performance memory test on recognition and recall are also used in the evaluation process. User performances on trained skills are automatically recorded by logging detailed information of "time on task" and "accuracy" of task performance. Their performance procedures are also recorded using video. A short post-test interview was also used to obtain insight into user perceptions of their interaction and experience with the VT. Memory test questionnaires (MTQ) were also used to aid assessment of engagement and immersion of user experience in VTs in previous studies [32]. In the present study, an online memory test is used to assess user ability of recognition and recall. By focusing on questions related to VT structure and characteristics, users reveal levels of spatial awareness, sense of presence, and attention to the VT.

13.4 CASE STUDY 1

13.4.1 Method

A total of 30 participants (4 female and 26 male) were involved in this study. They were recruited from Deakin University, Waurn Pond Campus. The age of the participant population was between 25 and 34 years. Of these participants, 7 were very experienced VE users (VE expert—VEP), 11 were experienced with object assembly tasks (task expert—TEP), and 12 were neither experienced users of VE nor experienced in object assembly tasks (non-expert—NEP). Table 13.1 illustrates the profiles of the participants. The classification of user group was achieved based on users' self-reporting of their prior experience with VEs, 3D object manipulation, and expertise of assembly tasks. It is worth noting that, in this study, VE experts are

TABLE 13.1 Participant Profiles

Expertise	VE expert—VEP ($N = 7$)	Task expert—TEP ($N = 11$)	Non-expert—NEP ($N = 12$)
Gender	Male (26)		Female (4)
Age	25–34 years		

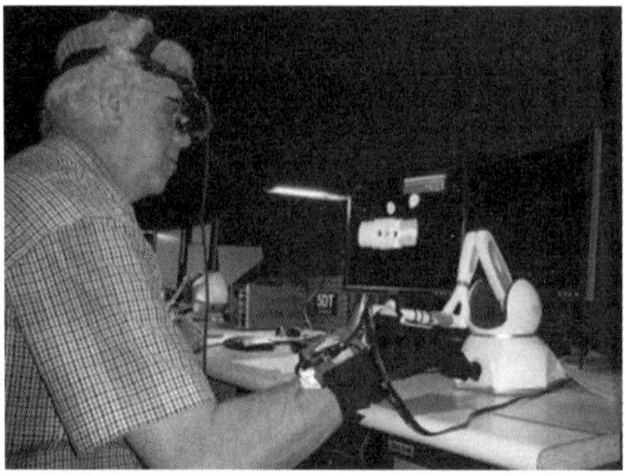

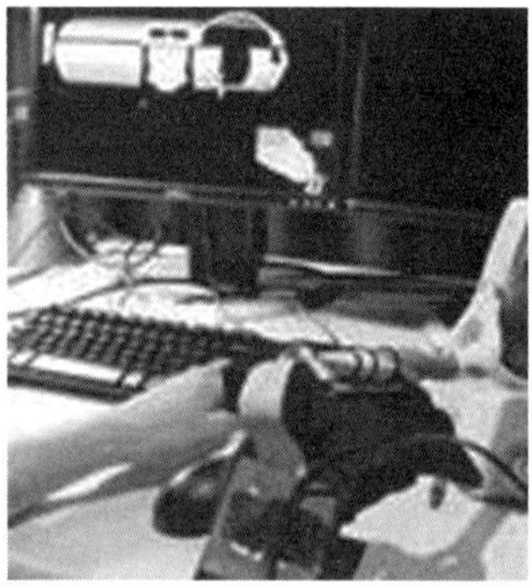

FIGURE 13.2 Experimental condition.

those participants who conduct research in the area of haptic interfaces or VE design and development. They were doctoral or postgraduate students or research staff with expertise in the VE field [39]. Domain/task experts were those experienced in object assembly procedures who frequently worked in either the engineering workshop or in real-life object assembly situations. Non-experts or novices were those who had no experience in VEs nor adequate knowledge of object assembly tasks in real life. This study was approved by the Deakin University Ethics Committee.

Seven object assembly tasks with various levels of difficulty were embedded in the VT system. The task involved selecting, rotating, releasing, inserting, and manipulating 3D objects. These tasks required users to utilize a data glove, a haptic device, a keyboard, and an HMD. To produce an easily controlled experiment and to be able to focus on examining the impact of users' prior experience on learning in VTs, the design of the experimental tasks required users to follow a prescribed sequence rather than allow them to freely move between training tasks. Each task needed to be performed within a specific order and each task was locked after successful completion. Participants are required to press "space button" after completion of each task to lock the task before they proceed to next one. Figure 13.2 illustrates the experimental condition.

The experiment proceeds with following procedures (as shown in Figure 13.3) upon the participant entering the experimental environment. Each participant is asked to complete a pre-test questionnaire (Pre-test Q). Each participant was then given a brief introduction to the system and performed a simple object assembly task, which served as a pre-test of the participant's ability to interact with, control, and use the various VE system control devices. An SE questionnaire (SEQ) was then filled out. Afterward, a training test was presented to each participant, which had 15 minutes to complete 7 object assembly tasks in the VT system. A PVE questionnaire (PVEQ) was then presented to participants. Lastly, an open-ended interview with each participant was carried out immediately following the test. Two weeks after the experimental test, participants were required to respond to the MTQ that required them to recall their learning tasks or procedures in the VE training system.

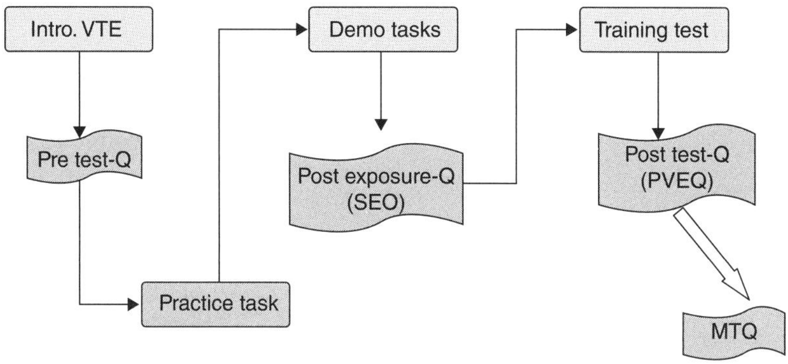

FIGURE 13.3 Experimental procedure.

13.4.2 Results

VT efficacy was hypothesized to be significantly affected by different levels of prior experience, in particular the users' prior experience in manipulating 3D objects in a gaming or computer environment (LOE3D). As VE efficacy was measured on a training test (TTS), SE, PVE, and memory (MMT), it was expected that people with a higher level of LOE3D have higher self-efficacy beliefs, achieve better outcomes in the training test, perceive the VT to be more effective, and have a higher achievement on the memory test.

13.4.2.1 Cognitive, Skill-Based Affective Learning Outcomes On average, participants achieved a task score of 64.67 (out of 100). As shown in Figure 13.4a, the mean score for TEP was 78.18, NEP achieved 38.33, and VEP, 88.57. Overall, VEP and TEP achieved similar levels of object assembly scores, although VEP outperformed TEP on each assembly task but task 6, and NEP achieved the lowest task score.

Only one participant received 0 task score as the user did not complete any task in the training test session. In addition, participants in both the task expert (TEP) and VE expert (VEP) groups achieved a higher task score than those in the non-expert (NEP) category. Both TEP and VEP groups had participants who successfully assembled all training tasks in the VE and thus received the full test score of 100. On the other hand, the highest task score in the NEP category was 80. A one-way between-groups analysis of variance (ANOVA) was performed. At the significant level of 0.05, an examination of the Levene test for homogeneity of variances suggested that this assumption had not been violated ($p > 0.05$). The F-ratio with an F-probability value less than 0.05 is significant, suggesting that a different type of expertise (VEP, TEP, and NEP) does significantly influence the task score in a VE training system. Post hoc analyses further showed that the difference lies between NEP and VEP ($p = 0.000$); and between NET and TEP ($p = 0.000$). No significant difference was found between VEP and TEP on the task score. Results also indicated, regardless of the users' prior

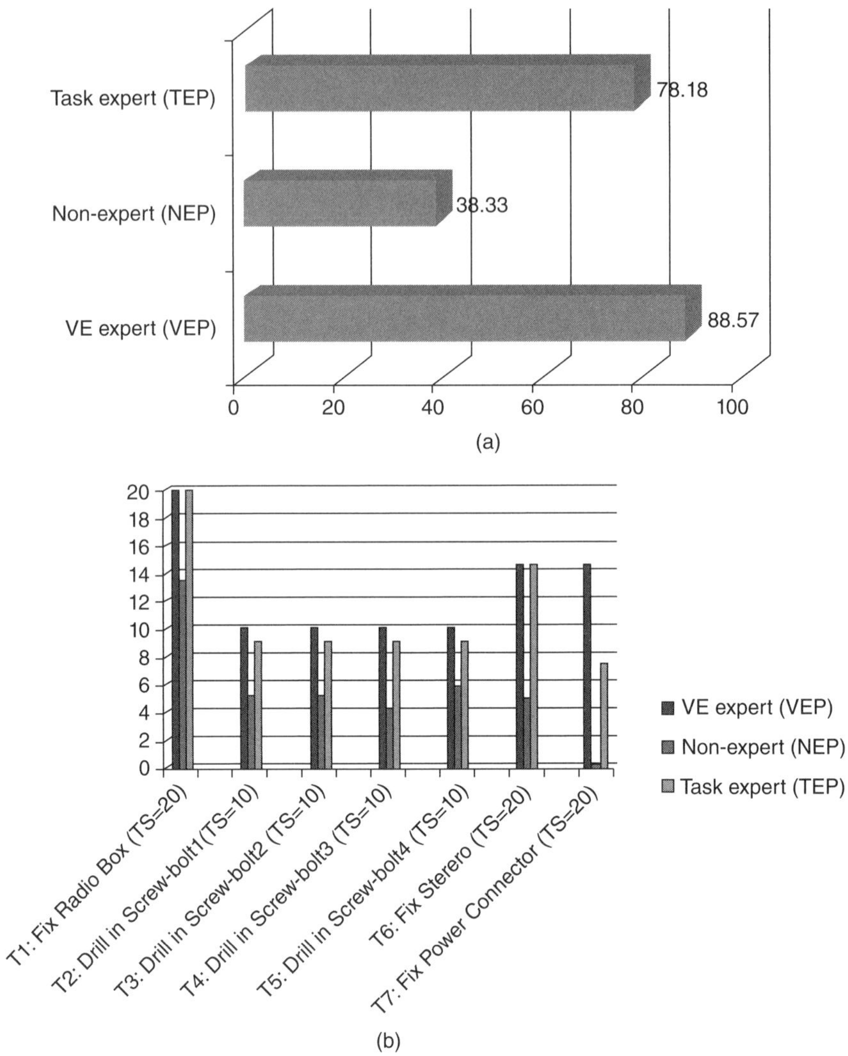

FIGURE 13.4 (a) Mean score of skill-based task outcomes according to users' prior expertise and (b) participants' skill-based outcomes on each assembly task.

expertise, all VEP, NEP, and TEP participants performed better on the task with a low LOD than those at moderate or high LOD. As Figure 13.4b illustrates, the highest task completion rate was achieved on Task 1 (T1), which had the least difficult level, among the VEP, NEP, and TEP user groups. None of the NEP participant was able to assemble the task at the highest LOD, that is, T5. Participants who had previous VE experience outperformed task/domain experts on task with moderate and highest LOD.

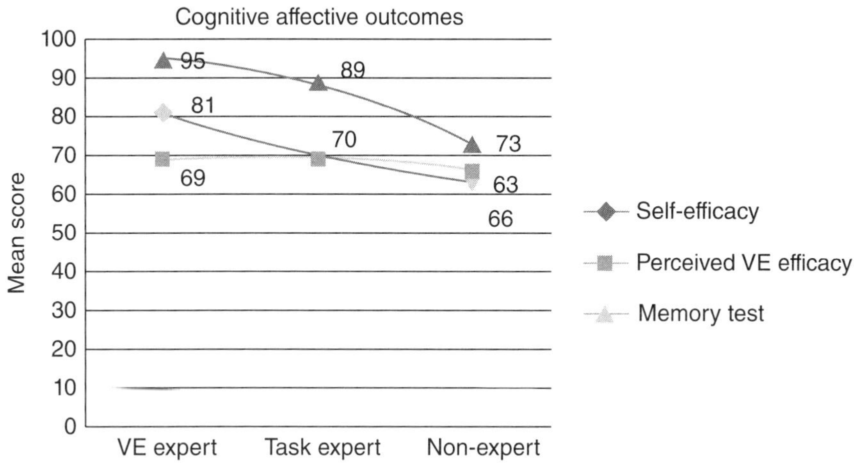

FIGURE 13.5 Cognitive and affective outcomes.

Cognitive and affective learning outcomes measured via SE, PVE, and MMT are illustrated in Figure 13.5. Interestingly, the mean score for self-efficacy (SE) between VEP and TEP differs widely compared with the similar results for NEP and TEP as shown in Figure 13.5. However, a one-way between-groups ANOVA analysis shows the difference is not significant ($F = 1.918, p = 0.170$). Therefore, individuals tend to have similar beliefs of self-efficacy, regardless of their differences in expertise. Mean scores of user perceptions of VT efficacy (PVE) show that subjects with different expertise perceive the VE to be equally effective: on a 100 scale, the mean subjective rating for all user groups was greater than 60 and closer to 70 for VEP and TEP. In addition, a one-way between-group ANOVA analysis supported the mean result and showed that no significant differences found among subjects on perceived VT efficacy ($F = 0.385, p > 0.05$). Twenty-six participants responded to the memory test (MMT). On average, participants achieved 84.80 out of 100, with 74.25 for NEP, 88.89 for TEP, and 94.64 for TEP. Moreover, ANOVA analysis shows that there is a significant difference on memory test score (MMT) across the different expertise ($F = 7.215$, $p = 0.004$). Post hoc testing further revealed that the difference exhibited between VEP and NEP ($p = 0.004$) and between NEP and TEP ($p = 0.030$). No significant difference was found between VEP and TEP ($p > 0.05$).

13.4.2.2 Individual's Characteristics Impact It is widely accepted that evaluators can make no assumptions about a novice user's ability to understand or use a given interaction technique or device. In situations where the VT evaluation is not aimed at a particular set of users, the best course of action is to evaluate the most diverse user population possible, in terms of age, gender, technical ability, physical characteristics, and so on and include these factors in any models of performance measurement [39]. To assess the utility of prior experience to explain task outcome, we use multiple predictors: computer use frequency (ComFreq), computer use history (CompHis), experience of manipulating 3D objects in a gaming or computer environment (LOE3D), and experience of manipulating 3D objects in a VE environment (ExpVE). These were included in a multiple linear regression (MLR) model to predict a training test score (TTS). Because the potential effect of one's object assembly skills in real life may have an influence on the performance one shows in the VE, experience of using electronic tools for object assembly tasks (ExpTool), and a perceived LOD of assembly task (PdifTask) were included as predictors in this model. Finally, due to the potential effects of age and gender on a TTS, and other response measures, these two variables were included in the model.

In general, the inclusion of these variables in the predictive model of TTS was aimed at avoiding biases in the parameter estimates; CompFreq, CompHis, LOE3D, and ExpVE that might have occurred if variance due to prior object assembly skills (ExpTool, PdifTask) or individual differences were not taken into account. However, it is anticipated that there are interrelationships among the variables. With this in mind, a standard approach of multiple regressions was performed, which allowed us to find out how the multiple predictors are combined to influence the TTS. The regression model used to assess the utility of multiple predictors on the TTS was structured as shown below:

$$TTS = \beta_0 + \beta_1\text{Age} + \beta_2\text{Gender} + \beta_3\text{CompFreq} + \beta_4\text{CompHis} \\ +\beta_5\text{LOE3D} + \beta_6\text{ExpVe} + \beta_7\text{ExpTool} + \beta_8\text{PdifTask}. \quad (13.1)$$

Results of the standardized regression coefficients analysis indicate that this regression model predicted TTSs well, that is, $F = 2.404$ and $p < 0.05$. Approximately 48% of the variability in TTSs was explained by this model ($R^2 = 0.478$). The results also showed that at the $\alpha = 0.05$ level, LOE3D is the most important predictor of TTSs ($\beta = 0.567$ and $p = 0.032$). More importantly, LOE3D alone accounts for 38% of the variance of a TTS, $F = 17.136$ and $p = 0.000$. Surprisingly, of the eight predictors, only the participants' prior experience of manipulating 3D objects in a gaming or computer environment (LOE3D) contributed significantly ($p = 0.001$) to the model (Table 13.2).

Correlation analysis (one-tailed) also confirmed that LOE3D was significantly and positively correlated with TTS, $r = 0.616$, $N = 30$, and $p = 0.000$. In other words, people who are more experienced in manipulating 3D objects in a gaming or computer environment tend to achieve higher TTSs. In addition, a moderate but significant linear relationship between gender and TTSs ($r = 0.321$, $N = 30$, $p = 0.042$) and between ExpVE and TTSs ($r = 0.358$, $N = 30$, $p = 0.026$) was found. These results show that males tend to outperform females, and people with more experience in manipulating 3D objects in VE achieve a higher TTS. In addition, younger people tend to have more experience in manipulating 3D objects in a gaming or computer environment than elder ones ($r = 0.508$, $N = 30$, and $p = 0.004$).

TABLE 13.2 Results of Standardized Regression Coefficients Analysis on Individual Parameter Estimates

Variables	df	t Value for H_0	$p > T$
Intercept	1	0.111	$p > 0.05$
Gender	1	0.606	$p > 0.05$
Age	1	0.127	$p > 0.05$
Comptreq	1	0.051	$p > 0.05$
CompHis	1	−0.699	$p > 0.05$
LOE3D	1	2.301	$p < 0.05^*$
ExpVE	1	0.403	$p > 0.05$
ExpTool	1	−0.314	$p > 0.05$
PdifTask	1	1.130	$p > 0.05$

*Significant at the $\alpha = 0.05$ level.

13.4.3 Discussion

The analysis results of the study provide strong evidence to support the hypothesis that LOE3D has a significant effect on user task performance. As expected, higher performance is associated with higher LOE3D, that is, participants with moderate and high LOE3D perform better than those with low LOE3D. However, at this stage it cannot be concluded that participants with high LOE3D outperform those with low LOE3D significantly; in terms of perception and memory, no significant effect of LOE3D was found on user perceptions of self-efficacy and perceived VE efficacy, as well as memory. In other words, it is not clear whether higher perception is associated with higher LOE3D or higher ability of recall (memory). On the other hand, the results demonstrate that objective task performance is positively and significantly associated with user perceived VE efficacy, as well as memory as shown in Table 13.3, which also has a positive and significant association with users' self-efficacy beliefs. These results suggest that not each subjective user perception measure is correlated well with each of the objective measures, but the associations are undeniable.

Furthermore, it has been found that users with a different range of LOE3D have little to no effect on SE and PVE. Prior experience in manipulating 3D objects was gathered based on subjective self-reporting of user expertise. The manipulation of experimental groups was affected by such information. With respect to LOE3D on TTS and VE efficacy, LOE3D of high, moderate, and low ranges may not have been substantial enough to affect a users' ability to predict their performance and to rate VE system efficacy. Besides, subjective perception may not be consistent with objective task performance measures. As in motivational/affect literature, self-efficacy and user attitude (perception) should be used as a supplement to objective measures in evaluating VT system performance [40, 41]. Supported by the results of this study, it is clear that user perception measures are equally important (if not superior) for the assessment of system efficacy. Although self-efficacy does not correlate with an objective measure of task performance significantly well, a positive relationship can be detected between PVE and VE efficacy; and both user perception measures of self-efficacy and PVE efficacy are positively and strongly correlated with VT efficacy.

13.5 CASE STUDY 2

13.5.1 Method

Seventy-six volunteers participated in an evaluation in a controlled laboratory environment. Participants were trained on a series of object assembly tasks. The participants had diverse backgrounds with their profile as shown in Table 13.3 and fell into four age groups: 18–24 ($N = 32$), 25–34 ($N = 33$), 35–45 ($N = 8$), and over 45 ($N = 3$). All participants were recruited from the School of Engineering, Deakin University, and the study was approved by the university's ethics committee. Most participants reported limited prior experience in manipulating 3D objects in a computer environment (LOE3D). Specifically, 46% ($N = 35$) participants indicated that they had no experience or only minimum LOE3DExp. About 37.3% ($N = 28$) reported moderate LOE3D and 16% ($N = 12$) with high LOE3D. In terms of prior experience with VEs (LOEveExp), most participants 88% ($N = 66$) indicated that they had little to no prior experience, 10.7% ($N = 8$) reported moderate LOEveExp, and only 1.3% ($N = 1$) reported high LOEveExp.

Based on an initial user-based evaluation (from Case study 1), a new version of the VT system was improved in terms of human–machine interface (HMI) (e.g., user interface, haptic interface, and interaction style) and training design (e.g., task procedure and training modes). The system architecture retained its key features, but induced more explicit multimodal feedback and had a more intuitive design, compared with earlier version. The following improvements were made in the second study, based on the evaluation results from Case study 1, with the aim of improving the usability and learning effectiveness of the VT:

- Improved hardware components: A 3D mouse, as shown in Figure 13.6b, was included instead of the keyboard to provide a more direct, enjoyable, and immersive interaction style for the user. This also allowed easy manipulation of the 3D virtual world. A haptic grappler was also specifically designed to attach to the end of the haptic device stylus, with the intention of providing the user with a better tactile sense in the real world when holding the object in the VE,

TABLE 13.3 Participant Profiles

Expertise	Low LOE3D ($N = 35$)	Moderate LOE3D ($N = 28$)		High LOE3D ($N = 12$)
Gender	Male (56)			Female (20)
Age	18–24 ($N = 32$)	25–34 ($N = 33$)	35–45 ($N = 8$)	Over 45 ($N = 3$)

LOE3D, level of prior experience in manipulating 3D objects in a computer environment.

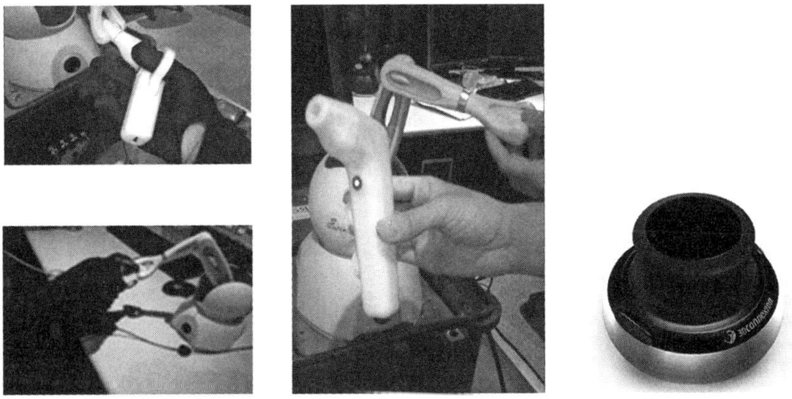

(a) (b)

FIGURE 13.6 (a) Haptic device connected with a grappler and (b) 3D mouse.

as Figure 13.6a illustrates. In addition, newer version of the HMD was also utilized, which gave users a better 3D-depth perception of the virtual objects and world, as well as better user comfort.

- Improved user interface design: Orientation arrows and more user interface buttons were added to improve the usability of the user interface. Users no longer need to press the space key each time after performing a certain task in order to block the task and proceed to the next one. Both the users' hands can be utilized—one to control the haptic device and one to control the 3D mouse.
- Fixed task procedure: Users need to follow a fixed performance procedure while performing object assembly tasks. As in Case study 1, problems became apparent when the task procedure was open and the first-time users/participants started to focus on exploring different options rather than on learning the assigned tasks according to LOD. A more structured training design was also derived to enhance training effectiveness in VT.

These changes enabled a solution to the system deficiency and better refined the training design. Overall, the HMI of the developed system can be defined by the I/O devices that are used by the user to interact with the VT environment, as shown in Figure 13.7. In general, a user wears a data glove attached to the haptic device to perform tasks. The data glove provides the visualization of a virtual hand within the VE, whereas the haptic device provides the force sensation to the user as well as the tracking information in terms of location and orientation of the hand. The user is able to grasp and manipulate the objects by selecting and making a predefined hand gesture, in particular, a fist to grasp and an open hand to let go. While the objects are in the user's grasp they can be maneuvered through the VE with constraints imposed by the haptics, that is, not being able to pass through other objects. The user is then supposed to assemble the objects by fitting them to appropriate locations. To

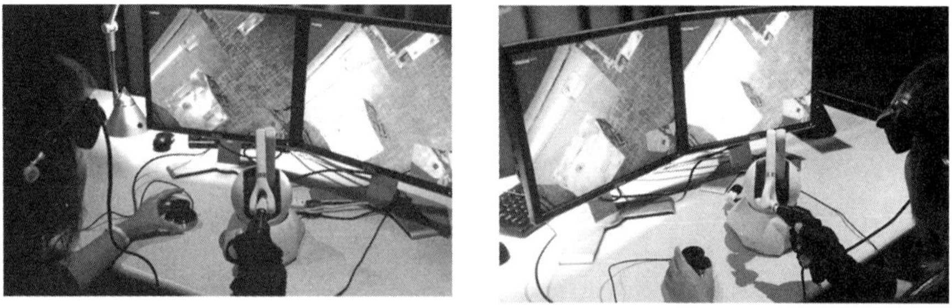

(a) (b)

FIGURE 13.7 Human–machine interface (HMI).

be able to assemble the object, the user has to perform an alignment of the objects according to the available fitting space, as the physical constraints imposed by the haptic device restrict the assembly operation. The user is also provided with visual and audio feedback to inform them of different events that occur during the operation such as completion of any specific assembly operation, wrong assembly sequence, or operation.

The VT was displayed by two 20-inch LCD monitors on a Windows PC (Core II duo, 2.6 GHz, 4 GB RAM, and nVidia Quadro 4600 GPU), an Emagin SVGA HMD by Emagin-Display (800 × 600, 26 degrees FoV diagonal), and a 3D Connexion mouse. Utilization of two LCD screens was intended to provide larger view of the visual interface. SensAble PHANToM Omni haptic device was used to provide 3DOF force feedback. It was placed at a comfortable distance from the participants' seating position. Participants wearing the data glove used their dominant hand and manipulated the virtual objects via a haptic stylus, which was attached to the specifically designed grappler. Participants were asked to close their hand and hold the grappler of the PHANToM device when they grabbed virtual objects in the VE, just like holding a real object in a real world. The two 17-inch LCD monitors were placed in front of participants' eyes at a comfortable distance. The 3D mouse was placed next to the participants' nondominant hand. These hardware components were used to provide users with force feedback, 3D object perception, and 3D environment manipulation. Software components include a user interface that consists of a series of user menus and a 3D visual model of assembly objects. Four training modes were included in the experimental process to better facilitate the learning of assembly tasks. The training process consisted of user-selectable difficulty levels and training modes as follows:

- **Mode I**—3D animated simulation explaining the task procedures and sequence of operations necessary in order to achieve successful assembly.
- **Mode II**—Experimental learning with a simple object assembly task, enabling first-person user experience with VT.

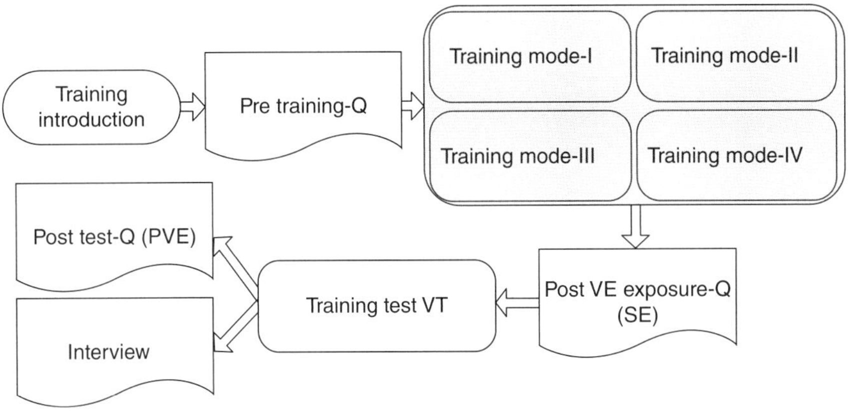

FIGURE 13.8 Flowchart of the experiment process.

- **Mode III**—Interactive simulation of assembly operations allowing the user to perform the training tasks in a predefined sequence. Each task needed to be performed in a specific order.
- **Mode IV**—Self-exploration of assembly tasks offering the participant the freedom to practice the required skills with no restriction on a task sequence.

The aforementioned training modes require different levels of interaction from the participant. In a general sense, the lower the LOD of a particular training mode, the less interaction or input required from the user/participant and the more audio-visual feedback provided. Similarly, as expected, as the LOD increased, interactivity increased and the assistive feedback decreased accordingly. Overall, the experiment follows the similar procedures, as shown in Figure 13.8, as the first case study except the four training modes.

13.5.2 Results

13.5.2.1 Cognitive, Skill-Based Affective Learning Outcomes A total of 75 participants successfully completed one or more assembly tasks of various LOD (low to high) within 15 minutes. One participant did not assemble any of the required objects with the VT system and withdrew from the testing. Nineteen participants responded to the memory test and majority of them performed well ($M = 72$, SD = 20, and $N = 19$). The mean score of the memory test was 72. In terms of subjective rating on self-efficacy and perceived VE efficacy, participants achieved moderate self-efficacy beliefs with a mean score of 59 (SD = 15 and $N = 75$), but achieved high level of perceived VE efficacy (PVE) ($M = 74$, SD = 16, and $N = 76$) (Figure 13.9).

Perceived VE efficacy (PVE) score ($M = 74$, SD = 16, and $N = 76$) was higher than self-efficacy (SE) score ($M = 59$, SD = 15, $N = 75$), but both were lower than participants' actual TTSs ($M = 77$, SD = 24, $N = 75$). This suggests that although

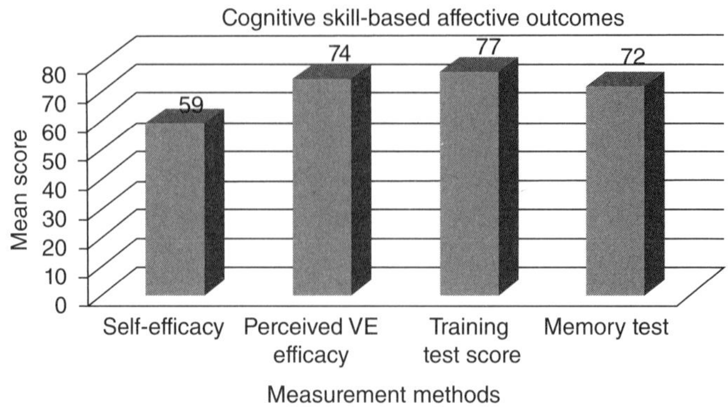

FIGURE 13.9 Mean score of cognitive, skill-based, affective learning outcomes.

users were not confident with their ability in performing learnt object assembly skills before commencing the training test in the VT, they possessed high perceptions of the system efficacy. Users' high level of real task performance outcomes in the training test provided further evidence to suggest that VT is well designed.

13.5.2.2 LOE3D Impact Case study 1 proved that LOE3D was the most important predictor of TTS. Therefore, due to the potential effects of LOE3D, a one-way ANOVA analysis with a Turkey post hoc test was performed to investigate LOE3D on task performance. Figure 13.10 illustrates mean TTS result by LOE3D. The ANOVA test showed significant effects of LOE3D on task performance outcomes ($F = 3.598$ $p < 0.05$), specifically between users with low LOE3D and high LOE3D, and between users with moderate LOE3D and high LOE3D. Furthermore, a Turkey post hoc test

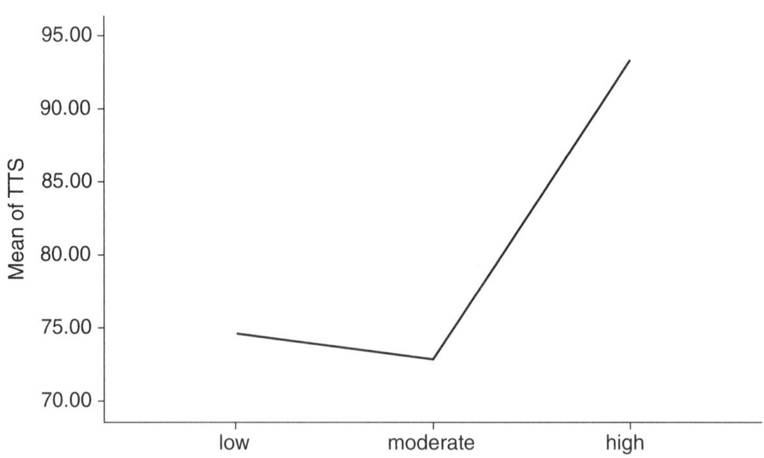

FIGURE 13.10 LOE effect on TTS.

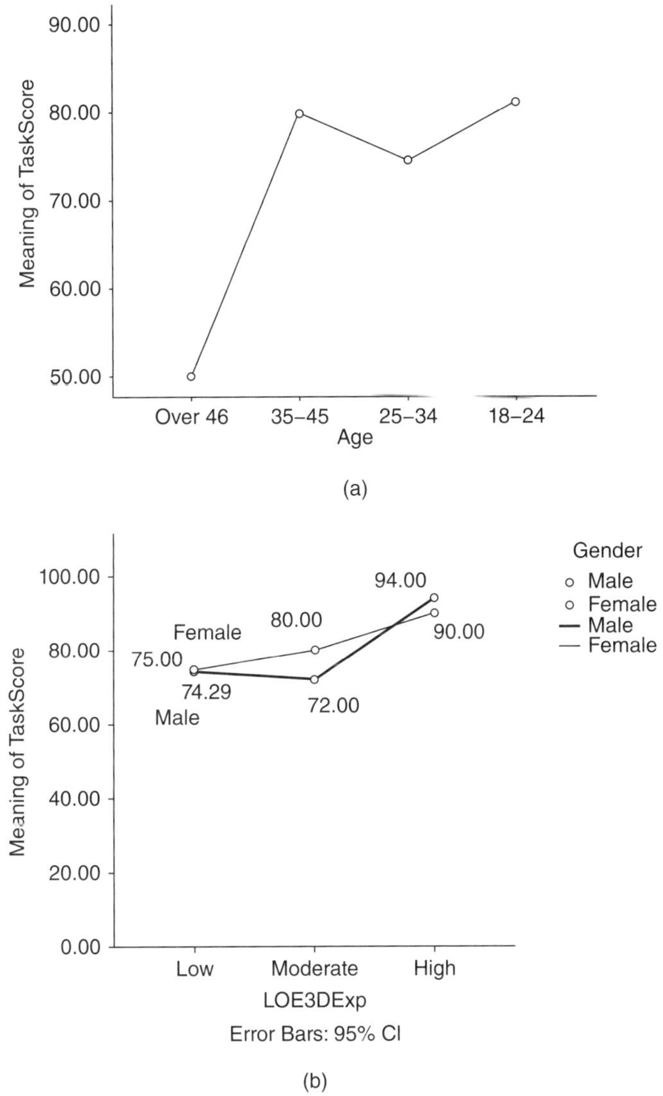

FIGURE 13.11 Task performance by age (a) and by gender (b).

revealed that participants had higher TTSs under the high LOE3D ($M = 93$) than under the moderate ($M = 73$) and low LOE3D ($M = 75$). No significant difference was found between users with low and moderate LOE3D ($p > 0.05$).

Mean task performance results by age were 50 (SD = 26), 80 (SD = 23), 75 (SD = 25), and 81 (SD = 22) for participants aged over 46, between 35 and 45, between 25 and 24, and between 18 and 24, respectively, as shown in Figure 13.11a. Clearly, the young adult group aged between 18 and 24 performed better ($M = 81$) than all the other older groups. Interestingly, participants aged between 35 and 45 outperformed (M = 80) younger people aged between 25 and 24 (M = 75). The oldest adults performed poorly (M = 50) compared with the others. In terms of gender, 56 males and 19 females completed a training test on object assembly tasks in the VT. Mean task performance results by gender showed 77 (SD = 25) for males, and 77 (SD = 22) for females, as Figure 13.11b illustrates. This suggests that overall males and females performed equally. Male and female participants were further categorized based on their LOE3D. In addition, gender by LOE3D interactions on performance were 74, 72, and 94 for males; and 75, 80, and 90 for females, of low, moderate, and high LOE3D, respectively. Interestingly, females performed better than males at low and moderate LOE3D levels, whereas males did better than females at the high LOE3D level.

13.5.2.3 System Feedback Feedback is considered a primary instructional strategy and a powerful component in learning. It can be defined as "the provision of knowledge of progress to be used in adjusting subsequent behaviour, and is a fundamental element for any adaptive system or organism" [40 p. 3]. Importantly, feedback acts both to inform the recipient of expected behaviors and to provide information for satisfactory performance outcome.

In the present study, training involving procedural knowledge and motor skill acquisition is delivered by a haptic–audio–visual environment, which requires high level of cognitive activities. Extrinsic feedback is provided by the VT and examined through gaining insight of user perceptions of the design efficacy and how well various system feedbacks facilitate or impede learning. In this sense, feedback relates to information that is sent back to the user about what action has been accomplished through the use of a system I/O device [42, 43]. Figure 13.12 indicates that visual feedback that induced from HMD received highest rating in terms of quality of feedback.

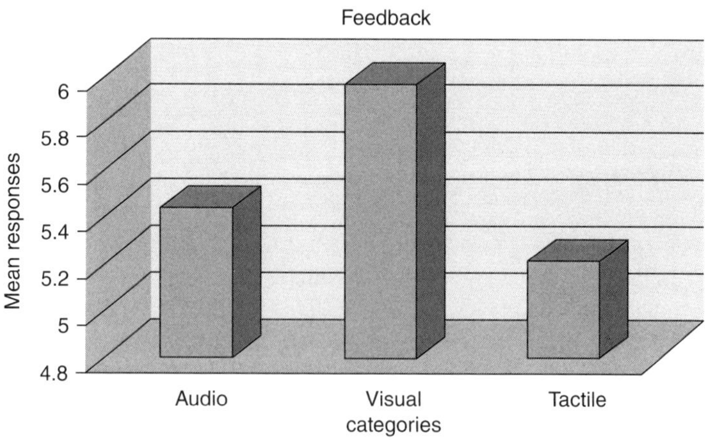

FIGURE 13.12 Audio-visual and tactile feedback in the VT.

Users' mean rating result ($M = 6$; strongly agree) suggests that majority of users feel visual feedback, for example, color change of the object, was appropriate and useful to assist them in manipulating virtual objects. Overall, users agree that both audio feedback and tactile or haptic feedback, for example, force sensation, were appropriate and did not distract them from performing the task. Past research showed that external feedback would not affect procedural knowledge acquisition in online learning situations, and argue that intrinsic feedback may affect procedural knowledge acquisition [44]. We do not know whether these findings hold in VE training situations. A regression test was performed to examine the effect of various feedbacks on performance and satisfaction. Results showed that the overall feedback had significant effect on performance, that is, $F\ (2.9, 1571) = 3$, and $p < 0.05$. Post hoc tests using Turkey's HSD revealed that visual feedback from HMD had significant effect on performance ($\beta = -0.357$, $p = 0.016$). On the other hand, tactile/force feedback ($\beta = -0.31$, $p = 0.824$) and audio feedback ($\beta = -0.068$, $p = 0.639$) had no significant influence on performance. In addition, no effect of cognitive load ($\beta = 0.033$, $p = 0.519$) on performance was found. Interestingly, satisfaction with the input tool (haptic device, data glove, and 3D mouse were comfortable to operate together in unison) lead to a significant effect on performance ($\beta = 0.313$, $p = 0.024$).

It was suggested in previous research [45] that as users gain experience, their intentions were more strongly influenced by affect or motivational factor and perceived behavioral control and less influenced by perceived usefulness. This was also found in this study. For example, through observing users' task performance and screening video interview transcripts, a pattern emerged showing that participants more experienced in manipulating 3D objects in a gaming or computer environment prior to their participation were the ones with a stronger intention of training in VEs. These are also the ones who enjoyed their VE training experience more, and had higher perceptions of system efficacy, compared with those less-experienced participants. On the other hand, more-experienced participants were also more critical of the design features, therefore less tolerable on usability problems. As one participant explained: "Feel good [after training], perhaps if [felt a bit] frustrated because I have far more experience in 3D object manipulation. What frustrated [me] is give me a day or two, I properly get used to it. It is really a matter of getting used to it. I can pretty much do everything."

Forty-eight participants responded to the questions regarding "what is the most positive aspect(s) of VE?" via the perceived VE efficacy questionnaire (PVE). Among these valid responses, "feeling of reality/realism" and "interesting, fun, and enjoyable [experience]" each accounts for 19% ($N = 9$) positive user experience with VT. Further, 8% ($N = 4$) participants reported that they were able to "manipulate 3D objects," and VE is "ease of use and learn." These positive aspects are best summarized by some of the comments made by the participants as illustrated below:

- "Display an environment similar to the real world."
- "The possibility of performing 3D manipulation."
- "It was simple and easy to use, the instructions on the screen were clear."

- "Feeling interested in doing assembly tasks in [the] VE, otherwise wouldn't feel like to do such task in real environment."

In terms of "system- or interface-related" user feedback, responses were diverse. Out of these, 46% ($N = 26$) respondents who had positive perceptions about learning in the VE, some positive feedback was related to the functionality of a particular system interface, for example, HMD (8%, $N = 4$), data glove (8%, $N = 4$), and some were related to the system feedback mechanism, for example, audio feedback (13%, $N = 6$) or force feedback (8%, $N = 4$). Others found overall interaction style of the system (19%, $N = 9$) as the most positive aspect of VT. Scanning user feedback illustrated some evidence to support such results, highlighted below:

- "The audio helped in the virtual training environment as the user could identify that they have correctly placed certain object in place."
- "Force feedback and accuracy was very good."
- "3D mouse, haptic device and data gloves can [enable me to have] excellent in control [of] virtual object."
- "Sound and interaction; force feedback [are useful and adequate]."
- "The whole task [environment] by itself [is impressive], I really enjoyed it, the whole thing."
- "No physical pressure included."

Overall, the subjective responses from the perceived VE efficacy questionnaire (PVEQ) provided rich information for quantifying VT efficacy. Not only did this subjective data have indicated usability problems of the VT design, but also illustrated potential areas for improvement to enhance VT efficacy.

13.5.3 Discussion

13.5.3.1 LOE3D Effects on TTS, SE, and PVE Efficacy In Case study 1, one-way ANOVA is performed and results show that there is significant effect of LOE3D on task performance, that is, $F = 7.586$ and $p < 0.05$. Turkey post hoc test revealed that participants performed task better ($p < 0.05$) under the moderate LOE3D ($M = 82$) than under the high ($M = 85$) and low LOE3D ($M = 43$). Contrary to our expectations, results on SE and PVE revealed no significant effect ($p > 0.05$) of LOE3D. SE and PVE were observed as dependent measures in this study. People with low LOE3D have similar self-efficacy beliefs and perceive VT to be effective as those with moderate and high LOE3D. Mean scores on TTS, SE, and PVE indicate LOE3D have effects on these measures that account for VE efficacy as shown in Figures 13.13a and 13.13b for case studies 1 and 2, respectively. Additionally,

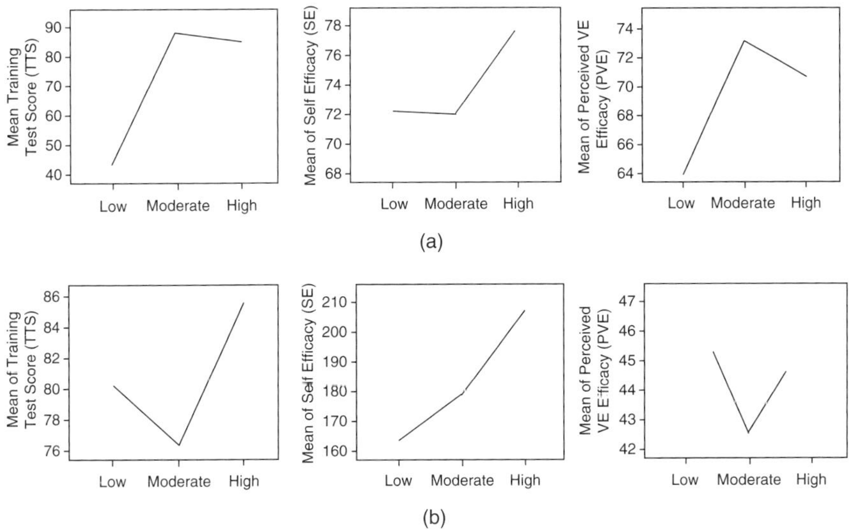

FIGURE 13.13 Effects of LOE3D on TTS, SE, and PVE from (a) Case study 1 and (b) Case study 2.

one-way ANOVA shows that there is no statistical significant difference on participants' ability of recall in memory test (MMT) across LOE3D ($F = 1.852$ and $p > 0.05$). All participants were able to recall learning tasks or procedures at high level, regardless different ranges of LOE3D ($M > 80$).

In Case study 2, Turkey post hoc test revealed that participants had higher self-efficacy beliefs under the high LOE3D level ($M = 12$) than under the moderate ($M = 28$) and low LOE3D ($M = 36$) levels. Results on PVE, TTS, and MMT revealed no significant effect ($p > 0.05$) of LOE3D. Likewise as in Case study 1, all participants were able to recall learning task or procedures at high level, regardless the different ranges of LOE3D. Mean score of learning outcomes measured via SE, PVE, TTS, and MMT are reported in Figure 13.14, which demonstrate high levels of skill-based learning outcomes and perceptions of VE efficacy from the collected user samples.

In addition, it is apparent that through the utilization of a haptic device, which provided realistic force feedback, users were able to engage in object assembly training with stronger sense of "reality." Unlike previous studies that put the entire focus on the design aspect, and overlooked how effective the design was, we assessed the design through two large-scale user testing. Overall, the above results indicate the mixed modalities evoked by the VE engaged users' perceptual, cognitive, motor skills and understanding about the virtual world, thus, providing empirical evidence of training effectiveness using this VT system. The results also highlight users' capability in identification of critical usability problems of the VT

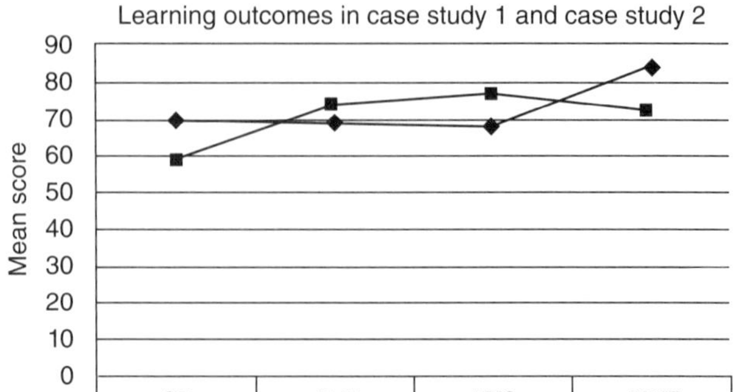

	SE	PVE	TTS	MMT
Case study 1	70	68	68	84
Case study 2	59	74	77	72

FIGURE 13.14 Comparison of learning outcomes from case studies 1 and 2.

system, and they possess a positive general perception of the utility of VEs for training.

13.5.3.2 Contributions The significance of this research is related to the increasingly important role of simulation in training, education, and in the everyday life of people. The studies revealed important and neglected aspects of design and evaluation to the world of simulation, VT design, and the human factor areas of research. For instance, the effective design of computer-based simulation requires designers to strive for high interactivity to facilitate user interaction with learning content, and to build links and to reflect on how the new knowledge integrates with their existing knowledge. Importantly, the simulation should provide "intrinsic" feedback—feedback that comes naturally from the system itself—to users upon their actions. Such form of feedback is very useful in procedural task learning.

It is evident through our studies that an implicit learning mode is included (e.g., using audio, visual, and haptic feedback) which proved to be effective to facilitate user learning of procedural tasks. On the other hand, explicit learning that can be characterized by the formation and refinement of mental models [46] is not easily assessed/obvious, but its association with users' prior experience in manipulating 3D environment (LOE3D) on task performance is apparent. The results reported above should be useful in the design and administration of VT training programs. For example, increasing VT experience (e.g., more time on training), enhancing the system and user interface to be more user-friendly (e.g., requiring little or no knowledge of mathematics), and increasing training support (e.g., encouragement during training) are all useful to maximize training outcomes. Increased experience with computers and the implementation of more user-friendly interfaces were also found to be helpful in improving users' self-efficacy beliefs [47]. The overall work is applicable to wide audience for designing or using simulations and advanced computer technologies that support interaction and learning, as well as to the decision-makers that make use of these technologies.

13.6 LESSONS LEARNT AND FUTURE WORK

13.6.1 Training Design and Method

The work by Popovici and Marhan [48] views learning and training as complex and complementary processes, especially when practical skills are to be acquired, such as for object assembly operations. They discuss that in order for learning to be effective, there should be continuous motivational factors (which most VTs are targeted toward) to engage learners and achieve high levels of focus. When practical skills are to be acquired, training is necessary. As a VT is not a common technology for novice users [39], there is the possibility of developing inaccurate assumptions about their understating of the technology and capability in performing tasks. In Case study 2, after participants underwent four training models, a mixed result for the utility of the haptic device and HMD was found, which may have influenced participants' perceptions of the VT efficacy. The data glove and 3D mouse achieved higher user acceptance, adoption, and rating, compared with the haptic device and HMD. Closer look at the visualization usability and interactivity of the VT confirmed such findings. Overall, user perceived visualization usability (as shown in Figure 13.15) of the VT was high, with a mean rating close to 6 (strongly agree) among five response categories. In particular, intuitive user interface design ($M = 5.9$), effective simulation of training tasks ($M = 5.8$), and clear information representation ($M = 5.8$) of the VT were found, which support such claims according to user feedback. With respect to the HMD for 3D image depth perception in VT, users also reported positive feedback. For instance, users claimed that "I am quite excited ... because this thing [grab the HMD at hand] kind of makes training easier. It tries to make training realistic with what you see ... " and "the 3 dimensional is very real, I think,

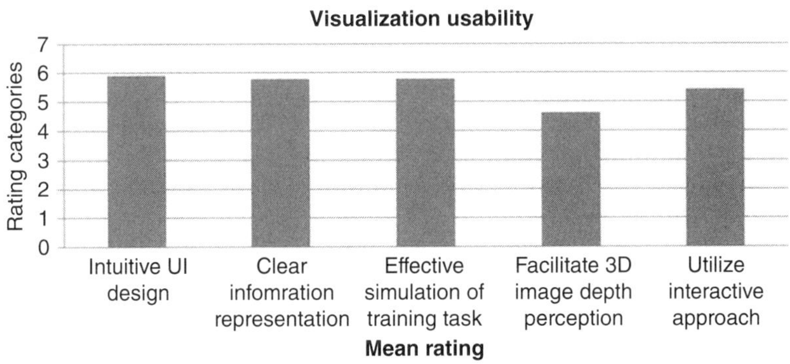

FIGURE 13.15 Visualization usability.

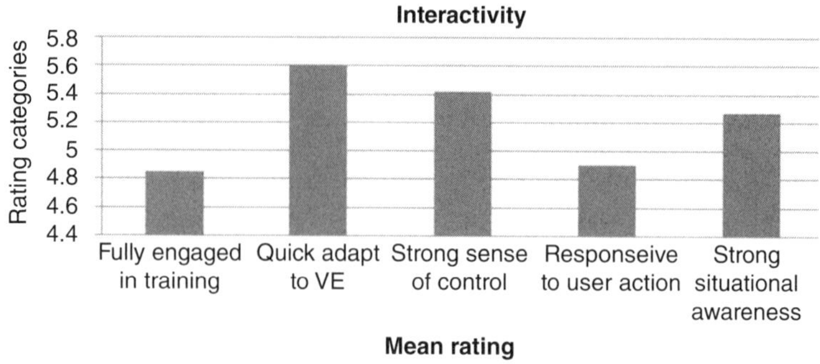

FIGURE 13.16 Interactivity/engagement and control.

just like the real world." Also, a user reported "It [user interface] was simple and easy to use, the instructions on the screen were clear."

In terms of interactivity, participants had positive perceptions toward the VT (as Figure 13.16 illustrates). A feeling of being in control is another indicator of one's confidence in performing a task effectively. Research implies that higher levels of feeling in control is associated with one's performance outcome [49], which was also found in this study. For instance, a user reported "3D mouse, haptic device and data gloves can provide excellent control in manipulating virtual objects." On the other hand, usability problems, such as "perhaps the movement of the glove could be more flexible, I felt a little restrictive with the gripper," that participants encounter in the VT which makes them feel less in control are likely to contribute to a poor performance outcome and deserve attention. High level ($M = 5.6$) of user perceptions on question, "I was able to quickly adapt to the virtual training environment," indicates the user interfaces were easy to use and learn.

Some indeed enjoyed the different experiences that the VT brought to them and indicated that this was due to the unique technological characteristics that made them experience a feeling of "reality." In terms of realism, a user reported "I mean, I like the fact it [training] does mimic the real life. You can pick something up and move it, you can open your hand and drop it, [it] make sense." Interview scripts also suggest that more time on training for users to enable them to be more familiar with the task environment and multimodal information displays may enhance their feeling of immersion and presence.

13.6.2 Measurement Methods

Two perceptually based subjective measures, self-efficacy and perceived VE efficacy, were used to gather user feedback on the existing design and suggestions for improvement. In contrast to recent findings, which have shown a significant relationship between self-efficacy and performance in computer training, this study shows that self-efficacy failed to demonstrate significant effect on task performance in VT. Participants' unfamiliarity with the VT technology may offer a possible explanation for this finding. As a VT is not a common technology for novice users, as Hasan [50] explained, self-efficacy is usually directly linked to performance in computer training, where users are familiar with specific training applications. However, performing VT tasks requires a specific set of skills and cross-domain tasks that it is not possible to capture through self-efficacy beliefs. Thus, it is possible that if the VT was more visible and available for trial or if longer training sessions were allowed, then a significant association between self-efficacy and task performance may be found.

Moreover, a user's perceived VT efficacy has also been shown to have no significant effect on task performance. Research on a technology acceptance model [51] suggests that user perceptions of technology often show an association with technology adaption behavior, whereas no clear evidence suggests an association between a user perception of technology and their ability to perform tasks using the technology. For example, if users perceive a technology to be useful and easy to use, they are more likely to adopt the technology, that is, use the technology to assist their work. However, positive perceptions of the technology may not be attributed to how well they are able to use the technology, especially when the technology is not readily assessable by its users, such as VT, and inaccurate judgments between user perception and performance are likely to occur. Therefore, it is important to incorporate both performance and perception measures when evaluating VTs.

This research has integrated perceptions of self-efficacy and VT efficacy with performance and memory in VT evaluation. Even though the study results illustrated no direct impact by perceived VE efficacy on memory ($r = 0.43$, $p > 0.95$, $N = 19$), a positive association between the two is apparent. Even though user perceptions are often used in VE evaluation accompanied by task performance, this study initiated a specific focus on building a connection between the two types of measures. A lack of studies on user perceptions in a VT system and memory made it difficult to compare our result with the existing literature. Nevertheless, there has been a noticeable increase in awareness of the utility of assessing user memory in terms of their training and interaction experience. For example, Sutcliffe [31] claimed that "post evaluation memory tests could be usefully incorporated into assessment of VEs as a check on the perceived severity of usability problems ($p = .324$)." From the past experience and this study, the inclusion of a memory test is critical in VT evaluation.

Further, this evaluation method also explores users' self-efficacy beliefs on the memory, which to the best of our knowledge, has not been studied in VT evaluation. Interestingly, a negative association was found between the two variables ($r = -0.231$, $p > 0.05$, and $N = 19$). It shows that people with higher self-efficacy beliefs performed poorly on the memory test, which contradicted our expectations. Regression analysis further indicated that self-efficacy could not be attributed as a key predictor of memory tests results, and only 5% of the variance of memory test was attributable to self-efficacy. However, we believe that it is useful to include self-efficacy measures in VT evaluation, and that memory tests could be used to check on the perceived self-efficacy severity of VE efficacy. According to Gist [52] and Mitchell [53], it needs to be acknowledged that many factors can influence actual performance in the time span between assessments of self-efficacy and performance. Thus, measuring learning performance for an extended period of time after training can facilitate better understanding of user perceptions on performance. In addition, the measurement of user-perceived VE efficacy (PVE) involves comprehensive measures of VT efficacy perceptions from the user perspective, and unlike previous research which measured only one or limited factors of user perceptions, the complexity and completeness of the PVE measure/scale were strong.

13.6.3 Prior Experience with a Force-Reflective Haptic Interface

Research has found that users with the prior experience of force-reflective haptic interfaces exhibit higher performance levels and were able to correctly identify three to four sphere sizes with a set of spheres ranging from 10 to 80 mm in radius, in a force-reflective haptic interface employing PHANToM [23]. The results indicated that by gradually increasing stimulus complexity yields more accurate measurements of information transmission at little additional cost in terms of the total number of experimental trials. In addition, Hale [54] found that tactile cues enhance performance when paired with head tracking, and suggested that the multimodal capacity of VE training systems may advance knowledge, skills, and attitudes of trainees. Similarly, Bicchi [55] found that cognitive and physical interaction tightly depends on each other [56] while using haptic interfaces. The authors explained that "physical interaction can help in setting rules for cognitive evaluation of the environment, while cognitive aspects may improve the physical interaction by setting suitable control interaction parameters." ECC [8] agrees and emphasizes that human cognition is embodied in or involves bodily activities and physical interaction between humans and artifacts, which potentially assist human cognition and learning. In this study, users who had experience with a haptic interface to some degree may have been assisted by this when performing tasks.

Furthermore, much work has been accomplished in the past in the area of using position displays to deliver speech information to people with hearing or visual impairment. A significant challenge in the general use of these devices is the amount of training that is required for the user to learn to associate the often unfamiliar haptic sensations with meaningful messages or events. Force-reflective devices have the advantage of delivering information that feels "natural" to novice users [23]. This is supported by this study, since it is possible that a novice user may be distracted by this "unnatural" effect—"feeling through an object" [57] (cited in Reference 23)—that they are not familiar with [23]. In this study, most participants were novices and the majority agreed that operating the haptic device, that is, PHANToM, did not distract them from performing the task, and believed that the VE system was well designed and intuitive to use. In addition, users reported that they not only engaged the task with a strong sense of being visually immersed in the training but they also experienced a "direct" interaction effect with the virtual objects by utilizing PHANToM that enabled a 3D force-vector. Results also suggested that not all participants comprehended the haptic feedback effectively, or use such feedback guide their task performance in the VT.

13.6.4 Future Work

In the future, the effective design of computer-based simulation requires designers to strive for high interactivity to facilitate user interaction with learning content, and to build links and to reflect on how the new knowledge integrates with their existing knowledge. Two main issues should be considered. The first one concerns the design of a learning activity. It must be clear and provide quick and unambiguous feedback in order to flow (an "optimal experience" in which a person enters a mental state of complete absorption or engagement in an interactivity) [49], as flow leads to focused and deep concentration. The second issue needs to be further explored is spatial aspects of interaction related to the instructions and activities conducted by the users provided by the VT. The VT, perceived as spatial surroundings, can be used to greatly enhance the interaction between the user and the world [58]. Spatial behavior, such as moving, orientation, and position in 3D space during locomotion in which human motor capabilities permit movement, is an important part of the HVEI experience [59]. Integration of VE system components such as handheld 3D mouse, data gloves, and 3D trackers can facilitate a user's spatial interaction experience. Overall, increasing training time, enhancing the system and its user interface toward to be more user-friendly (e.g., requiring little or no knowledge of mathematics), and increasing training support (e.g., encouragement during training) are useful for maximum training outcomes.

13.7 CONCLUSIONS

Case study 1 demonstrated that individual's characteristics and VT design features can influence user task performance and cognitive learning outcomes, regardless of levels of expertise. It was also found that individual characteristics such as age, gender, previous experience of computer use, usage frequency, computer use history, VTs, and task/domain expertise, all play a part in user performance and skill-based learning in VTs. Specifically, the experiment described in Case study 1 showed that participants achieved similar levels of object assembly in the VT environment. Clear effects of participants' expertise on task performance were found. Experienced participants of VTs outperformed those with minimal or no experience on the object assembly tasks; and these experienced participants of object assembly in real life achieved a better task performance than those with no such experience. In addition, a high level of task performance and memory test results indicated that the immersive and interactive quality of the VT environment was at a high level, and had positive effects on participants' cognitive and motor learning. Mixed results were gathered based on participants' response to the post-test questionnaire, in which they rated their perceived efficacy of the VT environment.

In Case study 2, a large sample of 76 representative users was involved in controlled experiment settings. The experiment suggested some of the complexities and problems encountered in applying individual difference research to the user-centered design of haptic–audio–visual training systems. This study found that users with a different range of LOE3D had little to no effect on the subjective perceptions of VT efficacy as well as objective task outcomes (TTS) and the ability for recall of learnt skills (MMT). As mentioned above, different levels of prior experience in manipulating 3D were gathered based on the subjective self-reporting of expertise. Even though the empirical study involved a procedure which has some control of a participant's ability to perform object assembly tasks in VT prior to their training, the manipulation of the experimental group may be affected by the self-reporting of demographical information. With respect to LOE3D on self-efficacy, LOE3D of high, moderate, and low ranges may not have been substantial enough to affect a participant's ability to perform training tasks, on the retention test or to rate the VT system efficacy, respectively. Further, VT system feedback has a significant effect on performance. In particular, visual feedback was shown to contribute performance the most, compared with tactile/force feedback and auditory feedback.

REFERENCES

[1] M. C. Salzman, C. Dede, R. B. Loftin, and J. Chen. A model for understanding how virtual reality aids complex conceptual learning. *Presence: Teleoperators and Virtual Environments, Special Issue on Education*, 8: 293–316, 1999.

[2] J. Isdale. Introduction to VRTechnology. In: IEEE Virtual Reality Conference 2003 (VR 2003), 2003, p. 302.

[3] J.-W. J. Lin. *Enhancement of user-experiences in immersive virtual environments that employ wide-field displays*. PhD Dissertation. Department of Industrial Engineering, University of Washington, Washington, DC, 2004, p. 207.

[4] A. Dunser, H. Kaufmann, K. Steinbugl, and J. Cgluck. Virtual and augmented reality as spatial ability training tools. In: 7th ACM SIGCHI, 2006, pp. 125–132.

[5] R. M. Satava. Metrics for objective assessment of surgical skills workshop. In: Defense Technical Information Center, Final Proceedings 20021114249, December 2001.

[6] A. G. Sutcliffe. *Multimedia and Virtual Reality: Designing Multisensory User Interfaces*. Mahwah, NJ: Lawrence Erlbaum Associates, 2003.

[7] M. Steffin. Virtual reality therapy of multiple sclerosis and spinal cord injury: design considerations for a haptic-visual interface. In G. Riva (ed). Virtual Reality in Neuro-Psycho-Physiology, 1997, pp. 185–208.

[8] W. Winn. Learning in artifical environments: embodiment, embeddedness and dynamic adaption. *Technology, Instruction, Cognition & Learning*, 1: 87–114, 2002.

[9] M. C. Salzman, C. Dede, and R. B. Loftin. Virtual reality's frames of reference: A visualization technique for mastering abstract information spaces. In: Proceedings of the Conference on Computer Human Interaction, Pittsburgh, USA, 1999, pp. 489–495.

[10] B. Dalgarno, J. Hedberg, and B. Harper. The contribution of 3D environments to conceptual understanding. In: Proceedings of the 19th Annual Conference of the Australasian Society for Computers in Learning in Tertiary Education, Auckland, NZ, 2002, pp. 145–158.

[11] K. M. Stanney, M. Mollaghasemi, L. Reeves, R. Breaux, and D. A. Graeber. Usability engineering of virtual environments (VEs): identifying multiple criteria that drive effective VE system design. *Journal of Human-Computer Studies*, 58: 447–481, 2003.

[12] T. Manninen. Rich interaction model for game and virtual environment design. PhD Dissertation. Department of Information Processing Science, University of Oulu, Oulu, Finland, 2004, p. 178.

[13] C. Coelho, J. Tichon, T. J. Hine, G. Wallis, and G. Riva. Media presence and inner presence: the sense of presence in virtual reality technologies. In: *From Communication to Presence: Cognition, Emotions and Culture Towards the Ultimate Communicative Experience*, edited by G. Riva, M. T. Anguera, B. K. Wiederhold, and F. Mantovani. Amsterdam, The Netherlands: IOS Press, 2006.

[14] H. K. Padda, A. Seffah, and S. Mudur. Towards a metrics-based framework for assessing comprehension of software visualization systems. In: *Human Interface and the Management of Information: Methods, Techniques and Tools in Information Design*, vol. 4557/2007. Berlin/Heidelberg: Springer, 2007, pp. 335–344.

[15] A. N. Tuch, J. A. Bargas-Avila, K. Opwis, and F. H. Willhelm. Visual complexity of websites: effects on users' experience, physiology, performance, and memory. *International Journal of Human-Computer Studies*, 67: 703–715, 2009.

[16] S. Haque. A meta-analysis of the training effectiveness of virtual reality surgical simulators. *IEEE Transactions on Information Technology in Biomedicine*, 10: 51–58, 2006.

[17] D. P. Ausubel. *Educational Psychology: A Cognitive View*. New York: Holt, Rinehart and Winston, Inc., 1968.

[18] C. Chen. Individual differences in spatial-semantic virtual environments. *Journal of the American Society of Information Science*, 51: 529–542, 2000.

[19] W. H. Hong, J. Y. L. Thong, W.-M. Wong, and K.-Y. Tam. Determinants of user acceptance of digital libraries: an empirical examination of individual differences and system characteristics. *Journal of Management Information Systems*, 18: 97–124, 2001.

[20] D. N. Rapp, H. A. Taylor, and G. R. Crane. The impact of digital libraries on cognitive processes: psychological issues of hypermedia. *Computers in Human Behavior*, 19: 609–628, 2003.

[21] B. Allen. Individual differences and conundrums of user-centered design: two experiments. *Journal of the American Society of Information Science*, 51: 508–520, 2000.

[22] X.-D. Yang, W. F. Bischof, and P. Boulanger. Validating the performance of haptic motor skill training. In: Symposium on Haptic Interfaces for Virtual Environments and Teleoperator Systems, Reno, NV, 2008, pp. 129–135.

[23] H. Z. Tan. Identification of sphere size using the PHANToM: towards a set of building blocks for rendering haptic environment. In: Proceedings of the ASME Dynamic Systems and Control Division, Dallas, TX, 1997.

[24] H. Z. Tan. Information transmission with a multi-finger tactual display. PhD Dissertation, Massachusetts Institute of Technology, Cambridge, MA, 1996.

[25] C. Conati and H. Maclaren. Exploring the role of individual differences in information visualization. In: AVI0'08, Napoli, Italy, 2008, pp. 199–206.

[26] M. Slater, V. Linakis, M. Usoh, and R. Kooper. Immersion, presence, and performance in virtual environments: an experiment with tri-dimensional chess. In: ACM Symposium on Virtual Reality Software and Technology, Hong Kong, 1996, pp. 163–172.

[27] D. Waller, E. Hunt, and D. Knapp. The transfer of spatial knowledge in virtual environment training. *Presence: Teleoperators and Virtual Environments*, 7: 129–143, 1998.

[28] R. Moreno. Does the modality principle hold for different media? A test of the method-affects-learning hypothesis. *Journal of Computer Assisted Learning*, 22: 149–158, 2006.

[29] R. Moreno and R. E. Mayer. Virtual reality and learning: cognitive and motivational effects of students' sense of presence. In: The Ninth International Conference on Human–Computer Interaction, 2001, pp. 65–67.

[30] A. G. Sutcliffe and K. D. Kaur. Evaluating the usability of virtual reality user interfaces. *Behavior and Information Technology*, 19: 415–426, 2000.

[31] A. Sutcliffe, B. Gault, and J.-E. Shin. Presence, memory and interaction in virtual environments. *International Journal of Human-Computer Studies*, 62: 307–327, 2005.

[32] J.-W. J. Lin. User experience modeling and enhancement for virtual environment that employ wide-field displays. In: *Digital Human Modeling*, edited by V. G. Duffy. Berlin/Heidelberg: Springer, 2007, pp. 423–433.

[33] D. A. Bowman, J. L. Gabbard, and D. Hix. A survey of usability evaluation in virtual environments: classification and comparison of methods. *Presence: Teleoperators and Virtual Environments*, 11: 404–424, 2002.

[34] J. F. Jensen. Film theory meets 3D: a film theoretic approach to the design and analysis of 3D spaces. In: *Virtual Interaction: Interaction in Virtual Inhabited 3D Worlds*, edited by L. Qvortrup, E. Granum, B. Holmqvist, S. Kolstrup, and K. Halskov Madsen. London: Springer, 2001, pp. 311–328.

[35] S. G. Schar. The influence of the user interface on solving well- and ill-defined problems. *International Journal of Human-Computer Studies*, 44: 1–18, 1996.

[36] T. Koschmann. Medical education and computer literacy: learning about, through, and with computers. *Academic Medicine*, 70: 818–821, 1995.

[37] J. Vince. *Virtual Reality Systems*. New Delhi, India: Pearson Education, 1995.

[38] D. Jia, A. Bhatti, and S. Nahavandi. MUSTe method for quantifying virtual environment training system efficacy. In: Annual Conference of the Australian Computer-Human Interaction Special Interest Group (CHISIG) of the Human Factors and Ergonomics Society of Australia (HFESA) (OZCHI09), Melbourne, Australia, 2009, pp. 305–308.

[39] D. A. Bowman, E. Kruijff, J. J. LaViola, and I. Poupyrev. *3D User Interfaces: Theory and Practice*. Boston, MA: Addison-Wesley/Pearson Education, 2004.

[40] F. Mantovani. VR learning: potential and challenges for the use of 3D environments in education and training. In: *Towards Cyber Psychology: Mind, Cognitions and Society in the Internet Age*, edited by R. Giuseppe and G. Carlo. Amsterdam, The Netherlands: IOS Press, 2001, pp. 207–226.

[41] J. M. Riley, D. B. Kaber, and J. V. Draper. Situation awareness and attention allocation measures for quantifying telepresence experiences in teleoperation. *Human Factors and Ergonomics in Manufacturing*, 14: 51–67, 2004.

[42] D. Stone, C. Jarrett, M. Woodroffe, and S. Minocha. *User Interface Design and Evaluation*. San Francisco, CA: Morgan Kaufmann Publishers, 2005.

[43] M. Virvou and G. Katsionis. On the usability and likeability of virtual reality games for education: the case of VR-ENGAGE. *Computers & Education*, 50: 154–178, 2006.

[44] L. Chai. To have or have not: an examination of feedback, learner control and knowledge type in online learning. In: The 36th Hawaii International Conference on System Sciences (HICSS'03), Hawaii, AK, 2003, p. 6a.

[45] R. Thompson, D. Compeau, C. Higgins, and N. Lupton. Intentions to use information technologies: an integrative model. In: *End User Computing Challenges and Technologies: Emerging Tools and Applications*, edited by S. Clarke. New York: Information Science Reference, 2008, pp. 79–101.

[46] S. G. Schar, F. Stoll, and H. Krueger. The effect of the interface on learning style in a simulation-based learning situation. *Human-Computer Interaction*, 9: 235–253, 2001.

[47] M. Igbaria and J. Iivari. The effects of self-efficacy on computer usage. *OMEGA International Journal of Management Science*, 23: 587–605, 1995.

[48] D. M. Popovici and A.-M. Marhan. Virtual reality-based environments for learning and training. In: *Product Engineering: Tools and Methods Based on Virtual Reality*, edited by D. Talaba and A. Amditis. Dordrecht, The Netherlands: Springer Netherlands, 2008, pp. 123–142.

[49] J. M. Pearce. An investigation of interactivity and flow: student behaviour during online instruction. PhD Dissertation,Department of Information Systems Department of Information Systems, University of Melbourne, Melbourne, Australia, 2004, p. 342.

[50] B. Hasan. The impact of multilevel computer self-efficacy on effectiveness of computer training. In: *End User Computing Challenges and Technologies: Emerging Tools and Applications*, edited by S. Clarke. Hershey, PA: Information Science Reference, 2008, pp. 33–47.

[51] F. D. Davis. User acceptance of information technology: system characteristics, user perceptions and behavioral impacts. *International Journal of Man-Machine Studies*, 38: 475–487, 1993.

[52] M. E. Gist. Training design and pedagogy: implications for skill acquisition, maintenance, and generalization. In: *Training for a Rapidly Changing Workplace: Applications of Psychological Research*, edited by A. E. M.A. Quinones. Washington, DC: American Psychological Association, 1997, pp. 201–222.

[53] T. R. Mitchell, H. Hopper, D. Daniels, J. George-Falvy, and L. R. James. Predicting self-efficay and performance during skill acquisition. *Journal of Applied Psychology*, 79: 506–517, 1994.

[54] K. S. Hale, K. M. Stanney, and L. Malone. Enhancing virtual environment spatial awareness training and transfer through tactile and vestibular cues. *Ergonomics*, 52: 187–203, 2009.

[55] A. Bicchi, A. D. Santis, B. Sicillian, and A. D. Luca. An atias of physical human–robot interaction. *Journal of Mechanism and Machine Theory*, 43: 253–270, 2008.

[56] S. Ricciardi, M. Nappi, L. Paolino, M. Sebillo, G. Vitiello, G. Gigante, D. Pascarella, L. Travascio, and A. Vozella. Dependability issues in visual-haptic interfaces. *Journal of Visual Languages & Computing*, 21: 33–40, 2010.

[57] T. H. Massie. Initial haptic explorations with the phantom: virtual touch through point interaction. PhD Dissertation, Department of Mechanical Engineering, Massachusetts Institute of Technology, Cambridge, MA, 1996.

[58] M. Nakao, R. Kitamura, T. Sato, and K. Minato. A model for sharing haptic interaction. *IEEE Transactions on Haptics*, 3: 292–296, 2010.

[59] Y. Visell, A. Law, and J. R. Cooperstock. Touch is everywhere: floor surfaces as ambient haptic displays. *IEEE Transactions on Haptics*, 2: 148–159, 2009.

CHAPTER 14

COMPUTER SUPPORTED COLLABORATIVE DESIGN: TECHNOLOGIES, SYSTEMS, AND APPLICATIONS

WEIMING SHEN[1], JEAN-PAUL BARTHÈS[2], and JUNZHOU LUO[3]
[1]National Research Council Canada, Ottawa, ON, Canada
[2]UMR CNRS 7253 Heudiasyc, Université de Technologie de Compiègne, Compiègne Cedex, France
[3]School of Computer Science and Engineering, Southeast University, Nanjing, PR China

14.1 INTRODUCTION

Industries and societies today require new technologies to address increasingly complex design issues for products, structures, buildings, systems, processes, and services while meeting the high expectations of customers. Computer supported collaborative design (CSCD) has emerged in response to this requirement. With the rapid advancement of Internet and Web-based technologies, CSCD was a very active research and development (R&D) area in the past two decades and progressed dramatically. To achieve its full potential, more and more research and commercial CSCD systems have been recently developed. The depth and breadth of these systems and applications are far beyond the traditional definition of concurrent engineering [1]. CSCD is carried out not only among multidisciplinary design teams within the same company but also across the boundaries of companies and time zones, with increased numbers of customers and suppliers involved in the process. This chapter presents a literature review of this R&D area, particularly CSCD technologies, systems, and applications. Research challenges and opportunities on CSCD are also discussed and highlighted.

Contemporary Issues in Systems Science and Engineering, First Edition.
Edited by MengChu Zhou, Han-Xiong Li and Margot Weijnen.

14.2 HISTORY OF COMPUTER SUPPORTED COLLABORATIVE DESIGN

14.2.1 CSCD

Traditional engineering design systems use a sequential mode of design generation, which breaks a design task into a number of sub-tasks that can be sequentially executed in a predefined workflow. Recently, such a sequential design mode has been found to be brittle and inflexible. It often requires numerous iterations, which make design expensive and time-consuming, and also limits the number of design alternatives that can be examined. On the other hand, sequential design is usually practiced with a downstream-wise information flow. Information feedback from downstream operations to the upstream design is usually performed by human interactions. It may thus cause insufficient design evaluation and optimization.

CSCD, also called cooperative design, concurrent design, or interdisciplinary design, is the process of designing a product, structure, system, process, or service through collaboration among multidisciplinary teams associated with the entire lifecycle. In the context of mechanical engineering design, this includes those functions such as preliminary design, detailed design, manufacturing, assembly, testing, quality control, and product service as well as those from suppliers and customers [2]. An important objective of CSCD is to address the insufficient or even absent manufacturability checks concurrently by detecting and considering conflicts and constraints at earlier design stages. To support collaborative design, information and communication technologies are used to augment the capabilities of the individual specialists, and enhance the ability of collaborators to interact with each other and with computational resources. CSCD is not only compulsory for complex products such as the development of the Boeing 777 airplane, which involves 130,000 parts, 6,800 internal people, and more than 10,000 external people, but also quite helpful for many middle- or even small-size products such as tooling and electronic products [3].

With the globalization of the manufacturing industry, CSCD is required to support distributed design. Members on a collaborative team often work in parallel and independently using different engineering tools distributed at remote locations, even across enterprise boundaries and across various time zones around the world. The resulting design process is then called distributed collaborative design [4].

Engineering design has some unique characteristics, for example, diverse and complex forms of information, interdisciplinary collaboration, and heterogeneous software tools, which make interactions difficult to support. Traditional approaches to sharing design information among collaborators and their tools include the development of integrated tools and the establishment of common data standards. These approaches are not good at supporting effective collaborative design because of the highly distributed nature of the design teams and engineering tools as well as the complexity and dynamics of design environments. A successful implementation of CSCD needs: (1) a series of new strategies, including an efficient communication strategy for a multidisciplinary group of people from the design and manufacturing departments to share and exchange ideas and comments; (2) an integration strategy to link heterogeneous software tools in product design, analysis, simulation, and manufacturing optimization to realize obstacle-free engineering information exchange and sharing; and (3) an interoperability strategy to manipulate downstream manufacturing applications as services to enable designers to evaluate manufacturability or assembleability as early as possible [5]. On the other hand, the objective of a design team has multiple facets, for example, optimizing the mechanical function of the product, minimizing the production or assembly costs, or ensuring that the product can be easily and economically serviced and maintained. Achieving global satisfaction, cooperative strategy, such as negotiation, optimization, and trade-off, is an important aspect of CSCD.

14.2.2 CSCD Eve: 1980s

This section provides an overview of the related research fields, including CSCW (computer supported cooperative work), concurrent engineering, and HCI (human–computer interaction), which triggered the emergence of CSCD.

14.2.2.1 Computer Supported Cooperative Work According to Schmidt and Bannon [6], the term *Computer Supported Cooperative Work* (CSCW) was first used by Greif and Cashman in 1984 to describe the topic of an interdisciplinary workshop that they were organizing on how to support people in their work arrangements with computers [7]. Subsequently, the term was abbreviated to CSCW. The definition of CSCW and the history of this research field are beyond the scope of this chapter. Readers are suggested to consult a well-established journal called *Computer Supported Cooperative Work* (CSCW). In fact, many people simply refer to this area by the term of *Groupware*, though others consider this to be too narrow. Generally speaking, *Groupware* is widely used in commercial software products, whereas CSCW is used more often in the research community.

14.2.2.2 Concurrent Engineering The concept of concurrent engineering was initially proposed in the late 1980s as a potential means to minimize product development time. It was defined as "a systematic approach to the integrated, concurrent design of products and their related processes, including manufacture and support" [8]. In a concurrent engineering environment, techniques, algorithms, and software tools are connected to allow product designers and developers to interact with each other. With concurrent engineering, more time and money are usually spent in the initial design stage to ensure the overall optimization of concept selection. Product design changes can be reduced at the late stages, leading to better engineered products with better total quality, time, and cost competitiveness. There are a number of implementation strategies, from the parallelization of product lifecycle functions to the upfront consideration of DFX activities such as design for manufacturability, assembleability, serviceability, and recycleability, to the cooperation and coordination of product design teams with different expertise [8, 9], all of which have laid a solid foundation for CSCD. To ensure the success of concurrent engineering, more emphasis is put on the establishment of team-work culture between design and manufacturing teams, and the enhancement of quick and effective communication. Balamuralikrishna *et al.* [10] summarized concurrent engineering as three T's: tools, training, and time. *Tools* refer to the communication facilities between the personnel in the multidisciplinary departments to address the information exchange that is obstructed by the complexity and wide range of specialized disciplinary areas and interdependent activities. *Training* provides a mechanism for employees to work collaboratively and concurrently, making the best use of the company's resources. *Time* means that corporations need time to carefully investigate and plan concurrent engineering as it involves many complex software tools and information infrastructures. Many reported cases have shown that a hurried implementation of concurrent engineering usually has a high probability of backfiring. In industry, more companies have realized the great benefits of concurrent engineering.

14.2.2.3 Human–Computer Interaction Research on human–computer interaction (HCI) was started as early as computers were invented. Myers [11] presented a brief HCI history. However, there is currently no widely agreed definition that covers a full range of topics that form the area of HCI, from computer graphics to ergonomics, and from virtual reality to digital human modeling.

Computer graphics was born from the use of CRT and pen devices very early in the history of computers. Work in computer graphics has continued to develop algorithms and hardware to allow the display and manipulation of ever more realistic-looking objects—which led to rapid developments of CAD/CAM tools in the 1980s.

There are many HCI-related international conferences with the most widely recognized one being the HCI International Conference Series [12], as well as scientific journals like *Human–Computer Interaction* [13]. Please refer to Reference 14 for more resources on the subject.

With its initial R&D focus on interaction between one user and one computer, HCI R&D was then extended to human–human interaction via networked computers, which is, in fact, the essence of CSCD.

14.2.2.4 Blackboard, DAI, and Software Agents The blackboard architecture was proposed in the HEARSAY project [15] as a means to organize and control large artificial intelligence (AI) systems. Its first version, HEARSAY I was used for speech recognition based on the idea of cooperating independent acoustic, lexical, syntactic, and semantic knowledge sources.

The introduction of the Contract-Net is a milestone in the history of distributed artificial intelligence (DAI). The Contract-Net protocol was developed by Smith [16] and demonstrated on a distributed sensing system. The Contract-Net implemented a negotiation-based approach for allocating tasks dynamically to nodes in the network. When a node has to solve a problem that is beyond its expertise, it broadcasts a task-announcement message that describes the task to be solved. Nodes that receive the message and wish to solve the problem then return a bid message. The node that issued the task-announcement message, called the *manager*, waits for bids for a certain period and then selects one (or more) bidder(s) to do the task, who is called the *contractor*. Thus, the choice of the contractor is done after the selection by the manager and by mutual agreement. To be able to function correctly, the system must include a high-level protocol that defines several types of messages with a structured content. Contract-Net has been widely used in various agent systems for negotiation among agents.

The Contract-Net protocol offered an early practical means for dealing with open systems from a software engineering point of view. Contrary to the blackboard approach, there is no shared memory where data and partial results are made available to the various knowledge sources. The Contract-Net approach has separate knowledge sources attached to distinct nodes in a computation network.

The actor model proposed by Hewitt [17] offers a model of computation for open systems at a finer grain than the Contract-Net approach. In the actor approach, problem solving is viewed as the result of the interaction of the activities of knowledge sources working independently and locally (with limited expertise and limited knowledge). Each node communicates with a limited number of other knowledge sources.

The concept of agents has evolved from the concepts of blackboard, Contract-Net, and actors. Separately, in such applied fields as manufacturing, object-oriented systems were developed with increasing intelligence being incorporated into the objects. What began as passive objects became "active objects" or "rule-based objects" or "intelligent objects," and finally "intelligent agent objects" as this stream of evolution merged with that of DAI [18]. All these technologies provide a good foundation for developing collaborative design systems. Under the CSCD context, an agent can be considered as a software system that communicates and cooperates with other software systems to solve a complex problem that is beyond the capability of each individual software system [19].

14.2.3 CSCD Emergence: 1990s

Modern design, particularly engineering design, is intrinsically multidisciplinary [20]. Various tools such as CAD/CAM/CAE (Computer-Aided Design/Computer-Aided Manufacturing/Computer-Aided Engineering) tools, developed and commercialized by different vendors without common specifications (or even with intentionally defined unique specifications for self-protection), do not address the needs of multidisciplinary design. On the other hand, large organizations, for example, Boeing, Airbus, or GM, must find a way to coordinate their R&D teams that are geographically distributed around the world in an effective way to carry out new product developments within a very limited time frame. Technologies like CSCW and intelligent agents have been investigated to meet this need, particularly to enhance communication, cooperation, and coordination among design team members as well as software tools. Some CSCW tools, like groupware, were directly used to facilitate communication among engineers and designers. Software agents were used to integrate different engineering tools. Examples of early applications of software agents in collaborative design include PACT [21], DIDE [22], and SiFAs [23].

With its emergence around 1993, the Web was quickly applied in the development of collaborative design systems, particularly for geographically distributed designers to share design documents. Along with the Web, a number of associated representation technologies have been developed, such as HTML (Hyper Text Mark-up Language), XML (eXtensible Mark-up Language), VRML (Virtual Reality Mark-up Language), to enable better cross-platform and cross-enterprise exchange of multimedia information and design models. In terms of system infrastructure, many early collaborative design systems were also developed using the Blackboard architecture and distributed object technologies, like CORBA and COM/DCOM.

14.2.4 CSCD: Today

During the past two decades, a large number of CSCD systems were developed and reported, especially on the applications of CSCW, Web, software agents, and recently Web Services, Semantic Web, and Computing Grids, for collaborative design. A few CAD vendors and other software firms also started developing and promoting collaborative design tools, for example, AutoDesk's Inventor™ [24], Buzzsaw™ [25], Streamline™ [26], ArchiCAD TeamWork™ [27], Creo/elements Direct (formerly CoCreate's OneSpace Solution) [28], Dassault Systèmes' ENOVIA (formerly Matrix PLM Platform) [29], or UGS' PLM solutions [30].

CSCD has been one of the most important applications of CSCW technologies. The most widely used CSCW techniques in collaborative design systems include *groupware* techniques for facilitating communication among design team members and *context awareness* techniques for enhancing coordination among team members.

Web-based and agent-based approaches were dominant in the implementation of CSCD technologies, systems, and applications during the past two decades, and will be further discussed in the following sections.

14.3 METHODS, TECHNIQUES, AND TECHNOLOGIES

14.3.1 Communication, Coordination, and Cooperation

Communication, coordination, and cooperation are three fundamental aspects of CSCD technologies. Communication enables sub-systems/software agents/human design team members to exchange information and to coordinate their activities. Communication is the most important way in which cooperation and coordination among sub-systems/agents/people takes place. A detailed discussion can be found in Reference 31.

14.3.1.1 Communication Communication is crucial in distributed collaborative design systems. Indeed, the complexity of building a single system has been traded for a reduced complexity of each sub-system or agent. However, the difficulty has been transferred in part to the level of communication and to the issue of combining the separately achieved sub-tasks through cooperation and coordination.

Communication can be implemented in very different ways, depending on the nature of the sub-systems or agents, the global architecture of the system, the timing of the exchanges, or the number of receivers for a message. Indeed, communication can occur between humans and/or machines. It can be direct by means of messages or indirect by posting; it can be synchronous or asynchronous; and it can concern a single sub-system/agent or several sub-systems/agents.

Communication can be direct or indirect. In multi-agent systems communication is direct between agents by using explicit messages. In client/server-based systems, communication is indirect by writing to the blackboard (which, however, is equivalent to a multicast communication strategy in the case of partitioned blackboards; broadcast otherwise). In the latter case, the communication mechanisms can be quite simple, like the SHARED workspace [32] in the DICE project.

Communication can be synchronous or asynchronous. Indeed, people work together via computer network in different modes: (1) co-located synchronous mode—they work in the same location and at the same time; (2) distributed synchronous mode—work in different sites but at the same time; (3) co-located asynchronous mode—they work in the same location but at different times; (4) distributed asynchronous mode—they work in different locations and at different times. Each mode requires hardware and software support for effective and efficient operations. It is important for users to recognize the distinctiveness of each mode, since their protocols, network, and storage requirements vary. In large design projects, both synchronous and asynchronous communications are required. Synchronous communication can be useful for teleconferencing (e.g., for design review meetings) [33]. However, most of the work will be done through asynchronous interactions.

The overall architecture of a design system has a strong influence on the way communications are organized. In client/server-based systems, the blackboard architecture has a tendency to impose indirect, synchronous, and general (broadcasted) communication. In multi-agent systems on the other hand, we find mostly direct, asynchronous communications with a variety in the number of receivers (point-to-point, multicast, and broadcast). The nature of the agents is somewhat orthogonal to the architecture of the system. Thus, we have several kinds of systems: (1) systems for supporting the communication between human designers and facilitating their access to various tools; (2) systems having mostly automated software agents; (3) systems having both human designers and automated software agents where communication occurs among the designers, among the agents, and between designers and agents. In the second category, human designers are of course present. However, the corresponding system usually has a single user, and the communication between the user and the system is traditional.

In order to support communication it is necessary to develop protocols. There are two levels of protocols: one for securing the communication process, another for supporting the exchange of information. They are supported by distinct communication languages. Among the protocols of communication proposed or developed, there exist some simple protocols such as Contract-Net [16] with a number of variations. The protocols supporting the exchange of information include a proposition from Gaspar [34] offering a protocol with four types of messages, each corresponding to the communication modes among agents; a more complex protocol like SANP [35], based on the results of linguistic researches; and KQML (Knowledge Query and Manipulation Language) [36].

14.3.1.2 Coordination According to Lesser and Corkill [37], "the objectives of the coordination process are to ensure that all necessary portions of the overall problem are included in the activities of at least one agent, that agents interact in a manner which permits their activities to be developed and integrated into an overall solution, that team members act in a purposeful and consistent manner, and that all of these objectives are achievable within the available computational and resource limitations."

Large design projects include complex processes distributed across time, involving many participants and functional perspectives. The design of a new commercial jet, for example, requires the integrated contribution of thousands of individuals spread over several continents and spanning possibly decades of time. Effective coordination is critical to the success of this cooperative process since the distributed activities are typically highly interdependent due to shared resources, input–output relationships, and so on. The sheer complexity of these interdependencies has begun to overwhelm traditional manual organizational schemes and paper-based coordination techniques, resulting in often huge rework costs, slowed schedules, and reduced product quality. As a result, while individual productivity may be high, the failure of existing coordination support practices and technologies has severe impact on the bottom line [38].

Coordination is central to the successful operation of CSCD systems, particularly to agent-based collaborative design systems whose complexity is high and stability is essential. Without coordination, a group of agents can quickly degenerate into a chaotic collection of individuals. The easiest way of ensuring coherent behavior would be to provide the group with an agent that has a wider perspective of the system. Such an agent then becomes the central controller of the system. This central controller could gather information from the agents in the group, create plans, and assign tasks to individual agents in order to ensure coherence. In the PACT project [21], the controller role is played by agents called facilitators, in CONDOR by collective agents [39], and in ABCDE by managers [40].

Using a single central controller for a large group of agents has an obvious problem. As the size of the group increases it becomes very difficult for the central controller to be informed of all the agents' beliefs and intentions. Also such a controller can become a severe communication bottleneck and would render the remaining components/agents unusable if it failed.

Based on the comprehensive review of the literature, we summarize a number of fundamental mechanisms and techniques that can be used to support multi-agent coordination [31]:

- *Mutual adjustment*: Agents share information and resources to achieve some common goal, by adjusting their behavior according to the behaviors of the other agents. In order to do so, agents usually need to exchange considerable information and make many adjustments. No agent has prior control over the others and the decision-making is a joint process. It is the simplest form of coordination. Coordination in peer groups and in markets is usually by mutual adjustment.

- *Direct supervision*: One agent has some degree of control over others and can control the use of shared resources by the subordinate agents and may also prescribe certain aspects of their behavior. Such a supervision relationship is often established through mutual adjustment (e.g., following acceptance of employment or a contract).
- *Coordination by standardization*: The supervisor agent coordinates other agents through standardization, by establishing standard procedures for subordinate agents to deal with situations they encounter. In mutual adjustments, such standard procedures are implemented by acceptance. While in direct supervision, they are implemented through (mandatory) requests. Routine procedures in companies or computer programs are examples of coordination by standardization.
- *Mediated coordination*: A mediator acts as a facilitator (e.g., finding or delivering information), and as a broker (a "go-between") and advisor on resource negotiations or supervisor (exercising some degree of direct supervision). The first role is mandatory, but the others are optional. A mediator facilitates or brokers mutual adjustment among agents and may also use direct supervision.
- *Coordination by reactive behavior*: Agents react to particular stimuli (situations) with specific behaviors (actions). With appropriately selected or evolved stimuli–behavior groupings and distributions, system-level patterns of coordinated behavior emerge so as to contribute to the achievement of common or system goals.

There are also other mechanisms for multi-agent coordination described in the literature such as Markov tracking [41], coordination techniques dealing with commitments and conventions [42], and knowledge-based reasoning techniques [43].

14.3.1.3 Cooperation Collaborative engineering design projects, particularly concerning those large and complex products and structures such as automobiles, locomotives, and airplanes, require the cooperation of multidisciplinary design teams using multiple sophisticated and powerful engineering tools such as commercial CAD/CAE tools, engineering database systems, and knowledge-based systems. Individuals or groups of multidisciplinary design teams work in parallel and independently with various engineering tools located at different sites. At any moment, individual members may be working on different versions of a design or viewing the design from various perspectives (such as profitability, manufacturability, resource capability, and capacity) at various levels of details. Effective cooperation is imperative.

In general collaborative systems, the cooperation can be fully cooperative, partly cooperative, and antagonistic. However, in CSCD systems there is usually no antagonistic situation, that is, it is either fully cooperative (particularly within the same company) or partly cooperative (usually among collaborative companies or stakeholders).

Fully cooperative sub-systems/agents that are able to resolve non-independent problems often pay a price in high communication costs. These sub-systems/agents may change their goals to suit the needs of other sub-systems/agents in order to ensure cohesion and coordination. Most real CSCD systems have a weak degree of cooperation, that is, they are partly cooperative, in order to reduce the communication cost and design time.

Fully cooperative agents exist in cooperative distributed problem solving (CDPS) systems where agents work together in a loosely coupled network to solve problems that are beyond their individual capabilities. In this network, each agent is capable of contributing to solve complex problems and can work independently, but the problems faced by the agents cannot be completed without full cooperation. This level of cooperation is necessary because no single agent has sufficient knowledge and resources to solve a given problem, although different agents might have the expertise to separately solve its different parts (i.e., sub-problems). CDPS agents cooperatively build a solution to a problem by using their local knowledge and resources to individually solve sub-problems, and then by integrating the solutions for these sub-problems into an overall solution. CDPS is particularly useful when distributed control becomes uncertain. The presence of uncertainties, such as incomplete and imprecise information, makes the need for full cooperation even more crucial. CDPS is also useful when each agent has specific knowledge and does not have a clear idea about what it might do or what information it might exchange with others. In this context, Smith's Contract-Net [16] provides a cooperative framework to minimize communication and allow load balancing and distributed control, while also maintaining coherent behavior.

While there are different cooperation methods [44] in the literature for general collaborative systems, cooperation methods widely used in CSCD systems include coordination as discussed above and conflict resolution through negotiation as described below.

14.3.2 Negotiation and Conflict Resolution

Conflicts occur in multidisciplinary collaborative design environments for two main reasons: individual participants in the cooperative design process lack the complete knowledge of global objectives necessary to ensure that their locally preferred solutions are in alignment with these higher-level objectives, and individual disciplines have individual ideas about what constitutes the best design. Even individual design requires trade-offs because of competing design criteria, such as safety, cost, and social acceptance, as well as requirements and specifications. The ability of human designers to avoid or minimize conflict through judicious trade-offs and other methods is one of their most valuable skills.

Conflict detection and resolution are especially difficult when the design task as well as the knowledge concerning competing factors are distributed among different actors with different perspectives. Methods like negotiation, hierarchical structures, constraint weakening, creation of new variables, or user interaction, can be used for conflict resolution. It is also possible to combine several such methods within the same system.

Researchers have developed conflict resolution strategies for various types of conflicts: introducing an approach to conflict resolution based on classifying the conflict and mapping it to a specific strategy [45]; resolving conflicts in resource allocation among cooperative agents [46]; resolving conflicts via negotiation [47–49]; assisting the designer by using expert systems that offer criticisms and suggestions concerning design decisions in SNEAKERS [50].

Two interesting examples can be found in the project NEXT-LINK [51] and in the project SHARED-DRIM [52]. NEXT-LINK uses a simple and ubiquitous notion of constraint to represent all conflicts and provides a support for human-guided conflict resolution. Next-Link architecture describes all conflicts in terms of constraint violations, using the Redux ontology [121]. Redux describes an AI search-based ontology that maps directly onto an engineering task-based ontology. A constraint violation in NEXT-LINK simply means that some combination of assignments has been declared to be inconsistent. This is not a logical notion of consistency, but rather it is dependent upon the domain semantics. Redux, as a domain-independent agent, has no notion of such semantics. So, it does not insist that every constraint violation should be fixed immediately, nor does it draw any inference from such violation. What Redux does do, given a conflict among a set of assignments, is to determine the AND/OR set of decisions supporting the set, and then notifies each of the agents involved in the constraint violation.

On the contrary, SHARED-DRIM [52] uses an automatic conflict-resolution technology for resolving the known conflicts when solutions are available in its knowledgebase. In SHARED-DRIM, human design agents use the constructs provided by DRIM to make their recommendation. Every time they make a recommendation, the system searches to see if there exists a recommendation that introduces/modifies the same object in the database. If there is no such object, the system creates the object. However, if the object exists in the database, the system retrieves it for comparison. If an inconsistency is found, the system informs all interested participants about the changes made to the object. In order to provide the causes of conflict, the system must isolate the model entities needed to understand the cause of change. This process is achieved by the hierarchy division function. Hierarchy division selects the part of an intent hierarchy from which the conflict under consideration can be explained. After the system has reached the level of detail or abstraction in which a source for the conflict can be signaled, it begins to explore the validity of the recommendation that is related to the source. To do this, the system uses a function called consistency checking. By detecting inconsistencies in any object and providing the causes of conflict, the system instantiates a negotiation process between all involved professionals. In this negotiation process, the computer may take an active or passive role. An active role implies that the system has access to knowledge that prescribes how to relax a specific type of constraint. On the other hand, the computer may take a passive role by providing design agents only with information on the existence of the conflict and the decision which led to the problem. The human design agents then make the decisions.

We believe that it is not possible to conduct large complex design projects while enforcing consistency at all levels. Thus, conflicts are inherent to a design process. Consequently, conflict resolution is a domain that will see many developments in the future.

14.3.3 Ontology and Semantic Integration

Collaborative engineering design projects are complex, involving many different stakeholders, and require sharing knowledge independently from the use of information and communication technologies. Ontological engineering is a new discipline that allows descriptions of the world to be handled within human–human, human–machine, and machine–machine communications, irrespective of the particular computing environments in which these occur.

Ontologies are a key element of CSCD. They both define and organize all concepts being used in the collaborative design process at different levels. Several ontologies are usually involved in the design process: (1) a domain ontology modeling all concepts to be found in the application domain; (2) a process ontology modeling concepts related to the process itself. A domain ontology helps different specialists to translate among the various professional jargons concerning the product domain. It allows indexing project documents semantically for easy retrieval. It can also be used to organize knowledge elements when capitalizing knowledge [53]. It also allows querying knowledge, either internally or externally through semantic Web Services [54]. A process ontology can be used as a backbone to organize the design process and provide context during different design phases. Combination of a domain ontology and a process ontology can be used to support natural language interaction, for example, between a human and his/her personal assistant agent.

Finally, because of globalization, ontologies need to be multilingual to support interaction in multi-cultural environments.

Unfortunately, it is difficult to build ontologies, to maintain them, and to use them correctly. The current popular formats like OWL produce structural information. Reasoners like RACER, Pellet, FACT++, Flora, JENA, JESS, Bossam, or SPARQL are defined outside the ontologies, which limits their efficiency. A comparison between reasoners was done by Huang *et al.* [55]. Significant R&D efforts are still required in this area.

14.3.4 Personal Assistance and Human–Machine Interaction

In the early 1970s, minicomputer-based CAD systems were composed of workstations including a graphics screen, a keyboard, a graphics tablet and stylus, and sometimes a large digitizer. At the end of the 1970s, engineers gathered around such workstations to work together, discussing new design around 3D wire mesh mock ups of a target product, for example, organizing the piping network of a power plant heat exchanger or organizing cranes on a construction site. Input could be done by means of graphical commands drawn on the tablet. Such small groups of designers were involved in co-located cooperative design, using the system as a recording device.

Later on, networks were introduced and one could envision distributed cooperative design. Today both organizations exist. They have different requirements.

14.3.4.1 Co-located Design Co-located design involves several people working together in the same environment. The tools must: (1) support the design process; (2) record design decisions; and, (3) prepare the final documentation [56]. Project review meetings are of this type.

First, CSCW human–machine interaction (HMI) benefits from all hardware novelties that have appeared on the market or are developed in the laboratories, starting from traditional screen/keyboard/mouse/menus interaction devices. Thus, we have seen the development of wall screens, virtual reality devices (CAVE, helmets, goggles), haptic devices, interactive boards, graphic surfaces, vocal inputs, tablets, smart phones, and 3D screens, and there will be clearly more to come.

The use of constructive solid geometry (CSG) and the possibility of displaying the designed product through 3D digital mock ups (DMU) using wall panels increased the quality of interaction for complex products. One step further was brought by collaborative augmented reality, allowing users to manipulate 3D objects in a virtual environment [57]. Today wall-size screens are replaced by large interactive touch panels. If different devices have been tested to interact with wall screens like magnetic rings or bracelets for pointing at the wall, an interesting approach consists in using laser pointers [58]. Haptic devices have been introduced to manipulate virtual 3D objects by means of gloves providing force feedback [59]. Sound and voice interfaces are proposed, for example, in the CALO project that tested a vocal interaction used in meetings during which participant interventions are recorded and the system is able to produce a summary of the discussion at the end of the meeting [60]. Voice interfaces and natural language interaction can also be used to select tasks, triggering specific dialogs before launching a particular task [61, 62]. The availability of graphics surfaces (tables) now offers the possibility of having multi-user multi-touch interaction [63]. Interaction is done through gestures [64] that still need to be standardized.

In the last 20 years, the development of the multi-agent technology brought major changes both to the architecture of the CSCW environments and to the HMI. The concept of personal assistant emerged slowly and now can be specialized to personal design assistants acting as specific technical secretaries that help doing the tedious repetitive work [65]. In some cases, personal assistant agents can be embodied in talking avatars [66]. The concept of digital butler proposed by Negroponte [67], consisting of giving a personal assistant a staff of specialized agents, allows an easier expansion of the system by adding more specialized agents as needed [68].

All the above devices need to be organized in consistent systems, and proposals have been done to develop systems containing wall screens and touch surfaces like the Roomware® approach [69] setting up a room combining the different devices, or the TATIN-PIC project combining multi-user multi-touch surfaces with vocal input and personal devices like tablets or smart phones for supporting preliminary cooperative design [63]. An interesting initiative is the PAL program proposed by DARPA with the goal of organizing a command center room for preparing (designing) missions cooperatively [70].

Second, interaction devices must be integrated and used in the context of design, meaning that supporting software has to be developed to this end. This includes 3D interaction, multimodal interaction, virtual reality, avatars, personal assistants, and much more.

As mentioned previously, CSG has been used to build 3D mock ups, which can be used to index design knowledge by attaching information [71]. Multimodal interaction includes now gestures that are charted to produce standards [64]. Multi-agent technology has been introduced as a supporting architecture, in particular with the definition of personal design assistants [65, 72].

14.3.4.2 Distributed Cooperative Design Distributing cooperation brings additional requirements with respect to co-located activities, namely exchange of documents, organization of the work, for example, version control, addressing the problem of awareness or coordination.

Distributed cooperative design was first developed around groupware and in particular document management (DM) systems like Lotus Notes® [73], as sharing documents is a prerequisite to collaborative work. Many DM products are today available including open source ones like Alfresco [74], Nuxeo [75], Zotero [76], or others. Another important requirement, in particular when developing software, is version control, which called for approaches like CVS [77], SVN [78], Git [79], and many others. Some researchers claim that such traditional systems are becoming inadequate and should be replaced by systems allowing real time collaborative editing like Google Wave [80]. Another problem when people are not co-located is that of awareness. Awareness of what other people are doing remotely needs be enforced [81, 82]. Finally, coordination is also more difficult to achieve when teams are distributed. This can be improved by the use of design assistant agents [83].

14.3.5 Collaborative Workflows

Large organizations must find a way to coordinate their R&D teams (distributed geographically around the world) in an effective way to carry out new product/process/system developments within very limited time framework. This requires not only simple integration of engineering tools but the seamless integration of the way they do their job, the efficiency of data sharing and transfer, and the awareness and consistency of design changes. Ambiguity in the description of roles, responsibilities, interactions, and processes makes collaborative design very difficult. Change management and product data coordination are two challenging issues as well.

Therefore we need to find a feasible technology to tackle the following requirements:

- integration of heterogeneous engineering tools installed in different environments at different locations;
- integration of multidisciplinary teams that may transverse organization boundaries;

- integration of segments of business logics of different engineering and manufacturing activities such that they can function as a whole business;
- effective data management and data exchange standards; and
- change management and version control.

We believe that three major technologies make up the collaborative workflow technology:

(1) web-related technology, such as WWW, Web Services, and Web Semantics, for a convenient media to publish, share, and exchange information and for providing unit engineering and manufacturing services over the Internet;
(2) intelligent agent technology as a semantic "glue" for adaptive process coordination and providing active assistance to a multidisciplinary team; and
(3) process modeling and coordination technologies—workflow.

Web is a convenient media to publish and share information relevant to the spectrum of the design process, from concept generation and prototyping to virtual manufacturing and product realization. Web-based infrastructure has been used in a number of collaborative design systems. In most cases, it is primarily used by multidisciplinary team members as a medium to share design data/information/knowledge and in some cases for product data management and project management through integration of the Web with the related technologies.

Web technology alone, however, is not the solution to collaborative design systems, although it makes communication physically viable through a common network. In order to work on a distributed project effectively, remote engineers and designers need active helps to coordinate their efforts. This coordination may involve the translation of terminology among disciplines or locating/providing engineering services (e.g., finite element analysis). Agent technology provides support to enhance the performance of collaborative design systems [31]. In agent-based collaborative design systems, agents have mostly been used for supporting cooperation among designers, providing a "semantic glue" between traditional tools, or for allowing better simulations. Shen *et al.* [31] provided a detailed discussion on issues in developing agent-based collaborative design systems and a review of significant, related projects or systems.

Both agent technology and Web technology are very useful in the integration of design processes. The collaborative workflow strategy still requires for a third technology called workflow. Workflow management is a fast evolving technology which is increasingly being exploited by businesses in a variety of industries. Its primary characteristic is the automation of processes involving combinations of human- and machine-based activities, particularly those involving interaction with information technology (IT) applications and tools. In recent years, the need for reorganization, continuous improvement of business processes, and the advances in IT have created a huge market for workflow management systems. Modern workflow management systems enable the design, enactment, and monitoring of workflow in a heterogeneous and distributed environment, allowing efficient process execution and management [84]. Very recently, Kim [85] presented a model-driven workflow fragmentation framework for collaborative workflow architectures and systems.

14.3.6 Collaborative Virtual Workspaces and Environments

With the integration of virtual reality, software agents, and Internet/Web-based technologies, collaborative virtual workspaces or virtual environments are being widely applied in almost all e-business and engineering domains for collaboration among distributed teams.

Rosenman *et al.* [86] presented a framework for collaborating in a virtual environment including an IFC-based database containing the various models and relationships, a virtual world environment for collaboration, and an agent-based society for handling communication among the users.

Aspin [87] proposed an interaction mechanism that enables a group of co-located users to collaboratively interact with a common visual environment through the use of lightweight remote computing devices. Applying an object-based distributed shared memory system enables the description of the active sessions to be distributed to both the collection of services, forming the design/review session configuration, and the remote interface applications that support individual user interaction. This distributed system then forms a synchronized, distributed description of the session content that both informs services of the session content and provides a centralized system for managing user interaction.

In an interesting experimental work, Hammond *et al.* [88] used a socio-technical theory as a framework to explore differences in engineering design team decision-making as a function of various communication media. Their results indicate that design teams communicating via an electronic medium perceive an increase in mental workload and interact less frequently, but for a greater total amount of time. These results brought interesting implications and suggestions for the management of distributed design teams or the management of human aspects [89].

14.3.7 New Representation Schemes for Collaborative Design

In a collaborative design process, product models need to be disseminated in a broader scope. Product models are the most important knowledge and properties of the product development companies. As a result, companies are usually reluctant to share these models directly to avoid the leakage of the commercial secrets to competitors. This consideration makes it difficult to realize the full potential and benefits of collaboration. On the other hand, a product model is proprietary to a CAD system. In a collaborative design environment with multiple users, it is infeasible or uneconomical to install a CAD system for every user to view or manipulate the product model. To address these concerns, research efforts have been made to develop new representation schemes of product models based on VRML, including X3D (eXtensible 3D) [90], Web 3D [91], U3D (Universal 3D) [92], JT [93], or OpenHSF [94]. These representation schemes retain the essential visualization information of proprietary product models to support display-based manipulations, such as rotation and zooming, annotation, and mark-up. Most of these schemes are open in formats and the features inside are neutral such that they have much broader acceptance than those of the proprietary product models. Major applications of these schemes for collaboration include customer surveys of product concepts and initial models, high-level project reviews among management, development and service departments, sales promotion, e-documents (e.g., Acrobat 3D), sharing catalogs, and visualization functions in product data management/product lifecycle management (PDM/PLM) systems. Since only the visualization information is included in these schemes, crucial design information is protected.

14.3.8 New Visualization Systems for Collaborative Design

In order to support the new representation schemes, some new visualization systems have been developed, for example, Cimmetry Systems AutoVue [95], Actify SpinFire [96], SolidWorks eDrawing [97], RealityWave ConceptStation [98], and Autodesk Streamline [26]. The visualization-based platforms are cost-effective solutions to replace CAD systems to facilitate collaborative activities for various users. With new representation schemes and visualization systems, teams can collaborate more effectively, such as by taking on design discussions, reviewing new products, and conducting customer surveys to get design feedback as early as possible. This may overcome some drawbacks of proprietary CAD product models that hinder collaborative activities. A visualization-based collaborative system uses a two-tier or three-tier client/server architecture. Java Applet and Microsoft ActiveX technologies are widely used for developing Web-based or specialized clients. Core functions or services are implemented in Java Servlet or Microsoft .Net ASP at the server side to provide system support and maintenance [99–101]. Recently, Java3D has been widely used to enable visualization-based manipulations of 3D objects and scenes, for example, to build, render, and control 3D objects for Web-based collaboration [102, 103].

14.3.9 Product Data Management and Product Lifecycle Management Systems

Product data management (PDM) and product lifecycle management (PLM) systems have been adopted by industry to facilitate engineering design. Such systems promise that the "right information" is provided to the "right person" in the "right time" according to the "right order." Mainstream solutions include UGS TeamCenter [104], PTC Windchill [105], ENOVIA VPLM, ENOVIA MatrixOne, and ENOVIA SmarTeam [106]. Actually, the systems can be regarded as the system-level integrated implementation of the current collaborative technologies to support engineering design [107]. These systems have distinguished characteristics, while sharing the following common functionalities:

- team management—to map the structure of a product development team to a hierarchical structure of organizations;
- product structure management—usually a bill of materials (BOM) structure root, to represent the physical structure of a developed product at different levels, which generally contains assemblies, sub-assemblies, and components;
- workflow and process management—to allow an organization to automate procedures in which information, tasks, and documents are passed among participants;
- design change management—to manage change information in design processes;
- visualization-based collaborative workspace—to retain the visualization information of product models based on light-weight visualization schemes to support multiple users to manipulate the product models, such as rotation, measurement, annotation, and mark-up; and
- integration interfaces with CAD, shop floor execution systems, legacy enterprise information systems, and other partners on the product value chain.

From a research perspective, important issues need to be solved for the better application of PDM/PLM systems. The design process, along with the product itself, should be considered as a crucial component of an engineering enterprise's intellectual capital [108]. Five aspects of design processes have been studied, including support for design information transformations, support for design decision-making, modeling and representation of design processes, analysis of design processes, and synthesis of design processes. Qiu and Wong [109] developed a dynamic workflow mechanism to accommodate the changes during design by minimizing the repetitive execution of finished workflow nodes. This approach can address the data integrity issue by managing various workflow data such as node properties and scripts. Concurrency control is the foremost mechanism to organize synchronous activities to avoid conflicts. Locking is a primary means in managing concurrency control to prevent people from colliding, and three types of locking, that is, non-optimistic, optimistic, and visual, have been developed and used in various applications. Negotiation can formalize and implement the mediation and facilitation functions among people to handle conflicts. Some research projects [107, 110, 111] have been carried out to enhance PDM systems to support pre- and post-design stages. Huang *et al.* [112] developed a Web-based system to manage engineering changes.

14.3.10 Security and Privacy

A major concern of implementing network-enabled collaborative design systems is the assurance that proprietary information about the intellectual property owned by an organization or information about the company operations is available to authorized individuals only. Collaborative design involves sharing intellectual property in the form of detailed design information as well as competitive enterprise information. For general acceptance of the collaborative design approach, the secrecy of the proprietary or competitive information must be maintained.

In addition to maintaining secrecy, collaborative design systems must also accommodate privacy of the individuals and organizations involved in collaborative design activities. Gathering and processing information about the activities of individuals or groups while managing or operating processes or machinery via computer networks can provide a great deal of details concerning the ways in which the individuals interact as well as process-related information. In a highly competitive business world, we must ensure that business intelligence information or the information provided by individuals or organizations is only shared in a fashion dictated by those involved.

14.4 COLLABORATIVE DESIGN SYSTEMS

14.4.1 System Architectures

In all kinds of collaborative design systems, two types of system architectures are dominant: the client/server architecture used mostly in web-based collaborative design systems as discussed in detail in Section 14.4.2, and federated system architectures used in most other collaborative design systems, including agent-based collaborative design systems and service-oriented collaborative design systems.

The client/server system architecture has been popular in traditional distributed computing systems, particularly web-based systems. A number of federated architectures have been proposed in the literature. Three approaches, facilitators, brokers, and mediators, are well known and widely used in collaborative design systems. The facilitator approach was proposed for the SHADE project [113] and demonstrated in the PACT project [21], where several facilitators (facilitation agents) were used to ease the communication and coordination among agents. CIIMPLEX [114] and some other agent based collaborative design systems also use this approach. Brokers (also called broker agents) have been used in a number of agent-based collaborative design and manufacturing systems [115, 116]. The mediator approach was initially proposed by Wiederhold [117] for distributed database management systems. It has also been successfully used in collaborative design and manufacturing system [118].

In addition to the three federated approaches described above, there are some other agent-based collaborative engineering design and manufacturing systems that could be classified as federated architectures. The First-Link project [119] was intended to test a specific software architecture where a central product model (*central node*) is accessed by independent software modules, somewhat like a blackboard architecture. The Next-Link project [120] was the successor to First-Link and was to study how to coordinate the independent agents, with a mechanism called the *Redux'* services developed by Petrie [121] being added to the system for this purpose. In the Process-Link project [122], this type of *central node* was replaced by an *agent manager* which is very similar to the previously described mediator. SiFAs [23] used a similar architecture with a *design board* that is visible to all agents participating in design. Lander *et al.* [123] proposed using a multi-blackboard architecture for managing agent interactions within a collaborative engineering design system. CAAD (Cooperative Agents and Applications in Design) used a similar approach to allow several designers to cooperate through a multi-blackboard system [124].

Other collaborative design system architectures include the autonomous agent approach in DIDE [22]. The DIDE project used this approach in implementing agent-based intelligent engineering design environments. DIDE is organized as a population of asynchronous cognitive agents integrating engineering tools and human specialists within an open environment. Each tool (or interface for a human specialist) is encapsulated as an agent. Engineering tools and human specialists are then connected through a local network and communicate via this network. Each can also communicate directly with other agents located in any other local networks, using the Internet. All agents are independent and autonomous. They exchange design data and knowledge via a local network or the Internet. DIDE does not use any facilitators or mediators. There is no static global control structure in the system. DIDE is specially designed to allow dynamic changes in the agent set (i.e., one can add new agents to or remove existing agents from the system without stopping and reinitializing the working environment). Some service-oriented collaborative design systems also use the autonomous agent approach [125–127].

One of the important features of autonomous agents is their independence. However, agents in federated multi-agent systems do not act completely independently. They communicate or interact with other agents through facilitators, brokers or other types of middle agents, and may be coordinated through agents such as mediators.

In the facilitator architecture, agents interact through facilitators that translate agent-specific knowledge into and out of a standard knowledge interchange language. Each agent can therefore reason (internally) in its own terms, asking other agents for information and providing other agents with information as needed through the facilitators. The facilitator model supports greater flexibility than direct communication used in the autonomous agent systems, because agents do not need to know detailed information about which other agents exist and what their capabilities are.

In the mediator approach, learning (where implemented) is at the group level, while in the autonomous agent approach, learning is at the individual level. Each autonomous agent has to have knowledge about its environment and the other agents, and has to learn in some sense if it is to update its knowledge.

Initially, one may think that the federation approach abandons the decentralization principle of distributed organizations by redefining centralized architectures. This is not so, since the ultimate goal of a federation organization is to release the agents from the burden of establishing their own communication links and assisting them to coordinate their activities with other agents. In such systems, individual agents are then free to concentrate their reasoning capabilities on autonomous planning and responding to the environment's stimuli. When clusters are formed, the organization at any instant may appear to be composed of "partial hierarchies." If these structures are dynamically created, however, and destroyed when their tasks are accomplished, it becomes evident that centralization is only being used locally, temporarily, and where it is advantageous.

In summary, the autonomous agent architecture is well suited for developing collaborative design systems when existing engineering tools are encapsulated as agents and connected to the system for providing special services, and the system consists of a small number of agents. Federated architectures are suitable for developing more complex collaborative design systems composed of a large number of problem-solving agents, data/information agents, and service agents. Implemented appropriately, these architectures can provide computational simplicity and manageability. The client/server approach will continue to be used for industrial applications, but they will be progressively replaced by the federated architectures and the autonomous system architectures.

14.4.2 Web-Based/Centralized Systems

The Web was originally developed for information sharing within internationally dispersed teams and the dissemination of information by support groups. Proposed and developed early in the 1990s, the Web has quickly become a convenient media to publish and share information relevant to the design process, from concept generation and prototyping to virtual manufacturing and product realization. It has been adopted as the most popular implementation architecture of a collaborative product development (including design and manufacturing) tool. A CSCD system developed with the Web as a backbone will primarily provide: (1) access to catalog and design information on components and sub-assemblies; (2) communication among multidisciplinary design team members (including customers, designers, and production engineers) in multimedia formats; and (3) authenticated access to design tools, services, and documents. However, since the Web is still fast evolving, particularly with the development of Web Services, Semantic Web, and Cloud Computing technologies, many researchers and working groups in and outside the World Wide Web Consortium are working hard to improve the current Web infrastructure and supporting tools. Web-based infrastructure has been used in a number of collaborative product design systems. In most cases, the Web is primarily used by multidisciplinary team members as a medium to share design data/information/knowledge; while in some cases, it is integrated with other related technologies and is used for product data management and project management.

A comprehensive review of some Web-based tools and systems can be found in References 4 and 128. Most Web-based collaborative design systems are developed using Java and CORBA [129–131], and some others are developed using Common Lisp, for example, WWDL [132], or Prolog, for example, WebCADET [133]. In addition to HTML and Java Applets for developing client-side user interfaces, ActiveX [134, 135] and VRML [131, 132] are widely used.

However, Web technology alone is not a complete solution to collaborative design systems, although it makes communication physically viable through a common network. In order to collaborate on a distributed design project, remote engineers and designers need active supports to coordinate their efforts. This coordination involves the translation of terminology among disciplines, locating/providing engineering analysis services, virtual prototyping services, and project management [136–138].

Web servers should not only be a repository of information but also provide intelligent services to help users to solve design problems. Such servers may be called software agents and will be discussed below.

14.4.3 Agent-Based/Distributed Systems

Application of software agents to collaborative design has been demonstrated by various research projects. PACT [21] might be one of the earliest successful projects in this area. The interesting aspects of PACT include its federation architecture using facilitators and wrappers for legacy system integration. SHARE [139] was concerned with developing open, heterogeneous, network-oriented environments for concurrent engineering, particularly for design information and data capturing and sharing through asynchronous communication. SiFAs [23] was intended to address the issues of patterns of interaction, communication, and conflict resolution using simple single-function agents. DIDE [22] was developed to study system openness, legacy systems integration, and distributed collaboration. ICM [140] developed a shared graphical modeling environment for collaborative design activities. Co-Designer [141] was intended to support localized design agents in the generation and management of conceptual design variants. Concept Database [142] described a strategic design support for version control, workflow management, and information gathering. A-Design [143] presented a new design generation methodology, which combines aspects of multi-objective optimization, multi-agent systems, and automated design synthesis. It provided designers with a new search strategy for the conceptual stages of product design that incorporates agent collaboration with an adaptive selection of design alternatives. Some projects also addressed the issue of integration and collaboration among product design, process planning, and manufacturing scheduling [118, 144, 145].

In agent-based collaborative design systems, software agents are mostly used for supporting cooperation among designers, enhancing interoperability between traditional computational tools, or allowing better simulations (particularly distributed simulations) [31].

14.4.4 Service-Oriented Systems

Both the Web and agent technologies are very useful in implementing collaborative design systems. The attractiveness of the Web for propagating information makes it appropriate to integrate with agents for accessing and manipulating information automatically. The challenge is to build a Web-based environment that enables and supports seamless interactions among human designers, software agents, and Web servers using the available emerging technologies [146].

A Web-based collaborative design system usually uses a client/server architecture in which the interaction between components is predefined and components are strongly coupled. This kind of approach is insufficient to support dynamic collaborative design environments, where tasks are usually involving complex and nondeterministic interactions, producing results that might be ambiguous and incomplete. An agent-based collaborative design system is a loosely coupled network of problem-solvers that work together to solve problems that are beyond their individual capabilities [31]. Software agents in such systems are communicative, collaborative, autonomous (or semi-autonomous), reactive (or even proactive), and intelligent. Different system architectures have been proposed and used to implement agent-based systems, as discussed in Section 14.4.1.

Although agent technology has been recognized as a promising approach for developing collaborative design systems, those agents that have so far been implemented in various prototype and industrial applications are actually not very "intelligent." In this view, agent applications in the Web-based collaborative design field are still facing many challenging questions. WebBlow [146] is an interesting attempt on the integration of the Web and software agents in implementing a multidisciplinary design optimization environment [147, 148]. Before the emergence of Web Services, the concept of an active Web server was proposed to integrate the Web and agent technologies [149]. Since the active Web servers have very similar features of Web Services, it is natural for the further work to implement collaborative design systems using Web Services [125].

During the past decade, more and more collaborative design systems have been developed using Web Services as well as Semantic Web and Grid Computing techniques.

14.4.5 Collaborative Design Over Supply Chain (Virtual Enterprise)

Design of a complex product, system, or structure involves multiple stakeholders over the supply chain. Early involvement of suppliers in the design process can significantly reduce design changes during the manufacturing/construction process, and therefore reduce the cost and time. The collaboration in this scenario is different from the collaborative design with one company, since companies involved in such collaboration are usually competing at least for profits. In fact, companies may even be competing to survive in the increasingly competitive global market and sometimes may be getting into legal confrontations. One of the major concerns on collaborative design over the supply chain is the product information sharing [150]. Therefore, there is a co-existence of competition and cooperation over supply chains [151]. Even though collaborative design over the supply chain is not a new research topic, there has not been much progress in this area.

Related R&D work has been focused on enterprise collaboration, either vertically along a supply chain or horizontally among peers (even competitors). Virtual enterprise (VE) is one of the most important types of enterprise collaboration. It has the highest demand for sophisticated ICT technologies. A virtual enterprise can be defined as "a network of independent organizations that jointly form an entity committed to provide a product or service" [152, 153]. Thus, from the customer's perspective, as far as that product/service is concerned, these independent organizations, for all practical and operational purposes, are virtually acting as a single entity/enterprise. A similar approach is used in MetaMorph II [118] where a hybrid agent-based mediator-centric architecture is used to integrate partners, suppliers, and customers dynamically with the lead enterprise through their respective mediators within a supply chain network via the Internet and Intranets. A detailed discussion on applications of agent technology to Virtual Enterprise can be found in Reference 154.

14.5 APPLICATIONS

Most CSCD applications mentioned above are related to product design in mechanical engineering [31], as did the CAD technologies. Almost all major companies producing mechanical products and systems are using CSCD technologies both internally with the companies and external over their supply chains. However, challenges still exist in terms of communication of design rationale, management of product data, collaboration and coordination of design teams, cultural and social aspects of design, information security and privacy, etc., which will be further discussed in the following section.

Due to the fragmented nature and adversarial behavior of the construction sector, the development and deployment of collaboration technologies and systems in architecture, engineering, and construction are behind the manufacturing sector [31]. Because the use of a single central repository to store the design information is not usually a viable option in such a unique industry [155], distributed loosely coupled integration and collaboration solutions using intelligent agents and Web Services technologies would be the most promising [156].

Recently, Building Information Modeling (BIM) [157] has been considered as an important enabling technology that drives the construction industry in improving productivity and efficiency. It combines some features of CAD, PDM, and PLM in mechanical engineering design. It can facilitate collaboration among stakeholders during the design, construction, and maintenance of buildings and facilities. It can also streamline the information integration throughout the lifecycle of buildings and facilities. A particular interesting application of BIM and collaboration technologies in architecture, engineering, and construction is change management [158]. However, BIM application in the construction industry is still at an early stage. According to Howard and Bjork [157], "the formal standards on BIM, such as the IFCs are complex and have not had the resources for rapid development and promotion that their potential deserved," and therefore it will take some time for this approach to be widely adopted.

While the focus of this chapter is on CSCD applications in the manufacturing and construction industries, CSCD technologies have been applied to many other domains including aerospace, automotive, logistics, transportation, power and energy, healthcare, infrastructure, administration, social networks, entertainment, and most recently to emergence response.

14.6 RESEARCH CHALLENGES AND OPPORTUNITIES

CSCD has been an active R&D area for about two decades. Some manufacturing and engineering companies have partially implemented in-house collaborative design systems. We expect a great future for CSCD and envision future ideal collaborative design systems as being:

- fully integrated with all necessary software tools connected through the network covering the full product lifecycle from conceptual design to detailed design (with detailed modeling, simulation, and optimization), virtual prototyping, manufacturing/construction, service and maintenance, and final disposal;
- integrated with physical testing and validation systems for "hardware-in-the-loop" simulations during the new product/process/system/service development;
- implemented as semi-automated interactive systems that involve human interventions;
- operated on a collaborative computing environment with automated computing load balancing, quick access, and fast transfer of large volumes of engineering data;
- secured with sophisticated security and privacy protection mechanisms;
- able to allow users to choose favorite software tools according to their experience and preference;
- able to provide different users (including engineers and managers, sales and services staff, as well as customers and suppliers) with different access privileges to the same data/information;
- able to manage knowledge gained during previous design projects;
- able to advise users about good practices, taking into account their personal profile; and
- able to display a proactive behavior.

To achieve this vision, a number of challenging issues have been identified for an academic research, further development, and wider deployment of collaborative design systems in industry. In fact, these challenges are also opportunities for the CSCD research community. Among others, the following areas are believed to be future research opportunities and challenges:

- Ontology and semantics-based integration: One of the most difficult tasks in collaborative design is to agree on the ontological commitments that enable knowledge-level communication among the distributed design parties. Another difficulty is the integration of the various available design tools. If the tool data and models are encapsulated, rather than using a standardized and unified formalism, each tool will be free to use the most appropriate internal representations and models for its intended tasks. This is not a new research topic, but the progress in this area has not been satisfactory. The emergence of the Semantic Web makes progress in this area more likely to occur.

- Interoperability among systems and/or models: Models help designers understand the nature of a design process by ignoring some of the not-so-important details. When deciding how to model a design process, determining the appropriate levels of abstraction is very critical for the model to be beneficial to its users. A key issue in collaborative design from a designer's perspective is how to bridge the multi-faceted models required to support a complex design project at various stages of the design process. The challenge is to use the relevant model for each task (the right abstraction and granularity) and to communicate the results in a suitable form to the various parties involved, whose needs are different and interests are diverse. One way to address this issue is through *collaborative design process modeling* which has been an active research topic recently, but significant efforts are still required.
- Reversibility and version control: A requirement is to record changes in a consistent manner through the various representations of the design product. This implies an extended form of version control as well as the possibility, may be limited, to propagate changes forward and backward. Also because the product is the result of changes that can be done by several groups in parallel, a reconciliation mechanism should be provided to integrate all changes at a given point.
- Product-centric design methodology: A product-centric design methodology is considered as a suitable approach for distributed collaborative product design. Featuring its self-learning ability, product-centric design fits well in a dynamically changing environment. Comprehensive care needs to be taken in modeling, collaboration, design, and development issues in the whole product lifecycle. Fundamental research is still required in this area.
- Data/information/knowledge management: Challenges in this area include knowledge discovery, support for natural language processing and information retrieval, the capturing of design intent in multimedia formats, dynamic knowledge management, self-learning, reasoning, and knowledge reuse. Based on the current Web infrastructure, users are allowed to access server resources primarily through HTTP and FTP. Using appropriate protocols to access the right data at the right locations is essential in collaborative design environments. This feature is particularly useful in large collaborative design and engineering projects where access to large volumes of data at different locations is frequent.
- Collaborative intelligent user interfaces: Human involvement in the collaborative engineering design processes is unavoidable. Designers need to interact with a design system and negotiate with peers via a user interface. The challenge is to make intelligent interfaces available to all resources such that the designers will have more flexibility for efficient and effective designing. The interfaces should be integrated, expressive, goal oriented, cooperative, easy to use, and customizable. In practice, the design process should be supported by a design environment in which the user is part of the system, rather than by a set of tools

for which the user is external to the system. This approach integrates the users' expertise and knowledge directly into the supporting design system.

- Distributed design project management: There must be some ways of managing all the resources involved, including people, organizations, software tools, and equipment. Relevant research issues are collaborative workflow, conflict management, cost and task management, activity scheduling, and computing resource management. In an interesting experimental work, Hammond *et al.* [88] used a socio-technical theory as a framework to explore differences in engineering design team decision-making as a function of various media of communication. Their results indicate that design teams communicating via an electronic medium perceive an increase in mental workload and interact less frequently, but for a greater total amount of time. These results brought interesting implications and suggestions for the management of distributed design teams. More research efforts are needed on how to increase people awareness when working in a distributed system.
- Drag and drop functionality: Drag and drop is a highly desired function in collaborative design using multiple computational tools. For example, a part designed under a CAD system may be moved to a CAE tool's graphical interface for analysis and simulation, and to a DFM tool's graphical interface for manufacturability analysis. It becomes more convenient if there is a drag and drop function that can copy or move a graphical object from one CAD/CAE system to another, particularly in a Web-based collaborative design environment. In fact, it is a type of communication between the two systems through the moving graphical object. The challenge is therefore to develop a common model or language for these related systems. A significant amount of research is needed to determinate standard geometric representations for features that can be used by different CAD and simulation tools. It also requires R&D of drag and drop type standards similar to OLE (object linking and embedding) which provides a protocol for organizing data in a standard format for exchange between different systems.
- Security and privacy: With the implementation and deployment of CSCD applications in industry, security and privacy issues become more and more important. The number of papers on this topic submitted to CSCWD conferences has increased significantly during the past few years. This will continue, particularly with more practical techniques and applications.
- Software self-management and self-healing: Since software self-management and self-healing have become an active research area, it would be natural to extend the research into CSCD systems.
- Social software and mass collaboration: Social software approaches and Wiki-style collaboration tools may be developed and used for knowledge-intensive collaborative design systems [159, 160].
- Cultural and social issues: With the industry globalization and the development of worldwide consortia, special attention is required with respect to cultural problems. Future collaborative design systems will need to integrate results

from social sciences in order to address the cultural differences of designers and users as well as be able to provide instant translation of natural language interactions.

- Intelligent knowledgeable environments should be developed in which a user is advised proactively with respect to the design process expertise, the domain expertise, or the knowledge acquired from previous projects. Such environments should be capable of acquiring knowledge on the fly and organizing it for future projects.
- Most systems are currently developed by engineers with the goal of filling technical needs. Usability studies need to be done extensively to make sure that the newly developed systems are easier to use and do not increase the cognitive load of the users. In general, human sciences have an increasingly important role to play in the future of CSCD.

14.7 CONCLUSIONS

CSCD has been recognized by industry as a way to address the requirements resulting from increasingly complex design of products, structures, buildings, systems, processes, and services, as well as to meet high customer expectations. With the fast development and advancement of Internet and Web-based technologies during the past two decades, various CSCD technologies and systems have been developed and applied to different domains including aerospace, automotive, manufacturing, logistics, transportation, power and energy, healthcare, infrastructure, administration, social networks, and entertainment. To achieve their full potential and the vision of fully integrated collaborative design systems, significant R&D efforts are still required. Some research challenges discussed above should be addressed within the next few years, though some of them may need a few decades to be thoroughly addressed.

REFERENCES

[1] J. Hartley. *Concurrent Engineering*. Cambridge, MA: Productivity Press, 1992.

[2] E. Sprow. Chrysler's concurrent engineering challenge. *Manufacturing Engineering*, 108(4): 35–42, 1992.

[3] K. T. Ulrich and S. D. Eppinger. *Product Design and Development*, 3rd edn. NewYork: McGraw-Hill, 2000.

[4] W. Shen and L. Wang. Web-based and agent-based approaches for collaborative product design: an overview. *International Journal of Computer Applications in Technology*, 16(2/3): 103–112, 2003.

[5] W. D. Li, S. K. Ong, and A. Y. C. Nee. *Integrated and Collaborative Product Development Environment—Technologies and Implementation*. World Scientific, Singapore, 2006.

[6] K. Schmidt and L. Bannon. Taking CSCW seriously. *Computer Supported Cooperative Work*, 1(1/2): 7–40, 1992.

[7] I. Greif (ed.). *Computer-Supported Cooperative Work: A Book of Readings*. San Mateo, CA: Morgan Kaufmann Publishers, 1988.

[8] J. Turino. *Managing Concurrent Engineering*. New York: Van Nostrand Reinhold, 1992.

[9] B. Prasad. *Concurrent Engineering Fundamentals: Integrated Product Development*, Vol. 2. One Saddle River, NJ: Prentice Hall, 1997.

[10] R. Balamuralikrishna, R. Athinarayanan, and X. S. Song. The relevance of concurrent engineering in industrial technology programs. *Journal of Industrial Technology*, 16(3): 1–5, 2000.

[11] B. A. Myers. A brief history of human computer interaction technology. *ACM Interactions*, 5(2): 44–54, 1998.

[12] HCI International Conference. Available at www.hci-international.org/ (accessed October 5, 2012).

[13] Human–Computer Interaction. Available at http://www.tandfonline.com/loi/hhci20 (accessed October 5, 2012).

[14] HCI Bibliography: Human–Computer Interaction Resources. Available at hcibib.org/ (accessed October 5, 2012).

[15] R. Reddy, L. Erman, R. Fennel, and R. Neely. The HEARSAY speech understanding system: an example of the recognition process. *IEEE Transactions on Computers*, C-25: 427–431, 1976.

[16] R. G. Smith. The contract net protocol: high-level communication and control in a distributed problem solver. *IEEE Transactions on Computers*, C-29(12): 1104–1113, 1980.

[17] C. Hewitt. Control structure as patterns of passing messages. In: *Artificial Intelligence: An MIT Perspective*, edited by P. H. Winston and R. H. Brown. The MIT Press, Cambridge, MA, USA, 1979, pp. 435–465.

[18] D. H. Norrie and A. D. Kwok. Object-oriented distributed artificial intelligence. In: *New Results and New Trends in Computer Science*, *LNCS 555*, edited by H. Maurer. Springer-Verlag, Berlin, Germany, 1991, pp. 225–242.

[19] W. Shen, Q. Hao, H. Yoon, and D. H. Norrie. Applications of agent systems in intelligent manufacturing: an update review. *International Journal of Advanced Engineering Informatics*, 20(4): 415–431, 2006.

[20] T. Tomiyama. Collaborative product development in ill-structured problem domains. In: Proceedings of the 10th International Conference on CSCW in Design, 2006, pp. 15–20.

[21] M. R. Cutkosky, R. S. Engelmore, R. E. Fikes, M. R. Genesereth, T. R. Gruber, W. S. Mark, J. M. Tenenbaum, and J. C. Weber. PACT: an experiment in integrating concurrent engineering systems. *IEEE Computer*, 26(1): 28–37, 1993.

[22] W. Shen and J. P. Barthès. DIDE: a multi-agent environment for engineering design. In: Proceedings of First International Conference on Multi-Agent Systems (ICMAS'95), San Francisco, CA, 1995, pp. 344–351.

[23] D. C. Brown, B. Dunskus, D. L. Grecu, and I. Berker. SINE: support for single function agents. In: Proceedings of Applications of AI in Engineering, Udine, Italy, 1995.

[24] Autodesk Inventor. Available at http://usa.autodesk.com/autodesk-inventor/ (accessed October 5, 2012).

[25] Autodesk Buzzsaw. Available at http://usa.autodesk.com/buzzsaw/ (accessed October 5, 2012).

[26] Autodesk Streamline. Available at http://usa.autodesk.com/adsk/servlet/pc/index?id=2164339&siteID=123112 (accessed October 5, 2012).

[27] Graphisoft ArchiCAD TeamWork™. Available at http://www.graphisoft.com/products/archicad/teamwork/ (accessed October 5, 2012).

[28] CoCreate, OneSpace. Available at www.cocreate.com/ (accessed October 5, 2012).

[29] Matrix PLM Platform. Available at http://www.3ds.com/products/enovia (accessed October 5, 2012).

[30] UGS, PLM Solutions. Available at http://www.ugs.com/solutions/ (accessed October 5, 2012).

[31] W. Shen, D. H. Norrie, and J. P. Barthes. *Multi-agent Systems for Concurrent Intelligent Design and Manufacturing*. London, UK: Taylor and Francis, 2001.

[32] A. Wong and D. Sriram. SHARED: an information model for cooperative product development. *Research in Engineering Design*, 5(1): 21–39, 1993.

[33] K. C. Lee, W. H. MansfieldJr., and A. P. Sheth. A framework for controlling cooperative agents. *IEEE Computer*, 26(7): 8–16, 1993.

[34] G. Gaspar. Communication and belief changes in a society of agents, towards a formal model of autonomous agents. *Decentralized Artificial Intelligence*, 2: 71–88, 1991.

[35] M. K. Chang and C. C. Woo. Speech-act-based negotiation protocol: design, implementation, and test use. *ACM Transactions on Information Systems*, 12(4): 360–382, 1991.

[36] T. Finin, D. McKay, and R. Fritzson. Specification of the KQML: agent-communication language. Technical Report EIT TR 92-04, Enterprise Integration Technologies, Palo Alto, CA, 1992.

[37] V. Lesser and D. Corkill. Functionally accurate, cooperative distributed systems. *IEEE Transactions on Systems, Man and Cybernetics*, 11(1): 81–96, 1981.

[38] M. Klein. iDCSS: integrating workflow, conflict and rationale-based concurrent engineering coordination technologies. *Concurrent Engineering: Research and Application*, 3(1): 21–27, 1995.

[39] C. Iffenecker. Un système multi-agents pour le support des activités de conception de produits. Thèse de l'Université Paris VI, 1994.

[40] S. Balasubramanian and D. H. Norrie. A multi-agent intelligent design system integrating manufacturing and shop-floor control. In: *Proceedings of First International Conference on Multi-agent Systems*. The AAAI Press/The MIT Press, Cambridge, MA, USA, 1995, pp. 3–9.

[41] R. Washington. Markov tracking for agent coordination. In: *Proceedings of the Second International Conference on Autonomous Agents*. The ACM Press, New York, NY, USA, 1998, pp. 70–77.

[42] N. R. Jennings. Coordination techniques for distributed artificial intelligence. In: *Foundations of DAI*, edited by G. M. P. O'Hare and N. R. Jennings. John Wiley & Sons, New York, NY, USA, 1996, pp. 187–210.

[43] M. S. Mazer. Reasoning about knowledge to understand distributed AI systems. *IEEE Transactions on Systems, Man and Cybernetics*, 21(6): 1333–1346, 1991.

[44] J. Ferber. *Les Systèmes Multi-Agents*. Paris: InterEditions, 1995.

[45] M. Klein. Supporting conflict resolution in cooperative design systems. *IEEE Transactions on Systems, Man and Cybernetics*, 21(6): 1379–1390, 1991.

[46] S. E. Conry, R. A. Meyer, and V. R. Lesser. Multistage negotiation in distributed planning. In: *Readings in Distributed Artificial Intelligence*, edited by A. H. Bond and L. Gasser. Morgan Kaufmann Publishers, Inc., San Francisco, CA, USA, 1988, pp. 367–383.

[47] K. P. Sycara. Cooperative negotiation in concurrent engineering design. In: *Computer-Aided Cooperative Product Development*, LNCS 492. Springer-Verlag, Berlin, Germany, 1991, pp. 269–297.

[48] B. Dunskus, D. L. Grecu, D. C. Brown, and I. Berker. Using single function agents to investigate negotiation. *AIEDAM: Artificial Intelligence for Engineering Design, Analysis and Manufacturing*, 9(4): 299–312, 1995.

[49] D. Bahler, C. Dupont, and J. Bowen. Anaxiomatic approach that supports negotiated resolution of design conflicts in concurrent engineering. In: *Artificial Intelligence in Design*, edited by J. S. Gero and D. Sudweeks. Kluwer Academic Publishers, Dordrecht, The Netherlands, 1994, pp. 363–379.

[50] R. E. Douglas, D. C. Brown, and D. C. Znger. A concurrent engineering demonstration and training system for engineers and managers. *International Journal of CAD/CAM and Computer Graphics*, 8(3): 263–301, 1993.

[51] C. Petrie, M. Cutkosky, T. Webster, A. Conru, and H. Park. Next-Link: an experiment in coordination of distributed agents. In: Proceedings of AID-94 Workshop on Conflict Resolution, Lausanne, Switzerland, 1994.

[52] F. Pena, D. Sriram, and R. Logcher. SHARED-DRIMS: SHARED design recommendation—intent management system. In: Proceedings of 2nd IEEE Workshop on Enabling Technologies Infrastructure for Collaborative Enterprises (WET ICE), 1993, pp. 213–221.

[53] C. A. Tacla and J-P. A. Barthès. A multi-agent architecture for evolving memories. In: AMKM 2003: Agent-Mediated Knowledge Management: LNCS 2926, 2004, pp. 388–404.

[54] M. L. Sbodio, D. Martin, and C. Moulin. Discovering semantic web services using SPARQL and intelligent agents. *Web Semantics Science Services and Agents on the World Wide Web*, 8(4): 310–328, 2010.

[55] T. Huang, W. Li, and C. Yang. Comparison of ontology reasoners: Racer, Pellet, Fact++. In: American Geophysical Union, Fall Meeting, 2008.

[56] C. A. Tacla, A. R. Freddo, E. C. Paraiso, M. P. Ramos, and G. Y. Sato. Supporting small teams in cooperative building application domain models. *Expert System Applications*, 38(2): 1160–1170, 2011.

[57] A. Butz, T. Höllerer, S. Feiner, B. MacIntyre, and C. Bescher. Enveloping users and computers in a collaborative 3D augmented reality. In: Proceedings of IWAR'99 (International Workshop on Augmented Reality), San Francisco, CA, 1999, pp. 35–44.

[58] M. Wissen. Implementation of a laser-based interaction technique for projection screens. *ERCIM News*, 46: 31–32, 2001.

[59] I. A. Twombly, K. Montgomery, J. Smith, and R. Boyle. The virtual glovebox (VGX): a semi-immersive virtual environments for training astronauts in life science experiments. *Journal of Systemics, Cybernetics and Informatics*, 2(3): 30–34, 2004.

[60] G. Tur, A. Stolcke, L. Voss, S. Peters, D. Hakkani-Tur, J. Dowding, B. Favre, R. Fernandez, M. Frampton, M. Frandsen, C. Frederickson, M. Graciarena, D. Kintzing, K. Leveque, S. Mason, J. Niekrasz, M. Purver, K. Riedhammer, E. Shriberg, J. Tein, D. Vergyri, and F. Yang. The CALO meeting assistant system. *IEEE Transactions on Audio, Speech, and Language Processing*, 18(6): 1601–1611, 2010.

[61] A. Nguyen. An agent-based approach to dialogue management in personal assistants. PhD Dissertation, University of New South Wales, Australia, 2007.

[62] J.-P. A. Barthès. Flexible communication based on linguistic and ontological cues. In: Proceedings of the 5th International MCETECH Conference on eTechnologies, Les Diablerets, Switzerland, 2011, pp. 131–145.

[63] A. Jones, A. Kendira, D. Lenne, T. Gidel, and C. Moulin. The TATIN-PIC project: a multi-modal collaborative work environment for preliminary design. In: Proceedings of 15th International Conference on Computer Supported Cooperative Work in Design—CSCWD 2011, Lausanne, Switzerland, 2011, pp. 154–161.

[64] M. R. Morris, J. O. Wobbrock, and A. D. Wilson. Understanding users' preferences for surface gestures. In: Proceedings of Graphics Interface (GI'10), Ottawa, Ontario, 2010, pp. 261–268.

[65] K. Sugawara, S. Ben Yaala, Y. Manabe, C. Moulin, N. Shiratori, and J.-P. A. Barthes. Conversation-based support for requirement definition by a personal design assistant. In: Proceedings of 10th IEEE International Conference on Cognitive Informatics & Cognitive Computing, 2011, pp. 262–267.

[66] E. C. Paraiso, Y. Campbell, and C. A. Tacla. Webanima: a web based embodied conversational assistant to interface users with multi-agent based CSCW application. In: Proceedings of CSCWD 2008, 2008, pp. 337–342.

[67] N. Negroponte. *Being Digital*. Westminister, MD: Alfred a Knopf, Inc., 1995.

[68] J.-P. A. Barthès. OMAS—a flexible multi-agent environment for CSCWD. *Future Generation Computer Systems*, 27(1): 78–87, 2011.

[69] T. Prante, N. Streitz, and P. Tandler. Roomware: computers disappear and interaction evolves. *Computer*, 37(12): 47–54, 2004.

[70] DARPA PAL Program. Available at https://pal.sri.com/Plone/framework (accessed October 5, 2012).

[71] Thouvenin, I., Lenne, D., Guenand, A., and Aubry, S. Knowledge integration in early design stages for collaboration on a virtual mock up. In: Proceedings of the Ninth International Conference on Computer Supported Cooperative Work in Design, vol. 2, 2005, pp. 1141–1145.

[72] Y. Zhang, H. Ghenniwa, and W. Shen. Agent-based personal assistance in collaborative design environments. In: *CSCW in Design II*, LNCS 3865, edited by W. Shen, K. M. Chao, Z. Lin, J. P. Barthes, and A. James. Springer-Verlag, Berlin, Germany, 2005, pp. 284–293.

[73] IBM Lotus Software. Available at http://www-01.ibm.com/software/lotus/ (accessed October 5, 2012).

[74] Alfresco. Available at www.alfresco.com/ (accessed October 5, 2012).

[75] Nuxeo. Available at http://www.nuxeo.com/en (accessed October 5, 2012).

[76] Zotero. Available at www.zotero.org/ (accessed October 5, 2012).

[77] Concurrent Versions System. Available at http://cvs.nongnu.org/ (accessed October 5, 2012).

[78] Apache Subversion. Available at http://subversion.apache.org/ (accessed October 5, 2012).

[79] Git. Available at http://git-scm.com/ (accessed October 5, 2012).

[80] Google Wave. Available at http://wave.google.com/ (accessed October 5, 2012).

[81] M. Kirsch-Pinheiro, J. Valdeni de Lima, and M. R. S. Borges. A framework for awareness support in groupware systems. In: Proceedings of CSCWD 2002, 2002, pp. 13–18.

[82] M. P. Locatelli and G. Vizzari. Awareness in collaborative ubiquitous environments: the multilayered multi-agent situated system approach. *ACM Transactions on Autonomous and Adaptive Systems (TAAS)*, 2(4), 2007, Article no.13.

[83] F. Enembreck, I. Thouvenin, M.-H. Abel, and J.-P. A. Barthes. An ontology-based multi-agent environment to improve collaborative design. In: Proceedings of 6th International Conference on the Design of Cooperative Systems, Coop04, French Riviera, France, 2004, pp. 81–89.

[84] F. Casati, S. Ceri, B. Pernici, and G. Pozzi. Workflow evolution. *Data & Knowledge Engineering*, 24: 211–238, 1998.

[85] K. Kim. A model-driven workflow fragmentation framework for collaborative workflow architectures and systems. *Journal of Network and Computer Applications*, 35(1): 97–110, 2012.

[86] M. A. Rosenman, G. Smith, M. L. Maher, L. Ding, and D. Marchant. Multidisciplinary collaborative design in virtual environments. *Automation in Construction*, 16(1): 37–44, 2007.

[87] R. Aspin. Supporting collaboration, in co-located 3D visualization, through the use of remote personal interfaces. *Journal of Computing in Civil Engineering*, 21(6): 393–401, 2007.

[88] J. M. Hammond, C. M. Harvey, R. J. Koubek, W. D. Compton, and A. Darisipudi. Distributed collaborative design teams: media effects on design processes. *International Journal of Human–Computer Interaction*, 18(2): 145–165, 2005.

[89] T. Hartmann, M. Fischer, and J. Haymaker. Implementing information systems with project teams using ethnographic–action research. *Advanced Engineering Informatics*, 23(1): 57–67, 2009.

[90] X3D (eXtensible 3D). Available at www.x3d.com (accessed October 5, 2012).

[91] W3D (Web 3D). Available at www.macromedia.com (accessed October 5, 2012).

[92] U3D. Available at www.intel.com/technology/systems/u3d/ (accessed October 5, 2012).

[93] JT Open. Available at www.jtopen.com (accessed October 5, 2012).

[94] OpenHSF. Available at www.openhsf.org (accessed October 5, 2012).

[95] Cimmetry AutoVue. Available at www.cimmetry.com/ (accessed October 5, 2012).

[96] Spinfire. Available at http://autoweb.net/web2006/products-spinfire.shtml (accessed October 5, 2012).

[97] Solidworks eDrawings. Available at http://www.solidworks.com/edrawings (accessed October 5, 2012).

[98] RealityWave ConceptStation. Available at http://products.datamation.com/e-business/groupware/982097698.html (accessed October 5, 2012).

[99] N. Shyamsundar and R. Gadh. Collaborative virtual prototyping of product assemblies over the internet. *Computer-Aided Design*, 34(10): 755–768, 2002.

[100] L. Chen, Z. J. Song, and L. Feng. Internet-based real-time collaborative assembly modelling via an e assembly system: status and promise. *Computer-Aided Design*, 36(9): 835–847, 2004.

[101] S. S. Zhang, W. Shen, and H. Ghenniwa. A review of internet-based product information sharing and visualization. *Computers in Industry*, 54(1): 1–15, 2004.

[102] L. Wang, S. Lang, and W. Shen. A Java3D enabled cyber workspace. *Communication of the ACM*, 45(11): 45–49, 2002.

[103] W. D. Li, J. H. Fuh, and Y. S. Wong. An Internet-enabled integrated system for co-design and concurrent engineering. *Computers in Industry*, 55(1): 87–103, 2004.

[104] UGS TeamCenter. Available at http://wwwprd.ugs.com/products/teamcenter/ (accessed October 5, 2012).

[105] PTC Windchill. Available at http://www.ptc.com/appserver/mkt/products/home.jsp?k=37 (accessed October 5, 2012).

[106] ENOVIA VPLM, MatrixOne and SmarTeam. Available at http://www.3ds.com/products-solutions/plm-solutions/enovia/products/ (accessed October 5, 2012).

[107] X. W. Xu and T. Liu. A web-enabled PDM system in a collaborative design environment. *Robotics and Computer Integrated Manufacturing*, 19(4): 315–328, 2003.

[108] J. H. Panchal, M. G. Fernández, C. J. J. Paredis, J. K. Allen, and F. Mistree. Leveraging design process related intellectual capital—a key to enhancing enterprise agility. In: *Collaborative Product Design and Manufacturing Methodologies and Applications*, edited by W. D. Li, S. K. Ong, A. Y. C. Nee, and C. A. McMahon. Springer-Verlag, Berlin, Germany, 2007, pp. 211–243.

[109] Z. M. Qiu and Y. S. Wong. Dynamic workflow change in PDM systems. *Computers in Industry*, 58(5): 453–463, 2007.

[110] M. Saad and M. L. Maher. Shared understanding in computer-supported collaborative design. *Computer-Aided Design*, 28(3): 183–192, 1996.

[111] S. R. Fussell, R. E. Kraut, F. J. Lerch, W. L. Scherlis, M. W. McNally, and J. J. Cadiz. Coordination, overload and team performance: effects of team communication strategies. In: Proceedings of the ACM Conference on Computer-Supported Cooperative Work, Seattle, Washington, DC, November 14–18, 1998, pp. 275–284.

[112] G. Q. Huang, W. Y. Yee, and K. L. Mak. Development of a web-based system for engineering change management. *Robotics and Computer Integrated Manufacturing*, 17(3): 255–267, 2001.

[113] J. McGuire, D. Huokka, J. Weber, J. Tenenbaum, T. Gruber, and G. Olsen. SHADE: technology for knowledge-based collaborative engineering. *Journal of Concurrent Engineering: Applications and Research*, 1(3): 137–146, 1993.

[114] Y. Peng, T. Finin, Y. Labrou, B. Chu, J. Long, W. J. Tolone, and A. Boughannam. A multi-agent system for enterprise integration. In: Proceedings of PAAM'98, London, UK, 1998, pp. 213–229.

[115] H. Park, J. Tenenbaum, and R. Dove. Agile infrastructure for manufacturing systems: a vision for transforming the US manufacturing base. In: Proceedings of Defense Manufacturing Conference, 1993.

[116] P. M. D. Gray, S. M. Embury, K. Hui, and A. Prce. An agent-based system for handling distributed design constraints. In: Working Notes of the Agent-Based Manufacturing Workshop, Minneapolis, MN, 1998.

[117] G. Wiederhold. Mediators in the architecture of future information systems. *IEEE Computer*, 25(3): 38–49, 1992.

[118] W. Shen, F. Maturana, and D. H. Norrie. MetaMorph II: an agent-based architecture for distributed intelligent design and manufacturing. *Journal of Intelligent Manufacturing*, 11(3): 237–251, 1999.

[119] H. Park, M. Cutkosky, A. Conru, and S. H. Lee. An agent-based approach to concurrent cable harness design. *Artificial Intelligence for Engineering Design, Analysis and Manufacturing*, 8(1), 45–61, 1994.

[120] C. Petrie, M. Cutkosky, T. Webster, A. Conru, and H. Park. Next-Link: an experiment in coordination of distributed agents. In: AID-94 Workshop on Conflict Resolution, Lausanne, Switzerland, 1994.

[121] C. Petrie. The Redux' server. In: Proceeding of International Conference on Intelligent and Cooperative Information Systems (ICICIS), Rotterdam, The Netherlands, 1993.

[122] Goldmann, S. Procura: a project management model of concurrent planning and design. In: Proceeding of WET ICE'96, Stanford, CA, 1996.

[123] S. E. Lander, S. M. Staley, and D. D. Corkill. Designing integrated engineering environment: blackboard-based integration of design and analysis tools. *Concurrent Engineering: Research and Applications*, 4(1): 59–72, 1996.

[124] N. E. Branki and A. Bridges. An architecture for cooperative agents and applications in design. In: *Advanced Technologies*, edited by M. R. Beheshi and K. Zreik. Elsevier Science Inc., New York, NY, USA, 1993, pp. 221–230.

[125] W. Shen and Q. Hao. A service oriented framework for blow molded automotive parts design and optimization. In: Proceedings of SAE 2004 Congress, Detroit, MI, SAE 2004-01-1244, 2004.

[126] R. Quadrel, R. Woodbury, S. Fenves, and S. Talukdar. Controlling asynchronous team design environment by simulated annealing. *Research in Engineering Design*, 5(2): 88–101, 1993.

[127] W. T. Tsai. Service-oriented system engineering: a new paradigm. In: Proceedings of the 2005 IEEE International Workshop on Service-Oriented System Engineering (SOSE'05), 2005, pp. 3–6.

[128] W. D. Li and Z. Qiu. State-of-the-art technologies and methodologies for collaborative product development systems. *International Journal of Production Research*, 44(13): 2525–2559, 2006.

[129] V. Jagannathan, G. Almasi, and A. Suvaiala. Collaborative infrastructures using the WWW and CORBA-based environments. In: Proceedings of the IEEE Workshops on Enabling Technologies Infrastructure for Collaborative Enterprises (WET ICE'96), 1996, pp. 292–297.

[130] G. Q. Huang and K. L. Mak. Web-based morphological charts for concept design in collaborative product development. *Journal of Intelligent Manufacturing*, 10: 267–278, 1999.

[131] A. Wallis, Z. Haag, and R. Foley. A multi-agent framework for distributed collaborative design. In: Proceedings of the IEEE Workshops on Enabling Technologies Infrastructure for Collaborative Enterprises (WET ICE'98), 1998, pp. 282–287.

[132] Zdrahal, Z. and Domingue, J. The world wide design lab: an environment for distributed collaborative design. In: Proceedings of 1997 International Conference on Engineering Design, Tampere, Finland, 1997.

[133] N. H. M. Caldwell and P. A. Rodgers. WebCADET: facilitating distributed design support. In: Proceedings of IEE Colloquium on Web-Based Knowledge Servers, London, UK, 1998, pp. 9/1–9/4.

[134] G. Q. Huang, S. W. Lee, and K. L. Mak. Web-based product and process data modelling in concurrent 'design for X'. *Robotics and Computer-Integrated Manufacturing*, 15(1): 53–63, 1999.

[135] G. Q. Huang and K. L. Mak. Design for manufacture and assembly on the Internet. *Computer in Industry*, 38(1): 17–30, 1999.

[136] R. H. Allen, S. Nidamarthi, S. P. Regalla, and R. D. Sriram. Enhancing collaboration using an Internet integrated workbench. In: Proceedings of 1999 ASME Design Engineering Technical Conference, Las Vegas, NV, 1999.

[137] X. Q. Liu, S. Raorane, and M. C. Leu. A web-based intelligent collaborative system for engineering design. In: *Collaborative Product Design and Manufacturing Methodologies and Applications*, edited by W. D. Li, S. K. Ong, A. Y. C. Nee, and C. A. McMahon. Springer-Verlag, Berlin, Germany, 2007, pp. 37–58.

[138] F. Mervyn, A. Senthil Kumar, and A. Y. C. Nee. A 'plug-and-play' computing environment for collaborative product design and manufacturing across an extended enterprise. In: *Collaborative Product Design and Manufacturing Methodologies and Applications*, edited by W. D. Li, S. K. Ong, A. Y. C. Nee, and C. A. McMahon. Springer-Verlag, Berlin, Germany, 2007, pp. 71–92.

[139] G. Toye, M. Cutkosky, L. Leifer, J. Tenenbaum, and J. Glicksman. SHARE: a methodology and environment for collaborative product development. In: *Proceeding of 2nd Workshop on Enabling Technologies: Infrastructure for Collaborative Enterprises*. IEEE Computer Society Press, 1993, pp. 33–47.

[140] R. Fruchter, K. A. Reiner, G. Toye, and L. J. Leifer. Collaborative mechatronic system design. *Concurrent Engineering: Research and Applications*, 4(4): 401–412, 1996.

[141] M. J. Hague and A. Taleb-Bendiab. Tool for management of concurrent conceptual engineering design. *Concurrent Engineering: Research and Applications*, 6(2): 111–129, 1998.

[142] A. Varma, A. Dong, B. Chidambaram, A. Agogino, and W. Wood. Web-based tool for engineering design. In: Proceedings of AID'96 Workshop on Agents and Web-Based Design Environments, 1996.

[143] M. I. Campbell, J. Cagan, and K. Kotovsky. A-design: an agent-based approach to conceptual design in a dynamic environment. *Research in Engineering Design*, 11: 172–192, 1999.

[144] S. Balasubramanian, F. Maturana, and D. H. Norrie. Multi-agent planning and coordination for distributed concurrent engineering. *International Journal of Cooperative Information Systems*, 5(2–3): 153–179, 1996.

[145] M. Mahesh, S. K. Ong, and A. Y. C. Nee. A web-based framework for distributed and collaborative manufacturing of engineering parts. In: *Collaborative Product Design and Manufacturing Methodologies and Applications*, edited by W. D. Li, S. K. Ong, A. Y. C. Nee, and C. A. McMahon. Springer-Verlag, Berlin, Germany, 2007, pp. 141–154.

[146] Y. Wang, W. Shen, and H. Ghenniwa. WebBlow: a web/agent-based multidisciplinary design optimization environment. *Computers in Industry*, 52(1): 17–28, 2003.

[147] W. Shen and H. H. Ghenniwa. Multidisciplinary design optimization: a framework for technology integration. In: Proceedings of the First International Conference on Multidisciplinary Design Optimization, London, ON, 2001, pp. 22–28.

[148] W. Shen and H. H. Ghenniwa. A distributed multidisciplinary design optimization framework based on web and agents. In: Proceedings of the 2002 ASME DETC/CIE Conference, Montreal, Canada, September 29–October 2, 2002, DETC2002/CIE-34461, 2002.

[149] W. Shen. Web-based infrastructure for collaborative product design: an overview. In: Proceedings of 5th International Conference on CSCW in Design, Hong Kong, 2000, pp. 239–244.

[150] G. Q. Huang, J. S. K. Lau, and K. L. Mak. The impacts of sharing production information on supply chain dynamics: a review of the literature. *International Journal of Production Research*, 41(7): 1483–1517, 2003.

[151] A. Surana, S. Kumara, M. Greaves, and U. N. Raghavan. Supply-chain networks: a complex adaptive systems perspective. *International Journal of Production Research*, 43(2): 4235–4265, 2005.

[152] H. S. Jagdev and K. D. Thoben. Anatomy of enterprise collaboration. *Production Planning and Control*, 12(5): 437–451, 2001.

[153] L. M. Camarinha-Matos and H. Afsarmanesh. *Infrastructure for Virtual Enterprise*. Norwell, MA: Kluwer Academic Publisher, 1999.

[154] V. Marik and M. Pechoucek. Agent technology for virtual organization. In: Proceedings of PRO-VE'03, Lugano, Switzerland, 2003, pp. 243–252.

[155] N. Bakis, G. Aouad, and M. Kagioglou. Towards distributed product data sharing environments—progress so far and future challenges. *Automation in Construction*, 16(6): 586–595, 2007.

[156] W. Shen, Q. Hao, H. Mak, J. Neelamkavil, H. Xie, J. Dickinson, J. R. Thomas, A. Pardasani, and H. Xue. Systems integration and collaboration in architecture, engineering, construction and facilities management: a review. *Advanced Engineering Informatics*, 24(2): 196–207, 2010.

[157] R. Howard and B. C. Bjork. Building information modelling—experts' views on standardisation and industry deployment. *Advanced Engineering Informatics*, 22(2): 271–280, 2008.

[158] M. Sun, S. Senaratne, A. Fleming, I. Motowa, and M. L. Yeoh. A change management toolkit for construction projects. *Architectural Engineering and Design Management*, 2(4): 261–271, 2006.

[159] D. Tapscott and A. D. Williams. *Wikinomics: How Mass Collaboration Changes Everything*. New York, NY: Penguin Group, 2007.

[160] D. Richards. Collaborative knowledge engineering: socialising expert systems. In: Proceedings of CSCWD 2007, 2007, pp. 635–640.

CHAPTER 15

SUPPORT COLLABORATION WITH ROLES

HAIBIN ZHU[1], MENGCHU ZHOU[2], and MING HOU[3]
[1]Department of Computer Science and Mathematics, Nipissing University, North Bay, ON, Canada
[2]Department of Electrical and Computer Engineering, New Jersey Institute of Technology, Newark, NJ, USA
[3]Defence Research and Development of Canada, Toronto, ON, Canada

15.1 INTRODUCTION

Collaboration occurred early in human history, when humankind began to thrive for survival. They collaborated to build shelters to protect themselves from natural disasters. They collaborated to defend themselves from attacks by ferocious animals, such as, wolves, tigers, and lions. They collaborated to kill animals to obtain enough food. Therefore, collaboration is a fundamental activity for the human world. No person lives totally alone.

People collaborate when one single person cannot accomplish a complex or difficult task, or if one single person can complete a task but it is not economical to do so. Collaboration is necessary when the task is so difficult and complex that one person cannot accomplish it, or if the task is too costly. For example, to move a piano into a living room is a difficult task for one person to accomplish; one person may cook tofu in a restaurant, but it is too costly for him/her to make tofu by himself/herself. He/she must collaborate with a local grocery store and purchase tofu from the store.

Collaboration requires people in a group to fulfill their obligations and respect the rights of others. To collaborate, people generally participate in an organization or a group. To collaborate, the related parties must establish common goals by negotiation, divide the whole task into subtasks, distribute subtasks to related parties, and finally, integrate all complete subtasks to a whole accomplishment. We can determine the requirements for roles by examining collaboration in a group. All individuals should have clear positions within a group and their roles should be related but not interfered.

Contemporary Issues in Systems Science and Engineering, First Edition.
Edited by MengChu Zhou, Han-Xiong Li and Margot Weijnen.

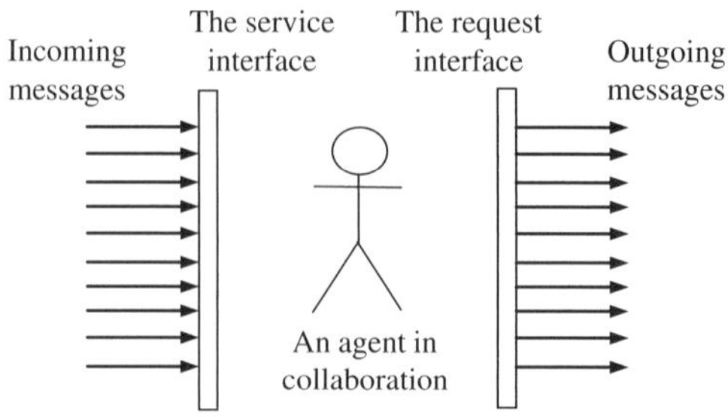

FIGURE 15.1 A role is a wrapper of an agent.

To form a successful organization, we need to have the members in the organization play roles. With roles, collaboration is made easy and more fruitful.

Roles are fundamental tools to support collaboration activities. They are also appropriate trade-off abstraction tools between objects and classes. A role can be applied when describing concepts at a position between a class and an object, that is, separation of concerns [1, 2].

In collaboration, people face environments and other people who are involved. To collaborate with other people, they must know what they can do and what they would like to do. To work efficiently, they must know what objects in the environment they can access. To be cooperative, they must tell others what they can do and what they would like to do for them. When collaborating, working efficiently and being cooperative involve playing roles. Therefore, in general, a person has two kinds of existence: one as a server and the other as a client. When a person plays a role, he/she provides specific services and possesses specific rights to ask for services. With this common sense, a role can be considered as a view of people to the collaborative environment. When they play a specific role, they have a special view of their surroundings. Their role in a collaborative environment is actually as a *wrapper* with a *service interface* and a *request interface* as shown in Figure 15.1. Therefore, we can separate a person's role into two parts:

(1) The service interface, including incoming messages.
(2) The request interface, including outgoing messages.

In collaboration without role mechanisms, if people want to cooperate with others, they simply send messages to other people. In applying a system without role mechanisms, people send messages to objects in the system. Other people may ask the system for services, get the states of objects, and send messages to the objects. The system and its objects are actually the collaboration media of people. In a system with role mechanism, people play roles when they interact with collaborative systems. Roles include two major aspects: *rights* and *responsibilities*. A role is a wrapper for a player in collaboration (Figure 15.1).

In this chapter, we consider "collaboration" as a generalized concept. We denote collaboration in the following categories:

(1) Among people, that is, natural collaboration.
(2) Among people through systems, that is, computer-supported cooperative work (CSCW).
(3) Among people and systems, that is, human–computer interaction (HCI).
(4) Among systems, that is, distributed and collaborative systems.

Therefore, in the following descriptions, we use *people, agents, or members* as role players: when we use people, we mean people collaborate naturally or with a system; when we use agents, we mean agents or systems collaboration; and when we use members, we emphasize the relationships between a part (members) and the whole (the group).

15.2 BENEFITS OF ROLES IN COLLABORATION

Collaboration involves *trust, dynamics, interaction, adaptation, coordination, and information sharing*. Roles can be made to facilitate an organizational structure, provide orderly system behavior, and consolidate system security for both human and nonhuman entities that collaborate and coordinate their activities with or within systems [3, 4].

15.2.1 Establishing Trust in Collaboration

Trust of an agent is an indicator of "being honest and reliable" that gives confidence for other agents to cooperate with the agent. A high degree of trust in an agent means that the agent will very likely be chosen as a collaboration partner.

If an agent has trust, it is easy for it to be assigned tasks to do, to earn credits, and earn money. Trust in collaborative systems means that an agent situates in an environment where each agent tries to choose the most reliable partner from potential agents and establish a strategy to adopt with the environment [5].

Trust may increase with positive experience and decrease with negative experience or, over time, without any experience. This also makes it clear that trust is changeable. With roles, agents are regulated for their behaviors. Agents that lose trust will definitely lose roles by regular agent evaluation and dynamic role assignment. The most trustable agents hold the most important and valuable roles. Agents that gradually lose trust will be finally discarded by the system, that is, no roles are assigned to them.

Roles are commonly accepted concepts that present enough information for trust. With roles, agents or people become role players. The trust of a player is in fact reflected by the roles he/she is playing. Playing a role means that the player possesses enough trust. In collaboration with roles, agents actually do not have to care about the trust of the other agents because they mainly interact with role players (indicating trust). It is the system that ensures the trust of agents.

In fact, roles are an equivalent mechanism to reputation model in multiagent systems to establish trust [6]. In e-commerce systems, reputation enables buyers to choose the best sellers in the system. Moreover, reputation encourages sellers to behave well, because they will be discarded by future buyers if their reputation goes down due to bad behavior. Roles assist to build reputations in the following aspects:

- Providing methods to gather information that defines the trustworthiness of an agent, using relationships among roles and agents.
- Providing reliable evaluation methods for the qualifications of agents to play roles.
- Providing mechanisms to promote qualifications that truly describe the trustworthiness of an agent.
- Building agents' beliefs, that is, a role played by agent a tells that a is capable of carrying out the delegated tasks as expected for the role.

With roles, agents cooperate with other agents through an environment that includes roles that dictate the rules of agent behaviors. Role mechanisms make agents not self-interested and therefore make sure that the rules of the system prevent lying and collusion between agents. Compared with traditional trust models in multiagent systems, roles apply a centralized evaluation methodology, thus enabling better aggregation and evaluation of qualifications, and hence trust. In fact, roles establish a system-level trust environment that provides truth-eliciting interaction protocols; develop reputation mechanisms that foster trustworthy behavior; and set up security mechanisms that ensure new entrants can be trusted [7, 5].

15.2.2 Establishing Dynamics

In human communities, even though individual interests are emphasized and personal property is protected, we still pursue a better group interest through teamwork. That is to say, to pursue the highest group performance is a stronger requirement than the individual's interest. The social axiom followed by adaptive collaboration is that *every person hopes to become a great member in a great group*. For example, to be a member of a National Basketball Association (NBA) team is an ideal for most basketball players in the world, and likewise, to be a member of a National Hockey League (NHL) team is an ideal for most hockey players in the world.

The process of collaboration can affect members' satisfaction, motivation, and productivity. The design of the collaboration process largely affects the success of collaboration among a group of people [8]. Many people collaborate passively and only act as requested. To encourage people to work more is a general aim for all regulations or policies of societies or organizations. To build a system that can encourage people to actively contribute is exciting and challenging. To be a good group, dynamics is required. Group dynamics is defined as a force or impetus for agents to join, advance, progress, and work in an environment, a community of agents, or a system. From this point of view, group dynamics is considered to be an important property of both a multiagent system and its agents.

A rational agent is one that does the right things. According to Russell and Norvig [9], "for each possible percept sequence, it should select an action that is expected to maximize its performance measure, given the evidence provided by the percept sequence and whatever built-in knowledge the agent has". Rationality is one of the dynamics of agent performance in a collaborative system [10]. However, the expression of rationality has not been comprehensively investigated.

Roles are, in fact, the rationale for agents to live. Computer systems are added to the Internet so as to perform certain roles and to provide services on the Internet. To design a system, roles are made to match requests and services. If there are no more requests and services, role production should cease. There may be a period of time when roles are not claimed by agents. After this period, role production should also stop.

Roles can encourage members' participation in group activities by specially designing facilities such as credits. The basic assumptions for establishing dynamics by their roles are based on Maslow's hierarchy of needs [11]. We can say that diligent agents hope to become great members in a great group, and that those agents that never contribute to a team will be finally discarded. Based on this assertion, we assume that most agents would like to improve themselves by actively participating in collaboration. We can divide agents into different types [12]:

(1) Diligent agents try to do as many jobs as possible. They want to take on as many responsibilities as possible and only require the least rights.
(2) Ordinary agents only want to complete the regular jobs assigned to them. They would like to take responsibilities but require matching rights.
(3) Lazy agents try to do as few jobs as possible. They want to grasp as many rights as possible and take as few responsibilities as possible.

To restrain type (3), encourage type (1), and support type (2), coupling both rights and responsibilities tightly with the help of roles is required. That is to say, more rights mean more responsibilities. If agents hope to grasp more rights, they must play more roles. As a result, they must take more responsibilities. For those who do not care about rights, there would be still enough rights for them to take on more responsibilities [12].

With roles, agents can express progress by telling how many significant roles they are playing and have played. That is to say, roles can be taken as a flag of advances for agents in collaboration. With roles, lazy agents will be finally restricted and discarded by the team and diligent agents will finally be the majorities of the team and be promoted to play more important roles that give them opportunities to contribute more [13].

15.2.3 Facilitating Interaction

Interaction is essential for collaboration. People may fear interaction, as in society, misunderstanding can produce conflict, hate, and even war. That is to say, the major problem of interaction is misunderstanding. Therefore, interaction requires common needs, common backgrounds, common views, and common goals. All these constitute a shared model that is required for a successful interaction.

- A shared model makes interaction *concise* and *simple*. For example, even though mathematicians interact with one another when trying to solve a very complicated mathematical problem, their interactions are concise because they share a lot of mathematical models.
- A shared model makes interaction *understandable*. Again, from the example of mathematicians, a difficult problem is understandable between two parties because they share the same models.
- A shared model makes interaction *comfortable*. For example, musicians interact with one another comfortably through the use of musical symbols, that is, a shared model. Mathematicians interact comfortably through the use of formal systems, that is, a shared model.
- A successful outcome may result from a model that is shared by both sides of an interaction. For example, when one person asks another person for directions, it is easy to make this interaction successful if there is a map, that is, the shared model. However, if there is no map at hand, the individuals involved lose the shared model and their interaction may be unsuccessful (Figure 15.2).
- Based on Norman's cognitive model [14] (Figure 15.3), there should be a general shared model in a computer system for HCI. Therefore, the computer icon should be replaced by the cognitive process as shown in Figure 15.4.

Ideal interactions should have a symmetric sharing model, that is, both parties share the same model, and information flow is easily regulated (Figure 15.4). In

FIGURE 15.2 The different mental models between two persons.

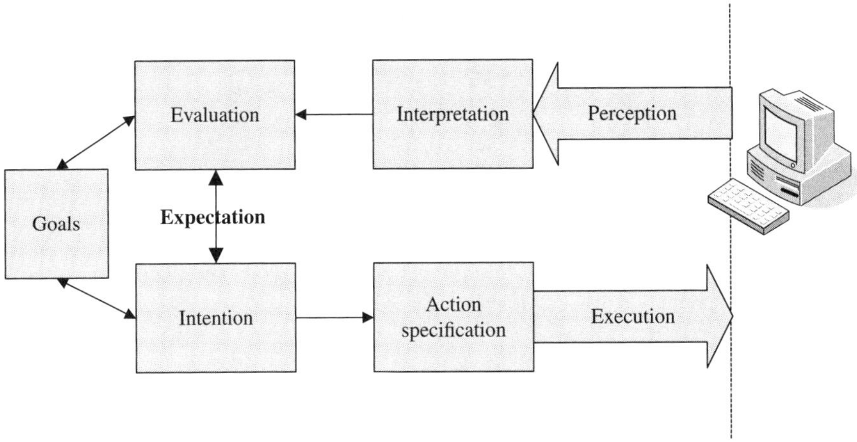

FIGURE 15.3 Norman's cognitive model [14].

Figure 15.4, we hope to express an ideal way for two parties to interact, that is, the two parties should have the same models not only for action specification and interpretation, but also for internal processing. For example, two normal people with the same five senses (which are physical models for perceptions, executions, actions, and interpretations) may fail in an interaction because they have different models for processing knowledge as based on each individual's unique prior training.

Roles are promising shared models for interaction. Based on roles, the scenario of human interaction through computers is as follows (Figure 15.5):

(1) An individual user classifies his/her knowledge and ability into roles.
(2) The computer holds different roles corresponding to the presumed user roles with its role engine [3, 15].
(3) The user logs in by selecting his/her role.
(4) The computer assumes a compatible role to interact with the human user.
(5) If the user role is compatible with the role presented by the computer, interaction continues.

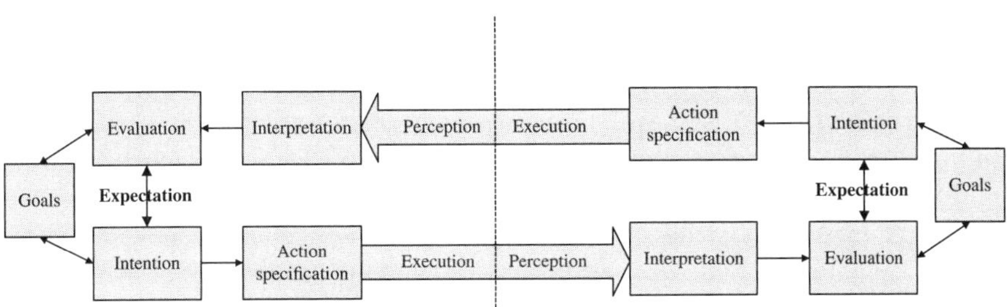

FIGURE 15.4 Ideal interaction based on Norman's model.

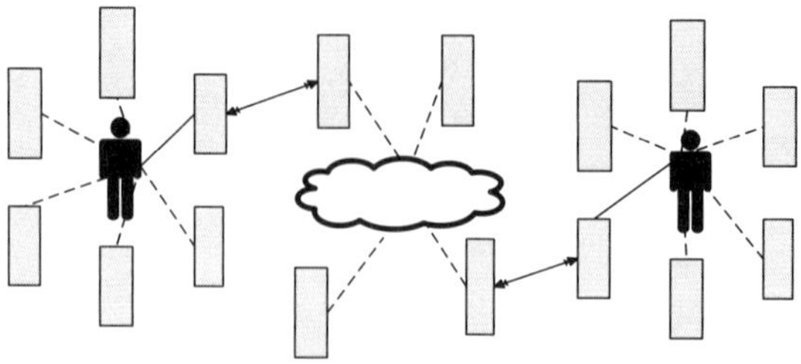

FIGURE 15.5 Interaction through shared models, i.e., roles. Note: The cloud icon expresses a collaborative system. The bars are roles.

(6) If there are some incompatibilities between their roles:
   (a) The person may change his/her role in order to interact with the computer.
   (b) The computer may also adjust its role so that it is more compatible with that of the person, and this makes the interaction successful.
   (c) The user may instruct the computer to change its role.
   (d) The computer may suggest that the user change his/her role.

15.2.4 Support Adaptation

Change is an eternal phenomenon of the world. Adaptability is a special property which enables people to sustain themselves in dynamic environments. The often quoted "survival of the fittest" is a phrase that captures this law of nature. Collaboration is required in situations where one individual is unable to accomplish a given task. In collaboration, the structures, members, relationships among members, and environments are always changing. To maintain good collaborative performance, adaptability is required.

After members join a group, their common goal is maintaining the significance of the group. In other words, they must adapt to and pursue the highest performance of their group. In reality, however, it is not the group members who do not want to adapt to the group. In many cases, it is the system that does not allow individuals to adapt since they must follow a set of static rules of the system.

With the help of computers, group work or adaptability of the group may be made easy to implement. That is to say, computer systems should assist people to adapt to the group, leading to higher group performance, thus serving the interests of the initiating individual [16]. It is essential for a person to be adaptive within a group, since optimal group performance can only be achieved when everybody within the group contributes at their highest potential. Ultimately this will require individuals to adapt and change existing behavior. This is why traditional research on adaptation focuses on the adaptability of individual agents [17–19] and the adaptability of machines to individual users [20–25]. However, true adaptive collaboration concentrates on the adaptability of a group as a whole.

Group adaptation includes many aspects, such as group structure and the adaptabilities of individuals. It reflects the abilities of the group to work, live, socialize, and compete. The adaptability of a group is dependent on its organization, structure, culture, and regulations. There are many factors affecting collaborative performance in a group. For example, culture, interests, personalities, health, equipment, hardware resources, benefits, abilities, powers, motivations, and situations are all relevant factors. How to optimize the group performance in a timely fashion is a big challenge to researchers of man–machine systems and collaborative technologies when considering the number of factors.

The fundamental way to implement adaptive collaboration [3] is to model the group and the factors affecting group performance using computer-based algorithms to timely recommend a group state, in order to obtain the highest performance. Therefore, adaptive collaboration with roles works using the following procedure (where, $\mathcal{A}$ is the agent set and $\mathcal{R}$ the role set):

AdaptiveCollaboration ($\mathcal{A}$, $\mathcal{R}$){

- *Step 1*: Collect information about the group.
- *Step 2*: Evaluate the agents on each role to get all the qualification indicators of agents.
- *Step 3*: Check the group performance.
- *Step 4*: Role assignment.
- *Step 5*: Let agents ($\mathcal{A}$) play roles ($\mathcal{R}$) according to the assignment until the group work is done:
  - *Step 5-1*: Collect information about the group.
  - *Step 5-2*: If the group structure changes significantly, go to *Step 2*.

}

15.2.5 Information Sharing

Information sharing is the foundation for collaboration. Swift and secure information sharing is required in collaboration. Roles provide a new trade-off abstraction among classes, processes, and objects. Therefore, roles also provide a new way for information sharing. Information sharing in the context of collaboration is concerned with two major problems: protection and accessibility.

15.2.5.1 Protection Database management systems are the most widely applied systems that support the sharing of data. Roles are used as a tool that allows administrators to collect users into a single unit against which one can apply permissions. In role-based access control (RBAC) [26], accessibility, and protection are implemented mainly by roles. RBAC aims to apply roles to simplify the tasks of security administrators in order to enforce access control policies. A role is described as a set of transactions that a user or set of users can perform within the context of an organization. Research on RBAC has developed rapidly since its inception, especially in the

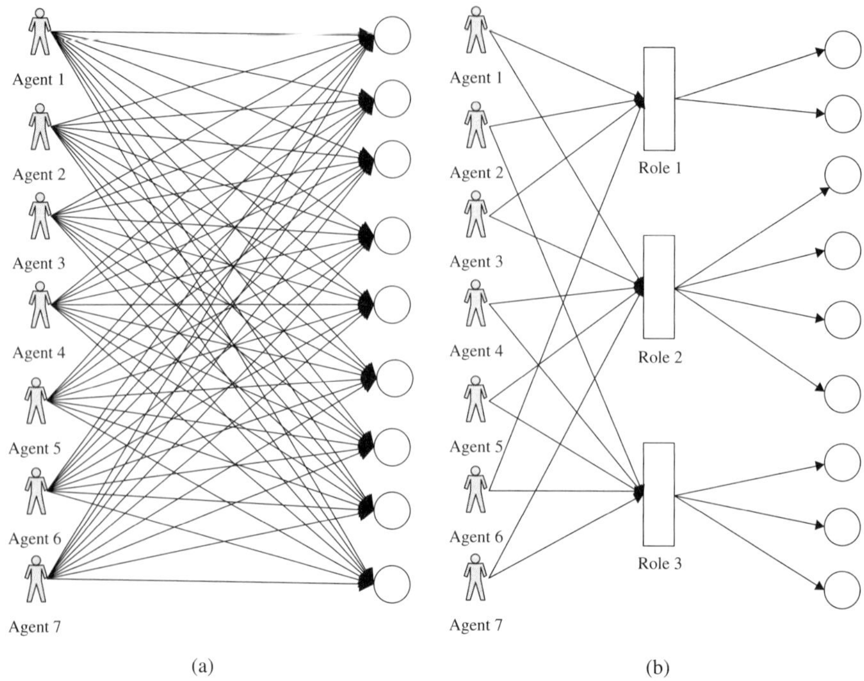

FIGURE 15.6 Information sharing with roles. (a) Access objects without roles. (b) Access objects with roles.

computer security and protection fields. In particular, many researchers discuss the architectures and mechanisms of RBAC [26–28]. Their research demonstrates that roles are excellent underlying mechanisms in dealing with access control and system security. The effort on RBAC has a mature, consistent, and standardized definition. The role concept in RBAC actually comes from the idea used in operating systems. A role is a tag that can be used by a system to perform protection on its resources. The system grants users (based on their roles) the access rights to files in order to accomplish protection.

15.2.5.2 Accessibility As for accessibility, conflict avoidance is the major concern. When there is an intention to access some objects, no conflict is the basic requirement for successful access. Without roles, agents access objects directly with operations. Agents experience conflict when they access critical shared objects. With roles, agents are offered rights to access objects by roles (Figure 15.6). In Figure 15.6a, there are no roles. Suppose at a time, each of the nine objects is accessed by one of the seven agents and the accesses are random and even. The probability of conflict when accessing each object (i.e., two agents access the same object) is $(1/2)^2 \times C(7,2)/9 = 0.5833$. In Figure 15.6b, with three roles, agents are assigned two roles, and each role is designed to access some of the objects. Then, the probability of conflict on each object is at most (Role 2 in Figure 15.6b) $(1/2)^2 \times C(5,2)/3/4 = 0.2083$. Therefore, with roles, occasions of conflict in information sharing can be significantly decreased.

15.2.6 Other Benefits

It is easy for a manager to distribute tasks using roles, because it is easier to find a technician than to find a specific person [15]. Based on roles, managers have ways to evaluate their staff. Roles provide a balanced means for anonymity and credibility. Managers can more easily find qualified players (people or agents) in collaborations. On the other hand, one person can play different roles and serve many clients at the same time period to save the human resources of a company. Two or more people can play the same role to serve a client, in order to improve the efficiency and quality of a specific service and make the client feel more comfortable and more satisfied. Roles can also help players:

- Identify the role player "self" [29].
- Avoid irrelevant interruptions that harm working performance [30].
- Enforce independence by hiding people under roles [31].
- Remove ambiguities to resolve expectation conflicts [8, 32, 33].
- Work with personalized user interfaces [31].
- Distribute tasks based on the overall requirements of a group [34, 4].
- Decrease the workload of system administrators [4, 34].
- Implement separation of concerns [4, 34].
- Decrease the search space of knowledge [4, 34].

15.3 ROLE-BASED COLLABORATION

Because roles have inherent advantages to facilitate collaboration, a role-based collaboration (RBC) methodology was proposed [4, 35]. RBC means that people (agents) collaborate in an environment where they can have clear role specifications and role structures. That is to say, we can also carry out RBC in normal life. RBC is defined as: *A computational methodology that mainly uses roles as underlying mechanisms to facilitate all the aspects of collaboration, including: abstraction, trust, interaction, classification, separation of concerns, dynamics, and sharing.*

The properties of RBC are as follows:

- *Clear role specification*—It is easy for people to understand their responsibilities and rights. Their specified roles are taken as the basis for collaboration. All problems relevant to collaboration can be solved with the help of roles, such as coordination, information sharing, view sharing, and conflict resolution.
- *Flexible role transition*—It is flexible and easy for players to transfer from one role to another. Players may have many roles attached to them but they can

play only one role at a time. This will help players concentrate on the tasks constrained by their current roles.

- *Flexible role facilitation*—It is easy for role facilitators to specify roles. Because the system is developing, even existing roles might be required to adjust to reflect the development of the system.
- *Flexible role negotiation*—It is easy to negotiate the specification between a player and a role facilitator. For example, Mr. Obama may need to negotiate with the US Congress for some special responsibilities or rights to adjust the role of the President of the United States of America.

From these properties, we find that roles are the key media for people to interact or to collaborate. People are allowed to concentrate on interacting with other roles, but not other people, in order to make collaboration more operable. Role specification and role negotiation are the major tasks to form an environment for collaboration. The specification of the interactions among roles defines the processes or procedures of collaboration. Based on the properties of RBC, role specification is the fundamental and key mechanism.

RBC means that people use (agents work in) a role-based collaborative system to cooperate and to obtain an ideal collaboration result. To build a collaborative system, we must understand the fundamental principles with which we need to comply. Many principles we learnt from modeling methodologies, software engineering, and collaborative systems development are the "shoulders of giants."

Roles are themselves not an isolated concept. Without the support of object technology, it would be impossible to specify a role. With object-oriented principles, we can conceptually construct the underlying components of a role-based collaborative system including classes, objects, messages, interfaces, agents, roles, groups, and people. In this way, we can express a role's incoming messages with class mechanisms. Therefore, we can use a special object, such as an agent, to express a player that provides services in a collaborative system. Here, we need to emphasize that agents are representatives of people in a system and that they can do some programmable jobs for people.

A common idea used in object modeling methodologies is that a role is specified from the viewpoint of a server. That is to say, when an object plays a role, it responds to incoming messages [4]. However, no traditional object modeling methodology can specify a role from the viewpoint of a client. This viewpoint considers roles with certain outgoing messages to facilitate the rights of a player. Therefore, we need to introduce the ideas from roles in RBAC that accommodate the rights of people. That is to say, it is a must to combine both the role concepts of object modeling and those of RBAC to support the role models in collaborative systems. Hence, RBC complies with *four sets* of principles, that is, *object, agent, role, and group principles* [4, 35].

Object Principles

- O1: Everything in the world is an object. An object can be used to express everything in a collaborative system.

- O2: Every system is composed of objects and a system is also an object.
- O3: The evolution and development of a system is caused by the interactions among the objects inside or outside the system.
- O4: Objects can be created, modified, and deleted.
- O5: A message is a way to activate the services of an object.
- O6: An interface is a list of message patterns.
- O7: Interactions among objects are expressed by sending messages that are requests to invoke objects' actions.
- O8: Each object is an instance of a class which shows the commonality of a group of objects.
- O9: A class X may inherit another class Y. Y is called a superclass while X is called a subclass.
- O10: Classes can be taken as templates of objects.

Agent Principles
According to References 36–45, we have:

- A1: Agents are special objects that simulate the behavior of people.
- A2: Agents can be created, modified, and deleted.
- A3: Agents are autonomous. They should be able to reply to incoming messages and send outgoing messages based on their situations.
- A4: Agents are adaptive. They should be able to understand their environment and take actions to change the environment and make it better for them to live.
- A5: Agents are social. They should be able to interact with other agents.
- A6: Agents are collaborative. They may join a group to work for a common goal or quit a group if they do not want to cooperate more.
- A7: Agents are flexible. Not all agent actions are predicted. They can dynamically choose which actions to invoke, and in what sequence, in response to the state of its environment.
- A8: Agents are mobile. They can move from one site to another in a system.

Role Principles

- R1: A role is independent of agents.
- R2: A role can be created, changed, and deleted.
- R3: A role includes both responsibilities (the service interface) when an agent is taken as a server and rights (the request interface) when the agent is taken as a client.
- R4: Roles can be interface roles.
- R5: Roles can be process roles. In such roles, what to do, how to do, and what to access are all rigidly specified.

- R6: Roles are taken as media for interactions.
- R7: Playing a role means that the agent is attached to a role. A role can be played by one or more agents at the same time.
- R8: An agent may play one or more roles but can play only one role at a time.
- R9: Roles can be used to support indirect and direct interactions.
- R10: Roles can have hierarchical relationships. Higher level roles can be taken as goals for agents playing roles at lower levels.

Group Principles

- G1: A group is a fundamental structure in a collaborative system.
- G2: A group can be created, changed, and deleted.
- G3: Before specifying a group, we must specify all the roles in it.
- G4: To form a group means letting agents join the group and play roles. They are named as the members of this group.
- G5: A group can be embedded, that is, one group may be an object in another.
- G6: A group can be overlapped with other groups, that is, the members may belong to two or more groups.
- G7: A group can be public or private.
- G8: A group can be open or closed.

We could describe the process of RBC as follows (Figure 15.7):

- *Step 1: Negotiate roles*—People discuss or negotiate to specify the roles relevant to collaboration. If a compromise or an agreement is obtained then the collaboration continues to Step 2, else it aborts.
- *Step 2: Assign roles*—Every person is assigned one or more roles. If agreement is obtained then the collaboration continues to Step 3, else it aborts.
  - *Step 2.1: Agent evaluation*—Rate the qualification of an agent for a role. It requires checking the capabilities, experiences, and credits of agents based on role specifications. In the business world, a person is required to have new knowledge, skills, and habits to be qualified for a new position. It is a fundamental yet difficult problem that requires advanced methodologies, such as information classification, data mining, pattern searching, and matching, to produce high quality evaluations.
  - *Step 2.2: Group role assignment*—Initiate a group by assigning roles to agents (the group members) to obtain the group's highest performance based on the agents' qualifications, which are the results of agent evaluation.
  - *Step 2.3: Role transfer (also called dynamic role assignment)*—Reassign roles to agents or transfer roles of agents to meet the requirements of the system changes.

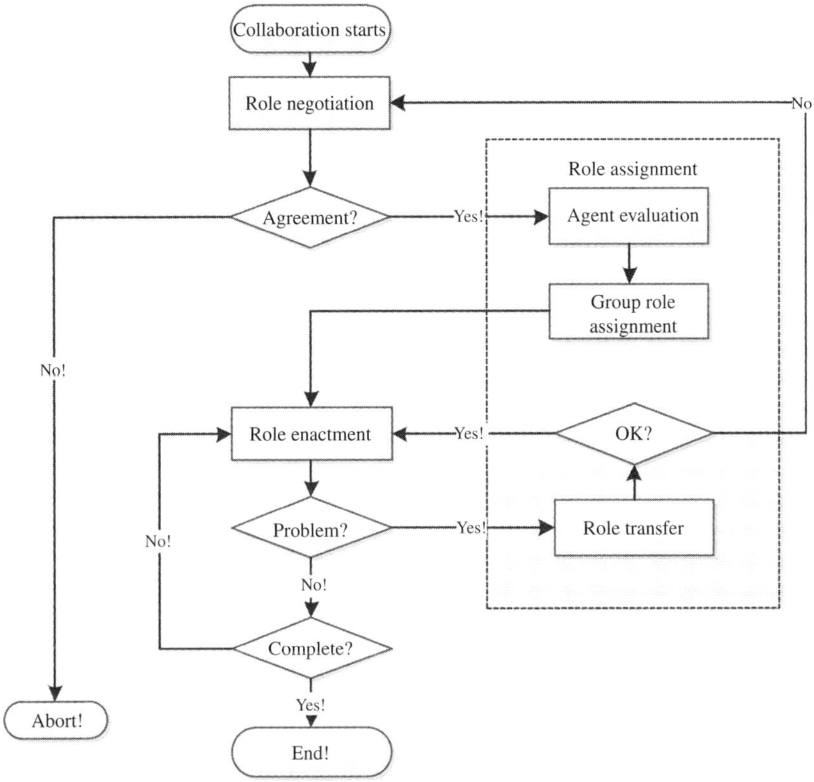

FIGURE 15.7 The process of RBC.

- *Step 3: Play roles*—People work according to their roles until collaboration completes successfully or some conflicts or discontents occur.
  - *Step 3.1: Check incoming messages*—People understand what they need to do at this time. Incoming messages are confined by role responsibilities (the service interface). If conflicts or discontents occur, the collaboration goes to Step 1.
  - *Step 3.2: Issue outgoing messages*—To provide services, people need to access and interact with the environment by sending messages or asking for others' services. If there are no incoming messages, people could think and issue messages as they want. Outgoing messages are confined by role rights (the request interface). If conflicts or discontents occur, the collaboration goes to Step 1.

Role assignments need to keep the following properties:

- *Fairness*—Starvation or overloading should be avoided. Starvation means that an agent has not received role assignments for a time longer than a limit, while overloading means an agent is assigned too many roles in a limited time.

- *Balanced workload*—Agents should be transferred to roles according to their workload.
- *Least conflict*—Agents should be transferred to roles with the least opportunity for agents to have conflicts in sharing information when they are playing roles.

Based on the above criteria, we can judge if a collaborative system is role based or not. In fact, many traditional systems that apply role concepts cannot be called role-based systems because they only support some role views but not the underlying role mechanisms as the central collaborative mechanisms.

15.4 E-CARGO MODEL

The E-CARGO model is an abstract structure based on roles for a group of people and agents who are involved in collaboration or team work. With the E-CARGO model [4, 46, 47], collaboration is based on roles. In E-CARGO, a system $\sum$ can be described as a 9-tuple $\sum ::= <C, O, \mathcal{A}, \mathcal{M}, \mathcal{R}, \mathcal{E}, \mathcal{G}, s_0, \mathcal{H}>$, where C is a set of classes, O is a set of objects, $\mathcal{A}$ is a set of agents, $\mathcal{M}$ is a set of messages, $\mathcal{R}$ is a set of roles, $\mathcal{E}$ is a set of environments, $\mathcal{G}$ is a set of groups, s_0 is the initial state of a collaborative system, and $\mathcal{H}$ is a set of users. In such a system, $\mathcal{A}$ and $\mathcal{H}$, $\mathcal{E}$ and $\mathcal{G}$ are tightly coupled sets. In RBC, a player and his/her agent play a role together. Every group should work within an environment. An environment regulates a group. With this tight coupling, a role-based collaborative system is composed of both computers and human beings. With the participation of people $\mathcal{H}$, such as joining in a team $\sum$, accessing objects of the team, sending messages through roles, and forming a group in an environment, $\sum$ evolves, develops, and functions. The results of team work are a new state of $\sum$ that is expressed by the values of C, O, $\mathcal{A}$, $\mathcal{M}$, $\mathcal{E}$, $\mathcal{G}$, and $\mathcal{H}$. To apply RBC methodologies into autonomic computing, agents $\mathcal{A}$ will be incorporated with some functions of human beings $\mathcal{H}$.

Roles are considered abstract interface specifications, or a process to describe a specific behavior as proposed in the E-CARGO model. Agents are role players. In the formalizations, if χ is a set, $|\chi|$ is its cardinality; $a.b$ means b of a or a's b; "$a=b$" means a is equivalent to b; "$a:= b$" means that a is assigned with b; and "$a::= b$" means that a is defined as b. In E-CARGO, agents are considered to have only one central processing unit and they can only play one role and process related messages at one time.

Definition 1 *class*. $c ::= <n, \mathcal{D}, \mathcal{F}, X>$ where

- n is the identification of the class;
- $\mathcal{D}$ is a data structure description for storing the state of an object including pairs of classes and an external identification;
- $\mathcal{F}$ is a set of the function definitions or implementations; and

- $\mathcal{X}$ is a unified interface of all the objects of this class. It is a set of all the message patterns relevant to the functions of this class. A message pattern tells how to send a message to invoke a function.

We use c to express a specific class and C the set of all classes. Next, we define an *object* based on a class. □

Definition 2 *object.* $o ::= < n, c, s >$ where

- n is the identification of the object;
- c is the object's class identified by the class identification or name; and
- s is a data structure whose values are called attributes, properties, or states.

We use o to express a specific object and O the set of all the objects. The above two definitions comply with object principles. □

Definition 3 *message.* $m ::= < n, v, \ell, \mathcal{P}, t >$ where

- n is the identification of m;
- v is null or the receiver of the message expressed by an identification of a role;
- ℓ is the pattern of m, specifying the types, sequence, and number of parameters;
- $\mathcal{P}$ is a set of objects taken as parameters with the message pattern ℓ, where $\mathcal{P} \subset O$; and
- t is a tag that expresses any, some, or all message. □

Definition 4 *role.* A role is defined as $r ::= <n, I, ⓠ, ⓡ, \mathcal{A}_c, \mathcal{A}_p, \mathcal{A}_o, \mathcal{R}_i, O_r>$ where

- n is the identification of the role;
- $I ::= < \mathcal{M}_{in}, \mathcal{M}_{out} >$ denotes a set of messages, where $\mathcal{M}_{in}$ expresses incoming messages to the relevant agents, and $\mathcal{M}_{out}$ expresses a set of outgoing messages or message templates to roles, that is, $\mathcal{M}_{in}, \mathcal{M}_{out} \subset \mathcal{M}$;
- ⓠ is a set of qualification requirements for an agent to play r;
- ⓡ is a set of rights offered to the agent when playing r;
- $\mathcal{A}_c$ is a set of agents who are currently playing r;
- $\mathcal{A}_p$ is a set of agents who are qualified to play r;
- $\mathcal{A}_o$ is a set of agents who used to play r;
- $\mathcal{R}_i$ is a set of roles interrelated with r [31]; and
- O_r is a set of objects that can be accessed by the agents playing this role. □

Definition 5 *agent.* An agent is defined as $a ::= < n, c_a, ⓠ, ⓡ, r_c, \mathcal{R}_p, \mathcal{R}_o, \mathcal{N}_g>$, where

- n is the identification of a;
- c_a is a special class that describes the common properties of agents or people;
- Ⓠ is the set of qualifications possessed by a;
- Ⓡ is the set of rights requested by a;
- r_c means a role that a is currently playing. If it is empty, then this agent is free;
- $\mathcal{R}_p$ means a set of roles that a is qualified to play ($r_c \notin a.\mathcal{R}_p$);
- $\mathcal{R}_o$ means a set of roles that a played before; and
- $\mathcal{N}_g$ means a set of groups that a belongs to.

All current roles and potential roles of agent a (i.e., $a.\mathcal{R}_p \cup \{a.r_c\}$) form its repository role set, denoted as $\mathcal{R}_r$. □

Definition 6 *environment.* $e ::= <n, \mathcal{R}_e, \mathcal{B}>$ where

- n is the identification of the environment;
- $\mathcal{R}_e$ is a finite set of roles; and
- $\mathcal{B} = \{ <r, q> \}$ is a set of tuples of role (or more exactly, a role class) and role range. The cardinality range q tells how many agents may play this role in this environment and is expressed by $[l, u]$. □

For example, q might be [1, 1], [2, 2], [1, 10], [3, 50], ... It states how many agents may play the same role r in the group.

Definition 7 *group.* $g = < n, e, \mathcal{A}_g, \mathcal{J}>$ where

- n is the identification of the group;
- e is an environment for the group to work;
- $\mathcal{A}_g$ is a finite set of agents; and
- $\mathcal{J}$ is a set of tuples of identifications of an agent and role, that is, $\mathcal{J} =\{< a, r>| \exists q \in Q, o_s \in O \ni (<r, q> \in e.\mathcal{B})\}$, where, Q is used to express the whole set of pairs of positive integers, such as, $[l, u]$. □

When an agent joins a group, it must play a role and attach with (put aside a part of its memory for) a role instance. With E-CARGO, a role-based system is composed of *groups* of *agents* playing *roles* situated in an *environment* to access *classes* of *objects*.

15.5 A CASE STUDY WITH RBC AND E-CARGO

A soccer team is composed of soccer players who trust each other (Table 15.1) [47]. There are 20 soccer players (a_0–a_{19}), we have $\mathcal{A}_g = \{a_0, a_1, \ldots, a_{19}\}$, where, $a_0.n$ = "Adam", $a_1.n$ = "Bob", ...

TABLE 15.1 The Players' Training Cost Indices

Players \ Positions	Goalkeeper	Back	Midfield	Forward
Adam	0.18	0.82	0.29	0.01
Bob	0.35	0.80	0.58	0.35
Chris	0.84	0.85	0.86	0.36
Doug	0.96	0.51	0.45	0.64
Edward	0.22	0.33	0.68	0.33
Fred	0.18	0.75	0.47	0.73
George	0.62	0.21	0.03	0.30
Harry	0.96	0.50	0.10	0.73
Ice	0.25	0.18	0.23	0.39
Joe	0.56	0.35	0.80	0.62
Kris	0.49	0.09	0.33	0.58
Larry	0.38	0.54	0.72	0.20
Matt	0.91	0.31	0.34	0.15
Nolan	0.85	0.34	0.43	0.18
Oak	0.44	0.06	0.66	0.37
Peter	0.05	0.53	0.71	0.11
Quick	0.74	0.32	0.36	0.75
Rick	0.08	0.06	0.83	0.78
Sam	0.64	0.79	0.29	0.30
Ted	0.61	0.10	0.74	0.52

Step 1: Role Negotiation—In the soccer game, how many roles should be set up is simple, because on a soccer field there are normally four roles: goalkeeper (r_0), backs (r_1), midfields (r_2), and forwards (r_3). Therefore, we have $\mathcal{R}_e = \{r_0, r_1, r_2, r_3\}$; The formation is a major concern of the coach, that is, the environment should be carefully planned and arranged. At the beginning, the coach sets up an initial formation, For example, 4-3-3: 1 goalkeeper (r_0), 4 backs (r_1), 3 midfields (r_2), and 3 forwards (r_3), that is.,:

$\mathcal{B}_0 = \{<r_0, [1, 1]>, <r_1, [4, 4]>, <r_2, [3, 3]>, <r_3, [3, 3]>\}$; and

$e_0 ::= <formation0, \mathcal{R}_e, \mathcal{B}_0>$.

Step 2: Role Assignment

Step 2.1: Agent evaluation—The coach evaluates each player in the team with a different qualification index value for each role (Table 15.1) with historical data such as emotion, health, training performance, etc. [48].

Step 2.2: Group Role Assignment—An optimization assignment can be obtained through a specific assignment algorithm [47] that assigns the most qualified players among all. The assignment $\mathcal{T}_0$ is shown as a matrix in Figure 15.8a.

Step 3: Role Enactment—The game starts, and players play the game by interacting with messages (Steps 3.1 and 3.2). During the game, the coach finds that a formation needs to be adjusted, For example, $\mathcal{B}_1 = \{<r_0, [1, 1]>, <r_1, [3, 3]>, <r_2, [5, 5]>, <r_3, [2, 2]>\}$, $e_1 ::= <formation1, \mathcal{R}_e, \mathcal{B}_1>$. Now, go to Step 2.3: Role transfer is required. Because of the availability of an efficient algorithm, we redo Step 2.2 with Table 15.1, and obtain a new assignment $\mathcal{J}_1$ (Figure 15.8b). Note: In real situations, the coach can actually modify the qualification index table according to the performance of each player in the field and obtain a better assignment.

With similar reasons, two more formations are required, that is,

$\mathcal{B}_2 = \{<r_0, [1, 1]>, <r_1, [2, 2]>, <r_2, [5, 5]>, <r_3, [3, 3]>\}$, and
$\mathcal{B}_3 = \{<r_0, [1, 1]>, <r_1, [1, 1]>, <r_2, [3, 3]>, <r_3, [6, 6]>\}$. Then, two new environments are:

$e_2 = <formation2, \mathcal{R}_e, \mathcal{B}_2>$ and $e_3 = <formation3, \mathcal{R}_e, \mathcal{B}_3>$.

The corresponding role assignments ($\mathcal{J}_2$ and $\mathcal{J}_3$ in Figure 15.8) are processed by the coach.

Finally, the team obtains the best collaboration performance [16] with a group in an adaptive organization, where n = "Well-Organized Soccer Team":

$g_0 = <n, e_0, \mathcal{A}_g, \mathcal{J}_0>; g_1 = <n, e_1, \mathcal{A}_g, \mathcal{J}_1>; g_2 = <n, e_2, \mathcal{A}_g, \mathcal{J}_2>$; and $g_3 = <n, e_3, \mathcal{A}_g, \mathcal{J}_3>$.

$$
\text{(a)}\ \begin{bmatrix} 0&0&0&1\\0&0&0&0\\0&0&0&0\\0&0&0&0\\0&0&0&0\\0&0&0&0\\0&0&1&0\\0&0&1&0\\0&0&1&0\\0&0&0&0\\0&1&0&0\\0&0&0&0\\0&0&0&1\\0&0&0&1\\0&1&0&0\\1&0&0&0\\0&0&0&0\\0&1&0&0\\0&0&0&0\\0&1&0&0 \end{bmatrix}
\quad
\text{(b)}\ \begin{bmatrix} 0&0&0&1\\0&0&0&0\\0&0&0&0\\0&0&0&0\\0&0&0&0\\0&0&0&0\\0&0&1&0\\0&0&1&0\\0&0&1&0\\0&0&0&0\\0&0&1&0\\0&0&0&0\\0&0&0&1\\0&0&0&0\\0&1&0&0\\1&0&0&0\\0&0&0&0\\0&1&0&0\\0&0&1&0\\0&1&0&0 \end{bmatrix}
\quad
\text{(c)}\ \begin{bmatrix} 0&0&0&1\\0&0&0&0\\0&0&0&0\\0&0&0&0\\0&0&0&0\\0&0&0&0\\0&0&1&0\\0&0&1&0\\0&0&1&0\\0&0&0&0\\0&0&1&0\\0&0&0&0\\0&0&0&1\\0&0&0&0\\0&1&0&0\\0&0&0&1\\0&0&0&0\\1&0&0&0\\0&0&1&0\\0&1&0&0 \end{bmatrix}
\quad
\text{(d)}\ \begin{bmatrix} 0&0&0&1\\0&0&0&0\\0&0&0&0\\0&0&0&0\\0&0&0&0\\0&0&0&0\\0&0&1&0\\0&0&1&0\\0&0&1&0\\0&0&0&0\\0&0&0&0\\0&0&0&1\\0&0&0&1\\0&0&0&1\\0&1&0&0\\0&0&0&1\\0&0&0&0\\1&0&0&0\\0&0&0&1\\0&0&0&0 \end{bmatrix}
$$

FIGURE 15.8 The role assignments for different environments: (a) $\mathcal{J}_0$ (b) $\mathcal{J}_1$ (c) $\mathcal{J}_2$ (d) $\mathcal{J}_3$.

15.6 CONCLUSIONS

In summary, roles are promising mechanisms to support collaboration. They help establish trust among members in a group; create dynamics in a group for members to actively work; facilitate interaction with shared models; enable adaptation of members in a group; and support information sharing with protection. They have been widely applied in many different fields for many years and there will be many more applications to come [34]. This chapter concludes that:

(1) Roles are important in information systems design but are often introduced intuitively and then applied.
(2) RBC is an emerging computational methodology that facilitates well-defined organizational structures, conceptualization, abstraction, classification, interaction, separation of concerns, and orderly system behavior, and consolidates system security for both human and nonhuman entities that collaborate and coordinate their activities with or within systems.
(3) The E-CARGO model is a promising abstract template for organizing team work and evaluating the performance of a team.

The research and practice on roles still faces many challenges. Therefore, we can foresee more interesting and successful activities in the research and practice of roles.

REFERENCES

[1] E. A. Kendall. Role model designs and implementations with aspect-oriented programming. *ACM SIGPLAN Notices*, 34(10): 353–369, 1999.
[2] B. Pernici. Objects with roles. *ACM SIGOIS Bulletin, The Conference on Office Information Systems*, 11(2–3): 205–215, 1990.
[3] H. Zhu. Role-based autonomic systems. *International Journal of Software Science and Computational Intelligence*, 2(3): 32–51, 2010.
[4] H. Zhu and M. C. Zhou. Role-based collaboration and its kernel mechanisms. *IEEE Transactions on Systems, Man and Cybernetics, Part C*, 36(4): 578–589, 2006.
[5] S. D. Ramchurn, D. Huynh, and N. R. Jennings. Trust in multi-agent systems. In: *The Knowledge Engineering Review*, 19(1). Cambridge University Press, pp. 1–25, 2004.
[6] J. Sabater, and C. Sierra. REGRET: a reputation model for gregarious societies. In: Proceedings of the 1st International Joint Conference on Autonomous Agents and Multi-Agent Systems, Bologna, Italy, 2002, pp. 475–482.
[7] P. Dasgupta. Trust as a commodity. In: *Trust: Making and Breaking Cooperative Relations*, edited by D. Gambetta. Blackwell, 1998, pp. 49–72.
[8] D. Hellriegel, J. W. SlocumJr., and R. W. Woodman. *Organizational Behavior*. St. Paul, MN: West Publishing Co., 1983.
[9] D. Russell and P. Norvig. *Artificial Intelligence: A Modern Approach*, 2nd edn. Upper Saddle River, NJ: Pearson Education, Inc., 2003.

[10] H. Zhu. Role as dynamics of agents in multi-agent systems. *System and Informatics Science Notes*, 1(2): 165–171, 2007.

[11] A. Maslow. *Motivation and Personality*, 2nd edn. Harper & Row, 1970.

[12] H. Zhu. Encourage participants' contributions by roles. In: Proceedings of The IEEE International Conference on Systems, Man and Cybernetics, Big Island, HI, 2005, pp. 1574–1579.

[13] H. Zhu. Roles versus classes. In: Proceedings of the International Conference on Software Engineering Theory and Practice, Orlando, FL, 2007, pp. 89–95.

[14] D. A. Norman. Cognitive engineering. In: *User Centered System Design: New Perspectives on Human–Computer Interaction*, edited by D. A. Norman and S. W. Draper. Hillsdale, NJ: Lawrence Erlbaum Associates, 1986, pp. 32–65.

[15] H. Zhu, R. Alkins, and M. Grenier. Role-based chatting. *Journal of Software*, 3(6): 69–78, 2008.

[16] N. L. Kerr and R. S. Tindale. Group performance and decision making. *Annual Review of Psychology*, 55: 623–655, 2004.

[17] P. de Wilde, M. Chli, L. Correia, R. Ribeiro, P. Mariano, V. Abramov, and J. Goossenaerts. Adapting populations of agents. In: *Adaptive Agents and Multi-agent Systems, LNAI 2636*, edited by E. Alonso, D. Kudenko, and D. Kazakov. Berlin: Springer, 2003, pp. 110–124.

[18] S. van Splunter, N. J. E. Wijngaards, and F. M. T. Brazier. Structuring agents for adaptation. In: *Adaptive Agents and Multi-agent Systems, LNAI 2636*, edited by E. Alonso, D. Kudenko, and D. Kazakov. Berlin: Springer, 2003, pp. 174–186.

[19] G. Picard and M.-P. Gleizes. An agent architecture to design self-organizing collectives: principles and application. In: *Adaptive Agents and Multi-agent Systems, LNAI 2636*, edited by E. Alonso, D. Kudenko, and D. Kazakov. Berlin: Springer, 2003, pp. 110–124.

[20] R. Atterer, M. Wnuk, and A. Schmidt. Knowing the user's every move: user activity tracking for website usability evaluation and implicit interaction. In: Proceedings of the 15th International Conference on World Wide Web, Cambridge, UK, 2006, pp. 203–212.

[21] P. Brusilovsky and E. Millán. User models for adaptive hypermedia and adaptive educational systems. In: *The Adaptive Web: Methods and Strategies of Web Personalization, LNCS*, Vol. 4321, edited by P. Brusilovsky, A. Kobsa, and W. Nejdl. Springer, 2007, pp. 3–53.

[22] G. Fischer. User modeling in human–computer interaction. *User Modeling and User-Adapted Interaction*, 11: 65–86, 2001.

[23] C. Gena. Methods and techniques for the evaluation of user-adaptive systems. *The Knowledge Engineering Review*, 20(1): 1–37, 2005.

[24] M. Hou, R. D. Kobierski, and M. Brown. Intelligent adaptive interfaces for the control of multiple UAVs. *Journal of Cognitive Engineering and Decision Making*, 1(3): 327–362, 2007.

[25] A. Jameson. Adaptive interface and agents. In: *The Human Computer Interaction Handbook*, edited by J. A. Jacko and A. Sears. Erlbaum, 2003, pp. 305–330.

[26] D. F. Ferraiolo and D. R. Kuhn. Role-based access control. In: Proceedings of the NIST-NSA National (USA) Computer Security Conference, 1992, pp. 554–563.

[27] D. F. Ferraiolo, J. A. Cugini, and D. R. Kuhn. Role-based access control (RBAC): features and motivations. In: Proceedings of the 11th Annual Computer Security Applications Conference (CSAC '95), Los Alamitos, CA, 1995, pp. 241–248.

[28] D. F. Ferraiolo, R. Sandhu, S. Gavrila, D. R. Kuhn, and R. Chandramouli. Proposed NIST standard: role-based access control. *ACM Transactions on Information and System Security*, 4(2): 224–274, 2001.

[29] B. E. Ashforth. *Role Transitions in Organizational Life: An Identity-Based Perspective*. Lawrence Erlbaum Associates, Inc., 2001.

[30] M. Czerwinski, E. Cutrell, and E. Horvitz. Instant messaging: effects of relevance and timing. In: *People and Computers XIV: Proceedings of HCI 2000*, vol. 2, edited by S. Turner and P. Turner. British Comput. Soc., 2000, pp. 71–76.

[31] H. Zhu and M. Hou. Role-based human–computer interaction. *International Journal of Cognitive Informatics and Natural Intelligence*, 5(2): 37–57, 2011.

[32] R. P. Bostrom. Role conflict and ambiguity: critical variables in the MIS user-designer relationship. In: Proceedings of the 17th Annual Computer Personnel Research Conference, Miami, FL, 1980, pp. 88–115.

[33] D. I. Hawkins, R. J. Best, and K. A. Coney. *Consumer Behavior*. Plano, TX: Business Publications, Inc., 1983.

[34] H. Zhu and M. C. Zhou. Roles in information systems: a survey. *IEEE Transactions on Systems, Man and Cybernetics, Part C*, 38(3): 377–396, 2008.

[35] H. Zhu. Role mechanisms in collaborative systems. *International Journal of Production Research*, 44(1): 181–193, 2006.

[36] N. R. Jennings and M. Wooldridge. Software agents. *IEE Review*, 42(1): 17–20, 1996.

[37] N. R. Jennings, K. Sycara, and M. Wooldridge. A roadmap of agent research and development. *Autonomous Agents and Multi-agent Systems*, 42(1): 7–38, 1998.

[38] P. Maes. Modeling adaptive autonomous agents. *Artificial Life*, 1(1): 135–162, 1994.

[39] H. S. Nwan. Software agents: an overview. *Knowledge Engineering Review*, 11(3): 205–244, 1996.

[40] H. S. Nwan, L. Lee, and N. R. Jennings. Coordination in software agent systems. *BT Technology Journal*, 14(4): 79–89, 1996.

[41] J. Odell, M. Nodine, and R. Levy. A metamodel for agents, roles, and groups. In: *Agent-Oriented Software Engineering (AOSE), Lecture Notes on Computer Science*, Vol. 3382, edited by J. Odell, P. Giorgini, and J. Müller. Berlin: Springer, pp. 78–92, 2005.

[42] J. Odell, H. van Dyke Parunak, and M. Fleischer. The role of roles in designing effective agent organizations. In: *Software Engineering for Large-Scale Multi-agent Systems, Lecture Notes on Computer Science*, Vol. 2603, edited by A. Garcia, C. Lucena, F. Zambonelli, A. Omicini, and J. Castro. Berlin: Springer, pp. 27–38, 2003.

[43] M. Wooldridge and N. R. Jennings. Intelligent agents: theory and practice. *The Knowledge Engineering Review*, 10(2): 115–152, 1995.

[44] M. Wooldridge, N. R. Jennings, and D. Kinny. The gaia methodology for agent-oriented analysis and design. *Journal of Autonomous Agents and Multi-agent Systems*, 3(3): 285–312, 2000.

[45] F. Zambonelli, N. R. Jennings, and M. Wooldridge. Developing multiagent systems: the gaia methodology. *ACM Transactions on Software Engineering Methodology*, 12(3): 317–370, 2003.

[46] H. Zhu, M. Hou, and M. C. Zhu. Establishing the foundations of adaptive collaboration. In: Proceedings of the ACM/IEEE International Symposium on Collaborative Technologies and Systems (CTS'10), Chicago, IL, May 2010, pp. 546–554.

[47] H. Zhu, M. C. Zhou, and R. Alkins. Group role assignment via a Kuhn-Munkres algorithm-based solution. *IEEE Transactions on Systems, Man and Cybernetics, Part A: Systems and Humans*, 42(3): 739–750, 2012.

[48] H. Zhu and M. Grenier. Agent evaluation for role assignment. In: Proceedings of the IEEE 8th International Conference on Cognitive Informatics (ICCI'09), Hong Kong, China, 2009, pp. 405–411.

PART IV

CLOUD AND SERVICE-ORIENTED COMPUTING

CHAPTER 16

CONTROL-BASED APPROACHES TO DYNAMIC RESOURCE MANAGEMENT IN CLOUD COMPUTING

PENGCHENG XIONG[1], CALTON PU[1], ZHIKUI WANG[2], and GUEYOUNG JUNG[3]
[1]College of Computing, Georgia Institute of Technology, Atlanta, GA, USA
[2]HP Labs, Palo Alto, CA, USA
[3]Xerox Research Center, Norwalk, CT, USA

16.1 INTRODUCTION

Cloud computing is the delivery of computing as a service whereby shared resources, software, and information are provided to computers and other devices as a utility over a network [1]. Cloud computing paradigm provides a range of attractive features such as resource elasticity, cost efficiency, and ease of management. For example, Amazon Elastic Compute Cloud (Amazon EC2) is a web service that provides resizable compute capacity in the cloud [2]. It enables you to increase or decrease computing resource within minutes, for example, you can commission one, hundreds, or even thousands of server instances simultaneously. In another example, cloud computing enables sharing of physical resources and allows multiple services to run in different virtual machines (VMs) in a single physical server. Because enterprise applications can run on consolidated and shared hardware infrastructure, the overall utilization of the shared cloud infrastructure increases while hardware, software, power, cooling, real estate, and also management costs reduce. In traditional data centers, the average resource utilization is about 5–20% [3, 4]. In contrast, shared cloud computing environments are likely to achieve much higher utilization levels.

Contemporary Issues in Systems Science and Engineering, First Edition.
Edited by MengChu Zhou, Han-Xiong Li and Margot Weijnen.

16.1.1 Public Cloud Computing

There are many types of public cloud computing: infrastructure as a service (IaaS), platform as a service (PaaS), and software as a service (SaaS) [1]. Along with SaaS and IaaS, PaaS providers offer a computing platform and a solution stack as a service. The service is able to facilitate the deployment of applications without the cost and complexity of buying and managing the underlying hardware and software and provisioning hosting capabilities. From a PaaS provider's business point of view, profit is determined based on two factors: revenues and resource expenditures. PaaS providers' revenues are determined through the delivery of client request service under Service-Level Agreements (SLAs). The actual revenue model depends on the chosen performance metrics (e.g., round trip time, availability, bandwidth, or throughput) and the statistics defined on them. PaaS providers may rent their resources from IaaS providers in order to offer their service. Computing resources are regarded as services and billed similarly to other commodities.

Although cloud computing brings in advantages over traditional computing paradigms, PaaS providers often face a resource sharing dilemma under workload fluctuation where statically configured virtual resources are often either overprovisioned or overloaded. On the one hand, PaaS providers always pursue high resource utilization, because the higher the utilization, the lower the hardware cost, operating cost, and maintenance cost. On the other hand, resource utilization should not be pushed too high or the revenue could be jeopardized due to bad service-level compliance under time-varying workload.

16.1.2 Dynamic Resource Management: Control-Based Approaches

To deal with the time-varying demand of the modern workloads, at least three resource management strategies have been considered for consolidated environments, that is, capacity planning [5], virtual machine migration [6], and dynamic resource allocation [3, 4, 7–9]. These techniques are complementary to one another because they typically operate at different timescales and different scopes of a data center. Different techniques have been applied for sharing of the resources, for instance, statistical multiplexing, optimization, and recently, control theory. Different metrics may be utilized to evaluate the efficiency of resource sharing, for instance, the resource utilization of the servers in the infrastructure level, and the end-to-end performance such as response times and throughputs of the services/applications that can be perceived by the end users.

As a powerful mechanism for dealing with unpredictable changes, uncertainties, and disturbances in systems, control-based approaches have been used for automatic resource management in cloud computing environments [7]. Control-based approaches are natural choices to tackle the time-varying workloads and systems under control. For example, feedback controllers can not only react promptly to the changes of the systems and the workloads, but also tune the resource allocation to the application in fine time granularity. In many cases that have been reported [8, 9], the performance management problem is usually formulated as a tracking problem. The performance of the application is maintained at the target level and the resource allocation is usually pushed to the limit. As a concrete example [8], the mean end-to-end response time of user requests is taken as the Service-Level Objective (SLO) metric, the value of which has to be maintained below some threshold.

However, although much progress has been made in the dynamic resource management using control-based approaches in cloud computing environments, there are two significant research challenges. First, only single metric like mean response time (MRT) [8, 9] has been considered in the study of quality of service in individual applications. Second, to evaluate the performance of dynamic controllers, a single benchmark application is usually utilized in the previous work although the application may be driven by different synthetic workloads or utilization traces collected from production environment. Little attention has been paid to the behavior of the applications, including that of the workload generators [10]. No insights have been provided into the relationship between performance of the closed-loop system and the behavior of the applications either.

In this chapter, we study these two issues through experiments on a test bed with multi-tier applications hosted on multiple virtual machines. Our study can be summarized in three parts.

First, we show that, when the resource utilization of the applications is maintained at constant levels, the end-to-end performance metrics of the applications that can be perceived by the users can be very different. This is because the metrics such as MRTss depend on many factors, for example, the workload intensity, the target utilization levels, and especially the behavior of the workload generators. Merely using a utilization controller is far less than enough to control the end-to-end performance.

Second, we show that, even when the MRT of the applications is maintained at constant levels, the other end-to-end performance metrics of the application that can be perceived by the users can still be different. The same MRT can come with different response time variations and throughputs. More than one performance metric, for instance, MRT, percentile response time, and throughput have to be used to characterize the performance the user will receive.

Third, our study provides insights into different controller performance when the same feedback controller is applied to applications with different workload generators. What we have found implies that the evaluation of the controller has to be done on multiple applications that can represent different practical scenarios. And in reality, the behavior of the applications has to be carefully considered when the controller parameters are configured.

16.2 EXPERIMENTAL SETUP AND APPLICATION MODELS

To evaluate the quality of service provided by applications hosted in virtualized environment, we set up a test bed based on Xen technology [11], and implement utilization controllers and performance controllers that can allocate the CPU resources dynamically to the components of multi-tier applications. The proposed test bed,

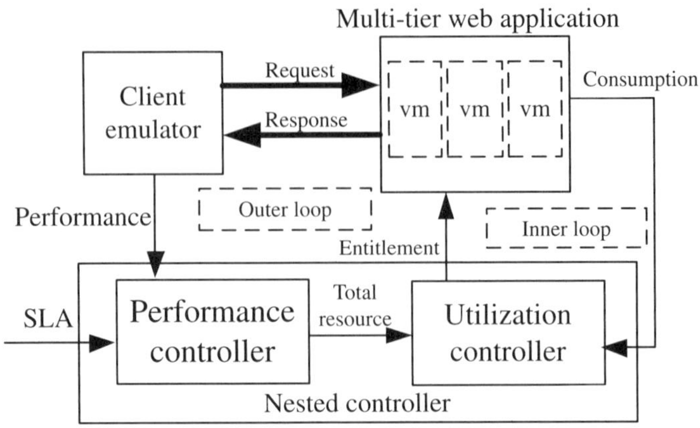

FIGURE 16.1 Nested controller for a multi-tier web application in virtualized environment.

control architecture and algorithms, and the two workload generators that can generate requests following closed or open system models are described as follows.

16.2.1 Test Bed and Control Architecture for a Multi-Tier Application

Figure 16.1 shows a three-tier web application composed of a web tier, an application tier, and a database tier. We use RUBiS as the benchmark application [12]. It is an online auction benchmark comprised of a front-end Apache Web server, a Tomcat application server, and back-end MySQL database server. Each tier of the application is hosted in one Xen virtual machine. Our test bed consists of three machines as shown in Figure 16.1. One of them is used for hosting three VMs, one for a client emulator, and the last one for a nested controller.

Each machine is an Intel Pentium 4 1.80GHz, 1 GB RAM PC with Gigabit Ethernet connected with the switch. All machines run Linux kernel 2.6.16.29. The hosting machine runs Xen 3.0.3. We use Apache v2.0.54 as the web server, Jakarta Tomcat v5.0.28 as the application server, and MySQL v3.23.58 as the database server.

For our experimental evaluation purpose, we implement the nested controllers as shown in Figure 16.1. In the inner loop, there is one utilization controller for each VM. In the outer loop, a performance controller is used for end-to-end performance guarantee of the whole application. We use "entitlement" (u) and "consumption/usage" (v) to refer to the CPU shares (in percentage of total CPU capacity) allocated to a virtual machine and the CPU share actually used by the virtual machine, respectively. We use "utilization" (r) to refer to the ratio between consumption and entitlement, that is, $r = v/u$.

16.2.1.1 Utilization Controller Utilization control has been utilized for workload management in product systems such as HP's gWLM [13] and IBM's Enterprise Workload Manager [14]. The CPU allocation to each VM is tuned dynamically to meet a given, and usually constant utilization target r_{ref}, for instance, 75%. In our approach, we use the simple utilization controller

$$u(k) = v(k-1)/r_{ref}.$$

That is, the new given CPU shares are to satisfy the utilization requirement if the CPU consumption is not going to vary. In our experiments, the CPU consumption is collected using "xentop" command. The CPU shares are entitled using "xm sched-credit" command. The new CPU shares are changed in the inner loop control interval k, that is, every 10 seconds.

16.2.1.2 Performance Controller We choose the end-to-end MRT as the controlled performance metric. The performance controller works every outer loop control interval j, that is, every 90 seconds, larger than that of an inner loop. For simplicity, we use an integrator controller for response time tracking. The controller reads the mean response time $mrt\,(j-1)$ in the previous control interval $j-1$ and then compute the utilization setting $r(j)$ for the current control interval j as

$$r(j) = r(j-1) + \Delta r$$

where $\Delta r = G^{*}(MRT_{ref} - mrt(j-1))/\ MRT_{ref}$.

Here G is the integrator gain. We set $G = 0.1$ in our experiments. The utilization setting is then sent to the utilization controller through TCP/IP-based connection. In our experiments, the utilization settings of all the three tiers are set to the same, although they can be different.

Many controllers have been developed for dynamic resource allocation in virtualized servers. We chose the nested control architecture for the following two reasons. First, it is intuitive and easy to implement in a scalable way. For instance, each VM is associated with one utilization controller that can be implemented inside a local server since only local data are sampled. On the other hand, each performance controller is designed for one application, components of which may span multiple servers. No interaction is needed between the performance controllers of different applications. The second, and the more important reason is that, this design is very simple. Even for multi-tier applications, the resource allocation can still be driven by simple feedback controllers instead of MIMO controllers for end-to-end performance control. Compared with the complexity of the more advanced controllers, the simplicity of this architecture makes it easier to analyze and understand the effect of factors such as application behavior on the performance of the closed-loop system.

The nested control architecture has been utilized in References 8 and 9. In our experiments, the total CPU requests from the utilization controllers are always less than the CPU capacity to avoid resource contention, for which arbitration is needed, and the other mechanism such as VM migration and workload consolidation can help [6].

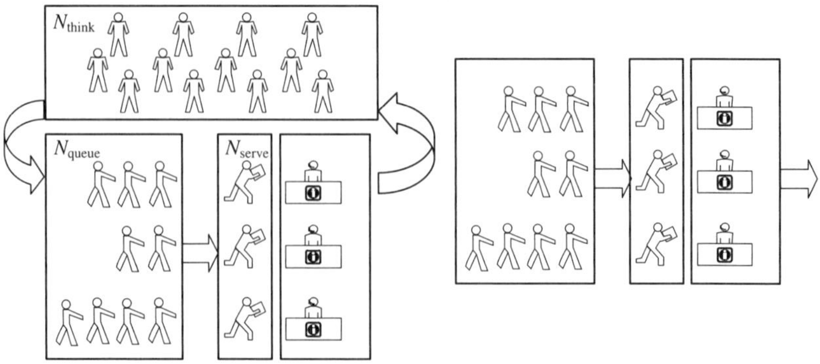

(a) (b)

FIGURE 16.2 (a) Closed and (b) open models.

16.2.2 System Models for the Application: Open or Closed

As shown in Reference 10, the standard web-related workload generators, or the benchmark applications can be classified as closed, open, or semi-open systems. Similar to the system designers, the controller designers pay little attention to the application behaviors, which can have significant effect not only on the performance of the controller itself, but also on the performance of the applications that can be perceived by the end users. We describe how RUBiS workload generator works as a closed system, and how we modify it such that it can work as an open system as well.

16.2.2.1 RUBiS and Its Close-Queue Model A workload generator is called a closed one if new user requests are triggered only after previous requests have been completed or timeout, as shown in Figure 16.2a. The Original Client Emulator, called "OCE" for RUBiS is closed and session oriented. Each session represents a virtual user. Each virtual user connects to the front-end Apache server using a persistent HTTP connection. Within each session, the client generates new request after the previous request is replied, and waits for an exponentially distributed "thinking time." Each session has three possible statuses: (a) waiting in a queue; (b) being served by a server, or (c) "think" for some amount of time. The action sequence of each session follows these steps: (a) to (b), (b) to (c), and (c) back to (a). If we denote the number of sessions in these three statues with N_{queue}, $N_{serve,}$ and N_{think}, respectively, then we have the total number of sessions $N = N_{think} + N_{system} + N_{serve}$. N is also called multiprogramming level (MPL), a parameter that represents the intensity of the workload.

There are 26 transaction types in RUBiS. The types of a new request generated by the virtual users are defined by a transition table. In our experiments, we used "Browsing mix" that has 10 transaction types, for example, Home, Browse, and ViewItem. These transactions have different service demand.

16.2.2.2 Modified RUBiS and Its Open-Queue Model In an open system as shown in Figure 16.2b, new requests are generated independently on completion of the previous requests. To quantify the difference between open and closed systems, we modify the original RUBiS client emulator such that it can work as an open system. A signal producer is created that produces a signal from interval to interval. The length of the intervals follows an exponential distribution. The signal is then put into a queue which is shared by all the sessions that can trigger the generation of new requests in each session other than waiting for the reply of previous requests. We create a large enough number of sessions to compete for the signal. It is guaranteed that that once a signal is produced in the shared queue, it can be immediately consumed by one of the sessions to send a request. The rate of signal defines the request rates. Thus, the behavior of the modified RUBiS application can be modeled as an open system. We call the Modified Client Emulator for RUBiS "MCE."

Note that, there is another alternative to the open and closed systems, that is, semi-open model which is more practical as argued in [10]. In the semi-open model, users arrive separately as in an open system. However, after users receive a response for the previous requests, they will stay and make a follow-up request with some probability p and will leave the system with probability $1 - p$. We do not evaluate this type of systems, but leave that for future work.

16.3 DYNAMIC RESOURCE ALLOCATION THROUGH UTILIZATION CONTROL

The relationship between resource utilization and MRT has been studied and modeled extensively. However, no much work has been done from the viewpoint of dynamic resource allocation. We evaluate this relationship for a three-tier application when the CPU shares of the VMs are dynamically tuned through the utilization controllers.

16.3.1 Design of Experiments

To evaluate the performance of the RUBiS application, we run experiments driven by the two workload generators but under similar workload conditions.

In the case with OCE, the default workload generator, the approximate sending rate is defined by two parameters, the MPL and the thinking time. We tried two MPL values. In one case, MPL = 10, and the mean of the thinking time is 350 ms. In the second case, MPL = 100, and the mean of the thinking time is 3500 ms. In both cases, the average sending rate is approximately 28 req/s when the response times of the requests are much less than the thinking time.

In the case with MCE, that is, the modified open workload generator, the signal producer produces signal every 35 ms on average to the shared signal queue. The producing interval follows exponential distribution with a mean of 35 ms, which means an average sending rate of about 28 req/s.

In all the three cases, the same transition tables are applied. As stated before, we use 10 "Browsing mix" transaction types in our experiment. Each session begins with

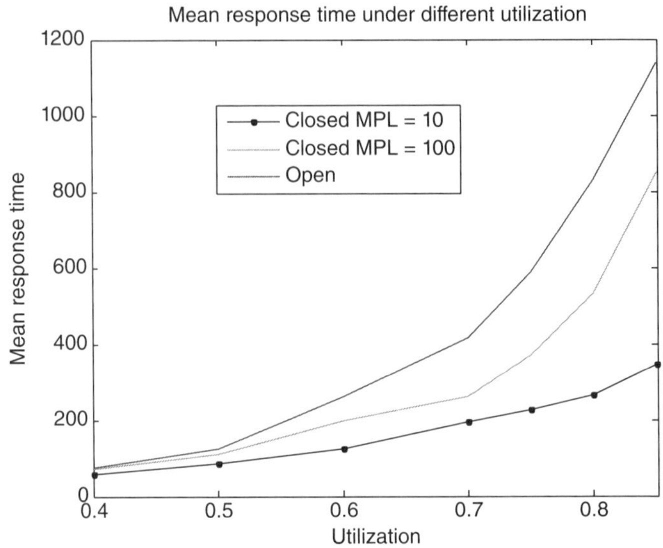

FIGURE 16.3 Relationship between utilization and mean response time.

transaction "Home." After a certain time, the transaction type distribution becomes stable. Because MCE has a much larger number of sessions than OCE, before we collect data for response time, we warm up the experiments for 2 minutes to make both MCE and OCE enter a status where transaction type distribution is stable. Thus, the distribution of transaction types should be the same for OCE and MCE during experiments.

To evaluate the relationship between MRT and utilization, the utilization settings of the VMs are kept unchanged in each experiment that lasts for 10 minutes, maintained by the utilization controllers. All the application response times, the application throughputs, resource entitlement, and resource consumption are collected in each experiment for offline analysis as shown in Figures 16.3, 16.4, 16.5, and 16.6. We change r_{ref} from 0.4 to 0.85, and for each setting, run three experiments driven by the three workloads, respectively.

16.3.2 Performance of the Application Under Control

Figure 16.3 shows the MRTs and the average utilization levels in all the experiments. From which we can find that, the relationship can be very different between the cases with open or closed models. Moreover, even with the same closed models, the curves can be affected by the values of the parameter MPL.

From Figure 16.3, it is not surprising to see that the MRT increases exponentially along with the utilization, which is consistent with the results in traditional queuing theory. However, those two curves for the cases with closed system models are much smoother. This difference is mainly due to the "self-tuning" capability of the OCE.

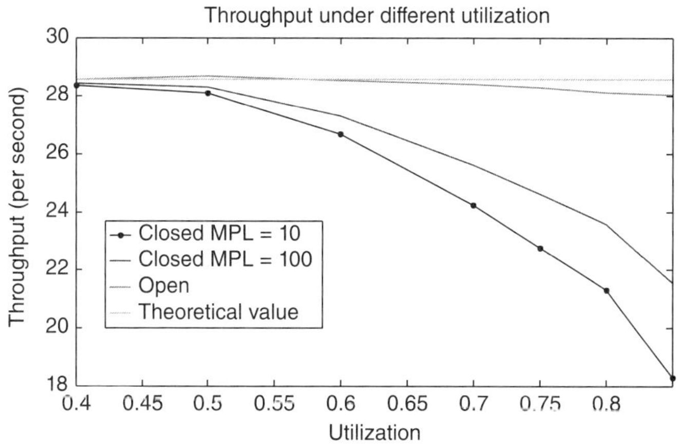

FIGURE 16.4 Relationship between utilization and throughput.

In the closed case, new requests can only be generated after the previous one is replied. This means that, the workload generator can tune the sending rate of the request corresponding to the congestion conditions in the servers along the path. When the servers are more loaded, more requests are queued, and the end-to-end response times increase which slow down the generation rate of the new requests. On the other hand, the same amount of requests is generated in the open cases; even

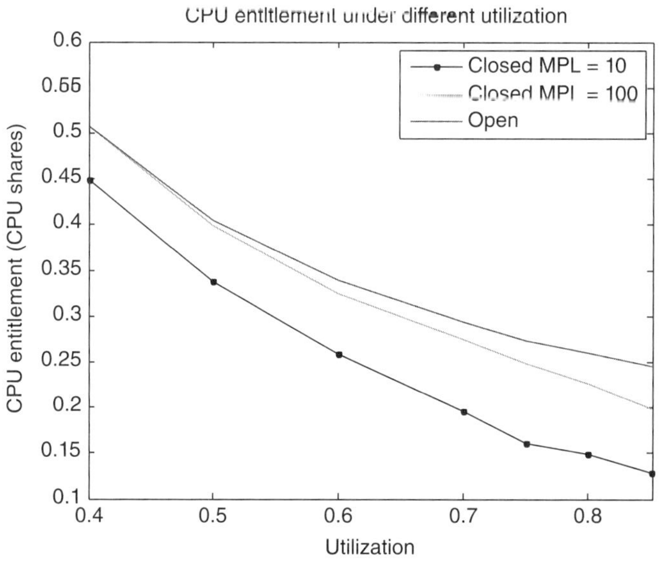

FIGURE 16.5 Relationship between utilization and CPU entitlement.

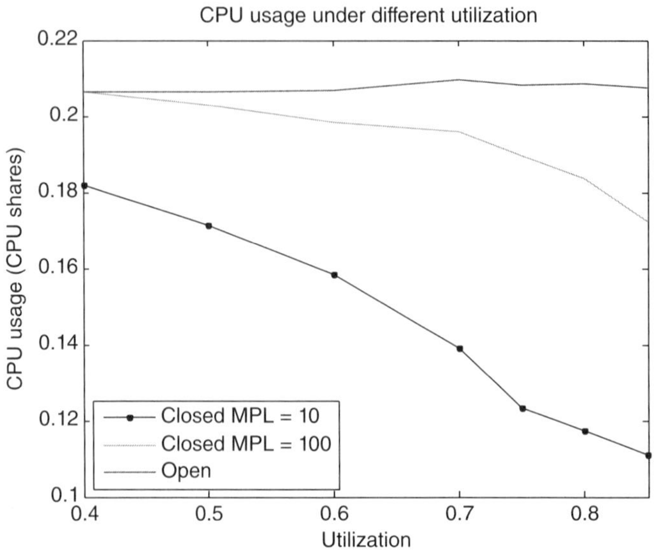

FIGURE 16.6 Relationship between utilization and CPU consumption.

though the capacity available to the servers is reduced as the utilization is increased. The increased utilization and the service times (due to reduction of CPU capacity) both drive up the response times.

The difference between the two closed cases with different MPL settings is not as obvious as that between closed and open ones. On the one hand, this is because the thinking times in the two cases are set up such that the maximum sending rates are the same even if the MPL settings are significantly different. On the other hand, the average queue size with larger MPL values is potentially larger than that with smaller MPL values since at the same time, more requests may be in the pipes.

The effect of the "self-tuning" capability of the closed system can be further observed from Figure 16.4, which shows the throughputs and resource consumption of the VMs in the experiments. In the open cases, the sending rates are the same, and capacity is always available for the VMs. The throughputs are at 28 req/s in all experiments. In the closed case, the throughput decreases along with the increase of utilization due to the increase of the response time.

Note that, in the experiments for the results shown in Figures 16.3 and 16.4, the CPU resources are allocated dynamically. The increase of the utilization is due to changes of CPU shares available to the VMs. It is interesting to investigate further what those differences mean if we would like to maintain application performance through dynamic resource allocation. In the following arguments, we assume that the response time can be maintained at target levels through dynamic resource allocation as we show in Figure 16.1.

The first conclusion is that, application performance cannot be guaranteed at all if only utilization is under control. As we can see from Figure 16.3, for the

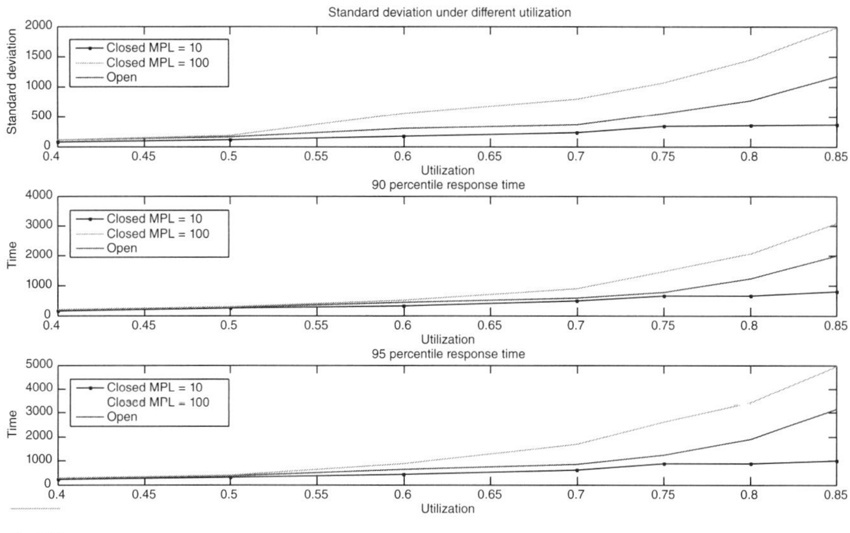

FIGURE 16.7 Illustration for the relationship between utilization and standard deviation of the percentile response time.

same utilization settings, for instance, 80%, the MRTs perceived by the users can be very different, depending on the models followed by the application/workload generator.

Second, the MRT may not be controllable for the open system when the utilization can be pushed too high. As we can see from Figure 16.3 again, the gain from utilization to MRT increases exponentially along with the utilization. For high response time target, the actual response time can be very sensitive with respect to changes of the utilization, or the resource allocation. On the other hand, the MRT for the closed systems can be much more controllable due to the smooth relationship between the metric and resource allocation.

Third, the same MRT settings can result in very different throughputs, depending again on the models the applications follow, as seen in Figure 16.4. Surprisingly, in the closed case, higher response time settings can result in lower throughputs, which is not intuitive. For instance, for an MRT setting of 440 ms, the throughput is around 18 req/s, compared with 28 req/s for a response time setting of 50 ms. The CPU resource consumption and entitlement are also reduced.

Fourth, the same MRT settings can result in very different response time distributions as well due to different workload generators. It has been well known that the MRT is not a very reasonable metric to measure the quality of service provided by the web applications. In Figure 16.7, we show the statistics of the response times of the individual requests during the experiments, including the standard deviation, the 90-percentile and 95-percentile. As we can see, for the open one, the variance of the response times can be very different with different MRT settings. When the utilization is pushed too high, the variance can be much larger than the MRT. Large variance also exists with the open systems, although in general it is much smaller than that in the open case. Between the two types of workload generators, we can see that, the performance that can be perceived by the end users can be very different even with the same MRT settings.

In summary, the experimental results and the analysis provide us insights into the application behaviors, and the effect of the application behaviors on the end-to-end performance. Moreover, it can provide us guidance on design and evaluation of feedback controllers for dynamic resource allocation and performance guarantee.

First, the problem may not be always formulated as a tracking problem. In the closed case, pushing the MRT to a high threshold can compromise the throughput significantly.

Second, the metrics and the target of the metrics have to be carefully chosen. MRT may not be good enough. More metrics, for instance, throughput, and percentile response time may be used that can better characterize the performance the user will experience.

Third, the behavior of the system under control can be very different along with factors like the workload generator and the operation region. Careful consideration has to be taken to configure the controller parameters such as the gains. When we evaluate the performance of controllers (not the application under control), different benchmarks have to be applied before we conclude the evaluation.

16.4 PERFORMANCE GUARANTEE THROUGH DYNAMIC RESOURCE ALLOCATION

We studied the relationship between the utilization, response time, throughput, and resource allocation when the application is under a utilization controller. Arguments have been made on the implication of the workload generators and workloads on the performance. For a validation purpose, we run experiments with the utilization controller as shown in Figure 16.1.

In another set of experiments, we use a real workload trace as shown in Figure 16.8 to evaluate the performance of the applications when under the nested controllers, that is, performance controller and utilization controller. The workload trace is generated based on the web traces from the 1998 World Cup site [15]. We have scaled down the request rates to fit our experimental environment. For the closed system, we change the number of sessions to mimic the change of request rate. For example, we use 10 sessions to mimic 10k request rate and 20 sessions to mimic 20k request rate. Note that this is only an approximation since the actual rates may be tuned by the workload generator itself. For the open system, we change the rates of signal production to mimic the change of request rates. For example, we set the mean interval of signal production to 70 ms to mimic 10k request rate and set the mean interval of signal production to 35 ms to mimic 20k request rate. In the experiments, we set MRT_{ref} = 350 ms. The initial utilization for both models is set to 0.3. Each experiment runs

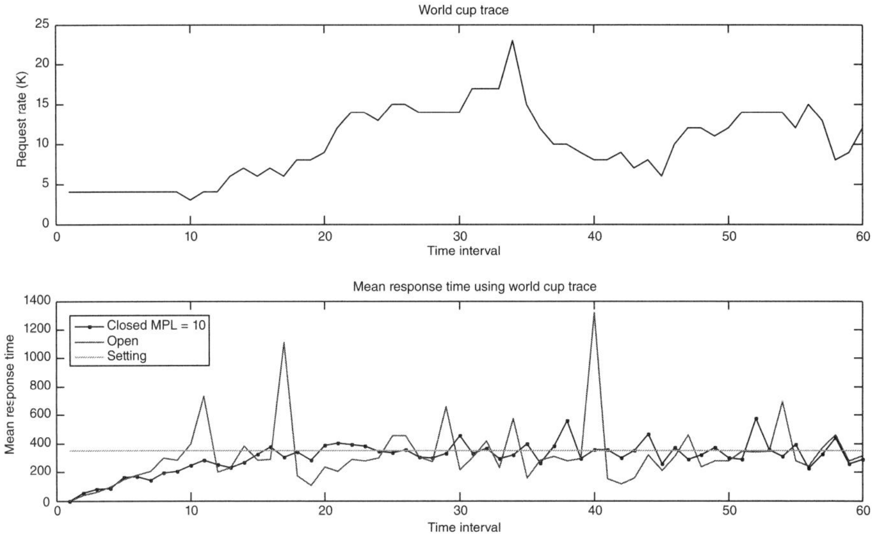

FIGURE 16.8 Request rate and mean response time for 1998 World Cup Trace.

90 minutes. The result from two experiments is shown in Figure 16.8, one for the open system, and the other one for the closed system with MPL = 10.

To evaluate the steady-state statistics, we collect the response time of each transaction during intervals from 20 to 60. The MRTs from the open and closed systems during that time period are 348.7 ms and 342.26 ms, respectively; both are close to the reference. However, the distributions of the individual response times are actually different between the two cases. The cumulative distribution functions of the

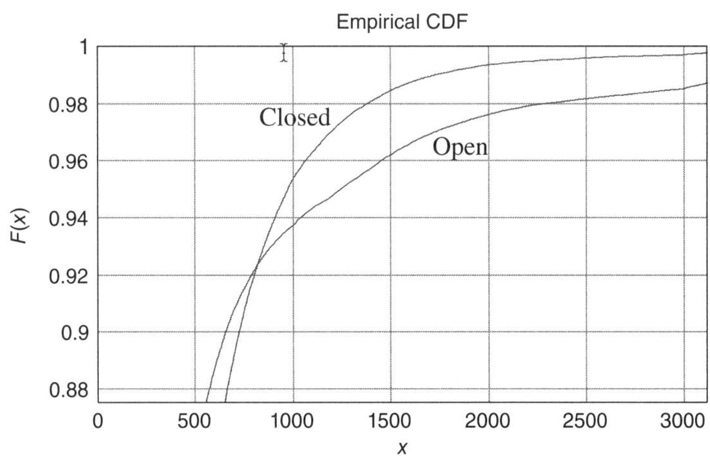

FIGURE 16.9 CDF of response times.

TABLE 16.1 Statistical Summary of Response Times (ms)

	Min	Max	Mean	Median	Std
Open	1	21,803	348.70	167	718
Closed	1	5043	343.26	244	385.47

TABLE 16.2 Percentile Response Times (ms)

	50%	90%	95%	97.5%
Open	167	714	1337	2013
Closed	244	750	1000	1319

response times are shown in Figure 16.9. Overall, the variance of the response time in the closed case is less than that of the open one.

More detailed statistical numbers are shown in Table 16.1 including min, max, mean, median and std (standard deviation), The std value of the open case is almost doubled than that of the closed one. This is consistent with what we can estimate from Figures 16.3 and 16.4. From Figure 16.3, the utilization should be kept within 0.8–0.85 for the closed case and around 0.7 for the open case if the setting response time is 350 ms. Then we check the standard deviation from Figure 16.4. We can find that, for the closed one the standard deviation is around 350 and that for the open one is around 800.

Table 16.2 shows multiple percentile numbers of the response times. We can see that, 95% of the time, the closed model can guarantee the response time within 1000 ms, compared with 1337 ms for the open one.

16.5 CONCLUSIONS

Although a number of studies have proposed solutions to address dynamic resource allocations, most of them do not explicitly account for the impact of various workload models, that is, the behaviors of applications and their workload generators in their approaches. By deploying closed-loop-based benchmarks only, they have dealt with somewhat less aggressive workload variations in their approaches. For instance, Xu *et al.* have developed a two-level control model by using fuzzy control and continuous learning to determine resource allocation for a given workload [16]. They have evaluated their approach by deploying Java Pet Store benchmark only with a closed-loop-based client emulator. Urgaonkar *et al.* have evaluated their queuing model for multi-tier applications by deploying RUBiS and RUBBoS, which are based on closed-loop systems, but do not show any results with open systems [17]. Schroeder *et al.* have introduced the impact of choosing workload models on the dynamic resource allocation by deploying open, closed, and semi-open models and comparing those throughputs and response times [10]. While they have focused on the choice of appropriate scheduling schemes against workload models, we have rather focused on how those models affect the implementation of control theoretic methodologies that are widely employed for regulating the performance characteristics of computer systems.

As shown in our experiments, using only MRT metric to solve the problem could be risky due to the lack of consideration of the distribution of the metric. By employing analytic queuing models and solvers based on mean values, some approaches have limitations in minimizing the number of SLO violations [18–21]. For example, Bennani *et al.* have proposed open and closed queuing network models to address the problem of resource allocation, but used MRT metric in their utility function [18]. Jung *et al.* have used a layered queuing network model, but used a solver, which is based on MRT for a given allocation and workload [19]. Urgaonkar *et al.* have addressed the resource allocation problem by using closed queuing model and RUBiS benchmark, but utilized mean value analysis technique to solve their models [20]. Cunha *et al.* have integrated the tail distribution of response times in their problem formulations. However, they have considered only MRT as a threshold to determine violations [21]. A few researches have been done on the management of the response time distribution such as percentile response time for dynamic resource allocation. For instance, in Reference 22, they have used a realistic SLA scheme that offers a percentile threshold as a guideline to how many requests must meet a given response time over an interval. Meanwhile, our work has proposed that MRT itself is not enough to address the resource allocation problems by providing insights of percentile and MRT-based solutions through extensive experiments.

Control theoretic techniques were applied to address the problem of resource allocation in the past few years [3, 4, 7–9, 23, 24]. However, most of them limit their solutions to closed-loop systems and/or mean-value-based SLOs. For example Padala *et al.* have evaluated their approaches by using three different benchmarks in which only MRT is used in their problem formulation [3]. Although most of the current work uses control theory to manage the resources in virtualized environment, little attention has been paid to the behaviors of the applications themselves, and most of the time, only the MRT is tracked.

In this chapter, we show the different static and dynamic behaviors between closed and open models through extensive experiments. We found that the behaviors of the applications, especially those of the workload generators, have significant effect on the performance of the application when the physical resources of the systems are shared by the applications. We have also showed that multiple metrics have to be considered to characterize the application performance that the user will receive. As ongoing and future work, we are studying other benchmarks, and developing new control problem definition, new control architectures and algorithms.

REFERENCES

[1] http://en.wikipedia.org/wiki/Cloud_computing (accessed January 19, 2015).

[2] http://aws.amazon.com/ec2/ (accessed January 19, 2015).

[3] P. Padala, K. Shin, X. Zhu, M. Uysal, Z. Wang, S. Singhal, A. Merchant, and K. Salem. Adaptive control of virtualized resources in utility computing environments. In: Proceedings of the 2nd ACM SIGOPS/EuroSys European Conference on Computer Systems, Lisbon, Portugal, 2007, pp. 289–302.

[4] P. Padala, K. Hou, X. Zhu, M. Uysal, Z. Wang, S. Singhal, A. Merchant, and K. Shin. Automated control of multiple virtualized resources. In: Proceedings of the European Conference on Computer Systems, March 2009.

[5] J. Rolia, L. Cherkasova, M. Arlitt, and A. Andrzejak. A capacity management service for resource pools. In: The 5th International Workshop on Software and Performance (WOSP'05), Spain, 2005

[6] C. Clark, K. Fraser, S. Hand, J. G. Hansenf, E. Julf, C. Limpach, I. Pratt, and A. Warfield. Live migration of virtual machines. In: Proceedings of NSDI, 2005.

[7] X. Zhu, M. Uysal, Z. Wang, S. Singhal, A. Merchant, P. Padala, and K. Shin. What does control theory bring to systems research? *Operating Systems Review* 43(1): 62–69 (2009)

[8] X. Zhu, Z. Wang, and S. Singhal. Utility-driven workload management using nested control design. In: Proceedings of American Control Conference (ACC), June, 2006.

[9] Z. Wang, X. Zhu, and S. Singhal. Utilization and SLO-based control for dynamic sizing of resource partitions. In: Proceedings of the 16th International Workshop on DSOM, 2005, pp. 133–144.

[10] B. Schroeder, A. Wierman, and M. Harchol-Balter. Open versus closed: a cautionary tale. In: Proceedings of 3rd Conference on Networked Systems Design & Implementation, San Jose, CA, 2006.

[11] P. Barham, B. Dragovic, K. Fraser, S. Hand, T. Harris, A. Ho, R. Neugebauer, I. Pratt, and A. Warfield. Xen and the art of virtualization. In: Proceedings of the 19th ACM SOSP, October, 2003, pp. 164–177.

[12] http://rubis.ow2.org/ (accessed January 19, 2015).

[13] HP Global Workload Manager (gWLM). https://h20392.www2.hp.com/portal/swdepot/displayProductInfo.do?productNumber=gwlm (accessed January 19, 2015).

[14] IBM Enterprise Workload Manager, http://www.redbooks.ibm.com/redbooks/pdfs/sg246785.pdf (accessed January 19, 2015).

[15] M. Arlitt and T. Jin. Workload characterization of the 1998 World Cup web site. HP Technical Report, HPL-99-35, 1999.

[16] J. Xu, M. Zhao, J. Fortes, R. Carpenter, and M. Yousif. On the use of fuzzy modeling in virtualized data center management. In: Proceedings of 4th International Conference on Autonomic Computing, June, 2007.

[17] B. Urgaonkar, P. Shenoy, A. Chandra, and P. Goyal. Dynamic provisioning of multi-tier internet applications. In: Proceedings of 4th International Conference on Autonomic Computing, 2005, pp. 217–228.

[18] M. Bennani and D. Manesce. Resource allocation for autonomic data centers using analytic performance models. In: Proceedings of 2nd International Conference on Autonomic Computing, 2005, pp. 217–228.

[19] G. Jung, M. Hiltunen, K. Joshi, R. Schlichting, and C. Pu. Generating adaptation policies for multi-tier applications in consolidated server environments. In: Proceedings of IEEE International Conference on Autonomic Computing, Chicago, IL, June, 2008.

[20] B. Urgaonkar, G. Pacifici, P. Shenoy, M. Spreitzer, and A. Tantawi. An analytical model for multi-tier internet services and its applications. In: Proceedings of the 2005 ACM SIGMETRICS, pp. 291–302.

[21] I. Cunha, J. Almeida, V. Almeida, and M. Santos. Self-adaptive capacity management for multi-tier virtualized environments. In: Proceedings of 10th IFIP/IEEE International Symposium on Integrated Network Management, 2007, pp. 129–138.

[22] K. Xiong and H. Perros. Resource optimization subject to a percentile response time SLA for enterprise computing. In: IEEE Global Telecommunications Conference, 2006, pp. 1–6.

[23] Y. Diao, N. Gandhi, J. L. Hellerstein, S. Parekh, and D. M. Tilbury. MIMO control of an Apache web server: modeling and controller design. In: Proceedings of American Control Conference, Anchorage, AK, 2002.

[24] C. Lu, X. Wang, and X. Koutsoukos. Feedback utilization control in distributed real-time systems with end-to-end tasks. *IEEE Transactions on Parallel and Distributed Systems*, 16(6): 550–561, 2005.

CHAPTER 17

A PETRI NET SOLUTION TO PROTOCOL-LEVEL MISMATCHES IN SERVICE COMPOSITION

PENGCHENG XIONG[1], MENGCHU ZHOU[2], CALTON PU[1], and YUSHUN FAN[3]

[1]College of Computing, Georgia Institute of Technology, Atlanta, GA, USA
[2]Department of Electrical and Computer Engineering, New Jersey Institute of Technology, Newark, NJ, USA
[3]Department of Automation, Tsinghua University, Beijing, PR China

17.1 INTRODUCTION

Since the early of this century, service-oriented computing has become the representative direction of distributed computing. The core part of service-oriented computing is web services. They are considered as self-contained, self-describing, modular applications that can be published, registered, and invoked across the web. Web Services Description Language (WSDL), Universal Description, Discovery, and Integration (UDDI), and Simple Object Access Protocol (SOAP) are the three standards for web service description, registration, and binding, respectively. A provider offers a service by publishing its description at UDDI. A requester queries UDDI in order to discover a suitable service to perform a certain task. Service-oriented computing and Cloud computing have a reciprocal relationship—the former provides computing of services and the latter provides services of computing.

Most of the time, a single web service cannot satisfy the complex functionality required by a user. Multiple web services need to be composed together in order to fulfill the request. This trend has triggered a significant number of research efforts on the web services' composition both in academia and industry. Web Services Business Process Execution Language (BPEL) [1, 2] is an OASIS [3] standard

Contemporary Issues in Systems Science and Engineering, First Edition.
Edited by MengChu Zhou, Han-Xiong Li and Margot Weijnen.

TABLE 17.1 Taxonomy of Service Composition Mismatches

Service composition mismatches	Interface mismatches	Message signature mismatches
		Message split/merge mismatches
		Message missing/extra mismatches
	Protocol-level mismatches	Unspecified reception
		Mutual waiting mismatches
		Nonlocal choice mismatches

executable language for specifying interactions with web services. It is becoming the *de facto* standard for implementing business processes using web service technology. However, it lacks a formal semantics, which makes the web service composition mismatches hard to analyze. In order to avoid this disadvantage, several attempts have been made to translate BPEL into formal models. By using them, it is able to detect inconsistencies in the BPEL specification and formally analyze BPEL processes using techniques of computer-aided verification. For example, Fisteus *et al.* use finite state machines [4], Ferrara [5] uses process algebra, Fahland and Reisig [6] use abstract state machines, Ouyang *et al.* [7], Hinz *et al.* [8], and Lohmann *et al.* [9] use Petri nets. Other methods, like pi-calculus [10], message sequence charts (MSC) [11], and conversations [12] are also used. Different researchers propose different Petri net semantics for BPEL. Ouyang *et al.* [7] transform BPEL into Petri nets represented in the Petri Net Markup Language (PNML) [13] by BPEL2PNML. They model a BPEL process as a workflow net (WFN) [14]. Lohmann *et al.* [15] adopts open workflow nets (oWFNs) for modeling BPEL processes and developed a compiler BPEL2oWFN. Both BPEL2PNML and BPEL2oWFN implement reduction techniques to reduce the model after their generation.

Despite all these efforts on formal models, as one of the most important problems in service composition, protocol-level mismatch is still a highly complicated problem to solve. The state-of-the-art method to deal with it is to generate adaptors to check deadlock freeness based on a reachability graph. When this property is violated, the generation process will repeat itself until no deadlock state is found. The main drawback of this method is that it does not take into account the future deadlock state and requires many interactions with a developer. Thus it suffers from low efficiency. In this chapter, we model multiple web service interaction with a Petri net called Composition net (C-net). The protocol-level mismatch problem is transformed into the empty siphon problem of a C-net. By this transformation, we not only take future deadlock states into consideration but also find the optimal solution that involves the fewest interactions with a developer. The proposed method is proved to achieve higher efficiency for resolving protocol-level mismatch issues than traditional ones.

In web service composition, when multiple web services are developed by different groups or vendors, they often fail to invoke each other because of mismatches. Service composition mismatches can be divided into interface- and protocol-level ones [16]. Table 17.1 gives a detailed taxonomy of already known service composition mismatches.

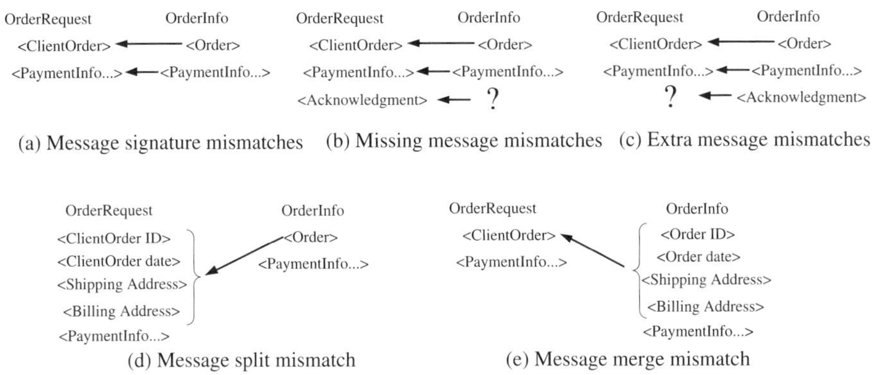

FIGURE 17.1 Illustration for interface mismatches.

17.1.1 Interface Mismatches

Interface mismatches include message signature mismatches, message split/merge, and message missing/extra mismatches [17, 18] as shown in Figure 17.1.

Message signature mismatches can be defined as: a service client's message has a different name and/or data type and/or data range from that of a service provider. For example, the message with the name "ClientOrder" on the service client's site may correspond to the message with the name "Order" on the provider's site. Note that, the case when the provider's message has a different data range from that of the client is also defined as parameter constraint mismatch in [17].

Message split mismatch can be defined as: a service client requires multiple messages to achieve certain functionality while a service provider can offer only a single message. For example, a provider's message *mp* contains *n* parts, that is , mp_1, mp_2, ... , and mp_n while a client considers mp_1, mp_2, ... , and mp_n as *n* separate messages.

Message merge mismatch can be defined as: a service client requires a single message to achieve certain functionality while a service provider offers multiple messages. This is the reverse case of message split mismatch. For example, a client requires message *mc* that is a combination of messages mc_1, mc_2, ... , and mc_n from a provider.

Missing message mismatches can be defined as: a service client expects a message that a service provider does not issue. For example, a client expects an acknowledgment message while although a provider receives the previous message, it does not issue an acknowledgment. Note that the missing message should not affect the semantics of the client. Extra message mismatches are defined as: a provider issues an extra message that a client does not expect. This is the opposite case of missing message mismatches. For example, a provider is sending an acknowledgment message that a client does not expect. Figure 17.1 shows the cases for interface mismatches.

Services can be composed if there are no interface mismatches. There are significant research results toward service interface level mismatch problems such as service-adaptation-based method [17, 18], schema-matching-based method [16], information retrieval techniques [19], and clustering-based approach [20]. Service adaptation generates a service (the adaptor) that mediates the interactions between two services with interface mismatches such that two services are interoperable. For example, Nezhad *et al.* [16] propose a service adaptor that is based on matching of schemas of messages to solve the service interface mismatch problem. They firstly obtain the XML schemas of all the messages on a service provider and client. Secondly, they calculate the similarity between all the possible message pairs considering message types and message orders. Finally, they pick the pair with the largest similarity. This method is proved to be useful to solve the message signature mismatch problems. To resolve the other kinds of interface mismatch problems, Benatallah *et al.* [17] propose a method based on a service adaptor. To resolve message split mismatch issues, the adaptor extracts mp_1, mp_2, ... , and mp_n from mp and then sends them to a service client. To avoid message merge mismatch, it integrates mc_1, mc_2, ... , and mc_n to generate a single message mc and sends it to a service client. To avoid missing message mismatch, it generates a new message by using the template of the expected message and sends it to the service client. Finally, to avoid extra message mismatch, it simply intercepts the extra message and discards it.

17.1.2 Protocol-Level Mismatches

Even if the service interface level matches perfectly, there may be protocol-level mismatch causing problems such as unspecified reception and deadlock [21]. Figure 17.2 shows the mismatch cases for unspecified reception, mutual waiting, and nonlocal choice, respectively. Unspecified reception mismatches can be defined

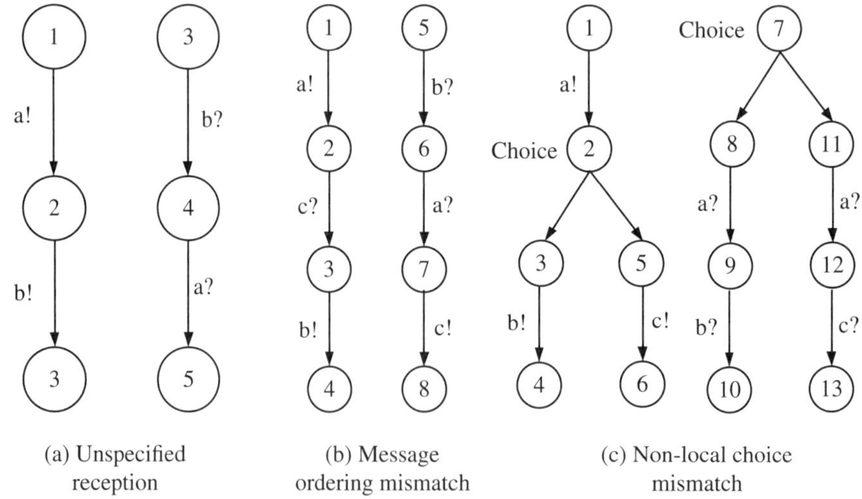

FIGURE 17.2 Illustration for protocol mismatches.

as: a service client's sending message order is different from a service provider's receiving message order. For example, as shown in Figure 17.2a, the client is sending message *a* first (denoted as *a*!), then message *b* (denoted as *b*!). However, the service provider is expecting to receive message *b* first (denoted as *b*?), then *a* (denoted as *a*?). Their interaction will result in an unspecified reception mismatch for the provider.

Deadlock mainly comes from mutual waiting mismatches [16] and nonlocal choice mismatches [22, 23]. Mutual waiting mismatches can be defined as: both of a service client and provider are expecting to receive a message from each other. For example, as shown in Figure 17.2b, after the client sends message *a* (denoted as *a*!), it is expecting to receive *c* (denoted as *c*?) from a provider. However, the provider is expecting to receive *b* (denoted as *b*?) from the client. Both of them are waiting for messages from each other.

Nonlocal choice mismatches can be defined as: a service client and provider make different local choices that should actually be synchronized. For example, as shown in Figure 17.2c, after the client sends message *a* (denoted as *a*!), it makes a local choice, that is, either to send a message *b* (denoted as *b*!) or send a message *c* (denoted as *c*!). However, the provider also makes a local choice, that is, after it receives message *a* (denoted as *a*?), it waits for message *b* (denoted as *b*?) or message *c* (denoted as *c*?). If the service client and provider happen to make the same choices, for example, the client is sending *b* and the provider is expecting to receive *b*, then there is no problem. However, if the client is sending message *b* and the provider is expecting to receive *c*, then there is a nonlocal choice mismatch.

To resolve unspecified reception mismatch issues, Nezhad *et al.* [16] propose a service-adaptor-based method. The adaptor first buffers the pending message, for example, message *a* in Figure 17.2a and then send it to the provider when it is expected. Actually, some commercial products, for example, the Oracle BPEL engine, already have the function to automatically buffer the pending messages for web service interactions. To solve deadlock mismatches, previously proposed methods [16–18] mainly take the following steps: (1) check the deadlock using a state-space-based method; and (2) provide some message and repeat testing the state space until there is no deadlock. In other words, they offer quite limited help in resolving deadlock mismatches. Finding deadlocks involves intense interactions with developers. When protocols become complicated, it is very hard for developers to find the best solutions. For example, it is nontrivial [16] to find the best solution for the mutual waiting mismatches outlined in Figure 17.2b.

We present a Petri-net-based method to find protocol-level mismatches and then generate solutions to fix the mismatch problems. The approach is based on an observation of Petri net objects called siphons. A siphon is a subset of Petri net nodes with a property analogous to program safety properties. The number of tokens in a siphon never increases and an empty siphon always remains empty. Protocol-level mismatch happens if and only if there is an empty siphon in Petri net models derived from BPEL composition.

Technically, the approach consists of three steps. First, we adopt BPEL for web services as the web service composition language. In the first step, the BPEL description of a composite service is translated into a Petri net model. Second, a mixed-integer programming formulation is used to detect the maximal empty siphons, which are then used to find protocol-level mismatches. Third, an algorithm is proposed to find siphon-based solutions for protocol-level mismatches by adding tokens in siphons to prevent them from becoming empty. Finally, time complexity comparisons are made between the approach and existing ones. The proposed one can achieve higher efficiency for resolving protocol-level mismatching issues.

17.2 MODELING SERVICE INTERACTION WITH PETRI NETS

Basic Petri net concepts, simple sequential workflow net (SSN), workflow module net (WMN) and C-net are first discussed. Modeling service interaction with a C-net is then performed.

17.2.1 Basic Petri Nets

The following definitions and discussions are mostly based on References 14, 24, and 25.

Definition 1 A Petri net is a 3-tuple, $N = (P, T, F)$ where:

(i) $P = \{p_1, p_2, \ldots, p_m\}$, $m > 0$, is a finite set of places pictured by circles;
(ii) $T = \{t_1, t_2, \ldots, t_n\}$, $n > 0$, is a finite set of transitions pictured by bars, with $P \cup T \neq \emptyset$ and $P \cap T = \emptyset$;
(iii) $F \subseteq (P \times T) \cup (T \times P)$ is the incidence relation. Based on F, we can derive the input and output functions that are $m \times n$ matrices. $\boldsymbol{I} : P \times T \rightarrow \{0, 1\}$ is an input function. $\boldsymbol{I}(p, t) = 1$ if $(p, t) \in F$; and 0 otherwise. $\boldsymbol{O} : P \times T \rightarrow \{0, 1\}$ is an output function. $\boldsymbol{O}(p, t) = 1$ if $(t, p) \in F$; and 0 otherwise. □

Post-set of t is the set of its output places, that is, $t^{\bullet} = \{p | \boldsymbol{O}(p, t) \neq 0\}$. Pre-set of t is the set of its input places, that is, ${}^{\bullet}t = \{p | \boldsymbol{I}(p, t) \neq 0\}$. Post (Pre)-set of p is the set of its output (input) transitions, denoted by $p^{\bullet}$ and ${}^{\bullet}p$ respectively. $M : P \rightarrow \mathbb{N}^0$, is a marking where $M(p)$ represents the number of tokens in place p and $\mathbb{N}^0 = \{0, 1, 2, \ldots\}$. An initial marking is denoted by M_0. Tokens are pictured by dots. (N, M) is called a net system or marked net. p is marked by M iff $M(p) > 0$. A place subset $S \subseteq P$ is marked by M iff at least one place in S is marked. The sum of tokens in S is denoted by $M(S) = \sum_{p \in S} M(p)$. A transition $t \in T$ is enabled under M if $\forall p \in {}^{\bullet}t$: $M(p) > 0$ holds, denoted as $M[t>$. If $M[t>$ holds, t may fire, resulting in a new marking M', denoted as $M[t > M'$, with $M'(p) = M(p) - I(p,t) + O(p,t)$.

M' is reachable from M iff there exists a firing sequence $\sigma = t_{i1}t_{i2} \ldots t_{ik}$, such that $M[t_{i1} > M_1[t_{i2} > \ldots M_{n\text{-}1}[t_{ik} > M'$ holds. The set of markings reachable from M_0 in N is denoted as $R(N, M_0)$. Given a marked net (N, M_0) and $N = (P, T, F)$, a transition $t \in T$ is live under M_0 if $\forall M \in R(N, M_0)$, $\exists M' \in R(N, M)$, $\ni M'[t >$ holds. N is dead

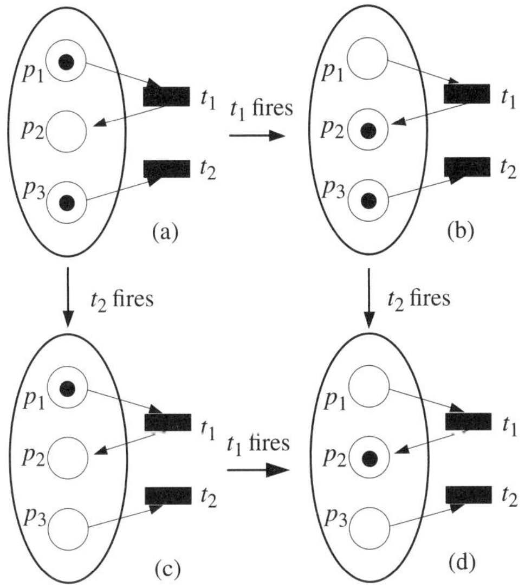

FIGURE 17.3 Illustration of a Petri net siphon.

under M_0 if $\forall t \in T$, $M_0[t>$ cannot hold. (N, M_0) is live if $\forall t \in T$: t is live under M_0. A Petri net N is said to be deadlock free if at least one transition is enabled at every reachable marking.

$N_X = (P_X, T_X, F_X)$ is a subnet of the Petri net $N = (P, T, F)$ iff $P_X \subseteq P$, $T_X \subseteq T$ and $F_X = F \cap ((P_X \times T_X) \cup (P_X \times T_X))$. N_X is generated by P_X iff $T_X = {}^\bullet P_X \cup P_X{}^\bullet$ (where the pre-sets and post-sets are taken w.r.t. F). Note that $\forall Q \subseteq P, {}^\bullet Q = \bigcup_{p \in Q} {}^\bullet p$ and $Q^\bullet = \bigcup_{p \in Q} p^\bullet$. A Petri net is called a state machine if $\forall t \in T$, $|{}^\bullet t| = |t^\bullet| = 1$. An asymmetric choice net or AC net is a Petri net such that $\forall p_1, p_2 \subseteq P$:$p_1^\bullet \cap p_2^\bullet \neq \emptyset \Rightarrow p_1^\bullet \subseteq p_2^\bullet$or $p_1^\bullet \supseteq p_2^\bullet$.

A P-vector is a column vector $X : P \rightarrow \mathbb{Z}$ indexed by P and a T-vector is a column vector $Y : T \rightarrow Z$ indexed by T, where $\mathbb{Z}$ is the set of integers. The incidence matrix of N is a matrix $[N] = O\text{-}I$. X is a P-invariant (place invariant) if $X \neq 0$ and $X^T[N] = \mathbf{0}^T$ hold. Y is a T-invariant (transition invariant) if $Y \neq 0$ and $[N]Y = \mathbf{0}$ hold.$||X|| = \{p \in P | X(p) \neq 0\}$and $||Y|| = \{t \in T | Y(t) \neq 0\}$are called the support of X and Y, respectively.

A nonempty place set $S \subseteq P$ is called a siphon if ${}^\bullet S \subseteq S^\bullet$ holds as shown in Figure 17.3. A siphon is minimal if there does not exist a siphon such that $S' \subset S$and $S' \neq S$. A siphon has a property: a siphon free of tokens at a marking remains token free. Marking M is a dead marking if no transition is enabled at M. A siphon S that eventually becomes empty is called potential deadlock. A Petri net is called potential deadlock free if all the siphons are always marked.

Definition 2 A simple sequential workflow net (SSN) is a Petri net $N = \{P \cup \{\alpha, \beta\}, T, F\}$ if:

(i) N has two special places: α and β where α is a source place, that is, ${}^\bullet\alpha = \emptyset$, and β is a sink place, that is, $\beta^\bullet = \emptyset$.

(ii) If we add a new transition t to N which connects α with β, that is, ${}^\bullet t = \beta$ and $t^\bullet = \alpha$, then the resulting extended net $\overline{N} = (\overline{P}, \overline{T}, \overline{F})$ where $\overline{P} = P, \overline{T} = T \cup \{t\}$, and $\overline{F} = F \cup \{(\beta, t), (t, \alpha)\}$, is a strongly connected state machine.

(iii) Every circuit of N contains t.

(iv) $M_0(\alpha) = 1$ and $M_0(p) = 0, \forall p \neq \alpha$. □

State α is defined as $M(\alpha) = 1$ and $M(p) = 0, \forall p \in P \backslash \{\alpha\}$ while state β is defined as $M(\beta) = 1$ and $M(p) = 0, \forall p \in P \backslash \{\beta\}$. Note that SSN is a special kind of workflow net [14] and is close to the definition of simple sequential processes in Reference 26. SSN has a soundness property:

(i) For every state M reachable from state α, there exists a firing sequence leading from M to β.

(ii) State β is the only state reachable from α with at least one token inβ.

(iii) There are no dead transitions in it.

Definition 3 A workflow module net (WMN) is an extended Petri net $N = \{P \cup P_I \cup P_O, T, F\}$, where:

(i) the subnet generated by P is an SSN;

(ii) $M_0(\alpha) = 1$ and $M_0(p) = 0, \forall p \neq \alpha$; and

(iii) P_I and P_O denote the input and output interfaces for the workflow module respectively, satisfying

   (a) $P_I \neq \emptyset$, $P_O \neq \emptyset$, $P_I \cap P_O = \emptyset$, and $(P_I \cup P_O) \cap P = \emptyset$;

   (b) $\forall p \in P_I$, $p^{\bullet\bullet} \cap P \neq \emptyset$, and $|p^\bullet| = 1$;

   (c) $\forall p \in P_O$, ${}^{\bullet\bullet}p \cap P \neq \emptyset$, and $|{}^\bullet p| = 1$; and

   (d) $\forall p \in P_I \cup P_O$, ${}^\bullet p \cap p^\bullet = \emptyset$. □

We define P as the set of process places and $P_I \cup P_O$ as the set of interface ones. The initial marking of a WMN follows that of its SSN according to item (ii) of Definition 3.

Definition 4 Two WMNs $N_j = \{P_j \cup P_{I_j} \cup P_{O_j}, T_j, F_j\}$, $j \in \{1, 2\}$ are composable, if $P_1 \cap P_2 = \emptyset$, $T_1 \cap T_2 = \emptyset$and $(P_{I_1} \cap P_{O_2}) \cup (P_{I_2} \cap P_{O_1}) = P_E \neq \emptyset$. For every $p \in P_E$, if $x \in {}^\bullet p$ and $y \in p^\bullet$, we call (x, p, y) an information channel. The two-member C-net denoted as $N = N_1 \oplus N_2$ is defined as follows:

(i) $P = P_1 \cup P_2$,

(ii) $P_I = \{(P_{I_1} \cup P_{I_2}) \backslash P_E\}$, $P_O = \{(P_{O_1} \cup P_{O_2}) \backslash P_E\}$,

(iii) $T = T_1 \cup T_2$, and

(iv) $F = F_1 \cup F_2$. □

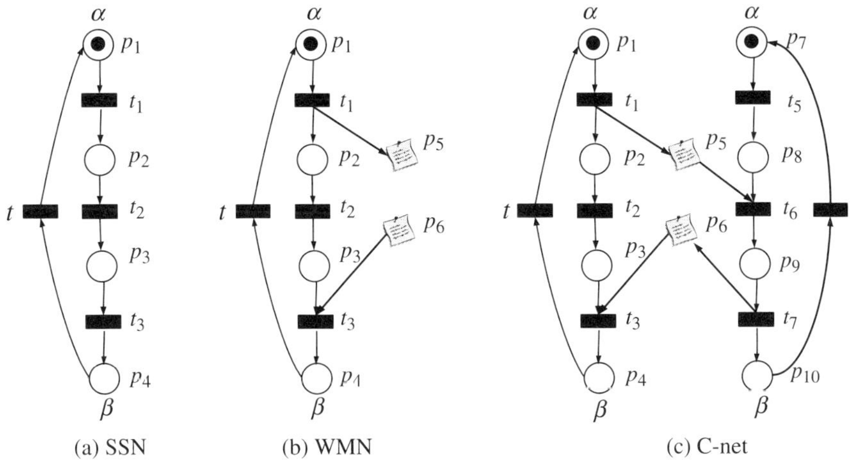

FIGURE 17.4 Illustration for Petri net siphon.

Two WMN N_1 and N_2 are composable when they interact through a set of common places. For example, if $P_{I_2} \cap P_{O_1} = P' \neq \emptyset$, N_1 sends information through the set of interfaces P' that is received by N_2.

Definition 5 An n-member C-net denoted as $N = \oplus_{i=1}^{n} N_i$ is defined recursively following Definition 4. A C-net is a complete net iff $P_I = \emptyset$ and $P_O = \emptyset$. □

For example, Figure17.4a–17.4c show an SSN, WMN, and C-net respectively. In an SSN, place α (i.e., p_1) and β (i.e., p_4) denote the start and end status, respectively and p_{2-3} are internal places. A WMN is formed by adding two interface places and their corresponding arcs to an SSN. In Figure 17.4b, p_5 and p_6 are output and input places, respectively. A WMN can be considered as a single-member C-net. Two or more WMNs can be composed to form a C-net. A complete C-net does not have additional input and output interfaces for external interaction. Figure 17.4c shows a complete C-net composed of two WMNs. One of the WMN is shown in Figure 17.4b which has an output and an input place, denoted as p_5 and p_6' respectively. The other one has an input and output place, denoted as p_5 and p_6' respectively. Thus, they can be composed to form a complete C-net.

17.2.2 Model Web Service Interaction with C-Net

We then show how we model web service interaction with C-nets. We assume that no service interface level mismatches exist, that is, the message signature and number of interfaces in both parties match. We divide the basic structures in BPEL, that is, *receive, reply, invoke, assign, throw, terminate, wait, empty,* and *link* into two categories. The first category is internal control logic that includes *assign, terminate, wait,* and *empty*. The second category is external control logic that includes *receive,*

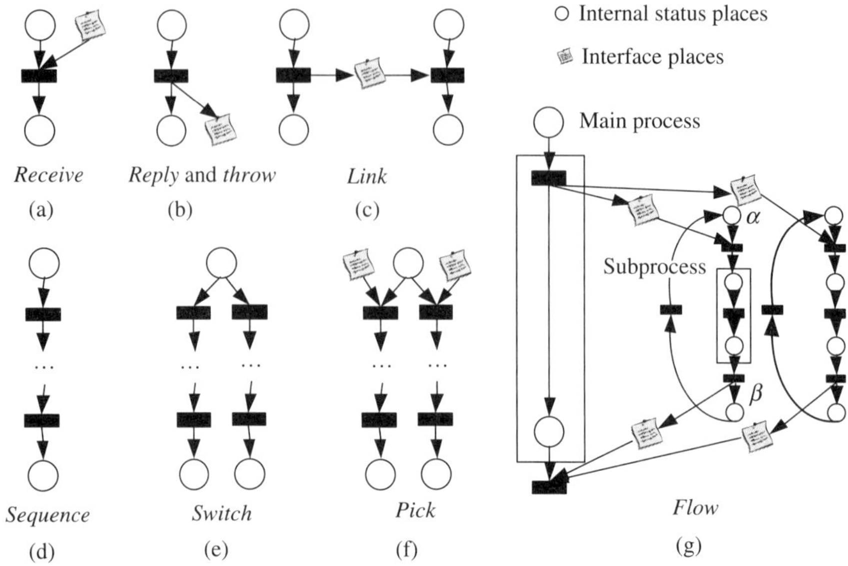

FIGURE 17.5 Transforming BPEL into WMNs.

reply, *invoke*, *throw,* and *link*. Basic structures in the first category are not related to the interaction between different web services and we model them as internal places and transitions in a WMN. Basic structures in the second category are related to the interaction between different web services and we model them as transitions connected with internal and interface places as shown in Figure 17.5. Note that, *invoke* is a combination of *reply* and *receive*, and *link* is modeled as an information channel.

There are *sequence, flow, pick, switch,* and *while* structured activities in a BPEL process. Based on basic structures, a WMN can cover the *sequence, pick,* and *switch* structures. The semantics of *while* structure are similar to while loop in programming languages like *Java*. Here we approximate the number of loops in a finite *while* structured activity and transform the activity to a *sequence* activity by expanding cycles [26]. We can transform the processes that are executed in parallel in the *flow* structure into the same processes that are invoked simultaneously in the *invoke* structure while maintaining the business logic. For example, we can divide the processes that are executed in parallel into separate BPEL processes while maintaining the business logic as shown in Figure 17.5g.

For example, assume that we have two web services as shown in Figure 17.6, that is, a customer service and an online shop service. The customer service first sends an order message, waits for delivery, and then sends a payment message; while the online shop service waits for the order and payment, and then delivers the product.

We model the action of sending order information of a customer service in Figure 17.7a. Here we model the order information message as p_{11} and the customer service

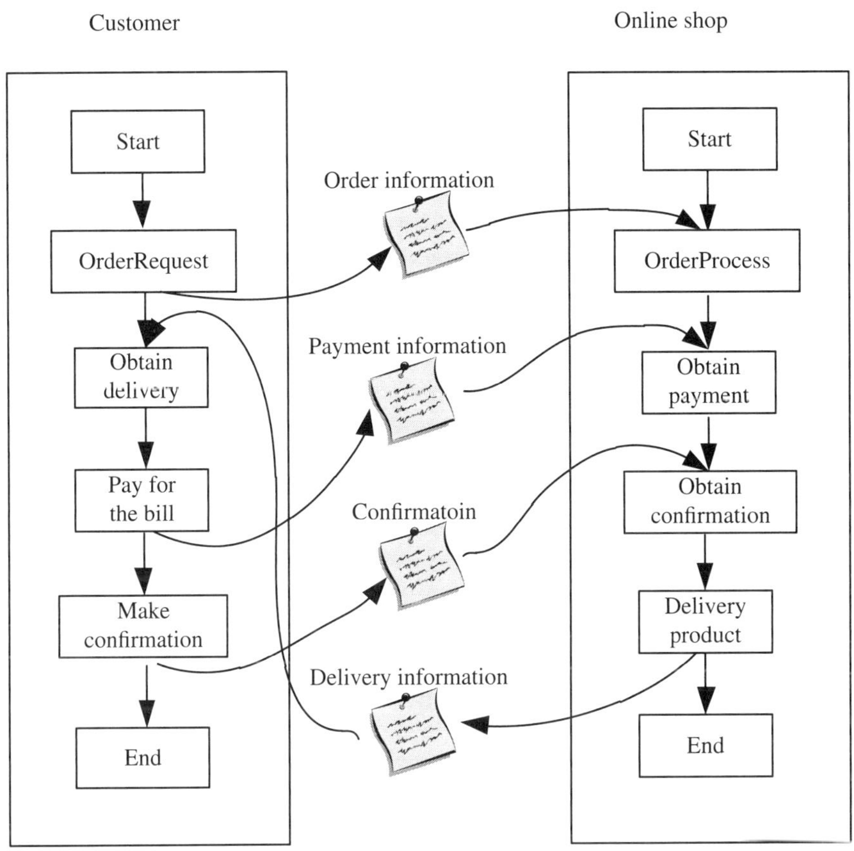

FIGURE 17.6 A customer service and an online shop service.

status before and after sending the message as p_1 and p_2, respectively. We also model the action of receiving order information of an online shop service in Figure 17.7b. We model the order information message as the same p_{11} and the online shop service status before and after receiving the message as p_6 and p_7, respectively.

Note that the interface places do not have tokens initially because no message is created. They can have tokens if and only if some transition wants to send a message through the information channel while they can lose a token if and only if some transition wants to receive a message through the information channel. A token in them models the situation when the required message is ready. We assume that the maximum number of tokens that an interface place can hold is one. Otherwise a BPEL process is not correct.

The firing of transitions in a C-net simulates the interaction of web services. For example, order information is ready in Figure 17.8a. After t_6 fires, the order information is received by the online shop service in Figure 17.8b. Following the modeling method, the final C-net for two web services is shown in Figure 17.9a.

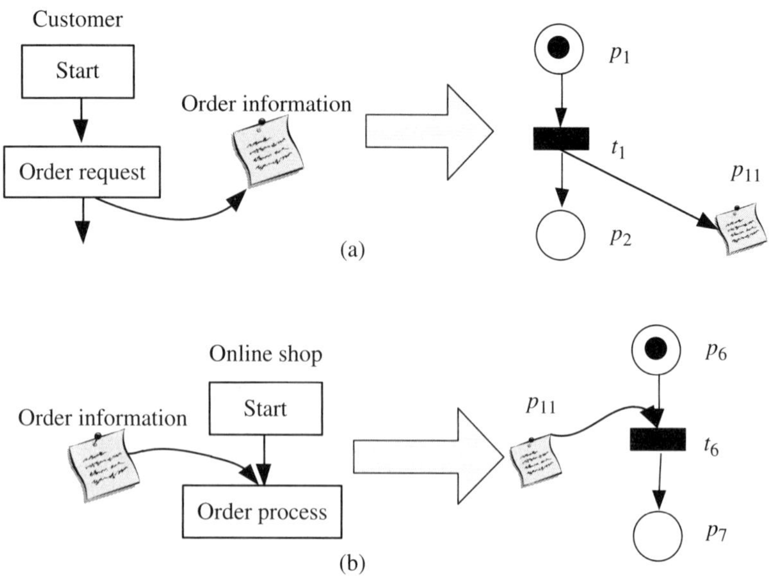

FIGURE 17.7 Modeling the case in Figure 17.6 as C-net.

17.3 PROTOCOL-LEVEL MISMATCH ANALYSIS

Mismatches in protocol level mainly come from mutual waiting mismatches and nonlocal choice mismatches. We propose a detection method, a core algorithm, as well as a comprehensive solution method for protocol-level mismatch.

17.3.1 Protocol-Level Mismatch Detection

Definition 6 A C-net $N = \oplus_{i=1}^{n} N_i$ matches at protocol level if N is live. □

Note that, according to Theorem 4 in Reference 25, a state machine N is live iff N is strongly connected and M_0 has least one token. Following this theorem, since an

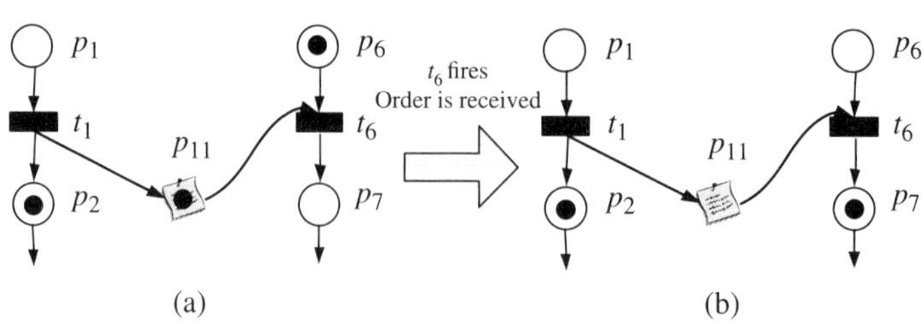

FIGURE 17.8 Modeling the interaction of web services.

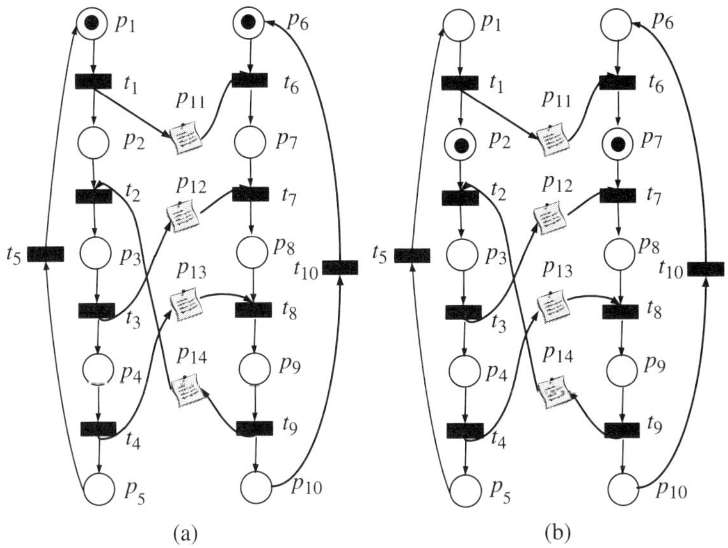

FIGURE 17.9 Modeling the interaction of web services in Figure 17.6

SSN is a strongly connected state machine with one token, it is live. However, this theorem is not true for C-nets.

Theorem 1 A C-net is live iff it is potential deadlock free.
Proof:

(1) A C-net is an asymmetric choice net.
$\forall p_1, p_2 \subseteq$C-net and $p_1^{\bullet} \cap p_2^{\bullet} \neq \emptyset$, there are four cases.
Case 1: $p_1 \in P_E$ and $p_2 \in P_E$. According to Definition 3, we have $p_1^{\bullet} \cap p_2^{\bullet} = \{t\}$.
Case 2: $p_1 \in P_E$ and $p_2 \notin P_E$. We have $p_1^{\bullet} \subseteq p_2^{\bullet}$.
Case 3: $p_2 \in P_E$ and $p_1 \notin P_E$. We have $p_1^{\bullet} \supseteq p_2^{\bullet}$.
Case 4: $p_1 \notin P_E$ and $p_2 \notin P_E$. However, since the extended SSN is a strongly connected state machine, we have $\forall t \in T$, $|{}^{\bullet}t| = |t^{\bullet}| = 1$. This case is impossible.
For all the cases, we have $p_1^{\bullet} \cap p_2^{\bullet} \neq \emptyset \Rightarrow p_1^{\bullet} \subseteq p_2^{\bullet}$or $p_1^{\bullet} \supseteq p_2^{\bullet}$. Then a C-net is an asymmetric choice net.

(2) According to Reference 27, an asymmetric choice net is live iff it is potential deadlock free.
Based on (1) and (2), a C-net is live iff it is potential deadlock free. □

According to the definition of potential deadlock, a siphon S eventually becomes empty when a C-net is not live and all the siphons are always marked when a C-net is live.

Theorem 2 A C-net $N = \oplus_{i=1}^{n} N_i$ matches at protocol level iff all the siphons are always marked, that is, $\forall M \in R(N, M_0)$, $\forall$(minimal) siphon S, $M(S) \neq 0$.

Proof: Obvious from Definition 6 and Theorem 1. □

For instance, there are five minimum siphons in Figure 17.6a, that is, $S_1 = \{p_{1\text{–}5}\}$, $S_2 = \{p_{6\text{–}10}\}$, $S_3 = \{p_1, p_{3\text{–}5}, p_{7\text{–}9}, p_{11}, p_{14}\}$, $S_4 = \{p_3, p_{8\text{–}9}, p_{12}, p_{14}\}$, $S_5 = \{p_{3\text{–}4}, p_9, p_{13\text{–}14}\}$. $M_0(S_{1\text{–}3}) = 1$ and $M_0(S_{4\text{–}5}) = 0$. Since there are empty siphons at the initial marking, that is, S_4 and S_5, there exists protocol-level mismatching. This is true because after t_1 and t_6 fire, there is a deadlock as shown in Figure 17.9b.

Thus the problem of protocol-level mismatching of web service interaction is transformed to the problem of empty minimal siphons in a C-net. We can use the mixed-integer programming algorithm to detect the maximal empty siphon [27]. Other efficient methods can be found in References 28, 29, 30, 31.

In the aspect of detecting protocol-level mismatch, neither of the reachability analysis-based method nor mixed-integer programming has clear computational advantage over the other because both of them have exponential complexity [32].

Suppose that we have a C-net $N = \oplus_{i=1}^{n} N_i$, where some minimal siphons can become empty. Our main goal is to introduce into the system a solution to guarantee that no empty minimal siphons are reachable during the evolution of the new C-net, that is , the new C-net at the protocol-level matches.

17.3.2 Core Algorithm

As stated in Theorem 2, a C-net $N = \oplus_{i=1}^{n} N_i$ matches at protocol level iff $\forall M \in R(N, M_0)$, $\forall$(minimal) siphon S, $M(S) \neq 0$. According to the property of a siphon, the number of tokens in a siphon will never increase and an empty siphon will always remain empty.

From a modeling perspective, the most intuitive way is to add tokens to the C-net such that all of the empty siphons become marked. For example, adding one token to p_{11} in Figure 17.9 means that the order information is ready while adding to p_{12} means that the payment information is ready. From a service perspective, the customer is waiting for the delivery information while the online shop is waiting for the payment information. Thus, developers must be involved to provide additional information at the deadlock point, that is, to ask the customer to provide the payment information or to ask the online shop to provide the delivery information.

In the above cases, there are at least four choices for a developer:

Choice 1(C1): provide nothing

Choice 2(C2): provide the payment information

Choice 3(C3): provide the delivery information

Choice 4(C4): provide the payment and delivery information

But developers may not always make a good decision. A good decision implies the correctness and efficiency of a solution.

Its correctness means that a solution resolves the current protocol-level mismatch. C1 is obviously not a correct solution since both the customer and online shop web services are still waiting. C2 is also not a correct solution although it can solve the current waiting status for the online shop. After choosing C2, payment information is provided and t_7 can fire. However, the system goes to another mutual waiting state, that is, the online shop is waiting for confirmation and the customer is waiting for delivery. This is because siphon S_5 is empty.

Its efficiency means that a solution requires the smallest amount of information to solve as many future protocol-level mismatches as possible. C4 is not an efficient solution because it needs 2 kinds of messages, that is, payment and delivery information. Note that choices 1–4 require 0, 1, 1, and 2 kinds of messages, respectively. Although C3 and C4 are all correct solutions, C3 is more efficient than C4 because it requires fewer kinds of messages.

We propose an algorithm to choose a correct and the most efficient solution by linear programming.

Core Algorithm:

INPUT: (1) n-member C-net N with a minimum siphon set $\Omega = \Omega_N \cup \Omega_Y$. Ω_N denotes the nonempty siphon set, and $\Omega_Y = \{S_1, S_2, \ldots S_i\}$ denotes the empty siphon set when a protocol mismatch happens. (2) Interface place set $P_E = \{p_1, p_2, \ldots p_j\}$

OUTPUT: A list of messages that should be provided. We denote the list as a $j \times 1$vector L where $L(j) = 1$ if $p_j \in L$; and 0 otherwise.

BEGIN:

Step 1. /* Calculate the contribution matrix of every message to the siphon*/

Constitute an $i \times j$ matrix A, where $A(i, j) = 1$ if $p_j \in S_i$; and 0 otherwise.

Step 2. /* Optimization*/

Compute the following linear programming problem:

Minimize $\mathbf{1}^* L$

s.t. $A^* L = \mathbf{1}^{\mathrm{T}}$

Step 3. /* Return result*/

Return L

END

The time complexity for the algorithm is linear. We explain the idea underlying this algorithm as follows.

Firstly, meeting the constraint $A^* L = \mathbf{1}^{\mathrm{T}}$ can return correct solutions. As shown in Figure 17.10, we have $A(i, j) = 1$ if $p_j \in S_i$; and 0 otherwise, and $L(j) = 1$ if $p_j \in L$; and 0 otherwise. It is clear that the solutions of the constraint guarantee that each empty siphon is marked by exactly one token. Moreover, if the constraint is not satisfied, there is at least one empty siphon.

Secondly, the objective function of the linear programming formulation can return the best solution. In the contribution matrix, the more siphons the message p_j is involved in, the more 1s it has in the jth column. Since the objective function calculates the sum of messages, the solution has the smallest total number of messages if the proposed objective function is minimized.

Finally, we can claim that, through the algorithm, we can provide a correct and efficient solution to protocol-level mismatch problems.

$$\begin{array}{c} \\ S_1 \\ S_2 \\ S_3 \\ \vdots \\ S_i \end{array} \begin{array}{c} p_1\ p_2\ p_3 \cdots p_j \\ \begin{pmatrix} a_{11} & \cdots & a_{1j} \\ \vdots & \ddots & \vdots \\ a_{i1} & \cdots & a_{ij} \end{pmatrix} \end{array} \begin{pmatrix} L_1 \\ L_2 \\ L_3 \\ \vdots \\ L_j \end{pmatrix} = \begin{pmatrix} 1 \\ 1 \\ 1 \\ \vdots \\ 1 \end{pmatrix}$$

FIGURE 17.10 Illustration of linear programming

17.3.3 Comprehensive Solution to Protocol-Level Mismatch

Based on the core algorithm, we propose a comprehensive solution to protocol-level mismatch as follows. The input of the algorithm is web services and their interaction described in BPEL. We assume that there is no interface mismatches for these web services' interaction.

Comprehensive solution:
BEGIN:

Step 1. Transform web service interaction into Petri net models as simple sequential workflow net (SSN), workflow module net (WMN) and Composition net (C-net).

Step 2. Check if there will be empty siphon by using mixed-integer programming according to "protocol-level mismatch detection." If none of them can become empty, go to end. Otherwise denote the set of minimal siphons in the C-net as Ω. Start the web service interaction as well as monitoring all the minimal siphons.

Step 3. During the monitoring period, if one or more of the minimal siphons become empty, denote the empty siphon set as Ω_Y and go to Step 4. The monitoring ends when the web service interaction terminates properly, that is, for each SSN, state β is reached from state α. Go to end.

Step 4. Take Ω_Y as the input of core algorithm. Run the core algorithm. Provide the list of messages that are returned by the algorithm. Go back to Step 3.

END

We explain the idea underlying the algorithm as follows:

Firstly, according to the taxonomy of service composition mismatches, we may have both the interface and protocol mismatches. Since interface matches are a prerequisite to achieve protocol matches, before using the solution, we have to guarantee that there is no interface mismatch.

Secondly, we model web service interaction by Petri net models such that we can identify and check the minimal siphons. If we cannot detect any empty minimal siphon, then there are no protocol mismatches and we finish our task. Otherwise, we have to monitor those minimal siphons until all the web services terminate properly. Before they terminate, when there is an empty siphon, we can use the core algorithm to obtain the list of messages that should be provided.

Finally, we provide those messages and solve the protocol mismatch problem.

We make an overview of the solution. As mentioned previously, there are three kinds of protocol-level mismatches, that is, unspecified reception, mutual waiting, and nonlocal choice. The cause for unspecified reception is a wrong sequence of sending messages and it can be solved by buffering previous messages. The cause for mutual waiting mismatches is that the web services follow a conflict "send and receive" sequence and they are waiting for each other to send messages. The cause for nonlocal choice mismatches is that the web services follow a conflict "send and receive" choice and wait for each other to send messages. In both the cases, the missing messages are responsible for the mismatch. By using our core algorithm, we not only try to find those missing messages but also find the smallest set of those missing ones. Thus, after we provide those messages, we are able to resolve the protocol mismatches.

We outline the complexity analysis of the solution step by step.

Step 1 is doing transformation tasks. Thus its complexity linearly depends on the size of BPEL.

Step 2 is answering an "Yes or No" question, that is, whether or not there is an empty siphon at any reachable markings. Then its complexity depends on the time when the algorithm will find an empty siphon. In one case, if there is a mutual waiting mismatch, generally speaking, there is at least one empty siphon at the initial marking. For example, there is a mutual waiting mismatch in Figure 17.6 and there is an empty siphon at the initial marking, that is, $S_4 = \{p_3, p_{8\text{–}9}, p_{12}, p_{14}\}$. Using the method to detect empty siphon at the initial marking, the maximal empty siphon that contains S_4 can be detected in polynomial time [27]. In another case, if there is a nonlocal choice mismatch, generally speaking, there is no empty siphon at the initial marking but there will be an empty one at a future marking. This is because a nonlocal choice mismatch will happen only when services make conflicting choices. In other words, it may be "hidden" when services happen to make the same choices. To detect an empty siphon for a nonlocal choice mismatch at a future marking has exponential complexity in the worst case. Fortunately, since we already propose an approach to resolve the mismatch, the answer to whether or not there is an empty siphon at any reachable markings is not very important.

Step 3 is generating the minimal siphons. It is a prerequisite for Step 4. The generation of minimal siphons has exponential complexity in the worst case. Fortunately, since siphons are a kind of special structures of Petri nets, they are not related with the initial marking and the number of all markings and we can generate them offline.

Step 4 is trying to control the empty siphons. Because it is using linear programming, the complexity is polynomial.

17.4 ILLUSTRATING EXAMPLES

As previously mentioned, protocol mismatch includes mutual waiting and nonlocal choice. We use two concrete examples to illustrate how to use the method mentioned above, especially the core algorithm.

The scenario in Figure 17.6 can be classified as mutual waiting mismatch. After t_1 and t_6 fire, there is a deadlock. We have $\Omega_N = \{S_{1\text{–}3}\}$, $\Omega_Y = \{S_{4\text{–}5}\}$ where $S_4 = \{p_3, p_{8\text{–}9}, p_{12}, p_{14}\}$, $S_5 = \{p_{3\text{–}4}, p_9, p_{13\text{–}14}\}$, and $P_E = \{p_{11\text{–}14}\}$. Because $p_{12} \in S_4$, $p_{14} \in S_4$, $p_{13} \in S_5$, $p_{14} \in S_5$, we have $A = ((0\ 1\ 0\ 1)\ (0\ 0\ 1\ 1))$. The result is $L = (0\ 0\ 0\ 1)^T$. It means that a developer should ask the online shop to provide the delivery information (the token in the interface place p_{14} denotes the delivery information). Moreover, if we check the method that the developer asks the customer to provide the payment information, that is, $L' = (0\ 1\ 0\ 0)^T$, we find that this method will fail. This is simply because L' is not a solution of the linear programming problem. Because $A * L' = (1\ 0)^T$, although S_4 is marked, S_5 is still empty.

We use a scenario mentioned in References 22, 23 as shown in Figure 17.11 to illustrate the nonlocal choice mismatches. There are three web services, denoted as

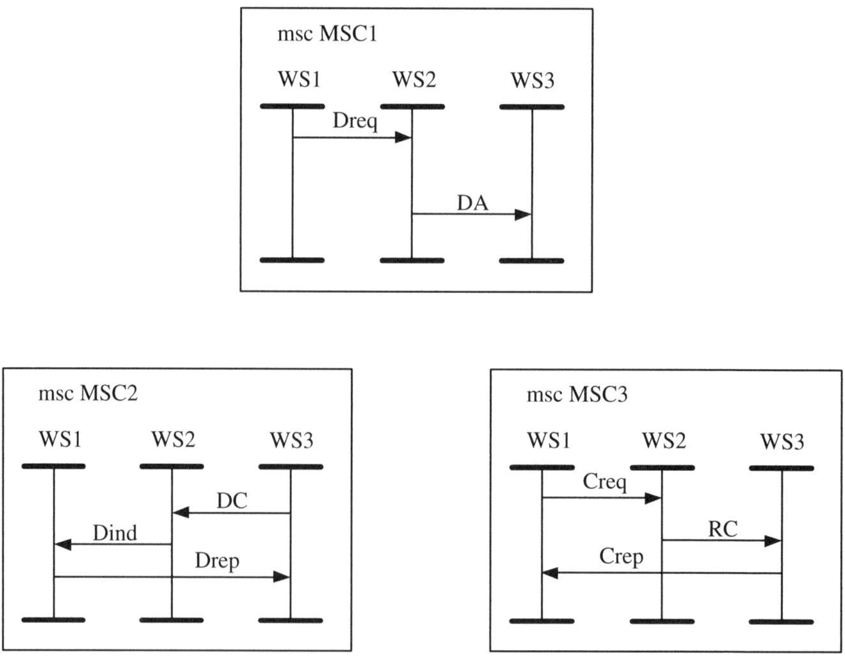

FIGURE 17.11 Nonlocal choice protocol mismatch.

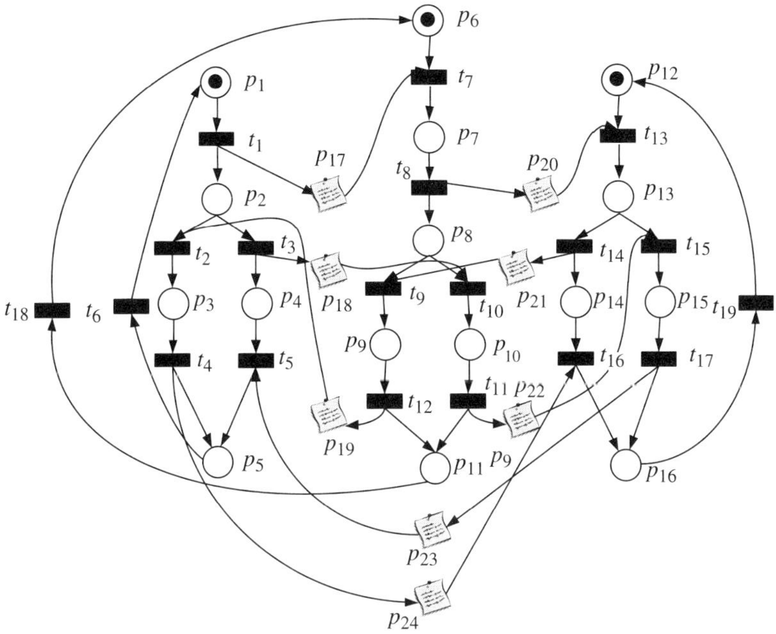

(a)

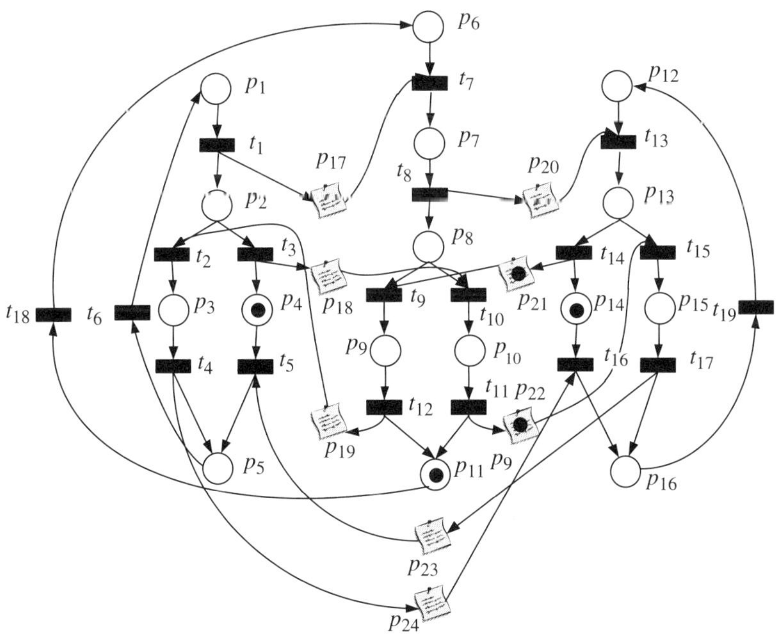

(b)

FIGURE 17.12 The C-nets for the nonlocal choice protocol mismatch in Figure 11.

WS1–WS3. We first denote their interaction through MSC. WS1 first sends "Dreq" to WS2, and then WS2 sends "DA" to WS3. After WS1 first sends "Dreq" to WS2, it is the first web service to decide whether to MSC2 or MSC3. For example, if it decides to proceed by following MSC3, it will send "Creq" to WS2, WS2 sends "RC" to WS3, and then WS1 waits for WS3 to send back "Crep".

We model the messages "Dreq," "DA," "Creq," "RC," "Crep," "DC," "Dind," and "Drep" with p_{17}, p_{20}, p_{18}, p_{22}, p_{23}, p_{21}, $p_{19,}$ and p_{24}, respectively. Then we build the C-net for the scenario as shown in Figure 17.12a. There will be no nonlocal choice mismatches if WS1 and WS3 can make the same choices, that is, following MSC2 or MSC3. However, if they follow different MSCs, then there is a nonlocal choice mismatch as shown in Figure 17.12b. For example, if WS1 decides to proceed following MSC3 and WS3 decides to proceed following MSC2, then after t_1, t_3, t_7, t_8, t_{10}, t_{13}, t_{14}, t_{11}, and t_{18} fire, there will be a deadlock. The situation will be: WS1 is waiting for "Crep" to fire t_5; WS2 is waiting to fire t_9 because "DC" arrives; WS3 is waiting for "Drep" to fire t_{16}.

The three web services are waiting for each other. There are 15 minimal siphons in the C-nets. We have $\Omega_N = \{S_{1-11}\}$, $\Omega_Y = \{S_{12-15}\}$, and $P_E = \{p_{17-24}\}$. For simplicity, we only list the siphons in Ω_Y, that is , $S_{12} = \{p_1, p_3, p_5, p_{7-9}, p_{13}, p_{15}, p_{17}, p_{19-20}, p_{23}\}$, $S_{13} = \{p_1, p_3, p_5, p_{7-9}, p_{12-13}, p_{15-17}, p_{19}, p_{23-24}\}$, $S_{14} = \{p_{1-3}, p_5, p_7, p_{13}, p_{15}, p_{17}, p_{20}, p_{23}\}$, and $S_{15} = \{p_{1-3}, p_5, p_{12-13}, p_{15-16}, p_{23-24}\}$. The result is $L = (0\ 0\ 0\ 0\ 0\ 0\ 1\ 0)^T$. Here p_{23} denotes "Crep." It means that the developer should ask WS3 to provide "Crep."

17.5 CONCLUSIONS

In service composition, services interact with each other through a sequence of messages to achieve a certain business goal. However, mismatches at the interface and protocol levels may render the composite service unusable. There are a plethora of works on service interface level mismatches and most of the interface level mismatch can be solved by using adaptors. However, the mismatches at protocol levels are not well addressed. Existing studies based on adaptors and reachability analyses fails to provide an efficient solution.

For instance, after transforming BPEL into Petri nets represented in PNML, Ouyang *et al.* [7] use WofBPEL to support three types of analysis, that is , reachability analysis, competing message-consuming activities, and garbage collection of queued messages by generating the full state space. Similarly, König *et al.* [33] use Fiona to automatically analyze the interactional behavior of a given oWFN. Martens [34] proposes a BPEL annotated Petri nets (BPN) and presents a decision algorithm for the controllability of a BPN model based on the communication graph (c-graph). The examination of interactions between the composed BPEL processes is transformed into the verification of deadlock freeness of a BPN. Nezhad *et al.* [16] generate a mismatch tree to handle deadlock situations. The basis of such tree is similar to the reachability graph in Petri nets. Although the current methods provide useful insights into the problem by adopting Petri-net-based analysis methods, for example,

WofBPEL, Fiona, c-graph, and mismatch tree, their analysis is mainly based on a reachable state space and they do not propose an effective solution to resolve the protocol-level mismatch issues.

For example, following the mismatch tree approach that is proposed by Nezhad *et al.*, we have choices 1–4 for the scenarios in Figure 17.4. For each choice, we have to search all the reachable state space and find out if it results in a future deadlock. Note that, the complexity for generating the reachable state space is exponential. Normally, assume that there are w web services and each of them is waiting for a message, the number of solutions to consider should be $O(2^w)$. Assume that there are k nodes in a Petri net. Then the complexity for generating the reachable state space is $O(Ce^k)$ for each solution. Thus the total complexity to find the best solution would be $O(C2^w e^k)$.

Compared with their work, the proposed one is more efficient. For example, to solve the same problem in Figure 17.4, after generating the siphons, it simply uses the linear programming to find the best solution. Assume that there are k nodes in the Petri net. Then the complexity for generating the minimal siphons is exponential, that is, $O(Ce^k)$. Because the complexity to solve linear programming is polynomial time, the total complexity is still $O(Ce^k)$.

Moreover, because siphons depend on only the structure of Petri nets, they can be generated offline. Thus, the generation process can run in parallel with web services' interaction. That is to say, we can generate the minimal siphons before the protocol mismatch appears and we can run the linear programming right after the protocol mismatch appears. However, the reachable state space method is strongly based on the input status. Thus, only when the protocol mismatch really happens, can we obtain the input status and then test the candidate solutions in reachable state space method. A detailed comparison is shown in Figure 17.13.

Our previous work [35] also proposes a method to solve nonlocal choice mismatches. There are two major differences between this work and Reference 35. (1) The protocol mismatches that this work can solve include both mutual waiting and nonlocal choice mismatches. Our previous work [35] can only solve nonlocal choice mismatches. Compared with Reference 35, this work can solve more general protocol mismatches. (2) Although both of the works can solve nonlocal choice mismatches, they are using different methods. The previous work uses a method based on adding information links and needs to modify BPEL source code. It does not allow nonlocal choice mismatches to happen. Compared with it, this work proposes a method based on providing missing messages. It can allow nonlocal choice mismatches to happen and then it provides those messages to resolve protocol mismatches.

The main contribution of this chapter is to propose a Petri net siphon-based solution to protocol-level mismatches. The whole approach follows the steps of modeling, analysis, and solution. Firstly, the web service composition is modeled by a set of Petri nets, that is, SSN, WMN, and C-net. Secondly, the protocol-level mismatch problem is analyzed by Petri net siphons. We make a useful conclusion that, there should be a protocol-level mismatch if and only if there is an empty siphon in the C-net. Finally, based on this conclusion, we use a linear programming method to find the best solution. Our approach provides an optimized and also automatic solution

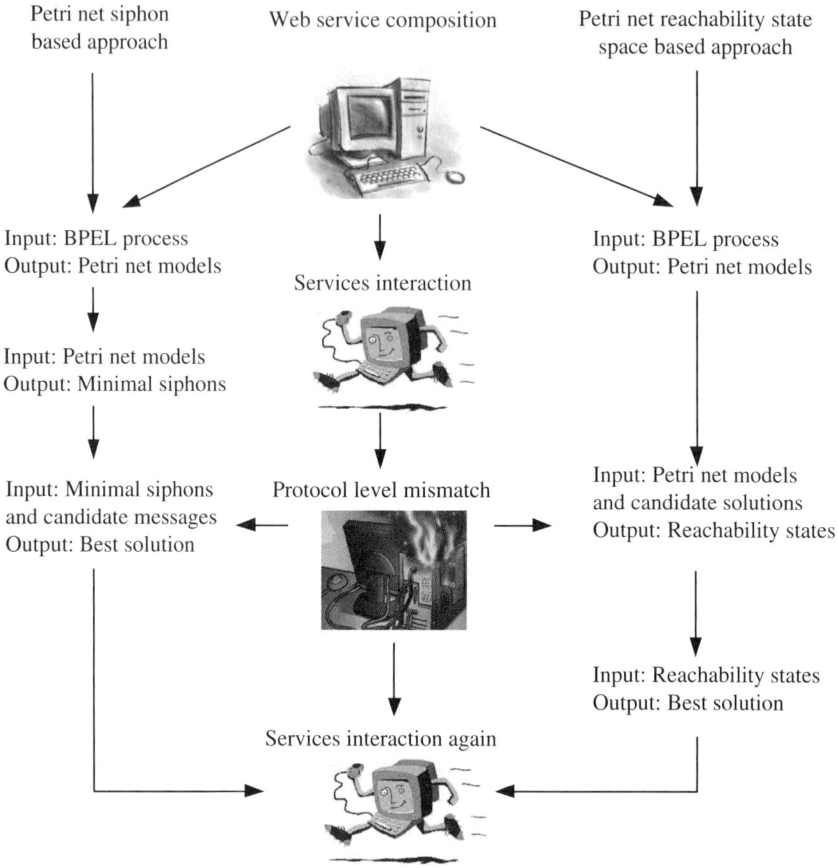

FIGURE 17.13 The comparison between Petri net siphon-based and reachability state-space-based methods.

for correcting protocol mismatches. This approach greatly reduces the amount of interactions with developers.

In future, first, we would like to extend our approach to address the protocol discovery from message logs. Our current approach is based on the already known BPEL. We would like to extend our approach to deal with the situation when BPEL is unknown and conversations between two or more services should be discovered by message logs. Data mining and machine learning technique may help to accomplish this goal [36]. Second, we would like to extend our approach to address the dynamically evolving service protocols. As mentioned in References 37 and 38, service protocols can dynamically adapt to changes in business needs and/or regulations. For example, Ryu *et al.* [37] propose a comprehensive approach to dynamic protocol evolution management using a finite state machine model. We only target static service protocols. We plan to adapt our approach to deal with dynamic protocol evolution and find out how changes in protocol can impact the analysis and solution. Finally, although the search for siphons can be performed offline and the computation of minimum siphons is simple, in some complex structured C-net, such computation can be expensive. Some polynomial complex algorithms to find and control only the problematic siphons should be explored for C-nets by making full use of their special structural information and considering the linear relationship among all siphons. Some recent advances [28–31] may provide good help along this direction.

REFERENCES

[1] F. Curbera, N. Mukhi, W. Nagy, and S. Weerawarana. Implementing BPEL4WS: the architecture of a BPEL4WS implementation. *Concurrency and Computation: Practice & Experience*, 18(10): 1219–1228, 2006.

[2] P. Wohed, W. M. P. van der Aalst, M. Dumas, and A. H. M. ter Hofstede. Analysis of web services composition languages: the case of BPEL4WS. In: Proceedings of 22nd International Conference on Conceptual Modeling, Chicago, IL, 2003, pp. 200–215.

[3] https://www.oasis-open.org/ (accessed January 19, 2015).

[4] J. Fisteus, L. Fernández, and C. Kloos. Formal verification of BPEL4WS business collaborations. In: Proceedings of the 5th International Conference on Electronic Commerce and Web Technologies, Zaragoza, Spain, August, 2004, pp. 76–85.

[5] A. Ferrara. Web services: a process algebra approach. In: Proceedings of the 2nd International Conference on Service Oriented Computing, New York City, 2004, pp. 242–251.

[6] D. Fahland and W. Reisig. ASM-based semantics for BPEL: the negative control flow. In: Proceedings of the 12th International Workshop on Abstract State Machines, Paris, France, March, 2005, pp. 131–151.

[7] C. Ouyang, E. Verbeek, W. M. P. van der Aalst, S. Breutel, M. Dumas, and A. H. M. ter Hofstede. WofBPEL: a tool for automated analysis of BPEL processes. In: Proceedings of the 3rd International Conference on Service Oriented Computing, Amsterdam, The Netherlands, December, 2005, pp. 484–489.

[8] S. Hinz, K. Schmidt, and C. Stahl. Transforming BPEL to Petri nets. In: Proceedings of the 3rd International Conference on Business Process Management, Nancy, France, September, 2005, pp. 220–235.

[9] N. Lohmann. A feature-complete Petri net semantics for WS-BPEL 2.0. In: Proceedings of the 4th International Workshop on Web Services and Formal Methods, Brisbane, Australia, September, 2007, pp. 77–91.

[10] R. Lucchia and M. Mazzara. A pi-calculus-based semantics for WS-BPEL. *Journal of Logic and Algebraic Programming*, 70(1): 96–118, 2007.

[11] H. Foster, S. Uchitel, J. Magee, and J. Kramer. Tool support for model-based engineering of web service compositions. In: Proceedings of the 2005 IEEE International Conference on Web Services, Orlando, FL, July, 2005, Vol. 1, pp. 95–102.

[12] X. Fu, T. Bultan, and J. Su. Analysis of interacting BPEL web services. In: Proceedings of the 13th International World Wide Web Conference, New York, May, 2004, pp. 621–630.

[13] http://www.pnml.org/ (accessed January 19, 2015).

[14] W. M. P. van der Aalst. The application of Petri nets to workflow management. *The Journal of Circuits, Systems and Computers*, 8(1): 21–66, 1998.

[15] N. Lohmann, P. Massuthe, C. Stahl, and D. Weinberg. Analyzing interacting BPEL processes. In: Proceedings of the 4th International Conference on Business Process Management, Vienna, Austria, September, 2006, pp. 17–32.

[16] H. Nezhad, B. Benatallah, A. Martens, F. Curbera, and F. Casati. Semi automated adaptation of service interactions. In: Proceedings of the 16th International World Wide Web Conference, Banff, AB, Canada, May, 2007, pp. 993–1002.

[17] B. Benatallah, F. Casati, D. Grigori, H. Nezhad, and F. Toumani. Developing adaptors for web services integration. In: Proceedings of CAiSE, Porto, Portugal, 2005, pp. 415–429.

[18] B. Benatallah, F. Casati, and F. Toumani. Representing, analysing and managing web service protocols. *Data & Knowledge Engineering*, 58(3): 327–357, 2006.

[19] Y. Wang and E. Stroulia. Flexible interface matching for web-service discovery. In: Proceedings of the 4th International Conference on Web Information System Engineering, Rome, Italy, December, 2003, pp. 147–156.

[20] X. Dong, A. Halevy, J. Madhavan, E. Nemes, and J. Zhang. Similarity search for web services In: Proceedings of the 30th International Conference on Very Large Data Bases, Toronto, ON, Canada, September, 2004, pp. 372–383.

[21] D. M. Yellin and R. E. Strom. Protocol specifications and component adaptors. *ACM Transactions on Programming Languages and Systems*, 19(2): 292–333, March, 1997.

[22] H. Ben-Abdallah and S. Leue. Syntactic detection of process divergence and non-local choice in message sequence charts. In: Proceedings of the 2nd International Workshop on Tools and Algorithms for the Construction and Analysis of Systems, Enschede, The Netherlands, April, 1997, pp. 259–274.

[23] A. Martens. Usability of web services. In: Proceedings of the 4th International Conference on Web Information Systems Engineering Workshops, Rome, Italy, December, 2003, pp. 182–190.

[24] M. C. Zhou and K. Venkatesh. *Modeling, Simulation and Control of Flexible Manufacturing Systems: A Petri Net Approach.* Singapore: World Scientific, 1998.

[25] T. Murata. Petri nets: properties, analysis and applications. *Proceedings of the IEEE*, 77(4): 541–580, 1989.

[26] J. Ezpeleta, J. Colom, and J. Martinez. A Petri net based deadlock prevention policy for flexible manufacturing systems. *IEEE Transactions on Robotics and Automation*, 11(2): 173–184, 1995.

[27] F. Chu and X. L. Xie. Deadlock analysis of Petri nets using siphons and mathematical programming. *IEEE Transactions on Robotics and Auto*, 13(6): 793–804, 1997.

[28] A. R. Wang, Z. W. Li, J. Y. Jia, and M. C. Zhou. An effective algorithm to find elementary siphons in a class of Petri nets. *IEEE Transactions on Systems, Man, and Cybernetics, Part A*, 39(4): 912–923, 2009.

[29] Z. W. Li and M. C. Zhou. *Deadlock Resolution in Automated Manufacturing Systems: A Novel Petri Net Approach*. New York: Springer, 2009.

[30] H. Hu, M. C. Zhou, and Z. W. Li. Liveness enforcing supervision of video streaming systems using non-sequential Petri nets. *IEEE Transactions on Multimedia*, 11(8): 1446–1456, 2009.

[31] N. Q. Wu, M. C. Zhou, and Z. W. Li. Resource-oriented Petri net for deadlock avoidance in flexible assembly systems. *IEEE Transactions on Systems, Man, and Cybernetics: Part A*, 38(1): 56–69, 2008.

[32] Z. W. Li and M. C. Zhou. Elementary siphons of Petri nets and their applications to deadlock prevention in flexible manufacturing systems. *IEEE Transactions on Systems, Man, and Cybernetics, Part A*, 34(1): 38–51, 2004.

[33] D. König, N. Lohmann, S. Moser, C. Stahl, and K. Wolf. Extending the compatibility notion for abstract WS-BPEL processes. In: Proceedings of the 17th International World Wide Web Conference, Beijing, China, April, 2008, pp. 785–794.

[34] A. Martens, R. Hamadi, and B. Benatallah. A Petri net based model for web service composition. In: Proceedings of the 14th Australian Database Conference, Adelaide, Australia, February, 2003, pp. 191–200.

[35] P. C. Xiong, Y. S. Fan, and M. C. Zhou. A Petri net approach to analysis and composition of web services. *IEEE Transactions on Systems, Man, and Cybernetics, Part A*, 40(2): 376–387, 2010.

[36] W. M. P. van der Aalst and H. Verbeek. Process mining in web services: the WebSphere case. *IEEE Data Engineering Bulletin*, 31(3): 45–48, 2008.

[37] S. Ryu, F. Casati, H. Skogsrud, B. Benatallah, and R. Saint-Paul. Supporting the dynamic evolution of web service protocols in service-oriented architectures. *ACM Transactions on the Web*, 2(2), April 2008.

[38] S. Ryu, R. Saint-Paul, B. Benatallah, and F. Casati. A framework for managing the evolution of business protocols in web services. In: Proceedings of the 4th Asia-Pacific Conference on Conceptual Modelling, Ballarat, Victoria, Australia, January, 2007, pp. 49–59.

CHAPTER 18

SERVICE-ORIENTED WORKFLOW SYSTEMS

WEI TAN[1] and MENGCHU ZHOU[2]
[1]IBM T. J. Watson Research Center, Yorktown Heights, NY, USA
[2]Department of Electrical and Computer Engineering, New Jersey Institute of Technology, Newark, NJ, USA

18.1 INTRODUCTION

From a technical point of view, with the emergence and maturation of service-oriented architecture (SOA), web services have begun to play a key role in both business and scientific workflows. In these applications, web services provide a uniform interface to various resources, data, and appliances. In business domain, web services are a key technology to glue various components; in scientific domain, grid middleware such as Globus [1] are adopting web services standards. From an economic point of view, web services are now gaining momentum in the so-called network economics. Many companies such as eBay, Google, and Amazon all use web services to extend their business and make profit out of it.

Service composition, or service-based workflow, offers enhanced reusability, agility, and dynamicity. Nowadays, many resources including data, computational resources, and others are wrapped with service interfaces to enable uniform access. Many applications, both business and scientific, are achieved through complex and distributed procedures. Many studies have been conducted on how to tailor the traditional workflow technology in order to address the new challenges brought about by such a service-oriented computing (SOC) paradigm. Workflow systems are increasingly being developed to enable users to integrate, orchestrate, and monitor various local or remote service resources to perform tasks like transaction processing, batch job execution, and scientific discovery. In this chapter we use *web service* and *service* interchangeably, and *service workflow* and *service composition* interchangeably.

Contemporary Issues in Systems Science and Engineering, First Edition.
Edited by MengChu Zhou, Han-Xiong Li and Margot Weijnen.

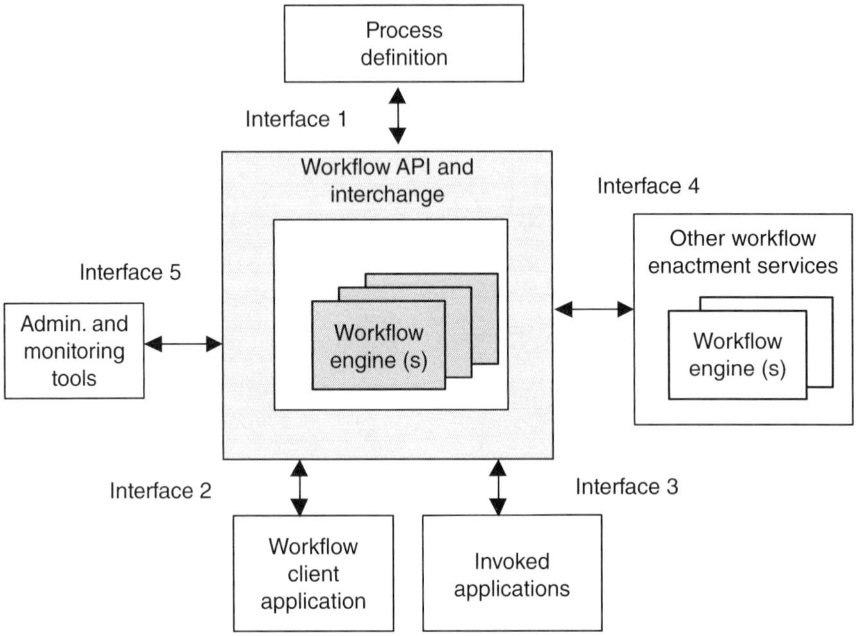

FIGURE 18.1 The workflow reference model (by WfMC [2]).

Workflow technologies can be traced back to the late 1970s. They originated from office automation (OA) and flourish in business process management (BPM) area. The requirements and characteristics of workflows in SOC are partially overlapping those of traditional workflows. For example, they both need to define tasks and dependencies, and the topology of a service workflow system is similar to the workflow reference model [2] proposed by WfMC, as shown in Figure 18.1. On the other hand, workflows in a SOC paradigm, have their unique features, and therefore raise new challenges to be addressed by academia as well as industry. Here we classify these features and challenges into five categories which correspond to five interfaces as defined in the workflow reference model and shown in Figure 18.1.

Interface 1: Process Definition

- A challenge for a workflow orchestrating web-scale resources is that, how to derive a composite workflow that fulfills a given user requirement both functionally and quality of service (QoS)wise. Typically in the web there are many services (i.e., resources) that provide similar functionality. Hence, techniques to facilitate the automatic service matchmaking and composition are highly needed. On the contrary, traditional OA workflows use a limited set of components in a relatively closed environment. Thus, their resource matchmaking is a relatively trivial task.

- Given the heterogeneity of various service resources, an approach to glue them together with the least cost and engineering effort, and the maximum QoS is also desired.

Interface 2: Workflow Client Application

The client application interface allows users to retrieve the tasks together with the related data, and submit the execution results. In SOC, user authentication and authorization is of great importance to ensure the integrity of workflows that span multiple organizations.

Interface 3: Invoked Applications

In SOC, Simple Object Access Protocol (SOAP) and REpresentational State Transfer (REST) [3] protocols are the industry standard to access a service. Fault handling is crucial because SOC workflows are often long-running, cross-organization, and expensive. The workflow system should be capable of identifying and handling failures to guarantee reliable execution in such cases.

Interface 4: Other Workflow Enactment Services

When multiple workflows are involved in an interaction, some technique is needed to guarantee the correctness of workflow(s). To take part in the collaboration with the workflows controlled by other enactment services, the flow engine should overcome heterogeneity at both process level (e.g., in data format, data semantics, and process logic) and system level (e.g., in security mechanisms).

Interface 5: Administration and Monitoring Tools

Interface 5 provides administration and monitoring interfaces such as user management, auditing, and resource management. In a SOC workflow system without a central control, it is challenging to guarantee the correctness and quality of an execution across multiple organizations.

18.2 WORKFLOW IN SOC: STATE OF THE ART

In this section we summarize the research of workflow in SOC into five topics, that is, languages for service composition, automatic service composition, mediation-aided service composition, verification of service workflow, and decentralized execution of workflow.

18.2.1 Languages for Service Composition

Industrial community has proposed many service orchestration specifications, including Web Service Business Process Execution Language (WS-BPEL, or BPEL for short) [4], Web Service Choreography Interface (WSCI) [5], Web Service Choreography Description Language (WS-CDL) [6], and Ontology Web Language-Service (OWL-S) [7]. Business Process Model and Notation (BPMN) [8] as a graphical representation specifying business processes, although not exclusively designed for SOC, also has a tight relation with it.

18.2.1.1 Business Process Execution Language Among these specification languages, BPEL [4] is becoming dominant not only because it has been proposed by Organization for the Advancement of Structured Information Standards (OASIS) as an industry standard, but also because it is execution oriented and supported by major software companies as well as the open-source community.

BPEL defines a metamodel and an XML-based grammar for describing the behavior of a business process that is composed of web services as well as exposed as a web service. The BPEL process defines the execution structure on how multiple web Service invocations are coordinated to achieve a business goal. BPEL also includes other model elements like data manipulation, fault/exception handling, and event processing. One favorite feature of BPEL is that, it utilizes many XML specifications like XML Schema, WSDL, XPath, and Extensible Stylesheet Language Transformation (XSLT). WSDL messages and XML Schemas are used as data type definitions; XPath and XSLT provide support for data assignment; WSDL is used to model all external partner services.

18.2.1.2 Business Process Model and Notation The motivation of BPMN is to provide a uniformed, intuitive, and graphical notation for business process models. It can be used as a common language among multiple stakeholders to share the knowledge of business processes: business analysts can use it to express the business goal and function to be achieved; IT experts can use it to implement a workflow that automates the process; managers can use it to monitor the execution and examine the performance of business. Although not originally designed for SOC, it is closely related to the concepts in it. For example, it has a model element called *service task* to depict a service; BPMN to BPEL mapping is also a part of its specification [9].

18.2.1.3 WSCI and WS-CDL Different from BPEL, WSCI and WS-CDL are not targeted at describing executable workflows. They, from a global viewpoint, define the peer-to-peer collaborations of services by specifying the ordered message exchanges between them. These two specifications are both abstract, and can describe the global observable behavior of multiple parties in collaboration. While BPEL is called a service orchestration language, WSCI and WS-CDL are usually referred to as service choreography languages.

18.2.1.4 Scientific Workflow Systems that Embrace SOC Scientific community also developed many workflow systems to orchestrate the data processing routines in scientific experiments. Various scientific workflow systems such as Kepler [10], Taverna [11], and OMII-UK [12] have developed tools to compose services and other components used by the e-science community.

18.2.2 Automatic Service Composition

Automatic service composition methods can be further classified into two categories, that is, planning based [13–15] and optimization based [16, 17]. Planning-based methods mainly concern the functional properties (i.e., the behavior) of the composite service, while the optimization-based ones concern the nonfunctional properties.

A common problem description for planning-based approaches is given in [18]: the automatic service composition problem can be described as a 5-tuple $< S, S_0, G, A, \Gamma >$, in which S is the set of all possible system states; S_0 and G are the initial and goal states, respectively, for the service composition; A is the set of all possible actions (i.e., all service operations); and $\Gamma = S \times A \times S$ defines the precondition and effect for each action. Then, if we start from the initial (or goal) state, with all the possible actions in Γ, various AI-planning methods can be used to derive a feasible plan to the goal (or initiate) state. Research in this category usually uses an ontology modeling language like OWL-S [7] to add semantic annotations to service descriptions. Then situation calculus [19, 20], PDDL [21], or rule-based approaches such as SWORD [22] and HTN [23] can be used to derive feasible solution(s).

Optimization-based approaches are used to derive composite services which satisfy certain QoS requirement. QoS constraints include cost, time, availability, reliability; usually there are local constraints (constraints imposed on single service) and global ones (constraints imposed on multiple services). Optimization methods like linear programming [16] and genetic algorithm [17] are used to solve the QoS constraints and yield feasible solution(s).

18.2.3 Mediation-Aided Service Composition

The service orchestration and choreography specifications, and most of the automatic service composition methods, assume the direct composition between services. It is made based on the assumptions that:

- The incoming messages of one service are the exact ones provided by its partner(s); the outgoing messages of one service are the exact ones consumed by its partner(s).
- Two services in composition consent to the message format and exchange sequence, such that their composition process always executes in a logically correct way (e.g., terminates properly).

If these two assumptions do not hold, we say that services are *partially compatible*. Briefly, there are two methods to make partially compatible services work with each other, that is, configuration [24] and mediation [25]. Configuration is a heavyweight approach that changes the original service within some predefined variable points to make it work smoothly with other services. Casati and Shan [24] propose a configuration-based approach that enables the dynamic and adaptive composition of e-services. Its adaptability is achieved through dynamic service discovery, dynamic conversation selection, multiservice nodes, and dynamic service node creation.

Recently the mediation-aided approach is attracting more attention. Benatallah *et al.* [26] provide a summary of the mismatch patterns in web Service composition. These patterns include Message Ordering Mismatch, Extra Message Mismatch, Missing Message Mismatch, Message Split Mismatch, and Message Merge Mismatch. Based on the work in Reference 26, Kongdenfha *et al.* [27] propose the use of an aspect-oriented programming (AOP) approach to weave adaptor code into the partial compatible services. Brogi and Popescu [28] present a method to automatically generate a mediator between two BPEL services. Nezhad *et al.* [29] also propose an automata-based method to model the protocol of service interaction and to identify mismatches automatically. They use a schema matchmaking technique to handle the issue of message mapping, and they also propose some heuristic methods for deadlock resolution between two services.

18.2.4 Verification of Service Workflows

The verification of workflows in an SOC paradigm has its unique features, compared to the verification of traditional workflows. First, the model elements in specifications like BPEL, are much more complicated than those in former workflow specifications like WfMC's XPDL [30]. BPEL concepts such as correlation set, death path elimination, compensation, and fault handling are unique, which brings complexity in verification. Second, because workflows in SOC usually interact and interrelate to each other by message exchange, the correctness of a workflow not only relies on its own internal logic, but also relies on how its partners collaborate with it. Even if a workflow is correct from a traditional single-process point of view, its composition with another one may still fail because these two workflows do not agree on their interactions.

Based on the formal method used, the researches in this area can be classified into several categories, that is, Petri-net-based, automata-based, and process-algebra-based methods.

18.2.4.1 Petri-Net-Based Methods Ouyang *et al.* [31] build up a comprehensive Petri net formalism for various BPEL model elements, including basic activities, structured activities, event handler, control link, and fault handling. This model covers the major part of BPEL specification, and is ideal for the verification of a single BPEL process. Martens *et al.* [32] try to verify the choreography of multiple BPEL processes. The properties verified include usability (give a BPEL process, whether a proper environment exists to interact with it), equivalence (whether one BPEL process can safely replace another in a choreography), and compatibility (whether the composition of BPEL processes can be used by an environment). Hinz *et al.* [33] transform BPEL into Petri nets, and then use Computational Tree Logic (CLT) and model-checking tool to verify various temporal properties.

18.2.4.2 Automata-Based Methods Su *et al.* [34, 35] focus on the automata model for services and apply model checking via Linear Temporal Logic (LTL). A special point of their research is a technique called synchronizability analysis to tackle the problem of state space explosion brought about by asynchronous messaging. Their result shows that, if a composite web service is synchronizable, its conversation set remains the same when asynchronous communication is replaced with synchronous communication. Thus a synchronous communication model can be used in LTL model checking. Kazhamiakin *et al.* [36] develop a set of parametric communication models in service compositions. These models range from synchronous communications to asynchronous ones with complex buffer structures. In addition, they develop a technique to associate with a service composition, the most adequate communication model that is sufficient to capture all the behaviors of the composition. Using this model, the analysis before the actual verification can bring about an improved performance in verification.

18.2.4.3 Process-Algebra-Based Methods Process algebra [37] is an algebraic approach to the modeling and analysis of concurrent processes. Its advantage is that it provides not only temporal logic model checking, but also bisimulation analysis through which whether two processes have equivalent behaviors can be determined. Foster *et al.* transform BPEL into a kind of process algebra called Finite State Process (FSP), and then use a model-checking tool to verify properties like whether the implementation satisfies the abstract design specifications [38], whether the composition of two services is compatible [39], and whether the composition of BPEL services satisfies the properties defined in WS-CDL [40]. A formal BPEL model based on Pi-calculus (a kind of process algebra based on Calculus of Communicating System) can be found in Reference 41; a Pi-calculus-based technique to analyze the behavioral substitution of web services is proposed in Reference 42.

18.2.5 Decentralized Execution of Workflows

Workflow systems are often built on the client/server architecture in which a single workflow engine takes the responsibility for the operation of a whole process. In many circumstances this sort of centralized systems may not fully meet the requirements. The services involved in a grid workflow come from different organizations, and probably these services can only be controlled by the workflow engines inside their own domain boundary; therefore sometimes a cross-organizational workflow cannot be orchestrated by one single engine. Secondly, the reliability and performance of a workflow can be increased when multiple workflow engines collaborate to execute it.

Partitioning an integrated workflow into small fragments each of which is orchestrated by one engine is a preliminary requirement for decentralized execution. A team from IBM India Research Lab has conducted a series of studies in the decentralized execution of composite BPEL services [43–46]. They investigated: how to partition a BPEL into multiple parts, especially the partition of fault-handling code; the model partition policies to improve execution performance; how to partition the model when dataflow is constrained. Recently a process-mining-based model fragmentation technique has been proposed for distributed workflow execution [47].

18.3 OPEN ISSUES

There are still many challenges to tackle when bringing workflow and service computing technology together, especially in a rapidly evolving IT environment. Here we discuss some open issues in this field and highlight a few exciting topics to pursue.

18.3.1 Social Network Meets Service Computing

The proliferation of social network web sites such as Twitter, Facebook, and Wikipedia, has provided a great means for connecting people who share interests and activities across various borders. These success stories make business and scientific community regard social network (SN) as a means for further improvement of their computing infrastructure. Advances in social network techniques, has gaining momentum to impact the major internet-based computing paradigms, including service computing. Currently, social network connects people or groups who expose similar interests or features. In service computing, social networks can connect not only people and organizations, but also other entities such as software components, web services, data, and workflows. More importantly, SN in service computing can record, trace, and visualize complex relations among people and nonhuman artifacts, as well as accumulate the wisdom of crowds which has profound technical and social impact. For example, by examining interaction patterns among services, users, and providers, users can identify trusted service providers or partners to share resources. myExperiment [48] is such a social network encouraging the sharing of bioinformatics web services and processes, and eventually inspiring collaborative scientific exploration. We have analyzed the workflows stored at myExperiment applying social network analysis techniques [49]. Through this work we reveal the current usage pattern of services in scientific workflows, and propose how this knowledge can be extracted to facilitate the reuse of best practices.

A recent work [50, 51] has collected over 1 million web service invocation results on 100 real-world web services from service users all over the world. It then develops a collaborative filtering technique such that when a user request a QoS value of a service, it will first identify similar users, then systematically combine their observation of that particular service's QoS and return the combined value. This work sheds a light on the use of many social recommendation techniques to the area of service composition.

18.3.2 More Practical and Flexible Service Composition

Current service composition techniques mainly assume that (1) services are well annotated with the metadata needed by composition algorithms, and (2) services can perfectly work together as a service chain. These assumptions do not hold in real-life scenarios, because services are developed and hosted in a distributed manner. However they need to interact with one another in the ways not necessarily foreseen when they are separately developed.

Firstly, different services hardly share a common terminology and documents that describe their input/output and functionalities are sometimes incomplete or inaccurate.

Secondly, as presented earlier in this chapter, partial compatibility is common in real-life service composition. It refers to the situation that, two (or more) services provide complementary functionality and can be linked together in principle; however, their interfaces and interaction patterns do not fit each other exactly. Hence, they cannot be directly composed.

A couple of techniques are promising to address this challenge. Data mashup is a lightweight technique that can aggregate web feeds, web pages, and other contents. Based on the REST [52] standard, for developers it alleviates the pain to annotate services, while for end users it offers a relatively easier alternative to do service composition.

As we have discussed, a mediation-aided approach is promising to address the issue of partial compatibility. In our previous work we have used mediator to glue services and resolve the logic (i.e., task sequence) conflict between them [53]. More work is needed to address other aspects such as temporal conflicts and even legal ones.

18.3.3 Workflow as a Service

With the emergence of cloud computing [54], researchers start thinking of how workflow can benefit from it. We argue that hosting workflow systems in cloud computing and providing workflow as a service has several advantages. First, like any other middleware, workflow management systems are complex to design, deploy, and manage; moreover, they need to scale out to cope with sudden transaction increase. These features make it desirable to provision a workflow system in a software-as-a-service manner.

Another benefit that is specific to workflow is that, workflow-as-a-service can enable the sharing of best practices. Current approaches to service composition are generally used in isolated stand-alone applications. A typical scenario to design a service composition is that, developers use a graphical tool to compose the services that they are aware of, with the business logic they have in mind; after the composition is completed, it is uploaded to a workflow engine to execute. In this case, developers do not leverage any existing knowledge, and the knowledge they gained through the composition practice is neither accumulated nor shared among people who may undertake similar tasks. Think of this alternate: in a hosted workflow platform enhanced by social network technology, when users are to compose a new workflow, they can see similar workflows already completed and contributed by others. During the process of composition they can obtain context-aware assistance such as which services to use, in what sequence, and even who to consult for questions. The platform will also record their behavior in this composition process, such as which services they have eventually used, which action they have taken when been recommended some services. This behavior log along with the workflow they have just completed, are contributed back to the platform and benefit others. We have done some preliminary work in workflow recommendation in a hosted platform [55], but several issues remain open, such as how to accurately capture users' requirement with least interruption to their work, and how to measure the effectiveness of the recommendation technique.

18.3.4 Novel Applications

With the advancement of service computing and workflow technology, they are mature enough to be applicable to many emerging applications, like health care [36], data-driven science [56], and mobile industry [57].

In recent years, medical information systems play an increasingly important role in supporting doctors and nurses, enhancing the quality of medical services, reducing medical expense, and improving the care of chronic patients. Therefore, medical information technology has drawn much attention in various countries, for example, the research and development of Hospital Information Systems (HIS), Electronic Medical Record (EMR) systems, Picture Archiving and Communication Systems (PACS), and the National Health Information Network in the United States of America, the EMR systems, online choosing of medical institutions and appointment systems, health network infrastructure, and PACS in the United Kingdom, the electronic health record (EHR) system in Canada, the National E-Health Technology Architecture in Australia, the Red System in Denmark, the Grenoble Integrated HIS of No. 8 health center in France, and the "Jin Wei" Project in China [58].

Majority of previous work focused on information standardization and integration for hospitals or medical institutions. The health care community has recognized the need to transform from the current hospital-centralized, treatment-based mode to prevention-oriented comprehensive health care mode in which hospitals, communities, families, and individuals are closely involved. The new mode needs to provide individuals with intelligent health information management and health care services. It allows them to enjoy medical prevention and health care services in their daily life. The advancement of SOC is promising to facilitate the construction of service systems with higher reusability, flexibility, extensibility, and robustness. Based on SOC various medical functions can be packaged and used in the form of services which can in turn be customized and composed to meet user needs that are diverse, uncertain, and personalized.

The second author of this chapter is involved in an ongoing effort named Public-oriented Health care Information Service Platform (PHISP) that can provide customizable and personalized health care services for individuals. In PHISP there are three layers, that is, body sensor networks (BSN), cloud platform, and health care service system, as described as follows and illustrated in Figure 18.2.

(1) *BSN*. According to different circumstances and personalized needs, appropriate health information collection terminals (i.e., sensors) are configured for different individuals. BSN is used to realize the multimode acquisition, integration, encryption, and real-time transmission of personal health information in living or working environment.

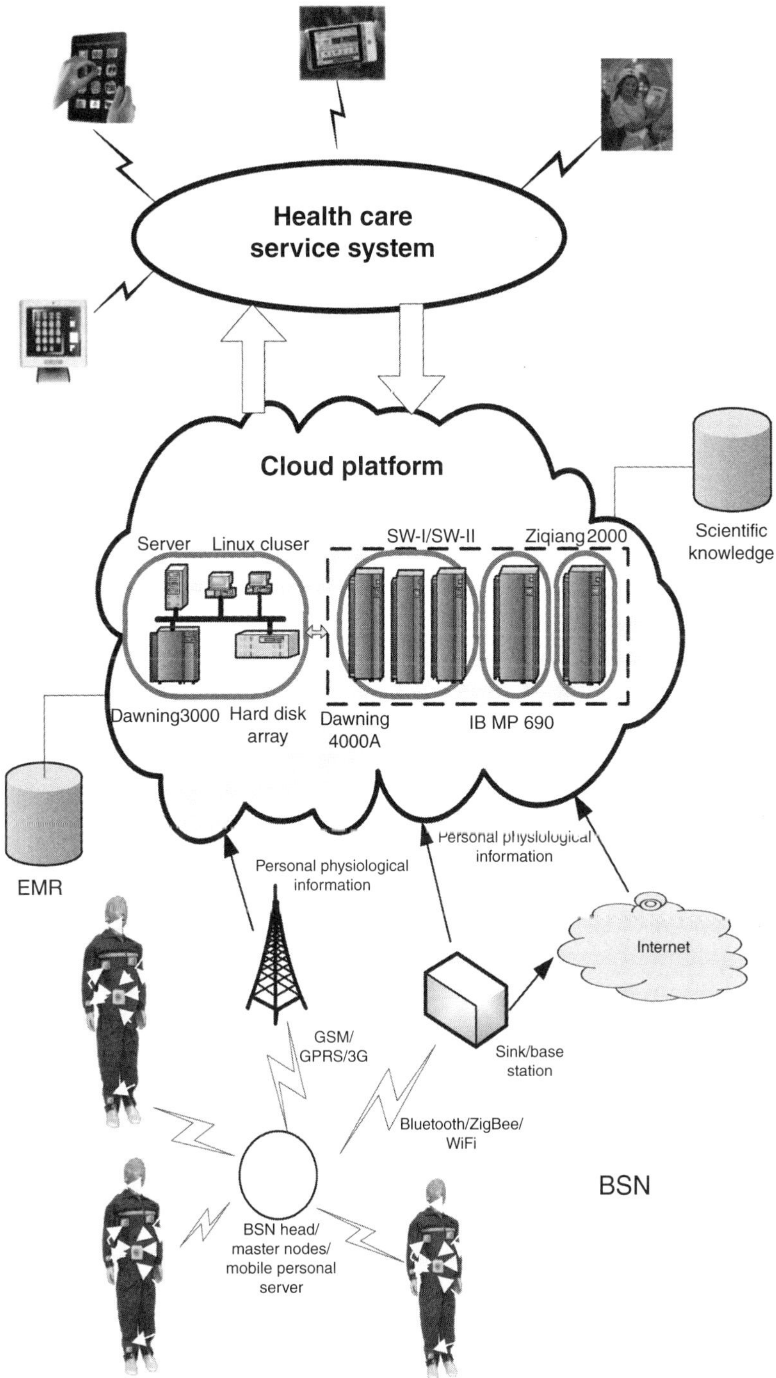

FIGURE 18.2 The architecture of the PHISP [36].

(2) *Cloud platform.* Based on cloud computing technology, it achieves the rapid storage, management, retrieval, and analysis of massive health data, which mainly includes EMR repository, scientific knowledge base of health care, and personal health data acquired from BSN.

(3) *Health care service system.* It includes a personal health information management system, dynamic personal health monitoring and real-time early warning system, personal health risk assessment and guidance system, seasonal disease early warning system, decision-making library for various diseases, and typical disease care services.

PHISP is in collaboration with many universities and medical institutions in China. It has been used to provide teaching, research, medical treatment, and health care services for college students and teachers, researchers, medical professionals, and ordinary people. PHISP supports personal health information management, personal health risk assessment and guidance, dynamic personal health monitoring and real-time early warning, active recommendation of personalized medical treatment, active seasonal disease warning and other health care services for individuals.

Another novel application is the cancer Grid (caGrid) Workflow Toolkit [56], an open-source tool suite for data-centric cancer research. Cancer Biomedical Informatics Grid (caBIG) [59] is an NIH-sponsored information network enabling the collaboration of cancer researchers and caGrid [60] is its underlying service-based computation infrastructure. CaGrid requires that services are composed and orchestrated in a given sequence to realize data pipelines, which are often called scientific workflows. CaGrid selected Taverna as its workflow system due to its integration with web service technology and extensible plug-in framework. By extending Taverna, the caGrid Workflow Toolkit (or the toolkit for short) provided a tool suite to compose and orchestrate web services in caGrid. Using it users can access more than 100 services and are offered with a rich set of features including (1) service discovery; (2) data query and transfer; (3) security configuration and enforcement; (4) state management in service interactions; (5) invocation of data and analytical services.

With the proliferation of web services and software tools to compose them, we look forward to more novel application which we do not usually see at this moment.

18.4 CONCLUSIONS

Building workflows using service composition has become an important method to build composite applications and reuse existing resources. Therefore, workflow-based service composition and orchestration is now a hot topic in both academia and industry.

This chapter summarized research activities in the field of workflow in SOC. Five major research topics, that is, languages for service orchestration, automatic service composition, mediation-aided service composition, verification of service workflows, and decentralized execution of workflows are discussed.

There are still many challenges to tackle when bringing workflow and service computing technology together, especially in a rapidly evolving IT environment. We have laid out several topics that are potentially appealing, that is, to combine the power of social network and service computing, more practical and flexible service composition, workflow as a service, and novel applications.

REFERENCES

[1] I. Foster and C. Kesselman. Globus: a metacomputing infrastructure toolkit. *International Journal of High Performance Computing Applications*, 11: 115–128, 1997.

[2] WfMC. The workflow reference model. Available at http://www.wfmc.org/standards/docs/tc003v11.pdf, 2005 (accessed January 24, 2015).

[3] C. Pautasso, O. Zimmermann, and F. Leymann. Restful web services vs. "big" web services: making the right architectural decision. In: 17th International Conference on World Wide Web, Beijing, China, 2008, pp. 805–814.

[4] Web services business process execution language version 2.0. Available at http://docs.oasis-open.org/wsbpel/2.0/OS/wsbpel-v2.0-OS.html (accessed July 31, 2011).

[5] W3C. Web service choreography interface (WSCI) 1.0. Available at http://www.w3.org/TR/wsci/, 2002 (accessed January 24, 2015).

[6] W3C. Web services choreography description language version 1.0. Available at http://www.w3.org/TR/2004/WD-ws-cdl-10-20041217/, 2004 (accessed January 24, 2015).

[7] W3C. OWL-S: Semantic markup for web services. Available at http://www.w3.org/Submission/OWL-S/, 2004 (accessed January 24, 2015).

[8] S. A. White and D. Miers. *Bpmn Modeling and Reference Guide*. Future Strategies Inc, 2008.

[9] OMG. Business process model and notation (BPMN) version 2.0. Available at http://www.omg.org/spec/BPMN/2.0/, 2010 (accessed January 24, 2015).

[10] B. Ludäscher, I. Altintas, C. Berkley, D. Higgins, E. Jaeger, M. Jones, E. A. Lee, J. Tao, and Y. Zhao. Scientific workflow management and the Kepler system. *Concurrency and Computation: Practice & Experience*, 18: 1039–1065, 2005.

[11] T. Oinn, P. Li, D. B. Kell, C. Goble, A. Goderis, M. Greenwood, D. Hull, R. Stevens, D. Turi, and J Zhao. Taverna/myGrid: aligning a workflow system with the life sciences community. In: *Workflows for E-science: Scientific Workflows for Grids*, edited by I. J. Taylor, E. Deelman, D. B. Gannon, and M. Shielsds. Guildford: Springer, 2007, pp. 300–319.

[12] L. Chen, W. Emmerich, and W. Wasseman. Grid services orchestration with OMII-BPEL. In: *Grid Computing: Infrastructure, Service, and Applications*, edited by L. Wang, W. Jie, and J. Chen. CRC Press, 2009, pp. 191–222.

[13] D. Berardi, D. Calvanese, G. D. Giacomo, M. Lenzerini, and M. Mecella. Automatic composition of e-services that export their behavior. In: Proceedings of the 1st International Conference on Service Oriented Computing (ICSOC), 2003, pp. 43–58.

[14] Z. Dang, O. H. Ibarra, and J. Su. On composition and lookahead delegation of e-services modeled by automata. *Theoretical Computer Science*, 341: 344–363, 2005.

[15] C. E. Gerede, R. Hull, O. H. Ibrra, and J. Su. Automated composition of e-services: lookaheads. In: 2nd International Conference on Service Oriented Computing, New York City, 2004, pp. 252–262.

[16] D. Ardagna and B. Pernici. Dynamic web service composition with QoS constraints. *International Journal of Business Process Integration and Management*, 1: 233–243, 2006.

[17] G. Canfora, M. Di Penta, R. Esposito, and M. L. Villani. An approach for QoS-aware service composition based on genetic algorithms. In: Conference on Genetic and Evolutionary Computation, Washington DC, 2005, pp. 1069–1075.

[18] J. Rao and X. Su. A survey of automated web service composition methods. In: Semantic Web Services and Web Process Composition, San Diego, CA, 2004, pp. 43–54.

[19] S. Narayanan and S. A. McIlraith. Simulation, verification and automated composition of web services. In: Eleventh International World Wide Web Conference, WWW2002, Honolulu, HI, 2002, pp. 77–88.

[20] S. McIlraith and T. Son. Adapting golog for composition of semantic web services. In: Eighth International Conference on Principles and Knowledge Representation and Reasoning (KR-02), Toulouse, France, 2002, pp. 482–493.

[21] D. McDermott. Estimated-regression planning for interactions with web services. In: AI Planning Systems Conference (AIPS'02), 2002, pp 204–211.

[22] S. R. Ponnekanti and A. Fox. SWORD: a developer toolkit for web service composition. In: Proceedings of the Eleventh International World Wide Web Conference, Honolulu, HI, 2002.

[23] E. Sirin, B. Parsia, D. Wu, and D. Nau. HTN planning for web service composition using SHOP2. *Journal of Web Semantics*, 1: 377–396, 2004.

[24] F. Casati and M. C. Shan. Dynamic and adaptive composition of e-services. *Information Systems*, 26: 143–163, 2001.

[25] D. Fensel and C. Bussler. The web service modeling framework WSMF. *Electronic Commerce Research and Applications*, 1(2): 113–137, 2002.

[26] B. Benatallah, F. Casati, D. Grigori, M. Nezhad, and F. Toumani. Developing adapters for web services integration. In: Proceedings of the International Conference on Advanced Information Systems Engineering (CAiSE), 2005, pp. 415–429.

[27] W. Kongdenfha, R. Saint-Paul, B. Bentallah, and F. Casati. An aspect-oriented framework for service adaptation. In: 4th International Conference on Service-Oriented Computing—ICSOC 2006, Chicago, IL, 2006, pp. 15–26.

[28] A. Brogi and R. Popescu. Automated generation of BPEL adapters. In: 4th International Conference on Service-Oriented Computing—ICSOC 2006, Chicago, IL, 2006, pp. 27–39.

[29] H. R. M. Nezhad, B. Bentallah, A. Martens, F. Curbera, and F. Casati. Semiautomated adaptation of service interactions. In: 16th International World Wide Web Conference (WWW2007), Banff, AB, Canada, 2007, pp. 993–1002.

[30] WfMC. Process Definition Interface—XML Process Definition Language. Available at http://www.wfmc.org/standards/docs/TC-1025_xpdl_2_2005-10-03.pdf, 2005 (accessed January 24, 2015).

[31] H. M. W. Verbeek and W. M. P. V. D. Aalst. Analyzing BPEL processes using Petri nets. In: Second International Workshop on Applications of Petri Nets to Coordination,

Workflow and Business Process Management, Florida International University, Miami, FL, 2005, pp. 59–78.

[32] A. Martens. Analyzing web service based business processes. In: 8th International Conference on Fundamental Approaches to Software Engineering (FASE 2005), Edinburgh, UK, 2005, pp. 19–33.

[33] S. Hinz, K. Schmidt, and C. Stahl. Transforming BPEL to Petri nets. In: 3rd International Conference on Business Process Management, Nancy, France, 2005, pp. 220–235.

[34] T. Bultan, J. Su, and X. Fu. Analyzing conversations of Web services. *IEEE Internet Computing*, 10: 18–25, 2006.

[35] X. Fu, T. Bultan, and J. Su. Synchronizability of conversations among web services. *IEEE Transactions on Software Engineering*, 31: 1042–1055, 2005.

[36] P. Wang, Z. Ding, C. Jiang, and M. Zhou. Web service composition techniques in a health care service platform. In: 2011 IEEE International Conference on Web Services (ICWS), 2011, pp. 355–362.

[37] J. C. M. Baeten and W. P. Weijland. *Process Algebra*. New York: Cambridge University Press, 1991.

[38] H. Foster, S. Uchitel, J. Magee, and J. Kramer. Model-based verification of web service compositions. In: 18th IEEE International Conference on Automated Software Engineering, 2003, pp. 152–161.

[39] H. Foster, S. Uchitel, J. Magee, and J. Kramer. Compatibility verification for web service choreography. In: IEEE International Conference on Web Services 2004, 2004, pp. 738–741.

[40] H. Foster, S. Uchitel, J. Magee, and J. Kramer. Model-based analysis of obligations in web service choreography. In: Advanced International Conference on Telecommunications and International Conference on Internet and Web Applications and Services (AICT/ICIW 2006), Guadeloupe, French 2006, p. 149.

[41] R. Lucchi and M. Mazzara. A pi-calculus based semantics for WS-BPEL. *Journal of Logic and Algebraic Programming*, 70: 96–118, 2007.

[42] L. Kuang, Y. Xia, S. Deng, and J. Wu. Analyzing behavioral substitution of web services based on Pi-calculus. In: IEEE International Conference on Web Services (ICWS), 2010, pp. 441–448.

[43] G. Chafle, S. Chandra, V. Mann, and M. G. Nanda. Orchestrating composite web services under data flow constraints. In: 2005 IEEE International Conference on Web Services, vol. 1, 2005, pp. 211–218.

[44] G. B. Chafle, S. Chandra, V. Mann, and M. G. Nanda. Decentralized orchestration of composite web services. In: 13th International World Wide Web Conference, New York, 2004, pp. 134–143.

[45] M. G. Nanda, *et al.* Decentralizing execution of composite web services. In: 19th Annual ACM SIGPLAN Conference on Object-Oriented Programming, Systems, Languages, and Applications Table of Contents, Vancouver, BC, Canada, 2004, pp. 170–187.

[46] M. G. Nanda and N. Karnik. Synchronization analysis for decentralizing composite web services. In: 2003 ACM Symposium on Applied Computing Table of Contents, Melbourne, FL, 2003, pp. 407–414.

[47] S. X. Sun, Q. Zeng, and H. Wang. Process-mining-based workflow model fragmentation for distributed execution. *IEEE Transactions on Systems, Man and Cybernetics, Part A: Systems and Humans*, 41: 294–310, 2011.

[48] D. De Roure, C. Goble, and R. Stevens. The design and realisation of the my experiment virtual research environment for social sharing of workflows. *Future Generation Computer Systems*, 25: 561–567, 2009.

[49] W. Tan, J. Zhang, I. Foster, D. De Roure, and C. Goble. Network analysis of scientific workflows: a gateway to reuse. *IEEE Computer*, 43: 54–61, 2010.

[50] Z. Zheng, H. Ma, M. R. Lyu, and I. King. QoS-aware web service recommendation by collaborative filtering. *IEEE Transactions on Services Computing*, 4: 140–152, 2011.

[51] L. Chen, Y. Feng, J. Wu, and Z. Zheng. An enhanced QoS prediction approach for service selection. In: 2011 IEEE International Conference on Services Computing (SCC), 2011, pp. 727–728.

[52] E. M. Maximilien, A. Ranabahu, and K. Gomadam. An online platform for web APIs and service mashups. *IEEE Internet Computing*, 12: 32–43, 2008.

[53] W. Tan, Y. Fan, and M. Zhou. A Petri net-based method for compatibility analysis and composition of web services in business process execution language. *IEEE Transactions on Automation Science and Engineering*, 6: 94–106, 2009.

[54] M. Armbrust, A. Fox, R. Griffith, A. D. Joseph, R. Katz, A. Konwinski, G. Lee, D. Patterson, A. Rabkin, I. Stoica, and M. Zaharia. Above the clouds: a berkeley view of cloud computing. Technical Report UCB/EECS-2009-28, EECS Department, University of California, Berkeley, 2009.

[55] W. Tan, J. Zhang, R. Madduri, and I. Foster. Providing map and GPS assistance to service composition in bioinformatics. In: IEEE International Conference on Services Computing (SCC), Washington, DC, 2011.

[56] W. Tan, R. Madduri, A. Nendic, S. Soiland-Reyes, D. Sulakhe, I. Foster, and C. A. Goble. CaGrid workflow toolkit: a taverna based workflow tool for cancer grid. *BMC Bioinformatics*, 11: 542, 2010.

[57] Y. Natchetoi, V. Kaufman, and A. Shapiro. Service-oriented architecture for mobile applications. In: Proceedings of the 1st International Workshop on Software Architectures and Mobility, Leipzig, Germany, 2008, pp. 27–32.

[58] P. Wang, Z. Ding, C. J. Jiang, and M. C. Zhou. A web service based public-oriented personalized health care platform. Tongji University Technical Report 01-12-2010, 2010.

[59] A. C. Von Eschenbach and K. Buetow. Cancer informatics vision: caBIG^{TM}. *Cancer informatics*, 2: 22, 2006.

[60] J. Saltz, T. Kurc, S. Hastings, S. Langella, S. Oster, D. Ervin, A, Sharma, T. Pan, M. Gurcan, J. Permar *et al.* e-Science, caGrid, and translational biomedical research. *IEEE Computer*, 41: 58–66, 2008.

PART V

SENSING, NETWORKING, AND OPTIMIZATION IN ROBOTICS AND MANUFACTURING

CHAPTER 19

REHABILITATION ROBOTIC PROSTHESES FOR UPPER EXTREMITY

HAN-PANG HUANG[1], YI-HUNG LIU[2], WEI-CHEN LEE[3], JIUN-YIH KUAN[1], and TZU-HAO HUANG[1]

[1]Department of Mechanical Engineering, National Taiwan University, Taiwan
[2]Department of Mechanical Engineering, National Taipei University of Technology, Taiwan
[3]Department of Mechanical Engineering, National Taiwan University of Science and Technology, Taiwan

19.1 INTRODUCTION

One of the most valuable applications for medical mechatronics is the robotic prosthesis for it not only plays a crucial role in rehabilitation but also helps much in improving life quality of patients. Recently, various rehabilitation robots have been developed. Yet, there are still critical issues to be addressed, for example, the design complexity for a rehabilitation robot arm. The method to reduce the complexity is first introduced in this chapter. The method is based on a robust control law and an independent joint model, and is derived in detail in this chapter.

In addition to rehabilitation robot arms, rehabilitation robot hands have also been studied for more than 30 years. With the recent development of Touch Bionic's i-LIMB^{TM} Hand, Defense Advanced Research Projects Agency (DARPA)'s RP 2009, and other commercial powered prosthetic hands, they have regained the focus in both engineering and medical fields. The current problems associated with the rehabilitation robot hands include, but are not limited to, mechanism design, electroneurographic (ENG)-based control, power system, and training. Therefore, this chapter also emphasizes on the mechanical mechanisms of the rehabilitation robot hand and the associated power systems.

Contemporary Issues in Systems Science and Engineering, First Edition.
Edited by MengChu Zhou, Han-Xiong Li and Margot Weijnen.

Myoelectric control plays a key role in rehabilitation robot arm/hand. A robot arm/hand with a myoelectric control system is known as a neuroprosthesis. Previously, most myoelectric control systems concentrated on the improvement of electromyography (EMG) classification accuracy. However, few studies have discussed the stability of neuroprostheses, a critical challenge in the field. Accuracy and stability are two totally different concepts for myoelectric control systems. The former refers to the classification performance based on EMG signals, whereas the latter refers to the ability to online reject the EMG patterns that do not belong to the classes (i.e., types of postures) defined in the off-line training stage. Multi-class classification has been a common strategy used to improve EMG classification accuracy. This strategy, however, fails to improve stability of neuroprostheses (or myoelectric control systems), as the two concepts are different in nature. In this chapter, a novel method that can improve stability of neuroprostheses will also be introduced.

19.2 REHABILITATION ROBOT ARM AND CONTROL

From a mechanical perspective, a rigid rehabilitation robot arm should be kinematically redundant and composed of at least seven degrees of freedom (DOFs) from the shoulder to the wrist to imitate the human arm behaviors [1]. Since a rehabilitation robot arm, consisting of multiple DOFs, is designed to execute complex and dexterous motions, how to construct a fast and precise control system is an important issue. Due to high complexity of the high-order dynamic model arisen from the coupling effects and payload variations, normal model-based controllers are difficult to be implemented in practical applications. Therefore, in this chapter, we focus on robust control law design based on the independent joint model to reduce the complexity of the design process.

The dynamic model of a rehabilitation robot arm can be described by a set of nonlinear and highly coupled differential equations. It is known that the model-based controllers [2, 3] can be used to enhance tracking capability with respect to the simple joint controller, such as proportional-derivative (PD) or proportional-integral-derivative (PID) type. However, the design of such a control algorithm based on a complete system model is generally complicated due to the presence of nonlinearity and dynamic coupling [4]. Thus, a more realistic controller design problem is to determine a required torque of each motor to achieve the tracking control of a robot joint, as an independent joint control scheme.

For a rehabilitation robot, each joint motor is connected to the links through gear trains with high reduction ratios, where the link dynamics appears as dynamic loads. Fortunately, these dynamic loads apparently reduced by the high gear ratios implies the interactions of the nonlinearity and coupling effects of robot links with joint driving systems largely reduced [5]. Thus, the joint driving systems' tendency that dominates the system dynamics essentially allows us to design the independent joint controller with treating the dynamic loads as disturbance torques and directly dealing with its effect by computation. However, as pointed out in Reference 6, it suffers from many problems. For example, the robot model and its parameters are in general unknown due to uncertainties. A robust controller, such as a sliding mode controller, is another alternative.

Sliding mode control is robust with respect to matched internal and external disturbances, guaranteeing robust tracking performance in a decoupling way with respect to system uncertainties and disturbances through the use of switching control or variable structure control during sliding motion [7]. However, the undesired chattering phenomenon of the control signal produced by the switch mechanism has become the major obstacle and great difficulty in some practical applications, particularly for the control of mechanical systems, such as rehabilitation robot systems, possessing a large range of dynamic responses.

In the literature, to overcome the preceding drawback, the first approach for chattering reduction is boundary layer control (BLC) [7]. In the BLC design, however, an antagonistic result of zero tracking error and smooth control law at the same time establishes a compromise between the allowable tracking error and control activity [8, 9]. Especially when the measurement noise is of a level larger than the boundary layer width, the high-frequency oscillations in the noise are reflected and amplified in the control signal [9]. Another way is to use higher-order sliding mode control (HOSMC) [10, 11]. However, a drawback of continuous approximation methods is the decrease of the accuracy of the system and the sliding mode stability. Furthermore, observer-based sliding mode control (OBSMC), introducing observers to the modeled part of the system with a sliding manifold defined in terms of the observer states, has been introduced for its effectiveness in the presence of unmodeled dynamics. The almost ideal high-frequency observer control signal is filtered by the high-gain fast dynamical part of the system such that a smooth control may be actually applied [12].

Another approach for chattering reduction is the dynamic sliding mode control (DSMC) or second-order sliding mode control [10–13], where an integrator or any low-pass filter is placed in front of the original system. The time derivative of the control input treated as the control variable for the augmented system, a system with the integrator, makes it higher in order compared with the original system. However, the designed filter may improve the stability of the sliding system and keep the controlled system with a chattering-free control input to the real system. In particular, it also keeps the system with desired performance and characteristics.

Despite its superiority to BLC, OBSMC, and HOSMC, the design of the DSMC is challenging for the DSMC designed system being one order larger than the original system. As a result, the new sliding variable in DSMC contains an uncertainty term due to the external disturbance and/or parametric uncertainty. Evaluation of the new sliding variable in the DSMC becomes difficult. Several variable structure estimators are introduced to estimate the sliding variable in the DSMC, but they have to assume a priori that the system state is uniformly bounded before proving the system stability, or that stability is guaranteed only if a differential inequality with bounded coefficients is satisfied. To overcome the problem of sliding variable estimation in the DSMC to stabilize a control system, Chen *et al.* have proposed to use a two-dimensional LTR observer [9, 14]. All assumptions on the bounded trajectory, the differential inequality condition, and the detection of sign change of the derivative of an accessible signal are

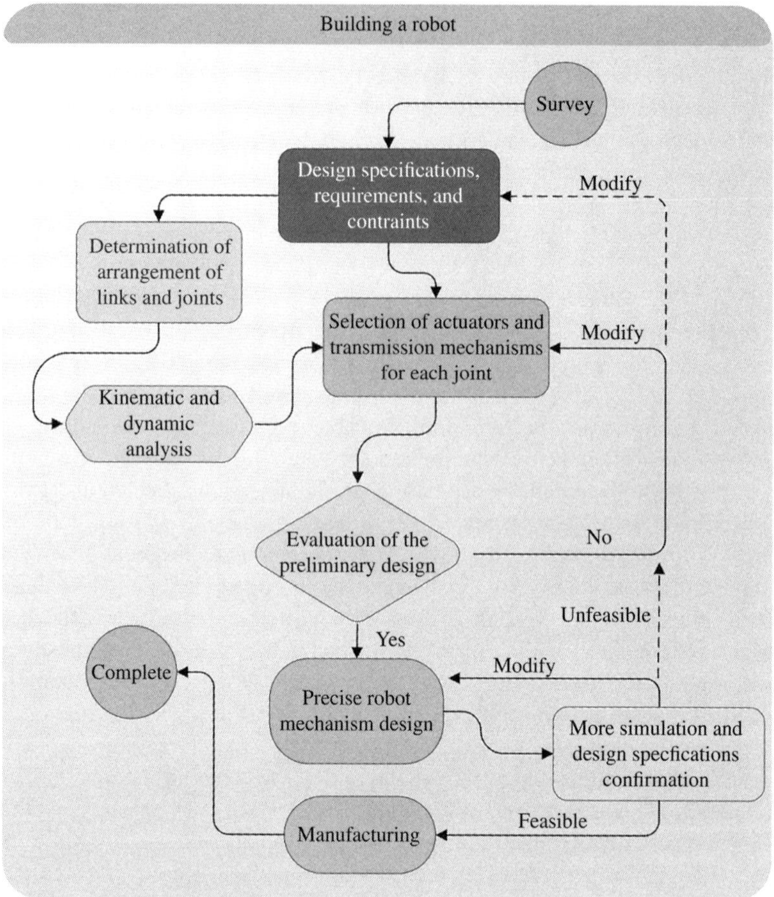

FIGURE 19.1 Design flow chart of building a robot.

not necessary in their controller design. This work extends their controller, designed for a regulation control problem, to the tracking control of the robot arm.

19.2.1 Mechanism Design

Mechanism design is crucial to the performance of a rehabilitation robot arm, and involves arrangement of actuators and sensors, distribution of masses, analyses of the kinematics and dynamics, and a compromise between safety and performance. Therefore, we introduce a design procedure to build a robot, as shown in Figure 19.1.

To imitate a human, a rehabilitation robot arm is about 60 centimeters in length to resemble a normal human, and possesses seven DOFs. In its mechanical design, the SOLIDWORKS software is utilized to design a robot, and the ADAMS software

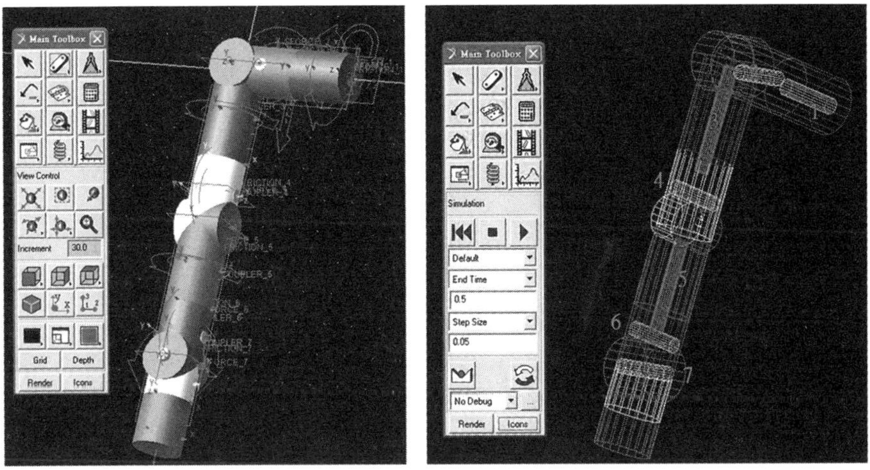

FIGURE 19.2 ADAMS model of the preliminary design.

is used to calculate and determine the required motor torque for the robot operated in a critical situation with maximum allowable speed, and to verify the experimental results. After simulation, the dynamic information of a robot can be obtained easily.

Estimating a required rated torque of each joint precisely can reduce the cost and weight of chosen motors, and thus two steps are carried out in the following, that is, estimating a maximum required static torque of each joint by calculation and then a maximum required dynamic torque by simulation.

To estimate both required static and dynamic rated torques, we assume that (1) the weight of the structure of the upper arm or lower arm is about 1000 g, and (2) the nominal payload held by the gripper is 900 g. Moreover, the critical posture of an arm should be horizontal against gravity, that is, the arm extending to reach the maximum reachable distance. To estimate a required dynamic rated torque of each joint, we use the ADAMS to build maximum velocity of each joint that reaches 270° per second at 2.5 s within the total simulation time of 5 s, and then test each joint independently when the other joints are fixed. The simulation results are that the estimated required rated torques of joints 1 and 2 are similar; since they have similar mechanical properties, so do joints 3 and 4, and joints 5, 6, and 7, as shown in Figure 19.2.

After finishing the preceding work, we design a rehabilitation robot arm, the National Taiwan University (NTU) Rehabilitation Robot Arm II, as shown in Figures 19.3 and 19.4, which can be assembled to a 120–150 cm tall rehabilitation robot or any robot platform. This new 7-DOF rehabilitation robot arm includes a set of mechanisms linked with a series of revolute joints, seven servo motors, and one additional gripper to locate an end-effector within certain workspace. The six revolute joints contribute to DOFs necessary for vertical, horizontal, and rotational motion

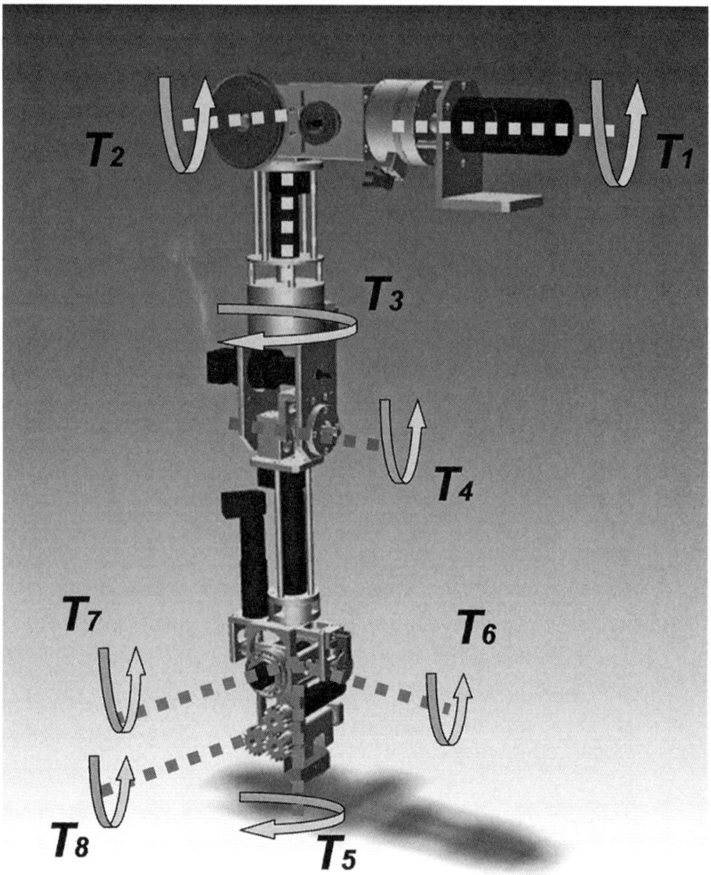

FIGURE 19.3 Mechanical structure of the NTU Rehabilitation Robot Arm II.

in the three dimensional space. In addition, it can achieve singularity avoidance and obstacle avoidance, fulfill complicated tasks, and perform dexterous motion by one more redundant revolute joint and design of the first three axes intersection and last three axes intersection together. The mechanical structure of the NTU Rehabilitation Robot Arm II is shown in Figure 19.3. The end-effector link of the robot arm is not only able to integrate with the designed one DOF gripper but also with different end-effectors to achieve various functions.

However, there are still some drawbacks of the robot, that is, the only one bevel gear used to perform the pitch motion to reduce the width of the wrist induces some backlash playing between loosely fitting parts, and single bearing support for joints 2 and 4 may cause slight vibration for high-frequency manipulation. On the other hand, this rehabilitation robot arm is compact enough to allow it easily to integrate into any platform. The dexterity and manipulability of this 7-DOF robot arm are so high

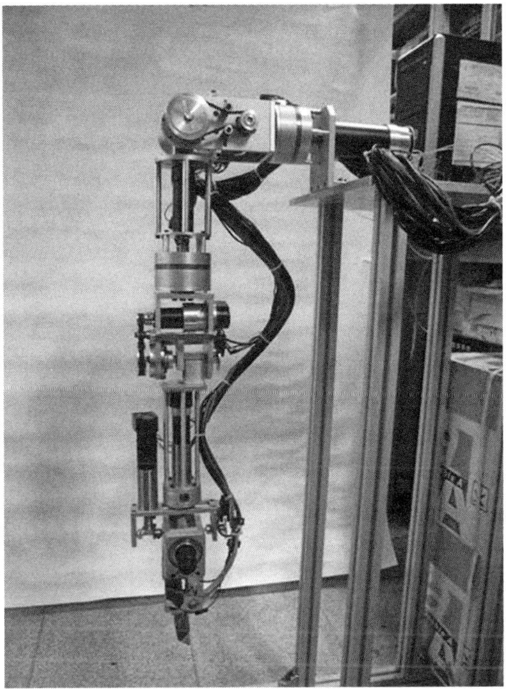

FIGURE 19.4 NTU Rehabilitation Robot Arm II.

TABLE 19.1 Specifications of the NTU Rehabilitation Robot Arm II

Total weight	About 6.5 kg (including the shoulder)
Maximum payload	Exceeds 1 kg
Maximum joint speed	Exceeds 270°/s
Maximum reachable distance	650 mm
Maximum width	150 mm
Motor	Brush DC × 7 and Brushless DC × 1, Brake × 3
Reduction device	Harmonic drive and gearhead
Transmission	Timing belt, bevel gear, and gear

that it can fulfill different complicated tasks easily. Table 19.1 shows more detailed specifications of the NTU Rehabilitation Robot Arm II.

19.2.2 Dynamic Model of an Individual Joint

The problem of controlling a rehabilitation robot arm with multiple joints is highly nonlinear and suffers from variable system uncertainty for a serial structure, payload variations, and so forth. It can be simplified by using a set of individual joint models with a robust control strategy to overcome the system uncertainty. To develop robust independent joint sliding mode controllers with gravity compensation, we have to consider a modified model, the dynamics of joint motors, and then transform this complete coupling dynamic model with the link effect into a suitable and simple independent joint model.

Using the Newton's principle, the dynamic model of the motor i can be expressed as

$$J_i\ddot{\phi}_i + D_i\dot{\phi}_i = u_i - n_i\tau_i - f_i + k_i, \tag{19.1}$$

where J_i, D_i, u_i, and f_i are the mass inertia, viscous coefficient, input torque, and Coulomb friction of the motor i, respectively; $n_i(\leq 1)$, the reduction ratio of gear train i; τ_i, the applied torque acted on link i; ϕ_i, the motor angle; and k_i is defined as

$$k_i = -J_i\left(\dot{\vec{\Omega}}^{l}_{i-1} \cdot \vec{\lambda}_i\right), \tag{19.2}$$

where $\vec{\Omega}^{l}_{i-1}$ denotes the absolute angular velocity vector of the robot link $i-1$, and $\vec{\lambda}_i$ denotes the unit vector of the motor axis. Note that the term k_i in the above equation represents the dynamic coupling effect on the joint motor i induced from motion of the front links, which has not been considered in general approaches [6].

Most of robot arms are driven by the DC servomotors of which the motor shaft is coupled with a high reduction ratio gear train and the load of the link. The equations of motion of the robot arm can be derived from Lagrange-Euler equations [4] as

$$\sum_{j=1}^{n} M_{ij}(q_1, q_2, \ldots, q_n)\ddot{q}_j + N_i(q_1, q_2, \ldots, q_n, \dot{q}_1, \dot{q}_2, \ldots, \dot{q}_n) + G_i(q_1, q_2, \ldots, q_n) = \tau_1, \quad i = 1, 2, \ldots, n \tag{19.3}$$

where $q_i = n_i\phi_i$ is the output link angle; τ_i, the applied torque supported by the transmission gear train i; M_{ij}, the inertia element; N_i, the centrifugal and Coriolis element; and G_i, the gravity element. By substituting Equation (19.3) into Equation (19.1), we obtain

$$J_i\ddot{\phi}_i + D_i\dot{\phi}_i = u_i - n_i\left(\sum_{j=1}^{n} M_{ij}\ddot{q}_j + N_i + G_i\right) - f_i + k_i. \tag{19.4}$$

The gravity term G_i can be rewritten as

$$G_i = G_i^0 + \Delta G_i, \tag{19.5}$$

where G_i^0 is the nominal value, and ΔG_i is the uncertain value induced by payload variations. Therefore, Equation (19.4) can be modified as

$$J_i\ddot{\phi}_i + D_i\dot{\phi}_i = u_i - n_iG_i^0 + d_i(t), \tag{19.6}$$

where $d_i(t)$ is the matched noise of which the direction coming in the equation is the same as that of the input torque u_i and defined as

$$d_i(t) = -f_i + k_i - n_i\left(\sum_{j=1}^{n} M_{ij}\ddot{q}_j + N_i + \Delta G_i\right). \tag{19.7}$$

Note that the matched noise $d_i(t)$ is relatively small because of the high reduction ratio, and assumed to be differentiable. Thus, it is bounded as

$$\left|d_i(t)\right| \le \delta_i(t), \tag{19.8}$$

where $\delta_i(t)$ is a known positive function with an upper bound to be roughly estimated by the maximum value of the designed acceleration of each joint.

19.2.3 LTR-Observer-Based Individual Joint Dynamic Sliding Mode Control with Gravity Compensation

A tracking controller design problem can be defined as follows: given the desired motor trajectory ϕ_i^d with some system uncertainty, design a simple robust tracking controller for the motor driving torque u_i such that the motor position ϕ_i can track the desired motor trajectory ϕ_i^d. Let the tracking error of the joint motor i be

$$e_i = \phi_i - \phi_i^d. \tag{19.9}$$

Then the tracking error equation derived from Equation (19.6) can be expressed as

$$J_i\ddot{e}_i + D_i\dot{e}_i = u_i - n_iG_i^0 + d_i(t) - J_i\ddot{\phi}_i^d - D_i\dot{\phi}_i^d. \tag{19.10}$$

A DSMC tracking controller design with a two-dimensional LTR observer is derived in the following section and depicted in Figure 19.5.

19.2.3.1 Sliding Variable Design in Individual Joint DSMC Considering the individual joint Equation (19.6) and defining the new state $x_1^i = \phi_i$ and $x_2^i = \dot{\phi}_i$, one can find out that the relative degree of the system is 2, and $\ddot{y}_i$ is unknown for the occurrence of the uncertainty. In the DSMC, the control design for the original system becomes an augmented system, one dimension higher than the original one; namely, the relative degree of the controlled system is $2 + 1 = 3$. Therefore, we define a sliding variable of joint i as

$$s_i = \ddot{e}_i + \lambda_1^i \cdot \dot{e}_i + \lambda_0^i \cdot e_i, \tag{19.11}$$

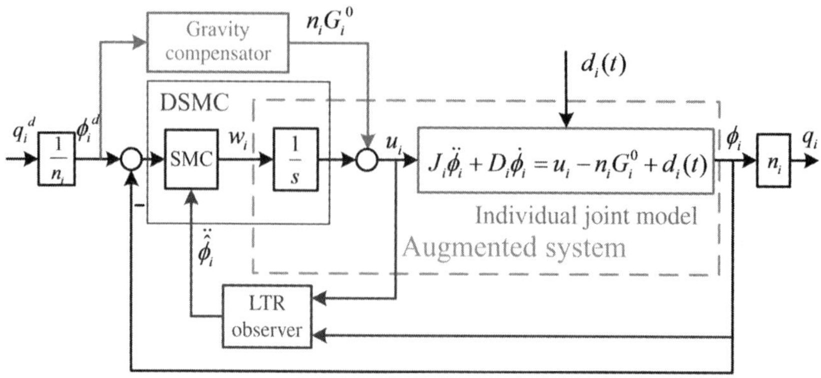

FIGURE 19.5 Control scheme of the IJDSMC.

where

$$\begin{aligned} e_i &= y_i - y_i^d = x_1^i - \phi_i^d, \\ \dot{e}_i &= \dot{y}_i - \dot{y}_i^d = x_2^i - \dot{\phi}_i^d, \\ \ddot{e}_i &= \ddot{y}_i - \ddot{y}_i^d = -\frac{D_i}{J_i}x_2^i + \frac{1}{J_i}\left(u_i - n_iG_i^0 + d_i\right) - \ddot{\phi}_i^d. \end{aligned}$$

The control driving the defined sliding variable s_i to zero is

$$u_i = n_iG_i^0 + J_i\ddot{\phi}_i^d + \int_{t_0}^{t} w_i dt \Rightarrow w_i = \dot{u}_i - J_{i i^d} - n_i\dot{G}_i^0, \tag{19.12}$$

where w_i is sliding mode control for the augmented system with uncertainty and the third derivative of the error is

$$\begin{aligned} e_i^{(3)} &= y_i^{(3)} -{}_i^d \\ &= \left\{ \left(\frac{D_i}{J_i}\right)^2 x_2^i - \frac{D_i}{J_i^2}\left(u_i - n_iG_i^0\right) + \frac{1}{J_i}w_i + \Delta d_i \right\} -{}_i^d, \end{aligned} \tag{19.13}$$

where $\Delta d_i = \frac{D_i}{J_i^2} \cdot d_i + \frac{1}{J_i} \cdot \dot{d}_i$. Note that λ_1^i and λ_0^i are positive and can be chosen to stabilize the ODE of the tracking error such that $e_i \to 0 \Rightarrow y_i \to y_i^d$ asymptotically [7].

19.2.3.2 LTR Observer Design in Individual Joint DSMC

The control variable $w_i = \dot{u}_i - J_{i i^d} - n_i\dot{G}_i^0$ is aimed to drive the new sliding variable s_i to (almost) zero in the occurrence of uncertainty. Since $\ddot{y}_i$ component of the defined sliding variable cannot be evaluated from the occurrence of the uncertainty, an observer is implemented in the control system. Define a two-dimensional state vector of the new augmented system $q^i = [\dot{y}_i \quad \ddot{y}_i]^T \in R^2$ such that the new state q^i can satisfy the following model:

$$\begin{aligned} \dot{q}^i &= A_i' q^i + B_i'(r_i + \Delta d_i), \\ \dot{y}_i &= C_i' q^i, \end{aligned} \tag{19.14}$$

where $\Delta d_i = \frac{D_i}{J_i^2} \cdot d_i + \frac{1}{J_i} \cdot \dot{d}_i$, $A_i' = \begin{bmatrix} 0 & 1 \\ 0 & 0 \end{bmatrix}$, $B_i' = \begin{bmatrix} 0 \\ 1 \end{bmatrix}$, $C_i' = [\,1\ 0\,]$, and $r_i = \left(\frac{D_i}{J_i}\right)^2 x_2^i - \frac{D_i}{J_i^2}\left(u_i - n_i G_i^0\right) + \frac{1}{J_i} w_i +_i^d$.

An LTR observer can be constructed as

$$\begin{aligned} \dot{\hat{q}}^i &= A_i' \, \hat{q}^i + B_i' \, r_i + L_i(\dot{y}_i - \dot{\hat{y}}_i), \\ \dot{\hat{y}}_i &= C_i' \, \hat{q}^i, \end{aligned} \tag{19.15}$$

where the gain of the LTR observer $L_i = \frac{1}{\mu_i} Q_i C\prime_i^T$ is subject to the algebraic Riccati equation:

$$Q_i(A_i' + I)^{\mathrm{T}} + (A_i' + I)Q_i - Q_i \frac{C_i' \mathrm{T} C_i'}{\mu_i} Q_i + \pi_i B_i' B_i' \mathrm{T} = 0, \tag{19.16}$$

where Q_i is a positive definite matrix gained by solving the algebraic Riccati equation, with $u_i = 1$ and $\pi_i > 0$ being sufficiently large. The following result is due to Reference 9.

Theorem 1 The LTR observer in Equation (19.15) satisfying Equation (19.16) achieves a small estimation error $\tilde{q}_i = q_i - \hat{q}_i$ in the sense that $\lim_{t\to\infty} \|\tilde{q}_i(t)\| \le \varepsilon_1^i \, \|x^i\| + \varepsilon_2^i \, \|u_i\| + \varepsilon_3^i$, where $\lim_{t\to\infty} \varepsilon_j^i = 0$ for j = 1, 2, and 3. □

From Theorem 1 and the estimated error of $\ddot{y}_i$ defined as $\ddot{\tilde{y}}_i = \ddot{y}_i - \ddot{\hat{y}}_i$, one can conclude that $\ddot{\tilde{y}}_i$ obtained from the LTR observer in Equation (19.15) achieves a small value whose magnitude can be controlled by the design parameter π_i in the observer Riccati Equation (19.16), such that

$$\ddot{\tilde{y}}_i| \le \varepsilon_1^i ||x^i|| + \varepsilon_2^1 |u_i| + \varepsilon_3^i. \tag{19.17}$$

We can also infer that after some finite time, the above equation approaches zero as π_i approaches infinity in the observer Riccati Equation (19.16), as follows:

$$\ddot{\tilde{y}}_i(t) = o\left(||x^i||\right) + o\left(|u_i|\right) + o(1). \tag{19.18}$$

In addition, one can obtain the estimated sliding variable $\hat{s}_i$ by utilizing the estimated states $\hat{\ddot{y}}_i$, such that $\hat{s}_i = \ddot{\hat{e}}_i + \lambda_1^i \cdot \dot{e}_i + \lambda_0^i \cdot e_i$, where$\ddot{\hat{e}}_i = \hat{\ddot{y}}_i - \ddot{\phi}_i^d$. Define the estimated error of the sliding variable as $\tilde{s}_i = s_i - \hat{s}_i$ such that the estimated error of the sliding variable becomes

$$\tilde{s}_i = \tilde{\ddot{y}}_i, \tag{19.19}$$

which is useful in the individual joint DSMC law.

19.2.3.3 *LTR Observer Design in Individual Joint DSMC*

The introduced individual joint DSMC law is

$$\begin{aligned} u_i &= n_i G_i^0 + J_i \ddot{\phi}_i^d + \int_{t_0}^t w_i dt, \\ &= n_i G_i^0 + J_i \ddot{\phi}_i^d + \int_{t_0}^t \left[\psi_i \left(\hat{s}_i \right) + v_i \left(\hat{s}_i \right) \right] dt, \end{aligned} \tag{19.20}$$

where $\psi_i(\hat{s}_i)$ is nominal control for the system without uncertainty

$$\begin{aligned} \psi_i(\hat{s}_i) = -J_i \Bigg\{ & \left[\left(\frac{D_i}{J_i} \right)^2 + \lambda_0^i \right] \cdot \dot{\phi}_i \\ & - \frac{D_i}{J_i^2} \cdot (u_i - n_i G_i^0) + \lambda_1^i \cdot \hat{e}_i - \lambda_0^i \cdot \dot{\phi}_i^d + \sigma_i \hat{s}_i \Bigg\} \end{aligned} \tag{19.21}$$

and $v_i(\hat{s}_i)$ is switching control to reduce the adverse effects of system uncertainty

$$v_i \left(\hat{s}_i \right) = -J_i \cdot \Delta \left(x^i, u_i \right) \cdot \text{sgn} \left(\hat{s}_i \right), \tag{19.22}$$

where $\Delta_i \left(x^i, u_i \right) = \rho_1^i \left\| x^i \right\| + \rho_2^i \left| u_i \right| + \rho_3^i$ is chosen large enough such that $\Delta_i \left(x^i,\ u_i \right)$ is an upper bound of Δd_i.

Theorem 2 Given the individual joint model Equation (19.6), the introduced DSMC system with bounded control u_i in Equation (19.20) can globally and asymptotically drive the tracking errors to zero, in that the tracking errors are asymptotically driven into a residual set around the origin, with the size of the residual set approaching zero as the design parameter π_i in the observer Riccati Equation (19.16) approaches infinity.

Proof: To examine the stability of the controlled system, choose the Lyapunov function of the joint motor i as

$$V_i = \frac{s_i^2}{2} > 0. \tag{19.23}$$

Its time derivative is

$$\dot{V}_i = s_i \cdot \left\{ \left[\left(\frac{D_i}{J_i} \right)^2 x_2^i - \frac{D_i}{J_i^2} \left(u_i - n_i G_i^0 \right) + \frac{1}{J_i} (\psi_i(\hat{s}_i) + v_i(\hat{s}_i)) + \Delta d_i \right] \right.$$
$$\left. + \lambda_1^i \cdot \ddot{e}_i + \lambda_0^i \cdot \left[x_2^i - \dot{\phi}_i^d \right] \right\}. \tag{19.24}$$

Substituting control law Equation (19.20) with Equations (19.21), and (19.22) into Equation (19.6), and using Equations (19.13) and (19.19), we have

$$\begin{aligned} \dot{V}_i &= s_i \cdot \left\{ -\sigma_i \hat{s}_i + \lambda_1^i \cdot \ddot{\tilde{y}}_i \right\} + s_i \cdot \left(-\Delta \left(x^i, u_i \right) \cdot \operatorname{sgn}\left(\hat{s}_i \right) + \Delta d_i \right), \\ &= -\sigma_i \cdot s_i^2 + s_i \cdot \left(\sigma_i + \lambda_1^i \right) \cdot \ddot{y}_i \\ &\quad + s_i \cdot \left[-\Delta \left(x^i, u_i \right) \cdot \operatorname{sgn}\left(\hat{s}_i \right) + \Delta d_i \right]. \end{aligned} \tag{19.25}$$

There may be two possible cases for the square brackets in the above equation:

Case 1: $|s_i| \leq |\tilde{s}_i|$.

In this case, using Equation (19.18) and Theorem 1, one can show that

$$|s_i| \leq \varepsilon_1^i \left\| x^i \right\| + \varepsilon_2^i |u_i| + \varepsilon_3^i. \tag{19.26}$$

Case 2: $|s_i| > |\tilde{s}_i|$

Using Equation (19.18) and $\operatorname{sgn}(\hat{s}_i) = \operatorname{sgn}(s_i)$ for $\hat{s}_i = s_i - \tilde{s}_i$, Equation (19.25) becomes

$$\begin{aligned} \dot{V}_i &\leq -\sigma_i \cdot s_i^2 + s_i \cdot \left(\sigma_i + \lambda_1^i \right) \cdot \ddot{\tilde{y}}_i - \left(\Delta \left(x^i, u_i \right) - \Delta d_i \right) \left\| s_i \right\|, \\ &\leq -\sigma_i \cdot s_i^2 + s_i \cdot \left(\sigma_i + \lambda_1^i \right) \cdot \ddot{\tilde{y}}_i, \\ &\leq -\sigma_i \cdot \left\| s_i \right\| \left\{ \left\| s_i \right\| - \left(1 + \frac{\lambda_1^i}{\sigma_i} \right) \cdot \left[\varepsilon_1 \left\| x \right\| + \varepsilon_2 \left\| u \right\| + \varepsilon_3 \right] \right\}. \end{aligned} \tag{19.27}$$

By the Lyapunov stability theorem and Theorem 1, it guarantees that $s_i \to 0$ asymptotically, while $|s_i| > \left(1 + \frac{\lambda_1^i}{\sigma_i} \right) \cdot \left[\varepsilon_1^i \left\| x^i \right\| + \varepsilon_2^i |u_i| + \varepsilon_3^i \right]$ where $\lim_{\pi \to \infty} \varepsilon_j^i = 0$ for $j = 1, 2$, and 3. After some finite time, we have

$$|s_i| \leq \left(1 + \frac{\lambda_1^i}{\sigma_i} \right) \cdot \left[\varepsilon_1^i \left\| x^i \right\| + \varepsilon_2^i |u_i| + \varepsilon_3^i \right]. \tag{19.28}$$

We conclude from cases 1 and 2 that

$$s_i = o\left(\left\| x^i \right\| \right) + o\left(|u_i| \right) + o(1). \tag{19.29}$$

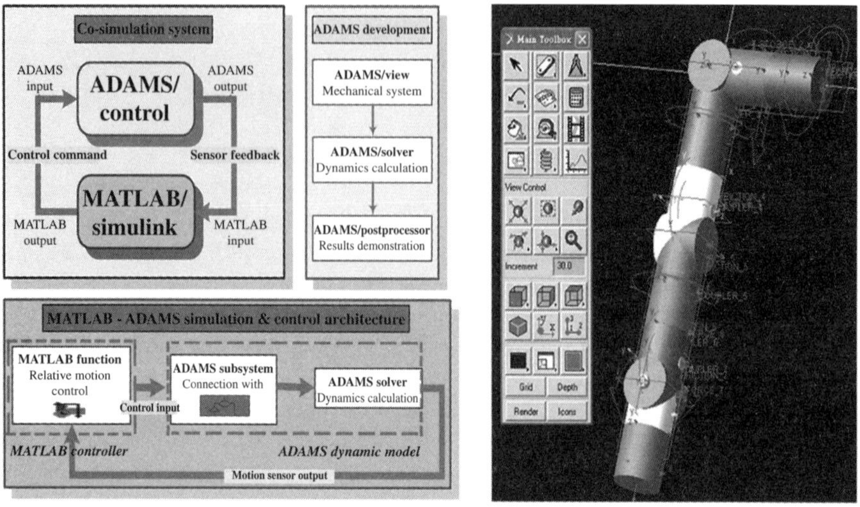

FIGURE 19.6 The co-simulation architecture of ADAMS/MATLAB and a dynamic model of the 7-DOF NTU Rehabilitation Robot Arm II built in ADAMS.

The tracking error leads to $e_i \to 0 \Rightarrow y_i \to y_i^d$ as $\lim_{\pi \to \infty} \varepsilon_j^i = 0$ after some finite time. □

19.2.4 Simulation of the NTU Rehabilitation Robot Arm II

Based on the dynamic model of the 7-DOF NTU Rehabilitation Robot Arm II built in ADAMS, as shown in Figure 19.6, the closed-loop co-simulations can be achieved by realizing different controllers in MATLAB/Simulink exchanging data and parameters with each other. With the initial state $q_i^0 = 0$ $(i = 1, 2, \ldots, 7)$, the designed cubic spline trajectories of all seven links are chosen as $q_i^f = \left[-45°, -60°, -70°, 0°, -30°, -45°, 90°\right]$ $(\mathrm{i} = 1, 2, 3, \ldots, 7)$ within 2 s. In addition, the individual joint SMC and the observer-based individual joint SMC are also carried out under high system parameter variations (50%), including the random payload, weight, and viscous parameter variations of each link. It is worthy to notice that a sudden stop in the given trajectories is used to examine sensitivity of the controlled system with respect to discontinuous reference command inputs, especially for an individual joint DSMC system using a planned jerk command for the designed LTR observer. The simulation results as shown in Figures 19.7–19.9 demonstrate that robot arm can achieve the tracking target as we expect while showing the different levels of chattering alleviation of each control input.

As mentioned earlier, for a high-order rehabilitation robot arm, the simulations show that the chattering phenomenon inherited from the switching mechanism is hard to be conquered by utilizing BLC or an estimated sliding variable, in view of the fact that the measurement noise, produced by utilizing a set of the independent joint

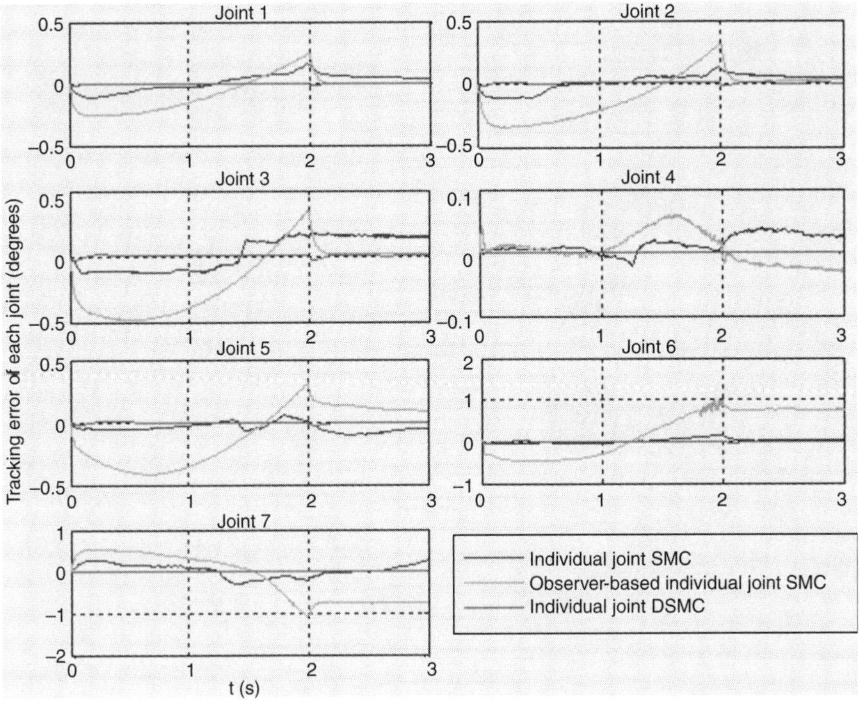

FIGURE 19.7 Simulated tracking errors of the NTU Rehabilitation Robot Arm II by individual joint SMC, observer-based individual joint SMC, and individual joint DSMC with 50% system parameter variations.

models during the manipulation of rehabilitation robot arms, is of a level larger than the boundary layer width within the acceptable tracking errors. In other words, the acceptable tracking error and the chattering phenomenon are hard to be accomplished simultaneously in the BLC-based control. Instead, the individual joint DSMC can effectively eliminate the chattering phenomenon and achieve good tracking performances everywhere, except where a discontinuous command input is used, as shown in Figures 19.8 and 19.9.

19.2.5 Experimental Results for the NTU Rehabilitation Robot Arm II

Practical experiments of the NTU Rehabilitation Robot Arm II are conducted to verify the performance and robustness of the individual joint SMC and individual joint DSMC tracking controllers under high system parameter variations (over 20%) and the dead zone phenomenon of the motor drivers. Similar to the simulation results, the experimental results as shown in Figure 19.10 demonstrate that the robot arm can achieve the tracking target as expected. In addition, the acceptable tracking errors and the chattering phenomenon are hard to be accomplished simultaneously in a

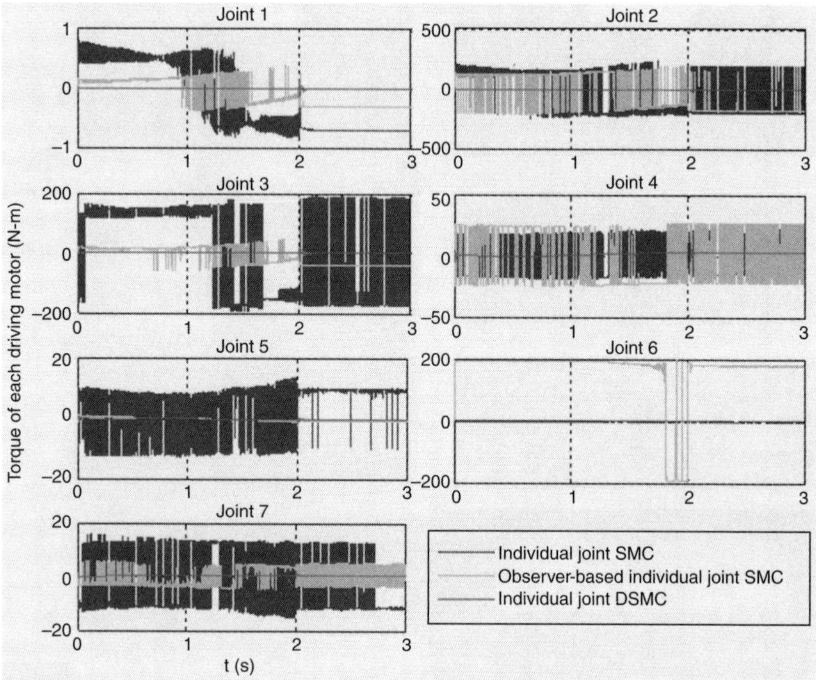

FIGURE 19.8 Simulated driving torques of the NTU Rehabilitation Robot Arm II by individual joint SMC, observer-based individual joint SMC, and individual joint DSMC with 50% system parameter variations.

high-order rehabilitation robot arm system with the BLC-based control. Instead, the individual joint DSMC can effectively diminish the chattering phenomenon with the better tracking performances.

19.3 REHABILITATION ROBOT HAND

Rehabilitation robotic hands, or prosthetic hands, have been developed for several decades. The Vaduz hand, one of the first electrically powered prosthetic hands, was developed around 1960 by Wilms. With the advance of modern mechatronics, it is expected that a robotic hand can fully replace a real human hand such that amputees can regain their life of quality. However, even with today's technology, there is no robotic hand that can function like a human hand. Among the developed robotic hands, most of them can be categorized into two types: prehensors (or grippers) and hands. Prehensors, such as commercially available VASI™ Hand and Otto Bock's SensorHand™, usually have one or two DOF and act like pliers. Prehensors are easy to control and can provide a large grip force, and thus it is one of the most popular robotic hands. One of the disadvantages of prehensors is that they usually

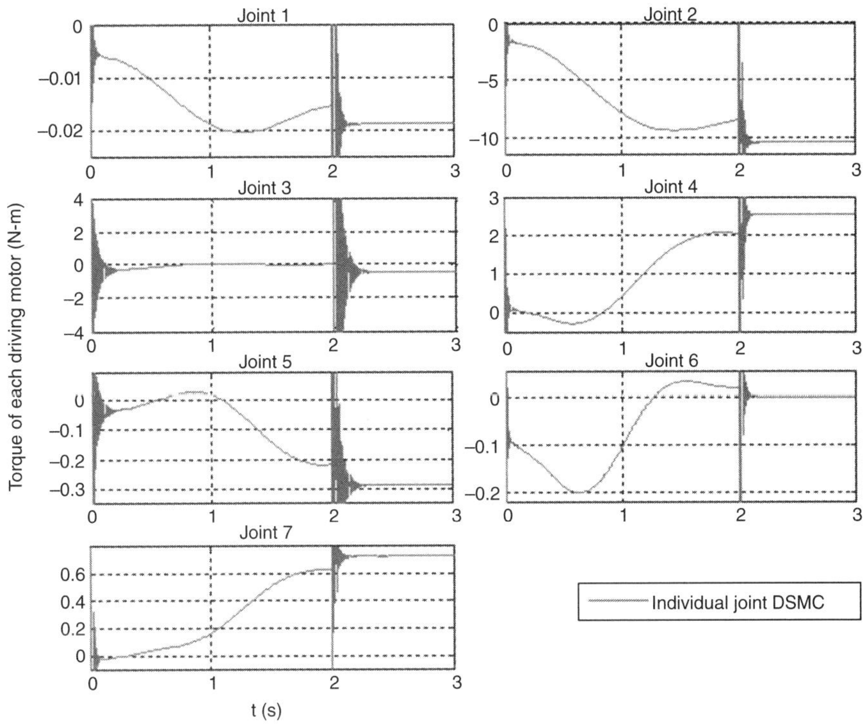

FIGURE 19.9 Simulated driving torques of the NTU Rehabilitation Robot Arm II by individual joint DSMC with 50% system parameter variations (in larger scale).

do not look like a real hand with multiple fingers, and, hence, cosmetic skin or gloves to cover them are needed. In addition, the lack of multiple fingers limits their functions. Therefore, many researchers have been pursuing the development of more sophisticated multi-fingered robotic hands. For the remaining part of this section, we will mainly focus on these hands.

Figure 19.11 shows the common components of a robotic hand. A robotic hand receives commands/signals most likely from users' muscle, neural systems, or brain directly. The signals are sent to a microprocessor or central processing unit and then identified in different ways. Finally, the microprocessor sends the identified signals to the corresponding actuators to achieve the desired hand motion. Many robotic hands are equipped with various sensors, such as force sensors, slip sensors, and temperature sensors, to detect the outside environment and feedback to the microprocessor to make the hand motion more precise or to respond to the environment more quickly. To function properly, all the electronics and actuators require adequate power. Almost all the robotic hands use batteries as the power source because of their portability. With the progress of nickel hydride batteries and lithium ion batteries, the robotic hands can be used longer than before. However, how long a robot hand can be used per charge is still a compromise between the battery capacity and the battery weight.

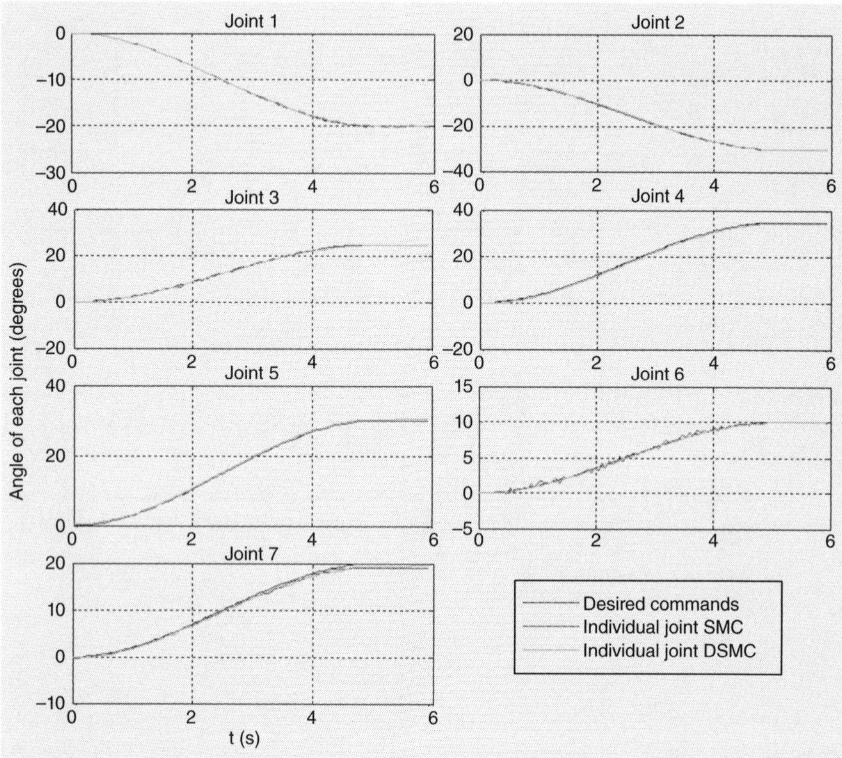

FIGURE 19.10 Tracking results of the NTU Rehabilitation Robot Arm II by individual joint SMC with 20% system parameter variations.

In the following sections, some background technology and recent development of the rehabilitation robotic hands are presented, and the current issues and challenges in this field are discussed briefly.

A popular research topic in the robotic hand field is the hand control. There are various ways to control the robotic hands in terms of the control input. The Vaduz hand detects the shape change of muscle due to muscle contraction. Thus, this method to control the multi-DOF robotic hand is seldom used now. Currently, most robotic hands are controlled by users' EMG signals. Motion Control Hand™ (Motion Control), SensorHand™, Greifer Hand™, Transcarpal Hand™ (Otto Bock), Centri Ultralite Hand™ (Fillauer), and i-LIMB Hand (Touch Bionics) are all examples of the EMG-controlled hands. Neuro-controlled hands are getting attraction in recent years. Revolutionizing Prosthesis 2009 (RP 2009), which is funded by DARPA of the United States, is an example. Currently, the injectable or implantable electrodes are under development such that the low-noise signals from the amputees' remaining muscles or nerves can be directly accessed.

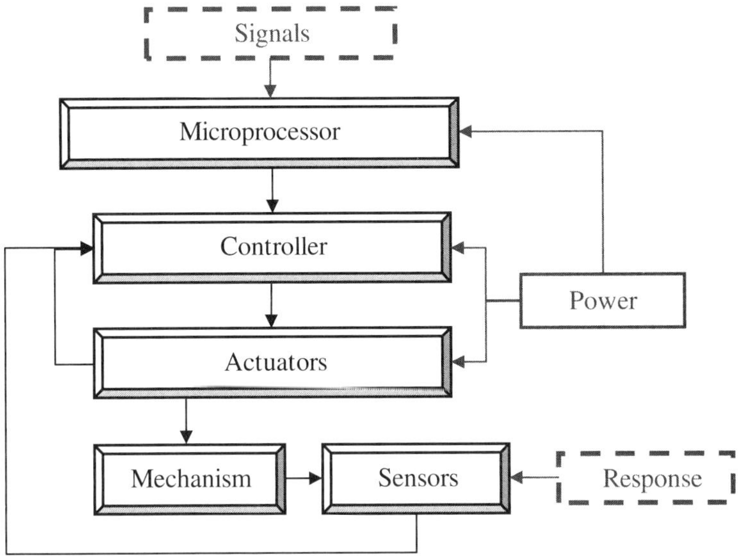

FIGURE 19.11 The major components of a robotic hand.

There are several control strategies for robotic hands. One intuitive approach (or so-called low-level control) is to allow users to control each finger directly, but this method usually requires a long period of training. Another approach (or so-called high-level control) is to define certain grip patterns that can cover most of our activities of daily life, and then use these patterns as a quick start to perform the specific task. According to the research, about 90% of the hand capability can be realized by four basic modes: cylindrical, hook, tip, and lateral [15]. Carrozza *et al.* [16] developed the low-level control scheme and high-level control scheme, and performed the high-level control on the robotic hand by foot movements [17]. The most obvious advantage of this approach is the short learning curve. Touch Bionic's i-LIMB Hand adopts a similar control methodology that receives only "open" and "close" signals from the two electrodes attached to the forearm of an amputee [18]. Kyberd *et al.* [19] also states that the control of a robotic hand should be realized in a hierarchical manner such that the amputees can send simple commands to control the hand and do not need to concentrate on the detailed control of each digit.

Various control schemes, such as classical PID, adaptive, robust, neural, fuzzy, sliding-mode, and their combinations, are used to control the fingers' torque, force, velocity, and trajectory. Several excellent reviews of the control strategies for prosthetic hands can be found in References 20–23.

Another popular research topic in the field of robotic hands is the mechanism design. Because of the cosmetic consideration, the sizes of most of the robotic hands are limited to that of a human hand. To design a multi-DOF robotic hand with multiple actuators within the limited space is a challenge. Many researchers use DC motors as the actuators. However, even the smallest DC motor that can output adequate torque to drive the finger of a robotic hand can occupy substantial space. To resolve this design problem, many fingers, such as the CyberHand [16], SPRING hand [24], HIT/DLR hand II [25], MANUS-HAND [15], DART hand [26], and i-LIMB Hand adopt the underactuated design, which uses one motor to pull tendons (or cables, belts) or linkages connected to two or three joints so that two or three phalanges can flex together by using one motor only.

Some researchers also use the motor to drive linkages or even gears [27] connected to phalanges to move them together. To simply use linkages or gears may not achieve adaptive grasp, and thus custom-designed springs are employed to allow certain compliance [28, 29]. The backlash and wear of the gears may cause control difficulty for robotic hands.

To fit all the electronic and mechanical components into a space no larger than a human hand usually requires the components to be small, which creates reliability issues. Some researchers have tried to avoid the complicated finger design by seeking a simplified hand that may not look like a real hand but can achieve similar functions with simpler and more reliable solutions. Early on, Hirose *et al.* [30] developed a chain-like mechanism which can be controlled by pulling two cables and possesses a self-adaption capability. Another innovative "hand" is developed by Brown *et al.* [31]. They use a balloon-like elastic bag filled with granular material as the hand. By contacting the object to be gripped and then evacuating the air out of the bag, it is reported that the hand can hold the object reliably. This hand may inspire us to develop hands of different design that can fit amputees' need but also can avoid the mechanical complexity.

Another topic of research in the field of robotic hands is the actuators. Instead of using common DC motors or ultrasonic motors, some researchers [32] drive the robotic hands hydraulically or pneumatically. The hydraulics can provide large force and possess certain self-adaption capabilities. The pneumatics usually requires larger space for its tank and piston, and Utah/MIT hand [33] is an example. However, the leakage of oils is a problem associated with the use of hydraulics, and the difficulty of controlling the compressible air is associated with the use of pneumatics [34].

To eliminate the backlash and problems associated with the gears and to reduce the weight of the motors used in the robotic hands, some other actuations have been introduced. One actuation that can be used in robotic hands is the shape memory alloy (SMA). Bergamasco *et al.* [35] and Bundhoo *et al.* [36] built SMA actuators and used them as human's tendons to drive the fingers. However, the cooling seems problematic, causing the response time to be unacceptably long. It is shown by O'Toole *et al.* [37] that the efficiency of cooling still needs to be increased to improve the dynamic response of SMA. An unconventional actuator used for robotic hands is the twist drive [38], which generates pulling force by using a motor to twist a pair of strings and is still at the early stage of development.

To implement various sensors on robotic hands for sensation is also an active research topic. Many robotic hands are equipped with sensors including contact sensors, pressure sensors, tri-axial force sensors [16], position sensors [39], torque sensors [40], and temperature sensors. These sensors can give the amputees a certain degree of feedback, which is needed according to the investigation among users of the robotic hands.

So far, no robotic hands can be made as dexterous as human hands. There are still many issues and challenges to make an artificial robotic hand with functions similar to a human's hand. These issues include cost, weight, volume, appearance, function, human interface, signal identification, sensation, training, reliability, response time, grasping force, self-adaptation capability, and power management. It is believed that a robotic hand with current technology cannot perform all the functions that a human hand can and it is difficult to resolve all the issues on one robotic hand. Consequently, the developed robotic hand is always a compromise among all these issues. Researchers may find part of the solutions through the newly developed compact-sized motors, light-weight materials with high strength, neural control with implantable electrodes, artificial skin with embedded sensors, etc.

19.4 STABILITY OF NEUROPROSTHESIS

EMG signal recognition plays a key role in neural–machine/human–robot interface. The most commonly seen application is the myoelectric prosthesis control. The input of a multifunction prosthetic hand is the EMG signals measured at different sites on the skin surface of a forearm using electrodes. The EMG signals from those recording sites within a timeframe (window) form an EMG pattern representing a specific type of muscle contraction, that is, a specific type of movement. To execute multiple functions, one has to construct a classifier that receives the EMG patterns and in real time outputs their corresponding movement labels into the control unit of the prosthetic hand. Various classifiers have been tested, including Bayes classifier [41], *K*-nearest neighbor classifiers [42, 43], linear classifiers [44–46], fuzzy logic systems [47, 48], fuzzy clustering-based maximum membership decision criterion [49], neuro-fuzzy systems [50, 51], artificial neural networks (ANN) [47, 52–56], Gaussian mixture models [57], and support vector machines (SVM) [58].

While the existing classifiers have shown different levels of success, there is still a problem that has never been discussed in the research field of EMG recognition. This problem is related to the "stability" of an EMG recognition system. Before the problem statement, we define the so-called target pattern and the nontarget pattern first.

Definition 1 (*Target pattern and nontarget pattern*): Supposing that an EMG pattern classifier (EMG-PC) has N outputs (N classes, i.e., there are N kinds of movements to be classified), if an input EMG pattern X belongs to one of the N classes, then X is called a target pattern; otherwise it is a nontarget pattern. □

For example, if an EMG-PC has two output classes: hook grasp (HG) and wrist flexion (WF). We need to collect a set of training patterns for each of the two classes in order to train the EMG-PC, because a classifier needs to be trained with a supervised learning manner. After training, the EMG-PC can be used to discriminate test EMG patterns and assign a class label to each of them. We now consider two different situations:

Situation 1: If an input EMG pattern X enters the EMG-PC, and if X belongs to the class "WF," then there will be two classification results: (1) if X is classified as the class "WF," it is classified correctly, and (2) if X is classified as the class "HG," it is misclassified (an error).

Situation 2: If another input EMG pattern Y enters the EMG-PC, and if Y belongs to the class "*lateral pinch*," then there will be only one classification result: this EMG pattern Y must be a misclassified error because the class "*lateral pinch*" is not predefined in the EMG-PC.

From the previous works, it can be seen from References 41 to 58 that no matter what classifier is used, the classifier can only assign a class label to an EMG pattern, and this label must belong to one of the classes that have been included in the training phase. Also, all the experiments are performed on the test EMG patterns that originally belong to these defined classes. Those works do not conduct any experiment on the EMG patterns that do not belong to the defined classes, because it can be expected that the EMG classification rate will be much lower. That is, none of the existing EMG recognition systems considers the second situation (Situation 2). If Situation 2 happens, the prosthesis equipped on an amputee will be out of control: whenever the amputee suddenly thinks of a movement that does not belong to the classes included in the classifier's training, his muscles may start to contract unwittingly. Then, the EMG recognition system would start to perform the classification task on the acquired EMG raw data. As a result, an undesired movement command will be sent to the controller of the prosthesis.

According to the above analysis, we see that a stable EMG recognition system should be able to reject all EMG nontarget patterns and accept all EMG target patterns; otherwise the prosthesis will be uncontrollable. To solve this critical problem, we need to design a nontarget EMG pattern filter in which a "watch dog" function should be embedded, and the watch-dog function should be capable of rejecting all nontarget patterns and accepting all target patterns. In other words, only the target EMG patterns can be authorized to enter the EMG-PC such that the latter prosthesis will only make the desired movements that have been defined. To realize this function, this section introduces a nontarget pattern filtering (NTPF) scheme.

The NTPF scheme is developed based on the concept of one-class classification (also called novelty detection [59, 60]), which drew increasing attention in the pattern recognition filed in the past decade. One-class classifiers are to find a compact description for a specific class (usually being referred to as *target class*), and unlike two-class classifiers, such as SVM [61], one-class classifiers can be built on just the target class. Some promising one-class classifiers include the one-class SVM [62], the single-class minimax probability machine [63], the support vector data description (SVDD) [64, 65], and the kernel principal component analysis–based method [66]. This paper adopted the SVDD as the basic component of the NTPF scheme, as its success has recently been shown in various applications requiring novelty/anomaly detections [67–70].

SVDD aims at looking for a hypersphere with minimum volume that can enclose all or most of target data. The hypersphere boundary is then used to distinguish target data from outliers (nontarget data). What makes SVDD attractive, like SVM, is also the use of the kernel trick, by which the hypersphere boundary can be flexible to fit the shape of an arbitrary target set in the input space. More importantly, training SVDD is easy because only target data are needed in the training phase. We just need to collect EMG target patterns as the training set for training the SVDD-based NTPF scheme. We do not need to make any effort to collect all possible EMG nontarget patterns. This can save us much time. For an amputee, this advantage is particularly significant in practical use.

19.4.1 SVDD-Based Target EMG Pattern Estimation

Given a target training set $T = \{\mathbf{x}_i \in \mathbf{R}^d\}_{i=1}^n$, SVDD aims at finding a minimum-volume sphere with center $\mathbf{a}_F$ and radius R in the feature space F such that all or most of the target EMG training patterns are enclosed by the hypersphere, which can be formulated as the constrained optimization problem as

$$\begin{aligned} &\text{Minimize} \quad O_p(R, \mathbf{a}_F, \xi) = R^2 + C \sum_{i=1}^{n} \xi_i \\ &\text{subject to} \quad \left\| \varphi(\mathbf{x}_i) - \mathbf{a}_F \right\|^2 \leq R^2 + \xi_i, \\ &\qquad\qquad\quad \xi_i \geq 0, \quad \forall i \in \{1, 2, ..., n\} \end{aligned} \tag{19.30}$$

where C is the penalty weight, $\xi_i' s$ are slack variables, and ϕ is a nonlinear mapping: $\phi\ .\ \mathbf{x}_i \in \mathbf{R}^d \rightarrow \phi(\mathbf{x}_i) \subset F$. The primal problem is usually solved by the dual problem as

$$\begin{aligned} &\text{Maximize} \quad O_d(\alpha) = 1 - \sum_{i=1}^{n} \sum_{j=1}^{n} \alpha_i \alpha_j K(\mathbf{x}_i, \mathbf{x}_j) \\ &\text{subject to} \quad \sum_{i=1}^{n} \alpha_i = 1, \\ &\qquad\qquad\quad 0 \leq \alpha_i \leq C, \quad \forall i \in \{1, 2, ..., n\} \end{aligned} \tag{19.31}$$

where α_i are positive weights, and K is the kernel function defined as the inner dot product of feature vectors in feature space F: $K(\mathbf{x}_i, \mathbf{x}_j) = \phi(\mathbf{x}_i) \cdot \phi(\mathbf{x}_j)$. In this chapter, we consider the radial basis function (RBF) kernel, defined as

$$K(\mathbf{x}_i, \mathbf{x}_j) = \exp\left(-\frac{\left\| \mathbf{x}_i - \mathbf{x}_j \right\|}{2\sigma^2} \right), \tag{19.32}$$

where σ is the kernel parameter specified by users. According to the Kuhn–Tucker (KT) conditions, the target training data can be classified into the three categories: (1) the data points with $\alpha_i = 0$ are inside of the sphere, (2) the data points with $0 < \alpha_i < C$ lie on the boundary of the sphere, and (3) the remaining data points (with $\alpha_i = C$) fall outside the sphere. The data points satisfying $0 < \alpha_i \leq C$ are called support vectors (SVs). We call the SVs with $0 < \alpha_i < C$ the unbounded SVs (UBSVs), and call SVs with $\alpha_i = C$ the bounded SVs (BSVs). The center $\mathbf{a}_F$ is expanded by the images of SVs:

$$\mathbf{a}_F = \sum_{i=1}^{n_s} \alpha_i \phi\left(\mathbf{x}_i\right), \tag{19.33}$$

where n_s is the number of SVs. The radius R can be determined by taking any $\mathbf{x}_k \in$ UBSVs and calculating the distance from its image $\phi(\mathbf{x}_k)$ to the center $\mathbf{a}_F$:

$$R = \left(1 - 2 \sum_{\mathbf{x}_i \in \mathrm{SVs}} \alpha_i K\left(\mathbf{x}_i, \mathbf{x}_k\right) + \sum_{\mathbf{x}_i \in \mathrm{SVs}} \sum_{\mathbf{x}_j \in \mathrm{SVs}} \alpha_i \alpha_j K\left(\mathbf{x}_i, \mathbf{x}_j\right)\right)^{1/2}. \tag{19.34}$$

The output for a test EMG pattern $\mathbf{x}$ is determined by comparing its distance to $\mathbf{a}_F$ with radius R:

$$D_{\mathrm{SVDD}}(\mathbf{x}) = \left\|\phi(\mathbf{x}) - \mathbf{a}_{\mathrm{F}}\right\|^2 - R^2. \tag{19.35}$$

The kernel expression of the SVDD decision function is given by

$$D_{\mathrm{SVDD}}(\mathbf{x}) = K(\mathbf{x}, \mathbf{x}) - 2 \sum_{i=1}^{n} \alpha_i K\left(\mathbf{x}, \mathbf{x}_i\right) + \sum\nolimits_{i,j=1}^{n} \alpha_i \alpha_j K\left(\mathbf{x}_i, \mathbf{x}_j\right) - R^2. \tag{19.36}$$

If K is RBF kernel, we obtain the SVDD decision function:

$$D_{\mathrm{SVDD}}(\mathbf{x}) = c - 2 \sum_{\mathbf{x}_i \in \mathrm{SVs}} \alpha_i K\left(\mathbf{x}, \mathbf{x}_i\right), \tag{19.37}$$

where $c = \left(1 - R^2\right) + \sum_{i,j=1}^{n} \alpha_i \alpha_j K\left(\mathbf{x}_i, \mathbf{x}_j\right)$ is a constant. If $D_{\mathrm{SVDD}}(\mathbf{x}) \leq 0$, $\mathbf{x}$ is accepted as a target EMG pattern; otherwise it is rejected as nontarget EMG pattern.

19.4.2 Nontarget EMG Pattern Filtering Scheme

The NTPF scheme is a generic form. Its flow chart is depicted in Figure 19.12. Supposing that there are M channels (M surface electrodes at different recording sites), the signal processing and feature extraction tasks perform on each channel independently. Supposing that there are totally d features extracted from the M channels, the

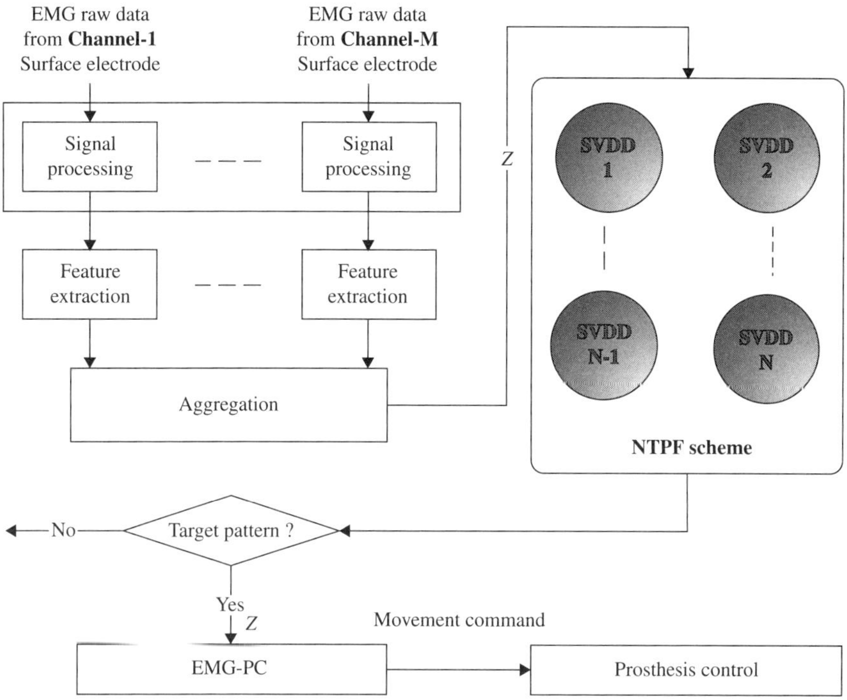

FIGURE 19.12 The introduced NTPF scheme in a generic EMG recognition system.

d features form a feature vector of d dimension, called an EMG pattern. After the EMG pattern Z arrives, it will not directly go to the EMG-PC for classification, but enters the introduced NTPF scheme first.

Traditionally, the SVDD is trained to enclose the patterns of a specific class. In EMG classification, there are usually many movements (classes) to be classified. The performance of the SVDD may not be the optimum if using only one hypersphere to enclose all the training patterns belonging to these classes, because these patterns would not form a compact distribution but several disjoint clusters in the input space. One remedy is to decrease the value of the kernel parameter σ to sufficiently small. By doing so, all the patterns can be enclosed by the single SVDD hypersphere tightly. However, the generalization performance of the SVDD will drop greatly because a too small σ results in a large number of support vectors. It has been proven that the larger the number of the support vectors, the poorer the generalization performance of the SVDD [64, 65]. Therefore, this paper adopts a "grouping" strategy. Instead of using one single SVDD hypersphere, we use multiple SVDD hyperspheres to learn the data descriptions for the EMG patterns originated from different target classes. Since there are more than one SVDD, we call the SVDDs an SVDD ensemble.

The SVDD ensemble is composed of N SVDDs (SVDDs 1–N), where N is the number of target classes. If the EMG-PC has N kinds of output movements, the SVDD ensemble has N SVDDs. Each SVDD is responsible for enclosing the EMG training patterns of a target class. Therefore, there will be N hyperspheres in the SVDD ensemble after training. The decision-making strategy used in the NTPF scheme follows the simple rules as:

(1) *If the EMG pattern Z falls inside any of the N hyperspheres, then it is a target pattern. It will be sent to the EMG-PC for further classification.*
(2) *The EMG pattern Z is classified as a nontarget pattern if it does not fall inside any of the N hyperspheres. Since it is a nontarget EMG pattern, it is ignored and thus not fed into the EMG-PC.*

Using the simple rules can easily achieve the goal of NTPF, and as a result, the EMG recognition system will be stable and will not send wrong motion commands to the prosthesis, unless the classification rate of the EMG-PC itself is not satisfactory. One may ask the question: since each SVDD in the ensemble is able to describe the distribution of a specific target class, why don't we use the SVDD ensemble to perform the classification task directly? (because as long as Z falls inside the hypersphere of the ith SVDD, it belongs to the ith class). In theory, it may work. But, using the SVDD ensemble to accomplish the classification task may cause an ambiguous situation in practice. For example, two different hyperspheres may have an overlapping region. Consequently, the pattern Z may fall inside the two hyperspheres at the same time. In this ambiguous situation, the class label of the input EMG pattern Z cannot be determined. Hence, it is still necessary to use a multi-class classifier to accomplish the task of multi-class EMG classification because the SVDD is not designed for multi-class classification, but for outlier/novelty detection.

19.4.3 Illustrative Example

In order to obtain the EMG patterns, we have the hardware and software settings as follows.

Three EMG surface electrodes are placed on *palmaris longus*, *extensor digitorum*, and *flexor carpi ulnaris* as shown in Figure 19.13 and therefore three channels are used. EMG signals are acquired via the EMG surface electrodes. A 60 Hz notch filter is used to reduce the effect of noise to the EMG signals. A 30–400 Hz band pass filter is also used to obtain meaningful EMG signals. After filtering, the EMG signals are sampled by an AD converter with sampling frequency 2.5 kHz and stored in the memory. The sampled EMG raw data are then transferred from the memory to a computer (PC) by parallel ports. The above process is common for all channels, and has been implemented with a digital signal processor (DSP)-based module, as shown in Figure 19.14. The details can be referred to the work in Reference 58.

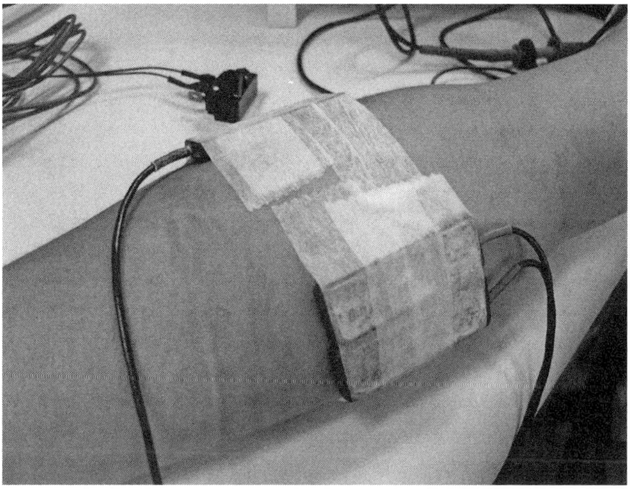

FIGURE 19.13 EMG signal recording sites.

The feature extraction methods include two different methods: autoregressive model (ARM) [71], and the histogram of EMG (HEMG) [72]. For ARM, a fourth-order model is adopted. By using a fourth-order ARM, each channel extracts four feature components from EMG raw data. By using HEMG, a feature vector containing nine feature components is obtained from each channel. Therefore, a 13-dimensional feature vector is obtained from each channel. Since there are 3 surface electrodes, there are totally 39 features. In other words, each EMG pattern is a 39-dimensional

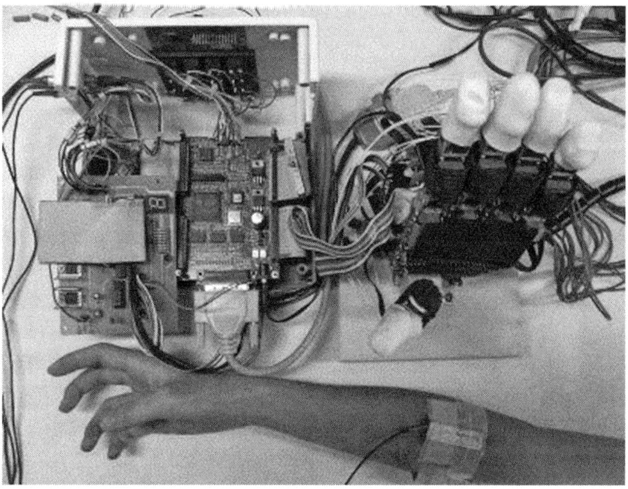

FIGURE 19.14 The DSP-based EMG processing module and the multi-finger prosthesis.

vector. Notice that the raw data are received by a window with size of 1000. That is, there would be 1000 EMG raw data collected in 400 ms for each channel since the sampling frequency is set to 2.5 kHz.

In this example, we are interested in several types of postures, including *power grasp (PG)*, *hook grasp (HG)*, *wrist flexion (WF)*, *lateral pinch (LP)*, *flattened hand (FH)*, *centralized grip (CEG)*, *three-jaw chuck (TJC)*, and *cylindrical grasp (CYG)*. For each of the eight kinds of postures, we collect 20 EMG patterns from an amputee. There are totally 160 EMG patterns.

Two kinds of circumstances are described as follows:

The first circumstance. In this circumstance, we follow the traditional EMG classification schemes. That is, training a classifier that can recognize the classes of the eight kinds of postures. First, for each class, we randomly select 10 EMG patterns as training samples. The remaining 10 EMG patterns are used as the test samples. Hence, we have 80 training samples and 80 test samples, respectively. Next, we use the 80 training patterns to train the EMG pattern classifier. The SVM [61] is adopted as the classifier in this work for its high generalization performance. However, SVM is a two-class classifier. Hence, we construct a multi-class classifier by using the one-against-one method and the voting strategy [58]. As a result, there are 28 SVMs in total. The Gaussian kernel is used for each SVM. The optimal hyperparameters for the 28 SVMs, the penalty weight and the kernel parameter, are obtained by taking twofold cross-validation. After testing, the classification rate is 87.5%. It is not surprising, because the class labels of the 80 test samples have been included in the training. Next, we consider a more practical circumstance.

The second circumstance. Here we first define the target class. The target class includes the following six postures: *PG*, *HG*, *WF*, *LP*, *FH*, and *CEG*. Similarly, we randomly select 10 EMG patterns from each of the posture class as the training samples, and the remaining ones are test samples. There are totally 60 training samples. We then use the 60 training samples to train the multi-class SVMs. There are 15 SVMs in total. In addition, all the EMG patterns belonging to the postures TJC and CYG are defined as the nontarget EMG patterns, and used as the test samples. Therefore, there are $60 + 40 = 100$ test samples in total. We then feed the 100 test samples into two different EMG classification systems: one has only the SVM classifier (SYSTEM 1), and the other (SYSTEM 2) involves the introduced nontarget EMG pattern filtering (NTPF) scheme, which is placed prior to the SVM classifier. The final classification rates of the two systems are reported in Table 19.2.

The advantage of the NTPF is clear, which can be seen from Table 19.2 that the SYSTEM 1 obtains a poor classification performance of 51%. In fact, it is not surprising because for SYSTEM 1, the 40 nontarget EMG patterns must be misclassifications. On the contrary, SYSTEM 2 achieves a relatively high classification rate (87%) because the NTPF scheme rejects almost all the nontarget EMG patterns (in the experiment, there are only four nontarget EMG patterns that are accepted

TABLE 19.2 Comparisons of the Classification Rates Among Two EMG Classification Systems, Without (SYSTEM 1) and With (SYSTEM 2), in the NTPF Scheme

	SYSTEM 1	SYSTEM 2
Classification rate (%)	51	87

by the NTPF scheme). In other words, four nontarget EMG patterns enter the SVM classifier. Another nine classification errors are contributed by the test EMG patterns belonging to the target class.

19.5 CONCLUSIONS

In general, two different kinds of rehabilitation robots, active assistive robots and passive constrained robots, are developed as successful rehabilitation tools. An active assistive robot serves as a force source, transmitting power to a body segment to actively guide a patient passively following the desired rehabilitative motions. On the other hand, a passive constrained robot serves as a passive constraint. The robot can actively change preset constraints of a patient's motions to allow the patient to move spontaneously according to the given motion pattern of the passive treatment. Because the introduced rehabilitation arm has high DOFs, it can be a passive and active rehabilitation robot simultaneously and provide an efficient and gentle way to accomplish various common rehabilitation exercises in order to recover the patient's motor function or assist human motion.

Both simulation and experimental results of a tracking problem for the 7-DOF NTU Rehabilitation Robot Arm II have been presented to show the intrinsic advantages and disadvantages of different controllers. Additionally, the individual joint DSMC is able to achieve the robust tracking control of a multi-DOF rehabilitation robot arm with parameter variations, external disturbances, and payload variations. In comparison with other control algorithms, the individual joint DSMC shows better performance.

In this chapter, we have also introduced a method which can improve stability of neuroprostheses, the nontarget EMG pattern filtering (NTPF) scheme. The results have shown that the introduced NTPF scheme is able to maintain a high EMG classification rate. And as a result, the latter prosthesis can perform in more stable manner. The success of the scheme should be attributed to the use of the one-class classifier, the SVDD.

However, the decision function of SVDD is expanded by the images of the training data points, which may result in a great run-time complexity if the size of the target set is large. To improve the usability of a neuroprosthesis, the run-time complexity of SVDD needs to be reduced further. In practice, speeding up decision functions expressed by the kernel expansion has been considered an important research topic in the SV community, and there has been a large body of works proposed to address this issue, for example, References 73 to 79. These works were mainly designed to improve the classification speed of SVM by approximating the SVM hyperplane with a subset of SVs. When using the methods developed in these works, both positive and negative training sets are needed in the procedure of the hyperplane approximation, because SVM is essentially a binary classifier. However, as aforementioned, SVDD is a one-classifier: the premise for one to use SVDD is that only the training data of one of the two classes are available. Hence, though the existing methods are effective in reducing the run-time complexity of SVM, unfortunately they cannot be applied to solve the problem of SVDD run-time complexity. Recently, Liu *et al.* [80] have proposed a method called fast SVDD to address this critical issue. Their method can greatly reduce the run-time complexity of SVDD. But, their method has limitations. For example, the generalization loss of fast SVDD is significant when faced with complex data sets, as pointed out in the discussion part in Reference 80. Thus, to avoid the delay between the EMG classification system and the prosthesis control, a more robust method to reduce the run-time complexity of SVDD such that it can be used more practically is required, which is a worth studying issue in the future.

REFERENCES

[1] T. Asfour and R. Dillmann. Human-like motion of a humanoid robot arm based on a closed-form solution of the inverse kinematics problem. In: Proceedings of IEEE/RSJ International Conference on Intelligent Robots and Systems, 2003, pp. 1407–1412.

[2] G. G. Elmer and H. In Joong. An approach to nonlinear feedback control with applications to robotics. In: Proceedings of IEEE Conference on Decision Control, 1983, pp. 134–138.

[3] T. Hsia. Adaptive control of robot manipulators—a review. In: Proceedings of 1986 IEEE International Conference on Robotics and Automation, 1986, vol. 3, pp. 183–189.

[4] R. C. G. K.S. Fu and C.S.G. Lee. *Robotics: Control, Sensing, Vision, and Intelligence*. New York: McGraw-Hill, 1987.

[5] C. T. Chang, H. P. Huang, and J. Y. Kuan. Independent joint sliding mode control of a humanoid robot arm. In: Proceedings of 33rd IEEE Annual Conference on Industrial Electronics Society, 2007, pp. 2754–2759.

[6] T. C. S. Hsia, T. A. Lasky, and Z. Guo. Robust independent joint controller design for industrial robot manipulators. *IEEE Transactions on Industrial Electronics*, 38(1): 21–25, 1991.

[7] J. J. Slotine and W. Li. *Applied Nonlinear Control*. Englewood Cliffs, New Jersey: Prentice Hall, 1991.

[8] V. Parra-Vega and G. Hirzinger. Chattering-free sliding mode control for a class of nonlinear mechanical systems. *International Journal of Robust and Nonlinear Control*, 11: 1161–1178, 2001.

[9] M. S. Chen, C. H. Chen, and F. Y. Yang. An LTR-observer-based dynamic sliding mode control for chattering reduction. *Automatica*, 43: 1111–1116, 2007.

[10] A. J. Koshkouei, K. J. Burnham, and A. S. I. Zinober. Dynamic sliding mode control design. In: Proceedings of IEE Conference on Control Theory and Applications, 2005, pp. 392–396.

[11] S. Purwar. Higher order sliding mode controller for robotic manipulator. In: Proceedings of 22nd IEEE International Symposium on Intelligent Control, Singapore, 2008, pp. 556–561.

[12] G. Bartolini, A. Ferrara, and E. Usani. Chattering avoidance by second-order sliding mode control. *IEEE Transactions on Automatic Control*, 43(2): 241–246, 1998.

[13] L. M. Capisani, A. Ferrara, and L. Magnani. Second order sliding mode motion control of rigid robot manipulators. In: Proceedings of IEEE Conference on Decision Control, New Orleans, LA, 2008, pp. 3691–3696.

[14] J. Pratt, P. Dilworth, and G. Pratt. Virtual model control of a bipedal walking robot. In: Proceedings of IEEE International Conference on Robotics and Automation, 1997, vol. 1, pp. 193–198.

[15] J. L. Pons, E. Rocon, R. Ceres, D. Reynaerts, B. Saro, S. Levin, and W. Van Moorleghem. The MANUS-HAND Dextrous robotics upper limb prosthesis: mechanical and manipulation aspects. *Autonomous Robots*, 16: 143–163, 2004.

[16] M. Carrozza, G. Cappiello, S. Micera, B. Edin, L. Beccai, and C. Cipriani. Design of a cybernetic hand for perception and action. *Biological Cybernetics*, 95(6): 629–644, 2006.

[17] M. C. Carrozza, A. Persichetti, C. Laschi, F. Vecchi, R. Lazzarini, V. Tamburrelli, P. Vacalebri, and P. Dario. A novel wearable interface for robotic hand prostheses. In: Proceedings of 2005 IEEE International Conference on Rehabilitation Robotics, Chicago, IL, 2005, pp. 109–112.

[18] C. Connolly. Prosthetic hands from touch bionics. *Industrial Robot*, 35: 290–293, 2008.

[19] P. J. Kyberd, O. E. Holland, P. H. Chappell, S. Smith, R. Tregidgo, P. J. Bagwell, and M. Snaith. MARCUS: a two degree of freedom hand prosthesis with hierarchical grip control. *IEEE Transactions on Rehabilitation Engineering*, 3(1): 70–76, 1995.

[20] D. S. Naidu, C.-H. Chen, A. Perez, and M. P. Schoen. Control strategies for smart prosthetic hand technology: an overview. In: Proceedings of 30th Annual International IEEE EMBS Conference, Vancouver, Canada, 2008, pp. 4314–4317.

[21] M. Zecca, S. Micera, M. C. Carrozza, and P. Dario. Control of multifunctional prosthetic hands by processing the electromyographic signal. *Critical Reviews in Biomedical Engineering*, 30: 459–485, 2002.

[22] P. Kyberd, P. Chappell, and D. Gow. Advances in the control of prosthetic arms. *Technology and Disability*, 15: 57–61, 2003.

[23] D. P. J. Cotton, A. Cranny, P. H. Chappell, N. M. White, and S. P. Beeby. Control strategies for a multiple degree of freedom prosthetic hand. In: Mini Symposia UKACC Control, 2006, pp. 211–218.

[24] M. C. Carrozza, C. Suppo, F. Sebastiani, B. Massa, F. Vecchi, R. Lazzarini, M. R. Cutkosky, and P. Dario. The SPRING hand: development of a self-adaptive prosthesis for restoring natural grasping. *Autonomous Robots*, 16: 125–141, 2004.

[25] H. Liu, K. Wu, P. Meusel, G. Hirzinger, M. H. Jin, Y. W. Liu, S. W. Fan, T. Lan, and Z. P. Chen. A dexterous humanoid five-fingered robotic hand. In: Proceedings of 17th IEEE International Symposium on Robot and Human Interactive Communication, RO-MAN, 2008, pp. 371–376.

[26] N. Thayer and S. Priya. Design and implementation of a dexterous anthropomorphic robotic typing (DART) hand. *Smart Materials and Structures*, 20: 035010, 2011.

[27] L. R. Lin and H. P. Huang. NTU hand: a new design of dexterous hands. *Transactions of the ASME Journal of Mechanical Design*, 120: 282–292, 1998.

[28] N. Dechev, W. L. Cleghorn, and S. Naumann. Multiple finger, passive adaptive grasp prosthetic hand. *Mechanism and Machine Theory*, 36: 1157–1173, 2001.

[29] S. Ge, H. Li, J.-J. Cabibihan, Y. Tan, J. Sun, and W. Zhang. COSA finger: a coupled and self-adaptive under-actuated unit for humanoid robotic hand. In: *Social Robotics*, vol. 6414. Berlin/Heidelberg: Springer, 2010, pp. 172–181.

[30] S. Hirose and Y. Umetani. The development of soft gripper for the versatile robot hand. *Mechanism and Machine Theory*, 13: 351–359, 1978.

[31] E. Brown, N. Rodenberg, J. Amend, A. Mozeika, E. Steltz, M. R. Zakin, H. Lipson, and H. M. Jaeger. Universal robotic gripper based on the jamming of granular material. *Proceedings of the National Academy of Sciences of the United States of America*, 107: 18809–18814, 2010.

[32] A. Kargov, T. Werner, C. Pylatiuk, and S. Schulz. Development of a miniaturised hydraulic actuation system for artificial hands. *Sensors and Actuators, A: Physical*, 141: 548–557, 2008.

[33] S. Jacobsen, E. Iversen, D. Knutti, R. Johnson, and K. Biggers. Design of the Utah/M.I.T. dextrous hand. In: IEEE International Conference on Robotics and Automation, 1986, pp. 1520–1532.

[34] L. W. Tsai. *Robot Analysis: The Mechanics of Serial and Parallel Manipulators*. New York: John Wiley & Sons, 1999.

[35] M. Bergamasco, F. Salsedo, and P. Dario. Shape memory alloy micromotors for direct-drive actuation of dexterous artificial hands. *Sensors and Actuators*, 17: 115–119, 1989.

[36] V. Bundhoo and E. J. Park. Design of an artificial muscle actuated finger towards biomimetic prosthetic hands. In: Proceedings of 2005 International Conference on Advanced Robotics, ICAR'05, Seattle, WA, 2005, pp. 368–375.

[37] K. T. O'Toole, M. M. McGrath, and E. Coyle. Analysis and evaluation of the dynamic performance of SMA actuators for prosthetic hand design. *Journal of Materials Engineering and Performance*, 18: 781–786, 2008.

[38] T. Sonoda and I. Godler. Multi-fingered robotic hand employing strings transmission named Twist Drive. In: 2010 IEEE/RSJ International Conference on Intelligent Robots and Systems, 2010, pp. 2733–2738.

[39] M. C. Carrozza, B. Massa, S. Micera, R. Lazzarini, M. Zecca, and P. Dario. The development of a novel prosthetic hand—ongoing research and preliminary results. *IEEE/ASME Transactions on Mechatronics*, 7: 108–114, 2002.

[40] H. Huang, L. Jiang, Y. W. Liu, L. Q. Hou, H. G. Cai, and H. Liu. The mechanical design and experiments of HIT/DLR prosthetic hand. In: ROBIO'06 IEEE International Conference on Robotics and Biomimetics, 2006, pp. 896–901.

[41] D. Graupe, J. Salahi, and K. H. Kohn. Multifunction prosthesis and orthosis control via micro-computer identification of temporal pattern differences in single-site myoelectric signals. *Journal of Biomedical Engineering*, 4: 17–22, 1982.

[42] D. Peleg, E. Braiman, E. Yom-Tov, and G. F. Inbar. Classification of finger activation for use in a robotic prosthesis arm. *IEEE Transactions on Neural Systems and Rehabilitation Engineering*, 10(4): 290–293, 2002.

[43] H. P. Huang, Y. H. Liu, and C. S. Wong. Automatic EMG feature evaluation for controlling a prosthetic hand using supervised feature mining method: An intelligent approach. In:

Proceedings of 2003 IEEE International Conference on Robotics and Automation, Taipei, Taiwan, 2003, vol. 1, pp. 220–225.

[44] K. Englehart, B. Hudgins, and P. A. Parker. A wavelet-based continuous classification scheme for multifunction myoelectric control. *IEEE Transactions on Biomedical Engineering*, 48(3): 302–311, 2001.

[45] S. M. ElBasiouny, A. M. El-Bialy, M. F. Taher, A. H. Kandil, and M. E. Rasmy. A myoelectric prosthesis controller. In: Proceedings of IEEE 29th Annual Bioengineering Conference, 2003, pp. 140–141.

[46] K. Englehart and B. Hudgins. A robust, real-time control scheme for multifunction myoelectric control. *IEEE Transactions on Biomedical Engineering*, 50(7): 848–854, 2003.

[47] F. H. Y. Chan, Y. S. Yang, F. K. Lam, Y. T. Zhang, and P. A. Parker. Fuzzy EMG classification for prosthesis control. *IEEE Transactions on Rehabilitation Engineering*, 8(3): 305–311, 2000.

[48] A. B. Ajiboye and R. F. Weir. A heuristic fuzzy logic approach to EMG pattern recognition for multifunctional prosthesis control. *IEEE Transactions on Neural Systems and Rehabilitation Engineering*, 13(3): 280–291, 2005.

[49] K. Momen, S. Krishnan, and T. Chau. Real-time classification of forearm electromyographic signals corresponding to user-selected intentional movements for multifunction prosthesis Control. *IEEE Transactions on Neural Systems and Rehabilitation Engineering*, 15(4): 535–542, 2007.

[50] X. Zhang, Y. Yang, X. Xu, and M. Zhang. Wavelet based neuro-fuzzy classification for EMG control. In: 2002 International Conference on Communications, Circuits and Systems and West Sino Exposition Proceedings, Chengdu, China, 2002, vol. 2, pp. 1087–1089.

[51] S. E. Hussein and M. H. Granat. Intention detection using a neurofuzzy EMG classifier. *IEEE Engineering in Medicine and Biology Magazine*, 21(6): 123–129, 2002.

[52] B. Hudgins, P. A. Parker, and R. N. Scott. A new strategy for multi-function myoelectric control. *IEEE Transactions on Biomedical Engineering*, 40(1): 82–94, 1993.

[53] B. Karlik, M. O. Tokhi, and M. Alci. A fuzzy clustering neural network architecture for multifunction upper-limb prosthesis. *IEEE Transactions on Biomedical Engineering*, 50(11): 1255–1261, 2003.

[54] H. P. Huang, Y. H. Liu, L. W. Liu, and C. S. Wong. EMG classification for prehensile postures using cascaded architecture of neural networks with self-organizing maps. In: Proceedings of 2003 IEEE International Conference on Robotics and Automation, Taipei, Taiwan, 2003, vol. 1, pp. 1497–1502.

[55] Y. Matsumura, Y. Mitsukura, M. Fukumi, and N. Akamatsu. Recognition of EMG signal patterns by neural networks. In: Proceedings of 9th International Conference on Neural Information Processing, 2002, vol. 2, pp. 750–754.

[56] J. G. Hincapie and R. F. Kirsch. Feasibility of EMG-based neural network controller for an upper extremity neuroprosthesis. *IEEE Transactions on Neural Systems and Rehabilitation Engineering*, 17(1): 80–90, 2009.

[57] Y. Huang, K. B. Englehart, B. Hudgins, and A. D. C. Chan. A Gaussian mixture model based classification scheme for myoelectric control of powered upper limb prostheses. *IEEE Transactions on Biomedical Engineering*, 52: 1801, 2005.

[58] Y. H. Liu, H. P. Huang, and C. H. Weng. Recognition of electromyographic signals using cascaded kernel learning machine. *IEEE/ASME Transactions on Mechatronics*, 12(3): 253–264, 2007.

[59] M. Markou and S. Singh. Novelty detection: a review, part I: statistical approaches. *Signal Processing*, 83: 2481–2497, 2003.

[60] M. Markou and S. Singh. Novelty detection: a review, part II: neural network based approaches. *Signal Processing*, 83: 2499–2521, 2003.

[61] V. N. Vapnik. *Statistical Learning Theory*. New York: John Wiley & Sons, 1998.

[62] B. Schölkopf, J. C. Platt, J. Shawe-Taylor, A. J. Smola, and R. C. Williamson. Estimating the support of a high-dimensional distribution. *Neural Computation*, 13: 1443–1471, 2001.

[63] G. R. G. Lanckriet, L. El Ghaoui, and M. I. Jordan. Robust novelty detection with single-class MPM. In: *Advances in Neural Information Processing Systems*, vol. 15, Cambridge, MA: MIT Press, 2003.

[64] D. Tax and R. Duin. Support vector domain description. *Pattern Recognition Letters*, 20: 1191–1199, 1999.

[65] D. Tax and R. Duin. Support vector data description. *Machine Learning*, 54: 45–66, 2004.

[66] H. Hoffmann. Kernel PCA for novelty detection. *Pattern Recognition*, 40: 863–874, 2007.

[67] A. Banerjee, P. Burlina, and C. Diehl. A support vector method for anomaly detection in hyperspectral imagery. *IEEE Transactions on Geoscience and Remote Sensing*, 44: 2282–2291, 2006.

[68] L. Nanni. Machine learning algorithms for T-cell epitopes prediction. *Neurocomputing*, 69: 866–868, 2006.

[69] J. Park, D. Kang, J. Kim, J. T. Kwok, and I. W. Tsang. SVDD-based pattern denoising. *Neural Computation*, 19: 1919–1938, 2007.

[70] Y. H. Liu, S. H. Lin, Y. L. Hsueh, and M. J. Lee. Automatic target defect identification for TFT-LCD array process inspection using kernel FCM based fuzzy SVDD ensemble. *Expert Systems with Applications*, 36: 1978–1998, 2009.

[71] D. Graupe, J. Magnussen, and A. M. Beex. A microprocessor system for multifunctional control of upper-limb prosthesis via myoelectric signal identification. *IEEE Transactions on Automatic Control*, 23(4): 538–544, 1978.

[72] Z. K. Mahyar, W. C. Bruce, B. Kambiz, and M. H. Reza. EMG feature evaluation for movement control of upper extremity prostheses. *IEEE Transactions on Rehabilitation Engineering*, 3(4): 324–333, 1995.

[73] E. Osuna and F. Girosi. Reducing the run-time complexity of support vector machines. In: *Advance in Kernel Methods—Support Vector Learning*, edited by B. Schölkopf, C. J. C. Burges, and A. Smola. Cambridge, MA: MIT Press, 1999, pp. 271–283.

[74] B. Tang and D. Mazzoni. Multiclass reduced-set support vector machines. In: Proceedings of 23rd International Conference on Machine Leaning, Pittsburgh, PA, 2006, pp. 921–928.

[75] Y. M. Wen and B. L. Lu. A cascade method for reducing training time and the number of support vectors. *Lecture Notes in Computer Science*, 3173: 480–486, 2004.

[76] T. Wu and H. Zhao. Constructing support vector classifiers with unlabeled data. In: *Advances in Neural Networks*, Part I. Berlin/Heidelberg: Springer, 2004, pp. 494–499.

[77] Y. Zhan and D. Shen. Design efficient support vector machine for fast classification. *Pattern Recognition*, 38: 157–161, 2005.

[78] J. Guo, N. Takahashi, and T. Nishi. A learning algorithm for improving the classification speed of support vector machines. In: Proceedings of 2005 European Conference on Circuit Theory and Design, Cork, Ireland, 2005, vol. 3, pp. 381–384.

[79] A. Navia-Vázquez. Compact multi-class support vector machines. *Neurocomputing*, 71: 400–405, 2007.

[80] Y. H. Liu, Y. C. Liu, and Y. J. Chen. Fast support vector data descriptions for novelty detection. *IEEE Transactions on Neural Networks*, 21(8): 1296–1313, 2010.

CHAPTER 20

ACCELEROMETER-BASED BODY SENSOR NETWORK (BSN) FOR MEDICAL DIAGNOSIS ASSESSMENT AND TRAINING

MING-YIH LEE[1,2], KIN FONG LEI[1,2], WEN-YEN LIN[1,3], WANN-YUN SHIEH[1,4], WEN-WEI TSAI[1,2], SIMON H. FU[1,5], and CHUNG-HSIEN KUO[6]

[1]Healthy Aging Research Center, Chang Gung University, Taiwan
[2]Graduate Institute of Medical Mechatronics, Chang Gung University, Taiwan
[3]Department of Electrical Engineering, Chang Gung University, Taiwan
[4]Department of Computer Science and Information Engineering, Chang Gung University, Taiwan
[5]IMS Technology and Consultant Inc., Taiwan
[6]Department of Electrical Engineering, National Taiwan University of Science and Technology, Taiwan

20.1 INTRODUCTION

According to the United Nations, 21.7% of the world population will be aged 60 or above in 2050, compared with 10% in 2000. Since health conditions tend to deteriorate with age, body sensor network (BSN) has increasingly become an attractive topic because of the significant demand for health services, as well as related equipment and supplies. BSN can provide real-time remote monitoring of the health situation of a particular person. In operation, wearable, miniaturized, and low-power consumption sensors are attached on or implanted in human body for collecting biological signals. A wide range of sensors has been utilized in the BSN. Recently, accelerometer can be fabricated by micro-electromechanical system (MEMS) technology that makes sensing of acceleration become simple and low in cost. Moreover, the sensor itself is miniaturized that can be integrated into various BSN applications. In this chapter,

Contemporary Issues in Systems Science and Engineering, First Edition.
Edited by MengChu Zhou, Han-Xiong Li and Margot Weijnen.

the use of accelerometers for BSN for monitoring the body motions is discussed. Recent advances in the computation of motion identification are described including tilting angle, muscle strength, and gait performance. Based on specific computing algorithms, different interpretations of the signal from accelerometers can be formulated. Moreover, several medical diagnosis assessment and training applications are introduced to demonstrate the capability of an accelerometer-based BSN. Furthermore, the concept of using biped humanoid robots to develop the BSN simulation system is discussed. The reasons for using such robots are their similar skeletal structures and desirable locomotion patterns such as walking and falls. In addition, the sensors such as accelerometers could be placed on the robot's body and limbs to collect the motion data for verifications with controllable locomotion. Finally, the BSN simulation system could generate helpful BSN information for evaluating the algorithms' performance in laboratories before the BSN systems are deployed to the human's body.

20.2 BODY SENSOR NETWORK

According to the United States Census Bureau, a significant birth rate between 1946 and 1964 was recorded. These persons who were born during the demographic post World War II baby boom are called baby boomers. They are nearly one-third of Americans, that is, 76 million people. With this statistical information, 2011 is the year of kicking off "Age Wave" according to the World Health Organization (WHO) regarding its definition of "old" people who are of age of 65 and above. Therefore, in the coming 10–20 years, there exists a significant demand and potential market on Age Wave associated with healthy aging and health care products.

Healthy aging is about staying physically and psychologically healthy in senior years. Early detection, early diagnosis, early warning, and early action (4E) and individual care, intelligent care, interactive care, and integrated care (4I) are proposed to provide development approaches for healthy aging products. Wearable remote sensing is one of the key technologies used to monitor health situation for the applications in 4E and 4I. Wearable electronics embodies the convergence of mobile, personal, and wireless communication technologies, with emerging textile components. Consequently, BSN and body area network (BAN) become hot research topics for the applications of personal medical care and remote home care.

The concept of BSN was first raised by Professor Guang-Zhong Yang, Imperial College, in 2002 [1] and the specification of a BSN node was also defined in 2004 [2]. BSN utilizes wearable, miniaturized, and low power consumption sensor devices, which are attached on or implanted in a human body, to collect biological signals through wireless communication for the monitoring of health situation. Therefore, the development of BSN includes the knowledge of electronics, computing, communication, and bioengineering technologies. The BSN consists of a number of BSN nodes as shown in Figure 20.1. A BSN node generally includes sensors, processor, transceiver, and power. The node captures human biological signals, such as ECG and

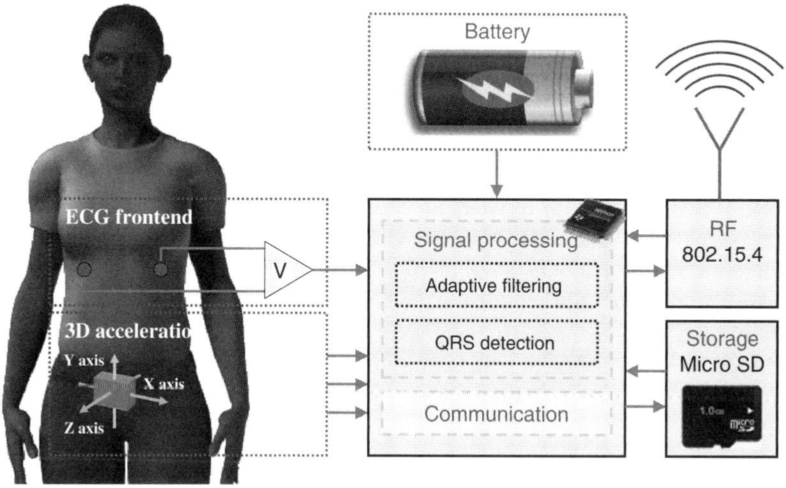

FIGURE 20.1 Architecture of a body sensor network node.

gesture, by using miniaturized sensors. Then, the signals are transferred from different BSN nodes and collected by the BSN. Various biological signals can be monitored by a BSN node including kinematic, kinetic, bio-physiological, and functional signal to represent the health situation, as shown in Figure 20.2. Kinetic signals cover the movement of joint, tremble of arms, trajectory of body movement, etc. Kinematic signals include articular twist force, foot pressure, etc. The bio-physiological signals

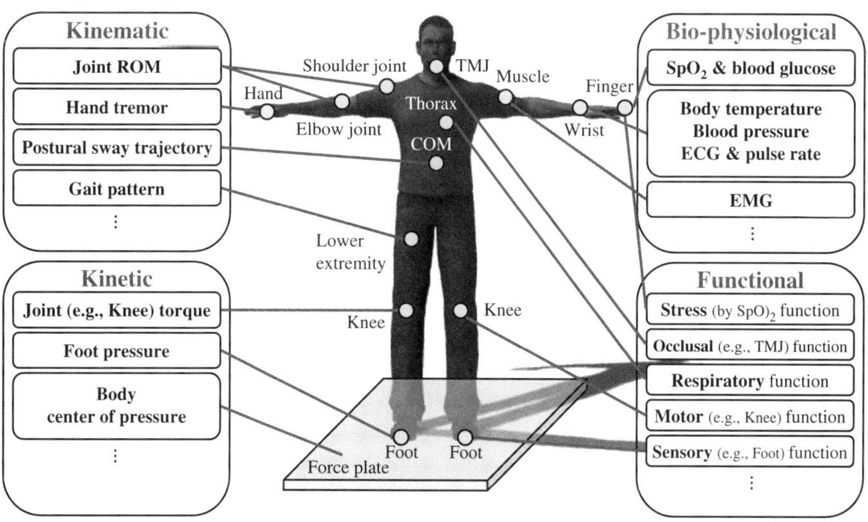

FIGURE 20.2 Categories of body sensors.

include blood sugar, blood oxygen, body temperature, blood pressure, ECG, heart beat rate, etc. Heart-lung, motion, and feeling are functional signals. Different signals require different sensors in proper position for monitoring biological signals. In addition, to satisfy the requirement of moving signal acquisition, the miniatured power source and low power consumption are the key elements of BSN.

Most recently, a wide range of BSN applications has been developed. Different sensors, such as ECG, body movement, and joint movement, are captured by a BSN and the signals are wirelessly transferred through smart phones. Medical centers can remotely monitor the health situation of their patients through Internet [3—5]. Moreover, a wireless body sensor node technique called Telos mote is developed for capturing the muscle electricity, body acceleration, and angular velocity [6]. These signals can wirelessly transfer to remote sites through a system called CodeBlue. The future applications include medical care for emergency patients going to and out of hospital and apoplexy patient recovery [6]. A wearable sensor platform called MIThril, that is a vest with sensors and transmitters, has been proposed to demonstrate the transformation of the biological information wirelessly and in real time [7]. Furthermore, optical linear encoder (OLE) technology is utilized to develop a miniaturized device for monitoring the angle of articulation. The demonstrated smart suit integrating a series of OLEs can sense the body motion and apply to virtual recovery scenario [8].

20.3 INFORMATION RETRIEVED FROM ACCELEROMETER

Nowadays, the inertial sensing has been widely used in many fields of applications. Indeed, the first four out of five modes of motion sensing, *acceleration*, *vibration*, *shock*, *tilt*, and *rotation*, are actually different manifestations of acceleration over different periods of time [9]. As a result, the accelerometer-based BSNs are developed and mainly targeting for medical applications driven by the increasing demands of tele-care due to the population aging of the societies and lifestyle changes.

The raw data measured by an accelerometer are the combination of dynamic acceleration of the object movement and the static earth gravity force projected on the single-, duo-, or tri-axial of the devices. The data are processed through certain algorithms and become more meaningful information: *acceleration*, *distanc*e, and *orientation*. Therefore, the object movement can be retrieved and these algorithms can be used for different applications.

- **Acceleration**: This is the direct meaning of the raw data carried. The magnitude of the data represents the *acceleration*. Fast back-and-forth changes of magnitude within a time period show *vibrations*. A significant change in magnitude within a short period is a *shock*. By such information, some biomedical parameters can be interpreted, such as heart rate, respiratory rate, and snoring rate, and used by BSNs for different applications, such as sleep apnea diagnosis [10], physical activity assessment [11], fall-risk estimation [12], and wearable automatic fall detectors [13–15].

- **Distance**: The distance of a linear movement can be computed by double integration of the acceleration, as shown in Equation (20.1). If the BSN is calibrated when it starts with a still position by setting its initial distance d_0 to zero, then the moving distance can be calculated. As a result, the information can be applied to some medical cases, for example, compression depth estimation for cardiopulmonary resuscitation (CPR) quality assessment [16], posture analysis for Parkinson's disease [17, 18], and walking distance estimation [19].

$$\mathrm{d}(t) = \int \int_0^t \mathrm{a}(\tau)\mathrm{d}\tau + \mathrm{d}_0 \tag{20.1}$$

- **Orientation**: When the net acceleration or force on the device node over time is gravity, the projections of the gravity vector on the axes of the accelerometer can be used to determine the static inclination or tilt angles of the node by calculating mathematical trigonometric functions from the projected gravity vector on the three axes of the accelerometer. The information is typically utilized in the posture monitoring applications, as in Reference 20, and some position monitoring applications, such as head's angle position monitoring after vitreoretinal surgery [21].

There are many other applications utilizing two or more kinds of information retrieved from accelerometers. For example, the wearable automatic fall detectors typically combine the shock sensing with tilt information to improve the accuracy as detected in References 13—15 and 22. Rissanen *et al.* [23] combine acceleration and the distance information for the analysis of dynamic voluntary muscle contractions in Parkinson's disease, and Khan *et al.* [24] propose an approach to physical activity recognition by combining acceleration data and tilt information.

20.4 RECENT ADVANCES IN ACCELEROMETER-BASED BSN

The signal output directly from an accelerometer is the measurement of the dynamic acceleration of the movement of the objective and the static earth gravity. However, by a specific algorithm, different interpretations of the biological identification can be obtained. Tilting angle, muscle strength, and gait performance are discussed next.

20.4.1 Tilting Angle Identification

When a three-axis accelerometer is used for inclination sensing, the orientation (inclination and tilt angles) of the device can be determined through domain transformation as illustrated in Figure 20.3. In the angular domain, ρ is the size of the gravity vector, that is, 1 **g**; θ, the angle between x-axis and the projected gravity vector on the x-y plane; and φ, the angle between z-axis and the gravity vector.

The computation of ρ and θ is too complicated to implement on a microcontroller. Hence, most current researches adopt offline computation [25], streaming the raw

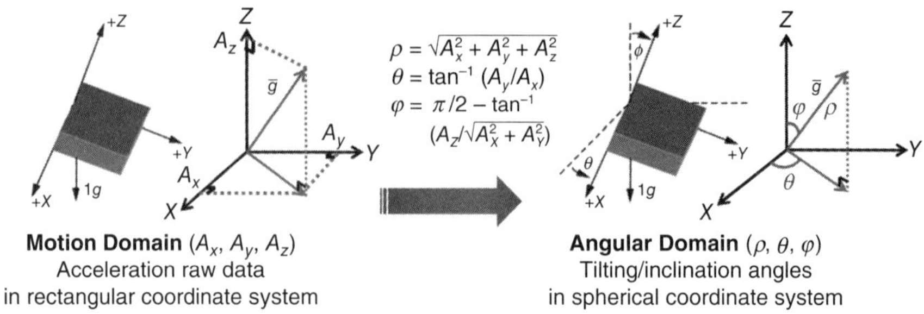

FIGURE 20.3 Angles of spherical coordinate system and coordinate transformation.

data to PC/PDA for computation [26–28] and table lookup [29]. However, the offline computation approach could not identify the tilting information in the system; as a result, the system does not have the capability for early detections for some critical behaviors and to generate real-time warnings. The streaming approach causes the concerns of excessive RF signal transmission thereby leading to higher system cost and higher power consumption. Table lookup approach requires huge memory space for table storage and produce higher system cost even though memory chips are more and more affordable.

20.4.1.1 Memory-Less and Low-Complexity Algorithm for Tilt Angle Identification The proposed memory-less and low-complexity algorithm called "CGU tilting angle identification algorithm" is based on the 2D coordinate rotation digital computer (CORDIC) [30] operations in system tilting angle identification. Using 2D CORDIC operations supplants the complex multiplication, division, and square root operations required in the original trigonometric functions with only basic adders and binary shifters [31] for rotation of a 2D vector through a sequence of simple elementary rotations. As a result, the tilt angle identification could be performed in a low-cost and high-portability closed embedded system. Consequently, early detection and prevention on some critical behaviors are feasible to implement. Real-time warnings and actions can be taken for some emergency situations in some medical applications.

The algorithm is conducted in two phases as shown in Figure 20.4. Phase I, as shown in Figure 20.4a, is to rotate $\vec{g}$ vector along z-axis and have its projected vector on x-y plane aligned to x-axis, and forming $\vec{g}'$ vector. The rotation is performed by 2D CORDIC on x-y plane with initial 2D vector (A_x, A_y) and final vector $(A'_x, 0)$. The iterative equations are shown in Figure 20.4a and the rotated angle θ and $A'_x = \sqrt{A_x^2 + A_y^2}$ are accrued. Phase II, as shown in Figure 20.4b, is to rotate $\vec{g}'$ vector on x-z plane and have it aligned to x-axis. The rotation is performed by 2D CORDIC on x-z plane with initial 2D vector (A'_x, A_z) and have it aligned to x-axis. The iterative equations are shown in Figure 20.4b and the rotated angle ψ is accrued through CORDIC operations and $\varphi = {}^{\pi}/_{2} - \psi$. Clearly, the inclination and tilt angles, θ and

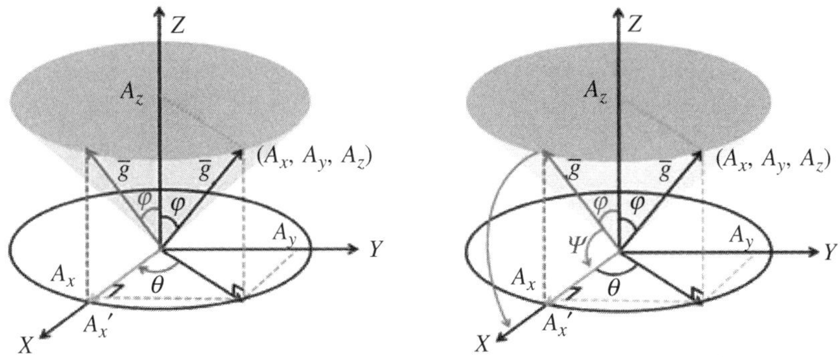

$x_{i+1} = x_i - y_i - d_i - 2^{-i}$

$y_{i+1} = y_i + x_i - d_i - 2^{\ i}$

$\theta_{i+1} = \theta_i + d_i - \arctan(2^{-i})$

Initially, $x_0 = A_x$, $y_0 = A_y$ and $\theta_0 = 0$

(a) Phase I: calculation of θ and $\sqrt{A_x^2 + A_y^2}$

$x_{i+1} = x_i - y_i - d_i - 2^{-i}$

$y_{i+1} = y_i + x_i - d_i - 2^{-i}$

$\Psi_{i+1} = \Psi_i + d_i - \arctan(2^{-i})$

Initially, $x_0 = A_x'$, $y_0 = A_z$ and $\Psi_0 = 0$

(b) Phase II: calculation of φ

FIGURE 20.4 CGU tilt angle identification algorithm: (a) Phase I and (b) Phase II.

φ, are obtained with simple shifting and addition operations only by this proposed algorithm.

20.4.1.2 Application of Head Positioning Study for Obstructive Sleep Apnea Syndrome The clinical studies show the transient spontaneous remission of patients with obstructive sleep apnea syndrome during diagnoses. However, how the transient spontaneous remission happens is still unclear. It might be caused by body position changes, arousal, and head position changes. Unfortunately, the existing polysomnography is incapable of measuring head position with high accuracy. To help in the study of how the head position relates to the transient spontaneous remission of obstructive sleep apnea syndrome, an accelerometer-based sensor is attached on the forehead of the patient, as shown in Figure 20.5. The measured acceleration raw data are transformed into angular domain to retrieve the orientation information of a patient's head.

The proposed algorithm described above can identify the head position with the representation of angles: θ and φ as shown in Figure 20.5. The patient is instructed to rotate her head toward right of the body and it could be observed that angle φ starts to change in a positive range since the angle, φ, between z-axis and gravity vector increases as the head rotates toward right. In this position, the projected vector of the gravity vector on x-y plane is almost aligned to y-axis; therefore, angle θ is almost equal to 90°. Then, the patient is instructed to rotate her head back and toward the left of the body. It is observed that the angle φ starts to change in a positive range again while θ is almost equal to negative 90°, as in this position the projected vector of the gravity vector on x-y plane is almost aligned to the opposite direction of y-axis.

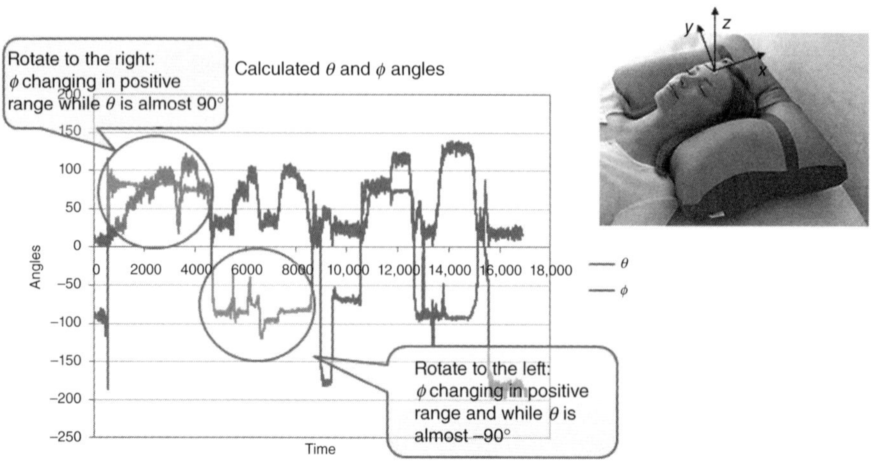

FIGURE 20.5 Accuracy comparison of θ and φ with a varied number of iterations.

The proposed algorithm can be used in the implementation of in-system tilting angle identification so that the output data can be integrated into the polysomnography for overall diagnoses. This is just a case of how the algorithm helps in in-system tilting angle identification, and there are other medical applications where the algorithm can be used.

20.4.2 Muscle Strength Identification

Vibration can be explained as fast back-and-forth changes of magnitude within a time period. With this property, muscle strength can be identified by an accelerometer after a specific algorithm. The muscle signals including electromyography (EMG) [32–37] and mechanomyography (MMG) [38–42] have been studied for many years. The level of muscular contraction can be estimated by either EMG or MMG signal. This relation can work as a noninvasive approach for musculoskeletal assessment in many applications, for example, rehabilitation equipment, athletic training, and prosthetic control. MMG is recognized as a low-frequency signal produced by oscillations of contracting skeletal muscle fibers and relates to the number and firing rate of the recruited motor units [38–42]. Therefore, MMG signal can be picked up by various sensors, such as condenser microphones [43, 44], piezoelectric transducers [44–46], MEMS-based accelerometers [47, 48], and strain sensor [49], attached on the muscle surface during contraction. In this section, the estimation of muscular contraction by the use of a MEMS-based accelerometer is described to illustrate an example of using inertia-based sensors for the biomedical applications.

20.4.2.1 Muscle Strength Identification by the Use of Accelerometer

An experimental setup shown in Figure 20.6 is utilized to perform the muscle strength identification. During the experiment, the subject is comfortably seated and the right

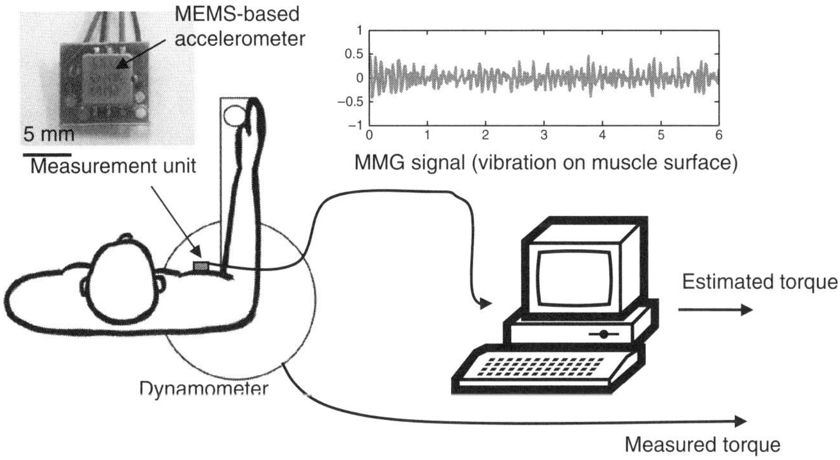

FIGURE 20.6 Illustration of the experiment setup for the identification of the muscle torque estimated by MMG signal.

arm is fixed horizontally on an in-house dynamometer. The shoulder abduction is at 0° and elbow angle is at 90°. The torque produced by biceps brachii is applied to the dynamometer and measured by a torque transducer (AKC-205A, 0–100 Nm, China Academy of Aerospace Aerodynamics) installed on the shaft of the dynamometer. A measurement unit consists of a dual-axis MEMS-based accelerometer (ADXL213, Analog Devices, Norwood, MA, USA; 5 mm × 5 mm × 2 mm LCC package) that is attached on the surface of the biceps brachii to measure the MMG signal. Then, the signal is input to a signal processing unit. A fourth order Butterworth filter with bandpass of 2–40 Hz is utilized to shape the signal, and the sampling rate of 200 Hz is set. Hence, two characteristics of MMG signal are extracted including root mean square (RMS) amplitude and the frequency variance. The RMS amplitude is the average amplitude of MMG signal within a preset time window. The frequency variance, *var*, can be computed as follows [50]:

$$\text{var} = \frac{\int_0^{f_s/2} \left(f - f_{\text{mean}}\right)^2 S(f)\,df}{\int_0^{f_s/2} S(f)\,df}, \tag{20.2}$$

where f_{mean} is the mean power frequency (MPF), $S(f)$ is the power density spectrum (PDS) of the signal, and f_s is the sampling frequency. MPF represents the moment of the PDS, and frequency variance measures the scale or degree of being spread out. A two-layer neural network that can adapt the characteristic of MMG signal for a particular muscle is utilized as the MMG-torque estimator. The RMS and frequency variance of MMG signal are the inputs of the estimator and the output is the estimated torque.

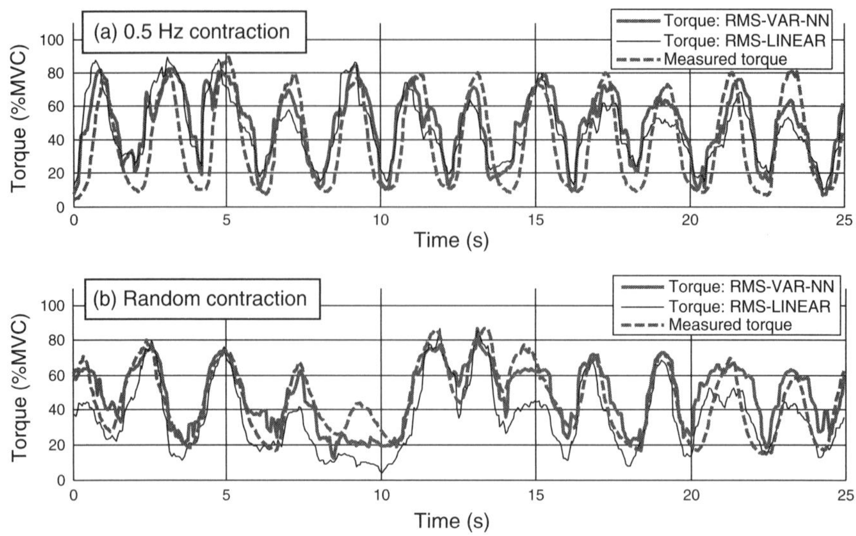

FIGURE 20.7 Examples of the MMG-torque estimation. Subjects were instructed to perform contraction at (a) 0.5 Hz and (b) random frequency.

The MMG-torque estimation is demonstrated by a dynamic (constant posture and varying torque) tracking contraction experiment. A constant frequency contraction and a random frequency contraction are performed to evaluate the MMG-torque estimator. Before the estimation, the MMG-torque estimator is trained by input-output data using a back-propagation algorithm. The input-output data are MMG signal (input) and measured torque (output) under sinusoidal contractions at 1, 0.5, 0.25, and 0.125 Hz. Because the estimator has learnt from the input-output data of a particular muscle, the estimator is muscle-dependent. After training, the subject is instructed to perform contractions under 0.5 Hz and random frequency. The comparison results of the estimated torques from MMG-torque estimator (MMG-VAR-NN) and MMG RMS amplitude linear mapping (RMS-LINEAR), which are commonly used, are shown in Figure 20.7. Also, the measured torque from the dynamometer is plotted in it. It is observed that the proposed MMG-torque estimator achieves better accuracy than linear mapping for both experiments. The error decreases 6% for the 0.5 Hz contraction and 30% for random contraction. From this study, the MMG-torque estimator can be potentially applied to many applications and research investigations that need musculoskeletal assessment.

20.4.3 Gait Performance Identification

The processes of walking or running are called the “gait.” Human gait is a complex movement that includes activities in many muscles and joints of the body [51, 52]. In clinical research, gait analysis has been used to investigate the features of normal or abnormal gait for demonstrating a change from treatment or from disease progression.

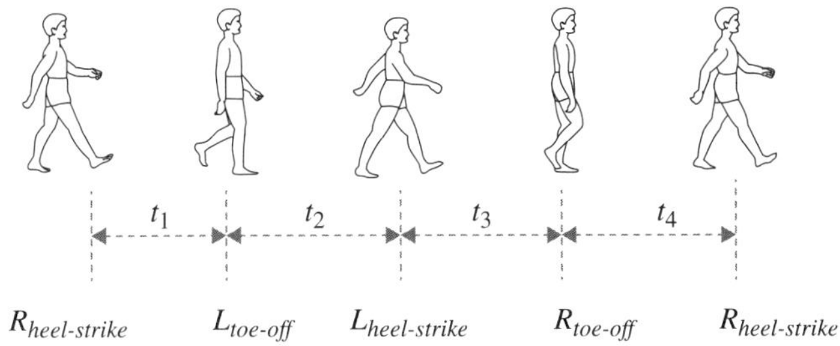

FIGURE 20.8 The process of a gait cycle.

By analyzing the gait of elderly people, it can be observed whether they have an abnormality or degradation in their walking ability [53]. For the patients with leg or foot injury, the gait analysis can be used to evaluate the effectiveness of rehabilitation or assess balance and mobility monitoring after operation [53].

To analyze the gait, we can break the repetitive body movements into similar cycles, usually referred to as gait cycles. Each gait cycle can be characterized by several gait events. Take Figure 20.8 for example. Figure 20.8 shows a process from the posture of the right foot touching the ground (at the same time the left foot preparing to leave the ground) to the same posture of the right foot touching the ground again. If the posture of any one foot touching the ground is defined as the "heel-strike," and the posture of any one foot leaving the ground is defined as the "toe-off," then a gait cycle can be broken into five gait events: $R_{heel\text{-}strike}$, $L_{toe\text{-}off}$, $L_{heel\text{-}strike}$, $R_{toe\text{-}off}$, $R_{heel\text{-}strike}$, as shown in Figure 20.8 [51, 54], where "*R*" denotes the right foot and "*L*" denotes the left foot.

By the definition of a gait cycle, several gait parameters can be quantified as follows:

- **Gait cycle time (GCT)**: The period of the time between the heel-strike of one foot to the next heel-strike of the same foot, that is, $\text{GCT} = t_1 + t_2 + t_3 + t_4$, where t_i is the time interval between two sequential gait events in Figure 20.8.
- **Step number (Nstep)**: The number of steps in a sample distance. This parameter is counted for each leg separately, for example, we can count the number of heel-strikes as the step number for each foot separately.

In the past, many gait analysis techniques have been proposed [55, 56, 57]. Most of them are video-based approaches, which pasted many reflective markers on the body and used multiple cameras from different directions to record the process of walking. This kind of approaches can provide very accurate measurement for gait analysis. However, they usually need expensive gait laboratory equipment. Also, the testing is restricted to a laboratory environment. An alternative approach is to use the accelerometers attached to the body for the purpose of examining segmental accelerations during walking.

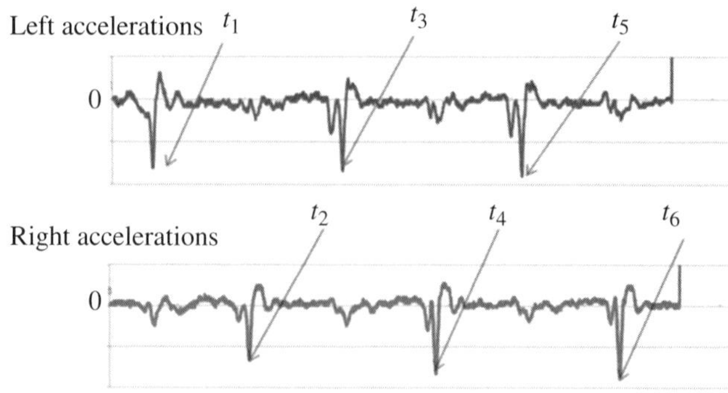

FIGURE 20.9 The results of the continuous accelerations on the left and right feet.

Figure 20.9 shows the signals of the continuous accelerations on the left foot and the right foot during a short period of walking measured by the accelerometers. If we assume:

k: The total step number,
t_n: The time of the *n*th heel-strike (left or right), where $n = 1, 2, \ldots, k$, and
X(m): The position of the *m*th time point, where $m = t_1, t_2, \ldots, t_n$.

Then from Figure 20.9, it is found that *k* can be obtained by counting the total number of peak points from each accelerometer, and t_n can be obtained by recording the time of each peak point. Assume that the position of the starting point is zero. By performing the integration operation on those acceleration functions in double times, the relative distance of each peak point away from the starting point, that is, *X(m)*, can be obtained. Note that the accelerometers should be calibrated first before testing. Based on *k*, t_n, and *X(m)*, many other spatio-temporal parameters for gait analysis [54] can be derived as shown in Table 20.1. Compared with conventional video-based approaches, using the accelerometers for gait performance identification has the benefits from their small size, low cost, and low power, which can be applied to lightweight or wearable devices brought by the subjects without any spatial or temporal restrictions. The development of gait-related clinical applications such as a diagnostic tool for long-term abnormal gait detection, an outcome evaluation after

TABLE 20.1 Expressions of Spatio-Temporal Gait Parameters

Gait parameters	T_n	T_{step}	F_{step}	D_n	S_{step}	V_{step}
Equation	$t_n - t_{n-1}$	$\frac{\sum_{n=1}^{K}(t_n - t_{n-1})}{K}$	$\frac{1}{T_{step}}$	$X(t_n) - X(t_{n-1})$	$\frac{\sum_{n=1}^{K}[X(t_n) - X(t_{n-1})]}{K}$	$\frac{S_{step}}{T_{step}}$

T_n, the period time of the *n*th step; T_{step}, average period time of a step; F_{step}, average step frequency; D_n, the distance of the *n*th step; S_{step}, average distance of a step; V_{step}, average speed of a step.

gait-related treatments, or rehabilitation would become a very important research issue in the future.

20.5 APPLICATIONS OF ACCELEROMETER-BASED BSN FOR REHABILITATION

The ability to control postural steadiness is an important determinant of overcoming the internal and external environmental stimulus and disturbances across the falls associated with the loss of balance. Traditionally, assessments of postural steadiness are based on patient's ability to control the angular deviation of the center of mass (COM) within the base of support (BOS). The imbalance is defined by deviations of COM that lies outside normal age-matched reference limits [58]. The deviation of COM depends on the angular movement of a body and height of COM. COM can be figured out at approximately 55% of the body height and the position is around trunk at L3-L5 [59]. It is much easier to mount accelerometer at COM on the body than to use a 3D motion analysis system.

In addition, changes in stance conditions can result in the degradation of sensory input. For example, it is well established that both visual feedback and task specificity, such as reducing stance width, increase the range and speed of body movement [60, 61]. Various rehabilitation approaches have been used to improve skill reacquisition of the impaired balance. However, different types of biofeedback, such as knowledge of performance (KP) and knowledge of results (KR), provided to the subject are the key factors in motor skill reacquisition [62]. KP requires concurrent feedback to evaluate goal-related performance, which requires motion analysis systems and significant post-processing to retrieve relevant body referenced motion data. In contrast, KR displays only result of overall performance at the end of task. Besides, the KR or terminal feedback about goal-related movement outcomes has become the focus of many researches in healthy [63, 64] as well as stroke [65, 66] subjects, mostly because of the ease with which it can be quantified.

20.5.1 Human Stability Evaluation System

A human stability evaluation system (HSES) is developed for the analysis of movement and postural control [67]. The HSES consists of a 3D acceleration measurement module (MMA7260, Freescale, Inc.), a microcontroller unit (MSP430, Texas Instruments), a RF module, a status display module, a power module, and a human interface unit, as shown in Figure 20.10. The sway movements of subjects are calculated by the tilting angles and the height of COM. The tilting angles are calculated by the gravitational vector that is directly measured by the acceleration measurement module. The module is mounted on a belt that the subject wears around the waist. This makes the system portable, quick, and easy to use, allowing the registration of how the trunk is moving during a tandem stance (TS) real-time task without significantly limiting the subject's sway movements. The data from the acceleration measurement module are sampled in every 10 ms, stored temporarily on the HSES, and transferred to the personal computer via bluetooth. The microcontroller unit utilizes an ultralow power

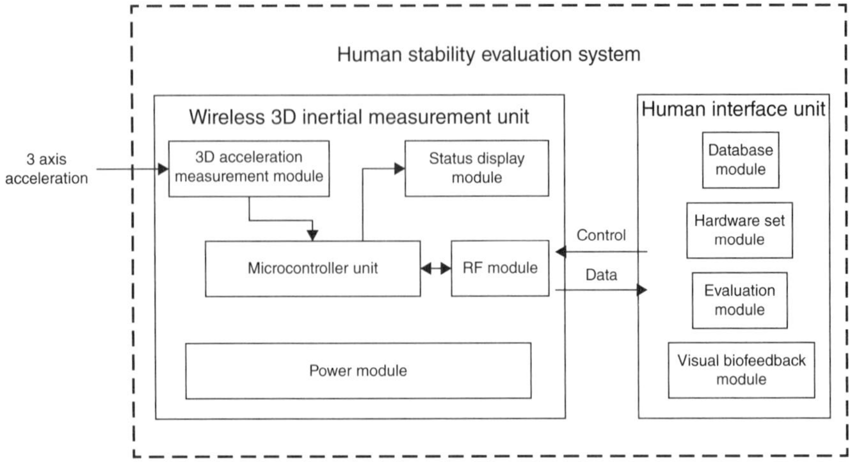

FIGURE 20.10 Architecture of human stability evaluation system.

microprocessor that achieves extended battery life for the portable measurement applications. The functions of the microcontroller unit include receiving the analog signals from the acceleration measurement module and uploading sway movements information for stability assessment and biofeedback training.

20.5.2 Postural Stability Evaluation for Stroke Patients

Twelve stroke patients (3 females and 9 males; mean age: 54.8 ± 12.2 years) have been recruited for this study. Six patients are with right-hand affected side and others with left-hand affected side. All participant patients are evaluated by therapists as Brunnstorm stage IV or V based on neurologic injury, motor control disorder, and Berg balance scale. Both Berg balance scale and HSES are used to assess the postural stability under four static standing tasks: standing unsupported with eye open, standing unsupported with eye closed, tandem stance with eye open, and standing feet together with eye open as shown in Figure 20.11. The anterior–posterior sway (pitch) and medial–lateral sway (roll) angle are monitored by HSES.

The results of Berg balance scale show that all patients are assessed at top marks via their performances under four static standing tasks. It means that the Berg balance scale could not be used to differentiate patients between Brunnstorm stage IV and Brunnstorm stage V. Moreover, the other postural stability results, outputs of HSES, are used as nonparametric statistics to prove the hypothesis that these data can be used to identify the patients' Brunnstorm stage IV or V from static standing tests. In the statistic results, there is no significant difference in both pitch angle and roll angle between patients with Brunnstorm stage IV and V under standing unsupported with eye open, standing unsupported with eye closed, and standing feet together with eye open postures; nevertheless, Brunnstorm stage IV and V patients show significant difference in roll angle under tandem stance with eye open posture (Brunnstorm stage

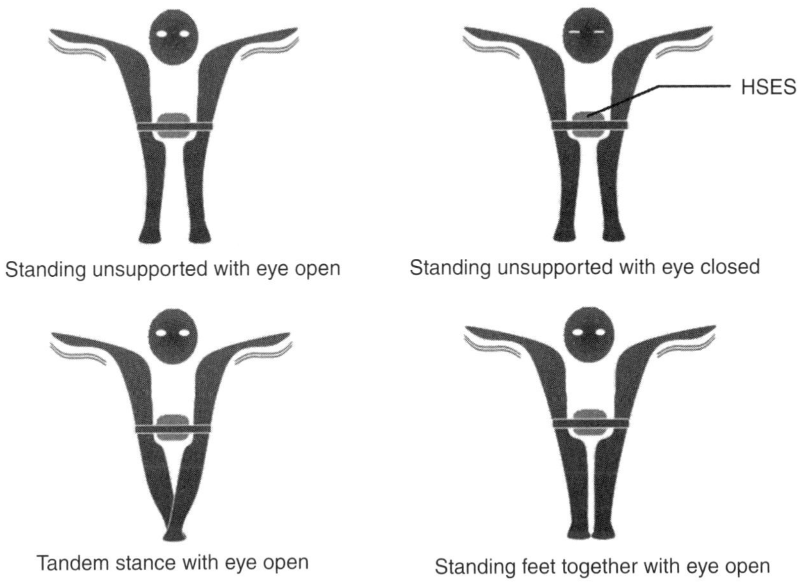

FIGURE 20.11 Four postures of static standing tasks.

IV: 1.36 ± 0.26, Brunnstorm stage V: 0.85 ± 0.05, $p = 0.002$). In conclusion, it is noticed that the HSES could provide more precisely quantitative evaluation results to classify patients than traditional Berg balance scale.

20.5.3 Postural Stability Training for Stroke Patients

The proposed HSES is integrated with an augmented biofeedback training unit, contains KP and KR strategies, for stability assessment and biofeedback training in clinical applications (Figure 20.12).

Eighteen subjects (13 males and 5 females, age 31–70 years, mean age 58.15 ± 13.28 years) with Brunnstorm stage IV and above who previously suffered stroke for at least 6 months are recruited for this 3 months trial study from Chung Shan Medical University Hospital. All subjects are randomly assigned into three groups: (1) KP ($n = 6$) practices a TS task involving three repetitions per day and 3 days per week, with real-time visual feedback about movement precision; (2) KR ($n = 6$) are trained on the same task and schedule as KP but with faded TS task scores; (3) C ($n = 6$) practices the same task and schedule as KP without any feedback. Initially, the subjects are standing relaxed with eyes open and weight evenly distributed on both feet. Thereafter, the subjects are instructed to stand in a TS position, a clinical measure of postural steadiness, with one foot in front of the other. The collected data of HSES in the study are peak-to-peak excursions in trunk angular displacement in both the roll and pitch planes. Parameters within the group of stoke patients are studied with the paired-sample t-test. The level of significance is set at $p < 0.05$.

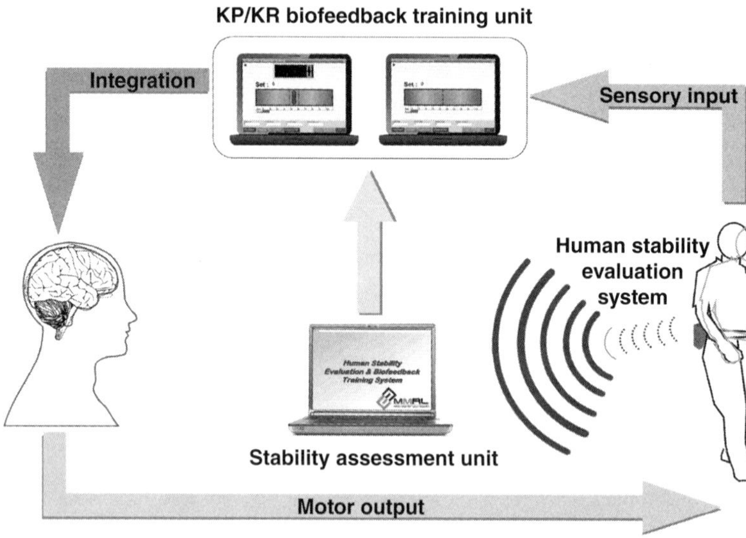

FIGURE 20.12 Illustration of conceptual design of the proposed human stability evaluation system integrated with KP/KR biofeedback for postural stability training.

The intervention has a significant influence on balance during a TS task. The results reveal that biofeedback appears to be an important variable in TS. Across all participants, the KP group has significantly reduced average trunk angular displacements in both the roll (pretest: 1.25 ± 0.06°, posttest: 0.92 ± 0.02°, $p = 0.02$) and pitch (pretest: 0.92 ± 0.03°, posttest: 0.81 ± 0.02°, $p = 0.04$) direction after biofeedback training intervention during TS but not in the KR group in roll (pretest: 1.01 ± 0.01°, posttest: 0.98 ± 0.02°, $p = 0.77$) and pitch (pretest: 0.86 ± 0.00°, posttest: 0.84 ± 0.01°, $p = 0.68$) direction. For the control group, there is no significant difference between pretest and posttest in the roll (pretest: 0.99 ± 0.08°, posttest: 1.00 ± 0.02°, $p = 0.97$) and pitch (pretest: 0.79 ± 0.05°, posttest: 0.76 ± 0.01°, $p = 0.84$) direction, as shown in Table 20.2. Interestingly, motor improvements depend on the type of feedback

TABLE 20.2 Population Means and Standard Error of Trunk Roll and Pitch Measures in Stroke Patients During Tandem Stance Tasks

	Roll			Pitch		
	Pretest	Posttest	*p*	Pretest	Posttest	*p*
	Average trunk angular displacement (degrees)					
KP	1.25 ± 0.06	0.92 ± 0.02	0.02*	0.92 ± 0.03	0.81 ± 0.02	0.04*
KR	1.01 ± 0.01	0.98 ± 0.02	0.77	0.86 ± 0.00	0.84 ± 0.01	0.68
Control	0.99 ± 0.08	1.00 ± 0.02	0.97	0.79 ± 0.05	0.76 ± 0.01	0.84

*Significant difference from respective before and after biofeedback training intervention.

provided in training. Compared with the control group, subjects who received KP biofeedback improved average trunk angular displacement in trunk sway during TS, whereas those who had KR biofeedback had no significant improvement. However, differences may be attributed to different motor control processes associated with different types of biofeedback. From this study, accelerometer-based system could be potentially applied to balance evaluation and postural stability training.

20.6 BSN SIMULATION SYSTEM

In this section, a BSN simulation system is discussed based on the uses of biped humanoid robot techniques. In general, biped humanoid robots are designed to perform similar locomotion with human beings by considering similar skeletal structure design approaches. Therefore, the locomotion of regular walking or falling down of human beings could be simulated by using biped humanoid robots. Practically, the sensors such as accelerometers could be placed on the robot's body and limbs to collect the BSN data for verifications with controllable locomotion.

In recent years, BSNs are successfully developed for fall detection systems. Falls are serious problems for elderly [68–70]. They may cause permanent injuries. Hence, it is important to implement a fall detection system [71] for elderly in an aging society. For example, Song *et al.* [72] propose a living aid robot (RoLA) capable of detecting falls. According to human's reflexes, it is difficult to collect realistic sensor's signals directly from human experimentation of falling down suddenly or in high-risk situations such as stair falls.

Due to the difficulties of collecting signals of falling down in a human's experiment, the BSN simulation system is proposed to collect the falling down signal by using a small-size biped humanoid robot [73, 74] with similar sensor setups and configurations of subjects. Hence, a BSN simulation system is employed to collect the signal of falls of biped humanoid robots, instead of a real person.

In order to evaluate the falling down sensor signal correlations with the biped humanoid robot and the human beings, the skeletal structures of the biped humanoid robot and popular human beings are compared first. Based on similar skeletal structures, the fall behaviors between them can be properly evaluated with similar conditions. First of all, the degree-of-freedom between the biped humanoid robot and human beings is discussed. The structure of the biped humanoid robot of a case study is shown in the *left*-hand side of Figure 20.13.

Moreover, the heights of the small-size biped humanoid robot and human beings are quite different. A scaled down skeletal model is used for the BSN simulation system. The scaled down skeletal model is examined by considering the skeletal structures of the male students in our laboratory. Form the practical measurements, the height of the biped humanoid robot is 43 cm; the averaged height of the male students in our laboratory is 173 cm. In this case, the height scale of the robot and the male students in our laboratory is 1:4. In addition, the length of a robot's arm is 19 cm; the length of average student's arm is 75 cm. The arm's length scale of the robot and students is 1:3.8. The length of the robot's leg is 22.6 cm; while the

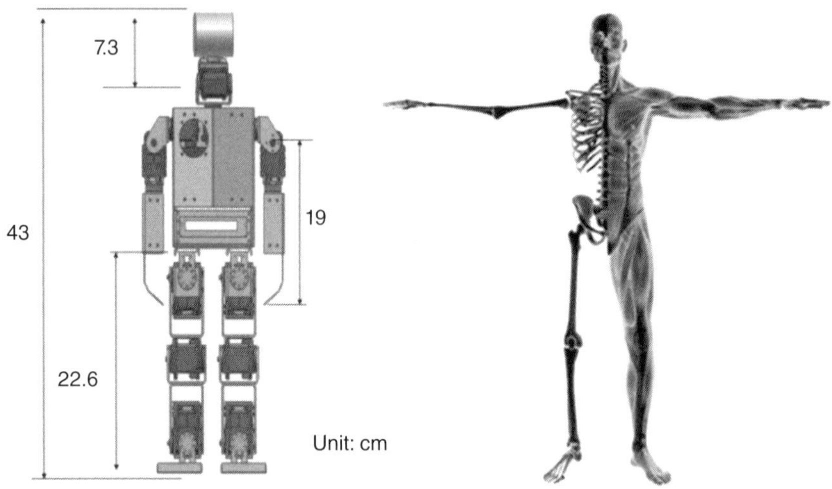

FIGURE 20.13 Skeletal structures of biped robot (*left*) and human beings (*right*).

length of student's leg is 104 cm. The leg's scale of the robot and students is 1:4.6. The aforementioned scaled down parameter is around 0.25 for the height and four extremities; hence, such scale down assumptions are feasible.

Based on the structure similarity between the biped humanoid robot and the human beings, three fall scenarios are discussed as follows:

1. *Scenario I*: The biped robot as shown in Figure 20.14 is stumbled over the protuberance in the straight line. As it kicks the protuberance, it falls down immediately. Therefore, this test platform receives the fall signals, and then

FIGURE 20.14 Fall scenario of stumbling over protuberance in straight line.

FIGURE 20.15 Fall scenario of slips on a wet floor.

records them in the database. They will be used to determine whether people fall down or not.

2. *Scenario II*: The biped humanoid robot as shown in Figure 20.15 slips instantaneously on the wet road. In this situation, the robot walks in the straight line on the carpet. As the robot steps on the carpet, we suddenly pull the carpet, and it

FIGURE 20.16 Fall scenario of falling down from stairs.

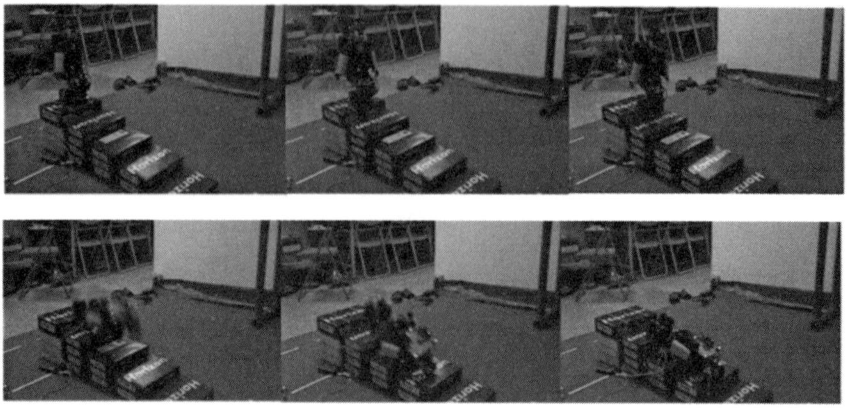

FIGURE 20.17 Stair fall experiments of a biped robot.

falls down instantly. Then the test platform receives the fall signals, and then records them in the database. They will be used to determine whether people fall down or not.

3. *Scenario III*: Figure 20.16 shows the situation of a biped robot being falling down from stairs. In this situation, the robot walks straight until it falls down from first stair. As it falls, the test platform receives the fall signals immediately, and then records them in the database. They will be used to determine whether people fall down or not.

The stair fall is also examined with a biped humanoid robot, as shown in Figure 20.17. A stair environment is arranged by considering the robot's dimension. In this figure, the robot walks forward without considering the gap existing in its way. As a consequence, the robot falls after crossing the first step. The tri-axis accelerometer sensors are also applied in this experiment. The experimental data and details are found in Reference 75.

BSN simulation system provides a safe and reproducible testing platform for the posture or fall detection. Robot-based BSN simulation system can be a human-like skeletal structure, and applied to simulate accidental scenarios without any human hurt and repeat tests. For this reason, when more and more accelerometer-based algorithms will be developed, the specific issues of how to design various robot-based BSN simulation systems with high mimicking reality will be addressed.

20.7 CONCLUSIONS

BSN can provide real-time remote monitoring of human health situation. That is excellent for the applications of early detection, early diagnosis, early warning, and early action (4E), and individual care, intelligent care, interactive care, and integrated care (41). In BSN, various sensors have been utilized for collecting the biological signals. MEMS-based accelerometers are one of the popular devices for measuring the body motion because of their miniaturized size, light weight, low cost, and low power consumption. This chapter introduced different interpretations of the signal of an accelerometer. Tilting angle, muscle strength, and gait performance can be formulated by using specific algorithms. Moreover, several medical diagnosis assessment and training applications are discussed to demonstrate the accelerometer-based BSN. The use of accelerometers in BSN can have a number of variations based on different algorithms. As the need of healthy aging is increasing, accelerometer-based BSN has great potential for various rehabilitation applications. Furthermore, BSN simulation system is also discussed in this chapter, and it exhibits an innovative and a promising research direction for effectively verifying the algorithms in laboratories before the BSN system is deployed.

BSN is an emerging technology for providing continuous, ubiquitous, and free-motion health monitoring. The open research issues for accelerometer-based BSN in clinical applications are suggested as follows: (1) head tilting angle identification for the diagnostics of polysomnography; (2) multi-muscle strength evaluation for the diagnostics of muscle coordination; (3) gait analysis for the long-term diagnostics of abnormal gait; (4) power-assisted exoskeletal prosthesis development for amputee and elderly; (5) human-like robotic fall simulation for the estimation of accidental scenarios without any human hurt. Moreover, integrating with other technologies, for example, textile, a concept of "Smart Clothing" or "Intelligent Clothing" for collecting human's information unconsciously is also suggested to achieve real-time health monitoring. The development of the BSN becomes a multidisciplinary technological issue. Integration of sensing, computing, communication, power management, and textile is also the technical challenge.

REFERENCES

[1] Imperial College London. Biosketch of Professor Guang-Zhong Yang. Available at http://ubimon.doc.ic.ac.uk/gzy/m365.html, 2009 (accessed January 25, 2012).

[2] Body Sensor Networks. About body sensor networks (BSN). Available at http://ubimon.doc.ic.ac.uk/bsn/m185.html, 2009 (accessed January 25, 2012).

[3] E. Jovanov, A. Milenkovic, C. Otto, and P. C. de Groen. A wireless body area network of intelligent motion sensors for computer assisted physical rehabilitation. *Journal of NeuroEngineering and Rehabilitation*, 2(6): 2–6, 2005.

[4] University of Oulu. Optoelectronics and measurement unit (OPME), Annual Report 2007. Available at http://www.infotech.oulu.fi/Annual/2007/opme.html, 2007 (accessed January 25, 2012).

[5] Michigan Technological University. Integrated Microsystems Enterprise (IME) Program. Available at http://www.enterprise.mtu.edu/im/index.html, 2010 (accessed January 25, 2012).

[6] Harvard University. CodeBlue: wireless sensors for medical care. Available at http://fiji.eecs.harvard.edu/CodeBlue, 2009 (accessed January 25, 2012).

[7] Massachusetts Institute of Technology, MIT Media Lab. MIThril Home. Available at http://www.media.mit.edu/wearables/mithril/index.html, 2009 (accessed January 25, 2012).

[8] K. D. Nguyen, I. M. Chen, Z. Luo, S. H. Yeo, and H. B.-L. Duh. A wearable sensing system for tracking and monitoring of functional arm movement. *IEEE/ASME Transactions on Mechatronics*, 16(2): 213–220, 2011.

[9] R. O'Reilly and H. Weinberg. The five motion senses: MEMS inertial sensing to transform applications. Analog Devices, Inc., White Papers, 2010.

[10] D. Morillo, J. Ojeda, L. Foix, and A. Jimenez. An accelerometer-based device for sleep apnea screening. *IEEE Transactions on Information Technology in Biomedicine*, 14(2): 491–499, 2010.

[11] D. Curone, A. Tognetti, E. L. Secco, G. Anania, N. Carbonaro, D. De Rossi, and G. Magenes. Heart rate and accelerometer data fusion for activity assessment of rescuers during emergency interventions. *IEEE Transactions on Information Technology in Biomedicine*, 14(3): 702–710, 2010.

[12] M. R. Narayanan, S. J. Redmond, M. E. Scalzi, S. R. Lord, B. G. Celler, and N. H. Lovell. Longitudinal falls-risk estimation using triaxial accelerometry. *IEEE Transactions on Biomedical Engineering*, 57(3): 534–541, 2010.

[13] A. Diaz, M. Prado, L. M. Roa, J. Reina-Tosina, and G. Sanchez. Preliminary evaluation of a full-time falling monitoring for the elderly. In: Proceedings of 26th Annual International Conference Engineering in Medicine Biology Society, San Francisco, CA, 2004, vol. 1, pp. 2180–2183.

[14] A. K. Bourke, J. V. O'Brien, and G. M. Lyons. Evaluation of a threshold-based tri-axial accelerometer fall detection algorithm. *Gait & Posture*, 26(2): 194–199, 2007.

[15] G. Williams, K. Doughty, K. Cameron, and D. A. Bradley. A smart fall and activity monitor for telecare applications. In: Proceedings 20th Annual International Conference on IEEE Engineering in Medicine Biology Society, Hong Kong, China, 1998, vol. 3, pp. 1151–1154.

[16] S. Aase and H. Myklebust. Compression depth estimation for CPR quality assessment using DSP on accelerometer signals. *IEEE Transactions on Biomedical Engineering*, 49(3): 263–268, 2002.

[17] L. Palmerini, L. Rocchi, S. Mellone, F. Valzania, and L. Chiari. Feature selection for accelerometer-based posture analysis in Parkinson's disease. *IEEE Transactions on Information Technology in Biomedicine*, 15(3): 481–490, 2011.

[18] S. Mellone, L. Palmerini, A. Cappello, and L. Chiari. Hilbert-Huang-based tremor removal to assess postural properties from accelerometers. *IEEE Transactions on Biomedical Engineering*, 58(6): 1752–1761, 2011.

[19] A. Sabatini, C. Martelloni, S. Scapellato, and F. Cavallo. Assessment of walking features from foot inertial sensing. *IEEE Transactions on Biomedical Engineering*, 52(3): 486–494, 2005.

[20] R. Nevins, N. G. Durdle, and V. J. Raso. A posture monitoring system using accelerometers. In: Proceedings of the 2002 IEEE Canadian Conference on Electrical & Computer Engineering (CCECE 2002), 2002, vol. 2, pp. 1087–1092.

[21] P. Čech, J. Dlouhý, M. Čižek, I. Vícha, and J. Rozman. Head position monitoring system design. In: 2010 20th International Conference on Radioelektronika, Brno, Czech Republic, 2010, pp. 1–4.

[22] Q. Li, J. A. Stankovic, M. A. Hanson, and A. T. Barth. Accurate, fast fall detection using gyroscopes and accelerometer-derived posture information. In: 6th International Workshop on Wearable and Implantable Body Sensor Network (BSN 2009), Berkeley, CA, 2009, pp. 138–143.

[23] S. M. Rissanen, M. Kankaanpaa, M. P. Tarvainen, A. Yu. Meigal, J. Nuutinen, I. M. Tarkka, O. Airaksinen, and P. A. Karjalainen. Analysis of dynamic voluntary muscle contractions in Parkinson's disease. *IEEE Transactions on Biomedical Engineering*, 56(9): 2280–2288, 2009.

[24] A. Khan, Y. K. Lee, S. Y. Lee, and T. S. Kim. A triaxial accelerometer-based physical-activity recognition via augmented-signal features and a hierarchical recognizer. *IEEE Transactions on Information Technology in Biomedicine*, 14(5): 1166–1172, 2010.

[25] R. J. Nevins, N. G. Durdle, and V. J. Raso. A posture monitoring system using accelerometers. In: IEEE Canadian Conference on Electrical and Computer Engineering, (CCECE 2002), Manitoba, Canada, 2002, Vol. 2, pp. 1087–1092.

[26] E. Farella, A. Pieracci, L. Benini, and A. Acquaviva. A wireless body area sensor network for posture detection. In: Proceedings of 11th IEEE Symposium on Computers and Communications (ISCC'06), Sardinia, Italy, 2006, pp. 454–459.

[27] W. T. Huang, C. N. Hung, C. C. Tuan, C. C. Chen, Y. J. Chang, and Y. Y. Chen. Novel design of interactive caption and recognition postures within the wearable and non-vision system over wireless sensor network. In: 2010 International Conference on Complex, Intelligent and Software Intensive Systems (CISIS), Krakow, Poland, 2010, pp. 752–757.

[28] H. R. Silva, J. A. Afonso, P. C. Morim, P. M. Oliveira, J. H. Correia, and L. A. Rocha. Wireless hydrotherapy smart-suit network for posture monitoring. In: IEEE International Symposium on Industrial Electronics (ISIE 2007), Vigo, Spain, 2007, pp. 2713–2717.

[29] P. Cech, J. Dlouhy, M. Cizek, I. Vicha, and J. Rozman. Software for automatic head position monitoring. In: 19th International Conference on Radioelektronika (RADIOELEKTRONIKA'09), Bratislava, Slovak Republic, 2009, pp. 31–34.

[30] J. E. Volder. The CORDIC trigonometric computing technique. *IRE Transactions on Electronic Computers*, EC-8(3): 330–334, 1959.

[31] Y. H. Hu. CORDIC-based VLSI architectures for digital signal processing. *IEEE Signal Processing Magazine*, 9(3): 16–35, 1992.

[32] A. L. Hof and J. W. Van den Berg. EMG to force processing I: an electrical analogue of the hill muscle model. *Journal of Biomechanics*, 14(11): 747–758, 1981.

[33] U. P. Wyss and V. A. Pollak. Surface electromyogram (EMG)/muscle force: a muscle model based on RMG peaks. *Engineering in Medicine*, 13(1): 27–33, 1984.

[34] E. N. Zuniga and D. G. Simons. Nonlinear relationship between averaged electromyogram potential and muscle tension in normal subjects. *Archives of Physical Medicine and Rehabilitation*, 50: 613–620, 1969.

[35] E. A. Clancy and N. Hogan. Multiple site electromyograph amplitude estimation. *IEEE Transactions on Biomedical Engineering*, 42(2): 203–211, 1995.

[36] E. A. Clancy and K. A. Farry. Adaptive whitening of the electromyogram to improve amplitude estimation. *IEEE Transactions on Biomedical Engineering*, 47(6): 709–771, 2000.

[37] E. A. Clancy, O. Bida, and D. Rancourt. Influence of advanced electromyogram (EMG) amplitude processors on EMG-to-torque estimation during constant-posture, force-varying contractions. *Journal of Biomechanics*, 39: 2690–2698, 2006.

[38] D. T. Barry. Vibrations and sounds from evoked twitches. *Electroencephalography and Clinical Neurophysiology*, 32(1–2): 35–40, 1992.

[39] D. T. Barry. Acoustic signals from frog skeletal muscles. *Biophysical Journal*, 51(5): 769–773, 1987.

[40] D. T. Barry and N. M. Cole. Muscle sounds are emitted at the resonant frequencies of skeletal muscle. *IEEE Transactions on Biomedical Engineering*, 37(5): 525–531, 1990.

[41] J. V. Frangioni, T. S. Kwan-Gett, L. E. Dobrunz, and T. A. McMahon. The mechanism of low frequency sound production in muscle. *Biophysical Journal*, 51(5): 775–783, 1987.

[42] C. Orizio. Muscle sound: bases for the introduction of a mechanomyographic signal in muscle studies. *Critical Reviews in Biomedical Engineering*, 21(3): 201–243, 1993.

[43] P. Madeleine, H. Y. Ge, A. Jaskolska, D. Farina, A. Jaskolski, and L. Arendt-Nielsen. Spectral moments of mechanomyographic signals recorded with accelerometer and microphone during sustained fatiguing contractions. *Medical & Biological Engineering & Computing*, 44: 290–297, 2006.

[44] A. Jaskolska, P. Madeleine, A. Jaskolski, K. Kisiel-Sajewicz, and L. Arendt-Nielsen. A comparison between mechanomyographic condenser microphone and accelerometer measurements during submaximal isometric, concentric and eccentric contractions. *Journal of Electromyography & Kinesiology*, 17: 336–347, 2007.

[45] M. Shinohara, M. Kouzaki, T. Yoshihisa, and T. Fukunaga. Mechanomyography of the human quadriceps muscle during incremental cycle ergometry. *European Journal of Applied Physiology*, 76: 314–319, 1997.

[46] D. B. Smith, T. J. Housh, J. R. Stout, G. O. Johnson, T. K. Evetovich, and K. T. Ebersole. Mechanomyographic responses to maximal eccentric isokinetic muscle actions. *Journal of Applied Physiology*, 82: 1003–1007, 1997.

[47] P. Madeleine, C. Cescon, and D. Farina. Spatial and force dependency of mechanomyographic signal features. *Journal of Neuroscience Methods*, 158: 88–99, 2006.

[48] K. Akataki, K. Mita, M. Watakabe, and K. Itoh. Mechanomyographic responses during voluntary ramp contractions of the human first dorsal interosseous muscle. *European Journal of Applied Physiology*, 89: 520–525, 2003.

[49] M. Ouamer, M. Boiteux, M. Petitjean, L. Travens, and A. Sales. Acoustic myography during voluntary isometric contraction reveals non-propagative lateral vibration. *Journal of Biomechanics*, 32: 1279–1285, 1999.

[50] E. Kwatny, D. H. Thomas, and H. G. Kwatny. An application of signal processing techniques to the study of myoelectric signals. *IEEE Transactions on Biomedical Engineering*, 17(4): 303–313, 1970.

[51] M. Salehi and M. Razzazi. A sensor-based framework for detecting human gait cycles using acceleration signals. In: Proceedings of the 17th International Conference on Software, Telecommunications & Computer Networks, Split/Hvar/Korcula, 2009, pp. 328–332.

[52] J. A. Lee, S. H. Cho, Y. J. Lee, H. K. Yang, and J. W. Lee. Portable activity monitoring system for temporal parameters of gait cycles. *Journal of Medical Systems*, 34(5): 959–966, 2009.

[53] B. Dijkstra, W. Zijlstra, E. Scherder, and Y. Kamsma. Detection of walking periods and number of steps in older adults and patients with Parkinson's disease: accuracy of a pedometer and an accelerometry-based method. *Oxford Journal: Age and Ageing*, 37(4): 436–441, 2008.

[54] W. Zijlstra and A. Hof. Assessment of spatio-temporal gait parameters from trunk accelerations during human walking. *Gait & Posture*, 18(2): 1–10, 2003.

[55] R. Davis, S. Õunpuu, D. Tyburski, and J. Gage. A gait analysis data collection and reduction technique. *Human Movement Science*, 10(5): 575–587, 1991.

[56] D. Robertson. *Research Methods in Biomechanics*. Champaign, IL: Human Kinetics Pubs., 2004.

[57] R. Best and R. Begg. Overview of movement analysis and gait features, Chapter 1. In: *Computational Intelligence for Movement Sciences: Neural Networks and Other Emerging Techniques*, edited by M. Palaniswami. Hershey PA, USA: Idea Group, 2006, pp. 11–18.

[58] J. Gill, J. H. J. Allum, M. G. Carpenter, M. Held-Ziolkowska, A. L. Adkin, F. Honegger, and K. Pierchala. Trunk sway measures of postural stability during clinical balance tests: effects of age. *Journals of Gerontology Series A: Biological Sciences and Medical Sciences*, 56(7): M438–M447, 2001.

[59] D. A. Winter, A. E. Patla, and J. S. Frank. Assessment of balance control in humans. *Medical Progress Through Technology*, 16(1): 31–51, 1990.

[60] W. M. Paulus, A. Straube, and T. Brandt. Visual stabilization of posture. Physiological stimulus characteristics and clinical aspects. *Brain*, 107(4): 1143–1163, 1984.

[61] B. L. Day, M. J. Steiger, P. D. Thompson, and C. D. Marsden. Effect of vision and stance width on human body motion when standing: implications for afferent control of lateral sway. *Journal of Physiology*, 469: 479–499, 1993.

[62] R. A. Schmidt and T. D. Lee. *Motor and Learning: A Behavioral Emphasis*, 2nd edn. Champaign, IL: Human Kinetics, 1988.

[63] C. J. Winstein. Knowledge of results and motor learning-implications for physical therapy. *Physical Therapy*, 71(2): 140–149, 1991.

[64] C. J. Winstein and R. Schmidt. Sensorimotor feedback. In *Human Skills*, 2nd edn, edited by D. H. Holding. Chichester, England: John Wiley & Sons, 1989, pp. 17–47.

[65] C. Bütefisch, H. Hummelsheim, P. Denzler, and K. H. Mauritz. Repetitive training of isolated movements improves the outcome of motor rehabilitation of the centrally paretic hand. *Journal of the Neurological Sciences*, 130(1): 59–68, 1995.

[66] M. C. Cirstea, A. Ptito, and M. F. Levin. Arm reaching improvements with short-term practice depend on the severity of the motor deficit in stroke. *Experimental Brain Research*, 152(4): 476–488, 2003.

[67] K. S. Soon, M. Y. Lee, W. W. Tsai, and C. F. Lin. A new trunk sway assessment protocol using biofeedback inertial-based sensing modality for stroke patients. In: 2011 International Conference on System Science and Engineering, Macau, June 8–10, 2011, pp. 675–678.

[68] C. Cooper, G. Campion, and L. J. Melton. Hip fractures in the elderly: a world-wide projection. *Osteoporosis International*, 2(6): 285–289, 1992.

[69] A. J. Blake and K. Morgan. Falls by elderly people at home: prevalence and associated factors. *Age Ageing*, 17: 365–372, 1988.

[70] R. W. Sattin. Falls among older persons: a public health perspective. *Annual Review of Public Health*, 13: 489–508, 1992.

[71] Q. Li. Accurate, fast fall detection using gyroscopes and accelerometer-derived posture information. In: Sixth International Workshop on Wearable and Implantable Body Sensor Networks, 2009, pp. 138–143.

[72] K. T. Song, C. Y. Tsai, F. S. Huang, J. W. Hong, C. Y. Lin, C. W. Chen, and Z. S Lin. Development of the robot of living aid: RoLA. In: IEEE International Conference on Automation and Logistics, 2008, pp. 443–448.

[73] K. Ogata, K. Terada, and Y. Kuniyoshi. Falling motion control for humanoid robots while walking. In: IEEE-RAS International Conference on Humanoid Robots, Paris, France, 2009, pp. 306–311.

[74] K. Kaneko, K. Harada, F. Kanehiro, G. Miyamori, and K. Akachi. Humanoid robot HRP-3. In: IEEE/RSJ International Conference on Intelligent Robots and Systems, Nice, France, 2008, pp. 2471–2478.

[75] P. C. Chia, C. H. Lee, T. S. Chen, C. H. Kuo, and M. Y Lee. Correlations of falling signals between biped robots and humans with 3-axis accelerometers. In: International Conference on System Science and Engineering, Macau, 2011, pp. 509–514.

CHAPTER 21

TELEPRESENCE ROBOTS FOR MEDICAL AND HOMECARE APPLICATIONS

JUN-MING LU[1] and YEH-LIANG HSU[2,3]
[1]Department of Industrial Engineering and Engineering Management, National Tsing Hua University, Taiwan
[2]Department of Mechanical Engineering, Yuan Ze University, Taiwan
[3]Gerontechnology Research Center, Yuan Ze University, Taiwan

21.1 INTRODUCTION

Over the past few decades, robotics made tremendous progress in saving, protecting, and improving human lives, and is now adopted across a broad range of applications in medicine and homecare services. While technologies are developed at a rapid pace, people are expecting robots to be part of their lives in a more natural way. Among a wide variety of robots, telepresence robots, which allow the user to experience the virtual presence in another place, offer the potential to meet this rising demand better and is thus of great concern. This chapter will explore the up-to-date research findings and industry practices in telepresence robots for medical and homecare applications. Moreover, the key contributing factors to the success of telepresence robots will be discussed as well to address the future trends and opportunities.

Tracing back to the first use of the term "robot" by Čapek in his play entitled "Rossum's Universal Robots" in 1920, robots were originally regarded as the artificial people produced to work as servants. From then on, the term "robot" began to be widely adopted to describe the human-like machines that assist human beings. In the beginning, most robots were developed to facilitate repetitive works for industrial applications [1]. With further advancement of robotics, more people realized the great potential of robots. Hence, robots were also expected to enrich the daily lives of human beings more naturally and directly by performing all kinds of services

Contemporary Issues in Systems Science and Engineering, First Edition.
Edited by MengChu Zhou, Han-Xiong Li and Margot Weijnen.

[2]. Nowadays, robots are generally categorized by the application fields into two main types: industrial robots and service robots. More specifically, either category of robots can include both autonomous and teleoperated robots, depending on how the robots are controlled.

Robots have the advantages of high precision, strong consistency, and reliable stability. Thus, in the field of medical applications, the use of robots exactly helps to overcome the technical limitations of conventional surgery performed by physicians. The first robot-assisted surgery was performed in 1985. The Unimation Puma 200 robot, which was equipped with a computerized tomographic scanner and a probe guide, was used for stereotactic brain surgery [3]. With the advancement of technologies, medical robotics was further developed to extend human capabilities in surgery. Intuitive Surgical [4] introduced the da Vinci® Surgical System with advanced supersensory for telepresence. By integrating a surgeon's console, a patient-side cart with robotic arms, and a high-performance vision system, the surgeon's hand movements can be seamlessly translated into precise and minimally invasive movements. In addition, robots also play an increasingly important role in modern medicine, ranging from training the medical and nursing staff, assisting diagnosis, to facilitating patients' rehabilitation and care. Some medical schools make use of human patient simulator (HPS), a robot that mimics human's feelings of pain or discomfort, to help the soon-to-be doctors and nurses prepare to treat real patients [5]. Besides, the InTouch Health [6] developed a mobile robot called RP-7 to enable physicians to be remotely present for diagnosis. It helps to remove time and distance barriers and effectively extend the physician's reach to manage patient care. In this way, patients feel more satisfied because physicians seem to spend more time with them [7]. Further, Kaczmarski and Granosik [8] presented the rehabilitation robot RRH1. By helping the patients replay trained exercises such as hip and knee flexion/extension and leg abduction/adduction, the rehabilitation for the lower extremities can be easily performed with safety. Moreover, robotic assistive limbs that enhance the caregiver's strength for patient handling [9] and the robotic wheelchair with the function of automatic navigation [10] can provide much help in homecare. Toward a higher quality of life, interactive robots serve as a new type of communication tools for medical or homecare use. Seal robot Paro is an example of robot-assisted therapy for improving mental health [11].

As shown in Table 21.1, the robots for medical and homecare applications are summarized according to the participants and activities involved. Generally speaking, the use of robots has made revolutionary changes by greatly helping the medical community in various ways to serve patients, improve quality of life, and prevent health problems for patients.

Among the wide variety of robots, telepresence robots do have the benefits of providing closer connections between the two ends of users, which is often emphasized and demanded in healthcare. Thus, in the next three sections, recent advances of telepresence robots for medical and homecare applications, including rehabilitation and therapy, monitoring and assistance, and communication, are introduced for better understanding. Subsequently, key factors contributing to the success of telepresence robots are discussed to reveal the real needs from the users' perspective.

TABLE 21.1 Categories of Robots for Medical and Homecare Applications

	Participants						
Activities	Medical staff	Nursing staff	Patient	Caregiver	Caretaker	Family and friends	Others
Training	[5]	[5]					
Surgery	[3, 4]	[3, 4]	[3, 4]				
Diagnosis	[7]		[7]				
Consultation	[7, 12]	[7, 12, 13]			[7, 12, 13]		
Monitoring	[7]	[7, 12]	[7, 12]	[12, 14]	[12, 14–16]	[12, 14–16]	
Rehabilitation	[8, 17]	[8, 17]	[8, 17]				
Therapy		[18]	[11, 18–20]		[11, 18–20]		
Assistance		[21]	[10, 21–23]	[21]	[10, 21–23]	[21]	
Communication	[12]	[12]		[15, 16, 23, 24]	[12, 15, 16, 23–25]	[12, 15, 16, 23–25]	
Pharmacy							[26]
Prosthetics	[27]	[27]	[27]				

21.2 SURGERY, DIAGNOSIS, AND CONSULTATION

As mentioned in the previous section, the da Vinci® Surgical System makes use of telepresence technology to enable surgeons to perform delicate and complex operations with increased vision, precision, dexterity, and control [4]. However, the system costs over 1 million dollars, and it takes longer time for surgical procedures than traditional ones. These are the major challenges to be overcome. In addition to serving as an assistant on the first line of medical services, telepresence robots can assist the doctors in diagnosis and consultation as well. InTouch Health [6] has released the mobile robot RP-7 that enables a physician to extend his/her reach to manage healthcare by making oneself remotely present near the patients. The robot doctor also allows direct connection to Class II medical devices, such as electronic stethoscopes, otoscopes, and ultrasound, for transmitting medical data to the remote physician. Medical personnel can thus discuss treatment plans and interact with patients remotely, which helps improve the efficiency of medical diagnosis and treatment for nonlife-threatening emergencies. In order to expand the range of use, Brière *et al.* [12] present Telerobot, an in-home telehealth robot for clinical applications. Telerobot is controlled via two screens. One is for the clinical information system and the other displays the control interface with a virtual joystick and the video stream. Designed with good mobility, Telerobot is able to serve as the agent of the family doctor in home environments. In addition, Luo *et al.* [13] develop a telemedicine robot that allows the medical staff from long distance to provide consultation for the elderly people living at home. Combined with the wearable sensors, the robot can detect the emergencies such as falls and immediately inform the patient-related family members.

21.3 REHABILITATION AND THERAPY

Telepresence helps extend not only human vision and hearing but also the sense of touch, which is important for physical rehabilitation. Mouri *et al.* [17] propose a novel hand telerehabilitation system comprising a hand rehabilitation support system for the patient, an anthropomorphic robot hand for the therapist, and a remote monitoring system for diagnosing the degree of recovery. A therapist applies the force to the robot hand, and the force is then transmitted to the patient via the rehabilitation support system. This makes both participants experience the face-to-face rehabilitation even though they are in fact far way from each other. Besides, the remote monitoring system provides quantitative data in real time, resulting in higher efficiency of treatments. Psychologically, therapeutic robots based on telepresence also have the potential for improving people's mental health. The EU project IROMEC (Interactive Robotic Social Mediators as Companions) has developed a therapeutic robot for children with minor motor disabilities or communication deficiencies. Since autonomous robots pose the particular hazard to handicapped children who are not able to react properly to a moving robot, the IROMEC robot makes use of telepresence and is controlled by the remote therapeutic personnel to play with the children by following them or by dancing [18].

21.4 MONITORING AND ASSISTANCE

The use of a robot can be an alternative to locating cameras everywhere in the house. Instead of creating the feeling of being watched all the time, the robot will only look around when there is any possible risk detected. iRobot LE is developed for people to use it as a security guard to monitor a house as if the remote user actually goes around in the home environment. With the telepresence capability, it gives the remote user access to not only security in monitoring house or investigating household but also checking the conditions of the elderly people living alone [14]. To be more active while living with people, telepresence robots can provide assistance in various ways. By extending the concept of smart homes, Helal and Abdulrazak [21] develop TeCaRob, a telecare robot, to provide physical assistance for people with special needs in the healthcare center. Caregivers stay in the remote operation center and wait for the senior people's needs. The robots can assist the elderly in many ways such as transferring and moving them, feeding them, giving medications, or doing some tasks for them.

21.5 COMMUNICATION

Derived from the idea of a mobile robot with videophone embedded, Michaud *et al.* [16] present a teleoperated robot with wheels. Telepresence is provided for both ends with auditory and visual information, but the feeling of "staying with the person at the same place" is however limited due to the machine-like appearance. Tsai *et al.* [24] develop a telepresence robot called TRIC for interpersonal communication for the daily use of the elderly in the home environment. With the human-like appearance, the robot can better serve as the avatar of the children or grandchildren for expressing their care. Given high mobility by means of omnidirectional wheels and ultrasonic sensors, it is able to move in all directions, turn around, and avoid collisions with the environment. Toward improved convenience of home use, it is then redesigned into a compact size with more plentiful presentation of the remote user's emotions and feelings, by means of eye contact, facial expression, and body language. The physical face-to-face interaction among people can be thus rebuilt to provide a more natural communication as if both users are being together with each other [15].

21.6 KEY FACTORS CONTRIBUTING TO THE SUCCESS OF TELEPRESENCE ROBOTS

Although technologies have made great contribution to the development of telepresence robots, the most important concern remains the user acceptance. In other words, since a telepresence robot is intended to serve as the avatar or agent of a human for interacting with the environment or other people, it is necessary to provide the realistic sense of "being there" for the remote controller, as well as the experience of "acting like a real person" for the local user who stays with the robot. Broadbent *et al.* [28] discuss the user acceptance of social robots in terms of robot factors and human factors. The two categories provide totally different views for the requirements of a robot. The former focuses on the functions or utility of the robot, while the latter highlights the relationships between the user characteristics and their feelings toward the robot. Based on this concept, the key affecting factors to the success of telepresence robots are further discussed and summarized hereafter. These findings can provide practical guidelines for researchers, professionals, and practitioners in this field.

21.6.1 Robot Factors of Acceptance

Robots are regarded as the products of technology and scientific innovation. In addition to the advanced functions that benefit human beings, how the robot interacts with people actually influences the user acceptance more greatly. This is especially critical for telepresence robots, in which the feeling of presence is highlighted. It is not only about making the remote user experience exactly what the robot perceives and where it travels but also about whether the local user considers the robot as a realistic one. Generally, these robot factors include its anthropomorphism, physical characteristics, and personality. As the robot acts or reacts more naturally as real humans do, users will show greater interest and be more willing to interact with it.

21.6.1.1 *Anthropomorphism* Anthropomorphism, or human likeness, refers to projecting human characteristics to nonhuman animals or nonliving things. Generally, it involves various attributes of the robot, such as appearance, facial expression, and body motion. As the level of anthropomorphism goes higher, the interaction performance can be further improved [29]. For example, Goetz *et al.* [30] indicate that the appearance of a robot influences people's perceptions of a robot, as well as their willingness to follow the instructions given by the robot. However, the humanoid robots are not always the preferred ones. As shown in the 2000-people survey conducted by Arras and Cerqui [31], only 19% prefer a humanoid appearance. In fact, the user acceptance depends on whether the level of anthropomorphism matches the sociability required in the jobs. More specifically, people would prefer human-like robots as office clerk or hospital message carrier, while machine-like robots are expected to be lab assistant, inspector, or guardian [30]. Considering the task involved, people also tend to cooperate with human-like robots rather than machine-like robots [32]. Further, the user perception also relates to the dimension of the robot's head and its facial expression. Powers and Kiesler [23] find that a shorter chin contributes to the perception of higher sociability and higher intentions to follow the robot's medical advice. In addition, large smiles with slow transitions are seen as more appealing by the users [33].

21.6.1.2 *Physical Characteristics* As a robot becomes closer to a real human, the physical characteristics such as gender, age, height, and weight will then have impacts on the user acceptance. Powers *et al.* [33] report that participants would say fewer words to the female robot than to the male robot in a human–robot dialog. This phenomenon might be explained by the traditional role stereotypes. Besides, the age of a robot can influence its role that people experience. For example, if the robot has an adult humanoid appearance, people will expect it to be able to converse more naturally than the robot with a younger appearance [25]. Moreover, the preference of the robot size is determined by the tasks it involves. Robots for home use are expected to have a smaller size [34], whereas robots for patient handling require a larger size to support the weight and thus increase a user's confidence toward the robot [22].

21.6.1.3 *Personality* In addition to the physical characteristics, the personality of a robot, including its emotional, attitudinal, and behavioral response patterns, also plays an important role in user acceptance. For efficient use, it has to match either the user's characteristics or its own role. Obviously, a caring and empathic personality will encourage interaction between the user and the robot [35]. Besides, Heerink *et al.* [36] report that a more socially communicative robot would be more likely to be accepted as a conversational partner. Further, perceptions of knowledge and sociability are found to be able to change people's intention to follow the robot's advice [23]. Moreover, Ţăpuş *et al.* [37] demonstrate that a robot's adaptability to the user's personality is important for the user to gain improvements via rehabilitation exercises.

21.6.2 Human Factors of Acceptance

There is a great diversity in human beings. People are with quite different physical characteristics, backgrounds, and experiences. In addition to the robot factors, the user acceptance is also affected by these many human factors. For telepresence robots, from the remote user's point of view, the feeling of presence may vary among different people even though the condition remains the same. As for a local user, one unique activity that the robot performs may produce completely different perception or responses among different people. Thus, it is necessary to investigate these factors for a better robot design.

21.6.2.1 Physical Characteristics Arras and Cerqui [31] investigated the relationship between age and the willingness of living on a daily basis with robots. The results show that young adults tend to give more positive responses than older adults (over the age of 65 years). Nevertheless, under the assumption that one is with impaired mobility and is unable to handle the daily activities, the older adults are more willing to accept a robot to help them gain independence. Besides, the gender of users also makes difference. While interacting with a robot, males wonder more about the technical aspects, whereas females are more interested in its name [38]. Further, in Nomura *et al.*'s study [39], the experimental results imply a gender difference in relationships between negative attitudes and anxiety, and behavior toward robots. Among people who have high negative attitudes and anxiety toward interaction with robots, males tend to avoid touching or talking with robots, while females still talk to the robot but not engage in much self-disclosure with it.

21.6.2.2 Backgrounds and Experiences With different backgrounds and experiences, people may have varying attitudes toward technologies or robots. Giuliani *et al.* [34] report that as one's educational level gets higher, he/she tries to make connections with technological solutions more frequently. Besides, lack of familiarity with technologies can be a major reason for people feeling uncertain about robots [40]. As people stay longer with a robot, it will also change their attitudes toward it. Over a 2-month study conducted by Wada and Shibata [20], residents in a care center finally develop much better personal relationships with the therapy robot. Moreover, cultural differences also make impacts on people's attitudes toward robots. Bartneck *et al.* [41] report that Americans are more positive in their attitudes toward robots than other cultures across Asia, Europe, and North America. Differences are also found between European groups, in which French-speaking people accept a human-like robot more than Germans do [31].

21.6.2.3 Roles and Needs People have different attitudes toward robots while they are playing different roles. For example, robots are generally accepted by patients and their families as a powerful assistant. However, robots often produce negative attitudes among the medical and nursing staff. Wasen [42] indicates that robots sometimes make assistant surgeons feel isolated in surgery, and they are also annoyed because of the difficulty in moving the robots around. As for nurses' reactions, many of them are distrustful of the technology, worrying that their job security is threatened [43], as well as feeling stressed while working with a robot [19].

21.6.3 SUMMARY

From the robot's perspective, since a telepresence robot is intended to serve as an avatar of a real human, it is expected to look and act as human beings do. Besides, in order to meet a local user's expectations, it would be better to make the robot's physical characteristics or personality match its own role or those of the remote controller. From the user's perspective, human-centered design is definitely critical to the success of a telepresence robot. Once the real demands can be explored and realized, telepresence robots will eventually enter our lives as new roles for modern medicine and homecare.

21.7 CONCLUSIONS

Robots have greatly changed people's lives by contributing to the advancement of modern medicine and healthcare. Telepresence especially assists in realizing the remote medicine and healthcare with realistic senses and feedbacks. Unlike autonomous robots, telepresence robots are never expected to totally take over the medical and nursing staff's jobs. Instead, they cooperate with the professionals toward a higher quality of life without borders of time and distance. From this point of view, the technological challenges for telepresence robots can be easier to be overcome. Nevertheless, no matter how technologies are accelerating the development of robotics, the most important thing is to meet the real demands of human beings. From both robot and human perspectives, the principle of using telepresence robots is to rebuild the face-to-face experiences among people in medical treatment or homecare services. Thus, the ease of use becomes a fundamental issue in the future development. Besides, customized considerations can further improve the acceptance for a wide range of users. Furthermore, it is also necessary for the providers of telepresence robots to clearly identify the user scenarios and develop adequate business models. This will be critical for the survival of such applications and services.

REFERENCES

[1] Robotic Institute of America. *RIA Worldwide Robotics Survey and Directory*, Dearborn, MI: Robotic Institute of America, 1979.

[2] International Federation of Robotics. United Nations/Economic Commission for Europe, World Robotics 1998, Geneva, 1998.

[3] Y. S. Kwoh, J. Hou, E. A. Jonckheere, and S. Hayall. A robot with improved absolute positioning accuracy for CT guided stereotactic brain surgery. *IEEE Transactions of Biomedical Engineering*, 35(2): 153–161, 1988.

[4] Intuitive Surgical. da Vinci® Surgical System. Available at http://www.intuitivesurgical.com/products/davinci_surgical_system/, 2010 (accessed January 16, 2015).

[5] METI. The human patient simulator. Available at http://www.meti.com/eng/patient-simulators, 2011 (accessed January 16, 2015).

[6] InTouch Health. Products and Services: RP-7i ROBOT. Available at http://www.intouchhealth.com/products-and-services/products/rp-7i-robot/, 2011 (accessed January 16, 2015).

[7] A. Gerrard, J. Tusia, B. Pomeroy, A. Dower, and J. Gillis. On-call physician-robot dispatched to remote. Labrador, News in Health, Media Centre, Dalhousie University, Canada, 2010.

[8] M. Kaczmarski and G. Granosik. Rehabilitation robot RRH1. *Archive of Mechanical Engineering*, 57(1): 103–113, 2011.

[9] H. Satoh, T. Kawabata, and Y. Sankai. Bathing care assistance with robot suit HAL. In: Proceedings of IEEE International Conference on Robotics and Biomimetics, Guilin, China, 2009, pp. 498–503.

[10] J. Pineau and A. Atrash. SmartWheeler: a robotic wheelchair test-bed for investigating new models of human–robot interaction. In: AAAI Spring Symposium on Multidisciplinary Collaboration for Socially Assistive Robotics, 2007, pp. 59–64.

[11] K. Wada, T. Shibata, T. Musha, and S. Kimura. Robot therapy for elders affected by dementia. *IEEE Engineering in Medicine and Biology Magazine*, 27(4): 53–60, 2008.

[12] S. Brière, P. Boissy, and F. Michaud. In-home telehealth clinical interaction using a robot. In: Proceedings of the 4th ACM/IEEE international conference on Human robot interaction, La Jolla, CA, 2009.

[13] R. C. Luo, C.T. Chen, and Y. J. Pu. Internet based remote supervisory system for telemedicine robotics application. In: Proceedings of IEEE Workshop on Advanced Robotics and its Social Impacts, Tokyo, Japan, 2009, pp. 153–158.

[14] iRobot Corporation. iRobot fot the home. Available at http://www.irobot.com/For-the-Home.aspx, 2011 (accessed January 16, 2015).

[15] J. M. Lu, C. H. Lu, Y. W. Chen, J. A. Wang, and Y. L. Hsu. TRiCmini—a telepresence robot towards enriched quality of life of the elderly. In: Proceedings of the 1st Asia Pacific eCare and TeleCare Congress, Hong Kong, China, 2011, vol. 18.

[16] F. Michaud, P. Boissy, D. Labonté, H. Corriveau, A. Grant, M. Lauria, R. Cloutier, M. A. Roux, D. Iannuzzi, and M. P. Royer. A telementoring robot for home care. In: Proceedings of the 2nd International Conference on Technology and Aging, Toronto, Canada, 2007, vol. 21, pp. 138–145.

[17] T. Mouri, H. Kawasaki, T. Aoki, Y. Nishimoto, S. Ito, and S. Ueki. Telerehabilitation for fingers and wrist using a hand rehabilitation support system and robot hand. In: Proceedings of the 9th International Symposium on Robot Control (SYROCO'09), Gifu, Japan, 2009, pp. 751–756.

[18] P. Böhm and T. Gruber. A novel HAZOP study approach in the RAMS analysis of a therapeutic robot for disabled children. In: Proceedings of the 29th International Conference on Computer Safety, Reliability and Security, Vienna, Austria, 2010, pp. 15–27.

[19] T. Saito, T. Shibata, K. Wada, and K. Tanie. Relationship between interaction with the mental commit robot and change of stress reaction of the elderly. In: Proceedings of the 2003 IEEE International Symposium on Computational Intelligence and Automation, Kobe, Japan, 2003, pp. 119–124.

[20] K. Wada and T. Shibata. Robot therapy in a care house—change of relationship among the residents and seal robot during a 2-month long study. In: Proceedings of the 16th IEEE International Symposium on Robot and Human Interactive Communication RO-MAN, Jeju Island, Korea, 2007, pp. 107–112.

[21] A. Helal and B. Abdulrazak. TeCaRob: tele-care using telepresence and robotic technology for assisting people with special needs. *International Journal of Assistive Robotics and Mechatronics*, 7(3): 46–53, 2006.

[22] T. Mukai, M. Onishi, T. Odashima, S. Hirano, and L. Zhiwei. Development of the tactile sensor system of a human-interactive robot "RI-MAN". *IEEE Transactions on Robotics*, 24(2): 505–512, 2008.

[23] A. Powers and S. Kiesler. The advisor robot: tracing people's mental model from a robot's physical attributes. In: Proceedings of the 1st ACM SIGCHI/SIGART Conference on Human–Robot Interaction, Salt Lake City, UT, 2006, pp. 218–225.

[24] T. C. Tsai, Y. L. Hsu, A. I. Ma, T. King, and C. H. Wu. Developing a telepresence robot for interpersonal communication with the elderly in a home environment. *Telemedicine and e-Health*, 13(4): 407–424, 2006.

[25] C. L. Breazeal. Sociable machines: expressive social exchange between humans and robots. Doctor Thesis, Massachusetts Institute of Technology, 2000.

[26] Intelligent Hospital Systems. Product—RIVA (Robotic IV Automation System). Available at http://www.intelligenthospitals.com/product.php, 2011 (accessed January 16, 2015).

[27] A. Tsoli and O. C. Jenkins. Robot grasping for prosthetic applications. In: *Robotics Research, STAR66*, edited by M. Kaneko and Y. Nakamura. Berlin/Heidelberg: Springer, 2010, pp. 1–12.

[28] E. Broadbent, R. Stafford, and B. MacDonald. Acceptance of healthcare robots for the older population: review and future directions. *International Journal of Social Robotics*, 1: 319–330, 2009.

[29] D. Li, P. L. Rau, and Y. Li. A cross-cultural study: effect of robot appearance and task. *International Journal of Social Robotics*, 2(2): 175–186, 2010.

[30] J. Goetz, S. Kiesler, and A. Powers. Matching robot appearance and behavior to tasks to improve human–robot cooperation. In: The 12th IEEE International Workshop on Robot and Human Interactive Communication (RO-MAN 2003), Millbrae, CA, 2003, vol. IXX, pp. 55–60.

[31] K. O. Arras and D. Cerqui. Do we want to share our lives and bodies with robots? A 2000-people survey. Technical Report 0605-001. Swiss Federal Institute of Technology (EPFL), Lausanne, Switzerland, 2005.

[32] P. J. Hinds, T. L. Roberts, and H. Jones. Whose job is it anyway? A study of human-robot interaction in a collaborative task. *Human–Computer Interaction*, 19(1–2): 151–181, 2004.

[33] A. Powers, A. D. I. Kramer, S. Lim, J. Kuo, S. Lee, and S. Kiesler. Eliciting information from people with a gendered humanoid robot. In: Proceedings of the 2005 IEEE

International Workshop on Robots and Human Interactive Communication, Nashville, TN, 2005, pp. 158–163.

[34] M. V. Giuliani, M. Scopelliti, and F. Fornara. Elderly people at home: technological help in everyday activities. In: Proceedings of the 2005 IEEE International Workshop on Robots and Human Interactive Communication, Nashville, TN, 2005, pp. 365–370.

[35] T. W. Bickmore and R. W. Picard. Towards caring machines. In: CHI'04 Extended Abstracts on Human Factors in Computing Systems, Vienna, Austria, 2004, pp. 1489–1492.

[36] M. Heerink, B. Krose, V. Evers, and B. Wielinga. The influence of a robot's social abilities on acceptance by elderly users. In: Proceedings of the 15th IEEE International Symposium on Robot and Human Interactive Communication RO-MAN'06, Hateld, UK, 2006, pp. 521–526.

[37] A. Ţăpuş, C. Ţăpuş, and M. J. Matarić. Hands-off therapist robot behavior adaptation to user personality for post-stroke rehabilitation therapy. In: Proceedings of the IEEE International Conference on Robotics and Automation, Rome, Italy, 2007, pp. 1547–1553.

[38] W. Taggart, S. Turkle, and C. D. Kidd. An interactive robot in a nursing home: preliminary remarks. In: Proceedings of Toward Social Mechanisms of Android Science, A Cognitive Science Society Workshop, Stresa, Italy, 2005, pp. 56–61.

[39] T. Nomura, T. Kanda, T. Suzuki, and K. Kato. Prediction of human behavior in human–robot interaction using psychological scales for anxiety and negative attitudes toward robots. *IEEE Transactions on Robotics*, 24: 442–451, 2008.

[40] M. I. Dijkers, P. C. deBear, R. F. Erlandson, K. Kristy, D. M. Geer, and A. Nichols. Patient and staff acceptance of robotic technology in occupational therapy: a pilot study. *Journal of Rehabilitation Research & Development*, 28: 33–44, 1991.

[41] C. Bartneck, T. Suzuki, T. Kanda, and T. Nomura. The influence of people's culture and prior experiences with Aibo on their attitude towards robots. *AI and Society*, 21: 217–230, 2006.

[42] K. Wasen. Person-friendly robot interaction: social, psychological and technical impacts in health care work. In: Proceedings of the 2005 IEEE International Workshop on Robots and Human Interactive Communication, Nashville, TN, 2005, pp. 643–648.

[43] J. Novek, S. Bettes, K. Burke, and P. Johnston. Nurses' perceptions of the reliability of an automated medication dispensing system. *Journal of Nursing Care Quality*, 14: 1–13, 2000.

CHAPTER 22

ADVANCES IN CLIMBING ROBOTS

JIZHONG XIAO[1,2] and HONGGUANG WANG[2]
[1]The City College, City University of New York, New York, USA
[2]Shenyang Institute of Automation, Chinese Academy of Sciences, Shenyang, PR China

22.1 INTRODUCTION

An increasing interest in the development of special climbing robots was witnessed in the last decades. Motivations are typically to increase the operation efficiency in dangerous environments or difficult-to-access places, and to protect human health and safety in hazardous tasks. Climbing robots with the ability to maneuver on vertical surfaces have been strongly requested by various industries in order to perform dangerous operations such as inspection of high-rise buildings, spray painting and sand blasting of gas tanks, maintenance of nuclear facilities, aircraft inspection, surveillance and reconnaissance in counter-terrorism and urban warfare, and assistance in fire fighting and search and rescue operations. Such capabilities of climbing robots would not only allow them to replace human workers in those dangerous duties but also eliminate costly scaffolding.

Since Nishi developed the first climbing robot [1, 2] in Japan in the late 1980s, many research projects on climbing robots have been carried out around the world. For example, CLAWAR (Climbing and Walking Robots) association, the oldest Robotics Network of Excellence funded by European Union [3] over the period of 1997 to 2005, was established to promote the advancement of climbing and walking robot technologies. Many impressive results that enhance the mobility and working capability of climbing robots were achieved over the time. More recently, the *RiSE* (Robotics in Scansorial Environments) consortium in the United States has carried out a large multidisciplinary effort to develop a family of biologically inspired climbing robots capable of agile maneuvers in vertical terrestrial settings, resulting in very successful Stickybot and Spinybot [4, 5] produced by Stanford University. In China,

Contemporary Issues in Systems Science and Engineering, First Edition.
Edited by MengChu Zhou, Han-Xiong Li and Margot Weijnen.

(a) (b) (c)

FIGURE 22.1 Some typical application scenarios of climbing robots: (a) inspection of a storage tank; (b) solar panel cleaning robot (Reprinted with permission from Serbot AG. Copyright © 2014); and (c) surveillance robot.

the national "Hi-Tech Research and Development Program of China" has supported many research projects on climbing robots since 1990s, resulting in some impressive glass-wall cleaning robots [6, 7] developed by Beihang University, and miniature wall-climbing robots [8] for surveillance by Harbin Institute of Technology.

The climbing robots have been used for different types of applications mainly in the following areas: (1) Inspection and maintenance in *hazardous* environments. The chemical and nuclear industries are the two primary fields where climbing robots are expected to perform remote inspection and maintenance tasks in highly contaminated areas or radioactive environments. Application examples include the retrieval of irradiated material samples [9], the inspection of storage tanks [10, 11], the handling and manipulation of nuclear fuel [12, 13]. (2) Welding and manipulation in construction and shipbuilding industries, where the climbing robots are designed for the painting or sanding of ship hulls, the inspection of the metallic skeletons of bridges [14]. (3) Cleaning and inspection for high-rise buildings in the service industry [15, 16]. (4) Urban search and rescue in military and civilian applications [17–19]. Figure 22.1 shows some typical application scenarios where the climbing robots are used for (1) inspection of a storage tank, (2) cleaning of solar panels, and (3) surveillance and reconnaissance. Employing climbing robots in these scenarios will relieve human workers from dangerous tasks, make automatic inspection and manipulation possible, and increase the productivity.

22.2 TECHNOLOGIES FOR ADHERING TO SURFACES

To successfully accomplish a specific task, the climbing robots should have reliable adhesion, maneuverability, high payload/weight ratio, and adaptability on a variety of wall surfaces. The most challenging work in climbing robot design is the adhesion method to ensure that the robots grip the target surface firmly without sacrifice of maneuverability. So far, researchers around the world have developed adhesion mechanisms that mainly fall in six categories: (1) magnetic adhesion devices for climbing ferrous surfaces; (2) vacuum suction techniques for operation on smooth and

FIGURE 22.2 A wall-climbing robot with permanent magnet adhesion [11].

nonporous surfaces; (3) attraction force generators based on aerodynamic principles; (4) grippers; (5) bio-mimetic approaches inspired by climbing animals; and (6) other emerging techniques such as compliant electroadhesion. This section presents the six existing adhering technologies with emphasis on latter four categories which were the most recent advancement in last several years.

22.2.1 Magnetic Adhesion

Magnetic adhesion devices are the most promising attachment mechanism for robots navigating on steel structures. Robots using permanent magnets or electromagnets are presented in References 10, 11, 20, and 21 for climbing large steel structures and in References 22 and 23 for internal inspection of iron pipes. Usually, magnetic adhesion combines with round wheels [22], tracked wheels [10, 11, 21, 24], or legs [20, 23] to move on steel surfaces. Figure 22.2 shows a representative tracked wheel wall-climbing robot with permanent magnet adhesion mechanism designed by Dalhousie University, Canada, for inspecting oil tanks with nondestructive sensors. Each track consists of a roller chain, two sprockets, and some evenly placed permanent magnetic units. When the robot moves, there are always a certain number of units in good contact with wall surface, which enables reliable adhesion while carrying a payload of over 30 kg.

The most salient advantage of magnet-based climbing robots is their energy efficiency—it requires little or no additional energy to adhere on ferrous surfaces. However, their applications are limited to steel walls due to the nature of magnets. One innovative variance to this limitation is the Windoro robot [25]—a commercial window-cleaning robot from south Korea unveiled in 2010 as shown in Figure 22.3. The robot uses two modules that sandwich the glass by permanent magnets. One module works as the navigation unit and the other module is the cleaning unit, which has four spinning microfiber pads and a reservoir that dispenses detergent. The robot is designed for cleaning windows 6 to 25 mm thick at homes and stores, but not suitable for cleaning glass panels on skyscrapers.

FIGURE 22.3 Windoro: window-cleaning robot using magnets.

22.2.2 Vacuum Suction Techniques

In applications on nonferromagnetic wall surfaces, climbing robots most generally use vacuum suctions to produce the adhesion force on smooth and nonporous surfaces. The vacuum can be generated through the Venturi principle, or a vacuum pump that is either on-board or external to the robot, or passive suctions [26, 27].

A Venturi tube [28], illustrated in Figure 22.4, consists of convergent entrance, cylindrical throat, and divergent outlet. When high-speed air flows into this tube, its shape causes the air to speed up and creates a vacuum. This low-pressure area will cause air to be drawn through any instruments that require a vacuum or partial vacuum, which can produce the adhesion force for a climbing robot. The Venturi tube is one of the easiest and not expensive apparatus for climbing robot since it has no moving parts. University of Portmouth, UK, developed a Bigfoot climbing robot based on a Venturi tube [29], which is able to operate on concrete wall surfaces and overcome steps of 2 cm height. The main problem for this robot is how to balance sealing effect and mobility.

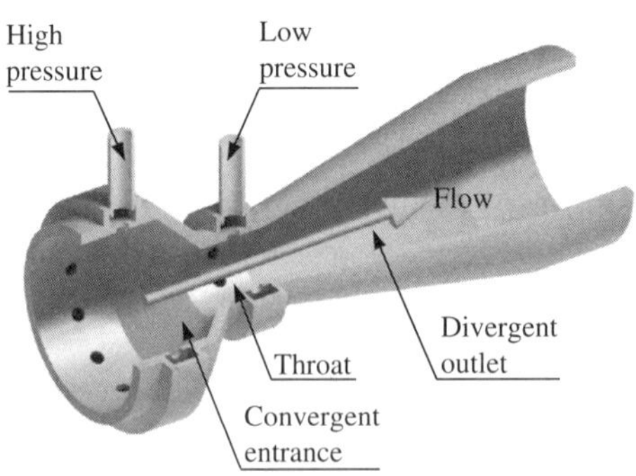

FIGURE 22.4 The Venturi tube principle.

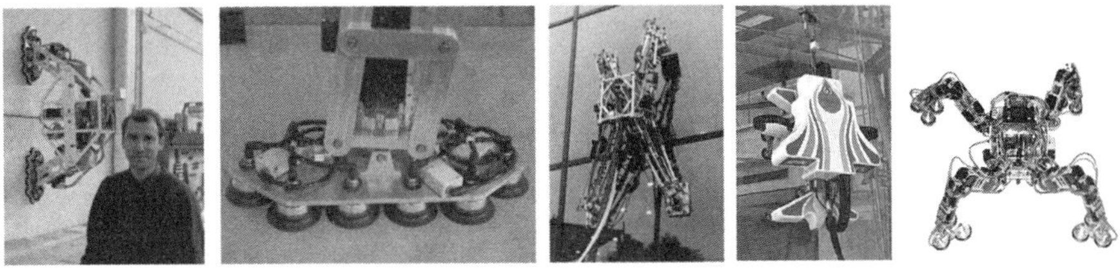

(a) (b) (c) (d) (e)

FIGURE 22.5 Climbing robots based on pneumatic actuators and vacuum suction cups: (a) ROMA II; (b) vacuum cups of ROMA II; (c) NINJA-I; (d) Sky Cleaner 3; and (e) MRWALLSPECT-III.

The early works on climbing robots generally use vacuum suction and a pneumatic pump to produce a strong adhesive force. The robots adopt biped, quadruped, or sliding frames for locomotion and are normally bulky. Figure 22.5a shows the ROMA II, which is a pneumatic climbing robot designed at University Carlos III of Madrid, Spain, for inspection and maintenance tasks in complex 3D environments [14, 30]. It has two legs and a pneumatically driven grasping mechanism with 10 vacuum cups, which are connected in pairs as shown in Figure 22.5b. The vacuum system is able to produce a strong grasping force that supports an overall weight of 20 kg. NINJA-I is a quadruped climbing robot developed at Tokyo Institute of Technology, Japan [31, 32], whose foot is a valve-regulated multiple sucker. Figure 22.5d shows the Sky Cleaner 3 that is designed by Beihang University, China, to clean glass-wall [6, 7] of the Shanghai Science and Technology Museum. Sky Cleaner 3 is a pneumatic climbing robot employed with 14 suction pads and sliding-frame locomotion that can carry a payload of approximately 60 kg with its body weight (45 kg). The system uses a supporting vehicle that is stationed on the ground and provides electricity, air, and cleaning liquid for glass-wall cleaning task. Sungkyunkwan University, Korea, developed a quadruped walking and climbing robot MRWALLSPECT III, as shown in Figure 22.5e. This robot has better ability to overcome various obstacles than most of previous quadruped wall-climbing robots because of its terrain adaptability [33]. Some other examples of vacuum suction– based robots include the ROBUG series climbing robots developed [34, 35] at University of Portsmouth, UK; and ROBIN [36] at Vanderbilt University, USA. Besides those robots built in academic institutes, some robots have been put into practical use. For example, MACS robots [37] at the Jet Propulsion Laboratory (JPL) use suction cups for surface adherence when inspecting the exterior of large military aircraft; Robicen robots [38] use pneumatic actuators and suction pads for remote inspection in nuclear power plants; and SADIE robots [39] use a sliding-frame mechanism and vacuum gripper feet for the weld inspection of gas duct internals at nuclear power stations.

In the past decade, there was a strong need to develop miniature climbing robots for inspection, surveillance, and reconnaissance purposes. Researchers at Michigan State University have designed two types of light-weight, self-contained miniature climbing robots using bipedal articulated structure and active suction. The robots

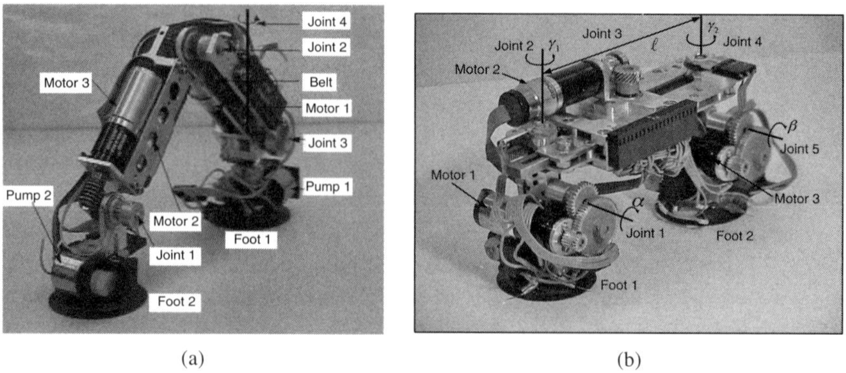

FIGURE 22.6 Miniature climbing robots with active suction feet: (a) FLIPPER and (b) CRAWLER.

are called FLIPPER and CRAWLER, with two suction feet located at the robot's extremities as shown in Figure 22.6. The foot weighs only 35 grams with a 40 mm diameter suction cup and uses a micro-pump to generate active suction [40]. The FLIPPER is a biped robot and its middle joint is of revolute type. The motion of the FLIPPER is achieved by sticking one foot on the surface and "flipping" its body as shown in Figure 22.6a. The CRAWLER robot achieves linear motion through extension and contraction of the robot legs. It is used to travel through inside pipes. The size and weight of the robot are minimized through under-actuated mechanism, wherein five joints are driven by only three motors. The dimension of the robot is approximately 80 mm in height and 50 mm in width. It weighs about 450 grams.

Aforementioned climbing robots are legged robots using suction cups for adhesion. Seoul National University developed a tracked climbing robot [41] with 24 suction pads installed on each of the timing belt, achieving a locomotion speed of 15 m/min as shown in Figure 22.7. City University of Hong Kong, China, also developed a tracked wheel climbing robot with similar functions [42].

An alternative way to realizing climbing without a vacuum pump is to use "passive suction cups" that are made of elastic material and are evacuated simply by pressing them to the surface. The cups are normally released by pulling them away from the surface through kinematic structure. Since the vacuum is generated only by utilizing the robot's locomotive system, no energy is consumed for adhesion. The climbing robot DEXTER (DEXTerous Electric Rover) [43] developed at University of Lubeck, Germany, is a practical example. A Swiss company, Serbot AG, produced a commercial GEKKO robot [44] for the cleaning of glass facade or solar panels using multiple suction cups as shown in Figures 22.8a and 22.8b. The innovation is based on the suction feet that rotate on two trapezoid-shaped geared belt drives, enabling the robot to push and peel off cups and achieve flexible movement in every chosen direction.

The great defect of the suction foot is that even a very small gap between the suction cup and the surface may cause the robot to drop off the wall, thereby it

FIGURE 22.7 Tracked wheel climbing robot developed by Seoul National University. (Reprinted with permission from Springer Science and Business Media from Reference 42)

only works well on flat and smooth surfaces such as the glass walls and ceramic tiles. Some researchers have attempted to overcome this problem. A wall-climbing robot with scanning type suction cups is reported in Reference 45. In recent years, Beihang University, China, has developed a robotic foot based on vibration suction as shown in Figure 22.9 [46] and demonstrated that the regular vibration can improve the suction cup's adsorption capacity. Multiple suction cups increase the robustness of the adhesion mechanism, as any single failure is compensated by adhesion force generated by other suction cups. Besides, Kawagwa University, Japan, has presented a wheeled climbing robot with a vacuum-based wet adhesion system, which allows low-friction performance when a suction cup adheres on rough surface as shown in Figure 22.10 [47]. Other examples of climbing robots using vacuum suction techniques can be found in References 48–53.

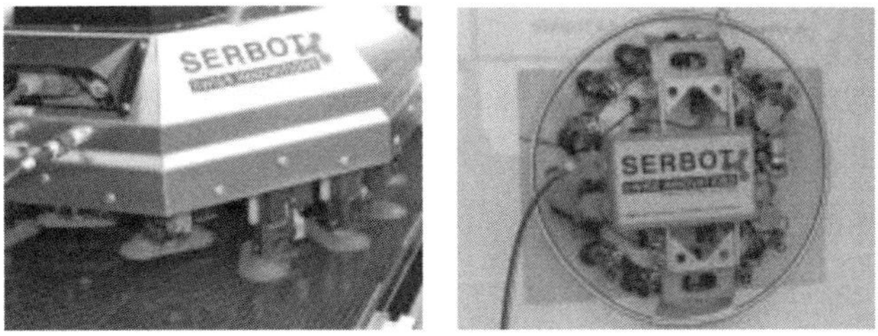

(a) (b)

FIGURE 22.8 Climbing robots with passive suction feet: (a) GEKKO Junior, *side view*, developed by Serbot AG (Swiss Innovations); and (b) GEKKO Junior, *bottom view*. (Reprinted with permission from Serbot AG. Copyright © 2014)

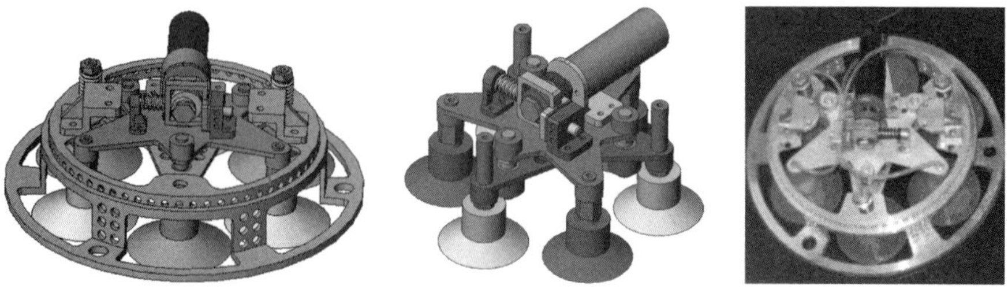

(a) (b) (c)

FIGURE 22.9 Robotic foot based on vibration suction: (a) CAD model; (b) assemble of air-relief mechanism and vibration mechanism; and (c) real prototype. (Reprinted with permission from Elsevier from Reference 46)

The vacuum suction-based adhesion mechanism is easy to control and has good payload/weight ratio. It has been widely adopted in climbing robots for operating on smooth surfaces. However, the major limitation of the vacuum suction is that it requires perfect sealing of the chamber, and thus it is not suitable for nonsmooth surfaces such as brick walls with cracks. Any gap in the seal will cause the vacuum to break down in time, resulting in unreliable functioning of the adhesion mechanism, or in the worst case, causing the robot to drop off the wall. Second, the suction mechanism requires time to evacuate the cup to build up enough adhesion force or to release suction for peeling off the cup from the surface. This reduces the locomotion speed of a climbing robot. Third, the vacuum suction cannot be used in outer space because there is no air to generate a suction force.

22.2.3 Aerodynamic Attraction

The third choice is to create adhesion force based on aerodynamic principles including the use of propeller [54–56], and controlled vortex to generate negative pressure for aerodynamic attraction [18, 19]. The negative pressure chamber is not sensitive to

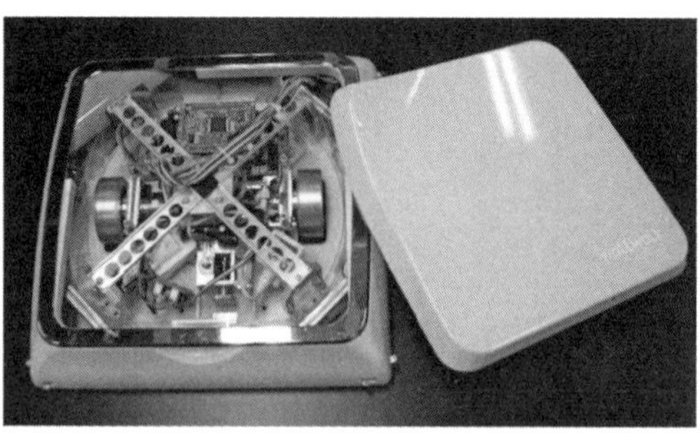

FIGURE 22.10 WallWalker, a wheeled climbing robot with vacuum-based wet adhesion.

(a) (b)

FIGURE 22.11 (a) TITO robot hanged by a crane for wall-cleaning operation and (b) an industrial fan is used in TITO to generate propulsive force for surface attachment. (Reprinted with permission from Reference 56. Copyright © 2009 Emerald Group Publishing Limited)

air leakage, but requires a good balance between adhesion and mobility to ensure the reliable attachment of the climbing robots to rough surfaces.

The early work by Nishi uses propellers to produce the thrust force and press the climbing robot onto the wall surface [54, 55]. TITO, a propulsion-based wall-cleaning robot, is a recent implementation of the technology [56] as shown in Figure 22.11. TITO has adopted a powerful fan to create a propulsive force for the surface attachment and attained the ability to clean a building facade of 10,000 m^2 in 12 hours.

The City College of New York (CCNY) has developed three generations of wall-climbing robot named City-Climber (Figures 22.12, 22.13, 22.14). The adhesion mechanism adopted in City-Climbers is based on the aerodynamic attraction produced by a vacuum rotor package, which generates a low-pressure zone enclosed by a chamber. The technology takes advantage of the merits of both the vacuum suction

(a) **(b)** **(c)**

FIGURE 22.12 (a) Vacuum rotor package to generate aerodynamic attraction; (b) exploded view of City-Climber prototype-I; and (c) City-Climber prototype-I with on-board camera, operating on brick wall.

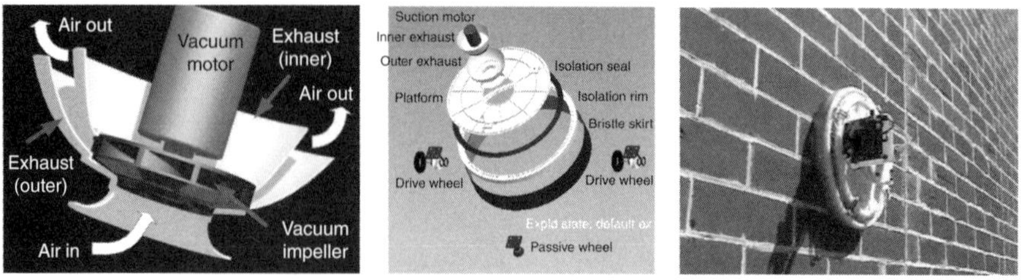

FIGURE 22.13 (a) Exploded view of City-Climber prototype-II; (b) two climbing robot modules connected by a hinge being able to make ground-to-wall, and wall-to-ceiling transitions; and (c) the City-Climber prototype-II rests on a brick wall.

and vortex technology, thus making good balance between strong adhesion force and high mobility. As the City-Climber robots do not require perfect sealing as the vacuum suction technique does, they can move on essentially any kinds of smooth or rough surfaces.

As shown in Figure 22.12a, the vacuum rotor package consists of a vacuum motor with impeller and exhaust cowling to direct air flow. It is essentially a radial flow device that combines two types of air flow. The high-speed rotation of the impeller causes the air to be accelerated toward the outer perimeter of the rotor, away from the center radically. Air is then pulled along the spin axis toward the device creating a low-pressure region, or partial vacuum region if sealed adequately, in front of the device. With the exhaust cowling, the resultant exhaust of air is directed toward the rear of the device, actually helping to increase the adhesion force by thrusting the device forward. Figure 22.12b shows the exploded view of the City-Climber prototype-I that consists of the vacuum rotor package, an isolation rim, a vacuum chamber with flexible bristle skirt seal, and internal three-wheel drive. The entire bristle surface is covered in a thin sheet of plastic to keep a good sealing, while the flexing of bristle allows the device to slide on rough surfaces. A pressure force isolation rim connecting the platform and the bristle skirt seal is made of re-foam. The rim improves the robot mobility, and also enhances sealing by reducing the deformation of the skirt. The

(a) (b)

FIGURE 22.14 City-Climber prototype-III: (a) one module resting on a stone wall and (b) two modules making ground-to-wall transition.

FIGURE 22.15 CROMSCI robot, *bottom view* of the construction drawing. (Reprinted with permission from Reference 62. Copyright © 2008 World Scientific Publishing Company)

driving system and the payload are mounted on the platform, thus the re-foam makes the skirt and the robot system adaptable to the curve of rough surfaces. Technical details can be found in References 18 and 19. The City-Climber prototype-II adopts the modular design that combines wheeled locomotion and articulated structure to achieve both quick motion of an individual module on planar surfaces and smooth wall-to-wall transition by a set of two modules as illustrated in Figure 22.13. The most important improvements in City-Climber prototype-III are the redesign of transition mechanism and the adoption of a six-wheel driving system to increase the contact friction and avoid wheel slippage while climbing vertical walls. Note that the wheels are outside of the robot frame, making it possible for each module to make ground-to-wall transition with ease. The two modules are closely coupled to reduce the torque required to lift up other module, as shown in Figure 22.14. Other variations of climbing robots using the same adhesion principle as that of City-Climber can be found in References 17 and 57–61.

University of Kaiserslautern, Germany, has produced a climbing robot prototype CROMSCI featuring the same attachment principle with an improved adhesion mechanism that is able to inspect concrete walls with cracks [62, 63]. The round shape with a complete diameter of 80 cm is divided into seven single vacuum chambers that are supported by one large reservoir chamber at the top of the robot as shown in Figures 22.15 and 22.16. Each chamber receives its negative pressure from the reservoir which is evacuated by a suction engine. The air pressure sensors in each chamber and the reservoir provide information to a close-loop controller that opens and closes valves to evacuate the chambers separately depending on the actual leak tightness. If one or more working chambers are losing negative pressure, they can be isolated from the vacuum system by closing the valve to avoid the propagation of normal pressure to the other chambers. The movement of the CROMSCI robot is driven by three steerable harmonic-driven wheels. A movable manipulator is installed on an outer ring to carry inspection sensors. The overall weight of the system is about 50 kg with an additional payload of up to 10 kg.

FIGURE 22.16 CROMSCI robot, real prototype adhered to a vertical wall. (Reprinted with permission from Reference 62, Copyright © 2008 World Scientific Publishing Company)

The climbing robots based on negative pressure and areodynamic attraction have demonstrated the capability of moving on brick and concrete walls with considerable success. They are not sensitive to air leakage and do not require perfect sealing as the vacuum suction technique does. However, the major disadvantage of the areodynamic attraction method is the requirement of constant rotation of electric motor to maintain negative pressure no matter the robots are moving or stationary, which increases power consumption significantly. The noise caused by the propeller or imperfect assembly of the suction engine is another issue to be addressed for some surveillance tasks. The adhesion method cannot work in outer space either.

22.2.4 Grasping Grippers

The three aforementioned adhesion mechanisms enable the climbing robots to move on large flat walls or ceilings. For operation on irregular curved surfaces or trusswork, for example, tree trunks, construction scaffolds, trussed frames of bridges, towers and space structures, many climbing robots are developed by the use of grippers.

Ilmenau University of Technology, Germany, has designed a modular climbing robot named RAUPI as shown in Figure 22.17a. It can climb on a pipe-like structure and transit between bars. Grippers of RAUPI are coated by a special polymer. It can produce a friction coefficient of up to $\mu = 1.6$ between gripper and a wooden surface [64]. University of Coimbra, Portugal, has developed a pole-climbing robot named 3D CLIMBER with the capability of climbing and manipulating across 3D structures as shown in Figure 22.17b. The climbing module consists of a 3-DOF planar serial arm and a *Z*-axis rotating mechanism. Each gripper includes two unique multi-fingered V-shaped bodies that can grasp circular profiles with diameters ranging from 200 to 350 mm. The 3D CLIMBER can overcome bends, T-junctions, flanges, and sharp changes on the pole's diameter [65]. The Chinese University of

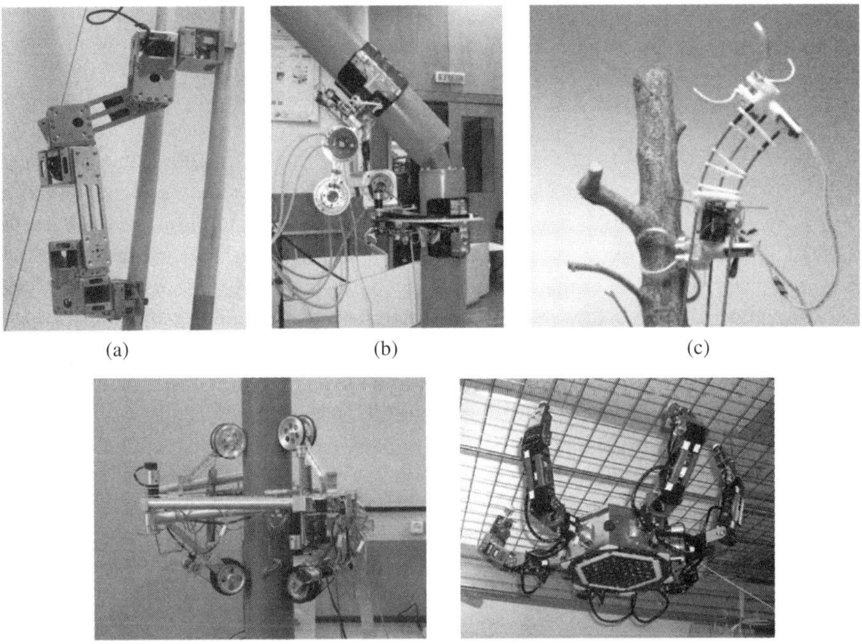

(a) (b) (c) (d) (e)

FIGURE 22.17 (a) RAUPI; (b) 3D CLIMBER (Reprinted with permission from Reference 65. Copyright © 2011 Elsevier); (c) TREEBOT; (d) UT-PCR robot; and (e) ASTERISK robot hanging from grid-like structure.

Hong Kong has developed a novel tree-climbing robot named TREEBOT as shown in Figure 22.17c. TREEBOT adopts an extensible continuum maneuver body and a pair of omnidirectional grippers that enable the robot to grasp tightly on large tree trunks and small branches in a wide range of gripping curvature and maneuver on a complex tree environment. TREEBOT is compact and light weighted and has payload capability of 1.75 kg, which is nearly three times of its own weight [66]. University of Tehran, Iran, has developed a wheel-based climbing robot to climb up tubular structure, which is named as UT-PCR (pole-climbing robot) and shown in Figure 22.17d [67]. Osaka University, Japan, has developed a climbing robot ASTERISK with six limbs as shown in Figure 22.17e. ASTERISK can realize the following operations: omnidirectional gait on flat and irregular terrains, climbing stairs, passing through narrow tunnels, and manipulating objects using two limbs [68]. The work in References 69–74 introduces other climbing robots by using grasping grippers.

22.2.5 Bio-Mimetic Approaches Inspired by Climbing Animals

In the past decade, many researchers focused on investigating the sticking abilities of climbing animals such as geckos, beetles, and snails. Significant progress has been made to mimic the behavior of these climbing animals. Inspired by these animals' adhesion mechanism, two types of adhesion methods have recently been applied in climbing robots: gecko-inspired synthetic dry adhesives, and beetle-inspired microspines.

22.2.5.1 *Gecko-Inspired Synthetic Dry Adhesives* According to References 75, 77, and 78, geckos have the ability to climb on almost any surface, whether wet or dry, smooth or rough, which has attracted scientists' attention for decades. It has been found that geckos rely on their feet to adhere to surfaces using patches of compliant micro-scale hairs that provide a mechanism for dry adhesion through van der Waals forces. The gecko feet have very fine keratinous hairs or setae on its toes. Each seta consists of a hair that is approximately 5 μm in diameter, and atop each of these setae sit hundreds of nano-fibers (spatulae) that are 200 nm in diameter. There are billions of tiny setae that make contact with the surface and generate a significant collective surface area of contact. The dry adhesion comes from molecular forces such as van der Waals forces that act between all materials in contact.

The Nanorobotics Lab at Carnegie Mellon University (CMU), USA, has put tremendous effort for producing synthetic gecko foot hairs with nano/micro hair hierarchy for dry adhesion and has achieved commendable success. Figures 22.18a and 22.18b show two-level synthetic polymer microfibers with controlled geometry

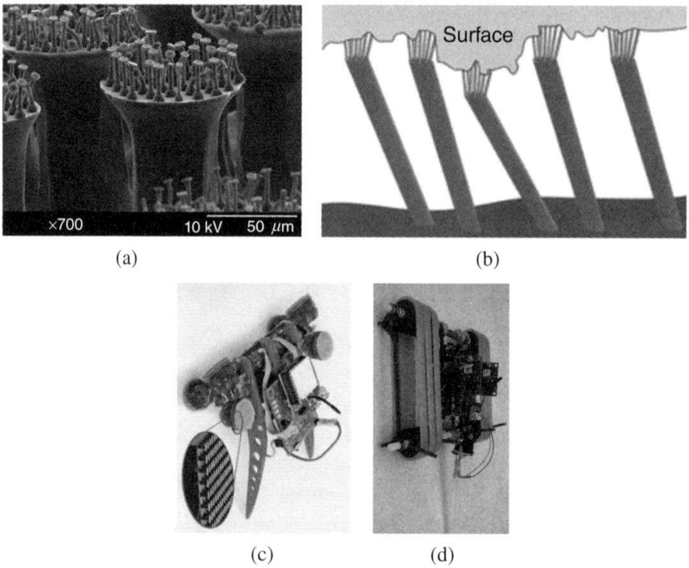

FIGURE 22.18 Dry adhesion and climbing robot prototypes at CMU. (a) Synthetic angled fibers with various tip geometry; (b) hierarchical fibers can conform to various surface roughnesses (Reprinted with permission from Reference 78. Copyright © 2009 American Chemical Society); (c) Waalbot-6 with 50 μm diameter fibrillar adhesive footpads; and (d) Tankbot using a tacky polymer as treads.

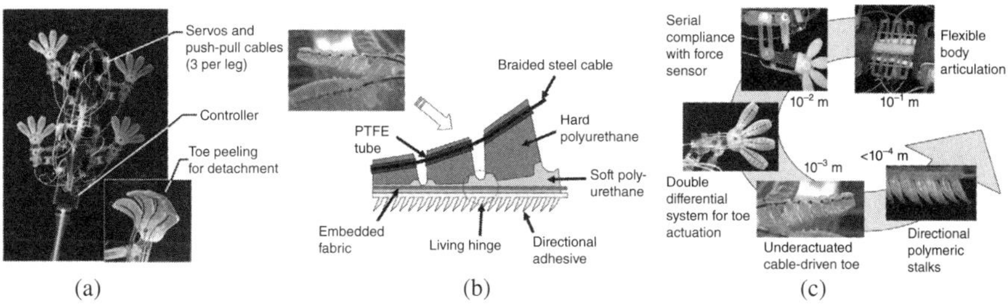

FIGURE 22.19 Stickybot, a novel bio-inspired robot capable of climbing smooth surfaces. (a) Overall structure; *inset*: details of toes curling to facilitate detachment; (b) schematic of cross-section view of Stickybot toe fabricated via shape deposition manufacturing; and (c) whole-body compliance over a range of length scale.

and saucer-like flat mushroom shapes at the tips for increased adhesion [79]. The increased compliance of the hierarchical structures allows the microfibers to conform to various surface roughnesses. Utilizing the synthetic gecko hair material for footpads, the group has produced many climbing robot prototypes such as the Waalbot, Wormbot, and Tankbot as shown in Figures 22.18c–22.18e. The tri-leg design of Waalbot in Figure 22.18c uses two motors for rotary legged motion to overcome small obstacles and low-angle surface curvatures, and includes passive joint and elastic flexures to provide the preload and peeling forces necessary for the footpads to climb using dry adhesives [80, 81, 82]. The Tankbot [83] uses tacky polymer as treads to climb a steep surface with van der Waals forces as shown in Figure 22.18d. Readers are referred to References 76, 83, 84, and 85 for details of the dry-adhesion-based climbing robots.

The most notable bio-inspired climbing robot as shown in Figure 22.19 is Stickybot developed at Stanford University, which can climb glass and other smooth surfaces using directional adhesive pads on its toes. The Stickybot combines many innovative design principles inspired from the geckos including directional adhesion, a hierarchy of compliant structures, and control of tangential contact forces [5, 86]. As shown in Figure 22.19b, the Stickybot's toes are covered with arrays of small, angled polymer stalks that induce directional adhesive when pulled tangentially from the tips toward the ankles and release when pulled in the opposite direction. As illustrated in Figure 22.19c, the Stickybot employs whole body compliance that conforms at multiple levels, including flexible body articulation, a double differential system for toe actuation, under-actuated cable-driven toe, and directional polymeric stalks. It also employs force feedback control in the tangential (fore-aft) direction coupled with a grasp-space stiffness controller to balance forces among the feet and achieve smooth engagement and disengagement of the toes. Combination of the three innovations empowers the Stickybot with the capability of climbing a variety of vertical surfaces including glass, glossy ceramic tile, acrylic, and polished granite at speeds up to 4.0 cm/s.

The gecko-inspired climbing robots using dry/directional adhesives have many advantages over the conventional climbing robots using other adhesion methods.

First, they can operate on a large variety of surfaces, for example, steel walls and nonferromagnetic surfaces, and are suitable for applications in outer space since the dry adhesive does not rely on surface materials or the atmosphere. Second, no energy is required to maintain attachment. Therefore, it does not generate noise. Third, fast locomotion speeds can be realized, because attachment of dry adhesion is nearly instantaneous as is the detachment. However, the major limitation of the dry adhesives is that the synthetic microfibers or stalk tips are made of a polyurethane that degrades with time, and can be contaminated in dirty environments or twisted on themselves, that are thereby degrading some of the adhesion performance. It is a challenging work to synthesize gecko foot hairs that are more durable and dirt-resistant, and can produce dry adhesive force strong enough for practical use, especially when large payload is required.

22.2.5.2 Beetle-Inspired Microspines Another successful bio-inspired climbing robot developed at Stanford University as part of the research effort of RiSE consortium is called Spinybot, which is based on microspines observed on beetles. Beetles and arthropods climb well on natural surfaces using legs equipped with a large number of small, sharp spines. Unlike the claws of a cat, the small spines or claws do not need to penetrate the surface. Instead, they exploit small asperities (bumps or pits) on the surfaces. A large number of spines are needed because each contact sustains only a limited force. As spines become smaller, it is possible to ascend smoother surfaces because the density of useable spine–asperity contacts increase rapidly. Spinybot as shown in Figure 22.20a is designed to climb rough surfaces such as brick and concrete and can achieve current locomotion speed of 2.3 cm/s. If encountering a very smooth surface, the robot either fails to proceed or falls. Moreover, the robot lacks sufficient degrees of freedom to negotiate corners and transitions from vertical to horizontal surfaces [4]. Figure 22.20b shows the RiSE V3 robot designed by Boston Dynamic, working with researchers at the University of

(a) (b)

FIGURE 22.20 Climbing robots using microspines. (a) Spinybot on a concrete wall and detailed view of microspines engaging asperities on the concrete surface and (b) RiSE V3 robot climbing a pole.

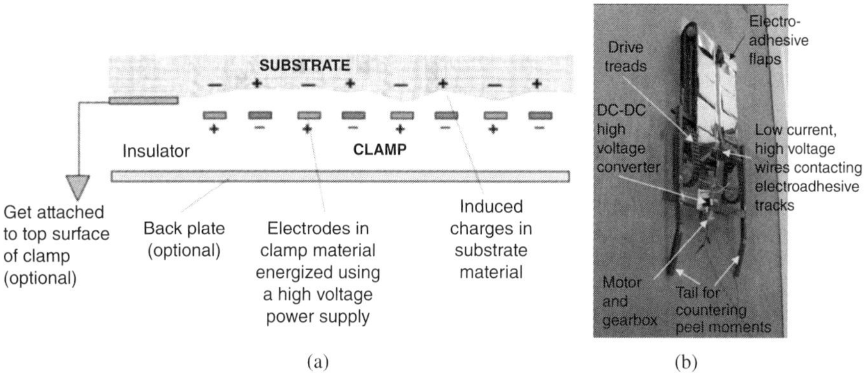

FIGURE 22.21 (a) Basic structure for an electroadhesive clamp and (b) tracked robot with electroadhesive flaps.

Pennsylvania. The quadruped robot uses claws to grip vertical surfaces and makes its way up vertical terrains at a rate of 21 cm/s. The work in References 87 and 88 introduces some other results in the microspines robot field.

22.2.6 Emerging Technologies for Climbing Robots

In recent years, some innovative ideas and emerging technologies for climbing robots have been proposed.

Stanford Research Institute (SRI) International has applied a novel clamping technology called compliant electroadhesion to climbing robots [89]. As shown in Figure 22.21, electroadhesion uses electrostatic forces between the substrate material (wall surface) and the electroadhesive pads. These pads are composed of conductive electrodes that are deposited on the surface of a polymer. When alternate positive and negative charges are induced on adjacent electrodes, the electric fields set up opposite charges on the substrate and thus cause electrostatic adhesion between the electrodes and the induced charges on the substrate. These charges do not neutralize themselves to those on the clamp because of the trapped air gaps as well as insulator material on the clamp. Electroadhesion includes inducing electrostatic charges on a wall substrate using a power supply connected to compliant pads situated on the climbingrobot. This technology produces high clamping forces of 0.2–1.4 N/cm^2 depending on different substrate. Moreover, the technology can operate on a wide variety of common building substrates, both rough and smooth, both electrically conductive and insulating. The technology consumes a very little power (about 20 mW/N) and shows the ability to repeatedly stick to wall substrates that are heavily covered by dust or other debris. Using this technology, SRI International has designed a variety of wall-climbing robots including legged and tracked robots. Figure 22.21b shows a tracked robot that is demonstrated to climb various surfaces with speeds of up to 15 cm/s [89].

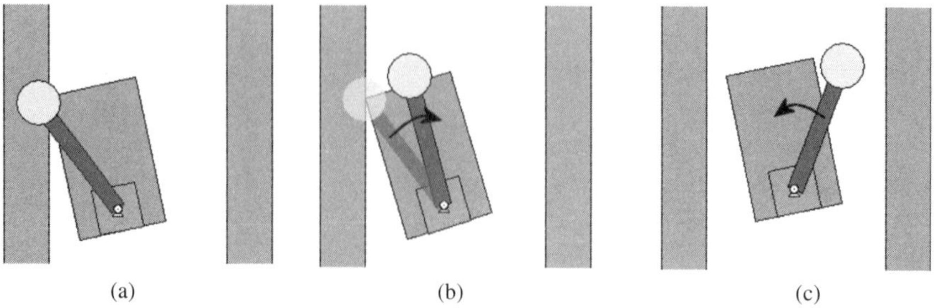

FIGURE 22.22 Single actuator dynamic climbing between two parallel walls, three phases: (a) impact; (b) stance; and (c) flight.

Researchers at CMU have proposed a climbing mechanism using a single actuator and dynamic movements to climb between two parallel vertical walls [90]. The proposed mechanism consists of two links: the first is the main body which contacts the walls and the second is a pendulum which is connected to the main body through an actuated revolute joint. The pendulum swings back and forth, causing the main body to rotate. While doing so, the main body hits the walls and uses the dynamic movement of the pendulum to push itself upward as shown in Figure 22.22. The simulation and experimental studies have provided some insights into the mechanism for design considerations of new climbing robots.

University of Utah has developed an Oscillating Climbing Robot with a recursive name ROCR as shown in Figure 22.23. ROCR is a pendulam two-link, serial chain

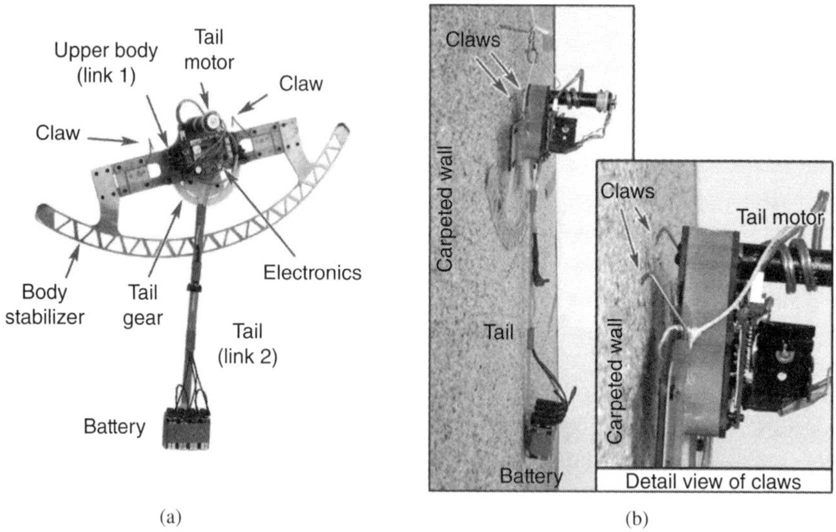

FIGURE 22.23 ROCR: oscillating climbing robot: (a) ROCR prototype with dactyl claw-gripping mechanisms and a pivoting tail attached to the center of its first link and (b) detailed view of ROCR's claw attachment to the wall.

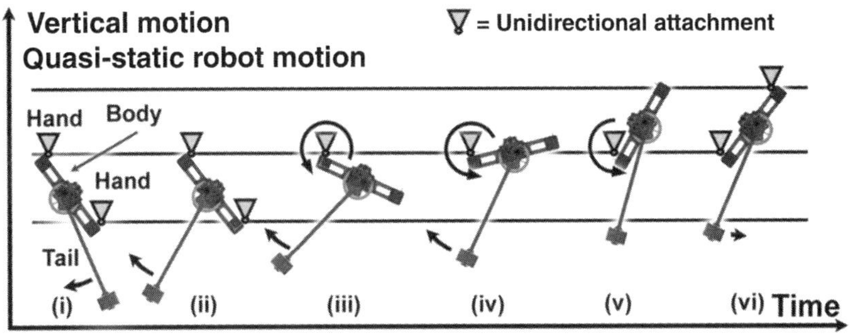

(a)

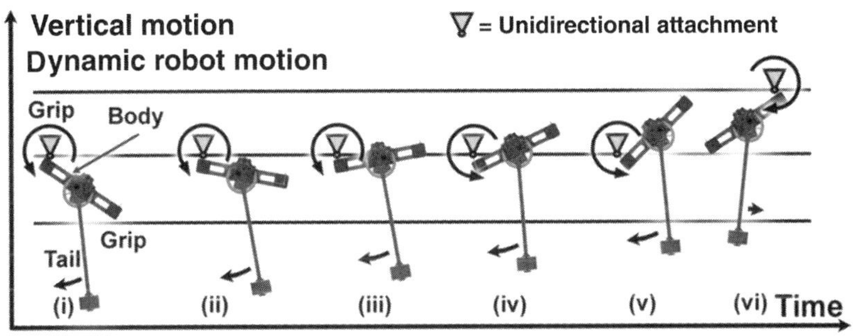

(b)

FIGURE 22.24 Example climbing gaits for ROCR: (a) quasi-static mass-shifting, equilibrium-based gait and (b) oscillating dynamic gait achieves higher efficiency. *Arrows* indicate tail motion relative to robot body and body rotation about an engaged gripping mechanism.

robot that utilizes alternating handholds and an actuated tail to propel itself upward in a climbing style inspired by human climbers and brachiating gibbons [91]. The two dactyl claw grippers are made from 1.0 mm diameter spring steel that is bent such that they provide directional adhesion, that is, ROCR's claws bear weight when pulled downward, but automatically release when the force on the claw is reversed. As portrayed in Figure 22.24, ROCR alternately grips the wall with one hand at a time and swings its tail, causing a center of gravity shift that raises its free hand, which then grips the climbing surface. ROCR's bio-inspired oscillating climbing strategy is efficient, requiring a minimum of input energy in order to climb vertical walls.

22.3 LOCOMOTION TECHNIQUES OF CLIMBING ROBOTS

Besides the adhesion mechanism, mechanical structure and mobility are other design factors determined by specific applications of climbing robots. Climbing large,

(a) (b) (c)

FIGURE 22.25 Climbing robots using wheels or tracks: (a) Alicia II, using wheeled locomotion (Reprinted with permission from Reference 92. Copyright © 2004 Emerald Group Publishing Limited); (b) TRIPILLAR, using tracked locomotion; and (c) TRIPILLAR cruises on ferromagnetic surfaces from floor to ceiling. (Pictures b and c, reprinted with permission from Reference 93. Copyright © 2002 Elsevier).

nearly flat surfaces demands simple machines fewer degrees of freedom in their motion. Wheeled vehicles or robots using sliding-frame systems may satisfy this requirement. On the other hand, in more complicated tasks where robots are required to overcome obstacles or make floor-to-wall, wall-to-ceiling transitions, articulated limbed structures with multiple legs are predominant. This section highlights some representative locomotion technologies implemented in climbing robots that fall in the following five major categories: the wheel-driven locomotion, tracked locomotion, legged locomotion, wheel–leg hybrid locomotion, and locomotion based on a sliding frame. Climbing robots using wheels or chain-tracks have a similarity that both move with a rotational mechanism. They can achieve fast locomotion on flat surfaces as long as the adhesion mechanisms (i.e., magnet, suction, or aerodynamic attraction) provide reliable attachment to the surfaces. Climbing robots using wheeled or tracked locomotion and suction mechanism normally are not able to move and cross obstacles higher than a few centimeters since they will cause leakage of vacuum chamber. The tracked locomotion increases the contact area and allows the robot to better adhere to surfaces. Figure 22.25a shows the Alicia II that is a representative wheeled climbing robot [92], and Figures 22.25b and 22.25c show the TRIPILLAR robot developed at EPFL, Switzerland, which uses magnetic caterpillars in a triangular shape and is able to cruise on planar ferromagnetic surfaces from floor to ceiling [93]. Other climbing robots using wheeled locomotion introduced in previous sections include Windoro (Figure 22.3), WallWalker (Figure 22.10), City-Climbers (Figures 22.12–22.15), and CROMSCI (Figure 22.15). The tracked climbing robots using magnet (Figure 22.2), suction (Figure 22.3), and dry adhesive (Figure 22.18d) are also reported in the previous sections.

Legged locomotion is used to transit between surfaces, or cross over large obstacles, or to operate in trussed frames. The bipedal climbing robots with the ability to

FIGURE 22.26 Alicia 3, bipod robot using three Alicia II modules.

move between surfaces include ROMA II in Figure 22.5a, FLIPPER and CRAWLER robots in Figure 22.6. University of Catania, Italy, has developed Alicia 3 robot using three Alicia II modules linked by two legs as shown in Figure 22.26a, which allows the whole system to pass over obstacles of 10 cm in a few steps by detaching the three modules one by one [94]. For increased safety and load capacity, NINJA-I and MRWALLSPECT-III are quadruped robots using four legs protruding from a central body and carrying suction feet as shown in Figures 22.5c and 22.5d. A hexapod robot with the capability for welding ship hull is reported in Reference 95. Figure 22.17 shows some legged climbing robots with multiple limbs and grippers, which are commonly used in trussed frames. Stickybot (Figure 22.19) and Spinybot (Figure 22.20) both use legs and feet to mimic the climbing behavior of gecko and beetle. More limbs typically provide redundant support and often increase load capacity and safety. However, these benefits are achieved at the cost of increased complexity, size, and weight.

Wheel–leg hybrid locomotion allows the robot to overcome small obstacles or cracks on vertical walls. Figure 22.27a shows a climbing robot using wheel–leg hybrid locomotion with adhesive feet, developed at Case Western Reserve University, USA [96–98]. The robot is based on Mini-Whegs and its legs are redesigned to enable its feet to better attach and peel off from the surface in a manner similar to that of climbing animals. Waalbot in Figure 22.18c adopts same locomotion which includes passive joint and elastic flexures to provide the preload and peeling forces necessary for the footpads to climb using dry adhesives. Figures 22.27b and 22.27c show another wheel–leg hybrid locomotion in large scale. It consists of a mobile base with three wheels inside a big suction cup and a mechanical leg with three degrees of freedom and small sucker. The robot has the potential to achieve quick motion on a wall surface, as well as smooth wall-to-wall transitions [58].

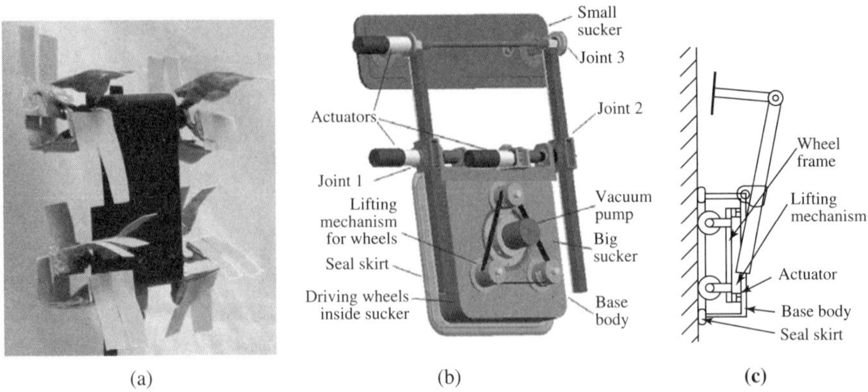

FIGURE 22.27 Hybrid locomotion: (a) Mini-Whegs on vertical glass with adhesive polymer feet and a tail; (b) climbing robot with a mobile base and a lifting leg; and (c) sketch of wheel–leg hybrid locomotion.

Sliding-frame locomotion uses simple sticking–moving–sticking process and is easy to control. Sky Cleaner 3 in Figure 22.5d and its ancestors are representative climbing robots using sliding-frame locomotion for glass cleaning. They are actuated by pneumatic cylinders for translation and vacuum grippers for attachment. Their sliding mechanism is shown in Figure 22.28a. Recently, a climbing robot using a double sliding platform and permanent magnets for attachment has been developed for grit blasting of ship hull as shown in Figure 22.28b. The robot adopts two modules that can move relatively to each other and can move up and along the shipside with any inclination during grit blasting [99]. It can also rotate to compensate for the hull curvature and avoid obstacles while performing the grit blasting task. The sliding-frame locomotion has two disadvantages: (1) the movement is discontinuous and the operational speed is low, and (2) the frames are normally large, which prevent it from being used in a narrow space.

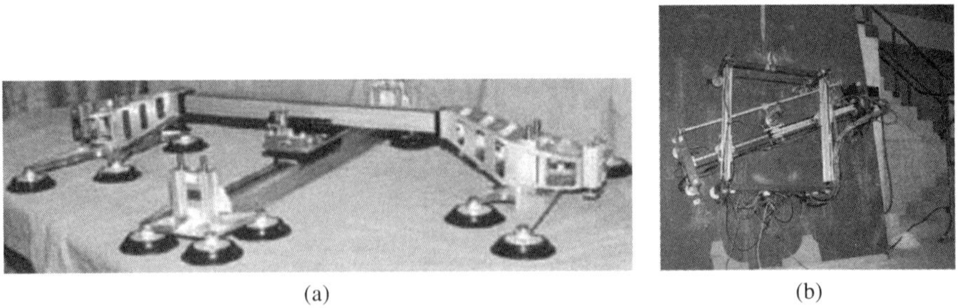

FIGURE 22.28 (a) Sliding mechanism of Sky Cleaner robots and (b) grit blasting robot using sliding-frame locomotion.

TABLE 22.1 Feature Comparison of Various Adhesion Techniques

Adhesion techniques		Applicable situations	Advantages	Limitations
Magnetic adhesion		Steel walls	Does not need energy to adhere; great suction force	Only can be used on steel walls
Vacuum suction techniques		Seamless and smooth surfaces	Great suction force; easy to control	Noise; slow speed; cannot be used in vacuum space
Aerodynamic attraction		Both smooth and rough surfaces	Low requirement for motion mechanism	High power consumption and noise; cannot work in vacuum space
Grasping grippers		Bars or irregular surfaces	Low noise and power consumption	Low security
Bio-inspired climbing robots	Gecko-inspired synthetic dry adhesives	Smooth surfaces	No noise; needs no energy to maintain; can be used in vacuum space and can realize fast locomotion speeds	Limited payload; and degrades with time, gets contaminated in dirty environments
	Beetle-inspired microspines	Rough surfaces		Cannot work on smooth surfaces
Compliant electroadhesion		All kinds of surfaces	No noise; low power consumption; can operate on walls that are heavily covered by dust or debris	Low motion speed; need to be improved

22.4 CONCLUSIONS

The research and development of climbing robot technologies have received long-term worldwide attention since late 1980s. This chapter presents a comprehensive survey of the adhesion mechanisms and locomotion techniques used in climbing robots. It highlights some representative climbing robots with most successful implementation of these technologies. Some innovative ideas and emerging technologies for climbing robots are also introduced. The article classifies the adhesion mechanism into six categories and summarizes their advantages and limitations in Table 22.1.

The future research and development directions of climbing robots will mainly focus on the improvement of adhesion reliability and locomotion maneuverability, and the transformation of these innovative technologies into real-world applications. For any climbing robot to be used as a commercial product, reliability is the most critical and determining factor for commercial success. No matter what adhesion mechanism is used, it must be able to supply the necessary attraction force for the robot to operate on targeted wall surfaces reliably at all times. High maneuverability is required to improve the operational efficiency. We can foresee more and more successful use of climbing robots in practical applications in the coming years.

ACKNOWLEDGMENT

This work is supported in part by the U.S. Army Research Office under grant W911NF-05-1-0011 and National Science Foundation of USA under grants CNS-0551598, CNS-0619577, and IIS-0644127.

REFERENCES

[1] A. Nishi, Y. Wakasugi, and K. Watanabe. Design of a robot capable of moving on a vertical wall. *Advanced Robotics*, 1: 33–45, 1986.

[2] A. Nishi. A biped walking robot capable of moving on a vertical wall. *Mechatronics*, 2(6): 543–554, 1992.

[3] G. Virk. The CLAWAR project-developments in the oldest robotics thematic network. In: *2005 IEEE Robotics & Automation Magazine*, 2005, pp. 14–20.

[4] S. Kim, A. T. Asbeck, M. R. Cutkosky, and W. R. Provancher. SpinybotII: climbing hard walls with compliant microspines. In: Proceedings of the 12th International Conference on Advanced Robotics (ICAR 05), Seattle, WA, 2005, pp. 601–606.

[5] S. Kim, M. Spenko, S. Trujillo, B. Heyneman, D. Santos, and M. R. Cutkosky. Smooth vertical surface climbing with directional adhesion. *IEEE Transactions on Robotics*, 2008, 24(1): 65–74.

[6] H. X. Zhang, J. W. Zhang, G. H. Zong, W. Wang, and R. Liu. Sky Cleaner 3: a real pneumatic climbing robot for glass-wall cleaning. *IEEE Robotics & Automation Magazine*, 13(1): 32–41, 2006.

[7] H. X. Zhang, W. Wang, and J. W. Zhang. High stiffness pneumatic actuating scheme and improved position control strategy realization of a pneumatic climbing robot. In: IEEE International Conference on Robotics and Biomimetics (ROBIO 2008), Bangkok, Thailand, 2008, pp. 1806–1811.

[8] S. Q. Wu. *Study on Low Noise Wall climbing Robot Based on Negative Pressure Suction (in Chinese)*. Harbin, China: Harbin Institute of Technology, 2007.

[9] S. Galt, B. L. Luk, D. S. Cooke, and A. A. Collie. A tele-operated semi-intelligent climbing robot for nuclear applications. In: Proceedings of the Fourth Annual Conference on Mechatronics and Machine Vision in Practice, Toowoomba, Australia, 1997, pp. 118–123.

[10] Y. Wang, S. L. Liu, D. G. Xu, Y. Z. Zhao, H. Shao, and X. S. Gao. Development and application of wall-climbing robots. In: Proceedings of the 1999 IEEE International Conference on Robotics and Automation, Detroit, MI, 1999, pp. 1207–1212.

[11] L. P. Kalra, J. Gu, and M. Max. Wall climbing robot for oil tank inspection. In: Proceedings of the IEEE International Conference on Robotics and Biomimetics (ROBIO 2006), Kunming, China, 2006, pp. 1523–1528.

[12] J. Savall, A. Avello, and L. Briones. Two compact robots for remote inspection of hazardous areas in nuclear power plants. In: Proceedings of the 1999 IEEE International Conference on Robotics and Automation, Detroit, MI, 1999, pp. 1993–1998.

[13] L. Briones, P. Bustamante, and M. A. Serna. Wall-climbing robot for inspection in nuclear power plants. In: Proceedings of the 1994 IEEE International Conference on Robotics and Automation, San Diego, CA, 1994, vol. 2, pp. 1409–1414.

[14] M. Abderrahim, C. Balaguer, A. Gimenez, J. M. Pastor, and V. M. Padron. Roma: a climbing robot for inspection operations. In: Proceedings of the 1999 IEEE International Conference on Robotics and Automation, Detroit, MI, pp. 2303–2308.

[15] N. Elkmann, T. Felsch, M. Sack, J. Saenz, and J. Hortig. Innovative service robot systems for facade cleaning of difficult-to-access areas. In: Proceedings of the 2002 IEEE/RSJ International Conference on Intelligent Robots and Systems, Lausanne, Switzerland, 2002, pp. 756–762.

[16] H. X. Zhang, J. W. Zhang, R. Liu, and G. H. Zong. Mechanical design and dynamics of an autonomous climbing robot for elliptic half-shell cleaning. *International Journal of Advanced Robotic Systems*, 4(4): 437–446, 2007.

[17] S. Wu, M. Li, S. Xiao, and Y. Li. A wireless distributed wall climbing robotic system for reconnaissance purpose. In: Proceedings of the 2006 IEEE International Conference on Mechatronics and Automation, June 25–28, 2006, Luoyang, China, 2006, pp. 1308–1312.

[18] J. Z. Xiao, A. Sadegh, M. Elliot, A. Calle, A. Persad, and H. M. Chiu. Design of mobile robots with wall climbing capability. In: Proceedings of the 2005 IEEE/ASME International Conference on Advanced Intelligent Mechatronics, Monterey, CA, 2005, pp. 438–443.

[19] J. Z. Xiao and A. Sadegh. City-Climber: a new generation of wall-climbing robots, Chapter 18. In: *Climbing and Walking Robots—Towards New Applications*, edited by H. Zhang. Vienna, Austria: I-Tech Education and Publishing, 2007, pp. 383–402.

[20] J. C. Grieco, M. Prieto, M. Armada, and P. Santo. A six-legged climbing robot for high payloads. In: Proceedings of the 1998 IEEE International Conference on Control Applications, Trieste, Italy, 1998, vol. 1, pp. 446–450.

[21] W. M. Shen, J. Gu, and Y. J. Shen. Proposed wall climbing robot with permanent magnetic tracks for inspecting oil tanks. In: Proceedings of the IEEE International Conference on Mechatronics & Automation, Niagara Fall, Canada, 2005, vol. 4, pp. 2072–2078.

[22] Y. Kawaguchi, I. Yoshida, H. Kurumatani, T. Kikuta, and Y. Yamada. Internal pipe inspection robot. In: Proceedings of the 1995 IEEE International Conference on Robotics and Automation, Nagoya, Japan, 1995, vol. 1, pp. 857–862.

[23] L. Z. Sun, P. Sun, X. J. Qin, and C. M. Wang. Micro robot in small pipe with electromagnetic actuator. In: Proceedings of the 1998 International Symposium on Micromechatronics and Human Science (MHS 98), Nagoya, Japan, 1998, pp. 243–248.

[24] Z. C. Gui, Q. Chen, and Z. G. Sun. Wall climbing robot employing multi-body flexible permanent magnetic adhesion system (in Chinese). *Chinese Journal of Mechanical Engineering*, 44(6): 177–182, 2008.

[25] E. Guizzo. Windoro window-cleaning robot demo, IEEE Spectrum robotics blog. Available at http://spectrum.ieee.org/automaton/robotics/home-robots/windoro-window-cleaning-robot-demo (accessed January 17, 2011).

[26] M. F. Silva, J. Machado, and J. K. Tar. A survey of technologies for climbing robots adhesion to surfaces. In: IEEE International Conference on Computational Cybernetics (ICCC 2008), Stara Lesna, Slovakia, 2008, pp. 127–132.

[27] B. Chu, K. Jung, C.S. Han, and D. Hong. A survey of climbing robots: locomotion and adhesion. *International Journal of Precision Engineering and Manufacturing*, 11(4): 633–647, 2010.

[28] http://en.wikipedia.org/wiki/Venturi_effect (accessed January 18, 2015).

[29] F. Weise, J. Kohnen, and H. Wiggenhauser. Non-destructive sensors for inspection of concrete structures with a climbing robot. In: Proceedings of the Fourth International Conference on Climbing and Walking Robots, CLAWAR 2001, Karlsruhe, Germany, 2001, pp. 945–952.

[30] C. Balaguer, A. Giménez, and A. Jardón. Climbing robots' mobility for inspection and maintenance of 3D complex environments. *Autonomous Robots*, 18(3): 157–169, 2005.

[31] S. Hirose, A. Nagakubo, and R. Toyama. Machine that can walk and climb on doors, walls and ceilings. In: Proceedings of the Fifth International Conference on Advanced Robotics, Piscataway, NJ, 1991, pp. 753–758.

[32] A. Nagakubo and S. Hirose. Walking and running of the quadruped wall-climbing robot. In: Proceedings of the 1994 IEEE International Conference on Robotics and Automation, San Diego, CA, 1994, pp. 1005–1012.

[33] T. Kang, H. Kim, T. Son, and H. Choi. Design of quadruped walking and climbing robot. In: IEEE/RSJ International Conference on Intelligent Robots and Systems (IROS 2003), Las Vegas, NV, 2003, vol. 1, pp. 619–624.

[34] B. L. Luk, A. A. Collie, and Billingsley. Robug II: An intelligent wall climbing robot. In: Proceedings of the 1991 IEEE International Conference on Robotics and Automation, Sacramento, CA, 1991, pp. 2342–2347.

[35] B. L. Luk, D. S. Cooke, S. Galt, A. A. Collie, and S. Chen. Intelligent legged climbing service robot for remote maintenance applications in hazardous environments. *Robotics and Autonomous Systems*, 53(2): 142–152, 2005.

[36] R. T. Pack, J. L. Christopher, and K. Kawamura. A rubbertuator-based structure climbing inspection robot. In: Proceedings of the 1997 IEEE International Conference on Robotics and Automation, Albuquerque, NM, 1997, pp. 1869–1874.

[37] P. G. Backes, Y. Bar-Cohen, and B. Joffe. The multifunction automated crawling system (MACS). In: Proceedings of the 1997 IEEE International Conference on Robotics and Automation, Albuquerque, NM, 1997, pp. 335–340.

[38] M. A. Serna, A. Avello, L. Briones, and P. Bustamante. ROBICEN: a pneumatic climbing robot for inspection of pipes and tanks. *Lecture Notes in Control and Information Sciences, Experimental Robotics*, 232: 325–334, 1998.

[39] T. S. White, N. Hewer, B. L. Luk, and J. Hazel. Design and operational performance of a climbing robot used for weld inspection in hazardous environments. In: Proceedings of the 1998 IEEE International Conference on Control Applications, Trieste, Italy, 1998, pp. 451–455.

[40] L. Tummala, R. Mukherjee, N. Xi, D. Aslam, H. Dulimarta, J. Z. Xiao, M. Minor, and G. Dang. Climbing the walls. *IEEE Robotics & Automation Magazine*, 9(4): 10–19, 2002.

[41] H. Kim, D. Kim, H. Yang, K. Lee, K. Seo, D. Chang, and J. Kim. Development of a wall-climbing robot using a tracked wheel mechanism. *Journal of Mechanical Science and Technology*, 22(8): 1490–1498, 2008.

[42] J. Zhu, D. Sun, and S. K. Tso. Development of a tracked climbing robot. *Journal of Intelligent and Robotic Systems Archive*, 35(4): 427–443, 2002.

[43] W. Brockmann. Concept for energy-autarkic, autonomous climbing robots. In: Proceedings of the 9th International Conference on Climbing and Walking Robots, 2006, pp. 107–114.

[44] http://www.serbot.ch (accessed January 18, 2015).

[45] T. Yano, S. Numao, and Y. Kitamura. Development of a self-contained wall climbing robot with scanning type suction cups. In: Proceedings of the 1998 IEEE/RSJ International Conference on Intelligent Robots and Systems, Victoria, Canada, 1998, pp. 249–254.

[46] W. Wang, K. Wang, G. H. Zong, and D. Z. Li. Principle and experiment of vibrating suction method for wall-climbing robot. *Vacuum*, 85: 107–112, 2010.

[47] T. Miyake, H. Ishihara, and T. Tomino. Vacuum-based wet adhesion system for wall climbing robots–lubricating action and seal action by the liquid. In: IEEE International Conference on Robotics and Biomimetics (ROBIO 2008), Bangkok, Thailand, 2008, pp. 1824–1829.

[48] D. Sun, J. Zhu, C. Lai, and S. K. Tso. A visual sensing application to a climbing cleaning robot on the glass surface. *Mechatronics*, 14(10): 1089–1104, 2004.

[49] H. Choi, J. Park, and T. Kang. A self-contained wall climbing robot with closed link mechanism. *KSME International Journal*, 18(4): 573–581, 2004.

[50] Y. Jiang, H. G. Wang, L. J. Fang, and M. Y. Zhao. A novel approach to fault detection and identification in suction foot control of a climbing robot. In: IEEE/RSJ International Conference on Intelligent Robots and Systems (IROS 2006), Beijing, China, 2006, pp. 3423–3428.

[51] B. L. Luk, A. A. Collie, V. Piefort, and G. S. Virk. Robug III: a tele-operated climbing and walking robot. In: UKACC International Conference on CONTROL 96, 1996, vol. 1, pp. 347–352.

[52] Y. Ota, T. Kuga, and K. Yoneda. Deformation compensation for continuous force control of a wall climbing quadruped with reduced-DOF. In: Proceedings of the 2006 IEEE International Conference on Robotics and Automation (ICRA 2006), Orlando, FL, 2006, pp. 468–474.

[53] M. T. Li, W. Guo, and L. Sun. Study of an inchworm-like micro-robot. *High Technology Letters*, 12(2): 74–78, 2006.

[54] A. Nishi and H. Miyagi. A wall climbing robot using propulsive force of propeller. In: Proceedings of the Fifth International Conference on Advanced Robotics, Piscataway, NJ, 1991, pp. 320–325.

[55] A. Nishi and H. Miyagi. Mechanism and control of propeller type wall-climbing robot. In: Proceedings of the IEEE/RSJ International Conference on Intelligent Robots and Systems, Munich, Germany, 1994, pp. 1724–1729.

[56] T. Akinfiev, M. Armada, and S. Nabulsi. Climbing cleaning robot for vertical surfaces. *Industrial Robot: An International Journal*, 36(4): 352–357, 2009.

[57] S. Y. Liu, X. S. Gao, K. J. Li, J. Li, and X. G. Duan. A small-sized wall-climbing robot for anti-terror scout. In: IEEE International Conference on Robotics and Biomimetics (ROBIO 2007), Sanya, China, 2007, pp. 1866–1870.

[58] Y. L. Fu, Z. H. Li, and S. G. Wang. A wheel-leg hybrid wall climbing robot with multi-surface locomotion ability. In: IEEE International Conference on Mechatronics and Automation (ICMA 2008), Takamatsu, Japan, 2008, pp. 627–632.

[59] B. Ma, R. Liu, R. Zhang, L. Tian, and Nashunbuhe. Design of wall climbing robots with transition capability. In: IEEE International Conference on Robotics and Biomimetics (ROBIO 2007), Sanya, China, 2007, pp. 1871–1875.

[60] Y. K. Song, C. M. Lee, L. M. Koo, D. T. Tran, H. Moon, and H. R. Choi. Development of wall climbing robotic system for inspection purpose. In: IEEE/RSJ International Conference on Intelligent Robots and Systems (IROS 2008), Nice, France, 2008, pp. 1990–1995.

[61] J. Li, X. S. Gao, N. J. Fan, W. Zhu, J. Yin, and Y. J. Jia. Wall climbing robot based on negative pressure-thrust suction method. In: IEEE International Conference on Mechatronics and Automation (ICMA 2008), Takamatsu, Japan, 2008, pp. 604–609.

[62] C. Hillenbrand, D. Schmidt, and K. Berns. Cromsci—a climbing robot with multiple sucking chambers for inspection tasks. In: 11th International Conference on Climbing and Walking Robots (CLAWAR), Coimbra, Portugal, September 8–10, 2008, pp. 311–318.

[63] T. Leichner, T. Gastauer, B. Sauer, C. Hillenbrand, D. Schmidt, and K. Berns. Design and development of a sealing-system for climbing-robots, 6th National Conference on Mechanical Engineering, Budapest, Hungary, May 29–30, 2008. G-2008-B-06.

[64] J. Mämpel, K. Gerlach, C. Schilling, and H. Witte. A modular robot climbing on pipe-like structures. In: 4th International Conference on Autonomous Robots and Agents (ICARA 2009), Wellington, New Zealand, 2009, pp. 87–91.

[65] M. Tavakoli, L. Marques, and A. Almeida. 3DCLIMBER: climbing and manipulation over 3D structure. *Mechatronics*, 21(1): 48–62, 2011.

[66] T. L. Lam and Y. S. Xu. A flexible tree climbing robot: Treebot—design and implementation. In: 2011 IEEE International Conference on Robotics and Automation, Shanghai, China, May 9–13, 2011.

[67] A. Baghani, M. N. Ahmadabadi, and A. Harati. Kinematics modeling of a wheel-based pole climbing robot (UT-PCR). In: Proceedings of the 2005 IEEE International Conference on Robotics and Automation (ICRA 2005), Barcelona, Spain, 2005, pp. 2099–2104.

[68] K. Inoue, T. Tsurutani, T. Takubo, and T. Arai. Omni-directional gait of limb mechanism robot hanging from grid-like structure. In: Proceeding of the IEEE/RSJ International Conference on Intelligent Robots and Systems (IROS 2006), Beijing, China, 2006, 1732–1737.

[69] Y. S. Guan, L. Jiang, X. M. Zhang, J. Qiu, and X. F. Zhou. 1-DoF robotic joint modules and their applications in new robotic systems. In: IEEE International Conference on Robotics and Biomimetics (ROBIO 2008), Bangkok, Thailand, 2008, pp. 1905–1910.

[70] F. Y. Xu and X. S. Wang. A wheel-based cable climbing robot with descending speed restriction. In: Control and Decision Conference (CCDC 2008), Yantai, Shandong, China, 2008, pp. 1570–1575.

[71] S. Mahdavi, E. Noohi, and M. N. Ahmadabadi. Basic movements of a nonholonomic wheel-based pole climbing robot. In: 2007 IEEE/ASME International Conference on Advanced Intelligent Mechatronics, Zurich, Switzerland, 2007, pp. 1–6.

[72] H. Kawasaki, S. Murakami, H. Kachi, and S. Ueki. Novel climbing method of pruning robot. In: SICE Annual Conference, Tokyo, Japan, 2008, pp. 160–163.

[73] R. Aracil, R. Saltarén, and O. Reinoso. Parallel robots for autonomous climbing along tubular structures. *Robotics and Autonomous Systems*, 42(2): 125–134, 2003.

[74] Y. Yoon and D. Rus. Shady3D: a robot that climbs 3D trusses. In: 2007 IEEE International Conference on Robotics and Automation (ICRA 2007), Rome, Italy, 2007, pp. 4071–4076.

[75] K. Autumn, Y. Liang, T. Hsieh, W. Zesch, W. P. Chan, T. Kenny, R. Fearing, and R. J. Full. Adhesive force of a single gecko foot-hair. *Nature*, 405(6787): 681–685, 2000.

[76] C. Menon, Y. Li, D. Sameoto, and C. Martens. Abigaille-I: towards the development of a spider-inspired climbing robot for space use. In: 2nd IEEE RAS & EMBS International Conference on Biomedical Robotics and Biomechatronics (BioRob 2008), Scottsdale, AZ, 2008, pp. 384–389.

[77] C. Menon, M. Murphy, and M. Sitti. Gecko inspired surface climbing robots. In: IEEE International Conference on Robotics and Biomimetics (ROBIO 2004), Shenyang, China, 2004, pp. 431–436.

[78] M. Sitti and R. S. Fearing. Synthetic gecko foot-hair micro/nano-structures for future wall-climbing robots. In: Proceedings of the 2003 IEEE In-ternational Conference on Robotics and Automation (ICRA 2003), Taipei, China, 2003, vol. 1, pp. 1164–1170.

[79] M. P. Murphy, S. Kim, and M. Sitti. Enhanced adhesion by gecko-inspired hierarchical fibrillar adhesives. *ACS Applied Materials & Interfaces*, 1(4): 849–855, 2009.

[80] M. P. Murphy, W. Tso, M. Tanzini, and M. Sitti. Waalbot: an agile small-scale wall climbing robot utilizing pressure sensitive adhesives. In: Proceeding of the IEEE/RSJ International Conference on Intelligent Robots and Systems (IROS 2006), Beijing, China, 2006, pp. 3411–3416.

[81] M. P. Murphy and M. Sitti. Waalbot: an agile small-scale wall-climbing robot utilizing dry elastomer adhesives. *IEEE/ASME Transactions on Mechatronics*, 12(3): 330–338, 2007.

[82] B. Aksak, M. P. Murphy, and M. Sitti. Gecko inspired micro-fibrillar adhesives for wall climbing robots on micro/nanoscale rough surfaces. In: Proceedings of 2008 IEEE International Conference on Robotics and Automation (ICRA 2008), Pasadena, CA, 2008, pp. 3058–3063.

[83] O. Unver and M. Sitti. Tankbot: a miniature, peeling based climber on rough and smooth surfaces. In: Proceedings of the 2009 IEEE International Conference on Robotics and Automation (ICRA 2009), Kobe, Japan, May 19–23, 2009.

[84] M. Carlo and S. Metin. A biomimetic climbing robot based on the gecko. *Journal of Bionic Engineering*, 3(3): 115–125, 2006.

[85] C. Menon and M. Sitti. Biologically inspired adhesion based surface climbing robots. In: Proceedings of the 2005 IEEE International Conference on Robotics and Automation (ICRA 2005), Barcelona, Spain, 2005, pp. 2715–2720.

[86] S. Kim, M. Spenko, S. Trujillo, B. Heyneman, V. Mattoli, and M. R. Mattoli. Whole body adhesion: hierarchical, directional and distributed control of adhesive forces for a climbing robot. In: IEEE International Conference on Robotics and Automation (ICRA 2007), Rome, Italy, 2007, pp. 1268–1273.

[87] J. E. Clark, D. I. Goldman, P. Lin, G. Lynch, T. S. Chen, H. Komsuoglu, R. J. Full, and D. Koditschek. Design of a bio-inspired dynamical vertical climbing robot. In: Proceedings of Robotics: Science and Systems 2007, Atlanta, GA, 2007. Online proceedings.

Available at http://www.roboticsproceedings.org/rss03/p02.html (accessed January 18, 2015).

[88] X. W. Guan, H. Zhang, A. H. Ji, and Z. D. Dai. Development of claw-inspired foot structure for wall-climbing robot. *Mechanical & Electrical Engineering Magazine*, 26(2): 1–3, 2009.

[89] H. Prahlad, R. Pelrine, S. Stanford, J. Marlow, and R. Kornbluh. Electroadhesive robots-wall climbing robots enabled by a novel, robust, and electrically controllable adhesion technology. In: IEEE International Conference on Robotics and Automation (ICRA 2008), Pasadena, CA, 2008, pp. 3028–3033.

[90] A. Degani, A. Shapiro, H. Choset, and M. T. Mason. A dynamic single actuator vertical climbing robot. In: Proceedings of the 2007 IEEE/RSJ International Conference on Intelligent Robots and Systems, San Diego, CA, October 29–November 2, 2007.

[91] W. R. Provancher, S. Jensen-Segal, and M. E. Fehlberg. ROCR: an energy-efficient dynamic wall-climbing robot. *IEEE/ASME Transactions on Mechatronics*, 16(5): 897–906, 2011.

[92] D. Longo and G. Muscato. A modular approach for the design of the Alicia 3 climbing robot for industrial inspection. *Industrial Robot: An International Journal*, 31(2): 148–158, 2004.

[93] G. L. Rosa, M. Messina, G. Muscato, and R. Sinatra. A low-cost lightweight climbing robot for the inspection of vertical surfaces. *Mechatronics*, 12(1): 71–96, 2002.

[94] D. Longo and G. Muscato. The Alicia 3 climbing robot, a three-module robot for automatic wall inspection. *IEEE Robotics and Automation Magazine*, 13(1): 42–50, 2006.

[95] M. Armada, M. Prieto, T. Akinfiev, R. Fernandez, P. Gonzalez, E. Garcia, H. Montes, S. Nabulsi, R. Ponticelli, J. Sarria, J. Estremera, S. Ros, J. Grieco, and G. Fernandez. On the design and development of climbing and walking robots for the maritime industries. *Maritime Research*, 2(1): 9–32, 2005.

[96] K. A. Daltorio, A. D. Horchler, S. Gorb, R. E. Ritzmann, and R. D. Quinn. A small wall-walking robot with compliant, adhesive feet. In: Proceedings of IEEE/RSJ International Conference on Intelligent Robots and Systems (IROS 2005), Edmonton, AB, Canada, 2005, pp. 4018–4023.

[97] K. A. Daltorio, S. Gorb, A. Peressadko, A. D. Horchler, R. E. Ritzmann, and R. D. Quinn. A robot that climbs walls using micro-structured polymer feet. In: Proceedings of the 8th International Conference on Climbing and Walking Robots and the Support Technologies for Mobile Machines, London, UK, 2006, pp. 131–138.

[98] K. A. Daltorio, T. E. Wei, S. N. Gorb, R. E. Ritzmann, and R. D. Quinn. Passive foot design and contact area analysis for climbing mini-whegs. In: IEEE International Conference on Robotics and Automation (ICRA 2007), Rome, Italy, 2007, pp. 1274–1279.

[99] A. Faina, D. Souto, A. Deibe, F. Lopez-Pena, R. J. Duro, and X. Fernandez. Development of a climbing robot for grit blasting operations in shipyards. In: Proceedings of the 2009 IEEE International Conference on Robotics and Automation, Kobe, Japan, May 19–23, 2009, pp. 200–205.

CHAPTER 23

DATA PROCESSING IN CURRENT 3D ROBOTIC PERCEPTION SYSTEMS

CANG YE
Department of Systems Engineering, University of Arkansas at Little Rock, Little Rock, AR, USA

23.1 INTRODUCTION

The past decade has witnessed tremendous advancement in robotic autonomy. One of the driving forces making this happen is the advancement in 3D robotic vision. New 3D imaging sensors and data processing methods have drastically changed the way a robot interacts with its environment and improved its intelligence. The most popular 3D perception sensors in robotics include stereovision system and light detection and ranging (LIDAR). Recently a new class of 3D imaging sensor—flash LIDAR camera (FLC)—has been developed and applied to robotic vision. The commercially available products include the SwissRanger SR4000 [1] and the TigerEye [2] FLCs. Since Microsoft introduced Kinect for Xbox 360 in 2010 as a remote free user interface, the Kinect sensor has attracted much attention from the robotics community [3]. The sensor projects light pattern on the scene and measures the range through triangulation. It produces both intensity and range image just like the FLC does. The data processing methods developed for the FLC may be adapted to the Kinect with ease. The details of the Kinect sensor are beyond the scope of this chapter. This section introduces three representative sensing systems: stereovision, LIDAR, and FLC.

23.1.1 Stereovision

Stereovision has been widely used in robotics. The representative applications of stereovision have been demonstrated in NASA's Mars Exploration Rover [4, 5] program and the DARPA's Learning Applied to Ground Robots [6–9] program. The

Contemporary Issues in Systems Science and Engineering, First Edition.
Edited by MengChu Zhou, Han-Xiong Li and Margot Weijnen.

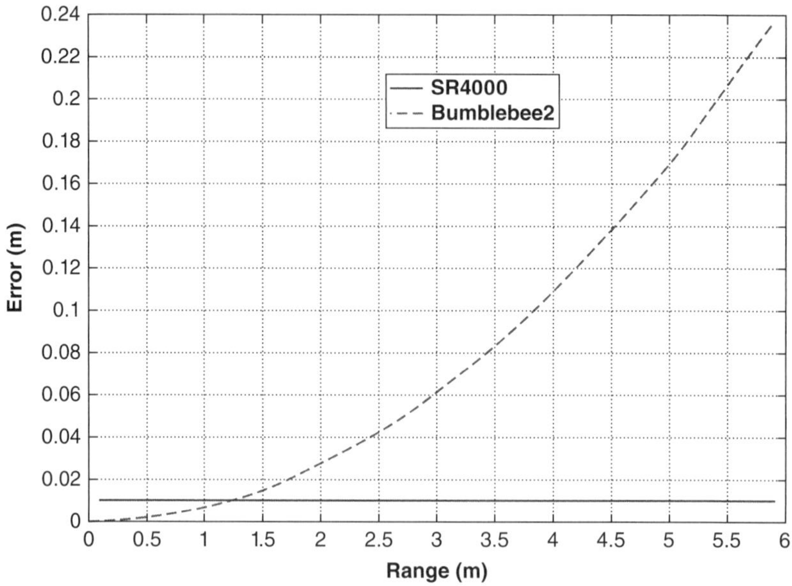

FIGURE 23.1 Range measurement errors of the SR4000 and Bumblebee2 stereovision system.

projects in both programs use stereovision as the major/single-perception sensor for robot navigation including robot pose estimation and obstacle detection/avoidance. A stereovision resembles human eye's depth perception. It performs stereo matching between features in the left and right images of the sensor and determines the depth (range) information of the matched features by using their disparity information. A stereovision system is cost-effective and has low power dissipation. However, it has the following disadvantages: (1) its range measurement accuracy drops quadratically with the true distance; and (2) it cannot produce complete range data of the environment. These properties are illustrated by Figures 23.1 and 23.2. The dotted curve in Figure 23.1 depicts the Bumblebee2 stereovision system's range measurement error that increases quadratically with the true range [3]. The measurement error is 1.6 cm

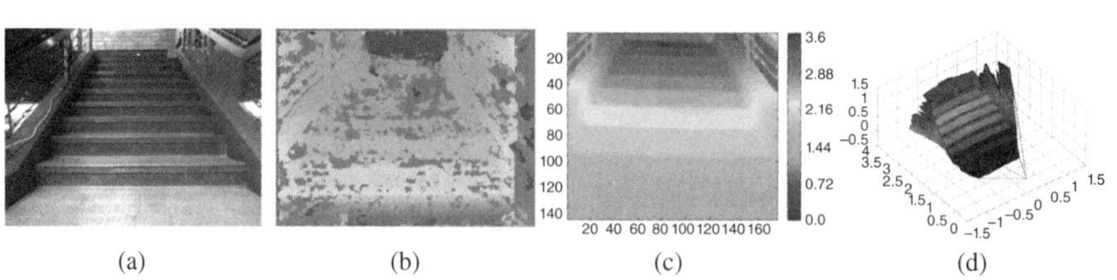

FIGURE 23.2 Comparison of the range images of the SwissRanger and the Bumblebee2 system: (a) left image of the Bumblebee 2; (b) range image of the Bumblebee2; (c) range image of the SR3000; and (d) 3D plot of the range data in part c.

(~1%) at 1.5 m and 17 cm (3.4%) at 5 m. This means that the system can maintain good measurement accuracy only for a short range. Figure 23.2b shows the range image of the Bumblebee2 for an indoor scene with a stairway. It can be seen that there are quite a lot of missing data that may affect the usefulness of the range data.

Robot pose estimation is a process of determining a robot's position (*X*, *Y*, *Z*) and Euler angles (roll, ϕ; pitch, θ; yaw, ψ angles) in the world coordinate. The representative stereovision-based pose estimation approach is the visual odometry (VO) algorithm [4]. The VO method estimates the robot's ego-motion (i.e., pose change) by detecting features in a stereo image pair and tracking them from one frame to the next. The features' 3D positions in each frame are determined by stereo matching. Feature tracking between frames is performed by selecting features in an image and locating them in the subsequent image by spatial correlation search. The VO method is only effective in a short range, that is, it may incur large pose estimation errors for distant features. This is because a stereovision system's depth measurement error quadratically increases with the true distance. To retain certain pose estimation accuracy, the robot may need to point the stereovision system to the terrain at its immediate vicinity. This reduces the robot's look-ahead distance and may affect the dual use of the stereovision data for obstacle avoidance. In principle, the VO method assumes a static environment for feature matching. This assumption does not hold when the robot's shadow is within the field of view of the stereovision system. The robot shadow may produce moving features which sometimes may cause failure of the VO method [10].

Obstacle detection/avoidance consists of a procedure of collecting 3D data from the sensor, building a terrain map, analyzing the traversal property of the map, and determining proper action for a robot. Usually an elevation map is used to describe terrain and the traversal property of each cell of the map can be described by its drivability (drivable, occupied, or unknown) [11] or the so-called traversability index (TI) [12]. The TI of a cell is calculated using the slope of the least-square plane (LSP) of the terrain patch surrounding the cell and the LSP fitting error. Terrain data captured by stereovision are not reliable for obstacle avoidance as they may miss excessive range data. This is because stereo-matching can only be performed on feature-rich points and thus featureless points generate missing data. Incomplete range data may affect the effectiveness of obstacle avoidance. Dense stereo [4, 5] may improve data density at the cost of a longer processing time and thus a lower frame rate. However, the range data are still incomplete and the density of range data depends on scene.

23.1.2 LIDAR

A LIDAR system [13, 14] outperforms a stereovision system in terms of the completeness and accuracy of range data. It has a consistent range measurement error of a few millimeters and is less prone to the missing data problem. As a result, it has been widely used in robotic perception. For example, the DARPA grand challenge vehicle [11] uses a sensor suite comprising cameras, RADAR sensors, and 2D LIDARs for terrain mapping and obstacle avoidance. In the most recent Urban Challenge competition, the Velodyne 3D LIDARs [15] were used for autonomous navigation of the robotic vehicles [16, 17]. Although these robotic systems have successfully demonstrated much higher autonomy, the size, weight, and power consumption of the sensor suites make such a multi-sensor approach unsuitable for small robots.

The Velodyne LIDAR may generate dense laser data (133,333 points/frame with a 10 Hz frame rate) for terrain mapping. However, it is too expensive and bulky for a small robot. To overcome this problem, the author developed a 3D terrain mapping and navigation method using a single 2D LIDAR in Reference 12. In the mapping system, the robot movement sweeps the 2D LIDAR's laser scan-line over the terrain and produces 3D terrain data over time. These laser scans are registered into a 3D map using the robot pose information computed from the data of wheel-odometry and a three-axis gyro. The system requires accurate robot pose information for laser data registration. Its efficacy has been validated on much smaller robotic platforms—the Segway Robotic Mobility Platform (SRMP) [9] and the Pioneer robot [18].

23.1.3 Flash LIDAR Camera (FLC)

An FLC illuminates the entire scene with a single laser pulse (or modulated infrared light) and focuses the image onto the 3D sensor's focal plane array (FPA). Each pixel on the FPA measures the time of flight (TOF) and thus the object distance to produce a cloud of points. The FLC also produces an intensity image simultaneously. Currently, the commercially available systems include SwissRanger SR4000 and TigerEye 3D FLCs. The former uses modulated infrared light and computes the TOF through phase shift measurement, whereas the latter uses laser pulse and direct TOF measurement. An FLC has the following advantages over stereovision: (1) it measures depth by TOF and therefore has consistent measurement accuracy (approximately ± 1 cm) in its full measurement range; (2) it is able to produce complete range data of the scene; and (3) as an active sensor it can be used for both daytime and nighttime driving and is not affected by the self-shadow problem. Figure 23.1 compares the range measurement accuracies of the SwissRanger SR4000 and the Bumblebee2 stereovision system. The SR4000 (solid line) has a consistent measurement error (approximately ± 1 cm) for a range of up to 5 m, whereas the stereovision's error (dotted curve) increases quadratically with range (approximately 1 cm at 1.25 m, and approximately 17 cm at 5 m). The data completeness of the SR4000 is demonstrated in Figures 23.2c and 23.2d.

SwissRanger SR4000 (Figure 23.3) uses a CMOS imaging sensor. It is small in size ($65 \times 65 \times 68$ mm^3) and has a power dissipation of 9.6 W (even lower dissipation in a trigger mode). The sensor illuminates its entire environment with modulated infrared light. Based on phase shift measurement, it detects the range values up to 5 m (with ± 1 cm accuracy) for every point sensed by the imaging pixels ($176 \times 144 = 25{,}344$ pixels). The sensor produces a 3D range image and an intensity image simultaneously at a rate up to 54 frames per second. The small size and the capability of producing both visual and range data have made it an ideal imaging sensor for small robot autonomy. Simultaneous visual and range data processing allows for robot pose estimation, obstacle detection, and scene recognition by using a single sensor modality.

FIGURE 23.3 The SR4000.

The remainder of the chapter is organized as follows: Section 23.2 presents a complete LIDAR-based terrain mapping and navigation system. Section 23.3 introduces robot pose estimation and 3D range data segmentation methods in an FLC-based perception system. Section 23.4 concludes the chapter and discusses some future research directions.

23.2 AN LIDAR-BASED TERRAIN MAPPING AND NAVIGATION SYSTEM

A terrain mapping and navigation system is depicted in Figure 23.4. It uses a Pioneer P3DX robot as the mobile platform. The sensors used for mapping and navigation

(a) (b)

FIGURE 23.4 The P3DX robot equipped with sensors for mapping and navigation: (a) side view and (b) front view.

include a 2D LIDAR (Sick LMS 200) and a three-axis fiber optic gyro. The gyro measures the robot's Euler angular rate that is combined with the wheel-odometry data for robot pose estimation. The on-board computer uses a 2.0 GHz Pentium 4 processor. It acquires the laser range data from the LIDAR and the robot Euler rate from the gyro through an RS-422 card and the wheel-odometry data through an RS-232 port. A high-speed RS-422 serial card (Quatech DSC-200/300 PCI board) is used to communicate with the LIDAR at 500k baud. The gyro and odometry data are used to estimate the robot's pose, based on which the laser data are registered into a terrain map. The navigation software then analyzes the terrain map and determines suitable motion commands to control the robot. The onboard computer also transmits the robot's pose and laser data to an off-board laptop computer via the wireless router. The process of terrain mapping and navigation is thus visualized on the laptop computer in real time.

23.2.1 Overview of the Mapping and Navigation System

As depicted in Figure 23.5, the terrain mapping and navigation system consists of three main modules: terrain mapping, terrain traversability analysis (TTA), and path planning. The perception sensor (the LIDAR) is mounted on the front-end of the robot. It looks forward and downward at the terrain with an angle of $-10°$ from the robot base. While the robot is in motion, the fanning laser beams profile the terrain ahead of the robot and produce a terrain map. In the system, a $2\frac{1}{2}$D grid-type terrain map (elevation map) is used for navigation.

The TTA module transforms the terrain map into a traversability map where each cell holds a value representing the degree of difficulty for the robot to move across that cell. This value is called traversability index (TI) as mentioned before. The path planning module converts the traversability map into a number of polar traversability indexes (PTIs) each of which represents the level of difficulty for the robot to move along the corresponding direction. Based on the PTIs, the path planning module

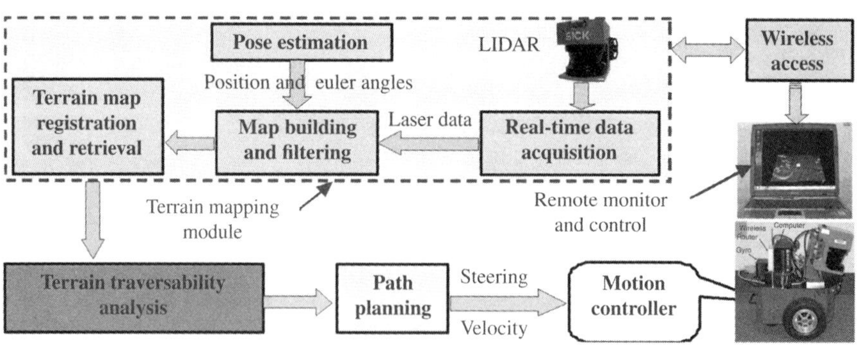

FIGURE 23.5 Diagram of the navigation system: the module within the dashed lines is the terrain mapping module.

determines the steering and velocity commands of the robot and sends these motion commands to control the robot.

23.2.2 Terrain Mapping

Terrain mapping is a process of transforming laser range measurements to 3D points in the world coordinate (also called the navigational frame) and registering these data points into a map. In this work, an extended terrain map (ETM) is first built. Each ETM consists of an elevation map and a so-called certainty map. Both are 2D grid-type maps. Each cell in the elevation map holds a value representing the height of the object at that cell; while each cell in the certainty map holds a value representing the certainty that the corresponding cell in the elevation map is occupied by the object.

Homogeneous coordinate transformation is used to convert a range measurement into a point in the navigational frame. Figure 23.6 depicts the coordinate systems used for terrain mapping. The left figure depicts the diagram of the robot and the right figure details the coordinate systems fixed on the robot body. In Figure 23.6, $x_b y_b z_b$ is the robot body frame fixed to the midpoint of the wheel axel. Frames $x_s y_s z_s$ and $x_l y_l z_l$ are the coordinate systems fixed to the LRF's mounting bracket and the LRF, respectively. The origin of frame $x_s y_s z_s$ locates at $(0, m, g + h)$ in frame $x_b y_b z_b$. The angle between z_s and z_l is α, which is the LRF's initial tilt-down angle, that is, $-10°$. Frame $x_r y_r z_r$ is the coordinate system attached to the laser receiver and its origin locates at $(0, b, p)$ in frame $x_l y_l z_l$. The navigational frame $x_n y_n z_n$ is aligned with the robot body frame when a navigation task is specified and it remains constant throughout the entire navigation task. The origin of the robot body frame is at (u, v, w) in the navigational frame. The three Euler angles are defined as follows: roll (ϕ), pitch (θ), and yaw (ψ), which are the rotation angles around y_b, x_b, and z_b axes, respectively.

The homogeneous transformation matrix T_r^n can be derived through successive rotations and translations from frame $x_r y_r z_r$ to frame $x_n y_n z_n$. The coordinates of a point in frame $x_b y_b z_b$ can be converted into the coordinates of a point in frame $x_n y_n z_n$

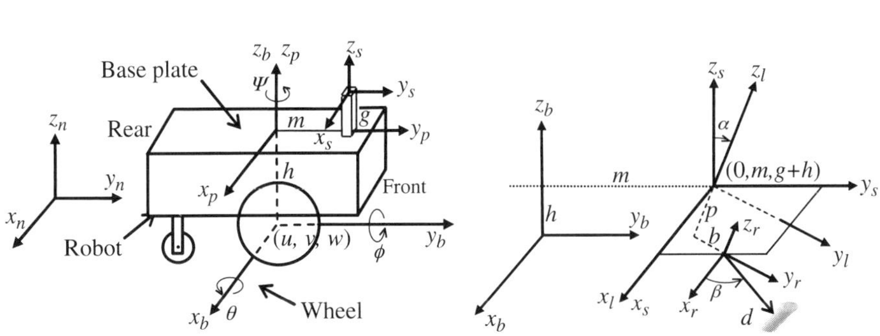

FIGURE 23.6 The coordinate systems for terrain mapping and navigation.

by applying a roll rotation *Roty(ϕ)* followed by a pitch rotation *Rotx(θ)*, a yaw rotation *Roty(ψ)*, and a translation *Trans(u, v, w)*. The corresponding transformation matrix T_b^n is given by

$$
\begin{aligned}
T_b^n &= Trans(u, v, w)\ \ Rotz(\psi)\ \ Rotx(\theta)\ \ Roty(\phi) \\
&= \begin{bmatrix} (c\psi c\phi - s\psi s\theta s\phi) & -s\psi c\theta & (c\psi s\phi + s\psi s\theta c\phi) & u \\ (s\psi c\phi + c\psi s\theta s\phi) & c\psi c\theta & (s\psi s\phi - c\psi s\theta c\phi) & v \\ -c\theta s\phi & s\theta & c\theta c\phi & c + w \\ 0 & 0 & 0 & 1 \end{bmatrix},
\end{aligned} \tag{23.1}
$$

where cx and sx stand for $cos(x)$ and $sin(x)$, respectively. Similarly, the transformation matrix that converts the coordinates of a point in frame $x_r y_r z_r$ into the coordinates of a point in frame $x_b y_b z_b$ is

$$
\begin{aligned}
T_r^b &= Trans(0, m, g + h) Rotx(\alpha) Trans(0, b, p) \\
&= \begin{bmatrix} 1 & 0 & 0 & 0 \\ 0 & c\alpha & -s\alpha & bc\alpha - ps\alpha + m \\ 0 & s\alpha & c\alpha & bs\alpha + pc\alpha + g + h \\ 0 & 0 & 0 & 1 \end{bmatrix}.
\end{aligned} \tag{23.2}
$$

Therefore, T_r^n is computed by

$$
T_r^n = T_b^n T_r^b = \begin{bmatrix} q_{11} & q_{12} & q_{13} & q_{14} \\ q_{21} & q_{22} & q_{23} & q_{24} \\ q_{31} & q_{32} & q_{33} & q_{34} \\ 0 & 0 & 0 & 1 \end{bmatrix}. \tag{23.3}
$$

A range measurement l is converted into a 3D point $\bar{d}^r = (l\cos\beta, l\sin\beta, 0)^T$ in frame $x_r y_r z_r$, where $\beta = n\pi/180$ is the scanning angle of the nth laser measurement. Transforming $\bar{d}^r$ into a point $(x_n, y_n, z_n)^{\mathrm{T}}$ in the navigational frame produces

$$
\bar{d}^n = T_r^n \bar{d}^r = \begin{bmatrix} q_{11} lc\beta + q_{12} ls\beta + q_{14} \\ q_{21} lc\beta + q_{22} ls\beta + q_{24} \\ q_{31} lc\beta + q_{32} ls\beta + q_{34} \\ 1 \end{bmatrix} = \begin{bmatrix} x_n \\ y_n \\ z_n \\ 1 \end{bmatrix}. \tag{23.4}
$$

For every two consecutive measurements (at time steps t and $t + 1$) on the object at the same location (x_n, y_n), the expected change in height value Δz should not be bigger than

$$\Delta z^t_{\max} = \Delta u \tan\phi - \Delta v \tan(\theta + \alpha) + \Delta w + \frac{\partial z_n}{\partial \psi}\Delta\psi + \frac{\partial z_n}{\partial \theta}\Delta\theta + \frac{\partial z_n}{\partial \phi}\Delta\phi + \frac{\partial z_n}{\partial l}\Delta l, \tag{23.5}$$

where Δl is the maximum measurement error of the LRF (32 mm according to Reference 10); Δu, Δv, Δw, $\Delta\phi$, $\Delta\theta$, and $\Delta\psi$ are the changes in the robot's pose from t to $t + 1$. $\partial z_n/\partial\psi$, $\partial z_n/\partial\theta$, $\partial z_n/\partial\phi$, and $\partial z_n/\partial l$ are the partial derivatives that can be derived from Equation (23.4).

In this work, a grid cell size of 80 × 80 mm is used. The height value of cell (i, j) in the elevation map is denoted by $h_{i,j}$, while the certainty value of this cell in the certainty map is denoted by $c_{i,j}$. Using the above transformation, range data are translated into a point (x_n, y_n, z_n). In order to register this point into the map at the associated location, the coordinate values x_n and y_n are converted to the grid indices i and j, respectively. Then the certainty value $c_{i,j}$ and height value $h_{i,j}$ of the cell in the terrain map are updated. Specifically, $c_{i,j}$ is updated by

$$c^{t+1}_{i,j} = \begin{cases} c^t_{i,j} + a & \text{if } \left|z^{t+1}_n - h^t_{i,j}\right| \le \Delta z^t_{\max} \text{ or } c^t_{i,j} = 0 \\ c^t_{i,j} & \text{otherwise} \end{cases}, \tag{23.6}$$

where $a \ge 1$ is the increment of the certainty value and it can be any positive integer. In this study, the certainty value is represented by 1 byte and $a = 3$ is used. As a result, all cells illuminated more than 84 times have the same certainty value. The height value $h_{i,j}$ of the cell in the elevation map is updated by

$$h^{t+1}_{i,j} = \begin{cases} z^{t+1}_n & \text{if } z^{t+1}_n > h^t_{i,j} \text{ or } c^t_{i,j} = 0 \\ h^t_{i,j} & \text{otherwise} \end{cases}. \tag{23.7}$$

$c_{i,j}$ and $h_{i,j}$ are initially zero and updated by Equations (23.6) and (23.7), respectively. In every 13.3 ms, the 181 range measurements of the LIDAR are mapped into both maps. In Equation (23.6), the condition $\left|z^{t+1}_n - h^t_{i,j}\right| \le \left|\Delta z^t_{\max}\right|$ is called motion continuity constraint [19]. Every two consecutive measurements on a real object at the same location should satisfy this constraint. According to Equation (23.6), cells occupied by real object are assigned continuous certainty increments and result in large certainty values. Therefore, they can be identified easily. Mixed pixels and random noise, on the other hand, do not satisfy the motion continuity constraint and thus result in small certainty values. In addition, they are spatially isolated in the elevation map. These characteristics can be used to filter out mixed pixels and random noises. Details of the filtering method are referred to in Reference 13.

23.2.3 Terrain Traversability Analysis

The task of TTA is to transform a terrain map into a traversability one by assigning a TI value to each cell in the terrain map. This process is divided into two steps: estimating terrain slopes and roughness, and computing the TI value for each cell.

A square terrain patch P centering at the robot geometric center (RGC) is formed in the terrain map. In this paper, P is chosen in such a way that it exactly envelops the robot regardless of the robot's orientation. The number of data points in the patch is $N = (2L + 1) \times (2L + 1)$, where $2L + 1$ is the number of cells each side length of P has. We then fit a LSP to P. The normal to the LSP $\boldsymbol{n} = (n_x, n_y, n_z)$ and the residual of the fit σ are found by using the singular value decomposition (SVD) method [20].

23.2.3.1 *Scheme 1 for TTA* For this scheme, the slope of the terrain patch is estimated by

$$\varphi = \cos^{-1}(n_z), \tag{23.8}$$

and the TI value of cell (i, j) is calculated using σ and φ as follows:

$$\tau_{i,j} = F_1\varphi + F_2\sigma/N. \tag{23.9}$$

23.2.3.2 *Scheme 2 for TTA* Scheme 1 has been implemented in an SRMP recently [12]. The disadvantage of using Equation (23.9) is that it does not reflect the fact that the level of difficulty for a robot to traverse a slope is related to the robot's yaw angle. To account for this, an alternative method of computing TI is proposed. As the method computes a TI along a specific direction, the robot yaw angle ψ is known. Considering that the robot is on the fitted plane whose normal vector is $\boldsymbol{n}$, the roll and pitch of the robot can be estimated by

$$\begin{cases} \phi = \sin^{-1}(n_x \cos\psi + n_y \sin\psi) \\ \theta = \sin^{-1}((n_x \sin\psi - n_y \cos\psi)/\cos\phi) \end{cases}. \tag{23.10}$$

The TI value of cell (i, j) is then calculated by

$$\tau_{i,j} = F_1 \max(|\phi|, |\theta|) + F_2\sigma/N. \tag{23.11}$$

Since the first term depends upon the yaw angle ψ, the use of the TI in Equation (23.11) allows the robot to steer into a direction that minimizes $\tau_{i,j}$. This enables the robot to snake up/down a steep ramp. F_1 and F_2 in Equations (23.9) and (23.11) are chosen by simulation runs in typical urban environments. In this work, $F_1 = 3.7$ and $F_2 = 0.15$ are used and this set of parameters results in the roughness having a larger contribution than the slope.

Apparently, the TTA process expands an obstacle's boundaries in order to account for the dimensions of a robot. This allows us to treat the robot as a single point in planning the robot motion.

23.2.4 PTI Histogram for Path Planning

The path planning algorithm determines the robot's next heading direction and velocity based on the traversal property of the local terrain map surrounding the robot. In this work, the traversal property is described by the form of PTI.

In order to compute PTIs, a square-shaped local traversability map S^* obtained by the above TTA method is formed at the RGC. There are $w_s \times w_s$ cells in S^* and the RGC is at the center of S^*. Figure 23.7 is the traversability map obtained by applying the TTA method to a computer-generated elevation map. A white grid in Figure 23.7 represents a cell with zero TI, whereas the other grids represent cells with nonzero TIs. A cell with nonzero TI in the traversability map generates an imaginary vector field, which exerts a virtual repulsive force on the robot and pushes it away from the

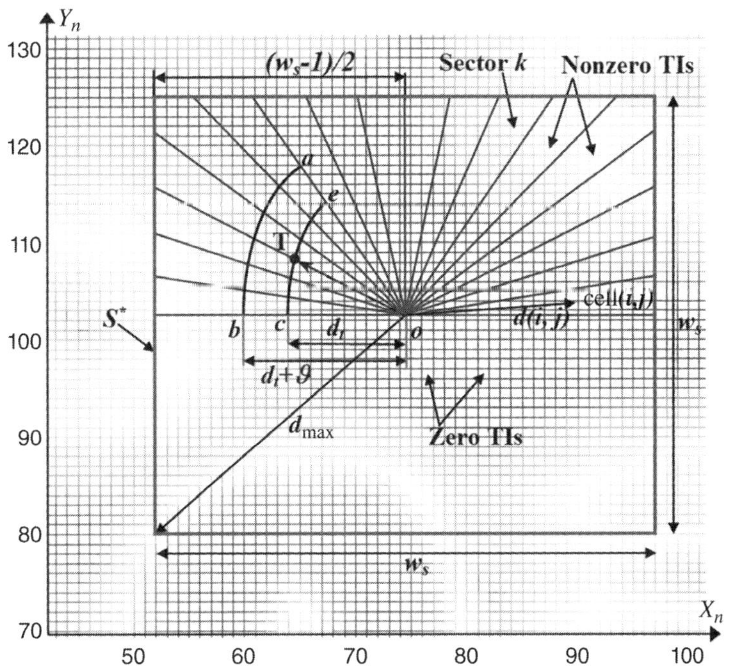

FIGURE 23.7 Transformation of a traversability map into a histogram: X_n and Y_n axes represent the world coordinates; o is the RGC. For simplicity, each sector is drawn in 10° and only half of the sectors are shown.

cell. This is called the "traversability field" that is defined in terms of a direction and a magnitude computed in the same way as in Reference 21.

The direction of the traversability field generated by cell (i, j) is given by

$$\varepsilon_{i,j} = \tan^{-1}\frac{y_j - y_o}{x_i - x_o} \tag{23.12}$$

and the magnitude of the traversability field is

$$m_{i,j} = \tau_{i,j}^2(d_{\max} - d_{i,j}), \tag{23.13}$$

where x_i and y_j are the coordinates of cell (i, j); x_0 and y_0 are the present coordinates of the RGC; $\tau_{i,j}$ is the TI value of cell (i, j); $d_{\max}$ is the distance between the vertices (the four farthest cells) of S^* and the RGC; and $d_{i,j}$ is the distance between cell (i, j) and the RGC. In Equation (23.13), $\tau_{i,j}$ is squared such that the impact of a cell with a small/large TI value is diminished/magnified. Obviously, $m_{i,j}$ decreases with increasing $d_{i,j}$ and it is zero for each of the four vertices of S^*.

In order to evaluate the overall difficulty of traversing along a direction, the magnitude of the traversability field produced by all cells in the same direction should be summed up. For this reason, S^* is divided into n sectors, each of which has an angular resolution of $\xi = 360°/n$. Each sector k, for $k = 0, \ldots, n-1$, has a discrete angle $\rho = k\xi$. Cell (i, j) in S^* is assigned to the kth sector according to

$$k = \mathrm{int}(\varepsilon_{i,j}/\xi), \tag{23.14}$$

where int(.) rounds the enclosed value to the nearest integer. For sector k, the magnitude of the traversability field produced by all cells in this sector is calculated by

$$h_k = \sum_{i,j} m_{i,j}, \tag{23.15}$$

where h_k is the distance-weighted sum of the squared TIs of all cells in sector k. Apparently, a larger value of h_k means that direction $\rho = k\xi$ is harder for the robot to move along with. In other words, h_k is a TI representing the overall difficulty of traversing the terrain in the corresponding direction. Therefore, it is called the polar traversability index (PTI). In this chapter $\xi = 5°$, that is, there are 72 sectors. The PTIs are represented in a form of histogram. The path planning method then clusters the histogram into candidate valleys (comprising consecutive sectors with PTIs below a threshold) and hills (comprising consecutive sectors with PTIs above the threshold). One of the candidate valleys that is the closest to the target direction and incurs the smallest heading change is selected as the wining valley. Sector k_h located inside the wining valley is then chosen as the robot's next heading direction. The details on how the path planning method determines the robot's next heading direction are referred to in Reference 18.

Since the controller of the robot accepts steering rate and velocity as the control commands, the robot's next heading direction is used to compute the steering rate as follows:

$$\dot{\psi}' = \left(k_h(t+1) - k_h(t)\right)\xi/\Delta T, \tag{23.16}$$

where $k_h(t)$ and $k_h(t+1)$ are the robot's heading directions at time steps t and $t+1$, respectively; and $\Delta T = 93.3$ ms is the control interval that equals to the time for the navigation system to acquire seven sets of laser data. When the robot heads on an obstacle, a reduction in velocity is required since this gives the robot more time for ON. It also allows the robot to obtain more range data on the potential hazard. For this reason, the robot velocity is calculated by

$$v' = v_{\rm max}\left(1 - \min(h_{\rm c}, h_{\rm m})/h_{\rm m}\right), \tag{23.17}$$

where $v_{\rm max}$ is the robot's maximum velocity, $h_{\rm c}$ is the PTI in the robot's current heading direction, and $h_{\rm m}$ is a constant that is empirically chosen to produce a sufficient reduction in velocity. A reduction in velocity is also required when the robot approaches its target. In this work, the velocity is further reduced by

$$v'' = v'\min(d_t, d_m)/d_m, \tag{23.18}$$

where $d_{\rm t}$ is the distance between the target and the RGC, and $d_{\rm m}$ is a constant (1.5 m in this work). Note that $\dot{\psi}'$ and v'' are actually the projection of the steering rate $\dot{\psi}$ and velocity V on the $x_n y_n$ plane. They are converted to $\dot{\psi}$ and V. A low pass filter is used to smooth the steering and velocity commands before being sent to the motion controller. In the system, the smoothed command is computed by

$$\Theta_a(t+1) = \lambda\Theta(t) + (1-\lambda)\Theta(t+1), \tag{23.19}$$

where $\Theta_a(t) = \left(V_a(t), \dot{\psi}_a(t)\right)^{\rm T}$ and $\Theta(t) = (V(t), \dot{\psi}(t))^{\rm T}$. $V_a(t)$ and $\dot{\psi}_a(t)$ are the smoothed velocity and steering commands that are actually applied to the robot. A larger value of λ produces smoother control commands but a slower system response. $\lambda = 0.75$ is used in the system to provide a relatively smooth control.

23.2.5 Experimental Results

The complete terrain mapping and navigation system was tested with the pioneer robot on an outdoor terrain outside the Fribourgh Hall at the University of Arkansas at Little Rock. As shown in Figure 23.8, the terrain is uneven and contains curbs and a ramp. A number of cardboard boxes were used to construct obstacle courses. A navigation task was specified to the robot by giving the robot its target location. The system was initialized using the robot's current position as its start point (the robot pose was set to zero).

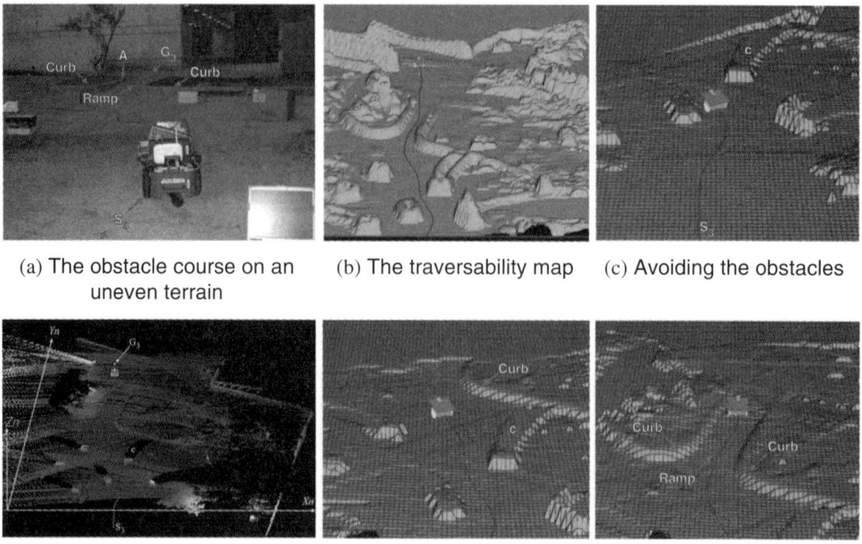

(a) The obstacle course on an uneven terrain (b) The traversability map (c) Avoiding the obstacles

(d) 3D terrain map in point cloud (e) Avoiding the curb (f) Climbing up the ramp

FIGURE 23.8 Experiment on an outdoor terrain outside the Fribourgh Hall at UALR: the robot ran from S_3 to G_3. The curve represents the robot trajectory. The robot in part (b) is just for illustration. It is actually a point in the traversability map. (a) The obstacle course on an uneven terrain; (b) the traversability map; (c) avoiding the obstacles; (d) 3D terrain map in point-cloud; (e) avoiding the curb; and (f) climbing up the ramp.

The terrain mapping and navigation system was implemented on the onboard computer that ran RTLinux as the OS. The LIDAR's angular resolution was set to 1°. Each laser scan takes 13.3 ms and contains 181 range data that covers a 180° field of view. The Real-time Data Acquisition (RDA) section of the terrain mapping module (Figure 23.5) resides in the real-time Linux kernel as a real-time thread. The RDA thread stores the laser, gyroscopic, and wheel-odometry data in a circular buffer. Each time a set of seven laser scans and the associated gyroscopic and odometry data are fetched and sent into a shared memory, from which the Map Building and Filtering (MBF) section reads the data set, computes the robot pose, and registers the laser data into a terrain map. MBF also sends the data set to a remote laptop computer for visualization of navigation. The TTA module then obtains the local terrain map (55 × 55 cells) and transforms it into a traversability map. Then the path planning module converts the traversability map into a PTI histogram, determines the motion commands, and sends the commands to the robot. The above process repeats every 93.3 ms (7 × 13.33 ms) until the robot reaches its target.

An RTLinux driver for the high-speed RS422 serial card was written in C code for communication with the LIDAR and gyro. The serial driver runs in the real-time Linux kernel. The RDA thread reads data from the serial driver and communicates with the MBF through the shared memory. The MBF is a process in the regular

Linux kernel, which is treated as a process in the RTLinux kernel. To maintain data correctness in the interprocess communication, a data flag is created in the shared memory to indicate the state of data set, that is, ready for the MBF to read or empty for the RDA to write. The RDA's circular buffer may accommodate 10 laser scans and the associated gyroscopic and odometry data. This arrangement will prevent data loss if the shared memory is temporarily not available for RDA writing.

Figure 23.8d shows one of the autonomous runs from S_3 to G_3 on the terrain. S_3 has a higher elevation than G_3. The area immediately before the ramp is a valley that is lower than the neighboring area, including S_3 and G_3. The 3D terrain map (rendered on the remote laptop computer) is depicted in Figure 23.8d in a form of point-cloud where the dark-gray points represent data with negative elevation (below S_3). The black spots are unperceived areas (no range data). In this map a brighter point has a bigger elevation value. Figure 23.8c displays the elevation map at the point where the robot is avoiding the first obstacle it has encountered. Figure 23.8e shows the elevation map at the time when the robot is avoiding the curb and aligning itself with the ramp. Figure 23.8f shows the elevation map after the robot successfully traverses the ramp. In the elevation maps, the unperceived areas (e.g., area c) of the lower terrain surface look like obstacles because their Z coordinate values (zero) are bigger than that of the surrounding terrain surfaces whose elevation values are negative.

Data collected from all experimental runs have demonstrated that the roll, pitch, and yaw angles, and the velocity of the robot change smoothly. This means that the navigation method produces smooth motion commands for the robot. Figure 23.9 depicts the robot's roll, pitch, and yaw angles during the experimental run from S_3 to G_3 in Figure 23.8. The roll and pitch angles are limited in $(-1°, 8°)$ and $(-3°, 7°)$, respectively. This means that the navigator guides the robot to traverse moderate terrain. Both roll and pitch angles have changed smoothly during the entire navigation task except for the points around "S" in Figures 23.9a and 23.9b where relatively larger changes in roll and pitch are observed. This is because the robot runs over the left curb at the upper end of the ramp as labeled by *A* in Figure 23.8a. The curb is low-profiled and considered traversable. However, it generates relatively larger changes in the robot's roll and pitch angles. As shown in Figure 23.9c, the robot's yaw angle changes very smoothly during the whole process. This indicates that the navigation method generates smooth steering commands for the navigation of the robot.

23.3 FLC-BASED SYSTEMS

As indicated earlier, an FLC is ideal for the autonomy of a small robot or robotic device. A robotic device is a portable robotic system with navigational functions such as positioning, obstacle detection, and scene recognition. A typical example is a computer vision enhanced white cane for the visually impaired [22]. For indoor navigation, the representative functions include feature-based robot/device pose estimation, object detection, and scene recognition. Feature-based pose estimation (FPE)

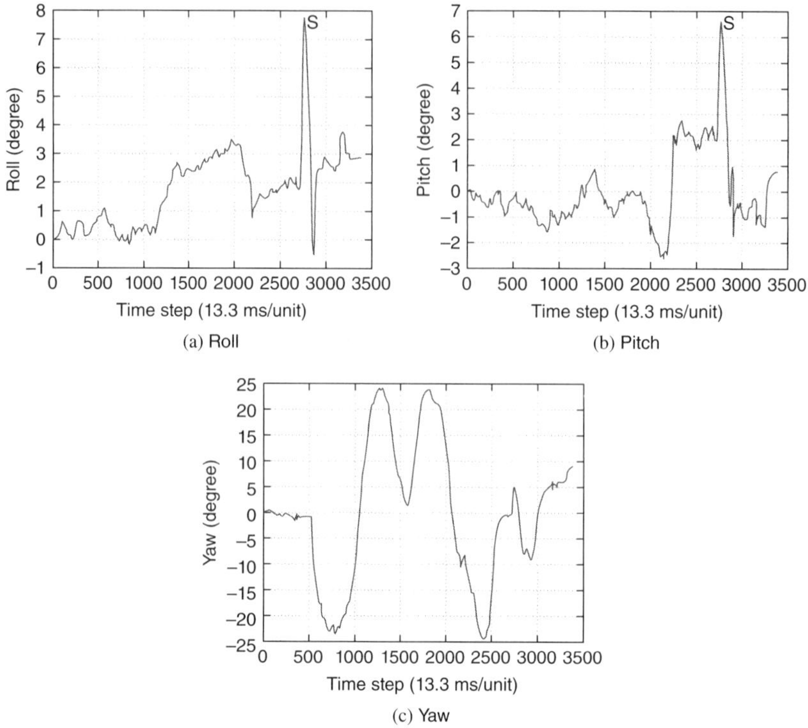

FIGURE 23.9 (a) Roll; (b) pitch; and (c) yaw angles of the robot during the run from S_3 to G_3 as shown in Figure 23.8.

involves a procedure of visual and range data coprocessing, called VR-odometry, while the key to object detection and scene recognition is 3D data segmentation.

23.3.1 VR-Odometry

23.3.1.1 Operating Principle The proposed FPE method is to perform feature detection and matching in the SR4000's intensity images and use the matched features' range data to determine the pose change in two consecutive image frames. The method is termed VR-odometry as it uses both visual and range data to determine the pose change.

In this work, the scale-invariant feature transform (SIFT) feature detector [23] is used to extract features in an intensity image and match them to the SIFT features in the next intensity image. As the 3D coordinates of the matched features are known from the range data, the feature-matching process in fact solves the 3D point correspondence (data association) problem between each two sampling steps, and result in two 3D data sets, $\{p_i\}$ and $\{p_i'\}$; $i = 1, \ldots, N$. N is the number of matched

SIFT features in the two image frames. The pose estimation problem is then to find rotation and translation matrices, R and T, that minimize the error residual

$$e^2 = \sum_{i=1}^{i=N} \|p_i' - Rp_i - T\|^2. \tag{23.20}$$

This least-square data sets fitting problem can be solved by the SVD method in Reference 24. As feature-matching in intensity images may result in incorrect data association (outliers), a random sample consensus (RANSAC) process is implemented to reject the outliers. The entire method has the following steps:

(1) Extract the SIFT features in two consecutive images, find the matched features, and locate the corresponding 3D data sets $\{p_i\}$ and $\{p_i'\}$.

(2) Randomly select four associated points from the two data sets and form $\{p_k\}$ and $\{p_k'\}$; $k = 1, \ldots, 4$. Then, find the least-square rotation and translation matrices ($\hat{R}_k$ and $\hat{T}_k$) for $\{p_k\}$ and $\{p_k'\}$.

(3) Project the entire data set $\{p_i\}$ onto $\{p_i'\}$ using the found transformation ($\hat{R}_k$ and $\hat{T}_k$) and compute the error

$$e_i^2 = \|p_i' - \hat{R}_k p_i - \hat{T}_k\|^2 \tag{23.21}$$

for each data-pair $i = 1, \ldots, N$. A threshold ε is used to score S_k for this transformation: S_k is incremented once for each $e_{i^2} < \varepsilon$.

(4) Steps 2 and 3 are repeated for a fixed number of iterations or until all combination of point set selections are exhausted, whichever is smaller. The transformation with the highest score is recorded. The corresponding data sets $\{p_j\}$ and $\{p_j'\}, j = 1, \ldots, S_k$, where each data-pair satisfies the threshold test in Step 3, are selected and used to compute the maximum likelihood transformation estimate $\hat{R}$ and $\hat{T}$ by the SVD least-square fitting method. The sensor's rotation can be computed from $\hat{R}$ and its translation is determined by $\hat{T}$. The pose change is thus determined.

23.3.1.2 Experimental Results A rudimentary implementation of the VR-odometry has been performed in Matlab environment to validate the idea. The implementation uses SIFT feature descriptors as they are invariant to translations, rotations, and scaling and robust to residual small distortions. Figure 23.10 illustrates how the VR-odometry finds the correct correspondences between features in two successive images. The images were captured when a human handheld an SR4000 and moved in a hallway. Figure 23.10a depicts the detected SIFT features. They were matched with the features in the next frame based on the SIFT features' scales and orientations. Figure 23.10b shows the initial matches that exhibit some outliers (mismatches). The RANSAC process removes the outliers and the results are shown in Figure 23.10c.

A series of experiments was carried out in an office environment with objects ranging 1.5–4.5 m from the SR4000. The sensor was installed on a pan-tilt unit.

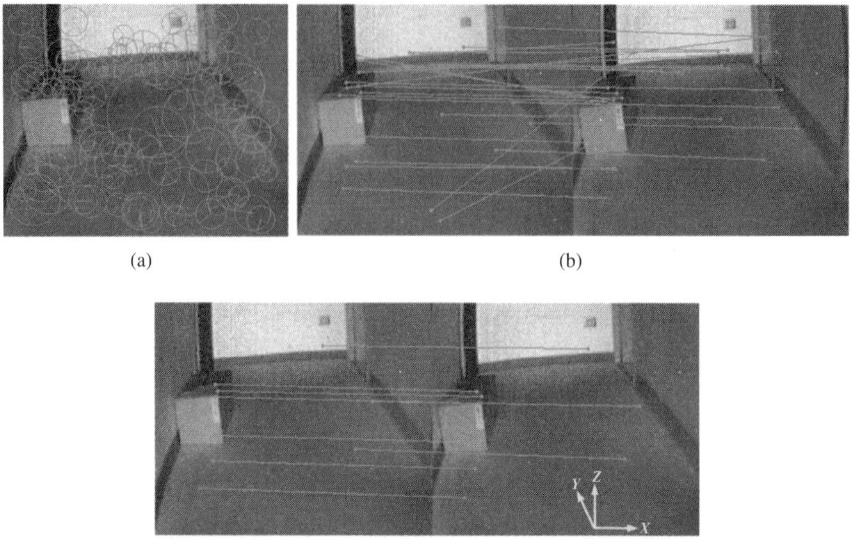

(c)

FIGURE 23.10 SIFT feature matching and RANSAC for outlier removal: (a) SIFT features of an image frame (each circle and the straight line represent the scale and orientation of the SIFT feature located at the center of the circle); (b) initial matched features in two consecutive image frames; and (c) matched features after RANSAC. Intensity images were captured in a hallway. The time interval between the right and left images in part (b) is 100 ms. Therefore, the objects in the right image are closer to the sensor.

Experimental runs, with various combinations of pitch, yaw rotation, and X, Y translations, were performed to quantify the measurement accuracy and repeatability. In all experiments, roll ϕ and Z translation are always zero. Total 1000 images for each pose were taken to compute pose changes and the error statistics. It is noted that in all experiments, we used raw sensor data for pose estimation.

Distribution of Pose Measurement Errors Two experiments were carried out to examine the distribution of pose measurement errors. In the first experiment, the SR4000's pose change is zero. The experimental result demonstrates that the measurement error of pose ($\phi, \theta, \psi, x, y, z$) is zero-mean Gaussian with standard deviation (0.1°, 0.2°, 0.2°, 7 mm, 3 mm, 6 mm). The result indicates that the VR-odometry's inherent error is a white Gaussian noise. Also, the measurement accuracy (mean error) and repeatability (standard deviation) are excellent, meaning that the sensor's noise has limited effect on the VR-odometry's performance in pose estimation.

In the second experiment, the SR4000 had a combination of rotation ($\theta = -5.9°$, $\psi = 5.0°$) and translation ($X = -80$ mm, $Y = 130$ mm). The distribution of the pose measurement error follows a normal distribution whose mean and standard deviation are (−0.1°, 0.2°, −0.3°, 8 mm, −2 mm, 11 mm) and (0.5°, 0.4°, 0.4°, 13 mm, 5 mm, 11 mm), respectively. Compared with the first experiment, the results exhibit a bias in the mean error and relatively larger standard deviation.

Both experiments show a decent accuracy in orientation measurement. Considering that the SR4000's average angular resolution is about 0.25°, the accuracy of orientation measurement is reasonable. The VR-odometry also demonstrates a very good accuracy and repeatability in measuring motion along Y-axis due to the sensor's high accuracy and repeatability in depth measurement. The sensor's resolution in measuring motion along X and Z axes drops proportionally with the depth (Y) value. At $Y = 5$ m, the resolution is about 23 mm. The VR-odometry's measurement accuracy and repeatability in X and Z axes seem reasonable. However, proper camera calibration and data filtering may improve the performance.

Accuracy and Repeatability of Pose Measurements Experiments were performed to inspect the pose estimation performance for each individual motion. Due to the constraint of the test facility, only experiments with pitch and yaw rotations and X and Y translations were performed. In the first experiment, the sensor underwent a pitch rotation in the range [3°, 21°] (increment: 3°/step). Total 1000 images were captured before and after each pitch rotation for computing the pose change. It is observed that the accuracy (mean error) and the repeatability of the pitch measurement are quite decent for a pitch rotation in the range [3°, 18°]: the mean errors are in the range [0°, 0.21°], and the standard deviations are in [0.53°, 0.86°]. The relative mean errors are between 0.0% and 2.0%.

The second experiment is to test the VR-odometry's performance in measuring yaw rotation in the range [−21°, −3°]. The mean errors are in the range [−0.36°, 0.03°] and the standard deviations are in [0.39°, 0.88°]. The relative mean errors are between −1.0% and 2.7%.

Table 23.1 shows the statistics of the roll, pitch, and yaw measurements of both experiments. It can be observed that the measurements are accurate if the pitch/yaw

TABLE 23.1 Measurement Accuracy of Rotation

MV: (μ, σ) / TV: (ϕ,θ,ψ)	Roll ϕ (°)	Pitch θ (°)	Yaw ψ (°)
(0, 3, 0)	(0.06, 0.27)	(3.06, 0.53)	(−0.16, 0.57)
(0, 6, 0)	(0.07, 0.16)	(6.00, 0.64)	(0.02, 0.66)
(0, 9, 0)	(0.01, 0.21)	(9.03, 0.59)	(0.24, 0.65)
(0, 12, 0)	(0.22, 0.21)	(12.14, 0.86)	(0.24, 0.69)
(0, 15, 0)	(0.11, 0.33)	(15.21, 0.63)	(−0.12, 0.91)
(0, 18, 0)	(0.00, 0.31)	(18.04, 0.77)	(−0.46, 0.80)
(0, 21, 0)	(0.29, 0.44)	(22.09, 1.54)	(0.14, 1.06)
(0, 0, −3)	(−0.03, 0.31)	(0.10, 0.52)	(−2.97, 0.39)
(0, 0, −6)	(−0.03, 0.19)	(−0.02, 0.50)	(−6.15, 0.45)
(0, 0, −9)	(0.02, 0.25)	(0.20, 0.60)	(−9.24, 0.61)
(0, 0, −12)	(−0.02, 0.26)	(0.15, 0.66)	(−12.22, 0.64)
(0, 0, −15)	(−0.09, 0.32)	(0.36, 0.68)	(−15.18, 0.76)
(0, 0, −18)	(−0.13, 0.30)	(0.15, 0.69)	(−18.33, 0.78)
(0, 0, −21)	(0.04, 0.39)	(0.20, 0.77)	(−21.36, 0.88)

MV, measured values; TV, true values; μ, mean error; σ, standard deviation; camera's angular resolution: pitch = 0.25°, yaw = 0.24°. Total 1000 samples were used to compute the statistics.

TABLE 23.2 Measurement Accuracy of Translation

MV: (μ, σ) \ TV: (X, Y, Z)	X (mm)	Y (mm)	Z (mm)
(305, 0, 0)	(297, 16)	(−2, 7)	(−8, 24)
(610, 0, 0)	(579, 64)	(−9, 11)	(−4, 75)
(0, 305, 0)	(−15, 73)	(303, 12)	(35, 55)
(0, 610, 0)	(−0, 98)	(607, 17)	(83, 98)

MV, measured values; TV, true values; μ, mean error; σ, standard deviation; camera's resolution at 5 m: $X = 22$ mm, $Z = 23$ mm, $Y = 10$ mm. Total 1000 samples were used to compute the statistics.

rotation is within ±18°, a large image motion between image frames compared with the camera's field of view. The mean errors are mostly within the camera's angular resolution (approximately 0.25°). There are a few exceptions that the mean errors go beyond ±0.25° (the worst case: −0.46°). This can be improved if proper data filtering and sensor calibration are performed. For repeatability, the standard deviations of the roll measurements are consistently small. This is probably because that there was no roll rotation in the experiments. The repeatability of pitch and yaw measurements need to be improved in our future work.

The third experiment is to test the measurements of translation when the sensor moves along X or Y directions. A step size of 305 mm was used in the experiments. We found that the VR-odometry did not produce satisfactory results if the translation was bigger than 610 mm. This suggests that a smaller step size should be used in the future study. The results are tabulated in Table 23.2, from which we can observe that the VR-odometry has highly consistent accuracy and repeatability in measuring movement along Y-axis (depth). This is attributed to the sensor's consistent measurement accuracy in Y-axis (±1 cm). This is its apparent advantage over the stereovision-based approach. However, it has relatively larger mean errors and/or standard deviations in X/Z measurements (compared with the camera's resolution). These need to be improved in our future work.

The results in Tables 23.1 and 23.2 are computed by using raw sensor data. Proper data filtering and camera calibration should improve the measurement accuracy and repeatability in the future work. It should be noted that in some cases the VR-odometry does not find a solution to the pose estimation. The error states were recorded and the data were discarded, meaning that the samples for computing some of the statistics in Tables 23.1 and 23.2 are slightly smaller than 1000. Some of the failures resulted because the RANSAC process did not find a sufficient number of matched features (possibly due to the use of a too small threshold ε). We have not yet looked into the cause of the other failure cases.

In our current implementation, we use the SIFT feature descriptors and the RANSAC method for the sake of the accuracy and reliability of pose estimation. Both approaches are computationally expensive. In our future work, we will develop a more efficient method with real-time performance. This may be achieved through the following efforts: (1) investigate other feature descriptors and adopt one with less computational cost such as speeded up robust features (SURF) [25] for the VR-odometry; and (2) use spatial invariants between the detected features' 3D points (e.g., distances between the 3D points) for outlier rejection (or inlier detection) to accelerate or remove the need of the RANSAC process. In terms of improving the method's accuracy and repeatability, we will calibrate the camera, develop a filtering method to filter out data with low confidence levels, and develop a feature sampling method to select features with more accurate depth values.

23.3.2 Three-Dimensional Data Segmentation

Segmentation of an indoor scene into planar surfaces and grouping them into structures are the essential steps for object/obstacle detection and scene recognition. In this work, a range data frame is represented as a tri-band color image where each pixel's RGB values represent the x, y components of the surface normal and the depth information (y value) of the corresponding point. This scheme encodes a 3D point's local geometric information (surface normal) and global information (location) into a pixel in the color image. This RGB image is called an enhanced range image (ERI), as it enhances the edges and surfaces of objects in the original range image.

With this image enhancement, one may simply apply the normalized cuts (NC) method to the ERI and segment the range data. However, this process is computationally expensive as the number of pixels in the ERI is huge. To reduce the computational cost, the mean shift method is first used to segment the ERI into a number of regions, called super-pixels (SPs), each of which contains a group of homogeneous pixels. A graph is then constructed by treating each SP as a node and the NC method is employed to partition the graph into a number of subgraphs, each of which represents a segment of the range data. The spatial location and color of each SP are computed as the centroid and mean color value of its ERI pixels, respectively. Finally, neighboring planar segments with the same orientation are merged. The above segmentation method is termed an extended normalized cuts (ENC) method. In the ENC, a node of the graph corresponds to a region rather than a single ERI pixel. This lowers the graph's node count and thus the computational time. The proposed segmentation method is less sensitive to noise because: (1) the mean color computation of an SP smoothes the noise of range data in each SP; and (2) the NC method takes both color and spatial information in clustering SPs.

Extended Normalized Cuts Method

The ENC method is explained in this section.

23.3.2.1 *Statistics of Plannar Segments* To classify a segment (an SP or a region containing a number of SPs) as planar or nonplanar, an LSP is fitted to the data points of the segment and the normal $\vec{N}$ and the plane fit error (PFE) Δ is computed by the SVD method [26]. A segment with a sufficiently small Δ, that is, $\Delta < \delta$, is labeled as a planar segment. The angle between planar segments i and j is computed by

$$\theta_{i,j} = \cos^{-1}(\vec{N}_i \bullet \vec{N}_j). \tag{23.22}$$

They are considered as parallel if $\theta_{i,j}$ is sufficiently small, that is, $\theta_{i,j} < \varepsilon$.

23.3.2.2 *Graph Construction and Partitioning* A graph $G = (V, E)$ is constructed by treating each planar SP as a node. The edge weight, which is a measure of similarity between two nodes V_i and V_j, is calculated by

$$w_{i,j} = e^{-\frac{\|F(i)-F(j)\|_2^2}{\sigma_F}} \begin{cases} e^{-\frac{d_{i,j}}{\sigma_D}} & \text{if } V_i \text{ and } V_j \text{ are neighbors} \\ 0 & \text{otherwise} \end{cases}, \tag{23.23}$$

where σ_F and σ_D are positive constants, $F(p)$ is the color vector of node p for $p = i, j$, and $d_{i,j}$ is the Euclidean distance between the LSPs of SPs i and j (i.e., nodes i and j). If $\theta_{i,j} < \varepsilon$ is satisfied, then $d_{i,j}$ is the distance from the centroid of the data in SP i to the LSP of the data in SP j. Otherwise, $d_{i,j} = 0$. Two SPs are considered as neighbors if they have at least two neighboring pixels in the ERI space. The weight computation in Equation (23.23) takes into account the statistics of 3D data points. This may result in a better segmentation result. We then apply the NC algorithm to the graph and cluster the SPs into a set of segments $S = \{s_1, s_2, \ldots, s_N\}$.

23.3.2.3 *Labeling and Merging of Planar Segments* In the last step of the ENC method, the segments in S are labeled as planar and nonplanar according to Section 23.3.2.1. Two neighboring planar segments, s_i and s_j, are merged if $\theta_{i,j} \le \varepsilon$ and $d_{i,j}$ is less than a predetermined threshold.

The ENC plane extraction method is summarized as follows:

(a) Construct ERI from the range data and apply the mean shift algorithm to obtain a number of SPs.

(b) Obtain planar SP_i; for $i = 1, \mathrm{L}\ , m$ from the resulted SPs according to Section 23.3.2.1.

(c) Construct a graph G on SP_i for $i = 1, \mathrm{L}\ , m$ and compute the similarity matrix $\boldsymbol{W}$ of order $n \times n$ by Equation (23.23).

(d) Apply the NC algorithm to graph G with $\boldsymbol{W}$ as the input and obtain N segments, s_i for $i = 1, \ldots, N$, each of which contains a number of SPs. Each segment r in s_i is further classified to form a set of planar segments $P = \{p_1, p_2, \ldots, p_t\}$; $t \le N$.

(e) Construct a binary matrix $K = \{k_{i,j}|_{i=1,\ldots,t;\, j=1,\ldots,t}\}$ to record the neighborhood relationship among segments in P, where

$$k_{i,j} = \begin{cases} 1 & \text{if } p_i \text{ and } p_j \text{ are neighbors and } \theta_{\mathrm{i,j}} \le \varepsilon, d_{i,j} \le \tau \\ 0 & \text{otherwise.} \end{cases} \tag{23.24}$$

It is noted that a segment is treated as its own neighbor. Therefore, $k_{i,i} = 1$.

(f) In the final step, the entire planar surfaces are extracted by merging those segments whose k values equal zero. This is done by using the depth-first search algorithm.

Experimental Results

The segmentation method has been validated with range data captured from the SR4000 in a number of representative indoor environments. In all experiments, a prespecified segment number $N = 75$ is used for the NC method. For the results shown in this section, an unlabeled segment is represented in black and a labeled segment (a planar segment) with a random nonblack color (a random gray value when displayed in black and white). The labeled segments are then overlaid on ERI.

To justify the use of the NC algorithm, the proposed method without the NC component is also run on each data. This method is referred to as mean-shift dominated (MSD) segmentation method. The segmentation performances of the two methods are compared using the following segmentation quality index (SQI):

$$\text{SQI} = \frac{(R - A)}{A}, \tag{23.25}$$

where A is the actual number of planes (hand labeled in the experiments) and R is the number of resulting planes by the segmentation method. The sign of SQI indicates over-segmentation (positive sign) or under-segmentation (negative sign) and the magnitude of SQI represents the segmentation quality. A SQI value closer to 0 indicates a better segmentation result.

The first experiment is to segment the range data of a hallway as shown in Figure 23.11a. Figures 23.11b and 23.11c display the range image and ERI with four prominent planes manually labeled. After applying the mean shift algorithm to ERI, we obtained 270 SPs as shown in Figure 23.11d. The use of SPs causes a reduction of node count from 25344 to 270. As the edge count of an n-node graph is $n \times (n - 1)/2$, the number of edge-weight computations is reduced from 321,146,496 to 36,315, about 10^4 times smaller. This computational reduction will be referred to as computational reduction factor (CRF) further on. The initial grouping of SPs by the NC method is shown in Figure 23.11e. The extracted planar segments by the ENC method (after merging homogenous planar segments) are shown in Figure 23.11f whose corresponding labeled point-cloud is depicted in Figure 23.11g. The result of MSD is shown in Figure 23.11h.

From a qualitative perspective, the MSD results in fragmented walls (in planes labeled as 2 and 3), whereas the ENC method is able to extract the walls in their entirety. There are some misclassifications at the intersection of the front wall and floor in Figure 23.11f. This is most likely due to the use of a small N. However, it should be noted that their impact on navigating a robot may be ignored as they occur at a faraway location from the sensor.

The second experiment was carried out to examine the method's performance on a scene with a stairway as shown in Figure 23.12a. As can be observed, the ENC

FIGURE 23.11 Segmentation of the range data of a hallway: (a) actual scene; (b) range image; (c) ERI with prominent planes marked in black; (d) SPs extracted from the ERI; (e) results after applying the NC algorithm on part (d); (f) extracted planes by the ENC method; (g) extracted planes in point-cloud; and (h) extracted planes by the MSD method.

method extracts most of the horizontal (tread) and vertical (riser) surfaces almost in their entirety.

The SQIs of the two methods in the first experiment are $SQI_{MSD} = 6$ and $SQI_{ENC} = 1$ and that of the second experiment are $SQI_{MSD} = 4.9$ and $SQI_{ENC} = 0.4$. These results indicate that the ENC method has a much better segmentation performance. We ran the USF segmentation method [27] on the above two cases and found a very large number of unclassified pixels in each case, indicating a useless segmentation result.

Additional experiments with various configurations of objects were performed and similar results were obtained. This demonstrates that inclusion of the NC method can substantially improve the segmentation result. All of our experiments have also indicated that the proposed method performs much better than the USF method in the cases of noisy range data of SR4000. More detailed information on the comparative study and the runtime performance of the ENC method can be found in Reference 28.

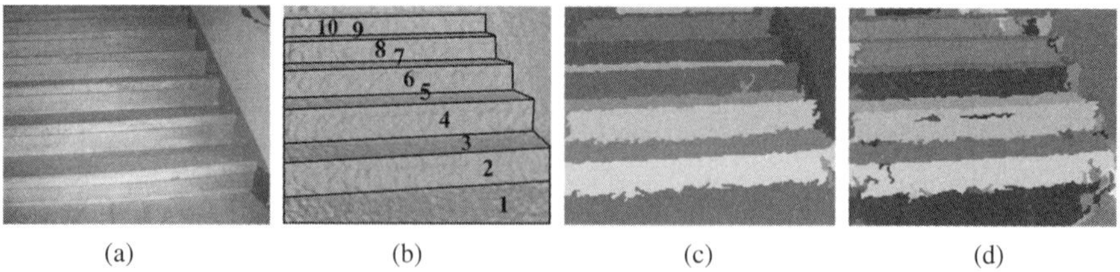

FIGURE 23.12 Segmentation of the range data of a stairway: SPs = 217, CRF=1.31×10^4: (a) actual scene; (b) ERI with the marked prominent planes; (c) extracted planes by the ENC method; and (d) extracted planes by the MSD method.

23.4 CONCLUSIONS

This chapter introduces the data processing methods for two typical robotic perception systems: LIDAR-based and FLC-based systems. The former employs a 2D LIDAR and utilizes the robot motion to perform in-motion mapping of the environment. Since each of the LIDAR's data frames contains only 181 range data, accurate robot pose information is needed to register the laser data frames captured at different locations over time. The robot pose information is computed from the data of a 3D gyro and wheel-odometry. While the mapping method is cost-effective compared with a 3D LIDAR-based system, it lacks reliability in a dynamic environment, that is, an environment with moving object(s). For the navigation method, the use of a $2\frac{1}{2}$D map in TTA incurs a disadvantage that the system cannot handle 3D objects (e.g., overhang object) with accuracy. A point-cloud processing capability will enable true 3D navigation capability. For an FLC-based system, the VR-odometry has good accuracy in estimating pose change between two image frames. This accuracy may be improved with a refined VR-odometry method. For estimating long-duration pose change, the VR-odometry output has to be integrated over time. This is usually achieved by using a state filter to track the robot pose. The ENC plane extraction method segments 3D point-cloud into planar surfaces with connectivity information. The information may be used to group neighboring surfaces into objects for object/obstacle detection and scene recognition that may support robot localization, symbolic map-building, and autonomous navigation.

At the time of writing this chapter, we were developing a robotic navigational device for the visually impaired. The device, called smart cane, is a conventional white cane enhanced by computer vision technology. It uses a SwissRanger SR4000 for sensing the environment. By simultaneously processing the sensor's visual and range data, the smart cane is expected to provide positioning, way-finding, and object-detection functions for the blind user. The positioning function is achieved by obtaining and tracking the pose information of the VR odometery. The way-finding and object-detection functions will be achieved by grouping the planar surfaces extracted by the ENC method into structures and applying pattern recognition methods to the structures. In the near future, the system will be able to detect indoor structures such as a stairway, doorway, and hallway and allow the blind traveler to use them as navigational waypoints. It will also be able to detect an overhang obstacle or drop-off and provide hazard warning to the user or guide the user to walk around the hazard. The pose information of the device may also be used to stitch point-cloud data obtained at multiple locations to form a larger view/map of the environment. The computer vision methods we have been developing for the smart cane may be applied to small robots or other portable robotic devices. For instance, the object level understanding of the surrounding will allow a robot to perform object manipulation. Further development in advanced data processing methods for the Flash LIDAR sensors has great potential in revolutionizing the autonomy of small robots and robotic devices and making them much smarter in the future.

It is worthwhile to mention that the Kinect sensor offers a very cost-effective way to produce range and visual data for robotic application. Currently, it has attracted much attention from researchers in computer vision and robotics. Compared with a Flash LIDAR, the Kinect has lower range measurement accuracy for a distant object due to the triangulation measurement principle and simultaneous use of multiple Kinect sensors which may cause interference. In addition, the Kinect's larger footprint makes it unsuitable for a small-sized robot or robotic device (e.g., our smart cane system). However, its low cost (1/33 of the cost of a SwissRanger SR4000) make it very attractive for indoor robotic applications. Because of the data similarity, developments in data processing methods for Kinect will also advance the state of the art of Flash LIDAR-based perception systems.

ACKNOWLEDGMENTS

The work on the LIDAR-based system was funded by Arkansas Space Grant Consortium under grant number UALR16800, a NASA EPSCOR Planning and Preparatory Award, and a matching fund from the Office of Research and Sponsored Programs of the University of Arkansas at Little Rock. The work on the FLC-based system was funded by the National Science Foundation (IIS-1017672), NASA (NNX09A072A), and Office of Naval Research Summer Faculty Research Fellow program. The author would like to thank Guru Prasad Hegde for his work in validating the data segmentation method. The development of the VR-odometry and the data collection were carried out at the Space and Naval Warfare Systems Center Pacific. The author also thanks Michael Bruch for hosting the research.

REFERENCES

[1] SwissRanger SR4000 Specification [online] http://www.mesa-imaging.ch/products/sr4000/ (accessed January 23, 2015).

[2] TigerEye 3D Flash LIDAR Camera Kit [online] http://www.advancedscientificconcepts.com/products/older-products/tigereye.html (accessed January 23, 2015).

[3] P. Henry, M. Krainin, E. Herbst, X. Ren, and D. Fox. RGB-D mapping: using Kinect-style depth cameras for dense 3D modeling of indoor environments. In: Proceedings of International Symposium on Experimental Robotics, Delhi, India, 2010.

[4] Y. Cheng, M. W. Maimone, and L. Matthies. Visual odometry on the Mars exploration rovers—a tool to ensure accurate driving and science imaging. *IEEE Robotics and Automation Magazine*, 13(2): 54–62, 2006.

[5] M. Maimone, Y. Cheng, and L. Matthies. Two years of visual odometry on the mars exploration rovers. *Journal of Field Robotics*, 24(3): 169–186, 2007.

[6] J. Sun, T. Mehta, D. Wooden, M. Powers, J. Rehg, T. Balch, and M. Egerstedt. Learning from examples in unstructured, outdoor environments. *Journal of Field Robotics*, 23(11–12): 1019–1036, 2006.

[7] D. Wooden. A guide to vision-based map building. *IEEE Robotics & Automation Magazine*, 13(2): 94–98, 2006.

[8] A. Howard, M. Turmon, L. Matthies, B. Tang, A. Angelova, and E. Mjolsness. Towards learned traversability for robot navigation: from underfoot to the far field. *Journal of Field Robotics*, 23(11–12): 1005–1017, 2006.

[9] M. Agrawal, K. Konolige, and R. Bolles. Localization and mapping for autonomous navigation in outdoor terrains: a stereo vision approach. In: Proceedings of IEEE Workshop on Applications of Computer Vision, Austin, TX, 2007, pp. 7–12.

[10] M. Deans, C. Kunz, R. Sargent, E. Park, and L. Perdersen. Combined feature based and shape based visual tracker for robot navigation. In: Proceedings of IEEE Aerospace Conference, Big Sky, MT, 2005, pp. 339–346.

[11] S. Thrun, M. Montemerlo, H. Dahlkamp, D. Stavens, A. Aron, J. Diebel, P. Fong, J. Gale, M. Halpenny, G. Hoffmann, K. Lau, C. Oakley, M. Palatucci, V. Pratt, P. Stang, S. Strohband, C. Dupont, L.-E. Jendrossek, C. Koelen, C. Markey, C. Rummel, J. v. Niekerk, E. Jensen, P. Alessandrini, G. Bradski, B. Davies, S. Ettinger, A. Kaehler, A. Nefian, and P. Mahoney. Stanley: the robot that won the DARPA grand challenge. *Journal of Field Robotics*, 23(9): 661–692, 2006.

[12] C. Ye. Navigating a mobile robot by a traversability field histogram. *IEEE Transactions on Systems, Man, and Cybernetics–Part B: Cybernetics*, 37(2): 361–372, 2007.

[13] C. Ye and J. Borenstein. Characterization of a 2-D laser scanner for mobile robot obstacle negotiation. In: Proceedings of IEEE International Conference on Robotics and Automation, Washington, DC, 2002, pp. 2512–2518.

[14] Y. Okubo1, C. Ye, and J. Borenstein. Characterization of the Hokuyo 2-D laser rangefinder for mobile robot obstacle negotiation. In: Proceedings of the Unmanned Systems Technology IX Conference at the 2009 SPIE Defense and Security Symposium, Orlando, FL, 2009.

[15] Velodyne HDL-64E Product Description, [online] http://velodynelidar.com/lidar/hdlproducts/hdl64e.aspx (accessed January 23, 2015).

[16] C. Urmson, J. Anhalt, D. Bagnell, C. Baker, R. Bittner, M. N. Clark, J. Dolan, D. Duggins, T. Galatali, C. Geyer, M. Gittleman, S. Harbaugh, M. Hebert, T. M. Howard, S. Kolski, A. Kelly, M. Likhachev, M. McNaughton, N. Miller, K. Peterson, B. Pilnick, R. Rajkumar, P. Rybski, B. Salesky, Y. W. Seo, S. Singh, J. Snider, A. Stentz, W. "Red" Whittaker, Z. Wolkowicki, J. Ziglar, H. Bae, T. Brown, D. Demitrish, B. Litkouhi, J. Nickolaou, V. Sadekar, W. Zhang, J. Struble, M. Taylor, M. Darms, and D. Ferguson. Autonomous driving in urban environments: boss and the urban challenge. *Journal of Field Robotics*, 25(8): 425–466, 2008.

[17] M. Montemerlo, J. Becker, S. Bhat, H. Dahlkamp, D. Dolgov, S. Ettinger, D. Haehnel, T. Hilden, G. Hoffmann, B. Huhnke, D. Johnston, S. Klumpp, D. Langer, A. Levandowski, J. Levinson, J. Marcil, D. Orenstein, J. Paefgen, I. Penny, A. Petrovskaya, M. Pflueger, G. Stanek, D. Stavens, A. Vogt, and S. Thrun. Junior: the Stanford entry in the urban challenge. *Journal of Field Robotics*, 25(9): 569–597, 2008.

[18] C. Ye. A method for mobile robot obstacle negotiation. *International Journal of Intelligent Control and Systems*, 10(3): 188–200, 2005.

[19] C. Ye and J. Borenstein. A novel filter for terrain mapping with laser rangefinders. *IEEE Transactions on Robotics*, 20(5): 913–921, 2004.

[20] D. H. Golub and C. F. Van Loan. *Matrix Computations*. The John Hopkins University Press, Baltimore, USA, 1989.

[21] J. Borenstein and Y. Koren. The vector field histogram−fast obstacle-avoidance for mobile robots. *IEEE Transactions on Robotics and Automation*, 7(3): 278–288, 1991.

[22] C. Ye. Navigating a portable robotic device by a 3D imaging sensor. In: Proceedings of IEEE Sensors Conference, Big Island, HI, 2010, pp. 1005–1010.

[23] D. G. Lowe. Distinctive image features from scale-invariant keypoints. *International Journal of Computer Vision*, 2(60): 91–110, 2004.

[24] K. S. Arun, T. S. Huang, and S. D. Blostein. Least square fitting of two 3-d point sets. *IEEE Transactions on Pattern Analysis and Machine Intelligence*, 9(5): 698–700, 1987.

[25] H. Bay, T. Tuytelaars, and L. V. Gool. SURF: speeded up robust features. *Computer Vision and Image Understanding*, 110(3): 346–359, 2008.

[26] C. Ye. A method for mobile robot obstacle negotiation. *International Journal of Intelligent Control and Systems*, 10(3): 188–200, 2005.

[27] A. Hoover, G. Jean-Baptiste, X. Jiang, P. J. Flynn, H. Bunke, D. B. Goldgof, K. Bowyer, D. W. Eggert, A. Fitzgibbon, and R. B. Fisher. An experimental comparison of range image segmentation algorithms. *IEEE Transactions on Pattern Analysis and Machine Intelligence*, 18(7): 673–689, 1996.

[28] G. M. Hegde, C. Ye, and G. Anderson. An extended normalized cuts method for real-time planar feature extraction from noisy range images. In: Proceedings of IEEE International Conference on Intelligent Robots and Systems, Taipei, China, 2010, pp. 1190–1195.

CHAPTER 24

HYBRID/ELECTRIC VEHICLE BATTERY MANUFACTURING: THE STATE-OF-THE-ART

CLAUDIA P. ARENAS GUERRERO[1], FENG JU[2], JINGSHAN LI[2], GUOXIAN XIAO[3], and STEPHAN BILLER[3]
[1]Cummins Inc., Columbus, IN, USA
[2]University of Wisconsin, Madison, WI, USA
[3]General Motors Research & Development Center, Warren, MI, USA

24.1 INTRODUCTION

The increasing demand for alternative and sustainable energy sources to replace gasoline powered engines has intrigued the development of battery technologies for hybrid and electric vehicles. This has brought both challenges and opportunities. Not only are new battery designs with advanced features needed, but also a high manufacturing capacity to satisfy the increasing market is required. The purpose of this chapter is to summarize the recent advances in automotive battery technologies and their manufacturing processes, as well as environmental and recycling issues.

The goal of automotive industry is to reduce fuel consumption, emissions and manufacturing cost, and enhance reliability, safety, and comfort [1]. The overall power system design, the cranking and starting systems, the load leveling, and the standby power also affect the vehicle's electrical system, and impact the characteristics of the battery. The increase in the power-consuming functions for electrical systems also requires the improvement of the battery system and operation conditions [2, 3]. Therefore, the selection of appropriate battery technology for an automotive application is critically important. The production of hybrid electric vehicles (HEVs) and electric vehicles (EVs) has created a need for higher voltage batteries with long life and increased reliability [4]. To meet these demands, Nickel–metal hydride (NiMH) and Lithium-ion (Li-ion) technologies are major contestants. The former is

Contemporary Issues in Systems Science and Engineering, First Edition.
Edited by MengChu Zhou, Han-Xiong Li and Margot Weijnen.

already commercially used in current HEVs with good results, while Li-ion batteries are awaited with high expectation. In this paper, we review the main aspects in technology and manufacturing processes for electric vehicle batteries, such as NiMH and Li-ion batteries.

The remainder of the chapter is structured as follows: Section 24.2 outlines the typical requirements for automobile batteries. Section 24.3 illustrates the different types of electric vehicles. The development of vehicle battery technology is introduced in Section 24.4. NiMH and Li-ion batteries are described in Sections 24.5 and 24.6 respectively. Finally, challenges and conclusions are given in Sections 24.7 and 24.8.

24.2 VEHICLE BATTERY REQUIREMENTS

The demand for electric power in vehicles is growing. In particular, using electrical energy for vehicle's propulsion presents important advantages in the attempt to reduce fuel consumption and CO_2 emissions. In addition to electrically controlled and powered systems for braking, steering, and stabilization, new functions have been introduced, such as an improved battery cycling for the internal combustion engine (ICE) stop–start operation mode, the torque assist/acceleration assist mode, electrical brake energy recovery, and energy management to detect the battery capability to accept charge [5]. These trends are imposing new requirements on the energy storage devices [6]. The new EV designs have to provide fuel economy, reduced emissions, longer cycle life, safety, comfort, reliability, and availability.

In brief, some common technical demands on batteries are identified as follows [3,7-10]:

- High discharge power to hold up new cranking technologies.
- High battery capacity and high cycling capability to cover increased energy demands for engine-off and idle periods.
- Good recharge capability, for quick response during such periods.
- Redundant design and battery monitoring and management to ensure high reliability for cranking and constant energy to safety and relevant components.
- Improved automatic stop/start features and driving strategies to minimize pollution and energy waste during engine idling.
- Higher power capacity. The demand, that started from less than 500 W in 1960s, was increased to more than 2 kW by the year 2000. Considerable increments have been experienced by actual EVs. The Chevrolet Volt pack contains 16 kWh of Li-ion cells of which 8 kWh are usable.

These functions may be realized independently or as a combination with each other. Satisfying all these requirements simultaneously and efficiently with respect to technical and cost restrictions is still in progress.

24.3 HYBRID, PLUG-IN HYBRID, AND ELECTRIC VEHICLE

24.3.1 Hybrid Electric Vehicle

An HEV is propelled by combining mechanical power from an ICE with electrical power from a battery (primary and secondary power sources). Normally, the vehicle works most of the time in the engine's maximum efficiency zone, where the output is maintained at an almost constant power level. The secondary power source is an energy storage device, usually a battery or a supercapacitor, that provides auxiliary power and takes advantage of regenerative braking and deceleration energy, which makes HEVs very efficient. The battery starts working when more power than the engine can provide is needed. Based on the percentage of battery use inside the vehicle's design, different levels of hybridization can be distinguished [4, 6, 11, 12]:

- Micro-HEVs combine automatic engine stop/start operation with regenerative braking, typical generator capacity in the range of 2–4 kW, and conventional 12 V battery technology.
- Mild-HEVs add electrically assisted launch from stop, and charge recuperation during regenerative braking. Typical battery capacities are 6–10 kW. The higher electromechanical power level also provides higher fuel saving benefits from regenerative braking.
- Full or power-assist HEVs offer energy for engine starting, idle loads, full electric launch, torque assistance, regenerative braking energy capture, and limited electric-only range. Used power is around 60 kW. Electric drive and battery typically operate at high voltages above 200 V.

All of these powertrain hybridization levels increase fuel economy by allowing the engine to shut down while idling and during deceleration. They capture electrical energy through regenerative braking. In all cases, the batteries have the advantage of self-recharging. In general, the full hybrid tends to reach its best mileage in city driving, while the medium hybrid tends to reach its best mileage on the highway [13].

24.3.2 Plug-In Hybrid Electric Vehicle

PHEVs provide typical full HEV capability, while at the same time, their batteries can theoretically be recharged from the typical residential AC power grid. This imposes large energy storage demands on the battery, which are not met by previous HEV batteries. The electrical power requirement depends on vehicle weight and is above 70 kW. Important improvements in powertrain configurations are also needed in order to handle the increased thermal management demands [6, 12].

PHEVs can work in two modes: charge depleting (CD) and charge sustaining (CS). When a PHEV is fully charged, it is driven in a CD mode, where the vehicle is powered by the energy stored in the battery. Once the state of charge (SoC) reaches its minimum level, the vehicle changes to the CS mode, and the fuel engine start driving the vehicle (like an HEV). After being fully charged, the range in a CD mode (or

All Electric Range-AER) is the distance a PHEV can work electrically with the heat engine off, before the engine turns on for the first time (switched to the CS mode). A common notation is used to describe the CD range: PHEV-*x*, where *x* means the CD range in miles (i.e., PHEV-10, is a vehicle with the CD range of 10 miles) [14, 15].

A PHEV operates like a HEV when the battery is discharged to a certain SoC (depleted). Most HEVs operate in a CS mode around a predefined SoC. During CS operation, the battery has to meet the discharge and regenerative power with available energy around this SoC. The United States Advanced Battery Consortium (USABC) and the FreedomCAR partnership have set battery requirements and targets for CS power-assist HEVs [14].

24.3.3 Electric Vehicle

EV, also called battery electric vehicle (BEV), is a fully electric vehicle with no supplementary fuel engine. The battery must be recharged by plugging into the electric grid every time the mileage capacity of the battery is exhausted. The main advantages of EVs are their independence of fuel use and their low emissions, which depend on the power source of the utility plant producing the electricity to recharge the EV. A disadvantage of EVs is their still limited driving range, which though may be several hundred miles, is still lower than driving demands. Thus, the principal challenge for the EV batteries is to provide a good driving range performance before recharging, comparable to that of the ICE (approximately 300 miles). Recharging the battery pack of an EV fully can take from 4 to 8 hours and an 80% recharge can be done within 10 minutes [16]. Since the need to recharge can occur on the road, all the road infrastructure must be changed to provide this service, and customers have to adjust their schedules around it. The idea of lacking a back-up system in EVs is a big limitation that implies more technical demands, and drives to the faster development of PHEV and HEV technologies [13, 17].

24.4 BATTERY TECHNOLOGY DEVELOPMENT

Developing high performance rechargeable batteries for EVs is a challenging task because of the need to fulfill multiple battery requirements: high energy, high power, safety, long life, low cost, resistance, and minimal environmental impact [18]. In recent years, automakers are focusing on battery for HEVs, including 42 V enhanced power generators/alternators, and integrated starter-generator (ISG) systems. US automakers have developed a set of general battery performance and price goals through the Partnership for a Next Generation of Vehicles (PNGV). At the same time, the charter of the USABC was created to develop advanced batteries for EV applications [4].

Lead–acid batteries were used in earlier EVs (e.g., GM EV1) [19, 20]. However, the limited driving range had limited its market acceptability and activated the development of higher specific energy batteries. NiMH batteries quickly became the technology of choice for the emerging electric vehicle market. Major automotive companies including Daimler Chrysler, Ford, GM, Honda and Toyota, have developed HEVs with NiMH batteries into the market [20]. Toyota introduced Prius, the first commercial HEV, using a 6.25 Ah cylindrical sealed NiMH D-size cell [21].

Recently, given the high cost of NiMH technology, battery researchers have turned their interest into Li-ion technology. Li-ion batteries have a higher energy density than any other type batteries. Even though NiMH battery holds about twice the energy per pound of the conventional lead–acid battery, Li-ion cells hold roughly twice as much energy per pound as NiMH batteries do. The ability of Li-ion batteries of carrying relatively high energy makes the development of EV possible. In addition, lithium is the lightest solid element, so it is easily portable, and cheaper than all the other battery raw materials [19]. To make Li-ion batteries practical for mass-produced EVs, new technologies must be developed to increase the energy storage and fast discharge capability of the batteries. In addition, the battery cycle life must be lengthened to the average vehicle life. Finally, the cost must be kept as low as possible [22].

The Toyota Vitz CVT 4, a small car sold only in Japan, was the first production car to use Li-ion battery. It used a four-cell, 12 Ah Li-ion battery pack to power its electric accessories and restart the engine after idle stop [22]. Recently, Tesla Motors, in San Carlos, CA, has offered the Tesla Roadster, an all-electric sport car that uses 6831 Li-ion cells. They give the car up to 250 miles of range, as well as the acceleration of 0–60 miles/h in less than 4 seconds [22]. But the major developments are just starting. In January 2010, GM launched the first Li-ion battery pack manufacturing plant operated by a major automaker in Brownstown, MI [23, 24]. The raw lithium cells are delivered from LG Chem's plant in Korea. After the packs are assembled at this facility they are then transported to Detroit-Hamtramack plant where the Volts will be assembled. Production began in March 2010. LG Chem also built a Li-ion cell factory in Michigan in April 2010 [10, 25].

Other battery manufacturers are also increasing their investments in Li-ion technology. A123 Systems has built a new world-class Li-ion battery manufacturing facilities in the United States, with the largest Li-ion automotive battery manufacturing plant located in southeast Michigan [26]. This plant enables the company to dramatically expand production capacity, with full production volumes designed to supply battery systems for five million HEVs or half million plug-in EVs per year by 2013 [27]. Similar efforts on new lithium battery developments have also been observed from many other manufacturers, in addition to the research that the major automakers themselves are conducting [22].

24.5 NICKEL-METAL HYDRIDE BATTERY

All currently commercially available HEVs use NiMH technology as the energy storage system (ESS). NiMH batteries have experienced great advances in the past 15 years, with an evident increase in energy and specific power. The NiMH battery used in today's HEVs is relatively small, has favorable life-cycle characteristics, high power, light weight, high level of safety, good thermal performance, and a configurable design [28]. In fact, NiMH batteries are the only tested and considered reliable electric power source for HEVs [13, 29–34]. However, the expensive raw materials make the cost a serious issue [21].

24.5.1 NiMH Battery Manufacturing

NiMH batteries are composed of nickel hydroxide on the positive electrode and an alloy consisting of vanadium, titanium, nickel, and other metals on the negative electrode [29]. In all cases, the materials have complex microstructures that allow the hydrogen storage alloys to operate in the aggressive environment within the battery. A critical design factor within the metal hydride surface oxide is to achieve a balance between passivation and corrosion. Passivation can cause problems for high rate discharge and cycle life, and corrosion is highly destructive. Oxidation and corrosion of the anode metals consume electrolyte, change the SOC balance, and create corrosion products capable of causing premature oxygen evolution. Porosity within the oxide is important to allow ionic access to the metallic catalysts and therefore promote high rate discharge [30].

The electrode groups consist of alternately stacked positive electrodes and negative electrodes interleaved with separators. Inserting these electrode groups into a resin battery case, and sealing with a cover equipped with a valve after filling with alkaline electrolyte, form the cell. A battery module consists of 10 cells connected in series by metal plates, and these are configured to permit airflow between the cells to ensure a uniform temperature distribution. A sealing plate is equipped with a valve to prevent bursting with an increase of internal pressure. This battery has been developed by optimizing the reaction area of the electrodes, reducing resistance for current collection and improving electrolyte composition to obtain high power characteristics [31].

24.5.2 NiMH Batteries in Commercial Vehicles

Today, several automakers have used the NiMH battery packs in commercial vehicles [19, 32–34].

24.5.2.1 Toyota Prius The battery pack of the second generation Toyota Prius consists of 28 Panasonic prismatic NiMH modules, each containing six 1.2 V cells, which are connected in series to produce a nominal voltage of 201.6 V through 168 cells. The weight of the complete battery pack is 53.3 kg. The power capability increases with higher temperatures and decreases at lower temperatures. A computer system is dedicated to keeping the Prius battery at the optimal temperature and charge level, and supply conditioned air from the cabin for cooling the batteries.

24.5.2.2 Toyota Highlander The NiMH battery used in Highlander Hybrid (also Lexus RX 400h) is packaged in a newly developed metal battery case. The 240 cells can deliver high voltage of 288 V, but the motor generators units can operate on variable voltage from 280 V to 650 V. The boost converter can change the 288 V supplied by battery pack to 500 V. This battery pack provides 40% more power than the Prius battery, and is 18% smaller. Each of the modules has its own monitoring and cooling control system, which can reduce efficiency losses due to excessive heat, and ensure the required supply of electric power to the motor. The battery-monitoring unit manages discharge and recharging by the generator and motors to keep the charge level constant.

24.5.2.3 Ford Escape The Ford Escape Hybrid's battery pack, made by Sanyo, consists of 250 individual NiMH cells, similar in shape to a size D flashlight battery. Each individual battery cell, contained in a stainless steel case, is 1.3 V. The cells are welded and wrapped together in groups of five to form a module, and there are 50 modules in the battery pack, resulting about 330 V total voltage of the battery pack.

24.5.2.4 Honda Insight The Honda Insight's battery pack is made of 120 Panasonic 1.2 V NiMH D cells. Total battery pack output is 144 V. The batteries are located under the cargo compartment floor, along with the Honda Integrated Motor Assist's power control unit. Honda used technology developed for its EV Plus electric car for the original development of the Insight's battery system.

24.5.2.5 Saturn Vue Saturn Vue uses a belt alternator starter (BAS) system, where a large electric motor assists the traditional gas engine during acceleration. The electric motor is powered by a 1.8 kWh, 300 V NiMH battery, which in turn is partially charged through regenerative braking. The battery pack consists of 22 NMH modules and weighs about 55 lb.

24.5.3 Cost

The biggest disadvantage of NiMH batteries is the cost. The battery cost consists of two parts, manufacturing cost and material expense [21]. The manufacturing cost depends on production volumes and manufacturing technology. It was reported that a dry powder roller processing technology had a lower manufacturing cost for NiMH cells. The material cost depends on the negative electrode, positive electrode, separator, cell hardware and others, where the negative electrode is the highest one, around 45% in total material cost. Therefore, developing a cheap electrode is important for cost reduction of NiMH batteries [35].

24.5.4 Recycling

To recycle the current generation of NiMH batteries, car makers destroy the battery to find the metals between the plastic, plates, steel case, and the wiring, to make sure the materials are processed for disposal. Some car makers offer a compensation to help ensure the battery is returned to a dealership and properly recycled [36].

After the market launch of NiMH batteries, a German company, NIREC, began work on the recycling of these batteries in order to put nickel back into the cycle of materials. The batteries first pass through a cutting chamber to open up the casing and release the stored hydrogen, then go into a collecting tank. After a stabilization period, the material can be taken out. After separation of the plastic content, a usable product is obtained with a high nickel content reusable in stainless steel production [37].

24.6 LITHIUM-ION (Li-Ion) BATTERY

The recent development in Li-ion batteries leads to a good anticipation in terms of performance and cost, potential application in EV, and customer acceptance [11]. In comparison with other rechargeable batteries, Li-ion provides very high specific energy and a large number of charge/discharge cycles. However, Li-ion batteries are only commercially available in small sizes at this time. Large numbers of cells have to be assembled in series/parallel configurations to achieve the desired battery size. In addition, the highly efficient and reliable battery packs for EV require high energy density, high charge–discharge efficiency, good load and temperature storage characteristics, accurate detection of battery SOC, long battery expected life, and good recyclability. All these requirements, combined with safety issues, vehicle autonomy, and low cost, present the challenges for the battery development [4, 38–40].

24.6.1 Lithium Technology

In addition to automotive companies, many other manufacturers have invested in the Lithium technology. SAFT, a leading designer and manufacturer of advanced technology batteries, has focused on Li-ion energy storage technology. To meet the different energy and voltage requirements of individual automakers, a six cell module is designed to provide three different voltages and capacities, by internally connecting the cells in different fixed, series and parallel connections. In addition, SAFT's design places individual cell monitoring and control at the module level, and uses liquid to cool the module and provide effective thermal management [41].

Another manufacturer, AVESTOR, has been oriented to the Lithium–Metal–Polymer battery (LMPB) development. The LMPB is an advanced battery technology under development for both EV and HEV applications. A lithium metal polymer cell is made by laminating together five thin materials including an insulator, a lithium foil anode, a solid conductive polymer electrolyte, a metal oxide cathode, and a current collector. The total thickness of this laminate is less than 100 μ (0.004 in) and it is formed into a prismatic shape to construct an electrochemical cell (EC). Different cells can be fabricated using the same manufacturing equipment. This flexibility allows LMPB product configurations to be adapted to different vehicles with little changes in the process and production cost. Because of the solid state nature of LMP, the cells can be wired into parallel and/or series arrangements within a single container to build the desired module capacity and voltage [41, 42].

24.6.2 Manufacturing Processes

24.6.2.1 Cell Design The main feature of Li-ion cells is that the current is carried by lithium ions, from the positive electrode (cathode) to the negative electrode (anode) during charging, and from negative to positive during discharging. The basic cell chemistry and design are the same for all types of Li-ion automotive cells. Thin layers of cathode, separator, and anode are rolled up and inserted into a cylindrical can. The gaps are filled with liquid electrolyte. The size of the cell and the quantities of different materials contained in the cell are the main difference between a cell optimized for high energy density (for EV) and the one for high power (for HEV). Both cells can utilize the same basic spiral-wound design, same materials, and even the same current collectors and separators. However, the high power cells must be smaller than the high energy cells in order to dissipate the higher heat load generated. In addition, high power batteries require less total energy, such that the battery packs for HEVs can be made smaller and lighter [43].

24.6.2.2 Electrodes Electrode manufacturing processes consist of coating metallic foils (Al for positive and Cu for negative) with a slurry containing the active material, an electronic conductor (for the positive mix), and a binder dissolved in a solvent. After drying, this coating is calendered to the desired thickness and density. The electrode thickness depends on the maximal power required. This allows a wide range of power and energy ratio designs with the same basic electrode manufacturing process [44, 45]. The material used for the cathode is a metal oxide. Sony is using manganese, while SAFT and Polystor use nickel, or mixed nickel/cobalt-based cathodes, and Varta and Shin-Kobe (Hitachi) are developing cells with manganese-based cathodes. The anode is generally made of graphite, coated on copper foil, but other forms of carbon can be used [43].

24.6.2.3 Electrolytes There is no significant difference in the nature of electrolytes, where LiPF6 is the only suitable salt, and cyclic or linear carbonates are the basis of the organic solvents [44]. Lately, there is great interest in developing flame-retardant electrolytes, using additives, rather than replacing the organic solvents [43].

24.6.2.4 Separators Separators, all in the form of thin microporous membranes, may be either in polypropylene or polyethylene, or mixed polymers [44]. Separators for either high energy or high power Li-ion batteries are typically made from polyolefin using layers with 50% porosity. The separators keep the positive and negative electrodes apart, and serve as a safety device: if a cell becomes too hot, the low melting polymers melt, closing off the pores through which the ions travel, and cutting off the cell current [43].

24.6.2.5 Cell Assembly There are multiple designs for assembling cells into a battery pack for electric and hybrid vehicles. The most commonly used one is the modular design, where cells are packaged together into a module. Given that Li-ion cells can suffer serious damage on overcharge or overdischarge, control devices are

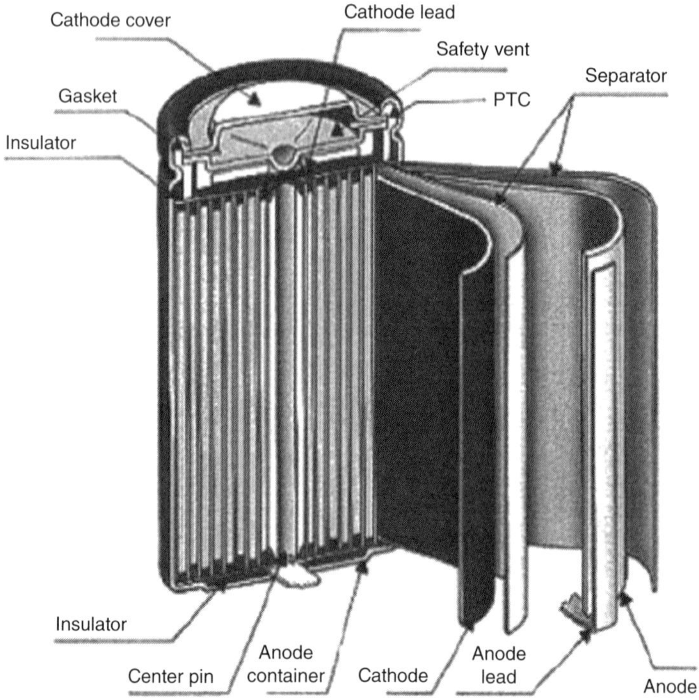

FIGURE 24.1 A schematic of the cylindrical cell [46].

included in each module. In batteries with organic electrolyte, the end of charge and discharge voltages must be monitored for each cell or parallel group, with specific battery management features. Depending on the requirements of the vehicle, the modules can be combined into different sizes of battery packs [44, 45]. Most automotive cell designs are cylindrical as shown in Figure 24.1, although some manufacturers produce prismatic cells as shown in Figure 24.2. Prismatic cells offer better volume filling in assembly, but they are more sensitive to deform on cycling or aging. Cylindrical cells have better mechanical characteristics and dimensional stability within the electrode stack. Because of the small electrode's thickness, the cell stack is built by rolling the electrodes in spiral. The coil may be cylindrical or flattened to make a rectangular-shaped cell [44, 45].

The process to assemble cylindrical cells is described as follows: as shown in Figure 24.3 the electrolytes are formed from pastes of active material powders, binders, solvents, and additives. The coating machines spread them on current collector foils, such as aluminum for the cathode side and copper for the anode side. After drying and calendering processes, and when homogeneous thickness and particle size are achieved, they are cut to the correct width. The stacks are formed by putting separator, anode, separator, cathode, and then are winded to cylindrical cells, inserted in

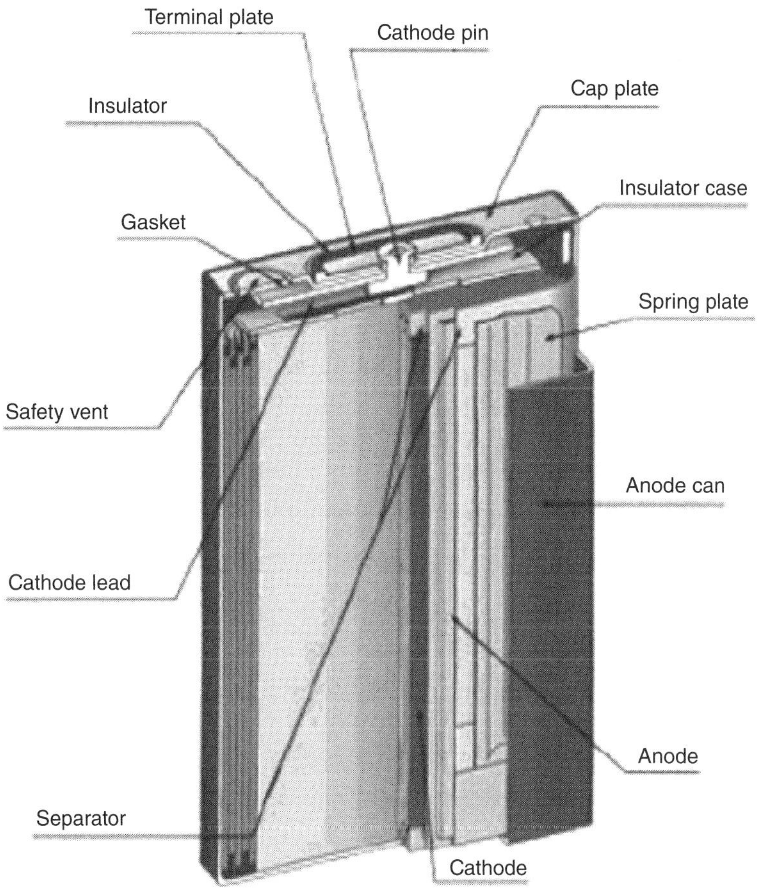

FIGURE 24.2 A schematic of the prismatic cell [46].

cylindrical cases, and welded to a conducting tab. After vacuum and heat drying, the cells are then filled with electrolyte. The electrolyte has to wet the separator, soak in, and wet the electrodes. As a part of the vacuum addition of an electrolyte process, the wetting and soaking processes are the slowest steps and determine the speed of the line. Some insulators, seals, and safety devices are also attached and connected. Finally, the cells are charged and tested to satisfy a manufacturer's requirements [45–50].

24.6.2.6 Packaging The packaging is a critical step after assembly. It is important that there is no reaction of the electrolyte with the packaging material or release of water from the material into the electrolyte [51]. Typically, aluminum cases for Li-ion batteries are used, since it is lighter and cheaper. Plastic is not used because it could be dissolved by the organic electrolytes [43].

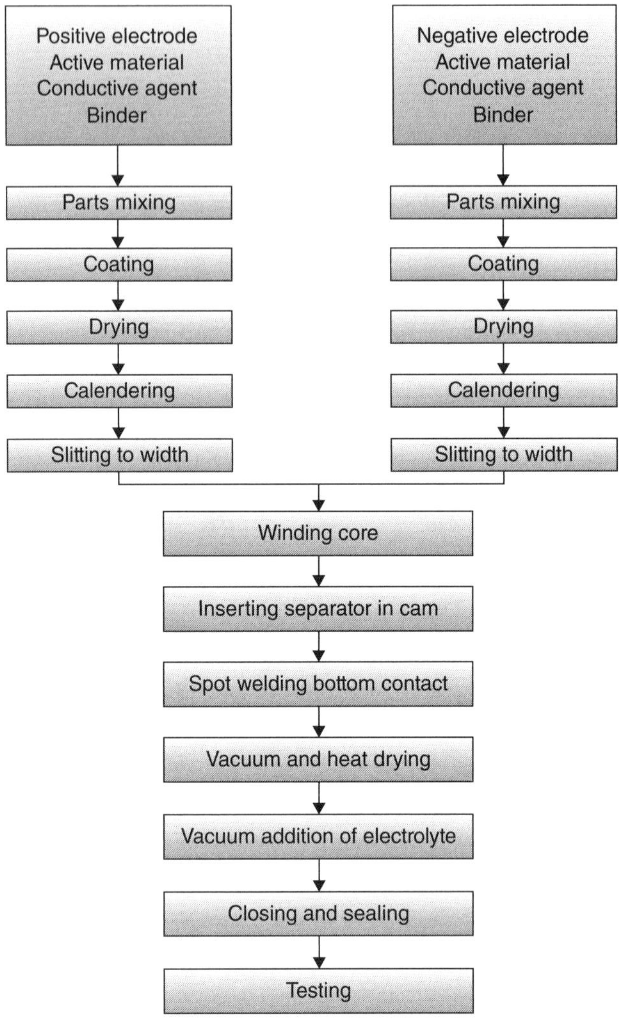

FIGURE 24.3 Process for the Li-ion battery cell fabrication.

24.6.2.7 Process Control This manufacturing process is based on tight tolerances and strict control through the process. Contamination or physical damage on the electrodes have to be identified, because they can cause penetration of the separator, resulting in internal short circuits in the cell. A protection method for prevention or control is needed [52].

24.6.2.8 Battery Assembly Once the individual cells are produced, they are stacked and assembled together into a module as shown in Figure 24.4. The individual cells are placed one on top of another, and isolated with foam, such that

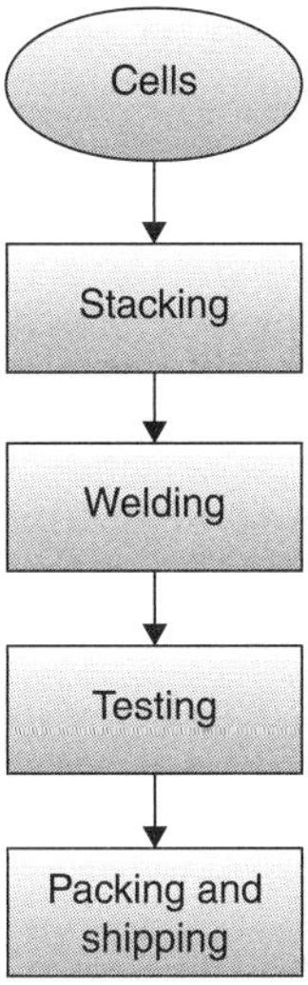

FIGURE 24.4 Li-ion battery assembly process.

they do not touch each other. Each module has a determined number of cells with total energy in the range of 1–2 kWh. Li-ion cell's flexibility allows direct parallel connection to get the desired capacity without other electronic components, and the ability to work in any position. Given that Li-ion cells can suffer serious damage on overcharge/overdischarge, each cell or module has an electronic board for specific battery management, monitoring, and control. Depending on the requirements of a vehicle, the modules can be combined into different sizes of battery packs. Finally, a battery system consists of assembled modules electrically connected in series, with a cooling system (air or liquid), a fuse for protection against short circuit, a contactor to insulate the battery from the vehicle, and a ground fault detector. These subsystems are usually managed by a controller that collects information and communicates with the automobile [44, 45].

Due to the structure of battery packs, manufacturing of Li-ion batteries requires welding of multiple layers within a tight product envelope. Ultrasonic welding technology is commonly applied in today's Li-ion battery manufacturing processes. Such a technology can deliver ultrasonic energy directly to the metals to be welded and create a durable metallurgical bond between workpieces without melting them. This bond is generated without using heat, current, fluxes, or fillers. Thus it requires no cleaning or consumable costs [53].

24.6.3 Li-Ion Batteries in Commercial Vehicles

Nowadays, many manufacturers build their own battery assembly plants and apply Li-ion batteries into commercial vehicles.

24.6.3.1 Chevy Volt In GM plant, three primary assembly processes are identified: the battery module pre-assembly, where cells are processed and installed into one of the three battery modules. The final assembly area is the place where the three modules are assembled and tested, which is required for each battery pack. Also in this area, the prismatic-shaped cells are processed and installed by flexible automated equipment into modules, which are then delivered to the battery pack main line. The area of battery pack main line has an Automated Guided Cart (ACG) system that enables operations for thermal and electrical assembly, quality, and dimensional checks. Battery pack's final testing, verification and packaging for shipment are introduced in this area too [10, 23, 24]. The Volt's battery pack is made up of multiple linked battery modules (more than 200 battery cells), storing 16 kWh in a 5-foot-long container weighing 400 lb as shown in Figure 24.5 [10, 24].

24.6.3.2 Nissan Leaf Nissan Leaf is powered by a 24 kWh Li-ion battery pack that can deliver power up to 120 hp. The battery pack contains 48 modules and each module consists of 4 cells, such that there are 192 cells in total, and the battery pack weighs 660 lb all together. Considering the battery's weight, Nissan places the battery pack under the seats and rear foot space in order to keep structural rigidity as shown in Figure 24.6 [48, 49].

24.6.4 Safety

Li-ion battery technology is not intrinsically safe. Short circuit, overcharge, overdischarge, and high temperature can lead to thermal runaway, fire, and explosion [46]. When an overvoltage affects the cells, overcharge can be avoided by using redundant voltage thresholds. If a short or overloading occurs, overheating can be avoided by limiting the maximum current, with fuses or thermal sensors [44]. A basic external

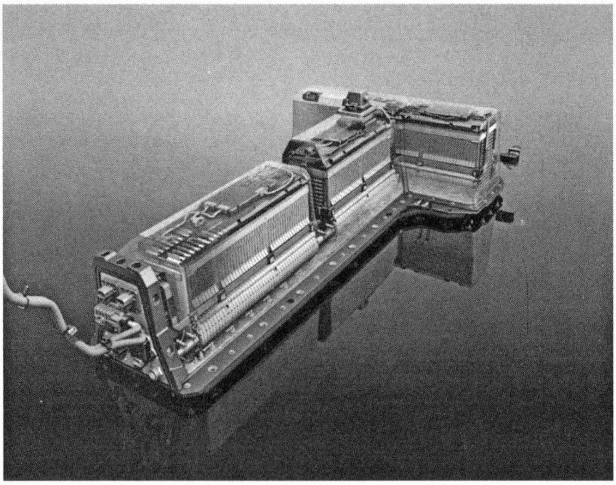

FIGURE 24.5 Battery pack in Chevy Volt [54].

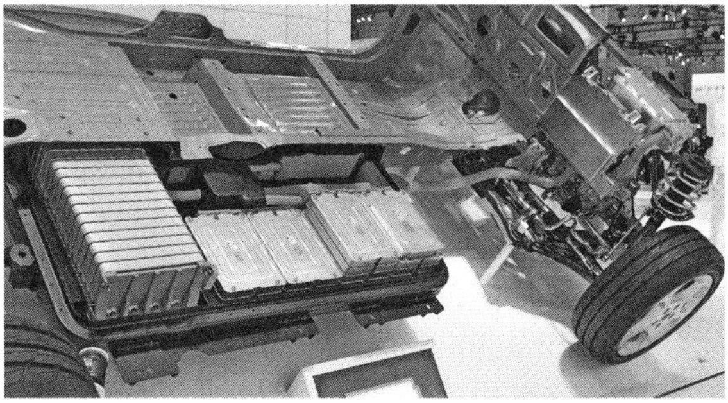

FIGURE 24.6 Battery pack in Nissan Leaf [55].

control system consists of a bypass circuit controlled by a microcontroller. When a cell reaches a given SOC (or discharge) before other cells, the bypass circuit is activated, discontinuing the charging (or discharging) process for that cell until balance is reached. In this way, balance is maintained between the cells and damage to individual cells is avoided [43]. Other safety mechanisms are also considered by manufacturers, such as pressure release valves, one-shot fuses, reversible and irreversible positive temperature coefficient elements, shutdown separators, chemical shuttles, nonflammable electrolytes and coatings [56].

24.6.5 Cost

Current li-ion battery cost per kWh is still high, that is, between $1000 and $1200 according to [57]. The main costs are due to the high cost of raw materials, materials processing, packaging and manufacturing of the cells [46].

24.6.6 Environmental Issues

The use of organic solvents in the electrode manufacture implies a high cost and multiple environmental concerns because of their toxicity, which may reduce the potential to take manufacturing cost down to the level required for wide adoption in many applications. Because of this, different approaches have been provided. For example, International Battery Company focuses on manufacturing Li-ion cells using a water soluble binder based process for both the cathode and anode, where water is used as a medium to dissolve and disperse the binders and the electrode materials. By eliminating the solvent from the manufacturing process, the material cost and capital investment can be reduced [58].

24.6.7 Recycling

Recycling of automotive batteries is easier than recycling of small portable cells. First, collection plans have been used for return of automotive batteries with each

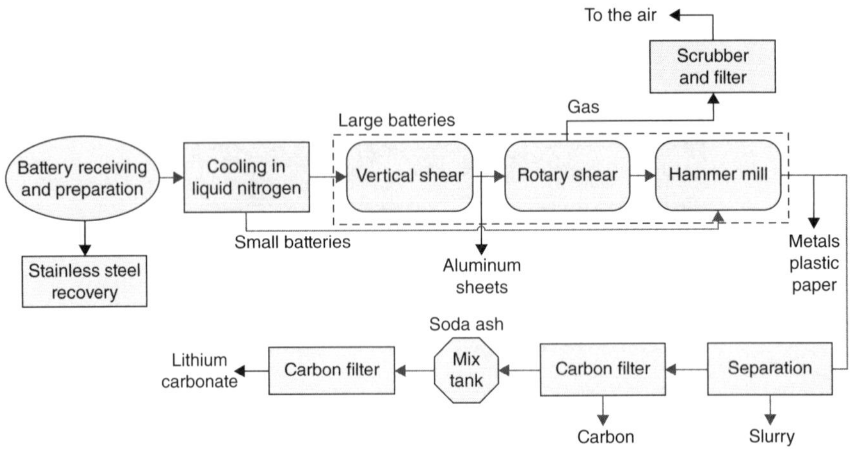

FIGURE 24.7 Recycling process for the Li-ion battery at Toxco (based on Reference[43]

new battery purchase and stripping batteries from wrecks before shredding. Second, the batteries are large enough to permit separation by type, to maximize the value of recovered materials. Third, there is enough material in large pieces to justify partial disassembly. The recycling approaches typically include discharging the cells, venting the electrolyte solvent, shredding the entire package, and recovering the solvent flashed off [43]. Recently, the US Department of Energy granted $9.5 million to Toxco to build the first recycling facility in America [36]. Sony, Accurec, SNAM, Nippon Mining & Metals Co., and GS Yuasa Corp., have also developed recycling processes for lithium batteries [36, 37, 43]. Toxco is a leading battery recycler and works on single-charge and rechargeable lithium batteries found in electronics devices and industrial applications [36]. Its process consists of chilling the batteries to −325°F in liquid nitrogen as shown in Figure 24.7. Large batteries are sheared into three pieces in a caustic bath, to dissolve the lithium salts, and neutralize the acidic elements. The salts are used to produce lithium carbonate. Hydrogen and organic burn off at the surface of the process bath. Cobalt can be recovered from the sludge. The large pieces remaining are set to a hammer mill, to recover ferrous and nonferrous metals. Plastics and paper float to the top and are recovered for disposal or recycling. The carbon sludge is filtered out, but it is not economical to reuse [36, 43].

24.7 CHALLENGES

Although Li-ion technology is the forefront of advanced battery design and offers multiple advantages in the electric vehicle application, it still faces challenges to meet the specific standards for performance, life span, safety, cost, etc. [59]. There are some issues yielding to progress, which are listed below.

- Joining manufacturing and its quality control.

  Since the battery pack consists of hundreds of cells, the joining process needs to be taken to firmly create electrical contacts from the positive tab to the negative one. Once the battery is manufactured, it is expected to operate reliably and safely for the whole life cycle stipulated by the manufacturers, which is typically more than 10 years. Therefore, it is crucial to ensure the high quality of joining process which can provide consistent and reliable joints.
- Battery design for manufacturing.

  Although the Li-ion battery has high energy density and a large number of charge/discharge cycles, the demand for batteries of higher energy and lower cost is still driving the research for improving the battery design. Recently, an innovative approach, referred to as a semisolid flow cell, to the design of Li-ion battery cell has been proposed [60]. It can potentially provide energy densities that are 10 times higher than those of conventional batteries. More importantly, the manufacturing process would be simplified and streamlined by eliminating many of the unit operations, such as electrode coating, calendering, and cell assembly. Thus, the equipment and production costs could be reduced.
- Battery remanufacturing.

  Remanufacturing not only implies reproduction of the original product, but also includes upgrading its performance by installing or substituting an advanced component or introducing new technology. A typical remanufacturing process includes disassembly, cleaning, sorting, checking, reconditioning, and reassembly. After remanufacturing, the battery is expected to serve another full life cycle. Therefore, remanufacturability should be taken into account in the battery design and manufacturing processes, as well as the material handling and production logistics. For example, the battery assembly needs to consider the possibility of disassembling and replacing cells in remanufacturing.
- Battery inspection and diagnostic testing methods.

  Manufacturers have to pay attention to battery charging and diagnostic testing in order to ensure the high level robustness. From a designer's perspective, additional hardware and software must be applied to maintain better performance, reliability, and safety [61].
- Information feedback of battery usage to manufacturing process.

  Battery manufacturing process should be considered as an integrated part within the battery life cycle. The design of manufacturing process and quality management need to take into account the potential battery usage. In other words, the prediction of battery life and performance should include the impacts from manufacturing specification and quality control, while the battery usage information needs to be fed back to the manufacturing and design process to improve battery quality [62].
- In addition, aspects such as assembly process safety improvement, assembly tooling durability and battery assembly productivity and quality improvement also need to be considered.

Obviously, all these improvements are closely related to the cost. Currently, the cost of battery pack for electric vehicles is relatively high due to facts such as high scrap rate and expensive equipment. With more experiences and scale effects, the individual battery pack will become less expensive [57]. However, how to deal with the tradeoff between principles such as performance, safety, life span, and cost will still be the main issue in battery manufacturing.

24.8 CONCLUSIONS

In this chapter, we review the recent development in hybrid/electric vehicle battery manufacturing. More specifically, we focus on Nickel-metal and Lithium-ion batteries which are used in most hybrid, plug-in hybrid, and electric vehicles. In particular, for Li-ion batteries, we introduce the manufacturing processes, including cells and their assembly, packaging, process control and finally, battery assembly. Moreover, the commercial applications (Chevy Volt and Nissan Leaf) and the related recycling and environmental issues are discussed. The challenges we are facing to improve performance, life span, and reduce cost are briefly outlined. It can be seen that substantial effort needs to be devoted to mass production of batteries for EVs and HEVs in the near future, which will provide this area with great challenges and promising research opportunities.

REFERENCES

[1] R. Friedrich and G. Richter. Performance requirements of automotive batteries for future car electrical systems. *Journal of Power Sources*, 78: 4–11, 1999.

[2] J. R. Pierson and R. T. Johnson. The battery designer's challenge—satisfying the ever-increasing demands of vehicle electrical systems. *Journal of Power Sources*, 33: 309–318, 1991.

[3] E. Meissner and G. Richter. Vehicle electric power systems are under change! Implications for design, monitoring and management of automotive batteries. *Journal of Power Sources*, 95: 13–23, 2001.

[4] T. J. Miller. Lithium ion battery automotive applications and requirements. In: Proceedings of the 17th Annual Battery Conference on Applications and Advances, Long Beach, CA, 2002, pp. 113–118.

[5] E. Meissner and G. Richter. The challenge to the automotive battery industry: the battery has to become an increasingly integrated component within the vehicle electric power system. *Journal of Power Sources*, 144: 438–460, 2005.

[6] M. Anderman. The challenge to fulfill electrical power requirements of advanced vehicles. *Journal of Power Sources*, 127: 2–7, 2004.

[7] E. Meissner and G. Richter. Battery monitoring and electrical energy management precondition for future vehicle electric power systems. *Journal of Power Sources*, 116: 79–98, 2003.

[8] B. Spier and G. Gutmann. 42-V battery requirements lead acid at its limits. *Journal of Power Sources*, 116,: 99–104, 2003.

[9] L. Webster. GM finally unveils production chevy volt: first look. Available at http://www.popularmechanics.com/automotive/new_cars/4283076.html (accessed September 16, 2008).

[10] GM begins mass production at volt battery assembly plant today. Available at http://gm-volt.com/2010/01/07/gm-begins-mass-production-at-volt-battery-assembly-plant-today/ (accessed January 7, 2010).

[11] O. Bitsche and G. Gutmann. Systems for hybrid cars. *Journal of Power Sources*, 127: 8–15, 2004.

[12] E. Karden, S. Ploume, B. Fricke, T. Miller, and K. Snyder. Energy storage devices for future hybrid electric vehicles. *Journal of Power Sources*, 168: 2–11, 2007.

[13] R. Ruegg and P. Thomas. Linkages of DOE's energy storage R&D to batteries and ultracapacitors for hybrid, plug-in hybrid, and electric vehicles. US Department of Energy – Energy Efficiency and Renewable Energy, February 2008.

[14] A. A. Pesaran, T. Markel, H. S. Tataria, and D. Howell. Battery requirements for plug-in hybrid electric vehicles analysis and rationale. In: 23rd International Electric Vehicle Symposium, Anaheim, California, December 25, 2007.

[15] J. Axsen, A. Burke, and K. Kurani. Batteries for plug-in hybrid electric vehicles (PHEVs): goals and the state of technology circa 2008. Institute of Transportation Studies, University of California Davis, CA, May 2008.

[16] C. D. Anderson and J. Anderson. *Electric and Hybrid Cars: A History*, McFarland, 2005.

[17] J. S. Cunningham. *An Analysis of Battery Electric Vehicle Production Projections*. Massachusetts Institute of Technology, 2009.

[18] F. McLarnon. *Lithium Batteries for Hybrid-Electric Vehicles*. Lawrence Berkeley National Laboratory *EETD News*, 1: 1–2, 2000.

[19] The hybrid car battery: a definitive guide. Available at http://www.hybridcars.com/hybrid-car-battery (accessed November 6, 2008).

[20] P. Gifford. Development of advanced nickel/metal hydride batteries for electric and hybrid vehicles. *Journal of Power Sources*, 80: 157–163, 1999.

[21] R. F. Nelson. Power requirements for batteries in hybrid electric vehicles. *Journal of Power Sources*, 91: 226, 2000.

[22] J. Voelcker. Lithium batteries for hybrid cars. Available at http://spectrum.ieee.org/green-tech/advanced-cars/lithium-batteries-for-hybrid-cars (accessed January 2007).

[23] S. Abuelsamid. GM lithium-ion battery plant goes to brownstown township, MI. Available at http://green.autoblog.com/2009/08/13/officially-official-gm-lithium-ion-battery-plant-goes-to-bro/(accessed August 13, 2009).

[24] GM builds first lithium-ion battery for chevrolet volt. Available at http://www.gm.com/corporate/responsibility/environment/news/2010/voltbattery_010710.jsp (accessed January 7, 2010).

[25] LG chem opens 'largest' EV battery plant, projects one-quarter world market share. Available at http://gm-volt.com/2011/04/15/lg-chem-opens-ochong-battery-plant-expects-major-market-share/ (accessed April 15, 2011).

[26] A123 systems opens the largest lithium ion automotive battery manufacturing plant in North America. Available at http://ir.a123systems.com/releasedetail.cfm?ReleaseID=506787 (accessed September 13, 2010).

[27] A123 systems seeks grant for hybrid car batteryproduction. Available at http://www.xhybridcar.com/index.php?option=com_content&task=view&id=697&Itemid=122 (accessed January 7, 2009).

[28] Plugging into the future. *The Economist*, June 10, 2006.

[29] S. M. Lukic, J. Cao, R. C. Bansal, F. Rodriguez, and A. Emadi. Energy storage systems for automotive applications. *IEEE Transactions on Industrial Electronics*, 55: 2258–2267, 2008.

[30] M. A. Fetcenko, S. R. Ovshinskya, B. Reichmana, K. Younga, C. Fierroa, J. Kocha, A. Zallena, W. Maysa, and T. Ouch. Recent advances in NiMH battery technology. *Journal of Power Sources*, 165: 544–551, 2007.

[31] A. Taniguchi, N. Fujioka, M. Ikoma, and A. Ohta. Development of nickel/metal-hydride batteries for EVs and HEVs. *Journal of Power Sources*, 100: 117–124, 2001.

[32] Nickel-metal hydride application manual. Available at http://data.energizer.com/PDFs/nickelmetalhydride_appman.pdf (accessed May 2011).

[33] G. Pistoia, *Battery Operated Devices and Systems - From Portable Electronics to Industrial Products*, Elsevier, 2009.

[34] Saturn Vue hybrid SUV | the two mode green line", Available at http://www.hybrid-suv.com/saturn/2009-vue-hybrid-twomode (accessed March 2008).

[35] T.-K. Ying, X.-P. Gao, W.-K. Huc, F. Wud, and D. Noreus. Studies on rechargeable NiMH batteries. *International Journal of Hydrogen Energy*, 31: 525–530, 2006.

[36] Birth of industry to recycle lithium auto batteries. Available at http://www.hybrid-cars.com/environment/birth-industry-recycle-lithium-auto-batteries-26047.html (accessed August 28, 2009).

[37] H. A. Kiehne, *Battery Technology Handbook*, 2nd edn, CRC Press, 2003.

[38] Y. Nishi, K. Katayama, J. Shigetomi, and H. Horie. The development of lithium-ion secondary battery systems for EV and HEV. In: Proceedings of 13th Annual Battery Conference on Applications and Advances, Long Beach, CA, 1998, pp. 31–36.

[39] B. Kennedy, D. Patterson, and S. Camilleri. Use of lithium-ion batteries in electric vehicles. *Journal of Power Sources*, 90: 156–162, 2000.

[40] A. Affanni, A. Bellini, G. Franceschini, P. Guglielmi, and C. Tassoni. Battery choice and management for new-generation electric vehicles. *IEEE Transactions on Industrial Electronics*, 52: 1343—1349, 2005.

[41] R. A. Sutula. Progress report for the electric vehicle battery research and development program. U.S. Department of Energy—Office of Advanced Automotive Technologies, January 2001.

[42] H. Wilkinson and S. Cornay. Avestor lithium-metal-polymer batteries deployed throughout North America. Proceedings of 27th IEEE International Telecommunication Conference, pp. 217–221, Berlin, Germany, 2005.

[43] L. Gaines and R. Cuenca. Costs of lithium-ion batteries for vehicles. Center for Transportation Research—Argonne National Laboratory, May 2000.

[44] G. Nazri and G. Pistoia, *Lithium Batteries: Science and Technology*, New York: Kluwer Academic Publisher, 2004.

[45] W. A. van Schalkwijk and B. Scrosati, *Advances in Lithium-Ion Batteries*, New York: Kluwer Academic Publisher, 2002.

[46] C. Daniel. Materials and processing for lithium-ion batteries. *Journal of Mechanics*, 60: 43–48, 2008.

[47] M. Wakihara. Recent developments in lithium ion batteries. *Materials Science and Engineering*, R33, 109–134, 2001.

[48] Details on Nissan leaf battery pack, including how recharging speed affects battery life. Available at http://green.autoblog.com/2010/05/27/details-on-nissan-leaf-battery-pack-including- how-recharging-sp/ (accessed May 27, 2010).

[49] 13 Key questions and answers about Nissan leaf battery pack and rrdering. Available at http://www.hybridcars.com/news/13-key-questions-and-answers-about-nissan-leaf-battery-pack-and-ordering-28007.html (accessed May 27, 2010).

[50] Battery grade. Available at http://www.megagraphite.com/products/byapplication/batterygrade (accessed May 2011).

[51] U. Heider, R. Oesten, and M. Jungnitz. Challenge in manufacturing electrolyte solutions for lithium and lithium ion batteries quality control and minimizing contamination level. *Journal of Power Sources*, 81–82: 119–122, 1999.

[52] Lithium battery manufacturing. Available at http://www.mpoweruk.com/battery_manufacturing.htm (accessed May 2011).

[53] For battery assembly, ultrasonic technology to weld multiple layers to tabs, terminals. Available at http://weldingdesign.com/processes/main/battery-assembly-ultrasonic-040-9/ (accessed April 1, 2009).

[54] http://media.gm.com/media/us/en/chevrolet/news.detail.html/ content/Pages/news/us/en/2011/Feb/0214_battery.html (accessed Febuary 14, 2011)

[55] http://en.wikipedia.org/wiki/File:Nissan_Leaf_012.JPG.

[56] P. G. Balakrishnan, R Ramesh, and T. Prem Kumar . Safety mechanisms in lithium-ion batteries. *Journal of Power Sources*, 155 : 401–414, 2006.

[57] The Boston Consulting Group. Batteries for electric cars, 2010.

[58] J. Muthu and J. Battaglini. Advancements in water-based processing for large format lithium ion cells. Available at http://www.internationalbattery.com/pdf/International Battery_ManufacturingWhitePaper.pdf (accessed May 2011).

[59] Lithium ion battery assembly challenges. http://www.iccnexergy.com/articles/1244/lithium-ion-battery-assembly-challenges/ (accessed January 2011).

[60] M. Duduta, B. Ho, V. C. Wood, P. Limthongkul, V. E. Brunini, W. C. Carter, and Y. Chiang, "Semi-solid lithium rechargeable flow battery. *Advanced Energy Materials*, 1: 511–516, 2011.

[61] L. M. Trumm. GM li-ion battery pack manufacturing. May 11, 2011.

[62] F. Ju, J. Wang, J. Li, S. Biller, and G. Xiao. Virtual battery: a simulation framework for batteries in electric vehicles. In: Proceedings of IEEE Conference on Automation Science and Engineering, Trieste, Italy, 2011, pp. 648–653.

CHAPTER 25

RECENT ADVANCES AND ISSUES IN FACILITY LOCATION PROBLEMS

FENG CHU[1], ZHANGUO ZHU[2], and SAÏD MAMMAR[1]
[1]Laboratoire d'IBISC EA 4526, Université d'Evry Val d'Essonne, Evry Cédex, France
[2]College of Economics and Management, Nanjing Agricultural University, Nanjing, PR China

25.1 INTRODUCTION

Due to the growing competition in economic activities, more and more firms focus on the integration aspect of a supply chain to improve their profitability and service level to customers. Locating facilities, for example, plants, depots/warehouses, and transshipment points, as an important strategic decision, impacts the efficiency of a supply chain observably. Improper locations of facilities that last for a considerable amount of time incur excessive cost throughout their whole life time, no matter how well other decisions such as production planning and inventory control are optimized [1]. The facilities location problem is well covered in the existing literature. Early models include median problems, covering problems, and center problems, the capacitated plant location problem (CPLP), the uncapacitated plant location problem, as detailed in Reference 2. With the increasing complexity of a supply chain researchers growingly focus on the variants of classical location problems that reflect the real world. Melo *et al.* [3] indicates that good location models are needed to support the supply chain network design phase and give a literature review of facility location models in the context of supply chain management. Multilayer facilities, location routing, competitive facility location, distance limitations, and service levels are aspects usually taken into consideration in a facility location model to fit into a complicated supply chain context.

Multilayer location problems involve different kinds of facilities in the system. A transportation process usually includes several parts between layers. An important

Contemporary Issues in Systems Science and Engineering, First Edition.
Edited by MengChu Zhou, Han-Xiong Li and Margot Weijnen.

type of it is two-stage capacitated facility location problem (TSCFLP) that involves flows of commodity shipped from capacitated origins (plants) to destinations (customers) via capacitated transshipment points (TPs). Related results are presented by Ro and Tcha [4], Tragantalerngsak *et al.* [5], Hinojosa *et al.* [6], Jayaraman and Pirkul [7], Romeijn *et al.* [8], Lu and Bostel [9], and Elhedhli and Gzara [10].

As an important research area within locational analysis, the location-routing problem (LRP) minimizes the overall cost including plant cost and routing cost satisfying the demand of the customers by limited vehicle and plant capacities in a potential set. It is a combination of two difficult problems: facility location and vehicle routing. Webb [11], Christofides and Eilon [12], Eilon *et al.* [13], Jacobsen and Madsen [14], Drezner and Wesolowsky [15], Laporte [16], etc. are among the first to study this topic. Then much other work including those by Albareda-Sambola *et al.* [17], Caballero *et al.* [18], and Aksen and Altinkemer [19] has been done. Albareda-Sambola *et al.* [17] investigate a combined LRP, define an auxiliary network and give a compact formulation of the problem in terms of finding a set of paths in the auxiliary network that fulfills additional constraints. Caballero *et al.* [18] have investigated a multiobjective LRP and solved it with a multiobjective metaheuristic procedure based on tabu search. Aksen and Altinkemer [19] have studied an LRP from the perspective of distribution logistics. For more details on LRP see the survey by Nagy and Salhi [20].

In facility location problems the distance limitation and quality of service level to customers become important consideration in recent years. Nozick [21] has developed a fixed charge facility location model with coverage restriction incorporating both cost and customer service responsiveness considerations. It tends to obtain good trade-off between them and two heuristics algorithms based on Lagrangian relaxation are proposed to solve the problem. Campbell [22] has studied a hub location problem considering time definite transportation where service levels are imposed by limiting the maximum travel distance for each origin–destination pair in the network. Numerical experiments based on a realistic case are given to test the effects of different service level constraints. Albareda-Sambola *et al.* [23] have studied a plant location problem that is different from a usual LRP. In their problem every potential plant has a fleet of vehicles. A vehicle visits one customer and goes back to the plant once and it has a total travel distance limitation. Hence the number of customers visited by a vehicle depends on the limitation. The authors have developed a tabu search algorithm to solve it. Villegas *et al.* [24] describes a biobjective (cost-coverage) facility location problem that is derived from the Colombian coffee supply network where reducing costs and providing high level of service in terms of coverage to affiliated customers are both considered together. Three algorithms are developed to solve the problem.

As described above, there exist many variants of location problems. This chapter discusses some most recent advances and issues in this field including a capacitated plant location problem with multicommodity flow (CPLPMF) [25], a multitype transshipment point location problem with multicommodity flow (MTTLMF) [26], a large scale new variant of capacitated clustering problem (VCCP) [27], a location problem with selective matching and vehicles assignment (LCSMVA) [28], and competitive facility location and design with reactions of competitors already in the market (CFL-DRCM) [29]. Their problem descriptions and formulations are introduced next.

25.2 A CAPACITATED PLANT LOCATION PROBLEM WITH MULTICOMMODITY FLOW

CPLPMF is a combination of the CPLP and multicommodity min-cost flow problem (MMCF). Since a customer's demand is assumed to be much larger than the capacity of one vehicle, details about how vehicles make tours of several customers to meet their demands as investigated in Vehicle Routing Problem (VRP) can be ignored. Thus, the decisions to be made are where to open a set of plants and how to move flows of different commodity types from open plants to customers through several transshipment points. There usually exist two types of transshipment points. One, known as bulk shipment, is to combine small shipments from different origins but the same destination into a large shipment, or to consolidate shipments for different destinations but with the same origin. The other one, acted as intermediary nodes, usually occurs in the situation where no vacant vehicles are available directly from open plants to customers or that one transportation mode cannot be employed for the entire journey. For the former one, transshipment points can provide reductions in overall shipment cost owing to economies of scale in the transshipment, and for the latter one, commodities have to be transported through some transshipment points and finally reach their destinations.

25.2.1 Problem Description

In a system, there exists a company needing to build some new plants to produce new different types of commodities to serve customers in different regions according to its transshipment network and the transshipment information offered by its shipment partners. The production capacity of each plant and capacities on arcs linking plants, transshipment points, and customers are limited. Generally, the capacity on each arc can be viewed as the tonnage of vehicles, for example, trains, ships, and trucks, proposed to the company by the transportation companies. For example, the number of vacant coaches of a train between regions available to transport commodities can be obtained by consulting shipment partners/a railway company. Each plant has a fixed opening cost and the unit transshipment cost of each commodity on each arc is also known in advance. The decisions to be made by the company in this problem are the locations of new plants and the flows of commodities on different arcs.

25.2.2 Problem Formulation

A directed network $G(N, A)$ is composed of a set N of nodes and a set A of directed arcs (i,j) connecting node i and node j. The set of nodes N is composed of three categories including the set of potential plants I, the set of transshipment points J, and the set of customers K. The set of different commodity types is denoted by C.

The cost of moving a unit flow of commodity $c \in C$ on each arc $(i,j) \in A$ is h^c_{ij} and the capacity of arc (i,j) is l_{ij}. The customer k's demand for commodity c is d^c_k, $k \in K$, $c \in C$. Each potential plant $i \in I$ with a limited capacity S_i has a fixed cost g_i. It is assumed that $\sum_{i \in I} S_i \geqslant d(C) \triangleq \sum_{k \in K} \sum_{c \in C} d^c_k$. x^c_{ij} as a decision variable indicating flows of commodity c moving on arc (i,j). Another decision variable y_i is equal to 1 if and only if a plant i is opened, and otherwise equal to 0.

The problem can be formulated as follows:

$$Z = min\left(\sum_{(i,j)\in A}\sum_{c\in C} h^c_{ij}x^c_{ij} + \sum_{i\in I} g_i y_i\right) \tag{25.1}$$

such that

$$\sum_{j\in J}\sum_{c\in C} x^c_{ij} \leqslant d(C)y_i, \quad \forall i \in I \tag{25.2}$$

$$\sum_{j\in J}\sum_{c\in C} x^c_{ij} \leqslant S_i, \quad \forall i \in I \tag{25.3}$$

$$\sum_{i\in I} S_i y_i \geqslant d(C) \tag{25.4}$$

$$\sum_{j\in N:(i,j)\in A} x^c_{ij} - \sum_{j\in N:(j,i)\in A} x^c_{ji} = 0, \quad \forall i \in J, \forall c \in C \tag{25.5}$$

$$\sum_{i\in J} x^c_{ik} = d^c_k, \quad \forall k \in K, \forall c \in C \tag{25.6}$$

$$\sum_{c\in C} x^c_{ij} \leqslant l_{ij}, \quad \forall (i,j) \in A \tag{25.7}$$

$$x^c_{ij} \geqslant 0, \quad \forall (i,j) \in A, \forall c \in C \tag{25.8}$$

$$y_i = \{0, 1\}, \quad \forall i \in I. \tag{25.9}$$

The objective (25.1) is to minimize the total cost of opening plants and transportation from plants to customers. Constraints (25.2) ensure flows on arcs to be connected with only open plants. Constraints (25.3) guarantee that the total flows moving out of a plant do not exceed its production capacity with the assumption that each plant is connected with at least one transshipment point. Constraints (25.2) and (25.3) can be integrated into (25.10) and it is split to ease the Lagrangian relaxation detailed in this paper. Constraint (25.4) is a redundant constraint but it can be used to obtain the lower bounds. The flow conservation is ensured by constraints (25.5) and the demands for each customer are due to constraints (25.6). Constraints (25.7) mean that the arc capacities are limited:

$$\sum_{j \in J} \sum_{c \in C} x_{ij}^{c} \leqslant S_i y_i \quad \forall i \in I. \tag{25.10}$$

When all of the customers are connected directly with the plants and the arc capacity between each customer and each plant is sufficient, the CPLPMF corresponds to the special case, CFLP, which is known to be NP-hard. Therefore, CPLPMF is NP-hard. Commercial optimization software packages can only solve small instances of CPLPMF exactly and a solving process becomes time consuming with a large scale size of instances. It is necessary to develop an effective algorithm to solve it. Hence, an algorithm consisting of a Lagrangian method and a Tabu Search is developed. Computational tests on random instances under different parameters are carried out and the computational results shows that the solution method is effective since gaps between the upper and lower bound are on average around 2%.

25.3 A MULTITYPE TRANSSHIPMENT POINT LOCATION PROBLEM WITH MULTICOMMODITY FLOW

MTTLMF is an extension of TSCFLP considering both MMCF and several TP types for the goal to locate TPs with appropriate types to ship different commodities from plants to customers while minimizing the total cost for opening TPs and shipment. For TSCFLP in the literature, there exists a basic assumption that the flows among facilities in the same layer are not allowed implying that each path from an origin to a destination may traverse only one transshipment point (TP) although commodities are usually shipped via several TPs (commodities are unloaded from one vehicle and temporarily stored in a TP before being sent to another TP using another vehicle in a long-haul transportation) in practice. TPs in this problem play the role of bulk shipment consolidating flows from different origins and distributing them to different destinations. The property that makes no restrictions on the number of TPs in the transportation between plants and customers in MMCF is kept in MTTLMF. Different types of TPs at each potential site are also considered, which makes the problem under study more realistic.

25.3.1 Problem Description

Facing increasing global competition, a multinational company needs to relocate some plants in developing countries to take advantage of the low cost in materials and manpower. Different types of commodities produced in plants are provided to customers. Due to high cost, direct shipments from plants to customers are not adopted. Instead, the company builds TPs to collect commodities from different origins and transits them together to obtain the economy of scale. There are different types of TPs at a potential site but at most one type can be selected. The capacities for each plant and each type of TPs are limited. The opening cost of TPs can be estimated and the corresponding unit transportation cost is obtained from the shipment partners. The decisions to be made by the company include the sites and types of new open TPs and flows of different commodities.

25.3.2 Problem Formulation

N, A, and C are identical to those in CPLPMF. Only those new ones must be defined. I is the set of potential plants, and S_i^c is the capacity of plant $i \in I$ for commodity $c \in C$. J is used to denote the set of potential TP sites and T_j is the set of possible types of TP at site $j \in J$. The capacity for any TP $j \in J$ with type $t \in T_j$ is v_{jt}. Since the capacity of TP is the total volume of all commodities it can hold, v_c is used to represent the volume of one unit commodity c for each commodity $c \in C$. K denotes the set of customers and d_k^c means the demand of customer k for commodity c. h_{ij}^c denotes the shipment cost for moving one unit flow of commodity c on arc (i,j) and f_{jt} is used to specify the fixed cost caused by opening TP j with type $t \in T_j$. x_{ij}^c is a real-number variable denoting the flow of commodity c moving on arc $(i,j) \in A$. y_{jt} is a binary variable. y_{jt} is equal to 1 if and only if TP j with type t is open, otherwise 0. Moreover, flows moving directly from plants to customers are not allowed. The MTTLMF can be formulated as follows:

$$Z = min\left(\sum_{j\in J}\sum_{t\in T_j} f_{jt}y_{jt} + \sum_{(i,j)\in A}\sum_{c\in C} h_{ij}^c x_{ij}^c\right), \tag{25.11}$$

such that

$$\sum_{j\in J} x_{ij}^c \leqslant S_i^c, \quad \forall i \in I, \forall c \in C \tag{25.12}$$

$$\sum_{j\in J} x_{jk}^c = d_k^c, \quad \forall k \in K, \forall c \in C \tag{25.13}$$

$$\sum_{k\in K} x_{jk}^c + \sum_{j'\in J\setminus\{j\}} x_{jj'}^c - \sum_{j'\in J\setminus\{j\}} x_{j'j}^c - \sum_{i\in I} x_{ij}^c = 0, \quad \forall j \in J, \forall c \in C \tag{25.14}$$

$$\sum_{c\in C} v^c \left(\sum_{j'\in J\setminus\{j\}} x^c_{j'j} + \sum_{i\in I} x^c_{ij} \right) \leqslant \sum_{t\in T_j} v_{jt} y_{jt}, \quad \forall j \in J \tag{25.15}$$

$$\sum_{t\in T_j} y_{jt} \leqslant 1, \quad \forall j \in J \tag{25.16}$$

$$y_{jt} = \{0, 1\}, \quad \forall j \in J, \forall t \in T_j \tag{25.17}$$

$$x^c_{ij} \geqslant 0, \quad \forall (i,j) \in A, \forall c \in C. \tag{25.18}$$

Objective (25.11) is to minimize the fixed cost of locating TPs with different types and shipment cost incurred by transporting commodities from plants to customers via TPs. Constraints (25.12) ensure that each plant does not supply more than its capacity. Constraints (25.13) mean the demand of each customer must be satisfied. The flow conservation is ensured by constraints (25.14). Constraints (25.15) mean that flows just move via open TP and do not exceed its capacity of certain type. Constraints (25.16) represent that at most one type of TP can be opened at each site.

As presented above, it is a mixed integer programming problem and it is difficult to solve it because of the mixture of binary variables and continuous decision variables. Motivated by Keskin and Üster [30, 31], a two-stage method is used. Firstly, the values of $y_{jt} \forall j \in J, \forall t \in T_j$ are given to be $0/1$. Secondly, for the TPs with $y_{jt} = 1$, MTTLMF becomes a classical MMCF that can be solved easily. It is necessary to note that set $\hat{J}$ represents the open TPs.

The reduced MTTLMF (named R-MMCF) can be modeled as follows:

$$Z = min \sum_{(i,j)\in A} \sum_{c\in C} h^c_{ij} x^c_{ij} \tag{25.19}$$

such that

$$\sum_{j\in \hat{J}} x^c_{ij} \leqslant S^c_i, \quad \forall i \in I, \forall c \in C \tag{25.20}$$

$$\sum_{j\in \hat{J}} x^c_{jk} = d^c_k, \quad \forall k \in K, \forall c \in C \tag{25.21}$$

$$\sum_{k\in K} x^c_{jk} + \sum_{j'\in \hat{J}\setminus\{j\}} x^c_{jj'} - \sum_{j'\in \hat{J}\setminus\{j\}} x^c_{j'j} - \sum_{i\in I} x^c_{ij} = 0, \quad \forall j \in \hat{J}, \forall c \in C \tag{25.22}$$

$$\sum_{c\in C} v^c \left(\sum_{j'\in \hat{J}\setminus\{j\}} x^c_{j'j} + \sum_{i\in I} x^c_{ij} \right) \leqslant v_{jt'}, \quad \forall j \in \hat{J}, y_{jt'} = 1, t' \in \hat{T}_j \tag{25.23}$$

$$x^c_{ij} \geqslant 0, \quad \forall (i,j) \in A, \forall c \in C. \tag{25.24}$$

In this way, MTTLMF is divided into two subproblems including the location subproblem and the transportation subproblem. The locations and types of TPs can be obtained by deciding variables $Y_{jt} \forall j \in J, \forall t \in T_j$ with scatter search (SS) initiated by Glover [32]. R-MMCF that is a real-number linear programming problem can be solved optimally by CPLEX 9.0 based on open TPs generated with SS. If the solution obtained is feasible, the total cost can be calculated by adding both the cost of opening TPs (the first stage) and the transportation cost (the second stage).

SS is a population-based metaheuristic and the general template of scatter search mainly consists of five parts: diversification generation, improvement, reference set update, subset generation, and solution combination. Details about them are referred to in Reference 33. In this problem, the overall algorithm includes seven parts: seed solution generation method, diversification generation method, local search, reference set update method, subset generation method, solution combination method, and improvement method.

Several sets of instances including the ratio of TP/plant number, the ratio of customer/TP number, the number of commodities, and the number of potential types for selection are generated to demonstrate the performance of SS. Since it is difficult to obtain the lower bound of MTTLMF, the performance of SS is evaluated by comparing the solution to the lower bound provided by CPLEX 9.0. It is showed that the scatter search performs efficiently over different kinds of instances generated.

25.4 A LARGE SCALE NEW VARIANT OF CAPACITATED CLUSTERING PROBLEM

VCCP is a new variant of the capacitated clustering problem, which describes a location-replenishment problem in upstream supply network design. In VCCP, plants that procure certain raw materials from a set of suppliers are to be opened in different sites. Each open plant has a minimum capacity requirement that must be satisfied and each supplier can be assigned to at most one plant. The total supply volume of all suppliers exceeds the sum of all requirements from the plants. The number of plants to be opened is also fixed. It is very different from most classical location problems considering the location of upstream plants and the distribution of final products to satisfy the demand of downstream customers.

25.4.1 Problem Description

In a green energy investment project of Champagne-Ardenne region of France, the local government plans to build four homogenous plants in six potential sites collecting the wheat straw to produce the ethanol with the minimum total cost of opening the plants and assigning suppliers to the plants. Under the consideration of potential demand of ethanol and economies of scale, each plant is subject to a minimum capacity requirement of wheat straw satisfied by enough suppliers. Moreover, each supplier is assigned to one plant at most and the supplier must supply all its volume available once it is assigned to an open plant. There exist about 2000 suppliers in all in the region, and the total supply volume of all suppliers exceeds the total requirement of all plants to be built. The decisions to be made are the location of plants and the assignment of suppliers to each open plant. VCCP is motivated by this project.

25.4.2 Problem Formulation

I denotes the set of potential plants and J denotes the set of suppliers. For plant $i \in I$, S_i is an associated minimum capacity requirement and g_i is a fixed opening cost. w_j is the supply volume of supplier $j \in J$. h_{ij} denotes the cost of allocation from supplier $j \in J$ to plant $i \in I$. y_j is a binary variable representing the status of a plant. y_j is equal to 1 if and only if plant i is open. x_{ij} is also a binary variable and equal to 1 if and only if supplier j is assigned to plant i.

VCCP can then be formulated as follows:

$$Z = min \sum_{j \in J} \sum_{i \in I} h_{ij} x_{ij} + \sum_{i \in I} g_i y_i \tag{25.25}$$

such that

$$\sum_{i \in I} x_{ij} \leq 1, \quad \forall j \in J \tag{25.26}$$

$$\sum_{j \in J} w_j x_{ij} \geq S_i y_i, \quad \forall i \in I \tag{25.27}$$

$$\sum_{j \in P} y_j = p \tag{25.28}$$

$$y_j \in \{0, 1\}, \quad \forall j \in P \tag{25.29}$$

$$x_{ij} \in \{0, 1\}, \quad \forall i \in S, \forall j \in P. \tag{25.30}$$

The objective function (25.25) is to minimize the total cost of plant opening cost and allocation cost from suppliers to plants. The assignment constraints (25.26) guarantee that each supplier is assigned to at most one plant. Constraints (25.27) ensure that the minimum supply requirements of each plant is satisfied. Constraint (25.28) means that the number of plants to be located is p.

It is not difficult to prove that VCCP is NP-hard. Considering its complexity, it is necessary to develop heuristics to handle large scale instances of the problem. A Lagrangian approach with two phases of dual optimization, the subgradient deflection method in the first phase and the standard subgradient method in the second phase, is developed to approximately solve VCCP.

The proposed approach has been tested on random instances that are generated with reference to the facility location problem in the green energy investment project of the Champagn-Ardenne region of France and the way used in Reference 34 for the CFLP. Numerical tests on 200 random instances have proved the effectiveness of the method. It is showed that tight lower bounds and high quality feasible solutions have been obtained for all tested instances.

25.5 A LOCATION PROBLEM WITH SELECTIVE MATCHING AND VEHICLES ASSIGNMENT

LCSMVA is a location problem which involves many characteristics such as selective matching between customers and suppliers, especially analyze the role of vehicles.

25.5.1 Problem Description

In a supply chain, a company needs to locate some plants to transport material/parts from suppliers to produce products and distribute products to serve customers with vehicles of different capacities. The vehicles depart and return to the same plant. In real life, the travel distance of a vehicle is usually limited because of the driver's working time or other factors. The maximal travel distance leads to different shipment modes: selective matching shipment or direct shipment. If the distance of the trigonal trip from plant to customer, from customer to supplier, and back to plant is less than the maximal travel distance, the customers and suppliers are possible to be served in the way of matching. Otherwise, they are just possible to be served in direct shipment. All the entities except customers have limited capacities. Each plant has a fixed opening cost. The cost of each vehicle contains two parts: the fixed cost and the variable cost. The decisions to be made include locating the potential plants, selecting an appropriate shipment mode, and determining the number of different vehicles serving different customers and suppliers.

25.5.2 Problem Formulation

The nodes sets include the set of potential plants I, the set of customers K, the set of potential suppliers J, and the set of vehicle types M. v_{ijk} means the distance from the plant $i \in I$ visiting customer $k \in K$, from customer $k \in K$ to supplier $j \in J$, from supplier $j \in J$ back to plant $i \in I$. v_{ik} is the distance between plant i and customer k, and v_{ij} is the distance between plant i and supplier j. S_i is the capacity of plant i and

g_i is the fixed opening cost. w_j is the limited capacity of supplier j. d_k is the demand of customer k. h_m is the cost per unit distance per unit demand for the mth types of vehicles. q_{ijk} is equal to 1 if and only if the distance from plant i to customer k and supplier j and back to plant i is less than L, and otherwise, 0, where L is the distance limitation parameter. e denotes the volume ratio between products and material/parts. r_{im} is the fixed cost of the mth type of vehicles in plant i. o_{im} is the loading capacity of the mth type vehicles of plant i. π_{im} is a parameter that denotes the upper bound of the number of the mth type of vehicles of plant i.

LCSMVA can be modeled as follows:

$$Z = min\left(\sum_i g_i y_i + \sum_i \sum_k \sum_j \sum_m h_m q_{ijk} v_{hij} x_{ijkm} + \sum_i \sum_k \sum_m 2h_m v_{ik} z_{ikm} + \sum_i \sum_j \sum_m 2h_m v_{ij} t_{ijm} + \sum_i \sum_m r_{im} u_{im}\right) \tag{25.31}$$

such that

$$\sum_i \sum_j \sum_m q_{ijm} x_{ijkm} + \sum_i \sum_m z_{ikm} = d_k, \quad \forall k \in K \tag{25.32}$$

$$e\left(\sum_k \sum_j \sum_m x_{ijkm} + \sum_k \sum_m z_{ikm}\right) = \sum_j \sum_m t_{ijm} + \sum_k \sum_j \sum_m x_{ijkm}, \quad \forall i \in I \tag{25.33}$$

$$\sum_k \sum_j \sum_m q_{ikj} x_{ijkm} + \sum_k \sum_m z_{ikm} \leq S_i, \quad \forall i \in I \tag{25.34}$$

$$\sum_k \sum_j q_{ikj} x_{ijkm} + \sum_k z_{ikm} \leq o_{im} u_{im}, \quad \forall m \in M, i \in I \tag{25.35}$$

$$\sum_i \sum_k \sum_m q_{ikj} x_{ijkm} + \sum_i \sum_m t_{ijm} \leq w_j, \quad \forall j \in J \tag{25.36}$$

$$\sum_m o_{im} u_{im} \leq S_i y_i, \quad \forall i \in I \tag{25.37}$$

$$\sum_k d_k \leq \sum_i S_i y_i \tag{25.38}$$

$$x_{ijkm} \geq 0, z_{ikm} \geq 0, t_{ijm} \geq 0, 0 \leq u_{im} \leq \pi_{im}, \quad \forall i \in I, \forall k \in K, \forall j \in J, \forall m \in M \tag{25.39}$$

$$x_{ijkm}, z_{ikm}, t_{ijm}, u_{im} \in integer, \quad \forall i \in I, \forall k \in K, \forall j \in J, \forall m \in M \tag{25.40}$$

$$y_i \in \{0, 1\}, \quad \forall i \in I. \tag{25.41}$$

The objective function (25.31) is to minimize the total opening cost of plants, allocation cost including the cost of trigonal shipments, direct shipments from plants to customers and from suppliers to plants, and the fixed cost of vehicles. Constraints (25.32) guarantee that the demands of all customers can be satisfied. The flow equilibrium in every open plant is ensured by constraints (25.33). Constraints (25.34) show that the number of products provided by each plant does not exceed its capacity. Constraints (25.35) show that the amount of products/materials shipped by different kinds of vehicles in each plant is limited. Constraints (25.36) ensure that the material supplied by each supplier does not exceed its capacity. Constraints (25.37) ensure that the total amount shipped by all vehicles in each plant is not more than the production capacity. y_i denotes whether a plant is open at site i. x_{ijkm} is the amount shipped by the mth type of vehicles with the matched method including customer k, supplier j and plant i in the same trigonal trip. z_{ikm} is the amount of customer k satisfied by the way of direct shipment from plant i with the mth type of vehicles. t_{ijm} is the amount of material supply from supplier j to plant i by the way of direct shipment with the kth type of vehicles. u_{im} denotes the number of the mth type of vehicles used in plant i and has an upper bound π_{im}. y_i is 0-1 binary variables. x_{ijkm}, z_{ikm}, and t_{ijm} are integer variables. u_{im} is an integer variable.

The model of LCSMVA is discussed, but no solution is found yet. Future work is required to design an efficient algorithm to solve this problem. Incorporating more practical items such as multicommodity, inventory, and multiperiod into the above formulation will also be needed.

25.6 COMPETITIVE FACILITY LOCATION AND DESIGN WITH REACTIONS OF COMPETITORS ALREADY IN THE MARKET

CFLDRCM is a problem of location and design of a new retail facility competing with other facilities already in the market under the consideration of the reactions of the competitors. Most studies in the literature just focus on the cases that no competitor exists or competitors exist but with no reactions. Obviously, it is not realistic in real life. Facing the possible loss of market share, the existing facilities in the market usually take into account some reactions, for example, improving their qualities. The decision-making for CFLDRCM includes two stages: quality decision and location decision, since in real world the location of a facility is always decided before its service quality. The competitive decision occurring only in the quality decisions is modeled as a game. The solution of this game is given by its Nash equilibrium, which is represented as functions of the location of the new facility. The best location of the new facility, based on the quality decisions in the quality decision stage, can be found by using an interval-based global optimization algorithm.

25.6.1 Problem Description

A retail firm plans to enter a market where customers and some retail facilities providing similar products are present. Customers are represented by demand points and each demand point has a buying power representing the combined purchase potential. As described above, facing the possible loss of market share incurred by a new retail firm, the existing retail firms in the market will generate some reactions. The attraction of each facility to a given demand point is proportional to the service quality of the facility and inversely proportional to the distance between the demand point and the facility. Since the relocation of existing retails is very expensive, the best way to compete with the new retail is to improve their qualities. The decisionis to seek for the location and the service quality of the new facility that maximize its profit.

25.6.2 Problem Formulation

The notations are as follows. I is the set of demand point. J is the set of existing facility. x, α_0, α_j are variables, which means location of the new facility ($x = (x_1, x_2)$), quality level of the new facility ($\alpha_0 > 0$), and new quality level of the jth existing facility ($\alpha_j > 0, \forall j = 1, \ldots, m$), respectively. $\alpha_j^{'}$ is the former quality level of the jth existing facility, $j = 1, \ldots, m$. p_i is the location of the ith demand point, $p_i = (p_i^1, p_i^2)$, f_j is the location of the jth existing facility, $f_j = (f_j^1, f_j^2)$, where $p_i^1, p_i^2, f_j^1, f_j^2$ denote the coordinates of corresponding points i and j. w_i is the buying power of the ith demand point. β_0 is the cost for increasing one unit of the quality level for the new facility. β_j is the cost for increasing one unit of the quality level for the jth facility. γ_i is the weight for the quality of a facility as perceived by the demand point i. λ_i is the weight for the distance to a facility as perceived by demand point i. $d_{\min}$ is the minimal distance between a demand point and a facility. $\alpha_{\min}$ is the minimum level of quality for all facilities. $\alpha_{\max}$ is the maximum level of quality for all facilities. d_{ij} is the distance between the ith demand point and the jth facility. d_{i0} is the distance between the ith demand point and the new facility. $g(.)$ is a nonnegative nondecreasing function and it is assumed that $g_i(d_{ij}) = \lambda_i d_{ij}$ for each $i = 1, \ldots, n$. $\frac{\alpha_{ij}}{g_i d_{ij}}$ is the attraction that demand point i has for facility j. $\frac{\alpha_{i0}}{g_i d_{i0}}$ is the attraction that demand point i has for the new facility. c is the income per unit of good sold. The Euclidian distance is used for the distance between the demand point i and the facility j.

The market share captured by the new facility can be given by

$$M_0(x, \alpha_0, \alpha_1, ..., \alpha_m) = \sum_{i=1}^{n} w_i \frac{\frac{\alpha_0}{d_{i0}}}{\frac{\alpha_0}{d_{i0}} + \sum_{j=1}^{m} \frac{\alpha_j}{d_{ij}}}. \tag{25.42}$$

The market shares captured by the existing facilities in the market become:

$$M_j(x, \alpha_0, \alpha_1, ..., \alpha_m) = \sum_{i=1}^{n} w_i \frac{\frac{\alpha_j}{d_{ij}}}{\frac{\alpha_0}{d_{i0}} + \sum_{k=1}^{m} \frac{\alpha_k}{d_{ik}}}, j = 1, 2, ..., m. \tag{25.43}$$

The profit obtained by existing facility j can be formulated as

$$\pi_j(x, \alpha_0, \alpha_1, ..., \alpha_m) = F(M_j(x, \alpha_0, \alpha_1, ..., \alpha_m)) - G(\alpha_j), \tag{25.44}$$

where F is used to transform the market share into expected sales and can be formulated as: $F(M_j(x, \alpha_0, \alpha_1, ..., \alpha_m)) = cM_j(x, \alpha_0, \alpha_1, ..., \alpha_m)$, c is the sales of unit market share. $G(\alpha_j) = \beta_j \alpha_j$ is the operating cost of a facility with service quality α_j.

Hence, the profit of the new facility is

$$\pi_0(x, \alpha_0, \alpha_1, ..., \alpha_m) = c \sum_{i=1}^{n} w_i \frac{\frac{\alpha_0}{d_{i0}}}{\frac{\alpha_0}{d_{i0}} + \sum_{j=1}^{m} \frac{\alpha_j}{d_{ij}}} - \beta_0 \alpha_0. \tag{25.45}$$

The profit of each facility already in the market is

$$\pi_j(x, \alpha_0, \alpha_1, ..., \alpha_m) = c \sum_{i=1}^{n} w_i \frac{\frac{\alpha_j}{d_{ij}}}{\frac{\alpha_0}{d_{i0}} + \sum_{k=1}^{m} \frac{\alpha_k}{d_{ik}}} - \beta_j(\alpha_j - \alpha_j'), j = 1, ..., m. \tag{25.46}$$

From the view of the new facility, CFLDRCM is formulated as

$$Max\ \pi_0(x, \alpha_0, \alpha_1, ..., \alpha_m) = c \sum_{i=1}^{n} w_i \frac{\frac{\alpha_0}{d_{i0}}}{\frac{\alpha_0}{d_{i0}} + \sum_{j=1}^{m} \frac{\alpha_j}{d_{ij}}} - \beta\alpha_0, \tag{25.47}$$

such that

$$d_{i0} \geq d_{\min}, \quad \forall i \in I$$
$$\alpha_0, \alpha_j \geq \alpha_{\min}, \quad \forall j \in J$$
$$\alpha_0, \alpha_j \leq \alpha_{\max}, \quad \forall j \in J$$

$x \in S \subset \mathbb{R}^2$, where S denotes the region of the market.

A two-stage method including quality decision and location decision is proposed. In the first stage, the objective of each facility is to make its best quality decision by considering the quality reactions of its competitors to maximize its profit. It is a competitive decision process that can be modeled as a noncooperative game. The solution in this stage is Nash equilibrium that can be formed as functions of the location of the new facility. In this stage, two cases including the case with only one facility already in the market and the case with multiple facilities already in the market are discussed. The objective of the second stage is to decide the best location of the new facility which can be obtained by replacing the quality variables of all facilities by their corresponding quality functions found in the first stage. An interval branch and bound algorithm can be used to obtain the solution of the second stage.

Computational experiments are proposed to evaluate the performance of the proposed method on different types of instances by varying the number of demand points and the number of existing facilities. The results show that the method can find a near-optimal solution for all tested instances with a reasonable computation time.

25.7 CONCLUSIONS AND FUTURE RESEARCH DIRECTIONS

Much work in this field was done by many researchers during the past few years. This chapter mainly focuses on those problems with which the authors have dealt. This section summarizes the research discussed in the previous sections and outlines some directions for future research. Five main conclusions are presented as follows.

The CPLPMF is addressed by introducing multicommodity flow into the distribution. A mixed integer linear model of the problem is formulated and a Lagrangian-based method is used to provide good lower and upper bounds of the model. Significant improvements of the quality of feasible solutions can be achieved by using a tabu search method. The result of the numerical analysis implies that the proposed solution method can generate high quality solutions to large-sized instances in a reasonable time within 2% difference from the optimal one on average.

The MTTLMF is also a multicommodity flow related location problem. Different from CPLPMF, its aim is to locate transshipment points (TPs) and determine the type for each open TP. MTTLMF is a realistic extension of TSCFLP considering both MMCF and several TP types. A mixed integer linear model is formulated to describe MTTLMF and a clustering-based scatter search is proposed to solve it. Computational experiments indicate that the scatter search is very effective even on large-sized problems (75 TPs and 225 customers).

As a new realistic variant of a capacitated clustering problem encountered in upstream supply chain network design, VCCP is solved by the proposed Lagrangian relaxation approach with two phases of dual optimization. Computational tests on random instances indicate that the proposed method generates a tight lower bound and a high quality feasible solution for all instances with up to 4000 suppliers, 200 potential sites, and 100 plants to locate.

Location issues under some different conditions including selective supplier–customer matching and specific vehicles assignment are addressed. A linear programming model is proposed to describe them although the solution method is not found yet.

In the last section, a practical type of location problems, CFLDRCM, is studied. After the model is formulated, a two-stage method including quality decision and location decision is proposed to solve it. Computational results indicate that the method can find a near-optimal solution on all randomly generated instances with a reasonable computation time.

Some ideas for future research are presented as follows. Considering the context of a supply chain, a natural extension is to integrate the production process and inventory into the location problems that will be more complicated and realistic than the problems discussed in this chapter. Another interesting extension can be done by a more detailed analysis of TPs/depots. A promising line would be to investigate what happens within TPs/depots, for example, the order of processing cargos, the workforce scheduling, the temporary storage, and the capacities of different platform(machines) within TPs. In CFLDRCM, a new facility entering the market is investigated. Therefore, its further extension is to consider the case of multiple facilities entering the market at the same time as usually happens in real life. In addition, a location problem in discrete space with similar settings to CFLDRCM is also a promising direction. As described above, integrating production process and depot location into CFLDRCM would be of significant benefit. It is necessary to note that adding new features probably invalidates existing approaches. Hence, the development of effective new methods for the extended problems is also important future work.

REFERENCES

[1] Q. Meng, Y. Huang, and R. L. Cheu. Competitive facility location on decentralized supply chains. *European Journal of Operational Research*, 196: 487–499, 2009.

[2] S. H. Owen and M. S. Daskin. Strategic facility location: a review. *European Journal of Operational Research*, 111: 423–447, 1998.

[3] M. T. Melo, S. Nickel, and F. Saldanha-da-Gama . Facility location and supply chain management—a review. *European Journal of Operational Research*, 196: 401–412, 2009.

[4] H. Ro and D. Tcha A branch and bound algorithm for the two-level uncapacitated facility location problem with some side constraints. *European Journal of Operational Research*, 18: 349–358, 1984.

[5] S. Tragantalerngsak, J. Holt, and J. Ronnqvist. Lagrangian heuristics for the two-echelon, single-source, capacitated facility location problem. *European Journal of Operational Research*, 102: 611–625, 1997.

[6] Y. Hinojosa, J. Puerto, and F. R. Fernández. A multiperiod two-echelon multicommodity capacitated plant location problem. *European Journal of Operational Research*, 123: 271–291, 2000.

[7] V. Jayaraman and H. Pirkul. Planning and coordination of production and distribution facilities for multiple commodities. *European Journal of Operational Research*, 133: 394–408, 2001.

[8] H. E. Romeijn, J. Shu, and C. Teo. Designing two-echelon supply networks. *European Journal of Operational Research*, 178: 449–462, 2007.

[9] Z. Lu and N. Bostel. A facility location model for logistics systems including reverse flows: The case of remanufacturing activities. *Computers & Operations Research* 34: 299–323, 2007.

[10] S. Elhedhli and F. Gzara. Integrated design of supply chain networks with three echelons, multiple commodities and technology selection. *IIE Transactions*, 40: 31–44, 2008.

[11] M. H. J. Webb. Cost functions in the location of depots for multiple-delivery journeys. *Operational Research Quarterly*, 19: 311–320, 1968.

[12] N. Christofides and S. Eilon. Expected distances in distribution problems. *Operational Research Quarterly*, 20: 437–443, 1969.

[13] S. Eilon, C. D. T. Watson-Gandy, and N. Christofides. *Distribution Management, Mathematical Modeling and Practical Analysis*, London: Griffin, 1971.

[14] S. K. Jacobsen and O. B. G. Madsen. A comparative study of heuristics for a two-level routing-location problem. *European Journal of Operational Research*, 5: 378–387, 1980.

[15] Z. Drezner and G. O. Wesolowsky. A trajectory approach for the round-trip location problem. *Transportation Science*, 16: 56–66, 1982.

[16] G. Laporte. A survey of algorithms for location-routing problems. *Investigación Operativa*, 1: 93–123, 1989.

[17] M. Albareda-Sambola, J. A. Diza, and E. Fernández. A compact model and tight bounds for a combined location-routing problem. *Computers & Operations Research*, 32: 407–428, 2005.

[18] R. Caballero, M. González, F. M. Guerrero, J. Molina, and C. Paralera. Solving a multiobjective location routing problem with a metaheuristic based on tabu search. Application to a real case in Andalusia. *European Journal of Operational Research*, 177: 1751–1763, 2007.

[19] D. Aksen and K . Altinkemer. A location-routing problem for the conversion to the "click-and-mortar" retailing: the static case. *European Journal of Operational Research*, 186: 554–575, 2008.

[20] S. Nagy, and S. Salhi. Location-routing: issues, models and methods. *European Journal of Operational Research*, 177: 649–672, 2007.

[21] L. K. Nozick. The fixed charge facility location problem with coverage restrictions. *Transportation Research Part E* , 37: 281–296, 2001.

[22] J. F. Campbell. Hub location for time definite transportation. *Computers & Operations Research*, 36: 3107–3116, 2009.

[23] M. Albareda-Sambola, E. Fernandez, and G. Laporte. The capacity and distance constrained plant location problem. *Computers & Operations Research*, 36: 597–611, 2009.

[24] J. G. Villegas, F. Palacios, and A. L. Medaglia. Solution methods for the bi-objective (cost-coverage) unconstrained facility location problem with an illustrative example. *Annual of Operation Research*, 147: 109–141, 2006.

[25] J. Li, F. Chu, and C. Prins. Lower and upper bounds for a capacitated plant location problem with multicommodity flow. *Computers & Operations Research*, 36: 3019–3030, 2009.

[26] J. Li, C. Prins, and F. Chu. A scatter search for a multi-type transshipment point location problem with multicommodity flow. *Journal of Intelligent Manufacturing*, 2010, doi:10.1007/s10845-010-0403-6

[27] Z. Yang, H. Chen, and F. Chu. A Lagrangian relaxation approach for a large scale new variant of capacitated clustering problem. *Computers & Industrial Engineering*, 61: 430–435, 2011.

[28] Z. Zhu, L. Sun, and F. Chu Location problem with selective matching and vehicles assignment. In: *International Conference on Management Science and Information Engineering*, Zhengzhou, China, 3, 534–536, 2010.

[29] N. Saidani, F. Chu, and H. Chen. Competitive facility location and design with reactions of competitors already in the market. *European Journal of Operational Research*, 219: 9–17, 2012.

[30] B. B. Keskin and H. Üster. A scatter search-based heuristic to locate capacitated transshipment points. *Computers and Operations Research*, 34(10): 3112–3125, 2007.

[31] B. B. Keskin and H. Üster. Meta-heuristic approaches with memory and evolution for a multi-product production/distribution system design problem. *European Journal of Operational Research*, 182(2): 663–682, 2007.

[32] F. Glover. Heuristics for integer programming using surrogate constraints. *Decision Science*, 8: 156–166, 1977.

[33] M. Laguna and R. Martí. *Scatter Search: Methodology and Implementations in C*. Boston: Kluwer Academic, 2003.

[34] G. Cornuèjols, R. Sridharan, and T. M. Thizy. A comparison of heuristics and relaxations for the capacitated plant location problem. *European Journal of Operational Research*, 50: 280–297, 1991.

INDEX

Contemporary Issues in Systems Science and Engineering, First Edition.
Edited by MengChu Zhou, Han-Xiong Li and Margot Weijnen.

IEEE PRESS SERIES ON SYSTEMS SCIENCE AND ENGINEERING

The focus of this series is to introduce the advances in theory and applications of systems science and engineering to industrial practitioners, researchers, and students. This series seeks to foster system-of-systems multidisciplinary theory and tools to satisfy the needs of the industrial and academic areas to model, analyze, design, optimize and operate increasingly complex man-made systems ranging from control systems, computer systems, discrete event systems, information systems, networked systems, production systems, robotic systems, service systems, and transportation systems to Internet, sensor networks, smart grid, social network, sustainable infrastructure, and systems biology.

Reinforcement and Systemic Machine Learning for Decision Making
Parag Kulkarni

Remote Sensing and Actuation Using Unmanned Vehicles
Haiyang Chao and YangQuan Chen

Hybrid Control and Motion Planning of Dynamical Legged Locomotion
Nasser Sadati, Guy A. Dumont, Kaveh Akbari Hamed, and William A. Gruver

Modern Machine Learning: Techniques and Their Applications in Cartoon Animation Research
Jun Yu and Dachen Tao

Design of Business and Scientific Workflows: A Web Service-Oriented Approach
Wei Tan and MengChu Zhou

Operator-based Nonlinear Control Systems: Design and Applications
Mingcong Deng

System Design and Control Integration for Advanced Manufacturing
Han-Xiong Li and XinJiang Lu

Sustainable Solid Waste Management: A Systems Engineering Approach
Ni-Bin Chang and Ana Pires

Contemporary Issues in Systems Science and Engineering
Mengchu Zhou, Han-Xiong Li, and Margot Weijnen